Intermediate
ALGEBRA

INTERMEDIATE ALGEBRA WITH P.O.W.E.R. LEARNING

Published by McGraw-Hill, a business unit of The McGraw-Hill Companies, Inc., 1221 Avenue of the Americas, New York, NY 10020. Copyright © 2014 by The McGraw-Hill Companies, Inc. All rights reserved. Printed in the United States of America. No part of this publication may be reproduced or distributed in any form or by any means, or stored in a database or retrieval system, without the prior written consent of The McGraw-Hill Companies, Inc., including, but not limited to, in any network or other electronic storage or transmission, or broadcast for distance learning.

Some ancillaries, including electronic and print components, may not be available to customers outside the United States.

This book is printed on acid-free paper.

1 2 3 4 5 6 7 8 9 0 DOW/DOW 1 0 9 8 7 6 5 4 3

ISBN 978–0–07–340627–5
MHID 0–07–340627–9

ISBN 978–0–07–748378–4 (Annotated Instructor's Edition)
MHID 0–07–748378–2

Senior Vice President, Products & Markets: *Kurt L. Strand*
Vice President, General Manager, Products & Markets: *Marty Lange*
Vice President, Content Production & Technology Services: *Kimberly Meriwether David*
Managing Director: *Ryan Blankenship*
Director, Developmental Mathematics: *Dawn R. Bercier*
Director of Development: *Rose Koos*
Director of Digital Content: *Nicole Lloyd*
Development Editors: *Liz Recker / Elizabeth O'Brien*
Market Development Manager: *Kim Leistner*
Marketing Director: *Alex Gay*
Director, Content Production: *Terri Schiesl*
Lead Project Manager: *Peggy J. Selle*
Buyer: *Nicole Baumgartner*
Senior Media Project Manager: *Sandra M. Schnee*
Senior Designer: *David W. Hash*
Cover/Interior Designer: *Rokusek Design, Inc.*
Cover Image: *Power button icon © tkemot*
Lead Content Licensing Specialist: *Carrie K. Burger*
Compositor: *Aptara®, Inc.*
Typeface: *10/13 Times New Roman MT Std*
Printer: *R. R. Donnelley*

All credits appearing on page or at the end of the book are considered to be an extension of the copyright page.

Library of Congress Cataloging-in-Publication Data

Cataloging-in-Publication Data has been requested from the Library of Congress.

www.mhhe.com

Intermediate
ALGEBRA

SHERRI MESSERSMITH
College of DuPage

LAWRENCE PEREZ
Saddleback College

ROBERT S. FELDMAN
University of Massachusetts Amherst

With contributions from William C. Mulford, *The McGraw-Hill Companies*

About the Authors

Sherri Messersmith
Professor of Mathematics, College of DuPage

Sherri Messersmith began teaching at the College of DuPage in Glen Ellyn, Illinois in 1994 and has over 25 years of experience teaching many different courses from developmental mathematics through calculus. She earned a Bachelor of Science degree in the Teaching of Mathematics at the University of Illinois at Urbana-Champaign and taught at the high school level for two years. Sherri returned to UIUC and earned a Master of Science in Applied Mathematics and stayed on at the university to teach and coordinate large sections of undergraduate math courses as well as teach in the Summer Bridge program for at-risk students. In addition to the P.O.W.E.R. Math Series, she is the author of a hardcover series of textbooks and has also appeared in videos accompanying several McGraw-Hill texts.

Sherri and her husband are recent empty-nesters and live in suburban Chicago. In her precious free time, she likes to read, cook, and travel; the manuscripts for her books have accompanied her from Spain to Greece and many points in between.

Lawrence Perez
Professor of Mathematics, Saddleback College

Larry Perez has fifteen years of classroom experience teaching math and was the recipient of the 2010 Community College Professor of the Year Award in Orange County, California. He realized early on that students bring to the classroom different levels of attitude, aptitude, and motivation sometimes accompanied by a tremendous fear of taking math. Confronted by this, he developed a passion for engaging students, demanding him to innovate traditional and online pedagogical techniques using architecture created with student feedback as the mechanism of design. He is the creator of the award-winning online learning environment Algebra2go® and has presented his work and methodology at conferences around the country.

Larry is a Veteran of the United States Navy Submarine Force and is a graduate of California State University Fullerton earning degrees in Electrical Engineering and Applied Mathematics. In his spare time he enjoys mountain biking and the great outdoors.

Robert S. Feldman
Dean and Professor of Psychology, University of Massachusetts Amherst

Bob Feldman still remembers those moments of being overwhelmed when he started college at Wesleyan University. "I wondered whether I was up to the challenges that faced me," he recalls, "and although I never would have admitted it then, I really had no idea what it took to be successful at college."

That experience, along with his encounters with many students during his own teaching career, led to a life-long interest in helping students navigate the critical transition that they face at the start of their own college careers. Bob, who went on to receive a doctorate in psychology from the University of Wisconsin-Madison, teaches at the University of Massachusetts Amherst, where he is the Dean of the College of Social and Behavioral Sciences and Professor of Psychology. He also directs a first-year experience course for incoming students.

Bob is a Fellow of both the American Psychological Association and the Association for Psychological Science. He has written more than 200 scientific articles, book chapters, and books, including P.O.W.E.R. Learning: *Strategies for Success in College and Life,* 6e and *Understanding Psychology,* 11e. He is president-elect of the FABBS Foundation, an umbrella group of societies promoting the behavioral and brain sciences.

Bob loves travel, music, and cooking. He and his wife live near the Holyoke mountain range in western Massachusetts.

Table of Contents

CHAPTER 1 — Real Numbers and Algebraic Expressions 1

Study Strategies: The P.O.W.E.R. Framework 2
Section 1.1 Set of Numbers 4
Section 1.2 Operations on Real Numbers 12
Section 1.3 Exponents, Roots, and Order of Operations 20
Section 1.4 Algebraic Expressions and Properties of Real Numbers 26
Group Activity 30
emPOWERme: Why Am I Going to College? 31
Chapter 1 Summary 32
Chapter 1 Review Exercises 34
Chapter 1 Test 35

CHAPTER 2 — Linear Equations in One Variable 36

Study Strategies: Reading Math (and Other) Textbooks 37
Section 2.1 Linear Equations in One Variable 39
Section 2.2 Applications of Linear Equations 52
Section 2.3 Geometry Applications and Solving Formulas 66
Section 2.4 More Applications of Linear Equations 76
Group Activity 85
emPOWERme: Organize Your Memory 86
Chapter 2 Summary 88
Chapter 2 Review Exercises 90
Chapter 2 Test 92
Cumulative Review for Chapters 1 and 2 93

CHAPTER 3 — Linear Inequalities and Absolute Value 94

Study Strategies: Time Management 95
Section 3.1 Linear Inequalities in One Variable 97
Section 3.2 Compound Inequalities in One Variable 109
Section 3.3 Absolute Value Equations and Inequalities 117
Group Activity 129
emPOWERme: Identify the Black Holes of Time Management 129
Chapter 3 Summary 131
Chapter 3 Review Exercises 132
Chapter 3 Test 133
Cumulative Review for Chapters 1–3 134

CHAPTER 4 — Linear Equations in Two Variables and Functions 135

Study Strategies: Taking Notes in Class 136
Section 4.1 Introduction to Linear Equations in Two Variables 138
Section 4.2 Slope of a Line and Slope-Intercept Form 155
Section 4.3 Writing an Equation of a Line 170
Section 4.4 Linear and Compound Linear Inequalities in Two Variables 186
Section 4.5 Introduction to Functions 196
Group Activity 213
emPOWERme: Checklist for Effective Notes 214
Chapter 4 Summary 215
Chapter 4 Review Exercises 221
Chapter 4 Test 225
Cumulative Review for Chapters 1–4 227

CHAPTER 5 — Solving Systems of Linear Equations 228

Study Strategies: Taking Math Tests 229
Section 5.1 Solving Systems of Linear Equations in Two Variables 232
Section 5.2 Solving Systems of Linear Equations in Three Variables 250
Section 5.3 Applications of Systems of Linear Equations 259
Section 5.4 Solving Systems of Linear Equations Using Matrices 273
Group Activity 281
emPOWERme: Studying Smart 281
Chapter 5 Summary 282
Chapter 5 Review Exercises 286
Chapter 5 Test 288
Cumulative Review for Chapters 1–5 289

CHAPTER 6 — Polynomials and Polynomial Functions 290

Study Strategies: Doing Math Homework 291
Section 6.1 The Rules of Exponents 293
Section 6.2 More on Exponents and Scientific Notation 303
Section 6.3 Addition and Subtraction of Polynomials and Polynomial Functions 313
Section 6.4 Multiplication of Polynomials and Polynomial Functions 324

Section 6.5	Division of Polynomials and Polynomial Functions 335
	Group Activity 344
	emPOWERme: The Right Time and Place for Homework 335
	Chapter 6 Summary 346
	Chapter 6 Review Exercises 348
	Chapter 6 Test 351
	Cumulative Review for Chapters 1–6 352

CHAPTER 7 Factoring Polynomials 353

Study Strategies: Working with a Study Group 354

Section 7.1	The Greatest Common Factor and Factoring by Grouping 356
Section 7.2	Factoring Trinomials 366
Section 7.3	Special Factoring Techniques 378
	Putting It All Together 388
Section 7.4	Solving Quadratic Equations by Factoring 392
Section 7.5	Applications of Quadratic Equations 402
	Group Activity 412
	emPOWERme: Switch "You" to "I" 413
	Chapter 7 Summary 414
	Chapter 7 Review Exercises 416
	Chapter 7 Test 418
	Cumulative Review for Chapters 1–7 419

CHAPTER 8 Rational Expressions, Equations, and Functions 420

Study Strategies: The Writing Process 421

Section 8.1	Simplifying, Multiplying, and Dividing Rational Expressions and Functions 423
Section 8.2	Adding and Subtracting Rational Expressions 436
Section 8.3	Simplifying Complex Fractions 450
Section 8.4	Solving Rational Equations 460
	Putting It All Together 471
Section 8.5	Applications of Rational Equations 478
Section 8.6	Variation 488
	Group Activity 496
	emPOWERme: Mad, Mad, Mad Math 496
	Chapter 8 Summary 497
	Chapter 8 Review Exercises 502
	Chapter 8 Test 505
	Cumulative Review for Chapters 1–8 506

CHAPTER 9 Radicals and Rational Exponents 507

Study Strategies: Working with Technology 508

Section 9.1	Radical Expressions and Functions 511
Section 9.2	Rational Exponents 525
Section 9.3	Simplifying Expressions Containing Square Roots 534
Section 9.4	Simplifying Expressions Containing Higher Roots 545
Section 9.5	Adding, Subtracting, and Multiplying Radicals 553
Section 9.6	Dividing Radicals 561
	Putting It All Together 573
Section 9.7	Solving Radical Equations 578
Section 9.8	Complex Numbers 587
	Group Activity 596
	emPOWERme: Information Please! 597
	Chapter 9 Summary 598
	Chapter 9 Review Exercises 603
	Chapter 9 Test 605
	Cumulative Review for Chapters 1–9 606

CHAPTER 10 Quadratic Equations and Functions 607

Study Strategies: Developing Financial Literacy 608

Section 10.1	The Square Root Property and Completing the Square 610
Section 10.2	The Quadratic Formula 623
	Putting It All Together 632
Section 10.3	Equations in Quadratic Form 636
Section 10.4	Formulas and Applications 645
Section 10.5	Quadratic Functions and their Graphs 653
Section 10.6	Application of Quadratic Functions and Graphing Other Parabolas 667
Section 10.7	Quadratic and Rational Inequalities 680
	Group Activity 690
	emPOWERme: Determine Your Saving Style 691
	Chapter 10 Summary 692
	Chapter 10 Review Exercises 696
	Chapter 10 Test 699
	Cumulative Review for Chapters 1–10 700

CHAPTER 11 Exponential and Logarithmic Functions 701

Study Strategies: Coping with Stress 702

Section 11.1	Composite and Inverse Functions 704
Section 11.2	Exponential Functions 719

Section 11.3	Logarithmic Functions 730
Section 11.4	Properties of Logarithms 743
Section 11.5	Common and Natural Logarithms and Change of Base 753
Section 11.6	Solving Exponential and Logarithmic Equations 765

Group Activity 776
emPOWERme: Progressive Relaxation 777
Chapter 11 Summary 778
Chapter 11 Review Exercises 783
Chapter 11 Test 786
Cumulative Review for Chapters 1–11 787

CHAPTER 12 Nonlinear Functions, Conic Sections, and Nonlinear Systems 788

Study Strategies: Improving Your Memory 789

Section 12.1	Graphs of Other Useful Functions 791
Section 12.2	The Circle 802
Section 12.3	The Ellipse 810
Section 12.4	The Hyperbola 817

Putting It All Together 828

| Section 12.5 | Nonlinear Systems of Equations 832 |
| Section 12.6 | Second-Degree Inequalities and Systems of Inequalities 839 |

Group Activity 844
emPOWERme: Memory Devices 845
Chapter 12 Summary 846
Chapter 12 Review Exercises 849
Chapter 12 Test 850
Cumulative Review for Chapters 1–12 851

APPENDIX

Section A.1	Review of Fractions A-1
Section A.2	Geometry Review A-12
Section A.3	Determinants and Cramer's Rule A-24
Section A.4	Synthetic Division and the Remainder Theorem A-33

This appendix is available online at www.connectmath.com and www.mcgrawhillcreate.com.

Student Answer Appendix SA-1
Instructor Answer Appendix (AIE only) IA-1
Credits C-1
Index I-1

Consistent Integration of Study Skills

In *Intermediate Algebra,* strategies for learning are presented alongside the math content, making it easy for students to learn math *and* study skills at the same time. The P.O.W.E.R. framework aligns with the math learning objectives, providing instructors with a resource that has been consistently integrated throughout the text.

A **STUDY STRATEGIES** feature begins each chapter. Utilizing the P.O.W.E.R. framework, these boxes present steps for mastering the different skills students will use to succeed in their developmental math course. For example, these boxes will contain strategies on time management, taking good notes and, as seen in the sample below, taking a math test.

CHAPTER AND SECTION POWER PLANS Before getting started on reading the chapter, a student will focus on Preparation and Organization skills in the **POWER Plans.** These tools give practical suggestions for setting and achieving goals. The steps revolve around best practices for student success and then apply P.O.W.E.R. toward learning specific concepts in math.

Chapter 5 POWER Plan

P Prepare | O Organize

What are your goals for Chapter 5?	How can you accomplish each goal?
1 Be prepared before and during class.	• Don't stay out late the night before, and be sure to set your alarm clock! • Bring a pencil, notebook paper, and textbook to class. • Avoid distractions by turning off your cell phone during class. • Pay attention, take good notes, and ask questions. • Complete your homework on time, and ask questions on problems you do not understand. • Plan ahead for tests by preparing many days in advance.
2 Understand the homework to the point where you could do it without needing any help or hints.	• Read the directions, and show all of your steps. • Go to the professor's office for help. • Rework homework and quiz problems, and find similar problems for practice. • Review old material that you have not mastered yet.
3 Use the P.O.W.E.R. framework to help you take tests: *Studying Smart*.	• Read the Study Strategy that explains how to study effectively for tests. • Do a "practice run" the night before the test by doing a practice test without notes. • Complete the emPOWERme that appears before the Chapter Summary.
4 Write your own goal. _____	• _____

What are your objectives for Chapter 5?	How can you accomplish each objective?
1 Be able to solve a system of linear equations in two variables by using the graphing, substitution, or elimination methods. Know when to use each method.	• Learn the procedures for each of these methods. • Know the terminology associated with the solutions such as independent and consistent. • Know how to check each answer.
2 Be able to determine when the solution to a system of equations is *no solution* or *infinite solutions*. Know what these solutions look like on a graph and how to write the answer.	• Learn the procedures for solving a system of equations and the possible answers when variables "drop out." • Learn the terminology associated with the solutions such as *inconsistent* and *dependent*. • Know what these results look like on a graph. • Know how to check your solutions.
3 Be able to solve a system of linear equations in three variables, including systems where there are missing terms.	• Learn the procedure for **Solving a System of Linear Equations in Three Variables.** • Know how to check your solutions.

www.mhhe.com/messersmith   CONSISTENT INTEGRATION OF STUDY SKILLS

W Hint
Be sure you are writing out each step as you are reading the example.

b) To solve $t^2 - 20 = 0$, begin by getting t^2 on a side by itself.

$$t^2 - 20 = 0$$
$$t^2 = 20 \qquad \text{Add 20 to each side.}$$
$$t = \pm\sqrt{20} \qquad \text{Square root property}$$
$$t = \pm\sqrt{4} \cdot \sqrt{5} \qquad \text{Product rule for radicals}$$
$$t = \pm 2\sqrt{5} \qquad \sqrt{4} = 2$$

Check:

$t = 2\sqrt{5}: \quad t^2 - 20 = 0$
$(2\sqrt{5})^2 - 20 \stackrel{?}{=} 0$
$(4 \cdot 5) - 20 \stackrel{?}{=} 0$
$20 - 20 = 0$ ✓

$t = -2\sqrt{5}: \quad t^2 - 20 = 0$
$(-2\sqrt{5})^2 - 20 \stackrel{?}{=} 0$
$(4 \cdot 5) - 20 \stackrel{?}{=} 0$
$20 - 20 = 0$ ✓

The solution set is $\{-2\sqrt{5}, 2\sqrt{5}\}$.

WORK HINTS provide additional explanation and point out common places where students might go wrong when solving a problem. Along with the ***Be Careful*** boxes, these tools act as a built-in tutor, helping students navigate the material and learn concepts even outside of class.

IN-CLASS EXAMPLES are available only in the Annotated Instructor Edition. These examples offer instructors additional problems to work through in class. In-class example problems align with the Guided Student Notes resource available with this package.

EXAMPLE 1

Solve using the square root property.

a) $x^2 = 9$ b) $t^2 - 20 = 0$ c) $2a^2 + 21 = 3$

Solution

a)
$$x^2 = 9$$
$$x = \sqrt{9} \quad \text{or} \quad x = -\sqrt{9} \qquad \text{Square root property}$$
$$x = 3 \quad \text{or} \quad x = -3$$

The solution set is $\{-3, 3\}$. The check is left to the student.

An equivalent way to solve $x^2 = 9$ is to write it as

$$x^2 = 9$$
$$x = \pm\sqrt{9} \qquad \text{Square root property}$$
$$x = \pm 3$$

The solution set is $\{-3, 3\}$. We will use this approach when solving equations using the square root property.

PUTTING IT ALL TOGETHER One of the challenges students struggle with is putting all of the steps they've learned together and *applying* that knowledge to a problem. *Putting It All Together* sections will help students understand the big picture and work through the toughest challenge when solving applications—*problem recognition,* or knowing *when* to use *what* method or thought process. These sections include a summary and several problems that help students reason through a problem using conversational, yet mathematically correct, language.

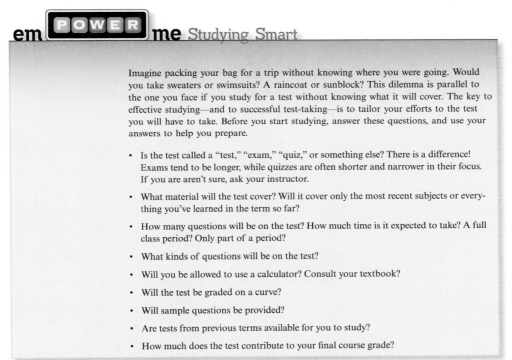

em**POWER**me Studying Smart

Imagine packing your bag for a trip without knowing where you were going. Would you take sweaters or swimsuits? A raincoat or sunblock? This dilemma is parallel to the one you face if you study for a test without knowing what it will cover. The key to effective studying—and to successful test-taking—is to tailor your efforts to the test you will have to take. Before you start studying, answer these questions, and use your answers to help you prepare.

- Is the test called a "test," "exam," "quiz," or something else? There is a difference! Exams tend to be longer, while quizzes are often shorter and narrower in their focus. If you are aren't sure, ask your instructor.
- What material will the test cover? Will it cover only the most recent subjects or everything you've learned in the term so far?
- How many questions will be on the test? How much time is it expected to take? A full class period? Only part of a period?
- What kinds of questions will be on the test?
- Will you be allowed to use a calculator? Consult your textbook?
- Will the test be graded on a curve?
- Will sample questions be provided?
- Are tests from previous terms available for you to study?
- How much does the test contribute to your final course grade?

em**POWER**me boxes circle back to the opening **Study Strategies** and give students a checklist to evaluate how well they followed through on all of the positive habits recommended to successfully master a skill.

The Messersmith/Perez/Feldman Series offers instructors a robust digital resources package to help you with all of your teaching needs.

> **Resources in your P.O.W.E.R. tool kit include:**
>
> - Connect Hosted by Aleks*
> - ALEKS 360*
> - Instructor Solutions Manual
> - Student Solutions Manual
> - Guided Student Notes*
> - Classroom Worksheets*
> - Instructor Resource Manual
> - Test Bank Files
> - Computerized Test Bank
> - Faculty Development and Digital Training*
> - PowerPoint Presentations
> - Extensive Video Package*
>
> *Details of these resources are included in the following pages!

Videos

Hundreds of videos are available to guide students through the content, offering support and instruction even outside your classroom.

Exercise Videos – These 3–5-minute clips show students how to solve various exercises from the textbook. With around thirty videos for every chapter, your students are supported even outside the classroom.

Lecture Videos – These 5–10-minute videos walk students through key learning objectives and problems from the textbook.

P.O.W.E.R. Videos – These engaging segments guide students through the P.O.W.E.R. framework and the study skills for each chapter.

Perform the operations and simplify: $4 - \sqrt{13} + 8 - 6\sqrt{13}$

Like radicals have the same index and the same radicand.

$$4 - \sqrt{13} + 8 - 6\sqrt{13}$$
$$12 - \sqrt{13} - 6\sqrt{13}$$
$$12 + (-1 - 6)\sqrt{13}$$
$$12 + (-7)\sqrt{13}$$
$$12 - 7\sqrt{13}$$

12 - 7 times square root (13).

Faculty Development and Digital Training

McGraw-Hill is excited to partner with our customers to ensure success in the classroom with our course solutions.

Looking for ways to be more effective in the classroom? Interested in learning how to integrate student success skills in your developmental math courses?

Workshops are available on these topics for anyone using or considering the Messersmith/Perez/Feldman P.O.W.E.R. Math Series. Led by the authors, contributors, and McGraw-Hill P.O.W.E.R. Learning consultants, each workshop is tailored to the needs of individual campuses or programs.

New to McGraw-Hill Digital Solutions? Need help setting up your course, using reports, and building assignments?

No need to wait for that big group training session during faculty development week. The McGraw-Hill Digital Implementation Team is a select group of advisors and experts in Connect Hosted by ALEKS™. The Digital Implementation Team will work one-on-one with each instructor to make sure you are trained on the program and have everything you need to ensure a good experience for you and your students.

Are you redesigning a course or expanding your use of technology? Are you interested in getting ideas from other instructors who have used ALEKS™ or Connect Hosted by ALEKS in their courses?

Digital Faculty Consultants (DFCs) are instructors who have effectively incorporated technology such as ALEKS and Connect Hosted by ALEKS Corp. in their courses. Discuss goals and best practices and improve outcomes in your course through peer-to-peer interaction and idea sharing.

Contact your local representative for more information about any of the faculty development, training, and support opportunities through McGraw-Hill. http://catalogs.mhhe.com/mhhe/findRep.do

Need a tool to help your students take better notes?

GUIDED STUDENT NOTES

By taking advantage of Guided Student Notes, your students will have more time to learn the material and participate in solving in-class problems while, at the same time, becoming better note takers. Ample examples are included for appropriate coverage of a topic that will not overwhelm students. Use them as they are or download and edit the Guided Student Notes according to your teaching style.

Guided Student Notes
MPF – Intermediate Algebra

Rules for Dividing Rational Expressions

Divide. Write each rational expression in lowest terms.

16) $\dfrac{42b^6}{c^3} \div \dfrac{2b}{c^5}$

17) $\dfrac{r^2 - 13r + 36}{2r + 10} \div \dfrac{12r - 3r^2}{16}$

18) $\dfrac{3n^2 - 22n - 16}{n^2} \div (3n + 2)^2$

Guided Student Notes
MPF – Intermediate Algebra

Name:_____

8.1 Simplifying, Multiplying, and Dividing Rational Expressions and Functions

Prepare What are my goals for this section?

Organize What am I going to do to accomplish these goals?

Work

Definition of a Rational Expression   **Definition of a Rational Function**

Determining the Domain of a Rational Function

1) If $f(x) = \dfrac{x^2 - 16}{x + 3}$,

   a) find $f(5)$

   b) find x so that $f(x) = 0$

   c) determine the domain of the function.

Develop your students' basic skills with a ready-made resource.

WORKSHEETS FOR STUDENT AND INSTRUCTOR USE

Worksheets for every section are available as an instructor supplement. These author-created worksheets provide a quick, engaging way for students to work on key concepts. They save instructors from having to create their own supplemental material and address potential stumbling blocks in student understanding. Classroom tested and easy to implement, they are also a great resource for standardizing instruction across a mathematics department.

The worksheets fall into three categories: Worksheets to Improve Basic Skills; Worksheets to Help Teach New Concepts; and Worksheets to Tie Concepts Together.

The worksheets are available in an instructor edition, with answers, and in a student edition, without answers.

Worksheet 5A
Messersmith – Intermediate Algebra

Evaluate.

1) $\sqrt{36}$ __6__
2) $\sqrt{144}$ __12__
3) $\sqrt{25}$ __5__
4) $\sqrt[3]{8}$ __2__
5) $\sqrt[3]{125}$ __5__
6) $\sqrt{81}$ __9__
7) $\sqrt[3]{27}$ __3__
8) $\sqrt[4]{16}$ __2__
9) $\sqrt[3]{1000}$ __10__
10) $\sqrt{121}$ __11__
11) $\sqrt{169}$ __13__
12) $\sqrt[3]{64}$ __2__
13) $\sqrt[4]{81}$ __3__
14) $\sqrt{4}$ __2__
15) $\sqrt[5]{1}$ __1__

Worksheet 5A
Messersmith – Intermediate Algebra   Name: _____

Evaluate.

1) $\sqrt{36}$ _____
2) $\sqrt{144}$ _____
3) $\sqrt{25}$ _____
4) $\sqrt[3]{8}$ _____
5) $\sqrt[3]{125}$ _____
6) $\sqrt{81}$ _____
7) $\sqrt[3]{27}$ _____
8) $\sqrt[4]{16}$ _____
9) $\sqrt[3]{1000}$ _____
10) $\sqrt{121}$ _____
11) $\sqrt{169}$ _____
12) $\sqrt[3]{64}$ _____
13) $\sqrt[4]{81}$ _____
14) $\sqrt{4}$ _____
15) $\sqrt[5]{1}$ _____

16) $\sqrt[3]{64}$ _____
17) $\sqrt[5]{32}$ _____
18) $\sqrt{49}$ _____
19) $\sqrt{100}$ _____
20) $\sqrt[3]{8}$ _____
21) $\sqrt[4]{1}$ _____
22) $\sqrt[3]{27}$ _____
23) $\sqrt{16}$ _____
24) $\sqrt[5]{32}$ _____
25) $\sqrt{121}$ _____
26) $\sqrt[4]{81}$ _____
27) $\sqrt[3]{125}$ _____
28) $\sqrt[4]{16}$ _____
29) $\sqrt{9}$ _____
30) $\sqrt{1}$ _____

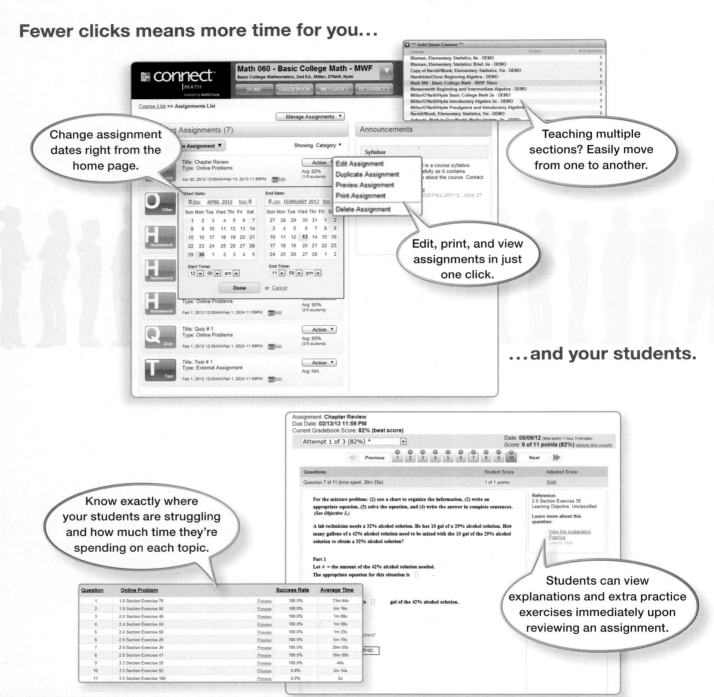

Quality Content For Today's Online Learners

Online Exercises were carefully selected and developed to provide a seamless transition from textbook to technology.

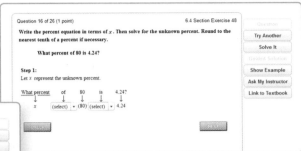

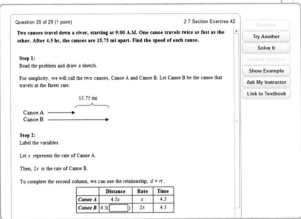

For consistency, the guided solutions match the style and voice of the original text as though the author is guiding the students through the problems.

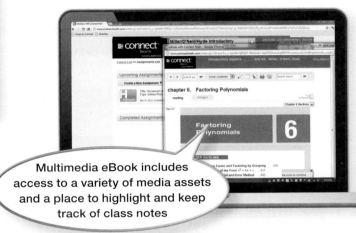

Multimedia eBook includes access to a variety of media assets and a place to highlight and keep track of class notes

ALEKS Corporation's experience with algorithm development ensures a commitment to accuracy and a meaningful experience for students to demonstrate their understanding with a focus towards online learning.

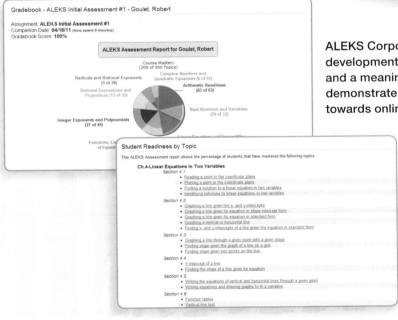

The ALEKS® Initial Assessment is an artificially intelligent (AI), diagnostic assessment that identifies precisely what a student knows. Instructors can then use this information to make more informed decisions on what topics to cover in more detail with the class.

ALEKS is a registered trademark of ALEKS Corporation.

www.successinmath.com

Hosted by ALEKS Corp.

ALEKS is a unique, online program that significantly raises student proficiency and success rates in mathematics, while reducing faculty workload and office-hour lines. ALEKS uses artificial intelligence and adaptive questioning to assess precisely a student's knowledge, and deliver individualized learning tailored to the student's needs. With a comprehensive library of math courses, ALEKS delivers an unparalleled adaptive learning system that has helped millions of students achieve math success.

ALEKS Delivers a Unique Math Experience:

- **Research-Based, Artificial Intelligence** precisely measures each student's knowledge
- **Individualized Learning** presents the exact topics each student is most **ready to learn**
- **Adaptive, Open-Response Environment** includes comprehensive tutorials and resources
- **Detailed, Automated Reports** track student and class progress toward course mastery
- **Course Management Tools** include textbook integration, custom features, and more

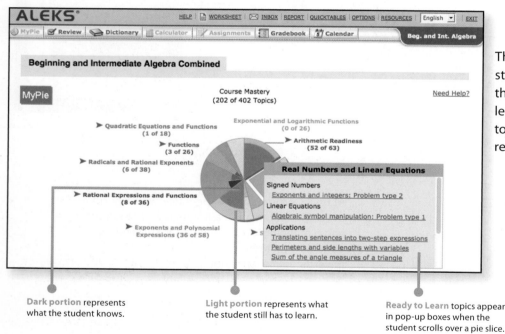

The ALEKS Pie summarizes a student's current knowledge, then delivers an individualized learning path with the exact topics the student is most ready to learn.

Dark portion represents what the student knows.

Light portion represents what the student still has to learn.

Ready to Learn topics appear in pop-up boxes when the student scrolls over a pie slice.

> "My experience with ALEKS has been effective, efficient, and eloquent. **Our students' pass rates improved from 49 percent to 82 percent with ALEKS.** We also saw student retention rates increase by 12% in the next course. Students feel empowered as they guide their own learning through ALEKS."
>
> —Professor Eden Donahou, *Seminole State College of Florida*

To learn more about ALEKS, please visit: **www.aleks.com/highered/math**

ALEKS is a registered trademark of ALEKS Corporation.

ALEKS® Prep Products

ALEKS Prep products focus on prerequisite and introductory material, and can be used during the first six weeks of the term to ensure student success in math courses ranging from Beginning Algebra through Calculus. ALEKS Prep quickly fills gaps in prerequisite knowledge by assessing precisely each student's preparedness and delivering individualized instruction on the exact topics students are most ready to learn. As a result, instructors can focus on core course concepts and see improved student performance with fewer drops.

> "ALEKS is wonderful. It is a professional product that takes very little time as an instructor to administer. Many of our students have taken Calculus in high school, but they have forgotten important algebra skills. ALEKS gives our students an opportunity to review these important skills."
>
> —Professor Edward E. Allen, *Wake Forest University*

A Total Course Solution

A cost-effective total course solution: fully integrated, interactive eBook combined with the power of ALEKS adaptive learning and assessment.

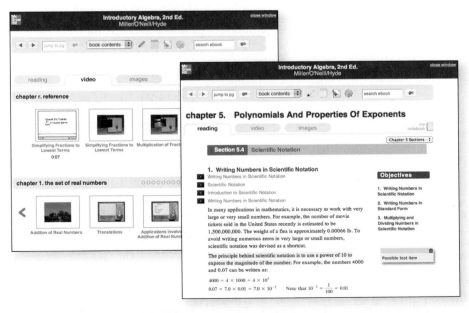

Students can easily access the full eBook content, multimedia resources, and their notes from within their ALEKS Student Accounts.

To learn more about ALEKS, please visit: **www.aleks.com/highered/math**

Acknowledgments

Manuscript Reviewers and Focus Group Participants

Thank you to all of the dedicated instructors who reviewed manuscript, participated in focus groups, and provided thoughtful feedback throughout the development of the *P.O.W.E.R.* series.

Darla Aguilar, *Pima Community College;* Scott Albert, *College of DuPage;* Bhagirathi Anand, *Long Beach City College;* Raul Arana, *Lee College;* Jan Archibald, *Ventura College;* Morgan Arnold, *Central Georgia Technical College;* Christy Babu, *Laredo Community College;* Michele Bach, *Kansas City Kansas Community College;* Kelly Bails, *Parkland College;* Vince Bander, *Pierce College, Pullallup;* Kim Banks, *Florence Darlington Technical College;* Michael Bartlett, *University of Wisconsin—Marinette;* Sarah Baxter, *Gloucester County College;* Michelle Beard, *Ventura College;* Annette Benbow, *Tarrant County College, Northwest;* Abraham Biggs, *Broward College;* Leslie Bolinger Horton, *Quinsigamond Community College;* Jessica Bosworth, *Nassau Community College;* Joseph Brenkert, *Front Range Community College;* Michelle Briles, *Gloucester County College;* Kelly Brooks, *Daytona State College (and Pierce);* Connie Buller, *Metropolitan Community College;* Rebecca Burkala, *Rose State College;* Gail Burkett, *Palm Beach State College;* Gale Burtch, *Ivy Tech Community College;* Jennifer Caldwell, *Mesa Community College;* Edie Carter, *Amarillo College;* Allison Cath, *Ivy Tech Community College of Indiana, Indianapolis;* Dawn Chapman, *Columbus Tech College;* Christopher Chappa, *Tyler Junior College;* Chris Chappa, *Tyler Junior College;* Charles Choo, *University of Pittsburgh at Titusville;* Patricia Clark, *Sinclair Community College;* Judy Kim Clark, *Wayne Community College;* Karen Cliffe, *Southwestern College;* Sherry Clune, *Front Range Community College;* Ela Coronado, *Front Range Community College;* Heather Cotharp, *West Kentucky Community & Tech College;* Danny Cowan, *Tarrant County College, Northwest;* Susanna Crawford, *Solano College;* George Daugavietis, *Solano Community College;* Joseph De Guzman, *Norco College;* Michaelle Downey, *Ivy Tech Community College;* Dale Duke, *Oklahoma City Community College;* Rhonda Duncan, *Midlands Technical College;* Marcial Echenique, *Broward College;* Sarah Ellis, *Dona Ana Community College;* Onunwor Enyinda, *Stark State College;* Chana Epstein, *Sullivan County Community College;* Karen Ernst, *Hawkeye Community College;* Stephen Ester, *St. Petersburg College;* Rosemary Farrar, *Southern West Virginia Community & Technical College;* John Fay, *Chaffey College;* Stephanie Fernandes, *Lewis and Clark Community College;* James Fiebiger, *Front Range Community College;* Angela Fipps, *Durham Technical Community College;* Jennifer Fisher, *Caldwell Community College & Technical Institute;* Elaine Fitt, *Bucks County Community College;* Carol Fletcher, *Hinds Community College;* Claude Fortune, *Atlantic Cape Community College;* Marilyn Frydrych, *Pikes Peak Community College;* Robert Fusco, *Broward College;* Jared Ganson, *Nassau Community College;* Kristine Glasener, *Cuyahoga Community College;* Ernest Gobert, *Oklahoma City Community College;* Linda Golovin, *Caldwell College;* Suzette Goss, *Lone Star College Kingwood;* Sharon Graber, *Lee College;* Susan Grody, *Broward College;* Leonard Groeneveld, *Springfield Tech Community College;* Joseph Guiciardi, *Community College of Allegheny County;* Susanna Gunther, *Solano College;* Lucy Gurrola, *Dona Ana Community College;* Frederick Hageman, *Delaware Technical & Community College;* Tamela Hanebrink, *Southeast Missouri State University;* Deborah Hanus, *Brookhaven College;* John Hargraves, *St. John's River State College;* Michael Helinger, *Clinton Community College;* Mary Hill, *College of DuPage;* Jody Hinson, *Cape Fear Community College;* Kayana Hoagland, *South Puget Sound Community College;* Tracey Hollister, *Casper College;* Wendy Houston, *Everett Community College;* Mary Howard, *Thomas Nelson Community College;* Lisa Hugdahl, *Milwaukee Area Tech College—Milwaukee;* Larry Huntzinger, *Western Oklahoma State College;* Manoj Illickal, *Nassau Community College;* Sharon Jackson, *Brookhaven College;* Lisa Jackson, *Black River Technical College;* Christina Jacobs, *Washington State University;* Gretta Johnson, *Amarillo College;* Lisa Juliano, *El Paso Community College, Northwest Campus;* Elias M. Jureidini, *Lamar State College/Orange;* Ismail Karahouni, *Lamar University;* Cliffe Karen, *Southwestern College;* David Kater, *San Diego City College;* Joe Kemble, *Lamar University;* Esmarie Kennedy, *San Antonio College;* Ahmed Khago, *Lamar University;* Michael Kirby, *Tidewater Community College VA Beach Campus;* Corrine Kirkbride, *Solano Community College;* Mary Ann Klicka, *Bucks County Community College;* Alex Kolesnik, *Ventura College;* Tatyana Kravchuk, *Northern Virginia Community College;* Randa Kress, *Idaho State University;* Julianne Labbiento, *Lehigh Carbon Community College;* Robert Leifson, *Pierce College;* Greg Liano, *Brookdale Community College;* Charyl Link, *Kansas City Kansas Community College;* Wanda Long, *Saint Charles County Community College;* Lorraine Lopez, *San Antonio College;* Luke Mannion, *St. John's University;* Shakir Manshad, *New Mexico State University;* Robert Marinelli, *Durham Technical Community College;* Lydia Matthews-Morales, *Ventura College;* Melvin Mays, *Metropolitan Community College (Omaha NE);* Carrie McCammon, *Ivy Tech Community College;* Milisa Mcilwain, *Meridian Community College;* Valerie Melvin, *Cape Fear Community College;* Christopher Merlo, *Nassau Community College;* Leslie Meyer, *Ivy Tech Community College/Central Indiana;* Beverly Meyers, *Jefferson College;* Laura Middaugh, *McHenry County College;* Karen Mifflin, *Palomar College;* Kris Mudunuri, *Long Beach City College;* Donald Munsey, *Louisiana Delta Community College;* Randall Nichols, *Delta College;* Joshua Niemczyk, *Saint Charles County Community College;* Katherine Ocker Stone, *Tusculum College;* Karen Orr, *Roane State;* Staci Osborn, *Cuyahoga Community College;* Steven Ottmann, *Southeast Community College, Lincoln Nebraska;* William Parker, *Greenville Technical College;* Joanne Peeples, *El Paso Community College;* Paul Peery, *Lee College;* Betty Peterson, *Mercer County Community College;* Carol Ann Poore, *Hinds Community College;* Hadley Pridgen, *Gulf Coast State College;* William Radulovich, *Florida State College @ Jacksonville;* Lakshminarayan Rajaram, *St. Petersburg College;* Kumars Ranjbaran, *Mountain View College;* Darian Ransom, *Southeastern Community College;* Nimisha Raval,

Central Georgia Technical College; Amy Riipinen, *Hibbing Community College;* Janet Roads, *Moberly Area Community College;* Marianne Roarty, *Metropolitan Community College;* Jennifer Robb, *Scott Community College;* Marie Robison, *McHenry County College;* Daphne Anne Rossiter, *Mesa Community College;* Anna Roth, *Gloucester County College;* Daria Santerre, *Norwalk Community College;* Kala Sathappan, *College of Southern Nevada;* Patricia Schubert, *Saddleback College;* William H. Shaw, *Coppin State University;* Azzam Shihabi, *Long Beach City College;* Jed Soifer, *Atlantic Cape Community College;* Lee Ann Spahr, *Durham Technical Community College;* Marie St. James, *Saint Clair County Community College;* Mike Stack, *College of DuPage;* Ann Starkey, *Stark State College of Technology;* Thomas Steinmann, *Lewis and Clark Community College;* Claudia Stewart, *Casper College;* Kim Taylor, *Florence Darlington Technical College;* Laura Taylor, *Cape Fear Community College;* Janet Teeguarden, *Ivy Tech Community College;* Janine Termine, *Bucks County Community College;* Yan Tian, *Palomar College;* Lisa Tolliver, *Brown Mackie South Bend;* David Usinski, *Erie Community College;* Hien Van Eaton, *Liberty University;* Theresa Vecchiarelli, *Nassau Community College;* Val Villegas, *Southwestern College;* David Walker, *Hinds Community College;* Ursula Walsh, *Minneapolis Community & Tech College;* Dottie Walton, *Cuyahoga Community College;* LuAnn Walton, *San Juan College;* Thomas Wells, *Delta College;* Kathryn Wetzel, *Amarillo College;* Marjorie Whitmore, *North West Arkansas Community College;* Ross Wiliams, *Stark State College of Technology;* Gerald Williams, *San Juan College;* Michelle Wolcott, *Pierce College, Puyallup;* Mary Young, *Brookdale Community College;* Loris Zucca, *Lone Star College, Kingwood;* Michael Zwilling, *University of Mount Union*

Student Focus Group Participants

Thanks to the students who reviewed elements of P.O.W.E.R. and talked candidly with the editorial team about their experiences in math courses.

Eire Aatnite, *Roosevelt University;* Megan Bekker, *Northeastern Illinois University;* Hiran Crespo, *Northeastern Illinois University;* John J. Frederick, Jr., *Harold Washington College;* Omar Gonzalez, *Wright College;* Yamizaret Guzman, *Western Illinois University;* Ashley Grayson, *Northeastern Illinois University;* Nathan Hurde, *University of Illinois at Chicago;* Zainab Khomusi, *University of Illinois at Chicago;* Amanda Koger, *Roosevelt University;* Diana Kotchounian, *Roosevelt University;* Adrana Martinez, *DePaul University;* Laurien Mosley, *Western Illinois University;* Jeffrey Moy, *University of Illinois at Chicago;* Jaimie O'Leary, *Northeastern Illinois University;* Trupti Patel, *University of Illinois at Chicago;* Pete Rodriguez, *Truman College;* Kyaw Sint Lay Wu, *University of Illinois at Chicago;* Shona L. Thomas, *Northeastern Illinois University;* Nina Turnage, *Roosevelt University;* Brittany K. Vernon, *Roosevelt University;* Kyaw Sint Lay Wu, *University of Illinois at Chicago*

Digital Contributors

Special thanks go to the faculty members who contributed their time and expertise to the digital offerings with *P.O.W.E.R.*

Jennifer Caldwell, *Mesa Community College*
Chris Chappa, *Tyler Junior College*
Tim Chappell, *MCC Penn Valley Community College*
Kim Cozean, *Saddleback College*
Katy Cryer
Cindy Cummins, *Ozarks Technical Community College*
Rob Fusco, *Bergen Community College*
Brian Huyvaert, *University of Oregon*
Sharon Jackson, *Brookhaven College*
Kelly Jackson, *Camden County College*

Theresa Killebrew, *Mesa Community College*
Corrine Kirkbride, *Solano Community College*
Brianna Kurtz, *Daytona State College*
Jamie Manche, *Southwestern Illinois College*
Amy Naughten
Christy Peterson, *College of DuPage*
Melissa Rossi, *Southwestern Illinois College*
Janine Termine, *Bucks County Community College*
Linda Schott, *Ozarks Technical Community College*

From the Authors

The authors would like to thank many people at McGraw-Hill. First, our editorial team: Elizabeth O'Brien, Liz Recker, and most of all, Dawn "Dawesome" Bercier, who believed in, championed, and put never-ending energy into our project from the beginning. To Ryan Blankenship, Marty Lange, Kurt Strand, and Brian Kibby: thank you for your continued support and vision that allows us to help students far beyond our own classrooms. Also, Kim Leistner, Nicole Lloyd, Peggy Selle, Peter Vanaria and Stewart Mattson have been instrumental in what they do to help bring our books and digital products to completion.

We offer sincere thanks to Vicki Garringer, Jennifer Caldwell and Sharon Bailey for their contributions to the series.

From Larry Perez: Thank you to my wife, Georgette, for your patience, support, and understanding throughout this endeavor. Thank you to Patrick Quigley and Candice Harrington for your friendship and support. Also, I must thank Dr. Harriet Edwards and Dr. Raghu Mathur for modeling inspirational and innovative pedagogy, examples which I still strive to emulate.

From Bob Feldman: I am grateful to my children, Jonathan, Joshua, and Sarah; my daughters-in-law Leigh and Julie; my smart, handsome, and talented grandsons Alex and Miles, and most of all to my wife, Katherine (who no longer can claim to be the sole mathematician in our family). I thank them all, with great love.

From Sherri Messersmith: Thank you to my daughters, Alex and Cailen, for being the smart, strong, supportive young women that you are; and to my husband, Phil, for understanding the crazy schedule I must keep that often does not complement your own. To Sue, Mary, Sheila, and Jill: everyone should have girlfriends like you. Thank you to the baristas at my hometown Starbucks for your always-smiling faces at 6 am and for letting me occupy the same seat for hours on end. Larry and Bob, thank you for agreeing to become my coauthors and for bringing your expertise to these books. Bill Mulford, we are immensely grateful for your hard work and creativity and for introducing Bob, Larry, and me in the first place. Working with our team of four has been a joy. And, finally, thank you Bill, for your friendship, your patience, and for working with me since the very beginning more than 8 years ago, without question the best student I've ever had.

Sherri Messersmith
Larry Perez
Bob Feldman

Application Index

BIOLOGY AND HEALTH

ages of mother and daughter, 227
ages of sisters, 268
antibiotic remaining in system, 729
bacteria population in culture, 763–764, 771–772, 775, 785
blood alcohol percentage, 153
cats and dogs treated per day, 287
distance person can see to horizon, 586
dog licenses issued per year, 742
drivers in fatal vehicle accidents, 152
foot length and shoe size, 184
generic *vs.* name brand drugs, 93
ibuprofen in bloodstream over time, 213
intravenous drip rate, 484
iodine in system, 776
length of hair, 147–149
medication dosage by weight, 484
peanut allergies treated per year, 739–740
spending on veterinary care, 91
steroid solution, 85

BUSINESS AND MANUFACTURING

billboard rentals, 184
dimensions of bulletin board, 409, 838
dimensions of desktop, 403
gold production by country, 267–268
hybrid vehicles sold, 132
manufacturing cost of notebooks, 495
market share of paper towel brands, 271
market share of tire brands, 287
production cost of clay pigeons, 690
production cost of purses, 690
profit function, 690, 699
profit on book sales, 319–320
profit on purse sales, 470
revenue during construction, 91
salary and commission over time, 168–169
Starbucks worldwide, 64
time to assemble conference notebooks, 486
time to put away clothes, 697
value of U.S. exports, 150

CONSTRUCTION AND WORK

area of Big Ben clock face, 72
area of ice rink, 678
area watered by sprinklers, 73
boards used for playhouse, 289
bridge arch, 816
cost to carpet room, 490
dimensions of cardboard for box, 647–648, 651, 698
dimensions of countertop, 417
dimensions of dog pen, 679
dimensions of door, 269
dimensions of garden, 403, 623, 678
dimensions of glass, 409

dimensions of lot, 261–262
dimensions of Parthenon foyer, 485
dimensions of playground, 269
dimensions of sheet metal, 651
dimensions of storage cube, 524
distance from ceiling of light fixture, 652
distance from ground of shirt on clothesline, 652
distance from ladder to wall, 410
farmland in county, 64
fencing for trapezoidal plot, 73
height of wall with ladder, 622
length of fence for animal pen, 406
length of gravel road, 91
length of pool, 72
length of room, 66–67
length of trapezoidal plot, 73
length of wire attached to pole, 407, 410
lengths of boards, 54–55, 65
lengths of cables, 65
lengths of chain, 55
lengths of pipes, 63
lengths of trim pieces, 63
lengths of wires, 62
maximum dimensions of outdoor café, 698
maximum fenced area, 670–671, 679
Oval Office equation, 816
radius of garden, 524
reinforcing the Leaning Tower of Pisa, 153
slope of a driveway, 167
slope of a highway, 157
slope of a parking garage ramp, 167
slope of a roof, 167
slope of a wheelchair ramp, 167
storage capacity of container, 809
time for ice to reach ground, 524
time to assemble swing set, 506
time to build tree house, 644
time to clean carpets, 504
time to clean pool, 486
time to fertilize lawn, 486
time to mow lawn, 484
time to paint bedroom, 482–484
time to paint billboard, 486
time to paint fence, 486
time to shovel snow, 486
time to trim bushes, 486
weight supported by beam, 496
width of pond border, 648–649
width of window shade, 129

CONSUMER APPLICATIONS

American *vs.* foreign cars, 91
appreciation of home value, 728–729
bouquet supply and demand, 698
break-even point for backpacks, 838
capacity of conical vase, 586
children at birthday party, 105–106

cost of attorney consultation, 801
cost of batteries, 484
cost of car rental, 248
cost of car washes, 270
cost of earrings and necklace, 263
cost of gasoline, 212, 479
cost of library rate postage, 797
cost of mailing large envelope, 797
cost of mailing small packages, 801
cost of metered parking, 801
cost of souvenirs, 270
cost of truck rental, 108, 248
depreciation of car value, 726, 728, 783
depreciation of truck value, 726
dimensions of rug, 409, 586
dimensions of storage box, 409
gallons of ethanol purchased, 79–80
hybrid vehicle registrations over time, 185
loudness of dishwasher, 763
maximum guests at inn, 678
moon jump rental, 152
motel rooms with queen size beds, 261
number of restaurants in U.S., 268
original price of backpack, 64
original price of book, 65
original price of calendar, 63
original price of coffee maker, 64
original price of refrigerator, 64
pages in book, 65
parking garage time limits, 108
personal consumption expenditures over time, 223
price of birthday gifts, 128–129
profit on dog house sales, 323
profit on toaster sales, 323
sale price of bathing suit, 63
sale price of clothing, 717
sale price of dress shirt, 58
sale price of jeans, 58
sale price of stroller, 63
sales of wine, 698
sales tax on clothing, 717
shovel supply and demand, 652
time to address invitations, 486
types of batteries purchased, 270
types of books sold, 480
types of stamps purchased, 83, 270
value of car over time, 167

DISTANCE AND TRAVEL

airport on-time departures, 286
cruise ships operating in North America, 349
distance between car and motorcycle, 418
distance between cyclists, 410
distance driven, 81
distance from home, 411
distance from LA to Chicago, 153
distance on bike, 64

xxiii

distance space shuttles travel, 312
distance to California, 62
distance traveled by jet, 225
distance traveled by sloth, 312
distance traveled by truck, 211
fuselage of Boeing 767, 816
height of dropped rock, 411
height of launched object, 411, 412, 418
height of rocket, 690
height of thrown object, 408
kinetic energy of car, 495
loudness of jet takeoff, 785
maximum height of ball, 669–670
maximum height of object, 670, 678, 698
passengers on New York flight, 62
speed during snowstorm, 644
speed of boat in current, 485
speed of boat in still water, 481–482, 485, 644
speed of car, 85
speed of current, 485, 486, 504
speed of driver at time of accident, 524
speed of plane in wind, 485, 486, 504, 644
speed of planes, 85
speed of walker, 486
speeds of car and bus, 287
speeds of car and train, 134, 271
speeds of car and truck, 270
speeds of cyclists, 265–266, 271
speeds of drivers, 83, 84
speeds of planes, 84, 271
speeds of trains, 84, 270
speeds of walker and cyclist, 271
students biking to class, 62
taxi charges per mile, 108
time for ball to reach ground, 629–630, 632
time for ball to reach height, 629–630, 632, 669–670
time for object to reach ground, 630, 632, 651, 698
time for object to reach height, 630, 632, 651, 670, 678, 698
time to catch up, 81–83, 84, 85, 92
time to meet, 93
time to travel distance, 495
time until distance apart, 84, 85
velocity of an object, 211
velocity of car, 544

EDUCATION

average salary for high school principals, 185
boys and girls in class, 53–54
children not in preschool, 91
freshman in class, 65
number of students over years, 160
per-pupil spending, 154
revenue from t-shirt fundraiser, 412
students giving speeches *vs.* writing papers, 269
students studying French and Spanish, 287
students taking notes in pen *vs.* pencil, 485
test average in class, 62, 108
time to grade tests, 644

ENTERTAINMENT

album downloads per year, 287
albums sold per artist, 65
American Idol viewers, 150
Broadway play attendees, 652
CD sales per artist, 91

cost of concert tickets, 262–263, 269
cost of movie tickets, 271
cost of theater seats, 287
Country Music Awards won, 269
dimensions of television screen, 651
earnings of two movies, 65
Emmy nominations by network, 260–261
movies nominated for Academy Awards, 268
original price of CD, 58–59
original price of video game, 59
profit on television sales, 323
revenue from comedy performance, 412
revenue from theater tickets, 490
revenue from ticket sales, 412
sale price of television, 63
types of movie tickets sold, 83, 85

ENVIRONMENT AND NATURE

altitude and barometric pressure, 153–154
area of oil spill, 717
average temperature in Tulsa, 717
carbon emissions per person, 312
cockroach population increase, 729
deer in wildlife refuge, 485
difference between elevations, 18
difference between temperatures, 15
elevation of city, 18
equivalent temperatures, 75
farms with milk cows, 185
gallons of water in water treatment plant, 213
garbage collected per year, 740
highest temperature in U.S., 18
lengths of rivers, 63
lowest temperature in Colorado, 18
pollution produced by population, 495
quills on porcupine, 349
radius of oil spill, 717
spread of magnetic stripes, 471
sulfur dioxide emissions over time, 180–181
temperatures in Anchorage, 54
weights of dogs, 63
wind chill, 533, 586

FINANCE AND INVESTMENT

average earnings for embalmers, 212
average salary for pharmacists over time, 169
compound interest on account, 759, 763, 785
compound interest on loan, 763
continuous compounding, 760, 763, 770–771, 775, 785
difference in median income, 18
earnings at part-time job, 212
earnings per week, 495
exchange rate between dollars and pesos, 169
income and hours worked, 169
interest earned on annuity, 729
interest earned on investment, 60, 64, 493
interest on two accounts, 64
interest rate needed, 775, 785
investments in three accounts, 65
investments in two accounts, 60–62, 64, 65, 91, 270, 485
net weekly pay, 783
salary over time, 64
salary per year, 65
value of stock over time, 222
wealthiest women in the world, 116

FOOD

caffeine in soda, 484
calories in ice cream, 227
calories in mayonnaise brands, 271
candy mixture, 287
chicken consumption per capita, 269
coffee blend, 84
cookie sales per month, 742
cost of cantaloupe and watermelon, 270
cost of chips and soda, 700
cost of granola, 484
cost of hamburger and fries, 270
cost of hot dog, fries, and soda, 271
cost of hot dog and soda, 287
cost of ice cream cone, shake, and sundae, 287
cost of meals, 269
cost of nut mixture, 270
cost of potatoes, 478–479
fat in Starbucks drink, 504
fruit juice mixture, 270
grams of protein in protein bars, 271
guacamole eaten during Super Bowl, 312
height of coffee can, 73
height of tomato sauce can, 73
length of candy bar, 419
milk consumption per capita, 169
nut mixture, 84
original price of dog food, 93
ounces in cereal box, 128
ounces in milk container, 128
ounces in soup can, 125
potato chips consumed per person, 152
profit on candy sales, 320
profit on salmon sales, 470
sales of dog food, 784
sales of hamburgers, 62
sandwich supply and demand, 652
sodium in drinks, 287
sugar in drinks, 62, 270
types of coffee ordered, 479–480
types of flour in mixture, 485
volume of soup can, 586

INTERNET AND TECHNOLOGY

dimensions of computer screen, 838
dimensions of iPod, 269
dimensions of laptop screen, 623
dimensions of monitor, 287
dimensions of mouse pad, 262
DVD data transfer rate, 212
Google quarterly revenue, 289
households with Internet access, 247
ink droplets per photo print, 312
length of copy machine paper, 72
original price of camera, 63
original price of cell phone, 65
profit on calculator sales, 323
profit on laptop sales, 323
sale price of cell phone, 63
sales of digital cameras, 700
samples of sound read from CD, 207–208, 212
surface area of CD, 809
text messages per month, 106
texts per person per day, 287
time for programming job, 486
time to print pictures, 486

time to set up alarm system, 486
USB data transfer rate, 225
width of printed area, 72
wireless communication subscribers over time, 223
YouTube video views, 289

SCIENCE AND CHEMISTRY

acid solution, 80–81, 84, 92, 265, 270
alcohol solution, 81, 84, 270, 287, 352, 506, 851
antifreeze solution, 84
cleaning solution, 485
current in circuit, 492
dimensions of aquarium, 409
focal length of lens, 470
force exerted on object, 495
force to stretch spring, 496
frequency of piano string, 495
height of firework shell, 411–412
height of tank, 72
hydrogen peroxide solution, 84, 264–265
illuminance, 652
impedance of circuits, 586
intensity of light, 492
loudness of space shuttle, 763
mass of water molecules, 349
period of pendulum, 524
pH of substances, 764, 785
power in electrical system, 495
power generated by Hoover Dam, 312
radioactive decay, 772–773, 775–776, 785
radius of water tank, 552
resistance of wire, 495
silver alloy, 85
speed of sound, 586
surface area of cube, 495
time to empty tank, 644
time to fill pool, 644

time to fill sink, 486
time to fill tub, 486
volume of box, 493
volume of candle wax, 552
volume of cylinder, 495, 586
volume of gas, 504
wave velocity, 586, 604
weight of ball, 504
weight of object above Earth, 496

SOCIOLOGY AND DEMOGRAPHICS

addresses of houses, 63
areas of countries, 62
babies born to teen mothers, 168, 678
change in housing starts, 18–19
females in Belgian workforce, 186
fingerprint comparisons per second, 212
households with pets, 485
increase in housing, 742
maximum traffic tickets written, 678
men in civilian labor force over time, 169–170
population change in Oakland, 19
population decrease, 775
population increase, 775, 785
population of Maine over time, 184–185
population of North Dakota over time, 185
tourism-related output in U.S., 649–650
tourists visiting per year, 742
violent crimes in U.S., 678
voters for each candidate, 63, 485

SPORTS AND HOBBIES

break-even point for basketballs, 838
colors in paint mixture, 485
cost of baseball tickets, 271
cost of football tickets and parking, 270
cost of soccer uniforms, 349
difference between golf scores, 15
difference in baseball attendance, 18

dimensions of bandana, 417
dimensions of fabric pieces, 651
dimensions of Ferris wheel, 809
dimensions of London Eye, 809
dimensions of painting, 409
dimensions of picture, 269
dimensions of sail, 651
female motocross spectators, 484
height of baseball, 407–408
height of bike ramp, 417, 651
length of kite string, 623
length of side of die, 524
lengths of jump ropes, 63
loudness of basketball game, 785
markup on fishing poles, 65
NBA championships won, 114
NCAA championships won, 269
NCAA championship viewers, 269
number of male runners, 62
Olympic medals per country, 63
Olympics participants, 64
pitchers on baseball team, 62
profit on bicycle sales, 323
racing winnings over time, 18
revenue of basketball teams, 271
schools in NCAA conferences, 717
snowboarding and ice skating participants, 246–247
soccer games played in season, 65
speed of baseball pitch, 133
speed of runner, 644
Super Bowl net yardage, 18
teams playing in Rose Bowl, 268
time to cut out shapes, 486
width of basketball lane, 73
width of picture frame border, 651
width of pillow sham border, 698
width of ping-pong table, 67
width of swimming pool border, 651

Real Numbers and Algebraic Expressions

CHAPTER 1

OUTLINE

Study Strategies: The P.O.W.E.R. Framework
1.1 Sets of Numbers
1.2 Operations on Real Numbers
1.3 Exponents, Roots, and Order of Operations
1.4 Algebraic Expressions and Properties of Real Numbers
Group Activity
emPOWERme: Why Am I Going to College?

Math at Work:

Computer Game Designer

Ever since he was a child, Dave Cantelmo has known what he wanted to do: create video games. "I still remember playing games on the classic systems I had growing up," Dave says. "I would spend hours and hours playing those games and always imagined the games I wanted to create myself one day."

In order to realize his ambitions, Dave put in the time and effort necessary to learn the design and computer programming skills involved in video game development. Early on in college, he was particularly focused on building up his math abilities, as math is critical to the technical side of video game creation. "I was always a little intimidated by math," Dave describes. "But with hard work and the help of my instructors, I was able to turn math into one of my strengths. Now, it's something I use in my job every day."

Creating a successful video game is a complex challenge. "Completing a video game takes the coordinated efforts of dozens of designers, writers, programmers, animators, and many other people, all working for years to turn an idea into a game people all over the world can play," Dave explains. He says the key to completing such a difficult task is taking a smart, organized approach, dividing the work into steps that will culminate in the finished product.

In this chapter, we will discuss the topic of real numbers. We will also introduce P.O.W.E.R., a framework that can help you succeed in the complex challenges you face, either in the classroom, on the job, or in your daily life.

Study Strategies — The P.O.W.E.R. Framework

The **P.O.W.E.R. Framework** is based on an acronym—a word formed from the first letters of a series of steps. P.O.W.E.R. stands for **P**repare, **O**rganize, **W**ork, **E**valuate, and **R**ethink. That's it. It's a simple framework, but an effective one. P.O.W.E.R. gives you a proven, ready-to-use approach to virtually any challenge you face, from studying for a math test to developing a presentation for your coworkers to writing the family grocery list. Think of its steps as a roadmap to success no matter what your task is.

Whether you are familiar with the P.O.W.E.R. framework already or are encountering it for the first time, take a moment now to review each of its steps in depth:

- Think about what you are trying to accomplish: Define both your short-term and long-term goals.
- **Long-term goals** are major accomplishments that take a significant amount of time and effort to achieve, such as graduating from college. **Short-term goals** are steps that are easier to accomplish and bring you closer to your long-term goals—for example, doing well on a math exam.

- Identify the tools you will need to complete your task.
- Effective organization involves gathering both the *physical* tools you will need to complete your task (for example, a textbook, pen, paper, and so forth) and doing the *mental* work (reviewing lecture notes or major concepts in your textbook, say) to ensure you are ready to succeed.

- With the previous steps as your foundation, do the work of completing your task.
- When doing math tasks in particular, it is important to work efficiently but patiently, neither trying to rush through the work nor becoming frustrated and giving up after the first difficulty you encounter.
- Stay motivated by keeping your goals in mind.

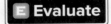

- Think back to your goal for the task. Did you meet your own expectations?
- Revise your work based on your assessment of it.
- In math courses, it's important to identify the specific obstacles that may be causing you to perform below your capabilities. Are there particular concepts you are struggling with? Would you benefit from working with a math tutor or your fellow students?

- Think critically about both the work you have done and the process you used to complete it. What did you do that worked well? Where do you see room for improvement?
- Take a step back and consider how the task you completed brought you closer to your long-term goals.

Chapter 1 POWER Plan

P Prepare | O Organize

What are your goals for Chapter 1?	How can you accomplish each goal?
1 Be prepared before and during class.	• Don't stay out late the night before and be sure to set your alarm clock! • Bring a pencil, notebook paper, and textbook to class. • Avoid distractions by turning off your cell phone during class. • Pay attention, take good notes, and ask questions. • Complete your homework on time and ask questions on problems you do not understand.
2 Understand the homework to the point where you could do it without needing any help or hints.	• Read the directions and show all of your steps. • Go to the professor's office for help. • Rework homework and quiz problems and find similar problems for practice.
3 Use the P.O.W.E.R. framework to help you organize your study: *Why Am I Going to College?*	• Read the Study Strategy that explains how to use P.O.W.E.R. • What does P.O.W.E.R. stand for? • Complete the emPOWERme that appears before the Chapter Summary.
4 Write your own goal. _____ _____	• _____ _____

What are your objectives for Chapter 1?	How can you accomplish each objective?
1 Learn the different sets of numbers.	• Learn the definitions in Section 1.1. • Take good notes in class.
2 Be able to add, subtract, multiply, and divide real numbers.	• Master the steps for these operations! Future sections build on this knowledge, so continually review these operations.
3 Understand how to evaluate expressions, roots, and exponents. Be able to simplify an expression using the order of operations.	• Understand the meaning of exponents and how to identify the base and exponent. • Memorize the powers of integers in Section 1.3. • Learn the **Order of Operations:** Use **P**lease **E**xcuse **M**y **D**ear **A**unt **S**ally to help you remember the order of operations.
4 Understand the properties of real numbers and how they can be used to simplify problems.	• Understand the meaning of each property based on the meaning of the word. (Commutative property = "commute.") • Be aware of when the properties can be used to help simplify a problem.
5 Write your own goal. _____ _____	• _____ _____

W Work	Read Sections 1.1 to 1.4, and complete the exercises.		
E Evaluate Complete the Chapter Review and Chapter Test. How did you do?		**R Rethink**	• How did you perform on the goals for the chapter? Which steps could be improved for next time? If you had the chance to do this chapter over what would you do differently? • Think of a job you might like to have and describe how you would need to use what you have just learned to effectively do that job. • How has the P.O.W.E.R. framework helped you master the objectives of this chapter? Where else could you use this framework? Make it a point to use P.O.W.E.R. to complete another task this week.

1.1 Sets of Numbers

P Prepare	**O Organize**
What are your objectives for Section 1.1?	How can you accomplish each objective?
1 Identify Numbers and Graph Them on a Number Line	• Know the definition of *natural numbers, whole numbers, integers, rational numbers,* and *irrational numbers.* • Determine when a number may belong to more than one number set. • Know how to draw/label a number line and determine where that number should be graphed. • Complete the given examples on your own. • Complete You Trys 1, 2, and 3.
2 Compare Numbers Using Inequality Symbols	• Write the inequality symbols in math notation and in words. • Know how to determine the larger of two given numbers. • Complete the given examples on your own. • Complete You Trys 4 and 5.
3 Find the Absolute Value of a Number	• Know the definitions of *additive inverse* and *absolute value.* • Be able to show, on a number line, the *additive inverse* and the *absolute value* of a given number. • Complete the given examples on your own. • Complete You Trys 6 and 7.

 Read the explanations, follow the examples, take notes, and complete the You Trys.

4   CHAPTER 1   **Real Numbers and Algebraic Expressions**   www.mhhe.com/messersmith

1 Identify Numbers and Graph Them on a Number Line

Why should we review sets of numbers and arithmetic skills? Because the manipulations done in arithmetic are precisely the same set of skills needed to learn algebra. Let's begin by defining some numbers used in arithmetic:

The set of **natural numbers** is {1, 2, 3, 4, 5, ...}.

The set of **whole numbers** is {0, 1, 2, 3, 4, 5, ...}.

Natural numbers are often thought of as the counting numbers. Whole numbers consist of the natural numbers and zero. Let's look at other sets of numbers. We begin with integers. Remember that, on a number line, positive numbers are to the right of zero, and negative numbers are to the left of zero.

> ### Definition
> The set of **integers** includes the set of natural numbers, their negatives, and zero. The set of *integers* is {..., −3, −2, −1, 0, 1, 2, 3, ...}.

EXAMPLE 1 Graph each number on a number line: 5, 1, −2, 0, −4.

Solution

$$\xleftarrow{\quad}\underset{-6\ -5\ -4\ -3\ -2\ -1\ \ 0\ \ 1\ \ 2\ \ 3\ \ 4\ \ 5\ \ 6}{\overset{-4\ \ \ -2\ \ 0\ 1\ \ \ \ \ \ \ \ 5}{\bullet\ \ \bullet\ \ \bullet\ \bullet\ \ \ \ \ \ \ \ \bullet}}\xrightarrow{\quad}$$

5 and 1 are to the right of zero since they are positive. −2 is two units to the left of zero, and −4 is four units to the left of zero.

YOU TRY 1 Graph each number on a number line: 3, −1, 6, −5, −3.

W Hint
Notice that we use the set of *integers* to label the number line in Example 1.

Positive and negative numbers are also called *signed numbers*.

EXAMPLE 2 Given the set of numbers $\left\{-11, 0, 9, -5, -1, \frac{2}{3}, 6\right\}$, list the

a) whole numbers   b) natural numbers   c) integers

Solution

a) whole numbers: 0, 6, 9   b) natural numbers: 6, 9

c) integers: −11, −5, −1, 0, 6, 9

YOU TRY 2 Given the set of numbers $\left\{3, -2, -9, 4, 0, \frac{5}{8}, -\frac{1}{3}\right\}$, list the

a) whole numbers   b) natural numbers   c) integers

Notice in Example 2 that $\frac{2}{3}$ did not belong to any of these sets. That is because the whole numbers, natural numbers, and integers do not contain any fractional parts. $\frac{2}{3}$ is a *rational number*.

> **W Hint**
> Remember that whole numbers, natural numbers, and integers can all be written in the form $\frac{p}{q}$ where $q = 1$.

Definition

A **rational number** is any number of the form $\frac{p}{q}$, where p and q are integers and $q \neq 0$.

In other words, a rational number is any number that can be written as a fraction where the numerator and denominator are integers and the denominator does not equal zero.

Rational numbers include much more than numbers like $\frac{2}{3}$, which are already in fractional form. They also include numbers such as 3, 0.5, -7, $0.\overline{4}$, and the square root of 9 because each of these numbers can be written as a fraction.

To summarize, the set of rational numbers includes

1) integers, whole numbers, and natural numbers.
2) repeating decimals.
3) terminating decimals.
4) fractions and mixed numbers.

The set of rational numbers does *not* include nonrepeating, nonterminating decimals or radicals like $\sqrt{5}$ where the radicand is *not* a perfect square. These numbers cannot be written as the quotient of two integers. Numbers such as these are called *irrational numbers*.

> **W Hint**
> Note that it is not possible to write an irrational number as a fraction. Some irrational numbers are π, $\sqrt{5}$, 0.412738...

Definition

The set of numbers that cannot be written as the quotient of two integers is called the set of **irrational numbers**. Written in decimal form, an *irrational number* is a nonrepeating, nonterminating decimal.

If we put together the sets of numbers we have discussed up to this point, we get the *real numbers*.

Definition

The set of **real numbers** consists of the rational and irrational numbers.

We summarize the information next with examples of the different sets of numbers:

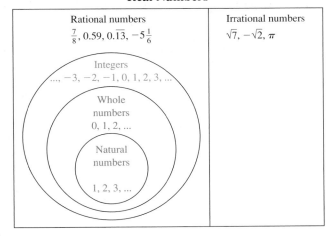

From the figure we can see, for example, that all whole numbers {0, 1, 2, 3, ...} are integers, but not all integers are whole numbers (−3, for example).

EXAMPLE 3

Given the set of numbers $\left\{0.\overline{2}, 37, -\dfrac{4}{15}, \sqrt{11}, -19, 8.51, 0, 6.149235...\right\}$, list the

a) integers b) natural numbers c) whole numbers
d) rational numbers e) irrational numbers f) real numbers

Solution

a) integers: −19, 0, 37
b) natural numbers: 37
c) whole numbers: 0, 37
d) rational numbers: $0.\overline{2}, 37, -\dfrac{4}{15}, -19, 8.51, 0$ Each of these numbers can be written as the quotient of two integers.
e) irrational numbers: $\sqrt{11}, 6.149235...$ These numbers *cannot* be written as the quotient of two integers.
f) real numbers: All of the numbers in this set are real.
$\left\{0.\overline{2}, 37, -\dfrac{4}{15}, \sqrt{11}, -19, 8.51, 0, 6.149235...\right\}$

YOU TRY 3

Given the set of numbers $\left\{-38, 0, \sqrt{15}, 6, \dfrac{3}{2}, 5.4, 0.\overline{8}, 4.981162...\right\}$, list the

a) whole numbers b) integers c) rational numbers
d) irrational numbers

SECTION 1.1 Sets of Numbers

2 Compare Numbers Using Inequality Symbols

Let's review the inequality symbols.

$<$ less than $\leq$ less than or equal to
$>$ greater than $\geq$ greater than or equal to
$\neq$ not equal to $\approx$ approximately equal to

We use these symbols to compare numbers as in $5 > 2$, $6 \leq 17$, $4 \neq 9$, and so on. How do we compare negative numbers?

W Hint
Think about why -8 is less than -7 on a number line. Draw a number line and notice that -8 is to the left of -7.

Note
As we move to the *left* on the number line, the numbers get smaller. As we move to the *right* on the number line, the numbers get larger.

EXAMPLE 4 Insert $>$ or $<$ to make the statement true. Look at the number line, if necessary.

a) $5 \_ 1$ b) $-4 \_ 3$ c) $-1 \_ -5$ d) $-5 \_ -2$

Solution

a) $5 > 1$ 5 is to the right of 1.   b) $-4 < 3$ -4 is to the left of 3.
c) $-1 > -5$ -1 is to the right of -5.   d) $-5 < -2$ -5 is to the left of -2.

[YOU TRY 4] Insert $>$ or $<$ to make the statement true.
a) $4 \_ 9$ b) $6 \_ -8$ c) $-3 \_ -10$

We use signed numbers in everyday situations.

EXAMPLE 5 Use a signed number to represent the change in each situation.

a) After a storm passed through Kansas City, the temperature dropped 18°.

b) Between 2006 and 2010, retail sales of gluten-free products rose by $1.3 billion. (www.celiac.com)

Solution

a) $-18°$ The negative number represents a decrease in temperature.
b) $1.3 billion   The positive number represents an increase in sales.

[YOU TRY 5] Use a signed number to represent the change.
After taking his last test, Julio raised his average by 3.5%.

CHAPTER 1 **Real Numbers and Algebraic Expressions**

3 Find the Absolute Value of a Number

Before we discuss absolute value, we will define *additive inverses*.

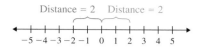

Notice that both -2 and 2 are a distance of 2 units from 0 but are on opposite sides of 0. We say that 2 and -2 are *additive inverses*.

W Hint
Notice that the additive inverse of a negative number is always positive. For a positive number, the additive inverse is always negative.

Definition

Two numbers are **additive inverses** if they are the same distance from 0 on the number line but on the opposite sides of 0. Therefore, if a is any real number, then $-a$ is its additive inverse.

Furthermore, $-(-a) = a$. We can see this on the number line.

EXAMPLE 6 Find $-(-4)$.

Solution

Beginning with -4, the number on the opposite side of zero and 4 units away from zero is 4. So, $-(-4) = 4$.

YOU TRY 6 Find $-(-11)$.

This idea of "distance from zero" can be explained in another way: *absolute value*.

Definition

If a is any real number, then the **absolute value of a,** denoted by $|a|$, is

    i) a if $a \geq 0$
    ii) $-a$ if $a < 0$

Remember, $|a|$ is never negative.

Note
The absolute value of a number is the distance between that number and 0 on the number line. It just describes the distance, *not* what side of zero the number is on. Therefore, the absolute value of a number is always positive or zero.

EXAMPLE 7

Evaluate each.

a) $|9|$ b) $|-7|$ c) $|0|$ d) $-|5|$ e) $|11 - 5|$

Solution

a) $|9| = 9$ 9 is 9 units from 0.

b) $|-7| = 7$ -7 is 7 units from 0.

c) $|0| = 0$

d) $-|5| = -5$ First, evaluate $|5|$: $|5| = 5$. Then, apply the negative symbol to get -5.

e) $|11 - 5| = |6|$ The absolute value symbols work like parentheses. First, evaluate what is inside: $11 - 5 = 6$.
   $= 6$ Find the absolute value.

[YOU TRY 7]

Evaluate each.

a) $|16|$ b) $|-4|$ c) $-|6|$ d) $|14 - 9|$

ANSWERS TO [YOU TRY] EXERCISES

1) [number line showing points at −5, −3, −1, 3, 6]

2) a) 0, 3, 4   b) 3, 4   c) −9, −2, 0, 3, 4

3) a) 0, 6   b) −38, 0, 6   c) −38, 0, 6, $\frac{3}{2}$, 5.4, $0.\overline{8}$ d) $\sqrt{15}$, 4.981162...

4) a) <   b) >   c) >

5) 3.5%   6) 11   7) a) 16   b) 4   c) −6   d) 5

E Evaluate 1.1 Exercises

Do the exercises, and check your work.

Objective 1: Identify Numbers and Graph Them on a Number Line

1) Given the set of numbers

$$\left\{-14, 6, \frac{2}{5}, \sqrt{19}, 0, 3.\overline{28}, -1\frac{3}{7}, 0.95\right\},$$

list the

a) whole numbers.
b) integers.
c) irrational numbers.
d) natural numbers.
e) rational numbers.
f) real numbers.

2) Given the set of numbers

$$\left\{5.2, 34, -\frac{9}{4}, -18, 0, 0.\overline{7}, \frac{5}{6}, \sqrt{6}, 4.3811275...\right\},$$

list the

a) integers.
b) natural numbers.

c) rational numbers.
d) whole numbers.
e) irrational numbers.
f) real numbers.

Determine whether each statement is true or false.

3) Every whole number is an integer.
4) Every rational number is a whole number.
5) Every real number is an integer.
6) Every natural number is a whole number.
7) Every integer is a rational number.
8) Every whole number is a real number.

Graph the numbers on a number line. Label each.

9) $6, -4, \frac{3}{4}, 0, -1\frac{1}{2}$

10) $5\frac{2}{3}, 1, -3, -4\frac{5}{6}, 2\frac{1}{4}$

11) $1.7, -\dfrac{4}{5}, 3\dfrac{1}{5}, -5, -2\dfrac{1}{2}$

12) $-5, 6\dfrac{1}{8}, 2\dfrac{3}{4}, -2\dfrac{5}{7}, 4.3$

13) $-6.8, -\dfrac{3}{8}, 0.2, 1\dfrac{8}{9}, -4\dfrac{1}{3}$

14) $-1, 5.9, 1\dfrac{7}{10}, -\dfrac{2}{3}, 0.61$

Objective 3: Find the Absolute Value of a Number
Evaluate.

15) $|-23|$

16) $|8|$

17) $\left|\dfrac{3}{2}\right|$

18) $|-13|$

19) $-|10|$

20) $-|6|$

21) $-|-19|$

22) $-\left|-1\dfrac{3}{5}\right|$

Find the additive inverse of each number.

23) 11

24) 5

25) -7

26) $-\dfrac{1}{2}$

27) -4.2

28) 2.9

Mixed Exercises: Objectives 2 and 3
Write each group of numbers from smallest to largest.

29) $7, -2, 3.8, -10, 0, \dfrac{9}{10}$

30) $-6, -7, 5.2, 5.9, 6, -1$

31) $-4\dfrac{1}{2}, \dfrac{5}{8}, \dfrac{1}{4}, -0.3, -9, 1$

32) $14, -5, 13.6, -5\dfrac{2}{3}, 1, \dfrac{6}{7}$

Determine whether each statement is true or false.

33) $9 \geq -2$

34) $-6 > 3$

35) $-7 \leq -4$

36) $10.8 \geq 10.2$

37) $\dfrac{1}{6} \leq \dfrac{1}{8}$

38) $-8.1 > -8.5$

39) $-5\dfrac{3}{10} < -5\dfrac{3}{4}$

40) $\dfrac{4}{5} \neq 0.8$

41) $|-9| \geq 9$

42) $-|-31| = 31$

Use a signed number to represent the change in each situation.

43) In 2001, Barry Bonds set the all-time home run record with 73. In 2002, he hit 46. That was a decrease of 27 home runs. (www.mlb.com)

44) In 2008, the median income for households in Texas was $50,043, and in 2009 it decreased by $1784 to $48,259. (www.census.gov)

45) In 2005, the U.S. unemployment rate was 5.1% and in 2006 it decreased by 0.5% to 4.6%. (www.census.gov)

46) During the 2010–2011 season, Kobe Bryant averaged 25.3 points per game (ppg). The following season he averaged 27.9 ppg, an increase of 2.6 over the previous year. (www.nba.com)

47) In Michael Jordan's last season with the Chicago Bulls (1997–1998) the average attendance at the United Center was 23,988. During the 2011–2012 season, the average attendance fell to 22,161. This was a decrease of 1827 people per game. (espn.go.com)

48) The 2004 Indianapolis 500 was won by Buddy Rice with an average speed of 138.518 mph. In 2012, Dario Franchitti won it with an average speed of 167.734 or 29.216 mph faster than in 2004. (www.indy500.com)

49) According to the 2000 census, the population of North Dakota was 642,200. In 2010, it increased by 30,391 to 672,591. (www.census.gov)

50) The 1999 Hennessey Viper Venom can go from 0 to 60 mph in 3.3 sec. The 2000 model goes from 0 to 60 mph in 2.7 sec, which is a decrease of 0.6 sec to go from 0 to 60 mph. (www.supercars.net)

R Rethink

R1) Think about why we call the numbers in this section *real numbers*. Why not just call them numbers? Does this mean that there are nonreal numbers?

R2) If the absolute value of a number is greater than the absolute value of another, what does this mean in terms of their location on a number line?

R3) In a computer programming language, what do you think "abs(−24)" means?

R4) Why is the additive inverse of a positive number always negative?

1.2 Operations on Real Numbers

P Prepare | O Organize

What are your objectives for Section 1.2?	How can you accomplish each objective?
1 Add Real Numbers	• Know the procedure for adding numbers with the same sign and adding numbers with different signs. • Know that the sum of a number and its *additive inverse* is 0. • Complete the given examples on your own. • Complete You Trys 1 and 2.
2 Subtract Real Numbers	• Know the definition of subtraction and how to rewrite subtraction problems using the definition. • Complete the given example on your own. • Complete You Try 3.
3 Solve Applied Problems Involving Addition and Subtraction	• Be able to determine if the problem requires subtraction or addition. • Be careful and use parentheses when necessary. • Learn to recognize whether your answer is logical. • Complete the given example on your own. • Complete You Try 4.
4 Multiply Real Numbers	• Know the rules for multiplying positive and negative numbers. • Complete the given example on your own. • Complete You Try 5.
5 Divide Real Numbers	• Know the rules for dividing positive and negative numbers. • Complete the given example on your own. • Complete You Try 6.

W Work

Read the explanations, follow the examples, take notes, and complete the You Trys.

In the previous section, we defined real numbers. In this section, we will discuss adding, subtracting, multiplying, and dividing real numbers.

1 Add Real Numbers

Recall that when we add two numbers with the same sign, the result has the same sign as the numbers being added.

$$2 + 5 = 7 \qquad -1 + (-4) = -5$$

Hint
The sum of two negative numbers is always negative. Similarly, the sum of two positive numbers is always positive.

Procedure Adding Numbers with the Same Sign

To add numbers with the same sign, find the absolute value of each number and add them. The sum will have the same sign as the numbers being added.

EXAMPLE 1

Add.

a) $-8 + (-2)$ b) $-35 + (-71)$

Solution

a) $-8 + (-2) = -(|-8| + |-2|) = -(8 + 2) = -10$

b) $-35 + (-71) = -(|-35| + |-71|) = -(35 + 71) = -106$

[YOU TRY 1]

Add.

a) $-2 + (-9)$ b) $-48 + (-67)$

Hint
Notice the use of parentheses when adding a negative number.

Let's review how to add numbers with different signs.

$$2 + (-5) = -3 \quad \text{and} \quad -8 + 12 = 4$$

Procedure Adding Numbers with Different Signs

To add two numbers with different signs, find the absolute value of each number. Subtract the smaller absolute value from the larger. The sum will have the sign of the number with the larger absolute value.

Hint
When adding numbers with different signs, the answer always has the sign of the number that is farthest from zero on the number line!

Let's apply this to $2 + (-5)$ and $-8 + 12$.

$2 + (-5)$: $|2| = 2 \quad |-5| = 5$
Since $2 < 5$, subtract $5 - 2$ to get 3. Since $|-5| > |2|$, the sum will be negative.
$$2 + (-5) = -3$$

$-8 + 12$: $|-8| = 8 \quad |12| = 12$
Subtract $12 - 8$ to get 4. Since $|12| > |-8|$, the sum will be positive.
$$-8 + 12 = 4$$

EXAMPLE 2

Add.

a) $-19 + 4$ b) $10.3 + (-4.1)$ c) $\dfrac{1}{4} + \left(-\dfrac{5}{9}\right)$ d) $-7 + 7$

Solution

a) $-19 + 4 = -15$ The sum will be negative because the number with the larger absolute value, $|-19|$, is negative.

b) $10.3 + (-4.1) = 6.2$ The sum will be positive because the number with the larger absolute value, $|10.3|$, is positive.

c) $\dfrac{1}{4} + \left(-\dfrac{5}{9}\right) = \dfrac{9}{36} + \left(-\dfrac{20}{36}\right)$ Get a common denominator.

$= -\dfrac{11}{36}$ The sum will be negative because the number with the larger absolute value, $\left|-\dfrac{20}{36}\right|$, is negative.

d) $-7 + 7 = 0$

Note

The sum of a number and its additive inverse is always 0. That is, if a is a real number, then $a + (-a) = 0$. Notice in part d) of Example 2 that -7 and 7 are additive inverses.

[YOU TRY 2]

Add.

a) $11 + (-8)$ b) $-17 + (-5)$ c) $-\dfrac{5}{8} + \dfrac{2}{3}$ d) $59 + (-59)$

2 Subtract Real Numbers

Subtraction of numbers can be defined in terms of the additive inverse. We'll begin by looking at a basic subtraction problem.

Represent $6 - 4$ on a number line.

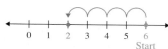

Begin at 6. To subtract 4, move 4 units to the left to get 2: $6 - 4 = 2$.

We use the same process to find $6 + (-4)$. This leads us to a definition of subtraction:

Definition

If a and b are real numbers, then $a - b = a + (-b)$.

The definition tells us that to subtract $a - b$,

1) change subtraction to addition.
2) find the additive inverse of b.
3) add a and the additive inverse of b.

EXAMPLE 3

Subtract.

a) $5 - 16$ b) $-19 - 4$ c) $20 - 5$ d) $7 - (-11)$

Solution

a) $5 - 16 = 5 + (-16) = -11$

　　　　　　↗　　　↖
　Change to　　Additive inverse
　addition　　　of 16

b) $-19 - 4 = -19 + (-4) = -23$

　　　　　　↗　　　↖
　Change to　　Additive inverse
　addition　　　of 4

c) $20 - 5 = 20 + (-5)$
　　　　　$= 15$

d) $7 - (-11) = 7 + 11 = 18$

　　　　　　↗　↖
　　Change to　Additive inverse
　　addition　　of -11

Hint
Notice in part d), that parentheses are used to separate the *subtraction* symbol and the *negative* sign.

YOU TRY 3

Subtract.

a) $3 - 10$ b) $-6 - 12$ c) $18 - 7$ d) $9 - (-16)$

In part d) of Example 3, $7 - (-11)$ changed to $7 + 11$. This shows that *subtracting a negative number is equivalent to adding a positive number.* Therefore, $-2 - (-6) = -2 + 6 = 4$.

3 Solve Applied Problems Involving Addition and Subtraction

Sometimes, we use signed numbers to solve real-life problems.

EXAMPLE 4

The lowest temperature ever recorded was $-129°F$ in Vostok, Antarctica. The highest temperature on record is $136°F$ in Al'Aziziyah, Libya. What is the difference between these two temperatures? (www.ncdc.noaa.gov)

Solution

$$\text{Difference} = \text{Highest temperature} - \text{Lowest temperature}$$
$$= 136 - (-129)$$
$$= 136 + 129$$
$$= 265$$

The difference between the temperatures is $265°F$.

Hint
Don't forget to include the correct units when needed.

YOU TRY 4

The best score in a golf tournament was -12, and the worst score was $+17$. What is the difference between these two scores?

4 Multiply Real Numbers

Let's review the rules for multiplying real numbers.

SECTION 1.2　Operations on Real Numbers　15

Procedure Multiplying Real Numbers

1) The product of two positive numbers is positive.
2) The product of two negative numbers is positive.
3) The product of a positive number and a negative number is negative.
4) The product of any real number and zero is zero.

EXAMPLE 5

Multiply.

a) $-7 \cdot (-3)$ b) $-2.5 \cdot 8$ c) $-\dfrac{4}{5} \cdot \left(-\dfrac{1}{6}\right)$ d) $-3 \cdot (-4) \cdot (-5)$

Solution

a) $-7 \cdot (-3) = 21$ The product of two negative numbers is positive.

b) $-2.5 \cdot 8 = -20$ The product of a negative number and a positive number is negative.

c) $-\dfrac{4}{5} \cdot \left(-\dfrac{1}{6}\right) = -\dfrac{\overset{2}{\cancel{4}}}{5} \cdot \left(-\dfrac{1}{\underset{3}{\cancel{6}}}\right) = \dfrac{2}{15}$ The product of two negatives is positive.

d) $\underbrace{-3 \cdot (-4)}_{12} \cdot (-5) = 12 \cdot (-5) = -60$ Multiply from left to right.

[YOU TRY 5]

Multiply.

a) $-2 \cdot 8$ b) $-\dfrac{10}{21} \cdot \dfrac{14}{15}$ c) $-2 \cdot (-3) \cdot (-1) \cdot (-4)$

W Hint
Notice how the parentheses are used when multiplying negative numbers!

Note
It is helpful to know that

1) an **even number** of negative factors in a product gives a positive result.

   $-3 \cdot 1 \cdot (-2) \cdot (-1) \cdot (-4) = 24$ Four negative factors

2) an **odd number** of negative factors in a product gives a negative result.

   $5 \cdot (-3) \cdot (-1) \cdot (-2) \cdot (3) = -90$ Three negative factors

5 Divide Real Numbers

Next we will review the rules for dividing signed numbers.

Procedure Dividing Real Numbers

1) The quotient of two positive numbers is a positive number.
2) The quotient of two negative numbers is a positive number.
3) The quotient of a positive and a negative number is a negative number.

EXAMPLE 6

Divide.

a) $-48 \div 6$ b) $-\dfrac{1}{12} \div \left(-\dfrac{4}{3}\right)$ c) $\dfrac{-6}{-1}$ d) $\dfrac{-27}{72}$

Solution

a) $-48 \div 6 = -8$

b) $-\dfrac{1}{12} \div \left(-\dfrac{4}{3}\right) = -\dfrac{1}{12} \cdot \left(-\dfrac{3}{4}\right)$ When dividing by a fraction, multiply by the reciprocal.

$= -\dfrac{1}{\underset{4}{12}} \cdot \left(-\dfrac{\overset{1}{3}}{4}\right)$

$= \dfrac{1}{16}$

W Hint
Remember, write all answers in lowest terms.

c) $\dfrac{-6}{-1} = 6$ The quotient of two negative numbers is positive, and $\dfrac{6}{1} = 6$.

d) $\dfrac{-27}{72} = -\dfrac{27}{72}$ The quotient of a negative number and a positive number is negative.

$= -\dfrac{3}{8}$ Divide 27 and 72 by 9.

It is important to note here that there are three ways to write the answer: $-\dfrac{3}{8}, \dfrac{-3}{8}$, or $\dfrac{3}{-8}$. These are equivalent. However, we usually write the negative sign in front of the entire fraction as in $-\dfrac{3}{8}$.

[YOU TRY 6]

Divide.

a) $-\dfrac{4}{21} \div \left(-\dfrac{2}{7}\right)$ b) $\dfrac{72}{-8}$ c) $\dfrac{-19}{-1}$

ANSWERS TO [YOU TRY] EXERCISES

1) a) -11 b) -115 2) a) 3 b) -22 c) $\dfrac{1}{24}$ d) 0   3) a) -7 b) -18 c) 11 d) 25

4) 29   5) a) -16 b) $-\dfrac{4}{9}$ c) 24   6) a) $\dfrac{2}{3}$ b) -9 c) 19

E Evaluate 1.2 Exercises

Do the exercises, and check your work.

Mixed Exercises: Objectives 1 and 2

1) Explain, in your own words, how to add two negative numbers.

2) Explain, in your own words, how to add a positive and a negative number.

3) Explain, in your own words, how to subtract two negative numbers.

4) Explain, in your own words, how to add two positive numbers.

Add or subtract as indicated.

5) $9 + (-13)$ 6) $-7 + (-5)$

7) $-2 - 12$ 8) $-4 + 11$

9) $-25 + 38$ 10) $-1 - (-19)$

11) $10 - (-17)$ 12) $-40 - (-6)$

13) $-794 - 657$
14) $380 + (-192)$
15) $-\dfrac{3}{10} + \dfrac{4}{5}$
16) $\dfrac{2}{9} - \dfrac{5}{6}$
17) $-\dfrac{5}{8} - \dfrac{2}{3}$
18) $\dfrac{3}{7} - \left(-\dfrac{1}{8}\right)$
19) $-\dfrac{11}{12} - \left(-\dfrac{5}{9}\right)$
20) $-\dfrac{3}{4} + \left(-\dfrac{1}{6}\right)$
21) $7.3 - 11.2$
22) $-14.51 + 20.6$
23) $-5.09 - (-12.4)$
24) $8.8 - 19.2$
25) $-1 - 4.2$
26) $288.11 - 1.367$
27) $18 - |-12|$
28) $-14 + |-11|$
29) $|13| + 9$
30) $|-7| - 19$
31) $|-5.2| - 4.8$
32) $\left|-\dfrac{1}{4}\right| + \dfrac{1}{4}$

Determine whether each statement is true or false. For any real numbers a and b,

33) $|a - b| = |b - a|$
34) $|a + b| = a + b$
35) $|a| + |b| = a + b$
36) $|a + b| = |a| + |b|$
37) $a + (-a) = 0$
38) $-b - (-b) = 0$

Objective 3: Solve Applied Problems Involving Addition and Subtraction

Applications of Signed Numbers: Write an expression for each and simplify.

39) The world's tallest mountain, Mt. Everest, reaches an elevation of 29,035 ft. The Mariana Trench in the Pacific Ocean has a maximum depth of 36,201 ft below sea level. What is the difference between these two elevations? (www.britannica.com)

40) In 2010, the total attendance at Major League Baseball games was 73,054,407. In 2011, the figure was 73,425,568. What was the difference in the number of people who went to ballparks from 2010 to 2011? (www.mlb.com)

41) The median income for a male with a bachelor's degree in 2008 was $57,278. In 2009, the median income was $54,091. What was the difference in the median income from 2008 to 2009? (www.census.gov)

42) Mt. Washington, New Hampshire, rises to an elevation of 6288 ft. New Orleans, Louisiana, lies 6296 ft below this. What is the elevation of New Orleans? (www.infoplease.com)

43) The lowest temperature ever recorded in Hawaii was 12°F while the lowest temperature in Colorado was 73° less than that. What was the coldest temperature ever recorded in Colorado?

44) The lowest temperature ever recorded in the United States was $-79.8°$F in the Endicott Mountains of Alaska. The highest U.S. temperature on record was 213.8° more than the lowest and was recorded in Death Valley, California. What is the highest temperature on record in the United States? (www.infoplease.com)

45) During one offensive drive in the first quarter of Super Bowl XXXVIII, the New England Patriots ran for 7 yd, gained 4 yd on a pass play, gained 1 yd on a running play, gained another 6 yd on a pass by Tom Brady, then lost 10 yd on a running play. What was the Patriots' net yardage on this offensive drive? (www.superbowl.com)

46) The bar graph shows Dale Earnhardt Jr.'s total winnings from races for the years 2004–2007. Use a signed number to represent the change in his winnings over the given years. (www.dalejrpitstop.com)

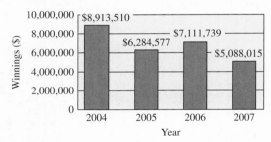

a) 2004–2005
b) 2005–2006
c) 2006–2007

47) The bar graph shows the number of housing starts (in thousands) during five months in 2003 in the Northeastern United States. Use a signed number to represent the change over the given months. (www.census.gov)

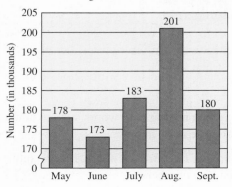

Housing Starts in Northeastern U.S.

a) May–June   b) June–July
c) July–August   d) August–September

48) The bar graph shows the population of Oakland, California from 2003 to 2006. Use a signed number to represent the change in Oakland's population over the given years. (U.S. Census Bureau)

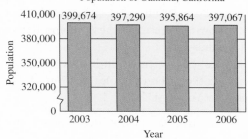

a) 2003–2004   b) 2004–2005
c) 2005–2006

Objective 4: Multiply Real Numbers
Fill in the blank with *positive* or *negative*.

49) The product of two negative numbers is _____.

50) The product of a positive number and a negative number is _____.

Multiply.

51) $-5 \cdot 9$

52) $3 \cdot (-11)$

53) $-14 \cdot (-3)$

54) $-16 \cdot (-31)$

55) $-1 \cdot (-6) \cdot (-7)$

56) $-2 \cdot 5 \cdot (-3)$

57) $\dfrac{7}{9} \cdot \left(-\dfrac{6}{5}\right)$

58) $-\dfrac{15}{32} \cdot \left(-\dfrac{8}{25}\right)$

59) $(-0.25)(1.2)$

60) $(-3.8)(-7.1)$

61) $8 \cdot (-2) \cdot (-4) \cdot (-1)$

62) $-5 \cdot (3) \cdot (-2) \cdot (-1) \cdot (-4)$

63) $(-8) \cdot (-9) \cdot 0 \cdot \left(-\dfrac{1}{4}\right) \cdot (-2)$

64) $(-6) \cdot \left(-\dfrac{2}{3}\right) \cdot 2 \cdot (-5)$

Objective 5: Divide Real Numbers

65) The quotient of a negative number and a positive number is _____.

66) The quotient of two negative numbers is _____.

Divide.

67) $-42 \div (-6)$

68) $-108 \div 9$

69) $\dfrac{-32}{-4}$

70) $\dfrac{56}{-7}$

71) $\dfrac{12}{-0.5}$

72) $\dfrac{-3.6}{0.9}$

73) $-\dfrac{12}{13} \div \left(-\dfrac{6}{5}\right)$

74) $-14 \div \left(-\dfrac{10}{3}\right)$

75) $\dfrac{0}{-4}$

76) $-\dfrac{0}{9}$

77) $\dfrac{360}{-280}$

78) $\dfrac{-84}{-210}$

79) $\dfrac{\dfrac{20}{21}}{\dfrac{5}{7}}$

80) $\dfrac{-\dfrac{3}{5}}{\dfrac{3}{4}}$

81) $\dfrac{-0.5}{10}$

82) $\dfrac{-11}{0.11}$

R Rethink

R1) Which is easier for you, $3 - 8$ or $3 + (-8)$? Why?

R2) When finding the sum of two integer values, when will the result be positive and when will it be negative?

R3) Why are some differences positive, and why are some negative?

R4) If the absolute value of a number is greater than the absolute value of another, how do you determine if their difference is positive or negative?

R5) Think about the similarities between multiplying and dividing signed numbers. Notice that the product of a negative and positive number is negative, and the quotient of a negative and positive number is also negative!

1.3 Exponents, Roots, and Order of Operations

P Prepare	**O Organize**
What are your objectives for Section 1.3?	How can you accomplish each objective?
1 Use Exponents	• Understand what an *exponent* represents. • Know how to determine the *base* and the *exponent* in an expression. • Memorize the powers given in this section. • Complete the given examples on your own. • Complete You Trys 1 and 2.
2 Find Square Roots	• Know what the $\sqrt{}$ symbol represents and what $-\sqrt{}$ represents. • Be able to identify the *radicand* and the type of solution you will obtain. • Complete the given example on your own. • Complete You Try 3.
3 Use the Order of Operations	• Memorize the **Order of Operations.** Write out the procedure before you begin each problem. • Use a memory aid, such as **P**lease **E**xcuse **M**y **D**ear **A**unt **S**ally, to help you remember the steps. • Don't skip steps, and copy the entire line as you perform each step. • Complete the given examples on your own. • Complete You Trys 4 and 5.

W Work   Read the explanations, follow the examples, take notes, and complete the You Trys.

1 Use Exponents

W Hint
Remember that
$2 \cdot 5 = 2 + 2 + 2 + 2 + 2$,
but two raised to the fifth power means
$2 \cdot 2 \cdot 2 \cdot 2 \cdot 2$.

Remember that exponents can be used to represent repeated multiplication. For example,

$$2 \cdot 2 \cdot 2 \cdot 2 \cdot 2 = 2^5 \leftarrow \text{exponent (or power)}$$
$$\uparrow$$
$$\text{base}$$

The *base* is 2, and 2 is a *factor* that appears five times. 5 is the *exponent* or *power*. We read 2^5 as "2 to the fifth power." 2^5 is called an **exponential expression**.

EXAMPLE 1   Rewrite each product in exponential form.

a) $6 \cdot 6 \cdot 6 \cdot 6$ b) $(-3) \cdot (-3)$

Solution

a) $6 \cdot 6 \cdot 6 \cdot 6 = 6^4$ b) $(-3) \cdot (-3) = (-3)^2$

[YOU TRY 1] Rewrite each product in exponential form.

a) $4 \cdot 4 \cdot 4 \cdot 4 \cdot 4 \cdot 4$ b) $\frac{3}{8} \cdot \frac{3}{8} \cdot \frac{3}{8} \cdot \frac{3}{8}$ c) $(-7) \cdot (-7) \cdot (-7) \cdot (-7) \cdot (-7)$

We can also evaluate an exponential expression.

EXAMPLE 2 Evaluate.

a) 3^2 b) $\left(-\frac{4}{5}\right)^3$ c) -2^4

Solution

a) $3^2 = 3 \cdot 3 = 9$

b) $\left(-\frac{4}{5}\right)^3 = \left(-\frac{4}{5}\right) \cdot \left(-\frac{4}{5}\right) \cdot \left(-\frac{4}{5}\right)$ The negative sign is included in the parentheses.
$= -\frac{64}{125}$ $-\frac{4}{5}$ appears as a factor 3 times.

W Hint
In Example 2, notice the bases in both parts a) and c) are positive!

c) $-2^4 = -1 \cdot 2 \cdot 2 \cdot 2 \cdot 2$ The negative sign is not included in the parentheses.
$= -16$ The base is 2, so the exponent of 4 applies only to the 2.

W Hint
Think about why $-2^4 = -16$, while $(-2)^4 = 16$.

Note
1 raised to any natural number power is 1 since 1 multiplied by itself equals 1.

[YOU TRY 2] Evaluate.

a) -3^4 b) 7^2 c) $\left(\frac{2}{5}\right)^3$ d) $\left(-\frac{1}{2}\right)^5$ e) -2^3

It is generally agreed that there are some skills in arithmetic that everyone should have in order to be able to acquire other math skills. Knowing the basic multiplication facts, for example, is essential for learning how to add, subtract, multiply, and divide fractions as well as how to perform many other operations in arithmetic and algebra. Similarly, memorizing powers of certain bases is necessary for learning how to apply the rules of exponents (Chapter 6) and for working with radicals (Chapter 9). Therefore, the powers listed here must be memorized in order to be successful in the previously mentioned, as well as other, topics. Throughout this book, it is assumed that students know the powers in the table:

W Hint
Create a set of flash cards with these "Powers to Memorize" and quiz yourself on a daily basis.

Powers to Memorize

$2^1 = 2$	$3^1 = 3$	$4^1 = 4$	$5^1 = 5$	$6^1 = 6$	$8^1 = 8$	$10^1 = 10$
$2^2 = 4$	$3^2 = 9$	$4^2 = 16$	$5^2 = 25$	$6^2 = 36$	$8^2 = 64$	$10^2 = 100$
$2^3 = 8$	$3^3 = 27$	$4^3 = 64$	$5^3 = 125$			$10^3 = 1000$
$2^4 = 16$	$3^4 = 81$					
$2^5 = 32$				$7^1 = 7$	$9^1 = 9$	$11^1 = 11$
$2^6 = 64$				$7^2 = 49$	$9^2 = 81$	$11^2 = 121$
						$12^1 = 12$
						$12^2 = 144$
						$13^1 = 13$
						$13^2 = 169$

2 Find Square Roots

In Example 2, we saw that $3^2 = 3 \cdot 3 = 9$. Here we will list and use the rules for the opposite procedure, finding **roots** of numbers.

The $\sqrt{}$ symbol represents the *positive* square root, or **principal square root**, of a number. For example,
$$\sqrt{9} = 3.$$

 $\sqrt{9} = 3$ but $\sqrt{9} \neq -3$. The $\sqrt{}$ symbol represents *only* the positive square root.

To find the negative square root of a number we must put a negative symbol in front of the $\sqrt{}$. For example, $-\sqrt{9} = -3$.

$\sqrt{}$ is the **square root symbol** or the **radical sign**. The number under the radical sign is the **radicand**.

$$\text{radical sign} \rightarrow \sqrt{9}$$
$$\uparrow$$
$$\text{radicand}$$

The entire expression, $\sqrt{9}$, is called a **radical**.

W Hint
Be familiar with the different types of problems involving radicals. This will help you do your homework more easily.

Property Radicands and Square Roots

1) If the radicand is a perfect square, the *square* root is a *rational* number.

   Example: $\sqrt{16} = 4$ 16 is a perfect square.

   $\sqrt{\dfrac{100}{49}} = \dfrac{10}{7}$ $\dfrac{100}{49}$ is a perfect square.

2) If the radicand is a negative number, the square root is *not* a real number.

   Example: $\sqrt{-25}$ is not a real number.

3) If the radicand is positive and *not* a perfect square, then the square root is an *irrational* number.

   Example: $\sqrt{13}$ is irrational.   13 is not a perfect square.

The square root of such a number is a real number that is a nonrepeating, nonterminating decimal. It is important to be able to approximate such square roots because sometimes it is necessary to estimate their places on a number line or on a Cartesian coordinate system when graphing.

For the purposes of graphing, approximating a radical to the nearest tenth is sufficient. A calculator with a √ key will give a better approximation of the radical.

EXAMPLE 3

Find each square root, if possible.

a) $\sqrt{100}$ b) $-\sqrt{\dfrac{81}{49}}$ c) $\sqrt{-9}$

Solution

a) $\sqrt{100} = 10$ since $(10)^2 = 100$.

b) $-\sqrt{\dfrac{81}{49}}$ means $-1 \cdot \sqrt{\dfrac{81}{49}}$. Therefore, $-\sqrt{\dfrac{81}{49}} = -1 \cdot \sqrt{\dfrac{81}{49}} = -1 \cdot \left(\dfrac{9}{7}\right) = -\dfrac{9}{7}$

c) *There is no such real number* since $3^2 = 9$ and $(-3)^2 = 9$. Therefore, $\sqrt{-9}$ is not a real number.

[YOU TRY 3]

Find each square root, if possible.

a) $-\sqrt{144}$ b) $\sqrt{\dfrac{25}{36}}$ c) $-\sqrt{\dfrac{1}{64}}$ d) $\sqrt{-36}$

3 Use the Order of Operations

We will begin this topic with a problem for the student:

[YOU TRY 4]

Evaluate $36 - 12 \div 4 + (6 - 1)^2$.

What answer did you get? 31? or 58? or 8? Or, did you get another result?
  Most likely you obtained one of the three answers just given. Only one is correct, however. If we do not have rules to guide us in evaluating expressions, it is easy to get the incorrect answer. Here are the rules we follow. This is called the **order of operations.**

Hint
Write down the procedure for the order of operations in your own words.

Procedure The Order of Operations

Simplify expressions in the following order:

1) If parentheses or other grouping symbols appear in an expression, simplify what is in these grouping symbols first.
2) Simplify expressions with exponents.
3) Perform multiplication and division from left to right.
4) Perform addition and subtraction from left to right.

Think about the "You Try" problem. Did you evaluate it using the order of operations? Let's look at that expression:

EXAMPLE 4

Evaluate $36 - 12 \div 4 + (6 - 1)^2$.

Solution

$36 - 12 \div 4 + (6 - 1)^2$ — First, perform the operation in the parentheses.
$36 - 12 \div 4 + 5^2$
$36 - 12 \div 4 + 25$ — Exponents are done before division, addition, and subtraction.
$36 - 3 + 25$ — Perform division before addition and subtraction.
$33 + 25$ — When an expression contains only addition and subtraction,
58 — perform the operations starting at the left and moving to the right.

W Hint
Write out all of the steps of each example as you read it. That way you are more likely to retain the information!

A good way to remember the order of operations is to remember the sentence, "**P**lease **E**xcuse **M**y **D**ear **A**unt **S**ally." (**P**arentheses, **E**xponents, **M**ultiplication, and **D**ivision from left to right, **A**ddition and **S**ubtraction from left to right) Don't forget that multiplication and division are at the same "level" in the process of performing operations and that addition and subtraction are at the same "level."

EXAMPLE 5

Evaluate.

a) $-2[20 - (40 \div 5)] + 7$ b) $\dfrac{(7-5)^3 \cdot 6}{30 - 9 \cdot 2}$

Solution

a) $-2[20 - (40 \div 5)] + 7$

This expression contains two sets of grouping symbols: **brackets** [ ] and **parentheses** ( ). Perform the operation in the **innermost** grouping symbol first, which is the parentheses in this case.

$-2[20 - (40 \div 5)] + 7 = -2[20 - 8] + 7$ — Innermost grouping symbol
$= -2[12] + 7$ — Brackets
$= -24 + 7$ — Perform multiplication before addition.
$= -17$ — Add.

b) $\dfrac{(7-5)^3 \cdot 6}{30 - 9 \cdot 2}$

The fraction bar in this expression acts as a grouping symbol. Therefore, simplify the numerator, simplify the denominator, then simplify the resulting fraction, if possible.

$\dfrac{(7-5)^3 \cdot 6}{30 - 9 \cdot 2} = \dfrac{2^3 \cdot 6}{30 - 18}$ — Parentheses / Multiply.

$= \dfrac{8 \cdot 6}{12}$ — Exponent / Subtract.

$= \dfrac{48}{12}$ — Multiply.

$= 4$

[YOU TRY 5]

Evaluate:

a) $5 + 3[15 - 2(3 + 1)]$ b) $\dfrac{7^2 - 3 \cdot 3}{5(12 - 8)}$

ANSWERS TO [YOU TRY] EXERCISES

1) a) 4^6 b) $\left(\frac{3}{8}\right)^4$ c) $(-7)^5$ 2) a) -81 b) 49 c) $\frac{8}{125}$ d) $-\frac{1}{32}$ e) -8
3) a) -12 b) $\frac{5}{6}$ c) $-\frac{1}{8}$ d) not a real number   4) 58 5) a) 26 b) 2

E Evaluate 1.3 Exercises

Do the exercises, and check your work.

Objective 1: Use Exponents
Write in exponential form.

1) $9 \cdot 9 \cdot 9 \cdot 9$

2) $2 \cdot 2 \cdot 2 \cdot 2 \cdot 2 \cdot 2 \cdot 2 \cdot 2$

Fill in the blank with *positive* or *negative*.

3) If a is a positive number, then $-a^6$ is _____.

4) If a is a positive number, then $(-a)^6$ is _____.

5) If a is a negative number, then $-a^5$ is _____.

6) Explain the difference between how you would evaluate -3^4 and $(-3)^4$. Then, evaluate each.

Write each expression using exponents. Do not evaluate.

7) $11 \cdot 11 \cdot 11 \cdot 11$ 8) $4 \cdot 4 \cdot 4$

9) $3.2 \cdot 3.2 \cdot 3.2$ 10) $0.2 \cdot 0.2 \cdot 0.2 \cdot 0.2 \cdot 0.2$

11) $\frac{1}{5} \cdot \frac{1}{5} \cdot \frac{1}{5} \cdot \frac{1}{5}$ 12) $\frac{3}{4} \cdot \frac{3}{4}$

13) $(-7)(-7)(-7)(-7)(-7)(-7)$

14) $(-1)(-1)(-1)$

Evaluate

15) 2^5 16) 9^2

17) $(11)^2$ 18) 4^3

19) $(-2)^4$ 20) $(-5)^3$

21) -7^2 22) -6^2

23) -2^3 24) -3^4

25) $\left(\frac{1}{5}\right)^3$ 26) $\left(\frac{3}{2}\right)^4$

27) Evaluate $(0.5)^2$ two different ways.

28) Explain why $1^{200} = 1$.

Objective 2: Find Square Roots
Decide if each statement is true or false. If it is false, explain why.

29) $\sqrt{49} = 7$ and $\sqrt{49} = -7$

30) $\sqrt{121} = 11$

31) $-\sqrt{4} = -2$

32) The square root of a negative number is a negative number.

Find all square roots of each number.

33) 64 34) 25

35) 400 36) 8100

37) $\frac{25}{16}$ 38) $\frac{49}{144}$

Find each square root, if possible.

39) $\sqrt{36}$ 40) $\sqrt{169}$

41) $-\sqrt{900}$ 42) $-\sqrt{1}$

43) $\sqrt{-25}$ 44) $\sqrt{-36}$

45) $\sqrt{\frac{100}{121}}$ 46) $\sqrt{\frac{4}{9}}$

47) $-\sqrt{\frac{1}{64}}$ 48) $-\sqrt{\frac{1}{25}}$

Objective 3: Use the Order of Operations

49) In your own words, summarize the order of operations.

Evaluate.

50) $16 - 2 \cdot 3$ 51) $23 + 4 \cdot 5$

52) $3[-1 - (-3)]$ 53) $4[9 - (-3)]$

54) $3 - (-4)6 + 12\sqrt{100}$

55) $2\sqrt{9} + 3 - 1 \cdot 4$

56) $35 - 7 + 8 - 3$

57) $-50 \div 10 + 15$

58) $6 \cdot (-4) - 2$

59) $20 - 3 \cdot 2 + 9$

60) $22 + 10 \div 2 - 1$

61) $\dfrac{1}{2} \cdot \dfrac{4}{5} - \dfrac{2}{5} \cdot \dfrac{3}{10}$

62) $\left(\dfrac{3}{2}\right)^2 - \left(-\dfrac{5}{4}\right)^2$

63) $15 - 3(6 - 4)^2$

64) $7 + 2(9 - 5)^2$

65) $-6[21 \div (3 + 4)] - 9$

66) $2[23 + (9 - 11)^3] + 3$

67) $4 + 3[(3 - 7)^3 \div (10 - 2)]$

68) $(8 + 2)^2 - 5[9 \cdot (3 + 1) - 5^2]$

69) $\dfrac{12(5 + 1)}{2 \cdot 5 - 1}$

70) $\dfrac{(14 - 4)^2 - 4^3}{4 \cdot 9 - 3 \cdot 11}$

71) $\dfrac{4(7 - 2)^2}{(12)^2 - 8 \cdot 3}$

72) $\dfrac{6(3 - 5)^2}{10 + 12 \div 2 + 4}$

R Rethink

R1) Which objective is the most difficult for you?

R2) Where have you encountered exponents other than this math course?

R3) Where in a 12 × 12 multiplication table are the perfect squares?

R4) Where are square roots in the order of operations?

1.4 Algebraic Expressions and Properties of Real Numbers

P Prepare	O Organize
What are your objectives for Section 1.4?	How can you accomplish each objective?
1 Evaluate Algebraic Expressions	• Know how to identify an *algebraic expression*. • Always substitute variable values into the expression by using parentheses. • Use the order of operations when simplifying. • Complete the given examples on your own. • Complete You Trys 1 and 2.
2 Properties of Real Numbers	• Know the definition of each property. Try to relate the name of the property to what the property is doing (i.e., *commutative* is "commuting" the numbers). • Notice that some properties work with addition and multiplication only. • Complete the given examples on your own. • Complete You Trys 3 and 4.

W Work   Read the explanations, follow the examples, take notes, and complete the You Trys.

1 Evaluate Algebraic Expressions

Here is an algebraic expression: $5x^3 - 9x^2 + \frac{1}{4}x + 7$. The *variable* is x. A **variable** is a symbol, usually a letter, used to represent an unknown number. The *terms* of this algebraic expression are $5x^3, -9x^2, \frac{1}{4}x,$ and 7. A **term** is a number or a variable or a product or quotient of numbers and variables. 7 is the **constant** or **constant term**. The value of a constant does not change. Each term has a **coefficient**.

> **W Hint**
> Remember, to evaluate an algebraic expression means to find its numerical value given the value(s) of the variable(s). We are not solving for anything, we are evaluating.

Definition

An **algebraic expression** is a collection of numbers, variables, and grouping symbols connected by operation symbols such as $+, -, \times,$ and $\div$.

Here are more examples of algebraic expressions:

$$10k + 9, \quad 3(2t^2 + t - 4), \quad 6a^2b^2 - 13ab - 2a + 5$$

We can **evaluate** an algebraic expression by substituting a value for a variable and simplifying. The value of an algebraic expression changes depending on the value that is substituted.

EXAMPLE 1

Evaluate $5x - 3$ when (a) $x = 4$ and (b) $x = -2$.

Solution

a) $5x - 3$ when $x = 4$ Substitute 4 for x.
     $= 5(4) - 3$ Use parentheses when substituting a value for a variable.
     $= 20 - 3$ Multiply.
     $= 17$ Subtract.

b) $5x - 3$ when $x = -2$ Substitute -2 for x.
     $= 5(-2) - 3$ Use parentheses when substituting a value for a variable.
     $= -10 - 3$ Multiply.
     $= -13$

[YOU TRY 1] Evaluate $10x + 7$ when $x = -4$.

EXAMPLE 2

Evaluate $4c^2 - 2cd + 1$ when $c = -3$ and $d = 5$.

Solution

$4c^2 - 2cd + 1$ when $c = -3$ and $d = 5$ Substitute -3 for c and 5 for d.
$= 4(-3)^2 - 2(-3)(5) + 1$ Use parentheses when substituting.
$= 4(9) - 2(-15) + 1$ Evaluate exponent and multiply.
$= 36 - (-30) + 1$ Multiply.
$= 36 + 30 + 1$
$= 67$

> **W Hint**
> Notice how the parentheses are being used when substituting in the numerical values for the variables.

[YOU TRY 2] Evaluate $b^2 + 7ab - 4a - 5$ when $a = \dfrac{1}{2}$ and $b = -6$.

2 Properties of Real Numbers

Like the order of operations, the properties of real numbers guide us in our work with numbers and variables.

W Hint
Commutative: the numbers are "commuting" or changing places.
Associative: the numbers are "associating" (or grouping) differently.
Inverse: an "opposite" operation is being applied to the number.

Summary Properties of Real Numbers

If a, b, and c are real numbers, then

Commutative Properties: $a + b = b + a$ and $ab = ba$
Associative Properties: $(a + b) + c = a + (b + c)$ and $(ab)c = a(bc)$
Identity Properties: $a + 0 = 0 + a = a$
$a \cdot 1 = 1 \cdot a = a$
Inverse Properties: $a + (-a) = -a + a = 0$
$b \cdot \dfrac{1}{b} = \dfrac{1}{b} \cdot b = 1 \ (b \neq 0)$
Distributive Properties: $a(b + c) = ab + ac$ and $(b + c)a = ba + ca$
$a(b - c) = ab - ac$ and $(b - c)a = ba - ca$

EXAMPLE 3

Use the stated property to rewrite each expression. Simplify, if possible.

a) $x \cdot 9$; commutative

b) $5 + (2 + n)$; associative

c) $-(-10m + 3n - 8)$; distributive

Solution

a) $x \cdot 9 = 9 \cdot x$ or $9x$ Commutative property

b) $5 + (2 + n) = (5 + 2) + n = 7 + n$ Associative property

c) $-(-10m + 3n - 8) = -1(-10m + 3n - 8)$
$= -1(-10m) + (-1)(3n) - (-1)(8)$ Apply the distributive property.
$= 10m + (-3n) - (-8)$ Multiply.
$= 10m - 3n + 8$ Simplify.

[YOU TRY 3] Use the stated property to rewrite each expression. Simplify, if possible.

a) $y \cdot 8$; commutative   b) $\left(-\dfrac{5}{12} \cdot \dfrac{4}{7}\right)\dfrac{7}{4}$; associative

c) $3(8x - 5y + 11z)$; distributive

EXAMPLE 4

Which property is illustrated by each statement?

a) $0 + 9 = 9$
b) $-1.3 + 1.3 = 0$
c) $\frac{1}{12} \cdot 12 = 1$
d) $7(1) = 7$

Solution

a) $0 + 9 = 9$ Identity property
b) $-1.3 + 1.3 = 0$ Inverse property
c) $\frac{1}{12} \cdot 12 = 1$ Inverse property
d) $7(1) = 7$ Identity property

[YOU TRY 4]

Which property is illustrated by each statement?

a) $4 \cdot \frac{1}{4} = 1$
b) $-8 + 8 = 0$
c) $-7.4(1) = -7.4$
d) $5 + 0 = 5$

ANSWERS TO [YOU TRY] EXERCISES

1) -33 2) 8   3) a) $8y$ b) $-\frac{5}{12}$ c) $24x - 15y + 33z$
4) a) inverse property   b) inverse property   c) identity property   d) identity property

E Evaluate 1.4 Exercises

Do the exercises, and check your work.

Objective 1: Evaluate Algebraic Expressions

1) Evaluate $2j^2 + 3j - 7$ when
   a) $j = 4$
   b) $j = -5$

2) Evaluate $6 - t^3$ when
   a) $t = -3$
   b) $t = 3$

Evaluate each expression when $x = -2$, $y = 7$, and $z = -3$.

3) $8x + y$
4) $10z - 3x$
5) $x^2 - 2z^2 + 4xy$
6) $x^2 + xy + 10$
7) $\frac{2x}{y + z}$
8) $\frac{x - y}{3z}$
9) $\frac{x^2 - y^2}{2z^2 + y}$
10) $\frac{5 + 3(y + 4z)}{x^2 - z^2}$

Objective 2: Properties of Real Numbers

11) What is the identity element for addition?

12) What is the identity element for multiplication?

13) What is the multiplicative inverse of 6?

14) What is the additive inverse of -9?

Which property of real numbers is illustrated by each example? Choose from the commutative, associative, identity, inverse, or distributive property.

15) $(-11 + 4) + 9 = -11 + (4 + 9)$

16) $5 \cdot 7 = 7 \cdot 5$

17) $20 + 8 = 8 + 20$

18) $16 + (-16) = 0$

19) $3(8 \cdot 4) = (3 \cdot 8) \cdot 4$

20) $5(3 + 7) = 5 \cdot 3 + 5 \cdot 7$

21) $(10 + 2)6 = 10 \cdot 6 + 2 \cdot 6$

22) $\dfrac{3}{4} \cdot 1 = \dfrac{3}{4}$

 23) $-24 + 0 = -24$

24) $\left(\dfrac{8}{15}\right)\left(\dfrac{15}{8}\right) = 1$

25) $9(a - b) = 9a - 9b$

26) $-3(c + d) = -3c - 3d$

Rewrite each expression using the indicated property.

27) $7(u + v)$; distributive

28) $12 + (3 + 7)$; associative

29) $k + 4$; commutative

30) $-8(c + 5)$; distributive

31) $-4z + 0$; identity

32) $9 + 11r$; commutative

33) Is $10c - 3$ equivalent to $3 - 10c$? Why or why not?

34) Is $8 + 5n$ equivalent to $5n + 8$? Why or why not?

Rewrite each expression using the distributive property. Simplify if possible.

35) $5(4 + 3)$

36) $8(1 + 5)$

37) $-2(5 + 7)$

38) $6(9 - 4)$

39) $-7(2 - 6)$

40) $-(9 - 5)$

41) $-(6 + 1)$

42) $(8 - 2)4$

43) $(-10 + 3)5$

44) $2(-6 + 5 + 3)$

45) $9(g + 6)$

46) $4(t - 5)$

47) $-5(z + 3)$

48) $-2(m + 11)$

49) $-8(u - 4)$

50) $-3(h - 9)$

51) $-(v - 6)$

52) $-(y - 13)$

53) $10(m + 5n - 3)$

54) $12(2a - 3b + c)$

55) $-(-8c + 9d - 14)$

56) $-(x - 4y + 10z)$

R Rethink

R1) Which exercises do you need help mastering?

R2) Which property of real numbers is the most difficult for you to understand?

R3) Why are addition and multiplication commutative while subtraction and division are not?

R4) Think about how the distributive property allows us to rewrite $14 \cdot 11$ as $14(10 + 1)$. Now apply the distributive property! Does this make the problem easier?

Group Activity — The Group Activity can be found online on Connect.

emPOWERme: Why Am I Going to College?

As you probably know by now, success in college takes a great deal of time, effort, and dedication. In order to stay motivated, it is important to focus on your goals for attending college. For many people, such as computer game designer Dave Cantelmo, college is an essential step toward achieving career aspirations. Other people go to college in order to build a better life for their families. Have you thought about why *you* are attending college?

Place a 1, 2, or 3 by the three *most* important reasons that you have for attending college:

_____ I want to get a good job when I graduate.
_____ I want to make my family proud.
_____ I couldn't find a job.
_____ I want to try something new.
_____ I want to get ahead at my current job.
_____ I want to pursue my dream job.
_____ I want to improve my reading and critical thinking skills.
_____ I want to become a more cultured person.
_____ I want to make more money.
_____ I want to learn more about things that interest me.
_____ A mentor or role model encouraged me to go.
_____ I want to prove to others that I can succeed.

Now consider the following:

- What do your answers tell you about yourself?
- What reasons besides these did you think about when you were applying to college?
- How do you think your reasons compare with those of other students who are starting college with you?
- How can you use your understanding of why you are in college to help you achieve college success?

Chapter 1: Summary

Definition/Procedure	Example
1.1 Sets of Numbers	
Natural numbers: {1, 2, 3, 4, …} **Whole numbers:** {0, 1, 2, 3, 4, …} **Integers:** {…, −3, −2, −1, 0, 1, 2, 3, …} A **rational number** is any number of the form $\frac{p}{q}$, where p and q are integers and $q \neq 0$. (p. 6)	The following numbers are rational: $-1, 2, \frac{3}{4}, 3.\overline{6}, 4.5$
An **irrational number** cannot be written as the quotient of two integers. (p. 6)	The following numbers are irrational: $\sqrt{7}, 5.1948…$
The set of **real numbers** includes the rational and irrational numbers. (p. 6)	Any number that can be represented on the number line is a real number.
The **additive inverse** of a is $-a$. (p. 9)	The additive inverse of 13 is −13.
Absolute Value $\|a\|$ is the distance of a from zero. (p. 9)	$\|-9\| = 9$
1.2 Operations on Real Numbers	
Adding Real Numbers To add numbers with the *same sign*, add the absolute value of each number. The sum will have the same sign as the numbers being added. (p. 13)	$-7 + (-4) = -11$
To add two numbers with *different signs*, subtract the smaller absolute value from the larger. The sum will have the sign of the number with the larger absolute value. (p. 13)	$-16 + 10 = -6$
Subtracting Real Numbers To subtract $a - b$, change subtraction to addition and add the additive inverse of b: $a - b = a + (-b)$. (p. 14)	$5 - 9 = 5 + (-9) = -4$ $-14 - (-6) = -14 + 6 = -8$ $11 - 4 = 11 + (-4) = 7$
Multiplying Real Numbers The product of two real numbers with the *same sign* is *positive*.	$9 \cdot 4 = 36 \qquad -6 \cdot (-5) = 30$
The product of a positive number and a negative number is *negative*.	$-3 \cdot 7 = -21 \qquad 8 \cdot (-1) = -8$
An *even number* of negative factors in a product gives a *positive* result.	$(-2)(-1)(3)(-4)(-5) = 120$ 4 negative factors
An *odd number* of negative factors in a product gives a *negative* result. (p. 16)	$(4)(-3)(-2)(-1)(3) = -72$ 3 negative factors
Dividing Real Numbers The quotient of two numbers with the *same* sign is positive.	$\frac{100}{4} = 25 \qquad -63 \div (-7) = 9$
The quotient of two numbers with *different* signs is negative. (p. 16)	$\frac{-20}{4} = -5 \qquad 32 \div (-8) = -4$

Definition/Procedure	Example
1.3 Exponents, Roots, and Order of Operations	
Exponents An **exponent** represents repeated multiplication. (p. 20)	Write $8 \cdot 8 \cdot 8 \cdot 8 \cdot 8$ in exponential form. $8 \cdot 8 \cdot 8 \cdot 8 \cdot 8 = 8^5$
Evaluating Exponential Expressions (p. 21)	Evaluate $(-2)^4$. $(-2)^4 = (-2)(-2)(-2)(-2) = 16$ Evaluate -2^4. $-2^4 = -1 \cdot 2^4 = -1 \cdot 2 \cdot 2 \cdot 2 \cdot 2 = -16$
Finding Roots If the *radicand* is a perfect square, then the square root is a *rational* number. If the *radicand* is a negative number, then the square root is *not* a real number. If the *radicand* is positive and not a perfect square, then the square root is an *irrational* number. (p. 22)	$\sqrt{49} = 7$ since $7^2 = 49$. $\sqrt{-36}$ is not a real number. $\sqrt{7}$ is irrational because 7 is not a perfect square.
Order of Operations **P**arentheses, **E**xponents, **M**ultiplication, **D**ivision, **A**ddition, **S**ubtraction. Remember that multiplication and division are at the same "level" when performing operations and that addition and subtraction are at the same level. (p. 24)	Evaluate $10 + (2 + 3)^2 - 8 \cdot 4$ $\begin{aligned} 10 + (2+3)^2 - 8 \cdot 4 & \\ = 10 + 5^2 - 8 \cdot 4 & \quad \text{Parentheses} \\ = 10 + 25 - 8 \cdot 4 & \quad \text{Exponents} \\ = 10 + 25 - 32 & \quad \text{Multiply.} \\ = 35 - 32 & \quad \text{Add.} \\ = 3 & \quad \text{Subtract.} \end{aligned}$
1.4 Algebraic Expressions and Properties of Real Numbers	
An **algebraic expression** is a collection of numbers, variables, and grouping symbols connected by operation symbols such as $+, -, \times,$ and $\div$. (p. 27)	$4y^2 - 7y + \dfrac{3}{5}$

Important Terms
     Variable   Constant
     Term   Coefficient
 	
 We can evaluate expressions for different values of the variables. (p. 27)
	Evaluate $2xy - 5y + 1$ when $x = -3$ and $y = 4$.
	
	 Substitute -3 for x and 4 for y and simplify.
	 $\begin{aligned} 2xy - 5y + 1 &= 2(-3)(4) - 5(4) + 1 \\ &= -24 - 20 + 1 \\ &= -24 + (-20) + 1 \\ &= -43 \end{aligned}$

Properties of Real Numbers If $a, b,$ and c are real numbers, then the following properties hold. **Commutative Properties:** $\quad a + b = b + a$ $\quad ab = ba$ **Associative Properties:** $\quad (a + b) + c = a + (b + c)$ $\quad (ab)c = a(bc)$	 **Commutative Properties:** $\quad 9 + 2 = 2 + 9$ $\quad (-4)(7) = (7)(-4)$ **Associative Properties:** $\quad (4 + 1) + 7 = 4 + (1 + 7)$ $\quad (2 \cdot 3)10 = 2(3 \cdot 10)$

Definition/Procedure	Example
Identity Properties: $$a + 0 = 0 + a = a$$ $$a \cdot 1 = 1 \cdot a = a$$	**Identity Properties:** $$\frac{3}{4} + 0 = \frac{3}{4}, \quad 5 \cdot 1 = 5$$
Inverse Properties: $$a + (-a) = -a + a = 0$$ $$b \cdot \frac{1}{b} = \frac{1}{b} \cdot b = 1$$	**Inverse Properties:** $$6 + (-6) = 0, \quad 8 \cdot \frac{1}{8} = 1$$
Distributive Properties: $a(b + c) = ab + ac$ and $(b + c)a = ba + ca$ $a(b - c) = ab - ac$ and $(b - c)a = ba - ca$ **(p. 28)**	**Distributive Properties:** $$9(3 + 4) = 9 \cdot 3 + 9 \cdot 4$$ $$= 27 + 36$$ $$= 63$$ $$4(n - 7) = 4n - 4 \cdot 7$$ $$= 4n - 28$$

Chapter 1: Review Exercises

(1.1)

1) Given the set of numbers,

$$\left\{ \sqrt{23},\ -6,\ 14.38,\ \frac{3}{11},\ 2,\ 5.\overline{7},\ 0,\ 9.21743819\ldots \right\},$$

list the

a) whole numbers.
b) natural numbers.
c) integers.
d) rational numbers.
e) irrational numbers.

2) Graph and label these numbers on a number line.

$$-2,\ 5\frac{1}{3},\ 0.8,\ -4.5,\ 3,\ -\frac{3}{4}$$

3) Evaluate $|-10|$.

(1.2) Add or subtract as indicated.

4) $-18 + 4$
5) $60 - (-15)$
6) $-\frac{5}{8} + \left(-\frac{2}{3}\right)$
7) $0.8 - 5.9$

8) The lowest temperature on record in the state of Wyoming is $-66°$F. Georgia's record low is $49°$ higher than Wyoming's. What is the lowest temperature ever recorded in Georgia? (www.usatoday.com)

Multiply or divide as indicated.

9) $(-10)(-7)$
10) $\left(-\frac{2}{3}\right)(15)$
11) $(3.7)(-2.1)$
12) $(-3)(-5)(-2)$
13) $(-1)(6)(-4)\left(-\frac{1}{2}\right)(-5)$
14) $-54 \div 6$
15) $\dfrac{-24}{-12}$
16) $\dfrac{38}{-44}$
17) $-\dfrac{20}{27} \div \dfrac{8}{15}$
18) $-\dfrac{8}{9} \div (-4)$

(1.3) Evaluate.

19) -5^2
20) $(-5)^2$
21) $(-3)^4$
22) $(-1)^9$
23) -2^6
24) $\sqrt{16}$
25) $\sqrt{49}$
26) $-\sqrt{4}$
27) $-\sqrt{36}$
28) $\sqrt{-64}$

Use the order of operations to simplify.

29) $64 \div (-8) + 6$
30) $15 - (3 - 7)^3$
31) $-11 - 3 \cdot 9 + (-2)^1$
32) $\dfrac{6 - 2(5 - 1)}{(-3)(-4) + 7 - 3}$
33) $\dfrac{3}{4} \cdot \left|-\dfrac{5}{7}\right|$
34) $2[3 - 4 - (-2)^2] \div 5$
35) $\dfrac{4^2 - (3 \cdot 5)}{|-4 - 2|}$
36) $\dfrac{-\left|-\dfrac{2}{3}\right|}{\dfrac{19}{9}}$
37) $12 + \sqrt{16} - 2^3 + 1$
38) $-\sqrt{4} + (-5) - |-9|$

(1.4)

39) List the terms and coefficients of
$c^4 + 12c^3 - c^2 - 3.8c + 11$.

40) Evaluate $-3m + 7n$ when $m = 6$ and $n = -2$.

41) Evaluate $\dfrac{t - 6s}{s^2 - t^2}$ when $s = -4$ and $t = 5$.

Which property of real numbers is illustrated by each example? Choose from the commutative, associative, identity, inverse, or distributive property.

42) $0 + 12 = 12$

43) $(8 + 1) + 5 = 8 + (1 + 5)$

44) $\left(\dfrac{4}{7}\right)\left(\dfrac{7}{4}\right) = 1$

45) $35 + 16 = 16 + 35$

46) $-6(3 + 8) = (-6)(3) + (-6)(8)$

Rewrite each expression using the distributive property. Simplify if possible.

47) $3(10 - 6)$

48) $(3 + 9)2$

49) $-(12 + 5)$

50) $-7(2c - d + 4)$

Chapter 1: Test

1) Given the set of numbers,

$$\left\{41, -8, 0, 2.\overline{83}, \sqrt{75}, 6.5, 4\dfrac{5}{8}, 6.37528861...\right\},$$

list the
a) integers.
b) irrational numbers.
c) natural numbers.
d) rational numbers.
e) whole numbers.

2) Graph the numbers on a number line. Label each.

$$6, \dfrac{7}{8}, -4, -1.2, 4\dfrac{3}{4}, -\dfrac{2}{3}$$

Perform the indicated operation(s). Write all answers in lowest terms.

3) $\dfrac{9}{14} \cdot \dfrac{7}{24}$

4) $\dfrac{1}{3} + \dfrac{4}{15}$

5) $5\dfrac{1}{4} - 2\dfrac{1}{6}$

6) $\dfrac{12}{13} \div 6$

7) $\dfrac{4}{7} - \dfrac{5}{6}$

8) $-11 - |-19|$

9) $25 + 15 \div 5$

10) $\dfrac{9}{10} \cdot \left(-\dfrac{2}{5}\right)$

11) $-8 \cdot (-6)$

12) $-13.4 + 6.9$

13) $30 - 5[-10 + (2 - 6)^2]$

14) $\dfrac{(50 - 26) \div 3}{3 \cdot 5 - 7}$

Evaluate.

15) 2^5

16) -3^4

17) $|-92|$

18) $|2 - 12| - 4|6 - 1|$

19) $-2(7 + \sqrt{25})$

20) $4(3 - 2)^2 + 19$

21) $\dfrac{(-\sqrt{36} \div 2)^3 + 7}{-|-11 + 3| \cdot (-4)}$

For Exercises 22–25, determine whether each statement is true or false.

22) $|b - a| = |a - b|$

23) $|a + b| = |a| + |b|$

24) If a is a positive number, then $-a^2$ is positive.

25) The square root of a negative number is a negative number.

26) Both the highest and lowest points in the continental United States are in California. Badwater, Death Valley, is 282 ft below sea level while, 76 mi away, Mount Whitney reaches an elevation of 14,505 ft. What is the difference between these two elevations?

27) Evaluate $9g^2 + 3g - 6$ when $g = -1$.

28) Which property of real numbers is illustrated by each example? Choose from the commutative, associative, identity, inverse, or distributive property.

a) $9(5 - 7) = 9 \cdot 5 - 9 \cdot 7$

b) $-6 + 6 = 0$

c) $8 \cdot 3 = 3 \cdot 8$

29) Rewrite each expression using the distributive property. Simplify if possible.

a) $-2(5 + 3)$

b) $5(t + 9u + 1)$

30) Is $x - 8$ equivalent to $8 - x$? Why or why not?

CHAPTER 2
Linear Equations in One Variable

OUTLINE

Study Strategies: Reading Math (and Other) Textbooks

2.1 Linear Equations in One Variable

2.2 Applications of Linear Equations

2.3 Geometry Applications and Solving Formulas

2.4 More Applications of Linear Equations

Group Activity

emPOWERme: Organize Your Memory

Math at Work:
Tailor

It has only been a few years since Elana Nicks opened her own tailoring shop, but she already feels like she's seen it all. She's done everything for customers from sewing up holes in their pockets to creating custom-fitting suits to altering a wedding dress—the day before the wedding! "I think my customers appreciate that I take my work seriously," Elana says. "I want every piece of clothing that leaves my shop to be perfect."

In order to maintain this level of excellence, Elana relies on the precision that math provides, using it to determine the exact details of how she will manipulate the fabric she is working with. "Some tailors are comfortable just eyeing things," Elana says. "I use the math I learned in college to keep things exact."

When one of Elana's customers has a tailoring need she's never handled before, she consults a small library of books and magazines she keeps in the back of her shop. "No one can do it all," Elana explains. "The key is being able to find the information to fill in the gaps in your knowledge." Elana says she learned to do this in college, working with textbooks and other assigned reading materials.

This chapter covers linear equations in one variable, and you will also find strategies you can use when you are reading math and other types of textbooks.

 **Study Strategies** Reading Math (and Other) Textbooks

Some students find textbooks intimidating. This attitude is understandable. Textbooks are typically long, heavy books, filled from cover to cover with information you'll need to master. Remember, though: when it comes to succeeding in a math course or in other college classes, a textbook is often your most valuable ally. If you use it effectively, it will equip you with the knowledge you'll need to do well on exams, problem sets, and in class. The strategies below will help you get the most out of your textbooks.

- When you get a new textbook, start by reading the preface and/or the introduction to understand the goals and thinking behind the book.
- Before you start reading a particular section, get a preview of the material by reviewing outlines, overviews, or section objectives.

- Before you start reading, make sure you have the appropriate tools handy, such as a pencil and paper, a highlighter, and the reading assignment.
- Find a quiet place to do your reading, one that is free of distractions.

- Read actively: this involves taking notes as you read (both in the text and on a separate piece of paper) and thinking about the material you are reading.
- The best way to read a math textbook is to write down all of the steps in the examples as you are reading them. Be sure you understand the math being done going from step to step.
- Highlight and/or underline only the essential information. One guideline: No more than 10% of what you read should be highlighted or underlined.
- To help you remember, connect the material to information you already know.
- In this book, do all the examples and You Try problems in your reading. These are opportunities to practice and apply the concepts you are learning about.

- Look over the reading again once you have finished, ensuring you have marked and/or made notes on all the essential material.
- In your own words, explain the major concepts of the reading, either aloud to yourself or to a fellow student.
- In a math class, do the homework exercises to determine how well you have learned the material in the section.

- Within 24 hours of reading the assignment, review it, along with your notes.
- Ask yourself what worked well and what did not in learning new material. Do you need to make any adjustments?

Chapter 2 P O W E R Plan

P Prepare | O Organize

What are your goals for Chapter 2?	How can you accomplish each goal?
1 Be prepared before and during class.	• Don't stay out late the night before, and be sure to set your alarm clock! • Bring a pencil, notebook paper, and textbook to class. • Avoid distractions by turning off your cell phone during class. • Pay attention, take good notes, and ask questions. • Complete your homework on time, and ask questions on problems you do not understand. • Plan ahead for tests by preparing many days in advance.
2 Understand the homework to the point where you could do it without needing any help or hints.	• Read the directions, and show all of your steps. • Go to the professor's office for help. • Rework homework and quiz problems, and find similar problems for practice. • Review old material that you have not mastered yet.
3 Use the P.O.W.E.R. framework to help you organize your study: *Organize Your Memory*.	• Read the Study Strategy that explains how to read a textbook. • How can you use P.O.W.E.R. to get the most out of this book? • Complete the emPOWERme that appears before the Chapter Summary.
4 Write your own goal. _____ _____	• _____

What are your objectives for Chapter 2?	How can you accomplish each objective?
1 Be able to solve equations, including those that contain fractions and decimals.	• Learn the **Properties of Equality**. • Be very organized with your steps, and write out each example as you read it. • Know how to check each answer. • Learn the procedure for canceling decimals and fractions in equations.
2 Be able to determine when an equation has *no solution* and when an equation has an *infinite number of solutions*.	• Be able to determine when variables "drop out" of an equation. • Recognize when the remaining numerical statement is "true" or "false."
3 Solve applied problems that include lengths, integers, percent changes, and interest calculations.	• Learn the procedure for solving applied problems. • Make sure you have answered the question, and not just solved the equation. Make sure your answer is logical.
4 Solve applied problems that involve geometry formulas and angle measurements.	• Learn the procedure for solving applied problems. • Draw diagrams, when needed. • Make sure your answer is logical. Check to make sure your angle measures sum to the proper total.
5 Solve applied problems that involve money, mixtures, and distance.	• Learn the procedure for solving applied problems. • Draw charts to help you organize your information. • Check to make sure your answer is logical.
6 Write your own goal. _____ _____	• _____

	W Work	Read Sections 2.1 to 2.4, and complete the exercises.	
E Evaluate Complete the Chapter Review and Chapter Test. How did you do?		**R Rethink**	• How did you perform on the goals for the chapter? Which steps could be improved for next time? If you had the chance to do this chapter over, what would you do differently? • Think of a job you might like to have and describe how you would need to use what you have just learned to effectively do that job. • How has the P.O.W.E.R. framework helped you master the objectives of this chapter? Where else could you use this framework? Make it a point to use P.O.W.E.R. to complete another task this week.

2.1 Linear Equations in One Variable

P Prepare	**O Organize**
What are your objectives for Section 2.1?	How can you accomplish each objective?
1 Define a Linear Equation in One Variable	• Be able to recognize an *equation*. • Know what it means to solve an equation, and be able to check a given solution.
2 Use the Properties of Equality	• Be familiar with the **Properties of Equality** and when they should be used to help solve equations. • After solving each equation, be sure to check your solution into the original equation. • Complete the given example on your own. • Complete You Try 1.
3 Combine Like Terms to Solve a Linear Equation	• Know the steps for solving a *linear equation*. • Be able to recognize *like terms* and combine them when they are on the same side of the equal sign. • Recognize that the **Properties of Equality** must be used when *like terms* are on opposite sides of an equation. • Complete the given example on your own. • Complete You Try 2.

(*continued*)

What are your objectives for Section 2.1?	How can you accomplish each objective?
4 Solve Equations Containing Fractions or Decimals	• Know how to find the LCD for all the fractions in the equation. • Review the rules for multiplying fractions. • Use the **Multiplication Property of Equality** to multiply both sides of the equation by the LCD. • For equations with decimals, be able to determine the smallest power of 10 that can be used to cancel decimals. • Use the **Multiplication Property of Equality** to multiply both sides of the equation by that power of 10. • Complete the given examples on your own. • Complete You Trys 3 and 4.
5 Solve Equations with No Solution or an Infinite Number of Solutions	• Be aware that when solving an equation, the variables may "drop out." • Understand when an equation has *no solution* and when the solution is *all real numbers*. • Complete the given example on your own. • Complete You Try 5.
6 Use the Five Steps for Solving Applied Problems	• Read the problem more than once. • Follow the procedure for solving applied problems and do not jump to setting up the equation before you choose a variable and define your unknown(s). • Make sure your answer is logical and that you have answered the question that is being asked. • Complete the given examples on your own. • Complete You Trys 6, 7, and 8.

Read the explanations, follow the examples, take notes, and complete the You Trys.

1 Define a Linear Equation in One Variable

What is an equation? It is a mathematical statement that two expressions are equal. $5 + 1 = 6$ is an equation.

> An equation contains an "=" sign and an expression does not.

$5x + 4$ is an *equation*. $9y + 2y$ is an *expression*.

We can **solve** equations, and we can **simplify** expressions.

In this section, we will begin our study of solving algebraic equations. Examples of algebraic equations include:

$$a + 6 = 11, \quad t^2 + 7t + 12 = 0, \quad \sqrt{n-4} = 16 - n$$

The first equation is an example of a linear equation; the second is a quadratic equation, and the third is a radical equation. In this section we will learn how to solve a linear equation. We will work with the other equations later in this book.

To **solve an equation** means to find the value or values of the variable that make the equation true.

$a + 6 = 11$: The **solution** is $a = 5$ since we can substitute 5 for the variable and the equation is true:

$$a + 6 = 11$$
$$5 + 6 = 11 \checkmark$$

> **Hint**
> Notice that there are no squared variable terms in a linear equation.

Definition

A **linear equation in one variable** is an equation that can be written in the form

$$ax + b = 0$$

where a and b are real numbers and $a \neq 0$.

Notice that the exponent of the variable, x, is 1 in a linear equation. For this reason, these equations are also known as first-degree equations. Here are other examples of linear equations in one variable:

$$5y - 8 = 19, \quad 2(c + 7) - 3 = 4c + 1, \quad \frac{2}{3}k + \frac{1}{4} = k - 5$$

2 Use the Properties of Equality

The properties of equality will help us solve equations.

> **Hint**
> Note that dividing both sides of an equation by c, is the same as multiplying both sides by $\frac{1}{c}$.

Property The Properties of Equality

Let a, b, and c be expressions representing real numbers. Then,

1) If $a = b$, then $a + c = b + c$ Addition property of equality
2) If $a = b$, then $a - c = b - c$ Subtraction property of equality
3) If $a = b$, then $ac = bc$ Multiplication property of equality
4) If $a = b$, then $\frac{a}{c} = \frac{b}{c} (c \neq 0)$ Division property of equality

These properties tell us that we can add, subtract, multiply, or divide both sides of an equation by the same real number without changing the solutions to the equation.

EXAMPLE 1

Solve and check each equation.

a) $w + 9 = 2$ b) $4k = -24$ c) $\frac{4}{7}a - 5 = 1$

Solution

Remember, to solve the equation means to find the value of the variable that makes the statement true. To do this, we want to *isolate* the variable; that is, we need to get the variable by itself.

a) $w + 9 = 2$: Here, 9 is being *added to* w. To get the w by itself, *subtract* 9 from each side.

$$w + 9 = 2$$
$$w + 9 - 9 = 2 - 9 \quad \text{Subtract 9.}$$
$$w = -7$$

Check: $w + 9 = 2$
$-7 + 9 = 2$
$2 = 2$ ✓

The solution set is $\{-7\}$.

b) $4k = -24$: On the left-hand side of the equation, the k is being *multiplied* by 4. So, we will perform the "opposite" operation and *divide* each side by 4.

> **W Hint**
> Be sure to check your answer. This is always a good practice!

$$4k = -24$$
$$\frac{4k}{4} = \frac{-24}{4} \quad \text{Divide by 4.}$$
$$k = -6$$

Check: $4k = -24$
$4(-6) = -24$
$-24 = -24$ ✓

The solution set is $\{-6\}$.

c) $\frac{4}{7}a - 5 = 1$: On the left-hand side, the a is being multiplied by $\frac{4}{7}$, and 5 is being *subtracted* from the a-term. To solve the equation, begin by eliminating the number being subtracted from the a-term.

$$\frac{4}{7}a - 5 = 1$$
$$\frac{4}{7}a - 5 + 5 = 1 + 5 \quad \text{Add 5 to each side.}$$
$$\frac{4}{7}a = 6 \quad \text{Combine like terms.}$$
$$\frac{7}{4} \cdot \frac{4}{7}a = \frac{7}{4} \cdot 6 \quad \text{Multiply each side by the reciprocal of } \frac{4}{7}.$$
$$1a = \frac{7}{\underset{2}{4}} \cdot \overset{3}{6} \quad \text{Simplify.}$$
$$a = \frac{21}{2}$$

Check: $\frac{4}{7}a - 5 = 1$
$\frac{\overset{2}{4}}{\underset{1}{7}}\left(\frac{\overset{3}{21}}{\underset{1}{2}}\right) - 5 = 1$
$6 - 5 = 1$ ✓

The solution set is $\left\{\frac{21}{2}\right\}$.

[YOU TRY 1] Solve and check each equation.

a) $b - 8 = 5$ b) $-9w = 36$ c) $20 = 13 + \frac{1}{6}n$

3 Combine Like Terms to Solve a Linear Equation

Sometimes it is necessary to combine like terms before we apply the properties of equality. Here are the steps we use to solve a linear equation in one variable.

> **Procedure** How to Solve a Linear Equation
>
> **Step 1:** **Clear parentheses** and **combine like terms** on each side of the equation.
>
> **Step 2:** **Get the variable on one side of the equal sign and the constant on the other side of the equal sign** (isolate the variable) using the addition or subtraction property of equality.
>
> **Step 3:** **Solve for the variable** using the multiplication or division property of equality.
>
> **Step 4:** **Check the solution** in the original equation.

EXAMPLE 2

Solve $8n + 5 - 3(2n + 9) = n + 2(2n + 7)$.

Solution

We will follow the steps to solve this equation.

Step 1: Clear the parentheses and combine like terms.

$$8n + 5 - 6n - 27 = n + 4n + 14 \quad \text{Distribute.}$$
$$2n - 22 = 5n + 14 \quad \text{Combine like terms.}$$

> **W Hint**
> Write out each step as you are reading the example.

Step 2: Isolate the variable. (The variable can be on either side of the equal sign.)

$$2n - 22 = 5n + 14$$
$$2n - 2n - 22 = 5n - 2n + 14 \quad \text{Get the variable on the right side of the } = \text{ sign by subtracting } 2n.$$
$$-22 = 3n + 14$$
$$-22 - 14 = 3n + 14 - 14 \quad \text{Get the constants on the left side of the } = \text{ sign by subtracting } 14.$$
$$-36 = 3n$$

Step 3: Solve for n using the division property of equality.

$$\frac{-36}{3} = \frac{3n}{3} \quad \text{Divide each side by 3.}$$
$$-12 = n$$

Step 4: Check $n = -12$ in the original equation.

$$8n + 5 - 3(2n + 9) = n + 2(2n + 7)$$
$$8(-12) + 5 - 3[2(-12) + 9] = -12 + 2[2(-12) + 7] \quad \text{Substitute } -12 \text{ for } n.$$
$$-96 + 5 - 3(-24 + 9) = -12 + 2(-24 + 7) \quad \text{Distribute.}$$
$$-91 - 3(-15) = -12 + 2(-17) \quad \text{Add.}$$
$$-91 + 45 = -12 + (-34) \quad \text{Multiply.}$$
$$-46 = -46 \checkmark$$

The solution set is $\{-12\}$.

[YOU TRY 2] Solve $5(3 - 2a) + 7a - 2 = 2(9 - 2a) + 15$.

4 Solve Equations Containing Fractions or Decimals

Some equations contain several fractions or decimals that make them appear more difficult to solve. Here are two examples:

$$\frac{2}{9}x - \frac{1}{2} = \frac{1}{18}x + \frac{2}{3} \quad \text{and} \quad 0.05c + 0.4(c - 3) = -0.3$$

Before applying the steps for solving a linear equation, we can eliminate the fractions and decimals from the equations.

Procedure Eliminating Fractions from an Equation

To eliminate the fractions, determine the least common denominator (LCD) for all of the fractions in the equation. Then, multiply both sides of the equation by the LCD.

EXAMPLE 3 Solve $\frac{2}{9}x - \frac{1}{2} = \frac{1}{18}x + \frac{2}{3}$.

Solution

The least common denominator of all of the fractions in the equation is 18. **Multiply both sides of the equation by 18 to eliminate the fractions.**

W Hint
Notice how the LCD is used with the distributive property to eliminate the fractions.

$18\left(\frac{2}{9}x - \frac{1}{2}\right) = 18\left(\frac{1}{18}x + \frac{2}{3}\right)$ Multiply by 18 to eliminate the denominators.

$18 \cdot \frac{2}{9}x - 18 \cdot \frac{1}{2} = 18 \cdot \frac{1}{18}x + 18 \cdot \frac{2}{3}$ Distribute.

$4x - 9 = x + 12$ Multiply.

$4x - x - 9 = x - x + 12$ Get the variable on the left side of the = sign by subtracting x.

$3x - 9 = 12$

$3x - 9 + 9 = 12 + 9$ Get the constants on the right side of the = sign by adding 9.

$3x = 21$

$\frac{3x}{3} = \frac{21}{3}$ Divide by 3.

$x = 7$ Simplify.

The check is left to the student. The solution set is {7}.

[YOU TRY 3] Solve $\frac{1}{5}y + 1 = \frac{3}{10}y + \frac{1}{4}$.

Just as we can eliminate the fractions from an equation to make it easier to solve, we can eliminate decimals from an equation.

> **Procedure** Eliminating Decimals from an Equation
>
> To eliminate the decimals from an equation, multiply both sides of the equation by the smallest power of 10 that will eliminate all decimals from the problem.

EXAMPLE 4

Solve $0.05c + 0.4(c - 3) = -0.3$.

Solution

We want to eliminate the decimals. The number containing a decimal place farthest to the right is 0.05. The 5 is in the *hundredths* place. Therefore, **multiply both sides of the equation by 100 to eliminate all decimals in the equation.**

$$100[0.05c + 0.4(c - 3)] = 100(-0.3)$$
$$100 \cdot (0.05c) + 100 \cdot [0.4(c - 3)] = 100(-0.3) \quad \text{Distribute.}$$

Now we will distribute the 100 to eliminate the decimals.

$$\begin{aligned} 5c + 40(c - 3) &= -30 &&\text{Distribute.} \\ 5c + 40c - 120 &= -30 &&\text{Distribute again.} \\ 45c - 120 &= -30 &&\text{Combine like terms.} \\ 45c - 120 + 120 &= -30 + 120 &&\text{Add 120 to each side.} \\ 45c &= 90 \\ \frac{45c}{45} &= \frac{90}{45} &&\text{Divide.} \\ c &= 2 \end{aligned}$$

The check is left to the student. The solution set is {2}.

YOU TRY 4

Solve $0.1d = 0.5 - 0.02(d - 5)$.

5 Solve Equations with No Solution or an Infinite Number of Solutions

Does every equation have a solution? Consider Example 5.

EXAMPLE 5

Solve $9a + 2 = 6a + 3(a - 5)$.

Solution

$$\begin{aligned} 9a + 2 &= 6a + 3(a - 5) \\ 9a + 2 &= 6a + 3a - 15 &&\text{Distribute.} \\ 9a + 2 &= 9a - 15 &&\text{Combine like terms.} \\ 9a - 9a + 2 &= 9a - 9a - 15 &&\text{Subtract } 9a. \\ 2 &= -15 &&\text{False} \end{aligned}$$

Notice that the variable has "dropped out." Is $2 = -15$ a true statement? No! This means that the equation has *no solution*. We can say that the solution set is the **empty set,** or **null set,** denoted by $\emptyset$.

We have seen that a linear equation may have one solution or no solution. There is a third possibility—a linear equation may have an infinite number of solutions.

EXAMPLE 6 Solve $p - 3p + 8 = 8 - 2p$.

Solution

$$p - 3p + 8 = 8 - 2p$$
$$-2p + 8 = 8 - 2p \quad \text{Combine like terms.}$$
$$-2p + 2p + 8 = 8 - 2p + 2p \quad \text{Add } 2p.$$
$$8 = 8 \quad \text{True}$$

W Hint
Check to see if $p = -2$, $p = 0$, and $p = 2$ are solutions to the equation given in Example 6. Notice that they all work!

Here, the variable has "dropped out," and we are left with an equation, $8 = 8$, that is true. This means that any real number we substitute for p will make the original equation true. Therefore, this equation has an *infinite number of solutions*. The solution set is **{all real numbers}**.

> **Summary** Outcomes When Solving Linear Equations
>
> There are three possible outcomes when solving a linear equation. The equation may have
>
> 1) **one solution.** Solution set: {a real number}. An equation that is true for some values and not for others is called a **conditional equation.**
>
> or
>
> 2) **no solution.** In this case, the variable will drop out, and there will be a false statement such as $2 = -15$. Solution set: $\emptyset$. An equation that has no solution is called a **contradiction.**
>
> or
>
> 3) **an infinite number of solutions.** In this case, the variable will drop out, and there will be a true statement such as $8 = 8$. Solution set: {all real numbers}. An equation that has all real numbers as its solution set is called an **identity.**

[YOU TRY 5] Solve.

a) $6 + 5x - 4 = 3x + 2 + 2x$ b) $3x - 4x + 9 = 5 - x$

6 Use the Five Steps for Solving Applied Problems

Equations can be used to describe events that occur in the real world. Therefore, we need to learn how to translate information presented in English into an algebraic equation. We will begin slowly, then throughout the chapter we will work our way up to more challenging problems. Yes, it may be difficult at first, but with patience and persistence, you can do it!

While no single method will work for solving all applied problems, the following approach is suggested to help in the problem-solving process.

> **W Hint**
> Rewrite the steps for solving applied problems in your own words.

Procedure Steps for Solving Applied Problems

Step 1: Read the problem carefully, more than once if necessary, until you understand it. Draw a picture, if applicable. Identify what you are being asked to find.

Step 2: Choose a variable to represent an unknown quantity. If there are any other unknowns, define them in terms of the variable.

Step 3: Translate the problem from English into an equation using the chosen variable. Some suggestions for doing so are:
- Restate the problem in your own words.
- Read and think of the problem in "small parts."
- Make a chart to separate these "small parts" of the problem to help you translate to mathematical terms.
- Write an equation in English, then translate it to an algebraic equation.

Step 4: Solve the equation.

Step 5: Check the answer in the original problem, and **interpret** the solution as it relates to the problem. Be sure your answer makes sense in the context of the problem.

EXAMPLE 7

Write an equation and solve.
Five more than twice a number is nineteen. Find the number.

Solution

How should we begin?

Step 1: Read the problem carefully. We must find an unknown number.

Step 2: Choose a variable to represent the unknown.

$$\text{Let } x = \text{the number}$$

Step 3: Translate the information that appears in English into an algebraic equation by reading the problem slowly and "in parts."

Statement:	Five more than	twice a number	is	nineteen
Meaning:	Add 5 to	2 times the unknown	equals	19
	↓	↓	↓	↓
Equation:	5 +	2x	=	19

The equation is $5 + 2x = 19$.

Step 4: Solve the equation.

$$5 + 2x = 19$$
$$5 - 5 + 2x = 19 - 5 \quad \text{Subtract 5 from each side.}$$
$$2x = 14 \quad \text{Combine like terms.}$$
$$x = 7 \quad \text{Divide each side by 2.}$$

Step 5: Check the answer. Does it make sense? Five more than twice seven is $5 + 2(7) = 19$. The answer is correct. The number is 7.

[YOU TRY 6] Write an equation and solve.
Nine more than twice a number is seventeen.

Sometimes, dealing with subtraction in a word problem can be confusing. So, let's look at an arithmetic problem first.

EXAMPLE 8 What is four less than ten?

Solution

To solve this problem, do we subtract $10 - 4$ or $4 - 10$? "Four less than ten" is written as $10 - 4$, and $10 - 4 = 6$. Six is four less than ten. The 4 is *subtracted from* the 10. Keep this problem in mind as you read the next example.

[YOU TRY 7] What is eight less than twelve?

EXAMPLE 9 Write the following statement as an equation, and find the number.
Eleven less than three times a number is the same as the number increased by nine. Find the number.

Solution

Step 1: **Read** the problem carefully. We must find an unknown number.

Step 2: **Choose a variable** to represent the unknown.

$$\text{Let } x = \text{the number}$$

Step 3: **Translate** the information that appears in English into an algebraic equation by reading the problem slowly and "in parts."

Statement:	Eleven less than	three times a number	is the same as	the number	increased by 9
Meaning:	Subtract 11 from	3 times the unknown	equals	the unknown	add 9

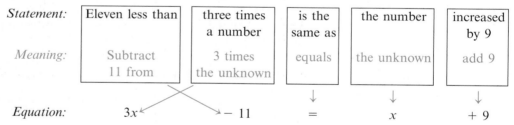

Equation: $3x \quad - 11 \quad = \quad x \quad + 9$

The equation is $3x - 11 = x + 9$.

Step 4: **Solve** the equation.

$$3x - 11 = x + 9$$
$$3x - x - 11 = x - x + 9 \qquad \text{Subtract } x \text{ from each side.}$$
$$2x - 11 = 9 \qquad \text{Combine like terms.}$$
$$2x - 11 + 11 = 9 + 11 \qquad \text{Add 11 to each side.}$$
$$2x = 20 \qquad \text{Combine like terms.}$$
$$x = 10 \qquad \text{Divide each side by 2.}$$

Step 5: **Check** the answer. Does it make sense? Eleven less than three times 10 is $3(10) - 11 = 19$. The number increased by nine is $10 + 9 = 19$. The answer is correct. The number is 10.

[**YOU TRY 8**] Write the following statement as an equation, and find the number.
Three less than five times a number is the same as the number increased by thirteen.

Using Technology

We can use a graphing calculator to solve a linear equation in one variable. First enter the left side of the equation in Y_1 and the right side of the equation in Y_2. Then graph the equations. The x-coordinate of the point of intersection is the solution to the equation.

We will solve $3x + 5 = 6$ algebraically and by using a graphing calculator, and then compare the results. First, use algebra to solve $3x + 5 = 6$. You should get $\left\{\dfrac{1}{3}\right\}$.

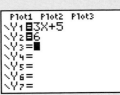

Next, use a graphing calculator to solve $3x + 5 = 6$.

1) Enter $3x + 5$ in Y_1 by pressing the [Y=] key and typing $3x + 5$ to the right of $\setminus Y_1 =$. Then press [ENTER].

2) Enter 6 in Y_2 by pressing the [Y=] key and typing 6 to the right of $\setminus Y_2 =$. Then press [ENTER].

3) Press [ZOOM] and select 6:ZStandard to graph the equations.

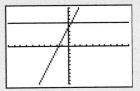

4) To find the intersection point, press [2nd] [TRACE] and select 5:intersect. Press [ENTER] three times. The x-coordinate of the intersection point is shown on the left side of the screen shown to the right, and is stored in the variable x.

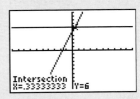

5) Return to the home screen by pressing [2nd] [MODE]. Enter [X,T,Θ,n] [ENTER] to display the solution. Since the result in this case is a decimal value, we can convert it to a fraction by entering [X,T,Θ,n] [MATH] [ENTER] to convert the result to a fraction as shown on the screen to the right.

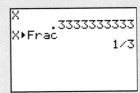

The calculator then gives us a solution set of $\left\{\dfrac{1}{3}\right\}$.

Solve each equation algebraically, and then verify your answer using a graphing calculator.

1) $2x - 3 = 5$ 2) $2x - 6 = 5$ 3) $3x - 6 = 7$
4) $4x - 8 = 3$ 5) $4x - 5 = x + 2$ 6) $3x + 7 = 2 - 5x$

ANSWERS TO [**YOU TRY**] **EXERCISES**

1) a) $\{13\}$ b) $\{-4\}$ c) $\{42\}$ 2) $\{20\}$ 3) $\left\{\dfrac{15}{2}\right\}$ 4) $\{5\}$ 5) a) {all real numbers} b) $\emptyset$
6) $9 + 2x = 17$; 4   7) 4   8) $5x - 3 = x + 13$; 4

SECTION 2.1 **Linear Equations in One Variable**

ANSWERS TO TECHNOLOGY EXERCISES

1) $\{4\}$ 2) $\left\{\dfrac{11}{2}\right\}$ 3) $\left\{\dfrac{13}{3}\right\}$ 4) $\left\{\dfrac{11}{4}\right\}$ 5) $\left\{\dfrac{7}{3}\right\}$ 6) $\left\{-\dfrac{5}{8}\right\}$

E Evaluate 2.1 Exercises

Do the exercises, and check your work.

Objective 1: Define a Linear Equation in One Variable

Identify each as an expression or an equation.

1) $7t - 2 = 11$
2) $\dfrac{3}{4}k + 5(k - 6) = 2$
3) $8 - 10p + 4p + 5$
4) $9(2z - 7) + 3z$
5) Can we solve $3(c + 2) + 5(2c - 5)$? Why or why not?
6) Can we solve $3(c + 2) + 5(2c - 5) = -6$? Why or why not?
7) Which of the following are linear equations in one variable?
   a) $y^2 + 8y + 15 = 0$
   b) $\dfrac{1}{2}w - 5(3w + 1) = 6$
   c) $8m - 7 + 2m + 1$
   d) $0.3z + 0.2 = 1.5$
8) Which of the following are linear equations in one variable?
   a) $-7p = 0$
   b) $-2 = 5g - 4 + g + 10 + 3g - 1$
   c) $9x + 4y = 3$
   d) $10 - 6(4n - 1) + 7$

Determine if the given value is a solution to the equation.

9) $-8p = 12; p = -\dfrac{3}{2}$
10) $2d + 1 = 13; d = -6$
11) $2(t - 5) + 7 = 3(2t - 9) - 2; t = 4$
12) $5 - 3(2k + 1) = 4k - 3; k = \dfrac{1}{2}$

Objective 2: Use the Properties of Equality

Solve and check each equation.

13) $r - 6 = 11$
14) $c + 2 = -5$
15) $-16 = k - 12$
16) $8 = t + 1$
17) $a + \dfrac{5}{8} = \dfrac{1}{2}$
18) $w - \dfrac{3}{4} = -\dfrac{1}{6}$
19) $3y = 30$
20) $-56 = -7v$
21) $-6 = \dfrac{k}{8}$
22) $30 = -\dfrac{x}{2}$
23) $\dfrac{2}{3}g = -10$
24) $\dfrac{7}{4}r = 42$
25) $-\dfrac{5}{3}d = -30$
26) $-\dfrac{5}{6} = -\dfrac{4}{9}x$
27) $0.5q = 6$
28) $0.3t = 3$
29) $3x - 7 = 17$
30) $5g + 19 = 4$
31) $8d - 15 = -15$
32) $4 = 7j - 8$
33) $\dfrac{4}{9}w - 11 = 1$
34) $\dfrac{5}{3}a + 6 = 41$
35) $\dfrac{10}{7}m + 3 = 1$
36) $\dfrac{9}{10}x - 4 = 11$
37) $5 - 0.4p = 2.6$
38) $1.8 = 1.2n - 7.8$

Mixed Exercises: Objectives 3 and 5

Solve and check each equation.

39) $10v + 9 - 2v + 16 = 1$
40) $-8g - 7 + 6g + 1 = 20$
41) $5 = -12p + 7 + 4p - 12$
42) $12 = 9y + 11 - 3y - 7$
43) $-12 = 7(2a - 3) - (8a - 9)$
44) $20 = 5r - 3 + 2(9 - 3r)$
45) $2y + 7 = 5y - 2$
46) $8n - 21 = 3n - 1$
47) $6 - 7p = 2p + 33$
48) $z + 19 = 5 - z$
49) $-8x + 6 - 2x + 11 = 3 + 3x - 7x$
50) $10 - 13a + 2a - 16 = -5 + 7a + 11$
51) $4(2t + 5) - 7 = 5(t + 5)$
52) $3(2m + 10) = 6(m + 4) - 8m$

53) $-9r + 4r - 11 + 2 = 3r + 7 - 8r + 9$

54) $3(4b - 7) + 8 = 6(2b + 5)$

55) $j - 15j + 8 = -3(4j - 3) - 2j - 1$

56) $n - 16 + 10n + 4 = 2(7n - 6) - 3n$

57) $8(3t + 4) = 10t - 3 + 7(2t + 5)$

58) $2(9z - 1) + 7 = 10z - 14 + 8z + 2$

59) $8 - 7(2 - 3w) - 9w = 4(5w - 1) - 3w - 2$

60) $4m - (6m + 5) + 2 = 8m + 3(4 - 3m)$

61) $7y + 2(1 - 4y) = 8y - 5(y + 4)$

Objective 4: Solve Equations Containing Fractions or Decimals

62) How can you eliminate the fractions from the equation $\frac{1}{6}x + \frac{5}{4} = \frac{1}{2}x - \frac{5}{12}$?

Solve each equation by first clearing fractions or decimals.

63) $\frac{1}{6}x + \frac{5}{4} = \frac{1}{2}x - \frac{5}{12}$

64) $\frac{3}{4}n + \frac{1}{2} = \frac{1}{2}n + \frac{1}{4}$

65) $\frac{2}{3}d - 1 = \frac{1}{5}d + \frac{2}{5}$

66) $\frac{1}{5}c + \frac{2}{7} = 2 - \frac{1}{7}c$

67) $\frac{m}{3} + \frac{1}{2} = \frac{2m}{3} + 3$

68) $\frac{a}{8} - 1 = \frac{a}{3} - \frac{7}{12}$

69) $\frac{1}{3} + \frac{1}{9}(k + 5) - \frac{k}{4} = 2$

70) $\frac{1}{2} = \frac{2}{9}(3x - 2) - \frac{x}{9} - \frac{x}{6}$

71) $0.05(t + 8) - 0.01t = 0.6$

72) $0.2(y - 3) + 0.05(y - 10) = -0.1$

73) $0.2(12) + 0.08z = 0.12(z + 12)$

74) $0.1x + 0.15(8 - x) = 0.125(8)$

75) $0.04s + 0.03(s + 200) = 27$

76) $0.06x + 0.1(x - 300) = 98$

Objective 6: Use the Five Steps for Solving Applied Problems

Write each statement as an equation, and find the number.

77) Four more than a number is fifteen.

78) Thirteen more than a number is eight.

79) Seven less than a number is twenty-two.

80) Nine less than a number is eleven.

81) Twice a number is -16.

82) The product of six and a number is fifty-four.

83) Seven more than twice a number is thirty-five.

84) Five more than twice a number is fifty-three.

85) Three times a number decreased by eight is forty.

86) Twice a number decreased by seven is -13.

87) Half of a number increased by ten is three.

88) One-third of a number increased by four is one.

89) Twelve less than five times a number is the same as the number increased by sixteen.

90) Three less than twice a number is the same as the number increased by eight.

91) Ten more than one-third of a number is the same as the number decreased by two.

92) A number decreased by nine is the same as seven more than half the number.

93) Twice the sum of a number and five is sixteen.

94) Twice the sum of a number and -8 is four.

95) Three times a number is fifteen more than half the number.

96) A number increased by fourteen is nine less than twice the number.

97) A number decreased by six is five more than twice the number.

98) A number divided by four is nine less than the number.

R Rethink

R1) What is the name of the quantity we use to clear the fractions from an equation?

R2) How do you know what number to use to clear the decimals from an equation?

R3) What symbol do we use to represent the solution of an equation that has no solution?

R4) Which types of equations are harder for you to solve?

2.2 Applications of Linear Equations

P Prepare | **O Organize**

What are your objectives for Section 2.2?	How can you accomplish each objective?
1 Solve Problems Involving General Quantities	• Read the problem more than once. • Follow the procedure for solving applied problems; do not jump to setting up the equation before you choose a variable and define your unknown(s) using that variable. • Make sure your answer makes sense and that you have answered the question that is being asked. • Be aware that the solution to the equation may not be the final answer to the question in all cases! • Complete the given example on your own. • Complete You Try 1.
2 Solve Problems Involving Lengths	• After reading the problem more than once, draw a diagram and label it with the given information. • Follow the procedure for solving applied problems. • Be aware that many problems require lengths to be added together. • Complete the given example on your own. • Complete You Try 2.
3 Solve Consecutive Integer Problems	• Understand the meaning of *consecutive integers, consecutive even integers,* and *consecutive odd integers.* • Follow the procedure for solving applied problems. • When setting up an equation, translate from the word problem into the equation. • Make sure your answer makes sense and that you answer the question that is being asked. • Complete the given examples on your own. • Complete You Trys 3 and 4.
4 Solve Problems Involving Percent Change	• Change a percent into its decimal form when solving an equation. • Follow the procedure for solving applied problems. • When setting up an equation, translate from the word problem into the equation. • Complete the given examples on your own. • Complete You Trys 5 and 6.
5 Solve Problems Involving Simple Interest	• Know the formula for calculating *simple interest.* • Be able to identify the *principal, interest earned, annual interest rate,* and the *time* to be substituted into the formula. One of those values is usually not given. • Don't forget to convert all rates to their decimal form when using them in an equation. • Follow the procedure for solving applied problems when needed. • Check to make sure your answer makes sense. • Complete the given examples on your own. • Complete You Trys 7, 8, and 9.

Read the explanations, follow the examples, take notes, and complete the You Trys.

In the previous section, you were introduced to solving applied problems. In this section, we will build on these skills so that we can solve other types of problems.

1 Solve Problems Involving General Quantities

EXAMPLE 1 Mrs. Ramirez has 26 students in her third-grade class. There are four more boys than girls. How many boys and girls are in her class?

Solution

Step 1: **Read** the problem carefully, and identify what we are being asked to find.

We must find the number of boys and number of girls in the class.

Step 2: **Choose a variable** to represent an unknown, and define the other unknown in terms of this variable.

In the statement "There are four more boys than girls," the number of boys is expressed *in terms of* the number of girls. Therefore, let

$$x = \text{the number of girls}$$

Define the other unknown (the number of boys) in terms of x. Since there are four *more* boys than girls,

$$\text{The number of girls} + 4 = \text{the number of boys}$$
$$x + 4 = \text{the number of boys}$$

Step 3: **Translate** the information that appears in English into an algebraic equation. One approach is to restate the problem in your own words.

We can think of the situation in this problem as:

The number of girls plus the number of boys is 26.

Let's write this as an equation.

Statement:	Number of girls	plus	Number of boys	is	26
	↓	↓	↓	↓	↓
Equation:	x	$+$	$(x + 4)$	$=$	26

The equation is $x + (x + 4) = 26$.

Step 4: **Solve** the equation.

$$x + (x + 4) = 26$$
$$2x + 4 = 26$$
$$2x + 4 - 4 = 26 - 4 \qquad \text{Subtract 4 from each side.}$$
$$2x = 22 \qquad \text{Combine like terms.}$$
$$\frac{2x}{2} = \frac{22}{2} \qquad \text{Divide each side by 2.}$$
$$x = 11 \qquad \text{Simplify.}$$

Step 5: **Check** the answer, and **interpret** the meaning as it relates to the problem.

Since x represents the number of girls, there are 11 girls in the class.

The expression $x + 4$ represents the number of boys, so there are $x + 4 = 11 + 4 = 15$ boys.

The answer makes sense because the total number of children in this third-grade class is $11 + 15 = 26$.

> **[YOU TRY 1]** The record low temperature for the month of January in Anchorage, Alaska, was (in °F) 84° less than January's record high temperature. The sum of the record low and record high is 16°. Find the lowest and highest temperatures ever recorded in Anchorage, Alaska, in January. (www.weather.com)

2 Solve Problems Involving Lengths

EXAMPLE 2 A carpenter has a board that is 15 ft long. He needs to cut it into two pieces so that one piece is half as long as the other. What will be the length of each piece?

Solution

Step 1: **Read** the problem carefully, and identify what we are being asked to find.

We must find the length of each of two pieces of a board.

A picture will be very helpful in this problem.

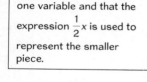

Step 2: **Choose a variable** to represent an unknown, and define the other unknown in terms of this variable.

One piece of the board must be half the length of the other piece. Therefore, let

$$x = \text{the length of one piece}$$

Define the other unknown in terms of x.

$$\frac{1}{2}x = \text{the length of the second piece}$$

> **W Hint**
> Notice that we only use one variable and that the expression $\frac{1}{2}x$ is used to represent the smaller piece.

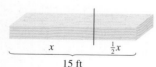

Step 3: **Translate** the information that appears in English into an algebraic equation. Let's label the picture on the left with the expressions representing the unknowns and then restate the problem in our own words.

From the picture we can see that the

length of one piece plus the length of the second piece equals 15 ft.

Let's write this as an equation.

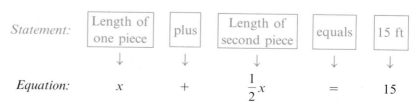

The equation is $x + \frac{1}{2}x = 15$.

Step 4: **Solve** the equation.

$$x + \frac{1}{2}x = 15$$

$$\frac{3}{2}x = 15 \qquad \text{Add like terms.}$$

$$\frac{2}{3} \cdot \frac{3}{2}x = \frac{2}{3} \cdot 15 \qquad \text{Multiply by the reciprocal of } \frac{3}{2}.$$

$$x = 10$$

Step 5: **Check** the answer, and **interpret** the solution as it relates to the problem.

One piece of board is 10 ft long.

The expression $\frac{1}{2}x$ represents the length of the other piece of board, so the length of the other piece is $\frac{1}{2}x = \frac{1}{2}(10) = 5$ ft.

The answer makes sense because the length of the original board was 10 ft + 5 ft = 15 ft.

[**YOU TRY 2**] A 24-ft chain must be cut into two pieces so that one piece is twice as long as the other. Find the length of each piece of chain.

3 Solve Consecutive Integer Problems

What *are* consecutive integers? "Consecutive" means one after the other, in order. In this section, we will look at consecutive integers, consecutive even integers, and consecutive odd integers.

W Hint
Notice that consecutive integers are always one unit apart from each other.

Consecutive integers differ by 1. For example, look at the consecutive integers 5, 6, 7, and 8. If $x = 5$, then $x + 1 = 6$, $x + 2 = 7$, and $x + 3 = 8$. Therefore, to define the unknowns for consecutive integers, let

$x =$ first integer
$x + 1 =$ second integer
$x + 2 =$ third integer
$x + 3 =$ fourth integer

and so on.

EXAMPLE 3   The sum of three consecutive integers is 126. Find the integers.

Solution

Step 1: **Read** the problem carefully, and identify what we are being asked to find.

We must find three consecutive integers with a sum of 126.

Step 2: **Choose a variable** to represent an unknown, and define the other unknowns in terms of this variable.

There are three unknowns. We will let x represent the first consecutive integer and then define the other unknowns in terms of x.

$$x = \text{the first integer}$$

Define the other unknowns in terms of x.

$$x + 1 = \text{the second integer}$$
$$x + 2 = \text{the third integer}$$

Step 3: Translate the information that appears in English into an algebraic equation. What does the original statement mean?

"The sum of three consecutive integers is 126" means that when the three numbers are *added* together the sum is 126.

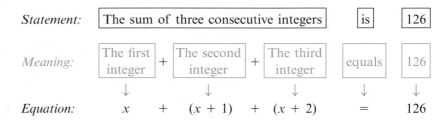

The equation is $x + (x + 1) + (x + 2) = 126$.

Step 4: Solve the equation.

$$x + (x + 1) + (x + 2) = 126$$
$$3x + 3 = 126$$
$$3x + 3 - 3 = 126 - 3 \qquad \text{Subtract 3 from each side.}$$
$$3x = 123 \qquad \text{Combine like terms.}$$
$$\frac{3x}{3} = \frac{123}{3} \qquad \text{Divide each side by 3.}$$
$$x = 41 \qquad \text{Simplify.}$$

Step 5: Check the answer, and **interpret** the solution as it relates to the problem.

The first integer is 41. The second integer is 42 since $x + 1 = 41 + 1 = 42$, and the third integer is 43 since $x + 2 = 41 + 2 = 43$.

The answer makes sense because their sum is $41 + 42 + 43 = 126$.

> **[YOU TRY 3]** The sum of three consecutive integers is 177. Find the integers.

Next, let's look at **consecutive even integers,** which are even numbers that differ by two such as $-10, -8, -6,$ and -4. If x is the first even integer, we have

-10	-8	-6	-4
x	$x + 2$	$x + 4$	$x + 6$

Therefore, to define the unknowns for consecutive even integers, let

$$x = \text{the first even integer}$$
$$x + 2 = \text{the second even integer}$$
$$x + 4 = \text{the third even integer}$$
$$x + 6 = \text{the fourth even integer}$$

and so on.

Will the expressions for **consecutive odd numbers** be any different? No! When we count by consecutive odds, we are still counting by twos. Look at the numbers 9, 11, 13, and 15 for example. If x is the first odd integer, we have

9	11	13	15
x	$x + 2$	$x + 4$	$x + 6$

> **Hint**
> Notice that both consecutive even integers and consecutive odd integers are always two units apart from each other.

To define the unknowns for consecutive odd integers, let

$$x = \text{the first odd integer}$$
$$x + 2 = \text{the second odd integer}$$
$$x + 4 = \text{the third odd integer}$$
$$x + 6 = \text{the fourth odd integer.}$$

EXAMPLE 4

The sum of two consecutive odd integers is 47 more than three times the larger integer. Find the integers.

Solution

Step 1: **Read** the problem carefully, and identify what we are being asked to find.

We must find two consecutive odd integers.

Step 2: **Choose a variable** to represent an unknown, and define the other unknowns in terms of this variable.

There are two unknowns. We will let x represent the first consecutive odd integer and then define the other unknown in terms of x.

$$x = \text{the first odd integer}$$
$$x + 2 = \text{the second odd integer}$$

Step 3: **Translate** the information that appears in English into an algebraic equation. Read the problem slowly and carefully, breaking it into small parts.

Statement:	The sum of two consecutive odd integers		is	47 more than	three times the larger integer
Meaning:	The first odd integer	+	The second odd integer	equals	add 47 to	3 times the larger integer
	↓	↓	↓	↓	↓
Equation:	x	+	$(x + 2)$	=	$47 +$	$3(x + 2)$

The equation is $x + (x + 2) = 47 + 3(x + 2)$.

Step 4: **Solve** the equation.

$$x + (x + 2) = 47 + 3(x + 2)$$
$$2x + 2 = 47 + 3x + 6 \quad \text{Combine like terms; distribute.}$$
$$2x + 2 = 3x + 53 \quad \text{Combine like terms.}$$
$$2x + 2 - 2 = 3x + 53 - 2 \quad \text{Subtract 2 from each side.}$$
$$2x = 3x + 51 \quad \text{Combine like terms.}$$
$$2x - 3x = 3x - 3x + 51 \quad \text{Subtract } 3x \text{ from each side.}$$
$$-x = 51 \quad \text{Combine like terms.}$$
$$x = -51 \quad \text{Divide each side by } -1.$$

Step 5: **Check** the answer, and **interpret** the solution as it relates to the problem.

The first odd integer is -51. The second integer is -49 since $x + 2 = -51 + 2 = -49$.

Check these numbers in the original statement of the problem. The sum of -51 and -49 is -100. Then, 47 more than three times the larger integer is $47 + 3(-49) = 47 + (-147) = -100$. The numbers are -51 and -49.

[YOU TRY 4] Twice the sum of three consecutive even integers is 18 more than five times the largest number. Find the integers.

4 Solve Problems Involving Percent Change

Percents pop up everywhere. "Earn 4% simple interest on your savings account." "The unemployment rate increased 1.2% this year." We've seen these statements online, or we have heard about them on television. In this section, we will introduce applications involving percents.

Here's another type of percent problem we might see in a store: "Everything in the store is marked down 30%." Before tackling an algebraic percent problem, let's look at an arithmetic problem. Relating an algebra problem to an arithmetic problem can make it easier to solve an application that requires the use of algebra.

EXAMPLE 5 Jeans that normally sell for $28.00 are marked down 30%. What is the sale price?

Solution

Concentrate on the *procedure* used to obtain the answer. This is the same procedure we will use to solve algebra problems with percent increase and percent decrease.

$$\text{Sale price} = \text{Original price} - \text{Amount of discount}$$

How much is the discount? It is 30% *of* $28.00.

Change the percent to a decimal. The amount of the discount is calculated by multiplying:

$$\text{Amount of discount} = (\text{Rate of discount})(\text{Original price})$$
$$\text{Amount of discount} = (0.30) \cdot (\$28.00) = \$8.40$$

$$\begin{aligned}\text{Sale price} &= \text{Original price} - \text{Amount of discount}\\ &= \$28.00 - (0.30)(\$28.00)\\ &= \$28.00 - \$8.40\\ &= \$19.60\end{aligned}$$

The sale price is $19.60.

W Hint
Remember that to convert a percent to a decimal, you move the decimal two places to the left.

[YOU TRY 5] A dress shirt that normally sells for $39.00 is marked down 25%. What is the sale price?

Next, let's solve an algebra problem involving a markdown or percent decrease.

EXAMPLE 6 The sale price of a Lil Wayne CD is $14.80 after a 20% discount. What was the original price of the CD?

Solution

Step 1: Read the problem carefully, and identify what we are being asked to find.

We must find the original price of the CD.

Step 2: Choose a variable to represent the unknown.

$$x = \text{the original price of the CD}$$

Hint

Remember that a *sale price* will always be less than the *original price*.

Step 3: Translate the information that appears in English into an algebraic equation. One way to figure out how to write an algebraic equation is to relate this problem to the arithmetic problem in Example 5. To find the sale price of the jeans in Example 5 we found that

Sale price = Original price − Amount of discount

where we found the amount of the discount by multiplying the rate of the discount by the original price. We will write an algebraic equation using the same procedure.

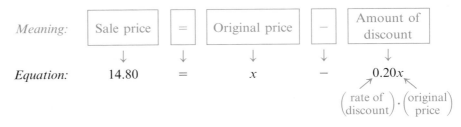

The equation is $14.80 = x - 0.20x$.

Step 4: Solve the equation.

$$14.80 = x - 0.20x$$
$$14.80 = 0.80x \quad \text{Combine like terms.}$$
$$\frac{14.80}{0.80} = \frac{0.80x}{0.80} \quad \text{Divide each side by 0.80.}$$
$$x = 18.5 \quad \text{Simplify.}$$

Step 5: Check the answer, and **interpret** the solution as it relates to the problem.

The original price of the CD was $18.50.

The answer makes sense because the amount of the discount is $(0.20)(\$18.50) = \$3.70$ which makes the sale price $\$18.50 - \$3.70 = \$14.80$.

[YOU TRY 6] A video game is on sale for $35.00 after a 30% discount. What was the original price of the video game?

5 Solve Problems Involving Simple Interest

When customers invest their money in bank accounts, their accounts earn interest. Why? Paying interest is a way for financial institutions to get people to deposit money.

There are different ways to calculate the amount of interest earned from an investment. In this section, we will discuss *simple interest*. **Simple interest** calculations are based on the initial amount of money deposited in an account. This initial amount of money is known as the **principal.**

The formula used to calculate simple interest is $I = PRT$, where

I = interest earned (simple)
P = principal (initial amount invested)
R = annual interest rate (expressed as a decimal)
T = amount of time the money is invested (in years)

We will begin with two arithmetic problems. The procedures used will help you understand more clearly how we arrive at the algebraic equation in Example 9.

EXAMPLE 7

If $500 is invested for 1 year in an account earning 6% simple interest, how much interest will be earned?

Solution

Use $I = PRT$ to find I, the interest earned.

$$P = \$500, \quad R = 0.06, \quad T = 1$$

$$I = PRT$$
$$I = (500)(0.06)(1)$$
$$I = 30$$

Hint
Notice that I represents a dollar amount and R represents a percent expressed as a decimal.

The interest earned will be $30.

YOU TRY 7

If $3500 is invested for 2 years in an account earning 4.5% simple interest, how much interest will be earned?

EXAMPLE 8

Tom invests $2000 in an account earning 7% interest and $9000 in an account earning 5% interest. After 1 year, how much interest will he have earned?

Solution

Tom will earn interest from two accounts. Therefore,

Total interest earned = Interest from 7% account + Interest from 5% account

$$\begin{array}{cccc} & P \quad R \quad T & & P \quad R \quad T \\ \text{Total interest earned} = & (2000)(0.07)(1) & + & (9000)(0.05)(1) \\ \text{Total interest earned} = & 140 & + & 450 \\ & = \$590 & & \end{array}$$

Tom will earn a total of $590 in interest from the two accounts.

YOU TRY 8

Abebe invests $1500 in an account earning 4% interest and $4000 in an account earning 6.7% interest. After 1 year, how much interest will she have earned?

Note
When money is invested for 1 year, $T = 1$. Therefore, the formula $I = PRT$ can be written as $I = PR$.

This idea of earning interest from different accounts is one we will use in Example 9.

EXAMPLE 9

Last year, Neema Reddy had $10,000 to invest. She invested some of it in a savings account that paid 3% simple interest, and she invested the rest in a certificate of deposit that paid 5% simple interest. In 1 year, she earned a total of $380 in interest. How much did Neema invest in each account?

60   CHAPTER 2   Linear Equations in One Variable

Solution

Step 1: **Read** the problem carefully, and identify what we are being asked to find.

We must find the amounts Neema invested in the 3% account and in the 5% account.

Step 2: **Choose a variable** to represent an unknown, and define the other unknown in terms of this variable.

Let x = amount Neema invested in the 3% account.

How do we write an expression, in terms of x, for the amount invested in the 5% account?

Total invested   Amount invested in 3% account
$\downarrow$ $\downarrow$
10,000 $-$ x = amount invested in the 5% account

We define the unknowns as

x = amount Neema invested in the 3% account
$10{,}000 - x$ = amount Neema invested in the 5% account

Step 3: **Translate** the information that appears in English into an algebraic equation. Use the "English equation" we used in Example 8. Since $T = 1$, we can compute the interest using $I = PR$.

Total interest earned = Interest from 3% account + Interest from 5% account

                              P R P R
380   = $(x)(0.03)$ + $(10{,}000 - x)(0.05)$

The equation is $380 = 0.03x + 0.05(10{,}000 - x)$.

We can also get the equation by organizing the information in a table:

	Amount Invested (in dollars) P	Interest Rate R	Interest Earned After 1 Year I
3% account	x	0.03	$0.03x$
5% account	$10{,}000 - x$	0.05	$0.05(10{,}000 - x)$

Total interest earned = Interest from 3% account + Interest from 5% account
       380   = $0.03x$ + $0.05(10{,}000 - x)$

The equation is $380 = 0.03x + 0.05(10{,}000 - x)$.

Either way of organizing the information will lead us to the correct equation.

Step 4: **Solve** the equation. Begin by multiplying both sides of the equation by 100 to eliminate the decimals.

$380 = 0.03x + 0.05(10{,}000 - x)$
$100(380) = 100[0.03x + 0.05(10{,}000 - x)]$ Multiply each side by 100.
$38{,}000 = 3x + 5(10{,}000 - x)$ Multiply.
$38{,}000 = 3x + 50{,}000 - 5x$ Distribute.
$38{,}000 = -2x + 50{,}000$ Combine like terms.
$-12{,}000 = -2x$ Subtract 50,000.
$6000 = x$ Divide by -2.

Step 5: **Check** the answer, and **interpret** the solution as it relates to the problem.

Neema invested $6000 at 3% interest. The amount invested at 5% is $10,000 − x or $10,000 − \$6000 = \$4000$.

Check:

Total interest earned = Interest from 3% account + Interest from 5% account
380 = 6000(0.03) + 4000(0.05)
   = 180 + 200
= 380

[YOU TRY 9] Christine received an $8000 bonus from work. She invested part of it at 6% simple interest and the rest at 4% simple interest. Christine earned a total of $420 in interest after 1 year. How much did she deposit in each account?

ANSWERS TO [YOU TRY] EXERCISES

1) record low: −34°F, record high: 50°F 2) 8 ft and 16 ft 3) 58, 59, 60 4) 26, 28, 30
5) $29.25 6) $50.00 7) $315 8) $328 9) $5000 at 6% and $3000 at 4%

E Evaluate 2.2 Exercises

Do the exercises, and check your work.

1) On a baseball team, there are 5 more pitchers than catchers. If there are c catchers on the team, write an expression for the number of pitchers.

2) On Wednesday, the Snack Shack sold 23 more hamburgers than hot dogs. Write an expression for the number of hamburgers sold if h hot dogs were sold.

3) There were 31 fewer people on a flight from Chicago to New York than from Chicago to Los Angeles. Write an expression for the number of people on the flight to New York if there were p people on the flight to L.A.

4) The test average in Mr. Muscari's second-period class was 3.8 points lower than in his first-period class. If the average test score in the first-period class was a, write an expression for the test average in Mr. Muscari's second-period class.

5) A survey of adults aged 20–29 revealed that, of those who exercise regularly, three times as many men run as women. If w women run on a regular basis, write an expression for the number of male runners.

6) At Roundtree Elementary School, s students walk to school. One-third of that number ride their bikes. Write an expression for the number of students who ride their bikes to class.

7) An electrician cuts a 14-ft wire into two pieces. If one is x ft long, how long is the other piece?

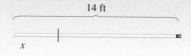

8) Jorie drives along a 142-mi stretch of highway from Oregon to California. If she drove m mi in Oregon, how far did she drive in California?

Objective 1: Solve Problems Involving General Quantities

Solve using the Five Steps for Solving Applied Problems. See Example 1.

9) A 12-oz serving of Pepsi has 6.5 more teaspoons of sugar than a 12-oz serving of Gatorade. Together they contain 13.1 teaspoons of sugar. How much sugar is in each 12-oz drink? (www.dentalgentlecare.com)

10) Two of the smallest countries in the world are the Marshall Islands and Liechtenstein. The Marshall Islands covers 8 mi^2 more than Liechtenstein. Find the area of each country if together they encompass 132 mi^2. (www.infoplease.com)

11) In the 2012 Summer Olympics in London, Germany won half as many medals as China. If they won a total of 132 medals, how many were won by each country? (www.london2012.com)

12) Latisha's Golden Retriever weighs twice as much as Janessa's Border Collie. Find the weight of each dog if they weigh 96 lb all together.

13) The Columbia River is 70 miles shorter than the Ohio River. Determine the length of each river if together they span 2550 miles. (ga.water.usgs.gov)

14) In the 2010 Alabama gubernatorial election, Ron Sparks had 235,220 fewer votes than Robert Bentley. If they received a total of 1,485,324 votes, how many people voted for each man? (www.sos.state.al.us)

Objective 2: Solve Problems Involving Lengths
Solve using the Five Steps for Solving Applied Problems. See Example 2.

15) A plumber has a 36-in.-long pipe. He must cut it into two pieces so that one piece is 14 in. longer than the other. How long is each piece?

16) A builder has to install some decorative trim on the outside of a house. The piece she has is 75 in. long, but she needs to cut it so that one piece is 27 in. shorter than the other. Find the length of each piece.

17) Calida's mom found an 18-ft-long rope in the garage. She will cut it into two pieces so that one piece can be used for a long jump rope and the other for a short one. If the long rope is to be twice as long as the short one, find the length of each jump rope.

18) A 55-ft-long drainage pipe must be cut into two pieces before installation. One piece is two-thirds as long as the other. Find the length of each piece.

Objective 3: Solve Consecutive Integer Problems
Solve using the five "Steps for Solving Applied Problems." See Examples 3 and 4.

19) The sum of two consecutive integers is 77. Find the integers.

20) The sum of three consecutive integers is 195. Find the integers.

21) Find two consecutive even integers such that twice the smaller is 10 more than the larger.

22) Find three consecutive odd integers such that four times the smallest is 56 more than the sum of the other two.

23) Find three consecutive odd integers such that their sum is five more than four times the largest integer.

24) The sum of two consecutive even integers is 52 less than three times the larger integer. Find the integers.

25) Two consecutive page numbers in a book add up to 345. Find the page numbers.

26) The addresses on the east side of Arthur Ave. are consecutive odd numbers. Two consecutive house numbers add up to 36. Find the addresses of these two houses.

Objective 4: Solve Problems Involving Percent Change
Find the sale price of each item.

27) A cell phone that regularly sells for $75.00 is marked down 15%.

28) A baby stroller that retails for $69.00 is on sale at 20% off.

29) At the end of the summer, the bathing suit that sold for $29.00 is marked down 60%.

30) An advertisement states that a TV that regularly sells for $399.00 is being discounted 25%.

Solve using the Five Steps for Solving Applied Problems. See Example 6.

31) A digital camera is on sale for $119 after a 15% discount. What was the original price of the camera?

32) Marie paid $15.13 for a hardcover book that was marked down 15%. What was the original selling price of the book?

33) In February, a store discounted all of its calendars by 60%. If Ramon paid $4.38 for a calendar, what was its original price?

34) An appliance store advertises 20% off all of its refrigerators. Mr. Kotaris paid $399.20 for the fridge. Find its original price.

35) The sale price of a coffeemaker is $22.75. This is 30% off of the original price. What was the original price of the coffeemaker?

36) Katrina paid $20.40 for a backpack that was marked down 15%. Find the original retail price of the backpack.

37) One hundred forty countries participated in the 1984 Summer Olympics in Los Angeles. This was 75% more than the number of countries that took part in the Summer Olympics in Moscow 4 years earlier. How many countries participated in the 1980 Olympics in Moscow? (www.olympics.org)

38) In 2010 there were about 1224 acres of farmland in Custer County. This is 32% less than the number of acres of farmland in 2000. Calculate the number of acres of farmland in Custer County in 2000.

39) In 2004, there were 7569 Starbucks stores worldwide. This is approximately 1681% more stores than 10 years earlier. How many Starbucks stores were there in 1994? (Round to the nearest whole number.) (www.starbucks.com)

40) Liu Fan's salary this year is 14% higher than it was 3 years ago. If he earns $37,050 this year, what did he earn 3 years ago?

Objective 5: Solve Problems Involving Simple Interest
Solve. See Examples 7 and 8.

41) Jenna invests $800 in an account for 1 year earning 4% simple interest. How much interest was earned from this account?

42) If $3000 is deposited into an account for 1 year earning 5.5% simple interest, how much money will be in the account after 1 year?

43) Rachel Levin has a total of $5500 to invest for 1 year. She deposits $4000 into an account earning 6.5% annual simple interest and the rest into an account earning 8% annual simple interest. How much interest did Rachel earn?

44) Maurice plans to invest a total of $9000 for 1 year. In the account earning 5.2% simple interest he will deposit $6000, and in an account earning 7% simple interest he will deposit the rest. How much interest will Maurice earn?

Solve using the Five Steps for Solving Applied Problems. See Example 9.

45) Amir Sadat receives a $15,000 signing bonus on accepting his new job. He plans to invest some of it at 6% annual simple interest and the rest at 7% annual simple interest. If he will earn $960 in interest after 1 year, how much will Amir invest in each account?

46) Lisa Jenkins invested part of her $8000 inheritance in an account earning 5% simple interest and the rest in an account earning 4% simple interest. How much did Lisa invest in each account if she earned $365 in total interest after 1 year?

47) Enrique's money earned $164 in interest after 1 year. He invested some of his money in an account earning 6% simple interest and $200 more than that amount in an account earning 5% simple interest. Find the amount Enrique invested in each account.

48) Saori Yamachi invested some money in an account earning 7.4% simple interest and twice that amount in an account earning 9% simple interest. She earned $1016 in interest after 1 year. How much did Saori invest in each account?

49) Last year, Clarissa invested a total of $7000 in two accounts earning simple interest. Some of it she invested at 9.5%, and the rest was invested at 7%. How much did she invest in each account if she earned a total of $560 in interest last year?

50) Ted has $3000 to invest. He deposits a portion of it in an account earning 5% simple interest and the rest at 6.5% simple interest. After 1 year he has earned $175.50 in interest. How much did Ted deposit in each account?

Mixed Exercises
Write an equation and solve. Use the Five Steps for Solving Applied Problems.

51) Irina was riding her bike when she got a flat tire. Then, she walked her bike 1 mi more than half the distance she rode it. How far did she ride her bike and how far did she walk if the total distance she traveled was 7 mi?

52) A book is open to two pages numbered consecutively. The sum of the page numbers is 373. What are the page numbers?

53) In August 2013, there were 2600 entering freshmen at a state university. This is 4% higher than the number of freshmen entering the school in August of 2012. How many freshmen were enrolled in August 2012?

54) In 2002, Mia Hamm played in eight fewer U.S. National Team soccer matches than in 2003. During those 2 years she appeared in a total of 26 games. In how many games did she play in 2002 and in 2003? (www.ussoccer.com)

55) A 53-in. board is to be cut into three pieces so that one piece is 5 in. longer than the shortest piece, and the other piece is twice as long as the shortest piece. Find the length of each piece of board.

56) Ivan has $8500 to invest. He will invest some of it in a long-term IRA paying 4% simple interest and the rest in a short-term CD earning 2.5% simple interest. After 1 year, Ivan's investments have earned $250 in interest. How much did Ivan invest in each account?

57) One-sixth of the smallest of three consecutive even integers is three less than one-tenth the sum of the other even integers. Find the integers.

58) A 58-ft-long cable must be cut into three pieces. One piece will be 6 ft longer than the shortest piece, and the longest portion must be twice as long as the shortest. Find the length of the three pieces of cable.

59) In 2011, *Harry Potter and the Deathly Hallows Part 2* grossed about $100 million more in theaters than *Twilight Saga: Breaking Dawn Part 1*. Together, they earned $662 million. How much did each movie earn in theaters in 2011? (boxofficemojo.com)

60) Henry marks up the prices of his fishing poles by 50%. Determine what Henry paid his supplier for his best-selling fishing pole if Henry charges his customers $33.75.

61) Tamara invests some money in three different accounts. She puts some of it in a CD earning 3% simple interest and twice as much in an IRA paying 4% simple interest. She also decides to invest $1000 more than what she has invested in the CD into a mutual fund earning 5% simple interest. Determine how much money Tamara invested in each account if she earned $290 in interest after 1 year.

62) The three top-selling albums of 2011 were Adele's *21*, Michael Buble's *Christmas,* and Lady Gaga's *Born This Way*. Adele sold 3.3 million more albums than Michael Buble, while Lady Gaga sold 0.4 million fewer albums than Buble. Find the number of albums sold by each artist if together they sold 10.4 million albums. (articles.latimes.com)

63) Zoe's current salary is $40,144. This is 4% higher than last year's salary. What was Zoe's salary last year?

64) Find the original price of a cell phone if it costs $63.20 after a 20% discount.

R Rethink

R1) Which objective is the most difficult for you?

R2) When solving an application problem, how do you determine which unknown to let the variable *x* represent?

R3) What kinds of professions do you think often deal with application problems involving lengths?

R4) What kinds of professions do you think often use percents?

R5) When was the last time you were given a percent, and what did it describe?

2.3 Geometry Applications and Solving Formulas

P Prepare	**O Organize**
What are your objectives for Section 2.3?	How can you accomplish each objective?
1 Solve Problems Using Formulas from Geometry	• Make a list of common geometry formulas. • Draw a diagram. • Follow the procedure for solving applied problems. • Be sure your answer makes sense. • Complete the given example on your own. • Complete You Try 1.
2 Solve Problems Involving Angle Measures	• Draw a diagram if one is not already given. • Follow the procedure for solving applied problems. • Make sure your answer is logical by verifying that the angle measures sum to the correct total (usually 90° or 180°). • Complete the given example on your own. • Complete You Try 2.
3 Solve Problems Involving Complementary and Supplementary Angles	• Know the definitions of a *complementary* and *supplementary angles* and how to define each ($90 - x$ or $180 - x$). • Follow the procedure for solving applied problems. • Complete the given example on your own. • Complete You Try 3.
4 Solve a Formula for a Specific Variable	• Solve for the given variable the same way you would solve an equation—use the **Properties of Equality.** • Be aware that you will not get a numerical answer, but rather a "rearrangement" of the given formula in terms of another variable. • Complete the given examples on your own. • Complete You Trys 4 and 5.

Read the explanations, follow the examples, take notes, and complete the You Trys.

In this section, we will make use of concepts and formulas from geometry to solve applied problems. We will also learn how to solve a formula for a specific variable. We begin with geometry.

1 Solve Problems Using Formulas from Geometry

EXAMPLE 1 The area of a rectangular room is 180 ft². Its width is 12 ft. What is the length of the room?

Solution

Step 1: Read the problem carefully, and identify what we are being asked to find.

We must find the length of the room.

A picture will be very helpful in this problem.

Step 2: **Choose a variable** to represent the unknown.

l = the length of the room

Label the picture with the width, 12 ft, and the length l.

Step 3: **Translate** the information that appears in English into an algebraic equation. We will use a known geometry formula. How do we know which formula to use? List the information we are given and what we want to find:

The room is in the shape of a rectangle, its area = 180 ft², and its width = 12 ft. We must find the width. Which formula involves the area, length, and width of a rectangle?

$$A = lw$$

Substitute the known values into the formula for the area of a rectangle, and solve for l.

$$A = lw$$
$$180 = l(12) \quad \text{Substitute the known values.}$$

Step 4: **Solve** the equation.

$$180 = 12l$$
$$\frac{180}{12} = \frac{12l}{12} \quad \text{Divide by 12.}$$
$$15 = l \quad \text{Simplify.}$$

Step 5: **Check** the answer, and **interpret** the solution as it relates to the problem.

If l = 15 ft, then $l \cdot w$ = 15 ft · 12 ft = 180 ft². Therefore, the length of the room is 15 ft.

> **Note**
> Remember to include the correct units in your answer!

[YOU TRY 1]

The area of a rectangular ping-pong table is 45 ft². Its length is 9 ft. What is the width of the ping-pong table?

2 Solve Problems Involving Angle Measures

EXAMPLE 2

Find the missing angle measures.

Solution

Step 1: **Read** the problem carefully, and identify what we are being asked to find.

Find the missing angle measures.

Step 2: The unknowns are already defined. We must find x, the measure of one angle, and then $3x + 5$, the measure of the other angle.

Hint

While working through this section, make a list of all the different geometric formulas that are used. This will make your homework easier.

> **Hint**
> Write down the equation in English before writing it using math.

Step 3: **Translate** the information into an algebraic equation. Since the sum of the angles in a triangle is 180°, we can write

English: | Measure of one angle | plus | Measure of second angle | plus | Measure of third angle | is | 180°
↓ ↓ ↓ ↓ ↓ ↓ ↓
Equation: $\quad x \quad + \quad 83 \quad + \quad (3x + 5) \quad = \quad 180$

The equation is $x + 83 + (3x + 5) = 180$.

Step 4: **Solve** the equation.

$$x + 83 + (3x + 5) = 180$$
$$4x + 88 = 180 \qquad \text{Combine like terms.}$$
$$4x + 88 - 88 = 180 - 88 \qquad \text{Subtract 88 from each side.}$$
$$4x = 92 \qquad \text{Combine like terms.}$$
$$x = 23 \qquad \text{Divide each side by 4.}$$

Step 5: **Check** the answer, and **interpret** the solution as it relates to the problem.

One angle, x, has a measure of 23°. The other unknown angle measure is $3x + 5 = 3(23) + 5 = 74°$.

The answer makes sense because the sum of the angle measures is $23° + 83° + 74° = 180°$.

[YOU TRY 2] Find the missing angle measures.

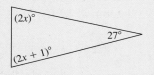

3 Solve Problems Involving Complementary and Supplementary Angles

Recall that two angles are **complementary** if the sum of their angles is 90°, and two angles are **supplementary** if the sum of their angles is 180°.

For example, if the measure of $\angle A$ is 52°, then

> **Hint**
> Notice that the measurement of a complement is $90 - x$, not $x - 90$. The measurement of a supplement is $180 - x$, not $x - 180$.

a) the measure of its complement is $90° - 52° = 38°$.

b) the measure of its supplement is $180° - 52° = 128°$.

Now, let's say an angle has a measure, $x°$. Using the same reasoning as above

a) the measure of its complement is $90 - x$.

b) the measure of its supplement is $180 - x$.

We will use these ideas to solve Example 3.

EXAMPLE 3

Twice the complement of an angle is 18° less than the supplement of the angle. Find the measure of the angle.

Solution

Step 1: **Read** the problem carefully, and identify what we are being asked to find.

We must find the measure of the angle.

Step 2: **Choose a variable** to represent an unknown, and define the other unknowns in terms of this variable.

This problem has three unknowns: the measures of the angle, its complement, and its supplement. Choose a variable to represent the original angle, then define the other unknowns in terms of this variable.

$$x = \text{the measure of the angle}$$

Define the other unknowns in terms of x.

$$90 - x = \text{the measure of the complement}$$
$$180 - x = \text{the measure of the supplement}$$

Step 3: **Translate** the information that appears in English into an algebraic equation.

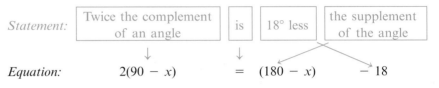

Equation: $2(90 - x) = (180 - x) - 18$

The equation is $2(90 - x) = (180 - x) - 18$.

Step 4: **Solve** the equation.

$2(90 - x) = (180 - x) - 18$	
$180 - 2x = 180 - x - 18$	Distribute.
$180 - 2x = 162 - x$	Combine like terms.
$180 - 162 - 2x = 162 - 162 - x$	Subtract 162 from each side.
$18 - 2x = -x$	Combine like terms.
$18 - 2x + 2x = -x + 2x$	Add $2x$ to each side.
$18 = x$	Simplify.

Step 5: **Check** the answer and **interpret** the solution as it relates to the problem.

The measure of the angle is 18°.

To check the answer, we first need to find its complement and supplement. The complement is 90° − 18° = 72°, and its supplement is 180° − 18° = 162°. Now we can check these values in the original statement: twice the complement is 2(72°) = 144°. Eighteen degrees less than the supplement is 162° − 18° = 144°.

[YOU TRY 3] Ten times the complement of an angle is 9° less than the supplement of the angle. Find the measure of the angle.

4 Solve a Formula for a Specific Variable

The formula $P = 2l + 2w$ allows us to find the perimeter of a rectangle when we know its length, l, and width, w. But, what if we were solving problems where we repeatedly needed to find the value of w? Then, we could rewrite $P = 2l + 2w$ so that it is solved for w:

$$w = \frac{P - 2l}{2}$$

Doing this means that we have *solved the formula $P = 2l + 2w$ for the specific variable, w.*

Solving a formula for a specific variable may seem confusing at first because the formula contains more than one letter. Keep in mind that we will solve for a specific variable the same way we have been solving equations up to this point.

We'll start by solving $2x + 5 = 13$ step by step for x and then applying the same procedure to solving $mx + b = y$ for x.

EXAMPLE 4

Solve $2x + 5 = 13$ and $mx + b = y$ for x.

Solution

Look at these equations carefully, and notice that they have the same form. Read the following steps in order.

Part 1 Solve $2x + 5 = 13$.

Don't quickly run through the solution of this equation. *The emphasis here is on the steps used to solve the equation and why we use those steps!*

$$2\boxed{x} + 5 = 13$$

We are solving for x. We'll put a box around it. What is the first step? "Get rid of" what is being added to the $2x$, that is "get rid of" the 5 on the left. Subtract 5 from each side.

$$2\boxed{x} + 5 - 5 = 13 - 5$$

Combine like terms.

$$2\boxed{x} = 8$$

Part 3 We left off needing to solve $2\boxed{x} = 8$ for x. We need to eliminate the "2" on the left. Since x is being multiplied by 2, we will *divide* each side by 2.

$$\frac{2\boxed{x}}{2} = \frac{8}{2}$$

Simplify.

$$x = 4$$

Part 2 Solve $mx + b = y$ for x.

Since we are solving for x, we'll put a box around it.

$$m\boxed{x} + b = y$$

The goal is to get the "x" on a side by itself. What do we do first? As in Part 1, "get rid of" what is being added to the "mx" term, that is "get rid of" the "b" on the left. Since b is being added to "mx," we will subtract it from each side. (We are performing the same steps as in Part 1!)

$$m\boxed{x} + b - b = y - b$$

Combine like terms.

$$m\boxed{x} = y - b$$

(We cannot combine the terms on the right, so the right remains $y - b$.)

Part 4 Now, we have to solve $m\boxed{x} = y - b$ for x. We need to eliminate the "m" on the left. Since x is being multiplied by m, we will *divide* each side by m.

$$\frac{m\boxed{x}}{m} = \frac{y - b}{m}$$

These are the same steps used in Part 3! Simplify.

$$\frac{\cancel{m}\boxed{x}}{\cancel{m}} = \frac{y - b}{m}$$

$$x = \frac{y - b}{m} \quad \text{or} \quad x = \frac{y}{m} - \frac{b}{m}$$

> **Note**
> To obtain the result $x = \dfrac{y}{m} - \dfrac{b}{m}$, we distributed the m in the denominator to each term in the numerator. Either form of the answer is correct.

When you are solving a formula for a specific variable, think about the steps you use to solve an equation in one variable.

[**YOU TRY 4**] Solve $an - pr = w$ for n.

EXAMPLE 5

$R = \dfrac{\rho L}{A}$ is a formula used in physics. Solve this equation for L.

Solution

$R = \dfrac{\rho \boxed{L}}{A}$ Solve for L. Put it in a box.

$AR = A \cdot \dfrac{\rho \boxed{L}}{A}$ Begin by eliminating A on the right.
Since ρL is being divided by A, we will *multiply* by A on each side.

$AR = \rho \boxed{L}$ Next, eliminate ρ.

$\dfrac{AR}{\rho} = \dfrac{\rho \boxed{L}}{\rho}$ Divide each side by ρ.

$\dfrac{AR}{\rho} = L$

EXAMPLE 6

$A = \dfrac{1}{2}h(b_1 + b_2)$ is the formula for the area of a trapezoid. Solve it for b_1.

Solution

$A = \dfrac{1}{2}h(\boxed{b_1} + b_2)$ There are two valid ways to solve this for b_1. We'll look at both of them.

Method 1
We will put b_1 in a box to remind us that this is what we must solve for. In Method 1, we will start by eliminating the fraction. Multiply both sides by 2.

$2A = 2 \cdot \dfrac{1}{2}h(\boxed{b_1} + b_2)$

$2A = h(\boxed{b_1} + b_2)$

> **Hint**
> Don't forget to put a box around the variable you are solving for.

We are solving for b_1, which is in the parentheses. The quantity in parentheses is being multiplied by h, so we can *divide both sides by h* to eliminate it on the right.

$\dfrac{2A}{h} = \dfrac{h(\boxed{b_1} + b_2)}{h}$ Divide each side by h.

$\dfrac{2A}{h} = \boxed{b_1} + b_2$ Simplify.

$\dfrac{2A}{h} - b_2 = \boxed{b_1} + b_2 - b_2$ Subtract b_2 from each side.

$\dfrac{2A}{h} - b_2 = b_1$ Simplify.

Method 2

To solve $A = \frac{1}{2}h(b_1 + b_2)$ for b_1, we can begin by distributing $\frac{1}{2}h$ on the right. We will put b_1 in a box to remind us that this is what we must solve for.

$$A = \frac{1}{2}h(\boxed{b_1} + b_2)$$

$$A = \frac{1}{2}h\boxed{b_1} + \frac{1}{2}hb_2 \qquad \text{Distribute.}$$

$$2A = 2\left(\frac{1}{2}h\boxed{b_1} + \frac{1}{2}hb_2\right) \qquad \text{Multiply by 2 to eliminate the fractions.}$$

$$2A = h\boxed{b_1} + hb_2 \qquad \text{Distribute.}$$

$$2A - hb_2 = h\boxed{b_1} + hb_2 - hb_2 \qquad \text{Since } hb_2 \text{ is being added to } hb_1, \text{ subtract } hb_2 \text{ from each side.}$$

$$2A - hb_2 = h\boxed{b_1} \qquad \text{Simplify.}$$

$$\frac{2A - hb_2}{h} = \frac{h\boxed{b_1}}{h} \qquad \text{Divide by } h.$$

$$\frac{2A - hb_2}{h} = b_1 \qquad \text{Simplify.}$$

Or, b_1 can be rewritten as $b_1 = \frac{2A}{h} - \frac{hb_2}{h} = \frac{2A}{h} - b_2$.

So, $b_1 = \dfrac{2A - hb_2}{h}$ or $b_1 = \dfrac{2A}{h} - b_2$. The forms are equivalent.

[YOU TRY 5] Solve for the indicated variable.

a) $t = \dfrac{qr}{s}$ for q 	b) $R = t(k - c)$ for c

ANSWERS TO [YOU TRY] EXERCISES

1) 5 ft   2) 76°, 77°   3) 81°   4) $n = \dfrac{w + pr}{a}$ 5) a) $q = \dfrac{st}{r}$ b) $c = \dfrac{kt - R}{t}$ or $c = k - \dfrac{R}{t}$

E Evaluate 2.3 Exercises   Do the exercises, and check your work.

Objective 1: Solve Problems Using Formulas from Geometry

Use a formula from geometry to solve each problem. See Example 1.

 1) The Torrence family has a rectangular, in-ground pool in their yard. It holds 1700 ft³ of water. If it is 17 ft wide and 4 ft deep, what is the length of the pool?

2) A rectangular fish tank holds 2304 in³ of water. Find the height of the tank if it is 24 in. long and 8 in. wide.

3) A computer printer is stocked with paper measuring 8.5 in. × 11 in. But, the printing covers only 48 in² of the page. If the length of the printed area is 8 in., what is the width?

4) The rectangular glass surface to produce copies on a copy machine has an area of 135 in². It is 9 in. wide. What is the longest length of paper that will fit on this copier?

5) The face of the clock on Big Ben in London has a radius of 11.5 ft. What is the area of this circular clock face? Use 3.14 for π. Round the answer to the nearest square foot. (www.bigben.freeservers.com)

6) A lawn sprinkler sprays water in a circle of radius 6 ft. Find the area of grass watered by this sprinkler. Use 3.14 for π. Round the answer to the nearest square foot.

7) The "lane" on a basketball court is a rectangle of length 19 ft. The area of the lane is 228 ft^2. What is the width of the lane?

8) A trapezoidal plot of land has the dimensions pictured below. If the area is 9000 ft^2,
   a) find the length of the missing side, x.
   b) how much fencing would be needed to completely enclose the plot?

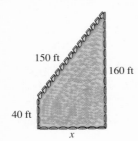

9) A large can of tomato sauce is in the shape of a right circular cylinder. If its radius is 2 in. and its volume is 24π in^3, what is the height of the can?

10) A coffee can in the shape of a right circular cylinder has a volume of 50π in^3. Find the height of the can if its diameter is 5 in.

Objective 2: Solve Problems Involving Angle Measures

Find the missing angle measures. See Example 2.

11)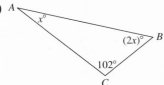

12)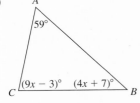

13)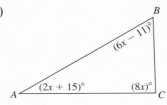

14)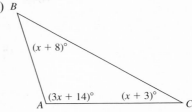

Find the measure of each indicated angle.

15)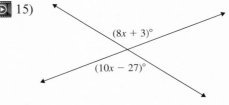

16)

17)

18)

19)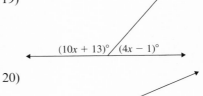

20)

21)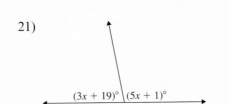

SECTION 2.3 Geometry Applications and Solving Formulas 73

22)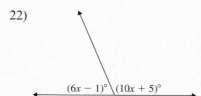

Objective 3: Solve Problems Involving Complementary and Supplementary Angles

23) If x = the measure of an angle, write an expression for its supplement.

24) If x = the measure of an angle, write an expression for its complement.

Solve each problem. See Example 3.

25) Ten times the measure of an angle is 7° more than the measure of its supplement. Find the measure of the angle.

26) The measure of an angle is 12° more than twice its complement. Find the measure of the angle.

27) Four times the complement of an angle is 40° less than twice the angle's supplement. Find the angle, its complement, and its supplement.

28) Twice the supplement of an angle is 24° more than four times its complement. Find the angle, its complement, and its supplement.

29) The sum of an angle and three times its complement is 55° more than its supplement. Find the measure of the angle.

30) The sum of twice an angle and its supplement is 24° less than five times its complement. Find the measure of the angle.

31) The sum of three times an angle and twice its supplement is 400°. Find the angle.

32) The sum of twice an angle and its supplement is 253°. Find the angle.

Objective 4: Solve a Formula for a Specific Variable

Substitute the given values into the formula. Then, solve for the remaining variable.

33) $I = Prt$ (simple interest); if $I = 240$ when $P = 3000$ and $r = 0.04$, find t.

34) $I = Prt$ (simple interest); if $I = 156$ when $P = 650$ and $t = 3$, find r.

35) $V = lwh$ (volume of a rectangular box); if $V = 96$ when $l = 8$ and $h = 3$, find w.

36) $V = \frac{1}{3}Ah$ (volume of a pyramid); if $V = 60$ when $h = 9$, find A.

37) $P = 2l + 2w$ (perimeter of a rectangle); if $P = 50$ when $w = 7$, find l.

38) $P = s_1 + s_2 + s_3$ (perimeter of a triangle); if $P = 37$ when $s_1 = 17$ and $s_3 = 8$, find s_2.

39) $V = \frac{1}{3}\pi r^2 h$ (volume of a cone); if $V = 54\pi$ when $r = 9$, find h.

40) $V = \frac{1}{3}\pi r^2 h$ (volume of a cone); if $V = 32\pi$ when $r = 4$, find h.

41) $S = 2\pi r^2 + 2\pi rh$ (surface area of a right circular cylinder); if $S = 120\pi$ when $r = 5$, find h.

42) $S = 2\pi r^2 + 2\pi rh$ (surface area of a right circular cylinder); if $S = 66\pi$ when $r = 3$, find h.

43) $A = \frac{1}{2}h(b_1 + b_2)$ (area of a trapezoid); if $A = 790$ when $b_1 = 29$ and $b_2 = 50$, find h.

44) $A = \frac{1}{2}h(b_1 + b_2)$ (area of a trapezoid); if $A = 246.5$ when $h = 17$ and $b_2 = 16$, find b_1.

45) Solve for x.
   a) $x + 12 = 35$ b) $x + n = p$
   c) $x + q = v$

46) Solve for t.
   a) $t - 3 = 19$ b) $t - w = m$
   c) $t - v = j$

47) Solve for y.
   a) $4y = 36$ b) $ay = x$
   c) $py = r$

48) Solve for n.
   a) $5n = 30$ b) $yn = c$
   c) $wn = d$

49) Solve for c.
   a) $\frac{c}{3} = 7$ b) $\frac{c}{u} = r$
   c) $\frac{c}{x} = t$

50) Solve for m.
   a) $\dfrac{m}{8} = 2$
   b) $\dfrac{m}{z} = p$
   c) $\dfrac{m}{q} = f$

51) Solve for d.
   a) $8d - 7 = 17$
   b) $kd - a = z$

52) Solve for g.
   a) $3g + 23 = 2$
   b) $cg + k = \pi$

53) Solve for z.
   a) $6z + 19 = 4$
   b) $yz + t = w$

54) Solve for p.
   a) $10p - 3 = 19$
   b) $np - r = d$

Solve each formula for the indicated variable.

55) $F = ma$ for m (physics)
56) $C = 2\pi r$ for r (geometry)
57) $n = \dfrac{c}{v}$ for c (physics)
58) $f = \dfrac{R}{2}$ for R (physics)
59) $E = \sigma T^4$ for σ (meteorology)
60) $p = \rho g y$ for ρ (geology)
61) $V = \dfrac{1}{3}\pi r^2 h$ for h (geometry)
62) $U = \dfrac{1}{2}LI^2$ for L (physics)
63) $R = \dfrac{E}{I}$ for E (electronics)
64) $A = \dfrac{1}{2}bh$ for b (geometry)
65) $I = PRT$ for R (finance)

66) $I = PRT$ for P (finance)
67) $P = 2l + 2w$ for l (geometry)
68) $A = P + PRT$ for T (finance)
69) $H = \dfrac{D^2 N}{2.5}$ for N (auto mechanics)
70) $V = \dfrac{AH}{3}$ for A (geometry)
71) $A = \dfrac{1}{2}h(b_1 + b_2)$ for b_2 (geometry)
72) $A = \pi(R^2 - r^2)$ for r^2 (geometry)

73) The perimeter, P, of a rectangle is $P = 2l + 2w$, where l = length and w = width.
   a) Solve $P = 2l + 2w$ for w.
   b) Find the width of the rectangle with perimeter 28 cm and length 11 cm.

74) The area, A, of a triangle is $A = \dfrac{1}{2}bh$, where b = length of the base and h = height.
   a) Solve $A = \dfrac{1}{2}bh$ for h.
   b) Find the height of the triangle that has an area of 30 in^2 and a base of length 12 in.

75) $C = \dfrac{5}{9}(F - 32)$ is a formula that can be used to convert from degrees Fahrenheit, F, to degrees Celsius, C.
   a) Solve this formula for F.
   b) The average high temperature in Mexico City, Mexico, in April is 25°C. Use the result in part a) to find the equivalent temperature in degrees Fahrenheit. (www.bbc.co.uk)

76) The average low temperature in Stockholm, Sweden, in January is −5°C. Use the result in Exercise 75a) to find the equivalent temperature in degrees Fahrenheit. (www.bbc.co.uk)

Rethink

R1) Which objective is the most difficult for you?
R2) What types of professions do you think often work with angle measurements?
R3) What kinds of professions do you think often deal with application problems involving lengths?
R4) When was the last time you had to solve a real-life application problem at home?
R5) What kinds of sports require a player to project a ball using some type of strategic angle measurement?

2.4 More Applications of Linear Equations

Prepare

What are your objectives for Section 2.4?	**Organize** How can you accomplish each objective?
1 Solve Problems Involving Money	• Be able to write the amount of money in *dollars* and *cents*. Be consistent with using one form when setting up your equation. • Know that the value of a quantity of coins is very different from the number of coins. • Follow the procedure for solving an applied problem. • Check to make sure your answer makes sense. • Complete the given examples on your own. • Complete You Trys 1 and 2.
2 Solve Mixture Problems	• Follow the procedure for solving an applied problem. • Use decimal form of the percent in the equation. • Remember that the concentration of a solution is very similar to the value of a coin—both are multiplied by a quantity in the equation. • Complete the given example on your own. • Complete You Try 3.
3 Solve Problems Involving Distance, Rate, and Time	• Learn the formula for calculating distance. • Follow the procedure for solving an applied problem. • Make a chart to keep track of your *distance, rate,* and *time.* • Set up the equation based on the information in the chart. • Make sure your answer is logical and answers the question asked. • Complete the given example on your own. • Complete You Try 4.

Work Read the explanations, follow the examples, take notes, and complete the You Trys.

1 Solve Problems Involving Money

Many application problems involve thinking about the number of coins or bills and their values. Let's look at how arithmetic and algebra problems involving these ideas are related.

EXAMPLE 1 Determine the amount of money you have in cents *and* in dollars if you have

a) 8 nickels   b) 7 quarters   c) 8 nickels and 7 quarters

Solution

You may be able to do these problems "in your head," but it is very important that we understand the *procedure* that is used to do this arithmetic problem so that we can apply the same procedure to algebra. So, read this carefully!

Parts a) and b): Let's begin with part a), finding the value of 8 nickels.

Here's how we find the value of 7 quarters:

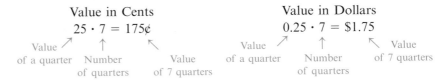

> **W Hint**
> On a sheet of paper, make a list of the different types of tables used in this section. This will make your homework easier!

A table can help us organize the information, so let's put both part a) and part b) in a table so that we can see a pattern.

Value of the Coins (in cents)

	Value of the Coin	Number of Coins	Total Value of the Coins
Nickels	5	8	$5 \cdot 8 = 40$
Quarters	25	7	$25 \cdot 7 = 175$

Value of the Coins (in dollars)

	Value of the Coin	Number of Coins	Total Value of the Coins
Nickels	0.05	8	$0.05 \cdot 8 = 0.40$
Quarters	0.25	7	$0.25 \cdot 7 = 1.75$

Notice that each time we want to find the total value of the coins we find it by multiplying.

$$\text{Value of the coin} \cdot \text{Number of coins} = \text{Total value of the coins}$$

c) Now let's write an equation in English to find the total value of the 8 nickels and 7 quarters.

English:	Value of 8 nickels	plus	Value of 7 quarters	equals	Total value of all the coins
	↓	↓	↓	↓	↓
Cents:	5(8) 40	+ +	25(7) 175	=	215¢
Dollars:	0.05(8) 0.40	+ +	0.25(7) 1.75	=	$2.15

We will use the same procedure that we just used to solve these arithmetic problems to write algebraic expressions to represent the value of a collection of coins.

EXAMPLE 2

Write expressions for the amount of money you have in cents *and* in dollars if you have

a) n nickels   b) q quarters   c) n nickels and q quarters

Solution

Parts a) and b) Let's use tables just like we did in Example 5. We will put parts a) and b) in the same table.

Value of the Coins (in cents)

	Value of the Coin	Number of Coins	Total Value of the Coins
Nickels	5	n	$5 \cdot n = 5n$
Quarters	25	q	$25 \cdot q = 25q$

Value of the Coins (in dollars)

	Value of the Coin	Number of Coins	Total Value of the Coins
Nickels	0.05	n	$0.05 \cdot n = 0.05n$
Quarters	0.25	q	$0.25 \cdot q = 0.25q$

If you have n nickels, then the expression for the amount of money in cents is $5n$. The amount of money in dollars is $0.05n$. If you have q quarters, then the expression for the amount of money in cents is $25q$. The amount of money in dollars is $0.25q$.

c) Write an equation in English to find the total value of n nickels and q quarters. It is based on the same idea that we used in Example 5.

English:	Value of n nickels	plus	Value of q quarters	equals	Total value of all the coins
	↓	↓	↓	↓	↓
Equation in cents:	$5n$	$+$	$2.5q$	$=$	$5n + 25q$
Equation in dollars:	$0.05n$	$+$	$0.25q$	$=$	$0.05n + 0.25q$

The expression in cents is $5n + 25q$. The expression in dollars is $0.05n + 0.25q$.

[YOU TRY 1]

Determine the amount of money you have in cents *and* in dollars if you have

a) 11 dimes b) 20 pennies c) 8 dimes and 46 pennies
d) d dimes e) p pennies f) d dimes and p pennies

Next, we'll apply this idea of the value of different denominations of money to an application problem.

EXAMPLE 3

Jamaal has only dimes and quarters in his piggy bank. When he counts the change, he finds that he has $18.60 and that there are twice as many quarters as dimes. How many dimes and quarters are in his bank?

Solution

Step 1: **Read** the problem carefully, and identify what we are being asked to find.

We must find the number of dimes and quarters in the bank.

Step 2: **Choose a variable** to represent an unknown, and define the other unknown in terms of this variable.

In the statement "there are twice as many quarters as dimes," the number of quarters is expressed *in terms of* the number of dimes. Therefore, let

$d =$ the number of dimes

Define the other unknown (the number of quarters) in terms of d:

$2d =$ the number of quarters

Step 3: **Translate** the information that appears in English into an algebraic equation.

Let's begin by making a table to write an expression for the value of the dimes and the value of the quarters. We will write the expression in terms of dollars because the total value of the coins, $18.60, is given in dollars.

	Value of the Coin	Number of Coins	Total Value of the Coins
Dimes	0.10	d	$0.10d$
Quarters	0.25	$2d$	$0.25 \cdot (2d)$

Hint

Remember, writing an equation in English will help you to write it using algebra.

Write an equation in English and substitute the expressions we found in the table and the total value of the coins to get an algebraic equation.

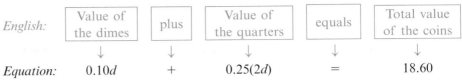

Step 4: Solve the equation.

$$0.10d + 0.25(2d) = 18.60$$
$$100[0.10d + 0.25(2d)] = 100(18.60) \quad \text{Multiply by 100 to eliminate the decimals.}$$
$$10d + 25(2d) = 1860 \quad \text{Distribute.}$$
$$10d + 50d = 1860 \quad \text{Multiply.}$$
$$60d = 1860 \quad \text{Combine like terms.}$$
$$\frac{60d}{60} = \frac{1860}{60} \quad \text{Divide each side by 60.}$$
$$d = 31 \quad \text{Simplify.}$$

Step 5: **Check** the answer, and **interpret** the meaning of the solution as it relates to the problem.

There were 31 dimes and $2(31) = 62$ quarters in the bank.

Check: The value of the dimes is $\$0.10(31) = \$3.10$, and the value of the quarters is $\$0.25(62) = \$15.50$. Their total is $\$3.10 + \$15.50 = \$18.60$.

[YOU TRY 2] A collection of coins consists of pennies and nickels. There are five fewer nickels than there are pennies. If the coins are worth a total of $4.97, how many of each type of coin is in the collection?

2 Solve Mixture Problems

Mixture problems involve combining two or more substances to make a mixture of them. We will begin with an example from arithmetic then extend this concept to be used with algebra.

EXAMPLE 4 The state of Illinois mixes ethanol (made from corn) in its gasoline to reduce pollution. If a customer purchases 15 gal of gasoline and it has a 10% ethanol content, how many gallons of ethanol are in the 15 gal of gasoline?

Solution

Write an equation in English first:

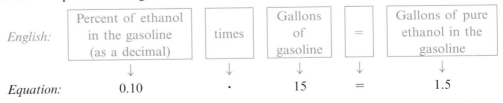

We *multiply* the percent of ethanol by the number of gallons of gasoline to get the number of gallons of ethanol in the gasoline.

The equation is $0.10(15) = 1.5$.

We can also organize the information in a table:

Percent of Ethanol in the Gasoline (as a decimal)	Gallons of Gasoline	Gallons of Ethanol in the Gasoline
0.10	15	0.10(15) = 1.5

Either way we find that there are 1.5 gal of ethanol in 15 gal of gasoline.

The idea above will be used to help us solve the next mixture problem.

EXAMPLE 5

A chemist needs to make 30 liters (L) of a 6% acid solution. She will make it from some 4% acid solution and some 10% acid solution that is in the storeroom. How much of the 4% solution and 10% solution should she use?

Solution

Step 1: **Read** the problem carefully, and identify what we are being asked to find.

We must find the amount of 4% acid solution and 10% acid solution she should use.

Step 2: **Choose a variable** to represent an unknown, and define the other unknown in terms of this variable. Let

$$x = \text{the number of liters of 4\% acid solution needed}$$

Define the other unknown (the amount of 10% acid solution needed) in terms of x. Since she wants to make a total of 30 L of acid solution,

$$30 - x = \text{the number of liters of 10\% acid solution needed}$$

Step 3: **Translate** the information that appears in English into an algebraic equation.

> **W Hint**
> Note that the percentage of the resulting solution will always lie between the percentages of two solutions that are being mixed.

Let's begin by arranging the information in a table. *Remember, to obtain the expression in the last column, multiply the percent of acid in the solution by the number of liters of solution to get the number of liters of acid in the solution.*

	Percent of Acid in Solution (as a decimal)	Liters of Solution	Liters of Acid in Solution
Mix these	0.04	x	$0.04x$
	0.10	$30 - x$	$0.10(30 - x)$
to make →	0.06	30	$0.06(30)$

Now, write an equation in English. Since we obtain the 6% solution by mixing the 4% and 10% solutions,

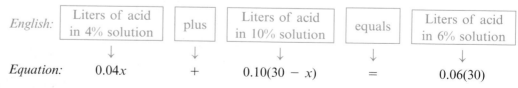

The equation is $0.04x + 0.10(30 - x) = 0.06(30)$.

Step 4: **Solve** the equation.

$$0.04x + 0.10(30 - x) = 0.06(30)$$
$$100[0.04x + 0.10(30 - x)] = 100[0.06(30)] \quad \text{Multiply by 100 to eliminate decimals.}$$
$$4x + 10(30 - x) = 6(30)$$
$$4x + 300 - 10x = 180 \quad \text{Distribute.}$$
$$-6x + 300 = 180 \quad \text{Combine like terms.}$$
$$-6x = -120 \quad \text{Subtract 300 from each side.}$$
$$x = 20 \quad \text{Divide by } -6.$$

Step 5: **Check** the answer, and **interpret** the meaning of the solution as it relates to the problem.

The chemist needs 20 L of the 4% solution.

Find the other unknown, the amount of 10% solution needed.

$$30 - x = 30 - 20 = 10 \text{ L of 10\% solution}$$

Check:

Acid in 4% solution + Acid in 10% solution = Acid in 6% solution
$$0.04(20) \quad + \quad 0.10(10) \quad = \quad 0.06(30)$$
$$0.80 \quad + \quad 1.00 \quad = \quad 1.80$$
$$1.80 = 1.80 \ \checkmark$$

[**YOU TRY 3**] How many milliliters (mL) of a 7% alcohol solution and how many milliliters of a 15% alcohol solution must be mixed to make 20 mL of a 9% alcohol solution?

3 Solve Problems Involving Distance, Rate, and Time

If you drive at 50 mph for 4 hr, how far will you drive? One way to get the answer is to use the formula

$$\text{Distance} = \text{Rate} \cdot \text{Time}$$
or
$$d = rt$$

$$d = (50 \text{ mph}) \cdot (4 \text{ hr})$$
Distance traveled = 200 mi

Notice that the rate is in miles per *hour* and the time is in *hours*. The units must be consistent in this way. If the time in this problem had been expressed in minutes, it would have been necessary to convert minutes to hours. The formula $d = rt$ will be used in Example 6.

EXAMPLE 6 Alex and Jenny are taking a cross-country road trip on their motorcycles. Jenny leaves a rest area first traveling at 60 mph. Alex leaves 30 min later, traveling on the same highway, at 70 mph. How long will it take Alex to catch Jenny?

Solution

Step 1: **Read** the problem carefully, and identify what we are being asked to find.

We must determine how long it takes Alex to catch Jenny.

> **Hint**
> Remember that it is always useful to sketch a picture to help you visualize these problems!

We will draw a picture to help us see what is happening in this problem.

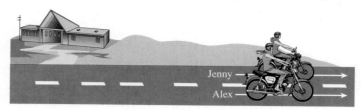

Since both girls leave the same rest area and travel on the same highway, when Alex catches Jenny they have driven the *same* distance.

Step 2: **Choose a variable** to represent an unknown, and define the other unknown in terms of this variable.

Alex's time is in terms of Jenny's time, so let

t = the number of hours Jenny has been riding when Alex catches her

Alex leaves 30 minutes ($\frac{1}{2}$ hour) after Jenny, so Alex travels $\frac{1}{2}$ hour *less than* Jenny.

$t - \frac{1}{2}$ = the number of hours it takes Alex to catch Jenny

Step 3: **Translate** the information that appears in English into an algebraic equation.

Let's make a table using the equation $d = rt$. Fill in the time and the rates first, then multiply those together to fill in the value for the distance.

	d	r	t
Jenny	$60t$	60	t
Alex	$70(t - \frac{1}{2})$	70	$t - \frac{1}{2}$

We will write an equation in English to help us write an algebraic equation. The picture shows that

English: Jenny's distance | is the same as | Alex's distance
↓ ↓ ↓
Equation: $60t$ = $70\left(t - \dfrac{1}{2}\right)$

The equation is $60t = 70\left(t - \dfrac{1}{2}\right)$.

Step 4: **Solve** the equation.

$$60t = 70\left(t - \frac{1}{2}\right)$$
$$60t = 70t - 35 \quad \text{Distribute.}$$
$$-10t = -35 \quad \text{Subtract } 70t.$$
$$\frac{-10t}{-10} = \frac{-35}{-10} \quad \text{Divide each side by } -10.$$
$$t = 3.5 \quad \text{Simplify.}$$

Step 5: **Check** the answer, and **interpret** the meaning of the solution as it relates to the problem.

Remember, Jenny's time is t. Alex's time is $t - \frac{1}{2} = 3\frac{1}{2} - \frac{1}{2} = 3$ hr.

It will take Alex 3 hr to catch Jenny.

Check to see that Jenny travels 60 mph · (3.5 hr) = 210 miles, and Alex travels 70 mph · (3 hr) = 210 miles. The girls travel the same distance.

[YOU TRY 4] The Kansas towns of Topeka and Voda are 230 mi apart. Alberto left Topeka driving west on Interstate 70 at the same time Ramon left Voda driving east toward Topeka on I-70. Alberto and Ramon meet after 2 hr. If Alberto's speed was 5 mph faster than Ramon's speed, how fast was each of them driving?

ANSWERS TO [YOU TRY] EXERCISES

1) a) 110¢, $1.10 b) 20¢, $0.20 c) 126¢, $1.26 d) $10d$ cents, $0.10d$ dollars e) $1p$ cents, $0.01p$ dollars
f) $10d + 1p$ cents, $0.10d + 0.01p$ dollars   2) 87 pennies, 82 nickels   3) 15 ml of 7%, 5 ml of 15%
4) Ramon: 55 mph, Alberto: 60 mph

E Evaluate 2.4 Exercises

Do the exercises, and check your work.

Objective 1: Solve Problems Involving Money

For Exercises 1–6, determine the amount of money a) in dollars and b) in cents given the following quantities.

1) 8 dimes
2) 32 nickels
3) 217 pennies
4) 12 quarters
5) 9 quarters and 7 dimes
6) 89 pennies and 14 nickels

For Exercises 7–12, write an expression which represents the amount of money in a) dollars and b) cents given the following quantities.

7) q quarters
8) p pennies
9) d dimes
10) n nickels
11) p pennies and q quarters
12) n nickels and d dimes

Solve using the Five Steps for Solving Applied Problems. See Example 3.

13) Gino and Vince combine their coins to find they have all nickels and quarters. They have 8 more quarters than nickels, and the coins are worth a total of $4.70. How many nickels and quarters do they have?

14) Danika saves all of her pennies and nickels in a jar. One day she counted them and found that there were 131 coins worth $3.43. How many pennies and how many nickels were in the jar?

15) Kyung Soo has been saving her babysitting money. She has $69.00 consisting of $5 bills and $1 bills. If she has a total of 25 bills, how many $5 bills and how many $1 bills does she have?

16) A bank employee is servicing the ATM after a busy Friday night. She finds the machine contains only $20 bills and $10 bills and that there are twice as many $20 bills remaining as there are $10 bills. If there is a total of $550.00 left in the machine, how many of the bills are twenties, and how many are tens?

17) A movie theater charges $9.00 for adults and $7.00 for children. The total revenue for a particular movie is $475.00. Determine the number of each type of ticket sold if the number of children's tickets sold was half the number of adult tickets sold.

18) At the post office, Ronald buys 12 more 45¢ stamps than 32¢ stamps. If he spends $11.56 on the stamps, how many of each type did he buy?

Objective 2: Solve Mixture Problems
Solve. See Example 4.

19) How many ounces of alcohol are in 40 oz of a 5% alcohol solution?

20) How many milliliters of acid are in 30 mL of a 6% acid solution?

21) Sixty milliliters of a 10% acid solution are mixed with 40 mL of a 4% acid solution. How much acid is in the mixture?

22) Fifty ounces of an 8% alcohol solution are mixed with 20 ounces of a 6% alcohol solution. How much alcohol is in the mixture?

Solve using the Five Steps for Solving Applied Problems. See Example 5.

23) How many ounces of a 4% acid solution and how many ounces of a 10% acid solution must be mixed to make 24 oz of a 6% acid solution?

24) How many milliliters of a 15% alcohol solution must be added to 80 mL of an 8% alcohol solution to make a 12% alcohol solution?

25) How many liters of a 40% antifreeze solution must be mixed with 5 L of a 70% antifreeze solution to make a mixture that is 60% antifreeze?

26) How many milliliters of a 7% hydrogen peroxide solution and how many milliliters of a 1% hydrogen peroxide solution should be mixed to get 400 mL of a 3% hydrogen peroxide solution?

27) Custom Coffees blends its coffees for customers. How much of the Aztec coffee, which sells for $6.00 per pound, and how much of the Cinnamon coffee, which sells for $8.00 per pound, should be mixed to make 5 lb of the Winterfest blend to be sold at $7.20 per pound?

28) All-Mixed-Up Nut Shop sells a mix consisting of cashews and pistachios. How many pounds of cashews, which sell for $7.00 per pound, should be mixed with 4 lb of pistachios, which sell for $4.00 per pound, to get a mix worth $5.00 per pound?

Objective 3: Solve Problems Involving Distance, Rate, and Time
Solve using the Five Steps for Solving Applied Problems. See Example 6.

29) Two cars leave Indianapolis, one driving east and the other driving west. The eastbound car travels 8 mph slower than the westbound car, and after 3 hr they are 414 mi apart. Find the speed of each car.

30) Two planes leave San Francisco, one flying north and the other flying south. The southbound plane travels 50 mph faster than the northbound plane, and after 2 hours they are 900 miles apart. Find the speed of each plane.

31) Maureen and Yvette leave the gym to go to work traveling the same route, but Maureen leaves 10 min after Yvette. If Yvette drives 60 mph and Maureen drives 72 mph, how long will it take Maureen to catch Yvette?

32) Vinay and Sadiva leave the same location traveling the same route, but Sadiva leaves 20 minutes after Vinay. If Vinay drives 30 mph and Sadiva drives 36 mph, how long will it take Sadiva to catch Vinay?

33) A passenger train and a freight train leave cities 400 mi apart and travel toward each other. The passenger train is traveling 20 mph faster than the freight train. Find the speed of each train if they pass each other after 5 hr.

34) A freight train passes the Old Towne train station at 11:00 A.M. going 30 mph. Ten minutes later a passenger train, headed in the same direction on an adjacent track, passes the same station at 45 mph. At what time will the passenger train catch the freight train?

35) A truck and a car leave the same intersection traveling in the same direction. The truck is traveling at 35 mph, and the car is traveling at 45 mph. In how many minutes will they be 6 mi apart?

36) At noon, a truck and a car leave the same intersection traveling in the same direction. The truck is traveling at 30 mph, and the car is traveling at 42 mph. At what time will they be 9 mi apart?

37) Ajay is traveling north on a road while Rohan is traveling south on the same road. They pass by each other at 3 P.M., Ajay driving 30 mph and Rohan driving 40 mph. At what time will they be 105 miles apart?

38) When Lance and Jan pass each other on their bikes going in opposite directions, Lance is riding at 22 mph, and Jan is pedaling at 18 mph. If they continue at those speeds, after how long will they be 100 mi apart?

39) At noon, a cargo van crosses an intersection at 30 mph. At 12:30 P.M., a car crosses the same intersection traveling in the opposite direction. At 1 P.M., the van and car are 54 miles apart. How fast is the car traveling?

40) A freight train passes the Naperville train station at 9:00 A.M. going 30 mph. Ten minutes later a passenger train, headed in the same direction on an adjacent track, passes the same station at 45 mph. At what time will the passenger train catch the freight train?

Mixed Exercises: Objectives 1–3
Solve.

41) At the end of her shift, a cashier has a total of $6.30 in dimes and quarters. There are 7 more dimes than quarters. How many of each of these coins does she have?

42) An alloy that is 30% silver is mixed with 200 g of a 10% silver alloy. How much of the 30% alloy must be used to obtain an alloy that is 24% silver?

43) A jet flying at an altitude of 35,000 ft passes over a small plane flying at 10,000 ft headed in the same direction. The jet is flying twice as fast as the small plane, and 30 minutes later they are 100 mi apart. Find the speed of each plane.

44) Tickets for a high school play cost $3.00 each for children and $5.00 each for adults. The revenue from one performance was $663, and 145 tickets were sold. How many adult tickets and how many children's tickets were sold?

45) A pharmacist needs to make 20 cubic centimeters (cc) of a 0.05% steroid solution to treat allergic rhinitis. How much of a 0.08% solution and a 0.03% solution should she use?

46) Geri is riding her bike at 10 mph when Erin passes her going in the opposite direction at 14 mph. How long will it take before the distance between them is 6 mi?

R Rethink

R1) Which objective is the most difficult for you?

R2) In which of your future courses, other than a math course, do you think you will need to solve applications problems similar to those in this section?

R3) Why is it useful to sometimes organize the given application data in a table?

Group Activity — The Group Activity can be found online on Connect.

emPOWERme Organize Your Memory

It's not enough to simply read assigned material in your math and other textbooks. To use the information inside and outside of class effectively, you need to *remember* it. Experts have found that the best way to remember new material is to connect it to information you have already learned or to mentally group similar kinds of material together. The key, in other words, is keeping your memory organized.

To see how this works, try this exercise, devised by critical thinking expert Diane Halpern. Read the following 15 words at a rate of approximately one per second:

girl
heart
robin
purple
finger
flute
blue
organ
man
hawk
green
lung
eagle
child
piano

Now, cover the list, and write down as many of the words as you can on a separate sheet of paper. How many words are on your list? _____

After you've done this, read the following list:

green
blue
purple
man
girl
child
piano
flute
organ
heart
lung
finger
eagle
hawk
robin

Now cover this second list and write down as many of the words as you can on the other side of the separate sheet of paper.

How many words did you remember this time? _____ Did you notice that the words on both lists are identical? Did you remember more the second time? (Most people do.) Why do you think most people remember more when the words are organized as they are in the second list? Discuss with your classmates ways in which you can organize material from this book to make it easier to remember.

Chapter 2: Summary

Definition/Procedure	Example
2.1 Linear Equations in One Variable	
The Properties of Equality 1) If $a = b$, then $a + c = b + c$ 2) If $a = b$, then $a - c = b - c$ 3) If $a = b$, then $ac = bc$ 4) If $a = b$, then $\dfrac{a}{c} = \dfrac{b}{c}$ ($c \neq 0$). (p. 41)	Solve $a + 4 = 19$. $a + 4 - 4 = 19 - 4$ Subtract 4 from each side. $a = 15$ The solution set is $\{15\}$. Solve. $\dfrac{3}{2}t = -30$ $\dfrac{2}{3} \cdot \dfrac{3}{2}t = \dfrac{2}{3} \cdot (-30)$ Multiply each side by $\dfrac{2}{3}$. $t = -20$ The solution set is $\{-20\}$.
How to Solve a Linear Equation *Step 1:* **Clear parentheses** and **combine like terms** on each side of the equation. *Step 2:* **Get the variable on one side of the equal sign and the constant on the other side of the equal sign** (isolate the variable) using the addition or subtraction property of equality. *Step 3:* **Solve for the variable** using the multiplication or division property of equality. *Step 4:* **Check the solution** in the original equation. (p. 43)	Solve $4(c + 1) + 7 = 2c + 9$. $4c + 4 + 7 = 2c + 9$ Distribute. $4c + 11 = 2c + 9$ Combine like terms. $4c - 2c + 11 = 2c - 2c + 9$ Get variable terms on one side. $2c + 11 = 9$ $2c + 11 - 11 = 9 - 11$ Get constants on one side. $2c = -2$ $\dfrac{2c}{2} = \dfrac{-2}{2}$ Divide by 2. $c = -1$ The solution set is $\{-1\}$.

Steps for Solving Applied Problems
 	
 Step 1: **Read** the problem carefully, more than once if necessary, until you understand it. Draw a picture, if applicable. Identify what you are being asked to find.
 Step 2: **Choose a variable** to represent an unknown quantity. If there are any other unknowns, define them in terms of the variable.
 Step 3: **Translate** the problem from English into an equation using the chosen variable.
 Step 4: **Solve** the equation.
 Step 5: **Check** the answer in the original problem, and **interpret** the solution as it relates to the problem. Be sure your answer makes sense in the context of the problem. (p. 47)
	The sum of a number and fifteen is eight. Find the number.
	 Step 1: **Read** the problem carefully.
	 Step 2: **Choose** a variable.
	      x = the number
	 Step 3: "The sum of a number and fifteen is eight" means
	      The number plus fifteen equals eight.
	      $x \quad + \quad 15 \quad = \quad 8$
	 Equation: $x + 15 = 8$
	 Step 4: **Solve** the equation.
	      $x + 15 = 8$
	      $x + 15 - 15 = 8 - 15$
	      $x = -7$
	 The number is -7.
	 Step 5: The **check** is left to the student.

2.2 Applications of Linear Equations	

Apply the **Five Steps for Solving Applied Problems** to solve this application involving consecutive odd integers. (p. 55)	The sum of three consecutive odd integers is 87. Find the integers.
	 Step 1: **Read** the problem carefully.
	 Step 2: **Choose** a variable to represent an unknown, and define the other unknowns.
	      x = the first odd integer
	      $x + 2$ = the second odd integer
	      $x + 4$ = the third odd integer

Definition/Procedure	Example
	Step 3: "The sum of three consecutive odd integers is 87" means First odd + Second odd + Third odd = 87 $x \quad + \quad (x+2) \quad + \quad (x+4) \quad = 87$ Equation: $x + (x+2) + (x+4) = 87$ **Step 4:** Solve $x + (x+2) + (x+4) = 87$. $3x + 6 = 87$ $3x + 6 - 6 = 87 - 6 \quad$ Subtract 6. $3x = 81$ $\dfrac{3x}{3} = \dfrac{81}{3} \quad$ Divide by 3. $x = 27$ **Step 5:** Find the values of all of the unknowns. $x = 27, \quad x + 2 = 29, \quad x + 4 = 31$ The numbers are 27, 29, and 31. The **check** is left to the student.

2.3 Geometry Applications and Solving Formulas

Formulas from geometry can be used to solve applications. (p. 66)	A rectangular bulletin board has an area of 180 in². It is 12 in. wide. Find its length. Use $A = lw$. Formula for the area of a rectangle $A = 180$ in², $w = 12$ in. Find l. $A = lw$ $180 = l(12) \quad$ Substitute values into $A = lw$. $\dfrac{180}{12} = \dfrac{l(12)}{12}$ $15 = l$ The length is 15 in.
To **solve a formula for a specific variable,** think about the steps involved in solving a linear equation in one variable. (p. 70)	Solve $C = kr - w$ for r. $C + w = k\boxed{r} - w + w \quad$ Add w to each side. $C + w = k\boxed{r}$ $\dfrac{C + w}{k} = \dfrac{k\boxed{r}}{k} \quad$ Divide each side by k. $\dfrac{C + w}{k} = r$

2.4 More Applications of Linear Equations

Use a table to organize the information given in a distance problem or a mixture problem. (p. 76)	Scott starts off on a 6-mile run. He runs part of the way at 6 mph, but then he walks the rest of the way at 3 mph. The time he walks is half of the time he runs. Determine the amount of time he walked. **Step 1:** Read the problem carefully. 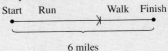 6 miles **Step 2:** Define the unknowns. t = amount of time Scott ran $\dfrac{1}{2}t$ = amount of time Scott walked

Definition/Procedure	Example
	Step 3: Make a table and fill in the values.

	$d =$	r	t
Run	$6t$	6	t
Walk	$\frac{3}{2}t$	3	$\frac{1}{2}t$

We can see from the picture in Step 1 that

Distance Run + Distance Walked = Total Distance

$$6t + \frac{3}{2}t = 6$$

Equation: $6t + \frac{3}{2}t = 6$

Step 4: Solve $6t + \frac{3}{2}t = 6$

$$\frac{12}{2}t + \frac{3}{2}t = 6 \quad \text{Get a common denominator to add like terms.}$$

$$\frac{15}{2}t = 6 \quad \text{Add.}$$

$$\frac{2}{15} \cdot \frac{15}{2}t = \frac{2}{15} \cdot 6$$

$$t = \frac{4}{5} \text{ hr}$$

Step 5: $t = \frac{4}{5}$ hr represents the amount of time Scott ran. We must determine how long he walked. According to the table, the time he spent walking was $\frac{1}{2}t$.

$$\frac{1}{2}t = \frac{1}{2}\left(\frac{4}{5}\right) = \frac{2}{5} \text{ hr}$$

Scott walked for $\frac{2}{5}$ hr or 24 minutes.

Chapter 2: Review Exercises

(2.1) Determine if the given value is a solution to the equation.

1) $2n + 13 = 10; n = -\frac{3}{2}$

2) $5 + t = 3t - 1; t = 4$

3) $\frac{3}{2}k - 5 = 1; k = -4$

4) $5 - 2(3p + 1) = 9p - 2; p = \frac{1}{3}$

5) How do you know that an equation has no solution?

6) What can you do to make it easier to solve an equation with fractions?

Solve each equation.

7) $-9z = 30$

8) $p - 11 = -14$

9) $21 = k + 2$

10) $56 = \frac{8}{5}m$

11) $-\frac{4}{9}w = -\frac{10}{7}$

12) $-c = 4$

13) $21 = 0.6q$

14) $8b - 7 = 57$

15) $6 = 15 + \frac{9}{2}v$

16) $2.3a + 1.5 = 10.7$

17) $\frac{2}{7} - \frac{3}{4}k = -\frac{17}{14}$

90 CHAPTER 2 Linear Equations in One Variable

18) $5(2z + 3) - (11z - 4) = 3$

19) $11x + 13 = 2x - 5$

20) $5(c + 3) - 2c = 4 + 3(2c + 1)$

21) $6 - 5(4d - 3) = 7(3 - 4d) + 8d$

22) $4k + 19 + 8k = 2(6k - 11)$

23) $0.05m + 0.11(6 - m) = 0.08(6)$

24) $1 - \dfrac{1}{6}(t + 5) = \dfrac{1}{2}$

25) $-0.78 = -0.6t$

26) $10 - 7b = 4 - 5(2b + 9) + 3b$

27) $0.18a + 0.1(20 - a) = 0.14(20)$

28) $3(r + 4) - r = 2(r + 6)$

29) $16 = -\dfrac{12}{5}d$

30) $\dfrac{1}{2}(n - 5) - 1 = \dfrac{2}{3}(n - 6)$

Solve using the Five Steps for Solving Applied Problems.

31) Twelve less than a number is five. Find the number.

32) A number increased by nine is one less than twice the number. Find the number.

(2.2)

33) Mr. Morrissey has 26 children in his kindergarten class, and c children attended preschool. Write an expression for the number of children who did not attend preschool.

34) In a parking lot, there were f foreign cars. Write an expression for the number of American cars in the lot if there were 14 more than the number of foreign cars.

Solve using the Five Steps for Solving Applied Problems.

35) In its first week, American Idol finalist Clay Aiken's debut CD sold 316,000 more copies than Idol winner Kelly Clarkson's debut CD. Together their debut CDs sold 910,000 copies. How many CDs did each of them sell during the first week? (www.top40.com)

36) The sum of three consecutive even integers is 258. Find the integers.

37) The road leading to the Sutter family's farmhouse is 500 ft long. Some of it is paved, and the rest is gravel. If the paved portion is three times as long as the gravel part of the road, determine the length of the gravel portion of the road.

38) Before road construction began in front of her ice cream store, Imelda's average monthly revenue was about $8200. In the month since construction began, revenue has decreased by 18%. What was the revenue during the first month of construction?

39) Americans spent about $11.0 billion on veterinary care for their pets in 2008. This is 8.5% more than they spent in 2007. How much did Americans spend on their pets' veterinary care in 2007? (Round to the nearest tenths place.) (http://articles.moneycentral.msn.com)

40) Jerome invested some money in an account earning 7% simple interest and $3000 more than that at 8% simple interest. After 1 yr, he earned $915 in interest. How much money did he invest in each account?

(2.3)

41) *Use a formula from geometry to solve:* The base of a triangle measures 9 cm. If the area of the triangle is 54 cm², find the height.

42) Find the missing angle measures.

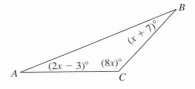

43) Find the missing angle measures.

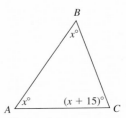

Find the measure of each indicated angle.

44)

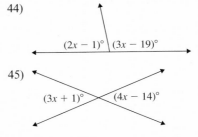

45)

46) *Solve:* Three times the measure of an angle is 12° more than the measure of its supplement. Find the measure of the angle.

Solve for the indicated variable.

47) $p - n = z$ for p
48) $y = mx + b$ for m
49) $pV = nRT$ for R
50) $C = \frac{1}{3}n(t + T)$ for t

(2.4) Solve.

51) A collection of coins contains 91 coins, all dimes and quarters. If the value of the coins is \$14.05, determine the number of each type of coin in the collection.

52) To make a 6% acid solution, a chemist mixes some 4% acid solution with 12 L of 10% acid solution. How much of the 4% solution must be added to the 10% solution to make the 6% acid solution?

53) Peter leaves Mitchell's house traveling 30 mph. Mitchell leaves 15 min later, trying to catch up to Peter, going 40 mph. If they drive along the same route, how long will it take Mitchell to catch Peter?

Chapter 2: Test

Solve.

1) $7p + 16 = 30$
2) $5 - 3(2c + 7) = 4c - 1 - 5c$
3) $\frac{5}{8}(3k + 1) - \frac{1}{4}(7k + 2) = 1$
4) $6(4t - 7) = 3(8t + 5)$
5) $8(3 + n) = 5(2n - 3)$
6) $0.06x + 0.14(x - 5) = 0.10(23)$

Solve using the Five Steps for Solving Applied problems.

7) Nine less than twice a number is 33. Find the number.

8) The sum of three consecutive even integers is 114. Find the numbers.

9) The tray table on the back of an airplane seat is in the shape of a rectangle. It is 5 in. longer than it is wide and has a perimeter of 50 in. Find the dimensions.

10) Motor oil is available in three types: regular, synthetic, and synthetic blend, which is a mixture of regular oil with synthetic. Bob decides to make 5 qt of his own synthetic blend. How many quarts of regular oil costing \$1.20 per quart and how many quarts of synthetic oil costing \$3.40 per quart should he mix so that the blend is worth \$1.86 per quart?

11) In 2012, Wisconsin had 10 drive-in theaters. This is 80% less than the number of drive-ins in 1967. Determine the number of drive-in theaters in Wisconsin in 1967. (www.drive-inthruwisconsin.com)

12) A contractor has 460 ft of fencing to enclose a rectangular construction site. If the length of the site is 160 ft, what is the width?

13) Two cars leave the same location at the same time, one traveling east and the other going west. The westbound car is driving 6 mph faster than the eastbound car. After 2.5 hr they are 345 miles apart. What is the speed of each car?

Solve for the indicated variable.

14) $R = \frac{kt}{5}$ for t
15) $S = 2\pi r^2 + 2\pi rh$ for h

Find the measure of each indicated angle.

16)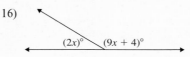

17) Find the measure of each indicated angle.

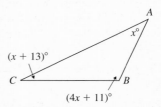

18) How much pure acid must be added to 6 gal of a 4% acid solution to make 20% acid solution?

Chapter 2: Cumulative Review for Chapters 1–2

Perform the operations and simplify.

1) $\dfrac{5}{12} - \dfrac{7}{9}$

2) $\dfrac{8}{15} \div 12$

3) $52 - 12 \div 4 + 3 \cdot 5$

4) $2^3 - 10(-9 + 4) + (-2)$

5) -3^4

6) $(-3)^2$

Given the set of numbers $\left\{-13.7, \dfrac{19}{7}, 0, 8, \sqrt{17}, 0.\overline{61}, \sqrt{81}, -2\right\}$ **identify**

7) the rational numbers

8) the integers

9) the whole numbers

10) the natural numbers

Determine which property is illustrated by each statement.

11) $9 + (4 + 1) = (9 + 4) + 1$

12) $3(n + 7) = 3n + 21$

13) $5 \times \dfrac{1}{5} = 1$

14) Rewrite the expression $8 + 5$ using the commutative property.

Solve each equation.

15) $-31 = \dfrac{4}{7}z + 9$

16) $12 - 4(2a - 5) = 7 + 3(a + 2) - 7a$

17) $\dfrac{1}{2}(t + 1) - \dfrac{1}{3} = \dfrac{17}{12} + \dfrac{1}{4}(2t - 5)$

18) Solve for R: $A = P + PRT$.

19) Find the supplement of $59°$.

20) Find the complement of $33°$.

21) Find the area and perimeter of the rectangle.

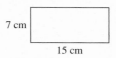

22) Find the missing angle and identify the triangle as acute, obtuse, or right.

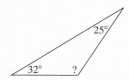

Write an equation and solve.

23) On Thursday, a pharmacist fills twice as many prescriptions with generic drugs as name brand drugs. If he filled 72 prescriptions, how many were with generic drugs, and how many were with name brand drugs?

24) Two friends start biking toward each other from opposite ends of a bike trail. One averages 8 mph, and the other averages 10 mph. If the distance between them when they begin is 9 miles, how long will it take them to meet?

25) Lorenzo purchased a bag of dog food for $28. This is 30% off the original price. What was the original price of the dog food?

CHAPTER 3
Linear Inequalities and Absolute Value

OUTLINE

Study Strategies: Time Management

3.1 Linear Inequalities in One Variable

3.2 Compound Inequalities in One Variable

3.3 Absolute Value Equations and Inequalities

Group Activity

emPOWERme: Identify the Black Holes of Time Management

Math at Work:
Air Traffic Controller

Kevin McCabe is responsible for managing the takeoffs, landings, and flight paths of dozens of airplanes a day. As an air traffic controller, he must ensure that all these planes get where they need to go safely and (hopefully!) on time. Obviously, monitoring the speed and trajectory of numerous airplanes in the sky and on the runway requires some pretty sophisticated math. "People don't often think about just how many planes are in the sky over the United States every single day," Kevin says. "It is complex mathematical calculations performed by air traffic controllers using computers that allow people to crisscross the country in the way we're accustomed to."

One unexpected result of his job, Kevin says, is that he has become much more organized in his daily life. "I never really kept a schedule before I started working in air traffic control," he describes. "But helping keep so many flights on schedule every day, I guess I just naturally wanted some more structure in my own time. Now, when I'm not working, I find that I get a lot more done, and actually feel like I have more free time."

In Chapter 3, we will learn about linear inequalities and absolute value. We will also discuss strategies to manage your time more effectively.

Study Strategies — Time Management

Even those of us who are not air traffic controllers may feel like we have a hundred urgent things to do to all at once. For college students in particular, schedules can become crowded with competing demands of work, study, family, and more. Yet there are strategies you can use to help prevent that dreaded feeling that there are simply not enough hours in the day. Use the framework below to improve your time management abilities.

P Prepare

- Start by creating a *time log*: a record of how you actually spend your time, including interruptions, noting blocks of time as short as 15 minutes.
- Review your time log, and consider whether you are spending time on what matters to you the most.
- To focus your schedule on the right areas, create a ranked list of your priorities.

O Organize

- Create a *master calendar* that shows every week of the term on a single page. Write in all your major class assignments and any significant events in your personal life.
- Create a *weekly timetable* with the days of the week across the top and the hours of the day along the side. Write in all your regularly scheduled activities, such as your classes, as well as one-time events.
- Finally, make a habit of creating a *daily to-do list* that you always have with you, checking off items you've accomplished and adding new ones that come up.

W Work

- Using the schedules you've created to help you, manage your time and stay focused on your priorities.
- Avoid procrastination and "time black holes"—activities that can unexpectedly take hours out of your day. (For help identifying your time black holes, see the emPOWERme on page 129.)

E Evaluate

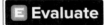

- Use your daily to-do list to assess how efficient you are in meeting your short-term obligations during the day.
- Use your weekly timetable and master calendar to ensure you are on track to meet your long-term obligations.

R Rethink

- Ask yourself how you feel during the day. Are you stressed? Constantly rushing? If so, try to think of ways to reduce your obligations.
- Reassess your priorities. Are you making time for the things that are most important to you?

Chapter 3 POWER Plan

P Prepare / O Organize

What are your goals for Chapter 3?	How can you accomplish each goal?
1 Be prepared before and during class.	• Don't stay out late the night before, and be sure to set your alarm clock! • Bring a pencil, notebook paper, and textbook to class. • Avoid distractions by turning off your cell phone during class. • Pay attention, take good notes, and ask questions. • Complete your homework on time, and ask questions on problems you do not understand. • Plan ahead for tests by preparing many days in advance.
2 Understand the homework to the point where you could do it without needing any help or hints.	• Read the directions and show all of your steps. • Go to the professor's office for help. • Rework homework and quiz problems and find similar problems for practice. • Review old material that you have not yet mastered.
3 Use the P.O.W.E.R. framework to help you manage your time: *Identify the Black Holes of Time Management.*	• Read the emPOWERme and do the checklist for time management issues. How many apply to you? • How many of the *controllable* items can you eliminate or reduce to make more time for studying? • Complete the emPOWERme that appears before the Chapter Summary.
4 Write your own goal.	•

What are your objectives for Chapter 3?	How can you accomplish each objective?
1 Be able to solve linear inequalities and write solutions in interval notation.	• Learn the **Properties of Equality.** • Understand *interval notation* and *set notation,* and learn how to write solutions using both.
2 Be able to solve compound inequalities in one variable containing the word *or* or the word *and.*	• Understand the meanings of *intersection* and *union.* • Learn the procedure for **Solving a Compound Inequality.**
3 Solve absolute value equations and inequalities.	• Learn the procedure for **Solving an Absolute Value Equation.** • Understand the different types of absolute value equations and inequalities and how to solve each. • Review interval notation.
4 Solve applied problems that involve inequalities or absolute value.	• Learn the procedure for solving applied problems. • Apply the correct procedure to solve the inequality or absolute value problem. • Make sure the answer is logical.
5 Write your own goal.	•

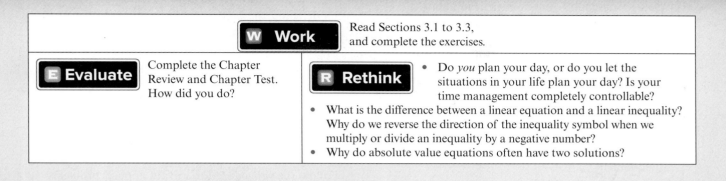

3.1 Linear Inequalities in One Variable

What are your objectives for Section 3.1?	How can you accomplish each objective?
1 Use Graphs and Set and Interval Notations	• Be able to recognize a *linear inequality*. • Understand how to write an answer in *set notation* and in *interval notation*. • Be able to determine when to use a bracket and when to use a parenthesis in *interval notation*. • Complete the given example on your own. • Complete You Try 1.
2 Solve Inequalities Using the Addition and Subtraction Properties of Inequality	• Learn the **Addition and Subtraction Properties of Inequality**, and know when they should be used to solve inequalities. • Complete the given example on your own. • Complete You Try 2.
3 Solve Inequalities Using the Multiplication Property of Inequality	• Learn the **Multiplication Property of Inequality**, and understand that we reverse the inequality symbol when we *multiply* or *divide* by a negative number. • Complete the given example on your own. • Complete You Try 3.
4 Solve Inequalities Using a Combination of the Properties	• Know all the properties of inequalities and when to use each property. • Remember to reverse the inequality symbol when multiplying or dividing by a negative number. • Complete the given example on your own. • Complete You Try 4.

(continued)

What are your objectives for Section 3.1?	How can you accomplish each objective?
5 Solve Compound Inequalities Containing Three Parts	• Be able to identify *compound inequalities*. • Understand the process for solving *three-part inequalities*. • Know how the solutions of *three-part inequalities* are graphed on a number line and written in *interval notation*. • Complete the given examples on your own. • Complete You Trys 5–7.
6 Solve Applications Involving Linear Inequalities	• Learn the phrases used with the different inequality symbols. • Follow the procedure for solving applied problems. • Make sure your answer is logical and that you have answered the question being asked. • Complete the given example on your own. • Complete You Try 8.

  Read the explanations, follow the examples, take notes, and complete the You Trys.

Recall the inequality symbols

   $<$ "is less than" $\leq$ "is less than or equal to"

   $>$ "is greater than" $\geq$ "is greater than or equal to"

We will use the symbols to form *linear inequalities in one variable*.

  While an equation states that two expressions are equal, an *inequality* states that two expressions are not necessarily equal. Here is a comparison of an equation and an inequality:

          Equation   **Inequality**

          $3x - 8 = 13$ $3x - 8 \leq 13$

> ### Definition
>
> A **linear inequality in one variable** can be written in the form $ax + b < c$, $ax + b \leq c$, $ax + b > c$, or $ax + b \geq c$ where a, b, and c are real numbers and $a \neq 0$.

The solution to a linear inequality is a set of numbers that can be represented in one of three ways:

    1) On a graph

    2) In *set notation*

    3) In *interval notation*

In this section, we will learn how to solve linear inequalities in one variable and how to represent the solution in each of those three ways.

1 Use Graphs and Set and Interval Notations

EXAMPLE 1 Graph each inequality and express the solution in set notation and interval notation.

a) $x \leq -1$ b) $t > 4$

Solution

a) $x \leq -1$:

Graphing $x \leq -1$ means that we are finding the solution set of $x \leq -1$. What value(s) of x will make the inequality true? The largest solution is -1. Then, any number *less than* -1 will make $x \leq -1$ true. We represent this **on the number line** as follows:

The graph illustrates that the solution is the set of all numbers less than and including -1.

Notice that the dot on -1 is shaded. This tells us that -1 is included in the solution set. The shading to the left of -1 indicates that *any* real number (not just integers) in this region is a solution.

We can express the solution set in **set notation** this way: $\{x \mid x \leq -1\}$.

$$\underset{\substack{\uparrow \\ \text{The set of} \\ \text{all values of } x}}{\{x} \quad \underset{\substack{\uparrow \\ \text{such that}}}{\mid} \quad \underset{\substack{\uparrow \\ x \text{ is less than} \\ \text{or equal to } -1.}}{x \leq -1\}}$$

In **interval notation** we write $(-\infty, -1]$

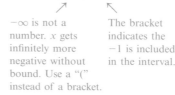

$-\infty$ is not a number. x gets infinitely more negative without bound. Use a "(" instead of a bracket.

The bracket indicates the -1 is included in the interval.

> **Note**
> The variable does not appear anywhere in interval notation.

b) $t > 4$:

We will plot 4 as an *open circle* on the number line because the symbol is ">" and *not* "$\geq$." The inequality $t > 4$ means that we must find the set of all numbers, t, greater than (but *not* equal to) 4. Shade to the right of 4.

 The graph illustrates that the solution is the set of all numbers greater than 4 but not including 4.

W Hint
Remember that we will never include ∞ or $-\infty$ in the solution set. Therefore, ∞ and $-\infty$ will always get a parenthesis when writing the solution set in interval notation.

We can express the solution set in *set notation* this way: $\{t \mid t > 4\}$
In *interval notation* we write

$(4, \infty)$

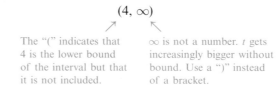

The "(" indicates that 4 is the lower bound of the interval but that it is not included.

∞ is not a number. t gets increasingly bigger without bound. Use a ")" instead of a bracket.

Hints for using interval notation:

1) The variable never appears in interval notation.
2) A number *included* in the solution set gets a bracket: $x \leq -1 \rightarrow (-\infty, -1]$
3) A number *not included* in the solution set gets a parenthesis: $t > 4 \rightarrow (4, \infty)$
4) The symbols $-\infty$ and ∞ *always* get parentheses.
5) The smaller number is always placed to the left. The larger number is placed to the right.
6) Even if we are not asked to graph the solution set, the graph may be helpful in writing the interval notation correctly.

[YOU TRY 1] Graph each inequality and express the solution in interval notation.

a) $k \geq -7$ b) $c < 5$

2 Solve Inequalities Using the Addition and Subtraction Properties of Inequality

The addition and subtraction properties of equality help us to solve equations. Similar properties hold for inequalities as well.

W Hint
Notice that whenever you *add* or *subtract* numbers to both sides of an inequality, the direction of the inequality is unchanged.

Property Addition and Subtraction Properties of Inequality

Let a, b, and c be real numbers. Then,

1) $a < b$ and $a + c < b + c$ are equivalent

   and

2) $a < b$ and $a - c < b - c$ are equivalent.

Adding the same number to both sides of an inequality or subtracting the same number from both sides of an inequality will not change the solution.

Note
The above properties hold for any of the inequality symbols.

EXAMPLE 2 Solve $y - 8 \geq -5$. Graph the solution set, and write the answer in interval and set notations.

Solution

$$y - 8 \geq -5$$
$$y - 8 + 8 \geq -5 + 8 \qquad \text{Add 8 to each side.}$$
$$y \geq 3$$

The solution set in interval notation is $[3, \infty)$. In set notation we write $\{y | y \geq 3\}$.

[YOU TRY 2] Solve $k - 10 \geq -4$. Graph the solution set, and write the answer in interval and set notations.

3 Solve Inequalities Using the Multiplication Property of Inequality

While the addition and subtraction properties for solving equations and inequalities work the same way, this is not true for multiplication and division. Let's see why.

Begin with an inequality we know is true: $2 < 5$. Multiply both sides by a *positive* number, say 3.

$$2 < 5 \quad \text{True}$$
$$3(2) < 3(5) \quad \text{Multiply by 3.}$$
$$6 < 15 \quad \text{True}$$

Begin again with $2 < 5$. Multiply both sides by a *negative* number, say -3.

$$2 < 5 \quad \text{True}$$
$$-3(2) < -3(5) \quad \text{Multiply by } -3.$$
$$-6 < -15 \quad \text{False}$$

To make $-6 < -15$ into a *true* statement, we must *reverse the direction of the inequality symbol*.

$$-6 > -15 \quad \text{True}$$

If you begin with a true inequality and *divide* by a positive number or by a negative number, the results will be the same as above since division can be defined in terms of multiplication. This leads us to the multiplication property of inequality.

Property Multiplication Property of Inequality

Let a, b, and c be real numbers.

1) If c is a *positive* number, then $a < b$ and $ac < bc$ are equivalent inequalities and have the same solutions.

2) If c is a *negative* number, then $a < b$ and $ac > bc$ are equivalent inequalities and have the same solutions.

It is also true that if $c > 0$ and $a < b$, then $\frac{a}{c} < \frac{b}{c}$. If $c < 0$ and $a < b$, then $\frac{a}{c} > \frac{b}{c}$.

For the most part, the procedures used to solve linear inequalities are the same as those for solving linear equations **except** *when you multiply or divide an inequality by a negative number, you must reverse the direction of the inequality symbol.*

EXAMPLE 3 Solve each inequality. Graph the solution set, and write the answer in interval and set notations.

a) $-5w \leq 20$ b) $5w \leq -20$

Solution

a) $-5w \leq 20$

First, divide each side by -5. *Since we are dividing by a negative number, we must remember to reverse the direction of the inequality symbol.*

$$-5w \leq 20$$
$$\frac{-5w}{-5} \geq \frac{20}{-5} \quad \text{Divide by } -5, \text{ so reverse the inequality symbol.}$$
$$w \geq -4$$

Interval notation: $[-4, \infty)$
Set notation: $\{w | w \geq -4\}$

b) $5w \leq -20$

First, divide by 5. Since we are dividing by a *positive* number, the inequality symbol remains the same.

$$5w \leq -20$$
$$\frac{5w}{5} \leq \frac{-20}{5} \quad \text{Divide by 5. Do } not \text{ reverse the inequality symbol.}$$
$$w \leq -4$$

Interval notation: $(-\infty, -4]$
Set notation: $\{w | w \leq -4\}$

> **[YOU TRY 3]** Solve $-\dfrac{1}{4}m < 3$. Graph the solution set, and write the answer in interval and set notations.

4 Solve Inequalities Using a Combination of the Properties

Often it is necessary to combine the properties to solve an inequality.

EXAMPLE 4 Solve $4(5 - 2d) + 11 < 2(d + 3)$. Graph the solution set, and write the answer in interval and set notations.

Solution

$$4(5 - 2d) + 11 < 2(d + 3)$$
$$20 - 8d + 11 < 2d + 6 \quad \text{Distribute.}$$
$$31 - 8d < 2d + 6 \quad \text{Combine like terms.}$$
$$31 - 8d - 2d < 2d - 2d + 6 \quad \text{Subtract } 2d \text{ from each side.}$$
$$31 - 10d < 6$$
$$31 - 31 - 10d < 6 - 31 \quad \text{Subtract 31 from each side.}$$
$$-10d < -25$$
$$\frac{-10d}{-10} > \frac{-25}{-10} \quad \text{Divide both sides by } -10. \text{ Reverse the inequality symbol.}$$
$$d > \frac{5}{2} \quad \text{Simplify.}$$

> **W Hint**
> When solving an inequality, remember that we must reverse the direction of the inequality symbol whenever we multiply or divide by a negative number!

> **W Hint**
> In this example, notice *when* the direction of the inequality symbol changed.

To graph the inequality, think of $\frac{5}{2}$ as $2\frac{1}{2}$.

Interval notation: $\left(\frac{5}{2}, \infty\right)$. Set notation: $\left\{d \mid d > \frac{5}{2}\right\}$

[YOU TRY 4] Solve $4(p + 2) + 1 > 2(3p + 10)$. Graph the solution set and write the answer in interval and set notations.

5 Solve Compound Inequalities Containing Three Parts

A **compound inequality** contains more than one inequality symbol. Some types of compound inequalities are

$$-5 < b + 4 < 1, \qquad t \leq \frac{1}{2} \text{ or } t \geq 3, \qquad \text{and} \qquad 2z + 9 < 5 \text{ and } z - 1 > 6$$

W Hint
When the variable term is between two numbers, both inequality signs are pointing in the same direction.

In this section, we will learn how to solve the first type of compound inequality, also called a **three-part inequality.** In Section 3.2 we will discuss the last two.

Consider the inequality $-2 \leq x \leq 3$. We can think of this in two ways:

1) x is *between* -2 and 3, and -2 and 3 are included in the interval.

   or

2) We can break up $-2 \leq x \leq 3$ into the two inequalities $-2 \leq x$ and $x \leq 3$.

Either way we think about $-2 \leq x \leq 3$, the meaning is the same. On a number line, the inequality would be represented as

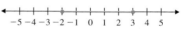

Notice that the **lower bound** of the interval on the number line is -2 (including -2), and the **upper bound** is 3 (including 3). Therefore, we can write the interval notation as

$$[-2, 3]$$

The endpoint, -2, is included in the interval, so use a bracket.   The endpoint, 3, is included in the interval, so use a bracket.

The set notation to represent $-2 \leq x \leq 3$ is $\{x \mid -2 \leq x \leq 3\}$.

Next, we will solve the inequality $-5 < b + 4 < 1$. To solve a three-part inequality you must remember that *whatever operation you perform on one part of the inequality must be performed on all three parts*. All properties of inequalities apply.

EXAMPLE 5 Solve $-5 < b + 4 < 1$. Graph the solution set, and write the answer in interval notation.

Solution

$$-5 < b + 4 < 1$$
$$-5 - 4 < b + 4 - 4 < 1 - 4 \qquad \text{Subtract 4 from each part of the inequality.}$$
$$-9 < b < -3$$

The graph of the solution set is 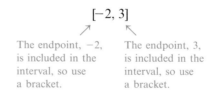 Every real number in the shaded region makes the original inequality true. In interval notation, we write $(-9, -3)$.

> **Note**
> Use parentheses in the interval notation because -9 and -3 are not included in the solution set.

[YOU TRY 5] Solve $-2 \leq 7k - 9 \leq 19$. Graph the solution set, and write the answer in interval notation.

We can eliminate fractions in an inequality by multiplying by the LCD of all of the fractions.

EXAMPLE 6 Solve $-\dfrac{7}{3} < \dfrac{1}{2}y - \dfrac{1}{3} \leq \dfrac{1}{2}$. Graph the solution set, and write the answer in interval notation.

Solution

The LCD of the fractions is 6. Multiply by 6 to eliminate the fractions.

$$-\frac{7}{3} < \frac{1}{2}y - \frac{1}{3} \leq \frac{1}{2}$$

$$6\left(-\frac{7}{3}\right) < 6\left(\frac{1}{2}y - \frac{1}{3}\right) \leq 6\left(\frac{1}{2}\right) \quad \text{Multiply all parts of the inequality by 6.}$$

$$-14 < 3y - 2 \leq 3$$

$$-14 + 2 < 3y - 2 + 2 \leq 3 + 2 \quad \text{Add 2 to each part.}$$

$$-12 < 3y \leq 5 \quad \text{Combine like terms.}$$

$$-\frac{12}{3} < \frac{3y}{3} \leq \frac{5}{3} \quad \text{Divide each part by 3.}$$

$$-4 < y \leq \frac{5}{3} \quad \text{Simplify.}$$

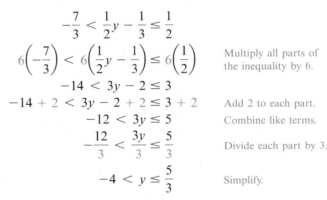

Interval notation: $\left(-4, \dfrac{5}{3}\right]$

W Hint
Remember, we can eliminate fractions in inequalities just like we do in equations!

[YOU TRY 6] Solve $-\dfrac{3}{4} < \dfrac{1}{3}z - \dfrac{3}{4} \leq \dfrac{5}{4}$. Graph the solution set, and write the answer in interval notation.

Remember, if we multiply or divide an inequality by a negative number, we reverse the direction of the inequality symbol. When solving a compound inequality like these, reverse *both* symbols.

EXAMPLE 7

Solve $11 < -3x + 2 < 17$. Graph the solution set, and write the answer in interval notation.

Solution

$$11 < -3x + 2 < 17$$
$$11 - 2 < -3x + 2 - 2 < 17 - 2 \quad \text{Subtract 2 from each part.}$$
$$9 < -3x < 15$$
$$\frac{9}{-3} > \frac{-3x}{-3} > \frac{15}{-3} \quad \text{When we divide by a negative number, reverse the direction of the inequality symbol.}$$
$$-3 > x > -5 \quad \text{Simplify.}$$

W Hint
In this example, notice *when* the direction of inequality changed.

Think carefully about what $-3 > x > -5$ means. It means "x is less than -3 *and* x is greater than -5." This is especially important to understand when writing the correct interval notation.

The graph of the solution set is ⟵—+—+—+—◇—+—◇—+—+—+—+—+—⟶
$\quad\quad\quad\quad\quad\quad\quad\quad\quad\quad\quad -7\ -6\ -5\ -4\ -3\ -2\ -1\ \ 0\ \ 1\ \ 2\ \ 3$

Even though we got $-3 > x > -5$ as our result, -5 is actually the lower bound of the solution set and -3 is the upper bound. The inequality $-3 > x > -5$ can also be written as $-5 < x < -3$.

The solution in interval notation is $(-5, -3)$.
$\quad\quad\quad\quad\quad\quad\quad\quad\quad\quad\quad\quad\quad\quad\quad\quad ↑\quad ↑$
$\quad\quad\quad\quad\quad\quad\quad\quad\quad\quad\text{Lower bound on the left\quad Upper bound on the right}$

[YOU TRY 7] Solve $4 < -2x - 4 < 10$. Graph the solution set, and write the answer in interval notation.

6 Solve Applications Involving Linear Inequalities

Certain phrases in applied problems indicate the use of inequality symbols:

$$\text{at least:}\ \geq \quad\quad \text{no less than:}\ \geq$$
$$\text{at most:}\ \leq \quad\quad \text{no more than:}\ \leq$$

W Hint
Make a note of these four phrases. This may help you remember which inequality symbol to use when doing your homework.

There are others. Next, we will look at an example of a problem involving the use of an inequality symbol. We will use the same steps that were used to solve applications involving equations.

EXAMPLE 8

Joe Amici wants to have his son's birthday party at Kiddie Fun Factory. The cost of a party is $175 for the first 10 children plus $3.50 for each additional child. If Joe can spend at most $200, find the greatest number of children who can attend the party.

Solution

Step 1: **Read** the problem carefully. We must find the greatest number of children who can attend the party.

Step 2: **Choose a variable** to represent the unknown quantity. We know that the first 10 children will cost $175, but we do not know how many *additional* guests Joe can afford to invite.

$\quad\quad\quad x$ = number of children **over** the first 10 who attend the party

Step 3: **Translate** from English to an algebraic inequality.

English: [Cost of first 10 children] + [Cost of additional children] [is at most] [$200]

Inequality: $175 + 3.50x \leq 200$

The inequality is $175 + 3.50x \leq 200$.

Step 4: **Solve** the inequality.

$$175 + 3.50x \leq 200$$
$$3.50x \leq 25 \quad \text{Subtract 175.}$$
$$x \leq 7.142\ldots \quad \text{Divide by 3.50.}$$

Step 5: **Check** the answer, and **interpret** the solution as it relates to the problem.

The result was $x \leq 7.142\ldots$, where x represents the number of additional children who can attend the party. Since it is not possible to have $7.142\ldots$ people and $x \leq 7.142\ldots$, in order to stay within budget, Joe can afford to pay for at most 7 additional guests *over* the initial 10.

Therefore, the greatest number of people who can attend the party is

The first 10 + additional = total
$$10 + 7 = 17$$

At most, 17 children can attend the birthday party. Does the answer make sense?

$$\text{Total Cost of Party} = \$175 + \$3.50(7)$$
$$= \$175 + \$24.50$$
$$= \$199.50$$

We can see that one more guest (at a cost of $3.50) would put Joe over budget.

[**YOU TRY 8**] For $4.00 per month, Van can send or receive 200 text messages. Each additional message costs $0.05. If Van can spend at most $9.00 per month on text messages, find the greatest number he can send or receive each month.

ANSWERS TO [YOU TRY] **EXERCISES**

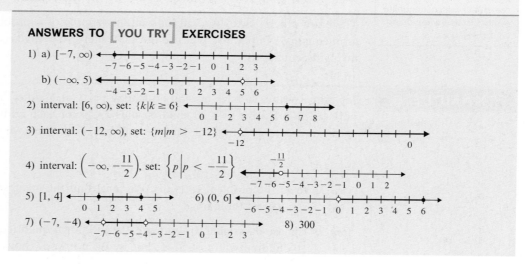

106 CHAPTER 3 Linear Inequalities and Absolute Value

Evaluate 3.1 Exercises

Do the exercises, and check your work.

Objective 1: Use Graphs and Set and Interval Notations

1) When do you use parentheses when writing a solution set in interval notation?

2) When do you use brackets when writing a solution set in interval notation?

Graph the inequality. Express the inequality in a) set notation and b) interval notation.

3) $x \geq 3$
4) $t \geq -4$
5) $c < -1$
6) $r < \dfrac{5}{2}$
7) $w > -\dfrac{11}{3}$
8) $p \leq 2$

Mixed Exercises: Objectives 2 and 3

Solve each inequality. Graph the solution set and write the answer in a) set notation and b) interval notation. See Examples 2 and 3.

9) $r - 9 \leq -5$
10) $p + 6 \geq 4$
11) $y + 5 \geq 1$
12) $n - 8 \leq -3$
13) $3c > 12$
14) $8v > 24$
15) $15k < -55$
16) $16m < -28$
17) $-4b \leq 32$
18) $-9a \geq 27$
19) $-14w > -42$
20) $-30t < -18$
21) $\dfrac{1}{5}z \geq -3$
22) $\dfrac{1}{3}x < -2$
23) $-\dfrac{9}{4}y < -18$
24) $-\dfrac{2}{5}p \geq 4$

Objective 4: Solve Inequalities Using a Combination of the Properties

Solve each inequality. Graph the solution set and write the answer in interval notation. See Example 4.

25) $8z + 19 > 11$
26) $5x - 2 \leq 18$
27) $12 - 7t \geq 15$
28) $-1 - 4p < 5$
29) $-23 - w < -20$
30) $16 - h \geq 9$
31) $6(7y + 4) - 10 > 2(10y + 13)$
32) $7a + 4(5 - a) \leq 4 - 5a$
33) $9c + 17 > 14c - 3$
34) $-11n + 6 \leq 16 - n$
35) $\dfrac{8}{3}(2k + 1) > \dfrac{1}{6}k + \dfrac{8}{3}$
36) $\dfrac{1}{2}(c - 3) + \dfrac{3}{4}c \geq \dfrac{1}{2}(2c + 3) + \dfrac{3}{8}$
37) $0.04x + 0.12(10 - x) \geq 0.08(10)$
38) $0.09m + 0.05(8) \leq 0.07(m + 8)$

Objective 5: Solve Compound Inequalities Containing Three Parts

Graph the inequality. Express the inequality in a) set notation and b) interval notation.

39) $1 \leq n \leq 4$
40) $-3 \leq g \leq 2$
41) $-2 < a < 1$
42) $-4 < d < 0$
43) $\dfrac{1}{2} < z \leq 3$
44) $-2 \leq y < 3$

Solve each inequality. Graph the solution set and write the answer in interval notation. See Examples 5–7.

45) $-8 \leq a - 5 \leq -4$
46) $1 \leq t + 3 \leq 7$
47) $9 < 6n < 18$
48) $-10 < 2x < 7$
49) $-19 \leq 7p + 9 \leq 2$
50) $-5 \leq 3k - 11 \leq 4$
51) $-6 \leq 4c - 13 < -1$
52) $-11 < 6m + 1 \leq -3$
53) $2 < \dfrac{3}{4}u + 8 < 11$
54) $2 \leq \dfrac{5}{2}y - 3 \leq 7$
55) $-\dfrac{1}{2} \leq \dfrac{5d + 2}{6} \leq 0$
56) $2 < \dfrac{2b + 7}{3} < 5$
57) $3 < 19 - 2j \leq 9$
58) $-13 \leq 14 - 9h < 5$
59) $0 \leq 4 - 3w \leq 7$
60) $-6 < -5 - z < 0$

Mixed Exercises: Objectives 2–5

Solve each inequality. Write the answer in interval notation.

61) $k + 11 > 4$
62) $5 < x + 9 < 12$
63) $-12p \geq -16$
64) $2w + 7 \geq 13$
65) $5(2b - 3) - 7b > 5b + 9$
66) $8 - m < 14$
67) $-12 < \dfrac{8}{5}t + 12 \leq 6$
68) $0.29 \geq 0.04a + 0.05$

69) $\dfrac{5}{4}(k+4) + \dfrac{1}{4} \geq \dfrac{5}{6}(k+3) - 1$

70) $-3 \leq 6c - 1 \leq 5$

71) $4 < 4 - 7y \leq 18$

72) $9z \leq -18$

Objective 6: Solve Applications Involving Linear Inequalities

Write an inequality for each problem and solve. See Example 8.

73) Carson's Parking Garage charges $4.00 for the first 3 hr plus $1.50 for each additional half-hour. Ted has only $11.50 for parking. For how long can Ted park his car in this garage?

74) Oscar makes a large purchase at Home Depot and plans to rent one of its trucks to take his supplies home. The most he wants to spend on the truck rental is $50.00. If Home Depot charges $19.00 for the first 75 min and $5.00 for each additional 15 min, for how long can Oscar keep the truck and remain within his budget? (www.homedepot.com)

75) A taxi charges $2.00 plus $0.25 for every $\dfrac{1}{5}$ of a mile. How many miles can you go if you have $12.00?

76) A taxi charges $2.50 plus $0.20 for every $\dfrac{1}{4}$ of a mile. How many miles can you go if you have $12.50?

77) Melinda's first two test grades in Psychology were 87 and 94. What does she need to make on the third test to maintain an average of at least 90?

78) Russell's first three test scores in Geography were 86, 72, and 81. What does he need to make on the fourth test to maintain an average of at least 80?

 Rethink

R1) Which exercises in this section do you find most challenging?

R2) When must you switch the direction of the inequality?

R3) If $a < b < c$, then why must it be true that $c > b > a$?

R4) What kind of paid services do you use that place a limit on the number of times you use the service? If there is a cost for going over the limit, how much is it?

3.2 Compound Inequalities in One Variable

Prepare

What are your objectives for Section 3.2?	How can you accomplish each objective?
1 Find the Intersection and Union of Two Sets	• Understand the meaning of *union* and *intersection*. • Know the symbols used to indicate *union* and *intersection*. • Complete the given example on your own. • Complete You Try 1.
2 Solve Compound Inequalities Containing the Word *And*	• Learn the procedure for **Solving a Compound Inequality Containing *and***. • Complete the given examples on your own. • Complete You Trys 2 and 3.
3 Solve Compound Inequalities Containing the Word *Or*	• Follow the procedure for **Solving a Compound Inequality Containing *or***. • Complete the given example on your own. • Complete You Try 4.
4 Solve Special Compound Inequalities	• Review the meaning of *union* and *intersection*. • Know the procedure for **Solving a Compound Inequality**. • Complete the given example on your own. • Complete You Try 5.
5 Application of Intersection and Union	• Understand the meaning of *and* and *or* in this type of application. • Complete the given example on your own. • Complete You Try 6.

Work Read the explanations, follow the examples, take notes, and complete the You Trys.

In Section 3.1, we learned how to solve a compound inequality like $-8 \leq 3x + 4 \leq 13$. In this section, we will discuss how to solve compound inequalities like these:

$$t \leq \frac{1}{2} \text{ or } t \geq 3 \quad \text{and} \quad 2z + 9 < 5 \text{ and } z - 1 > 6$$

But first, we must talk about set notation and operations.

1 Find the Intersection and Union of Two Sets

EXAMPLE 1 Let $A = \{1, 2, 3, 4, 5, 6\}$ and $B = \{3, 5, 7, 9, 11\}$.

The **intersection** of sets A and B is the set of numbers that are elements of A **and** of B. The *intersection* of A and B is denoted by $A \cap B$.

$A \cap B = \{3, 5\}$ because 3 and 5 are found in both A and B.

The **union** of sets A and B is the set of numbers that are elements of A **or** of B. The union of A and B is denoted by $A \cup B$. The set $A \cup B$ consists of the elements in A or in B or in *both*.

$$A \cup B = \{1, 2, 3, 4, 5, 6, 7, 9, 11\}$$

Note
Although the elements 3 and 5 appear in both set A and in set B, we do not write them twice in the set $A \cup B$.

Let $A = \{2, 4, 6, 8, 10\}$ and $B = \{1, 2, 5, 6, 9, 10\}$. Find $A \cap B$ and $A \cup B$.

Note
The word "*and*" indicates *intersection*, while the word "*or*" indicates *union*. This same principle holds when solving compound inequalities involving "*and*" or "*or*."

2 Solve Compound Inequalities Containing the Word *And*

EXAMPLE 2

Solve the compound inequality $c + 5 \geq 3$ and $8c \leq 32$. Graph the solution set, and write the answer in interval notation.

Solution

Step 1: Identify the inequality as "*and*" or "*or*" and understand what that means. These two inequalities are connected by "*and*." That means the solution set will consist of the values of c that make *both* inequalities true. The solution set will be the *intersection* of the solution sets of $c + 5 \geq 3$ and $8c \leq 32$.

Step 2: Solve each inequality separately.

$$\begin{array}{ccc} c + 5 \geq 3 & \text{and} & 8c \leq 32 \\ c \geq -2 & \text{and} & c \leq 4 \end{array}$$

Step 3: Graph the solution set to each inequality on its own number line even if the problem does not require you to graph the solution set. This will help you visualize the solution set of the compound inequality.

$c \geq -2$: ← +–+–+–+–+–+–+–+–+–+–+ →
 −5 −4 −3 −2 −1 0 1 2 3 4 5

$c \leq 4$: ← +–+–+–+–+–+–+–+–+–+–+ →
 −5 −4 −3 −2 −1 0 1 2 3 4 5

Step 4: Look at the number lines and think about where the solution set for the compound inequality would be graphed.

Since this is an "*and*" inequality, the solution set of $c + 5 \geq 3$ and $8c \leq 32$ consists of the numbers that are solutions to *both* inequalities. We can visualize it this way: if we take the number line above representing $c \geq -2$ and place it on top of the number line representing $c \leq 4$, what shaded areas would overlap (intersect)?

$c \geq -2$ and $c \leq 4$:

They intersect between -2 and 4, *including* those endpoints.

Step 5: Write the answer in interval notation.

The final number line illustrates that the solution to $c + 5 \geq 3$ and $8c \leq 32$ is $[-2, 4]$. The graph of the solution set is the final number line above.

Here are the steps to follow when solving a compound inequality.

W Hint
Even if you are not *asked* to graph the solution set of a compound inequality, graph it anyway. Looking at the graph will help you write the interval notation correctly.

Procedure Steps for Solving a Compound Inequality

Step 1: Identify the inequality as "*and*" or "*or*" and understand what that means.
Step 2: Solve each inequality separately.
Step 3: Graph the solution set to each inequality on its own number line even if the problem does not explicitly tell you to graph it. This will help you to visualize the solution to the compound inequality.
Step 4: Use the separate number lines to graph the solution set of the compound inequality.
    a) If it is an "*and*" inequality, the solution set consists of the regions on the separate number lines that would *overlap* (intersect) if one number line was placed on top of the other.
    b) If it is an "*or*" inequality, the solution set consists of the *total* (union) of what would be shaded if you took the separate number lines and put one on top of the other.
Step 5: Use the graph of the solution set to write the answer in interval notation.

[**YOU TRY 2**] Solve the compound inequality $y - 2 \leq 1$ and $7y > -28$. Graph the solution set, and write the answer in interval notation.

EXAMPLE 3 Solve the compound inequality $7y + 2 > 37$ and $5 - \frac{1}{3}y < 6$. Write the solution set in interval notation.

Solution

Step 1: This is an "*and*" inequality. The solution set will be the *intersection* of the solution sets of the separate inequalities $7y + 2 > 37$ and $5 - \frac{1}{3}y < 6$.

Step 2: We must solve each inequality separately.

$$7y + 2 > 37 \quad \text{and} \quad 5 - \frac{1}{3}y < 6$$

Subtract 2. $7y > 35$ and $-\frac{1}{3}y < 1$ Subtract 5, then multiply both sides by -3.

Divide by 7. $y > 5$ and $y > -3$ Reverse the direction of the inequality symbol.

W Hint

Remember that the solution set to an "and" compound inequality is found by looking at the *intersection* of the solution sets of two different inequalities.

Step 3: Graph the solution sets separately so that it is easier to find their intersection.

$y > 5$: ⟵—|—|—|—|—|—|—|—|—|—|—|—◇—|—⟶
 −6 −5 −4 −3 −2 −1 0 1 2 3 4 5 6

$y > -3$: ⟵—|—|—◇—|—|—|—|—|—|—|—|—|—⟶
 −6 −5 −4 −3 −2 −1 0 1 2 3 4 5 6

Step 4: If we were to put the number lines above on top of each other, where would they intersect?

$y > 5$ and $y > -3$: ⟵—|—|—|—|—|—|—|—|—|—|—|—◇—|—⟶
 −6 −5 −4 −3 −2 −1 0 1 2 3 4 5 6

Step 5: The solution, shown in the shaded region in Step 4, is $(5, \infty)$.

[YOU TRY 3] Solve each compound inequality and write the answer in interval notation.

a) $4x - 3 > 1$ and $x + 6 < 13$ b) $-\dfrac{4}{5}m > -8$ and $2m + 5 \leq 12$

---

3 Solve Compound Inequalities Containing the Word *Or*

Recall that the word "*or*" indicates the union of two sets.

EXAMPLE 4 Solve the compound inequality $6p + 5 \leq -1$ or $p - 3 \geq 1$. Write the answer in interval notation.

Solution

Step 1: These two inequalities are joined by "*or*." Therefore, the solution set will consist of the values of p that are in the solution set of $6p + 5 \leq -1$ *or* in the solution set of $p - 3 \geq 1$ *or* in *both* solution sets.

Step 2: Solve each inequality separately.

$$6p + 5 \leq -1 \quad \text{or} \quad p - 3 \geq 1$$
$$6p \leq -6$$
$$p \leq -1 \quad \text{or} \quad p \geq 4$$

W Hint

Remember that the solution set to an "or" compound inequality is found by looking at the *union* of the solution sets of two different inequalities.

Step 3: Graph the solution sets separately so that it is easier to find the *union* of the sets.

$p \leq -1$: ⟵—|—|—|—|—●—|—|—|—|—|—|—|—⟶
 −5 −4 −3 −2 −1 0 1 2 3 4 5 6

$p \geq 4$: ⟵—|—|—|—|—|—|—|—|—|—●—|—|—⟶
 −5 −4 −3 −2 −1 0 1 2 3 4 5 6

Step 4: The solution set of the compound inequality $6p + 5 \leq -1$ or $p - 3 \geq 1$ consists of the numbers which are solutions to the first inequality *or* the second inequality *or* both. We can visualize it this way: if we put the number lines on top of each other, the solution set of the compound inequality is the **total** (union) of what is shaded.

$p \leq -1$ or $p \geq 4$: ⟵—|—|—|—|—|—●—|—|—|—|—●—|—|—⟶
 −6 −5 −4 −3 −2 −1 0 1 2 3 4 5 6

Step 5: The solution, shown above, is written as $(-\infty, -1] \cup [4, \infty)$.
$ \uparrow$
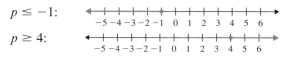
Use the *union* symbol for "or."

[YOU TRY 4] Solve $t + 8 \geq 14$ or $\dfrac{3}{2}t < 6$ and write the solution in interval notation.

4 Solve Special Compound Inequalities

EXAMPLE 5 Solve each compound inequality, and write the answer in interval notation.

a) $k - 5 < -2$ or $4k + 9 > 6$ b) $\dfrac{1}{2}w \geq 3$ and $1 - w \geq 0$

Solution

a) $k - 5 < -2$ or $4k + 9 > 6$

Step 1: The solution to this "*or*" inequality is the *union* of the solution sets of $k - 5 < -2$ and $4k + 9 > 6$.

Step 2: Solve each inequality separately.

$$k - 5 < -2 \quad \text{or} \quad 4k + 9 > 6$$
$$4k > -3$$
$$k < 3 \quad \text{or} \quad k > -\dfrac{3}{4}$$

W Hint
Do you see why it is helpful to graph each inequality separately?

Step 3: $k < 3$:

$k > -\dfrac{3}{4}$:

Step 4: $k < 3$ or $k > -\dfrac{3}{4}$:

If the number lines in Step 3 were placed on top of each other, the *total* (union) of what would be shaded is the entire number line. This represents all real numbers.

Step 5: The solution set of the compound inequality is $(-\infty, \infty)$.

b) $\dfrac{1}{2}w \geq 3$ and $1 - w \geq 0$

Step 1: The solution to this "*and*" inequality is the *intersection* of the solution sets of $\dfrac{1}{2}w \geq 3$ and $1 - w \geq 0$.

Step 2: Solve each inequality separately.

$$\dfrac{1}{2}w \geq 3 \quad \text{and} \quad 1 - w \geq 0$$
$$\qquad\qquad\qquad\qquad\qquad 1 \geq w \quad \text{Add } w.$$
$$\text{Multiply by 2.} \quad w \geq 6 \quad \text{and} \quad w \leq 1 \quad \text{Rewrite } 1 \geq w \text{ as } w \leq 1.$$

Step 3: $w \geq 6$:

$w \leq 1$:

Step 4: $w \geq 6$ and $w \leq 1$:

If the number lines in Step 3 were placed on top of each other, the shaded regions would *not* intersect. Therefore, the solution set is the empty set, $\varnothing$.

Step 5: The solution set of $\dfrac{1}{2}w \geq 3$ and $1 - w \geq 0$ is $\varnothing$.

[YOU TRY 5] Solve the compound inequalities, and write the solution in interval notation.

a) $-3w \leq w - 6$ and $5w < 4$ b) $9z - 8 \leq -8$ or $z + 7 \geq 2$

5 Application of Intersection and Union

EXAMPLE 6

The following table of selected NBA teams contains the number of times they have appeared in the play-offs as well as the number of NBA championships they have won through the 2009–2010 season.

Team	Play-Off Appearances	Championships
Boston Celtics	48	17
Chicago Bulls	29	6
Cleveland Cavaliers	18	0
Detroit Pistons	32	3
Los Angeles Lakers	46	11
New York Knicks	38	2

(www.basketball-reference.com)

List the elements of the set that satisfy the given information.

a) The set of teams with more than 20 play-off appearances and more than 5 championships

b) The set of teams with less than 30 play-off appearances or more than 5 championships

Solution

a) Because the two conditions in this statement are connected by *and*, we must find the team or teams that satisfy *both* conditions. The set of teams is

{Boston Celtics, Chicago Bulls, Los Angeles Lakers}

b) Because the two conditions in this statement are connected by *or*, we must find the team or teams that satisfy either the first condition, *or* the second condition, *or* both. The set of teams is

{Boston Celtics, Chicago Bulls, Cleveland Cavaliers, Los Angeles Lakers}

[YOU TRY 6] Use the table in Example 6, and list the elements of the set that satisfy the given information.

a) The set of teams with less than 40 play-off appearances and at least one championship

b) The set of teams with more than 30 play-off appearances or no championships

ANSWERS TO [YOU TRY] EXERCISES

1) $A \cap B = \{2, 6, 10\}$, $A \cup B = \{1, 2, 4, 5, 6, 8, 9, 10\}$ 2) ⟵|—+—+—◇—+—+—+—+—+—◆—+—|⟶ $-5\,-4\,-3\,-2\,-1\;0\;1\;2\;3\;4\;5$ $(-4, 3]$ 3) a) $(1, 7)$ b) $\left(-\infty, \dfrac{7}{2}\right]$ 4) $(-\infty, 4) \cup [6, \infty)$ 5) a) $\varnothing$ b) $(-\infty, \infty)$

6) a) {Chicago Bulls, Detroit Pistons, New York Knicks}
b) {Boston Celtics, Cleveland Cavaliers, Detroit Pistons, Los Angeles Lakers, New York Knicks}

E Evaluate 3.2 Exercises

Do the exercises, and check your work.

Objective 1: Find the Intersection and Union of Two Sets

1) Given sets A and B, explain how to find $A \cap B$.

2) Given sets X and Y, explain how to find $X \cup Y$.

Given sets $A = \{2, 4, 6, 8, 10\}$, $B = \{1, 3, 5\}$, $X = \{8, 10, 12, 14\}$, and $Y = \{5, 6, 7, 8, 9\}$ find

3) $A \cap X$
4) $X \cap Y$
5) $A \cup Y$
6) $B \cup Y$
7) $X \cap B$
8) $B \cap A$
9) $A \cup B$
10) $X \cup Y$

Each number line represents the solution set of an inequality. Graph the *intersection* of the solution sets and write the intersection in interval notation.

11) $x \geq -3$:
   $x \leq 2$:

12) $n \leq 4$:
   $n \geq 0$:

13) $t < 3$:
   $t > -1$:

14) $y > -4$:
   $y < -2$:

15) $c > 1$:
   $c \geq 3$:

16) $p < 2$:
   $p < -1$:

17) $z \leq 0$:
   $z \geq 2$:

18) $g \geq -1$:
   $g < -\dfrac{5}{2}$:

Mixed Exercises: Objectives 2 and 4

Solve each compound inequality. Graph the solution set, and write the answer in interval notation.

19) $a \leq 5$ and $a \geq 2$
20) $k > -3$ and $k < 4$
21) $b - 7 > -9$ and $8b < 24$

22) $3x \leq 1$ and $x + 11 \geq 4$

23) $5w + 9 \leq 29$ and $\dfrac{1}{3}w - 8 > -9$

24) $4y - 11 > -7$ and $\dfrac{3}{2}y + 5 \leq 14$

25) $2m + 15 \geq 19$ and $m + 6 < 5$

26) $d - 1 > 8$ and $3d - 12 < 4$

27) $r - 10 > -10$ and $3r - 1 > 8$

28) $2t - 3 \leq 6$ and $5t + 12 \leq 17$

29) $9 - n \leq 13$ and $n - 8 \leq -7$

30) $c + 5 \geq 6$ and $10 - 3c \geq -5$

Objective 1: Find the Intersection and Union of Two Sets

Each number line represents the solution set of an inequality. Graph the *union* of the solution sets and write the union in interval notation.

31) $p < -1$:
   $p > 5$:

32) $z < 2$:
   $z > 6$:

33) $a \leq \dfrac{5}{3}$:
   $a > 4$:

34) $v \leq -3$:
   $v \geq \dfrac{11}{4}$:

35) $y > 1$:
   $y > 3$:

36) $x \leq -6$:
   $x \leq -2$:

37) $c < \dfrac{7}{2}$:
   $c \geq -2$:

38) $q \leq 3$:
   $q > -2.7$:

SECTION 3.2 Compound Inequalities in One Variable 115

Mixed Exercises: Objectives 3 and 4
Solve each compound inequality. Graph the solution set, and write the answer in interval notation.

39) $z < -1$ or $z > 3$

40) $x \leq -4$ or $x \geq 0$

41) $6m \leq 21$ or $m - 5 > 1$

42) $a + 9 > 7$ or $8a \leq -44$

43) $3t + 4 > -11$ or $t + 19 > 17$

44) $5y + 8 \leq 13$ or $2y \leq -6$

45) $-2v - 5 \leq 1$ or $\dfrac{7}{3}v < -14$

46) $k - 11 < -4$ or $-\dfrac{2}{9}k \leq -2$

47) $c + 3 \geq 6$ or $\dfrac{4}{5}c \leq 10$

48) $\dfrac{8}{3}g \geq -12$ or $2g + 1 \leq 7$

49) $7 - 6n \geq 19$ or $n + 14 < 11$

50) $d - 4 > -7$ or $-6d \leq 2$

Mixed Exercises: Objectives 2–4
The following exercises contain *and* and *or* inequalities. Solve each inequality, and write the answer in interval notation.

51) $4n + 7 \leq 9$ and $n + 6 \geq 1$

52) $8t - 5 \geq 11$ or $-\dfrac{2}{5}t \geq 6$

53) $\dfrac{4}{3}x + 5 < 2$ or $x + 3 \geq 8$

54) $p + 10 < -3$ and $8p - 7 > 11$

55) $\dfrac{8}{3}w < -16$ or $w + 9 > -4$

56) $5y - 2 > 8$ and $\dfrac{3}{4}y + 2 < 11$

57) $7 - r > 7$ and $0.3r < 6$

58) $2c - 9 \leq -3$ or $10c + 1 \geq 7$

59) $3 - 2k > 11$ and $\dfrac{1}{2}k + 5 \geq 1$

60) $6 - 5a \geq 1$ or $0.8a > 8$

Objective 5: Application of Intersection and Union
The following table lists the net worth (in billions of dollars) of some of the wealthiest women in the world for the years 2004 and 2008.

Name	Net Worth in 2004	Net Worth in 2008
Liliane Bettencourt	17.2	22.9
Abigail Johnson	12.0	15.0
J. K. Rowling	1.0	1.0
Alice Walton	18.0	19.0
Oprah Winfrey	1.3	2.5

(www.forbes.com)

List the elements of the set that satisfy the given information.

61) The set of women with a net worth more than $15 billion in 2004 and in 2008.

62) The set of women with a net worth more than $10 billion in 2004 and less than $20 billion in 2008.

63) The set of women with a net worth less than $2 billion in 2004 or more than $20 million in 2008.

64) The set of women with a net worth more than $15 billion in 2004 or more than $2 billion in 2008.

R Rethink

R1) How does the *intersection* of two sets differ from the *union* of two sets?

R2) What is the mathematical symbol for the *union*, and what is the mathematical symbol for the *intersection* of two sets?

R3) How does the "and" compound inequality differ from the "or" compound inequality?

R4) Which exercises in this section do you find most challenging?

3.3 Absolute Value Equations and Inequalities

P Prepare | O Organize

What are your objectives for Section 3.3?	How can you accomplish each objective?
1 Understand the Meaning of an Absolute Value Equation	• Understand *absolute value* as it relates to a number line. • Learn the procedure for **Solving an Absolute Value Equation**. • Complete the given example on your own. • Complete You Try 1.
2 Solve an Equation of the Form $\|ax + b\| = k$	• Follow the procedure for **Solving an Absolute Value Equation** for $k > 0$. • Be sure to look at the value of k. • Complete the given examples on your own. • Complete You Trys 2 and 3.
3 Solve an Equation of the Form $\|ax + b\| = \|cx + d\|$	• Learn the procedure for **Solving an Absolute Value Equation** of this form. • Understand how to set up problems containing two absolute values. • Complete the given example on your own. • Complete You Try 4.
4 Solve Absolute Value Inequalities Containing $<$ or $\leq$	• Learn the procedure for **Solving Absolute Value Inequalities** of this form. • Know what the solution represents on the number line. • Complete the given examples on your own. • Complete You Trys 5 and 6.
5 Solve Absolute Value Inequalities Containing $>$ or $\geq$	• Learn the procedure for **Solving Absolute Value Inequalities** of this form. • Know what the solution represents on the number line and why these solutions are two parts. • Complete the given examples on your own. • Complete You Trys 7 and 8.
6 Solve Special Cases of Absolute Value Inequalities	• Review the meaning of *absolute value*. • Carefully interpret the meaning of the inequality symbol when solving these absolute value inequalities. • Complete the given example on your own. • Complete You Try 9.
7 Solve an Applied Problem Using an Absolute Value Inequality	• Read the problem carefully. • Choose the correct inequality symbol to use in your problem. • Follow the procedure for **Solving Absolute Value Inequalities**. • Complete the given example on your own.

Work — Read the explanations, follow the examples, take notes, and complete the You Trys.

In Section 1.1 we learned that the absolute value of a number describes its *distance from zero*.

$$|5| = 5 \quad \text{and} \quad |-5| = 5$$

5 units from zero   5 units from zero

$-7\ -6\ -5\ -4\ -3\ -2\ -1\ 0\ 1\ 2\ 3\ 4\ 5\ 6\ 7$

We use this idea of *distance from zero* to solve absolute value equations and inequalities.

1 Understand the Meaning of an Absolute Value Equation

EXAMPLE 1 Solve $|x| = 3$.

Solution

Since the equation contains an absolute value, **solve $|x| = 3$** means "*Find the number or numbers whose distance from zero is 3.*"

3 units from zero   3 units from zero

$-6\ -5\ -4\ -3\ -2\ -1\ 0\ 1\ 2\ 3\ 4\ 5\ 6$

Those numbers are 3 and -3. Each of them is 3 units from zero. The solution set is $\{-3, 3\}$.

Check: $|3| = 3, |-3| = 3$ ✓

[YOU TRY 1] Solve $|y| = 8$.

Hint
Write the procedure for solving an absolute value equation in your own words.

Procedure Solving an Absolute Value Equation for $k > 0$

If P represents an expression and k is a positive real number, then to solve $|P| = k$ we rewrite the absolute value equation as the *compound equation*

$$P = k \quad \text{or} \quad P = -k$$

and solve for the variable. P can represent expressions like x, $3a + 2$, $\frac{1}{4}t - 9$, and so on.

118 CHAPTER 3 Linear Inequalities and Absolute Value

2 Solve an Equation of the Form $|ax + b| = k$

EXAMPLE 2 Solve each equation.

a) $|m + 1| = 5$ b) $\left|\dfrac{3}{2}t + 7\right| + 5 = 6$

Solution

a) Solving $|m + 1| = 5$ means, "Find the number or numbers that can be substituted for m so that the quantity $m + 1$ is 5 units from 0."

$m + 1$ will be 5 units from zero if $m + 1 = 5$ or if $m + 1 = -5$, since both 5 and -5 are 5 units from zero. Therefore, we can solve the equation this way:

$|m + 1| = 5$

$m + 1 = 5$ or $m + 1 = -5$ Set the quantity inside the absolute value equal to 5 and -5.

$m = 4$ or $m = -6$ Solve.

Check: $m = 4$: $|4 + 1| \stackrel{?}{=} 5$ $\quad$ $m = -6$: $|-6 + 1| \stackrel{?}{=} 5$
$\qquad\qquad\quad |5| = 5$ ✓ $\qquad\qquad\qquad |-5| = 5$ ✓

The solution set is $\{-6, 4\}$.

b) Before we rewrite this equation as a compound equation, we must *isolate* the absolute value (get the absolute value on a side by itself).

$\left|\dfrac{3}{2}t + 7\right| + 5 = 6$

$\left|\dfrac{3}{2}t + 7\right| = 1$ Subtract 5 to get the absolute value on a side by itself.

$\dfrac{3}{2}t + 7 = 1$ or $\dfrac{3}{2}t + 7 = -1$ Set the quantities inside the absolute value equal to 1 and -5.

$\dfrac{3}{2}t = -6$ $\qquad\qquad$ $\dfrac{3}{2}t = -8$ Subtract 7.

$\dfrac{2}{3} \cdot \dfrac{3}{2}t = \dfrac{2}{3} \cdot (-6)$ $\qquad$ $\dfrac{2}{3} \cdot \dfrac{3}{2}t = \dfrac{2}{3} \cdot (-8)$ Multiply by $\dfrac{2}{3}$.

$t = -4$ or $t = -\dfrac{16}{3}$ Solve.

W Hint
Isolate the absolute value on one side of the equation before "splitting" the equation.

The check is left to the student. The solution set is $\left\{-\dfrac{16}{3}, -4\right\}$.

[YOU TRY 2] Solve each equation.

a) $|c - 4| = 3$ b) $\left|\dfrac{1}{4}n - 3\right| + 2 = 5$

Be sure to think about the value of k when solving $|ax + b| = k$.

EXAMPLE 3

Solve $|4y - 11| = -9$.

Solution

This equation says that the absolute value of the quantity $4y - 11$ equals *negative* 9. Can an absolute value be negative? No! This equation has *no solution*.

The solution set is $\emptyset$.

[YOU TRY 3] Solve $|d + 3| = -5$.

3 Solve an Equation of the Form $|ax + b| = |cx + d|$

Another type of absolute value equation involves two absolute values.

> **Procedure** Solve $|ax + b| = |cx + d|$
>
> If P and Q are expressions, then to solve $|P| = |Q|$ we rewrite the absolute value equation as the *compound equation*.
>
> $$P = Q \quad \text{or} \quad P = -Q$$
>
> and solve for the variable.

EXAMPLE 4

Solve $|2w - 3| = |w + 9|$.

Solution

This equation is true when the quantities inside the absolute values are the *same* or when they are *negatives* of each other.

$$|2w - 3| = |w + 9|$$

The quantities are the same   or   the quantities are negatives of each other.

$$2w - 3 = w + 9 \qquad\qquad 2w - 3 = -(w + 9)$$
$$w = 12 \qquad\qquad\qquad 2w - 3 = -w - 9$$
$$\qquad\qquad\qquad\qquad\qquad 3w = -6$$
$$\qquad\qquad\qquad\qquad\qquad w = -2$$

W Hint

Notice that the absolute value equation is split into two separate equations, and consider why the distributive property is necessary.

Check: $w = 12$: $|2(12) - 3| \stackrel{?}{=} |12 + 9|$ $w = -2$: $|2(-2) - 3| \stackrel{?}{=} |-2 + 9|$
$\qquad\qquad\qquad\qquad |24 - 3| \stackrel{?}{=} |21|$ $\qquad\qquad\qquad |-4 - 3| \stackrel{?}{=} |7|$
$\qquad\qquad\qquad\qquad |24| = 21$ ✓ $\qquad\qquad\qquad\qquad |-7| = 7$ ✓

The solution set is $\{-2, 12\}$.

BE CAREFUL In Example 4 and other examples like it, you *must* put parentheses around the expression with the negative as in $-(w + 9)$.

[YOU TRY 4] Solve $|c + 7| = |3c - 1|$.

Next, we will learn how to solve **absolute value inequalities.** Some examples of absolute value inequalities are

$$|t| < 6, \quad |n + 2| \leq 5, \quad |3k - 1| > 11, \quad \left|5 - \frac{1}{2}y\right| \geq 3$$

4 Solve Absolute Value Inequalities Containing $<$ or $\leq$

What does it mean to solve $|x| \leq 3$? It means to find the set of all real numbers whose distance from zero is 3 *units or less*.

3 is 3 units from 0.
−3 is 3 units from 0.

Any number *between* 3 and −3 is less than 3 units from zero. For example, if $x = 1$, $|1| \leq 3$. If $x = -2$, $|-2| \leq 3$. We can represent the solution set on a number line as

We can write the solution set in interval notation as $[-3, 3]$.

W Hint
Notice that a *three-part inequality* is used to solve absolute value inequalities of this form.

Procedure Solve $|P| \leq k$ or $|P| < k$

Let P be an expression and let k be a positive real number. To solve $|P| \leq k$, solve the three-part inequality $-k \leq P \leq k$. ($<$ may be substituted for $\leq$.)

EXAMPLE 5 Solve $|t| < 6$. Graph the solution set, and write the answer in interval notation.

Solution

We must find the set of all real numbers whose distance from zero is less than 6. We can do this by solving the three-part inequality $-6 < t < 6$.

We can represent this on a number line as

We can write the solution set in interval notation as $(-6, 6)$. Any number between −6 and 6 will satisfy the inequality.

[YOU TRY 5] Solve. Graph the solution set, and write the answer in interval notation.
$$|u| < 9$$

EXAMPLE 6

Solve each inequality. Graph the solution set, and write the answer in interval notation.

a) $|n + 2| \leq 5$ b) $|4 - 5p| < 16$

Solution

a) We must find the set of all real numbers, n, so that $n + 2$ is less than or equal to 5 units from zero. To solve $|n + 2| \leq 5$, we must solve the three-part inequality

$$-5 \leq n + 2 \leq 5$$
$$-7 \leq n \leq 3 \qquad \text{Subtract 2.}$$

The number line representation is

In interval notation, we write the solution as $[-7, 3]$. Any number between -7 and 3 (and including those endpoints) will satisfy the inequality.

b) Solve the three-part inequality.

$$-16 < 4 - 5p < 16$$
$$-20 < -5p < 12 \qquad \text{Subtract 4.}$$
$$4 > p > -\frac{12}{5} \qquad \text{Divide by } -5, \text{ and change the direction of the inequality symbols.}$$

This inequality means p is less than 4 and greater than $-\frac{12}{5}$. We can rewrite it as $-\frac{12}{5} < p < 4$.

The number line representation of the solution set is

In interval notation, we write $\left(-\frac{12}{5}, 4\right)$.

[YOU TRY 6]

Solve. Graph the solution set, and write the answer in interval notation.

$$|6k + 5| \leq 13$$

5 Solve Absolute Value Inequalities Containing > or ≥

To solve $|x| \geq 4$ means to find the set of all real numbers whose distance from zero is 4 *units or more.*

4 is 4 units from 0.
-4 is 4 units from 0.

Any number greater than 4 *or* less than -4 is more than 4 units from zero.

For example, if $x = 6$, $|6| \geq 4$. If $x = -5$, then $|-5| = 5$ and $5 \geq 4$. We can represent the solution set to $|x| \geq 4$ as

122   CHAPTER 3   Linear Inequalities and Absolute Value

The solution set consists of two separate regions, so we can write a compound inequality using *or*.

$$x \leq -4 \quad \text{or} \quad x \geq 4$$

In interval notation, we write $(-\infty, -4] \cup [4, \infty)$.

W Hint
Notice that an "*or*" compound inequality is used to solve absolute value inequalities of this form.

Procedure Solve $|P| \geq k$ or $|P| > k$

Let P be an expression and let k be a positive, real number. To solve $|P| \geq k$ ($>$ may be substituted for $\geq$), solve the compound inequality $P \geq k$ or $P \leq -k$.

EXAMPLE 7   Solve $|r| > 2$. Graph the solution set, and write the answer in interval notation.

Solution

We must find the set of all real numbers whose distance from zero is greater than 2. The solution is the compound inequality $r > 2$ or $r < -2$.

On the number line, we can represent the solution set as

In interval notation, we write $(-\infty, -2) \cup (2, \infty)$. Any number in the shaded region will satisfy the inequality. For example, to the right of 2, if $r = 3$, then $|3| > 2$. To the left of -2, if $r = -4$, then $|-4| > 2$.

[YOU TRY 7]   Solve $|d| \geq 5$. Graph the solution set, and write the answer in interval notation.

EXAMPLE 8   Solve each inequality. Graph the solution set, and write the answer in interval notation.

a) $|3k - 1| > 11$ b) $|c + 6| + 10 \geq 12$

Solution

a) To solve $|3k - 1| > 11$ means to find the set of all real numbers, k, so that $3k - 1$ is more than 11 units from zero on the number line. We will solve the compound inequality.

$$3k - 1 > 11 \quad \text{or} \quad 3k - 1 < -11$$

↑ $3k - 1$ is more than 11 units away from zero to the *right* of zero.   ↑ $3k - 1$ is more than 11 units away from zero to the *left* of zero.

$$\begin{array}{rcl} 3k - 1 > 11 & \text{or} & 3k - 1 < -11 \\ 3k > 12 & \text{or} & 3k < -10 \quad \text{Add 1.} \\ k > 4 & \text{or} & k < -\dfrac{10}{3} \quad \text{Divide by 3.} \end{array}$$

On the number line, we get

From the number line, we can write the interval notation $\left(-\infty, -\dfrac{10}{3}\right) \cup (4, \infty)$.

Any number in the shaded region will satisfy the inequality.

b) Begin by getting the absolute value on a side by itself.

$$|c + 6| + 10 \geq 12$$
$$|c + 6| \geq 2 \quad \text{Subtract 10.}$$
$$c + 6 \geq 2 \quad \text{or} \quad c + 6 \leq -2 \quad \text{Rewrite as a compound inequality.}$$
$$c \geq -4 \quad \text{or} \quad c \leq -8 \quad \text{Subtract 6.}$$

The graph of the solution set is

The interval notation is $(-\infty, -8] \cup [-4, \infty)$.

> **W Hint**
> Notice that we first isolate the absolute value on one side of the inequality before we write the "*or*" compound inequality.

[**YOU TRY 8**] Solve each inequality. Graph the solution set and write the answer in interval notation.

a) $|8q + 9| \geq 7$ b) $|k + 8| - 5 \geq 9$

Example 9 illustrates why it is important to understand what the absolute value inequality means before trying to solve it.

6 Solve Special Cases of Absolute Value Inequalities

EXAMPLE 9 Solve each inequality.

a) $|z + 3| < -6$ b) $|2s - 1| \geq 0$ c) $|4d + 7| + 9 \leq 9$

Solution

a) Look carefully at this inequality, $|z + 3| < -6$. It says that the absolute value of a quantity, $z + 3$, is *less than* a negative number. Since the absolute value of a quantity is always zero or positive, this inequality has *no solution*.

The solution set is $\emptyset$.

b) $|2s - 1| \geq 0$ says that the absolute value of a quantity, $2s - 1$, is greater than or equal to zero. An absolute value is *always* greater than or equal to zero, so *any* value of s will make the inequality true.

The solution set consists of all real numbers, which we can write in interval notation as $(-\infty, \infty)$.

> **W Hint**
> Do you see why it is important to carefully look at the constant once the absolute value expression has been isolated?

c) Begin by isolating the absolute value. $|4d + 7| + 9 \leq 9$
$$|4d + 7| \leq 0 \quad \text{Subtract 9.}$$

The absolute value of a quantity can *never be less than zero* but it *can equal zero*. To solve this, we must solve $4d + 7 = 0$.

$$4d + 7 = 0$$
$$4d = -7$$
$$d = -\frac{7}{4}$$

The solution set is $\left\{-\frac{7}{4}\right\}$.

[YOU TRY 9] Solve each inequality.

a) $|p + 4| \geq 0$ b) $|5n - 7| < -2$ c) $|6y - 1| + 3 \leq 3$

7 Solve an Applied Problem Using an Absolute Value Inequality

EXAMPLE 10

On an assembly line, a machine is supposed to fill a can with 19 oz of soup. However, the possibility for error is ± 0.25 oz. Let x represent the range of values for the amount of soup in the can. Write an absolute value inequality to represent the range for the number of ounces of soup in the can, then solve the inequality and explain the meaning of the answer.

Solution

If the *actual* amount of soup in the can is x and there is supposed to be 19 oz in the can, then the error in the amount of soup in the can is $|x - 19|$. If the possible error is ± 0.25 oz, then we can write the inequality

$$|x - 19| \leq 0.25$$
$$-0.25 \leq x - 19 \leq 0.25 \quad \text{Solve.}$$
$$18.75 \leq x \leq 19.25$$

The actual amount of soup in the can is between 18.75 and 19.25 oz.

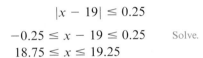

Using Technology

We can use a graphing calculator to solve an equation by entering one side of the equation as Y_1 and the other side as Y_2. Then graph the equations. Remember that absolute value equations like the ones found in this section can have 0, 1, or 2 solutions. *The x-coordinates of their points of intersection are the solutions to the equation.*

We will solve $|3x - 1| = 5$ algebraically and by using a graphing calculator, and then compare the results.

First, use algebra to solve $|3x - 1| = 5$. You should get $\left\{-\dfrac{4}{3}, 2\right\}$.

Next, use a graphing calculator to solve $|3x - 1| = 5$.
We will enter $|3x - 1|$ as Y_1 and 5 as Y_2. To enter $Y_1 = |3x - 1|$,

1) Press the $\boxed{Y=}$ key, so that the cursor is to the right of $\backslash Y_1=$.

2) Press $\boxed{\text{MATH}}$ and then press the right arrow, to highlight **NUM**. Also highlighted is 1:abs (which stands for *absolute value*).

3) Press $\boxed{\text{ENTER}}$ and you are now back on the $\backslash Y_1 =$ screen. Enter $3x - 1$ with a closing parentheses so that you have now entered $Y_1 = \text{abs}(3x - 1)$.

4) Press the down arrow to enter $\backslash Y_2 = 5$.

5) Press GRAPH.

 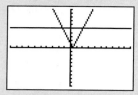

The graphs intersect at two points because there are two solutions to this equation. *Remember that the solutions to the equation are the x-coordinates of the points of intersection.*

To find these x-coordinates we will use the INTERSECT feature.

To find the left-hand intersection point, press 2nd TRACE and select 5:intersect. Press ENTER. Move the cursor close to the point on the left and press ENTER three times. You get the result in the screen below on the left.

To find the right-hand intersect point, press 2nd TRACE, select 5:intersect, and press ENTER. Move the cursor close to the point, and press ENTER three times. You will see the screen that is below on the right.

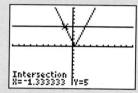 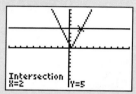

The screen on the left shows $x = -1.333333$. This is the calculator's approximation of $x = -1.\overline{3}$, the decimal equivalent of $x = -\frac{4}{3}$, one of the solutions found using algebra.

The screen on the right shows $x = 2$ as a solution, the same solution we obtained algebraically.

The calculator gives us a solution set of $\{-1.333333, 2\}$, while the solution set found using algebra is $\left\{-\frac{4}{3}, 2\right\}$.

Solve each equation algebraically, then verify your answer using a graphing calculator.

1) $|x - 1| = 2$ 2) $|x + 4| = 6$ 3) $|2x + 3| = 3$

4) $|4x - 5| = 1$ 5) $|3x + 7| - 6 = -8$ 6) $|6 - x| + 3 = 3$

ANSWERS TO [YOU TRY] EXERCISES

1) $\{-8, 8\}$ 2) a) $\{1, 7\}$ b) $\{0, 24\}$ 3) $\emptyset$ 4) $\left\{-\frac{3}{2}, 4\right\}$

5) $(-9, 9)$

6) $\left[-3, \frac{4}{3}\right]$

7) $(-\infty, -5] \cup [5, \infty)$

8) a) $(-\infty, -2] \cup \left[-\frac{1}{4}, \infty\right)$

b) $(-\infty, -22] \cup [6, \infty)$

9) a) $(-\infty, \infty)$ b) $\emptyset$ c) $\left\{\frac{1}{6}\right\}$

ANSWERS TO TECHNOLOGY EXERCISES

1) $\{-1, 3\}$ 2) $\{-10, 2\}$ 3) $\{-3, 0\}$ 4) $\{1, 1.5\}$ 5) $\emptyset$ 6) $\{6\}$

E Evaluate 3.3 Exercises

Do the exercises, and check your work.

Objective 1: Understand the Meaning of an Absolute Value Equation

1) In your own words, explain the *meaning* of the absolute value of a number.

2) Does $|x| = -8$ have a solution? Why or why not?

Objective 2: Solve an Equation of the Form $|ax + b| = k$
Solve.

3) $|q| = 6$
4) $|z| = 7$
5) $|q - 5| = 3$
6) $|a + 2| = 13$
7) $|4t - 5| = 7$
8) $|9x - 8| = 10$
9) $|12c + 5| = 1$
10) $|4 - 5k| = 11$
11) $\left|\dfrac{2}{3}b + 3\right| = 13$
12) $\left|\dfrac{3}{4}h + 8\right| = 7$
13) $\left|4 - \dfrac{3}{5}d\right| = 6$
14) $\left|\dfrac{3}{2}r + 5\right| = \dfrac{3}{4}$
15) $|m - 5| = -3$
16) $|2k + 7| = -15$
17) $|z - 6| + 4 = 20$
18) $|q + 3| - 1 = 14$
19) $|2a + 5| + 8 = 13$
20) $|6t - 11| + 5 = 10$
21) $|w + 14| = 0$
22) $|5h + 7| = -5$
23) $|8n + 11| = -1$
24) $|4p - 3| = 0$
25) $|5b + 3| + 6 = 19$
26) $1 = |7 - 8x| - 4$
27) $\left|\dfrac{5}{4}k + 2\right| + 9 = 7$
28) $|3m - 1| + 5 = 2$

Objective 3: Solve an Equation of the Form $|ax + b| = |cx + d|$
Solve the following equations containing two absolute values.

29) $|s + 9| = |2s + 5|$
30) $|j - 8| = |4j - 7|$
31) $|3z + 2| = |6 - 5z|$
32) $|1 - 2a| = |10a + 3|$
33) $\left|\dfrac{3}{2}x - 1\right| = |x|$
34) $|y| = \left|\dfrac{4}{7}y + 12\right|$
35) $\left|\dfrac{1}{4}t - \dfrac{5}{2}\right| = \left|5 - \dfrac{1}{2}t\right|$
36) $\left|k + \dfrac{1}{6}\right| = \left|\dfrac{2}{3}k + \dfrac{1}{2}\right|$

37) Write an absolute value equation that means x is 9 units from zero.

38) Write an absolute value equation that means y is 6 units from zero.

39) Write an absolute value equation that has a solution set of $\left\{-\dfrac{1}{2}, \dfrac{1}{2}\right\}$.

40) Write an absolute value equation that has a solution set of $\{-1.4, 1.4\}$.

Mixed Exercises: Objectives 4 and 5
Graph each inequality on a number line, and represent the sets of numbers using interval notation.

41) $-1 \leq p \leq 5$
42) $7 < t < 11$
43) $y < 2$ or $y > 9$
44) $a \leq -8$ or $a \geq \dfrac{1}{2}$
45) $n \leq -\dfrac{9}{2}$ or $n \geq \dfrac{3}{5}$
46) $-\dfrac{1}{4} \leq q \leq \dfrac{11}{4}$

Mixed Exercises: Objectives 4 and 6
Solve each inequality. Graph the solution set, and write the answer in interval notation.

47) $|m| \leq 7$
48) $|c| < 1$
49) $|3k| < 12$
50) $\left|\dfrac{5}{4}z\right| \leq 30$
51) $|w - 2| < 4$
52) $|k - 6| \leq 2$
53) $|3r + 10| \leq 4$
54) $|4a + 1| \leq 12$

55) $|7 - 6p| \leq 3$
56) $|17 - 9d| < 8$
57) $|5q + 11| < 0$
58) $|6t + 16| < 0$
59) $|8m - 15| \leq -5$
60) $|2x + 7| \leq -12$
61) $|2v + 5| + 3 < 14$
62) $|8c - 3| + 15 < 20$
63) $\left|\dfrac{3}{2}h + 6\right| - 2 \leq 10$
64) $7 + \left|\dfrac{8}{3}u - 9\right| < 12$

98) $|5b - 11| - 18 < -10$
99) $-\dfrac{3}{5} \geq \dfrac{5}{2}a - \dfrac{1}{2}$
100) $4 + 3(2r - 5) > 9 - 4r$
101) $|6k + 17| > -4$ 102) $|5 - w| \geq 3$
103) $5 \geq |c + 8| - 2$ 104) $0 \leq |4a + 1|$
105) $|5h - 8| > 7$
106) $\left|\dfrac{2}{3}y - 1\right| = \left|\dfrac{3}{2}y + 4\right|$

Mixed Exercises: Objectives 5 and 6
Solve each inequality. Graph the solution set, and write the answer in interval notation.

65) $|t| \geq 7$
66) $|p| > 3$
67) $|d + 10| \geq 4$
68) $|q - 7| > 12$
69) $|4v - 3| \geq 9$
70) $|6a + 19| > 11$
71) $|17 - 6x| > 5$
72) $|1 - 4g| \geq 10$
73) $|8k + 5| \geq 0$
74) $|5b - 6| \geq 0$
75) $|z - 3| \geq -5$
76) $|3r + 10| > -11$
77) $|w + 6| - 4 \geq 2$
78) $|2m - 1| + 4 > 5$
79) $-3 + \left|\dfrac{5}{6}n + \dfrac{1}{2}\right| \geq 1$
80) $\left|\dfrac{3}{2}y - \dfrac{5}{4}\right| + 9 \geq 11$

81) Explain why $|3t - 7| < 0$ has no solution.
82) Explain why $|4l + 9| \leq -10$ has no solution.
83) Explain why the solution to $|2x + 1| \geq -3$ is $(-\infty, \infty)$.
84) Explain why the solution to $|7y - 3| \geq 0$ is $(-\infty, \infty)$.

Mixed Exercises: Objectives 1–6
The following exercises contain absolute value equations, linear inequalities, and both types of absolute value inequalities. Solve each. Write the solution set for equations in set notation, and use interval notation for inequalities.

85) $|2v + 9| > 3$
86) $\left|\dfrac{5}{3}a + 2\right| = 8$
87) $3 = |4t + 5|$
88) $|4k + 9| \leq 5$
89) $9 \leq |7 - 8q|$
90) $|2p - 5| - 12 = 11$
91) $2(x - 8) + 10 < 4x$
92) $\dfrac{1}{2}n + 11 < 8$
93) $|6y + 5| \leq -9$
94) $8 \leq |5v + 2|$
95) $\left|\dfrac{4}{3}x + 1\right| = \left|\dfrac{5}{3}x + 8\right|$
96) $|7z - 8| \leq 0$
97) $|4 - 9t| + 2 = 1$

Objective 7: Solve an Applied Problem Using an Absolute Value Inequality

107) A gallon of milk should contain 128 oz. The possible error in this measurement, however, is ±0.75 oz. Let a represent the range of values for the amount of milk in the container. Write an absolute value inequality to represent the range for the number of ounces of milk in the container, then solve the inequality and explain the meaning of the answer.

108) Dawn buys a 27-oz box of cereal. The possible error in this amount, however, is ±0.5 oz. Let c represent the range of values for the amount of cereal in the box. Write an absolute value inequality to represent the range for the number of ounces of cereal in the box, then solve the inequality and explain the meaning of the answer.

109) Emmanuel spent $38 on a birthday gift for his son. He plans on spending within $5 of that amount on his daughter's birthday gift. Let b represent the range of values for the amount he will spend on his daughter's gift. Write an absolute value inequality to represent the range for the amount of money Emmanuel will spend on his daughter's birthday

gift, then solve the inequality and explain the meaning of the answer.

110) An employee at a home-improvement store is cutting a window shade for a customer. The customer wants the shade to be 32 in. wide. If the machine's possible error in cutting the shade is $\pm\frac{1}{16}$ in., write an absolute value inequality to represent the range for the width of the window shade, and solve the inequality. Explain the meaning of the answer. Let w represent the range of values for the width of the shade.

R Rethink

R1) Which objective is the most difficult for you?

R2) Which type of absolute value inequality requires the use of a three-part inequality to obtain the solution?

R3) Which type of absolute value inequality requires the use of a compound inequality to reach the solution?

Group Activity — The Group Activity can be found online on Connect.

emPOWERme — Identify the Black Holes of Time Management

"Where did my time go?" Do you ever ask yourself this question? Many of us feel that our weekend off from work or the evening set aside for study simply vanishes. One factor that contributes to this feeling is what are called "time black holes"—unexpected or unstructured activities that simply swallow up our precious hours. To identify the black holes in your schedule, complete the following exercise.

The first 19 items on this list are common problems that prevent us from getting things done. Check off the ones that are problems for you, and indicate whether you have control over them (controllable problems) or they are out of your control (uncontrollable problems).

	Big Problem for Me	Often a Problem	Seldom a Problem	Controllable (C) or Uncontrollable (U)?
1. Telephone interruptions				
2. Drop-in visitors				
3. Email interruptions				
4. Hobbies				
5. Mistakes				

(continued)

	Big Problem for Me	Often a Problem	Seldom a Problem	Controllable (C) or Uncontrollable (U)?
6. Inability to say "no"				
7. Socializing				
8. Snacking				
9. Errands and shopping				
10. Meals				
11. Children's interruptions				
12. Perfectionism				
13. Family appointments				
14. Looking for lost items				
15. Redoing mistakes				
16. Jumping from task to task				
17. Surfing the World Wide Web				
18. Reading newspapers, magazines, recreational books				
19. Car trouble				
20. Other				
21. Other				
22. Other				
23. Other				
24. Other				
25. Other				

Consider the following questions:

- Do your time management problems fall into any patterns?
- Are there problems that at first seem uncontrollable that can actually be controlled?
- What strategies might you use to deal with these problems?

Adapted, in part, from Ferner, J. D. (1980). *Successful Time Management*. NY: Wiley, p. 33.

Chapter 3: Summary

Definition/Procedure	Example

3.1 Linear Inequalities in One Variable

Solving Linear Inequalities in One Variable

We solve linear inequalities in very much the same way we solve linear equations *except when we multiply or divide by a negative number we must reverse the direction of the inequality symbol.*

We can graph the solution set, write the solution in set notation, or write the solution in interval notation. **(p. 101)**

Solve $x - 5 \leq -3$. Graph the solution set, and write the answer in both set notation and interval notation.

$$x - 5 \leq -3$$
$$x - 5 + 5 \leq -3 + 5 \quad \text{Add 5.}$$
$$x \leq 2$$

⟵+—+—+—+—+—⬤—+⟶
$-3\ -2\ -1\ \ 0\ \ 1\ \ 2\ \ 3$

Solution: $\{x | x \leq 2\}$ Set notation
Solution: $(-\infty, 2]$ Interval notation

A **compound inequality** contains more than one inequality symbol. **(p. 103)**

Some examples are
$$-8 \leq 3x + 1 \leq 5, \quad c < 1 \text{ and } c > 7,$$
$$4p \leq -6 \text{ or } p - 2 > 3$$

The compound inequality $-8 \leq 3x + 1 \leq 5$ is also called a **three-part inequality.**

3.2 Compound Inequalities in One Variable

The solution set of a compound inequality joined by **"and"** will be the **intersection** of the solution sets of the individual inequalities. **(p. 110)**

Solve the compound inequality $5x - 2 \geq -17$ and $x + 8 \leq 9$.

$$5x - 2 \geq -17 \quad \text{and} \quad x + 8 \leq 9$$
$$5x \geq -15$$
$$x \geq -3 \quad \text{and} \quad x \leq 1$$

⟵+—⬤—+—+—+—+—⬤—+—+⟶
$-4\ -3\ -2\ -1\ \ 0\ \ 1\ \ 2\ \ 3\ \ 4$

Solution in interval notation: $[-3, 1]$

The solution set of a compound inequality joined by **"or"** will be the **union** of the solution sets of the individual inequalities. **(p. 111)**

Solve the compound inequality $x - 3 < -1$ or $7x > 42$.

$$x - 3 < -1 \quad \text{or} \quad 7x > 42$$
$$x < 2 \quad \text{or} \quad x > 6$$

⟵+—+—○—+—+—+—+—○—+—+⟶
$0\ \ 1\ \ 2\ \ 3\ \ 4\ \ 5\ \ 6\ \ 7\ \ 8\ \ 9$

Solution in interval notation: $(-\infty, 2) \cup (6, \infty)$

3.3 Absolute Value Equations and Inequalities

Absolute Value Equations

If P represents an expression and k is a positive, real number, then to solve $|P| = k$ we rewrite the absolute value equation as the *compound equation* $P = k$ or $P = -k$ and solve for the variable. **(p. 118)**

Solve $|4a + 10| = 18$.

$$|4a + 10| = 18$$

$$4a + 10 = 18 \quad \text{or} \quad 4a + 10 = -18$$
$$4a = 8 \qquad\qquad 4a = -28$$
$$a = 2 \quad \text{or} \qquad a = -7$$

Check the solutions in the original equation.
The solution set is $\{-7, 2\}$.

Inequalities Containing $<$ or $\leq$

Let P be an expression and let k be a positive, real number. To solve $|P| \leq k$, solve the three-part inequality
$$-k \leq P \leq k$$
($<$ may be substituted for $\leq$.) **(p. 121)**

Solve $|x - 3| \leq 2$. Graph the solution set, and write the answer in interval notation.

$$-2 \leq x - 3 \leq 2$$
$$1 \leq x \leq 5 \quad \text{Add 3.}$$

⟵+—+—+—⬤—+—+—+—⬤—+—+⟶
$-2\ -1\ \ 0\ \ 1\ \ 2\ \ 3\ \ 4\ \ 5\ \ 6\ \ 7$

In interval notation, we write $[1, 5]$.

Definition/Procedure	Example
Inequalities Containing $>$ or $\geq$ Let P be an expression and let k be a positive, real number. To solve $\lvert P \rvert \geq k$, solve the compound inequality $\qquad P \geq k$ or $P \leq -k$ **(p. 122)** ($>$ may be substituted for $\geq$)	Solve $\lvert 2n - 5 \rvert > 1$. Graph the solution set, and write the answer in interval notation. $2n - 5 > 1$ or $2n - 5 < -1$ Solve. $2n > 6$ or $\quad\;\; 2n < 4$ Add 5. $\;\; n > 3$ or $\quad\;\;\; n < 2$ Divide by 2. $\begin{array}{c}\leftarrow\!\!+\!\!+\!\!+\!\!+\!\!+\!\!\circ\!\!+\!\!\circ\!\!+\!\!+\!\!\rightarrow\\ -3\,-2\,-1\;\;0\;\;1\;\;2\;\;3\;\;4\;\;5\;\;6\end{array}$ In interval notation, we write $(\infty, 2) \cup (3, \infty)$.

Chapter 3: Review Exercises

(3.1) Solve each inequality. Graph the solution set, and write the answer in interval notation.

1) $z + 6 \geq 14$
2) $-10y + 7 > 32$
3) $w + 8 > 5$
4) $-6k \leq 15$
5) $5x - 2 \leq 18$
6) $0.03c + 0.09(6 - c) > 0.06(6)$
7) $-15 < 4p - 7 \leq 5$
8) $-1 \leq \dfrac{5 - 3x}{2} \leq 0$
9) $3(3c + 8) - 7 > 2(7c + 1) - 5$
10) $-19 \leq 7p + 9 \leq 2$
11) $-3 < \dfrac{3}{4}a - 6 \leq 0$
12) $\dfrac{1}{2} < \dfrac{1 - 4t}{6} < \dfrac{3}{2}$

13) **Write an inequality and solve.** Gia's scores on her first three History tests were 94, 88, and 91. What does she need to make on her fourth test to have an average of at least 90?

(3.2)

14) $A = \{10, 20, 30, 40, 50\}$ $B = \{20, 25, 30, 35\}$
   a) Find $A \cup B$.   b) Find $A \cap B$.

Solve each compound inequality. Graph the solution set, and write the answer in interval notation.

15) $a + 6 \leq 9$ and $7a - 2 \geq 5$
16) $3r - 1 > 5$ or $-2r \geq 8$
17) $8 - y < 9$ or $\dfrac{1}{10}y > \dfrac{3}{5}$
18) $x + 12 \leq 9$ and $0.2x \geq 3$

The following table lists the number of hybrid vehicles sold in the United States by certain manufacturers in June and July of 2008. (www.hybridcars.com)

Manufacturer	Number Sold in June	Number Sold in July
Toyota	16,330	18,801
Honda	2,717	3,443
Ford	1,910	1,265
Lexus	1,476	1,562
Nissan	1,333	715

List the elements of the set that satisfy the given information.

19) The set of manufacturers who sold more than 3000 hybrid vehicles in June and July

20) The set of manufacturers who sold more than 5000 hybrid vehicles in June or fewer than 1500 hybrids in July

(3.3) Solve.

21) $\lvert m \rvert = 9$
22) $\left\lvert \dfrac{1}{2}c \right\rvert = 5$
23) $\lvert 7t + 3 \rvert = 4$
24) $\lvert 4 - 3y \rvert = 12$
25) $\lvert 8p + 11 \rvert - 7 = -3$
26) $\lvert 5k + 3 \rvert - 8 = 4$
27) $\left\lvert 4 - \dfrac{5}{3}x \right\rvert = \dfrac{1}{3}$
28) $\left\lvert \dfrac{2}{3}w + 6 \right\rvert = \dfrac{5}{2}$
29) $\lvert 7r - 6 \rvert = \lvert 8r + 2 \rvert$
30) $\lvert 3z - 4 \rvert = \lvert 5z - 6 \rvert$

31) $|2a - 5| = -10$
32) $|h + 6| - 12 = -20$
33) $|9d + 4| = 0$
34) $|6q - 7| = 0$

35) Write an absolute value equation which means *a is 4 units from zero*.

36) Write an absolute value equation which means *t is 7 units from zero*.

Solve each inequality. Graph the solution set, and write the answer in interval notation.

37) $|c| \le 3$
38) $|w + 1| < 11$
39) $|4t| > 8$
40) $|2v - 7| \ge 15$
41) $|12r + 5| \ge 7$
42) $|3k - 11| < 4$
43) $|4 - a| < 9$
44) $|2 - 5q| > 6$

45) $|4c + 9| - 8 \le -2$
46) $|3m + 5| + 2 \ge 7$
47) $|5y + 12| - 15 \ge -8$
48) $3 + |z - 6| \le 13$
49) $|k + 5| > -3$
50) $|4q - 9| < 0$
51) $|12s + 1| \le 0$

52) A radar gun indicated that a pitcher threw a 93 mph fastball. The radar gun's possible error in measuring the speed of a pitch is ± 1 mph. Write an absolute value inequality to represent the range for the speed of the pitch, and solve the inequality. Explain the meaning of the answer. Let *s* represent the range of values for the speed of the pitch.

Chapter 3: Test

Solve. Graph the solution set, and write the answer in interval notation.

1) $r + 7 \le 2$

2) $6m + 19 \le 7$

3) $9 - 3(2x - 1) < 4x + 5(x + 2) - 8$

4) $1 - 2(3x - 5) \le 2x + 5$

5) $-1 < \dfrac{w - 5}{4} \le \dfrac{1}{2}$

6) $3 < 3 - 2c < 9$

7) *Write an inequality and solve:* Rawlings Builders will rent a forklift for $46.00 per day plus $9.00 per hour. If they have at most $100.00 allotted for a one-day rental, for how long can they keep the forklift and remain within budget?

8) Given sets $A = \{1, 2, 3, 6, 12\}$ and $B = \{1, 2, 9, 12\}$, find each of the following.
   a) $A \cup B$
   b) $A \cap B$

Solve each compound inequality. Write the answer in interval notation.

9) $3n + 5 > 12$ or $\dfrac{1}{4}n < -2$

10) $y - 8 \le -5$ and $2y \ge 0$

11) $6 - p < 10$ or $p - 7 < 2$

Solve.

12) $|4y - 9| = 11$

13) $|d + 6| - 3 = 7$

14) $|3k + 5| = |k - 11|$

15) $\left|\dfrac{1}{2}n - 1\right| = -8$

16) Write an absolute value equation that means *x is 8 units from zero*.

Solve each inequality. Graph the solution set, and write the answer in interval notation.

17) $|2z - 7| \le 9$

18) $|4m + 9| - 8 \ge 5$

19) $|10 - 3w| < -2$

20) A scale in a doctor's office has a possible error of ± 0.75 lb. If Thanh's weight is measured as 168 lb, write an absolute value inequality to represent the range for his weight, and solve the inequality. Let *w* represent the range of values for Thanh's weight. Explain the meaning of the answer.

Chapter 3: Cumulative Review for Chapters 1–3

Perform the operations and simplify.

1) $\dfrac{3}{8} - \dfrac{5}{6}$

2) $\dfrac{5}{8} \cdot 12$

3) $26 - 14 \div 2 + 5 \cdot 7$

4) $-82 + 15 + 10(1 - 3)$

5) $-39 - |7 - 15|$

6) Find the area of a triangle with a base of length 9 cm and height of 6 cm.

Given the set of numbers $\left\{\dfrac{3}{4}, -5, \sqrt{11}, 2.5, 0, 9, 0.\overline{4}\right\}$ **identify**

7) the integers.

8) the rational numbers.

9) the whole numbers.

10) Which property is illustrated by $6(5 + 2) = 6 \cdot 5 + 6 \cdot 2$?

11) Does the commutative property apply to the subtraction of real numbers? Explain.

12) Combine like terms.
$11y^2 - 14y + 6y^2 + y - 5y$

Solve

13) $8t - 17 = 10t + 6$

14) $\dfrac{3}{2}n + 14 = 20$

15) $3(7w - 5) - w = -7 + 4(5w - 2)$

16) $\dfrac{x + 3}{10} = \dfrac{2x - 1}{4}$

17) $-\dfrac{1}{2}c + \dfrac{1}{5}(2c - 3) = \dfrac{3}{10}(2c + 1) - \dfrac{3}{4}c$

Solve using the five-step method.

18) Stu and Phil were racing from Las Vegas back to Napa Valley. Stu can travel 140 miles by train in the time it takes Phil to travel 120 miles by car. What are the speeds of the train and the car if the train is traveling 10 mph faster than the car?

Solve each inequality. Write the answer in interval notation.

19) $7 - k > 1$

20) $-10w + 3(2w - 5) \leq 7 + 2(w - 9)$

21) $-14 < 6y + 10 < 3$

22) $8x \leq -24$ or $4x - 5 \geq 6$

23) $\left|\dfrac{1}{2}m + 7\right| \leq 11$

24) $|3 - r| > 4$

25) Write an inequality and solve.

Anton has grades of 87 and 76 on his first two Biology tests. What does he need on the third test to keep an average of at least 80?

Linear Equations in Two Variables and Functions

CHAPTER 4

OUTLINE

Study Strategies: Taking Notes in Class

4.1 Introduction to Linear Equations in Two Variables

4.2 Slope of a Line and Slope-Intercept Form

4.3 Writing an Equation of a Line

4.4 Linear and Compound Linear Inequalities in Two Variables

4.5 Introduction to Functions

Group Activity

emPOWERme: Checklist for Effective Notes

Math at Work:
Surveyor

To most people, a neighborhood is a fairly straightforward collection of houses or other sorts of buildings, each constructed on its own clearly defined piece of land. Luis Aguilar, who works as surveyor for his city, knows that the reality is actually far more complex. "The boundaries between properties are usually difficult to determine," he explains. "Taking into account boundaries among shared spaces like roads and sidewalks only adds to the complexity." It is the job of surveyors like Luis to make all of the borders and boundaries in an area clear.

Surveyors measure distances, elevation, and other land features. These data are used to create maps, plan construction projects, and inform legal documents like deeds and leases. Usually, Luis works by finding a known reference point in a local area and uses this as the basis for calculations that tell him about all the land around it.

"A lot of my work is done in the field," Luis describes. "I'll visit a piece of land with a team of other surveyors, and we'll take careful, thorough notes, and then crunch those numbers back in the office."

In this chapter, we will discuss the topic of linear equations in two variables. We will also cover strategies you can use when taking notes in class.

 **Study Strategies** Taking Notes in Class

Many college students take at least three or four classes a term and may attend almost a dozen classes a week. In this context, it isn't realistic to try to remember everything that is said by an instructor in a particular class session. Hence, note-taking is essential to college success. In addition, taking notes helps process information, deepening your understanding. The strategies outlined below will help you take notes effectively, in this or in any class.

- Find a seat that lets you see and hear the instructor clearly.
- As you wait for the class to begin, review any assigned materials and your notes from the previous class.
- Turn off your cell phone!

- Plan to write your notes in a loose-leaf notepad, taking notes on *only one side of the page.*
- Always bring your textbook, paper, and something to write with to class. For a math class, you may need to bring a calculator, too.

- Listen actively. Think about what your instructor is saying, and ask questions about anything you don't understand.
- Focus on identifying and writing down key concepts, formulas, and insights.
- Keep in mind that instructors often write down key pieces of information on the board.

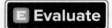
- Look over your notes as the class is finishing. Make sure you can read them and that they are thorough. (Use the emPOWERme checklist on page 214 to evaluate your notes in more detail.)
- If something was discussed that you didn't understand, plan to see your instructor during his or her office hours or talk about the concept with a fellow student or math tutor.

- Once the class is finished, *read over your notes.* This will help transfer the information into long-term memory.
- Try to put the day's lesson in the context of what you have been learning in the class throughout the term.

Chapter 4 POWER Plan

P Prepare | O Organize

What are your goals for Chapter 4?	How can you accomplish each goal?
1 Be prepared before and during class.	• Don't stay out late the night before, and be sure to set your alarm clock! • Bring a pencil, notebook paper, and textbook to class. • Avoid distractions by turning off your cell phone during class. • Pay attention, take good notes, and ask questions. • Complete your homework on time, and ask questions on problems you do not understand. • Plan ahead for tests by preparing many days in advance.
2 Understand the homework to the point where you could do it without needing any help or hints.	• Read the directions, and show all of your steps. • Go to the professor's office for help. • Rework homework and quiz problems, and find similar problems for practice. • Review old material that you have not mastered yet.
3 Use the P.O.W.E.R. framework to help you take notes in class: *Checklist for Effective Notes*	• Read the Study Strategy that explains how to become a better notetaker. • Be sure to personalize the strategy, and notice where you might be able to improve. • Complete the emPOWERme that appears before the Chapter Summary.
4 Write your own goal.	• _____

What are your objectives for Chapter 4?	How can you accomplish each objective?
1 Be able to graph linear equations, including vertical and horizontal lines.	• Be able to make a table of values that solves the linear equation. • Know how to plot points on the x- and y-axes. • Know what vertical and horizontal lines look like.
2 Be able to find the slope of a line, including slopes for vertical and horizontal lines. Be able to graph lines using slope and another point.	• Know the slope formula, given two points. • Know how to find the slope, given an equation. • Be able to use the slope to graph a line.
3 Be able to write an equation of a line, including vertical and horizontal lines. Be able to determine when lines are parallel or perpendicular, and be able to write the equation for a parallel or perpendicular line.	• Learn the **Point-Slope Formula**. • Know the different forms of a line, including slope-intercept, standard form, and vertical and horizontal lines. • Know when slopes are the same and when slopes are negative reciprocals.
4 Learn how to graph a linear inequality and find solutions to a compound linear inequality in two variables.	• Learn the procedure for **Graphing a Linear Inequality**. • Know the meaning of *union* and *intersection* when solving compound linear inequalities.

5 Know how to determine if a relation is a function. Be able to use function notation, evaluate function, and graph linear functions.	• Learn the definition of a function when applied to ordered pairs and graphs. • Know the vertical line test. • Be able to use function notation, and evaluate functions. • Use the slope and the y-intercept to graph linear functions.
6 Write your own goal. _____ _____	• _____ _____

	W Work	Read Sections 4.1 to 4.5, and complete the exercises.	
E Evaluate	Complete the Chapter Review and Chapter Test. How did you do?	**R Rethink**	• After completing the emPOWERme, *Checklist for Effective Notes,* see if a classmate could learn the material by reading your notes. • Can you explain the two ways to graph a linear equation? Can these two ways be used for vertical and horizontal lines? • Do all linear equations contain both an x and a y? If not, what type of lines would contain only one variable in the equation?

4.1 Introduction to Linear Equations in Two Variables

P Prepare   O Organize

What are your objectives for Section 4.1?	How can you accomplish each objective?
1 Plot Ordered Pairs	• Be able to draw and label the x-axis and y-axis. • Know how to determine the x-coordinate and the y-coordinate of an ordered pair, and understand what these values mean when compared to the origin. • Complete the given example on your own. • Complete You Try 1.
2 Decide Whether an Ordered Pair Is a Solution of an Equation	• Know how to substitute x- and y-values into an equation and simplify. • Complete the given example on your own. • Complete You Try 2.

What are your objectives for Section 4.1?	How can you accomplish each objective?
3 Graph a Linear Equation in Two Variables by Plotting Points	• Be able to recognize a *linear equation*. • Learn how to make a *table of values* that will satisfy the equation, and plot these points. • Complete the given example on your own. • Complete You Try 3.
4 Graph a Linear Equation in Two Variables by Finding the Intercepts	• Know what an *intercept* is and where it is located on the graph. • Learn the procedure for **Finding Intercepts,** and write it in your notes. • Complete the given example on your own. • Complete You Try 4.
5 Graph Linear Equations of the Forms $x = a$ and $y = b$	• Write the properties **The Graph of $x = a$** and **The Graph of $y = b$** in your own words, and learn them. • Be able to choose ordered pairs that will satisfy the equation, if necessary. • Complete the given examples on your own. • Complete You Trys 5 and 6.
6 Use the Midpoint Formula	• Learn the *midpoint* formula. • Understand what the *midpoint* is and where it is located. • Complete the given example on your own. • Complete You Try 7.
7 Solve Applications	• Read the problem carefully. • Be able to determine what the *x*-value represents and what the *y*-value represents before choosing ordered pairs. • Be able to make a *table of values*. • Make sure your answers make sense. • Complete the given example on your own.

 Read the explanations, follow the examples, take notes, and complete the You Trys.

We encounter graphs everywhere—in newspapers, books, and on the Internet. The graph in Figure 1 shows how many billions of dollars consumers spent shopping online for consumer electronics during the years 2001–2007.

Graphs like this one are based on the **Cartesian coordinate system,** or the **rectangular coordinate system.** Let's begin this section by reviewing how to work with a Cartesian coordinate system.

1 Plot Ordered Pairs

The Cartesian coordinate system has a horizontal number line, called the **x-axis**, and a vertical number line, called the **y-axis**. (See Figure 2.) The *x*-axis and *y*-axis determine a flat surface called a **plane**. The axes divide this plane into four **quadrants**. The point at which the *x*-axis and *y*-axis intersect is called the **origin**. The arrow at one end of the *x*-axis and one end of the *y*-axis indicates the positive direction on each axis.

Each point in the plane is represented by an **ordered pair** of real numbers (x, y). The first number in the ordered pair is called the **x-coordinate** (also called the **abscissa**), and it tells us the distance and direction of the point from the origin along the *x*-axis. The second number in the ordered pair is the **y-coordinate** (also called the **ordinate**), and it tells us the distance and direction of the point from the origin either on the *y*-axis or parallel to the *y*-axis. The origin has coordinates $(0, 0)$.

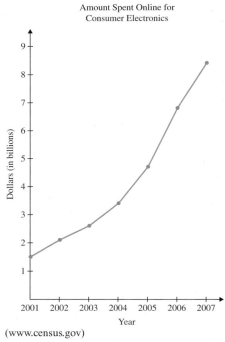

Amount Spent Online for Consumer Electronics

(www.census.gov)

Figure 1

> **Hint**
> Notice that from quadrant I, we move counterclockwise to move through the quadrants in order.

> **Hint**
> Beginning at the origin, notice that we first move horizontally, then move vertically to plot points.

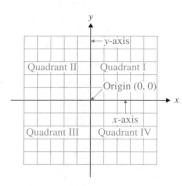

Figure 2

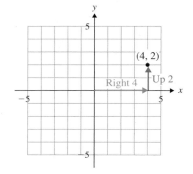

Figure 3

We have graphed the ordered pair $(4, 2)$ on the coordinate system in Figure 3. This is also called **plotting the point** $(4, 2)$. The *x*-coordinate of the point is 4. It tells us to move 4 units to the right from the origin along the *x*-axis. The *y*-coordinate of the point is 2. From the current position, move 2 units up, parallel to the *y*-axis.

We can use ordered pairs to represent the points on the line graph in Figure 1 just like we use them to represent points in the Cartesian coordinate system. In Figure 1, the years are on the horizontal axis, and the amount of money spent online on consumer electronics is on the vertical axis. If we move along the horizontal axis to the year 2005 and then move up, parallel to the vertical axis, we find the billions of dollars spent shopping online for consumer electronics during that year. This point can be represented by the ordered pair $(2005, 4.7)$, and it tells us that in 2005 shoppers spent approximately $4.7 billion online on consumer electronics.

EXAMPLE 1

Plot the points.

a) $(-2, 5)$ b) $(1, -4)$ c) $\left(\dfrac{7}{2}, 3\right)$

d) $(-5, -2)$ e) $(0, 1.5)$ f) $(-4, 0)$

Solution

To plot each point, move from the origin in the following ways:

a) $(-2, 5)$: Move left 2 units then up 5 units. This point is in quadrant II.

b) $(1, -4)$: Move right 1 unit then down 4 units. This point is in quadrant IV.

c) $\left(\dfrac{7}{2}, 3\right)$: Think of $\dfrac{7}{2}$ as $3\dfrac{1}{2}$. Move right $3\dfrac{1}{2}$ units then up 3 units. This point is in quadrant I.

d) $(-5, -2)$: Move left 5 units then down 2 units. This point is in quadrant III.

e) $(0, 1.5)$: The x-coordinate of 0 means that we don't move in the x-direction (horizontally). From the origin, move up 1.5 units on the y-axis. This point is not in any quadrant.

f) $(-4, 0)$: From the origin, move left 4 units. Since the y-coordinate is zero, we do not move in the y-direction (vertically). This point is not in any quadrant.

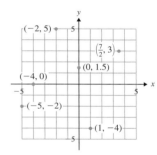

W Hint

Notice that negative x-values make you move left, and negative y-values make you move down.

YOU TRY 1

Plot the points.

a) $(3, 1)$ b) $(-2, 4)$ c) $(0, -5)$ d) $(2, 0)$

e) $(-4, -3)$ f) $\left(1, \dfrac{7}{2}\right)$

Note

The coordinate system should always be labeled to indicate how many units each mark represents.

A **solution of an equation in two variables** is written as an ordered pair so that when the values are substituted for the appropriate variables we obtain a true statement.

2 Decide Whether an Ordered Pair Is a Solution of an Equation

EXAMPLE 2

Determine whether each ordered pair is a solution of $4x + 5y = 11$.

a) $(-1, 3)$ b) $\left(\dfrac{3}{2}, 5\right)$

Solution

a) Solutions to the equation $4x + 5y = 11$ are written in the form (x, y) where (x, y) is called an *ordered pair*. Therefore, the ordered pair $(-1, 3)$ means that $x = -1$ and $y = 3$.

$$(-1, 3)$$
$$\quad \nearrow \quad \nwarrow$$
$$x\text{-coordinate} \quad y\text{-coordinate}$$

To determine whether $(-1, 3)$ is a solution of $4x + 5y = 11$, we substitute -1 for x and 3 for y. Remember to put these values in parentheses.

$$4x + 5y = 11$$
$$4(-1) + 5(3) = 11 \quad \text{Substitute } x = -1 \text{ and } y = 3.$$
$$-4 + 15 = 11 \quad \text{Multiply.}$$
$$11 = 11 \quad \text{True}$$

W Hint
Make sure you are substituting the variable x with the first quantity in the given ordered pair and the variable y with the second quantity of the ordered pair.

Since substituting $x = -1$ and $y = 3$ into the equation gives the true statement $11 = 11$, $(-1, 3)$ *is a solution* of $4x + 5y = 11$. We say that $(-1, 3)$ *satisfies* $4x + 5y = 11$.

b) The ordered pair $\left(\dfrac{3}{2}, 5\right)$ tells us that $x = \dfrac{3}{2}$ and $y = 5$.

$$4x + 5y = 11$$
$$4\left(\dfrac{3}{2}\right) + 5(5) = 11 \quad \text{Substitute } \dfrac{3}{2} \text{ for } x \text{ and } 5 \text{ for } y.$$
$$6 + 25 = 11 \quad \text{Multiply.}$$
$$31 = 11 \quad \text{False}$$

Since substituting $\left(\dfrac{3}{2}, 5\right)$ into the equation gives the false statement $31 = 11$, the ordered pair is *not* a solution to the equation.

[YOU TRY 2]

Determine whether each ordered pair is a solution of the equation $y = -\dfrac{3}{4}x + 5$.

a) $(12, -4)$ b) $(0, 7)$ c) $(-8, 11)$

If the variables in the equation are not x and y, then the variables in the ordered pairs are written in alphabetical order. For example, solutions to $2a + b = 7$ are ordered pairs of the form (a, b).

3 Graph a Linear Equation in Two Variables by Plotting Points

We saw in Example 2 that the ordered pair $(-1, 3)$ is a solution of the equation $4x + 5y = 11$. This is just one solution, however. The equation has an infinite number of solutions of the form (x, y), and we can represent these solutions with a graph on the Cartesian coordinate system.

The following table of values contains the solution we just verified as well as other solutions of $4x + 5y = 11$. Plot the points and connect them with a straight line. The line represents all solutions of the equation.

x	y
-1	3
4	-1
9	-5
0	$\frac{11}{5}$
$\frac{11}{4}$	0

$$4x + 5y = 11$$
$$4(-1) + 5(3) = -4 + 15 = 11$$
$$4(4) + 5(-1) = 16 + (-5) = 11$$
$$4(9) + 5(-5) = 36 + (-25) = 11$$
$$4(0) + 5\left(\frac{11}{5}\right) = 0 + 11 = 11$$
$$4\left(\frac{11}{4}\right) + 5(0) = 11 + 0 = 11$$

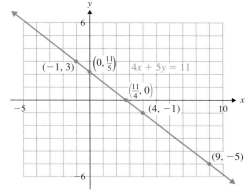

While equations like $x + 3 = 5$ and $2(t - 7) = 3t + 10$ are linear equations in one variable, the equation $4x + 5y = 11$ is an example of a linear equation in *two* variables. **The graph of a linear equation in two variables is a line, and every point on the line is a solution of the equation.**

Definition

A linear equation in two variables can be written in the form $Ax + By = C$, where A, B, and C are real numbers and where both A and B do not equal zero.

Other examples of linear equations in two variables are

$$3x - 5y = 10 \qquad y = \frac{1}{2}x + 7 \qquad -9s + 2t = 4 \qquad x = -8$$

(We can write $x = -8$ as $x + 0y = -8$, therefore it is a linear equation in two variables.) A solution to a linear equation in two variables is written as an ordered pair.

EXAMPLE 3   Graph $-x + 2y = 4$.

Solution

We will find three ordered pairs which satisfy the equation. Let's complete a table of values for $x = 0$, $x = 2$, and $x = -4$.

$x = 0$:
$-x + 2y = 4$
$-(0) + 2y = 4$
$2y = 4$
$y = 2$

$x = 2$:
$-x + 2y = 4$
$-(2) + 2y = 4$
$-2 + 2y = 4$
$2y = 6$
$y = 3$

$x = -4$:
$-x + 2y = 4$
$-(-4) + 2y = 4$
$4 + 2y = 4$
$2y = 0$
$y = 0$

Hint

If possible, use graph paper when graphing lines. Don't forget to label the x- and y-axes and to draw arrowheads on both ends of your lines.

We get the table of values

x	y
0	2
2	3
−4	0

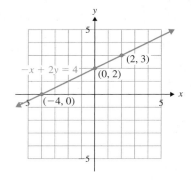

Plot the points (0, 2), (2, 3), and (−4, 0), and draw the line through them.

[YOU TRY 3] Graph each line.

a) $3x + 2y = 6$ b) $y = 4x - 8$

4 Graph a Linear Equation in Two Variables by Finding the Intercepts

In Example 3, the line crosses the x-axis at −4 and crosses the y-axis at 2. These points are called **intercepts**. What is the y-coordinate of any point on the x-axis? It is zero. Likewise, the x-coordinate of any point on the y-axis is zero.

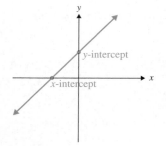

Definition

The **x-intercept** of the graph of an equation is the point where the graph intersects the x-axis.

The **y-intercept** of the graph of an equation is the point where the graph intersects the y-axis.

Hint

Notice that the y-intercept is where $x = 0$, and the x-intercept is where $y = 0$.

Procedure Finding Intercepts

To find the *x-intercept* of the graph of an equation, let $y = 0$ and solve for x.
To find the *y-intercept* of the graph of an equation, let $x = 0$ and solve for y.

Finding intercepts is very helpful for graphing linear equations in two variables.

EXAMPLE 4

Graph $y = -\frac{1}{2}x - 1$ by finding the intercepts and one other point.

Solution

We will begin by finding the intercepts.

x-intercept: Let $y = 0$, and solve for x.
$$0 = -\frac{1}{2}x - 1$$
$$1 = -\frac{1}{2}x$$
$$-2 = x \quad \text{Multiply both sides by } -2 \text{ to solve for } x.$$

The x-intercept is (−2, 0).

y-intercept: Let $x = 0$, and solve for y. $\quad y = -\frac{1}{2}(0) - 1$
$\qquad\qquad\qquad\qquad\qquad\qquad\qquad\qquad\quad y = 0 - 1 = -1$

The *y*-intercept is $(0, -1)$.

We must find another point. Let's look closely at the equation $y = -\frac{1}{2}x - 1$. The coefficient of x is $-\frac{1}{2}$. If we choose a value for x that is a multiple of 2 (the denominator of the fraction), then $-\frac{1}{2}x$ will not be a fraction.

Let $x = 2$. $\qquad y = -\frac{1}{2}x - 1$
$\qquad\qquad\qquad y = -\frac{1}{2}(2) - 1$
$\qquad\qquad\qquad y = -1 - 1$
$\qquad\qquad\qquad y = -2$

The third point is $(2, -2)$.

Plot the points, and draw the line through them.

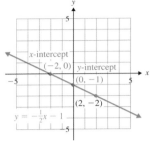

[**YOU TRY 4**] Graph $y = 3x + 6$ by finding the intercepts and one other point.

5 Graph Linear Equations of the Forms $x = a$ and $y = b$

The equation $x = a$ is a linear equation in two variables since it can be written in the form $x + 0y = a$. The same is true for $y = b$. It can be written as $0x + y = b$. Let's see how we can graph these equations.

EXAMPLE 5 Graph $x = 3$.

Solution

The equation is $x = 3$. (This is the same as $x + 0y = 3$.) $x = 3$ means that no matter the value of y, x always equals 3. We can make a table of values where we choose any value for y, but x is always 3.

x	y
3	0
3	1
3	-2

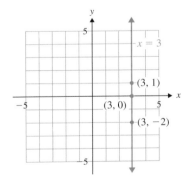

Plot the points, and draw a line through them. The graph of $x = 3$ is a *vertical line*.

W Hint
Notice that if x is constant, then the graphed line is vertical (parallel to the *y*-axis).

We can generalize the result as follows:

Property The Graph of $x = a$

If a is a constant, then the graph of $x = a$ is a *vertical line* going through the point $(a, 0)$.

[YOU TRY 5] Graph $x = -4$.

EXAMPLE 6 Graph $y = -2$.

Solution

The equation $y = -2$ is the same as $0x + y = -2$, therefore it is linear. $y = -2$ means that no matter the value of x, y always equals -2. Make a table of values where we choose any value for x, but y is always -2.

x	y
0	-2
2	-2
-2	-2

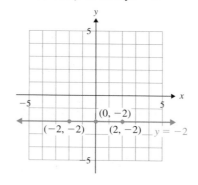

Plot the points, and draw a line through them. The graph of $y = -2$ is a *horizontal line*.

W Hint
Notice that if y is constant, then the graphed line is horizontal (parallel to the x-axis).

We can generalize the result as follows:

Property The Graph of $y = b$

If b is a constant, then the graph of $y = b$ is a *horizontal line* going through the point $(0, b)$.

[YOU TRY 6] Graph $y = 1$.

6 Use the Midpoint Formula

The **midpoint** of a line segment is the point that is exactly halfway between the endpoints of a line segment. We use the *midpoint formula* to find the midpoint.

W Hint
Notice that the midpoint formula calculates the arithmetic average of the given x- and y-values!

Definition The Midpoint Formula

If (x_1, y_1) and (x_2, y_2) are the endpoints of a line segment, then the **midpoint** of the segment has coordinates

$$\left(\frac{x_1 + x_2}{2}, \frac{y_1 + y_2}{2} \right)$$

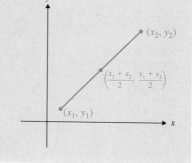

EXAMPLE 7

Find the midpoint of the line segment with endpoints $(-3, 4)$ and $(1, -2)$.

Solution

Begin by labeling the points: $\overset{(x_1, y_1)}{(-3, 4)}, \overset{(x_2, y_2)}{(1, -2)}$.
Substitute the values into the midpoint formula.

$$\text{Midpoint} = \left(\frac{x_1 + x_2}{2}, \frac{y_1 + y_2}{2}\right)$$
$$= \left(\frac{-3 + 1}{2}, \frac{4 + (-2)}{2}\right) \quad \text{Substitute values.}$$
$$= \left(\frac{-2}{2}, \frac{2}{2}\right)$$
$$= (-1, 1) \quad \text{Simplify.}$$

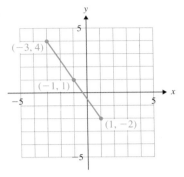

YOU TRY 7

Find the midpoint of the line segment with endpoints $(5, 2)$ and $(1, -3)$.

7 Solve Applications

Next we will see how a linear equation in two variables can be used to solve an application and how its graph can help us visualize the information that we get from the equation.

EXAMPLE 8

The length of an 18-year-old female's hair is measured to be 250 millimeters (mm) (almost 10 in.). The length of her hair after x days can be approximated by

$$y = 0.30x + 250$$

where y is the length of her hair in millimeters.

a) Find the length of her hair (i) 10 days, (ii) 60 days, and (iii) 90 days after the initial measurement, and write the results as ordered pairs.

b) Graph the equation.

c) How long would it take for her hair to reach a length of 274 mm (almost 11 in.)?

Solution

a) The problem states that in the equation $y = 0.30x + 250$,

x = number of days after the hair was measured
y = length of the hair (in millimeters)

We must determine the length of her hair after 10 days, 60 days, and 90 days. We can organize the information in a table of values, shown below.

x	y
10	
60	
90	

i) $x = 10$: $y = 0.30x + 250$
                 $y = 0.30(10) + 250$ Substitute 10 for x.
                 $y = 3 + 250 = 253$

After 10 days, her hair is 253 mm long. We can write this as the ordered pair (10, 253).

ii) $x = 60$: $y = 0.30x + 250$
                 $y = 0.30(60) + 250$ Substitute 60 for x.
                 $y = 18 + 250 = 268$

After 60 days, her hair is 268 mm long. (Or, after about 2 months, her hair is about 10.5 in. long. It has grown about half of an inch.) We can write this as the ordered pair (60, 268).

iii) $x = 90$: $y = 0.30x + 250$
                 $y = 0.30(90) + 250$ Substitute 90 for x.
                 $y = 27 + 250 = 277$

After 90 days, her hair is 277 mm long. Write this as the ordered pair (90, 277).

We can complete the table of values:

x	y
10	253
60	268
90	277

The ordered pairs are (10, 253), (60, 268), and (90, 277).

b) Graph the equation.

The x-axis represents the number of days after the hair was measured. Since it does not make sense to talk about a negative number of days, we will not continue the x-axis in the negative direction.

The y-axis represents the length of the female's hair. Likewise, a negative number does not make sense in this situation, so we will not continue the y-axis in the negative direction.

The scales on the x-axis and y-axis are different. This is because the size of the numbers they represent are quite different.

Here are the ordered pairs we must graph: (10, 253), (60, 268), and (90, 277).

The x-values are 10, 60, and 90, so we will let each mark in the x-direction represent 10 units.

The y-values are 253, 268, and 277. While the numbers are rather large, they do not actually differ by much. We will begin labeling the y-axis at 250, but each mark in the y-direction will represent 3 units. Because there is a large jump in values from 0 to 250 on the y-axis, we indicate this with "⌇" on the axis between the 0 and 250.

Notice, also, that we have labeled both axes. The ordered pairs are plotted on the graph shown at the left, and the line is drawn through the points.

c) We must determine how many days it would take for the hair to grow to a length of 274 mm. The length, 274 mm, is the y-value. We must find the value of x that corresponds to $y = 274$ since x represents the number of days.

The equation relating x and y is $y = 0.30x + 250$. We will substitute 274 for y and solve for x.

$$y = 0.30x + 250$$
$$274 = 0.30x + 250$$
$$24 = 0.30x$$
$$80 = x$$

It will take 80 days for her hair to grow to a length of 274 mm. This corresponds to the ordered pair (80, 274) and is consistent with what is illustrated by the graph.

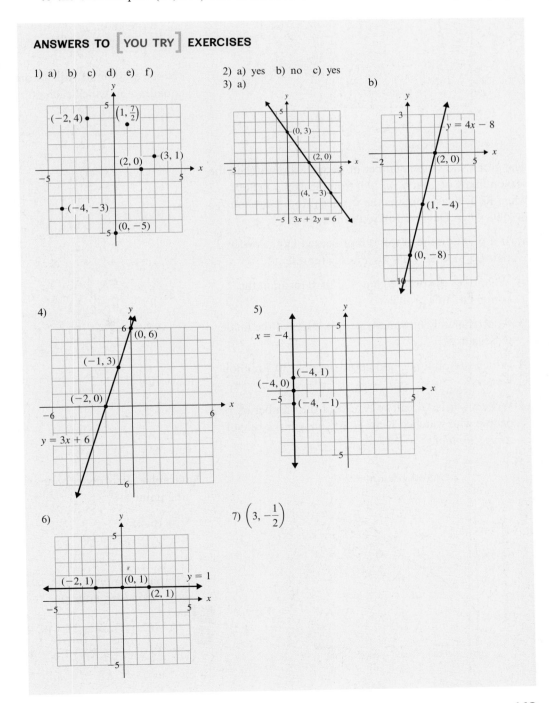

E Evaluate 4.1 Exercises

Do the exercises, and check your work.

Objective 1: Plot Ordered Pairs

1)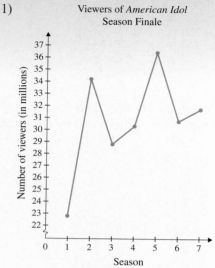

Viewers of *American Idol* Season Finale

(www.wikipedia.com)

The graph shows the number of people who watched the season finale of *American Idol* from Season 1 in 2002 when Kelly Clarkson was the winner through Season 7 in 2008 when David Cook took the crown.

a) If a point on the graph is represented by the ordered pair (x, y), then what do x and y represent?

b) What does the ordered pair (3, 28.8) mean in the context of this problem?

c) Approximately how many people watched the finale in Season 5?

d) Which season finale had approximately 30.7 million viewers?

e) Write an ordered pair to represent the number of people who watched Kelly Clarkson win the title of *American Idol*.

2)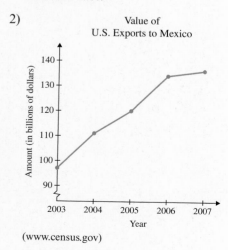

Value of U.S. Exports to Mexico

(www.census.gov)

The graph shows the value of U.S. exports to Mexico for the years 2003–2007.

a) If a point on the graph is represented by the ordered pair (x, y), then what do x and y represent?

b) What does the ordered pair (2004, 111) mean in the context of this problem?

c) During which year did exports total about $136 billion?

d) What was the value of exports to Mexico in 2005?

e) Write an ordered pair to represent the value of U.S. exports to Mexico in 2003.

Name each point with an ordered pair, and identify the quadrant in which each point lies.

3)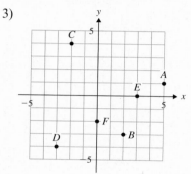

4)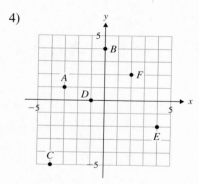

Graph each ordered pair, and explain how you plotted the points.

5) (6, 2)   6) (3, 4)

7) (−1, 4)   8) (−2, −3)

Graph the ordered pairs, labeling each point with the corresponding letter.

9) $A(2, -1)$, $B(5, 2)$, $C(0, 4)$, $D(-3, 0)$, $E\left(4, \dfrac{8}{3}\right)$, $F\left(-\dfrac{9}{2}, \dfrac{5}{4}\right)$, $G(-1.5, -4.2)$

150 CHAPTER 4 Linear Equations in Two Variables and Functions

10) $A(-3, -5)$, $B(-4, 5)$, $C(2, 0)$, $D(0, 3)$, $E\left(\frac{7}{2}, 1\right)$, $F\left(\frac{5}{2}, -\frac{5}{8}\right)$, $G(-0.7, 2.5)$

Fill in the blank with *positive, negative,* or *zero*.

11) The *x*-coordinate of every point in quadrant I is _____.

12) The *x*-coordinate of every point in quadrant II is _____.

13) The *y*-coordinate of every point in quadrant IV is _____.

14) The *y*-coordinate of every point in quadrant II is _____.

15) The *x*-coordinate of every point on the *y*-axis is _____.

16) The *y*-coordinate of every point on the *x*-axis is _____.

Objective 2: Decide Whether an Ordered Pair Is a Solution of an Equation

Determine whether each ordered pair is a solution of the given equation.

17) $7x + 2y = 4$; $(2, -5)$

18) $3x - 5y = 1$; $(-2, 1)$

19) $-4x + 7y = -4$; $(8, 4)$

20) $-2x - y = 13$; $(-8, 3)$

21) $y = 5x - 6$; $(3, 11)$

22) $x = -y + 9$; $(4, -5)$

23) $y = -\frac{3}{2}x - 19$; $(-10, -4)$

24) $5y = \frac{2}{3}x + 2$; $(12, 2)$

25) $x = 13$; $(5, 13)$

26) $y = -6$; $(-7, -6)$

Mixed Exercises: Objectives 3–5

27) Explain the difference between a linear equation in one variable and a linear equation in two variables. Give an example of each.

28) Is $x^2 + 6y = -5$ a linear equation in two variables? Explain your answer.

29) What do you get when you graph a linear equation in two variables? What does that graph represent?

30) Every linear equation in two variables has how many solutions?

31) What is the *y*-intercept of the graph of an equation? Explain how to find it.

32) What is the *x*-intercept of the graph of an equation? Explain how to find it.

Complete the table of values, and graph each equation.

33) $y = -2x + 5$

x	y
0	
-1	
2	
3	

34) $y = 3x - 1$

x	y
0	
1	
2	
-1	

35) $y = \frac{5}{2}x + 6$

x	y
0	
2	
-2	
-4	

36) $y = -\frac{2}{3}x + 4$

x	y
0	
-3	
3	
6	

37) $-3x + 6y = 9$

x	y
0	
	0
	4
-1	

38) $4x = 1 - y$

x	y
	0
0	
$\frac{5}{2}$	
	5

39) $y + 4 = 0$

x	y
0	
-3	
-1	
2	

40) $x = -\frac{3}{2}$

x	y
	5
	0
	-1
	-2

41) For $y = \frac{2}{3}x - 7$,

a) find *y* when $x = 3$, $x = 6$, and $x = -3$. Write the results as ordered pairs.

b) find *y* when $x = 1$, $x = 5$, and $x = -2$. Write the results as ordered pairs.

c) why is it easier to find the *y*-values in part a) than in part b)?

42) What ordered pair is a solution to every linear equation of the form $y = mx$, where *m* is a real number?

Graph each equation by finding the intercepts and at least one other point.

43) $y = -2x + 6$
44) $y = x - 3$
45) $3x - 4y = 12$
46) $5x + 2y = 10$
47) $x = -\frac{2}{3}y - 8$
48) $x = \frac{1}{4}y - 1$
49) $x - 4y = 6$
50) $2x + 3y = -6$
51) $y = -x$
52) $y = x$
53) $5y - 2x = 0$
54) $x + 3y = 0$
55) $x = 5$
56) $y = -1$
57) $y = 0$
58) $x = 0$
59) $y + 3 = 0$
60) $x - \frac{5}{2} = 0$
61) $x + 3y = 8$
62) $6x - y = 7$

63) a) What is the equation of the x-axis?
    b) What is the equation of the y-axis?

64) Let a and b be constants. If the lines $x = a$ and $y = b$ are graphed on the same axes, at what point will they intersect?

Objective 6: Use the Midpoint Formula

Find the midpoint of the line segment with the given endpoints.

65) $(1, 3)$ and $(7, 9)$
66) $(2, 10)$ and $(8, 4)$
67) $(-5, 2)$ and $(-1, -8)$
68) $(6, -3)$ and $(0, 5)$
69) $(-3, -7)$ and $(1, -2)$
70) $(-1, 3)$ and $(2, -9)$
71) $(4, 0)$ and $(-3, -5)$
72) $(-2, 4)$ and $(9, 3)$
73) $\left(\frac{3}{2}, -1\right)$ and $\left(\frac{5}{2}, \frac{7}{2}\right)$
74) $\left(\frac{9}{2}, \frac{3}{2}\right)$ and $\left(-\frac{7}{2}, -5\right)$
75) $(-6.2, 1.5)$ and $(4.8, 5.7)$
76) $(-3.7, -1.8)$ and $(3.7, -3.6)$

Objective 7: Solve Applications

Solve each application.

77) The number of drivers involved in fatal vehicle accidents in 2006 is given in the table. (www.census.gov)

Age	Number of Drivers
16	800
17	1300
18	1800
19	1900

a) Write the information as ordered pairs, (x, y), where x represents the age of the driver and y represents the number of drivers involved in fatal motor vehicle accidents.

b) Label a coordinate system, choose an appropriate scale, and graph the ordered pairs.

c) Explain the meaning of the ordered pair (18, 1800) in the context of the problem.

78) The number of pounds of potato chips consumed per person is given in the table. (U.S. Dept. of Agriculture)

Year	Pounds of Chips
2003	17.3
2004	16.6
2005	16.0
2006	16.2
2007	15.9

a) Write the information as ordered pairs, (x, y), where x represents the year and y represents the number of pounds of potato chips consumed per person.

b) Label a coordinate system, choose an appropriate scale, and graph the ordered pairs.

c) Explain the meaning of the ordered pair (2004, 16.6) in the context of the problem.

79) Horton's Party Supplies rents a "moon jump" for $100 plus $20 per hour. This can be described by the equation

$$y = 20x + 100$$

where x represents the number of hours and y represents the cost.

a) Complete the table of values, and write the information as ordered pairs.

x	y
1	
3	
4	
6	

b) Label a coordinate system, choose an appropriate scale, and graph the equation.

c) Explain the meaning of the ordered pair (4, 180) in the context of the problem.

d) For how many hours could a customer rent the moon jump if she had $280?

80) Kelvin is driving from Los Angeles to Chicago to go to college. His distance from L.A. is given by the equation

$$y = 62x$$

where x represents the number of hours driven, and y represents his distance from Los Angeles.

a) Complete the table of values and write the information as ordered pairs.

x	y
3	
8	
15	
20	

b) Label a coordinate system, choose an appropriate scale, and graph the equation.

c) Explain the meaning of the ordered pair (20, 1240) in the context of the problem.

d) What does the 62 in $y = 62x$ represent?

e) How many hours of driving time will it take for Kelvin to get to Chicago if the distance between L.A. and Chicago is about 2034 miles? (Round to the nearest hour.)

81) The blood alcohol percentage of a 180 lb male can be modeled by

$$y = 0.02x$$

where x represents the number of drinks consumed (1 drink = 12 oz of beer, for example) and y represents the blood alcohol percentage. (Taken from data from the U.S. Dept. of Health and Human Services)

a) Make a table of values using $x = 0, 1, 2,$ and 4, and write the information as ordered pairs.

b) Explain the meaning of each ordered pair in the context of the problem.

c) Graph the equation using the information in a). Use an appropriate scale.

d) If a 180-lb male had a blood alcohol percentage of 0.12, how many drinks did he have?

82) Concern about the Leaning Tower of Pisa in Italy led engineers to begin reinforcing the structure in the 1990s. The number of millimeters the Tower was straightened can be described by

$$y = 1.5x$$

where x represents the number of days engineers worked on straightening the Tower and y represents the number of millimeters (mm) the Tower was moved toward vertical. (In the end, the Tower will keep some of its famous lean.) (Reuters-9/7/2000)

a) Make a table of values using $x = 0, 10, 20,$ and 60, and write the information as ordered pairs.

b) Explain the meaning of each ordered pair in the context of the problem.

c) Graph the equation using the information in a). Use an appropriate scale.

d) Engineers straightened the Leaning Tower of Pisa a total of about 450 mm. How long did this take?

83) The relationship between altitude (in feet) and barometric pressure (in inches of mercury) can be modeled by

$$y = -0.001x + 29.86$$

for altitudes between sea level (0 feet) and 5000 feet, where x represents the altitude and y represents the barometric pressure. (From data taken from www.engineeringtoolbox.com)

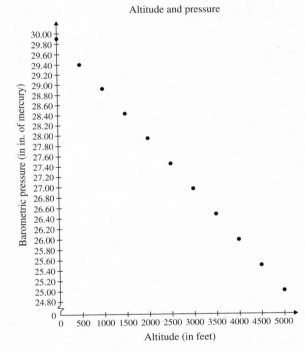

a) From the graph, estimate the pressure at the following altitudes: 0 ft (sea level), 1000 ft, 3500 ft, and 5000 ft.

b) Determine the barometric pressures at the same altitudes in a) using the equation. Are the numbers close?

c) Graph the line that models the data given on the original graph. Use an appropriate scale.

d) Can we use the equation $y = -0.001x + 29.86$ to determine the pressure at 10,000 ft? Why or why not?

84) The graph shows the actual per-pupil spending on education in the state of Washington for the school years 1994–2003.

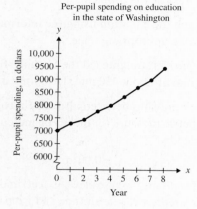

Algebraically, this can be modeled by

$$y = 293.3x + 6878.95$$

where x represents the number of years after the 1994–1995 school year. (So, $x = 0$ represents the 1994–1995 school year, $x = 1$ represents the 1995–1996 school year, etc.) y represents the per-pupil spending on education, in dollars. (www.effwa.org)

a) From the graph, estimate the amount spent per pupil during the following school years: 1994–1995, 1997–1998, and 2001–2002.

b) Determine the amount spent during the same school years as in a) using the equation. How do these numbers compare with the estimates from the graph?

R Rethink

R1) Which objective is the most difficult for you?

R2) When starting to graph a line, how do you determine which x or y values to start with?

R3) How can you tell if a line is going to be drawn vertically?

R4) How can you tell if a line is going to be drawn horizontally?

R5) In what other courses (besides math) have you had to use an equation of a line?

R6) When using the intercepts to graph a line, why should we always find a third point?

4.2 Slope of a Line and Slope-Intercept Form

Prepare / Organize

What are your objectives for Section 4.2?	How can you accomplish each objective?
1 Understand the Concept of Slope	• Understand the meaning of *slope*. • Write the property for **Slope of a Line** in your own words. • Complete the given example on your own. • Complete You Try 1.
2 Find the Slope of a Line Given Two Points on the Line	• Learn the formula for finding **The Slope of a Line**, and write it in your notes. • Understand the property that explains **Positive and Negative Slopes**. • Complete the given example on your own. • Complete You Try 2.
3 Use Slope to Solve Applied Problems	• Understand the meaning of points on a line. • Recognize what the slope of a line represents for an applied problem. • Complete the given example on your own.
4 Find the Slope of Horizontal and Vertical Lines	• Write the property for **Slopes of Horizontal and Vertical Lines** in your own words. • Recognize when the slope of a line is zero and when the slope of a line is undefined and why. • Complete the given example on your own. • Complete You Try 3.
5 Use Slope and One Point on a Line to Graph the Line	• Be able to plot the given ordered pair as a starting point. • Understand the meaning of slope to plot another point on the line. • Complete the given example on your own. • Complete You Try 4.
6 Graph a Line Using Slope and *y*-Intercept	• Learn the *slope-intercept form of a line*, $y = mx + b$. • Be able to determine the *slope* and the *y-intercept* for a given equation. • Complete the given example on your own. • Complete You Try 5.

Read the explanations, follow the examples, take notes, and complete the You Trys.

1 Understand the Concept of Slope

In Section 4.1, we learned to graph lines by plotting points. You may have noticed that some lines are steeper than others. Their "slants" are different too.

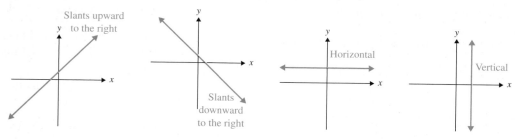

We can describe the steepness of a line with its *slope*.

> **Property** Slope of a Line
>
> The **slope** of a line measures its steepness. It is the ratio of the vertical change in y to the horizontal change in x. Slope is denoted by m.

We can also think of slope as a rate of change. *Slope* is the rate of change between two points. More specifically, it describes the rate of change in y to the change in x.

W Hint
Use two clearly defined points on the line to find the slope.

W Hint
Note that the numerator of the slope ratio will always represent the vertical change, and the denominator will always represent the horizontal change!

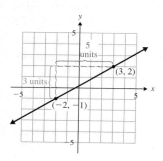

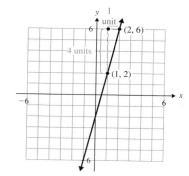

Slope = $\dfrac{3}{5}$ ← vertical change
← horizontal change

Slope = 4 or $\dfrac{4}{1}$ ← vertical change
← horizontal change

For example in the graph on the left, the line changes 3 units vertically for every 5 units it changes horizontally. Its slope is $\dfrac{3}{5}$. The line on the right changes 4 units vertically for every 1 unit of horizontal change. It has a slope of $\dfrac{4}{1}$ or 4.

Notice that the line with slope 4 is steeper than the line that has a slope of $\dfrac{3}{5}$.

Note
As the magnitude of the slope gets larger the line gets steeper.

Here is an application of slope.

EXAMPLE 1

A sign along a highway through the Rocky Mountains is shown on the left. What does this mean?

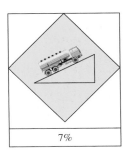

7%

Solution

Percent means "out of 100." Therefore, we can write 7% as $\frac{7}{100}$. We can interpret $\frac{7}{100}$ as the ratio of the vertical change in the road to horizontal change in the road.

The slope of the road is $\frac{7}{100}$. ← vertical change
$\phantom{\text{The slope of the road is }\frac{7}{100}.}$ ← horizontal change

The highway rises 7 ft for every 100 horizontal feet.

[YOU TRY 1]

The slope of a skateboard ramp is $\frac{7}{16}$ where the dimensions of the ramp are in inches. What does this mean?

2 Find the Slope of a Line Given Two Points on the Line

Here is line L. The points (x_1, y_1) and (x_2, y_2) are two points on line L. We will find the ratio of the vertical change in y to the horizontal change in x between the points (x_1, y_1) and (x_2, y_2).

To get from (x_1, y_1) to (x_2, y_2), we move *vertically* to point P then *horizontally* to (x_2, y_2). The x-coordinate of point P is x_1 and the y-coordinate of P is y_2.

When we moved *vertically* from (x_1, y_1) to point $P(x_1, y_2)$, how far did we go? We moved a vertical distance $y_2 - y_1$.

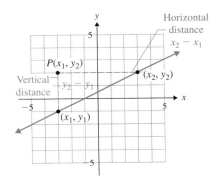

Note
The vertical change is $y_2 - y_1$ and is called the **rise**.

Then we moved *horizontally* from point $P(x_1, y_2)$ to (x_2, y_2). How far did we go? We moved a horizontal distance $x_2 - x_1$.

Note
The horizontal change is $x_2 - x_1$ and is called the **run**.

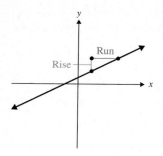

We said that the slope of a line is the ratio of the vertical change (rise) to the horizontal change (run). Therefore,

Formula The Slope of a Line

The **slope** (m) of a line containing the points (x_1, y_1) and (x_2, y_2) is given by

$$m = \frac{\text{vertical change}}{\text{horizontal change}} = \frac{y_2 - y_1}{x_2 - x_1}$$

W Hint
Notice that the slope of a line is a ratio! It is the ratio of $(y_2 - y_1)$ to $(x_2 - x_1)$ or rise to run.

We can also think of slope as:

$$\frac{\text{rise}}{\text{run}} \quad \text{or} \quad \frac{\text{change in } y}{\text{change in } x}$$

Let's look at some different ways to determine the slope of a line.

EXAMPLE 2

Determine the slope of each line.

a)
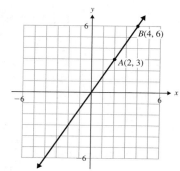

b)
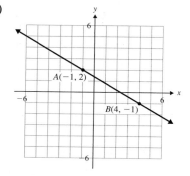

Solution

a) We will find the slope in two ways.

i) First, we will find the vertical change and the horizontal change by counting these changes as we go from A to B.

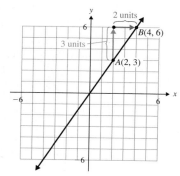

Vertical change (change in y) from A to B: 3 units

Horizontal change (change in x) from A to B: 2 units

$$\text{Slope} = \frac{\text{change in } y}{\text{change in } x} = \frac{3}{2} \quad \text{or} \quad m = \frac{3}{2}$$

ii) We can also find the slope using the formula.
Let $(x_1, y_1) = (2, 3)$ and $(x_2, y_2) = (4, 6)$.

$$m = \frac{y_2 - y_1}{x_2 - x_1} = \frac{6 - 3}{4 - 2} = \frac{3}{2}$$

You can see that we get the same result either way we find the slope.

b) i) First, find the slope by counting the vertical change and horizontal change as we go from A to B.

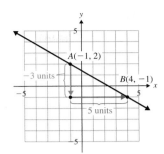

Vertical change (change in y) from A to B:
-3 units

Horizontal change (change in x) from A to B:
5 units

Slope $= \dfrac{\text{change in } y}{\text{change in } x} = \dfrac{-3}{5} = -\dfrac{3}{5}$ or $m = -\dfrac{3}{5}$

> **Hint**
> Always look at a line from left to right to determine if it is going up or down. If a line is going up, its slope is positive. If the line is going down, its slope is negative.

ii) We can also find the slope using the formula.
Let $(x_1, y_1) = (-1, 2)$ and $(x_2, y_2) = (4, -1)$.

$$m = \frac{y_2 - y_1}{x_2 - x_1} = \frac{-1 - 2}{4 - (-1)} = \frac{-3}{5} = -\frac{3}{5}$$

Again, we obtain the same result using either method for finding the slope.

Note

The slope of $-\dfrac{3}{5}$ can be thought of as $\dfrac{-3}{5}, \dfrac{3}{-5},$ or $-\dfrac{3}{5}$.

[YOU TRY 2] Determine the slope of each line by

a) counting the vertical change and horizontal change.   b) using the formula.

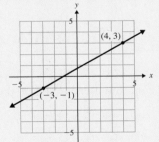
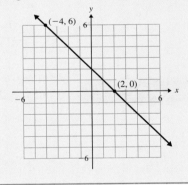

Notice that in Example 2a, the line has a positive slope and slants upward from left to right. As the value of x increases, the value of y increases as well. The line in 2b has a negative slope and slants downward from left to right. Notice, in this case, that as the line goes from left to right, the value of x increases while the value of y decreases. We can summarize these results with the following general statements.

Property Positive and Negative Slopes

A line with a **positive slope** slants upward from left to right. As the value of x increases, the value of y increases as well.

A line with a **negative slope** slants downward from left to right. As the value of x increases, the value of y decreases.

3 Use Slope to Solve Applied Problems

EXAMPLE 3 The graph models the number of students at DeWitt High School from 2006 to 2012.

a) How many students attended the school in 2006? in 2012?

b) What does the sign of the slope of the line segment mean in the context of the problem?

c) Find the slope of the line segment, and explain what it means in the context of the problem.

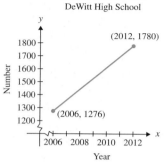

Number of Students at DeWitt High School

Solution

a) We can determine the number of students by reading the graph. In 2006, there were 1276 students, and in 2012 there were 1780 students.

b) The positive slope tells us that from 2006 to 2012 the number of students was increasing.

c) Let $(x_1, y_1) = (2006, 1276)$ and $(x_2, y_2) = (2012, 1780)$.

$$\text{Slope} = \frac{y_2 - y_1}{x_2 - x_1} = \frac{1780 - 1276}{2012 - 2006} = \frac{504}{6} = 84$$

The slope of the line is 84. Therefore, the number of students attending DeWitt High School between 2006 and 2012 increased by 84 per year.

Do all lines slant upward or downward? Let's look at Example 4.

4 Find the Slope of Horizontal and Vertical Lines

EXAMPLE 4 Find the slope of the line containing each pair of points.

a) $(-1, 2)$ and $(3, 2)$ b) $(-3, 4)$ and $(-3, -1)$

Solution

a) Let $(x_1, y_1) = (-1, 2)$ and $(x_2, y_2) = (3, 2)$.

$$m = \frac{y_2 - y_1}{x_2 - x_1} = \frac{2 - 2}{3 - (-1)} = \frac{0}{4} = 0$$

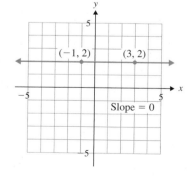

W Hint
Remember that if the numerator of a fraction is 0 and the denominator is nonzero, the fraction is equal to 0.

If we plot the points, we see that they lie on a horizontal line. Each point on the line has a y-coordinate of 2, so $y_2 - y_1$ always equals zero.

The slope of every horizontal line is zero.

W Hint
Horizontal lines are neither going up or down, and therefore they have a slope of zero.

160   CHAPTER 4   Linear Equations in Two Variables and Functions

Hint
Remember that if the denominator of a fraction is 0 it is said to be undefined.

Hint
We cannot define a vertical line as going up or down, and therefore we say its slope is undefined.

b) Let $(x_1, y_1) = (-3, 4)$ and $(x_2, y_2) = (-3, -1)$.

$$m = \frac{y_2 - y_1}{x_2 - x_1} = \frac{-1 - 4}{-3 - (-3)} = \frac{-5}{0} \text{ undefined}$$

We say that the slope is undefined. Plotting these points gives us a vertical line. Each point on the line has an x-coordinate of -3, so $x_2 - x_1$ always equals zero.

The slope of every vertical line is undefined.

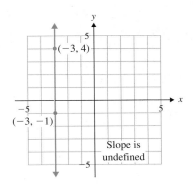

[**YOU TRY 3**] Find the slope of the line containing each pair of points.

a) $(5, 8)$ and $(-2, 8)$ b) $(4, 6)$ and $(4, 1)$

Property Slopes of Horizontal and Vertical Lines

The slope of a horizontal line, $y = b$, is **zero**. The slope of a vertical line, $x = a$, is **undefined**. (a and b are constants.)

5 Use Slope and One Point on a Line to Graph the Line

We have seen how we can find the slope of a line given two points on the line. Now, we will see how we can use the slope and *one* point on the line to graph the line.

EXAMPLE 5 Graph the line containing the point

a) $(2, -5)$ with a slope of $\frac{7}{2}$.   b) $(0, 4)$ with a slope of -3.

Solution

a) Plot the point $(2, -5)$.

Use the slope to find another point on the line.

$$m = \frac{7}{2} = \frac{\text{change in } y}{\text{change in } x}$$

To get from the point $(2, -5)$ to another point on the line, move up 7 units in the y-direction and right 2 units in the x-direction.

Plot this point, and draw a line through the two points.

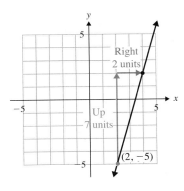

W Hint

In part b) we could have written $m = -3$ as $m = \dfrac{3}{-1}$. This would have given us a different point on the same line.

b) Plot the point $(0, 4)$.

What does the slope, $m = -3$, mean?

$$m = -3 = \frac{-3}{1} = \frac{\text{change in } y}{\text{change in } x}$$

To get from $(0, 4)$ to another point on the line, we will move *down* 3 units in the y-direction and *right* 1 unit in the positive x-direction. We end up at $(1, 1)$.

Plot this point, and draw a line through $(0, 4)$ and $(1, 1)$.

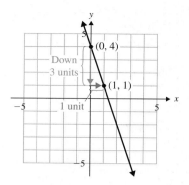

[**YOU TRY 4**] Graph the line containing the point

a) $(1, 1)$ with a slope of $-\dfrac{2}{3}$.

b) $(0, -5)$ with a slope of 3.

c) $(-2, 2)$ with an undefined slope.

We have already learned that a linear equation in two variables can be written in the form $Ax + By = C$ (this is called **standard form**), where A, B, and C are real numbers and where both A and B do not equal zero. Equations of lines can take other forms, too, and now we will learn about one of those forms.

In Example 5b, we graphed a line given its y-intercept, $(0, 4)$, and its slope, -3. This leads us into another good method for graphing a line: We can express a line in **slope-intercept** form and then graph it using its slope and y-intercept.

6 Graph a Line Using Slope and y-Intercept

We know that if (x_1, y_1) and (x_2, y_2) are points on a line, then the slope of the line is

$$m = \frac{y_2 - y_1}{x_2 - x_1}$$

Recall that to find the y-intercept of a line, we let $x = 0$ and solve for y. Let one of the points on a line be the y-intercept $(0, b)$, where b is a number. Let another point on the line be (x, y). See the graph on the left.

Substitute the points $(0, b)$ and (x, y) into the slope formula.

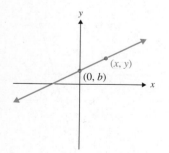

$$m = \frac{y_2 - y_1}{x_2 - x_1} = \frac{y - b}{x - 0} = \frac{y - b}{x}$$

Subtract y-coordinates ↑ Subtract x-coordinates

Solve $m = \dfrac{y - b}{x}$ for y.

$$mx = \frac{y - b}{x} \cdot x \quad \text{Multiply by } x \text{ to eliminate the fraction.}$$
$$mx = y - b$$
$$mx + b = y - b + b \quad \text{Add } b \text{ to each side to solve for } y.$$
$$mx + b = y$$
or
$$y = mx + b \quad \text{Slope-intercept form}$$

> **W Hint**
> Remember that the *b* value is the *y*-coordinate of the *y*-intercept ordered pair. The *y*-intercept is always written as the ordered pair (0, *b*).

Definition

The **slope-intercept form of a line** is $y = mx + b$, where m is the slope and $(0, b)$ is the *y*-intercept.

When an equation is in the form $y = mx + b$, we can quickly recognize the *y*-intercept and slope to graph the line.

EXAMPLE 6   Graph each line using its slope and *y*-intercept.

a) $y = 3x - 4$ b) $y = \frac{1}{2}x$ c) $5x + 3y = 12$

Solution

a) Graph $y = 3x - 4$.

Identify the slope, *m*, and *y*-intercept.

$$y = 3x - 4$$
$$m = 3, \quad y\text{-intercept is } (0, -4).$$

Plot the *y*-intercept first, then use the slope to locate another point on the line. Since the slope is 3, think of it as $\frac{3}{1}$. $\leftarrow$ change in *y* / change in *x*

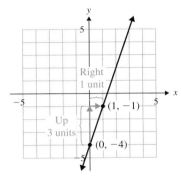

b) The equation $y = \frac{1}{2}x$ is the same as $y = \frac{1}{2}x + 0$.

Identify the slope, *m*, and *y*-intercept.

$$m = \frac{1}{2}, \quad y\text{-intercept is } (0, 0).$$

Plot the *y*-intercept, then use the slope to locate another point on the line.

Since $\frac{1}{2}$ is equivalent to $\frac{-1}{-2}$, we can use $\frac{-1}{-2}$ as the slope to locate yet another point on the line.

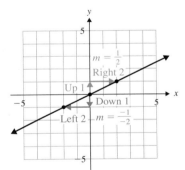

c) Rewrite $5x + 3y = 12$ in slope-intercept form, $y = mx + b$.

$$5x + 3y = 12$$
$$3y = -5x + 12 \quad \text{Add } -5x \text{ to each side.}$$
$$y = -\frac{5}{3}x + 4 \quad \text{Divide each side by 3.}$$
$$m = -\frac{5}{3}, \quad y\text{-intercept is } (0, 4).$$

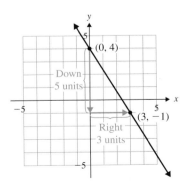

We can interpret the slope of $-\frac{5}{3}$ as either $\frac{-5}{3}$ or $\frac{5}{-3}$. Here we use $\frac{-5}{3}$. Using either form will give us a point on the line.

[**YOU TRY 5**] Graph each line using its slope and y-intercept.

    a) $y = \dfrac{2}{5}x - 3$ b) $y = -x$ c) $8x - 4y = 12$

Summary Methods for Graphing a Line

We have learned that we can use different methods for graphing lines. Given the equation of a line we can

1) make a table of values, plot the points, and draw the line through the points.
2) find the x-intercept by letting $y = 0$ and solving for x, and find the y-intercept by letting $x = 0$ and solving for y. Plot the points, then draw the line through the points.
3) put the equation into slope-intercept form, $y = mx + b$, identify the slope and y-intercept, then graph the line.

Using Technology

When we look at the graph of a linear equation, we should be able to estimate its slope. Use the equation $y = x$ as a guideline.

Step 1: Graph the equation $y = x$.

We can make the graph a thick line (so we can tell it apart from the others) by moving the arrow all the way to the left and hitting ENTER:

 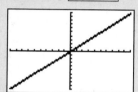

Step 2: Keeping this equation, graph the equation $y = 2x$:

 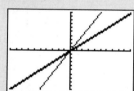

    a) Is the graph steeper or flatter than the graph of $y = x$?
    b) Make a guess as to whether $y = 3x$ will be steeper or flatter than $y = x$. Test your guess by graphing $y = 3x$.

Step 3: Clear the equation $y = 2x$, and graph the equation $y = 0.5x$:

 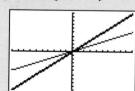

    a) Is the graph steeper or flatter than the graph of $y = x$?
    b) Make a guess as to whether $y = 0.65x$ will be steeper or flatter than $y = x$. Test your guess by graphing $y = 0.65x$.

Step 4: Test similar situations, except with negative slopes: $y = -x$.

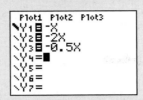 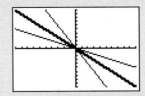

Did you notice that we have the same relationship, except in the opposite direction? That is, $y = 2x$ is steeper than $y = x$ in the positive direction, and $y = -2x$ is steeper than $y = -x$, but in the negative direction. And $y = 0.5x$ is flatter than $y = x$ in the positive direction, and $y = -0.5x$ is flatter than $y = -x$, but in the negative direction.

ANSWERS TO [YOU TRY] EXERCISES

1) The ramp rises 7 in. for every 16 horizontal inches.   2) a) $m = \dfrac{4}{7}$ b) $m = -1$

3) a) $m = 0$ b) undefined

4) a) 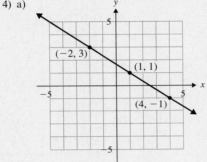   b)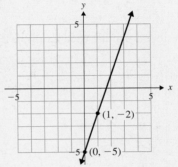

c) 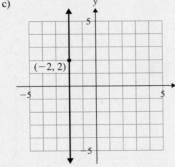   5 a)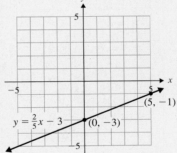

b) 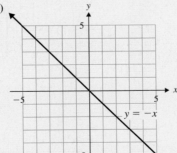   c)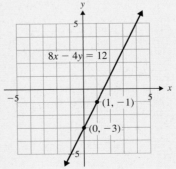

Evaluate 4.2 Exercises
Do the exercises, and check your work.

Objective 1: Understand the Concept of Slope

1) Explain the meaning of slope.
2) Describe the slant of a line with a positive slope.
3) Describe the slant of a line with a negative slope.
4) The slope of a horizontal line is _____.
5) The slope of a vertical line is _____.
6) If a line contains the points (x_1, y_1) and (x_2, y_2), write the formula for the slope of the line.

Mixed Exercises: Objectives 2 and 4

Determine the slope of each line by

a) counting the vertical change and the horizontal change as you move from one point to the other on the line
and
b) using the slope formula. (See Example 2.)

7)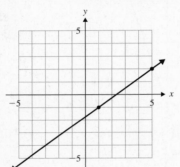

8)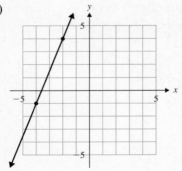

9)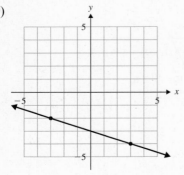

10)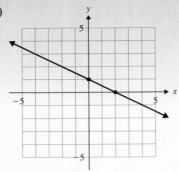

11)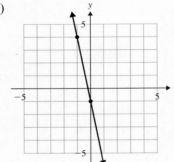

12)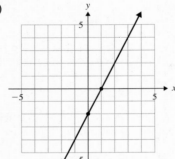

13)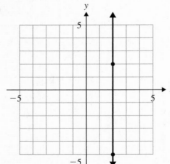

14)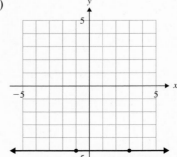

15) Graph a line with a positive slope and a negative *x*-intercept.

16) Graph a line with a negative slope and a positive *y*-intercept.

Use the slope formula to find the slope of the line containing each pair of points.

17) (3, 2) and (9, 5)

18) (4, 1) and (0, −5)

19) (−2, 8) and (2, 4)

20) (3, −2) and (−1, 6)

21) (9, 2) and (0, 4)

22) (−5, 1) and (2, −4)

23) (3, 5) and (−1, 5)

24) (−4, −4) and (−4, 10)

25) (3, 2) and (3, −1)

26) (0, −7) and (−2, −7)

27) $\left(\frac{3}{8}, -\frac{1}{3}\right)$ and $\left(\frac{1}{2}, \frac{1}{4}\right)$

28) $\left(\frac{3}{2}, \frac{7}{3}\right)$ and $\left(\frac{1}{3}, 4\right)$

29) (4.8, −1.6) and (6, 1.4)

30) (−1.7, −1.2) and (2.8, −10.2)

Objective 3: Use Slope to Solve Applied Problems

31) The slope of a roof is sometimes referred to as a *pitch*. A garage roof might have a 10-12 *pitch*. The first number refers to the rise of the roof, and the second number refers to how far over you must go to attain that rise (the run).

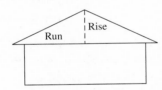

a) Find the slope of a roof with a 10-12 pitch.

b) Find the slope of a roof with an 8-12 pitch.

c) If a roof rises 8 in. in a 2-ft run, what is the slope of the roof? How do you write the slope in *x*-12 pitch form?

32) To make all buildings wheelchair accessible, the federal government has mandatory specifications for the slope of wheelchair ramps in new construction. A wheelchair ramp can have a maximum slope of $\frac{1}{12}$. Does the following ramp meet this requirement? (www.access-board.gov)

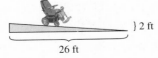

Use this information for Exercises 33 and 34.

Like many U.S. cities Evanston, Illinois, has building codes to regulate the steepness of driveways and parking garage ramps. A residential driveway can have a maximum slope of 12%, and the maximum slope of a ramp in an indoor parking garage is 15%. (www.cityofevanston.org)

33) A homeowner calls a landscape architect to tear out his old driveway and design a new one. The new driveway will have a different slope and will be constructed of brick pavers. In the new plans, the difference in height between the top and the bottom of the driveway will be 4 ft, and the linear distance between those two points will be 40 ft. Does the driveway meet the building code? Explain your answer.

34) The firm hired to build a parking garage in downtown Evanston wants to use plans from a garage it built in another city. In that structure, the ramps rose 1 ft for every 6.25 ft of horizontal distance. Can the construction firm use the same ramp plans in Evanston, or will it have to redesign them? Explain your answer.

35) Melissa purchased a new car in 2008. The graph shows the value of the car from 2008–2012.

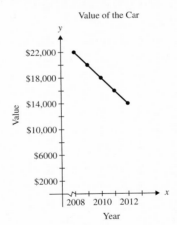

a) What was the value of the car in 2008?

b) Without computing the slope, determine whether it is positive or negative.

c) What does the sign of the slope mean in the context of this problem?

d) Find the slope of the line segment, and explain what it means in the context of the problem.

36) The graph shows the approximate number of babies (in thousands) born to teenage girls from 1997–2001. (www.census.gov)

a) Approximately how many babies were born to teen mothers in 1997? in 1998? in 2001?

b) Without computing the slope, determine whether it is positive or negative.

c) What does the sign of the slope mean in the context of the problem?

d) Find the slope of the line segment, and explain what it means in the context of the problem.

Objective 5: Use Slope and One Point on a Line to Graph a Line

Graph the line containing the given point and with the given slope.

37) $(-3, -2); m = \frac{5}{2}$ 38) $(1, 3); m = \frac{1}{4}$

39) $(1, -4); m = \frac{1}{3}$ 40) $(-4, 2); m = \frac{2}{7}$

41) $(4, 5); m = -\frac{2}{3}$ 42) $(2, -1); m = -\frac{3}{2}$

43) $(-5, 1); m = 3$ 44) $(-3, -5); m = 2$

45) $(0, -3); m = -1$ 46) $(0, 4); m = -3$

47) $(6, 2); m = -4$ 48) $(2, 5); m = -1$

49) $(-2, -1); m = 0$ 50) $(-3, 3); m = 0$

51) $(4, 0)$; slope is undefined   52) $(1, 6)$; slope is undefined

53) $(0, 0); m = 1$ 54) $(0, 0); m = -1$

Objective 6: Graph a Line Using Slope and y-Intercept

55) The slope-intercept form of a line is $y = mx + b$. What is the slope? What is the y-intercept?

56) How do you put an equation that is in standard form, $Ax + By = C$, into slope-intercept form?

Each of the following equations is in slope-intercept form. Identify the slope and the y-intercept, then graph each line using this information.

57) $y = \frac{2}{5}x - 6$ 58) $y = \frac{7}{4}x - 2$

59) $y = -\frac{5}{3}x + 4$ 60) $y = -\frac{1}{2}x + 5$

61) $y = \frac{3}{4}x + 1$ 62) $y = \frac{2}{3}x + 3$

63) $y = 4x - 2$ 64) $y = -3x - 1$

65) $y = -x + 5$ 66) $y = x$

67) $y = \frac{3}{2}x + \frac{1}{2}$ 68) $y = -\frac{3}{4}x - \frac{5}{2}$

69) $y = -2$ 70) $y = 4$

Put each equation into slope-intercept form, if possible, and graph.

71) $x + 3y = -6$ 72) $5x + 2y = 2$

73) $12x - 8y = 32$ 74) $y - x = 1$

75) $x + 9 = 2$ 76) $5 = x + 2$

77) $18 = 6y - 15x$ 78) $20x = 48 - 12y$

79) $y = 0$ 80) $y + 6 = 1$

81) Dave works in sales, and his income is a combination of salary and commission. He earns $34,000 per year plus 5% of his total sales. The equation $I = 0.05s + 34,000$ represents his total annual income, I, in dollars, when his sales total s dollars.

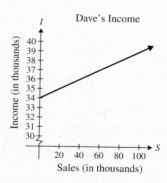

a) What is the I-intercept? What does it mean in the context of this problem?

b) What is the slope? What does it mean in the context of the problem?

c) Use the graph to find Dave's income if his total sales are $80,000. Confirm your answer using the equation.

82) Li Mei gets paid hourly in her after-school job. Her income is given by $I = 7.50h$, where I represents her income in dollars and h represents the number of hours worked.

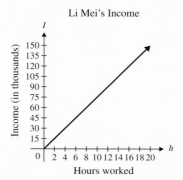

a) What is the I-intercept? What does it mean in the context of the problem?

b) What is the slope? What does it mean in the context of the problem?

c) Use the graph to find how much Li Mei earns when she works 14 hr. Confirm your answer using the equation.

83) The per capita consumption of whole milk in the United States since 1945 can be modeled by $y = -0.59x + 40.53$ where x represents the number of years after 1945, and y represents the per capita consumption of whole milk in gallons.
(U.S. Department of Agriculture)

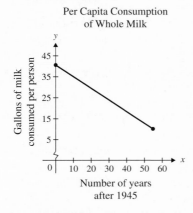

a) What is the y-intercept? What does it mean in the context of the problem?

b) What is the slope? What does it mean in the context of the problem?

c) Use the graph to estimate the per capita consumption of whole milk in the year 2000. Then, use the equation to determine this number.

84) On a certain day in 2011, the exchange rate between the American dollar and the Mexican peso was given by $p = 11.40d$, where d represents the number of dollars and p represents the number of pesos.

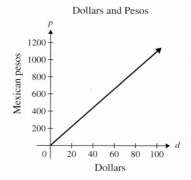

a) What is the p-intercept? What does it mean in the context of the problem?

b) What is the slope? What does it mean in the context of the problem?

c) Use the graph to estimate the value of 600 pesos in dollars. Then, use the equation to determine this number.

85) According to information taken from the U.S. Bureau of Labor Statistics, the average annual salary y, in dollars, of a pharmacist from 2000 to 2006 can be approximated by $y = 3986x + 68{,}613$ where x is the number of years after 2000. (www.bls.gov/oes)

a) What is the y-intercept? What does it mean in the context of this problem?

b) What is the slope? What does it mean in the context of this problem?

c) What was the average annual salary of a pharmacist in 2004?

d) Find the average annual salary of a pharmacist in 2014 if this equation continues to accurately model a pharmacist's average annual salary.

e) Using an appropriate scale, graph the equation for the years 2000–2006.

86) According to information taken from the U.S. Bureau of Labor Statistics, the percentage of men, y, ages 20–24 in the civilian labor force from 1986 to 2016 (projected) can be approximated by $y = -0.4x + 85.8$, where x is the number of years after 1986.
(http//www.bls.gov/opub/mlr/2007/11/art3full.pdf)

a) What is the y-intercept? What does it mean in the context of this problem?

b) What is the slope? What does it mean in the context of this problem?

c) What percentage of 20- to 24-year-old males were in the workforce in 2010?

d) What percentage of 20- to 24-year-old males are expected to be in the workforce in 2016?

e) Using an appropriate scale, graph the equation for the years 1986–2016.

Write the slope-intercept form for the equation of a line with the given slope and y-intercept.

87) $m = -4$; y-int: $(0, 7)$ 88) $m = 5$; y-int: $(0, 3)$

89) $m = \frac{8}{5}$; y-int: $(0, -6)$ 90) $m = \frac{4}{9}$; y-int: $(0, -1)$

91) $m = \frac{1}{3}$; y-int: $(0, 5)$ 92) $m = -\frac{1}{2}$; y-int: $(0, -3)$

93) $m = -1$; y-int: $(0, 0)$ 94) $m = 1$; y-int: $(0, 4)$

95) $m = 0$; y-int: $(0, -2)$ 96) $m = 0$; y-int: $(0, 0)$

R Rethink

R1) Given the graph of a line, how many different ways are there to determine the slope of the line?

R2) When looking at the graph of a line, how do you tell if the line has a positive slope or a negative slope?

R3) If all the ordered pairs on a line have the same x-value, what is the slope of the line?

R4) When you use the slope formula, does it matter which order you substitute in the x- and y-values?

R5) Which exercises in this section do you find most challenging?

4.3 Writing an Equation of a Line

Prepare / Organize

What are your objectives for Section 4.3?	How can you accomplish each objective?
1. Write an Equation of a Line Given Its Slope and y-Intercept	• Learn the procedure for **Writing an Equation of a Line Given Its Slope and y-Intercept,** and write it in your notes. • Complete the given example on your own. • Complete You Try 1.
2. Use the Point-Slope Formula to Write an Equation of a Line Given Its Slope and a Point on the Line	• Learn the **Point-Slope Formula,** and write it in your notes. • Be familiar with the final form of the equation, *standard form* or *slope-intercept* form. • Complete the given example on your own. • Complete You Try 2.
3. Use the Point-Slope Formula to Write an Equation of a Line Given Two Points on the Line	• Use the slope formula to first calculate slope given two points. • Use the **Point-Slope Formula,** with any given point and the slope, to write the equation. • Complete the given example on your own. • Complete You Try 3.

What are your objectives for Section 4.3?	How can you accomplish each objective?
4 Write Equations of Horizontal and Vertical Lines	• Learn the formulas for **Equations of Horizontal and Vertical Lines,** and write them in your notes. • Complete the given example on your own. • Complete You Try 4.
5 Use Slope to Determine if Two Lines Are Parallel or Perpendicular	• Be able to find the slope for a line if given an equation. • Understand the properties of **Parallel Lines** and **Perpendicular Lines,** and summarize them in your notes. • Complete the given example on your own. • Complete You Try 5.
6 Write an Equation of a Line That Is Parallel or Perpendicular to a Given Line	• Follow the examples, and create a step-by-step procedure for finding and writing the equations of parallel and perpendicular lines. • Complete the given examples on your own. • Complete You Trys 6 and 7.
7 Write a Linear Equation to Model Real-World Data	• Follow the example, then complete the example in your notes without looking at the example.

Read the explanations, follow the examples, take notes, and complete the You Trys.

The focus of Chapter 4, thus far, has been on graphing lines given their equations. In this section, we will switch gears. Given information about a line, we will write an equation of that line.

Recall the forms of lines we have discussed so far.

1) **Standard Form:** $Ax + By = C$, where A, B, and C are real numbers and where both A and B do not equal zero.

   We will now set an additional condition for when we write equations of lines in standard form:

   A, B, and C must be integers, and A must be positive.

2) **Slope-Intercept Form:** The slope-intercept form of a line is $y = mx + b$, where m is the slope, and the y-intercept is $(0, b)$.

It is common to express the equation of a line in one of these two forms. In the rest of this section, we will learn how to write equations of lines given information about their graphs.

1 Write an Equation of a Line Given Its Slope and y-Intercept

Procedure Write an Equation of a Line Given Its Slope and y-Intercept

If we are given the slope and y-intercept of a line, use $y = mx + b$ and substitute those values into the equation.

EXAMPLE 1

Find an equation of the line with slope $= -5$ and y-intercept $(0, 9)$.

Solution

Since we are told the slope and y-intercept, use $y = mx + b$.

$$m = -5 \quad \text{and} \quad b = 9$$

 Hint
Remember, a y-intercept point will always have an x-value of 0.

Substitute these values into $y = mx + b$ to get $y = -5x + 9$.

[YOU TRY 1] Find an equation of the line with slope $= \dfrac{2}{3}$ and y-intercept $(0, -6)$.

2 Use the Point-Slope Formula to Write an Equation of a Line Given Its Slope and a Point on the Line

When we are given the slope of a line and a point on that line, we can use another method to find its equation. This method comes from the formula for the slope of a line.

Let (x_1, y_1) be a given point on a line, and let (x, y) be any other point on the same line. See Figure 4. The slope of that line is

$$m = \frac{y - y_1}{x - x_1} \qquad \text{Definition of slope}$$
$$m(x - x_1) = y - y_1 \qquad \text{Multiply each side by } x - x_1.$$
$$y - y_1 = m(x - x_1) \qquad \text{Rewrite the equation.}$$

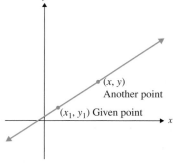

Figure 4

We have found the *point-slope form* of the equation of a line.

Formula Point-Slope Formula

The **point-slope form of a line** is $y - y_1 = m(x - x_1)$, where (x_1, y_1) is a point on the line and m is its slope.

Procedure Write an Equation of a Line Given Its Slope and a Point on the Line

If we are given the slope of the line and a point on the line, we can use the point-slope formula to find an equation of the line.

Note

The point-slope formula will help us write an equation of a line. We will not express our final answer in this form. We will write our answer in either slope-intercept form or in standard form.

EXAMPLE 2

A line has slope -3 and contains the point $(2, 1)$. Find the standard form for the equation of the line.

Solution

Although we are told to find the *standard form* for the equation of the line, we do not try to immediately "jump" to standard form. First, ask yourself, "*What information am I given?*"

We are given the slope and a point on the line. Therefore, we will begin by using the point-slope formula. Our *last* step will be to put it in standard form.

Use $y - y_1 = m(x - x_1)$. Substitute -3 for m. Substitute $(2, 1)$ for (x_1, y_1).

$$y - y_1 = m(x - x_1)$$
$$y - 1 = -3(x - 2) \quad \text{Substitute 2 for } x_1 \text{ and 1 for } y_1.$$
$$y - 1 = -3x + 6 \quad \text{Distribute.}$$

 Hint
Notice that the ordered pair (2, 1) is not the y-intercept because the x-value is not 0. Therefore you are not given b so we will not use y = mx + b in this example.

Since we are asked to express the answer in standard form, we must get the x- and y-terms on the same side of the equation.

$$3x + y - 1 = 6 \quad \text{Add } 3x \text{ to each side.}$$
$$3x + y = 7 \quad \text{Add 1 to each side.}$$

In standard form, the equation is $3x + y = 7$.

YOU TRY 2

A line has slope -5 and contains the point $(-1, 3)$. Find the standard form for the equation of the line.

3 Use the Point-Slope Formula to Write an Equation of a Line Given Two Points on the Line

We are now ready to discuss how to write an equation of a line when we are given two points on a line.

> **Procedure** Write an Equation of a Line Given Two Points on the Line
>
> To write an equation of a line given two points on the line,
>
> a) use the points to find the slope of line
>
> *then*
>
> b) use the slope and *either one* of the points in the point-slope formula.

EXAMPLE 3 Write an equation of the line containing the points (6, 4) and (3, 2). Express the answer in slope-intercept form.

Solution

We are given two points on the line, so first we will find the slope.

$$m = \frac{2-4}{3-6} = \frac{-2}{-3} = \frac{2}{3}$$

We will use the slope and *either one* of the points in the point-slope formula. (Each point will give the same result.) We will use (6, 4).

Substitute $\frac{2}{3}$ for m. Substitute (6, 4) for (x_1, y_1).

$$y - y_1 = m(x - x_1)$$
$$y - 4 = \frac{2}{3}(x - 6) \qquad \text{Substitute 6 for } x_1 \text{ and 4 for } y_1.$$
$$y - 4 = \frac{2}{3}x - 4 \qquad \text{Distribute.}$$
$$y = \frac{2}{3}x \qquad \text{Add 4 to each side to solve for } y.$$

The slope-intercept form of the equation is $y = \frac{2}{3}x$.

W Hint
In this example, substitute (3, 2) for (x, y) and verify that you get the same result.

[YOU TRY 3] Find the standard form for the equation of the line containing the points (−2, 5) and (1, 1).

4 Write Equations of Horizontal and Vertical Lines

In Section 4.2, we learned that the slope of a horizontal line is zero and that it has equation $y = b$, where b is a constant. The slope of a vertical line is undefined, and its equation is $x = a$, where a is a constant.

W Hint
Notice that the equation of a horizontal line has only a *y*-variable term, and a vertical line has only an *x*-variable term.

Formula Equations of Horizontal and Vertical Lines

Equation of a Horizontal Line: The equation of a horizontal line containing the point (a, b) is $y = b$.

Equation of a Vertical Line: The equation of a vertical line containing the point (a, b) is $x = a$.

EXAMPLE 4

Write an equation of the horizontal line containing the point $(5, -4)$.

Solution

The equation of a horizontal line has the form $y = b$, where b is the *y*-coordinate of the point. The equation of the line is $y = -4$.

[YOU TRY 4]

Write an equation of the horizontal line containing the point $(1, 7)$.

W Hint
Outline this summary, in your own words, in your notes.

Summary Writing Equations of Lines

If you are given

1) **the slope and *y*-intercept of the line,** use $y = mx + b$ and substitute those values into the equation.

2) **the slope of the line and a point on the line,** use the point-slope formula:

$$y - y_1 = m(x - x_1)$$

Substitute the slope for m and the point you are given for (x_1, y_1). Write your answer in slope-intercept or standard form.

3) **two points on the line,** find the slope of the line and then use the slope and *either one* of the points in the point-slope formula. Write your answer in slope-intercept or standard form.

The equation of a **horizontal line** containing the point (a, b) is $y = b$.
The equation of a **vertical line** containing the point (a, b) is $x = a$.

5 Use Slope to Determine if Two Lines Are Parallel or Perpendicular

Recall that two lines in a plane are *parallel* if they do not intersect. If we are given the equations of two lines, how can we determine if they are parallel?

Here are the equations of two lines, L_1 and L_2.

$$L_1: y = \frac{2}{3}x + 1 \qquad L_2: y = \frac{2}{3}x - 5$$

We will graph each line.

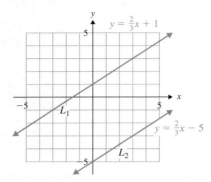

These lines are parallel. Their slopes are the same, but they have different y-intercepts. This is how we determine if two (nonvertical) lines are parallel. They have the same slope but different y-intercepts.

Property Parallel Lines

Parallel lines have the same slope but different y-intercepts. If two lines are vertical, they are parallel. However, their slopes are undefined.

The slopes of two lines can tell us about another relationship between the lines. The slopes can tell us if two lines are *perpendicular*. Recall that two lines are **perpendicular** if they intersect at 90° angles.

Here are the graphs of two perpendicular lines and their equations. We will see how their slopes are related.

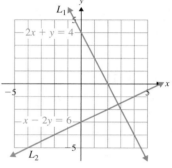

L_1 has equation $2x + y = 4$. L_2 has equation $x - 2y = 6$. Find the slopes of the lines by writing them in slope-intercept form.

L_1: $2x + y = 4$

$y = -2x + 4$

$m = -2$

L_2: $x - 2y = 6$

$-2y = -x + 6$

$y = \dfrac{-x}{-2} + \dfrac{6}{-2}$

$y = \dfrac{1}{2}x - 3$

$m = \dfrac{1}{2}$

How are the slopes related? They are **negative reciprocals.** That is, if the slope of one line is a, then the slope of a line perpendicular to it is $-\dfrac{1}{a}$. This is how we determine if two lines are perpendicular (where neither one is vertical).

> **Property** Perpendicular Lines
>
> When neither line is vertical, perpendicular lines have slopes that are negative reciprocals of each other.

EXAMPLE 5 Determine whether each pair of line is parallel, perpendicular, or neither.

a) $3x + 6y = 10$
   $x + 2y = -12$

b) $2x - 7y = 7$
   $21x + 6y = -2$

c) $y = -7x + 4$
   $7x - y = 9$

Solution

a) To determine if the lines are parallel or perpendicular, we must find the slope of each line.

Write each equation in slope-intercept form.

$$3x + 6y = 10 \qquad\qquad x + 2y = -12$$
$$6y = -3x + 10 \qquad\qquad 2y = -x - 12$$
$$y = -\frac{3}{6}x + \frac{10}{6} \qquad\qquad y = -\frac{x}{2} - \frac{12}{2}$$
$$y = -\frac{1}{2}x + \frac{5}{3} \qquad\qquad y = -\frac{1}{2}x - 6$$
$$m = -\frac{1}{2} \qquad\qquad m = -\frac{1}{2}$$

Each line has a slope of $-\frac{1}{2}$. Their y-intercepts are different. Therefore, $3x + 6y = 10$ and $x + 2y = -12$ are parallel lines.

b) Begin by writing each equation in slope-intercept form so that we can find their slopes.

$$2x - 7y = 7 \qquad\qquad 21x + 6y = -2$$
$$-7y = -2x + 7 \qquad\qquad 6y = -21x - 2$$
$$y = \frac{-2}{-7}x + \frac{7}{-7} \qquad\qquad y = -\frac{21}{6}x - \frac{2}{6}$$
$$y = \frac{2}{7}x - 1 \qquad\qquad y = -\frac{7}{2}x - \frac{1}{3}$$
$$m = \frac{2}{7} \qquad\qquad m = -\frac{7}{2}$$

The slopes are negative reciprocals, therefore the lines are perpendicular.

c) Again, we must find the slope of each line. $y = -7x + 4$ is already in slope-intercept form. Its slope is -7.

Write $7x - y = 9$ in slope-intercept form.

$$-y = -7x + 9 \qquad \text{Add } -7x \text{ to each side.}$$
$$y = \frac{-7}{-1}x + \frac{9}{-1} \qquad \text{Divide by } -1.$$
$$y = 7x - 9 \qquad \text{Simplify.}$$
$$m = 7$$

The slope of $y = -7x + 4$ is -7. The slope of $7x - y = 9$ is 7.

Since the slopes are different and not negative reciprocals, these lines are not parallel, and they are not perpendicular.

[YOU TRY 5] Determine whether each pair of lines is parallel, perpendicular, or neither.

a) $x + 6y = 12$
$y = 6x - 9$

b) $-4x + y = 1$
$2x - 3y = 6$

c) $3x - 2y = -8$
$15x - 10y = 2$

d) $x = 3$
$y = 2$

Let's learn how to write the equation of a line that is parallel or perpendicular to a given line.

6 Write an Equation of a Line That Is Parallel or Perpendicular to a Given Line

EXAMPLE 6

A line contains the point (4, 2) and is parallel to the line $y = \frac{3}{2}x + 2$. Write the equation of the line in slope-intercept form.

Solution

Let's look at the graph on the left to help us understand what is happening in this example. We must find the equation of the line in red. It is the line containing the point (4, 2) that is parallel to $y = \frac{3}{2}x + 2$.

The line $y = \frac{3}{2}x + 2$ has $m = \frac{3}{2}$. Therefore, the red line will have $m = \frac{3}{2}$ as well.

We know the slope, $\frac{3}{2}$, and a point on the line, (4, 2), so we use the point-slope formula to find its equation.

Substitute $\frac{3}{2}$ for m. Substitute (4, 2) for (x_1, y_1).

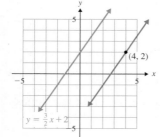

$$y - y_1 = m(x - x_1)$$
$$y - 2 = \frac{3}{2}(x - 4) \quad \text{Substitute 4 for } x_1 \text{ and 2 for } y_1.$$
$$y - 2 = \frac{3}{2}x - 6 \quad \text{Distribute.}$$
$$y = \frac{3}{2}x - 4 \quad \text{Add 2 to each side.}$$

The equation is $y = \frac{3}{2}x - 4$.

W Hint
When working with parallel lines, be sure to verify that your final equation indicates the same slope as in the given line.

[YOU TRY 6] A line contains the point (8, −5) and is parallel to the line $y = \frac{3}{4}x + \frac{2}{3}$. Write the equation of the line in slope-intercept form.

EXAMPLE 7

Find the standard form for the equation of the line that contains the point $(-5, 3)$ and is perpendicular to $5x - 2y = 6$.

Solution

Begin by finding the slope of $5x - 2y = 6$ by putting it into slope-intercept form.

$$5x - 2y = 6$$
$$-2y = -5x + 6 \qquad \text{Add } -5x \text{ to each side.}$$
$$y = \frac{-5}{-2}x + \frac{6}{-2} \qquad \text{Divide by } -2.$$
$$y = \frac{5}{2}x - 3 \qquad \text{Simplify.}$$
$$m = \frac{5}{2}$$

W Hint
Write down the steps as you are reading the example.

Then, determine the slope of the line containing $(-5, 3)$ by finding the *negative reciprocal* of the slope of the given line.

$$m_{\text{perpendicular}} = -\frac{2}{5}$$

The line for which we need to write an equation has $m = -\frac{2}{5}$ and contains the point $(-5, 3)$. Use the point-slope formula to find an equation of the line.

Substitute $-\frac{2}{5}$ for m. Substitute $(-5, 3)$ for (x_1, y_1).

$$y - y_1 = m(x - x_1)$$
$$y - 3 = -\frac{2}{5}[x - (-5)] \qquad \text{Substitute } -5 \text{ for } x_1 \text{ and } 3 \text{ for } y_1.$$
$$y - 3 = -\frac{2}{5}(x + 5)$$
$$y - 3 = -\frac{2}{5}x - 2 \qquad \text{Distribute.}$$

Since we are asked to write the equation in *standard form*, eliminate the fraction by multiplying each side by 5.

$$5(y - 3) = 5\left(-\frac{2}{5}x - 2\right)$$
$$5y - 15 = -2x - 10 \qquad \text{Distribute.}$$
$$5y = -2x + 5 \qquad \text{Add 15 to each side.}$$
$$2x + 5y = 5 \qquad \text{Add } 2x \text{ to each side.}$$

The equation is $2x + 5y = 5$.

[YOU TRY 7] Find the equation of the line perpendicular to $-3x + y = 2$ containing the point $(9, 4)$. Write the equation in standard form.

7 Write a Linear Equation to Model Real-World Data

As seen in previous sections of this chapter, equations of lines are used to describe many kinds of real-world situations. We will look at an example in which we must find the equation of a line given some data.

EXAMPLE 8

Since 1998, sulfur dioxide emissions in the United States have been decreasing by about 1080.5 thousand tons per year. In 2000, approximately 16,636 thousand tons of the pollutant were released into the air. (*Statistical Abstract of the United States*)

a) Write a linear equation to model this data. Let x represent the number of years after 1998, and let y represent the amount of sulfur dioxide (in thousands of tons) released into the air.

b) How much sulfur dioxide was released into the air in 1998? in 2004?

Solution

a) Ask yourself, "What information is given in the problem?"

i) ". . . emissions in the United States have been decreasing by about 1080.5 thousand tons per year" tells us the rate of change of emissions with respect to time. Therefore, this is the *slope*. It will be *negative* since emissions are decreasing.

$$m = -1080.5$$

ii) "In 2000, approximately 16,636 thousand tons . . . were released into the air" gives us a point on the line.

Let x = the number of years after 1998.

The year 2000 corresponds to $x = 2$.

y = amount of sulfur dioxide (in thousands of tons) released into the air.

Then, 16,636 thousand tons corresponds to $y = 16{,}636$.

A point on the line is **(2, 16,636)**.

Now that we know the slope and a point on the line, we can write an equation of the line using the point-slope formula:

$$y - y_1 = m(x - x_1)$$

Substitute -1080.5 for m. Substitute $(2, 16{,}636)$ for (x_1, y_1).

$y - y_1 = m(x - x_1)$
$y - 16{,}636 = -1080.5(x - 2)$ Substitute 2 for x_1 and 16,636 for y_1.
$y - 16{,}636 = -1080.5x + 2161$ Distribute.
$y = -1080.5x + 18{,}797$ Add 16,636 to each side.

The equation is $y = -1080.5x + 18{,}797$.

b) To determine the amount of sulfur dioxide released into the air in 1998, let $x = 0$ since x = the number of years *after* 1998.

$y = -1080.5(0) + 18{,}797$ Substitute $x = 0$.
$y = 18{,}797$

In 1998, 18,797 thousand tons of sulfur dioxide were released. Notice, the equation is in slope-intercept form, $y = mx + b$, and our result is b. That is because to find the y-intercept, let $x = 0$.

To determine how much sulfur dioxide was released in 2004, let $x = 6$ since 2004 is 6 years after 1998.

$$y = -1080.5(6) + 18,797 \quad \text{Substitute } x = 6.$$
$$y = -6483 + 18,797 \quad \text{Multiply.}$$
$$y = 12,314$$

In 2004, 12,314 thousand tons of sulfur dioxide were released into the air.

Using Technology

We can use a graphing calculator to explore what we have learned about perpendicular lines.

1) Graph the line $y = -2x + 4$. What is its slope?

2) Find the slope of the line perpendicular to the graph of $y = -2x + 4$.

3) Find the equation of the line perpendicular to $y = -2x + 4$ that passes through the point (6, 0). Express the equation in slope-intercept form.

4) Graph both the original equation and the equation of the perpendicular line:

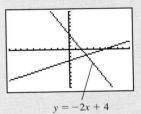

5) Do the lines above appear to be perpendicular?

6) Press ZOOM and choose 5:ZSquare:

7) Do the graphs look perpendicular now? Because the viewing window on a graphing calculator is a rectangle, *squaring* the window will give a more accurate picture of the graphs of the equations.

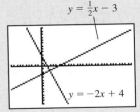

ANSWERS TO YOU TRY EXERCISES

1) $y = \frac{2}{3}x - 6$ 2) $5x + y = -2$ 3) $4x + 3y = 7$ 4) $y = 7$

5) a) perpendicular   b) neither   c) parallel   d) perpendicular

6) $y = \frac{3}{4}x - 11$ 7) $x + 3y = 21$

ANSWERS TO TECHNOLOGY EXERCISES

1) -2 2) $\frac{1}{2}$ 3) $y = \frac{1}{2}x - 3$ 5) No, because they do not meet at 90° angles.

7) Yes, because they meet at 90° angles.

Evaluate 4.3 Exercises

Do the exercises, and check your work.

Objective 1: Write an Equation of a Line Given Its Slope and y-Intercept

1) Explain how to find an equation of a line when you are given the slope and y-intercept of the line.

2) Is $\frac{2}{3}x - y = 5$ in standard form? Explain your answer.

Find an equation of the line with the given slope and y-intercept. Express your answer in the indicated form.

3) $m = -7$, y-int: $(0, 2)$; slope-intercept form
4) $m = 4$, y-int: $(0, -5)$; slope-intercept form
5) $m = 1$, y-int: $(0, -3)$; standard form
6) $m = -2$, y-int: $(0, 8)$; standard form

7) $m = -\frac{1}{3}$, y-int: $(0, -4)$; standard form
8) $m = \frac{5}{2}$, y-int: $(0, -1)$; standard form
9) $m = 1$, y-int: $(0, 0)$; slope-intercept form
10) $m = \frac{4}{9}$, y-int: $\left(0, -\frac{1}{6}\right)$; slope-intercept form

Objective 2: Use the Point-Slope Formula to Write an Equation of a Line Given Its Slope and a Point on the Line

11) a) If (x_1, y_1) is a point on a line with slope m, then the point-slope formula is _____.

b) Explain how to find an equation of a line when you are given the slope and a point on the line.

Find an equation of the line containing the given point with the given slope. Express your answer in the indicated form.

12) $(5, 8)$, $m = 3$; slope-intercept form
13) $(1, 6)$, $m = 5$; slope-intercept form
14) $(3, -2)$, $m = -2$; slope-intercept form
15) $(-9, 4)$, $m = -1$; slope-intercept form
16) $(-3, -7)$, $m = 1$; standard form

17) $(-2, -1)$, $m = 4$; standard form
18) $(2, 3)$, $m = -\frac{4}{5}$; slope-intercept form

19) $(-4, -5)$, $m = \frac{1}{6}$; slope-intercept form
20) $(-2, 0)$, $m = \frac{5}{8}$; standard form
21) $(6, 0)$, $m = -\frac{5}{9}$; standard form
22) $\left(\frac{1}{6}, -1\right)$, $m = 4$; slope-intercept form

Objective 3: Use the Point-Slope Formula to Write an Equation of a Line Given Two Points on the Line

23) Explain how to find an equation of a line when you are given two points on the line.

Find an equation of the line containing the two given points. Express your answer in the indicated form.

24) $(2, 5)$ and $(4, 1)$; slope-intercept form
25) $(3, 4)$ and $(7, 8)$; slope-intercept form
26) $(3, 2)$ and $(4, 5)$; slope-intercept form
27) $(-2, 4)$ and $(1, 3)$; slope-intercept form
28) $(5, -2)$ and $(2, -4)$; slope-intercept form
29) $(-1, -5)$ and $(3, -2)$; standard form
30) $(4, 1)$ and $(6, -3)$; standard form
31) $(-3, 0)$ and $(-5, 1)$; standard form
32) $(2, 1)$ and $(4, 6)$; standard form
33) $(2.5, 4.2)$ and $(3.1, 7.2)$; slope-intercept form
34) $(-3.4, 5.8)$ and $(-1.8, 3.4)$; slope-intercept form

Mixed Exercises: Objectives 1–4

Write the slope-intercept form of the equation of each line, if possible.

35)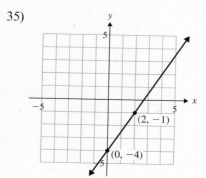

36)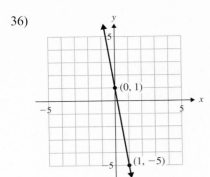

37)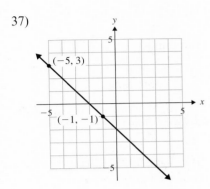

38)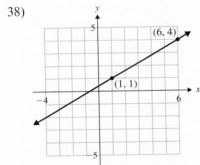

39)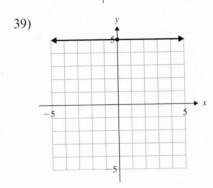

40)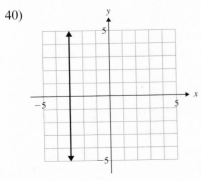

Write the slope-intercept form of the equation of the line, if possible, given the following information.

41) $m = 4$ and contains $(-4, -1)$

42) contains $(3, 0)$ and $(7, -2)$

43) $m = \dfrac{8}{3}$ and y-intercept $(0, -9)$

44) $m = 3$ and contains $(4, 5)$

45) contains $(-1, -2)$ and $(-5, 1)$

46) y-intercept $(0, 6)$ and $m = 7$

47) vertical line containing $(3, 5)$

48) vertical line containing $\left(-\dfrac{3}{4}, 2\right)$

49) horizontal line containing $(7, 4)$

50) horizontal line containing $(0, -8)$

51) contains $(0, 2)$ and $(6, 0)$

52) $m = 2$ and y-intercept $(0, -11)$

53) $m = 1$ and y-intercept $(0, 0)$

54) contains $(-3, -1)$ and $(2, -3)$

Objective 5: Use Slope to Determine if Two Lines Are Parallel or Perpendicular

55) How do you know if two lines are perpendicular?

56) How do you know if two lines are parallel?

Determine whether each pair of lines is parallel, perpendicular, or neither.

57) $y = -8x - 6$
$y = \dfrac{1}{8}x + 3$

58) $y = \dfrac{6}{11}x + 14$
$y = \dfrac{6}{11}x - 2$

59) $y = \dfrac{2}{9}x + 4$
$4x - 18y = 9$

60) $y = -\dfrac{5}{4}x - \dfrac{1}{3}$
$-4x + 5y = 10$

61) $-x + 5y = -35$
$y = 5x + 2$

62) $x - 4y = -12$
$2x - 6y = 9$

63) $y = x$
$x + y = 7$

64) $-3x + 2y = -10$
$3x - 4y = -2$

65) $4x - 3y = 18$
$-8x + 6y = 5$

66) $x + y = 4$
$y = -x$

67) $x = 5$
$x = -2$

68) $y = 4$
$x = 3$

SECTION 4.3 Writing an Equation of a Line

Lines L_1 and L_2 contain the given points. Determine if lines L_1 and L_2 are parallel, perpendicular, or neither.

69) L_1: (1, 2), (6, −13)
L_2: (−2, 5), (3, −10)

70) L_1: (3, −3), (1, 5)
L_2: (4, 3), (−12, −1)

71) L_1: (−1, −7), (2, 8)
L_2: (10, 2), (0, 4)

72) L_1: (1, 7), (−2, −11)
L_2: (0, −8), (1, −5)

73) L_1: (5, −1), (7, 3)
L_2: (−6, 0), (4, 5)

74) L_1: (−3, 9), (4, 2)
L_2: (6, −8), (−10, 8)

Objective 6: Write an Equation of a Line That Is Parallel or Perpendicular to a Given Line

Write an equation of the line *parallel* to the given line and containing the given point. Write the answer in slope-intercept form or in standard form, as indicated.

75) $y = 4x + 9$; (0, 2); slope-intercept form

76) $y = -3x - 1$; (0, 5); slope-intercept form

77) $y = \frac{1}{2}x - 5$; (4, 5); standard form

78) $y = 2x + 1$; (−2, −7); standard form

79) $x + 4y = 32$; (−8, 5); standard form

80) $4x + 3y = -6$; (−9, 4); standard form

81) $x + 5y = 10$; (15, 7); slope-intercept form

82) $18x - 3y = 9$; (2, −2); slope-intercept form

Write an equation of the line *perpendicular* to the given line and containing the given point. Write the answer in slope-intercept form or in standard form, as indicated.

83) $y = \frac{2}{3}x + 4$; (6, −3); slope-intercept form

84) $y = -\frac{4}{3}x + 2$; (8, 1); slope-intercept form

85) $y = -5x + 1$; (10, 0); standard form

86) $y = \frac{1}{4}x - 7$; (−2, 7); standard form

87) $x + y = 9$; (−5, −5); slope-intercept form

88) $y = x$; (10, −4); slope-intercept form

89) $24x - 15y = 10$; (16, −7); standard form

90) $2x + 5y = 11$; (4, 2); standard form

Write the slope-intercept form (if possible) of the equation of the line meeting the given conditions.

91) perpendicular to $2x - 6y = -3$ containing (2, 2)

92) parallel to $6x + y = 4$ containing (−2, 0)

93) parallel to $y = 2x + 1$ containing (1, −3)

94) perpendicular to $y = -x - 8$ containing (4, 11)

95) parallel to $x = -4$ containing (−1, −5)

96) parallel to $y = 2$ containing (4, −3)

97) perpendicular to $y = 3$ containing (2, 1)

98) perpendicular to $x = 0$ containing (5, 1)

99) perpendicular to $21x - 6y = 2$ containing (4, −1)

100) parallel to $-3x + 4y = 8$ containing (7, 4)

101) parallel to $y = 0$ containing $\left(-3, -\frac{5}{2}\right)$

102) perpendicular to $y = \frac{3}{4}$ containing (−2, 5)

Objective 7: Write a Linear Equation to Model Real-World Data

103) If a man's foot is 10 in. long, his U.S. shoe size is 8. A man's foot length of 10.5 in. corresponds to a shoe size of 9.5. Let L represent the length of a man's foot, and let S represent his shoe size.

  a) Write a linear equation that describes the relationship between the length of a man's foot in terms of his shoe size.

  b) If a man's foot is 11.5 in. long, what shoe size does he wear?

104) When a company charges customers $2000 to advertise on a billboard, 30% of its billboards are rented. If it cuts its rental fee to $1000, then 80% of its billboards are rented. Let C represent the cost of renting a billboard, and let P represent the percentage of the company's billboards that are being rented.

  a) Write a linear equation that describes the relationship of the percentage of billboards rented in terms of the cost.

  b) What percentage of its billboards would be rented if the company charges $1200?

105) Since 1998 the population of Maine has been increasing by about 8700 people per year. In 2001, the population of Maine was about 1,284,000. (*Statistical Abstract of the United States*)

  a) Write a linear equation to model this data. Let x represent the number of years after 1998, and let y represent the population of Maine.

  b) Explain the meaning of the slope in the context of the problem.

c) According to the equation, how many people lived in Maine in 1998? in 2002?

d) If the current trend continues, in what year would the population be 1,431,900?

106) Since 1997, the population of North Dakota has been decreasing by about 3290 people per year. The population was about 650,000 in 1997.
(*Statistical Abstract of the United States*)

a) Write a linear equation to model this data. Let x represent the number of years after 1997, and let y represent the population of North Dakota.

b) Explain the meaning of the slope in the context of the problem.

c) According to the equation, how many people lived in North Dakota in 1999? in 2002?

d) If the current trend holds, in what year would the population be 600,650?

107) The graph here shows the number of farms (in thousands) with milk cows between 1997 and 2002. x represents the number of years after 1997 so that $x = 0$ represents 1997, $x = 1$ represents 1998, and so on. Let y represent the number of farms (in thousands) with milk cows. (USDA, National Agricultural Statistics Service)

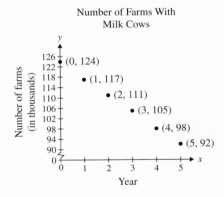

a) Write a linear equation to model this data. Use the data points for 1997 and 2002.

b) Explain the meaning of the slope in the context of the problem.

c) If the current trend continues, find the number of farms with milk cows in 2004.

108) The graph here shows the average salary for a public school high school principal for several years beginning with 1990. x represents the number of years after 1990 so that $x = 0$ represents 1990, $x = 1$ represents 1991, and so on. Let y represent the average salary of a high school principal.
(*Statistical Abstract of the United States*)

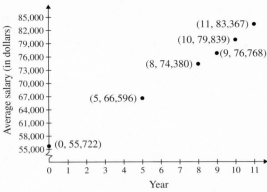

a) Write a linear equation to model this data. Use the data points for 1990 and 2001. (Round the slope to the nearest whole number.)

b) Explain the meaning of the slope in the context of the problem.

c) Use the equation to estimate the average salary in 1993.

d) If the trend continues, find the average salary in 2005.

109) The chart shows the number of hybrid vehicles registered in the United States from 2000 to 2003.
(*American Demographics*, Sept 2004, Vol 26, Issue 7)

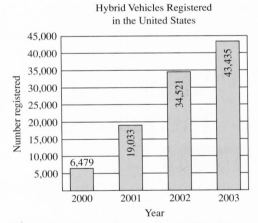

a) Write a linear equation to model this data. Use the data points for 2000 and 2003. Let x represent the number of years after 2000, and let y represent the number of hybrid vehicles registered in the United States. (Round the slope to the tenths place.)

b) Explain the meaning of the slope in the context of the problem.

c) Use the equation to determine the number of vehicles registered in 2002. How does it compare to the actual number on the chart?

d) If the current trend continues, approximately how many hybrid vehicles will be registered in 2010?

110) The chart shows the percentage of females in Belgium (15–64 yr old) in the workforce between 1980 and 2000. (*Statistical Abstract of the United States*)

Percentage of Females in the Workforce in Belgium

Year	Percentage
1980	47.0
1990	52.4
1995	56.1
2000	59.2

a) Write a linear equation to model this data. Use the data points for 1980 and 2000. Let x represent the number of years after 1980, and let y represent the percent of females in Belgium in the workforce.

b) Explain the meaning of the slope in the context of the problem.

c) Use the equation to determine the percentage of Belgian women in the workforce in 1990. How does it compare to the actual number on the chart?

d) Do the same as part c) for the year 1995.

e) In what year were 58% of Belgian women working?

R Rethink

R1) If two lines are perpendicular, how are their slopes related?

R2) How do you determine if an ordered pair represents a y-intercept point?

R3) Why do equations of vertical or horizontal lines only involve one variable term?

R4) Using the standard form $Ax + By = C$, can you determine the slope and y-intercept of a line by inspection? (*Hint:* Solve the equation $Ax + By = C$ for y).

R5) Which exercises in this section do you find most challenging?

4.4 Linear and Compound Linear Inequalities in Two Variables

P Prepare

What are your objectives for Section 4.4?

O Organize

How can you accomplish each objective?

What are your objectives for Section 4.4?	How can you accomplish each objective?
1 Graph a Linear Inequality in Two Variables	• Know the definition of a *linear inequality in two variables*. • Learn the two methods for graphing a linear inequality in two variables, and write them in your own words. • Understand what it means when a region is shaded and when a region is not shaded. • Complete the given examples on your own. • Complete You Trys 1 and 2.
2 Graph a Compound Linear Inequality in Two Variables	• Learn the procedure for **Graphing Compound Linear Inequalities in Two Variables.** • Understand the difference between the *intersection* of two solution sets and the *union* of two solution sets. • Choose a test point to check the shaded solution. • Complete the given examples on your own. • Complete You Trys 3 and 4.

Work — Read the explanations, follow the examples, take notes, and complete the You Trys.

In Chapter 3, we learned how to solve linear inequalities in *one variable* such as $2x - 3 \geq 5$.

We will begin this section by learning how to graph the solution set of linear inequalities in *two variables*. Then we will learn how to graph the solution set of *systems* of linear inequalities in two variables.

1 Graph a Linear Inequality in Two Variables

Definition

A **linear equality in two variables** is an inequality that can be written in the form $Ax + By \geq C$ or $Ax + By \leq C$ where A, B, and C are real numbers and where A and B are not both zero. ($>$ and $<$ may be substituted for $\geq$ and $\leq$.)

Here are some examples of linear inequalities in two variables.

$$x + y \geq 3, \quad y < \frac{1}{4}x + 3, \quad x \leq 2, \quad y > -4$$

Note
We can call $x \leq 2$ a linear inequality in two variables because we can write it as $x + 0y \leq 2$. Likewise, we can write $y > -4$ as $0x + y > -4$.

The solutions to linear inequalities in two variables, such as $x + y \geq 3$, are *ordered pairs* of the form (x, y) that make the inequality true. We graph a linear inequality in two variables on a rectangular coordinate system.

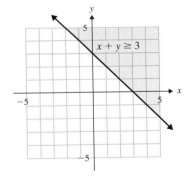

The points $(5, 2)$, $(1, 4)$, and $(3, 0)$ are some of the points that satisfy $x + y \geq 3$. There are infinitely many solutions. The points $(0, 0)$, $(-4, 1)$, and $(2, -3)$ are three of the points that do not satisfy $x + y \geq 3$. There are infinitely points that are not solutions.

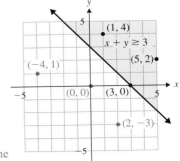

Points in the shaded region and on the line are in the solution set.

The points in the unshaded region are *not* in the solution set.

SECTION 4.4 Linear and Compound Linear Inequalities in Two Variables 187

> **Note**
> If the inequality had been $x + y > 3$, then the line would have been drawn as a *dotted line* and all points on the line would *not* be part of the solution set.

As we saw in the preceding graph, the line divides the *plane* into two regions or **half planes**. The line $x + y = 3$ is the **boundary line** between the two half planes. We will use this boundary line to graph a linear inequality in two variables. Notice that the boundary line is written as an equation; it uses an equal sign.

> **Procedure** Graph a Linear Inequality in Two Variables Using a Test Point
>
> 1) **Graph the boundary line.** If the inequality contains $\geq$ or $\leq$, make this boundary line *solid*. If the inequality contains $>$ or $<$, make it *dotted*.
> 2) **Choose a test point not on the line, and shade the appropriate region.** Substitute the test point into the inequality. If (0, 0) is not on the line, it is an easy point to test in the inequality.
>   a) If it *makes the inequality true,* shade the region *containing* the test point. All points in the shaded region are part of the solution set.
>   b) If the test point *does not satisfy the inequality,* shade the region on the *other* side of the line. All points in the shaded region are part of the solution set.

EXAMPLE 1 Graph $3x + 4y \leq -8$.

Solution

1) Graph the boundary line $3x + 4y = -8$ as a solid line.

2) Choose a test point not on the line and substitute it into the inequality to determine whether it makes the inequality true.

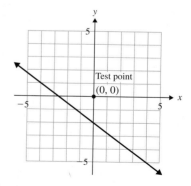

W Hint
If the inequality includes the *equals* condition, the boundary line is drawn solid. Points on this line make the inequality true.

Test Point	Substitute into $3x + 4y \leq -8$
(0, 0)	$3(0) + 4(0) \leq -8$
	$0 \leq -8$ False

188 CHAPTER 4 Linear Equations in Two Variables and Functions

Since the test point (0, 0) does *not* satisfy the inequality we will shade the region that does *not* contain the point (0, 0).

All points on the line and in the shaded region satisfy the inequality $3x + 4y \leq -8$.

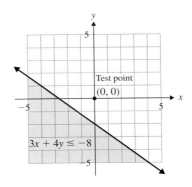

EXAMPLE 2

Graph $-x + 2y > -4$.

Solution

1) Since the inequality symbol is $>$, graph a *dotted* boundary line, $-x + 2y = -4$. (This means that the points *on* the line are not part of the solution set.)

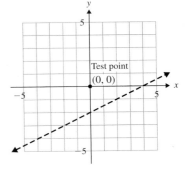

2) Choose a test point not on the line and substitute it into the inequality to determine whether it makes the inequality true.

> **Hint**
>
> If the inequality *does not* include the equals condition, the boundary line is dotted. Points on this line make the inequality false.

Test Point	Substitute into $-x + 2y > -4$
(0, 0)	$-(0) + 2(0) > -4$
	$0 > -4$ True

Since the test point (0, 0) satisfies the inequality, shade the region containing that point.

All points in the shaded region satisfy the inequality $-x + 2y > -4$.

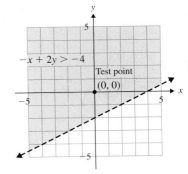

YOU TRY 1

Graph each inequality.

a) $2x + y \leq 4$ b) $x + 4y > 12$

If we write the inequality in *slope-intercept form*, we can decide which region to shade without using test points.

> **Procedure** Graph a Linear Inequality in Two Variables Using the Slope-Intercept Method
>
> 1) Write the inequality in the form $y \geq mx + b$ ($y > mx + b$) or $y \leq mx + b$ ($y < mx + b$), and graph the boundary line $y = mx + b$.
> 2) If the inequality is in the form $y \geq mx + b$ or $y > mx + b$, shade *above* the line.
> 3) If the inequality is in the form $y \leq mx + b$ or $y < mx + b$, shade *below* the line.

EXAMPLE 3   Graph each inequality using the slope-intercept method.

a) $y < -\dfrac{1}{3}x + 5$ b) $2x - y \leq -2$

Solution

a) The inequality $y < -\dfrac{1}{3}x + 5$ is already in slope-intercept form.

Graph the boundary line $y = -\dfrac{1}{3}x + 5$ as a *dotted line*.

Since $y < -\dfrac{1}{3}x + 5$ has a *less than* symbol, shade *below* the line. All points in the shaded region satisfy $y < -\dfrac{1}{3}x + 5$.

We can choose a point such as (0, 0) in the shaded region as a check. Substituting this point into $y < -\dfrac{1}{3}x + 5$ gives us $0 < -\dfrac{1}{3}(0) + 5$, or $0 < 5$, which is true.

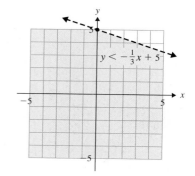

b) Solve $2x - y \leq -2$ for y.

$$2x - y \leq -2$$
$$-y \leq -2x - 2 \quad \text{Subtract } 2x.$$
$$y \geq 2x + 2 \quad \text{Divide by } -1, \text{ and change the direction of the inequality symbol.}$$

Graph $y = 2x + 2$ as a *solid line*.

Since $y \geq 2x + 2$ has a *greater than or equal to* symbol, shade *above* the line.

All points on the line and in the shaded region satisfy $2x - y \leq -2$.

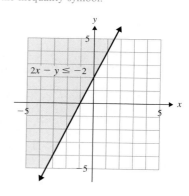

[YOU TRY 2] Graph each inequality using the slope-intercept method.

a) $y \geq -\dfrac{3}{4}x - 6$ b) $5x - 2y > 4$

2 Graph a Compound Linear Inequality in Two Variables

Linear inequalities in two variables are called *compound linear inequalities* if they are connected by the words *and* or *or*.

The solution set of a compound inequality containing *and* is the *intersection* of the solution sets of the inequalities. The solution set of a compound inequality containing *or* is the *union* of the solution sets of the inequalities.

> **Procedure** Graphing Compound Linear Inequalities in Two Variables
>
> 1) Graph each inequality separately on the same axes. Shade lightly.
> 2) If the inequality contains *and*, the solution set is the *intersection* of the shaded regions. Heavily shade this region.
> 3) If the inequality contains *or*, the solution set is the *union* (total) of the shaded regions. Heavily shade this region.

EXAMPLE 4

Graph $x \leq 2$ and $2x + 3y > 3$.

Solution

To graph $x \leq 2$, graph the boundary line $x = 2$ as a solid line. The x-values are *less than* 2 to the *left* of 2, so shade the region to the left of the line $x = 2$.

Graph $2x + 3y > 3$. Use a dotted boundary line.

The region shaded blue in the third graph is the *intersection* of the shaded regions and the solution set of the compound inequality. The part of the line $x = 2$ that is above the line $2x + 3y = 3$ is included in the solution set.

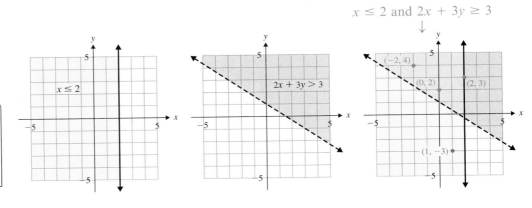

W Hint
Notice that the final solution is the set of ordered pairs that lie in the overlapping shaded region.

Any point in the solution set must satisfy *both* inequalities, and any point *not* in the solution set will not satisfy *both* inequalities. We check three test points next. (See the graph.)

Test Point	Substitute into $x \leq 2$	Substitute into $2x + 3y > 3$	Solution?
$(-2, 4)$	$-2 \leq 2$ True	$2(-2) + 3(4) > 3$ $8 > 3$ True	Yes
$(0, 2)$	$0 \leq 2$ True	$2(0) + 3(2) > 3$ $6 > 3$ True	Yes
$(1, -3)$	$1 \leq 2$ True	$2(1) + 3(-3) > 3$ $-7 > 3$ False	No

Although we show three separate graphs in Example 4, it is customary to graph everything on the same axes, shading lightly at first, then to heavily shade the region that is the graph of the compound inequality.

[YOU TRY 3] Graph the compound inequality $y \leq 3x - 1$ and $y + 2x \leq 4$.

EXAMPLE 5 Graph $y \leq \frac{1}{2}x$ or $2x + y \geq 2$.

Solution

Graph each inequality separately. The solution set of the compound inequality will be the *union* (total) of the shaded regions.

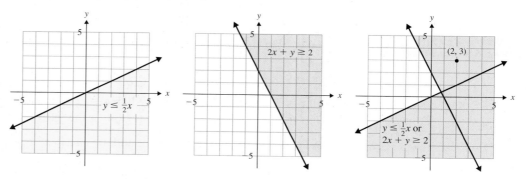

Any point in the shaded region of the third graph will be a solution to the compound inequality $y \leq \frac{1}{2}x$ or $2x + y \geq 2$. This means the point must satisfy $y \leq \frac{1}{2}x$ or $2x + y \geq 2$ or both. One point in the shaded region is $(2, 3)$.

Test Point	Substitute into $y \leq \frac{1}{2}x$	Substitute into $2x + y \geq 2$	Solution?
$(2, 3)$	$3 \leq \frac{1}{2}(2)$ $3 \leq 1$ False	$2(2) + 3 \geq 2$ $7 \geq 2$ True	Yes

Although $(2, 3)$ does not satisfy $y \leq \frac{1}{2}x$, it *does* satisfy $2x + y \geq 2$, so it *is* a solution of the compound inequality.

Choose a point in the region that is *not* shaded to verify that it does not satisfy either inequality.

[YOU TRY 4] Graph the compound inequality $x \geq -4$ or $x - 3y \leq -3$.

Using Technology

To graph a linear inequality in two variables using a graphing calculator, first solve the inequality for y. Then graph the boundary line found by replacing the inequality symbol with an = symbol. For example, to graph the inequality $2x - y \leq 5$, solve it for $y \geq 2x - 5$. Graph the boundary equation $y = 2x - 5$ using a solid line since the inequality symbol is $\leq$.

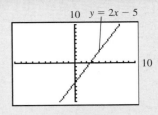

Press $\boxed{Y=}$, enter $2x - 5$ in Y_1, press $\boxed{ZOOM}$, and select 6:ZStandard to graph the equation as shown.

If the inequality symbol is $\leq$, shade below the boundary line. If the inequality symbol is $\geq$, shade above it. To shade above the line, press $\boxed{Y=}$ and move the cursor to the left of Y_1 using the left arrow key. Press $\boxed{ENTER}$ twice and then move the cursor to the next line as shown below left. Press $\boxed{GRAPH}$ to graph the inequality as shown below right.

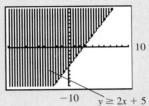

To shade below the line, press $\boxed{Y=}$ and move the cursor to the left of Y_1 using the left arrow key. Keep pressing $\boxed{ENTER}$ until you see ▲ next to Y_1, then move the cursor to the next line as shown below left. Press $\boxed{GRAPH}$ to graph the inequality $y \leq 2x - 5$ as shown below right.

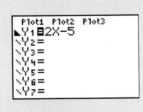
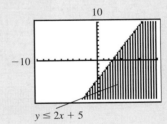

Graph the linear inequalities in two variables.

1) $y \leq 5x - 2$ 2) $y \geq x - 4$ 3) $x - 2y \leq 6$
4) $y - x \geq 5$ 5) $y \leq -4x + 1$ 6) $y \geq 3x - 6$

ANSWERS TO [YOU TRY] EXERCISES

1) a) 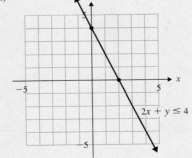   b)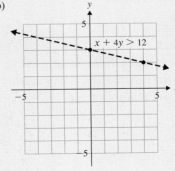

SECTION 4.4 Linear and Compound Linear Inequalities in Two Variables

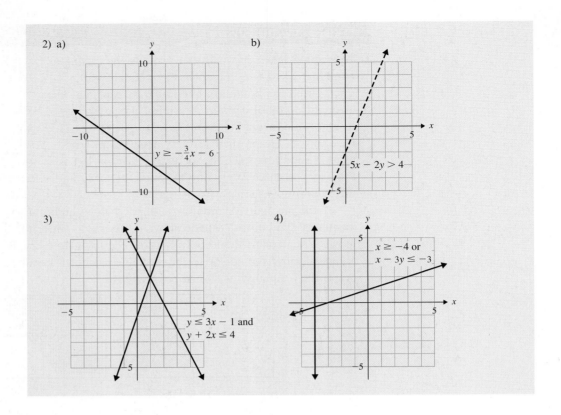

ANSWERS TO TECHNOLOGY EXERCISES

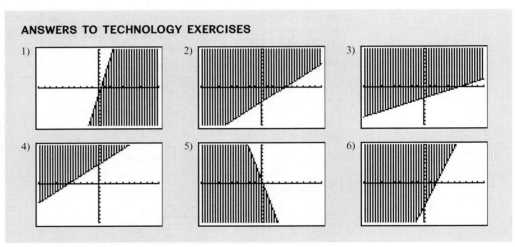

E Evaluate 4.4 Exercises

Do the exercises, and check your work.

Objective 1: Graph a Linear Inequality in Two Variables

The graphs of linear inequalities are given next. For each, find three points that satisfy the inequality and three that are not in the solution set.

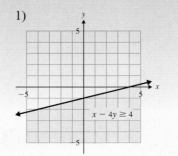
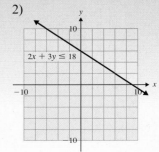

3)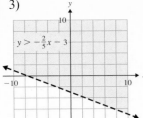

4)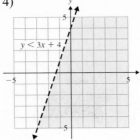

5)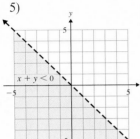

6)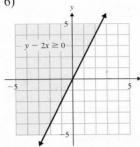

7) Will the boundary line you draw to graph $3x - 4y < 5$ be solid or dotted?

8) Are points on solid boundary lines included in the inequality's solution set?

Graph the inequalities. Use a test point.

9) $2x + y \geq 6$

10) $4x + y \leq 3$

11) $y < x + 2$

12) $y > \frac{1}{2}x - 1$

13) $2x - 7y \leq 14$

14) $4x + 3y < 15$

15) $y < x$

16) $y \geq 3x$

17) $y \geq -5$

18) $x < 1$

19) Should you shade the region above or below the boundary line for the inequality $y \leq 7x + 2$?

20) Should you shade the region above or below the boundary line for the inequality $y > 2x + 4$?

Use the slope-intercept method to graph each inequality.

21) $y \leq 4x - 3$

22) $y \geq \frac{5}{2}x - 8$

23) $y > \frac{2}{5}x - 4$

24) $y < \frac{1}{4}x + 1$

25) $6x + y > 3$

26) $2x + y > -5$

27) $9x - 3y \leq -21$

28) $3x + 5y < -20$

29) $x > 2y$

30) $x - y \leq 0$

31) To graph an inequality like $y \geq \frac{1}{3}x + 2$, would you rather use a test point or use the slope-intercept form?

32) To graph an inequality like $7x + 2y < 10$, would you rather use a test point or the slope-intercept form? Why?

Graph using either a test point or the slope-intercept method.

33) $y > -\frac{3}{4}x + 1$

34) $y \leq \frac{1}{3}x - 6$

35) $5x + 2y < -8$

36) $4x + y < 7$

37) $5x - 3y \geq -9$

38) $9x - 3y \leq 21$

39) $x > 2$

40) $y \leq 4$

41) $3x - 4y > 12$

42) $6x - y \leq 2$

Objective 2: Graph a Compound Linear Inequality in Two Variables

43) Is $(3, 5)$ in the solution set of the compound inequality $x - y \geq -6$ and $2x + y < 7$? Why or why not?

44) Is $(3, 5)$ in the solution set of the compound inequality $x - y \geq -6$ or $2x + y < 7$? Why or why not?

Graph each compound inequality.

45) $x \leq 4$ and $y \geq -\frac{3}{2}x + 3$

46) $y \leq \frac{1}{4}x + 2$ and $y \geq -1$

47) $y < x + 4$ and $y \geq -3$

48) $x < 3$ and $y > \frac{2}{3}x - 1$

49) $2x - 3y < -9$ and $x + 6y < 12$

50) $5x - 3y > 9$ and $2x + 3y \leq 12$

51) $y \leq -x - 1$ or $x \geq 6$

52) $y \leq 2$ or $y \leq \frac{4}{5}x + 2$

53) $y \leq 4$ or $4y - 3x \geq -8$

54) $x + 3y \geq 3$ or $x \geq -2$

55) $y > -\frac{2}{3}x + 1$ or $-2x + 5y \leq 0$

56) $y > x - 4$ or $3x + 2y \geq 12$

57) $x \geq 5$ and $y \leq -3$

58) $x \leq 6$ and $y \geq 1$

59) $y < 4$ or $x \geq -3$

60) $x \geq 2$ or $y \geq -6$

61) $2x + 5y < 15$ or $y \leq \frac{3}{4}x - 1$

62) $y - 2x \leq 1$ and $y \geq -\frac{1}{5}x - 2$

63) $y \geq \frac{2}{3}x - 4$ and $4x + y \leq 3$

64) $y < 5x + 2$ or $x + 4y < 12$

R Rethink

R1) Why do you need to use a test point that is not on the boundary line when solving a linear inequality in two variables?

R2) How do you represent the solution set of a system of linear inequalities in two variables?

R3) When do you draw a boundary line solid?

R4) When do you draw a boundary line dashed?

R5) Which objective is the most difficult for you?

4.5 Introduction to Functions

P Prepare	O Organize
What are your objectives for Section 4.5?	How can you accomplish each objective?
1 Define and Identify Relations, Functions, Domain, and Range	• Learn the definitions of a *relation, domain, range,* and *function,* and write them in your own words. • Be able to determine when a *relation* is also a *function.* • Understand **The Vertical Line Test.** • Complete the given examples on your own. • Complete You Trys 1 and 2.
2 Given an Equation, Determine Whether y Is a Function of x and Find the Domain	• Know the property for determining whether y is a function of x. • Know the procedure for **Finding the Domain of a Relation,** and write it in your own words. • Complete the given examples on your own. • Complete You Trys 3 and 4.
3 Use Function Notation	• Write the definition of *function notation* in your own words. • Be able to evaluate functions for given values of the independent variable. • Complete the given example on your own. • Complete You Try 5.
4 Find Function Values for Real-Number Values of the Variable	• Understand what it means to find a function value. • Be able to find the range for a given domain for functions represented by ordered pairs and graphs. • Complete the given examples on your own. • Complete You Trys 6 and 7.
5 Evaluate a Function for Variables or Expressions	• Know how to substitute variables or expressions to find a function value. • Complete the given example on your own. • Complete You Try 8.

What are your objectives for Section 4.5?	How can you accomplish each objective?
6 Define and Graph a Linear Function	• Learn the definition of a *linear function*. Write your own example in your notes. • Be able to graph a linear function using the slope and *y*-intercept. • Complete the given example on your own. • Complete You Try 9.
7 Solve Problems Using Linear Functions	• Know how to label the graph for a linear function. • Be able to choose a meaningful variable for the independent variable and the function. • Understand the meaning of function values. • Complete the given example on your own.

Read the explanations, follow the examples, take notes, and complete the You Trys.

If you are driving on a highway at a constant speed of 60 miles per hour, the distance you travel depends on the amount of time spent driving.

Driving Time	Distance Traveled
1 hr	60 mi
2 hr	120 mi
2.5 hr	150 mi
3 hr	180 mi

We can express these relationships with the ordered pairs

(1, 60)   (2, 120)   (2.5, 150)   (3, 180)

where the first coordinate represents the driving time (in hours), and the second coordinate represents the distance traveled (in miles).

We can also describe this relationship with the equation

$$y = 60x$$

where y is the distance traveled, in miles, and x is the number of hours spent driving.

The distance traveled *depends on* the amount of time spent driving. Therefore, the distance traveled is the **dependent variable,** and the driving time is the **independent variable.** In terms of x and y, since the value of y depends on the value of x, y is the *dependent variable* and x is the *independent variable*.

1 Define and Identify Relations, Functions, Domain, and Range

If we form a set of ordered pairs from the ones listed above, we get the *relation*
{(1, 60), (2, 120), (2.5, 150), (3, 180)}.

Definition

A **relation** is any set of ordered pairs.

Definition

The **domain** of a relation is the set of all values of the independent variable (the first coordinates in the set of ordered pairs). The **range** of a relation is the set of all values of the dependent variable (the second coordinates in the set of ordered pairs).

The domain of the given relation is {1, 2, 2.5, 3}. The range of the relation is {60, 120, 150, 180}.

The relation {(1, 60), (2, 120), (2.5, 150), (3, 180)} is also a *function* because every first coordinate corresponds to *exactly one* second coordinate. A function is a very important concept in mathematics.

> **Hint**
> Remember that only special types of relations can be classified as functions.

Definition

A **function** is a special type of relation. If each element of the domain corresponds to *exactly one* element of the range, then the relation is a function.

Relations and functions can be represented in another way—as a *correspondence* or a *mapping* from one set, the domain, to another, the range. In this representation, the domain is the set of all values in the first set, and the range is the set of all values in the second set. Our previously stated definition of a function still holds.

EXAMPLE 1 Identify the domain and range of each relation, and determine whether each relation is a function.

a) {(2, 0), (3, 1), (6, 2), (6, −2)}

b) $\left\{(-2, -6), (0, -5), \left(1, -\frac{9}{2}\right), (4, -3), \left(5, -\frac{5}{2}\right)\right\}$

c)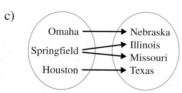

Solution

a) The *domain* is the set of first coordinates, {2, 3, 6}. (We write the 6 in the set only once even though it appears in two ordered pairs.) The *range* is the set of second coordinates, {0, 1, 2, −2}.

To determine whether or not this relation is a function ask yourself, *"Does every first coordinate correspond to exactly one second coordinate?"* No. In the ordered pairs (6, 2) and (6, −2), the same first coordinate, 6, corresponds to two different second coordinates, 2 and −2. Therefore, this relation is *not* a function.

b) The *domain* is {−2, 0, 1, 4, 5}. The *range* is $\left\{-6, -5, -\frac{9}{2}, -3, -\frac{5}{2}\right\}$.

Ask yourself, "Does every first coordinate correspond to *exactly one* second coordinate?" Yes. This relation *is* a function.

c) The *domain* is {Omaha, Springfield, Houston}. The *range* is {Nebraska, Illinois, Missouri, Texas}.

One of the elements in the domain, Springfield, corresponds to *two* elements in the range, Illinois and Missouri. Therefore, this relation is *not* a function.

[**YOU TRY 1**] Identify the domain and range of each relation, and determine whether each relation is a function.

a) {(−1, −3), (1, 1), (2, 3), (4, 7)}   b) {(−12, −6), (−12, 6), (−1, √3), (0, 0)}

c)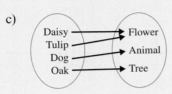

We stated earlier that a relation is a function if each element of the domain corresponds to *exactly one* element of the range.

If the ordered pairs of a relation are such that the first coordinates represent *x*-values and the second coordinates represent *y*-values (the ordered pairs are in the form (*x*, *y*)), then we can think of the definition of a function in this way:

Definition

A relation is a **function** if each *x*-value corresponds to exactly one *y*-value.

What does a function look like when it is graphed? Following are the graphs of the ordered pairs in the relations of Example 1a) and 1b).

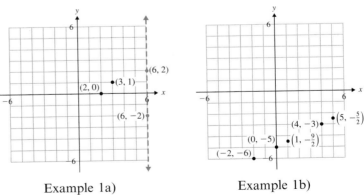

Example 1a)
not a function

Example 1b)
is a function

The relation in Example 1a) is *not* a function since the *x*-value of 6 corresponds to *two different y*-values, 2 and −2. Notice that we can draw a vertical line that intersects the graph in more than one point—the line through (6, 2) and (6, −2).

The relation in Example 1b), however, *is* a function—each *x*-value corresponds to only one *y*-value. Anywhere we draw a vertical line through the points on the graph of this relation, the line intersects the graph in *exactly one point*.

This leads us to the **vertical line test** for a function.

> **Procedure** The Vertical Line Test
>
> If there is no vertical line that can be drawn through a graph so that it intersects the graph more than once, then the graph represents a function.
>
> If a vertical line *can* be drawn through a graph so that it intersects the graph more than once, then the graph does *not* represent a function.

EXAMPLE 2   Use the vertical line test to determine whether each graph, in blue, represents a function. Identify the domain and range.

a) 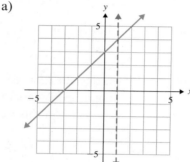   b)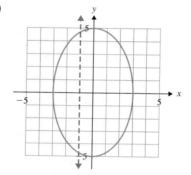

Solution

a) Anywhere a vertical line is drawn through the graph, the line will intersect the graph only once. *This graph represents a function.*

The arrows on the graph indicate that the graph continues without bound.

The domain of this function is the set of *x*-values on the graph. Since the graph continues indefinitely in the *x*-direction, the domain is the set of all real numbers. *The domain is* $(-\infty, \infty)$.

The range of this function is the set of *y*-values on the graph. Since the graph continues indefinitely in the *y*-direction, the range is the set of all real numbers. *The range is* $(-\infty, \infty)$.

b) This graph fails the vertical line test because we can draw a vertical line through the graph that intersects it more than once. *This graph does not represent a function.*

W Hint
Notice that the graph of a line, except for a vertical line, will always pass the vertical line test. Therefore, every nonvertical line represents a function.

The set of *x*-values on the graph includes all real numbers from −3 to 3. *The domain is* $[-3, 3]$.

The set of *y*-values on the graph includes all real numbers from −5 to 5. *The range is* $[-5, 5]$.

[YOU TRY 2] Use the vertical line test to determine whether each relation is also a function. Then, identify the domain and range.

a)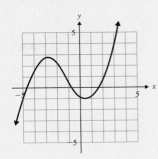

b)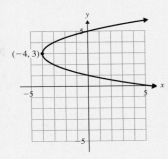

2 Given an Equation, Determine Whether y Is a Function of x and Find the Domain

We can also represent relations and functions with equations. The example given at the beginning of the section illustrates this.

The equation $y = 60x$ describes the distance traveled (y, in miles) after x hours of driving at 60 mph. If $x = 2$, $y = 60(2) = 120$. If $x = 3$, $y = 60(3) = 180$, and so on. For *every* value of x that could be substituted into $y = 60x$ there is *exactly one* corresponding value of y. Therefore, $y = 60x$ is a function.

Furthermore, we can say that *y is a function of x*. In the function described by $y = 60x$, the value of y *depends on* the value of x. That is, x is the independent variable and y is the dependent variable.

> **Property** *y* Is a Function of *x*
>
> If a function describes the relationship between x and y so that x is the independent variable and y is the dependent variable, then we say that *y is a function of x*.

EXAMPLE 3   Determine whether each relation describes y as a function of x.

a) $y = x + 2$ b) $y^2 = x$

Solution

a) To begin, substitute a couple of values for x and solve for y to get an idea of what is happening in this relation.

$x = 0$	$x = 3$	$x = -4$
$y = x + 2$	$y = x + 2$	$y = x + 2$
$y = 0 + 2$	$y = 3 + 2$	$y = -4 + 2$
$y = 2$	$y = 5$	$y = -2$

The ordered pairs (0, 2), (3, 5), and (−4, −2) satisfy $y = x + 2$. Each of the values substituted for x has *one* corresponding y-value. In this case, when *any* value is substituted for x there will be *exactly* one corresponding value of y. Therefore, $y = x + 2$ *is* a function.

b) Substitute a couple of values for x, and solve for y to get an idea of what is happening in this relation.

$x = 0$	$x = 4$	$x = 9$
$y^2 = x$	$y^2 = x$	$y^2 = x$
$y^2 = 0$	$y^2 = 4$	$y^2 = 9$
$y = 0$	$y = \pm 2$	$y = \pm 3$

The ordered pairs $(0, 0)$, $(4, 2)$, $(4, -2)$, $(9, 3)$, and $(9, -3)$ satisfy $y^2 = x$. Since $2^2 = 4$ and $(-2)^2 = 4$, $x = 4$ corresponds to two different y-values, 2 and -2. Likewise, $x = 9$ corresponds to the two different y-values of 3 and -3 since $3^2 = 9$ and $(-3)^2 = 9$. Finding one such example is enough to determine that $y^2 = x$ is *not* a function.

[YOU TRY 3] Determine whether each relation describes y as a function of x.

a) $y = 3x - 5$ b) $y^2 = x + 1$

Next, we will discuss how to determine the domain of a relation written as an equation.

Sometimes, it is helpful to ask yourself, "Is there any number that *cannot* be substituted for x?"

EXAMPLE 4   Determine the domain of each relation, and determine whether each relation describes y as a function of x.

a) $y = \dfrac{1}{x}$ b) $y = \dfrac{7}{x - 3}$ c) $y = -2x + 6$

Solution

a) To determine the domain of $y = \dfrac{1}{x}$ ask yourself, "Is there any number that *cannot* be substituted for x?" Yes. x cannot equal zero because a fraction is undefined if its denominator equals zero.

The domain contains all real numbers *except* 0. We can write the domain in interval notation as $(-\infty, 0) \cup (0, \infty)$.

$y = \dfrac{1}{x}$ *is a function* since each value of x in the domain will have only one corresponding value of y.

W Hint
Remember that a fraction with a denominator of 0 is undefined. Therefore, we must exclude any x-values from the domain of the function that make the denominators equal to 0.

b) Ask yourself, "Is there any number that *cannot* be substituted for x in $y = \dfrac{7}{x - 3}$?" Look at the denominator. When will it equal 0? Set the denominator equal to 0, and solve for x.

$x - 3 = 0$ Set the denominator $= 0$.
$x = 3$ Solve.

When $x = 3$, the denominator of $y = \dfrac{7}{x - 3}$ equals zero. The domain contains all real numbers *except* 3. Write the domain in interval notation as $(-\infty, 3) \cup (3, \infty)$.

$y = \dfrac{7}{x - 3}$ is a function. For every value that can be substituted for x there is only one corresponding value of y.

c) *Is there any number that cannot be substituted for x in $y = -2x + 6$?* No. Any real number can be substituted for x, and $y = -2x + 6$ will be defined.

The domain consists of all real numbers which can be written as $(-\infty, \infty)$.

Every value substituted for x will have exactly one corresponding y-value. $y = -2x + 6$ *is a function.*

Procedure Finding the Domain of a Relation

The domain of a relation that is written as an equation, where y is in terms of x, is the set of all real numbers that can be substituted for the independent variable, x. When determining the domain of a relation, it can be helpful to keep these tips in mind.

1) Ask yourself, "Is there any number that *cannot* be substituted for x?"
2) If x is in the denominator of a fraction, determine what value of x will make the denominator equal 0 by setting the expression equal to zero. Solve for x. This x-value is *not* in the domain.

The domain consists of all real numbers that can be substituted for x.

[YOU TRY 4] Determine the domain of each relation, and determine whether each relation describes y as a function of x.

a) $y = x - 9$ b) $y = -x^2 + 6$ c) $y = \dfrac{4}{x + 1}$

3 Use Function Notation

We can use *function notation* to name functions. If a relation is a function, then $f(x)$ can be used in place of y. In this case, $f(x)$ *is the same as y.*

For example, $y = x + 3$ is a function. We can also write $y = x + 3$ as $f(x) = x + 3$. *They mean the same thing.*

W Hint
Remember that only when a relation is classified as a function, do we replace y with $f(x)$.

Definition

$y = f(x)$ is called **function notation**, and it is read as "y equals f of x." $y = f(x)$ means that y is a function of x (y depends on x).

EXAMPLE 5

a) Evaluate $y = x + 3$ for $x = 2$.   b) If $f(x) = x + 3$, find $f(2)$.

Solution

a) To evaluate $y = x + 3$ for $x = 2$ means to substitute 2 for x and find the corresponding value of y.

$$y = x + 3$$
$$y = 2 + 3 \quad \text{Substitute 2 for } x.$$
$$y = 5$$

W Hint
Notice that $f(2)$ indicates that we replace the x-value with 2.

When $x = 2$, $y = 5$. We can also say that the ordered pair $(2, 5)$ satisfies $y = x + 3$.

b) To find $f(2)$ (read as "f of 2") means to find the value of the function when $x = 2$.

$$f(x) = x + 3$$
$$f(2) = 2 + 3 \quad \text{Substitute 2 for } x.$$
$$f(2) = 5$$

We can also say that the ordered pair $(2, 5)$ satisfies $f(x) = x + 3$ where the ordered pair represents $(x, f(x))$.

Note

Example 5 illustrates that evaluating $y = x + 3$ for $x = 2$ and finding $f(2)$ when $f(x) = x + 3$ is *exactly* the same thing. Remember, $f(x)$ is another name for y.

[YOU TRY 5] a) Evaluate $y = -2x + 4$ for $x = 1$.   b) If $f(x) = -2x + 4$, find $f(1)$.

Different letters can be used to name functions. $g(x)$ is read as "g of x," $h(x)$ is read as "h of x," and so on. Also, the function notation does *not* indicate multiplication; $f(x)$ does *not* mean f times x.

BE CAREFUL $f(x)$ does *not* mean f times x.

4 Find Function Values for Real-Number Values of the Variable

Sometimes, we call evaluating a function for a certain value *finding a function value*.

EXAMPLE 6

Let $f(x) = 6x - 5$ and $g(x) = x^2 - 8x + 3$. Find the following function values.

a) $f(3)$ b) $f(0)$ c) $g(-1)$

Solution

a) "Find $f(3)$" means to find the value of the function when $x = 3$. Substitute 3 for x.

$$f(x) = 6x - 5$$
$$f(3) = 6(3) - 5 = 18 - 5 = 13$$
$$f(3) = 13$$

We can also say that the ordered pair $(3, 13)$ satisfies $f(x) = 6x - 5$.

b) To find $f(0)$, substitute 0 for x in the function $f(x)$.

$$f(x) = 6x - 5$$
$$f(0) = 6(0) - 5 = 0 - 5 = -5$$
$$f(0) = -5$$

The ordered pair $(0, -5)$ satisfies $f(x) = 6x - 5$.

c) To find $g(-1)$, substitute -1 for every x in the function $g(x)$.

$$g(x) = x^2 - 8x + 3$$
$$g(-1) = (-1)^2 - 8(-1) + 3 = 1 + 8 + 3 = 12$$
$$g(-1) = 12$$

The ordered pair $(-1, 12)$ satisfies $g(x) = x^2 - 8x + 3$.

[**YOU TRY 6**] Let $f(x) = -4x + 1$ and $h(x) = 2x^2 + 3x - 7$. Find the following function values.

a) $f(5)$ b) $f(-2)$ c) $h(0)$ d) $h(3)$

We can also find function values for functions represented by a set of ordered pairs, a correspondence, or a graph.

EXAMPLE 7 Find $f(4)$ for each function.

a) $f = \{(-2, -11), (0, -5), (3, 4), (4, 7)\}$

b) Domain f Range c)

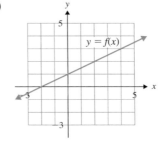

Solution

a) Since this function is expressed as a set of ordered pairs, finding $f(4)$ means finding the y-coordinate of the ordered pair with x-coordinate 4. The ordered pair with x-coordinate 4 is $(4, 7)$, so $f(4) = 7$.

b) In this function, the element 4 in the domain corresponds to the element -8 in the range. Therefore, $f(4) = -8$.

c) To find $f(4)$ from the graph of this function means to find the y-coordinate of the point on the line that has an x-coordinate of 4. Find 4 on the x-axis. Then, go straight up to the graph and move to the left to read the y-coordinate of the point on the graph where the x-coordinate is 4. That y-coordinate is 3. So, $f(4) = 3$.

W Hint
Remember that $f(4)$ indicates that 4 is in the domain of the function. $f(4)$ is equal to the range value that corresponds to 4.

[YOU TRY 7] Find $f(2)$ for each function.

a) $f = \{(-5, 8), (-1, 2), (2, -3), (6, -9)\}$

b)

c)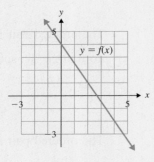

5 Evaluate a Function for Variables or Expressions

Functions can be evaluated for variables or expressions.

EXAMPLE 8

Let $h(x) = 5x + 3$. Find each of the following and simplify.

a) $h(c)$ b) $h(t - 4)$

Solution

a) Finding $h(c)$ (read as *h of c*) means to substitute c for x in the function h, and simplify the expression as much as possible.

$$h(x) = 5x + 3$$
$$h(c) = 5c + 3 \quad \text{Substitute } c \text{ for } x.$$

W Hint
Notice that $h(c)$ indicates that we replace the x-value with c.

b) Finding $h(t - 4)$ (read as *h of t minus 4*) means to substitute $t - 4$ for x in function h, and simplify the expression as much as possible. Since $t - 4$ contains two terms, we must put it in parentheses.

$$h(x) = 5x + 3$$
$$h(t - 4) = 5(t - 4) + 3 \quad \text{Substitute } t - 4 \text{ for } x.$$
$$h(t - 4) = 5t - 20 + 3 \quad \text{Distribute.}$$
$$h(t - 4) = 5t - 17 \quad \text{Combine like terms.}$$

W Hint
Notice that $h(t - 4)$ indicates that we replace the x-value with $t - 4$.

[YOU TRY 8] Let $f(x) = 2x - 7$. Find each of the following and simplify.

a) $f(k)$ b) $f(p + 3)$

6 Define and Graph a Linear Function

Earlier in this chapter, we learned that a linear equation can have the form $y = mx + b$. Similarly, a *linear function* has the form $f(x) = mx + b$.

W Hint
Remember that a vertical line has a undefined slope. Its equation cannot be written in $y = mx + b$ form. The graph of a vertical line does not represent a function.

Definition

A **linear function** has the form $f(x) = mx + b$, where m and b are real numbers, m is the *slope* of the line, and $(0, b)$ is the *y-intercept*.

EXAMPLE 9

Graph $f(x) = -\dfrac{1}{3}x - 1$ using the slope and y-intercept.

Solution

$$f(x) = -\dfrac{1}{3}x - 1$$

$m = -\dfrac{1}{3}$ y-int: $(0, -1)$

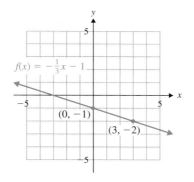

To graph this function, first plot the y-intercept, $(0, -1)$, then use the slope to locate another point on the line.

[YOU TRY 9]

Graph $f(x) = \dfrac{3}{4}x - 2$ using the slope and y-intercept.

7 Solve Problems Using Linear Functions

The independent variable of a function does not have to be x. When using functions to model real-life problems, we often choose a more "meaningful" letter to represent a quantity. For example, if the independent variable represents time, we may use the letter t instead of x. The same is true for naming the function.

No matter what letter is chosen for the independent variable, *the horizontal axis is used to represent the values of the independent variable, and the vertical axis represents the function values.*

EXAMPLE 10

A compact disc is read at 44.1 kHz (kilohertz). This means that a CD player scans 44,100 samples of sound per second on a CD to produce the sound that we hear. The function

$$S(t) = 44.1t$$

tells us how many samples of sound, $S(t)$, in *thousands* of samples, are read after t seconds. (www.mediatechnics.com)

a) How many samples of sound are read after 20 sec?
b) How many samples of sound are read after 1.5 min?
c) How long would it take the CD player to scan 1,764,000 samples of sound?
d) What is the smallest value t could equal in the context of this problem?
e) Graph the function.

Solution

a) To determine how much sound is read after 20 sec, let $t = 20$ and find $S(20)$.

$S(t) = 44.1t$
$S(20) = 44.1(20)$ Substitute 20 for t.
$S(20) = 882$ Multiply.

$S(t)$ is in thousands, so the number of samples read is $882 \cdot 1000 = 882{,}000$ samples of sound.

b) To determine how much sound is read after 1.5 min, do we let $t = 1.5$ and find $S(1.5)$? *No.* Recall that t is in *seconds*. Change 1.5 min to seconds before substituting for t. We must use the correct units in the function.

$$1.5 \text{ min} = 90 \text{ sec}$$

Let $t = 90$ and find $S(90)$.

$$S(t) = 44.1t$$
$$S(90) = 44.1(90)$$
$$S(90) = 3969$$

$S(t)$ is in thousands, so the number of samples read is $3969 \cdot 1000 = 3{,}969{,}000$ samples of sound.

c) Since we are asked to determine how *long* it would take a CD player to scan 1,764,000 samples of sound, we will be solving for t. What do we substitute for $S(t)$? $S(t)$ is in *thousands*, so substitute $1{,}764{,}000 \div 1000 = 1764$ for $S(t)$. Find t when $S(t) = 1764$.

$$S(t) = 44.1t$$
$$1764 = 44.1t \quad \text{Substitute 1764 for } S(t).$$
$$40 = t \quad \text{Divide by 44.1.}$$

It will take 40 sec for the CD player to scan 1,764,000 samples of sound.

d) Since t represents the number of seconds a CD has been playing, the smallest value that makes sense for t is 0.

e) Since $S(t)$ is in thousands of samples, the information we obtained in parts a), b), and c) can be written as the ordered pairs (20, 882), (90, 3969), and (40, 1764). In addition, when $t = 0$ (from part d) we obtain $S(0) = 44.1(0) = 0$. (0, 0) is an additional ordered pair on the graph of the function.

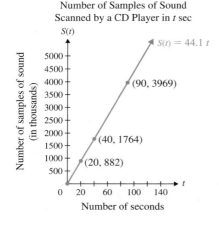

Number of Samples of Sound Scanned by a CD Player in t sec

ANSWERS TO [YOU TRY] EXERCISES

1) a) domain: $\{-1, 1, 2, 4\}$; range: $\{-3, 1, 3, 7\}$; yes b) domain: $\{-12, -1, 0\}$; range: $\{-6, 6, \sqrt{3}, 0\}$; no c) domain: {Daisy, Tulip, Dog, Oak}; range: {Flower, Animal, Tree}; yes
2) a) function; domain: $(-\infty, \infty)$; range: $(-\infty, \infty)$ b) not a function; domain: $[-4, \infty)$; range: $(-\infty, \infty)$ 3) a) yes b) no   4) a) $(-\infty, \infty)$; function b) $(-\infty, \infty)$; function c) $(-\infty, -1) \cup (-1, \infty)$; function   5) a) 2 b) 2   6) a) -19 b) 9 c) -7 d) 20
7) a) -3 b) 5 c) 1   8) a) $f(k) = 2k - 7$ b) $f(p + 3) = 2p - 1$
9) $m = \dfrac{3}{4}$, y-int: $(0, -2)$

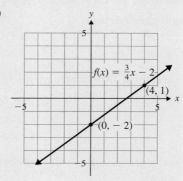

Evaluate 4.5 Exercises

Do the exercises, and check your work.

Objective 1: Define and Identify Relations, Functions, Domain, and Range

1) a) What is a relation?

   b) What is a function?

   c) Give an example of a relation that is also a function.

2) Give an example of a relation that is *not* a function.

Identify the domain and range of each relation, and determine whether each relation is a function.

3) $\{(5, 13), (-2, 6), (1, 4), (-8, -3)\}$

4) $\{(0, -3), (1, -4), (1, -2), (16, -5), (16, -1)\}$

5) $\{(9, -1), (25, -3), (1, 1), (9, 5), (25, 7)\}$

6) $\left\{(-4, -2), \left(-3, -\dfrac{1}{2}\right), \left(-1, -\dfrac{1}{2}\right), (0, -2)\right\}$

7)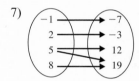

8)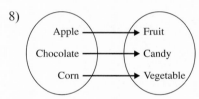

9)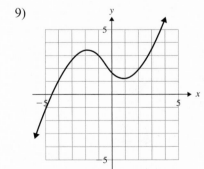

10)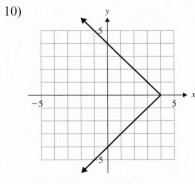

11)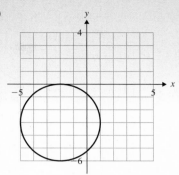

12)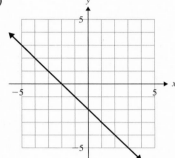

13)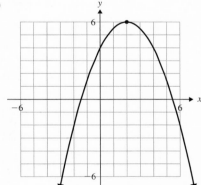

14)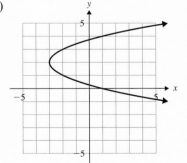

Objective 2: Given an Equation, Determine Whether y Is a Function of x and Find the Domain

Determine whether each relation describes y as a function of x.

15) $y = x - 9$

16) $y = x + 4$

17) $y = 2x + 7$

18) $y = \dfrac{2}{3}x + 1$

19) $x = y^4$

20) $x = y^2 - 3$

21) $y^2 = x - 4$

22) $y^2 = x + 9$

Determine the domain of each relation, and determine whether each relation describes y as a function of x.

23) $y = x - 5$

24) $y = 2x + 1$

25) $y = x^3 + 2$

26) $y = -x^3 + 4$

27) $x = |y|$

28) $x = y^4$

29) $y = -\dfrac{8}{x}$

30) $y = \dfrac{5}{x}$

31) $y = \dfrac{9}{x + 4}$

32) $y = \dfrac{2}{x - 7}$

33) $y = \dfrac{3}{x - 5}$

34) $y = \dfrac{1}{x + 10}$

35) $y = \dfrac{6}{5x - 3}$

36) $y = -\dfrac{4}{9x + 8}$

37) $y = \dfrac{15}{3x + 4}$

38) $y = \dfrac{5}{6x - 1}$

39) $y = -\dfrac{5}{9 - 3x}$

40) $y = \dfrac{1}{-6 + 4x}$

41) $y = \dfrac{x}{12}$

42) $y = \dfrac{x + 8}{7}$

Mixed Exercises: Objectives 3 and 4

43) Explain what it means when an equation is written in the form $y = f(x)$.

44) Does $y = f(x)$ mean "$y = f$ times x"? Explain.

45) a) Evaluate $y = 5x - 8$ for $x = 3$.
    b) If $f(x) = 5x - 8$, find $f(3)$.

46) a) Evaluate $y = -3x - 2$ for $x = -4$.
    b) If $f(x) = -3x - 2$, find $f(-4)$.

Let $f(x) = -4x + 7$ and $g(x) = x^2 + 9x - 2$. Find the following function values.

47) $f(5)$

48) $f(2)$

49) $f(0)$

50) $f\left(-\dfrac{3}{2}\right)$

51) $g(4)$

52) $g(1)$

53) $g(-1)$

54) $g(0)$

55) $g\left(-\dfrac{1}{2}\right)$

56) $g\left(\dfrac{1}{3}\right)$

57) $f(6) - g(6)$

58) $f(-4) - g(-4)$

For each function f in Exercises 59–64, find $f(-1)$ and $f(4)$.

59) $f = \left\{(-8, -1), \left(-1, \dfrac{5}{2}\right), (4, 5), (10, 8)\right\}$

60) $f = \{(-3, 16), (-1, 10), (0, 7), (1, 4), (4, -5)\}$

61)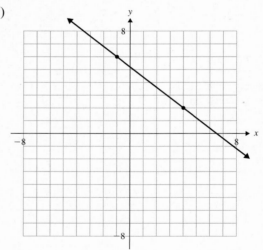

62)

63) Domain f Range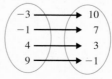

64) Domain f Range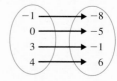

65) $f(x) = -3x - 2$. Find x so that $f(x) = 10$.

66) $f(x) = 5x + 4$. Find x so that $f(x) = 9$.

67) $g(x) = \frac{2}{3}x + 1$. Find x so that $g(x) = 5$.

68) $h(x) = -\frac{1}{2}x - 6$. Find x so that $h(x) = -2$.

Objective 5: Evaluate a Function for Variables or Expressions

Fill It In
Fill in the blanks with either the missing mathematical step or reason for the given step.

69) Let $f(x) = -9x + 2$. Find $f(n - 3)$.
    _____ Substitute $n - 3$ for x.
    $= -9n + 27 + 2$ _____
    _____ Simplify.

70) Let $f(x) = 4x - 5$. Find $f(k + 6)$.
    $f(k + 6) = 4(k + 6) - 5$ _____
    _____ Distribute.
    _____ Simplify.

71) $f(x) = -7x + 2$ and $g(x) = x^2 - 5x + 12$. Find each of the following, and simplify.
    a) $f(c)$ b) $f(t)$
    c) $f(a + 4)$ d) $f(z - 9)$
    e) $g(k)$ f) $g(m)$
    g) $f(x + h)$ h) $f(x + h) - f(x)$

72) $f(x) = 5x + 6$ and $g(x) = x^2 - 3x - 11$. Find each of the following, and simplify.
    a) $f(n)$ b) $f(p)$
    c) $f(w + 8)$ d) $f(r - 7)$
    e) $g(b)$ f) $g(s)$
    g) $f(x + h)$ h) $f(x + h) - f(x)$

Objective 6: Define and Graph a Linear Function

Graph each function by making a table of values and plotting points.

73) $f(x) = x - 4$

74) $f(x) = x + 2$

75) $f(x) = \frac{2}{3}x + 2$

76) $g(x) = -\frac{3}{5}x + 2$

77) $h(x) = -3$

78) $g(x) = 1$

Graph each function by finding the x- and y-intercepts and one other point.

79) $g(x) = 3x + 3$

80) $k(x) = -2x + 6$

81) $f(x) = -\frac{1}{2}x + 2$

82) $f(x) = \frac{1}{3}x + 1$

83) $h(x) = x$

84) $f(x) = -x$

Graph each function using the slope and y-intercept.

85) $f(x) = -4x - 1$

86) $f(x) = -x + 5$

87) $g(x) = -\frac{1}{4}x - 2$

88) $h(x) = \frac{3}{5}x - 2$

89) $g(x) = 2x + \frac{1}{2}$

90) $h(x) = 3x + 1$

Graph each function

91) $s(t) = -\frac{1}{3}t - 2$

92) $k(d) = d - 1$

93) $A(r) = -3r$

94) $N(t) = 3.5t + 1$

Objective 7: Solve Problems Using Linear Functions

95) A truck on the highway travels at a constant speed of 54 mph. The distance, D (in miles), that the truck travels after t hr can be defined by the function
$$D(t) = 54t$$
a) How far will the truck travel after 2 hr?
b) How far will the truck travel after 4 hr?
c) How long does it take the truck to travel 135 mi?
d) Graph the function.

96) The velocity of an object, v (in feet per second), of an object during free-fall t sec after being dropped can be defined by the function
$$v(t) = 32t$$
a) Find the velocity of an object 1 sec after being dropped.
b) Find the velocity of an object 3 sec after being dropped.
c) When will the object be traveling at 256 ft/sec?
d) Graph the function.

97) If gasoline costs $3.50 per gallon, then the cost, C (in dollars), of filling a gas tank with g gal of gas is defined by

$$C(g) = \$3.50g$$

a) Find $C(8)$, and explain what this means in the context of the problem.

b) Find $C(15)$, and explain what this means in the context of the problem.

c) Find g so that $C(g) = 42$, and explain what this means in the context of the problem.

98) Jenelle earns $7.50 per hour at her part-time job. Her total earnings, E (in dollars), for working t hr can be defined by the function

$$E(t) = 7.50t$$

a) Find $E(10)$, and explain what this means in the context of the problem.

b) Find $E(15)$, and explain what this means in the context of the problem.

c) Find t so that $E(t) = 210$, and explain what this means in the context of the problem.

99) A 16 × DVD recorder can transfer 21.13 MB (megabytes) of data per second onto a recordable DVD. The function $D(t) = 21.13t$ describes how much data, D (in megabytes), is recorded on a DVD in t sec. (www.osta.org)

a) How much data is recorded after 12 sec?

b) How much data is recorded after 1 min?

c) How long would it take to record 422.6 MB of data?

d) Graph the function.

100) The average hourly wage of an embalmer in Illinois in 2011 was $25.06. Seth's earnings, E (in dollars), for working t hr in a week can be defined by the function $E(t) = 25.06t$. (www.bls.gov)

a) How much does Seth earn if he works 30 hr?

b) How much does Seth earn if he works 27 hr?

c) How many hours would Seth have to work to make $877.10?

d) If Seth can work at most 40 hr per week, what is the domain of this function?

e) Graph the function.

101) Law enforcement agencies use a computerized system called AFIS (Automated Fingerprint Identification System) to identify fingerprints found at crime scenes. One AFIS system can compare 30,000 fingerprints per second. The function

$$F(s) = 30s$$

describes how many fingerprints, $F(s)$ in thousands, are compared after s sec.

a) How many fingerprints can be compared in 2 sec?

b) How long would it take AFIS to search through 105,000 fingerprints?

102) Refer to the function in Exercise 101 to answer the following questions.

a) How many fingerprints can be compared in 3 sec?

b) How long would it take AFIS to search through 45,000 fingerprints?

103) Refer to the function in Example 10 on p. 207 to determine the following.

a) Find $S(50)$, and explain what this means in the context of the problem.

b) Find $S(180)$, and explain what this means in the context of the problem.

c) Find t so that $S(t) = 2646$, and explain what this means in the context of the problem.

104) Refer to the function in Exercise 99 to determine the following.

a) Find $D(10)$, and explain what this means in the context of the problem.

b) Find $D(120)$, and explain what this means in the context of the problem.

c) Find t so that $D(t) = 633.9$, and explain what this means in the context of the problem.

105) The graph shows the amount, A, of ibuprofen in Sasha's bloodstream t hr after she takes two tablets for a headache.

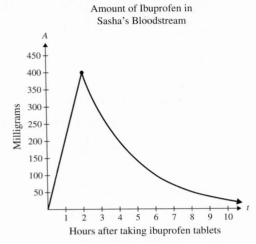

Amount of Ibuprofen in Sasha's Bloodstream

a) How long after taking the tablets will the amount of ibuprofen in her bloodstream be the greatest? How much ibuprofen is in her bloodstream at this time?

b) When will there be 100 mg of ibuprofen in Sasha's bloodstream?

c) How much of the drug is in her bloodstream after 4 hr?

d) Call this function A. Find $A(8)$, and explain what it means in the context of the problem.

106) The graph shows the number of gallons (in millions), G, of water entering a water treatment plant t hr after midnight on a certain day.

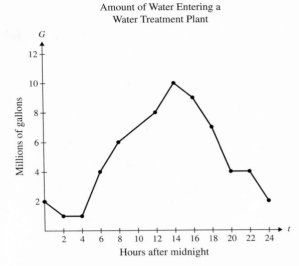

Amount of Water Entering a Water Treatment Plant

a) Identify the domain and range of this function.

b) How many gallons of water enter the facility at noon? At 10 P.M.?

c) At what time did the most water enter the treatment plant? How much water entered the treatment plant at this time?

d) At what time did the least amount of water enter the treatment plant?

e) Call this function G, find $G(18)$, and explain what it means in the context of the problem.

R Rethink

R1) What is the difference between a relation and a function?

R2) Are all functions considered to be relations?

R3) How do you determine if a graph represents a function?

R4) Is (a, b) an ordered pair of $f(b) = a$ or $f(a) = b$?

R5) If $h(3) = 7$, is 3 in the domain or range of the function?

R6) Which objective is the most difficult for you?

Group Activity — The Group Activity can be found online on Connect.

em POWER me Checklist for Effective Notes

Use the following checklist to evaluate your notes.

- ❑ My notes include the date, name of the class, and chapter number or section number, or topic being studied.
- ❑ My notes are written on one side of loose-leaf notepaper.
- ❑ My notes are legible and could be understood by someone other than myself.
- ❑ My notes do not include material unrelated to the class, such as things to pick up at the grocery store.
- ❑ My notes are focused on the major points the instructor discussed.
- ❑ My notes do not try to capture every single word the instructor said.
- ❑ My notes are organized and will be easy to study from.
- ❑ My notes include any material my instructor wrote on the board or featured in PowerPoint slides.
- ❑ My notes include any assignments the instructor gave at the end of class.
- ❑ My notes are in a notebook with my other notes from the course, so that I will be able to find them easily later.

Chapter 4: Summary

Definition/Procedure	Example

4.1 Introduction to Linear Equations in Two Variables

A **linear equation in two variables** can be written in the form $Ax + By = C$ where A, B, and C are real numbers and where both A and B do not equal zero.

To determine whether an ordered pair is a solution of an equation, substitute the values for the variables. **(p. 142)**

Is $(5, -3)$ a solution of $2x - 7y = 31$?
Substitute 5 for x and -3 for y.

$$2x - 7y = 31$$
$$2(5) - 7(-3) = 31$$
$$10 - (-21) = 31$$
$$10 + 21 = 31$$
$$31 = 31 \quad \text{True}$$

Yes, $(5, -3)$ is a solution.

Graphing by Plotting Points and Finding Intercepts

The graph of a linear equation in two variables, $Ax + By = C$, is a straight line. Each point on the line is a solution to the equation.

We can graph the line by plotting the points and drawing the line through them. **(p. 143)**

Graph $y = \frac{1}{2}x - 4$ by plotting points.

Make a table of values. Plot the points, and draw a line through them.

x	y
0	-4
2	-3
4	-2

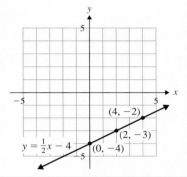

The **x-intercept** of an equation is the point where the graph intersects the x-axis. To find the x-intercept of the graph of an equation, let $y = 0$ and solve for x.

The **y-intercept** of an equation is the point where the graph intersects the y-axis. To find the y-intercept of the graph of an equation, let $x = 0$ and solve for y. **(p. 144)**

Graph $5x + 2y = 10$ by finding the intercepts and another point on the line.

x-intercept: Let $y = 0$ and solve for x.

$$5x + 2(0) = 10$$
$$5x = 10$$
$$x = 2$$

The x-intercept is $(2, 0)$.

y-intercept: Let $x = 0$, and solve for y.

$$5x + 2y = 10$$
$$5(0) + 2y = 10$$
$$2y = 10$$
$$y = 5$$

The y-intercept is $(0, 5)$. Another point on the line is $(4, -5)$.

Plot the points, and draw the line through them.

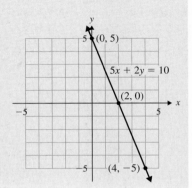

Definition/Procedure	Example
If a is a constant, then the graph of $x = a$ is a *vertical line* going through the point $(a, 0)$. If b is a constant, then the graph of $y = b$ is a *horizontal line* going through the point $(0, b)$. (p. 145)	Graph $x = 2$. Graph $y = 3$. 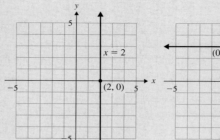
The Midpoint Formula If (x_1, y_1) and (x_2, y_2) are the endpoints of a line segment then the **midpoint** of the segment has coordinates $$\left(\frac{x_1 + x_2}{2}, \frac{y_1 + y_2}{2}\right)$$ (p. 146)	Find the midpoint of the line segment with endpoints $(-2, 5)$ and $(6, 3)$. $$\text{Midpoint} = \left(\frac{-2 + 6}{2}, \frac{5 + 3}{2}\right) = \left(\frac{4}{2}, \frac{8}{2}\right) = (2, 4)$$

4.2 Slope of a Line and Slope-Intercept Form

The **slope** of a line is the ratio of the vertical change in y to the horizontal change in x. Slope is denoted by m. The slope of a line containing the points (x_1, y_1) and (x_2, y_2) is $$m = \frac{y_2 - y_1}{x_2 - x_1}$$ The slope of a horizontal line is zero. The slope of a vertical line is undefined. (p. 158)	Find the slope of the line containing the points $(6, 9)$ and $(-2, 12)$. $$m = \frac{y_2 - y_1}{x_2 - x_1}$$ $$= \frac{12 - 9}{-2 - 6} = \frac{3}{-8} = -\frac{3}{8}$$ The slope of the line is $-\dfrac{3}{8}$.
If we know the slope of a line and a point on the line, we can graph the line. (p. 161)	Graph the line containing the point $(-5, -3)$ with a slope of $\dfrac{4}{7}$. Start with the point $(-5, -3)$, and use the slope to plot another point on the line. $$m = \frac{4}{7} = \frac{\text{change in } y}{\text{change in } x}$$ 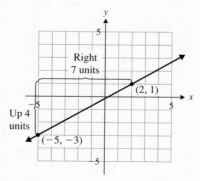

216   CHAPTER 4   Linear Equations in Two Variables and Functions

Definition/Procedure	Example

The Slope-Intercept Form of a Line

The **slope-intercept form of a line** is $y = mx + b$, where m is the slope and $(0, b)$ is the y-intercept.

If a line is written in slope-intercept form, we can use the y-intercept and the slope to graph the line. (p. 163)

Write the equation in slope-intercept form, and graph it.

$$6x + 4y = 16$$
$$4y = -6x + 16$$
$$y = -\frac{6}{4}x + \frac{16}{4}$$
$$y = -\frac{3}{2}x + 4 \quad \text{Slope-intercept form}$$

$m = -\dfrac{3}{2}$, y-intercept $(0, 4)$

Plot $(0, 4)$, then use the slope to locate another point on the line. We will think of the slope as $m = \dfrac{-3}{2} = \dfrac{\text{change in } y}{\text{change in } x}$.

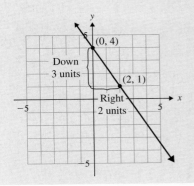

4.3 Writing an Equation of a Line

To write the equation of a line given its slope and y-intercept, *use $y = mx + b$ and substitute those values into the equation.* (p. 172)

Find an equation of the line with slope $= 3$ and y-intercept $(0, -8)$.

$$y = mx + b$$
$$y = 3x - 8 \quad \text{Substitute 3 for } m \text{ and } -8 \text{ for } b.$$

Point-Slope Form of a Line: The **point-slope form of a line** is $y - y_1 = m(x - x_1)$, where (x_1, y_1) is a point on the line and m is its slope.

Given the slope of the line and a point on the line, we can use the point-slope formula to find an equation of the line. (p. 173)

Find an equation of the line containing the point $(1, -4)$ with slope $= 2$. Express the answer in standard form.

Use $y - y_1 = m(x - x_1)$.

Substitute 2 for m. Substitute $(1, -4)$ for (x_1, y_1).

$$y - (-4) = 2(x - 1)$$
$$y + 4 = 2x - 2$$
$$-2x + y = -6$$
$$2x - y = 6 \quad \text{Standard form}$$

To write an equation of a line given two points on the line,

a) use the points to find the slope of the line

then

b) use the slope and *either one* of the points in the point-slope formula. (p. 174)

Find an equation of the line containing the points $(-2, 6)$ and $(4, 2)$. Express the answer in slope-intercept form.

$$m = \frac{2 - 6}{4 - (-2)} = \frac{-4}{6} = -\frac{2}{3}$$

We will use $m = -\dfrac{2}{3}$ and the point $(-2, 6)$ in the point-slope formula.

$$y - y_1 = m(x - x_1)$$

Definition/Procedure	Example
	Substitute $-\dfrac{2}{3}$ for m. Substitute $(-2, 6)$ for (x_1, y_1). $y - 6 = -\dfrac{2}{3}[x - (-2)]$ Substitute. $y - 6 = -\dfrac{2}{3}(x + 2)$ $y - 6 = -\dfrac{2}{3}x - \dfrac{4}{3}$ Distribute. $y = -\dfrac{2}{3}x + \dfrac{14}{3}$ Slope-intercept form
The equation of a *horizontal line* containing the point (a, b) is $y = b$. The equation of a *vertical line* containing the point (a, b) is $x = a$. **(p. 175)**	The equation of a horizontal line containing the point $(7, -4)$ is $y = -4$. The equation of a vertical line containing the point $(9, 1)$ is $x = 9$.
Parallel and Perpendicular Lines *Parallel lines* have the same slope. *Perpendicular lines* have slopes that are negative reciprocals of each other. **(p. 176)**	Determine whether the lines $5x + y = 3$ and $x - 5y = 20$ are parallel, perpendicular, or neither. Put each line into slope-intercept form to find their slopes. $\begin{array}{l\|l} 5x + y = 3 & x - 5y = 20 \\ y = -5x + 3 & -5y = -x + 20 \\ & y = \dfrac{1}{5}x - 4 \\ m = -5 & m = \dfrac{1}{5} \end{array}$ The lines are *perpendicular* because their slopes are negative reciprocals of each other.
To write an equation of the line parallel or perpendicular to a given line, we must find the slope of the given line first. **(p. 178)**	Write an equation of the line parallel to $2x - 3y = 21$ containing the point $(-6, -3)$. Express the answer in slope-intercept form. First, find the slope of $2x - 3y = 21$. $2x - 3y = 21$ $-3y = -2x + 21$ $y = \dfrac{2}{3}x - 7$ $m = \dfrac{2}{3}$ The slope of the parallel line is also $\dfrac{2}{3}$. Since this line contains $(-6, -3)$, use the point-slope formula to write its equation. $y - y_1 = m(x - x_1)$ $y - (-3) = \dfrac{2}{3}[x - (-6)]$ Substitute values. $y + 3 = \dfrac{2}{3}(x + 6)$ $y + 3 = \dfrac{2}{3}x + 4$ Distribute. $y = \dfrac{2}{3}x + 1$ Slope-intercept form

Definition/Procedure	Example

4.4 Linear and Compound Linear Inequalities in Two Variables

A **linear inequality in two variables** is an inequality that can be written in the form $Ax + By \geq C$ or $Ax + By \leq C$, where A, B, and C are real numbers and where A and B are not both zero. ($>$ and $<$ may be substituted for $\geq$ and $\leq$.) **(p. 187)**

Some examples of linear inequalities in two variables are

$$x + 3y \leq 2, \quad y > -\frac{2}{3}x + 5, \quad y \geq -1, \quad x < 4$$

Graph a Linear Inequality in Two Variables Using a Test Point

1) *Graph the boundary line.*
   a) If the inequality contains $\geq$ or $\leq$, make the boundary line *solid*.
   b) If the inequality contains $>$ or $<$, make the boundary line *dotted*.
2) *Choose a test point not on the line, and shade the appropriate region.* Substitute the test point into the inequality. If $(0, 0)$ is not on the line, it is an easy point to test in the inequality.
   a) If it *makes the inequality true,* shade the region *containing* the test point. All points in the shaded region are part of the solution set.
   b) If the test point *does not satisfy the inequality,* shade the region on the *other* side of the line. All points in the shaded region are part of the solution set. **(p. 188)**

Graph $2x + y > -3$.

1) Graph the boundary line as a *dotted* line.
2) Choose a test point not on the line, and substitute it into the inequality to determine whether it makes the inequality true.

Test Point	Substitute into $2x + y > -3$
$(0, 0)$	$2(0) + (0) > -3$
	$0 > -3$ True

Since the test point satisfies the inequality, shade the region containing $(0, 0)$.

All points in the shaded region satisfy $2x + y > -3$.

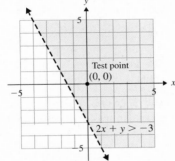

Graph a Linear Inequality in Two Variables Using the Slope-Intercept Method

1) Write the inequality in the form $y \geq mx + b$ ($y > mx + b$) or $y \leq mx + b$ ($y < mx + b$), and graph the boundary line $y = mx + b$.

2) If the inequality is in the form $y \geq mx + b$ or $y > mx + b$, shade *above* the line.

3) If the inequality is in the form $y \leq mx + b$ or $y < mx + b$, shade *below* the line. **(p. 190)**

Graph $-x + 3y \leq 6$ using the slope-intercept method.

Write the inequality in slope-intercept form by solving $-x + 3y \leq 6$ for y.

$$-x + 3y \leq 6$$
$$3y \leq x + 6$$
$$y \leq \frac{1}{3}x + 2$$

Graph $y = \frac{1}{3}x + 2$ as a solid line.

Since $y \leq \frac{1}{3}x + 2$ has a $\leq$ symbol, shade *below* the line.

All points on the line and in the shaded region satisfy $-x + 3y \leq 6$.

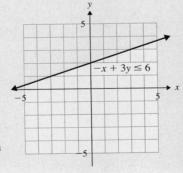

Definition/Procedure	Example
Graphing Compound Linear Inequalities in Two Variables 1) Graph each inequality separately on the same axes. Shade lightly. 2) If the inequality contains *and*, the solution set is the *intersection* of the shaded regions. Heavily shade this region. 3) If the inequality contains *or*, the solution set is the *union* (total) of the shaded regions. Heavily shade this region. (p. 191)	Graph the compound inequality $y \geq -4x + 3$ and $y \geq 1$. Since the inequality contains *and*, the solution set is the *intersection* of the shaded regions. Any point in the shaded area will satisfy *both* inequalities. 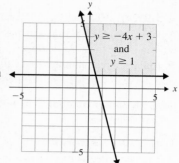

4.5 Introduction to Functions

A **relation** is any set of ordered pairs. A relation can also be represented as a correspondence or mapping from one set to another. (p. 197)	Relations: a) $\{(-4, -12), (-1, -3), (3, 9), (5, 15)\}$ b) 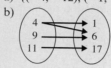
The **domain** of a relation is the set of values of the independent variable (the first coordinates in the set of ordered pairs). The **range** of a relation is the set of all values of the dependent variable (the second coordinates in the set of ordered pairs). (p. 198)	In a) above, the domain is $\{-4, -1, 3, 5\}$, and the range is $\{-12, -3, 9, 15\}$. In b) above, the domain is $\{4, 9, 11\}$, and the range is $\{1, 6, 17\}$.
A **function** is a relation in which each element of the domain corresponds to *exactly one* element of the range. (p. 198)	The relation above in a) *is* a function. The relation above in b) *is not* a function.
The Vertical Line Test (p. 200)	This graph represents a function. Anywhere a vertical line is drawn, it will intersect the graph only once. 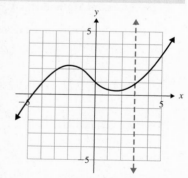 This is *not* the graph of a function. A vertical line can be drawn so that it intersects the graph more than once. 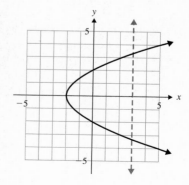

Definition/Procedure	Example
The **domain** of a relation that is written as an equation, where y is in terms of x, is the set of all real numbers that can be substituted for the independent variable, x. When determining the domain of a relation, it can be helpful to keep these tips in mind. 1) Ask yourself, "Is there any number that *cannot* be substituted for x?" 2) If x is in the denominator of a fraction, determine what value of x will make the denominator equal 0 by setting the denominator equal to zero. Solve for x. This x-value is *not* in the domain. (p. 203)	Determine the domain of $f(x) = \dfrac{9}{x+8}$. $x + 8 = 0$ Set the denominator = 0. $x = -8$ Solve. When $x = -8$, the denominator of $f(x) = \dfrac{9}{x+8}$ equals zero. The domain contains all real numbers *except* -8. The domain of the function is $(-\infty, -8) \cup (-8, \infty)$.
Function Notation If a function describes the relationship between x and y so that x is the independent variable and y is the dependent variable, then y is a function of x. $y = f(x)$ is called **function notation** and it is read as "y equals f of x." Finding a function value means evaluating the function for the given value of the variable. (p. 203)	If $f(x) = 9x - 4$, find $f(2)$. Substitute 2 for x and evaluate. $f(2) = 9(2) - 4 = 18 - 4 = 14$ $f(2) = 14$
A **linear function** has the form $$f(x) = mx + b$$ where m and b are real numbers, m is the *slope* of the line, and $(0, b)$ is the *y-intercept*. (p. 206)	Graph $f(x) = -3x + 4$ using the slope and y-intercept. The slope is -3 and the y-intercept is $(0, 4)$. Plot the y-intercept, and use the slope to locate another point on the line. 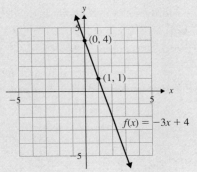

Chapter 4: Review Exercises

(4.1) Determine whether each ordered pair is a solution of the given equation.

1) $4x - y = 9$; $(1, -5)$
2) $3x + 2y = 20$; $(-4, 2)$
3) $y = \dfrac{5}{4}x + \dfrac{1}{2}$; $(2, 3)$
4) $x = 7$; $(7, -9)$

Complete the ordered pair for each equation.

5) $y = -6x + 10$; $(-3, \ \ )$
6) $y = \dfrac{2}{3}x + 5$; $(12, \ \ )$
7) $y = -8$; $(5, \ \ )$
8) $5x - 9y = 3$; $(\ \ , -2)$

Complete the table of values for each equation.

9) $y = x - 11$

x	y
0	
3	
-1	
-5	

10) $4x - 6y = 8$

x	y
	0
0	
3	
	-4

Plot the ordered pairs on the same coordinate system.

11) a) (5, 2) b) (−3, 0)
    c) (−4, 3) d) (6, −2)

12) a) (0, 1) b) (−2, −5)
    c) $\left(\frac{5}{2}, 1\right)$ d) $\left(4, -\frac{1}{3}\right)$

13) The fine for an overdue book at the Hinsdale Public Library is given by

$$y = 0.10x$$

where x represents the number of days a book is overdue and y represents the amount of the fine, in dollars.

a) Complete the table of values, and write the information as ordered pairs.

x	y
1	
2	
7	
10	

b) Label a coordinate system, choose an appropriate scale, and graph the ordered pairs.

c) Explain the meaning of the ordered pair (14, 1.40) in the context of the problem.

14) Fill in the blank with *positive, negative,* or *zero*.

a) The y-coordinate of every point in quadrant III is _____.

b) The x-coordinate of every point in quadrant IV is _____.

Complete the table of values, and graph each equation.

15) $y = -2x + 3$

x	y
0	
1	
2	
−2	

16) $3x + 2y = 4$

x	y
0	
	−2
1	
	4

Graph each equation by finding the intercepts and at least one other point.

17) $x - 2y = 6$
18) $5x + y = 10$
19) $y = -\frac{1}{6}x + 4$
20) $y = \frac{3}{4}x - 7$
21) $x = 5$
22) $y = -3$

Find the midpoint of the line segment with the given endpoints.

23) (3, 8) and (5, 2)
24) (−6, 1) and (−2, −1)
25) (7, −3) and (6, −4)
26) $\left(\frac{2}{3}, \frac{1}{4}\right)$ and $\left(-\frac{1}{6}, \frac{5}{8}\right)$

(4.2) Determine the slope of each line.

27)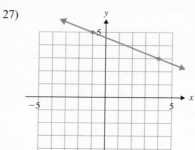

28)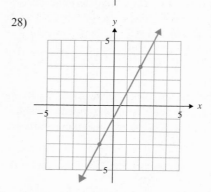

Use the slope formula to find the slope of the line containing each pair of points.

29) (1, 7) and (−4, 2)
30) (−2, −3) and (3, −1)
31) (−2, 5) and (3, −8)
32) (0, 4) and (8, −2)
33) $\left(\frac{3}{2}, -1\right)$ and $\left(-\frac{5}{2}, 7\right)$
34) (2.5, 5.3) and (−3.5, −1.9)
35) (9, 0) and (9, 4)
36) (−7, 4) and (1, 4)

37) Paul purchased some shares of stock in 2008. The graph shows the value of one share of the stock from 2008–2012.

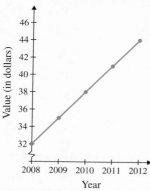

Value of One Share of Stock

a) What was the value of one share of stock the year that Paul made his purchase?

b) Is the slope of the line segment positive or negative? What does the sign of the slope mean in the context of this problem?

c) Find the slope. What does it mean in the context of this problem?

Graph the line containing the given point and with the given slope.

38) $(-3, -2)$; $m = 4$

39) $(1, 5)$; $m = -3$

40) $(-2, 6)$; $m = -\dfrac{5}{2}$

41) $(-3, 2)$; slope undefined

42) $(5, 2)$; $m = 0$

Identify the slope and y-intercept, then graph the line.

43) $y = x - 3$

44) $y = -2x + 7$

45) $y = -\dfrac{3}{4}x + 1$

46) $y = \dfrac{1}{4}x - 2$

47) $x - 3y = -6$

48) $2x - 7y = 35$

49) $x + y = 0$

50) $y + 3 = 4$

51) Personal consumption expenditures in the United States since 1998 can be modeled by $y = 371.5x + 5920.1$, where x represents the number of years after 1998, and y represents the personal consumption expenditure in billions of dollars. (Bureau of Economic Analysis)

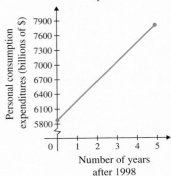

Personal Consumption Expenditures

a) What is the y-intercept? What does it mean in the context of the problem?

b) Has the personal consumption expenditure been increasing or decreasing since 1998? By how much per year?

c) Use the graph to estimate the personal consumption expenditure in the year 2002. Then, use the equation to determine this number.

(4.3)

52) Write the point-slope formula for the equation of a line with slope m and which contains the point (x_1, y_1).

Write the *slope-intercept form* of the equation of the line, if possible, given the following information.

53) $m = 7$ and contains $(2, 5)$

54) $m = -8$ and y-intercept $(0, -1)$

55) $m = -\dfrac{4}{9}$ and y-intercept $(0, 2)$

56) contains $(-6, -5)$ and $(4, 10)$

57) contains $(3, -6)$ and $(-9, -2)$

58) $m = \dfrac{1}{2}$ and contains $(8, -3)$

59) horizontal line containing $(1, 9)$

60) vertical line containing $(4, 0)$

Write the *standard form* of the equation of the line given the following information.

61) contains $(-2, 2)$ and $(8, 7)$

62) $m = -1$ and contains $(4, -7)$

63) $m = -3$ and contains $\left(\dfrac{4}{3}, 1\right)$

64) contains $(15, -2)$ and $(-5, -10)$

65) $m = 6$ and y-intercept $(0, 0)$

66) $m = -\dfrac{5}{3}$ and y-intercept $(0, 2)$

67) contains $(1, 1)$ and $(-7, -5)$

68) $m = \dfrac{1}{6}$ and contains $(17, 2)$

69) The chart shows the number of wireless communication subscribers worldwide (in millions) from 2001–2004. (Dell'Oro Group and Standard and Poor's Industry Surveys)

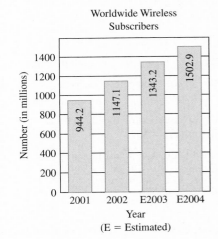

Worldwide Wireless Subscribers

a) Write a linear equation to model this data. Use the data points for 2001 and 2004. Let x represent the number of years after 2001, and let y represent the number of worldwide wireless subscribers, in millions. Round to the nearest tenth.

b) Explain the meaning of the slope in the context of the problem.

c) Use the equation to determine the number of subscribers in 2003. How does it compare to the value given on the chart?

Determine if each pair of lines is parallel, perpendicular, or neither.

70) $y = -\dfrac{1}{2}x + 7$
    $5x + 10y = 8$

71) $9x - 4y = -1$
    $-27x + 12y = 2$

72) $4x - 6y = -3$
    $-3x + 2y = -2$

73) $y = 6$
    $x = -2$

74) $x = 3$
    $x = 1$

75) $y = 6x - 7$
    $4x + y = 9$

76) $x + 2y = 22$
    $2x - y = 0$

Write an equation of the line *parallel* to the given line and containing the given point. Write the answer in slope-intercept form or in standard form, as indicated.

77) $y = 5x + 14$; $(-2, -4)$; slope-intercept form

78) $y = -3x + 1$; $(5, -19)$; slope-intercept form

79) $x - 4y = 9$; $(5, 3)$; standard form

80) $5x - 3y = 7$; $(4, 8)$; slope-intercept form

Write an equation of the line *perpendicular* to the given line and containing the given point. Write the answer in slope-intercept form or in standard form, as indicated.

81) $y = -\dfrac{1}{2}x + 9$; $(6, 5)$; slope-intercept form

82) $y = -x + 11$; $(-10, -8)$; slope-intercept form

83) $2x - 11y = 11$; $(2, -7)$; slope-intercept form

84) $-2x + 3y = 15$; $(-5, 7)$; standard form

85) Write an equation of the line parallel to $x = 7$ containing $(2, 3)$.

86) Write an equation of the line parallel to $y = -5$ containing $(-4, 9)$.

87) Write an equation of the line perpendicular to $y = 6$ containing $(-1, -3)$.

88) Write an equation of the line perpendicular to $x = 7$ containing $(6, 0)$.

(4.4) Graph each linear inequality in two variables.

89) $y \leq -2x + 7$

90) $y \geq -\dfrac{3}{2}x + 2$

91) $-3x + 4y > 12$

92) $5x - 2y \geq 8$

93) $y < x$

94) $x \geq 4$

Graph each compound inequality.

95) $y \geq \dfrac{3}{4}x - 4$ and $y \leq -5$

96) $y < \dfrac{5}{4}x - 5$ or $y < -3$

97) $4x - y < -1$ or $y > \dfrac{1}{2}x + 5$

98) $2x + 5y \leq 10$ and $y \geq \dfrac{1}{3}x + 4$

99) $4x + 2y \geq -6$ and $y \leq 2$

100) $2x + y \leq 3$ or $6x + y > 4$

(4.5) Identify the domain and range of each relation, and determine whether each relation is a function.

101) $\{(-3, 1), (5, 3), (5, -3), (12, 4)\}$

102)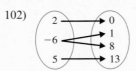

103)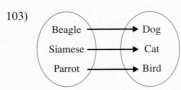

104)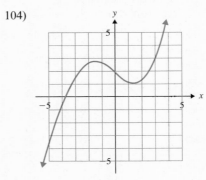

105)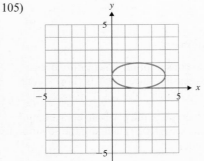

Determine the domain of each relation, and determine whether each relation describes *y* as a function of *x*.

106) $y = 4x - 7$

107) $y = \dfrac{8}{x + 3}$

108) $y = \dfrac{15}{x}$

109) $y^2 = x$

110) $y = x^2 - 6$

111) $y = \dfrac{5}{7x - 2}$

For each function, f, find f(3) and f(−2).

112) $f = \{(-7, -2), (-2, -5), (1, -10), (3, -14)\}$

113)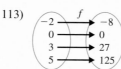

114)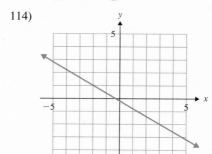

115) Let $f(x) = 5x - 12$, $g(x) = x^2 + 6x + 5$. Find each of the following and simplify.

  a) $f(4)$ b) $f(-3)$

  c) $g(3)$ d) $g(0)$

  e) $f(a)$ f) $g(t)$

  g) $f(k + 8)$ h) $f(c - 2)$

  i) $f(x + h)$ j) $f(x + h) - f(x)$

116) $h(x) = -3x + 7$. Find x so that $h(x) = 19$.

117) $f(x) = \frac{3}{2}x + 5$. Find x so that $f(x) = \frac{11}{2}$.

118) Graph $f(x) = -2x + 6$ by making a table of values and plotting points.

119) Graph each function using the slope and y-intercept.

  a) $f(x) = \frac{2}{3}x - 1$ b) $f(x) = -3x + 2$

120) Graph $g(x) = \frac{3}{2}x + 3$ by finding the x- and y-intercepts and one other point.

Graph each function.

121) $h(c) = -\frac{5}{2}c + 4$

122) $D(t) = 3t$

123) A USB 2.0 device can transfer data at a rate of 480 MB/sec (megabytes/second). Let $f(t) = 480t$ represent the number of megabytes of data that can be transferred in t sec. (www.usb.org)

  a) How many megabytes of a file can be transferred in 2 sec? in 6 sec?

  b) How long would it take to transfer a 1200 MB file?

124) A jet travels at a constant speed of 420 mph. The distance D (in miles) that the jet travels after t hr can be defined by the function

$$D(t) \, 5 = 420t$$

  a) Find $D(2)$, and explain what this means in the context of the problem.

  b) Find t so that $D(t) = 2100$, and explain what this means in the context of the problem.

Chapter 4: Test

1) Is $(9, -13)$ a solution of $5x + 3y = 6$?

2) Complete the table of values and graph $y = -2x + 4$.

x	y
0	
3	
−1	
	0

3) Fill in the blanks with *positive* or *negative*. In quadrant II, the x-coordinate of every point is _____ and the y-coordinate is _____.

4) For $2x - 3y = 12$,

  a) find the x-intercept.

  b) find the y-intercept.

  c) find one other point on the line.

  d) graph the line.

5) Graph $x = -4$.

6) Determine the midpoint of the line segment with endpoints $(2, 1)$ and $(10, -6)$.

7) Find the slope of the line containing the points

  a) $(-8, -5)$ and $(4, -14)$

  b) $(9, 2)$ and $(3, 2)$

8) Graph the line containing the point $(-3, 6)$ with slope $= -\frac{2}{5}$.

9) Graph the line containing the point $(4, 1)$ with an undefined slope.

10) Write the slope-intercept form for the equation of the line with slope -4 and y-intercept $(0, 5)$.

11) Write the standard form for the equation of a line with slope $\frac{1}{2}$ containing the point (3, 8).

12) Determine whether $4x - 7y = -7$ and $14x + 8y = 3$ are *parallel, perpendicular,* or *neither*.

13) Find the slope-intercept form of the equation of the line

   a) perpendicular to $y = -3x + 11$ containing (12, −5).

   b) parallel to $5x - 2y = 2$ containing (8, 14).

14) Mr. Kumar owns a computer repair business. The graph shows the annual profit since 2007. Let x represent the number of years after 2007, and let y represent the annual profit, in thousands.

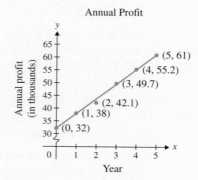
Annual Profit

   a) What was the profit in 2011?

   b) Write a linear equation (in slope-intercept form) to model this data. Use the data points for 2007 and 2012.

   c) What is the slope of the line? What does it mean in the context of the problem?

   d) What is the y-intercept? What does it mean in the context of the problem?

   e) If the profit continues to follow this trend, in what year can Mr. Kumar expect a profit of $90,000?

Graph each inequality.

15) $y \geq 3x + 1$ 16) $2x - 5y > 10$

Graph the compound inequality.

17) $-2x + 3y \geq -12$ and $x \leq 3$

18) $y < -x$ or $2x - y > 1$

Identify the domain and range of each relation, and determine whether each relation is a function.

19) $\{(-2, -5), (1, -1), (3, 1), (8, 4)\}$

20)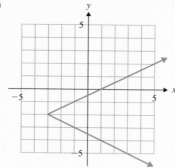

For each function, (a) determine the domain. (b) Is y a function of x?

21) $y = \frac{7}{3}x - 5$ 22) $y = \frac{8}{2x - 5}$

For each function, f, find $f(2)$.

23) $f = \{(-3, -8), (0, -5), (2, -3), (7, 2)\}$

24)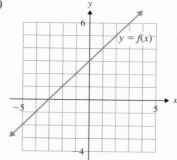

Let $f(x) = -4x + 2$ and $g(x) = x^2 - 3x + 7$. Find each of the following, and simplify.

25) $f(6)$ 26) $g(2)$

27) $g(t)$ 28) $f(h - 7)$

Graph the function.

29) $h(x) = -\frac{3}{4}x + 5$

30) A USB 1.1 device can transfer data at a rate of 12 MB/sec (megabytes/second). Let $f(t) = 12t$ represent the number of megabytes of data that can be transferred in t sec. (www.usb.org)

   a) How many megabytes of a file can be transferred in 3 sec?

   b) How long would it take to transfer 132 MB?

Chapter 4: Cumulative Review for Chapters 1–4

1) Write $\dfrac{252}{840}$ in lowest terms.

2) A rectangular picture frame measures 8 in. by 10.5 in. Find the perimeter of the frame.

Evaluate.

3) -2^6

4) $\dfrac{21}{40} \cdot \dfrac{25}{63}$

5) $3 - \dfrac{2}{5}$

6) Write an expression for "53 less than twice eleven" and simplify.

Solve each equation

7) $12 - 5(2n + 9) = 3n + 2(n + 6)$

8) $\dfrac{1}{3}p + \dfrac{1}{4} = \dfrac{2}{3}p - \dfrac{1}{6}(2p + 5)$

9) Solve for w: $t + zw = r$

10) $|7y + 2| = 16$

Solve. Write the solution in interval notation.

11) $19 - 4x \geq 25$

12) $3c - 10 < 2$ and $\dfrac{5}{2}c + 4 > -1$

For 13–15, write an equation and solve.

13) One serving of Ben and Jerry's Chocolate Chip Cookie Dough Ice Cream has 10% fewer calories than one serving of their Chunky Monkey® Ice Cream. If one serving of Chocolate Chip Cookie Dough has 270 calories, how many calories are in one serving of Chunky Monkey? (www.benjerry.com)

14) Find the missing angle measures.

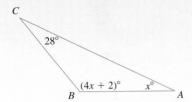

15) Lynette's age is 7 yr less than three times her daughter's age. If the sum of their ages is 57, how old is Lynette, and how old is her daughter?

16) Find the slope of the line containing the points $(-7, 8)$ and $(2, 17)$.

17) Graph $4x + y = 5$.

18) Write an equation of the line with slope $-\dfrac{5}{4}$ containing the point $(-8, 1)$. Express the answer in standard form.

19) Write an equation of the line perpendicular to $y = \dfrac{1}{3}x + 11$ containing the point $(4, -12)$. Express the answer in slope-intercept form.

20) Given $y = \dfrac{3}{x + 7}$,

a) is y a function of x?

b) determine the domain.

Let $f(x) = 8x + 3$. Find each of the following, and simplify.

21) $f(-5)$

22) $f(a)$

23) $f(t + 2)$

Graph each function.

24) $f(x) = 2$

25) $h(x) = -\dfrac{1}{4}x + 2$

CHAPTER

5 Solving Systems of Linear Equations

OUTLINE

Study Strategies: Taking Math Tests

5.1 Solving Systems of Linear Equations in Two Variables

5.2 Solving Systems of Linear Equations in Three Variables

5.3 Applications of Systems of Linear Equations

5.4 Solving Systems of Linear Equations Using Matrices

Group Activity

emPOWERme: Studying Smart

Math at Work:

Financial Planner

Bethany Wilson is used to handling a lot of responsibility. As a financial planner, she manages the savings, the stock portfolios, and the retirement accounts of dozens of clients. "I've gotten used to the pressure that comes with my job," she describes. "I'm proud that my clients trust me with such important aspects of their lives."

Financial planners like Bethany advise clients on what sorts of investments they should make, how much they need to save to meet their financial goals, how they can prepare for retirement, and so forth. "My job involves a great deal of financial forecasting," says Bethany. "I make calculations that show me how a client's finances will look in six months, in a year, even in twenty years."

Of course, life is unpredictable, and Bethany admits that even the best financial planner can't anticipate everything that might occur in the larger economy or in a client's life. "I don't have a crystal ball," Bethany says. "I've learned that the best I can do is to work hard and come to work prepared to meet my clients' needs. I focus on controlling what I can control, and so far, that strategy has served me well."

In this chapter, we will learn about solving systems of linear equations, as well as introduce some strategies you can use to prepare for the high-pressure situation of academic life: taking a math test.

Study Strategies Taking Math Tests

You've studied your class notes and your textbook. You've reviewed your homework and problem sets. You've even worked with your fellow students in a study group. Now it's time to sit down and actually take the test you've prepared for. When this moment comes, don't you owe it to yourself to perform your best? Use the following strategies to ensure that you get the grade you deserve.

- Study smart. Focus your preparation on the material that will be covered on the test. The emPOWERme on page 281 will help you.
- Practice, practice, practice! Repetition will help build your skills. Do problems using a timer to simulate the amount of time you will have to take the test.
- Get a good night's sleep before the test, as it is during sleep that short-term memory turns into long-term memory.
- Warm up for the test just like athletes do before they play a game. No matter how much you studied the night before, do several "warm-up" problems the same day as the test. This way, you will be in the groove of doing math and won't go into the test cold.

- Arrive at the test location early.
- Have multiple pens, pencils, scratch paper, and your textbook with you when you sit down to take the test.
- Warm-up your brain by reviewing your notes and working through problems like those that will appear on the test.

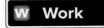

- Before you start the test, read all the instructions.
- Answer the easiest questions first, leaving time at the end to work on harder problems.
- Show all your work in a neat and organized way because your instructor may give you credit for this even if you don't finish the problem.
- Remain calm throughout the test. Take a deep breath and focus on doing your best if you start to feel anxious.

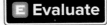

- Leave time at the end of the test to check over your work.
- Double-check your calculations to ensure you didn't make any careless errors.

- Review your test carefully when you get it back. You want to understand the source of any errors you made.
- Consider the process you used to study for the test, and think about ways it could be improved.

CHAPTER 5 Solving Systems of Linear Equations

Chapter 5 POWER Plan

P Prepare | O Organize

What are your goals for Chapter 5?	How can you accomplish each goal?
1 Be prepared before and during class.	• Don't stay out late the night before, and be sure to set your alarm clock! • Bring a pencil, notebook paper, and textbook to class. • Avoid distractions by turning off your cell phone during class. • Pay attention, take good notes, and ask questions. • Complete your homework on time, and ask questions on problems you do not understand. • Plan ahead for tests by preparing many days in advance.
2 Understand the homework to the point where you could do it without needing any help or hints.	• Read the directions, and show all of your steps. • Go to the professor's office for help. • Rework homework and quiz problems, and find similar problems for practice. • Review old material that you have not mastered yet.
3 Use the P.O.W.E.R. framework to help you take tests: *Studying Smart*.	• Read the Study Strategy that explains how to study effectively for tests. • Do a "practice run" the night before the test by doing a practice test without notes. • Complete the emPOWERme that appears before the Chapter Summary.
4 Write your own goal. _____ _____	• _____

What are your objectives for Chapter 5?	How can you accomplish each objective?
1 Be able to solve a system of linear equations in two variables by using the graphing, substitution, or elimination methods. Know when to use each method.	• Learn the procedures for each of these methods. • Know the terminology associated with the solutions such as independent and consistent. • Know how to check each answer.
2 Be able to determine when the solution to a system of equations is *no solution* or *infinite solutions*. Know what these solutions look like on a graph and how to write the answer.	• Learn the procedures for solving a system of equations and the possible answers when variables "drop out." • Learn the terminology associated with the solutions such as *inconsistent* and *dependent*. • Know what these results look like on a graph. • Know how to check your solutions.
3 Be able to solve a system of linear equations in three variables, including systems where there are missing terms.	• Learn the procedure for **Solving a System of Linear Equations in Three Variables.** • Know how to check your solutions.

4 Solve several applied problems that involve general quantities, geometry applications, mixture problems, and distance problems.	• Learn the procedure for **Solving an Applied Problem Using a System of Equations.** • Draw diagrams or charts when needed. • Be sure your answer makes sense.
5 Solve systems of linear equations using *Gaussian elimination*.	• Learn the vocabulary associated with *Gaussian elimination* and the definition of an *augmented matrix*. • Understand how to write a system of equations in an augmented matrix. • Learn matrix row operations, and perform them until the matrix is in *row echelon form*. • Write a system of equations, and solve. • Check your answer.
6 Write your own goal.	•

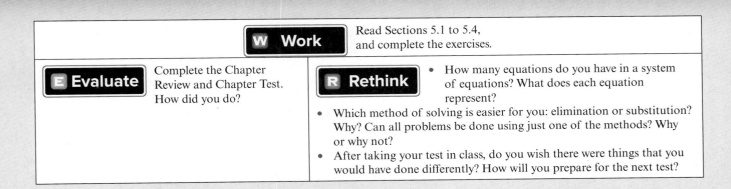

W Work	Read Sections 5.1 to 5.4, and complete the exercises.		
E Evaluate	Complete the Chapter Review and Chapter Test. How did you do?	**R Rethink**	• How many equations do you have in a system of equations? What does each equation represent? • Which method of solving is easier for you: elimination or substitution? Why? Can all problems be done using just one of the methods? Why or why not? • After taking your test in class, do you wish there were things that you would have done differently? How will you prepare for the next test?

5.1 Solving Systems of Linear Equations in Two Variables

P Prepare
What are your objectives for Section 5.1?

O Organize
How can you accomplish each objective?

1 Determine Whether an Ordered Pair Is a Solution of a System	• Be able to recognize a *system of linear equations*. • Know what it means when an ordered pair is a solution of a system of equations. • Be able to substitute an ordered pair into an equation. • Complete the given example on your own. • Complete You Try 1.
2 Solve a Linear System by Graphing	• Learn the definitions of a *consistent system*, *inconsistent system*, and *dependent system*. • Review how to graph a linear equation. • Learn the procedure for **Solving a System by Graphing**, and write it in your own words. • Be familiar with the different types of solutions and the terms associated with each of those solutions. • Complete the given examples on your own. • Complete You Trys 2, 3, and 4.
3 Solve a Linear System by Substitution	• Learn the procedure for **Solving a System by Substitution.** • Recognize when the substitution method is good to use. • Complete the given examples on your own. • Complete You Trys 5 and 6.
4 Solve a Linear System by Elimination	• Learn the procedure for **Solving a System of Two Linear Equations by the Elimination Method.** Summarize it in your notes in your own words. • Recognize when elimination is easier to use than the substitution method. • Complete the given examples on your own. • Complete You Trys 7 and 8.
5 Solve Special Systems	• Be aware that when solving using substitution or elimination, variables may "drop out," which may indicate *no solution* or *infinite solutions*. • In your notes, explain when a system has *no solution* or an *infinite number of solutions*. • Complete the given examples on your own. • Complete You Trys 9 and 10.

W Work   Read the explanations, follow the examples, take notes, and complete the You Trys.

What is a system of linear equations in two variables? A **system of linear equations** consists of two or more linear equations with the same variables. Three examples of such systems are

$$2x + 5y = 5 \qquad y = \frac{1}{3}x - 8 \qquad -3x + y = 1$$
$$x + 4y = -1 \qquad 5x - 6y = 10 \qquad x = -2$$

In the third system, we can think of $x = -2$ as an equation in two variables by writing it as $x + 0y = -2$. In this section we will learn how to solve a system of two equations in two variables.

1 Determine Whether an Ordered Pair Is a Solution of a System

We will begin our work with systems of equations by determining whether an ordered pair is a solution of the system.

Definition

A **solution of a system** of two equations in two variables is an ordered pair that is a solution of each equation in the system.

EXAMPLE 1

Determine whether $(-5, -1)$ is a solution of each system of equations.

a) $x - 3y = -2$
$-4x + y = 19$

b) $2x - 9y = -1$
$y = x + 8$

Solution

a) If $(-5, -1)$ is a solution of $\begin{array}{l} x - 3y = -2 \\ -4x + y = 19 \end{array}$ then when we substitute -5 for x and -1 for y, the ordered pair will make each equation true.

$$x - 3y = -2 \qquad\qquad -4x + y = 19$$
$$-5 - 3(-1) \stackrel{?}{=} -2 \quad \text{Substitute.} \qquad -4(-5) + (-1) \stackrel{?}{=} 19 \quad \text{Substitute.}$$
$$-5 + 3 \stackrel{?}{=} -2 \qquad\qquad 20 + (-1) \stackrel{?}{=} 19$$
$$-2 = -2 \quad \text{True} \qquad\qquad 19 = 19 \quad \text{True}$$

W Hint
Notice that we must substitute the x- and y-values of the ordered pair into both equations. Both results will be true if the ordered pair is a solution to the system.

Because $(-5, -1)$ is a solution of each equation, it is a solution of the system.

b) We will substitute -5 for x and -1 for y to see whether $(-5, -1)$ satisfies (is a solution of) each equation.

$$2x - 9y = -1 \qquad\qquad y = x + 8$$
$$2(-5) - 9(-1) \stackrel{?}{=} -1 \quad \text{Substitute.} \qquad -1 \stackrel{?}{=} (-5) + 8 \quad \text{Substitute.}$$
$$-10 + 9 \stackrel{?}{=} -1 \qquad\qquad -1 = 3 \quad \text{False}$$
$$-1 = -1 \quad \text{True}$$

Although $(-5, -1)$ is a solution of the first equation, it does *not* satisfy $y = x + 8$. Therefore, $(-5, -1)$ is *not* a solution of the system.

YOU TRY 1

Determine whether $(6, 2)$ is a solution of each system.

a) $-x + 4y = 2$
$3x - 5y = 6$

b) $y = x - 4$
$2x - 9y = -6$

Now we will learn how to *find* solutions to systems of linear equations using three methods: graphing, substitution, and elimination.

2 Solve a Linear System by Graphing

To **solve a system of equations in two variables** means to find the ordered pair or pairs that satisfy each equation in the system.

Recall from Chapter 4 that the graph of $Ax + By = C$ is a line. This line represents all solutions of the equation. If two lines intersect at a point, then that point is a solution of each equation.

Definition

When solving a system of equations by graphing, the point of intersection is the solution of the system. If a system has at least one solution, we say that the system is **consistent**. The equations are **independent** if the system has one solution.

EXAMPLE 2

Solve the system by graphing.

$$y = x + 1$$
$$x + 2y = 8$$

Solution

Graph each line on the same axes. The first equation is in slope-intercept form. Graph the second equation by writing it in slope-intercept form or by making a table of values and plotting points.

The point of intersection is (2, 3). Therefore, the solution of the system is (2, 3).

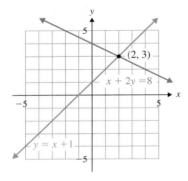

W Hint

Notice that when you are solving a system of two linear equations, you are trying to find the intersection of the two lines.

Note

Always use a straightedge to graph the lines. If the solution of a system contains numbers that are not integers, it may be impossible to accurately read the point of intersection. This is one reason why solving a system by graphing is not always the best way to find the solution. But it can be a useful method, and it is one that is used to solve problems not only in mathematics, but also in areas such as business, economics, and chemistry.

[YOU TRY 2]

Solve the system by graphing. $2x + 3y = 8$
$y = -4x - 4$

Do two lines *always* intersect? No! Let's see what that tells us about the solution to a system of equations.

EXAMPLE 3

Solve the system by graphing.

$$2y - x = 2$$
$$-x + 2y = -6$$

Solution

Graph each line on the same axes. In slope-intercept form, the first equation is $y = \frac{1}{2}x + 1$.

The second is $y = \frac{1}{2}x - 3$. Both lines have a slope of $\frac{1}{2}$. Their y-intercepts, (0, 1) and (0, −3), are different. They are parallel.

Because the lines do not intersect, the system has *no solution*. We write the solution set as ∅.

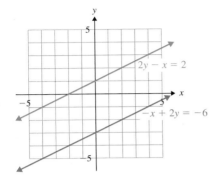

W Hint
Remember that the solution of a system of equations is the point of intersection. Since parallel lines do not intersect, there is no solution!

Definition

When solving a system of equations by graphing, if the lines are parallel, then the system has **no solution**. We write this as ∅. Furthermore, we say that a system that has no solution is **inconsistent**.

EXAMPLE 4

Solve the system by graphing.

$$y = -\frac{3}{2}x + 2$$
$$6x + 4y = 8$$

Solution

Graph each line on the same axes. In slope-intercept form, the second equation is $y = -\frac{3}{2}x + 2$. The equations are equivalent because the second equation is a multiple of the first. This means that the graph of each equation is the same line. Therefore, each point on the line satisfies each equation.

The system has an *infinite number of solutions* of the form $y = -\frac{3}{2}x + 2$. The solution set is $\left\{(x, y) \mid y = -\frac{3}{2}x + 2\right\}$, which we read as "the set of all ordered pairs (x, y) such that $y = -\frac{3}{2}x + 2$."

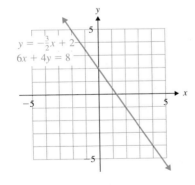

In Example 4, we could have used either equation to write the solution set. However, we will use an equation written in slope-intercept form or an equation written in standard form with integer coefficients that have no common factor other than 1.

Definition

When solving a system of equations by graphing, if the graph of each equation is the same line, then the system has an **infinite number of solutions**. Because the system has at least one solution, it is **consistent**.

The system in Example 4 contains equations that are equivalent, so the graphs of their lines are identical. We call such equations *dependent*. The systems in Examples 2 and 3, however, do not contain pairs of equivalent equations. The graphs of their lines intersect or are parallel. We call such equations *independent*.

Definition

The graphs of **dependent** linear equations are identical. The graphs of **independent** linear equations either intersect or are parallel.

> **W Hint**
> Write the procedure for **Solving a System by Graphing** in your own words.

Procedure Solving a System by Graphing

To solve a system by graphing, graph each line on the same axes.

1) If the lines intersect at a single point, then the point of intersection is the solution of the system. The system is *consistent,* and the equations are *independent.* (See Figure 5.1a.)

2) If the lines are parallel, then the system has *no solution*. We write the solution set as $\emptyset$. The system is *inconsistent,* and the equations are *independent.* (See Figure 5.1b.)

3) If the graphs are the same line, then the system has an *infinite number of solutions*. The system is *consistent,* and the equations are *dependent.* (See Figure 5.1c.)

> **W Hint**
> Use an acronym to help you remember the three cases: *CI* for one, *II* for none, *CD* for infinite.

Figure 5.1

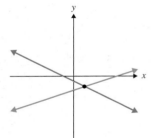
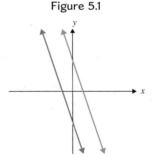
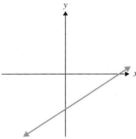

a) One solution
   Consistent system
   Independent equations

b) No solution
   Inconsistent system
   Independent equations

c) Infinite number of solutions
   Consistent system
   Dependent equations

[YOU TRY 3] Solve each system by graphing.

a) $6x - 8y = 8$
   $3x - 4y = -12$

b) $x + 2y = 8$
   $-6y - 3x = -24$

The graphs of lines can lead us to the solution of a system. But we can also learn something about the solution by looking at the equations of the lines *without* graphing them.

We saw in Example 4 that if a system's equations have the same slope and the same *y*-intercept, then they are the same line, so the system has an *infinite number of solutions*. Example 3 showed that if a system's equations have the same slope and different *y*-intercepts, then the lines are parallel and the system has *no solution*. We learned in Example 2 that if a system's equations have different slopes, then they will intersect and the system has *one solution*. In Example 5 we will use this information to determine whether a system has no solution, one solution, or an infinite number of solutions.

EXAMPLE 5

Without graphing, determine whether each system has no solution, one solution, or an infinite number of solutions.

a) $y = \dfrac{2}{3}x + 5$
   $-4x + 3y = 6$

b) $8x - 12y = 4$
   $-6x + 9y = -3$

c) $10x + 4y = -9$
   $5x + 2y = 14$

Solution

We will write each equation in slope-intercept form.

a) The first equation is already in slope-intercept form. Write the second equation, $-4x + 3y = 6$, as $y = \dfrac{4}{3}x + 2$. The slope in the first equation is $\dfrac{2}{3}$, and the second is $\dfrac{4}{3}$. The slopes are different, so this system has *one solution*.

> **W Hint**
> Remember, different slopes mean *one solution*. Same slope means either *no solution* or *infinite solutions* depending on the *y*-intercept!

b) Write each equation in slope-intercept form.

$8x - 12y = 4$
$-12y = -8x + 4$
$y = \dfrac{-8}{-12}x + \dfrac{4}{-12}$
$y = \dfrac{2}{3}x - \dfrac{1}{3}$

$-6x + 9y = -3$
$9y = 6x - 3$
$y = \dfrac{6}{9}x - \dfrac{3}{9}$
$y = \dfrac{2}{3}x - \dfrac{1}{3}$

These equations are the same. Therefore, this system has an *infinite number of solutions*.

c) Write each equation in slope-intercept form.

$10x + 4y = -9$
$4y = -10x - 9$
$y = \dfrac{-10}{4}x - \dfrac{9}{4}$
$y = -\dfrac{5}{2}x - \dfrac{9}{4}$

$5x + 2y = 14$
$2y = -5x + 14$
$y = \dfrac{-5}{2}x + \dfrac{14}{2}$
$y = -\dfrac{5}{2}x + 7$

The equations have the same slope but different *y*-intercepts. Their lines are parallel, so this system has *no solution*.

YOU TRY 4

Without graphing, determine whether each system has no solution, one solution, or an infinite number of solutions.

a) $10x + y = 4$
   $20x + 2y = -7$

b) $y = -\dfrac{2}{3}x + \dfrac{1}{2}$
   $4x + 6y = 3$

Using Technology

So far, we have learned that the solution of a system of equations is the point at which their graphs intersect. Let's see how we can solve a system by graphing using a graphing calculator. On the calculator we will solve the following system by graphing:

$$x + y = 5$$
$$y = 2x - 3$$

Begin by entering each equation using the $\boxed{Y=}$ key. Before entering the first equation, we must solve for y.

$$x + y = 5$$
$$y = -x + 5$$

Enter $-x + 5$ in Y_1 and $2x - 3$ in Y_2, press $\boxed{\text{ZOOM}}$, and select 6:ZStandard to graph the equations.

Since the lines intersect, the system has a solution. How can we find that solution? Once you see from the graph that the lines intersect, press $\boxed{\text{2nd}}$ $\boxed{\text{TRACE}}$. Select 5: intersect and then press $\boxed{\text{ENTER}}$ three times. The screen will move the cursor to the point of intersection and display the solution to the system on the bottom of the screen.

To obtain the exact solution to the system of equations, first return to the home screen by pressing $\boxed{\text{2nd}}$ $\boxed{\text{MODE}}$. To display the x-coordinate of the solution, press
$\boxed{\text{X, T, }\Theta\text{, n}}$ $\boxed{\text{MATH}}$ $\boxed{\text{ENTER}}$ $\boxed{\text{ENTER}}$ and to display the y-coordinate of the solution press $\boxed{\text{ALPHA}}$ $\boxed{\text{I}}$ $\boxed{\text{MATH}}$ $\boxed{\text{ENTER}}$ $\boxed{\text{ENTER}}$. The solution to the system is $\left(\dfrac{8}{3}, \dfrac{7}{3}\right)$.

Use a graphing calculator to solve each system.

1) $y = -2x + 5$
   $y = x - 4$

2) $y = -3x - 4$
   $y = x$

3) $2x - y = 3$
   $y - 4x = -8$

4) $x + 2y = -1$
   $x + 4y = 2$

5) $3x + 2y = 4$
   $4x + 3y = 7$

6) $6x - 3y = 10$
   $-2x + y = 4$

3 Solve a Linear System by Substitution

Another way to solve a system of equations is to use the *substitution method*. This method is especially good when one of the variables has a coefficient of 1 or -1.

EXAMPLE 6

Solve the system using substitution.

$$2x + 3y = 14$$
$$y = 3x + 1$$

Solution

The second equation is already solved for y; it tells us that y equals $3x + 1$. Therefore, we can substitute $3x + 1$ for y in the first equation, then solve for x.

$$2x + 3y = 14 \quad \text{First equation}$$
$$2x + 3(3x + 1) = 14 \quad \text{Substitute.}$$
$$2x + 9x + 3 = 14 \quad \text{Distribute.}$$
$$11x + 3 = 14$$
$$x = 1 \quad \text{Solve for } x.$$

> **Hint**
> Remember to write out the example as you are reading it!

We still need to find y. Substitute $x = 1$ in *either* equation, and solve for y. In this case, we will use the second equation since it is already solved for y.

$$y = 3x + 1 \quad \text{Second equation}$$
$$y = 3(1) + 1 \quad \text{Substitute.}$$
$$y = 4$$

Check $x = 1$, $y = 4$ in *both* equations.

$$2x + 3y = 14 \qquad\qquad y = 3x + 1$$
$$2(1) + 3(4) \stackrel{?}{=} 14 \quad \text{Substitute.} \qquad 4 \stackrel{?}{=} 3(1) + 1 \quad \text{Substitute.}$$
$$2 + 12 \stackrel{?}{=} 14 \quad \text{True} \qquad\qquad 4 \stackrel{?}{=} 3 + 1 \quad \text{True}$$

We write the solution as an ordered pair. The solution of the system is $(1, 4)$.

When we use the **substitution method,** we solve one of the equations for one of the variables in terms of the other. Then we substitute that expression into the other equation. We can do this because solving a system means finding the ordered pair (or pairs) that satisfies *both* equations.

Procedure Solving a System by Substitution

Step 1: Solve one of the equations for one of the variables. If possible, solve for a variable that has a coefficient of 1 or -1.

Step 2: Substitute the expression found in *Step 1* into the *other* equation. The equation you obtain should contain only one variable.

Step 3: Solve the equation in one variable from *Step 2*.

Step 4: Substitute the value found in *Step 3* into either of the equations to obtain the value of the other variable.

Step 5: Check the values in each of the original equations, and write the solution as an ordered pair.

EXAMPLE 7

Solve the system by substitution.

$$x + 4y = 3 \quad (1)$$
$$2x + 3y = -4 \quad (2)$$

Solution

We will follow the steps listed above.

Step 1: For which variable should we solve? The x in the first equation is the only variable with a coefficient of 1 or -1. Therefore, we will solve the first equation for x.

$$x + 4y = 3 \quad \text{Equation (1)}$$
$$x = 3 - 4y \quad \text{Subtract } 4y.$$

> **Hint**
> Notice that it is easy to solve the top equation for the variable x.

Step 2: Substitute $3 - 4y$ for x in equation (2).

$$2x + 3y = -4 \quad \text{Equation (2)}$$
$$2(3 - 4y) + 3y = -4 \quad \text{Substitute.}$$

Step 3: Solve the last equation for y.

$$2(3 - 4y) + 3y = -4$$
$$6 - 8y + 3y = -4 \quad \text{Distribute.}$$
$$-5y = -10 \quad \text{Subtract 6; combine like terms.}$$
$$y = 2$$

Step 4: Find x by substituting 2 for y in either equation. We will use equation (1).

$$x + 4y = 3 \quad (1)$$
$$x + 4(2) = 3 \quad \text{Substitute 2 for } y.$$
$$x = -5$$

Step 5: The check is left to the student. The solution of the system is $(-5, 2)$.

If no variable in the system has a coefficient of 1 or -1, solve for any variable.

[YOU TRY 5] Solve the system by substitution.

$$10x + 3y = -4$$
$$8x + y = 1$$

W Hint
Review the procedures for **Eliminating Fractions and Decimals from an Equation** in Section 2.1.

If a system contains an equation with fractions, first multiply the equation by the least common denominator to eliminate the fractions. Likewise, if an equation in the system contains decimals, begin by multiplying the equation by the power of 10 that will eliminate the decimals.

EXAMPLE 8 Solve the system by substitution.

$$\frac{2}{5}x - \frac{1}{3}y = 2 \quad (1)$$
$$-\frac{1}{6}x + \frac{1}{2}y = \frac{4}{3} \quad (2)$$

Solution

First, eliminate the fractions in each equation.

$$\frac{2}{5}x - \frac{1}{3}y = 2 \quad \text{Equation (1)} \qquad\qquad -\frac{1}{6}x + \frac{1}{2}y = \frac{4}{3} \quad \text{Equation (2)}$$
$$15\left(\frac{2}{5}x - \frac{1}{3}y\right) = 15 \cdot 2 \quad \text{Multiply by the LCD: 15} \qquad 6\left(-\frac{1}{6}x + \frac{1}{2}y\right) = 6 \cdot \frac{4}{3} \quad \text{Multiply by the LCD: 6}$$
$$6x - 5y = 30 \quad (3) \quad \text{Distribute.} \qquad\qquad -x + 3y = 8 \quad (4) \quad \text{Distribute.}$$

We now have an equivalent system of equations.

$$6x - 5y = 30 \quad (3)$$
$$-x + 3y = 8 \quad (4)$$

Step 1: The x in equation (4) has a coefficient of -1. Solve this equation for x.

$$-x + 3y = 8 \quad \text{Equation (4)}$$
$$-x = 8 - 3y \quad \text{Subtract } 3y.$$
$$x = -8 + 3y \quad \text{Divide by } -1.$$

Steps 2 and 3: Substitute $-8 + 3y$ for x in equation (3) and solve for y.

$$6x - 5y = 30 \quad \text{Equation (3)}$$
$$6(-8 + 3y) - 5y = 30 \quad \text{Substitute.}$$
$$-48 + 18y - 5y = 30 \quad \text{Distribute.}$$
$$13y = 78 \quad \text{Add 48; combine like terms.}$$
$$y = 6 \quad \text{Divide by 13.}$$

Step 4: Find x by substituting 6 for y in equation (3) or (4). Let's use equation (4) since it has smaller coefficients.

$$-x + 3y = 8 \quad \text{Equation (4)}$$
$$-x + 3(6) = 8 \quad \text{Substitute.}$$
$$-x + 18 = 8$$
$$-x = -10$$
$$x = 10 \quad \text{Divide by } -1.$$

Step 5: Check $x = 10$ and $y = 6$ in the *original* equations. The solution is $(10, 6)$.

[**YOU TRY 6**] Solve each system by substitution.

a) $\dfrac{2}{3}x - \dfrac{1}{9}y = -3$
$-\dfrac{5}{6}x + \dfrac{1}{2}y = \dfrac{1}{2}$

b) $0.01x + 0.04y = 0.15$
$0.1x - 0.1y = 0.5$

4 Solve a Linear System by Elimination

The next technique we will learn for solving a system of equations is the **elimination method**. (This is also called the **addition method**.) It is based on the addition property of equality, which says that we can add the *same* quantity to each side of an equation and preserve the equality.

If $a = b$, then $a + c = b + c$.

We can extend this idea by saying that we can add *equal* quantities to each side of an equation and still preserve the equality.

If $a = b$ and $c = d$, then $a + c = b + d$.

The objective of the elimination method is to add the equations (or multiples of one or both of the equations) so that one variable is eliminated. Then, we can solve for the remaining variable.

EXAMPLE 9

Solve the system using the elimination method.

$$3x - 2y = 2 \quad (1)$$
$$-5x + 2y = -10 \quad (2)$$

Solution

The left side of each equation is equal to the right side of each equation. Therefore, if we add the left sides together and add the right sides together,

W Hint

Notice that in this case, the coefficients of y are opposites. So when you add the two equations together, the y's will be eliminated.

we can set them equal. We will add these equations vertically, eliminate the y-terms, and solve for x.

$$\begin{aligned} 3x - 2y &= 2 \quad (1) \\ + \;\; -5x + 2y &= -10 \quad (2) \\ \hline -2x + 0y &= -8 \quad \text{Add equations (1) and (2).} \\ -2x &= -8 \quad \text{Simplify.} \\ x &= 4 \quad \text{Divide by } -2. \end{aligned}$$

To find y, we can substitute $x = 4$ into either equation. Here, we will use equation (1).

$$\begin{aligned} 3x - 2y &= 2 \quad (1) \\ 3(4) - 2y &= 2 \\ 12 - 2y &= 2 \\ -2y &= -10 \quad \text{Subtract 12.} \\ y &= 5 \quad \text{Divide by } -2. \end{aligned}$$

Check $x = 4$ and $y = 5$ in *both* equations.

$$\begin{array}{ll} 3x - 2y = 2 \quad (1) & -5x + 2y = -10 \quad (2) \\ 3(4) - 2(5) \stackrel{?}{=} 2 \quad \text{Substitute.} & -5(4) + 2(5) = -10 \quad \text{Substitute.} \\ 12 - 10 = 2 \quad \text{True} & -20 + 10 = -10 \quad \text{True} \end{array}$$

The solution is $(4, 5)$.

[YOU TRY 7] Solve the system using the elimination method.

$$\begin{aligned} x + 2y &= -6 \\ -x - 3y &= 13 \end{aligned}$$

In Example 9, we eliminated a variable by adding the equations. Sometimes eliminating a variable requires more steps, however. These steps are listed next.

> **Procedure** Solving a System of Two Linear Equations by the Elimination Method
>
> **Step 1:** Write each equation in the form $Ax + By = C$.
>
> **Step 2:** Determine which variable to eliminate. If necessary, multiply one or both of the equations by a number so that the coefficients of the variable to be eliminated are negatives of one another.
>
> **Step 3:** Add the equations, and solve for the remaining variable.
>
> **Step 4:** Substitute the value found in *Step 3* into either of the original equations to find the value of the other variable.
>
> **Step 5:** Check the solution in each of the original equations.

EXAMPLE 10 Solve the system using the elimination method.

$$\begin{aligned} 3x &= 7y + 5 \quad (1) \\ 2x - 3 &= 5y \quad (2) \end{aligned}$$

Solution

Step 1: Write each equation in the form $Ax + By = C$. We will number these new equations (3) and (4).

$3x = 7y + 5$ (1) $2x - 3 = 5y$ (2)
$3x - 7y = 5$ Subtract 7y. (3) $2x - 5y = 3$ Subtract 5y and add 3. (4)

Step 2: **Determine which variable to eliminate from equations (3) and (4).** Often, it is easier to eliminate the variable with the smaller coefficients, so we will eliminate x.

The least common multiple of 3 and 2 (the x-coefficients) is 6. We want one x-coefficient to be 6 and the other to be -6. Multiply equation (3) by 2 and equation (4) by -3.

Rewrite the System

$2(3x - 7y) = 2(5)$ 2 times (3) $\longrightarrow$ $6x - 14y = 10$
$-3(2x - 5y) = -3(3)$ -3 times (4) $-6x + 15y = -9$

> **W Hint**
> Notice that in this case, the coefficients of x are made to be opposites. This way, when you add the two equations together, the x's will be eliminated.

Step 3: Add the resulting equations to eliminate x. Solve for y.

$$6x - 14y = 10$$
$$+ \underline{-6x + 15y = -9}$$
$$y = 1$$

Step 4: Substitute $y = 1$ into equation (1) and solve for x.

$3x = 7y + 5$ (1)
$3x = 7(1) + 5$ Substitute.
$3x = 12$
$x = 4$

Step 5: **Check** to verify that (4, 1) satisfies each of the original equations. The solution is (4, 1).

[**YOU TRY 8**] Solve the system using the elimination method.

$$4x - 10 = -3y$$
$$5x = 4y - 3$$

Procedure Choosing Between the Substitution and Elimination Methods

If an equation contains fractions or decimals, begin by eliminating them. Then use the following guidelines.

1) If a variable has a coefficient of 1 or -1, solve for that variable and *use the substitution method.*

2) If none of the variables has a coefficient of 1 or -1, *use the elimination method.*

5 Solve Special Systems

As we saw in Examples 3 and 4, some systems have no solution, and some have an infinite number of solutions. How do the substitution and elimination methods illustrate these results?

EXAMPLE 11 Solve the system.

$$9y = 12x + 5 \quad (1)$$
$$6y - 8x = -11 \quad (2)$$

Solution

We will use the elimination method to solve this system.

Step 1: Write each equation in the form $Ax + By = C$.

$$9y = 12x + 5 \quad\longrightarrow\quad -12x + 9y = 5 \quad (3)$$
$$6y - 8x = -11 \quad\longrightarrow\quad -8x + 6y = -11 \quad (4)$$

Steps 2 and 3: Determine which variable to eliminate from equations (3) and (4). Eliminate y. The least common multiple of 9 and 6 is 18. One y-coefficient must be 18, and the other must be -18. Then add the equations.

$$-2(-12x + 9y) = -2(5) \quad\longrightarrow\quad 24x - 18y = -10$$
$$3(-8x + 6y) = 3(-11) \quad\quad\quad\quad + \underline{-24x + 18y = -33}$$
$$0 = -43 \quad \text{False}$$

Both variables are eliminated, and we get a false statement. The system is inconsistent, and the solution set is $\emptyset$. The graph of the system, at left, supports our work. The lines are parallel, so the system has no solution.

[YOU TRY 9] Solve the system.

$$24x + 6y = -7$$
$$4y + 3 = -16x$$

EXAMPLE 12 Solve the system.

$$2x - 8y = 20 \quad (1)$$
$$x = 4y + 10 \quad (2)$$

Solution

Step 1: Equation (2) is already solved for x, so we will use the substitution method and begin with Step 2.

Steps 2 and 3: Substitute $4y + 10$ for x in equation (1) and solve for y.

$$2x - 8y = 20 \quad \text{Equation (1)}$$
$$2(4y + 10) - 8y = 20 \quad \text{Substitute.}$$
$$8y + 20 - 8y = 20 \quad \text{Distribute.}$$
$$20 = 20 \quad \text{True}$$

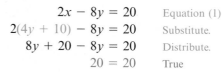

Because both variables are eliminated and we get a true statement, there are an infinite number of solutions to the system. The equations are dependent, and the solution set is $\{(x, y) | x = 4y + 10\}$, that is, all the points on the line formed by the equation $2x - 8y = 20$.

The graph to the left shows that the equations in the system are the same line, therefore the system has an infinite number of solutions.

[YOU TRY 10] Solve each system by substitution.

a) $6x + y = -8$
    $12x + 2y = -9$

b) $x - 3y = 5$
    $4x - 12y = 20$

Note
When you are solving a system of equations and both variables are eliminated,
1) if you get a *false statement*, like $0 = -43$, then the system has *no solution* and is *inconsistent*.
2) if you get a *true statement*, like $20 = 20$, then the system has an *infinite number of solutions*. The equations are *dependent*.

ANSWERS TO [YOU TRY] EXERCISES

1) a) no   b) yes
2) $(-2, 4)$

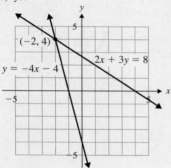

3) a) ∅

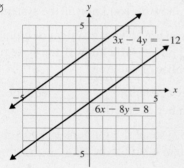

b) $\{(x, y) \mid x + 2y = 8\}$

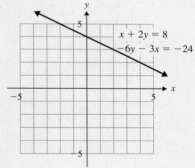

4) a) no solution   b) infinite number of solutions   5) $\left(\dfrac{1}{2}, -3\right)$ 6) a) $(-6, -9)$ b) $(7, 2)$

7) $(8, -7)$ 8) $(1, 2)$ 9) ∅   10) a) ∅   b) $\{x, y) \mid x - 3y = 5\}$

ANSWERS TO TECHNOLOGY EXERCISES

1) $(3, -1)$ 2) $(-1, -1)$ 3) $\left(\dfrac{5}{2}, 2\right)$ 4) $\left(-4, \dfrac{3}{2}\right)$ 5) $(-2, 5)$ 6) ∅

Evaluate 5.1 Exercises

Do the exercises, and check your work.

Objective 1: Determine Whether an Ordered Pair Is a Solution of a System

Determine whether the ordered pair is a solution of the system of equations.

1) $3x + 2y = 4$
   $4x - y = -3$
   $(-2, 5)$

2) $10x + 7y = -13$
   $-6x - 5y = 11$
   $\left(\frac{3}{2}, -4\right)$

3) $y = 5x - 7$
   $3x + 9 = y$
   $(-1, -2)$

4) $x - 5y = 7$
   $y = 2x + 13$
   $(-8, -3)$

Objective 2: Solve a Linear System by Graphing

5) If you are solving a system of equations by graphing, how do you know if the system has no solution?

6) If you are solving a system of equations by graphing, how do you know if the system has an infinite number of solutions?

7) A system with _____ equations has an infinite number of solutions.

8) If a system of equations has no solution, then the system is said to be _____.

Solve each system of equations by graphing. Identify any inconsistent systems or dependent equations.

9) $y = -\frac{2}{3}x + 3$
   $y = x - 2$

10) $y = \frac{3}{2}x - 4$
    $y = -2x + 3$

11) $y = -3x + 1$
    $x = 1$

12) $y - 4x = 1$
    $y = -3$

13) $\frac{3}{4}x - y = 0$
    $3x - 4y = 20$

14) $6x - 3y = 12$
    $-2x + y = -4$

15) $3x + y = 2$
    $y = 2x - 3$

16) $x = 2y - 6$
    $3x - 2y = 2$

17) $y = -3x + 1$
    $12x + 4y = 4$

18) $2x - y = 1$
    $-2x + y = -3$

Write a system of equations so that the given ordered pair is a solution of the system.

19) $(5, 1)$

20) $(-1, -4)$

21) $\left(-\frac{1}{2}, 3\right)$

22) $\left(0, \frac{2}{3}\right)$

For Exercises 23 and 24, determine which ordered pair could not be a solution to the system of equations that is graphed. Explain why you chose that ordered pair.

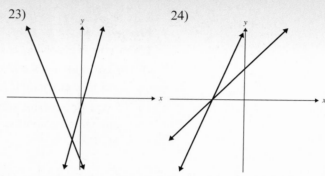

23)
A. $(-3, -3)$ C. $(-1, -8)$
B. $(-2, 1)$ D. $\left(-\frac{3}{2}, -\frac{9}{2}\right)$

24)
A. $(-6, 0)$ C. $(0, -5)$
B. $\left(-\frac{1}{2}, 0\right)$ D. $(-8.3, 0)$

25) How do you determine, *without graphing*, that a system of equations has exactly one solution?

26) How do you determine, *without graphing*, that a system of equations has no solution?

Without graphing, determine whether each system has no solution, one solution, or an infinite number of solutions. Do not solve the system.

27) $y = \frac{3}{2}x + \frac{7}{2}$
    $-9x + 6y = 21$

28) $y = 4x + 6$
    $8x - y = -7$

29) $5x - 2y = -11$
    $x + 6y = 18$

30) $5x - 8y = 24$
    $10x - 16y = -9$

31) $x + y = 10$
    $-9x - 9y = 2$

32) $x - 4y = 2$
    $3x + y = 1$

33) The graph shows the number of people, seven years of age and older, who have participated more than once in snowboarding and ice/figure skating from 1997–2003. (National Sporting Goods Association)

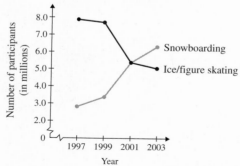

Snowboarders and Ice Skaters

246   CHAPTER 5   Solving Systems of Linear Equations

a) When were there more snowboarders than ice/figure skaters?

b) When did the number of snowboarders equal the number of skaters? How many people participated in each?

c) During which years did snowboarding see its greatest increase in participation?

d) During which years did skating see its greatest decrease in participation?

34) The graph shows the percentage of households in Delaware and Nevada with Internet access in various years from 1998 to 2003. (www.census.gov)

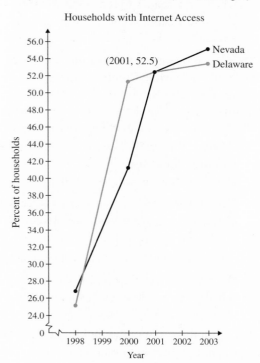

a) In 1998, which state had a greater percentage of households with Internet access? Approximately how many more were there?

b) In what year did both states have the same percentage of households with Internet access? What percentage of households had Internet access that year?

c) Between 2000 and 2001, which state had the greatest increase in the percentage of households with Internet access? How can this information be related to *slope*?

Solve each system using a graphing calculator.

35) $y = -2x + 4$
$y = x - 5$

36) $y = -x + 5$
$y = 3x + 1$

37) $-2x + y = 9$
$x - 3y = -2$

38) $4x + y = -6$
$x + 3y = 4$

39) $6x - 5y = 3.25$
$3x + y = -2.75$

40) $2x + 5y = 7.5$
$-3x + y = -11.25$

Objective 3: Solve a Linear System by Substitution

41) If you were asked to solve this system by substitution, why would it be easiest to begin by solving for y in the second equation?

$6x - 2y = -5$
$3x + y = 4$

Solve each system by substitution. Identify any inconsistent systems or dependent equations.

42) $y = 3x + 4$
$6x + y = -2$

43) $2x - y = 5$
$x = y + 6$

44) $x = y - 8$
$-3x - y = 12$

45) $2x + 5y = 8$
$x - 6y = 4$

46) $x + 3y = -12$
$3x + 4y = -6$

47) $9y - 18x = 5$
$2x - y = 3$

48) $2x + 30y = 9$
$x = 6 - 15y$

49) $5x - y = 8$
$4y = 10 - x$

50) $y - 4x = -1$
$8x + y = 2$

51) $10x + y = -5$
$-5x + 2y = 10$

52) $x + 2y = 6$
$x + 20y = -12$

53) $6x + y = -6$
$-12x - 2y = 12$

54) $x - 2y = 10$
$3x - 6y = 30$

55) $2x - y = 6$
$3y = -18 - x$

56) $2x - 9y = -2$
$6y - x = 0$

57) If an equation in a system contains fractions, what should you do first to make the system easier to solve?

58) If an equation in a system contains decimals, what should you do first to make the system easier to solve?

Solve each system by substitution. Identify any inconsistent systems or dependent equations.

59) $\frac{1}{4}x - \frac{1}{2}y = 1$
$\frac{2}{3}x + \frac{1}{6}y = \frac{25}{6}$

60) $\frac{2}{3}x + \frac{2}{3}y = 6$
$\frac{3}{2}x - \frac{1}{4}y = \frac{13}{2}$

61) $-\frac{2}{15}x - \frac{1}{3}y = \frac{2}{3}$
$\frac{2}{3}x + \frac{5}{3}y = \frac{1}{2}$

62) $y - \frac{5}{2}x = -2$
$\frac{3}{4}x - \frac{3}{10}y = \frac{3}{5}$

63) $\frac{3}{4}x + \frac{1}{2}y = 6$
$x = 3y + 8$

64) $0.01x + 0.10y = -0.11$
$0.02x - 0.01y = 0.20$

65) $0.2x - 0.1y = 0.1$
$0.01x + 0.04y = 0.23$

Mixed Exercises: Objectives 4 and 5
Solve each system using elimination. Identify any inconsistent systems or dependent equations.

66) $3x + 5y = -10$
$7x - 5y = 10$

67) $4x - 3y = -5$
$-4x + 5y = 11$

68) $7x + 6y = 3$
$3x + 2y = -1$

69) $-8x + 5y = -6$
$4x - 7y = 3$

70) $3x - y = 4$
$-6x + 2y = -8$

71) $5x - 6y = -2$
$10x - 12y = 7$

72) $2x - 9 = 8y$
$20y - 5x = 6$

73) $x - 6y = -5$
$-24y + 4x = -20$

74) $8x = 6y - 1$
$10y - 6 = -4x$

75) $9x - 7y = -14$
$4x + 3y = 6$

76) $6x + 5y = 13$
$5x + 3y = 5$

77) $7x + 2y = 12$
$24 - 14x = 4y$

78) $4 + 9y = -21x$
$14x + 6y = -1$

79) $\frac{x}{4} + \frac{y}{2} = -1$
$\frac{3}{8}x + \frac{5}{3}y = -\frac{7}{12}$

80) $\frac{1}{2}x + \frac{2}{3}y = -\frac{29}{6}$
$-\frac{1}{3}x + y = -4$

81) $\frac{x}{2} - \frac{y}{5} = \frac{1}{10}$
$\frac{x}{3} + \frac{y}{4} = \frac{5}{6}$

82) $x + \frac{y}{4} = \frac{7}{2}$
$\frac{2}{5}x + \frac{1}{2}y = -1$

83) $0.1x + 2y = -0.8$
$0.03x + 0.10y = 0.26$

84) $0.6x - 0.1y = 0.5$
$0.10x - 0.03y = -0.01$

85) $-0.4x + 0.2y = 0.1$
$0.6x - 0.3y = 1.5$

86) $x - 0.5y = 0.2$
$-0.3x + 0.15y = -0.06$

87) Noor needs to rent a car for one day while hers is being repaired. Rent-for-Less charges $0.40 per mile while Frugal Rentals charges $12 per day plus $0.30 per mile. Let x = the number of miles driven, and let y = the cost of the rental. The cost of renting a car from each company can be expressed with the following equations:

Rent-for-Less: $y = 0.40x$
Frugal Rentals: $y = 0.30x + 12$

a) How much would it cost Noor to rent a car from each company if she planned to drive 60 mi?

b) How much would it cost Noor to rent a car from each company if she planned to drive 160 mi?

c) Solve the system of equations using the substitution method, and explain the meaning of the solution.

d) Graph the system of equations, and explain when it is cheaper to rent a car from Rent-for-Less and when it is cheaper to rent a car from Frugal Rentals. When is the cost the same?

88) To rent a moving truck, Discount Van Lines charges $1.20 per mile while Comfort Ride Company charges $60 plus $1.00 per mile. Let x = the number of miles driven, and let y = the cost of the rental. The cost of renting a moving truck from each company can be expressed with the following equations:

Discount Van Lines: $y = 1.20x$
Comfort Ride: $y = 1.00x + 60$

a) How much would it cost to rent a truck from each company if the truck would be driven 100 mi?

b) How much would it cost to rent a truck from each company if the truck would be driven 400 mi?

c) Solve the system of equations using the substitution method, and explain the meaning of the solution.

d) Graph the system of equations, and explain when it is cheaper to rent a truck from Discount Van Lines and when it is cheaper to rent a truck from Comfort Ride. When is the cost the same?

Mixed Exercises: Objectives 3 and 4
Solve each system using *your choice* of the substitution or the elimination method. Identify any inconsistant systems or dependent equations. See the box about choosing between the substitution and elimination methods on page 243.

89) $-2x - y = -1$
$4x + y = -5$

90) $3x + y = 15$
$y = 4x - 6$

91) $3x + 4y = 9$
$5x + 6y = 16$

92) $0.02x + 0.07y = -0.24$
$0.05y - 0.04x = 0.10$

93) $\frac{x}{12} - \frac{y}{6} = \frac{2}{3}$
$\frac{x}{4} + \frac{y}{3} = 2$

94) $y = \frac{1}{3}x + 4$
$3y - x = 12$

95) $y = -3x - 20$
$6x + 2y = 5$

96) $\dfrac{5}{3}x - \dfrac{4}{3}y = -\dfrac{4}{3}$
$y = 2x + 4$

97) $0.1x + 0.5y = 0.4$
$-0.03x + 0.01y = 0.2$

98) $6x - 3y = -11$
$9x - 2y = 1$

Solve each system using *your choice* of the substitution or elimination method. Begin by combining like terms.

99) $5(2x - 3) + y - 6x = -24$
$8y - 3(2y + 3) + x = -6$

100) $8 + 2(3x - 5) - 7x + 6y = 2$
$9(y - 2) + 5x - 13y = -12$

101) $6(3x + 4) - 8(x + 2) = 5 - 3y$
$6x - 2(5y + 2) = -7(2y - 1) - 4$

102) $2(y - 6) = 3y + 4(x - 5)$
$2(4x + 3) - 5 = 2(1 - y) + 5x$

Objective 5: Solve Special Systems

103) When solving a system of linear equations, how do you know if the system has

a) no solution?

b) an infinite number of solutions?

104) Given the following system of equations,

$x + y = 8$
$x + y = c$

find c so that the system has

a) an infinite number of solutions.

b) no solution.

105) Given the following system of equations,

$x - y = 3$
$x - y = c$

find c so that the system has

a) an infinite number of solutions.

b) no solution.

106) Given the following system of equations,

$2x - 3y = 5$
$ax - 6y = 10$

find a so that the system has

a) an infinite number of solutions.

b) exactly one solution.

107) Find b so that $(2, -1)$ is a solution to the system

$3x - 4y = 10$
$-x + by = -7$

108) Find a so that $(4, 3)$ is a solution to the system

$ax + 5y = 3$
$2x - 3y = -1$

Extension

Let a, b, and c represent nonzero constants. Solve each system for x and y.

109) $ax + by = 5$
$2ax - by = 1$

110) $-3ax + 2by = 9$
$3ax - 4by = 1$

111) $x - 3by = 2$
$3x + by = -4$

112) $ax + 4y = 6$
$2ax + y = 5$

113) $ax + by = c$
$-ax + by = c$

114) $2ax - by = c$
$ax + 2by = 8c$

Systems like those in Exercises 115–118 can be solved by substituting one variable for $\dfrac{1}{x}$, another variable for $\dfrac{1}{y}$, and then using the elimination method. In these systems, substitute u for $\dfrac{1}{x}$, v for $\dfrac{1}{y}$, and solve for u and v using elimination. Since the original system is in terms of x and y, the last step is to solve for x and y.

In the first equation of Exercise 115, for example, $\dfrac{1}{x} = u$ and $\dfrac{2}{y} = 2 \cdot \dfrac{1}{y} = 2v$. Substitute u for $\dfrac{1}{x}$ and $2v$ for $\dfrac{2}{y}$ so that the equation becomes $u + 2v = -\dfrac{7}{8}$. Use this approach to solve each system.

115) $\dfrac{1}{x} + \dfrac{2}{y} = -\dfrac{7}{8}$
$\dfrac{1}{x} - \dfrac{1}{y} = \dfrac{5}{8}$

116) $\dfrac{1}{x} + \dfrac{1}{y} = \dfrac{4}{3}$
$\dfrac{4}{x} + \dfrac{1}{y} = \dfrac{13}{3}$

117) $-\dfrac{1}{x} + \dfrac{2}{y} = -\dfrac{7}{3}$
$\dfrac{2}{x} - \dfrac{3}{y} = 5$

118) $\dfrac{6}{x} + \dfrac{1}{y} = 0$
$-\dfrac{2}{x} + \dfrac{3}{y} = \dfrac{20}{3}$

Solve by graphing. Given the functions $f(x)$ and $g(x)$, determine the value of x for which $f(x) = g(x)$.

119) $f(x) = x - 3$, $g(x) = -\dfrac{1}{4}x + 2$

120) $f(x) = x + 4$, $g(x) = -2x - 2$

121) $f(x) = 3x + 3$, $g(x) = x + 1$

122) $f(x) = -\dfrac{2}{3}x + 1$, $g(x) = x - 4$

R Rethink

R1) This section has demonstrated three different methods for finding the intersection of two lines. What are the three methods? Which method do you prefer?

R2) What happens when you graph a system of linear equations that has no solution?

R3) Suppose you want to eliminate the *x*-values in a system of linear equations. What must you do with the equations to eliminate the *x*-values?

R4) Which objective is the most difficult for you?

5.2 Solving Systems of Linear Equations in Three Variables

P Prepare / O Organize

What are your objectives for Section 5.2?	How can you accomplish each objective?
1 Understand Systems of Three Equations in Three Variables	• Know how to recognize a *system of three equations in three variables*. • Know what a *three-dimensional coordinate system* looks like and what an *ordered triple* is. • Be aware of the different possible solutions of a system of three equations.
2 Solve Systems of Linear Equations in Three Variables	• Learn the procedure for **Solving a System of Linear Equations in Three Variables.** • Complete the given example on your own. • Complete You Try 1.
3 Solve Special Systems in Three Variables	• Be aware that when solving a system in three variables, variables may "drop out," which may indicate *no solution* or *infinite solutions*. • Be able to determine the difference between *no solution* and *infinite solutions* and know what those solutions mean. • Complete the given example on your own. • Complete You Try 2.
4 Solve a System with Missing Terms	• Notice that the procedure for **Solving a System of Linear Equations in Three Variables** is modified when equations are missing terms. • Complete the given example on your own. • Complete You Try 3.

 Read the explanations, follow the examples, take notes, and complete the You Trys.

In this section, we will learn how to solve a system of linear equations in *three* variables.

1 Understand Systems of Three Equations in Three Variables

Definition

A **linear equation in three variables** is an equation of the form $Ax + By + Cz = D$ where A, B, and C are not all zero and where A, B, C, and D are real numbers. Solutions of this type of an equation are **ordered triples** of the form (x, y, z).

An example of a linear equation in three variables is

$$2x - y + 3z = 12.$$

This equation has infinitely many solutions. Here are a few:

(5, 1, 1)   since $2(5) - (1) + 3(1) = 12$
(3, 0, 2)   since $2(3) - (0) + 3(2) = 12$
(6, −3, −1) since $2(6) - (-3) + 3(-1) = 12$

Ordered triples, like (1, 2, 3) and (3, 0, 2), are graphed on a three-dimensional coordinate system, as shown to the right. Notice that the ordered triples are *points*.

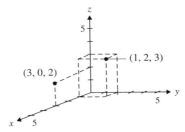

The graph of a linear equation in three variables is a *plane*.

A **solution of a system of linear equations in three variables** is an *ordered triple* that satisfies each equation in the system. Like systems of linear equations in two variables, systems of linear equations in *three* variables can have *one* solution, *no* solution, or *infinitely many* solutions.

Here is an example of a system of linear equations in three variables:

$$x + 4y + 2z = 10$$
$$3x - y + z = 6$$
$$2x + 3y - z = -4$$

In Section 5.1, we solved systems of linear equations in *two* variables by graphing. Since the graph of an equation like $x + 4y + 2z = 10$ is a *plane*, however, solving a system in three variables by graphing would not be practical. But let's look at the graphs of systems of linear equations in three variables that have one solution, no solution, or an infinite number of solutions.

One solution:

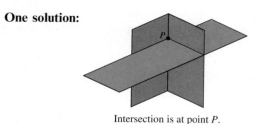

Intersection is at point P.

All three planes intersect at one point; this point is the solution of the system.

> **W Hint**
> Write down a description of each of these three cases using your own words.

No solution:

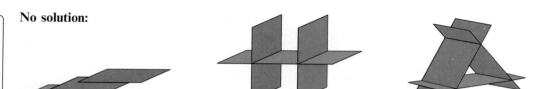

None of the planes may intersect or *two* of the planes may intersect, but if there is no solution to the system, *all three planes* do not have a common point of intersection.

Infinite number of solutions:

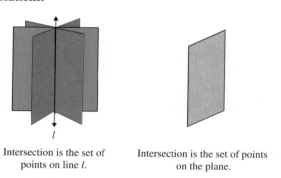

Intersection is the set of points on line *l*.

Intersection is the set of points on the plane.

The three planes may intersect so that they have a line or a plane in common. The solution to the system is the infinite set of points on the line or the plane, respectively.

2 Solve Systems of Linear Equations in Three Variables

First we will learn how to solve a system in which each equation has three variables.

Procedure Solving a System of Linear Equations in Three Variables

Step 1: **Label** the equations ①, ②, and ③.

Step 2: **Choose a variable to eliminate. Eliminate** this variable from *two* sets of *two* equations using the elimination method. You will obtain two equations containing the same two variables. Label one of these new equations Ⓐ and the other Ⓑ.

Step 3: Use the elimination method to **eliminate a variable from equations Ⓐ and Ⓑ**. You have now found the value of one variable.

Step 4: **Find the value of another variable** by substituting the value found in *Step 3* into equation Ⓐ or Ⓑ and solving for the second variable.

Step 5: **Find the value of the third variable** by substituting the values of the two variables found in *Steps 3* and *4* into equation ①, ②, or ③.

Step 6: **Check** the solution in each of the original equations, and **write the solution as an ordered triple.**

EXAMPLE 1

Solve ① $x + 2y - 2z = 3$
② $2x + y + 3z = 1$
③ $x - 2y + z = -10$

Solution

Steps 1 and 2: We have already **labeled** the equations. We'll **choose** to eliminate the variable y from *two* sets of *two* equations:

a) Add equations ① and ③ to eliminate y. Label the resulting equation Ⓐ.

$$\begin{array}{rl} ① & x + 2y - 2z = 3 \\ ③ + & x - 2y + z = -10 \\ \hline Ⓐ & 2x - z = -7 \end{array}$$

> **W Hint**
> This example has many steps. It will be very helpful to write it out as you are reading it!

b) Multiply equation ② by 2, and add it to equation ③ to eliminate y. Label the resulting equation Ⓑ.

$$\begin{array}{rl} 2 \times ② & 4x + 2y + 6z = 2 \\ ③ + & x - 2y + z = -10 \\ \hline Ⓑ & 5x + 7z = -8 \end{array}$$

> **Note**
> Equations Ⓐ and Ⓑ contain only *two* variables and they are the same variables, x and z.

Step 3: Use the elimination method to **eliminate a variable from equations** Ⓐ **and** Ⓑ. We will eliminate z from Ⓐ and Ⓑ. Multiply Ⓐ by 7, and add it to Ⓑ.

$$\begin{array}{rl} 7 \times Ⓐ & 14x - 7z = -49 \\ Ⓑ + & 5x + 7z = -8 \\ \hline & 19x = -57 \\ & \boxed{x = -3} \quad \text{Divide by 19.} \end{array}$$

Step 4: **Find the value of another variable** by substituting $x = -3$ into equation Ⓐ or Ⓑ. We will use Ⓐ since it has smaller coefficients.

$$\begin{array}{rl} Ⓐ & 2x - z = -7 \\ & 2(-3) - z = -7 \quad \text{Substitute } -3 \text{ for } x. \\ & -6 - z = -7 \quad \text{Multiply.} \\ & \boxed{z = 1} \quad \text{Add 6, and divide by } -1. \end{array}$$

Step 5: **Find the value of the third variable** by substituting $x = -3$ and $z = 1$ into equation ①, ②, or ③. We will use equation ①.

$$\begin{array}{rl} ① & x + 2y - 2z = 3 \\ & -3 + 2y - 2(1) = 3 \quad \text{Substitute } -3 \text{ for } x \text{ and } 1 \text{ for } z. \\ & -3 + 2y - 2 = 3 \quad \text{Multiply.} \\ & 2y - 5 = 3 \quad \text{Combine like terms.} \\ & \boxed{y = 4} \quad \text{Add 5, and divide by 2.} \end{array}$$

Step 6: **Check** the solution, $(-3, 4, 1)$, in each of the original equations, and **write the solution.**

① $\quad x + 2y - 2z = 3$
$\quad -3 + 2(4) - 2(1) \stackrel{?}{=} 3$
$\quad -3 + 8 - 2 \stackrel{?}{=} 3$
$\quad 3 = 3$
$\quad$ True

② $\quad 2x + y + 3z = 1$
$\quad 2(-3) + 4 + 3(1) \stackrel{?}{=} 1$
$\quad -6 + 4 + 3 \stackrel{?}{=} 1$
$\quad 1 = 1$
$\quad$ True

③ $\quad x - 2y + z = -10$
$\quad -3 - 2(4) + 1 \stackrel{?}{=} -10$
$\quad -3 - 8 + 1 \stackrel{?}{=} -10$
$\quad -10 = -10$
$\quad$ True

The solution is $(-3, 4, 1)$.

[YOU TRY 1] Solve $\quad x + 2y + 3z = -11$
$\qquad\qquad\quad 3x - y + z = 0$
$\qquad\qquad\ -2x + 3y - z = 4$

3 Solve Special Systems in Three Variables

Some systems in three variables have no solution, and some have an infinite number of solutions.

EXAMPLE 2 Solve ① $-3x + 2y - z = 5$
$\qquad\qquad$② $\quad x + 4y + z = -4$
$\qquad\qquad$③ $\ 9x - 6y + 3z = -2$

Solution

Steps 1 and 2: We have already *labeled* the equations. The *variable we choose to eliminate* is z, the easiest.

a) Add equations ① and ② to eliminate z. Label the resulting equation $\boxed{A}$.

$\quad$① $\quad -3x + 2y - z = 5$
$\quad$② $+\ \underline{x + 4y + z = -4}$
$\quad\boxed{A}\quad -2x + 6y \quad\ \ = 1$

b) Multiply equation ① by 3, and add it to equation ③ to eliminate z. Label the resulting equation $\boxed{B}$.

$\quad$① $-3x + 2y - z = 5 \longrightarrow 3 \times$ ① $\quad -9x + 6y - 3z = 15$
$\qquad\qquad\qquad\qquad\qquad\qquad\quad$③ $+\ \underline{9x - 6y + 3z = -2}$
$\qquad\qquad\qquad\qquad\qquad\qquad\ \boxed{B}\qquad\qquad\qquad\ 0 = 13\quad$ False

Since the variables are eliminated and we get the false statement $0 = 13$, equations ① and ③ have no ordered triple that satisfies each equation.

The system is inconsistent, so there is no solution. The solution set is $\emptyset$.

Note
If the variables are eliminated and you get a false statement, there is *no solution* to the system. The system is inconsistent, so the solution set is $\emptyset$.

EXAMPLE 3

Solve ① $-4x - 2y + 8z = -12$
② $2x + y - 4z = 6$
③ $6x + 3y - 12z = 18$

Solution

Steps 1 and 2: We label the equations and choose a variable, y, to eliminate.

a) Multiply equation ② by 2 and add it to equation ①. Label the resulting equation Ⓐ.

$$
\begin{array}{rl}
2 \times ② & 4x + 2y - 8z = 12 \\
① + & -4x - 2y + 8z = -12 \\
\hline
Ⓐ & 0 = 0 \quad \text{True}
\end{array}
$$

> **W Hint**
> Notice that just like in Section 5.1, Objective 5, if the variables drop out and you end up with a true statement, there are *an infinite number of solutions*.

The variables were eliminated, and we obtained the true statement $0 = 0$. This is because equation ① is a multiple of equation ②.

Notice, also, that equation ③ is a multiple of equation ②.

The equations in this system are dependent. There are an infinite number of solutions, and we write the solution set as $\{(x, y, z) | 2x + y - 4z = 6\}$. The equations all have the same graph.

[YOU TRY 2]

Solve each system of equations.

a) $8x + 20y - 4z = -16$
$-6x - 15y + 3z = 12$
$2x + 5y - z = -4$

b) $x + 4y - 3z = 2$
$2x - 5y + 2z = -8$
$-3x - 12y + 9z = 7$

4 Solve a System with Missing Terms

EXAMPLE 4

Solve ① $5x - 2y = 6$
② $y + 2z = 1$
③ $3x - 4z = -8$

Solution

First, notice that while this *is* a system of three equations in three variables, none of the equations contains three variables. Furthermore, each equation is "missing" a different variable.

> **Note**
> We will use many of the *ideas* outlined in the steps for solving a system of three equations, but we will use *substitution* rather than the elimination method.

Step 1: Label the equations ①, ②, ③.

Step 2: The goal of *Step 2* is to obtain two equations that contain the same two variables. We will modify this step from the way it was outlined on p. 252.

In order to obtain *two* equations with the same *two* variables, we will use *substitution*.

Since y in equation ② is the only variable in the system with a coefficient of 1, we will solve equation ② for y.

$$② \quad y + 2z = 1$$
$$y = 1 - 2z \quad \text{Subtract } 2z.$$

Substitute $y = 1 - 2z$ into equation ① to obtain an equation containing the variables x and z. Simplify. Label the resulting equation Ⓐ.

$$① \quad 5x - 2y = 6$$
$$5x - 2(1 - 2z) = 6 \quad \text{Substitute } 1 - 2z \text{ for } y.$$
$$5x - 2 + 4z = 6 \quad \text{Distribute.}$$
$$Ⓐ \quad 5x + 4z = 8 \quad \text{Add 2.}$$

Step 3: The goal of *Step 3* is to solve for one of the variables. Equations Ⓐ and ③ contain only x and z.

We will eliminate z from Ⓐ and ③. Add the two equations to eliminate z, then solve for x.

$$Ⓐ \quad 5x + 4z = 8$$
$$③ + 3x - 4z = -8$$
$$8x \quad\quad = 0$$
$$\boxed{x = 0} \quad \text{Divide by 8.}$$

Step 4: Find the value of another variable by substituting $x = 0$ into either Ⓐ, ①, or ③.

$$Ⓐ \quad 5x + 4z = 8$$
$$5(0) + 4z = 8 \quad \text{Substitute 0 for } x.$$
$$4z = 8$$
$$\boxed{z = 2} \quad \text{Divide by 4.}$$

Step 5: Find the value of the third variable by substituting $x = 0$ into ① or $z = 2$ into ②.

$$① \quad 5x - 2y = 6$$
$$5(0) - 2y = 6 \quad \text{Substitute 0 for } x.$$
$$-2y = 6$$
$$\boxed{y = -3} \quad \text{Divide by } -2.$$

Step 6: Check the solution $(0, -3, 2)$ in each of the original equations. The check is left to the student. The solution is $(0, -3, 2)$.

[YOU TRY 3] Solve $x + 2y = 8$
$\phantom{\text{Solve }} 2y + 3z = 1$
$\phantom{\text{Solve }} 3x - z = -3$

ANSWERS TO [YOU TRY] EXERCISES

1) $(2, 1, -5)$ 2) a) $\{(x, y, z) | 2x + 5y - z = -4\}$ b) $\varnothing$ 3) $(-2, 5, -3)$

E Evaluate 5.2 Exercises

Do the exercises, and check your work.

Objective 1: Understand Systems of Three Equations in Three Variables

Determine whether the ordered triple is a solution of the system.

1) $3x + y + 2z = 2$
   $-2x - y + z = 5$
   $x + 2y - z = -11$
   $(1, -5, 2)$

2) $4x + 3y - 7z = -6$
   $x - 2y + 5z = -3$
   $-x + y + 2z = 7$
   $(-2, 3, 1)$

3) $-x + y - 2z = 2$
   $3x - y + 5z = 4$
   $2x + 3y - z = 7$
   $(0, 6, 2)$

4) $6x - y + 4z = 4$
   $-2x + y - z = 5$
   $2x - 3y + z = 2$
   $\left(-\dfrac{1}{2}, -3, 1\right)$

5) Write a system of equations in x, y, and z so that the ordered triple $(4, -1, 2)$ is a solution of the system.

6) Find the value of c so that $(6, 0, 5)$ is a solution of the system
   $2x - 5y - 3z = -3$
   $-x + y + 2z = 4$
   $-2x + 3y + cz = 8$

Objective 2: Solve Systems of Linear Equations in Three Variables

Solve each system. See Example 1.

7) $x + 3y + z = 3$
   $4x - 2y + 3z = 7$
   $-2x + y - z = -1$

8) $x - y + 2z = -7$
   $-3x - 2y + z = -10$
   $5x + 4y + 3z = 4$

9) $5x + 3y - z = -2$
   $-2x + 3y + 2z = 3$
   $x + 6y + z = -1$

10) $-2x - 2y + 3z = 2$
    $3x + 3y - 5z = -3$
    $-x + y - z = 9$

11) $3a + 5b - 3c = -4$
    $a - 3b + c = 6$
    $-4a + 6b + 2c = -6$

12) $a - 4b + 2c = -7$
    $3a - 8b + c = 7$
    $6a - 12b + 3c = 12$

Objective 3: Solve Special Systems in Three Variables

Solve each system. Identify any systems that are inconsistent or that have dependent equations. See Examples 2 and 3.

13) $a - 5b + c = -4$
    $3a + 2b - 4c = -3$
    $6a + 4b - 8c = 9$

14) $-a + 2b - 12c = 8$
    $-6a + 2b - 8c = -3$
    $3a - b + 4c = 4$

15) $-15x - 3y + 9z = 3$
    $5x + y - 3z = -1$
    $10x + 2y - 6z = -2$

16) $-4x + 10y - 16z = -6$
    $-6x + 15y - 24z = -9$
    $2x - 5y + 8z = 3$

17) $-3a + 12b - 9c = -3$
    $5a - 20b + 15c = 5$
    $-a + 4b - 3c = -1$

18) $3x - 12y + 6z = 4$
    $-x + 4y - 2z = 7$
    $5x + 3y + z = -2$

Objective 4: Solve a System with Missing Terms

Solve each system. See Example 4.

19) $5x - 2y + z = -5$
    $x - y - 2z = 7$
    $4y + 3z = 5$

20) $-x + z = 9$
    $-2x + 4y - z = 4$
    $7x + 2y + 3z = -1$

21) $a + 15b = 5$
    $4a + 10b + c = -6$
    $-2a - 5b - 2c = -3$

22) $2x - 6y - 3z = 4$
    $-3y + 2z = -6$
    $-x + 3y + z = -1$

23) $x + 2y + 3z = 4$
    $-3x + y = -7$
    $4y + 3z = -10$

24) $-3a + 5b + c = -4$
    $a + 5b = 3$
    $4a - 3c = -11$

25) $-5x + z = -3$
    $4x - y = -1$
    $3y - 7z = 1$

26) $a + b = 1$
    $a - 5c = 2$
    $b + 2c = -4$

27) $4a + 2b = -11$
    $-8a - 3c = -7$
    $b + 2c = 1$

28) $3x + 4y = -6$
    $-x + 3z = 1$
    $2y + 3z = -1$

Mixed Exercises: Objectives 2–4

Solve each system. Identify any systems that are inconsistent or that have dependent equations.

29) $6x + 3y - 3z = -1$
    $10x + 5y - 5z = 4$
    $x - 3y + 4z = 6$

30) $2x + 3y - z = 0$
    $x - 4y - 2z = -5$
    $-4x + 5y + 3z = -4$

31) $7x + 8y - z = 16$
    $-\frac{1}{2}x - 2y + \frac{3}{2}z = 1$
    $\frac{4}{3}x + 4y - 3z = -\frac{2}{3}$

32) $3a + b - 2c = -3$
    $9a + 3b - 6c = -9$
    $-6a - 2b + 4c = 6$

33) $2a - 3b = -4$
    $3b - c = 8$
    $-5a + 4c = -4$

34) $5x + y - 2z = -2$
    $-\frac{1}{2}x - \frac{3}{4}y + 2z = \frac{5}{4}$
    $x - 6z = 3$

35) $-4x + 6y + 3z = 3$
    $-\frac{2}{3}x + y + \frac{1}{2}z = \frac{1}{2}$
    $12x - 18y - 9z = -9$

36) $x - \frac{5}{2}y + \frac{1}{2}z = \frac{5}{4}$
    $x + 3y - z = 4$
    $-6x + 15y - 3z = -1$

37) $a + b + 9c = -3$
    $-5a - 2b + 3c = 10$
    $4a + 3b + 6c = -15$

38) $2x + 3y = 2$
    $-3x + 4z = 0$
    $y - 5z = -17$

39) $x + 5z = 10$
    $4y + z = -2$
    $3x - 2y = 2$

40) $a + 3b - 8c = 2$
    $-2a - 5b + 4c = -1$
    $4a + b + 16c = -4$

41) $2x - y + 4z = -1$
    $x + 3y + z = -5$
    $-3x + 2y = 7$

42) $-2a + 3b = 3$
    $a + 5c = -1$
    $b - 2c = -5$

43) Given the following two equations, write a third equation to obtain a system of three equations in x, y, and z so that the system has no solution.

    $x + 3y - 2z = -9$
    $2x - 5y + z = 1$

44) Given the following two equations, write a third equation to obtain a system of three equations in x, y, and z so that the system has an infinite number of solutions.

    $9x - 12y + 3z = 21$
    $-3x + 4y - z = -7$

Extension

Extend the concepts of this section to solve each system. Write the solution in the form (a, b, c, d)

45) $a - 2b - c + d = 0$
    $-a + 2b + 3c + d = 6$
    $2a + b + c - d = 8$
    $a - b + 2c + 3d = 7$

46) $-a + 4b + 3c - d = 4$
    $2a + b - 3c + d = -6$
    $a + b + c + d = 0$
    $a - b + 2c - d = -1$

47) $3a + 4b + c - d = -7$
    $-3a - 2b - c + d = 1$
    $a + 2b + 3c - 2d = 5$
    $2a + b + c - d = 2$

48) $3a - 4b + c + d = 12$
    $-3a + 2b - c + 3d = -4$
    $a - 2b + 2c - d = 2$
    $-a + 4b + c + d = 8$

R Rethink

R1) When solving a system of linear equations in three variables, how do you know which variable to eliminate first?

R2) How does this section relate to Section 5.1 (Solving Linear Equations in Two Variables)?

R3) When solving a system of linear equations with three variables, what are the minimum number of equations needed to solve the system?

R4) Which objective is the most difficult for you?

5.3 Applications of Systems of Linear Equations

P Prepare

What are your objectives for Section 5.3?

O Organize

How can you accomplish each objective?

1 Solve General Two-Variable Problems	<ul><li>Learn the procedure for **Solving an Applied Problem Using a System of Equations,** and summarize it in your notes.</li><li>Use one of the methods learned in this chapter for solving the system.</li><li>Complete the given example on your own.</li><li>Complete You Try 1.</li></ul>
2 Solve Problems Involving Geometry	<ul><li>Make a list of common geometry formulas.</li><li>Draw a diagram if one is not already given.</li><li>Use the procedure for **Solving an Applied Problem Using a System of Equations.**</li><li>Complete the given example on your own.</li><li>Complete You Try 2.</li></ul>
3 Solve Problems Involving Cost	<ul><li>Use the procedure for **Solving an Applied Problem Using a System of Equations.**</li><li>Review the procedure for **Clearing Decimals Out of Equations.**</li><li>Complete the given example on your own.</li><li>Complete You Try 3.</li></ul>
4 Solve Mixture Problems	<ul><li>Use the procedure for **Solving an Applied Problem Using a System of Equations.**</li><li>Look back at Section 2.4 if you need to review the concepts involved in solving mixture problems.</li><li>Complete the given example on your own.</li><li>Complete You Try 4.</li></ul>
5 Solve Problems Involving Distance, Rate, and Time	<ul><li>Use the procedure for **Solving an Applied Problem Using a System of Equations.**</li><li>Refer to Section 2.4 to review the concepts of solving these problems using $d = rt$.</li><li>Complete the given example on your own.</li><li>Complete You Try 5.</li></ul>
6 Solve Problems Involving Three Variables	<ul><li>Extend the procedure for **Solving an Applied Problem Using a System of Equations in Two Variables** to include a third variable and third equation.</li><li>Complete the given example on your own.</li><li>Complete You Try 6.</li></ul>

 Read the explanations, follow the examples, take notes, and complete the You Trys.

1 Solve General Two-Variable Problems

In Section 2.1, we introduced the five-step problem-solving method. Throughout most of this section, we will modify the method for problems with *two* unknowns and *two* equations.

> **Procedure** Solving an Applied Problem Using a System of Equations
>
> *Step 1:* **Read** the problem carefully, more than once if necessary. Draw a picture, if applicable. Identify what you are being asked to find.
>
> *Step 2:* **Choose variables** to represent the unknown quantities. Label any pictures with the variables.
>
> *Step 3:* **Write a system of equations using two variables.** It may be helpful to begin by writing the equations in words.
>
> *Step 4:* **Solve** the system.
>
> *Step 5:* **Check** the answer in the original problem, and **interpret** the solution as it relates to the problem.

EXAMPLE 1

Write a system of equations, and solve.

In 2012, the prime-time TV shows of HBO received 21 more Emmy Award nominations than CBS. Together, their shows received a total of 141 nominations. How many Emmy nominations did HBO and CBS each receive in 2012? (www.emmys.org)

Solution

Step 1: **Read** the problem carefully, and identify what we are being asked to find. We must find the number of Emmy nominations received by HBO and by CBS.

Step 2: **Choose variables** to represent the unknown quantities.

x = number of nominations for HBO
y = number of nominations for CBS

Step 3: **Write a system of equations using two variables.**
Let's write the equations in English first. Then, we can translate them to algebraic equations.

To get one equation, use the information that says HBO and CBS received a total of 141 nominations.

Number of HBO nominations	+	Number of CBS nominations	=	Total number of nominations	
x	+	y	=	141	Equation (1)

To get the second equation, use the information that says that HBO received 21 more nominations than CBS.

Number of HBO nominations	was	21 more than	number of CBS nominations	
↓	↓	↓	↓	
x	=	21 +	y	Equation (2)

The system of equations is
$$x + y = 141$$
$$x = 21 + y.$$

Step 4: **Solve** the system. Let's use substitution.

$$x + y = 141$$
$$(21 + y) + y = 141 \quad \text{Substitute.}$$
$$21 + 2y = 141$$
$$2y = 120$$
$$y = 60$$

Find x by substituting $y = 60$ into $x = 21 + y$.

$$x = 21 + 60$$
$$x = 81$$

The solution to the system is $(81, 60)$.

Step 5: **Check** the answer in the original problem, and **interpret** the solution as it relates to the problem. *HBO received* 81 *nominations, and CBS received* 60. The total number of nominations was $81 + 60 = 141$. HBO received 21 more nominations than CBS, and $21 + 60 = 81$. The answer is correct.

[YOU TRY 1] Write a system of equations and solve.
The Turquoise Bay Motel has 51 rooms. Some of them have two double beds and half as many rooms have one queen-size bed. Determine how many rooms have two beds and how many rooms have one bed.

In Chapter 2 we solved problems involving geometry by defining the unknowns in terms of one variable. Now, we will use a system of equations to solve these applications.

2 Solve Problems Involving Geometry

EXAMPLE 2 Write a system of equations, and solve.
The Alvarez family purchased a rectangular lot on which they will build a new house. The lot is 70 ft longer than it is wide, and the perimeter is 460 ft. Find the length and width of the lot.

Solution

Step 1: **Read** the problem carefully. Draw a picture. We must find the length and width of the lot.

Step 2: **Choose variables** to represent each unknown quantity and label the picture.

Hint
Review the geometry formulas, if necessary.

$$l = \text{length of the lot}$$
$$w = \text{width of the lot}$$

Step 3: **Write a system of equations using two variables.**

We know the lot is 70 ft longer than it is wide, so

Length	is	70 ft more than	the width.
↓	↓	↓	↓
l	$=$	$70 +$	w Equation (1)

The perimeter of the lot is 460 ft. Write the second equation using the formula for the perimeter of a rectangle.

$$\text{Perimeter} = 2(\text{length}) + 2(\text{width})$$
$$460 = 2l + 2w \quad \text{Equation (2)}$$

The system of equations is $\begin{aligned} l &= 70 + w \\ 2l + 2w &= 460 \end{aligned}$

Step 4: **Solve** the system. Let's use substitution.

$$\begin{aligned} 2l + 2w &= 460 \\ 2(70 + w) + 2w &= 460 \quad &\text{Substitute.} \\ 140 + 2w + 2w &= 460 \quad &\text{Distribute.} \\ 4w + 140 &= 460 \\ 4w &= 320 \\ w &= 80 \end{aligned}$$

Find l by substituting $w = 80$ into $l = 70 + w$.

$$l = 70 + 80 = 150$$

The solution of the system is (150, 80). (The ordered pair is written as (l, w), in alphabetical order.)

Step 5: **Check** the answer in the original problem, and **interpret** the solution.

The length of the lot is 150 *ft, and the width of the lot is* 80 *ft.* Check the solution to verify that the numbers make sense.

> [YOU TRY 2] Write a system of equations, and solve.
> A rectangular mouse pad is 1.5 in. longer than it is wide. The perimeter of the mouse pad is 29 in. Find its dimensions.

Another common type of problem involves the cost of items and the number of items.

3 Solve Problems Involving Cost

EXAMPLE 3

Write a system of equations, and solve.

In 2004, Usher and Alicia Keys each had concerts at the Allstate Arena near Chicago. Kayla and Levon sat in the same section for each performance. Kayla bought four tickets to see Usher and four to see Alicia Keys for $360. Levon spent $220.50 on two Usher tickets and three Alicia Keys tickets. Find the cost of a ticket to each concert. (www.pollstaronline.com)

Solution

Step 1: **Read** the problem carefully. We must find the cost of an Usher ticket and the cost of an Alicia Keys ticket.

Step 2: **Choose variables** to represent each unknown quantity.

x = cost of one Usher ticket
y = cost of one Alicia Keys ticket

Step 3: **Write a system of equations using two variables.**

Use the information about Kayla's purchase.

$$\underbrace{\text{Cost of four Usher tickets}}_{4x} + \underbrace{\text{Cost of four Alicia Keys tickets}}_{4y} = \underbrace{\text{Total cost}}_{360.00} \qquad \text{Equation (1)}$$

where $4x$ = (Number of tickets)(Cost of each ticket) and $4y$ = (Number of tickets)(Cost of each ticket).

Use the information about Levon's purchase.

$$\underbrace{\text{Cost of two Usher tickets}}_{2x} + \underbrace{\text{Cost of three Alicia Keys tickets}}_{3y} = \underbrace{\text{Total cost}}_{220.50} \qquad \text{Equation (2)}$$

where $2x$ = (Number of tickets)(Cost of each ticket) and $3y$ = (Number of tickets)(Cost of each ticket).

The system of equations is
$$\begin{aligned} 4x + 4y &= 360.00 \\ 2x + 3y &= 220.50 \end{aligned}$$

Step 4: **Solve** the system. Let's use the elimination method. Multiply equation (2) by -2 to eliminate x.

$$\begin{aligned} 4x + 4y &= 360.00 \\ +\,\underline{-4x - 6y} &= \underline{-441.00} \\ -2y &= -81.00 \qquad \text{Add the equations.} \\ y &= 40.50 \end{aligned}$$

Find x. We will substitute $y = 40.50$ into $4x + 4y = 360.00$.

$$\begin{aligned} 4x + 4(40.50) &= 360.00 \qquad \text{Substitute.} \\ 4x + 162 &= 360.00 \\ 4x &= 198.00 \\ x &= 49.50 \end{aligned}$$

The solution to the system is $(49.50, 40.50)$.

Step 5: **Check** and **interpret** the solution.

The cost of a ticket to see Usher in concert was $49.50, *and the cost of a ticket to see Alicia Keys was* $40.50. Check the numbers to verify that they are correct.

[YOU TRY 3] Write a system of equations, and solve.

At Julie's Jewelry Box, all necklaces sell for one price and all pairs of earrings sell for one price. Cailen buys three pairs of earrings and a necklace for $19.00, while Marcella buys one pair of earrings and two necklaces for $18.00. Find the cost of a pair of earrings and the cost of a necklace.

Next, we will discuss how to solve mixture problems using two variables and a system of equations.

4 Solve Mixture Problems

EXAMPLE 4

Write a system of equations, and solve.

How many milliliters of an 8% hydrogen peroxide solution and how many milliliters of a 2% hydrogen peroxide solution should a pharmacist mix to get 300 ml of a 4% hydrogen peroxide solution?

Solution

Step 1: Read the problem carefully. We must find the amount of the 8% solution and the amount of the 2% solution she should use.

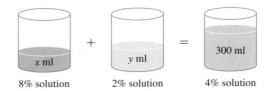

Step 2: Choose variables to represent each unknown quantity.

x = amount of 8% solution
y = amount of 2% solution

Step 3: Write a system of equations using two variables.

> **W Hint**
> Look at Section 2.4 if you need to review the concepts behind solving mixture problems.

Make a table to organize the information. To obtain the expression in the last column, multiply the percent of hydrogen peroxide in the solution by the amount of solution to get the amount of hydrogen peroxide in the solution.

	Percent of Hydrogen Peroxide in Solution (as a decimal)	Amount of Solution	Amount of Hydrogen Peroxide in Solution
Mix these	0.08	x	$0.08x$
	0.02	y	$0.02y$
to make →	0.04	300	$0.04(300)$

To get one equation, use the information in the *second column*:

English: Amount of 8% solution + Amount of 2% solution = Amount of 4% solution

Equation: $x \quad + \quad y \quad = \quad 300$ Equation (1)

> **W Hint**
> Note that the percent of the final solution is always *between* the percent of the two solutions being mixed. Note: 2% < 4% < 8%

To get the second equation, use the information in the *third column*:

English: Amount of hydrogen peroxide in 8% solution + Amount of hydrogen peroxide in 2% solution = Amount of hydrogen peroxide in 4% solution

Equation: $0.08x \quad + \quad 0.02y \quad = \quad 0.04(300)$ Equation (2)

The system of equations is $\begin{matrix} x + y = 300 \\ 0.08x + 0.02y = 0.04(300). \end{matrix}$

Step 4: **Solve** the system. Multiply the second equation by 100 to eliminate the decimals. We get $8x + 2y = 4(300)$.

$$x + y = 300$$
$$8x + 2y = 1200 \quad 4(300) = 1200$$

Use the elimination method. Multiply the first equation by -2 to eliminate y.

$$-2x - 2y = -600$$
$$+\ \underline{8x + 2y = 1200}$$
$$6x = 600 \quad \text{Add the equations.}$$
$$x = 100$$

To find y, substitute $x = 100$ into $x + y = 300$.

$$100 + y = 300 \quad \text{Substitute.}$$
$$y = 200$$

The solution to the system is (100, 200).

Step 5: **Check** and **interpret** the solution.

The pharmacist needs 100 *ml of an* 8% *hydrogen peroxide solution and* 200 *ml of a* 2% *solution to make* 300 *ml of a* 4% *hydrogen peroxide solution.* Check the answers in the equations to verify that they are correct.

[**YOU TRY 4**] Write a system of equations, and solve.

How many milliliters of a 3% acid solution must be added to 60 ml of an 11% acid solution to make a 9% acid solution?

5 Solve Problems Involving Distance, Rate, and Time

EXAMPLE 5 Write a system of equations, and solve.
Julia and Katherine start at the same point and begin biking in opposite directions. Julia rides 2 mph faster than Katherine. After 2 hr, the girls are 44 mi apart. How fast was each of them riding?

Solution

Step 1: **Read** the problem carefully. Draw a picture. We must find the speed at which each girl was riding.

Step 2: **Choose variables** to represent each unknown quantity.

$$x = \text{Julia's speed}$$
$$y = \text{Katherine's speed}$$

W Hint
Look at Section 2.4 if you need to review the concepts for solving problems using $d = rt$.

Step 3: **Write a system of equations using two variables.**

Let's make a table using the equation $d = rt$. Fill in their speeds (rates) x and y, and fill in 2 for the time since each rode for 2 hr. Multiply these together to fill in the values for the distance.

	d	r	t
Julia	$2x$	x	2
Katherine	$2y$	y	2

Hint
Notice that the two equations for the *system of linear equations* come from Rows 1 and 2.

Label the picture with expressions for the distances.

To get one equation, look at the picture and think about the distance between the girls after 2 hr.

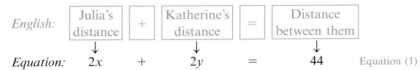

$$\text{Equation:} \quad 2x + 2y = 44 \quad \text{Equation (1)}$$

To get the second equation, use the information that says Julia rides 2 mph faster than Katherine.

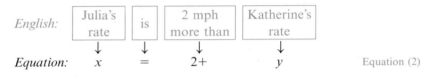

$$\text{Equation:} \quad x = 2 + y \quad \text{Equation (2)}$$

The system of equations is $\begin{aligned} 2x + 2y &= 44 \\ x &= 2 + y. \end{aligned}$

Step 4: **Solve** the system. Use substitution.

$$2x + 2y = 44$$
$$2(2 + y) + 2y = 44 \quad \text{Substitute } 2 + y \text{ for } x.$$
$$4 + 2y + 2y = 44 \quad \text{Distribute.}$$
$$4y + 4 = 44$$
$$4y = 40$$
$$y = 10$$

Find x by substituting $y = 10$ into $x = 2 + y$.

$$x = 2 + 10 = 12$$

The solution to the system is $(12, 10)$.

Step 5: **Check** and **interpret** the solution.

Julia rides her bike at 12 *mph, and Katherine rides hers at* 10 *mph.* Julia's speed is 2 mph faster than Katherine's, as stated in the problem. And, since each girl rode for 2 hr, the distance between them is

$\underbrace{2(12)}_{\text{Julia's distance}} + \underbrace{2(10)}_{\text{Katherine's distance}} = 24 + 20 = 44$ mi. The answer is correct.

YOU TRY 5

Write a system of equations, and solve.

Kenny and Kyle start at the same point and begin biking in opposite directions. Kenny rides 1 mph faster than Kyle. After 3 hr, the boys are 51 mi apart. How fast was each of them riding?

6 Solve Problems Involving Three Variables

To solve applications involving a system of three equations in three variables, we will extend the method used for two equations in two variables.

EXAMPLE 6

Write a system of equations, and solve.

The top three gold-producing nations in 2002 were South Africa, the United States, and Australia. Together, these three countries produced 37% of the gold during that year. Australia's share was 2% less than that of the United States, while South Africa's percentage was 1.5 times Australia's percentage of world gold production. Determine what percentage of the world's gold supply was produced by each country in 2002. (Market Share Reporter—2005, Vol. 1, "Mine Product" http://www.gold.org/value/market/supply-demand/min_production.html from World Gold Council)

Solution

Step 1: **Read** the problem carefully. We must determine the percentage of the world's gold produced by South Africa, the United States, and Australia in 2002.

Step 2: **Choose variables** to represent the unknown quantities.

x = percentage of world's gold supply produced by South Africa
y = percentage of world's gold supply produced by the United States
z = percentage of world's gold supply produced by Australia

Step 3: **Write a system of equations using the variables.**

To write one equation, we will use the information that says *together* the three countries produced 37% of the gold.

$$x + y + z = 37 \quad \text{Equation (1)}$$

To write a second equation, we will use the information that says Australia's share was 2% less than that of the United States.

$$z = y - 2 \quad \text{Equation (2)}$$

To write the third equation, we will use the statement that says South Africa's percentage was 1.5 times Australia's percentage.

$$x = 1.5z \quad \text{Equation (3)}$$

The system is
① $x + y + z = 37$
② $z = y - 2$
③ $x = 1.5z$

Step 4: **Solve** the system. Since two of the equations contain only two variables, we will modify our steps to solve the system.

Our plan is to rewrite equation ① in terms of a single variable, z, and solve for z.

Solve equation ② for y.

② $z = y - 2$
$z + 2 = y$ \quad Solve for y.

Now rewrite equation ① using the value for y from equation ② and the value for x from equation ③.

① $\quad x + y + z = 37 \quad$ Equation (1)
$\quad (1.5z) + (z + 2) + z = 37 \quad$ Substitute $1.5z$ for x and $z + 2$ for y.
$\quad 3.5z + 2 = 37 \quad$ Combine like terms.
$\quad 3.5z = 35 \quad$ Subtract 2.
$\quad \boxed{z = 10} \quad$ Divide by 3.5.

To solve for x, we can substitute $z = 10$ into equation ③.

③ $x = 1.5z$
$\quad x = 1.5(10) \quad$ Substitute 10 for z.
$\quad \boxed{x = 15} \quad$ Multiply.

To solve for y, we substitute $z = 10$ into equation ②.

② $z = y - 2$
$\quad 10 = y - 2 \quad$ Substitute 10 for z.
$\quad \boxed{12 = y} \quad$ Solve for y.

The solution of the system is (15, 12, 10).

Step 5: **Check** and **interpret** the solution.

In 2002, South Africa produced 15% *of the world's gold, the United States produced* 12%, *and Australia produced* 10%. The check is left to the student.

[**YOU TRY 6**] Write a system of equations, and solve.
Amelia, Bella, and Carmen are sisters. Bella is 5 yr older than Carmen, and Amelia's age is 5 yr less than twice Carmen's age. The sum of their ages is 48. How old is each girl?

ANSWERS TO [**YOU TRY**] **EXERCISES**

1) 34 rooms have two beds, 17 rooms have one bed.   2) 6.5 in. by 8 in.
3) A pair of earrings costs $4.00 and a necklace costs $7.00.   4) 20 ml
5) Kyle: 8 mph; Kenny: 9 mph   6) Amelia: 19; Bella: 17; Carmen: 12

E Evaluate 5.3 Exercises

Do the exercises, and check your work.

Objective 1: Solve General Two-Variable Problems

Write a system of equations, and solve.

1) The sum of two numbers is 36, and one number is two more than the other. Find the numbers.

2) One number is half another number. The sum of the two numbers is 108. Find the numbers.

3) Through 2012, the University of Southern California football team played in 13 more Rose Bowl games than the University of Michigan. All together, these teams have appeared in 53 Rose Bowls. How many appearances did each team make in the Rose Bowl? (www.tournamentofroses.com)

4) In 2012, *Hugo* was nominated for six more Academy Awards than the movie *The Girl with the Dragon Tatoo*. Together they received 16 nominations. How many Academy Award nominations did each movie receive? (oscar.go.com)

5) There were a total of 2626 IHOP and Waffle House restaurants across the United States at the end of 2004. There were 314 fewer IHOPs than Waffle Houses. Determine the number of IHOP and Waffle House restaurants in the United States.
(www.ihop.com, www.ajc.com)

6) Through 2012, the University of Kentucky men's basketball team won four fewer NCAA championships than UCLA. The two teams won a total of 19 championship titles. How many championships did each team win? (www.ncaasports.com)

7) The first NCAA Men's Basketball Championship game was played in 1939 in Evanston, Illinois. In 2011, 64,876 more people attended the final game in Houston than attended the first game in Evanston. The total number of people who attended these two games was 75,876. How many people saw the first NCAA championship in 1939, and how many were at the game in 2011? (www.ncaasports.com)

8) George Strait won 6 more Country Music Awards than Alan Jackson through 2011. They won a total of 38 awards. How many awards did each man win? (www.cmaawards.com)

9) Annual per capita consumption of chicken in the United States in 2001 was 6.3 lb less than that of beef. Together, each American consumed, on average, 120.5 lb of beef and chicken in 2001. Find the amount of beef and the amount of chicken consumed, per person, in 2001. (U.S. Department of Agriculture)

10) Mr. Chen has 27 students in his American History class. For their assignment on the Civil War, twice as many students chose to give a speech as chose to write a paper. How many students will be giving speeches, and how many will be writing papers?

Objective 2: Solve Problems Involving Geometry

11) The length of a rectangle is twice its width. Find the length and width of the rectangle if its perimeter is 78 in.

12) The width of a rectangle is 5 cm less than the length. If the perimeter is 46 cm, what are the dimensions of the rectangle?

13) Find the dimensions of a rectangular door that has a perimeter of 220 in. if the width is 50 in. less than the height of the door.

14) Yuki has a rectangular picture she wants to frame. Its perimeter is 42 in., and it is 4 in. longer than it is wide. Find the length and width of the picture.

15) An iPod Mini is rectangular in shape and has a perimeter of 28 cm. Its length is 4 cm more than its width. What are the dimensions of the iPod Mini?

16) Tiny Tots Day Care needs to put a new fence around their rectangular playground. They have determined they will need 220 ft of fencing. If the width of the playground is 30 ft less than the length, what are the playground's dimensions?

17) Find the measures of angles x and y if the measure of angle y is half the measure of angle x and if the angles are related according to the figure.

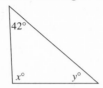

18) Find the measures of angles x and y if the measure of angle x is two-thirds the measure of angle y and if the angles are related according to the figure.

Objective 3: Solve Problems Involving Cost

19) Jennifer and Carlos attended two concerts with their friends at the American Airlines Arena in Miami. Jennifer and her friends bought five tickets to see Marc Anthony and two tickets to the Santana concert for $563. Carlos' group purchased three Marc Anthony tickets and six for Santana for $657. Find the cost of a ticket to each concert. (www.pollstaronline.com)

20) Both of the pop groups Train and Maroon 5 played at the House of Blues in North Myrtle Beach, SC in 2003. Three Maroon 5 tickets and two Train tickets would have cost $64, while four Maroon 5 tickets and four Train tickets would have cost $114. Find the cost of a ticket to each concert. (www.pollstaronline.com)

21) Every Tuesday, Stacey goes to Panda Express to buy lunch for herself and her colleagues. One day she buys 3 two-item meals and 1 three-item meal for $21.96. The next Tuesday she spends $23.16 on 2 two-item meals and 2 three-item meals. What is the cost of a two-item meal and of a three-item meal? (Panda Express menu)

22) Two families buy end-zone tickets to the same football game. One family buys four tickets and a parking pass for $313. The other family buys six tickets and a parking pass for $457. Find the price of a ticket and the cost of a parking pass.

23) At Sparkle Car Wash, two deluxe car washes and three regular car washes would cost $26.00. Four regular washes and one deluxe wash would cost $23.00. What is the cost of a deluxe car wash and of a regular wash?

24) On vacation, Wendell buys three key chains and five postcards for $10.00, and his sister buys two key chains and three postcards for $6.50. Find the cost of each souvenir.

25) Ella spends $7.50 on three cantaloupe and one watermelon at a farmers' market. Two cantaloupe and two watermelon would have cost $9.00. What is the price of a cantaloupe and the price of a watermelon?

26) One 12-oz serving of Coke and two 12-oz servings of Mountain Dew contain 31.3 tsp of sugar while three servings of Coke and one Mountain Dew contain 38.9 tsp of sugar. How much sugar is in a 12-oz serving of each drink? (www.dentalgentlecare.com)

27) Carol orders six White Castle hamburgers and a small order of fries for $5.05, and Momar orders eight hamburgers and two small fries for $7.66. Find the cost of a hamburger and the cost of an order of french fries at White Castle. (White Castle menu)

28) Six White Castle hamburgers and one small order of french fries contain 1150 calories. Eight hamburgers and two orders of fries contain 1740. Determine how many calories are in a White Castle hamburger and in a small order of french fries. (www.whitecastle.com)

Objective 4: Solve Mixture Problems

29) How many ounces of a 9% alcohol solution and how many ounces of a 17% alcohol solution must be mixed to get 12 oz of a 15% alcohol solution?

30) How many milliliters of a 4% acid solution and how many milliliters of a 10% acid solution must be mixed to obtain 54 ml of a 6% acid solution?

31) Raheem purchases 20 stamps. He buys some $0.44 stamps and some $0.28 stamps and spends $7.52. How many of each type of stamp did he buy?

32) How many pounds of peanuts that sell for $1.80 per pound should be mixed with cashews that sell for $4.50 per pound so that a 10-lb mixture is obtained that will sell for $2.61 per pound?

33) Sally invested $4000 in two accounts, some of it at 3% simple interest, the rest in an account earning 5% simple interest. How much did she invest in each account if she earned $144 in interest after one year?

34) Diego inherited $20,000 and puts some of it into an account earning 4% simple interest and the rest in an account earning 7% simple interest. He earns a total of $1130 in interest after a year. How much did he deposit into each account?

35) Josh saves all of his quarters and dimes in a bank. When he opens it, he has 110 coins worth a total of $18.80. How many quarters and how many dimes does he have?

36) Mrs. Kowalski bought nine packages of batteries when they were on sale. The AA batteries cost $1.00 per package and the C batteries cost $1.50 per package. If she spent $11.50, how many packages of each type of battery did she buy?

37) How much pure acid and how many liters of a 10% acid solution should be mixed to get 12 L of a 40% acid solution?

38) How many ounces of pure orange juice and how many ounces of a citrus fruit drink containing 5% fruit juice should be mixed to get 76 oz of a fruit drink that is 25% fruit juice?

Objective 5: Solve Problems Involving Distance, Rate, and Time

39) A passenger train and a freight train leave cities 400 mi apart and travel toward each other. The passenger train is traveling 20 mph faster than the freight train. Find the speed of each train if they pass each other after 5 hr.

40) A car and a truck leave the same location, the car headed east and the truck headed west. The truck's speed is 10 mph less than the speed of the car. After 3 hr, the car and truck are 330 mi apart. Find the speed of each vehicle.

41) Olivia can walk 8 mi in the same amount of time she can bike 22 mi. She bikes 7 mph faster than she walks. Find her walking and biking speeds.

42) A small, private plane can fly 400 mi in the same amount of time a jet can fly 1000 mi. If the jet's speed is 300 mph more than the speed of the small plane, find the speeds of both planes.

43) Nick and Scott leave opposite ends of a bike trail 13 mi apart and travel toward each other. Scott is traveling 2 mph slower than Nick. Find each of their speeds if they meet after 30 min.

44) Vashon can travel 120 mi by car in the same amount of time he can take the train 150 mi. If the train travels 10 mph faster than the car, find the speed of the car and the speed of the train.

Objective 6: Solve Problems Involving Three Variables

Write a system of equations, and solve.

45) Moe buys two hot dogs, two orders of fries, and a large soda for $9.00. Larry buys two hot dogs, one order of fries, and two large sodas for $9.50, and Curly spends $11.00 on three hot dogs, two orders of fries, and a large soda. Find the price of a hot dog, an order of fries, and a large soda.

46) A movie theater charges $9.00 for an adult's ticket, $7.00 for a ticket for seniors 60 and over, and $6.00 for a child's ticket. For a particular movie, the theater sold a total of 290 tickets, which brought in $2400. The number of seniors' tickets sold was twice the number of children's tickets sold. Determine the number of adults', seniors', and children's tickets sold.

47) A Chocolate Chip Peanut Crunch Clif Bar contains 4 fewer grams of protein than a Chocolate Peanut Butter Balance Bar Plus. A Chocolate Peanut Butter Protein Plus PowerBar contains 9 more grams of protein than the Balance Bar Plus. All three bars contain a total of 50 g of protein. How many grams of protein are in each type of bar? (www.clifbar.com, www.balance.com, www.powerbar.com)

48) A 1-tablespoon serving size of Hellman's Real Mayonnaise has 55 more calories than the same serving size of Hellman's Light Mayonnaise. Miracle Whip and Hellman's Light have the same number of calories in a 1-tablespoon serving size. If the three spreads have a total of 160 calories in one serving, determine the number of calories in one serving of each. (product labels)

49) The three NBA teams with the highest revenues in 2002–2003 were the New York Knicks, the Los Angeles Lakers, and the Chicago Bulls. Their revenues totaled $428 million. The Lakers took in $30 million more than the Bulls, and the Knicks took in $11 million more than the Lakers. Determine the revenue of each team during the 2002–2003 season. (*Forbes,* Feb. 16, 2004, p. 66)

50) The best-selling paper towel brands in 2002 were Bounty, Brawny, and Scott. Together they accounted for 59% of the market. Bounty's market share was 25% more than Brawny's, and Scott's market share was 2% less than Brawny's. What percentage of the market did each brand hold in 2002? (*USA Today,* Oct. 23, 2003, p. 3B from Information Resources, Inc.)

51) Ticket prices to a Cubs game at Wrigley Field vary depending on their dates. At the beginning of the 2012 season, Bill, Corrinne, and Jason bought tickets in the upper deck of the infield for several games. Bill spent $441 on four bronze dates, four silver dates, and three gold dates. Corrinne bought tickets for four bronze dates, three silver dates, and two gold dates for $341. Jason spent $262 on three bronze dates, three silver dates, and one gold date. How much did it cost to sit in the upper deck of the infield at Wrigley Field on a bronze date, silver date, and gold date in 2012? (http://chicago.cubs.mlb.com)

52) To see the Boston Red Sox play at Fenway Park in 2009, two field box seats, three infield grandstand seats, and five bleacher seats cost $530. The cost of four field box seats, two infield grandstand seats, and three bleacher seats was $678. The total cost of buying one of each type of ticket was $201. What was the cost of each type of ticket during the 2009 season? (http://boston.redsox.mlb.com)

53) The measure of the largest angle of a triangle is twice the middle angle. The smallest angle measures 28° less than the middle angle. Find the measures of the angles of the triangle. (Hint: Recall that the sum of the measures of the angles of a triangle is 180°.)

54) The measure of the smallest angle of a triangle is one-third the measure of the largest angle. The middle angle measures 30° less than the largest angle. Find the measures of the angles of the triangle. (Hint: Recall that the sum of the measures of the angles of a triangle is 180°.)

55) The smallest angle of a triangle measures 44° less than the largest angle. The sum of the two smaller angles is 20° more than the measure of the largest angle. Find the measures of the angles of the triangle.

56) The sum of the measures of the two smaller angles of a triangle is 40° less than the largest angle. The measure of the largest angle is twice the measure of the middle angle. Find the measures of the angles of the triangle.

57) The perimeter of a triangle is 29 cm. The longest side is 5 cm longer than the shortest side, and the sum of the two smaller sides is 5 cm more than the longest side. Find the lengths of the sides of the triangle.

58) The smallest side of a triangle is half the length of the longest side. The sum of the two smaller sides is 2 in. more than the longest side. Find the lengths of the sides if the perimeter is 58 in.

R Rethink

R1) When solving the application problems in Objectives 1–5, you end up with two linear equations. What would happen if you were to graph each of these lines?

R2) When solving the application problems in Objectives 1–5, there were always two unknowns. How did you determine which unknown to represent by x, and which unknown to represent by y?

R3) When solving these application problems, were any of your answers negative? If not, can application problems have negative answers?

R4) Which objective do you still need help mastering?

5.4 Solving Systems of Linear Equations Using Matrices

P Prepare	**O Organize**
What are your objectives for Section 5.4?	How can you accomplish each objective?
1 Learn the Vocabulary Associated with Gaussian Elimination	• Know the definitions of a *matrix* and an *augmented matrix* and the arrangement of numbers in the matrix from a given equation. • Understand the goal of the **Gaussian elimination method** and the final form of the matrix—*row echelon form*.
2 Solve a System of Two Equations Using Gaussian Elimination	• Learn the *matrix row operations*. • Label the row operation you will be doing before you begin each step. • Recognize the final matrix when there are 1's on the diagonal. • Learn how to write a system of equations from the final matrix. • Complete the given example on your own.
3 Solve a System of Any Number of Equations Using Gaussian Elimination	• Learn the procedure for **How to Solve a System of Equations Using Gaussian Elimination.** • Label the row operation you will be doing before you begin each step. • Recognize the final matrix when there are 1's on the diagonal. • Learn how to write a system of equations from the final matrix. • Complete the given example on your own. • Complete You Trys 1 and 2.

 Read the explanations, follow the examples, take notes, and complete the You Trys.

We have learned how to solve systems of linear equations by graphing, substitution, and the elimination method. In this section, we will learn how to use *row operations* and *Gaussian elimination* to solve systems of linear equations. We begin by defining some terms.

1 Learn the Vocabulary Associated with Gaussian Elimination

A **matrix** is a rectangular array of numbers. (The plural of *matrix* is *matrices*.) Each number in the matrix is an **element** of the matrix. An example of a matrix is

$$\begin{array}{c} \text{Column 1} \quad \text{Column 2} \quad \text{Column 3} \\ \downarrow \qquad\qquad \downarrow \qquad\qquad \downarrow \\ \begin{array}{c}\text{Row 1} \to \\ \text{Row 2} \to\end{array} \begin{bmatrix} 3 & -1 & 4 \\ 0 & 2 & -5 \end{bmatrix} \end{array}$$

We can represent a system of equations in an *augmented matrix*. An **augmented matrix** has a vertical line to distinguish between different parts of the equation. For example, we can represent the system below with the augmented matrix shown here:

$$\begin{array}{ll} 5x + 4y = 1 & \text{Equation (1)} \\ x - 3y = 6 & \text{Equation (2)} \end{array} \qquad \left[\begin{array}{cc|c} 5 & 4 & 1 \\ 1 & -3 & 6 \end{array}\right] \begin{array}{l} \text{Row 1} \\ \text{Row 2} \end{array}$$

Notice that the vertical line separates the system's coefficients from its constants on the other side of the = sign. The system needs to be in standard form, so the first column in the matrix represents the *x*-coefficients. The second column represents the *y*-coefficients, and the column on the right represents the constants.

Gaussian elimination is the process of using row operations on an augmented matrix to solve the corresponding system of linear equations. It is a variation of the elimination method and can be very efficient. Computers often use augmented matrices and row operations to solve systems.

The goal of Gaussian elimination is to obtain a matrix of the form $\left[\begin{array}{cc|c} 1 & a & b \\ 0 & 1 & c \end{array}\right]$ or $\left[\begin{array}{ccc|c} 1 & a & b & d \\ 0 & 1 & c & e \\ 0 & 0 & 1 & f \end{array}\right]$ when solving a system of two or three equations, respectively. Notice the 1's along the **diagonal** of the matrix and zeros below the diagonal. We say a matrix is in **row echelon form** when it has 1's along the diagonal and 0's below the diagonal. We get matrices in row echelon form by performing row operations. When we rewrite a matrix that is in row echelon form back into a system, its solution is easy to find.

2 Solve a System of Two Equations Using Gaussian Elimination

The row operations we can perform on augmented matrices are similar to the operations we use to solve a system of equations using the elimination method.

> **Definition** Matrix Row Operations
>
> Performing the following row operations on a matrix produces an equivalent matrix.
>
> 1) Interchanging two rows
> 2) Multiplying every element in a row by a nonzero real number
> 3) Replacing a row by the sum of it and the multiple of another row

Let's use these operations to solve a system using Gaussian elimination. Notice the similarities between this method and the elimination method.

EXAMPLE 1 Solve using Gaussian elimination. $\quad x + 5y = -1$
$ 2x - y = 9$

Solution

Begin by writing the system as an augmented matrix. $\begin{bmatrix} 1 & 5 & | & -1 \\ 2 & -1 & | & 9 \end{bmatrix}$

We will use the 1 in Row 1 to make the element below it a zero. If we multiply the 1 by -2 (to get -2) and add it to the 2, we get zero. We must do this operation to the entire row. Denote this as $-2R_1 + R_2 \rightarrow R_2$. (Read as, "$-2$ times Row 1 plus Row 2 makes the new Row 2.") We get a new Row 2.

> **W Hint**
> Write out the example as you are reading it!

Use this $\rightarrow \begin{bmatrix} \boxed{1} & 5 & | & -1 \\ \boxed{2} & -1 & | & 9 \end{bmatrix} \xrightarrow{-2R_1 + R_2 \rightarrow R_2} \begin{bmatrix} 1 & 5 & | & -1 \\ -2(1)+2 & -2(5)+(-1) & | & -2(-1)+9 \end{bmatrix}$

to make this 0. $= \begin{bmatrix} 1 & 5 & | & -1 \\ 0 & -11 & | & 11 \end{bmatrix}$ Multiply each element of Row 1 by -2, and add it to the corresponding element of Row 2.

> **Note**
> We are not *making* a new Row 1, so it stays the same.

We have obtained the first 1 on the diagonal with a 0 below it. Next we need a 1 on the diagonal in Row 2.

This column is in the correct form.
$\downarrow$

$\begin{bmatrix} 1 & 5 & | & -1 \\ 0 & \boxed{-11} & | & 11 \end{bmatrix} \xrightarrow{-\frac{1}{11}R_2 \rightarrow R_2} \begin{bmatrix} 1 & 5 & | & -1 \\ 0 & 1 & | & -1 \end{bmatrix}$ Multiply each element of Row 2 by $-\frac{1}{11}$ to get a 1 on the diagonal.

$\uparrow$ Make this 1.

We have obtained the final matrix because there are 1's on the diagonal and a 0 below. The matrix is in row echelon form. From this matrix, write a system of equations. The last row gives us the value of y.

> **W Hint**
> Notice on this matrix that 1's on the diagonal with a zero below means that you have eliminated the x in the linear equation represented by Row 2.

$\begin{bmatrix} 1 & 5 & | & -1 \\ 0 & 1 & | & -1 \end{bmatrix} \quad \begin{array}{l} 1x + 5y = -1 \\ 0x + 1y = -1 \end{array}$ or $\quad \begin{array}{ll} x + 5y = -1 & \text{Equation (1)} \\ y = -1 & \text{Equation (2)} \end{array}$

$\begin{array}{ll} x + 5(-1) = -1 & \text{Substitute } -1 \text{ for } y \text{ in equation (1).} \\ x - 5 = -1 & \text{Multiply.} \\ x = 4 & \text{Add 5.} \end{array}$

The solution is $(4, -1)$. Check by substituting $(4, -1)$ into both equations of the original system.

3 Solve a System of Any Number of Equations Using Gaussian Elimination

Here are the steps for using Gaussian elimination to solve a system of any number of equations. **Our goal is to obtain a matrix with 1's along the diagonal and 0's below—row echelon form.**

> **Procedure** How to Solve a System of Equations Using Gaussian Elimination
>
> **Step 1:** Write the system as an *augmented matrix*.
> **Step 2:** Use row operations to make the *first entry in column 1* be a 1.
> **Step 3:** Use row operations to make *all entries below the 1 in column 1* be 0's.
> **Step 4:** Use row operations to make the *second entry in column 2* be a 1.
> **Step 5:** Use row operations to make *all entries below the 1 in column 2* be 0's.
> **Step 6:** Continue this procedure until the matrix is in *row echelon form*—1's along the diagonal and 0's below.
> **Step 7:** Write the matrix in *Step 6* as a *system of equations*.
> **Step 8:** *Solve the system* from *Step 7*. The last equation in the system will give you the value of one of the variables; find the values of the other variables by using substitution.
> **Step 9:** *Check the solution* in each equation of the original system.

[YOU TRY 1] Solve the system using Gaussian elimination. $\quad x - y = -1$
$\qquad\qquad\qquad\qquad\qquad\qquad\qquad\qquad\qquad\qquad\quad -3x + 5y = 9$

Next we will solve a system of three equations using Gaussian elimination.

EXAMPLE 2 Solve using Gaussian elimination.

$$2x + y - z = -3$$
$$x + 2y - 3z = 1$$
$$-x - y + 2z = 2$$

Solution

Step 1: Write the system as an *augmented matrix*.

$$\begin{bmatrix} 2 & 1 & -1 & | & -3 \\ 1 & 2 & -3 & | & 1 \\ -1 & -1 & 2 & | & 2 \end{bmatrix}$$

W Hint
Are you writing out this example as you are reading it?

Step 2: To make the *first entry in column 1* be a 1, we *could* multiply Row 1 by $\frac{1}{2}$, but this would make the rest of the entries in the first row fractions. Instead, recall that we can interchange two rows. If we interchange Row 1 and Row 2, the first entry in column 1 will be 1.

$$\begin{array}{c} R_1 \leftrightarrow R_2 \\ \text{Interchange} \\ \text{Row 1 and Row 2.} \end{array} \quad \begin{bmatrix} ① & 2 & -3 & | & 1 \\ 2 & 1 & -1 & | & -3 \\ -1 & -1 & 2 & | & 2 \end{bmatrix}$$

Step 3: We want to make *all the entries below the 1 in column 1* be 0's. To obtain a 0 in place of the 2 in column 1, multiply the 1 by -2 (to get -2) and add it to the 2. Perform that same operation on the entire row to obtain the new Row 2.

$$\begin{array}{c} \text{Use this} \rightarrow \\ \text{to make} \rightarrow \\ \text{this zero.} \end{array} \begin{bmatrix} ① & 2 & -3 & | & 1 \\ \boxed{2} & 1 & -1 & | & -3 \\ -1 & -1 & 2 & | & 2 \end{bmatrix} \begin{array}{c} -2R_1 + R_2 \rightarrow R_2 \\ -2 \text{ times Row 1} + \\ \text{Row 2} = \text{new Row 2} \end{array} \begin{bmatrix} 1 & 2 & -3 & | & 1 \\ 0 & -3 & 5 & | & -5 \\ -1 & -1 & 2 & | & 2 \end{bmatrix}$$

To obtain a 0 in place of the -1 in column 1, add the 1 and the -1. Perform that same operation on the entire row to obtain a new Row 3.

$$\begin{array}{c}\text{Use this} \rightarrow \\ \text{to make} \\ \text{this zero.}\end{array} \begin{bmatrix} \boxed{1} & 2 & -3 & | & 1 \\ 0 & -3 & 5 & | & -5 \\ \boxed{-1} & -1 & 2 & | & 2 \end{bmatrix} \xrightarrow[\text{Row 1 + Row 3 = new Row 3}]{R_1 + R_3 \rightarrow R_3} \begin{bmatrix} 1 & 2 & -3 & | & 1 \\ 0 & -3 & 5 & | & -5 \\ 0 & 1 & -1 & | & 3 \end{bmatrix}$$

Step 4: Next, we want the *second entry in column 2* to be a 1. We *could* multiply Row 2 by $-\frac{1}{3}$ to get the 1, but the other entries would be fractions. Instead, interchanging Row 2 and Row 3 will give us a 1 on the diagonal and keep 0's in column 1. (Sometimes, though, fractions are unavoidable.)

$$\xrightarrow[\text{Interchange Rows 2 and 3.}]{R_2 \leftrightarrow R_3} \begin{bmatrix} 1 & 2 & -3 & | & 1 \\ 0 & \boxed{1} & -1 & | & 3 \\ 0 & -3 & 5 & | & -5 \end{bmatrix}$$

Step 5: We want to make *all the entries below the 1 in column 2* be 0's. To obtain a 0 in place of -3 in column 2, multiply the 1 above it by 3 (to get 3) and add it to -3. Perform that same operation on the entire row to obtain a new Row 3.

$$\begin{array}{c}\text{Use this} \\ \text{to make} \\ \text{this zero.}\end{array} \begin{bmatrix} 1 & 2 & -3 & | & 1 \\ 0 & \boxed{1} & -1 & | & 3 \\ 0 & \boxed{-3} & 5 & | & -5 \end{bmatrix} \xrightarrow[\text{3 times Row 2 + Row 3 = new Row 3}]{3R_2 + R_3 \rightarrow R_3} \begin{bmatrix} 1 & 2 & -3 & | & 1 \\ 0 & 1 & -1 & | & 3 \\ 0 & 0 & 2 & | & 4 \end{bmatrix}$$

We have completed *Step 5* because there is only one entry below the 1 in column 2.

Step 6: *Continue this procedure.* The last entry in column 3 needs to be a 1. (This is the last 1 we need along the diagonal.) Multiply Row 3 by $\frac{1}{2}$ to obtain the last 1.

$$\xrightarrow[\text{Multiply Row 3 by } \frac{1}{2}.]{\frac{1}{2}R_3 \rightarrow R_3} \begin{bmatrix} \boxed{1} & 2 & -3 & | & 1 \\ 0 & \boxed{1} & -1 & | & 3 \\ 0 & 0 & \boxed{1} & | & 2 \end{bmatrix}$$

> **W Hint**
> Notice on this matrix that 1's on the diagonal with zeros below means that you have eliminated both x and y in the linear equation that is represented by Row 3.

We are done performing row operations because there are 1's on the diagonal and zeros below.

Step 7: Write the matrix in *Step 6* as a *system of equations*.

$$\begin{array}{c} 1x + 2y - 3z = 1 \\ 0x + 1y - 1z = 3 \\ 0x + 0y + 1z = 2 \end{array} \quad \text{or} \quad \begin{array}{c} x + 2y - 3z = 1 \\ y - z = 3 \\ z = 2 \end{array}$$

Step 8: Solve the system in *Step 7*. The last row tells us that $z = 2$. Substitute $z = 2$ into the equation above it ($y - z = 3$) to get the value of y: $y - 2 = 3$, so $y = 5$.

Substitute $y = 5$ and $z = 2$ into $x + 2y - 3z = 1$ to solve for x.

$$\begin{array}{ll} x + 2y - 3z = 1 & \\ x + 2(5) - 3(2) = 1 & \text{Substitute values.} \\ x + 10 - 6 = 1 & \text{Multiply.} \\ x + 4 = 1 & \text{Subtract.} \\ x = -3 & \end{array}$$

The solution of the system is $(-3, 5, 2)$.

Step 9: *Check the solution* in each equation of the original system. The check is left to the student.

This procedure may seem long and complicated at first, but as you practice and become more comfortable with the steps, you will see that it is actually quite efficient.

YOU TRY 2 Solve the system using Gaussian elimination.
$$x + 3y - 2z = 10$$
$$3x + 2y + z = 9$$
$$-x + 4y - z = -1$$

If we are performing Gaussian elimination and obtain a matrix that produces a false equation as shown, then the system has *no solution*. The system is *inconsistent*.
$\begin{bmatrix} 1 & -6 & | & 9 \\ 0 & 0 & | & 8 \end{bmatrix}$ $0x + 0y = 8$ False

If, however, we obtain a matrix that produces a row of zeros as shown, then the system has an *infinite number of solutions*. The system is *consistent* with *dependent* equations. We write its solution as we did in previous sections.
$\begin{bmatrix} 1 & 5 & | & -1 \\ 0 & 0 & | & 0 \end{bmatrix}$ $0x + 0y = 0$ True

Using Technology

In this section, we have learned how to solve a system of three equations using Gaussian elimination. The row operations used to convert an augmented matrix to row echelon form can be performed on a graphing calculator.

Follow the nine-step method given in the text to solve the system using Gaussian elimination:
$$x + 2y - 3z = 1$$
$$y - z = 3$$
$$-2y + 4z = -4$$

Step 1: Write the system as an augmented matrix: $\begin{bmatrix} 1 & 2 & -3 & | & 1 \\ 0 & 1 & -1 & | & 3 \\ 0 & -2 & 4 & | & -4 \end{bmatrix}$

Store the matrix in matrix [A] using a graphing calculator. Press 2nd x^{-1} to select [A]. Press the right arrow key two times and press ENTER to select EDIT. Press 3 ENTER then 4 ENTER to enter the number of rows and number of columns in the augmented matrix. Enter the coefficients one row at a time as follows: 1 ENTER 2 ENTER (−) 3 ENTER 1 ENTER 0 ENTER 1 ENTER (−) 1 ENTER 3 ENTER 0 ENTER (−) 2 ENTER 4 ENTER (−) 4 ENTER. Press 2nd MODE to return to the home screen. Press 2nd x^{-1} ENTER ENTER to display matrix [A].

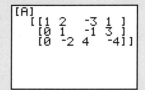

Notice that we can omit *Steps 2–4* because we already have two 1's on the diagonal and 0's below the first 1.

Step 5: Get the element in Row 3, column 2, to be 0. Multiply Row 2 by the opposite of the number in Row 3, column 2, and add to Row 3. The graphing calculator row operation used to multiply a row by a nonzero number and add to another row is ***row+(nonzero number, matrix name, first row, second row)**.

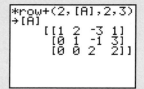

In this case, we have *row+(2, [A], 2, 3). To enter this row operation on your calculator, press 2nd x^{-1}, then press the right arrow to access the MATH menu. Scroll down to option F and press ENTER to display *row+(then enter 2 , 2nd x^{-1} ENTER , 2 , 3) as shown. Store the result back in matrix [A] by pressing STO> 2nd x^{-1} ENTER ENTER.

Step 6: To make the last number on the diagonal be 1, multiply Row 3 by $\frac{1}{2}$. The graphing calculator row operation used to multiply a row by a nonzero number is ***row(nonzero number, matrix name, row)**. In this case, we have *row(1/2, [A], 3). On your calculator, press 2nd x^{-1}, then press the right arrow to access the MATH menu. Scroll down to option E and press ENTER to display *row(then enter 1 ÷ 2 , 2nd x^{-1} ENTER , 3) as shown.

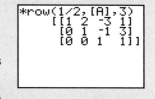

Step 7: Write the matrix from *Step 6* as:

$$\begin{bmatrix} 1 & 2 & -3 & | & 1 \\ 0 & 1 & -1 & | & 3 \\ 0 & 0 & 1 & | & 1 \end{bmatrix} \qquad \begin{matrix} 1x + 2y - 3z = 1 \\ 0x + 1y - 1z = 3 \\ 0x + 0y + 1z = 1 \end{matrix} \text{ or } \begin{matrix} x + 2y - 3z = 1 \\ y - z = 3 \\ z = 1 \end{matrix}$$

Step 8: Solve the system using substitution to obtain the solution $x = -4$, $y = 4$, $z = 1$ or $(-4, 4, 1)$.

Step 9: Check the solution.

Using Row Echelon Form

The row echelon form shown above is not unique. Another row echelon form can be obtained *in one step* using a graphing calculator. Given the original augmented matrix stored in [A], press 2nd x^{-1}, then press the right arrow, scroll down to option A, and press ENTER to display ref(which stands for row echelon form, and press ENTER. Press 2nd x^{-1} ENTER) ENTER to show the matrix in row echelon form.

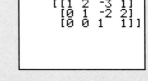

Using Reduced Row Echelon Form

The **reduced row echelon form** of an augmented matrix contains 1's on the diagonal and 0's *above* and *below* the 1's. We can find this using row operations as shown in the 9-step process, or directly in one step. Given the original augmented matrix stored in [A], press 2nd x^{-1}, then press the right arrow, scroll down to option B, and press ENTER to display ref(which stands for reduced row-echelon form, and press ENTER. Press 2nd x^{-1} ENTER) ENTER to show the matrix in reduced row echelon form.

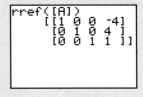

Write a system of equations from the matrix that is in reduced row echelon form.

$$\begin{bmatrix} 1 & 0 & 0 & | & -4 \\ 0 & 1 & 0 & | & 4 \\ 0 & 0 & 1 & | & 1 \end{bmatrix} \qquad \begin{matrix} 1x + 0y + 0z = -4 \\ 0x + 1y + 0z = 4 \\ 0x + 0y + 1z = 1 \end{matrix} \text{ or } \begin{matrix} x = -4 \\ y = 4 \\ z = 1 \end{matrix}$$

Use a graphing calculator to solve each system using Gaussian elimination.

1) $x + 2y = 1$
   $3x - y = 17$

2) $x - 5y = -3$
   $2x - 7y = 3$

3) $-5x + 2y = -4$
   $3x - y = 8$

4) $3x - 5y - 3z = 6$
   $-x + 3y + 2z = 1$
   $-2x + 7y + 5z = 6$

5) $3x + 2y + z = 9$
   $-5x - 2y - z = -7$
   $4x + y + z = 3$

6) $2x - y + 2z = -4$
   $-x + y - 2z = 7$
   $-3x + y - z = -1$

ANSWERS TO [YOU TRY] EXERCISES

1) (2, 3) 2) (4, 0, −3)

ANSWERS TO TECHNOLOGY EXERCISES

1) (5, −2) 2) (12, 3) 3) (12, 28) 4) (4, −3, 7) 5) (−1, 5, 2) 6) (3, 6, −2)

E Evaluate 5.4 Exercises

Do the exercises, and check your work.

Objective 1: Learn the Vocabulary Associated with Gaussian Elimination

Write each system in an augmented matrix.

1) $x - 7y = 15$
$4x + 3y = -1$

2) $x + 6y = 4$
$-5x + y = -3$

3) $x + 6y - z = -2$
$3x + y + 4z = 7$
$-x - 2y + 3z = 8$

4) $x + 2y - 7z = 3$
$3x - 5y = -1$
$-x + 2z = -4$

Write a system of linear equations in x and y represented by each augmented matrix.

5) $\begin{bmatrix} 3 & 10 & | & -4 \\ 1 & -2 & | & 5 \end{bmatrix}$

6) $\begin{bmatrix} 1 & -1 & | & 6 \\ -4 & 7 & | & 2 \end{bmatrix}$

7) $\begin{bmatrix} 1 & -6 & | & 8 \\ 0 & 1 & | & -2 \end{bmatrix}$

8) $\begin{bmatrix} 1 & 2 & | & 11 \\ 0 & 1 & | & 3 \end{bmatrix}$

Write a system of linear equations in x, y, and z represented by each augmented matrix.

9) $\begin{bmatrix} 1 & -3 & 2 & | & 7 \\ 4 & -1 & 3 & | & 0 \\ -2 & 2 & -3 & | & -9 \end{bmatrix}$

10) $\begin{bmatrix} 1 & 4 & -3 & | & -5 \\ -1 & 2 & 5 & | & 8 \\ 6 & -2 & -1 & | & 3 \end{bmatrix}$

11) $\begin{bmatrix} 1 & 5 & 2 & | & 14 \\ 0 & 1 & -8 & | & 2 \\ 0 & 0 & 1 & | & -3 \end{bmatrix}$

12) $\begin{bmatrix} 1 & 4 & -7 & | & -11 \\ 0 & 1 & 3 & | & -1 \\ 0 & 0 & 1 & | & 6 \end{bmatrix}$

Objective 2: Solve a System of Two Equations Using Gaussian Elimination

Solve each system using Gaussian elimination. Identify any inconsistent systems or dependent equations.

13) $x + 4y = -1$
$3x + 5y = 4$

14) $x - 3y = 1$
$-3x + 7y = 3$

15) $x + 4y = -6$
$2x + 5y = 0$

16) $x - 3y = 9$
$-6x + 5y = 11$

17) $4x - 3y = 6$
$x + y = -2$

18) $-4x + 5y = -3$
$x - 8y = -6$

Objective 3: Solve a System of Any Number of Equations Using Gaussian Elimination

19) $x + y - z = -5$
$4x + 5y - 2z = 0$
$8x - 3y + 2z = -4$

20) $x - 2y + 2z = 3$
$2x - 3y + z = 13$
$-4x - 5y - 6z = 8$

21) $x - 3y + 2z = -1$
$3x - 8y + 4z = 6$
$-2x - 3y - 6z = 1$

22) $x - 2y + z = -2$
$2x - 3y + z = 3$
$3x - 6y + 2z = 1$

23) $-4x - 3y + z = 5$
$x + y - z = -7$
$6x + 4y + z = 12$

24) $6x - 9y - 2z = 7$
$-3x + 4y + z = -4$
$x - y - z = 1$

25) $x - 3y + z = -4$
$4x + 5y - z = 0$
$2x - 6y + 2z = 1$

26) $x - y + 3z = 1$
$5x - 5y + 15z = 5$
$-4x + 4y - 12z = -4$

Extension

Extend the concepts of this section to solve these systems using Gaussian elimination.

27) $a + b + 3c + d = -1$
$-a + c - d = 7$
$2a + 3b + 9c - 2d = 7$
$a - 2b + c + 3d = -11$

28) $a - 2b - c + 3d = 15$
$2a - 3b + c + 4d = 22$
$-a + 4b + 6c + 7d = -3$
$3a + 2b - c - d = -7$

29) $w - 3x + 2y - z = -2$
$-3w + 8x - 5y + z = 2$
$2w - x + y + 3z = 7$
$w - 2x + y + 2z = 3$

30) $w + x - 4y + 2z = -21$
$3w + 2x + y - z = 6$
$-2w - x - 2y + 6z = -30$
$-w + 3x + 4y + z = 1$

R Rethink

R1) How is Gaussian elimination similar to the elimination method for solving a system of equations?

R2) If you had to solve a system of linear equations in two variables, would you prefer to use the *elimination method* shown in Section 5.1 or *Gaussian elimination*?

R3) Could you explain how to solve a system of two linear equations using Gaussian elimination to a friend?

R4) When solving a system of linear equations using *Gaussian elimination,* does it matter in which row you represent each equation? In other words, does the top equation have to be placed in Row 1?

Group Activity

The Group Activity can be found online on Connect.

emPOWERme Studying Smart

Imagine packing your bag for a trip without knowing where you were going. Would you take sweaters or swimsuits? A raincoat or sunblock? This dilemma is parallel to the one you face if you study for a test without knowing what it will cover. The key to effective studying—and to successful test-taking—is to tailor your efforts to the test you will have to take. Before you start studying, answer these questions, and use your answers to help you prepare.

- Is the test called a "test," "exam," "quiz," or something else? There is a difference! Exams tend to be longer, while quizzes are often shorter and narrower in their focus. If you are aren't sure, ask your instructor.
- What material will the test cover? Will it cover only the most recent subjects or everything you've learned in the term so far?
- How many questions will be on the test? How much time is it expected to take? A full class period? Only part of a period?
- What kinds of questions will be on the test?
- Will you be allowed to use a calculator? Consult your textbook?
- Will the test be graded on a curve?
- Will sample questions be provided?
- Are tests from previous terms available for you to study?
- How much does the test contribute to your final course grade?

Chapter 5: Summary

Definition/Procedure	Example

5.1 Solving Systems of Linear Equations in Two Variables

A **system of linear equations** consists of two or more linear equations with the same variables. A **solution of a system** of two equations in two variables is an ordered pair that is a solution of each equation in the system. (p. 233)

Determine whether $(6, 1)$ is a solution of the system
$$x + 3y = 9$$
$$-2x + 7y = -5.$$

$$x + 3y = 9 \qquad\qquad -2x + 7y = -5$$
$$6 + 3(1) \stackrel{?}{=} 9 \quad \text{Substitute.} \qquad -2(6) + 7(1) \stackrel{?}{=} -5 \quad \text{Substitute.}$$
$$6 + 3 \stackrel{?}{=} 9 \qquad\qquad -12 + 7 \stackrel{?}{=} -5$$
$$9 = 9 \quad \text{True} \qquad\qquad -5 = -5 \quad \text{True}$$

Since $(6, 1)$ is a solution of each equation in the system, *yes*, it is a solution of the system.

To **solve a system by graphing,** graph each line on the same coordinate axes.

a) If the lines intersect at a single point, then this point is the solution of the system. The system is **consistent** and the equations are **independent**.
b) If the lines are parallel, then the system has **no solution**. We write the solution set as $\emptyset$. The system is **inconsistent** and the equations are **independent**.
c) If the graphs are the same line, then the system has an **infinite number of solutions** consisting of the points on that line. The system is **consistent** and the equations are **dependent**. (p. 236)

Solve by graphing. $\quad y = \dfrac{1}{4}x - 3$
$$3x + 4y = 4$$

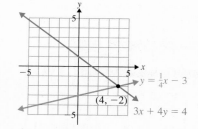

The solution of the system is $(4, -2)$. The system is consistent.

Steps for Solving a System by Substitution

Step 1: Solve one of the equations for one of the variables. If possible, solve for a variable that has a coefficient of 1 or -1.

Step 2: Substitute the expression found in *Step 1* into the *other* equation. The equation you obtain should contain only one variable.

Step 3: Solve the equation obtained in *Step 2*.

Step 4: Substitute the value found in *Step 3* into one of the equations to obtain the value of the other variable.

Step 5: Check the values in the original equations. (p. 238)

Solve by substitution. $\quad 2x - 7y = 2$
$$x - 2y = -2$$

Step 1: Solve for x in the second equation since its coefficient is 1.
$$x = 2y - 2$$

Step 2: Substitute $2y - 2$ for the x in the first equation.
$$2(2y - 2) - 7y = 2$$

Step 3: Solve the equation in *Step 2* for y.
$$4y - 4 - 7y = 2 \quad \text{Distribute.}$$
$$-3y - 4 = 2 \quad \text{Combine like terms.}$$
$$-3y = 6 \quad \text{Add 4.}$$
$$y = -2 \quad \text{Divide by } -3.$$

Step 4: Substitute $y = -2$ into the equation in *Step 1* to find x.
$$x = 2y - 2$$
$$x = 2(-2) - 2 \quad \text{Substitute } -2 \text{ for } y.$$
$$x = -4 - 2 \quad \text{Multiply.}$$
$$x = -6$$

Step 5: The solution is $(-6, -2)$. Verify this by substituting $(-6, -2)$ into each of the original equations.

Definition/Procedure	Example
Steps for Solving a System by the Elimination Method *Step 1:* Write each equation in the form $Ax + By = C$. *Step 2:* Determine which variable to eliminate. If necessary, multiply one or both of the equations by a number so that the coefficients of the variable to be eliminated are opposites. *Step 3:* Add the equations, and solve for the remaining variable. *Step 4:* Substitute the value found in *Step 3* into either of the original equations to find the value of the other variable. *Step 5:* Check the solution in each of the original equations. **(p. 241)**	Solve using the elimination method. $\quad 7x - 4y = 1$ $ -4x + 3y = 3$ Eliminate y. Multiply the first equation by 3, and multiply the second equation by 4 to rewrite the system with equivalent equations. Rewrite the system: $\quad \begin{array}{l} 3(7x - 4y) = 3(1) \\ 4(-4x + 3y) = 4(3) \end{array} \rightarrow \begin{array}{l} 21x - 12y = 3 \\ -16x + 12y = 12 \end{array}$ Add the equations: $\quad\quad 21x - 12y = 3$ $ + \underline{-16x + 12y = 12}$ $ 5x = 15$ $ x = 3$ Substitute $x = 3$ into either of the original equations and solve for y. $\quad\quad 7x - 4y = 1$ $\quad\quad 7(3) - 4y = 1$ $\quad\quad 21 - 4y = 1 \quad$ Multiply. $\quad\quad\quad -4y = -20$ $\quad\quad\quad\quad\; y = 5$ The solution is (3, 5). Verify this by substituting (3, 5) into each of the original equations.
If the variables are eliminated and a false equation is obtained, the system has **no solution. The system is inconsistent, and the solution set is** $\varnothing$. **(p. 244)**	Solve the system. $\quad 4x - 12y = 7$ $ x = 3y - 1$ We will solve by substitution. Substitute $3y - 1$ for x in the first equation. $\quad\quad 4(3y - 1) - 12y = 7$ Solve the equation for y. $\quad\quad 12y - 4 - 12y = 7 \quad$ Distribute. $\quad\quad\quad\quad\quad\quad -4 = 7 \quad$ False The system has no solution. The solution set is $\varnothing$.
If the variables are eliminated and a true equation is obtained, the system has an **infinite number of solutions.** The equations are **dependent. (p. 244)**	Solve the system. $\quad y = x + 4$ $ 2x - 2y = -8$ Substitute $x + 4$ for y in the second equation. $\quad\quad 2x - 2(x + 4) = -8$ Solve the equation for x. $\quad\quad 2x - 2x - 8 = -8 \quad$ Distribute. $\quad\quad\quad\quad\quad -8 = -8 \quad$ True The equations are dependent, and the solution set is $\{(x, y) \mid y = x + 4\}$.

Definition/Procedure	Example

5.2 Solving Systems of Linear Equations in Three Variables

A **linear equation in three variables** is an equation of the form $Ax + By + Cz = D$, where A, B, and C are not all zero and where A, B, C, and D are real numbers. Solutions of this type of an equation are **ordered triples** of the form (x, y, z). **(p. 251)**

$5x + 3y + 9z = -2$

One solution of this equation is $(-1, -2, 1)$ because substituting the values for x, y, and z satisfies the equation.

$$5x + 3y + 9z = -2$$
$$5(-1) + 3(-2) + 9(1) \stackrel{?}{=} -2$$
$$-5 - 6 + 9 \stackrel{?}{=} -2$$
$$-2 = -2 \quad \text{True}$$

Solving a System of Three Linear Equations in Three Variables

Step 1: Label the equations ①, ②, and ③.

Step 2: Choose a variable to eliminate. **Eliminate** this variable from *two* sets of *two* equations using the elimination method. You will obtain two equations containing the same two variables. Label one of these new equations Ⓐ and the other Ⓑ.

Step 3: Use the elimination method to **eliminate a variable from equations** Ⓐ **and** Ⓑ. You have now found the value of one variable.

Step 4: **Find the value of another variable** by substituting the value found in *Step 3* into equation Ⓐ or Ⓑ and solving for the second variable.

Step 5: **Find the value of the third variable** by substituting the values of the two variables found in *Steps 3* and *4* into equation ①, ②, or ③.

Step 6: **Check** the solution in each of the original equations, and **write the solution as an ordered triple. (p. 252)**

Solve
① $x + 2y + 3z = 5$
② $4x - 2y - z = -1$
③ $-3x + y + 4z = -12$

Step 1: Label the equations ①, ②, and ③.

Step 2: We will eliminate y from *two* sets of *two* equations.

a) Add equations ① and ② to eliminate y. Label the resulting equation Ⓐ.

① $x + 2y + 3z = 5$
② + $4x - 2y - z = -1$
Ⓐ $5x + 2z = 4$

b) To eliminate y again, multiply equation ③ by 2 and add it to equation ②. Label the resulting equation Ⓑ.

$2 \times $ ③ $-6x + 2y + 8z = -24$
② + $4x - 2y - z = -1$
Ⓑ $-2x + 7z = -25$

Step 3: Eliminate x from Ⓐ and Ⓑ. Multiply Ⓐ by 2 and Ⓑ by 5. Add the resulting equations.

$2 \times $ Ⓐ $10x + 4z = 8$
$5 \times $ Ⓑ + $-10x + 35z = -125$
$\phantom{5 \times \text{Ⓑ} + -10x + } 39z = -117$
$\phantom{5 \times \text{Ⓑ} + -10x + 39} \boxed{z = -3}$

Step 4: Substitute $z = -3$ into either Ⓐ or Ⓑ.

Ⓐ $5x + 2z = 4$
     $5x + 2(-3) = 4$ Substitute -3 for z.
     $5x - 6 = 4$ Multiply.
     $5x = 10$ Add 6.
     $\boxed{x = 2}$ Divide by 5.

Step 5: Substitute $x = 2$ and $z = -3$ into ① to solve for y.

① $x + 2y + 3z = 5$
     $2 + 2y + 3(-3) = 5$ Substitute.
     $2 + 2y - 9 = 5$ Multiply.
     $2y - 7 = 5$ Combine like terms.
     $2y = 12$ Add 7.
     $\boxed{y = 6}$ Divide by 2.

Step 6: The solution is $(2, 6, -3)$. The check is left to the student.

Definition/Procedure	Example

5.3 Applications of Systems of Linear Equations

Use the procedure outlined in the section to solve an applied problem. **Step 1: Read** the problem carefully. Draw a picture, if applicable. Identify what you are being asked to find. Label any pictures with the variables. **Step 2: Choose variables** to represent unknown quantities. **Step 3: Write a system of equations using the variables.** It may be helpful to begin by writing an equation in words. **Step 4: Solve** the system. **Step 5: Check** the answer in the original problem, and **interpret** the solution as it relates to the problem. **(p. 260)**	Natalia spent $23.80 at an office supply store when she purchased boxes of pens and paper clips. The pens cost $3.50 per box, and the paper clips cost $0.70 per box. How many boxes of each did she buy if she purchased 10 items all together? **Steps 1 and 2:** Define the variables. x = number of boxes of pens she bought y = number of boxes of paper clips she bought **Step 3:** One equation involves the *cost* of the items: Cost of pens + Cost of paper clips = Total cost $3.50x \quad + \quad 0.70y \quad = \quad 23.80$ The other equation involves the number of items: Number of pens + Number of paper clips = Total number of items $x \quad + \quad y \quad = \quad 10$ The system is $\quad 3.50x + 0.70y = 23.80$ $\quad x + y = 10$ **Step 4:** Multiply by 10 to eliminate the decimals in the first equation, and then solve the system using substitution. $10(3.50x + 0.70y) = 10(23.80)\quad$ Eliminate decimals. $35x + 7y = 238$ Solve the system $\begin{array}{l}35x + 7y = 238\\ x + y = 10\end{array}$ to determine that the solution is (6, 4). **Step 5:** Natalia bought 6 boxes of pens and 4 boxes of paper clips. Verify the solution.

5.4 Solving Systems of Linear Equations Using Matrices

An **augmented matrix** contains a vertical line to separate different parts of the matrix. **(p. 274)**	An example of an augmented matrix is $\begin{bmatrix} 1 & 4 & \vert & -9 \\ 2 & -3 & \vert & 8 \end{bmatrix}$.
Matrix Row Operations Performing the following row operations on a matrix produces an equivalent matrix. 1) Interchanging two rows 2) Multiplying every element in a row by a nonzero real number 3) Replacing a row by the sum of it and the multiple of another row **Gaussian elimination** is the process of performing row operations on a matrix to put it into *row echelon* form. A matrix is in **row echelon form** when it has 1's along the diagonal and 0's below. $\begin{bmatrix} 1 & a & \vert & b \\ 0 & 1 & \vert & c \end{bmatrix}\quad \begin{bmatrix} 1 & a & b & \vert & d \\ 0 & 1 & c & \vert & e \\ 0 & 0 & 1 & \vert & f \end{bmatrix}$ **(p. 274)**	Solve using Gaussian elimination. $\quad x - y = 5$ $2x + 7y = 1$ Write the system in an augmented matrix. Then, perform row operations to get it into row echelon form. $\begin{bmatrix} 1 & -1 & \vert & 5 \\ 2 & 7 & \vert & 1 \end{bmatrix} \xrightarrow{-2R_1 + R_2 \to R_2} \begin{bmatrix} 1 & -1 & \vert & 5 \\ 0 & 9 & \vert & -9 \end{bmatrix} \xrightarrow{\frac{1}{9}R_2 \to R_2} \begin{bmatrix} 1 & -1 & \vert & 5 \\ 0 & 1 & \vert & -1 \end{bmatrix}$ The matrix is in row echelon form since it has 1's on the diagonal and a 0 below. Write a system of equations from the matrix that is in row echelon form. $\begin{bmatrix} 1 & -1 & \vert & 5 \\ 0 & 1 & \vert & -1 \end{bmatrix} \quad \begin{array}{l} 1x - 1y = 5 \\ 0x + 1y = -1 \end{array} \text{ or } \begin{array}{l} x - y = 5 \\ y = -1 \end{array}$ Solving the system, we obtain the solution $(4, -1)$.

Chapter 5: Review Exercises

(5.1) Determine whether the ordered pair is a solution of the system of equations.

1) $-2x + y = 3$
   $3x - y = -17$
   $(-4, -5)$

2) $3x + 4y = 0$
   $9x + 2y = 5$
   $\left(\dfrac{2}{3}, -\dfrac{1}{2}\right)$

Solve each system by graphing.

3) $y = \dfrac{1}{2}x - 2$
   $y = 2x + 1$

4) $3x - y = 2$
   $x + y = 2$

5) $-2x + 3y = 15$
   $2x - 3y = -3$

6) $2x + 3y = 5$
   $y = \dfrac{1}{2}x + 4$

7) $4x + y = -4$
   $-2x - \dfrac{1}{2}y = 2$

Without graphing, determine whether each system has no solution, one solution, or an infinite number of solutions.

8) $x + 7y = -3$
   $4x - 9y = 1$

9) $y = -\dfrac{2}{3}x - 3$
   $4x + 6y = 5$

10) $5x - 4y = 2$
    $y = \dfrac{5}{4}x - \dfrac{1}{2}$

11) $15x - 10y = 4$
    $-9x + 6y = 1$

12) The graph shows the on-time departure percentages during the four quarters of 2003 at San Diego (Lindbergh) and Denver International Airports.

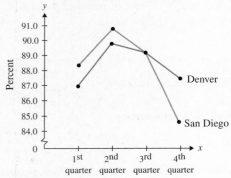

On-time Departures in 2003

(www.census.gov)

a) When was Denver's on-time departure percentage better than San Diego's?

b) When were their percentages the same and, approximately, what percentage of flights left these airports on time?

Solve each system by substitution. Identify any inconsistent systems or dependent equations

13) $x + 8y = -2$
    $2x + 11y = -9$

14) $y = \dfrac{5}{6}x - 2$
    $6y - 5x = -12$

15) $-2x + y = -18$
    $x + 7y = 9$

16) $6x - y = -3$
    $15x + 2y = 15$

17) $\dfrac{5}{2}x + \dfrac{9}{2}y = \dfrac{1}{2}$
    $\dfrac{1}{6}x + \dfrac{2}{3}y = -\dfrac{1}{3}$

18) $x + 8y = 2$
    $x = 20 - 8y$

Solve each system using the elimination method. Identify any inconsistent systems or dependent equations.

19) $x - y = -8$
    $3x + y = -12$

20) $-5x + 3y = 17$
    $x + 2y = 20$

21) $4x - 5y = -16$
    $-3x + 4y = 13$

22) $6x + 8y = 13$
    $9x + 12y = -5$

23) $0.12x + 0.01y = 0.06$
    $0.5x + 0.2y = -0.7$

24) $3(8 - y) = 5x + 3$
    $x + 2(y - 3) = 2(4 - x)$

Solve each system by your choice of the substitution or elimination method. Identify any inconsistent systems or dependent equations.

25) $2x - 3y = 3$
    $3x + 4y = -21$

26) $2(2x - 3) = y + 3$
    $8 + 5(y - 6) = 8(x + 3) - 9x - y$

27) $6x - 4y = 12$
    $15x - 10y = 30$

28) $\dfrac{3}{4}x - y = \dfrac{1}{2}$
    $-\dfrac{x}{3} + \dfrac{y}{2} = -\dfrac{1}{6}$

29) $15 - y = y + 2(4x + 5)$
    $2(x + 1) + 3 = 2(y + 4) + 10x$

30) $7x - y = -12$
    $-2x + 3y = 17$

(5.2) Determine whether the ordered triple is a solution of the system.

31) $x - 6y + 4z = 13$
    $5x + y + 7z = 8$
    $2x + 3y - z = -5$
    $(-3, -2, 1)$

32) $-4x + y + 2z = 1$
    $x - 3y - 4z = 3$
    $-x + 2y + z = -7$
    $(0, -5, 3)$

Solve each system using one of the methods in Section 5.2. Identify any inconsistent systems or dependent equations.

33) $2x - 5y - 2z = 3$
    $x + 2y + z = 5$
    $-3x - y + 2z = 0$

34) $x - 2y + 2z = 6$
    $x + 4y - z = 0$
    $5x + 3y + z = -3$

35) $5a - b + 2c = -6$
    $-2a - 3b + 4c = -2$
    $a + 6b - 2c = 10$

36) $2x + 3y - 15z = 5$
    $-3x - y + 5z = 3$
    $-x + 6y - 10z = 12$

37) $4x - 9y + 8z = 2$
    $x + 3y = 5$
    $6y + 10z = -1$

38) $-a + 5b - 2c = -3$
    $3a + 2c = -3$
    $2a + 10b = -2$

39) $x + 3y - z = 0$
    $11x - 4y + 3z = 8$
    $5x + 15y - 5z = 1$

40) $4x + 2y + z = 0$
    $8x + 4y + 2z = 0$
    $16x + 8y + 4z = 0$

41) $12a - 8b + 4c = 8$
    $3a - 2b + c = 2$
    $-6a + 4b - 2c = -4$

42) $3x - 12y - 6z = -8$
    $x + y - z = 5$
    $-4x + 16y + 8z = 10$

43) $5y + 2z = 6$
    $-x + 2y = -1$
    $4x - z = 1$

44) $2a - b = 4$
    $3b + c = 8$
    $-3a + 2c = -5$

45) $8x + z = 7$
    $3y + 2z = -4$
    $4x - y = 5$

46) $6y - z = -2$
    $x + 3y = 1$
    $-3x + 2z = 8$

(5.3) Write a system of equations, and solve.

47) One day at the Village Veterinary Practice, the doctors treated twice as many dogs as cats. If they treated a total of 51 cats and dogs, how many of each did the doctors see?

48) At Aurora High School, 183 sophomores study either Spanish or French. If there are 37 fewer French students than Spanish students, how many study each language?

49) At a Houston Texans football game in 2008, four hot dogs and two sodas cost $26.50, while three hot dogs and four sodas cost $28.00. Find the price of a hot dog and the price of a soda at a Texans game. (www.teammarketing.com)

50) The perimeter of a rectangular computer monitor is 66 in. The length is 3 in. more than the width. What are the dimensions of the monitor?

51) Find the measures of angles x and y if the measure of x is twice the measure of y.

52) A store owner plans to make 10 pounds of a candy mix worth $1.92/lb. How many pounds of gummi bears worth $2.40/lb, and how many pounds of jelly beans worth $1.60/lb must be combined to make the candy mix?

53) How many milliliters of pure alcohol and how many milliliters of a 4% alcohol solution must be combined to make 480 milliliters of an 8% alcohol solution?

54) A car and a tour bus leave the same location and travel in opposite directions. The car's speed is 12 mph more than the speed of the bus. If they are 270 mi apart after $2\frac{1}{2}$ hr, how fast is each vehicle traveling?

55) One serving (8 fl oz) of Powerade has 17 mg more sodium than one serving of Propel. One serving of Gatorade has 58 mg more sodium than the same serving size of Powerade. Together the three drinks have 197 mg of sodium. How much sodium is in one serving of each drink? (Product labels)

56) In 2003, the top highway truck tire makers were Goodyear, Michelin, and Bridgestone. Together, they held 53% of the market. Goodyear's market share was 3% more than Bridgestone's, and Michelin's share was 1% less than Goodyear's. What percent of this tire market did each company hold in 2003? (*Market Share Reporter*, Vol. I, 2005, p. 361: From: *Tire Business*, Feb. 2, 2004, p. 9)

57) One Friday, Serena, Blair, and Chuck were busy texting their friends. Together, they sent a total of 140 text messages. Blair sent 15 more texts then Serena, while Chuck sent half as many as Serena. How many texts did each person send that day?

58) The Recording Industry Association of America reports that in 2005 there were 14 million fewer downloads than in 2006, and in 2007 there were 14.9 million more downloads than the previous year. During all three years, 83.7 million albums were downloaded. How many albums were downloaded in 2005, 2006, and 2007? (www.riaa.com)

59) A family of six people goes to an ice cream store every Sunday after dinner. One week, they order two ice cream cones, three shakes, and one sundae for $13.50. The next week they get three cones, one shake, and two sundaes for $13.00. The week after that they spend $11.50 on one shake, one sundae, and four ice cream cones. Find the price of an ice cream cone, a shake, and a sundae.

60) An outdoor music theater sells three types of seats—reserved, behind-the-stage, and lawn seats. Two reserved, three behind-the-stage, and four lawn seats cost $360. Four reserved, two behind-the-stage, and five lawn seats cost $470. One of each type of seat would total $130. Determine the cost of a reserved seat, a behind-the-stage seat, and a lawn seat.

61) The measure of the smallest angle of a triangle is one-third the measure of the middle angle. The measure of the largest angle is 70° more than the measure of the smallest angle. Find the measures of the angles of the triangle.

62) The perimeter of a triangle is 40 in. The longest side is twice the length of the shortest side, and the sum of the two smaller sides is four inches longer than the longest side. Find the lengths of the sides of the triangles.

(5.4) Solve each system using Gaussian elimination.

63) $x - y = -11$
$2x + 9y = 0$

64) $x - 8y = -13$
$4x + 9y = -11$

65) $5x + 3y = 5$
$-x + 8y = -1$

66) $3x + 5y = 5$
$-4x - 9y = 5$

67) $x - 3y - 3z = -7$
$2x - 5y - 3z = 2$
$-3x + 5y + 4z = -1$

68) $x - 3y + 5z = 3$
$2x - 5y + 6z = -3$
$3x + 2y + 2z = 3$

Chapter 5: Test

1) Determine whether $\left(\dfrac{3}{4}, -5\right)$ is a solution of the system.

$8x + y = 1$
$-12x - 4y = 11$

In Exercises 2–11, when solving the systems, identify any inconsistent systems or dependent equations.

Solve each system by graphing.

2) $y = 2x - 3$
$3x + 2y = 8$

3) $x + y = 3$
$2x + 2y = -2$

Solve each system by substitution.

4) $5x + 9y = 3$
$x - 4y = -11$

5) $-9x + 12y = 21$
$y = \dfrac{3}{4}x + \dfrac{7}{4}$

Solve each system by elimination.

6) $4x - 3y = -14$
$x + 3y = 19$

7) $7x + 8y = 28$
$-5x + 6y = -20$

8) $-x + 4y + 3z = 6$
$3x - 2y + 6z = -18$
$x + y + 2z = -1$

Solve each system using any method.

9) $x - 8y = 1$
$-2x + 9y = -9$

10) $\dfrac{x}{6} + y = -\dfrac{1}{3}$
$-\dfrac{5}{8}x + \dfrac{3}{4}y = \dfrac{11}{4}$

11) $5(y + 4) - 9 = -3(x - 4) + 4y$
$13x - 2(3x + 2) = 3(1 - y)$

Write a system of equations, and solve.

12) A 6-in. Turkey Breast Sandwich from Subway has 390 fewer calories than a Burger King Whopper. If the two sandwiches have a total of 950 calories, determine the number of calories in the Subway sandwich and the number of calories in a Whopper. (www.subway.com, www.bk.com)

13) At a hardware store, three boxes of screws and two boxes of nails sell for $18 while one box of screws and four boxes of nails sell for $16. Find the price of a box of screws and the price of a box of nails.

14) The measure of the smallest angle of a triangle is 9° less than the measure of the middle angle. The largest angle is 30° more than the sum of the two smaller angles. Find the measures of the angles of the triangle.

Solve using Gaussian elimination.

15) $x + 5y = -4$
$3x + 2y = 14$

16) $-3x + 5y + 8z = 0$
$x - 3y + 4z = 8$
$2x - 4y - 3z = 3$

Chapter 5: Cumulative Review for Chapters 1–5

Perform the operations and simplify.

1) $\dfrac{3}{10} - \dfrac{7}{15}$

2) $5\dfrac{5}{6} \div 1\dfrac{13}{15}$

3) $(5 - 8)^3 + 40 \div 10 - 6$

4) Find the area of the triangle.

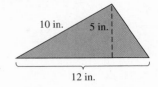

5) Simplify $-8(3x^2 - x - 7)$.

6) Solve $11 - 3(2k - 1) = 2(6 - k)$.

7) Solve $0.04(3p - 2) - 0.02p = 0.1(p + 3)$.

8) Solve $-47 \leq 7t - 5 \leq 6$. Write the answer in interval notation.

9) Write an equation, and solve.

   Google's second quarter revenue in 2012 was 35% more than in the second quarter of 2011. If Google's second-quarter revenue was $12.2 billion in 2012, what was it in the same quarter of 2011? Round to the nearest tenth of a billion. (www.eweek.com)

10) Solve $|8w - 3| > 7$. Write the answer in interval notation.

11) The area, A, of a trapezoid is $A = \dfrac{1}{2}h(b_1 + b_2)$

    where h = height of the trapezoid,
    b_1 = length of one base of the trapezoid, and
    b_2 = length of the second base of the trapezoid.

    a) Solve the equation for h.

    b) Find the height of the trapezoid that has an area of 39 cm² and bases of length 8 cm and 5 cm.

12) Graph $2x + 3y = 5$.

13) Graph $f(x) = 3x - 2$.

14) Find the x- and y-intercepts of the graph of $4x - 5y = 10$.

15) Write the slope-intercept form of the equation of the line containing $(-7, 4)$ and $(1, -3)$.

16) Determine whether the lines are *parallel, perpendicular,* or *neither.*

    $10x + 18y = 9$
    $9x - 5y = 17$

Determine the domain of each function, and write it in interval notation.

17) $f(x) = -6x + 11$

18) $g(x) = \dfrac{4}{7 - 2x}$

19) Let $f(x) = 4x + 9$. Find each of the following.

    a) $f(-5)$ b) $f(p)$ c) $f(n + 2)$

Solve each system of equations. Identify any inconsistent systems or dependent equations.

20) $9x + 7y = 7$
    $3x + 4y = -11$

21) $3(2x - 1) - (y + 10) = 2(2x - 3) - 2y$
    $3x + 13 = 4x - 5(y - 3)$

22) $\dfrac{5}{6}x - \dfrac{1}{2}y = \dfrac{2}{3}$
    $-\dfrac{5}{4}x + \dfrac{3}{4}y = \dfrac{1}{2}$

23) $4a - 3b = -5$
    $-a + 5c = 2$
    $2b + c = -2$

Write a system of equations, and solve.

24) Dhaval used twice as many 6-ft boards as 4-ft boards when he made a playhouse for his children. If he used a total of 48 boards, how many of each size did he use?

25) Pilar loves to watch animal videos on You Tube. Her favorite puppy video has twice as many views as her favorite pig video. Her favorite kitty video has 1.6 million more views than her favorite puppy video. All three videos have a total of 7.1 million views. How many views has each video had?

CHAPTER 6
Polynomials and Polynomial Functions

OUTLINE

Study Strategies: Doing Math Homework

6.1 The Rules of Exponents

6.2 More on Exponents and Scientific Notation

6.3 Addition and Subtraction of Polynomials and Polynomial Functions

6.4 Multiplication of Polynomials and Polynomial Functions

6.5 Division of Polynomials and Polynomial Functions

Group Activity

emPOWERme: The Right Time and Place for Homework

Math at Work:
Actuary

Nyesha Samuels has always been good at math, and coming out of college she wanted to find a career that would allow her to put her math skills to work on a daily basis. That is why she decided to be an actuary. Actuaries work for insurance companies, helping them determine the price of insurance policies.

Nyesha works for a car insurance company. "I help calculate the risk that a given person will end up in a car accident," Nyesha explains. "Then I help calculate how much that person needs to pay us for insurance based on that risk. The greater the risk of an accident, the more the person needs to pay."

Translating concepts like risk into dollars and cents obviously takes a lot of expertise. Actuaries have to take a series of tests in order to qualify themselves for the profession. "Right after college, I started working in another field, and after work I would come home and study for my actuary exams," Nyesha describes. "I learned that to build the career you want, you sometimes have to be willing to work in an office *and* at home."

In this chapter, we will discuss polynomials and polynomial functions and also introduce strategies you can apply to math homework.

Study Strategies — Doing Math Homework

Perhaps more than in any other academic field, homework is essential to learning math. Homework gives you the opportunity to apply and practice the skills you hear about from your instructor and read about in your textbook. It lets you develop mastery that you can use throughout your academic and your professional career. And, not least, it gives you valuable practice in the same sorts of problems you will see on exams. Use the strategies below to help improve your math homework performance.

P Prepare
- Make sure you are doing your homework in the right environment (see the emPOWERme on page 345.)
- Gather the materials you need to complete the homework: pens and pencils, scratch paper, your textbook, and the assignment itself.

O Organize
- Review the concepts in your homework, both in your textbook and in your class notes.
- Make sure you have everything you need to complete the homework.

W Work
- Always show your work neatly. This allows your instructor to see how you arrived at the solution to a problem, so he or she can point out any errors for you.
- Don't rely too much on your calculator, particularly if you won't be able to use one on exams.
- Don't give up if you get stuck. Remember that working to solve a difficult problem is a great way to learn.

E Evaluate
- Double-check your calculations to ensure you didn't make any careless errors.
- If there were topics you struggled with, make a point of asking about them in class or during your instructor's office hours.

R Rethink
- Review your homework when it is returned to see what you did right and what you did wrong.
- Keep all your returned assignments for each class organized and in the same place. Never throw away your homework!

Chapter 6 POWER Plan

P Prepare / O Organize

What are your goals for Chapter 6?	How can you accomplish each goal?
1 Be prepared before and during class.	• Don't stay out late the night before, and be sure to set your alarm clock! • Bring a pencil, notebook paper, and textbook to class. • Read the textbook section before coming to class. This will familiarize you with the material. • Avoid distractions by turning off your cell phone during class. • Pay attention, take good notes, and ask questions. • Complete your homework on time, and ask questions on problems you do not understand. • Plan ahead for tests by preparing many days in advance.
2 Understand the homework to the point where you could do it without needing any help or hints.	• Read the directions, and show all of your steps. • Go to the professor's office for help. • Rework homework and quiz problems, and find similar problems for practice. • Review old material that you have not mastered yet.
3 Use the P.O.W.E.R. framework to help you complete and understand your homework: *The Right Time and Place for Homework.*	• Read the Study Strategy about creating an environment for doing homework successfully. • How important do you think homework is for your success in this class? • Complete the emPOWERme that appears before the Chapter Summary.
4 Write your own goal.	• _____

What are your objectives for Chapter 6?	How can you accomplish each objective?
1 Be able to simplify all types of expressions containing exponents.	• Learn the **Rules for Exponents:** Power Rule, Product Rule, Quotient Rule, and Zero and Negative Exponent Rule. • Be able to identify the *base* and *exponent* in each problem. • Follow the **Order of Operations** when simplifying expressions.
2 Be able to write a number in scientific notation, and perform operations with numbers in scientific notation.	• Learn the definition of *scientific notation*. • Learn the procedure for **Writing a Number in Scientific Notation.** • Use the rules of exponents to perform operations on numbers in scientific notation.
3 Be able to add and subtract polynomials and polynomial functions. Be able to evaluate polynomial functions for a given value.	• Learn the procedure for **Adding and Subtracting Polynomials and Polynomial Functions.** • Review *like terms,* and be able to determine the *coefficient* of a number. • Know how to evaluate a polynomial function by using parentheses when substituting.
4 Be able to multiply polynomials and polynomial functions.	• Learn several different methods for multiplying different types of polynomials, including FOIL and the distributive property. • Review *combining like terms.*
5 Be able to divide polynomials and polynomial functions.	• Review long division of whole numbers. • Know the differences in dividing by a monomial versus dividing by a polynomial.
6 Write your own goal.	• _____

	W Work	Read Sections 6.1 to 6.5, and complete the exercises.	
E Evaluate	Complete the Chapter Review and Chapter Test. How did you do?	**R Rethink**	• If homework is not part of your grade in this class, how likely would you be to do your homework? How much does doing homework daily help your understanding of the material? • Why do you need to review exponent rules before multiplying polynomials? • Can you name the different types of polynomials? • Is there more than one way to multiply $(x + 3)(x + 7)$?

6.1 The Rules of Exponents

P Prepare	**O Organize**
What are your objectives for Section 6.1?	How can you accomplish each objective?
1 Use the Product Rule for Exponents	• Know the definition of an *exponential expression*. • Be able to identify the *base* and the *exponent*. • Learn the **Product Rule for Exponents**. • Complete the given example on your own. • Complete You Try 1.
2 Use the Power Rules for Exponents	• Learn the three **Power Rules for Exponents**. • Be sure to follow the **Order of Operations** when simplifying expressions containing exponents. • Complete the given examples on your own. • Complete You Trys 2 and 3.
3 Use 0 and Negative Exponents	• Know the definition for **Zero as an Exponent**. • Know how to identify the *base* and the *exponent*. • Learn the property for **Negative Exponents**. • Complete the given examples on your own. • Complete You Trys 4–6.
4 Use the Quotient Rule for Exponents	• Learn the property for the **Quotient Rule for Exponents**. • Be sure to follow the **Order of Operations** when simplifying expressions containing exponents. • Complete the given example on your own. • Complete You Try 7.

W Work   Read the explanations, follow the examples, take notes, and complete the You Trys.

W Hint
Review the powers of whole numbers in the table on p. 22, if necessary.

Recall from Section 1.3 that exponential notation is a shorthand way to represent a multiplication problem. For example, $3 \cdot 3 \cdot 3 \cdot 3$ can be written as 3^4, so $3^4 = 81$. The *base* is 3 and the *exponent*, or *power*, is 4. We read 3^4 as "3 to the fourth power" or as "3 to the fourth."

> **Definition**
>
> An **exponential expression** of the form a^n is $a^n = \underbrace{a \cdot a \cdot a \cdot \ldots \cdot a}_{n \text{ factors of } a}$, where a is any real number and n is a positive integer. The **base** is a, and n is the **exponent**.

1 Use the Product Rule for Exponents

Is there a rule we can use to *multiply* exponential expressions? Let's rewrite each of the following products as a single power of the base using what we already know:

1) $2^3 \cdot 2^4 = \underbrace{2 \cdot 2 \cdot 2}_{3 \text{ factors of } 2} \cdot \underbrace{2 \cdot 2 \cdot 2 \cdot 2}_{4 \text{ factors of } 2}$
$= \underbrace{2 \cdot 2 \cdot 2 \cdot 2 \cdot 2 \cdot 2 \cdot 2}_{7 \text{ factors of } 2}$
$2^3 \cdot 2^4 = 2^7$

2) $9^2 \cdot 9^3 = \underbrace{9 \cdot 9}_{2 \text{ factors of } 9} \cdot \underbrace{9 \cdot 9 \cdot 9}_{3 \text{ factors of } 9}$
$= \underbrace{9 \cdot 9 \cdot 9 \cdot 9 \cdot 9}_{5 \text{ factors of } 9}$
$9^2 \cdot 9^3 = 9^5$

Do you notice a pattern? *When we multiply expressions with the same base, keep the same base and add the exponents.* This is called the **product rule** for exponents.

W Hint
Notice that the base of each term must be the same to use the product rule.

> **Property** Product Rule
>
> Let a be any real number, and let m and n be integers. Then,
> $$a^m \cdot a^n = a^{m+n}$$

EXAMPLE 1

Find each product.

a) $2^2 \cdot 2^4$ b) $(-x)^8 \cdot (-x)^5$ c) $(-4p^7)(9p^3)$

d) $m^5 n^2$ e) $c^4 \cdot c \cdot c^{10}$ f) $(4a^2 b)(5a^3 b^6)$

Solution

a) $2^2 \cdot 2^4 = 2^{2+4} = 2^6 = 64$ The bases are the same, so add the exponents.

b) $(-x)^8 \cdot (-x)^5 = (-x)^{8+5} = (-x)^{13} = -x^{13}$

c) $(-4p^7)(9p^3) = (-4 \cdot 9)(p^7 \cdot p^3)$
$= -36p^{10}$ Associative and commutative properties

W Hint
Notice that parentheses are required when the base is a negative number.

d) $m^5 n^2$ This expression is in simplest form. We cannot use the product rule because the bases are different.

e) $c^4 \cdot c \cdot c^{10} = c^{4+1+10} = c^{15}$

f) $(4a^2 b)(5a^3 b^6) = (4 \cdot 5)(a^2 \cdot a^3)(b^1 \cdot b^6)$
$= 20a^5 b^7$

[YOU TRY 1] Find each product.

a) $3^2 \cdot 3^2$ b) $q^6 \cdot q^3$ c) $(-8d^4)(3d^5)$
d) $w^8 \cdot w^5 \cdot w$ e) $w^3 k^5$ f) $(8xy^6)(3x^2 y^8)$

BE CAREFUL Can the product rule be applied to $4^3 \cdot 5^2$? **No!** The bases are not the same, so we cannot simply add the exponents. To evaluate $4^3 \cdot 5^2$, we evaluate $4^3 = 64$ and $5^2 = 25$, then multiply:

$$4^3 \cdot 5^2 = 64 \cdot 25 = 1600$$

2 Use the Power Rules for Exponents

What does $(2^2)^3$ mean? We can rewrite $(2^2)^3$ as $2^2 \cdot 2^2 \cdot 2^2$. Then, using the product rule for exponents, we get $2^2 \cdot 2^2 \cdot 2^2 = 2^{2+2+2} = 2^6$.

Therefore $(2^2)^3 = 2^{2+2+2}$, or $2^{2 \cdot 3} = 2^6$. This result suggests the *basic power rule*. Two more power rules are derived from the basic power rule.

Property The Power Rules for Exponents

Let a be any real number and m and n be integers. Then,

1) **Basic Power Rule** $(a^m)^n = a^{mn}$

   To raise a power to another power, multiply the exponents.

2) **Power Rule for a Product** $(ab)^n = a^n b^n$

   To raise a product to a power, raise each factor to that power.

3) **Power Rule for a Quotient** $\left(\dfrac{a}{b}\right)^n = \dfrac{a^n}{b^n}$, where $b \neq 0$

   To raise a quotient to a power, raise the numerator and denominator to that power.

EXAMPLE 2 Simplify using the power rules.

a) $(3^8)^4$ b) $(n^3)^7$ c) $(-9y)^2$ d) $(5c^2)^3$ e) $\left(\dfrac{3}{8}\right)^2$ f) $\left(\dfrac{-2t}{u^4}\right)^5$

Solution

a) $(3^8)^4 = 3^{8 \cdot 4} = 3^{32}$ Basic power rule

b) $(n^3)^7 = n^{3 \cdot 7} = n^{21}$ Basic power rule

c) $(-9y)^2 = (-9)^2 y^2 = 81y^2$ Power rule for a product

d) $(5c^2)^3 = 5^3 \cdot (c^2)^3 = 125 c^{2 \cdot 3} = 125 c^6$ Begin with the power rule for a product.

e) $\left(\dfrac{3}{8}\right)^2 = \dfrac{3^2}{8^2} = \dfrac{9}{64}$ Power rule for a quotient

f) $\left(\dfrac{-2t}{u^4}\right)^5 = \dfrac{(-2t)^5}{(u^4)^5}$ Power rule for a quotient

$= \dfrac{(-2)^5 \cdot t^5}{u^{4 \cdot 5}}$ Power rule for a product

Basic power rule

$= \dfrac{-32t^5}{u^{20}}$ Simplify.

$= -\dfrac{32t^5}{u^{20}}, \quad u \ne 0$

[YOU TRY 2] Simplify using the power rules.

a) $(5^4)^3$ b) $(j^6)^5$ c) $(10p)^3$ d) $(2k^7)^4$ e) $\left(\dfrac{5}{12}\right)^2$ f) $\left(\dfrac{-4w}{y^9}\right)^3$

Now that we have learned the product and power rules for exponents, let's use the rules together. We must follow the order of operations. (See Section 1.3.)

EXAMPLE 3

Simplify.

a) $(2n)^3(3n^6)^2$ b) $2(4a^5b^3)^3$ c) $\dfrac{(6x^9)^2}{(2y^4)^3}$

Solution

a) Remember to perform the exponential operations before multiplication.

$(2n)^3(3n^6)^2 = (2^3n^3)(3)^2(n^6)^2$ Power rule
$= (8n^3)(9n^{12})$ Power rule and evaluate exponents
$= 72n^{15}$ Product rule

Hint

Notice how the rules for the *order of operations* are being used in these examples.

b) The order of operations tells us to perform exponential operations before multiplication.

$2(4a^5b^3)^3 = 2 \cdot (4)^3(a^5)^3(b^3)^3$ Order of operations and power rule
$= 2 \cdot 64a^{15}b^9$ Power rule
$= 128a^{15}b^9$ Multiply.

c) $\dfrac{(6x^9)^2}{(2y^4)^3}$

Here, the operations are division and exponents. Which comes first in the order of operations? **Exponents.**

$\dfrac{(6x^9)^2}{(2y^4)^3} = \dfrac{36x^{18}}{8y^{12}}$ Power rule

$= \dfrac{\overset{9}{\cancel{36}}x^{18}}{\underset{2}{\cancel{8}}y^{12}}$ Divide out the common factor of 4.

$= \dfrac{9x^{18}}{2y^{12}}$

BE CAREFUL When simplifying the expression $\dfrac{(6x^9)^2}{(2y^4)^3}$ in Example 3, you may want to simplify before applying the product rule, like this:

$$\dfrac{\overset{3}{\cancel{6}}x^9)^2}{\underset{1}{\cancel{(2}}y^4)^3} \neq \dfrac{(3x^9)^2}{(y^4)^3} = \dfrac{9x^{18}}{y^{12}} \quad \text{Incorrect!}$$

This method does not follow the rules for the order of operations, so it does *not* give us the correct answer.

[YOU TRY 3] Simplify.

a) $-5(2c^5d^3)^4$ b) $(11r^8t)^2(-r^{10}t^6)^3$ c) $\dfrac{(10a^7b^5)^3}{(8c^3)^2}$

Is it possible to have an exponent of zero or a negative exponent? If so, what do they mean?

3 Use 0 and Negative Exponents

Definition
Zero as an Exponent: If $a \neq 0$, then $a^0 = 1$.

Why does $a^0 = 1$? Let's evaluate $2^0 \cdot 2^3$. Using the product rule we get:
$$2^0 \cdot 2^3 = 2^{0+3} = 2^3 = 8$$
Since $2^0 \cdot 2^3 = 2^3$, it must be true that $2^0 = 1$. This is one way to understand that $a^0 = 1$.

Note
0^0 is undefined.

EXAMPLE 4 Evaluate each expression.

a) 4^0 b) -9^0 c) $(-3)^0$ d) $(8t)^0$

Solution

a) $4^0 = 1$ b) $-9^0 = -1 \cdot 9^0 = -1 \cdot 1 = -1$

c) $(-3)^0 = 1$ d) $(8t)^0 = 1,\ t \neq 0$

[YOU TRY 4] Evaluate.

a) 5^0 b) -7^0 c) $(-6)^0$ d) $(2w)^0$

So far we have worked with exponents that are zero or positive. What does a negative exponent mean? Let's use the product rule to find $2^3 \cdot 2^{-3}$.

$$2^3 \cdot 2^{-3} = 2^{3+(-3)} = 2^0 = 1.$$

Remember that a number multiplied by its reciprocal is 1, and here we have that a quantity, 2^3, times another quantity, 2^{-3}, is 1. Therefore, 2^3 and 2^{-3} are reciprocals! This leads to the definition of a negative exponent.

> **Property** Negative Exponent
>
> If n is any integer and a and b are not equal to zero, then $a^{-n} = \left(\dfrac{1}{a}\right)^n = \dfrac{1}{a^n}$ and $\left(\dfrac{a}{b}\right)^{-n} = \left(\dfrac{b}{a}\right)^n.$

Therefore, to rewrite an expression of the form a^{-n} or $\left(\dfrac{a}{b}\right)^{-n}$ with a positive exponent, *take the reciprocal of the base and make the exponent positive.*

EXAMPLE 5

Simplify each expression so that the result contains only positive exponents. Assume all variables do not equal zero.

a) 5^{-3} b) $\left(\dfrac{8}{7}\right)^{-2}$ c) $\left(\dfrac{1}{2}\right)^{-6}$ d) $(-6)^{-2}$

e) k^{-5} f) $\left(\dfrac{3}{w}\right)^{-4}$ g) $4c^{-2}$

Solution

a) 5^{-3}: The reciprocal of 5 is $\dfrac{1}{5}$, so $5^{-3} = \left(\dfrac{1}{5}\right)^3 = \dfrac{1^3}{5^3} = \dfrac{1}{125}.$

b) $\left(\dfrac{8}{7}\right)^{-2}$: The reciprocal of $\dfrac{8}{7}$ is $\dfrac{7}{8}$, so $\left(\dfrac{8}{7}\right)^{-2} = \left(\dfrac{7}{8}\right)^2 = \dfrac{7^2}{8^2} = \dfrac{49}{64}.$

 BE CAREFUL Notice that a negative exponent does not make the answer negative!

c) $\left(\dfrac{1}{2}\right)^{-6}$: The reciprocal of $\dfrac{1}{2}$ is 2, so $\left(\dfrac{1}{2}\right)^{-6} = 2^6 = 64.$

d) $(-6)^{-2} = \left(-\dfrac{1}{6}\right)^2 = \left(-1 \cdot \dfrac{1}{6}\right)^2 = (-1)^2 \left(\dfrac{1}{6}\right)^2 = 1 \cdot \dfrac{1^2}{6^2} = \dfrac{1}{36}.$

e) $k^{-5} = \left(\dfrac{1}{k}\right)^5 = \dfrac{1^5}{k^5} = \dfrac{1}{k^5}$

f) $\left(\dfrac{3}{w}\right)^{-4} = \left(\dfrac{w}{3}\right)^4 = \dfrac{w^4}{3^4} = \dfrac{w^4}{81}$

g) $4c^{-2} = 4 \cdot \left(\dfrac{1}{c}\right)^2$ Note that the base is c, *not* $4c$.

$= 4 \cdot \dfrac{1}{c^2} = \dfrac{4}{c^2}$

[YOU TRY 5] Simplify each expression so that the result contains only positive exponents. Assume all variables do not equal zero.

a) $(11)^{-2}$ b) $\left(\dfrac{1}{10}\right)^{-3}$ c) $\left(\dfrac{2}{3}\right)^{-3}$ d) -5^{-3}

e) $\left(\dfrac{3}{t}\right)^{-4}$ f) z^{-6} g) $9n^{-5}$

How could we rewrite $\dfrac{x^{-2}}{y^{-2}}$ with only positive exponents? One way would be to apply the power rule for exponents: $\dfrac{x^{-2}}{y^{-2}} = \left(\dfrac{x}{y}\right)^{-2} = \left(\dfrac{y}{x}\right)^{2} = \dfrac{y^2}{x^2}$.

Let's do the same for $\dfrac{a^{-5}}{b^{-5}}$: $\dfrac{a^{-5}}{b^{-5}} = \left(\dfrac{a}{b}\right)^{-5} = \left(\dfrac{b}{a}\right)^{5} = \dfrac{b^5}{a^5}$.

Notice that to rewrite the original expression with only positive exponents, the terms with the negative exponents "switch" their positions in the fraction. We can generalize this way:

> **W Hint**
> Notice that in this definition, only the exponents change signs.

Definition

If m and n are any integers and a and b are real numbers not equal to zero, then

$$\dfrac{a^{-m}}{b^{-n}} = \dfrac{b^n}{a^m}$$

EXAMPLE 6 Rewrite the expression with positive exponents. Assume the variables do not equal zero.

a) $\dfrac{m^{-9}}{n^{-4}}$ b) $\dfrac{7r^{-5}}{t^3}$ c) $h^{-2}k^{-1}$ d) $\dfrac{8ab^{-6}}{5c^{-3}}$ e) $\left(\dfrac{xy}{3z}\right)^{-4}$

Solution

a) $\dfrac{m^{-9}}{n^{-4}} = \dfrac{n^4}{m^9}$ To make the exponents positive, "switch" the positions of the terms in the fraction.

b) $\dfrac{7r^{-5}}{t^3} = \dfrac{7}{r^5 t^3}$ Since the exponent on t is positive, do not change its position in the expression.

c) $h^{-2}k^{-1} = \dfrac{h^{-2}k^{-1}}{1}$

> **W Hint**
> Write out the example as you are reading it.

$= \dfrac{1}{h^2 k}$ Move $h^{-2}k^{-1}$ to the denominator to write with positive exponents.

d) $\dfrac{8ab^{-6}}{5c^{-3}} = \dfrac{8ac^3}{5b^6}$ To make the exponents positive, switch the positions of the variables with the negative exponents.

e) $\left(\dfrac{xy}{3z}\right)^{-4} = \left(\dfrac{3z}{xy}\right)^{4}$ To make the exponent positive, take the reciprocal of the base.

$= \dfrac{3^4 \cdot z^4}{x^4 y^4}$ Use the power rule.

$= \dfrac{81 z^4}{x^4 y^4}$ Simplify.

[YOU TRY 6] Rewrite the expression with positive exponents. Assume the variables do not equal zero.

a) $\dfrac{a^{-2}}{b^{-7}}$ b) $\dfrac{d^{-8}}{2c^4}$ c) $6p^{-5}q^{-1}$ d) $\dfrac{10x^2 y^{-9}}{3z^{-8}}$ e) $\left(\dfrac{ab}{5c}\right)^{-3}$

4 Use the Quotient Rule for Exponents

Next, we will discuss how to simplify the quotient of two exponential expressions with the same base. Let's begin by simplifying $\dfrac{9^6}{9^4}$. One way we can do this is by writing the numerator and denominator without exponents:

$$\dfrac{9^6}{9^4} = \dfrac{\cancel{9}\cdot\cancel{9}\cdot\cancel{9}\cdot\cancel{9}\cdot 9 \cdot 9}{\cancel{9}\cdot\cancel{9}\cdot\cancel{9}\cdot\cancel{9}} \quad \text{Divide out common factors.}$$

$$= 9 \cdot 9 = 9^2 = 81$$

So, $\dfrac{9^6}{9^4} = 9^2$. Do you notice a relationship among the exponents? That's right. We *subtracted* the exponents: $6 - 4 = 2$.

> **Property** Quotient Rule for Exponents
>
> If m and n are any integers and $a \neq 0$, then
>
> $$\dfrac{a^m}{a^n} = a^{m-n}$$

Notice that the base in the numerator and denominator is a. *In order to apply the quotient rule, the bases must be the same. Subtract the exponent of the denominator from the exponent of the numerator.*

EXAMPLE 7 Simplify and write with positive exponents.

a) $\dfrac{2^{10}}{2^4}$ b) $\dfrac{w^{13}}{w^5}$ c) $\dfrac{3}{3^{-2}}$ d) $\dfrac{k^3}{k^{12}}$ e) $\dfrac{7^2}{2^3}$ f) $\dfrac{25m^{-8}n^7}{15m^{-5}n^2}$

Solution

a) $\dfrac{2^{10}}{2^4} = 2^{10-4} = 2^6 = 64$ The bases are the same, so subtract the exponents.

b) $\dfrac{w^{13}}{w^5} = w^{13-5} = w^8$ Because the bases are the same, subtract the exponents.

c) $\dfrac{3}{3^{-2}} = \dfrac{3^1}{3^{-2}} = 3^{1-(-2)}$ Since the bases are the same, subtract the exponents.

$\qquad\qquad = 3^3 = 27$ Be careful when subtracting the negative exponent!

d) $\dfrac{k^3}{k^{12}} = k^{3-12} = k^{-9}$ Same base, subtract the exponents.

$\qquad = \left(\dfrac{1}{k}\right)^9 = \dfrac{1}{k^9}$ Write with a positive exponent.

e) $\dfrac{7^2}{2^3} = \dfrac{49}{8}$ Because the bases are not the same, we cannot apply the quotient rule. Evaluate the numerator and denominator separately.

f) $\dfrac{25m^{-8}n^7}{15m^{-5}n^2} = \dfrac{\overset{5}{\cancel{25}}m^{-8}n^7}{\underset{3}{\cancel{15}}m^{-5}n^2}$ Simplify $\dfrac{25}{15}$.

$= \dfrac{5}{3}m^{-8-(-5)}n^{7-2}$ Apply the quotient rule.

$= \dfrac{5}{3}m^{-3}n^5$ Subtract.

$= \dfrac{5n^5}{3m^3}$ Write with only positive exponents.

[YOU TRY 7] Simplify and write with positive exponents.

a) $\dfrac{10^9}{10^6}$ b) $\dfrac{v^6}{v^{-2}}$ c) $\dfrac{x^3}{x^7}$ d) $\dfrac{2^8}{2^{11}}$ e) $\dfrac{16a^{11}b^{-5}}{40a^3b^{-2}}$

In Section 6.2 we will learn more about combining the rules of exponents.

ANSWERS TO [YOU TRY] EXERCISES

1) a) 81 b) q^9 c) $-24d^9$ d) w^{14} e) w^3k^5 f) $24x^3y^{14}$ 2) a) 5^{12} b) j^{30} c) $1000p^3$
d) $16k^{28}$ e) $\dfrac{25}{144}$ f) $-\dfrac{64w^3}{y^{27}}$ 3) a) $-80c^{20}d^{12}$ b) $-121r^{46}t^{20}$ c) $\dfrac{125a^{21}b^{15}}{c^6}$
4) a) 1 b) -1 c) 1 d) 1   5) a) $\dfrac{1}{121}$ b) 1000 c) $\dfrac{27}{8}$ d) $-\dfrac{1}{125}$ e) $\dfrac{t^4}{81}$ f) $\dfrac{1}{z^6}$ g) $\dfrac{9}{n^5}$
6) a) $\dfrac{b^7}{a^2}$ b) $\dfrac{1}{2c^4d^8}$ c) $\dfrac{6}{p^5q}$ d) $\dfrac{10x^2z^8}{3y^9}$ e) $\dfrac{125c^3}{a^3b^3}$ 7) a) 1000 b) v^8 c) $\dfrac{1}{x^4}$ d) $\dfrac{1}{8}$ e) $\dfrac{2a^8}{5b^3}$

E Evaluate 6.1 Exercises   Do the exercises, and check your work.

Objective 1: Use the Product Rule for Exponents

Identify the base and the exponent in each.

1) $(-7r)^5$

2) $-\dfrac{4}{5}h^3$

Rewrite each expression using exponents.

3) $9 \cdot 9 \cdot 9 \cdot 9 \cdot 9 \cdot 9$

4) $\left(-\dfrac{2}{7}m\right)\left(-\dfrac{2}{7}m\right)\left(-\dfrac{2}{7}m\right)\left(-\dfrac{2}{7}m\right)$

5) Evaluate $(3+4)^2$ and $3^2 + 4^2$. Are they equivalent? Why or why not?

6) For any values of a and b, does $(a+b)^2 = a^2 + b^2$? Why or why not?

7) Does $-3^4 = (-3)^4$? Why or why not?

8) Are $2k^4$ and $(2k)^4$ equivalent? Why or why not?

Evaluate the expression using the product rule, where applicable.

9) $2^2 \cdot 2^4$

10) $10^2 \cdot 10$

11) $3^3 \cdot 5^2$

12) $\left(\dfrac{1}{2}\right)^4 \cdot \left(\dfrac{1}{2}\right)$

Simplify the expression using the product rule. Leave your answer in exponential form.

13) $(-4)^2 \cdot (-4)^3 \cdot (-4)^2$ 14) $(8) \cdot (8)^7 \cdot (8)^2$

15) $a^2 \cdot a^3$ 16) $9w^4 \cdot w^3$

17) $k \cdot k^2 \cdot k^3$ 18) $-6z^5 \cdot z^6 \cdot z$

19) $(2p^2)(-12p^5)$ 20) $(7h^3)(8h^{12})$

21) $(5n^3)(-6n^7)(2n^2)$ 22) $\left(\dfrac{8}{5}y^3\right)(35y)(-2y^4)$

23) $\left(\dfrac{49}{24}t^5\right)(-4t^7)\left(-\dfrac{12}{7}t\right)$ 24) $(6v^9)\left(\dfrac{5}{8}v^3\right)\left(\dfrac{4}{15}v^4\right)$

www.mhhe.com/messersmith

Objective 2: Use the Power Rules for Exponents
Simplify the expression using one of the power rules.

25) $(a^9)^4$
26) $(w^5)^9$
27) $(2^3)^2$
28) $(3^2)^2$
29) $\left(\dfrac{1}{2}\right)^5$
30) $\left(\dfrac{3}{2}\right)^4$
31) $\left(\dfrac{4}{y}\right)^3$
32) $\left(\dfrac{w}{k}\right)^6$
33) $(-10r)^4$
34) $(2g)^5$
35) $(-4ab)^3$
36) $(-7mn)^2$

Mixed Exercises: Objectives 1 and 2
Simplify using the product and power rules.

37) $(k^9)^2(k^3)^2$
38) $(p^4)^2(p^7)^3$
39) $(5+3)^2$
40) $(11-8)^2$
41) $8(6k^7l^2)^2$
42) $2(-6a^5b)^2$
43) $-m^4(-5m^2)^3(-m^6)^2$
44) $4h^9(11h^5)^2(-h^4)^3$
45) $\dfrac{(4d^9)^2}{(-2c^5)^6}$
46) $\dfrac{(-5m^7)^3}{(5n^{12})^2}$
47) $\dfrac{6(a^8b^3)^5}{(2c)^3}$
48) $\dfrac{(3x^8)^3}{15(y^2z^3)^4}$
49) $(8u^3v^8)^2\left(-\dfrac{13}{4}uv^5\right)^2$
50) $\left(-\dfrac{3}{4}h^7k^2\right)^3\left(\dfrac{2}{9}h^4k\right)^2$
51) $\left(\dfrac{4r^6s^2}{t^3}\right)^3$
52) $\left(-\dfrac{5x^3z}{12y^8}\right)^2$
53) $\left(\dfrac{36n^4}{4m^8p^3}\right)^2$
54) $\left(\dfrac{21b^7}{14a^5c^2}\right)^4$

Objective 3: Use 0 and Negative Exponents
In Exercises 55–58, decide whether each statement is true or false.

55) Raising a positive base to a negative exponent will give a negative result. (Example: 5^{-2})
56) $9^0 = 0$
57) The reciprocal of 8 is $\dfrac{1}{8}$.
58) $4^{-3} + 2^{-3} = 6^{-3}$

Evaluate. Assume the variable does not equal zero.

59) 6^0
60) -12^0
61) $-(-7)^0$
62) 0^4
63) $(11)^0 + (-11)^0$
64) w^0
65) $-6y^0$
66) $x^0 - (9x)^0$
67) 6^{-2}
68) 2^{-4}
69) $\left(\dfrac{1}{7}\right)^{-2}$
70) $\left(\dfrac{1}{5}\right)^{-3}$
71) $\left(\dfrac{4}{3}\right)^{-3}$
72) $\left(\dfrac{3}{10}\right)^{-3}$
73) $\left(-\dfrac{1}{2}\right)^{-5}$
74) $\left(-\dfrac{6}{11}\right)^{-2}$
75) -2^{-6}
76) -4^{-3}
77) $2^{-3} - 4^{-2}$
78) $5^{-2} + 2^{-2}$
79) $-9^{-2} + 3^{-3} + (-7)^0$
80) $6^0 - 9^{-1} + 4^0 + 3^{-2}$

Rewrite each expression with only positive exponents. Assume the variables do not equal zero.

81) y^{-4}
82) c^{-1}
83) $\dfrac{a^{-10}}{b^{-3}}$
84) $\dfrac{u^{-6}}{v^{-2}}$
85) $\dfrac{x^4}{10y^{-5}}$
86) $\dfrac{7t^{-3}}{u^6}$
87) $8x^3y^{-7}$
88) $\dfrac{1}{2}r^{-5}t^3$
89) $\dfrac{8a^6b^{-1}}{5c^{-10}d}$
90) $\dfrac{1}{h^{-7}k^{-4}}$
91) $\left(\dfrac{a}{6}\right)^{-2}$
92) $\left(\dfrac{3}{q}\right)^{-4}$
93) $\left(\dfrac{12b}{cd}\right)^{-2}$
94) $\left(\dfrac{2xy}{z}\right)^{-6}$
95) $-6r^{-2}$
96) $2w^{-5}$
97) $-p^{-8}$
98) $-k^{-7}$
99) $\left(\dfrac{1}{x}\right)^{-1}$
100) $a^4\left(\dfrac{1}{b}\right)^{-9}$

Objective 4: Use the Quotient Rule for Exponents
Simplify using the quotient rule. Assume the variables do not equal zero.

101) $\dfrac{n^9}{n^4}$
102) $\dfrac{h^8}{h}$
103) $\dfrac{8^{11}}{8^9}$
104) $\dfrac{2^{15}}{2^{10}}$
105) $\dfrac{5^6}{5^9}$
106) $\dfrac{3^7}{3^{11}}$
107) $\dfrac{40w^8}{72w^3}$
108) $\dfrac{12d^6}{28d}$
109) $\dfrac{t^3}{t^5}$
110) $\dfrac{a^2}{a^9}$
111) $\dfrac{x^{-3}}{x^6}$
112) $\dfrac{h^{-10}}{h^{-2}}$
113) $\dfrac{-6k}{k^4}$
114) $\dfrac{15m^3}{m^{-1}}$
115) $\dfrac{6v^{-1}w}{54v^2w^{-5}}$
116) $\dfrac{54ab^{-2}}{36a^5b^{-7}}$
117) $\dfrac{20m^{-3}n^4}{4m^8n^6}$
118) $\dfrac{3c^{-4}d}{7c^{-1}d^4}$
119) $\dfrac{(x+y)^8}{(x+y)^3}$
120) $\dfrac{(a-3b)^{-4}}{(a-3b)^{-5}}$

R Rethink

R1) If a variable is raised to the zero power, the answer will always be what number? Assume the variable does not equal zero.

R2) Why is it useful to write expressions without negative exponents?

R3) When evaluating an expression with negative exponents, when do you end up with a negative answer?

R4) Which exercises in this section do you find most challenging?

6.2 More on Exponents and Scientific Notation

P Prepare / O Organize

What are your objectives for Section 6.2?	How can you accomplish each objective?
1 Combine the Rules of Exponents	• Review the **Rules of Exponents,** and summarize them in your notes. • Complete the given examples on your own. • Complete You Trys 1 and 2.
2 Understand Scientific Notation	• Learn the definition of *scientific notation*. • Be able to recognize when a number is in *scientific notation*. • Complete the given examples on your own. • Complete You Trys 3 and 4.
3 Convert a Number from Scientific Notation to Standard Form	• Make sure to be careful when moving the decimal place. • Know when to move the decimal point to the right or to the left. • Complete the given example on your own. • Complete You Try 5.
4 Convert a Number from Standard Form to Scientific Notation	• Learn the procedure on **How to Write a Number in Scientific Notation.** • Complete the given example on your own. • Complete You Try 6.
5 Perform Operations on Numbers in Scientific Notation	• Use the **Properties of Multiplication** and **Exponent Rules** to simplify numbers and expressions. • Complete the given example on your own. • Complete You Try 7.

W Work Read the explanations, follow the examples, take notes, and complete the You Trys.

1 Combine the Rules of Exponents

Let's review all the rules of exponents and then see how we can combine the rules to simplify expressions.

> **W Hint**
> Summarize these rules in your notes.

Summary Rules of Exponents

In the rules below, a and b are any real numbers and m and n are integers.

Product rule	$a^m \cdot a^n = a^{m+n}$
Basic power rule	$(a^m)^n = a^{mn}$
Power rule for a product	$(ab)^n = a^n b^n$
Power rule for a quotient	$\left(\dfrac{a}{b}\right)^n = \dfrac{a^n}{b^n},\ (b \neq 0)$
Quotient rule	$\dfrac{a^m}{a^n} = a^{m-n},\ (a \neq 0)$

Change from negative to positive exponents, where $a \neq 0$, $b \neq 0$, and m and n are any integers.

$$\dfrac{a^{-m}}{b^{-n}} = \dfrac{b^n}{a^m}, \quad \left(\dfrac{a}{b}\right)^{-n} = \left(\dfrac{b}{a}\right)^n$$

In the definitions below, $a \neq 0$ and n is any integer.

Zero as an exponent	$a^0 = 1$
Negative number as an exponent	$a^{-n} = \left(\dfrac{1}{a}\right)^n = \dfrac{1}{a^n}$

EXAMPLE 1 Simplify using the rules of exponents. Assume all variables represent nonzero real numbers. Write the answers with positive exponents.

a) $(5k^{-7})^3(2k^6)^2$ b) $\left(\dfrac{9a^8 b^5}{a^3 b^2}\right)^2$ c) $\dfrac{h^{-4} \cdot h^5}{h^6}$ d) $\left(\dfrac{8x^{-3}y^{10}}{20xy^{-3}}\right)^{-3}$

Solution

a) We must follow the order of operations. Therefore, evaluate the exponents first.

$$\begin{aligned}(5k^{-7})^3 \cdot (2k^6)^2 &= 5^3 k^{(-7)(3)} \cdot 2^2 k^{(6)(2)} &&\text{Apply the power rule.}\\ &= 125k^{-21} \cdot 4k^{12} &&\text{Simplify.}\\ &= 500k^{-21+12} &&\text{Multiply } 125 \cdot 4, \text{ and add the exponents.}\\ &= 500k^{-9} &&\text{Add the exponents.}\\ &= \dfrac{500}{k^9} &&\text{Write the answer using a positive exponent.}\end{aligned}$$

b) How can we begin this problem? We can use the quotient rule to simplify the expression before squaring it.

> **W Hint**
> Write out the example as you are reading it!

$$\begin{aligned}\left(\dfrac{9a^8 b^5}{a^3 b^2}\right)^2 &= (9a^{8-3}b^{5-2})^2 &&\text{Apply the quotient rule inside the parentheses.}\\ &= (9a^5 b^3)^2 &&\text{Simplify.}\\ &= 9^2 a^{(5)(2)} b^{(3)(2)} &&\text{Apply the product rule.}\\ &= 81a^{10} b^6\end{aligned}$$

304 CHAPTER 6 Polynomials and Polynomial Functions www.mhhe.com/messersmith

c) Let's begin by simplifying the numerator.

$$\frac{h^{-4} \cdot h^5}{h^6} = \frac{h^{-4+5}}{h^6} \quad \text{Add the exponents in the numerator.}$$

$$= \frac{h^1}{h^6}$$

Now, we can apply the quotient rule:

$$= h^{1-6} = h^{-5} \quad \text{Subtract the exponents.}$$

$$= \frac{1}{h^5} \quad \text{Write the answer using a positive exponent.}$$

d) Eliminate the negative exponent **outside** of the parentheses by taking the reciprocal of the base. Notice that we have **not** eliminated the negatives on the exponents **inside** the parentheses.

$$\left(\frac{8x^{-3}y^{10}}{20xy^{-3}}\right)^{-3} = \left(\frac{20xy^{-3}}{8x^{-3}y^{10}}\right)^{3}$$

We could apply the exponent of 3 to the quantity inside of the parentheses, but we could also simplify $\frac{20}{8}$ first and apply the quotient rule before cubing the quantity.

$$= \left(\frac{5}{2}x^{1-(-3)}y^{-3-10}\right)^3 \quad \text{Simplify } \frac{20}{8}, \text{ and subtract the exponents.}$$

$$= \left(\frac{5}{2}x^4 y^{-13}\right)^3$$

$$= \frac{125}{8}x^{12}y^{-39} \quad \text{Apply the power rule.}$$

$$= \frac{125x^{12}}{8y^{39}} \quad \text{Write the answer using positive exponents.}$$

[**YOU TRY 1**] Simplify using the rules of exponents. Assume all variables represent nonzero real numbers. Write the answers with positive exponents.

a) $\left(\dfrac{2r^{11}t^6}{r^5 t}\right)^5$ b) $(-w^{-3})^6(11w^4)^2$ c) $\dfrac{z^8 \cdot z^{-10}}{z^2}$ d) $\left(\dfrac{24a^{12}b^{-4}}{21a^7 b}\right)^{-2}$

It is possible for variables to appear in exponents. The same rules apply.

EXAMPLE 2

Simplify using the rules of exponents. Assume all variables represent nonzero integers. Write your final answer so that the exponents have positive coefficients.

a) $n^{3x} \cdot n^{2x}$ b) $\dfrac{a^{4b}}{a^{8b}}$

Solution

a) $n^{3x} \cdot n^{2x} = n^{3x+2x} = n^{5x}$ The bases are the same, so apply the product rule. Add the exponents.

b) $\dfrac{a^{4b}}{a^{8b}} = a^{4b-8b}$ The bases are the same, so apply the quotient rule. Subtract the exponents.

$= a^{-4b}$

$= \dfrac{1}{a^{4b}}$ Write the exponent with a positive coefficient.

[YOU TRY 2] Simplify using the rules of exponents. Assume all variables represent nonzero integers. Write your final answer so that the exponents have positive coefficients.

a) $6^{5p} \cdot 6^p \cdot 6^{8p}$ b) $(d^4)^{-5k}$

2 Understand Scientific Notation

The distance from the Earth to the Sun is approximately 150,000,000 km.

The gross domestic product of the United States in 2011 was $15,094,000,000,000.
(databank.worldbank.org)

A single rhinovirus (cause of the common cold) measures 0.00002 mm across.

Each of these is an example of a very large or very small number containing many zeros. Sometimes, performing operations with so many zeros can be difficult. This is why scientists and economists, for example, often work with such numbers in a different form called *scientific notation*.

Scientific notation is a short-hand method for writing very large and very small numbers. Writing numbers in scientific notation together with applying the rules of exponents can simplify calculations with very large and very small numbers.

Before discussing scientific notation further, we will review some principles behind the notation. Let's look at multiplying numbers by positive powers of 10.

EXAMPLE 3

Multiply.

a) 0.0538×10^3 b) $76 \cdot 10^2$

Solution

a) $0.0538 \times 10^3 = 0.0538 \times 1000 = 53.8$

b) $76 \cdot 10^2 = 76 \cdot 100 = 7600$

Notice that when we multiply a positive number by a positive power of 10, the result is *larger* than the original number. In fact, the exponent determines how many places to the *right* the decimal point is moved.

$0.0538 \times 10^3 = 53.8$ (3 places to right) $76 \cdot 10^2 = 76.00 \cdot 10^2 = 7600$ (2 places to right)

W Hint
When multiplying a number by a positive power of 10, the exponent tells you how many places to move the decimal to the right.

[YOU TRY 3] Multiply by moving the decimal point the appropriate number of places.

a) 6.44×10^5 b) $0.000937 \cdot 10^4$

What happens to a positive number when we multiply by a *negative* power of 10?

EXAMPLE 4

Multiply.

a) $59 \cdot 10^{-2}$ b) 138×10^{-4}

Solution

a) $59 \cdot 10^{-2} = 59 \cdot \dfrac{1}{100} = \dfrac{59}{100} = 0.59$

CHAPTER 6 Polynomials and Polynomial Functions

Hint
When multiplying a number by a negative power of 10, the negative exponent tells you how many places to move the decimal to the left.

b) $138 \times 10^{-4} = 138 \times \dfrac{1}{10,000} = \dfrac{138}{10,000} = 0.0138$

Is there a pattern? When we multiply a positive number by a negative power of 10, the result is *smaller* than the original number. The exponent determines how many places to the *left* the decimal point is moved:

$59 \cdot 10^{-2} = .59. \cdot 10^{-2} = 0.59$ $138 \times 10^{-4} = .0138. \times 10^{-4} = 0.0138$
2 places to the left   4 places to the left

[YOU TRY 4] Multiply.

a) $61 \cdot 10^{-3}$ b) 4.9×10^{-4}

Definition

A number is in **scientific notation** if it is written in the form $a \times 10^n$ where $1 \leq |a| < 10$ and n is an integer.

Multiplying $|a|$ by a *positive* power of 10 will result in a number that is *larger* than $|a|$. Multiplying $|a|$ by a *negative* power of 10 will result in a number that is *smaller* than $|a|$. $1 \leq |a| < 10$ means that a is a number that has *one* nonzero digit to the left of the decimal point.

Some examples of numbers written in scientific notation are

$9.15 \times 10^{-5}, \quad 8.1 \cdot 10^3, \quad 6 \times 10^{-2}.$

The following numbers are *not* in scientific notation:

26.31×10^5 0.87×10^{-4} $400 \cdot 10^2$
↑   ↑   ↑
2 digits to left   Zero is to left   3 digits to left
of decimal point   of decimal point.   of decimal point

3 Convert a Number from Scientific Notation to Standard Form

We will continue our discussion by converting from scientific notation to standard form, that is, a number without exponents.

EXAMPLE 5   Rewrite in standard form.

a) 7.094×10^4 b) 2.6×10^{-3} c) $8.1163 \cdot 10^3$

Solution

a) $7.094 \times 10^4 \to 7.0940. = 70,940$ Remember, multiplying a positive number by a positive
                     4 places to the right   power of 10 will make the result *larger* than 7.094.

b) $2.6 \times 10^{-3} \to .002.6 = 0.0026$ Multiplying 2.6 by a negative power of 10 will make
                    3 places to the left   the result *smaller* than 2.6.

c) $8.1163 \cdot 10^3 \to 8.1163. = 8116.3$
                  3 places to the right

[YOU TRY 5] Rewrite in standard form.

a) $6.09 \cdot 10^4$ b) $1.4 \cdot 10^{-5}$ c) 5.02147×10^3

4 Convert a Number from Standard Form to Scientific Notation

To write the number 25,000 in scientific notation, first locate its decimal point.

$$25,000.$$
↑
Decimal point is here.

Next, determine where the decimal point will be when the number is in scientific notation.

$$2500.$$
∧
Decimal point will be here.

In scientific notation, 25,000 must be 2.5×10^n, where n is an integer. Will n be positive or negative? The number 2.5 must be multiplied by a *positive* power of 10 to make it larger.

Count the number of places between the original and the final decimal place locations.

$$\underbrace{25000}_{4\,3\,2\,1}$$

Use the number of places, 4, as the exponent of 10.

$$25,000 = 2.5 \times 10^4 \text{ or } 2.5 \cdot 10^4$$

EXAMPLE 6

Write each number in scientific notation.

Solution

a) The distance from the Earth to the Sun is approximately 150,000,000 km.

150,000,000.   150,000,000   Move decimal point eight places.
∧   ↖
Decimal point   Decimal point
will be here.   is here.

$$150,000,000 \text{ km} = 1.5 \times 10^8 \text{ km or } 1.5 \cdot 10^8 \text{ km}.$$

b) A single rhinovirus measures 0.00002 mm across.

0.00002 mm
∧
Decimal point
will be here.

$$0.00002 \text{ mm} = 2 \times 10^{-5} \text{ mm or } 2 \cdot 10^{-5} \text{ mm}.$$

Procedure How to Write a Number in Scientific Notation

1) Locate the decimal point in the original number.
2) Determine where the decimal point needs to be when converting to scientific notation. Remember, there will be one nonzero digit to the left of the decimal point.
3) Count how many places you must move the decimal point to take it from its original place to its position for scientific notation.
4) If the absolute value of the resulting number is *smaller* than the absolute value of the original number, you will multiply the result by a *positive* power of 10. *Example:* $350.9 = 3.509 \times 10^2$ or $3.509 \cdot 10^2$.

    If the absolute value of the resulting number is *larger* than the absolute value of the original number, you will multiply the result by a *negative* power of 10. *Example:* $0.0000068 = 6.8 \times 10^{-6}$ or $6.8 \cdot 10^{-6}$.

[YOU TRY 6] Write each number in scientific notation.

a) The gross domestic product of the United States in 2011 was approximately $15,094,000,000,000.

b) The diameter of a human hair is approximately 0.001 in.

---

5 Perform Operations on Numbers in Scientific Notation

We use the rules of exponents to perform operations with numbers in scientific notation.

EXAMPLE 7 Perform the operations and simplify.

a) $(-2 \cdot 10^3)(3 \cdot 10^2)$ b) $\dfrac{9 \times 10^3}{2 \times 10^5}$

Solution

a) $(-2 \cdot 10^3)(3 \cdot 10^2) = (-2 \cdot 3)(10^3 \cdot 10^2)$ Commutative property
$= -6 \cdot 10^5$ Add the exponents.
$= -600,000$

b) $\dfrac{9 \times 10^3}{2 \times 10^5} = \dfrac{9}{2} \times \dfrac{10^3}{10^5}$

$= 4.5 \times 10^{-2} = 0.045$ Subtract the exponents.
         ↑
Write $\dfrac{9}{2}$ in decimal form.

[YOU TRY 7] Perform the operations and simplify.

a) $(4.1 \times 10^2)(3 \times 10^4)$ b) $\dfrac{5.2 \cdot 10^{-10}}{4 \cdot 10^{-6}}$

Using Technology

We can use a graphing calculator to convert a very large or very small number to scientific notation, or to convert a number in scientific notation to a number written without an exponent. Suppose we are given a very large number such as 35,000,000,000. If you enter any number with more than 10 digits on the home screen of your calculator and press ENTER the number will automatically be displayed in scientific notation as shown on the screen below. A small number with more than two zeros to the right of the decimal point (such as .000123) will automatically be displayed in scientific notation as shown below.

The E shown in the screen refers to a power of 10, so 3.5 E 10 is the number 3.5×10^{10} in scientific notation. 1.23 E-4 is the number 1.23×10^{-4} in scientific notation.

If a large number has 10 or fewer digits, or if a small number has fewer than three zeros to the right of the decimal point, then the number will not automatically be displayed in scientific notation. To display the number using scientific notation press MODE, select SCI, and press ENTER. When you return to the home screen, all numbers will be displayed in scientific notation as shown below.

 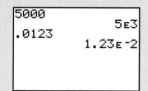

A number written in scientific notation can be entered directly into your calculator. For example, the number 2.38×10^7 can be entered directly on the home screen by typing 2.38 followed by 2nd , 7 ENTER as shown to the right. If you wish to display this number without an exponent, change the mode back to NORMAL and enter the number on the home screen as shown.

Write each number without an exponent using a graphing calculator.

1) 3.14×10^5 2) 9.3×10^7 3) 1.38×10^{-3}

Write each number in scientific notation using a graphing calculator.

4) 186,000   5) 5280   6) 0.0469

ANSWERS TO YOU TRY EXERCISES

1) a) $32r^{30}t^{25}$ b) $\dfrac{121}{w^{10}}$ c) $\dfrac{1}{z^4}$ d) $\dfrac{49b^{10}}{64a^{10}}$ 2) a) 6^{14p} b) $\dfrac{1}{d^{20k}}$ 3) a) 644,000 b) 9.37
4) a) 0.061 b) 0.00049   5) a) 60,900 b) 0.000014 c) 5021.47   6) a) 1.5094×10^{13} dollars b) 1.0×10^{-3} in.   7) a) 12,300,000 b) 0.00013

ANSWERS TO TECHNOLOGY EXERCISES

1) 314,000   2) 93,000,000   3) 0.00138   4) 1.86×10^5 5) 5.28×10^3 6) 4.69×10^{-2}

E Evaluate 6.2 Exercises

Do the exercises, and check your work.

Objective 1: Combine the Rules of Exponents

Simplify. Assume all variables represent nonzero real numbers. Write the answers with positive exponents.

1) $-10(-3g^4)^3$
2) $6(2d^3)^3$
3) $\dfrac{23t}{t^{11}}$
4) $\dfrac{r^{-6}}{r^{-2}}$
5) $\left(\dfrac{2xy^4}{3x^{-9}y^{-2}}\right)^3$
6) $\left(\dfrac{a^8 b^3}{10a^5}\right)^3$
7) $\left(\dfrac{7r^3}{s^8}\right)^{-2}$
8) $\left(\dfrac{3n^{-6}}{m^2}\right)^{-4}$
9) $(-k^4)^3$
10) $(t^7)^8$
11) $(-2m^5n^2)^5$
12) $(13yz^6)^2$
13) $(20w^2)\left(-\dfrac{3}{5}w^6\right)$
14) $\left(-\dfrac{9}{4}z^5\right)\left(\dfrac{2}{3}z^{-1}\right)$
15) $\left(\dfrac{a^5}{b^4}\right)^{-6}$
16) $\dfrac{x^{-4}}{y^{10}}$
17) $(-ab^3c^5)^2\left(\dfrac{a^4}{bc}\right)^3$
18) $\dfrac{(4v^3)^2}{(6v^8)^2}$
19) $\left(\dfrac{48u^{-7}v^2}{36u^3v^{-5}}\right)^{-3}$
20) $\left(\dfrac{xy^5}{5x^{-2}y}\right)^{-3}$
21) $\left(\dfrac{-3t^4u}{t^2u^{-4}}\right)^4$
22) $\left(\dfrac{k^7m^7}{12k^{-1}m^6}\right)^2$
23) $(h^{-3})^7$
24) $(-n^4)^{-5}$
25) $\left(\dfrac{h}{2}\right)^5$
26) $17m^{-2}$
27) $-7c^8(-2c^2)^3$
28) $5p^3(4p^7)^2$
29) $(12a^7)^{-1}(6a)^2$
30) $(9c^2d)^{-1}$
31) $\left(\dfrac{9}{20}d^5\right)(2d^{-3})\left(\dfrac{4}{33}d^9\right)$
32) $\left(\dfrac{f^8 \cdot f^{-3}}{f^2 \cdot f^9}\right)^5$
33) $\left(\dfrac{56m^4n^8}{21m^4n^5}\right)^{-2}$
34) $\dfrac{(2x^4y)^{-2}}{(5xy^3)^2}$
35) $\dfrac{(-r^{-1}t^2)^{-5}}{(2r^7t^{-1})^{-5}}$
36) $(-10p^{-14})\left(\dfrac{2}{3}p^5\right)\left(\dfrac{9}{20}p^3\right)$
37) $\left(\dfrac{c^5}{6a^4b^{-1}}\right)^{-2}\left(\dfrac{a^{-8}c}{2b^3}\right)^5\left(\dfrac{3b^5c^{-2}}{a}\right)^{-3}$
38) $\left(\dfrac{y^{-3}}{5xz^{-1}}\right)^{-1}\left(\dfrac{8x^{-3}}{z^5}\right)^{-2}\left(-\dfrac{2z^{-6}}{x^{-4}y}\right)^3$
39) $\dfrac{(x+y)^4}{(x+y)^9}$
40) $\dfrac{(s-t)^{10}}{(s-t)^3}$

Simplify. Assume that the variables represent nonzero integers. Write your final answer so that the exponents have positive coefficients.

41) $(p^{2n})^5$
42) $(3d^{4t})^2$
43) $y^m \cdot y^{10m}$
44) $t^{-6c} \cdot t^{9c}$
45) $x^{5a} \cdot x^{-8a}$
46) $b^{-2y} \cdot b^{-3y}$
47) $\dfrac{21c^{2x}}{35c^{8x}}$
48) $-\dfrac{5y^{-13a}}{8y^{-2a}}$

49) Find the area.

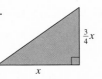

50) The area of a rectangle is $24n^5$ in^2. Its width is $6n^3$ in. Find the length.

Objective 2: Understand Scientific Notation

Determine whether each number is in scientific notation.

51) 4.73×10^3
52) -93×10^{-5}
53) 0.14×10^{-6}
54) 36.0×10^{-2}
55) $-4.3 \cdot 10^7$
56) $-2 \cdot 10^5$
57) 3.5×2^{-8}

58) Explain, in your own words, how to determine if a number is expressed in scientific notation.

59) Explain, in your own words, how to write 5.3×10^{-2} in standard form.

60) Explain, in your own words, how to write 8.76×10^6 in standard form.

Objective 3: Convert a Number from Scientific Notation to Standard Form

Write each number without an exponent.

61) -6.8×10^{-5}
62) 2.91×10^4
63) 3.45029×10^4
64) $-3.57 \cdot 10^3$
65) $-5 \cdot 10^{-5}$
66) $9 \cdot 10^{-7}$
67) 8.1×10^{-4}
68) 3×10^6
69) 2.645067×10^4
70) 7×10^2
71) $-3.921 \cdot 10^{-2}$
72) $4.1 \cdot 10^{-6}$

Objective 4: Convert a Number from Standard Form to Scientific Notation

Write each number in scientific notation.

73) 2110.5

74) 382.275

75) 0.0048

76) 0.000321

77) −400,000

78) 92,600

79) 11,000

80) −308,000

81) 0.0008

82) −0.00000089

83) −0.054

84) 9990

85) 6500

86) 0.00000002

Write each number in scientific notation.

87) In 1883 it cost approximately $15,000,000 to build the Brooklyn Bridge. (www.nycroads.com)

88) A typical music player may hold approximately 16,384,000 bytes of data.

89) The diameter of an atom is about 0.00000001 cm.

90) The oxygen-hydrogen bond-length in a water molecule is 0.000000001 mm.

Objective 5: Perform Operations on Numbers in Scientific Notation

Perform the operation as indicated. Write the final answer without an exponent.

91) $\dfrac{8 \cdot 10^6}{4 \cdot 10^2}$

92) $(9 \times 10^5)(3 \times 10^4)$

93) $(2.3 \times 10^3)(3 \times 10^2)$

94) $\dfrac{12 \cdot 10^7}{4 \cdot 10^4}$

95) $\dfrac{-8.8 \times 10^{-3}}{-2.2 \times 10^{-8}}$

96) $\dfrac{9.6 \times 10^{11}}{-4 \times 10^4}$

97) $(-1.5 \times 10^{-6})(6 \times 10^2)$

98) $(-4 \times 10^{-3})(-3.5 \times 10^{-1})$

99) $\dfrac{(0.004)(600,000)}{0.0003}$

100) $\dfrac{360}{(0.006)(2,000,000)}$

For each exercise, express each number in scientific notation, then solve.

101) When one of the U.S. space shuttles enters orbit, it travels at about 7800 m/s. How far does it travel in 24 hours? (Hint: Change hours to seconds, and write all numbers in scientific notation before doing the computations.) (hypertextbook.com)

102) According to Nielsen Media Research, over 92,000,000 people watched Super Bowl XLIII in 2009 between the Pittsburgh Steelers and the Arizona Cardinals. The California Avocado Commission estimates that about 46,000,000 pounds of avocados were eaten during that Super Bowl, mostly in the form of guacamole. On average, how many pounds of guacamole did each viewer eat during the Super Bowl?

103) In 2007, the United States produced about 6×10^9 metric tons of carbon emissions. The U.S. population that year was about 300 million. Find the amount of carbon emissions produced per person that year. (www.eia.doe.gov and www.census.gov)

104) A photo printer delivers approximately 1.1×10^6 droplets of ink per square inch. How many droplets of ink would a 4 in. × 6 in. photo contain?

105) The Hoover Dam generates approximately 4×10^9 kW-hours (kW-hr) of power per year. How much would it generate in 6 yr? (hooverdam.travelnevada.com)

106) A three-toed sloth moves at a rate of about 7.38×10^{-2} mph on the ground. How far could it go after 5 hr? (hypertextbook.com)

R Rethink

R1) Why is 24×10^{12} not considered to be written in scientific notation?

R2) When you express a very small number in scientific notation, will the exponent be positive or negative? Why?

R3) In what college courses other than math do you think scientific numbers are used?

R4) Which exercises in this section do you find most challenging?

6.3 Addition and Subtraction of Polynomials and Polynomial Functions

P Prepare	O Organize
What are your objectives for Section 6.3?	How can you accomplish each objective?
1 Learn Polynomial Vocabulary	• Learn the definition of a *polynomial*. • Be able to identify the *terms* of a polynomial, the *coefficients* of the terms, and the *degree* of the polynomial. • Be able to identify a *monomial, binomial,* and a *trinomial*. • Complete the given example on your own. • Complete You Try 1.
2 Add and Subtract Polynomials in One Variable	• Learn the procedures for **Adding and Subtracting Polynomials.** • Know how to identify *like terms*. • Complete the given examples on your own. • Complete You Trys 2 and 3.
3 Add and Subtract Polynomials in More Than One Variable	• Know how to identify *like terms* when the terms contain more than one variable. • Complete the given example on your own. • Complete You Try 4.
4 Define and Evaluate a Polynomial Function	• Learn the definition of a *polynomial function*. • Use parentheses when evaluating a polynomial function for a given value. • Follow the order of operations when simplifying. • Complete the given example on your own. • Complete You Try 5.
5 Add and Subtract Polynomial Functions	• Learn what it means to find the *sum and difference of two functions*. • Learn the two ways to evaluate the sum or difference of two functions for a given value to double check your answer. • Complete the given examples on your own. • Complete You Trys 6 and 7.

W Work — Read the explanations, follow the examples, take notes, and complete the You Trys.

1 Learn Polynomial Vocabulary

In Section 1.4 we defined an *algebraic expression* as a collection of numbers, variables, and grouping symbols connected by operation symbols such as $+$, $-$, $\times$, and $\div$. An example of an algebraic expression is $4x^3 + 6x^2 - x + \frac{5}{2}$. The *terms* of this algebraic expression are $4x^3$, $6x^2$, $-x$, and $\frac{5}{2}$. A **term** is a number or a variable or a product or quotient of numbers and variables. Not only is $4x^3 + 6x^2 - x + \frac{5}{2}$ an expression, it is also a *polynomial*.

Definition

A **polynomial in x** is the sum of a finite number of terms of the form ax^n, where n is a whole number and a is a real number.

Let's look more closely at the polynomial $4x^3 + 6x^2 - x + \frac{5}{2}$.

1) The polynomial is written in **descending powers of x** since the powers of x decrease from left to right. Generally, we write polynomials in descending powers of the variable.

2) Recall that the term without a variable is called a **constant**. The constant is $\frac{5}{2}$.

   The **degree of a term** equals the exponent on its variable. (If a term has more than one variable, the degree equals the *sum* of the exponents on the variables.) We will list each term, its coefficient, and its degree.

W Hint
In your notes and using your own words, write a description of each of the bold phrases in this three-step description of polynomials.

Term	Coefficient	Degree
$4x^3$	4	3
$6x^2$	6	2
$-x$	-1	1
$\frac{5}{2}$	$\frac{5}{2}$	0 $\left(\frac{5}{2} = \frac{5}{2}x^0\right)$

3) The **degree of the polynomial** equals the highest degree of any nonzero term. The polynomial above has degree 3. Or, we say that this is a **third-degree polynomial**.

EXAMPLE 1

Decide whether each expression *is* or *is not* a polynomial. If it is a polynomial, identify each term and the degree of each term. Then, find the degree of the polynomial.

a) $-7k^4 + 3.2k^3 - 8k^2 - 11$

b) $5n^2 - \frac{4}{3}n + 6 + \frac{9}{n^2}$

c) $x^3y^3 + 3x^2y + 3xy + 1$

d) $12t^5$

Solution

a) The expression $-7k^4 + 3.2k^3 - 8k^2 - 11$ is a polynomial in k. Its terms have whole number exponents and real coefficients. Its terms and degrees are in the table. *The degree of this polynomial is 4.*

Term	Degree
$-7k^4$	4
$3.2k^3$	3
$-8k^2$	2
-11	0

b) The expression $5n^2 - \frac{4}{3}n + 6 + \frac{9}{n^2}$ is *not* a polynomial because one of its terms has a variable in the denominator.

c) The expression $x^3y^3 + 3x^2y + 3xy + 1$ *is* a polynomial because the variables have whole number exponents and the coefficients are real numbers. Since this is a polynomial in two variables, we find the degree of each term by adding the exponents. *The degree of this polynomial is 6.*

Term	Degree
x^3y^3	6
$3x^2y$	3
$3xy$	2
1	0

d) The expression $12t^5$ is a polynomial even though it has only one term. The degree of the term is 5, and that is the degree of the polynomial as well.

[**YOU TRY 1**] Decide whether each expression *is* or *is not* a polynomial. If it is a polynomial, identify each term and the degree of each term. Then, find the degree of the polynomial.

a) $g^3 + 8g^2 - \frac{5}{g}$

b) $6t^5 - \frac{4}{9}t^4 + t + 2$

c) $a^4b^3 - 9a^3b^3 - a^2b + 4a - 7$

d) $x + 5x^{1/2} + 6$

Hint

In your notes, write the definitions for the bold-faced terms and give an example of each.

A polynomial with one term, like $12t^5$, is called a **monomial** ("mono" means one). Other examples of monomials are x^2, $-9j^4$, y, x^2y^2, and -2.

A **binomial** is a polynomial that consists of exactly two terms ("bi" means two). Some examples are $n + 6$, $5b^2 - 7$, $c^4 - d^4$, and $-12u^4v^3 + 8u^2v^2$.

A **trinomial** is a polynomial that consists of exactly three terms ("tri" means three). Here are some examples: $p^2 - 5p - 36$, $3k^5 + 24k^2 - 6k$, and $8r^4 + 10r^2s + 3s^2$.

It is important that you understand the meaning of these terms. We will use them throughout our study of algebra.

Recall from beginning algebra that **like terms** contain the same variables with the same exponents. For example, $4y^2$ and $\frac{2}{3}y^2$ are like terms, but $5x^6$ and $3x^4$ are not because their exponents are different. We add or subtract like terms by adding or subtracting the coefficients and leaving the variable(s) and exponent(s) the same. We use the same idea for adding and subtracting polynomials.

2 Add and Subtract Polynomials in One Variable

> **Procedure** Adding Polynomials
>
> To add polynomials, add like terms.

EXAMPLE 2 Add the polynomials.

a) $11c^2 + 3c - 9$ and $2c^2 + 5c + 1$

b) $(7y^3 - 10y^2 + y - 2) + (3y^3 + 4y^2 - 3)$

Solution

a) The addition problem can be set up horizontally or vertically. We will add these horizontally. Put the polynomials in parentheses since each contains more than one term. Use the associative and commutative properties to rewrite like terms together.

$$(11c^2 + 3c - 9) + (2c^2 + 5c + 1) = (11c^2 + 2c^2) + (3c + 5c) + (-9 + 1)$$
$$= 13c^2 + 8c - 8 \quad \text{Combine like terms.}$$

W Hint
When adding like terms, the exponents remain unchanged.

b) We will add these polynomials vertically. Line up like terms in columns and add.

$$\begin{array}{r} 7y^3 - 10y^2 + y - 2 \\ +\ 3y^3 + 4y^2 \quad\ \ - 3 \\ \hline 10y^3 - 6y^2 + y - 5 \end{array}$$

[YOU TRY 2] Add $(t^3 + 2t^2 - 10t + 1) + (3t^3 - 9t^2 - t + 6)$.

To subtract two polynomials such as $(6p^2 + 2p - 7) - (4p^2 - 9p + 3)$ we will use the distributive property to clear the parentheses in the second polynomial.

EXAMPLE 3 Subtract $(6p^2 + 2p - 7) - (4p^2 - 9p + 3)$.

Solution

$(6p^2 + 2p - 7) - (4p^2 - 9p + 3)$
$= (6p^2 + 2p - 7) - 1(4p^2 - 9p + 3)$
$= (6p^2 + 2p - 7) + (-1)(4p^2 - 9p + 3)$ Change -1 to $+(-1)$.
$= (6p^2 + 2p - 7) + (-4p^2 + 9p - 3)$ Distribute.
$= 2p^2 + 11p - 10$ Combine like terms.

In Example 3, notice that we changed the sign of each term in the second polynomial and then added it to the first.

Procedure Subtracting Polynomials

To subtract two polynomials, change the sign of each term in the second polynomial. Then, add the polynomials.

Let's see how we apply this rule to subtracting polynomials both horizontally and vertically.

EXAMPLE 4

Subtract $(-8k^3 - k^2 + 5k + 7) - (6k^3 - 3k^2 + k - 2)$.

a) horizontally   b) vertically

Solution

a) $(-8k^3 - k^2 + 5k + 7) - (6k^3 - 3k^2 + k - 2)$ Change the signs in the second polynomial, and add.
$= (-8k^3 - k^2 + 5k + 7) + (-6k^3 + 3k^2 - k + 2)$
$= -14k^3 + 2k^2 + 4k + 9$ Combine like terms.

W Hint
Do You Try 3 using both the vertical and horizontal methods. Decide which method works best for you.

b) To subtract vertically, line up like terms in columns.

$-8k^3 - k^2 + 5k + 7$ Change the signs in the $-8k^3 - k^2 + 5k + 7$
$-(6k^3 - 3k^2 + k - 2)$ second polynomial, and $+(-6k^3 + 3k^2 - k + 2)$
                                            add the polynomials. $-14k^3 + 2k^2 + 4k + 9$

You can see that adding and subtracting polynomials horizontally or vertically gives the same result.

[YOU TRY 3] Subtract $(9m^2 - 4m + 2) - (-m^2 + m - 6)$.

3 Add and Subtract Polynomials in More Than One Variable

To add and subtract polynomials in more than one variable, remember that like terms contain the same variables with the same exponents.

EXAMPLE 5

Perform the indicated operation.

a) $(x^2y^2 + 5x^2y - 8xy - 7) + (5x^2y^2 - 4x^2y - xy + 3)$

b) $(9cd - c + 3d + 1) - (6cd + 5c - 8)$

Solution

a) $(x^2y^2 + 5x^2y - 8xy - 7) + (5x^2y^2 - 4x^2y - xy + 3)$
$= 6x^2y^2 + x^2y - 9xy - 4$ Combine like terms.

b) $(9cd - c + 3d + 1) - (6cd + 5c - 8) = (9cd - c + 3d + 1) - 6cd - 5c + 8$
$= 3cd - 6c + 3d + 9$ Combine like terms.

W Hint
Notice where the distributive property is being used.

[YOU TRY 4] Perform the indicated operation.

a) $(-10a^2b^2 + ab - 3b + 4) - (-9a^2b^2 - 6ab + 2b + 4)$

b) $(5.8t^3u^2 + 2.1tu - 7u) + (4.1t^3u^2 - 7.8tu - 1.6)$

4 Define and Evaluate a Polynomial Function

In Chapter 4 we learned about linear functions of the form $f(x) = mx + b$. A linear function is a special type of polynomial function, which we will study now.

Definition

A **polynomial function of degree n** is given by
$f(x) = a_n x^n + a_{n-1} x^{n-1} + \cdots + a_1 x + a_0$, where $a_n, a_{n-1}, \ldots, a_1$, and a_0 are real numbers, $a_n \neq 0$, and n is a whole number.

Look at the polynomial $2x^2 - 5x + 7$. If we substitute 3 for x, the *only* value of the expression is 10:

$$2(3)^2 - 5(3) + 7 = 2(9) - 15 + 7 = 18 - 15 + 7 = 10$$

It is true that polynomials have different values depending on what value is substituted for the variable. It is also true that for any value we substitute for x in a polynomial like $2x^2 - 5x + 7$ there will be *only one value* of the expression. Since each value substituted for the variable produces *only one value* of the expression, we can use function notation to represent a polynomial like $2x^2 - 5x + 7$.

$f(x) = 2x^2 - 5x + 7$ is a *polynomial function* since $2x^2 - 5x + 7$ is a polynomial. Therefore, finding $f(3)$ when $f(x) = 2x^2 - 5x + 7$ is the same as evaluating $2x^2 - 5x + 7$ when $x = 3$.

EXAMPLE 6

If $f(x) = x^3 - 4x^2 + 3x + 1$, find $f(-2)$.

Solution

$f(x) = x^3 - 4x^2 + 3x + 1$
$f(-2) = (-2)^3 - 4(-2)^2 + 3(-2) + 1$ Substitute -2 for x.
$f(-2) = -8 - 4(4) - 6 + 1$
$f(-2) = -8 - 16 - 6 + 1$
$f(-2) = -29$

[YOU TRY 5] If $g(t) = 2t^4 + t^3 - 7t^2 + 12$, find $g(-1)$.

5 Add and Subtract Polynomial Functions

We have learned that we can add and subtract polynomials. These same operations can be performed with functions.

Definition

Given the functions $f(x)$ and $g(x)$, the **sum and difference of f and g** are defined by

1) $(f + g)(x) = f(x) + g(x)$
2) $(f - g)(x) = f(x) - g(x)$

The domain of $(f + g)(x)$ and $(f - g)(x)$ is the intersection of the domains of $f(x)$ and $g(x)$.

EXAMPLE 7

Let $f(x) = x^2 - 2x + 7$ and $g(x) = 4x - 3$. Find each of the following.

a) $(f + g)(x)$ b) $(f - g)(x)$ and $(f - g)(-1)$

Solution

a) $(f + g)(x) = f(x) + g(x)$
$= (x^2 - 2x + 7) + (4x - 3)$ Substitute the functions.
$= x^2 + 2x + 4$ Combine like terms.

b) $(f - g)(x) = f(x) - g(x)$
$= (x^2 - 2x + 7) - (4x - 3)$ Substitute the functions.
$= x^2 - 2x + 7 - 4x + 3$ Distribute.
$= x^2 - 6x + 10$ Combine like terms.

 Hint

Remember to use parentheses when you are substituting values!

Use the result above to find $(f - g)(-1)$.

$(f - g)(x) = x^2 - 6x + 10$
$(f - g)(-1) = (-1)^2 - 6(-1) + 10$ Substitute -1 for x.
$= 1 + 6 + 10$
$= 17$

We can also find $(f - g)(-1)$ using the rule this way:

$(f - g)(-1) = f(-1) - g(-1)$
$= [(-1)^2 - 2(-1) + 7] - [4(-1) - 3]$ Substitute -1 for x in $f(x)$ and $g(x)$.
$= (1 + 2 + 7) - (-4 - 3)$
$= 10 - (-7)$
$= 17$

YOU TRY 6

Let $f(x) = 3x^2 - 8$ and $g(x) = 2x + 1$. Find each of the following.

a) $(f + g)(x)$ and $(f + g)(-2)$ b) $(f - g)(x)$

We can use polynomial functions to solve real-world problems.

EXAMPLE 8

A publisher sells paperback romance novels to a large bookstore chain for $4.00 per book. Therefore, the publisher's revenue, in dollars, is defined by the function

$$R(x) = 4x$$

where x is the number of books sold to the retailer. The publisher's cost, in dollars, to produce x books is

$$C(x) = 2.5x + 1200$$

In business, profit is defined as revenue − cost. In terms of functions, this is written as $P(x) = R(x) - C(x)$, where $P(x)$ is the profit function.

a) Find the profit function, $P(x)$, that describes the publisher's profit from the sale of x books.

b) If the publisher sells 10,000 books to this chain of bookstores, what is the publisher's profit?

Solution

a) $P(x) = R(x) - C(x)$
   $= 4x - (2.5x + 1200)$ Substitute the functions.
   $= 1.5x - 1200$
   $P(x) = 1.5x - 1200$

b) Find $P(10,000)$.

$$P(10,000) = 1.5(10,000) - 1200$$
$$= 15,000 - 1200$$
$$= 13,800$$

The publisher's profit is $13,800.

[YOU TRY 7] A candy company sells its Valentine's Day candy to a grocery store retailer for $6.00 per box. The candy company's revenue, in dollars, is defined by $R(x) = 6x$, where x is the number of boxes sold to the retailer. The company's cost, in dollars, to produce x boxes of candy is $C(x) = 4x + 900$.

a) Find the profit function, $P(x)$, that defines the company's profit from the sale of x boxes of candy.

b) Find the candy company's profit from the sale of 2000 boxes of candy.

ANSWERS TO [YOU TRY] EXERCISES

1) a) not a polynomial   b) It is a polynomial of degree 5.   c) It is a polynomial of degree 7.

Term	Degree
$6t^5$	5
$-\dfrac{4}{9}t^4$	4
t	1
2	0

Term	Degree
a^4b^3	7
$-9a^3b^3$	6
$-a^2b$	3
$4a$	1
-7	0

d) not a polynomial   2) $4t^3 - 7t^2 - 11t + 7$ 3) $10m^2 - 5m + 8$
4) a) $-a^2b^2 + 7ab - 5b$ b) $9.9t^3u^2 - 5.7tu - 7u - 1.6$ 5) 6
6) a) $3x^2 + 2x - 7$; 1   b) $3x^2 - 2x - 9$ 7) a) $P(x) = 2x - 900$ b) $3100

E Evaluate 6.3 Exercises

Do the exercises, and check your work.

Objective 1: Learn Polynomial Vocabulary
Is the given expression a polynomial? Why or why not?

1) $-5z^2 - 4z + 12$

2) $9t^3 + t^2 - t + \dfrac{3}{8}$

3) $g^3 + 3g^2 + 2g^{-1} - 5$

4) $6y^4$

5) $m^{2/3} + 4m^{1/3} + 4$

6) $8c - 5 + \dfrac{2}{c}$

Determine whether each is a monomial, a binomial, or a trinomial.

7) $3x - 7$

8) $-w^3$

9) $a^2b^2 + 10ab - 6$

10) $16r^2 + 9r$

11) 1

12) $v^4 + 7v^2 + 6$

13) How do you determine the degree of a polynomial in one variable?

14) Write a fourth-degree polynomial in one variable.

15) How do you determine the degree of a term in a polynomial in more than one variable?

16) Write a fifth-degree monomial in x and y.

For each polynomial, identify each term in the polynomial, the coefficient and degree of each term, and the degree of the polynomial.

17) $4d^2 + 12d - 9$

18) $7y^3 + 10y^2 - y + 2$

19) $-9r^3s^2 - r^2s^2 + \frac{1}{2}rs + 6s$

20) $8m^2n^2 + 0.5m^2n - mn + 3$

Objective 2: Add and Subtract Polynomials in One Variable

Add.

21) $(11w^2 + 2w - 13) + (-6w^2 + 5w + 7)$

22) $(4f^4 - 3f^2 + 8) + (2f^4 - f^2 + 1)$

23) $(-p + 16) + (-7p - 9)$

24) $(y^3 + 8y^2) + (y^3 - 11y^2)$

25) $\left(-7a^4 - \frac{3}{2}a + 1\right) + \left(2a^4 + 9a^3 - a^2 - \frac{3}{8}\right)$

26) $\left(2d^5 + \frac{1}{3}d^4 - 11\right) + (10d^5 - 2d^4 + 9d^2 + 4)$

27) $\left(\frac{11}{4}x^3 - \frac{5}{6}\right) + \left(\frac{3}{8}x^3 + \frac{11}{12}\right)$

28) $\left(\frac{3}{4}c + \frac{1}{8}\right) + \left(\frac{3}{2}c - \frac{5}{6}\right)$

29) $(6.8k^3 + 3.5k^2 - 10k - 3.3) + (-4.2k^3 + 5.2k^2 + 2.7k - 1.1)$

30) $(0.6t^4 - 7.3t + 2.2) + (-1.8t^4 + 4.9t^3 + 8.1t + 7.1)$

Add.

31) $\begin{array}{r} 12x - 11 \\ + 5x + 3 \end{array}$

32) $\begin{array}{r} -6n^3 + 1 \\ + 4n^3 - 8 \end{array}$

33) $\begin{array}{r} 9r^2 + 16r + 2 \\ + 3r^2 - 10r + 9 \end{array}$

34) $\begin{array}{r} z^2 - 4z \\ +3z^2 + 9z + 4 \end{array}$

35) $\begin{array}{r} -2.6q^3 - q^2 + 6.9q - 1 \\ + 4.1q^3 - 2.3q + 16 \end{array}$

36) $\begin{array}{r} 9a^4 + 5.3a^3 - 7a^2 - 1.2a + 6 \\ + -8a^4 - 2.8a^3 + 4a^2 - 3.9a + 5 \end{array}$

Subtract.

37) $(8a^4 - 9a^2 + 17) - (15a^4 + 3a^2 + 3)$

38) $(16w^3 + 9w - 7) - (27w^3 - 3w - 4)$

39) $(j^2 + 18j + 2) - (-7j^2 + 6j + 2)$

40) $(-2m^2 + m + 5) - (3m^2 + m + 1)$

41) $(h^5 + 7h^3 - 8h) - (-9h^5 + h^4 + 7h^3 - 8h - 6)$

42) $(19s^5 - 11s^2) - (10s^5 + 3s^4 - 8s^2 - 2)$

43) $(-3b^4 - 5b^2 + b + 2) - (-2b^4 + 10b^3 - 5b^2 - 18)$

44) $(4t^3 - t^2 + 6) - (t^2 + 7t + 1)$

45) $\left(-\frac{5}{7}r^2 + \frac{4}{9}r + \frac{2}{3}\right) - \left(-\frac{5}{14}r^2 - \frac{5}{9}r + \frac{11}{6}\right)$

46) $\left(\frac{5}{6}y^3 + \frac{1}{2}y + 3\right) - \left(-\frac{1}{6}y^3 + y^2 - 3y\right)$

Subtract.

47) $\begin{array}{r} 17v + 3 \\ - 2v + 9 \end{array}$

48) $\begin{array}{r} 10q - 7 \\ - 4q + 8 \end{array}$

49) $\begin{array}{r} 2b^2 - 7b + 4 \\ - 3b^2 + 5b - 3 \end{array}$

50) $\begin{array}{r} -3d^2 + 16d + 2 \\ - 5d^2 + 7d - 3 \end{array}$

51) $\begin{array}{r} a^4 - 2a^3 + 6a^2 - 7a + 11 \\ - -2a^4 + 9a^3 - a^2 + 3 \end{array}$

52) $\begin{array}{r} 7y^4 + y^3 - 10y^2 + 6y - 2 \\ - -2y^4 + y^3 - 4y + 1 \end{array}$

53) Explain, in your own words, how to subtract two polynomials.

54) Do you prefer adding and subtracting polynomials vertically or horizontally? Why?

55) Will the sum of two trinomials always be a trinomial? Why or why not? Give an example.

56) Write a fourth-degree polynomial in x that does not contain a second-degree term.

Perform the indicated operations.

57) $(-3b^4 + 4b^2 - 6) + (2b^4 - 18b^2 + 4) + (b^4 + 5b^2 - 2)$

58) $(-7m^2 - 14m + 56) + (3m^2 + 7m - 6) + (9m^2 - 10)$

59) $\left(n^3 - \frac{1}{2}n^2 - 4n + \frac{5}{8}\right) + \left(\frac{1}{4}n^3 - n^2 + 7n - \frac{3}{4}\right)$

60) $\left(\frac{2}{3}z^4 + z^3 - \frac{3}{2}z^2 + 1\right) + \left(z^4 - 2z^3 - \frac{1}{6}z^2 + 8z - 1\right)$

61) $(u^3 + 2u^2 + 1) - (4u^3 - 7u^2 + u + 9) + (8u^3 - 19u^2 + 2)$

62) $(21r^3 - 8r^2 + 3r + 2) + (-4r^2 + 5) - (6r^3 - r^2 - 4r)$

63) $\left(\frac{3}{8}k^2 + k - \frac{1}{5}\right) - \left(2k^2 + k - \frac{7}{10}\right) + (k^2 - 9k)$

64) $\left(y + \frac{1}{4}\right) + \left(\frac{1}{2}y^2 - 3y + \frac{3}{4}\right) - \left(\frac{3}{4}y^2 - \frac{1}{2}y + 1\right)$

65) $(2t^3 - 8t^2 + t + 10) - [(5t^3 + 3t^2 - t + 8) + (-6t^3 - 4t^2 + 3t + 5)]$

66) $(x^2 - 10x - 6) - [(-8x^2 + 11x - 1) + (5x^2 - 9x - 3)]$

67) $(-12a^2 + 9) - (-9a^3 + 7a + 6) + (12a^2 - a + 10)$

68) $(5c + 7) - (c^2 + 4c - 2) - (-7c^3 - c + 4)$

Objective 3: Add and Subtract Polynomials in More Than One Variable

Each of the polynomials below is a polynomial in two variables. Perform the indicated operation(s).

69) $(4a + 13b) - (a + 5b)$

70) $(-2g - 3h) + (6g + h)$

71) $\left(5m + \frac{5}{6}n + \frac{1}{2}\right) + \left(-6m + n - \frac{3}{4}\right)$

72) $\left(-2c - \frac{2}{3}d + 1\right) - \left(2c + \frac{1}{9}d - \frac{4}{7}\right)$

73) $(-12y^2z^2 + 5y^2z - 25yz^2 + 16) + (17y^2z^2 + 2y^2z - 15)$

74) $(-8u^2v^2 + 2uv + 3) - (-9u^2v^2 - 14uv + 18)$

75) $(8x^3y^2 - 7x^2y^2 + 7x^2y - 3) + (2x^3y^2 + x^2y - 1) - (4x^2y^2 + 2x^2y + 8)$

76) $(r^3s^2 + r^2s^2 + 4) - (6r^3s^2 + 14r^2s^2 - 6) + (8r^3s^2 - 6r^2s^2 - 4)$

Write an expression for each, and perform the indicated operation(s).

77) Find the sum of $v^2 - 9$ and $4v^2 + 3v + 1$.

78) Add $11d - 12$ to $2d + 3$.

79) Subtract $g^2 - 7g + 16$ from $5g^2 + 3g + 6$.

80) Subtract $-9y^2 + 4y + 6$ from $2y^2 + y$.

81) Subtract the sum of $4n^2 + 1$ and $6n^2 - 10n + 3$ from $2n^2 + n + 4$.

82) Subtract $19x^3 + 4x - 12$ from the sum of $6x^3 + x^2 + x$ and $4x^3 - 3x - 8$.

Find the polynomial that represents the perimeter of each rectangle.

83)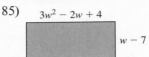
$3x + 8$, $x - 1$

84)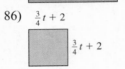
$a^2 + 5a - 3$, $a^2 - 2a + 2$

85) $3w^2 - 2w + 4$, $w - 7$

86) $\frac{3}{4}t + 2$, $\frac{3}{4}t + 2$

Objective 4: Define and Evaluate a Polynomial Function

87) If $f(x) = 5x^2 + 7x - 8$, find
  a) $f(-3)$
  b) $f(1)$

88) If $h(a) = -a^2 - 3a + 10$, find
  a) $h(5)$
  b) $h(-4)$

89) If $P(t) = t^3 - 3t^2 + 2t + 5$, find
  a) $P(4)$
  b) $P(0)$

90) If $G(c) = 3c^4 + c^2 - 9c - 4$, find
  a) $G(0)$
  b) $G(-1)$

91) If $f(x) = \frac{1}{3}x + 5$, find x so that $f(x) = 7$.

92) If $H(z) = -4z + 9$, find z so that $H(z) = 11$.

93) If $r(k) = \frac{2}{5}k - 3$, find k so that $r(k) = 13$.

94) If $Q(a) = 6a - 1$, find a so that $Q(a) = -9$.

Objective 5: Add and Subtract Polynomial Functions

For each pair of functions, find (a) $(f + g)(x)$, (b) $(f + g)(5)$, (c) $(f - g)(x)$, and (d) $(f - g)(2)$.

95) $f(x) = -3x + 1$, $g(x) = 2x - 11$

96) $f(x) = 5x - 9$, $g(x) = x + 4$

97) $f(x) = 4x^2 - 7x - 1$, $g(x) = x^2 + 3x - 6$

98) $f(x) = -2x^2 + x + 8$, $g(x) = 3x^2 - 4x - 6$

Let $f(t) = 4t - 1$, $g(t) = -t^2 + 6$, and $h(t) = 3t^2 - 4t$. Find each of the following.

99) $(g - h)(t)$

100) $(g + h)(t)$

101) $(g - h)(5)$

102) $(g + h)(0)$

103) $(f + g)(-1)$

104) $(f - g)(-2)$

105) $(h - f)\left(\dfrac{1}{2}\right)$

106) $(f - h)\left(\dfrac{2}{3}\right)$

107) Find two polynomial functions $f(x)$ and $g(x)$ so that $(f + g)(x) = 5x^2 + 8x - 2$.

108) Let $f(x) = 6x^3 - 9x^2 - 4x + 10$. Find $g(x)$ so that $(f - g)(x) = x^3 + 3x^2 + 8$.

109) A manufacturer's revenue, $R(x)$ in dollars, from the sale of x calculators is given by $R(x) = 12x$. The company's cost, $C(x)$ in dollars, to produce x calculators is $C(x) = 8x + 2000$.

 a) Find the profit function, $P(x)$, that defines the manufacturer's profit from the sale of x calculators.

 b) What is the profit from the sale of 1500 calculators?

110) $R(x) = 80x$ is the revenue function for the sale of x bicycles, in dollars. The cost to manufacture x bikes, in dollars, is $C(x) = 60x + 7000$.

 a) Find the profit function, $P(x)$, that describes the manufacturer's profit from the sale of x bicycles.

 b) What is the profit from the sale of 500 bicycles?

111) $R(x) = 18x$ is the revenue function for the sale of x toasters, in dollars. The cost to manufacture x toasters, in dollars, is $C(x) = 15x + 2400$.

 a) Find the profit function, $P(x)$, that describes the profit from the sale of x toasters.

 b) What is the profit from the sale of 800 toasters?

112) A company's revenue, $R(x)$ in dollars, from the sale of x dog houses is given by $R(x) = 60x$. The company's cost, $C(x)$ in dollars, to produce x dog houses is $C(x) = 45x + 6000$.

 a) Find the profit function, $P(x)$, that describes the company's profit from the sale of x dog houses?

 b) What is the profit from the sale of 300 dog houses?

For Exercises 113 and 114, let x be the number of items sold (in hundreds), and let $R(x)$ and $C(x)$ be in thousands of dollars.

113) A manufacturer's revenue, $R(x)$, from the sale of flat-screen TVs is given by $R(x) = -0.2x^2 + 23x$, while the cost, $C(x)$, is given by $C(x) = 4x + 9$.

 a) Find the profit function, $P(x)$, that describes the company's profit from the sale of x hundred flat-screen TVs.

 b) What is the profit from the sale of 2000 TVs?

114) A manufacturer's revenue, $R(x)$, from the sale of laptop computers is given by $R(x) = -0.4x^2 + 30x$, while the cost, $C(x)$, is given by $C(x) = 3x + 11$.

 a) Find the profit function, $P(x)$, that describes the company's profit from the sale of x hundred laptop computers.

 b) What is the profit from the sale of 1500 computers?

R Rethink

R1) How do you determine if a polynomial is written in descending powers?

R2) What are the differences between a binomial, a trinomial, and a polynomial?

R3) When combining like terms, do the exponents change?

R4) What is the difference between a coefficient and a variable?

R5) Which exercises in this section do you find most challenging?

6.4 Multiplication of Polynomials and Polynomial Functions

P Prepare	**O Organize**
What are your objectives for Section 6.4?	How can you accomplish each objective?
1 Multiply a Monomial and a Polynomial	• Use the distributive property. • Review the rules of exponents. • Complete the given example on your own. • Complete You Try 1.
2 Multiply Two Polynomials	• Learn the procedure for **Multiplying Polynomials,** and write it in your notes. • Understand that polynomials can be multiplied vertically as well as horizontally. • Complete the given examples on your own. • Complete You Trys 2 and 3.
3 Multiply Two Binomials Using FOIL	• Learn the procedure for multiplying using **FOIL,** and summarize it in your notes. • Complete the given example on your own. • Complete You Try 4.
4 Find the Product of More Than Two Polynomials	• Multiply using both methods to decide which you like best. • Complete the given example on your own. • Complete You Try 5.
5 Find the Product of Binomials of the form $(a + b)(a - b)$	• Learn the formula for the **Product of the Sum and Difference of Two Terms,** and write it in your notes. • Complete the given example on your own. • Complete You Try 6.
6 Square a Binomial	• Learn the formulas for the **Square of a Binomial,** and write them in your notes. • Complete the given example on your own. • Complete You Try 7.
7 Multiply Other Binomials	• Understand what the word *expand* means. • Multiply using methods already learned in this chapter. • Complete the given example on your own. • Complete You Try 8.
8 Multiply Polynomial Functions	• Learn the definition of the *product of two functions.* • Substitute using parentheses. • Use the multiplication methods learned in this chapter. • Complete the given example on your own. • Complete You Try 9.

W Work   Read the explanations, follow the examples, take notes, and complete the You Trys.

We have already learned that when multiplying two monomials, we multiply the coefficients and add the exponents of the same bases:

$$3x^4 \cdot 5x^2 = 15x^6 \qquad -2a^3b^2 \cdot 6ab^4 = -12a^4b^6$$

In this section, we will discuss how to multiply other types of polynomials.

1 Multiply a Monomial and a Polynomial

When multiplying a monomial and a polynomial, we use the distributive property.

EXAMPLE 1   Multiply $5n^2(2n^2 + 3n - 4)$.

Solution

$$5n^2(2n^2 + 3n - 4) = (5n^2)(2n^2) + (5n^2)(3n) + (5n^2)(-4) \qquad \text{Distribute.}$$
$$= 10n^4 + 15n^3 - 20n^2 \qquad \text{Multiply.}$$

[YOU TRY 1]   Multiply $6a^4(7a^3 - a^2 - 3a + 4)$.

2 Multiply Two Polynomials

W Hint
Notice how the distributive property is being used repeatedly.

To multiply two polynomials, we use the distributive property repeatedly. For example to multiply $(2r - 3)(r^2 + 7r + 9)$, we multiply each term in the second polynomial by $(2r - 3)$.

$$(2r - 3)(r^2 + 7r + 9) = (2r - 3)(r^2) + (2r - 3)(7r) + (2r - 3)(9) \qquad \text{Distribute.}$$

Next, we distribute again.

$$(2r - 3)(r^2) + (2r - 3)(7r) + (2r - 3)(9)$$
$$= (2r)(r^2) - (3)(r^2) + (2r)(7r) - (3)(7r) + (2r)(9) - (3)(9)$$
$$= 2r^3 - 3r^2 + 14r^2 - 21r + 18r - 27 \qquad \text{Multiply.}$$
$$= 2r^3 + 11r^2 - 3r - 27 \qquad \text{Combine like terms.}$$

This process of repeated distribution leads us to the following rule.

W Hint
Write out each procedure you learn in your notes.

Procedure Multiplying Polynomials

To multiply two polynomials, multiply each term in the second polynomial by each term in the first polynomial. Then, combine like terms. The answer should be written in descending powers.

Let's use this rule to multiply the polynomials in Example 2.

EXAMPLE 2 Multiply $(c^2 - 3c + 5)(2c^3 + c - 6)$.

Solution

Multiply each term in the second polynomial by each term in the first. Add like terms.

$(c^2 - 3c + 5)(2c^3 + c - 6)$
$= (c^2)(2c^3) + (c^2)(c) + (c^2)(-6) + (-3c)(2c^3) + (-3c)(c) + (-3c)(-6)$
$\quad + (5)(2c^3) + (5)(c) + (5)(-6)$ Distribute.
$= 2c^5 + c^3 - 6c^2 - 6c^4 - 3c^2 + 18c + 10c^3 + 5c - 30$ Multiply.
$= 2c^5 - 6c^4 + 11c^3 - 9c^2 + 23c - 30$ Combine like terms.

[YOU TRY 2] Multiply $(8z + 1)(7z^2 - z + 2)$.

Polynomials can be multiplied vertically as well. The process is similar to the way we multiply whole numbers. See the example at the right.

$$\begin{array}{r} 458 \\ \times\ 32 \\ \hline 916 \\ 13\ 74 \\ \hline 14{,}656 \end{array}$$

Multiply 458 by 2.
Multiply 458 by 3.
Add.

In the next example, we will find a product of polynomials by multiplying vertically.

EXAMPLE 3 Multiply vertically. $(3n^2 + 4n - 1)(2n + 7)$

Solution

Set up the multiplication problem like you would for whole numbers:

$$\begin{array}{r} 3n^2 + 4n - 1 \\ \times\ 2n + 7 \\ \hline 21n^2 + 28n - 7 \\ 6n^3 +\ 8n^2 - 2n \\ \hline 6n^3 + 29n^2 + 26n - 7 \end{array}$$

Multiply each term in $3n^2 + 4n - 1$ by 7.
Multiply each term in $3n^2 + 4n - 1$ by $2n$.
Line up like terms in the same column. Add.

[YOU TRY 3] Multiply vertically.

a) $(5x + 4)(7x^2 - 9x + 2)$ b) $\left(p^2 - \dfrac{3}{2}p - 6\right)(5p^2 + 8p - 4)$

3 Multiply Two Binomials Using FOIL

Multiplying two binomials is one of the most common types of polynomial multiplication used in algebra. A method called **FOIL** is one that is often used to multiply two binomials, and it comes from using the distributive property.

Let's use the distributive property to multiply $(x + 5)(x + 3)$.

$(x + 5)(x + 3) = (x + 5)(x) + (x + 5)(3)$ Distribute.
$\qquad\qquad\quad = x(x) + 5(x) + x(3) + 5(3)$ Distribute.
$\qquad\qquad\quad = x^2 + 5x + 3x + 15$ Multiply.
$\qquad\qquad\quad = x^2 + 8x + 15$ Combine like terms.

> **W Hint**
> Notice that FOIL is a useful acronym to help you remember how to multiply two binomials.

To be sure that each term in the first binomial has been multiplied by each term in the second binomial we can use FOIL. **FOIL** stands for **First Outer Inner Last**. Let's see how we can apply FOIL to the binomials we just multiplied.

$$(x + 5)(x + 3) = (x + 5)(x + 3) = x(x) + x(3) + 5(x) + 5(3)$$
$$= x^2 + 3x + 5x + 15 \quad \text{Multiply.}$$
$$= x^2 + 8x + 15 \quad \text{Combine like terms.}$$

You can see that we get the same result.

EXAMPLE 4

Use FOIL to multiply the binomials.

a) $(n + 9)(n - 4)$ b) $(5c - 8)(c - 1)$

c) $(x + 2y)(x - 6y)$ d) $(3a + 2)(b + 7)$

Solution

a) $(n + 9)(n - 4) = (n + 9)(n - 4) = n(n) + n(-4) + 9(n) + 9(-4)$ Use FOIL.
$$= n^2 - 4n + 9n - 36 \quad \text{Multiply.}$$
$$= n^2 + 5n - 36 \quad \text{Combine like terms.}$$

Notice that the middle terms are like terms, so we can combine them.

b) $(5c - 8)(c - 1) = 5c(c) + 5c(-1) - 8(c) - 8(-1)$ Use FOIL.
$$= 5c^2 - 5c - 8c + 8 \quad \text{Multiply.}$$
$$= 5c^2 - 13c + 8 \quad \text{Combine like terms.}$$

The middle terms are like terms, so we can combine them.

> **Hint**
> As you read each example, be sure you are writing every step on your own paper!

c) $(x + 2y)(x - 6y) = x(x) + x(-6y) + 2y(x) + 2y(-6y)$ Use FOIL.
$$= x^2 - 6xy + 2xy - 12y^2 \quad \text{Multiply.}$$
$$= x^2 - 4xy - 12y^2 \quad \text{Combine like terms.}$$

Like parts a) and b), we combined the middle terms.

d) $(3a + 2)(b + 7) = 3a(b) + 3a(7) + 2(b) + 2(7)$ Use FOIL.
$$= 3ab + 21a + 2b + 14 \quad \text{Multiply.}$$

In this case the middle terms were not like terms, so we could not combine them.

[YOU TRY 4]

Use FOIL to multiply the binomials.

a) $(a + 9)(a + 2)$ b) $(4t + 3)(t - 5)$

c) $(c - 6d)(c - 3d)$ d) $(5m + 2)(n + 6)$

With practice, you should be able to find the product of two binomials "in your head."

While the polynomial multiplication problems we have seen so far are the most common types we encounter in algebra, there are other products we will see as well.

4 Find the Product of More Than Two Polynomials

EXAMPLE 5

Multiply $5d^2(4d - 3)(2d - 1)$.

Solution

We can approach this problem a couple of ways.

Method 1
Begin by multiplying the binomials, *then* multiply by the monomial.

$$5d^2(4d - 3)(2d - 1) = 5d^2(8d^2 - 4d - 6d + 3) \quad \text{Use FOIL to multiply the binomials.}$$
$$= 5d^2(8d^2 - 10d + 3) \quad \text{Combine like terms.}$$
$$= 40d^4 - 50d^3 + 15d^2 \quad \text{Distribute.}$$

Method 2
Begin by multiplying $5d^2$ by $(4d - 3)$, then multiply *that* product by $(2d - 1)$.

$$5d^2(4d - 3)(2d - 1) = (20d^3 - 15d^2)(2d - 1) \quad \text{Multiply } 5d^2 \text{ by } (4d - 3).$$
$$= 40d^4 - 20d^3 - 30d^3 + 15d^2 \quad \text{Use FOIL to multiply.}$$
$$= 40d^4 - 50d^3 + 15d^2 \quad \text{Combine like terms.}$$

The result is the same. These may be multiplied by whichever method you prefer.

[YOU TRY 5] Multiply $-6x^3(x + 5)(3x - 4)$.

Special types of binomial products come up often in algebra. We will look at those next.

5 Find the Product of Binomials of the Form (a + b)(a − b)

Hint
Notice that when using FOIL to multiply binomials of this form, the middle terms of the product are always opposites of each other and cancel out.

Let's find the product $(p + 5)(p - 5)$. Using FOIL we get

$$(p + 5)(p - 5) = p^2 - 5p + 5p - 25$$
$$= p^2 - 25$$

Notice that the "middle terms," the p-terms, drop out. In the result, $p^2 - 25$, the first term (p^2) is the square of p and the last term (25) is the square of 5. They are subtracted. The resulting binomial is a *difference of squares*. This pattern always holds when multiplying two binomials of the form $(a + b)(a - b)$.

> **Formula** The Product of the Sum and Difference of Two Terms
> $$(a + b)(a - b) = a^2 - b^2$$

EXAMPLE 6

Multiply.

a) $(r + 7)(r - 7)$ b) $(3 + y)(3 - y)$

c) $(4a - 3b)(4a + 3b)$

Solution

a) The product $(r + 7)(r - 7)$ is in the form $(a + b)(a - b)$ because $a = r$ and $b = 7$.
$$(r + 7)(r - 7) = r^2 - 7^2$$
$$= r^2 - 49$$

b) $(3 + y)(3 - y) = 3^2 - y^2$
$$= 9 - y^2$$

 BE CAREFUL Notice that the answer to part (b) is $9 - y^2$ not $y^2 - 9$; subtraction is not commutative.

c) Since multiplication is commutative (the order in which we multiply does not affect the result), $(4a - 3b)(4a + 3b)$ is the same as $(4a + 3b)(4a - 3b)$. This is in the form $(a + b)(a - b)$, where $a = 4a$ and $b = 3b$.
$$(4a - 3b)(4a + 3b) = (4a)^2 - (3b)^2 = 16a^2 - 9b^2$$

[YOU TRY 6]

Multiply.

a) $(m + 9)(m - 9)$ b) $(4z + 3)(4z - 3)$ c) $(5c - 2d)(5c + 2d)$

6 Square a Binomial

W Hint
Notice that when you multiply binomials of this form using FOIL, the middle terms of the product always have the same sign and therefore give you two identical middle terms.

Another type of special binomial product is a **binomial square** such as $(x + 6)^2$. The expression $(x + 6)^2$ means $(x + 6)(x + 6)$. Therefore, we can use FOIL to multiply.
$$(x + 6)^2 = (x + 6)(x + 6) = x^2 + 6x + 6x + 36$$
$$= x^2 + 12x + 36$$

Notice that the outer and inner products, $6x$ and $6x$, are the same. When we add those terms, we see that the middle term of the result is *twice* the product of the terms in each binomial: $12x = 2(x)(6)$.

The *first* term in the result is the square of the *first* term in the binomial, and the *last* term in the result is the square of the *last* term in the binomial. We can express these relationships with these formulas:

> **Formula** The Square of a Binomial
> $$(a + b)^2 = a^2 + 2ab + b^2$$
> $$(a - b)^2 = a^2 - 2ab + b^2$$

We can think of the formulas in words as:

To square a binomial, you square the first term, square the second term, then multiply 2 times the first term times the second term and add.

Finding the products $(a + b)^2 = a^2 + 2ab + b^2$ and $(a - b)^2 = a^2 - 2ab + b^2$ is also called *expanding* the binomial squares $(a + b)^2$ and $(a - b)^2$.

$(a + b)^2 \neq a^2 + b^2$ and $(a - b)^2 \neq a^2 - b^2$.

EXAMPLE 7

Expand.

a) $(q + 4)^2$
b) $(u - 10)^2$
c) $(4x + 7y)^2$
d) $[(3x - y) + 1]^2$

Solution

a) $(q + 4)^2 = \underbrace{q^2}_{\text{Square the first term}} + \underbrace{2(q)(4)}_{\text{Two times first term times second term}} + \underbrace{4^2}_{\text{Square the second term}}$ $\quad a = q, b = 4$

$= q^2 + 8q + 16$

Notice, $(q + 4)^2 \neq q^2 + 16$. Do not "distribute" the power of 2 to each term in the binomial!

b) $(u - 10)^2 = \underbrace{u^2}_{\text{Square the first term}} - \underbrace{2(u)(10)}_{\text{Two times first term times second term}} + \underbrace{(10)^2}_{\text{Square the second term}}$ $\quad a = u, b = 10$

$= u^2 - 20u + 100$

W Hint
Using your own words, write down the pattern you see in these examples.

c) $(4x + 7y)^2 = (4x)^2 + 2(4x)(7y) + (7y)^2 \quad a = 4x, b = 7y$
$= 16x^2 + 56xy + 49y^2$

d) Although this expansion looks complicated, we use the same method. Here, $a = 3x - y$ and $b = 1$.

$[(3x - y) + 1]^2 = (3x - y)^2 + 2(3x - y)(1) + (1)^2 \quad (a + b)^2 = a^2 + 2ab + b^2$
$= (3x)^2 - 2(3x)(y) + (y)^2 + 2(3x - y)(1) + (1)^2 \quad \text{Expand } (3x - y)^2.$
$= 9x^2 - 6xy + y^2 + 6x - 2y + 1 \quad \text{Simplify.}$

YOU TRY 7

Expand.

a) $(t + 7)^2$
b) $(b - 12)^2$
c) $(2a + 5b)^2$
d) $[(3c - d) + 4]^2$

7 Multiply Other Binomials

To find other products of binomials, we use techniques we have already discussed.

EXAMPLE 8

Expand.

a) $(a + 2)^3$ b) $[(3c + d) + 2n][(3c + d) - 2n]$

Solution

a) Just as $x^2 \cdot x = x^3$, $(a + 2)^2 \cdot (a + 2) = (a + 2)^3$. So we can think of $(a + 2)^3$ as $(a + 2)^2(a + 2)$.

$$\begin{aligned}(a + 2)^3 &= (a + 2)^2(a + 2) \\ &= (a^2 + 4a + 4)(a + 2) &&\text{Square the binomial.} \\ &= a^3 + 2a^2 + 4a^2 + 8a + 4a + 8 &&\text{Multiply.} \\ &= a^3 + 6a^2 + 12a + 8 &&\text{Combine like terms.}\end{aligned}$$

b) Notice that $[(3c + d) + 2n][(3c + d) - 2n]$ has the form $(a + b)(a - b)$, where $a = 3c + d$ and $b = 2n$. So we can use $(a + b)(a - b) = a^2 - b^2$.

$$\begin{aligned}[(3c + d) + 2n][(3c + d) - 2n] &= (3c + d)^2 - (2n)^2 \\ &= (3c)^2 + 2(3c)(d) + (d)^2 - (2n)^2 &&\text{Expand } (3c + d)^2. \\ &= 9c^2 + 6cd + d^2 - 4n^2 &&\text{Simplify.}\end{aligned}$$

YOU TRY 8

Expand.

a) $(n - 3)^3$ b) $[(x - 2y) + 3z][(x - 2y) - 3z]$

8 Multiply Polynomial Functions

In Section 6.3 we learned how to add and subtract functions. Next we will learn how to multiply functions.

Definition

Given the functions $f(x)$ and $g(x)$, the **product of f and g** is defined by

$$(fg)(x) = f(x) \cdot g(x)$$

The domain of $(fg)(x)$ is the intersection of the domains of $f(x)$ and $g(x)$.

EXAMPLE 9

Let $f(x) = 5x - 2$ and $g(x) = 2x^2 + x - 7$. Find

a) $(fg)(x)$ b) $(fg)(-2)$

Solution

a) $$\begin{aligned}(fg)(x) &= f(x) \cdot g(x) \\ &= (5x - 2)(2x^2 + x - 7) &&\text{Substitute the functions.} \\ &= 10x^3 + 5x^2 - 35x - 4x^2 - 2x + 14 &&\text{Multiply.} \\ &= 10x^3 + x^2 - 37x + 14 &&\text{Combine like terms.}\end{aligned}$$

b) Since $(fg)(x) = 10x^3 + x^2 - 37x + 14$, substitute -2 for x to find $(fg)(-2)$.

$$(fg)(-2) = 10(-2)^3 + (-2)^2 - 37(-2) + 14 \quad \text{Substitute } -2 \text{ for } x.$$
$$= 10(-8) + 4 + 74 + 14 \quad \text{Simplify.}$$
$$= -80 + 92$$
$$= 12$$

Verify that the result in part b) is the same as $f(-2) \cdot g(-2)$.

[YOU TRY 9] Let $f(x) = -x + 4$ and $g(x) = 3x^2 - 8x - 6$. Find
a) $(fg)(x)$ b) $(fg)(3)$

ANSWERS TO [YOU TRY] EXERCISES

1) $42a^7 - 6a^6 - 18a^5 + 24a^4$ 2) $56z^3 - z^2 + 15z + 2$ 3) a) $35x^3 - 17x^2 - 26x + 8$
b) $5p^4 + \frac{1}{2}p^3 - 46p^2 - 42p + 24$ 4) a) $a^2 + 11a + 18$ b) $4t^2 - 17t - 15$ c) $c^2 - 9cd + 18d^2$
d) $5mn + 30m + 2n + 12$ 5) $-18x^5 - 66x^4 + 120x^3$ 6) a) $m^2 - 81$ b) $16z^2 - 9$
c) $25c^2 - 4d^2$ 7) a) $t^2 + 14t + 49$ b) $b^2 - 24b + 144$ c) $4a^2 + 20ab + 25b^2$
d) $9c^2 - 6cd + d^2 + 24c - 8d + 16$ 8) a) $n^3 - 9n^2 + 27n - 27$ b) $x^2 - 4xy + 4y^2 - 9z^2$
9) a) $-3x^3 + 20x^2 - 26x - 24$ b) -3

E Evaluate 6.4 Exercises

Do the exercises, and check your work.

Objective 1: Multiply a Monomial and a Polynomial

1) Explain how to multiply two monomials.
2) Explain how to multiply a monomial by a trinomial.

Multiply.

3) $(7k^4)(2k^2)$ 4) $(8p^6)(-p^4)$
5) $\left(\frac{7}{10}d^9\right)\left(\frac{5}{2}d^2\right)$ 6) $\left(-\frac{8}{9}c^5\right)\left(\frac{3}{10}c^7\right)$
7) $7y(4y - 9)$ 8) $-12m(11m - 4)$
9) $6v^3(v^2 - 4v - 2)$ 10) $3x^4(5x^3 + x - 7)$
11) $-3t^2(9t^3 - 6t^2 - 4t - 7)$
12) $-8u^5(9u^4 + 8u^3 + 12u - 1)$
13) $2x^3y(xy^2 + 8xy - 11y + 2)$
14) $5p^5q^2(-5p^2q + 12pq^2 - pq + 2q - 1)$
15) $-\frac{3}{4}t^4(20t^3 + 8t^2 - 5t)$
16) $\frac{2}{5}x^4(30x^2 - 15x + 7)$

Objective 2: Multiply Two Polynomials
Perform the indicated operations, and simplify.

17) $2(10g^3 + 5g^2 + 4) - (2g^3 - 14g - 20)$
18) $6(7m^2 + 7m + 9) - 11(4m^2 + 7m + 1)$
19) $7(a^3b^3 + 6a^3b^2 + 3) - 9(6a^3b^3 + 9a^3b^2 - 12a^2b + 8)$
20) $3(x^2y^2 + xy - 2) - 5(x^2y^2 + 6xy - 2) + 2(x^2y^2 + 6xy - 2)$
21) $(m + 9)(9m^2 + 4m - 7)$
22) $(q + 3)(5q^2 - 15q + 9)$
23) $(p - 6)(2p^2 + 3p - 5)$
24) $(s - 5)(7s^2 - 3s - 11)$
25) $(5y^3 - y^2 + 8y + 1)(3y - 4)$
26) $(4n^3 + 4n^2 - 5n - 7)(7n - 6)$
27) $\left(\frac{1}{2}k^2 + 3\right)(12k^2 + 5k - 10)$
28) $\left(\frac{2}{3}c^2 - 8\right)(6c^2 - 4c + 9)$

29) $(a^2 - a + 3)(a^2 + 4a - 2)$

30) $(r^2 + 2r + 5)(2r^2 - r - 3)$

31) $(3v^2 - v + 2)(-8v^3 + 6v^2 + 5)$

32) $(c^4 + 10c^2 - 7)(c^2 - 2c - 3)$

Multiply each horizontally and vertically. Which method do you prefer and why?

33) $(2x - 3)(4x^2 - 5x + 2)$

34) $(3n^2 + n - 4)(5n + 2)$

Objective 3: Multiply Two Binomials Using FOIL

35) What do the letters in the word FOIL represent?

36) Can FOIL be used to expand $(x + 9)^2$? Explain your answer.

Use FOIL to multiply.

37) $(w + 8)(w + 7)$ 38) $(k - 5)(k + 9)$

39) $(n - 11)(n - 4)$ 40) $(y - 6)(y - 1)$

41) $(4p + 5)(p - 3)$ 42) $(6t + 1)(t + 7)$

43) $(8n + 3)(3n + 4)$ 44) $(5b - 2)(8b + 5)$

45) $(0.4g - 0.9)(0.1g + 1.1)$

46) $(0.7 + 0.5m)(0.8m - 1.5)$

47) $(4a - 5b)(3a + 4b)$ 48) $(2x + 3y)(x - 6y)$

49) $(6p + 5q)(10p + 3q)$ 50) $(7m - 3n)(m - n)$

51) $\left(2a - \dfrac{1}{4}b\right)(2b + a)$ 52) $\left(w + \dfrac{3}{2}v\right)(2w - 3v)$

Write an expression for a) the perimeter of each figure and b) the area of each figure.

53)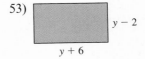

54)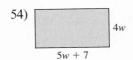

55)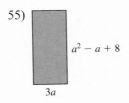

56)

$2x^2 - 3$ (square with sides $2x^2 - 3$)

Objective 4: Find the Product of More Than Two Polynomials

57) To find the product $2(n + 6)(n - 1)$, Raman begins by multiplying $2(n + 6)$ and then he multiplies that result by $(n - 1)$. Peggy begins by multiplying $(n + 6)(n - 1)$ and multiplies that result by 2. Who is right?

58) Find the product $(3a + 2)(a - 4)(a - 2)$

  a) by first multiplying $(3a + 2)(a - 4)$ and then multiplying that result by $(a - 2)$.

  b) by first multiplying $(a - 4)(a - 2)$ and then multiplying that result by $(3a + 2)$.

  c) What do you notice about the results?

Multiply. See Example 5.

59) $3(y + 4)(5y - 2)$

60) $4(7 - 3z)(2z - 1)$

61) $12g^2(2g + 5)(-g + 1)$

62) $-7r^2(r - 9)(r - 2)$

63) $(c + 3)(c + 4)(c - 1)$

64) $(x - 5)(x - 2)(x + 3)$

65) $10n\left(\dfrac{1}{2}n^2 + 3\right)(n^2 + 5)$

66) $12k\left(\dfrac{1}{4}k^2 - \dfrac{2}{3}\right)(k^2 + 1)$

67) $(r + t)(r - 2t)(2r - t)$

68) $(x + y)(x - 2y)(x + 3y)$

Objective 5: Find the Product of Binomials of the Form $(a + b)(a - b)$

Find the following special products. See Example 6.

69) $(3m + 2)(3m - 2)$ 70) $(5y - 4)(5y + 4)$

71) $(7a - 8)(7a + 8)$ 72) $(4x - 11)(4x + 11)$

73) $(2p + 7q)(2p - 7q)$ 74) $(6a - b)(6a + b)$

75) $\left(n + \dfrac{1}{2}\right)\left(n - \dfrac{1}{2}\right)$ 76) $\left(b - \dfrac{1}{5}\right)\left(b + \dfrac{1}{5}\right)$

77) $\left(\dfrac{2}{3} - k\right)\left(\dfrac{2}{3} + k\right)$ 78) $\left(\dfrac{4}{3} + z\right)\left(\dfrac{4}{3} - z\right)$

79) $(0.3x - 0.4y)(0.3x + 0.4y)$

80) $(1.2a + 0.8b)(1.2a - 0.8b)$

81) $(5x^2 + 4)(5x^2 - 4)$

82) $(9k^2 + 3l^2)(9k^2 - 3l^2)$

Objective 6: Square a Binomial
Expand.

83) $(y + 8)^2$

84) $(b + 6)^2$

85) $(t - 11)^2$

86) $(g - 5)^2$

87) $(4w + 1)^2$

88) $(7n + 2)^2$

89) $(2d - 5)^2$

90) $(3p - 5)^2$

91) $(6a - 5b)^2$

92) $(7x + 6y)^2$

93) Does $4(t + 3)^2 = (4t + 12)^2$? Why or why not?

94) Explain, in words, how to find the product $3(z - 4)^2$, then find the product.

Find the product.

95) $6(x + 1)^2$

96) $2(k + 5)^2$

97) $2a(a + 3)^2$

98) $-3(m - 1)^2$

99) $[(3m + n) + 2]^2$

100) $[(2c - d) + 7]^2$

101) $[(x - 4) - y]^2$

102) $[(3r + 2) - t]^2$

Objective 7: Multiply Other Binomials
Expand.

103) $(r + 5)^3$

104) $(w + 4)^3$

105) $(s - 2)^3$

106) $(q - 1)^3$

107) $(c^2 - 9)^2$

108) $\left(\dfrac{3}{8}x + 2\right)^2$

109) $(y + 2)^4$

110) $(b + 3)^4$

111) $[(v - 5w) + 4][(v - 5w) - 4]$

112) $[(4p + 3q) + 1][(4p + 3q) - 1]$

113) $[(2a + b) + c][(2a + b) - c]$

114) $[(x - 3y) - 2z][(x - 3y) + 2z]$

Mixed Exercises: Objectives 1, 3, 6, and 7

115) Does $(x + 5)^2 = x^2 + 25$? Why or why not?

116) Does $(y - 3)^3 = y^3 - 27$? Why or why not?

117) Express the volume of the cube as a polynomial.

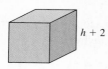

118) Express the area of the square as a polynomial.

119) Express the area of the shaded region as a polynomial.

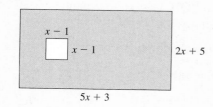

120) Express the area of the triangle as a polynomial.

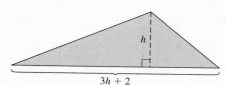

Objective 8: Multiply Polynomial Functions

Use the following functions for Exercises 121–129.
Let $f(x) = 3x - 5$, $g(x) = x^2$, and $h(x) = 2x^2 - 3x - 1$. Find each of the following.

121) $(fg)(x)$

122) $(gh)(x)$

123) $(gh)(-2)$

124) $(fg)(1)$

125) $(fh)(x)$

126) $(fg)(3)$

127) $(fh)\left(-\dfrac{1}{3}\right)$

128) $(gh)\left(\dfrac{1}{2}\right)$

129) Find $f(1)$, $g(1)$, and $f(1) \cdot g(1)$. How does this compare to the result in Exercise 124? Why is this true?

130) If $f(x) = x + 6$ and $(fg)(x) = 2x^2 + 12x$, find $g(x)$.

R Rethink

R1) What property are you using when you multiply polynomials?

R2) What does FOIL represent, and on what types of problems do you use it?

R3) How many special products did you encounter in this section?

R4) What is the most common mistake you make when multiplying polynomials?

R5) Which exercises in this section do you find most challenging?

6.5 Division of Polynomials and Polynomial Functions

P Prepare

What are your objectives for Section 6.5?

O Organize

How can you accomplish each objective?

What are your objectives for Section 6.5?	How can you accomplish each objective?
1 Divide a Polynomial by a Monomial	• Learn the procedure for **Dividing a Polynomial by a Monomial.** • Review the quotient rules for exponents. • Complete the given examples on your own. • Complete You Trys 1 and 2.
2 Divide a Polynomial by a Polynomial	• Review long division of whole numbers by writing out Example 3 in your notes. • Learn the long division process for polynomials. • Notice the similarities between whole number division and polynomial long division. • Complete the given examples on your own. • Complete You Trys 3–6.
3 Divide Polynomial Functions	• Learn the procedure for **Dividing Functions.** • Understand why certain values are not in the domain of the quotient. • Complete the given example on your own. • Complete You Try 7.

 Read the explanations, follow the examples, take notes, and complete the You Trys.

The last operation with polynomials we need to discuss is *division* of polynomials. We will consider this in two parts:

1) Dividing a polynomial by a monomial

$$\text{Examples:} \quad \frac{12a^2 - a + 15}{3}, \quad \frac{-48m^3 + 30m^2 - 8m}{8m^2}$$

and

2) Dividing a polynomial by a polynomial

$$\text{Examples:} \quad \frac{n^2 + 14n + 48}{n + 6}, \quad \frac{27z^3 - 1}{3z - 1}$$

1 Divide a Polynomial by a Monomial

The procedure for dividing a polynomial by a monomial is based on the procedure for adding or subtracting fractions.

To add $\frac{4}{15} + \frac{7}{15}$, we do the following:

$$\frac{4}{15} + \frac{7}{15} = \frac{4 + 7}{15} \quad \text{Add numerators, keep the common denominator.}$$
$$= \frac{11}{15}$$

Reversing the process above we can write $\frac{11}{15} = \frac{4 + 7}{15} = \frac{4}{15} + \frac{7}{15}$. We can generalize this result and say that $\frac{a + b}{c} = \frac{a}{c} + \frac{b}{c}$ ($c \neq 0$).

> **W Hint**
> Write this procedure in your notes, in your own words.

Procedure Dividing a Polynomial by a Monomial

To divide a polynomial by a monomial, divide *each term* in the polynomial by the monomial and simplify.

EXAMPLE 1

Divide.

a) $\dfrac{40a^2 - 25a + 10}{5}$ b) $\dfrac{9x^3 + 30x^2 + 3x}{3x}$

Solution

a) First, note that the polynomial is being divided by a *monomial*. That means we divide each term in the numerator by the monomial 5.

$$\frac{40a^2 - 25a + 10}{5} = \frac{40a^2}{5} - \frac{25a}{5} + \frac{10}{5}$$
$$= 8a^2 - 5a + 2$$

> **Hint**
> Notice that when dividing a trinomial by a monomial, the result is a trinomial.

Let's label the components of our division problem the same way as when we divide with integers.

$$\text{Dividend} \rightarrow \frac{40a^2 - 25a + 10}{5} = 8a^2 - 5a + 2 \leftarrow \text{Quotient}$$
$$\text{Divisor} \rightarrow$$

We can check our answer by multiplying the quotient by the divisor. The answer should be the dividend.

Check: $5(8a^2 - 5a + 2) = 40a^2 - 25a + 10$ ✓

The quotient $8a^2 - 5a + 2$ is correct.

b) $\dfrac{9x^3 + 30x^2 + 3x}{3x} = \dfrac{9x^3}{3x} + \dfrac{30x^2}{3x} + \dfrac{3x}{3x}$ Divide each term in numerator by $3x$.

$\phantom{b) \dfrac{9x^3 + 30x^2 + 3x}{3x}} = 3x^2 + 10x + 1$ Apply the quotient rule for exponents.

BE CAREFUL

Students will often incorrectly "cancel out" $\dfrac{3x}{3x}$ and get nothing. But $\dfrac{3x}{3x} = 1$ since a quantity divided by itself equals one.

Check: $3x(3x^2 + 10x + 1) = 9x^3 + 30x^2 + 3x$ ✓   The quotient is correct.

Note

In Example 1b), x cannot equal zero because then the denominator of $\dfrac{9x^3 + 30x^2 + 3x}{3x}$ would equal zero. Remember, a fraction is undefined when its denominator equals zero!

[YOU TRY 1]

Divide $\dfrac{24t^5 - 6t^4 - 54t^3}{6t^2}$.

EXAMPLE 2

Divide $(15m + 45m^3 - 4 + 18m^2) \div (9m^2)$.

Solution

Although this example is written differently, it is the same as the previous examples. Notice, however, the terms in the numerator are not written in descending powers. Rewrite them in descending powers before dividing.

$\dfrac{15m + 45m^3 - 4 + 18m^2}{9m^2} = \dfrac{45m^3 + 18m^2 + 15m - 4}{9m^2}$

$\phantom{\dfrac{15m + 45m^3 - 4 + 18m^2}{9m^2}} = \dfrac{45m^3}{9m^2} + \dfrac{18m^2}{9m^2} + \dfrac{15m}{9m^2} - \dfrac{4}{9m^2}$

$\phantom{\dfrac{15m + 45m^3 - 4 + 18m^2}{9m^2}} = 5m + 2 + \dfrac{5}{3m} - \dfrac{4}{9m^2}$ Apply the quotient rule, and simplify.

The quotient is *not* a polynomial since m and m^2 appear in denominators. The quotient of polynomials is not necessarily a polynomial.

[YOU TRY 2]

Divide $(6u^2 + 40 - 24u^3 - 8u) \div (8u^2)$.

2 Divide a Polynomial by a Polynomial

When dividing a polynomial by a polynomial containing two or more terms, we use *long division of polynomials*. This method is similar to long division of whole numbers. We will look at a long division problem here so that we can compare the procedure with polynomial long division.

EXAMPLE 3

Divide 4593 by 8.

Solution

$$\begin{array}{r} 5 \\ 8{\overline{\smash{)}4593}} \\ -40\downarrow \\ \hline 59 \end{array}$$

1) How many times does 8 divide into 45 evenly?   5
2) Multiply $5 \times 8 = 40$.
3) Subtract $45 - 40 = 5$.
4) Bring down the 9.

Start the process again.

$$\begin{array}{r} 57 \\ 8{\overline{\smash{)}4593}} \\ -40 \\ \hline 59 \\ -56\downarrow \\ \hline 33 \end{array}$$

1) How many times does 8 divide into 59 evenly?   7
2) Multiply $7 \times 8 = 56$.
3) Subtract $59 - 56 = 3$.
4) Bring down the 3.

Do the procedure again.

$$\begin{array}{r} 574 \\ 8{\overline{\smash{)}4593}} \\ -40 \\ \hline 59 \\ -56 \\ \hline 33 \\ -32 \\ \hline 1 \end{array}$$

1) How many times does 8 divide into 33 evenly?   4
2) Multiply $4 \times 8 = 32$.
3) Subtract $33 - 32 = 1$.
4) There are no more numbers to bring down, so the remainder is 1.

> **W Hint**
> Notice that when you have a remainder, your final answer is a mixed number.

Write the result.

$$4593 \div 8 = 574\frac{1}{8} \quad \begin{array}{l}\leftarrow \text{Remainder} \\ \leftarrow \text{Divisor}\end{array}$$

Check: $(8 \times 574) + 1 = 4592 + 1 = 4593$ ✓

[YOU TRY 3]   Divide 3827 by 6.

To divide two polynomials, we use a long division process similar to that of Example 3.

EXAMPLE 4

Divide $\dfrac{3x^2 + 19x + 20}{x + 5}$.

Solution

First, notice that we are dividing by a binomial. That tells us to use long division of polynomials.

We will work with the x in $x + 5$ like we worked with the 8 in Example 3.

> **W Hint**
> Write down the procedure for dividing a polynomial by a binomial in your notes, in your own words.

$$\begin{array}{r} 3x \\ x+5{\overline{\smash{)}3x^2 + 19x + 20}} \\ -(3x^2 + 15x)\downarrow \\ \hline 4x + 20 \end{array}$$

1) By what do we multiply x to get $3x^2$? $3x$
   Line up terms in the quotient according to exponents, so write $3x$ above $19x$.
2) Multiply $3x$ by $(x + 5)$: $3x(x + 5) = 3x^2 + 15x$.
3) Subtract $(3x^2 + 19x) - (3x^2 + 15x) = 4x$.
4) Bring down the $+20$.

338   CHAPTER 6   Polynomials and Polynomial Functions

Start the process again. Remember, work with the x in $x + 5$ like we worked with the 8 in Example 3.

$$\begin{array}{r} 3x + 4 \\ x + 5 \overline{)\smash{3x^2 + 19x + 20}} \\ -(3x^2 + 15x) \phantom{{}+20} \\ \hline 4x + 20 \\ -(4x + 20) \\ \hline 0 \end{array}$$

1) By what do we multiply x to get $4x$? 4
   Write $+4$ above $+20$.
2) Multiply 4 by $(x + 5)$: $4(x + 5) = 4x + 20$.
3) Subtract $(4x + 20) - (4x + 20) = 0$.
4) There are no more terms. The remainder is 0.

> **Hint**
> Notice that *like terms* are always lined up in the same columns.

Write the result.

$$\frac{3x^2 + 19x + 20}{x + 5} = 3x + 4$$

Check: $(x + 5)(3x + 4) = 3x^2 + 4x + 15x + 20 = 3x^2 + 19x + 20$ ✓

[**YOU TRY 4**] Divide.

a) $\dfrac{x^2 + 11x + 24}{x + 8}$ b) $\dfrac{2x^2 + 23x + 45}{x + 9}$

Next, we will look at a division problem with a remainder.

EXAMPLE 5 Divide $\dfrac{-11c + 16c^3 + 19 - 38c^2}{2c - 5}$.

Solution

When we write our long division problem, the polynomial in the numerator must be rewritten so that the exponents are in descending order. Then, perform the long division.

$$\begin{array}{r} 8c^2 \phantom{{}+ 19} \\ 2c - 5 \overline{)\smash{16c^3 - 38c^2 - 11c + 19}} \\ -(16c^3 - 40c^2) \phantom{{}- 11c + 19} \downarrow \\ \hline 2c^2 - 11c \phantom{{}+ 19} \end{array}$$

1) By what do we multiply $2c$ to get $16c^3$? $8c^2$
2) Multiply $8c^2(2c - 5) = 16c^3 - 40c^2$.
3) Subtract.
   $(16c^3 - 38c^2) - (16c^3 - 40c^2)$
   $= 16c^3 - 38c^2 - 16c^3 + 40c^2$
   $= 2c^2$
4) Bring down the $-11c$.

> **Hint**
> Notice how the parentheses and the distributive property are used in this example.

Repeat the process.

$$\begin{array}{r} 8c^2 + c \phantom{{}+ 19} \\ 2c - 5 \overline{)\smash{16c^3 - 38c^2 - 11c + 19}} \\ -(16c^3 - 40c^2) \phantom{{}- 11c + 19} \\ \hline 2c^2 - 11c \phantom{{}+ 19} \\ -(2c^2 - 5c) \phantom{{}+ 19} \downarrow \\ \hline -6c + 19 \end{array}$$

1) By what do we multiply $2c$ to get $2c^2$? c
2) Multiply $c(2c - 5) = 2c^2 - 5c$.
3) Subtract.
   $(2c^2 - 11c) - (2c^2 - 5c)$
   $= 2c^2 - 11c - 2c^2 + 5c$
   $= -6c$
4) Bring down the $+19$.

Hint
Are you writing out the example as you are reading it? Do you *understand* each step as it is being performed?

Continue.

$$\begin{array}{r} 8c^2 + c - 3 \\ 2c - 5 \overline{) 16c^3 - 38c^2 - 11c + 19} \\ \underline{-(16c^3 - 40c^2)} \\ 2c^2 - 11c \\ \underline{-(2c^2 - 5c)} \\ -6c + 19 \\ \underline{-(-6c + 15)} \\ 4 \end{array}$$

1) By what do we multiply $2c$ to get $-6c$? -3
2) Multiply $-3(2c - 5) = -6c + 15$.
3) Subtract.
$$(-6c + 19) - (-6c + 15)$$
$$= -6c + 19 + 6c - 15 = 4$$

We are done with the long division process. How do we know that? Because the degree of 4 (degree zero) is less than the degree of $2c - 5$ (degree one) we cannot divide anymore. *The remainder is* 4.

The answer is $\dfrac{16c^3 - 38c^2 - 11c + 19}{2c - 5} = 8c^2 + c - 3 + \dfrac{4}{2c - 5}$

Check: $(2c - 5)(8c^2 + c - 3) + 4 = 16c^3 + 2c^2 - 6c - 40c^2 - 5c + 15 + 4$
$= 16c^2 - 38c^2 - 11c + 19$ ✓

[YOU TRY 5] Divide $-23t^2 - 2 + 20t^3 - 11t$ by $5t + 3$.

As we saw in Example 5, we must write our polynomials so that the exponents are in descending order. We have to watch out for something else as well—missing terms. **If a polynomial is missing one or more terms, we put them into the polynomial with coefficients of zero.**

EXAMPLE 6 Divide $x^3 + 125$ by $x + 5$.

Solution

The degree of the polynomial $x^3 + 125$ is three, but it is missing the x^2-term and the x-term. We will insert these terms into the polynomial by giving them coefficients of zero.

$$x^3 + 125 = x^3 + 0x^2 + 0x + 125$$

Divide.
$$\begin{array}{r} x^2 - 5x + 25 \\ x + 5 \overline{) x^3 + 0x^2 + 0x + 125} \\ \underline{-(x^3 + 5x^2)} \\ -5x^2 + 0x \\ \underline{-(-5x^2 - 25x)} \\ 25x + 125 \\ \underline{-(25x + 125)} \\ 0 \end{array}$$

Hint
In your notes, summarize the procedures for dividing by a monomial and dividing by a polynomial with two or more terms. Include an example of each.

Therefore, $(x^3 + 125) \div (x + 5) = x^2 - 5x + 25$

Check: $(x + 5)(x^2 - 5x + 25) = x^3 - 5x^2 + 25x + 5x^2 - 25x + 125$
$= x^3 + 125$ ✓

[YOU TRY 6] Divide $\dfrac{2k^3 + 5k^2 + 91}{2k + 9}$.

3 Divide Polynomial Functions

Now that we've learned how to add, subtract, and multiply functions, let's learn how to divide functions.

> **Procedure** Dividing Functions
>
> Given the functions $f(x)$ and $g(x)$, the **quotient of f and g** is defined by
> $$\left(\frac{f}{g}\right)(x) = \frac{f(x)}{g(x)}, \text{ where } g(x) \neq 0.$$
> The domain of $\left(\frac{f}{g}\right)(x)$ is the intersection of the domains of $f(x)$ and $g(x)$ except for the values of x for which $g(x) = 0$.

EXAMPLE 7

Let $f(x) = 3x^2 + 19x + 20$ and $g(x) = x + 5$. Find

a) $\left(\frac{f}{g}\right)(x)$, and identify the value of x that is not in its domain.

b) $\left(\frac{f}{g}\right)(2)$.

Solution

a) $\left(\frac{f}{g}\right)(x) = \frac{f(x)}{g(x)}$

$= \frac{3x^2 + 19x + 20}{x + 5}$ Substitute the functions.

We can simplify this expression, but first let's look at the denominator of the quotient function. If $x = -5$, the denominator will equal zero. Therefore, -5 is not in the domain of $\left(\frac{f}{g}\right)(x)$.

W Hint
In your notes, summarize how you determine which values are not in the domain of $\frac{f(x)}{g(x)}$.

In Example 4 we used long division to find the quotient $\frac{3x^2 + 19x + 20}{x + 5} = 3x + 4$.

Therefore,
$$\left(\frac{f}{g}\right)(x) = 3x + 4, \text{ where } x \neq -5.$$

b) Since $\left(\frac{f}{g}\right)(x) = 3x + 4$, substitute 2 for x to find $\left(\frac{f}{g}\right)(2)$.

$\left(\frac{f}{g}\right)(2) = 3(2) + 4$ Substitute 2 for x.

$= 6 + 4$ Multiply.

$= 10$

Verify that this result is the same as $\frac{f(2)}{g(2)}$.

YOU TRY 7 Let $f(x) = 2x^2 + x - 3$ and $g(x) = x - 1$. Find

a) $\left(\dfrac{f}{g}\right)(x)$, and identify the value of x that is not in its domain.

b) $\left(\dfrac{f}{g}\right)(7)$.

ANSWERS TO YOU TRY EXERCISES

1) $4t^3 - t^2 - 9t$ 2) $-3u + \dfrac{3}{4} - \dfrac{1}{u} + \dfrac{5}{u^2}$ 3) $637\dfrac{5}{6}$ 4) a) $x + 3$ b) $2x + 5$

5) $4t^2 - 7t + 2 - \dfrac{8}{5t + 3}$ 6) $k^2 - 2k + 9 + \dfrac{10}{2k + 9}$ 7) a) $2x + 3$, where $x \neq 1$ b) 17

E Evaluate 6.5 Exercises

Do the exercises, and check your work.

Label the dividend, divisor, and quotient of each division problem.

1) $\dfrac{12c^3 + 20c^2 - 4c}{4c} = 3c^2 + 5c - 1$

2) $2p + 3 \overline{\smash{\big)}\, 10p^3 + p^2 - 25p - 6}$ with quotient $5p^2 - 7p - 2$

3) Explain, in your own words, how to divide a polynomial by a monomial.

4) When do you use long division to divide polynomials?

Objective 1: Divide a Polynomial by a Monomial

Divide.

5) $\dfrac{28k^4 + 8k^3 - 40k^2}{4k^2}$

6) $\dfrac{4a^5 - 10a^4 + 6a^3}{2a^3}$

7) $\dfrac{18u^7 + 18u^5 + 45u^4 - 72u^2}{9u^2}$

8) $\dfrac{-15m^6 + 10m^5 + 20m^4 - 35m^3}{5m^3}$

9) $(35d^5 - 7d^2) \div (-7d^2)$

10) $(-32q^6 - 8q^3 + 4q^2) \div (-4q^2)$

11) $\dfrac{9w^5 + 42w^4 - 6w^3 + 3w^2}{6w^3}$

12) $\dfrac{-54j^5 + 30j^3 - 9j^2 + 15}{9j}$

13) $(10v^7 - 36v^5 - 22v^4 - 5v^2 + 1) \div (4v^4)$

14) $(60z^5 + 3z^4 - 10z) \div (5z^2)$

Divide.

15) $\dfrac{90a^4b^3 + 60a^3b^3 - 40a^3b^2 + 100a^2b^2}{10ab^2}$

16) $\dfrac{24x^6y^6 - 54x^5y^4 - x^3y^3 + 12x^3y^2}{6x^2y}$

17) $(9t^5u^4 - 63t^4u^4 - 108t^3u^4 + t^3u^2) \div (-9tu^2)$

18) $(-45c^8d^6 - 15c^6d^5 + 60c^3d^5 + 30c^3d^3) \div (-15c^3d^2)$

19) Irene divides $16t^3 - 36t^2 + 4t$ by $4t$ and gets a quotient of $4t^2 - 9t$. Is this correct? Why or why not?

20) Kinh divides $\dfrac{15x^2 + 12x}{12x}$ and gets a quotient of $15x^2$. What was his mistake? What is the correct answer?

Objective 2: Divide a Polynomial by a Polynomial

Divide.

21) $\dfrac{g^2 + 9g + 20}{g + 5}$

22) $\dfrac{n^2 + 13n + 40}{n + 8}$

23) $\dfrac{p^2 + 8p + 12}{p + 2}$

24) $\dfrac{v^2 + 13v + 12}{v + 1}$

25) $\dfrac{k^2 + 4k - 45}{k + 9}$

26) $\dfrac{m^2 - 6m - 27}{m + 3}$

27) $\dfrac{h^2 + 5h - 24}{h - 3}$

28) $\dfrac{u^2 - 11u + 30}{u - 5}$

29) $\dfrac{4a^3 - 24a^2 + 29a + 15}{2a - 5}$

30) $\dfrac{28b^3 - 26b^2 + 41b - 15}{7b - 3}$

31) $(p + 45p^2 - 1 + 18p^3) \div (6p + 1)$

32) $(17z^2 - 10 - 12z^3 + 32z) \div (4z + 5)$

33) $(6t^2 - 7t + 4) \div (t - 5)$

34) $(7d^2 + 57d - 4) \div (d + 9)$

35) $\dfrac{61z + 12z^3 - 37 + 44z^2}{3z + 5}$

36) $\dfrac{23k^3 + 22k - 8 + 6k^4 + 44k^2}{6k - 1}$

37) $\dfrac{w^3 + 64}{w + 4}$

38) $\dfrac{a^3 - 27}{a - 3}$

39) $(16r^3 + 58r^2 - 9) \div (8r - 3)$

40) $(50c^3 + 7c + 4) \div (5c + 2)$

Mixed Exercises: Objectives 1 and 2
Divide.

41) $\dfrac{6x^4y^4 + 30x^4y^3 - x^2y^2 + 3xy}{6x^2y^2}$

42) $\dfrac{12v^2 - 23v + 14}{3v - 2}$

43) $\dfrac{-8g^4 + 49g^2 + 36 - 25g - 2g^3}{4g - 9}$

44) $(12c^2 + 6c - 30c^3 + 48c^4) \div (-6c)$

45) $\dfrac{6t^2 - 43t - 20}{t - 8}$

46) $\dfrac{-14u^3v^3 + 7u^2v^3 + 21uv + 56}{7u^2v}$

47) $(8n^3 - 125) \div (2n - 5)$

48) $(12a^4 - 19a^3 + 22a^2 - 9a - 20) \div (3a - 4)$

49) $(13x^2 - 7x^3 + 6 + 5x^4 - 14x) \div (x^2 + 2)$

50) $(18m^4 - 66m^3 + 39m^2 + 11m - 7) \div (6m^2 - 1)$

51) $\dfrac{-12a^3 + 9a^2 - 21a}{-3a}$

52) $\dfrac{64r^3 + 27}{4r + 3}$

53) $\dfrac{10h^4 - 6h^3 - 49h^2 + 27h + 19}{2h^2 - 9}$

54) $\dfrac{16w^2 - 3 - 7w + 15w^4 - 5w^3}{5w^2 + 7}$

55) $\dfrac{6d^4 + 19d^3 - 8d^2 - 61d - 40}{2d^2 + 7d + 5}$

56) $\dfrac{8x^4 + 2x^3 - 13x^2 - 53x + 14}{2x^2 + 5x + 7}$

57) $\dfrac{9c^4 - 82c^3 - 41c^2 + 9c + 16}{c^2 - 10c + 4}$

58) $\dfrac{15n^4 - 16n^3 - 31n^2 + 50n - 22}{5n^2 - 7n + 2}$

59) $\dfrac{k^4 - 81}{k^2 + 9}$

60) $\dfrac{b^4 - 16}{b^2 - 4}$

61) $\dfrac{49a^4 - 15a^2 - 14a^3 + 5a^6}{-7a^3}$

62) $\dfrac{9q^2 + 26q^4 + 8 - 6q - 4q^3}{2q^2}$

63) $\left(x^2 + \dfrac{13}{2}x + 3\right) \div (2x + 1)$

64) $\left(k^2 + \dfrac{11}{3}k + 2\right) \div (3k + 2)$

65) $\left(2w^2 + \dfrac{10}{3}w - 8\right) \div (3w - 4)$

66) $\left(3y^2 - \dfrac{41}{4}y + 9\right) \div (4y - 3)$

For each rectangle, find a polynomial that represents the missing side.

67) $y - 6$

Find the length if the area is given by $4y^2 - 23y - 6$.

68) $3x + 2$

Find the width if the area is given by $6x^2 + x - 2$.

69) $9a^3$

Find the width if the area is given by $18a^5 - 45a^4 + 9a^3$.

70)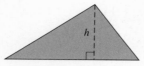

Find the length if the area is given by
$9w^3 + 6w^2 - 24w$.

71) Find the base of the triangle if the area is given by
$6h^3 + 3h^2 + h$.

72) Find the base of the triangle if the area is given by
$6n^3 - 2n^2 + 10n$.

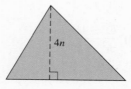

Objective 3: Divide Polynomial Functions

For each pair of functions, find $\left(\dfrac{f}{g}\right)(x)$ and identify any values of x that are not in the domain of the quotient function.

73) $f(x) = 5x^2 + 6x - 27$, $g(x) = x + 3$

74) $f(x) = 8x^2 - 22x + 15$, $g(x) = 4x - 5$

75) $f(x) = 12x^3 - 18x^2 + 2x$, $g(x) = 2x$

76) $f(x) = 24x^4 - 10x^2 + 9x$, $g(x) = 6x$

77) $f(x) = 3x^4 - 10x^3 + 9x^2 + 2x - 4$, $g(x) = x - 1$

78) $f(x) = 6x^4 + 11x^3 - 14x^2 - 27x - 4$, $g(x) = 3x + 4$

Use the following functions for Exercises 79–86.

Let $f(x) = 4x^2 - 1$, $g(x) = 2x + 1$, and $h(x) = 3x$. Find each of the following.

79) $\left(\dfrac{f}{g}\right)(x)$

80) $\left(\dfrac{h}{g}\right)(x)$

81) $\left(\dfrac{f}{g}\right)(5)$

82) $\left(\dfrac{h}{g}\right)(-1)$

83) $\left(\dfrac{g}{h}\right)(x)$

84) $\left(\dfrac{f}{g}\right)\left(\dfrac{1}{4}\right)$

85) $\left(\dfrac{g}{h}\right)\left(-\dfrac{2}{3}\right)$

86) $\left(\dfrac{h}{g}\right)(0)$

R Rethink

R1) What is the most common mistake you make when performing polynomial long division?

R2) How do you include the remainder in your final answer?

R3) When performing polynomial long division, why do you line up like terms in the same column?

R4) Which exercises in this section do you find most challenging?

Group Activity — The Group Activity can be found online on Connect.

emPOWERme: The Right Time and Place for Homework

Are you doing your homework in an environment that maximizes your chances for success? To find out, circle the number that best applies for each question using the following scale:

1) I keep my cell phone on my desk when I do my homework, in case anyone calls me.

    Highly accurate   4   3   2   1   Not true at all

2) The TV is on when I do my homework.

    Highly accurate   4   3   2   1   Not true at all

3) I do my homework in the same room my family uses to socialize.

    Highly accurate   4   3   2   1   Not true at all

4) The desk I use to do my homework is a total mess.

    Highly accurate   4   3   2   1   Not true at all

5) I often do my homework in bed.

    Highly accurate   4   3   2   1   Not true at all

6) I only have one pen or pencil handy with me when I start my homework.

    Highly accurate   4   3   2   1   Not true at all

7) I try to do my homework on the bus so that I can get it over with before I get home.

    Highly accurate   4   3   2   1   Not true at all

8) I have my e-mail account open while I am doing my homework.

    Highly accurate   4   3   2   1   Not true at all

9) If I haven't finished my homework, I do it right before class starts.

    Highly accurate   4   3   2   1   Not true at all

10) I play loud music in my headphones while I work on my homework.

    Highly accurate   4   3   2   1   Not true at all

Scoring: Total the numbers you have circled.

If the score is below 15, you do your homework in an environment that allows you to focus and perform your best. If your score is 16–25, you could make some changes to the place where you do your homework that would improve your performance.

If your score is above 25, the environment in which you do your homework is probably negatively affecting your performance. Consider doing your homework in the library or some other quiet place outside your home.

Chapter 6: Summary

Definition/Procedure	Example
6.1 The Rules of Exponents	
Let a and b be real numbers and m and n be integers. The following rules apply:	
Product Rule: $a^m \cdot a^n = a^{m+n}$ (p. 294)	$x^8 \cdot x^2 = x^{10}$
Basic Power Rule: $(a^m)^n = a^{mn}$ (p. 295)	$(t^3)^5 = t^{15}$
Power Rule for a Product: $(ab)^n = a^n b^n$ (p. 295)	$(2c)^4 = 2^4 c^4 = 16c^4$
Power Rule for a Quotient: $\left(\dfrac{a}{b}\right)^n = \dfrac{a^n}{b^n}$, where $b \neq 0$. (p. 295)	$\left(\dfrac{w}{5}\right)^3 = \dfrac{w^3}{5^3} = \dfrac{w^3}{125}$
Zero Exponent: If $a \neq 0$, then $a^0 = 1$. (p. 297)	$(-9)^0 = 1$
Negative Exponent: If n is a natural number and $a \neq 0$ and $b \neq 0$, then $a^{-n} = \left(\dfrac{1}{a}\right)^n = \dfrac{1}{a^n}$, and $\left(\dfrac{a}{b}\right)^{-n} = \left(\dfrac{b}{a}\right)^n$. (p. 298)	Evaluate. $\left(\dfrac{5}{2}\right)^{-3} = \left(\dfrac{2}{5}\right)^3 = \dfrac{2^3}{5^3} = \dfrac{8}{125}$
If $a \neq 0$ and $b \neq 0$, then $\dfrac{a^{-m}}{b^{-n}} = \dfrac{b^n}{a^m}$. (p. 299)	Rewrite each expression with positive exponents. a) $\dfrac{x^{-3}}{y^{-7}} = \dfrac{y^7}{x^3}$ b) $\dfrac{14m^{-6}}{n^{-1}} = \dfrac{14n}{m^6}$
Quotient Rule: If $a \neq 0$, then $\dfrac{a^m}{a^n} = a^{m-n}$. (p. 300)	Simplify. $\dfrac{4^9}{4^6} = 4^{9-6} = 4^3 = 64$
6.2 More on Exponents and Scientific Notation	
Combining the Rules of Exponents (p. 304)	Simplify. $\left(\dfrac{a^4}{2a^7}\right)^{-5} = \left(\dfrac{2a^7}{a^4}\right)^5 = (2a^3)^5 = 32a^{15}$
Scientific Notation A number is in **scientific notation** if it is written in the form $a \times 10^n$, where $1 \leq |a| < 10$ and n is an integer. That is, a is a number with one nonzero digit to the left of the decimal point. (p. 307)	Write in scientific notation. a) $78{,}000 \to 78{,}000. \to 7.8 \times 10^4$ b) $0.00293 \to 0.00293 \to 2.93 \times 10^{-3}$
Converting from Scientific Notation (p. 307)	Write without exponents. a) $5 \cdot 10^{-4} \to .0005. \to 0.0005$ b) $1.7 \times 10^6 = 1.700000 \to 1{,}700{,}000$
6.3 Addition and Subtraction of Polynomials and Polynomial Functions	
A *polynomial in x* is the sum of a finite number of terms of the form ax^n where n is a whole number and a is a real number. (p. 314)	Identify each term in the polynomial, the coefficient and degree of each term, and the degree of the polynomial. $5a^4b^2 - 16a^3b^2 - 4a^2b^3 + ab + 9b$
The **degree of a term** equals the exponent on its variable. If a term has more than one variable, the degree equals the *sum* of the exponents on the variables. The **degree of the polynomial** equals the highest degree of any nonzero term.	(see table below)

Term	Coefficient	Degree
$5a^4b^2$	5	6
$-16a^3b^2$	-16	5
$-4a^2b^3$	-4	5
ab	1	2
$9b$	9	1

The degree of the polynomial is 6.

Definition/Procedure	Example
To *add polynomials,* add like terms. Polynomials may be added horizontally or vertically. **(p. 315)**	Add the polynomials. $(6n^2 + 7n - 14) + (-2n^2 + 6n + 3)$ $= [6n^2 + (-2n^2)] + (7n + 6n) + (-14 + 3)$ $= 4n^2 + 13n - 11$
To *subtract two polynomials,* change the sign of each term in the second polynomial. Then, add the polynomials. **(p. 316)**	Subtract. $(3h^3 - 7h^2 + 8h + 4) - (12h^3 - 8h^2 + 3h + 9)$ $= (3h^3 - 7h^2 + 8h + 4) + (-12h^3 + 8h^2 - 3h - 9)$ $= -9h^3 + h^2 + 5h - 5$
$f(x) = 3x^2 + 8x - 4$ is an example of a *polynomial function* since $3x^2 + 8x - 4$ is a polynomial and since each real number that is substituted for x produces only one value for the expression. Finding $f(4)$ is the same as evaluating $3x^2 + 8x - 4$ when $x = 4$. **(p. 318)**	If $f(x) = 3x^2 + 8x - 4$, find $f(4)$. $f(4) = 3(4)^2 + 8(4) - 4$ Substitute 4 for x. $= 3(16) + 32 - 4$ $= 48 + 32 - 4$ $= 76$
Adding and Subtracting Polynomial Functions Given the functions $f(x)$ and $g(x)$, the **sum and difference of f and g** are defined by 1) $(f + g)(x) = f(x) + g(x)$ 2) $(f - g)(x) = f(x) - g(x)$ The domain of $(f + g)(x)$ and $(f - g)(x)$ is the intersection of the domains of $f(x)$ and $g(x)$. **(p. 318)**	Let $f(x) = x^2 - 5x + 3$ and $g(x) = 7x - 9$. Find $(f + g)(x)$ and $(f + g)(-5)$. $(f + g)(x) = f(x) + g(x)$ $= (x^2 - 5x + 3) + (7x - 9)$ $= x^2 + 2x - 6$ Use $(f + g)(x) = x^2 + 2x - 6$ to find $(f + g)(-5)$. $(f + g)(-5) = (-5)^2 + 2(-5) - 6$ $= 25 - 10 - 6$ $= 9$

6.4 Multiplication of Polynomials and Polynomial Functions

When multiplying a *monomial* and a *polynomial,* use the distributive property. **(p. 325)**	Multiply. $4a^3(-3a^2 + 7a - 2)$ $= (4a^3)(-3a^2) + (4a^3)(7a) + (4a^3)(-2)$ Distribute. $= -12a^5 + 28a^4 - 8a^3$
To *multiply two polynomials,* multiply each term in the second polynomial by each term in the first polynomial. Then, combine like terms. **(p. 325)**	Multiply. $(2c + 5)(c^2 - 3c + 6)$ $= (2c)(c^2) + (2c)(-3c) + (2c)(6) + (5)(c^2) + (5)(-3c) + (5)(6)$ $= 2c^3 - 6c^2 + 12c + 5c^2 - 15c + 30$ $= 2c^3 - c^2 - 3c + 30$
Multiplying Two Binomials We can use FOIL to multiply two binomials. **FOIL** stands for First Outer Inner Last. Multiply the binomials, then add like terms. **(p. 326)**	Use FOIL to multiply $(3k - 2)(k + 4)$. $ \text{F} \quad \text{O} \quad \text{I} \quad \text{L}$ $(3k - 2)(k + 4) = 3k \cdot k + 3k \cdot 4 - 2 \cdot k - 2 \cdot 4$ $= 3k^2 + 12k - 2k - 8$ $= 3k^2 + 10k - 8$
Special Products 1) $(a + b)(a - b) = a^2 - b^2$ 2) $(a + b)^2 = a^2 + 2ab + b^2$ 3) $(a - b)^2 = a^2 - 2ab + b^2$ **(p. 328)**	1) Multiply: $(y + 6)(y - 6) = y^2 - 6^2 = y^2 - 36$ 2) Expand: $(t + 9)^2 = t^2 + 2(t)(9) + 9^2 = t^2 + 18t + 81$ 3) Expand $(4u - 3)^2$. $(4u - 3)^2 = (4u)^2 - 2(4u)(3) + 3^2$ $= 16u^2 - 24u + 9$

Definition/Procedure	Example
Multiplying Functions Given the functions $f(x)$ and $g(x)$, the **product of f and g** is defined by $(fg)(x) = f(x) \cdot g(x)$. The domain of $(fg)(x)$ is the intersection of the domains of $f(x)$ and $g(x)$. (p. 331)	Let $f(x) = 3x - 7$ and $g(x) = 2x + 5$. Find $(fg)(x)$. $(fg)(x) = f(x) \cdot g(x)$ $= (3x - 7)(2x + 5)$ $= 6x^2 + x - 35$

6.5 Division of Polynomials and Polynomial Functions

To *divide a polynomial by a monomial*, divide *each term* in the polynomial by the monomial and simplify. (p. 336)	Divide $\dfrac{18r^4 + 2r^3 - 9r^2 + 6r - 10}{2r^2}$. $= \dfrac{18r^4}{2r^2} + \dfrac{2r^3}{2r^2} - \dfrac{9r^2}{2r^2} + \dfrac{6r}{2r^2} - \dfrac{10}{2r^2}$ $= 9r^2 + r - \dfrac{9}{2} + \dfrac{3}{r} - \dfrac{5}{r^2}$
To *divide a polynomial by another polynomial* containing two or more terms, use *long division*. (p. 337)	Divide $\dfrac{12m^3 - 32m^2 - 17m + 25}{6m + 5}$. $\;\;2m^2 - 7m + 3$ $6m+5\overline{)\,12m^3 - 32m^2 - 17m + 25}$ $-(12m^3 + 10m^2)$ $\;-42m^2 - 17m$ $-(-42m^2 - 35m)$ $\;18m + 25$ $-(18m + 15)$ Remainder ← 10 $\dfrac{12m^3 - 32m^2 - 17m + 25}{6m + 5} = 2m^2 - 7m + 3 + \dfrac{10}{6m + 5}$
Dividing Polynomial Functions Given the functions $f(x)$ and $g(x)$, the **quotient of f and g** is defined by $\left(\dfrac{f}{g}\right)(x) = \dfrac{f(x)}{g(x)}$, where $g(x) \neq 0$. The domain of $\left(\dfrac{f}{g}\right)(x)$ is the intersection of the domains of $f(x)$ and $g(x)$ except the values of x for which $g(x) = 0$. (p. 341)	Let $f(x) = 5x$ and $g(x) = x - 6$. Find $\left(\dfrac{f}{g}\right)(x)$. $\left(\dfrac{f}{g}\right)(x) = \dfrac{f(x)}{g(x)}$ $= \dfrac{5x}{x - 6}$, where $x \neq 6$

Chapter 6: Review Exercises

(6.1 and 6.2) Evaluate using the rules of exponents.

1) $\dfrac{3^{10}}{3^6}$

2) 8^{-2}

3) $\left(\dfrac{5}{4}\right)^{-3}$

4) $-4^0 + 7^0$

Simplify. Assume all variables represent nonzero real numbers. Write the answers with positive exponents.

5) $(z^6)^3$

6) $(4p^3)(-3p^7)$

7) $\dfrac{70r^9}{10r^4}$

8) $(-5c^4)^2$

9) $(-9t)(6t^6)$

10) $\dfrac{6m^{10}}{24m^6}$

11) $\dfrac{k^3}{k^{11}}$

12) $\dfrac{d^{-6}}{d^3}$

13) $(-2a^2b)^3(5a^{-12}b)$

14) $\dfrac{x^5 y^{-3}}{x^8 y^{-4}}$

15) $\left(\dfrac{3pq^{-10}}{2p^{-2}q^5}\right)^{-2}$

16) $(7c^{-8}d^2)(3c^{-2}d)^2$

17) $\left(\dfrac{40}{21}x^{10}\right)(3x^{-12})\left(\dfrac{49}{20}x^2\right)$

18) $\left(\dfrac{4r^{-3}t}{s^2}\right)^{-3}\left(\dfrac{3t^{-5}s}{r^2}\right)^{-2}\left(\dfrac{2rs^2}{t^3}\right)^4$

Simplify. Assume that the variables represent nonzero integers. Write the final answer so that the exponents have positive coefficients.

19) $x^{5t} \cdot x^{3t}$

20) $\dfrac{r^{9a}}{r^{3a}}$

21) $(y^{2p})^3$

22) $\dfrac{w^{-12a}}{w^{-3a}}$

23) True or False: $-x^2 = (-x)^2$ for every real number value of x. Explain your answer.

24) True or False: $(5y)^{-3} = \dfrac{-125}{y^3}$ if $y \ne 0$. Explain your answer.

Write each number without an exponent.

25) 9.38×10^5

26) -4.185×10^2

27) $1.05 \cdot 10^{-6}$

28) $2 \cdot 10^4$

Write each number in scientific notation.

29) 0.0000575

30) 36,940

31) 32,000,000

32) 0.0000004

Perform the operation as indicated. Write the final answer without an exponent.

33) $\dfrac{8 \cdot 10^6}{2 \cdot 10^{13}}$

34) $(9 \cdot 10^{-8})(4 \cdot 10^7)$

35) $\dfrac{-3 \times 10^{10}}{-4 \times 10^6}$

36) $(-4.2 \times 10^2)(3.1 \times 10^3)$

For Exercises 37 and 38, write each of the numbers in scientific notation, then solve the problem. Write the answer without exponents.

37) Eight porcupines have a total of about 2.4×10^5 quills on their bodies. How many quills would one porcupine have?

38) One molecule of water has a mass of $2.99 \cdot 10^{-23}$ g. Find the mass of 100,000,000 molecules.

(6.3) Identify each term in the polynomial, the coefficient and degree of each term, and the degree of the polynomial.

39) $4r^3 - 7r^2 + r + 5$

40) $x^3y + 6xy^2 - 8xy + 11y$

41) Evaluate $-x^2y^2 - 7xy + 2x + 5$ for $x = -3$ and $y = 2$.

Add or subtract as indicated.

42) $(5t^2 + 11t - 4) - (7t^2 + t - 9)$

43) $\begin{array}{r} 5.8p^3 - 1.2p^2 + p - 7.5 \\ + \underline{2.1p^3 + 6.3p^2 + 3.8p + 3.9} \end{array}$

44) $\left(\dfrac{4}{9}w^2 - \dfrac{3}{8}w + \dfrac{2}{5}\right) + \left(\dfrac{2}{9}w^2 + \dfrac{5}{8}w - \dfrac{9}{20}\right)$

45) Subtract $3a^2b^2 - 10a^2b + ab + 6$ from $a^2b^2 + 7a^2b - 3ab + 11$.

46) Find the sum of $6xy + 4x - y - 10$ and $-4xy + 2y + 3$, and subtract it from $-6xy - 7x + y + 2$.

47) Write a fifth-degree polynomial in x that does not contain a third-degree term.

48) Find the polynomial that represents the perimeter of the rectangle.

$d^2 + 3d + 5$

$d^2 - 5d + 2$

49) Let $f(x) = -2x^2 + 5x - 8$. Find each of the following functions values.

a) $f(-3)$

b) $f\left(\dfrac{1}{2}\right)$

50) Let $f(x) = 3x + 2$ and $g(x) = x^2 + 6x - 10$. Find each of the following.

a) $(f + g)(x)$

b) $(f + g)(0)$

c) $(f - g)(x)$

d) $(f - g)(-4)$

51) The number of cruise ships, $N(x)$, operating in North America from 2002–2006 can be modeled by the polynomial function $N(x) = 0.5x^3 - 4.357x^2 + 16.429x + 122.686$ where x is the number of years after 2002. (www.census.gov)

a) Approximately how many cruise ships were operating in 2002?

b) Approximately how many cruise ships were operating in 2006?

c) Find $N(1)$, and explain what it means in the context of the problem.

52) $R(x) = 20x$ is the revenue function for the sale of x children's soccer uniforms, in dollars. The cost to produce x soccer uniforms, in dollars, is

$C(x) = 14x + 400$

a) Find the profit function, $P(x)$, that describes the profit from the sale of x uniforms.

b) What is the profit from the sale of 200 uniforms?

(6.4) Find each product.

53) $-6m^3(9m^2 - 3m + 7)$

54) $7u^4v^2(-8u^2v + 7uv^2 + 12u - 3)$

55) $(2w + 5)(-12w^3 + 6w^2 - 2w + 3)$

56) $(x^2 + 4x - 11)(12x^4 - 7x^2 + 9)$

57) $(y - 7)(y + 8)$

58) $(3n - 7)(2n - 9)$

59) $(ab + 5)(ab + 6)$

60) $(9r + 2s)(r - s)$

61) $-(4d + 3)(6d + 7)$

62) $6c^3(4c - 5)(c - 2)$

63) $(p + 3)(p - 6)(p + 2)$

64) $(z + 4)(z + 1)(z + 5)$

65) $\left(\dfrac{3}{5}m + 2\right)\left(\dfrac{1}{3}m - 4\right)$

66) $\left(\dfrac{2}{9}t - 5\right)\left(\dfrac{1}{10}t - 3\right)$

67) $(z + 9)(z - 9)$

68) $\left(\dfrac{1}{5}n - 2\right)\left(\dfrac{1}{5}n + 2\right)$

69) $\left(\dfrac{7}{8} - r^2\right)\left(\dfrac{7}{8} + r^2\right)$

70) $\left(2a - \dfrac{1}{3}b\right)\left(2a + \dfrac{1}{3}b\right)$

Expand.

71) $(b + 7)^2$

72) $(x - 10)^2$

73) $(5q - 2)^2$

74) $(7 - 3y)^2$

75) $(x - 2)^3$

76) $(p + 10)^3$

77) $-2(3c - 4)^2$

78) $6w(w + 3)^2$

79) $[(m - 5) + n]^2$

80) $[(3r + 2t) + 1][(3r + 2t) - 1]$

81) Let $f(x) = 2x - 9$ and $g(x) = -5x + 4$. Find each of the following.

   a) $(fg)(x)$ b) $(fg)(2)$

82) If $f(x) = 3x - 1$ and $(fg)(x) = 1 - 3x$, find $g(x)$.

83) Write an expression for the a) area and b) perimeter of the rectangle.

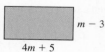

84) Express the volume of the cube as a polynomial.

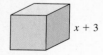

(6.5) Divide

85) $\dfrac{8t^5 - 14t^4 - 20t^3}{2t^3}$

86) $\dfrac{16p^4 + 56p^3 - 32p^2 + 8p}{-8p}$

87) $\dfrac{c^2 + 8c - 20}{c - 2}$

88) $\dfrac{y^2 - 15y + 56}{y - 8}$

89) $\dfrac{12r^3 - 13r^2 - 5r + 6}{3r + 2}$

90) $\dfrac{-66h^3 - 5h^2 + 45h - 4}{11h - 1}$

91) $(15x^4y^4 - 42x^3y^4 - 6x^2y + 10y) \div (-6x^2y)$

92) $(56a^6b^6 + 21a^4b^5 - 4a^3b^4 + a^2b - 7ab) \div (7a^3b^3)$

93) $(6q^2 + 2q - 35) \div (3q + 7)$

94) $(12r^2 - 16r + 11) \div (6r + 1)$

95) $\dfrac{23a - 7 + 15a^2}{5a - 4}$

96) $\dfrac{6m^4 + 2m^3 + 7m^2 + 5m - 20}{2m^2 + 5}$

97) $\dfrac{8t^4 + 32t^3 - 43t^2 - 44t + 48}{8t^2 - 11}$

98) $\dfrac{b^3 - 64}{b - 4}$

99) $\dfrac{f^3 + 125}{f + 5}$

100) $\dfrac{-23 - 46w + 32w^3}{4w + 3}$

101) $\dfrac{8k^2 - 8 + 15k^3}{3k - 2}$

102) $(7u^4 - 69u^3 + 15u^2 - 37u + 12) \div (u^2 - 10u + 3)$

103) $(6c^4 + 13c^3 - 21c^2 - 9c + 10) \div (2c^2 + 5c - 4)$

104) Find the base of the triangle if the area is given by $15y^2 + 12y$.

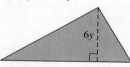

105) Find the length of the rectangle if the area is given by $6x^3 - x^2 + 13x - 10$.

106) Let $f(x) = x^2 + 6x + 8$, $g(x) = x + 4$, and $h(x) = 7x$. Find each of the following.

   a) $\left(\dfrac{f}{g}\right)(x)$

   b) $\left(\dfrac{f}{g}\right)(-9)$

   c) $\left(\dfrac{g}{h}\right)(x)$

   d) $\left(\dfrac{g}{h}\right)(3)$

Chapter 6: Test

1) Evaluate each expression.

   a) -3^4
   b) 2^{-5}
   c) $-6^0 - 9^0$
   d) $\left(\dfrac{3}{10}\right)^{-3}$
   e) $\dfrac{2^{11}}{2^{17}}$

Simplify. Assume all variables represent nonzero real numbers. Write the answers with positive exponents.

2) $(-3p^4)(10p^8)$

3) $\dfrac{a^5 b}{a^9 b^7}$

4) $(2y^{-4})^6 \left(\dfrac{1}{2}y^5\right)^3$

5) $\left(\dfrac{36xy^8}{54x^3 y^{-1}}\right)^{-2}$

6) $t^{10k} \cdot t^{3k}$

7) Write $7.283 \cdot 10^5$ without exponents.

8) Write 0.000165 in scientific notation.

9) Divide $\dfrac{-7.5 \times 10^{12}}{1.5 \times 10^8}$. Write the answer without exponents.

10) An electron is a subatomic particle with a mass of 9.1×10^{-28}g. What is the mass of 10,000,000,000 electrons? Write the answer in scientific notation.

11) Given the polynomial $5p^3 - p^2 + 12p + 9$,

    a) what is the coefficient of p^2?

    b) what is the degree of the polynomial?

12) Evaluate $-m^2 + 3n$ when $m = -5$ and $n = 8$.

In Exercises 13–23, perform the indicated operations.

13) $(10r^3 s^2 + 7r^2 s^2 - 11rs + 5) + (4r^3 s^2 - 9r^2 s^2 + 6rs + 3)$

14) Subtract $5j^2 - 2j + 9$ from $11j^2 - 10j + 3$.

15) $(c - 8)(c - 7)$

16) $(3y + 5)(2y + 1)$

17) $\left(u + \dfrac{3}{4}\right)\left(u - \dfrac{3}{4}\right)$

18) $(2a - 5b)(3a + b)$

19) $(7m - 5)^2$

20) $6(-n^3 + 4n - 2) - 3(2n^3 + 5n^2 + 8n - 1) + 4n(n^2 - 7n + 3)$

21) $(3 - 8m)(2m^2 + 4m - 7)$

22) $3x(x + 4)^2$

23) $[(5a - b) - 3]^2$

24) Expand $(s - 4)^3$.

Divide.

25) $\dfrac{r^2 + 10r + 21}{r + 7}$

26) $\dfrac{24t^5 - 60t^4 + 12t^3 - 8t^2}{12t^3}$

27) $(38v - 31 + 30v^3 - 51v^2) \div (5v - 6)$

28) Write an expression for the base of the triangle if the area is given by $12k^2 + 28k$.

29) A company's revenue, $R(x)$ in thousands of dollars, from the sale of x frozen pizzas (in thousands) is given by $R(x) = 5x + 9$. The cost, $C(x)$ in thousands of dollars, of producing these pizzas is given by $C(x) = 3x + 8$.

    a) Find the profit function, $P(x)$, that describes the profit from the sale of x pizzas.

    b) What is the profit from the sale of 8000 frozen pizzas?

30) Let $f(x) = x^2 - 5x - 24$, $g(x) = x - 8$, and $h(x) = x - 2$. Find each of the following.

    a) $(f + g)(x)$
    b) $(f + g)(-2)$
    c) $(gh)(x)$
    d) $(gh)(4)$
    e) $\left(\dfrac{f}{g}\right)(x)$ and any values not in its domain
    f) $\left(\dfrac{f}{g}\right)(7)$

Chapter 6: Cumulative Review for Chapters 1–6

1) Given the set of numbers
$$\left\{\frac{6}{11}, -14, 2.7, \sqrt{19}, 43, 0.\overline{65}, 0, 8.21079314\ldots\right\}$$
list the

   a) whole numbers.
   b) integers.
   c) rational numbers.

2) Evaluate $-2^4 + 3 \cdot 8 \div (-2)$.

3) Divide $2\frac{6}{7} \div 1\frac{4}{21}$.

Solve.

4) $-\dfrac{12}{5}c - 7 = 20$

5) $6(w + 4) + 2w = 1 + 8(w - 1)$

6) $|9 - 2n| + 5 = 18$

7) $A = \dfrac{1}{2}h(b_1 + b_2)$ for b_2

8) Solve the compound inequality, and write the answer in interval notation.
$$3y + 16 < 4 \quad \text{or} \quad 8 - y \geq 7$$

9) *Write an equation in one variable and solve.* How many milliliters of a 12% alcohol solution and how many milliliters of a 4% alcohol solution must be mixed to obtain 60 ml of a 10% alcohol solution?

10) Find the x- and y-intercepts of $5x - 2y = 10$ and sketch a graph of the equation.

11) Graph $x = -3$.

12) Write an equation of the line containing the points $(-5, 8)$ and $(1, 2)$. Express the answer in standard form.

13) What is the domain of $g(x) = \dfrac{1}{3}x + 5$?

14) Solve this system using the elimination method.
$$-7x + 2y = 6$$
$$9x - y = 8$$

15) *Write a system of two equations in two variables and solve.* The length of a rectangle is 7 cm less than twice its width. The perimeter of the rectangle is 76 cm. What are the dimensions of the figure?

Simplify. Assume all variables represent nonzero real numbers. The answers should not contain negative exponents.

16) $-5(3w^4)^2$

17) $\left(\dfrac{2n^{-10}}{n^{-4}}\right)^3$

18) $p^{10k} \cdot p^{4k}$

Perform the indicated operations.

19) $(4q^2 + 11q - 2) - 3(6q^2 - 5q + 4) + 2(-10q - 3)$

20) $(4g - 9)(4g + 9)$

21) $\dfrac{8a^4b^4 - 20a^3b^2 + 56ab + 8b}{8a^3b^3}$

22) $\dfrac{17v^3 - 22v + 7v^4 + 24 - 47v^2}{7v - 4}$

23) $(2p^3 + 5p^2 - 11p + 9) \div (p + 4)$

24) $(2c - 5)(c - 3)^2$

25) Let $f(x) = x^2 - 6$ and $g(x) = x^2 + 5x + 4$. Find

   a) $(fg)(x)$.
   b) $(fg)(2)$.

CHAPTER 7

Factoring Polynomials

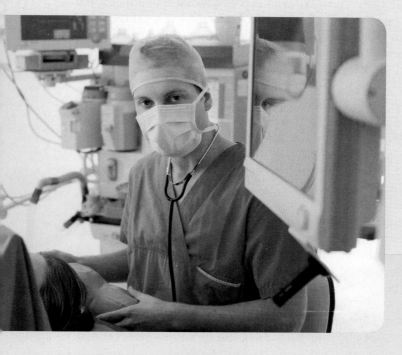

OUTLINE

Study Strategies: Working with a Study Group

7.1 The Greatest Common Factor and Factoring by Grouping

7.2 Factoring Trinomials

7.3 Special Factoring Techniques

Putting It All Together

7.4 Solving Quadratic Equations by Factoring

7.5 Applications of Quadratic Equations

Group Activity

emPOWERme: Switch "You" to "I"

Math at Work:
Nurse Anesthetist

Carl Burton plays an essential role in a team of medical professionals that performs surgery and other types of procedures. As a nurse anesthetist, he ensures that the patient is comfortable and pain-free during the procedure. "I try to imagine myself or a loved one on the operating table when I am doing my job," Carl says. "I want them to enjoy the same safety and comfort that I would expect."

When asked what the most important aspects of his job are, Carl points to two areas in which he has excelled since college: math and people-skills. "A nurse anesthetist has to have a lot of math capability at his or her fingertips," Carl describes. "Dosages need to be calculated and prepared, conversions need to be made between units, and patients' vital signs need be constantly measured."

"As for people-skills," Carl continues, "a team in an operating room should work together like a well-oiled machine. Everyone has to understand the information the other team-members need and be able to give it to them at the right time, and in a clear manner." Carl was raised in a small rural town. It was not until he went to college that he was exposed to many different types of people and learned the skills of collaboration and communication that are so vital to his career today.

In this chapter, we will cover the topic of factoring polynomials. We will also offer some advice on how to work effectively with your fellow students in the context of a study group.

 **Study Strategies** Working with a Study Group

You have numerous resources available for learning math in college: instructors, textbooks, tutors, math labs, and so on. However, in many ways the most valuable educational resource on campus is one that many students don't take full advantage of: other students. The students in your classes have the same responsibilities in terms of what they need to learn for assignments and tests that you do, and working together you can almost always accomplish course goals more easily than you could alone. Below are some strategies you can use to form and work effectively in a study group:

- Pick the members of your study group based on what they have to offer you and the other group members in terms of learning.
- Three to five people is typically a good size for a study group.
- Identify the specific goal the study group has. This goal can be short-term (to prepare for a specific test, for example) or long-term (to help one another with homework throughout the year, say).

- Set a regular time for your study group to meet.
- If you are preparing for a test, think about assigning each member of the study group a specific topic to teach the rest of the group.
- Each member of the group should bring his or her textbook, notes, and other course materials so that everyone can contribute equally.

- Find a quiet meeting place, but one that allows conversation.
- At the start of the session, set out your goals to ensure that everyone is on the same page.
- If you are studying for a test, try to make a game of asking one another questions.
- Be sure to communicate with one another respectfully. (The emPOWERme on page 413 can help with this.)

- At the end of the study session, take a moment to discuss how well you accomplished your goals for the session.
- Be open in discussing what worked and what didn't.

- Consider whether your study group is the right size and has the right mix of people to be effective.
- Ask yourself whether a study group is right for you. While group study is valuable, some people prefer studying on their own.

Chapter 7 POWER Plan

P Prepare | O Organize

What are your goals for Chapter 7?	How can you accomplish each goal?
1 Be prepared before and during class.	• _____ • _____ • _____ • _____
2 Understand the homework to the point where you could do it without needing any help or hints.	• _____ • _____ • _____
3 Use the P.O.W.E.R. framework to help you communicate effectively with a study group: Switch "*You*" to "*I*."	• _____ • _____ • _____
4 Write your own goal. _____	• _____

What are your objectives for Chapter 7?	How can you accomplish each objective?
1 Learn all the different methods for factoring polynomials.	• Learn the different procedures for factoring polynomials of different forms. • Make a summary list of when to use each method. • Review multiplication tables, if needed. • Check your answers using multiplication. • Do the *Putting It All Together* section more than once.
2 Be able to solve quadratic and higher degree equations by factoring.	• Learn the definition of *standard form*. • Learn the procedure for **Solving a Quadratic Equation by Factoring**. • Review multiplication tables, if needed. • Check solutions.
3 Be able to solve applied problems that involve quadratic equations.	• Learn the procedure for **Solving Applied Problems**. • Always draw a diagram, when appropriate. • Use the procedure for **Solving a Quadratic Equation by Factoring**. • Check your answers.
4 Write your own goal. _____	• _____

	W Work	Read Sections 7.1 to 7.5, and complete the exercises.
E Evaluate — Complete the Chapter Review and Chapter Test. How did you do?	**R** Rethink	• Have you ever formed a study group? If so, was it helpful? Why or why not? • What does it mean for a polynomial to be factored completely? Can you think of a problem that is not factored completely? • Can you name all the factoring methods off the top of your head? • Why do we have to discard some solutions when solving applied problems?

7.1 The Greatest Common Factor and Factoring by Grouping

P Prepare	**O** Organize
What are your objectives for Section 7.1?	How can you accomplish each objective?
1 Find the GCF of a Group of Monomials	• Know the definition of a *greatest common factor*. • Be able to identify *factors* of a number. • Complete the given examples on your own. • Complete You Trys 1 and 2.
2 Factoring vs. Multiplying Polynomials	• In your notes and in your own words, explain the difference between multiplying polynomials and factoring polynomials. • Learn the procedure for **Factoring Out the Greatest Common Factor.** • Be sure to check your answers. • Complete Example 3 on your own.
3 Factor Out the Greatest Common Monomial Factor	• Know how to determine the *greatest common monomial factor*. • Check your answers. • Complete the given examples on your own. • Complete You Trys 3 and 4.
4 Factor Out the Greatest Common Binomial Factor	• Know how to determine the *greatest common binomial factor*. • Check your answers. • Complete the given example on your own. • Complete You Try 5.
5 Factor by Grouping	• Be able to identify when to *factor by grouping*. • Learn the procedure for **Factoring by Grouping.** • Check your answers. • Complete the given examples on your own. • Complete You Trys 6–8.

 **Work** Read the explanations, follow the examples, take notes, and complete the You Trys.

Recall that we can write a number as the product of factors:

$$12 = 3 \cdot 4$$
$$\downarrow \quad \downarrow \quad \downarrow$$
Product Factor Factor

To **factor** an integer is to write it as the product of two or more integers. Therefore, 12 can also be factored in other ways:

$$12 = 1 \cdot 12 \qquad 12 = 2 \cdot 6 \qquad 12 = -1 \cdot (-12)$$
$$12 = -2 \cdot (-6) \qquad 12 = -3 \cdot (-4) \qquad 12 = 2 \cdot 2 \cdot 3$$

The last **factorization**, $2 \cdot 2 \cdot 3$ or $2^2 \cdot 3$, is called the **prime factorization** of 12 since all of the factors are prime numbers. The factors of 12 are 1, 2, 3, 4, 6, 12, -1, -2, -3, -4, -6, and -12. We can also write the factors as $\pm 1, \pm 2, \pm 3, \pm 4, \pm 6,$ and ± 12. (Read ± 1 as "plus or minus 1.")

In this chapter, we will learn how to factor polynomials, a skill that is used in many ways throughout algebra.

1 Find the GCF of a Group of Monomials

Definition
The **greatest common factor (GCF)** of a group of two or more integers is the *largest* common factor of the numbers in the group.

For example, if we want to find the GCF of 12 and 20, we can list their positive factors.

12: 1, 2, 3, 4, 6, 12
20: 1, 2, 4, 5, 10, 20

The greatest common factor of 12 and 20 is 4. We can also use prime factors.

We begin our study of factoring polynomials by discussing how to find the greatest common factor of a group of monomials.

EXAMPLE 1 Find the greatest common factor of x^5 and x^3.

Solution

We can write each monomial as the product of its prime factors. To find the GCF use each prime factor the *least* number of times it appears in any of the prime factorizations. Then, multiply.

We can write x^5 and x^3 as

$$x^5 = x \cdot x \cdot x \cdot x \cdot x$$
$$x^3 = x \cdot x \cdot x$$

In x^5, x appears as a factor *five times*. In x^3, x appears as a factor *three times*.

Hint
In this example, the term with the least number of factors of x determines the GCF.

The least number of times *x* appears as a factor is three. *There will be three factors of x in the GCF.*

$$\text{GCF} = x \cdot x \cdot x = x^3$$

In Example 1, notice that the power of 3 in the GCF is the smallest of the powers when comparing x^5 and x^3. This will always be true.

> **Note**
> The exponent on the variable in the GCF will be the *smallest* exponent appearing on the variable in the group of terms.

[YOU TRY 1] Find the greatest common factor of y^4 and y^7.

EXAMPLE 2 Find the greatest common factor for each group of terms.

a) $30k^4, 10k^9, 50k^6$ b) $-12a^8b, 42a^5b^7$ c) $63c^5d^3, 18c^3, 27c^2d^2$

Solution

a) The GCF of the coefficients, 30, 10, and 50, is 10. The smallest exponent on *k* is 4, so k^4 will be part of the GCF.

   The GCF of $30k^4$, $10k^9$, and $50k^6$ is $10k^4$.

b) The GCF of the coefficients, -12 and 42, is 6. The smallest exponent on *a* is 5, so a^5 will be part of the GCF. The smallest exponent on *b* is 1, so *b* will be part of the GCF.

   The GCF of $-12a^8b$ and $42a^5b^7$ is $6a^5b$.

c) The GCF of the coefficients is 9. The smallest exponent on *c* is 2, so c^2 will be part of the GCF. There is no *d* in the term $18c^3$, so there will be no *d* in the GCF.

   The GCF of $63c^5d^3$, $18c^3$, and $27c^2d^2$ is $9c^2$.

[YOU TRY 2] Find the greatest common factor for each group of terms.

a) $-16p^7, 8p^5, 40p^8$ b) $r^6s^5, 9r^8s^3, 12r^4s^4$

2 Factoring vs. Multiplying Polynomials

Earlier we said that to **factor an integer** is to write it as the product of two or more integers. To **factor a polynomial** is to write it as a product of two or more polynomials.

Throughout this chapter we will study different factoring techniques. We will begin by discussing how to factor out the greatest common factor.

Factoring a polynomial is the opposite of multiplying polynomials. Let's see how these procedures are related.

EXAMPLE 3

a) Multiply $2x(x + 5)$.   b) Factor out the GCF from $2x^2 + 10x$.

Solution

a) Use the distributive property to multiply.
$$2x(x + 5) = (2x)x + (2x)(5)$$
$$= 2x^2 + 10x$$

b) Use the distributive property to factor out the greatest common factor from $2x^2 + 10x$.
First, identify the GCF of $2x^2$ and $10x$: GCF $= 2x$.
Then, rewrite each term as a product of two factors with one factor being $2x$.
$$2x^2 = (2x)(x) \text{ and } 10x = (2x)(5)$$
$$2x^2 + 10x = (2x)(x) + (2x)(5)$$
$$= 2x(x + 5) \qquad \text{Distributive property}$$

When we factor $2x^2 + 10x$, we get $2x(x + 5)$. We can check our result by multiplying.
$$2x(x + 5) = 2x^2 + 10x \checkmark$$

W Hint
Write out this procedure using your own words.

Procedure Steps for Factoring Out the Greatest Common Factor

1) Identify the GCF of all of the terms of the polynomial.
2) Rewrite each term as the product of the GCF and another factor.
3) Use the distributive property to factor out the GCF from the terms of the polynomial.
4) Check the answer by multiplying the factors. The result should be the original polynomial.

3 Factor Out the Greatest Common Monomial Factor

EXAMPLE 4

Factor out the greatest common factor.

a) $12a^5 + 30a^4 + 6a^3$ b) $c^6 - 6c^2$ c) $4x^5y^3 + 12x^5y^2 - 28x^4y^2 - 4x^3y$

Solution

a) Identify the GCF of all of the terms: GCF $= 6a^3$
$$12a^5 + 30a^4 + 6a^3 = (6a^3)(2a^2) + (6a^3)(5a) + (6a^3)(1) \qquad \text{Rewrite each term using the GCF as one of the factors.}$$
$$= 6a^3(2a^2 + 5a + 1) \qquad \text{Distributive property}$$

Check: $6a^3(2a^2 + 5a + 1) = 12a^5 + 30a^4 + 6a^3 \checkmark$

b) The GCF of all of the terms is c^2.
$$c^6 - 6c^2 = (c^2)(c^4) - (c^2)(6) \qquad \text{Rewrite each term using the GCF as one of the factors.}$$
$$= c^2(c^4 - 6) \qquad \text{Distributive property}$$

Check: $c^2(c^4 - 6) = c^6 - 6c^2 \checkmark$

c) The GCF of all of the terms is $4x^3y$.

$$4x^5y^3 + 12x^5y^2 - 28x^4y^2 - 4x^3y$$
$$= (4x^3y)(x^2y^2) + (4x^3y)(3x^2y) - (4x^3y)(7xy) - (4x^3y)(1)$$ Rewrite each term using the GCF as one of the factors.
$$= 4x^3y(x^2y^2 + 3x^2y - 7xy - 1)$$ Distributive property

Check: $4x^3y(x^2y^2 + 3x^2y - 7xy - 1) = 4x^5y^3 + 12x^5y^2 - 28x^4y^2 - 4x^3y$ ✓

[YOU TRY 3]   Factor out the greatest common factor.

a) $56k^4 - 24k^3 + 40k^2$ b) $3a^4b^4 - 12a^3b^4 + 18a^2b^4 - 3a^2b^3$

Sometimes we need to take out a negative factor.

EXAMPLE 5   Factor out $-5d$ from $-10d^4 + 45d^3 - 15d^2 + 5d$.

Solution

$$-10d^4 + 45d^3 - 15d^2 + 5d$$
$$= (-5d)(2d^3) + (-5d)(-9d^2) + (-5d)(3d) + (-5d)(-1)$$ Rewrite each term using $-5d$ as one of the factors.
$$= -5d[2d^3 + (-9d^2) + 3d + (-1)]$$ Distributive property
$$= -5d(2d^3 - 9d^2 + 3d - 1)$$ Rewrite $+(-9d^2)$ as $-9d^2$ and $+(-1)$ as -1.

Check: $-5d(2d^3 - 9d^2 + 3d - 1) = -10d^4 + 45d^3 - 15d^2 + 5d$ ✓

 When taking out a negative factor, be very careful with the signs!

[YOU TRY 4]   Factor out $-p^2$ from $-p^5 - 7p^4 + 3p^3 + 11p^2$.

4 Factor Out the Greatest Common Binomial Factor

Until now, all of the GCFs have been monomials. Sometimes, however, the greatest common factor is a *binomial*.

EXAMPLE 6   Factor out the greatest common factor.

a) $x(y + 2) + 9(y + 2)$ b) $r(s + 4) - (s + 4)$

Solution

a) In the polynomial $\underbrace{x(y + 2)}_{\text{term}} + \underbrace{9(y + 2)}_{\text{term}}$, $x(y + 2)$ is a term and $9(y + 2)$ is a term. What do these terms have in common? $y + 2$

Hint

The GCF can also be a binomial. Notice how the distributive property is being used in this case.

The GCF of $x(y+2)$ and $9(y+2)$ is $(y+2)$. Use the distributive property to factor out $y+2$.

$$x(y+2) + 9(y+2) = (y+2)(x+9) \quad \text{Distributive property}$$

Check: $(y+2)(x+9) = (y+2)x + (y+2)9$ Distribute.

The result $(y+2)x + (y+2)9$ is the same as $x(y+2) + 9(y+2)$ because multiplication is commutative. ✓

Hint

Always write a "1" in front of the parentheses in a problem like Example 6b).

b) Let's begin by rewriting $r(s+4) - (s+4)$ as $\underbrace{r(s+4)}_{\text{term}} - \underbrace{1(s+4)}_{\text{term}}$.

The GCF is $s+4$.

$$r(s+4) - 1(s+4) = (s+4)(r-1) \quad \text{Distributive property}$$

The check is left to the student.

 BE CAREFUL It is important to write -1 in front of $(s+4)$. Otherwise, the following mistake is often made:

$$r(s+4) - (s+4) = (s+4)r \quad \text{THIS IS INCORRECT!}$$

The correct factor is $r-1$ not r.

[YOU TRY 5] Factor out the GCF.

a) $t(u-8) + 5(u-8)$ b) $z(z^2+2) - 6(z^2+2)$ c) $2n(m+7) - (m+7)$

Taking out a binomial factor leads us to our next method of factoring—factoring by grouping.

5 Factor by Grouping

When we are asked to factor a polynomial containing four terms, we often try to **factor by grouping**.

EXAMPLE 7 Factor by grouping.

a) $ab + 5a + 3b + 15$ b) $2pr - 5qr + 6p - 15q$

c) $x^3 + 6x^2 - 7x - 42$

Solution

a) Begin by grouping terms together so that each group has a common factor.

$$\underbrace{ab + 5a}_{\downarrow} + \underbrace{3b + 15}_{\downarrow}$$

Factor out a to get $a(b + 5)$. $= a(b + 5) + 3(b + 5)$ Factor out 3 to get $3(b + 5)$.
 $= (b + 5)(a + 3)$ Factor out $(b + 5)$.

Check: $(b + 5)(a + 3) = ab + 5a + 3b + 15$ ✓

> **W Hint**
> Write out all of the steps as you are reading the problem.

b) Group terms together so that each group has a common factor.

$$\underbrace{2pr - 5qr}_{\downarrow} + \underbrace{6p - 15q}_{\downarrow}$$

Factor out r to get $r(2p - 5q)$. $= r(2p - 5q) + 3(2p - 5q)$ Factor out 3 to get $3(2p - 5q)$.
 $= (2p - 5q)(r + 3)$ Factor out $(2p - 5q)$.

Check: $(2p - 5q)(r + 3) = 2pr - 5qr + 6p - 15q$ ✓

c) Group terms together so that each group has a common factor.

$$\underbrace{x^3 + 6x^2}_{\downarrow} - \underbrace{7x - 42}_{\downarrow}$$

Factor out x^2 to get $x^2(x + 6)$. $= x^2(x + 6) - 7(x + 6)$ Factor out -7 to get $-7(x + 6)$.
 $= (x + 6)(x^2 - 7)$ Factor out $(x + 6)$.

We **must** factor out -7 *not* 7 from the second group so that the binomial factors for both groups are the same! [If we had factored out 7, then the factorization of the second group would have been $7(-x - 6)$.]

Check: $(x + 6)(x^2 - 7) = x^3 + 6x^2 - 7x - 42$ ✓

> **[YOU TRY 6]** Factor by grouping.
> a) $2cd + 4d + 5c + 10$ b) $4k^2 - 36k + km - 9m$ c) $h^3 + 8h^2 - 5h - 40$

Sometimes we have to rearrange the terms before we can factor.

EXAMPLE 8   Factor $24y^2 - 3z + 4y - 18yz$ completely.

Solution

Group terms together so that each group has a common factor.

$$\underbrace{24y^2 - 3z}_{\downarrow} + \underbrace{4y - 18yz}_{\downarrow}$$

Factor out 3 to get $3(8y^2 - z)$. $= 3(8y^2 - z) + 2y(2 - 9z)$ Factor out $2y$ to get $2y(2 - 9z)$.

> **W Hint**
> Often, there is more than one way that the terms can be rearranged so that the polynomial can be factored by grouping.

The groups do not have common factors! Let's rearrange the terms in the original polynomial, and group the terms differently.

$$\underbrace{24y^2 + 4y}_{\downarrow} - \underbrace{18yz - 3z}_{\downarrow}$$

Factor out $4y$ to get $4y(6y + 1)$. $= 4y(6y + 1) - 3z(6y + 1)$ Factor out $-3z$ to get $-3z(6y + 1)$.
 $= (6y + 1)(4y - 3z)$ Factor out $(6y + 1)$.

Check: $(6y + 1)(4y - 3z) = 24y^2 - 3z + 4y - 18yz$ ✓

[YOU TRY 7] Factor $6a^2 - 5b + 15a - 2ab$ completely.

Often, we have to combine the two factoring techniques we have learned here. That is, we begin by factoring out the GCF and then we factor by grouping. Let's summarize how to factor a polynomial by grouping, and then look at another example.

> **Procedure** Steps for Factoring by Grouping
>
> 1) Before trying to factor by grouping, look at each term in the polynomial and ask yourself, *"Can I factor out a GCF first?"* If so, factor out the GCF from all of the terms.
> 2) Make two groups of two terms so that each group has a common factor.
> 3) Take out the common factor in each group of terms.
> 4) Factor out the common binomial factor using the distributive property.
> 5) Check the answer by multiplying the factors.

EXAMPLE 9 Factor completely. $4y^4 + 4y^3 - 20y^2 - 20y$

Solution

Notice that this polynomial has four terms. This is a clue for us to try factoring by grouping. *However,* look at the polynomial carefully and ask yourself, *"Can I factor out a GCF?"* Yes! Therefore, the first step in factoring this polynomial is to factor out $4y$.

$$4y^4 + 4y^3 - 20y^2 - 20y = 4y(y^3 + y^2 - 5y - 5) \quad \text{Factor out the GCF, } 4y.$$

The polynomial in parentheses has 4 terms. Try to factor it by grouping.

W Hint
Seeing a polynomial with 4 terms is your hint to try factoring by grouping.

$$4y(\underbrace{y^3 + y^2}_{} \; \underbrace{- 5y - 5}_{})$$
$$= 4y[y^2(y + 1) - 5(y + 1)] \quad \text{Take out the common factor in each group.}$$
$$= 4y(y + 1)(y^2 - 5) \quad \text{Factor out } (y+1) \text{ using the distributive property.}$$

Check: $4y(y + 1)(y^2 - 5) = 4y(y^3 + y^2 - 5y - 5)$
$$= 4y^4 + 4y^3 - 20y^2 - 20y \; \checkmark$$

[YOU TRY 8] Factor completely. $4ab + 14b + 8a + 28$

Remember, seeing a polynomial with four terms is a clue to try factoring by grouping. Not all polynomials will factor this way, however. We will learn other techniques later, and some polynomials must be factored using methods learned in later courses.

ANSWERS TO [YOU TRY] **EXERCISES**

1) y^4 2) a) $8p^5$ b) r^4s^3 3) a) $8k^2(7k^2 - 3k + 5)$ b) $3a^2b^3(a^2b - 4ab + 6b - 1)$
4) $-p^2(p^3 + 7p^2 - 3p - 11)$ 5) a) $(u - 8)(t + 5)$ b) $(z^2 + 2)(z - 6)$ c) $(m + 7)(2n - 1)$
6) a) $(c + 2)(2d + 5)$ b) $(k - 9)(4k + m)$ c) $(h + 8)(h^2 - 5)$ 7) $(2a + 5)(3a - b)$
8) $2(b + 2)(2a + 7)$

Evaluate 7.1 Exercises

Do the exercises, and check your work.

Objective 1: Find the GCF of a Group of Monomials

Find the greatest common factor of each group of terms.

1) $45m^3, 20m^2$
2) $18d^6, 21d^2$
3) $42k^5, 54k^7, 72k^9$
4) $25t^8, 55t, 30t^3$
5) $24r^3s^6, 56r^2s^5$
6) $27x^4y, 45x^2y^3$
7) $28u^2v^5, 20uv^3, -8uv^4$
8) $-6a^4b^3, 18a^2b^6, 12a^2b^4$
9) $21s^2t, 35s^2t^2, s^4t^2$
10) $p^4q^4, -p^3q^4, -p^3q$
11) $a(n-7), 4(n-7)$
12) $x^2(y+9), z^2(y+9)$
13) Explain how to find the GCF of a group of terms.
14) What does it mean to factor a polynomial?

Mixed Exercises: Objectives 2–4

Factor out the greatest common factor. Be sure to check your answer.

15) $30s + 18$
16) $14a + 24$
17) $24z - 4$
18) $63f^2 - 49$
19) $3d^2 - 6d$
20) $20m - 5m^2$
21) $30b^3 - 5b$
22) $42y^2 + 35y^3$
23) $r^9 + r^2$
24) $t^5 - t^4$
25) $\frac{1}{2}c^2 + \frac{5}{2}c$
26) $\frac{1}{8}k^2 + \frac{7}{8}k$
27) $10n^5 - 5n^4 + 40n^3$
28) $18x^7 + 42x^6 - 30x^5$
29) $2v^8 - 18v^7 - 24v^6 + 2v^5$
30) $12z^6 + 30z^5 - 15z^4 + 3z^3$
31) $8c^3 + 3d^2$
32) $m^5 - 5n^2$
33) $a^4b^2 + 4a^3b^3$
34) $20r^3s^3 - 14rs^4$
35) $50x^3y^3 - 70x^3y^2 + 40x^2y$
36) $21b^4d^3 + 15b^3d^3 - 27b^2d^2$
37) $m(n-12) + 8(n-12)$
38) $x(y+5) + 3(y+5)$

39) $a(9c+4) - b(9c+4)$
40) $p(8r-3) - q(8r-3)$
41) $y(z+11) + (z+11)$
42) $2u(v-7) + (v-7)$
43) $2k^2(3r+4) - (3r+4)$
44) $8p(3q+5) - (3q+5)$
45) Factor out -8 from $-64m - 40$.
46) Factor out -7 from $-14k + 21$.
47) Factor out $-5t^2$ from $-5t^3 + 10t^2$.
48) Factor out $-4v^3$ from $-4v^5 - 36v^3$.
49) Factor out $-a$ from $-3a^3 + 7a^2 - a$.
50) Factor out $-q$ from $-10q^3 - 4q^2 + q$.
51) Factor out -1 from $-b + 8$.
52) Factor out -1 from $-z - 6$.

Objective 5: Factor by Grouping

Factor by grouping.

53) $kt + 3k + 8t + 24$
54) $uv + 5u + 10v + 50$
55) $fg - 7f + 4g - 28$
56) $cd - 5d + 8c - 40$
57) $2rs - 6r + 5s - 15$
58) $3jk - 7k + 6j - 14$
59) $3xy - 2y + 27x - 18$
60) $4ab + 32a + 3b + 24$
61) $8b^2 + 20bc + 2bc^2 + 5c^3$
62) $8u^2 - 16uv^2 + 3uv - 6v^3$
63) $4a^3 - 12ab + a^2b - 3b^2$
64) $5x^3 - 30x^2y^2 + xy - 6y^3$
65) $kt + 7t - 5k - 35$
66) $pq - 2q - 9p + 18$
67) $mn - 8m - 10n + 80$
68) $hk + 6k - 4h - 24$
69) $dg - d + g - 1$
70) $qr + 3q - r - 3$
71) $5tu + 6t - 5u - 6$
72) $4yz + 7z - 20y - 35$
73) $36g^4 + 3gh - 96g^3h - 8h^2$
74) $40j^3 + 72jk - 55j^2k - 99k^2$
75) Explain, in your own words, how to factor by grouping.
76) What should be the first step in factoring $3xy + 6x + 15y + 30$?

Factor completely. You may need to begin by taking out the GCF first or by rearranging terms.

Fill It In

Fill in the blanks with either the missing mathematical step or reason for the given step.

77) $5mn + 15m + 10n + 30$

$5mn + 15m + 10n + 30$

= _____ Factor out the GCF.

= $5[m(n + 3) + 2(n + 3)]$ _____

= _____ Take out the binomial factor.

78) $3x^2y - 21x^2 - 6xy + 42x$

$3x^2y - 21x^2 - 6xy + 42x$

= _____ Factor out the GCF.

= $3x[x(y - 7) - 2(y - 7)]$ _____

= _____ Take out the binomial factor.

79) $7pq + 28q + 14p + 56$

80) $2ab + 8a + 6b + 24$

81) $8s^2t - 40st + 16s^2 - 80s$

82) $10hk^3 - 5hk^2 + 30k^3 - 15k^2$

83) $7cd + 12 + 28c + 3d$

84) $9rs + 12 + 2s + 54r$

85) $42k^3 + 15d^2 - 18k^2d - 35kd$

86) $12x^3 + 2y^2 - 3x^2y - 8xy$

87) $9f^2j^2 + 45fj + 9fj^2 + 45f^2j$

88) $n^3m - 4n^2 + mn^2 - 4n^3$

89) $4x^4y - 14x^3 + 28x^4 - 2x^3y$

90) $12a^2c^2 - 20ac - 4ac^2 + 60a^2c$

Mixed Exercises: Objectives 1–5

Factor completely.

91) $pq - 8p + 3q - 24$

92) $27d^3 + 36d^2 - 9d$

93) $a^4b^2 + 2a^3b^3 - 8a^2b^4$

94) $18rt^3 + 15t^3 - 18rt^2 - 15t^2$

95) $3h^3 - 8k^3 + 12h^2k^2 - 2hk$

96) $6yz - 8z - 3y + 4$

97) $2c^4 + 14c^2 + 84c + 12c^3$

98) $18w^3 + 45w^2$

99) Factor out $-8v$ from $-16v^3 - 56v^2 + 8v$.

100) Factor out $-3n^2$ from $-3n^4 + 27n^3 + 33n^2$.

R Rethink

R1) Why is the GCF for $x^4 + x^3 + x^2$ not equal to x^4, the term with the largest exponent power?

R2) Why are x and $(x - 4)$ considered to be factors of $x^2 - 4x$?

R3) How does multiplying two binomials together relate to factoring a trinomial?

R4) Can the factoring by grouping procedure be performed on a trinomial? Why or why not?

R5) Which exercises in this section do you find most challenging?

7.2 Factoring Trinomials

P Prepare

O Organize

What are your objectives for Section 7.2?	How can you accomplish each objective?
1 Factor a Trinomial of the Form $x^2 + bx + c$	• Review how to multiply polynomials using FOIL. • Learn the procedure for **Factoring a Polynomial of the Form $x^2 + bx + c$.** • Be able to determine when a trinomial is *prime*. • Complete the given example on your own. • Complete You Try 1.
2 More on Factoring a Trinomial of the Form $x^2 + bx + c$	• Ask yourself, *"Can I factor out a GCF?"* • Learn the procedure for **Factoring a Polynomial of the Form $x^2 + bx + c$.** • Know what it means for a polynomial to be *factored completely.* • Ask yourself, *"Can I factor again?"* • Complete the given example on your own. • Complete You Try 2.
3 Factor a Trinomial Containing Two Variables	• Learn the procedure for **Factoring a Polynomial of the Form $x^2 + bx + c$.** • Always begin by asking yourself, *"Can I factor out a GCF?"* and, after factoring, always ask yourself, *"Can I factor again?"* • Complete the given example on your own. • Complete You Try 3.
4 Factor $ax^2 + bx + c$ $(a \neq 1)$ by Grouping	• Learn the procedure for **Factoring by Grouping.** • Always begin by asking yourself, *"Can I factor out a GCF?"* and, after factoring, always ask yourself, *"Can I factor again?"* • Complete the given examples on your own. • Complete You Trys 4 and 5.
5 Factor $ax^2 + bx + c$ $(a \neq 1)$ by Trial and Error	• Review the FOIL process. • Know the procedure for **Factoring Trinomials by Trial and Error.** • Always begin by asking yourself, *"Can I factor out a GCF?"* and, after factoring, always ask yourself, *"Can I factor again?"* • Complete the given example on your own. • Complete You Trys 6 and 7.
6 Factor Using Substitution	• Understand the method of using *substitution*. • Complete the given example on your own. • Complete You Try 8.

W Work   Read the explanations, follow the examples, take notes, and complete the You Trys.

One of the factoring problems encountered most often in algebra is the factoring of trinomials. In this section, we will discuss how to factor trinomials like $x^2 + 11x + 18$, $3n^2 - n - 4$, $4a^2 - 11ab + 6b^2$, and many more. Let's begin with trinomials of the form $x^2 + bx + c$, where the coefficient of the squared term is 1.

1 Factor a Trinomial of the Form $x^2 + bx + c$

In Section 7.1 we said that the process of factoring is the opposite of multiplying. Let's see how this will help us understand how to factor a trinomial of the form $x^2 + bx + c$.

Multiply $(x + 4)(x + 7)$ using FOIL.

$$(x + 4)(x + 7) = x^2 + 7x + 4x + 4 \cdot 7 \quad \text{Multiply using FOIL.}$$
$$= x^2 + (7 + 4)x + 28 \quad \text{Use the distributive property, and multiply } 4 \cdot 7.$$
$$= x^2 + 11x + 28$$

$$(x + 4)(x + 7) = x^2 + 11x + 28$$

11 is the *sum* of 4 and 7.   28 is the *product* of 4 and 7.

So, if we were asked to *factor* $x^2 + 11x + 28$, we need to think of two integers whose *product* is 28 and whose *sum* is 11. Those numbers are 4 and 7. The *factored form* of $x^2 + 11x + 28$ is $(x + 4)(x + 7)$.

> **Procedure** Factoring a Polynomial of the Form $x^2 + bx + c$
>
> To factor a polynomial of the form $x^2 + bx + c$, find two integers m and n whose product is c and whose sum is b. Then,
>
> $$x^2 + bx + c = (x + m)(x + n).$$
>
> 1) If b and c are positive, then both m and n must be positive.
> 2) If c is positive and b is negative, then both m and n must be negative.
> 3) If c is negative, then one integer, m, must be positive and the other integer, n, must be negative.
>
> You can check the answer by multiplying the binomials. The result should be the original polynomial.

EXAMPLE 1

Factor, if possible.

a) $y^2 - 8y + 15$ b) $k^2 + k - 56$ c) $r^2 + 7r + 9$

Solution

a) To factor $y^2 - 8y + 15$, find the two integers whose *product* is 15 and whose *sum* is -8. Since 15 is positive and the coefficient of y is a negative number, -8, both integers will be negative.

Factors of 15	Sum of the Factors
$-1 \cdot (-15) = 15$	$-1 + (-15) = -16$
$-3 \cdot (-5) = 15$	$-3 + (-5) = -8$

W Hint
We are finding two things. The first is the product of the factors, and the second is the sum of the factors.

The numbers are -3 and -5: $y^2 - 8y + 15 = (y - 3)(y - 5)$.

Check: $(y - 3)(y - 5) = y^2 - 5y - 3y + 15 = y^2 - 8y + 15$ ✓

b) $k^2 + k - 56$

The coefficient of k is 1, so we can think of this trinomial as $k^2 + 1k - 56$.

Find two integers whose *product* is -56 and whose *sum* is 1. Since the last term in the trinomial is negative, one of the integers must be positive and the other must be negative.

Try to find these integers mentally. Two numbers with a product of *positive* 56 are 7 and 8. We need a product of -56, so either the 7 is negative or the 8 is negative.

Factors of -56	Sum of the Factors
$-7 \cdot 8 = -56$	$-7 + 8 = 1$

The numbers are -7 and 8: $k^2 + k - 56 = (k - 7)(k + 8)$.

Check: $(k - 7)(k + 8) = k^2 + 8k - 7k - 56 = k^2 + k - 56$ ✓

c) To factor $r^2 + 7r + 9$, find the two integers whose *product* is 9 and whose *sum* is 7. We are looking for two positive numbers.

Factors of 9	Sum of the Factors
$1 \cdot 9 = 9$	$1 + 9 = 10$
$3 \cdot 3 = 9$	$3 + 3 = 6$

There are no such factors! Therefore, $r^2 + 7r + 9$ does not factor using the methods we have learned here. We say that it is **prime**.

Note

We say that trinomials like $r^2 + 7r + 9$ are **prime** if they cannot be factored using the method presented here.

In later mathematics courses, you may learn how to factor such polynomials using other methods so that they are not considered prime.

[**YOU TRY 1**] Factor, if possible.

a) $s^2 + 5s - 66$ b) $d^2 - 6d - 10$ c) $x^2 - 12x + 27$

2 More on Factoring a Trinomial of the Form $x^2 + bx + c$

Sometimes it is necessary to factor out the GCF before applying this method for factoring trinomials.

Note

From this point on, the *first* step in factoring *any* polynomial should be to ask yourself, *"Can I factor out a greatest common factor?"*

And since some polynomials can be factored more than once, after performing one factorization, ask yourself, *"Can I factor again?"* If so, factor again. If not, you know that the polynomial has been completely factored.

EXAMPLE 2

Factor $5y^3 - 15y^2 - 20y$ completely.

Solution

Ask yourself, *"Can I factor out a GCF?"* Yes. The GCF is $5y$.

$$5y^3 - 15y^2 - 20y = 5y(y^2 - 3y - 4)$$

Look at the trinomial and ask yourself, *"Can I factor again?"* Yes. The integers whose product is -4 and whose sum is -3 are -4 and 1. Therefore,

$$5y^3 - 15y^2 - 20y = 5y(y^2 - 3y - 4)$$
$$= 5y(y - 4)(y + 1)$$

W Hint
When factoring a trinomial, we first look to see if the trinomial has a GCF. If it does, then we begin our factoring procedure by first factoring out the GCF.

We cannot factor again.

Check: $5y(y - 4)(y + 1) = 5y(y^2 + y - 4y - 4)$
$= 5y(y^2 - 3y - 4)$
$= 5y^3 - 15y^2 - 20y$ ✓

[YOU TRY 2] Factor $6g^4 + 42g^3 + 60g^2$ completely.

3 Factor a Trinomial Containing Two Variables

If a trinomial contains two variables and we cannot take out a GCF, the trinomial may still be factored according to the method outlined in this section.

EXAMPLE 3

Factor $a^2 + 9ab + 18b^2$ completely.

Solution

Ask yourself, *"Can I factor out a GCF?"* No. Notice that the first term is a^2. Let's rewrite the trinomial as

$$a^2 + 9ba + 18b^2$$

so that we can think of $9b$ as the coefficient of a. Find two expressions whose product is $18b^2$ and whose sum is $9b$. They are $3b$ and $6b$ since $3b \cdot 6b = 18b^2$ and $3b + 6b = 9b$.

$$a^2 + 9ab + 18b^2 = (a + 3b)(a + 6b)$$

We cannot factor $(a + 3b)(a + 6b)$ any more, so this is the complete factorization. The check is left to the student.

[**YOU TRY 3**] Factor completely.

  a) $x^2 + 15xy + 54y^2$ b) $3k^3 + 18ck^2 - 21c^2k$

If we are asked to factor each of these polynomials, $3x^2 + 18x + 24$ and $3x^2 + 10x + 8$, how do we begin? How do the polynomials differ?

The GCF of $3x^2 + 18x + 24$ is 3. To factor, begin by taking out the 3. We can factor using what we learned earlier.

$$3x^2 + 18x + 24 = 3(x^2 + 6x + 8) \quad \text{Factor out 3.}$$
$$= 3(x + 4)(x + 2) \quad \text{Factor the trinomial.}$$

In the second polynomial, $3x^2 + 10x + 8$, we *cannot* factor out the leading coefficient of 3. Next, we will discuss two methods for factoring a trinomial like $3x^2 + 10x + 8$ where we *cannot* factor out the leading coefficient.

4 Factor $ax^2 + bx + c$ $(a \neq 1)$ by Grouping

To factor $3x^2 + 10x + 8$, first find the product of 3 and 8. Then, find two integers (*Sum* is 10; *Product*: $3 \cdot 8 = 24$) whose *product* is 24 and whose *sum* is 10. The numbers are 6 and 4. Rewrite the middle term, $10x$, as $6x + 4x$, then factor by grouping.

W Hint
Notice that the two numbers we are trying to find are factors of the product of *a* and *c* whose sum is *b*.

$$3x^2 + 10x + 8 = 3x^2 + 6x + 4x + 8$$
$$= 3x(x + 2) + 4(x + 2) \quad \text{Take out the common factor from each group.}$$
$$= (x + 2)(3x + 4) \quad \text{Factor out } (x + 2).$$

$$3x^2 + 10x + 8 = (x + 2)(3x + 4)$$

Check: $(x + 2)(3x + 4) = 3x^2 + 4x + 6x + 8 = 3x^2 + 10x + 8$ ✓

EXAMPLE 4 Factor completely.

  a) $10p^2 - 13p + 4$ b) $5c^2 - 29cd - 6d^2$

  Solution

  a) *Sum* is -13; $10p^2 - 13p + 4$; *Product*: $10 \cdot 4 = 40$

Think of two integers whose *product* is 40 and whose *sum* is -13. (Both numbers will be negative.): -8 and -5

Rewrite the middle term, $-13p$, as $-8p - 5p$. Factor by grouping.

W Hint
Be mindful of the signs!

$$10p^2 - 13p + 4 = 10p^2 - 8p - 5p + 4$$
$$= 2p(5p - 4) - 1(5p - 4) \quad \text{Take out the common factor from each group.}$$
$$= (5p - 4)(2p - 1) \quad \text{Factor out } (5p - 4).$$

Check: $(5p - 4)(2p - 1) = 10p^2 - 13p + 4$ ✓

b)

$$5c^2 - 29cd - 6d^2$$

Product: $5 \cdot (-6) = -30$, Sum is -29

The integers whose *product* is -30 and whose *sum* is -29 are -30 and 1.

Rewrite the middle term, $-29cd$, as $-30cd + cd$. Factor by grouping.

$$\begin{aligned} 5c^2 - 29cd - 6d^2 &= \underbrace{5c^2 - 30cd}_{} + \underbrace{cd - 6d^2}_{} \\ &= 5c(c - 6d) + d(c - 6d) \quad \text{Take out the common factor from each group.} \\ &= (c - 6d)(5c + d) \quad \text{Factor out } (c - 6d). \end{aligned}$$

Check: $(c - 6d)(5c + d) = 5c^2 - 29cd - 6d^2$ ✓

[YOU TRY 4] Factor completely.

a) $2c^2 + 11c + 14$ b) $6n^2 - 23n - 4$ c) $8x^2 - 10xy + 3y^2$

EXAMPLE 5

Factor $24w^2 - 54w - 15$ completely.

Solution

It is tempting to jump right in and multiply $24 \cdot (-15) = -360$ and try to think of two integers with a product of -360 and a sum of -54. Before doing that, ask yourself, *"Can I factor out a GCF?"* Yes! We can factor out 3.

$$24w^2 - 54w - 15 = 3(\underbrace{8w^2 - 18w - 5}_{}) \quad \text{Factor out 3.}$$

Sum is -18; Product: $8 \cdot (-5) = -40$

Try to factor $8w^2 - 18w - 5$ by finding two integers whose *product* is -40 and whose *sum* is -18. The numbers are -20 and 2.

$$\begin{aligned} &= 3(\underbrace{8w^2 - 20w}_{} + \underbrace{2w - 5}_{}) \\ &= 3[4w(2w - 5) + 1(2w - 5)] \quad \text{Take out the common factor from each group.} \\ &= 3(2w - 5)(4w + 1) \quad \text{Factor out } 2w - 5. \end{aligned}$$

Check by multiplying: $3(2w - 5)(4w + 1) = 3(8w^2 - 18w - 5)$
$= 24w^2 - 54w - 15$ ✓

W Hint
Remember that the best way to read a math book is to write out the examples as you are reading them.

[YOU TRY 5] Factor completely.

a) $20m^3 - 8m^2 + 4m$ b) $6z^2 + 20z + 16$

5 Factor $ax^2 + bx + c$ ($a \neq 1$) by Trial and Error

Earlier, we factored $3x^2 + 10x + 8$ by grouping. Now we will factor it by trial and error, which is just reversing the process of FOIL.

EXAMPLE 6

Factor $3x^2 + 10x + 8$ completely.

Solution

Can we factor out a GCF? No. So try to factor $3x^2 + 10x + 8$ as the product of two binomials. Notice that all terms are positive, so all factors will be positive.

Begin with the squared term, $3x^2$. Which two expressions with integer coefficients can we multiply to get $3x^2$? $3x$ and x. Put these in the binomials.

$$3x^2 + 10x + 8 = (3x \quad)(x \quad) \qquad 3x \cdot x = 3x^2$$

Next, look at the last term, 8. What are the pairs of positive integers that multiply to 8? They are 8 and 1 as well as 4 and 2.

Try these numbers as the last terms of the binomials. The middle term, $10x$, comes from finding the sum of the products of the outer terms and inner terms.

First Try

$$3x^2 + 10x + 8 \stackrel{?}{=} (3x + 8)(x + 1) \quad \text{Incorrect}$$

These must both be $10x$. $\quad 8x + 3x = 11x$

> **W Hint**
> Notice that the *trial and error* method skips both the process of rewriting the middle term and factoring by grouping.

Switch the 8 and 1. $\quad 3x^2 + 10x + 8 \stackrel{?}{=} (3x + 1)(x + 8) \quad \text{Incorrect}$

These must both be $10x$. $\quad 1x + 24x = 25x$

Try using 4 and 2. $\quad 3x^2 + 10x + 8 \stackrel{?}{=} (3x + 4)(x + 2) \quad \text{Correct!}$

These must both be $10x$. $\quad 4x + 6x = 10x$

The factorization of $3x^2 + 10x + 8$ is $(3x + 4)(x + 2)$. Check by multiplying.

EXAMPLE 7

Factor $2r^2 - 13r + 20$ completely.

Solution

Can we factor out a GCF? No. To get a product of $2r^2$, we will use $2r$ and r.

$$2r^2 - 13r + 20 = (2r \quad)(r \quad) \qquad 2r \cdot r = 2r^2$$

Since the last term is positive and the middle term is negative, we want pairs of negative integers that multiply to 20. The pairs are -1 and -20, -2 and -10, and -4 and -5. Try these numbers as the last terms of the binomials. The middle term, $-13r$, comes from finding the sum of the products of the outer terms and inner terms.

$$2r^2 - 13r + 20 \stackrel{?}{=} (2r - 1)(r - 20) \quad \text{Incorrect}$$

These must both be $-13r$. $\quad -r + (-40r) = -41r$

Switch the -1 and -20: $\quad 2r^2 - 13r + 20 \stackrel{?}{=} (2r - 20)(r - 1)$

372 CHAPTER 7 Factoring Polynomials

Without multiplying we know that this choice is incorrect. How? In the factor $(2r - 20)$, a 2 can be factored out to get $2(r - 10)$. But, we said that we could not factor out a GCF from the original polynomial, $2r^2 - 13r + 20$. Therefore, it will not be possible to take out a common factor from one of the binomial factors.

> **Note**
> If you cannot factor out a GCF from the original polynomial, then you cannot take out a factor from one of the binomial factors either.

For the same reason, the pair -2 and -10 will not give us the correct factorization:

$$2r^2 - 13r + 20 \neq (2r - 2)(r - 10) \quad \text{2 can be factored out of } (2r - 2).$$
$$2r^2 - 13r + 20 \neq (2r - 10)(r - 2) \quad \text{2 can be factored out of } (2r - 10).$$

Try using -4 and -5: $\quad 2r^2 - 13r + 20 \neq (2r - 4)(r - 5) \quad$ 2 can be factored out of $(2r - 4)$.

Switch the -4 and -5: $\quad 2r^2 - 13r + 20 = (2r - 5)(r - 4) \quad$ Correct!

These must both be $-13r$. $\quad -5r + (-8r) \to -13r$

The factorization of $2r^2 - 13r + 20$ is $(2r - 5)(r - 4)$. Check by multiplying.

[YOU TRY 6] Factor completely.

a) $2m^2 + 11m + 12$ b) $3v^2 - 28v + 9$

EXAMPLE 8   Factor completely.

a) $12d^2 + 46d - 8$ b) $-2h^2 + 9h + 56$

Solution

a) Ask yourself, *"Can I take out a common factor?"* Yes! The GCF is 2.

$$12d^2 + 46d - 8 = 2(6d^2 + 23d - 4)$$

Now, try to factor $6d^2 + 23d - 4$. To get a product of $6d^2$, we can try either $6d$ and d or $3d$ and $2d$. Let's start by trying $6d$ and d.

$$6d^2 + 23d - 4 = (6d \quad)(d \quad)$$

List pairs of integers that multiply to -4: 4 and -1, -4 and 1, 2 and -2.

Try 4 and -1. Do not put 4 in the same binomial as $6d$ since then it would be possible to factor out 2. But, a 2 does not factor out of $6d^2 + 23d - 4$. Put the 4 in the same binomial as d.

$$6d^2 + 23d - 4 \stackrel{?}{=} (6d - 1)(d + 4)$$

$-d + 24d = 23d$ Correct!

Don't forget that the very first step was to factor out a 2. Therefore,
$$12d^2 + 46d - 8 = 2(6d^2 + 23d - 4) = 2(6d - 1)(d + 4).$$

Check by multiplying.

b) Since the coefficient of the squared term is negative, begin by factoring out -1. (There is no other common factor except 1.)
$$-2h^2 + 9h + 56 = -1(2h^2 - 9h - 56)$$

Try to factor $2h^2 - 9h - 56$. To get a product of $2h^2$ we will use $2h$ and h in the binomials.
$$2h^2 - 9h - 56 = (2h \quad)(h \quad)$$

We need pairs of integers so that their product is -56. The ones that come to mind quickly involve 7 and 8 and 1 and 56: -7 and 8, 7 and -8, -1 and 56, 1 and -56.

There are other pairs; if these do not work, we will list others.

Do *not* start with -1 and 56 or 1 and -56 because the middle term, $-9h$, is not very large. Using -1 and 56 or 1 and -56 would likely result in a larger middle term.

Try -7 and 8. Do not put 8 in the same binomial as $2h$ since then it would be possible to factor out 2.

$$2h^2 - 9h - 56 \stackrel{?}{=} (2h - 7)(h + 8)$$

$$\begin{array}{r} -7h \\ +16h \\ \hline 9h \end{array}$$ This must equal $-9h$. Incorrect!

Only the sign of the sum is incorrect. *Change the signs in the binomials to get the correct sum.*

$$2h^2 - 9h - 56 \stackrel{?}{=} (2h + 7)(h - 8)$$

$$\begin{array}{r} 7h \\ + (-16h) \\ \hline -9h \end{array}$$ Correct!

Remember that we factored out -1 to begin the problem.
$$-2h^2 + 9h + 56 = -1(2h^2 - 9h - 56) = -(2h + 7)(h - 8)$$

Check by multiplying.

[YOU TRY 7] Factor completely.

a) $15b^2 - 55b + 30$ b) $-4p^2 + 3p + 10$

We have seen two methods for factoring $ax^2 + bx + c$ ($a \neq 1$): factoring by grouping and factoring by trial and error. In either case, remember to begin by taking out a common factor from all terms whenever possible.

6 Factor Using Substitution

Some polynomials can be factored using a method called **substitution**. We will illustrate this method in Example 9.

EXAMPLE 9

Factor $3(x + 4)^2 + 11(x + 4) + 10$ completely.

Solution

Notice that the binomial $x + 4$ appears as a *squared quantity* and a *linear quantity*. We will use another letter to represent this quantity. We can use any letter except x. Let's use the letter u.

$$\text{Let } u = x + 4. \text{ Then, } u^2 = (x + 4)^2.$$

Substitute the u and u^2 into the original polynomial, and factor.

$$3(x + 4)^2 + 11(x + 4) + 10 =$$
$$3u^2 + 11u + 10 = (3u + 5)(u + 2)$$
$$= [3(x + 4) + 5][(x + 4) + 2] \quad \text{Since the original polynomial was in terms of } x, \text{ substitute } x + 4 \text{ for } u.$$
$$= (3x + 12 + 5)(x + 4 + 2) \quad \text{Distribute.}$$
$$= (3x + 17)(x + 6) \quad \text{Combine like terms.}$$

BE CAREFUL The final factorization is **not** the one containing the substitution variable, u. We must go back and replace u with the expression it represented so that the factorization is in terms of the original variable.

[YOU TRY 8]

Factor $2(x - 3)^2 + 3(x - 3) - 35$ completely.

Using Technology

We found some ways to narrow down the possibilities when factoring $ax^2 + bx + c (a \neq 1)$ using the trial and error method.

We can also use a graphing calculator to help with the process. Consider the trinomial $2x^2 - 9x - 35$. Enter the trinomial into Y_1 and press ZOOM and then enter 6 to display the graph in the standard viewing window.

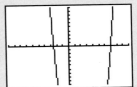

Look on the graph for the x-intercept (if any) that appears to be an integer. It appears that 7 is an x-intercept.

To check if 7 is an x-intercept, press TRACE then 7 and press ENTER. As shown on the graph, when $x = 7$, $y = 0$, so 7 is an x-intercept.

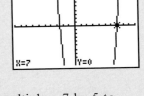

When an x-intercept is an integer, then x minus that x-intercept is a factor of the trinomial. In this case, $x - 7$ is a factor of $2x^2 - 9x - 35$. We can then complete the factoring as $(2x + 5)(x - 7)$, since we must multiply -7 by 5 to obtain -35.

Find an x-intercept using a graphing calculator, and factor the trinomial.

1) $3x^2 + 13x - 10$
2) $2x^2 + x - 36$
3) $5x^2 + 16x + 3$
4) $2x^2 - 13x + 21$
5) $3x^2 - 23x - 8$
6) $7x^2 + 11x - 6$

ANSWERS TO [YOU TRY] EXERCISES

1) a) $(s + 11)(s - 6)$ b) prime c) $(x - 9)(x - 3)$
2) $6g^2(g + 5)(g + 2)$ 3) a) $(x + 6y)(x + 9y)$ b) $3k(k + 7c)(k - c)$
4) a) $(2c + 7)(c + 2)$ b) $(6n + 1)(n - 4)$ c) $(4x - 3y)(2x - y)$
5) a) $4m(5m^2 - 2m + 1)$ b) $2(3z + 4)(z + 2)$
6) a) $(2m + 3)(m + 4)$ b) $(3v - 1)(v - 9)$
7) a) $5(3b - 2)(b - 3)$ b) $-(4p + 5)(p - 2)$
8) $(2x - 13)(x + 2)$

ANSWERS TO TECHNOLOGY EXERCISES

1) $(3x - 2)(x + 5)$ 2) $(2x + 9)(x - 4)$ 3) $(x + 3)(5x + 1)$ 4) $(x - 3)(2x - 7)$
5) $(x - 8)(3x + 1)$ 6) $(x + 2)(7x - 3)$

E Evaluate 7.2 Exercises

Do the exercises, and check your work.

Objective 1: Factor a Trinomial of the Form $x^2 + bx + c$

1) If $x^2 + bx + c$ factors to $(x + m)(x + n)$ and if c is positive and b is negative, what do you know about the signs of m and n?

2) If $x^2 + bx + c$ factors to $(x + m)(x + n)$ and if b and c are positive, what do you know about the signs of m and n?

3) When asked to factor a polynomial, what is the first question you should ask yourself?

4) What does it mean to say that a polynomial is prime?

5) After factoring a polynomial, what should you ask yourself to be sure that the polynomial is completely factored?

6) How do you check the factorization of a polynomial?

Factor completely, if possible. Check your answer.

7) $g^2 + 8g + 12$
8) $j^2 + 9j + 20$
9) $w^2 + 13w + 42$
10) $t^2 + 15t + 36$
11) $c^2 - 13c + 36$
12) $v^2 - 11v + 24$
13) $m^2 - m - 110$
14) $s^2 + 3s - 28$
15) $u^2 + u - 132$
16) $b^2 - 2b - 8$
17) $q^2 - 8q + 15$
18) $z^2 - 10z + 24$
19) $y^2 + 9y + 10$
20) $a^2 - 16a + 8$
21) $p^2 - 20p + 100$
22) $u^2 + 18u + 81$

Objective 2: More on Factoring a Trinomial of the Form $x^2 + bx + c$

Factor completely, if possible. Check your answer.

23) $3p^2 - 24p - 27$
24) $5n^2 + 40n + 60$
25) $2k^3 - 26k^2 + 80k$
26) $w^4 - 7w^3 + 12w^2$
27) $a^3b + 9a^2b - 36ab$
28) $4x^3y + 32x^2y + 24xy$

Factor completely by first taking out -1 and then by factoring the trinomial, if possible. Check your answer.

29) $-a^2 - 10a - 16$
30) $-y^2 - 9y - 18$
31) $-h^2 + 2h + 15$
32) $-j^2 - j + 56$
33) $-k^2 + 11k - 28$
34) $-b^2 + 17b - 66$
35) $-n^2 - 14n - 49$
36) $-z^2 + 4z - 4$

Objective 3: Factor a Trinomial Containing Two Variables

Factor completely. Check your answer.

37) $a^2 + 6ab + 5b^2$
38) $v^2 + 7vw + 6w^2$
39) $m^2 + 4mn - 21n^2$
40) $x^2 - 15xy + 36y^2$
41) $p^2 - 17pq + 72q^2$
42) $r^2 - 9rs + 20s^2$
43) $f^2 - 10fg - 11g^2$
44) $u^2 + 2uv - 48v^2$
45) $c^2 + 6cd - 55d^2$
46) $w^2 + 17wx + 60x^2$

Objective 4: Factor $ax^2 + bx + c$ ($a \neq 1$) by Grouping

Factor by grouping.

47) $2r^2 + 11r + 15$
48) $3a^2 + 10a + 8$
49) $5p^2 - 21p + 4$
50) $7j^2 - 30j + 8$
51) $11m^2 - 18m - 8$
52) $5b^2 + 9b - 18$
53) $6v^2 + 11v - 7$
54) $8x^2 - 14x + 3$
55) $10c^2 + 19c + 6$
56) $15n^2 + 22n + 8$
57) $6a^2 + ab - 5b^2$
58) $9x^2 - 13xy + 4y^2$

Objective 5: Factor $ax^2 + bx + c$ ($a \neq 1$) by Trial and Error

59) How do we know that $(2x - 4)$ cannot be a factor of $2x^2 + 13x - 24$?

60) How do we know that $(5c + 10)$ cannot be a factor of $5c^2 + 16c + 30$?

Factor by trial and error.

61) $5w^2 + 11w + 6$
62) $2g^2 + 13g + 18$
63) $3u^2 - 23u + 30$
64) $7a^2 - 17a + 6$
65) $7k^2 + 15k - 18$
66) $5z^2 - 18z - 35$
67) $8r^2 + 26r + 15$
68) $6t^2 + 23t + 7$
69) $6v^2 - 19v + 14$
70) $10m^2 + 47m - 15$
71) $10a^2 - 13ab + 4b^2$
72) $8x^2 - 19xy + 6y^2$
73) $6c^2 + 31cd + 18d^2$
74) $12m^2 - 16mn - 35n^2$

Objective 6: Factor Using Substitution

Use substitution to factor each polynomial.

75) $(n + 5)^2 + 6(n + 5) - 27$
76) $(p - 6)^2 + 11(p - 6) + 28$
77) $(k - 3)^2 - 9(k - 3) + 8$
78) $(t + 4)^2 - 10(t + 4) - 24$
79) $2(w + 1)^2 - 13(w + 1) + 15$
80) $3(c - 9)^2 + 14(c - 9) + 16$
81) $6(2y - 1)^2 - 5(2y - 1) - 4$
82) $10(3a + 2)^2 - 19(3a + 2) + 6$

Mixed Exercises: Objectives 1–6

Factor completely, if possible.

83) $4q^3 - 28q^2 + 48q$
84) $2y^2 - 19y + 24$
85) $6 + 7t + t^2$
86) $m^2n^2 - 5mn - 6$
87) $12c^2 + 15c - 18$
88) $-h^2 - 3h + 54$
89) $3(b + 5)^2 + 4(b + 5) - 20$
90) $a^3b + 10a^2b^2 + 24ab^3$
91) $7s^2 - 17st + 6t^2$
92) $(x + y)t^2 - 4(x + y)t - 21(x + y)$
93) $-10z^2 + 19z - 6$
94) $64p^2 - 112p + 49$
95) $c^2 + 6c - 5$
96) $4(2h + 1)^2 - 3(2h + 1) - 22$
97) $r^2 - 11r + 18$
98) $2k^2 + 13k + 21$
99) $12p^2(q - 1) - 49p(q - 1)^2 + 49(q - 1)^2$
100) $3w^2 - w - 6$

R Rethink

R1) How do you determine whether a trinomial is *prime* using the methods shown in this section?

R2) Have you noticed that you need quick recall of the multiplication facts from 1–12 in order to factor well? Do you need to review the facts?

R3) Which trinomials were the easiest for you to factor? Do they all have something in common?

7.3 Special Factoring Techniques

P Prepare

O Organize

What are your objectives for Section 7.3?	How can you accomplish each objective?
1 Factor a Perfect Square Trinomial	• Learn the formula for **Factoring a Perfect Square Trinomial.** • Review all of the *perfect squares.* • Always begin by asking yourself, *"Can I factor out a GCF?"* and, after factoring, always ask yourself, *"Can I factor again?"* • Complete the given examples on your own. • Complete You Try 1.
2 Factor the Difference of Two Squares	• Learn the formula for **Factoring the Difference of Two Squares.** • Always begin by asking yourself, *"Can I factor out a GCF?"* and, after factoring, always ask yourself, *"Can I factor again?"* • Recognize the difference between the *sum of two squares* and the *difference of two squares.* • Complete the given examples on your own. • Complete You Trys 2 and 3.
3 Factor the Sum and Difference of Two Cubes	• Learn the formula for **Factoring the Sum and Difference of Two Cubes.** • Review all the *perfect cubes.* • Always begin by asking yourself, *"Can I factor out a GCF?"* and, after factoring, always ask yourself, *"Can I factor again?"* • Complete the given examples on your own. • Complete You Trys 4 and 5.

W Work   Read the explanations, follow the examples, take notes, and complete the You Trys.

1 Factor a Perfect Square Trinomial

Recall that we can square a binomial using the formulas

$$(a + b)^2 = a^2 + 2ab + b^2$$
$$(a - b)^2 = a^2 - 2ab + b^2$$

For example, $(x + 5)^2 = x^2 + 2x(5) + 5^2 = x^2 + 10x + 25$.

Since factoring a polynomial means writing the polynomial as a product of its factors, $x^2 + 10x + 25$ factors to $(x + 5)^2$.

The expression $x^2 + 10x + 25$ is a *perfect square trinomial.* A **perfect square trinomial** is a trinomial that results from squaring a binomial.

We can use the factoring method presented in Section 7.2 to factor a perfect square trinomial, or we can learn to recognize the special pattern that appears in these trinomials. Above we stated that $x^2 + 10x + 25$ factors to $(x + 5)^2$. How are the terms of the trinomial and binomial related?

Compare $x^2 + 10x + 25$ to $(x + 5)^2$.

x^2 is the square of x, the first term in the binomial.

25 is the square of 5, the last term in the binomial.

Hint
Write out the definition of a *perfect square trinomial* using your own words.

We get the term $10x$ by doing the following:

$$10x = 2 \cdot x \cdot 5$$

Two times / First term in binomial / Last term in binomial

This follows directly from how we found $(x + 5)^2$ using the formula.

> **Formula** Factoring a Perfect Square Trinomial
> $$a^2 + 2ab + b^2 = (a + b)^2$$
> $$a^2 - 2ab + b^2 = (a - b)^2$$

> **Note**
> In order for a trinomial to be a perfect square, two of its terms must be perfect squares.

EXAMPLE 1

Factor $t^2 + 12t + 36$ completely.

Solution

We cannot take out a common factor, so let's see if this follows the pattern of a perfect square trinomial.

$$t^2 + 12t + 36$$

What do you square to get t^2? t → $(t)^2$ $(6)^2$ ← What do you square to get 36? 6

Does the middle term equal $2 \cdot t \cdot 6$? Yes.

$$2 \cdot t \cdot 6 = 12t$$

Therefore, $t^2 + 12t + 36 = (t + 6)^2$. Check by multiplying.

EXAMPLE 2

Factor completely.

a) $n^2 - 14n + 49$ b) $4p^3 + 24p^2 + 36p$

c) $9k^2 + 30k + 25$ d) $4c^2 + 20c + 9$

Solution

a) We cannot take out a common factor. However, since the middle term is negative and the first and last terms are positive, the sign in the binomial will be a minus ($-$) sign. Does this fit the pattern of a perfect square trinomial?

$$n^2 - 14n + 49$$

What do you square to get n^2? n → $(n)^2$ $(7)^2$ ← What do you square to get 49? 7

Does the middle term equal $2 \cdot n \cdot 7$? Yes: $2 \cdot n \cdot 7 = 14n$

W Hint
The coefficients of the first and last terms in the trinomials are both perfect squares.

Since there is a minus sign in front of $14n$, $n^2 - 14n + 49$ fits the pattern of $a^2 - 2ab + b^2 = (a - b)^2$ with $a = n$ and $b = 7$.

Therefore, $n^2 - 14n + 49 = (n - 7)^2$. Check by multiplying.

> **W Hint**
> Don't forget to ask yourself, "Can I factor out a GCF?"

b) From $4p^3 + 24p^2 + 36p$ we *can* begin by taking out the GCF of $4p$.

$$4p^3 + 24p^2 + 36p = 4p(p^2 + 6p + 9)$$

What do you square to get p^2? p → $(p)^2$ $(3)^2$ ← What do you square to get 9? 3

Does the middle term equal $2 \cdot p \cdot 3$? Yes: $2 \cdot p \cdot 3 = 6p$.

$$4p^3 + 24p^2 + 36p = 4p(p^2 + 6p + 9) = 4p(p + 3)^2$$

Check by multiplying.

c) We cannot take out a common factor. Since the first and last terms of $9k^2 + 30k + 25$ are perfect squares, let's see if this is a perfect square trinomial.

$$9k^2 + 30k + 25$$

What do you square to get $9k^2$? $3k$ → $(3k)^2$ $(5)^2$ ← What do you square to get 25? 5

Does the middle term equal $2 \cdot 3k \cdot 5$? Yes: $2 \cdot 3k \cdot 5 = 30k$.

Therefore, $9k^2 + 30k + 25 = (3k + 5)^2$. Check by multiplying.

d) We cannot take out a common factor. The first and last terms of $4c^2 + 20c + 9$ are perfect squares. Is this a perfect square trinomial?

$$4c^2 + 20c + 9$$

What do you square to get $4c^2$? $2c$ → $(2c)^2$ $(3)^2$ ← What do you square to get 9? 3

Does the middle term equal $2 \cdot 2c \cdot 3$? No: $2 \cdot 2c \cdot 3 = 12c$

This is *not* a perfect square trinomial. Applying a method from Section 7.2 we find that the trinomial does factor, however.

$4c^2 + 20c + 9 = (2c + 9)(2c + 1)$. Check by multiplying.

> **[YOU TRY 1]** Factor completely.
>
> a) $w^2 + 8w + 16$ b) $a^2 - 20a + 100$ c) $4d^2 - 36d + 81$

2 Factor the Difference of Two Squares

Another common type of factoring problem is a **difference of two squares**. Some examples of these types of binomials are

$$y^2 - 9, \quad 25m^2 - 16n^2, \quad 64 - t^2, \quad \text{and} \quad h^4 - 16.$$

Notice that in each binomial, the terms are being *subtracted*, and each term is a perfect square.

In Section 6.4, we saw that

$$(a + b)(a - b) = a^2 - b^2.$$

If we reverse the procedure, we get the factorization of the difference of two squares.

> **Formula** Factoring the Difference of Two Squares
> $$a^2 - b^2 = (a + b)(a - b)$$

Don't forget that we can check all factorizations by multiplying.

EXAMPLE 3 Factor completely.

a) $y^2 - 9$ b) $25m^2 - 16n^2$ c) $w^2 - \dfrac{9}{64}$ d) $c^2 + 36$

Solution

a) First, notice that $y^2 - 9$ is the difference of two terms *and* those terms are perfect squares. We can use the formula $a^2 - b^2 = (a + b)(a - b)$.

Identify a and b.

$$y^2 - 9$$
$$\downarrow \quad \downarrow$$

What do you square to get y^2? y $\quad (y)^2 \quad (3)^2 \quad$ What do you square to get 9? 3

Then, $a = y$ and $b = 3$. Therefore, $y^2 - 9 = (y + 3)(y - 3)$.

W Hint
Notice in this example that the coefficients of the first and last terms of the binomials are both perfect squares!

b) Look carefully at $25m^2 - 16n^2$. Each term *is* a perfect square, and they are being subtracted.

Identify a and b.

$$25m^2 - 16n^2$$
$$\downarrow \quad \downarrow$$

What do you square to get $25m^2$? $5m$ $\quad (5m)^2 \quad (4n)^2 \quad$ What do you square to get $16n^2$? $4n$

Then, $a = 5m$ and $b = 4n$. So, $25m^2 - 16n^2 = (5m + 4n)(5m - 4n)$.

c) Each term in $w^2 - \dfrac{9}{64}$ is a perfect square, and they are being subtracted.

$$w^2 - \dfrac{9}{64}$$
$$\downarrow \quad \downarrow$$

What do you square to get w^2? w $\quad (w)^2 \quad \left(\dfrac{3}{8}\right)^2 \quad$ What do you square to get $\dfrac{9}{64}$? $\dfrac{3}{8}$

So, $a = w$ and $b = \dfrac{3}{8}$. Therefore, $w^2 - \dfrac{9}{64} = \left(w + \dfrac{3}{8}\right)\left(w - \dfrac{3}{8}\right)$.

d) Each term in $c^2 + 36$ is a perfect square, but the expression is the *sum* of two squares. This polynomial does not factor.

$$c^2 + 36 \neq (c + 6)(c - 6) \text{ since } (c + 6)(c - 6) = c^2 - 36.$$
$$c^2 + 36 \neq (c + 6)(c + 6) \text{ since } (c + 6)(c + 6) = c^2 + 12c + 36.$$

So, $c^2 + 36$ is prime.

> **Note**
> If the sum of two squares does not contain a common factor, then it cannot be factored.

[YOU TRY 2] Factor completely.

a) $r^2 - 25$ b) $49p^2 - 121q^2$ c) $x^2 - \dfrac{25}{144}$ d) $h^2 + 1$

Remember that sometimes we can factor out a GCF first. And, after factoring once, ask yourself, *"Can I factor again?"*

EXAMPLE 4

Factor completely.

a) $128t - 2t^3$ b) $5x^2 + 45$ c) $h^4 - 16$

Solution

a) Ask yourself, *"Can I take out a common factor?"* Yes. Factor out $2t$.

$$128t - 2t^3 = 2t(64 - t^2)$$

Now ask yourself, *"Can I factor again?"* Yes. $64 - t^2$ is the difference of two squares. Identify a and b.

$$\begin{array}{cc} 64 & - \; t^2 \\ \downarrow & \downarrow \\ (8)^2 & (t)^2 \end{array}$$

So, $a = 8$ and $b = t$. $64 - t^2 = (8 + t)(8 - t)$.

Therefore, $128t - 2t^3 = 2t(8 + t)(8 - t)$.

 BE CAREFUL $(8 + t)(8 - t)$ is *not* the same as $(t + 8)(t - 8)$ because subtraction is not commutative. While $8 + t = t + 8$, $8 - t$ does not equal $t - 8$. You must write the terms in the correct order.

Another way to see that they are not equivalent is to multiply $(t + 8)(t - 8)$. $(t + 8)(t - 8) = t^2 - 64$. This is not the same as $64 - t^2$.

b) Ask yourself, *"Can I take out a common factor?"* Yes. Factor out 5.

$$5(x^2 + 9)$$

"Can I factor again?" No; $x^2 + 9$ is the *sum* of two squares. Therefore, $5x^2 + 45 = 5(x^2 + 9)$.

c) The terms in $h^4 - 16$ have no common factors, but they are perfect squares. Identify a and b.

$$\begin{array}{cc} h^4 & - \; 16. \\ \downarrow & \downarrow \\ (h^2)^2 & (4)^2 \end{array}$$

What do you square to get h^4? h^2 What do you square to get 16? 4

So, $a = h^2$ and $b = 4$. Therefore, $h^4 - 16 = (h^2 + 4)(h^2 - 4)$. Can we factor again?

$h^2 + 4$ is the *sum* of two squares. It will not factor.
$h^2 - 4$ is the difference of two squares, so it *will* factor.

$$h^2 - 4$$
$$\downarrow \quad \downarrow \qquad h^2 - 4 = (h + 2)(h - 2)$$
$$(h)^2 \quad (2)^2$$
$$a = h \text{ and } b = 2$$

Therefore, $h^4 - 16 = (h^2 + 4)(h^2 - 4) = (h^2 + 4)(h + 2)(h - 2)$.

[**YOU TRY 3**] Factor completely.

a) $12p^4 - 27p^2$ b) $y^4 - 1$ c) $2n^2 + 72$

3 Factor the Sum and Difference of Two Cubes

We can understand where we get the formulas for factoring the sum and difference of two cubes by looking at two products.

$(a + b)(a^2 - ab + b^2) = a(a^2 - ab + b^2) + b(a^2 - ab + b^2)$ Distributive property
$\qquad\qquad\qquad\qquad\quad = a^3 - a^2b + ab^2 + a^2b - ab^2 + b^3$ Distribute.
$\qquad\qquad\qquad\qquad\quad = a^3 + b^3$ Combine like terms.

So, $(a + b)(a^2 - ab + b^2) = a^3 + b^3$, the sum of two cubes.

Now, let's multiply $(a - b)(a^2 + ab + b^2)$.

$(a - b)(a^2 + ab + b^2) = a(a^2 + ab + b^2) - b(a^2 + ab + b^2)$ Distributive property
$\qquad\qquad\qquad\qquad\quad = a^3 + a^2b + ab^2 - a^2b - ab^2 - b^3$ Distribute.
$\qquad\qquad\qquad\qquad\quad = a^3 - b^3$ Combine like terms.

So, $(a - b)(a^2 + ab + b^2) = a^3 - b^3$, the difference of two cubes.

The formulas for factoring the sum and difference of two cubes, then, are as follows:

Hint
Write down any patterns you see in the formulas. This may help you remember them.

Formula Factoring the Sum and Difference of Two Cubes

$$a^3 + b^3 = (a + b)(a^2 - ab + b^2)$$
$$a^3 - b^3 = (a - b)(a^2 + ab + b^2)$$

Note

Notice that each factorization is the product of a binomial and a trinomial. To factor the sum and difference of two cubes

Step 1: Identify a and b.

Step 2: Place them in the binomial factor, and write the trinomial based on a and b.

Step 3: Simplify.

EXAMPLE 5

Factor completely.

a) $n^3 + 8$ b) $c^3 - 64$ c) $125r^3 + 27s^3$

Solution

a) Use Steps 1–3 to factor.

Step 1: Identify a and b.

$$n^3 + 8$$

What do you cube to get n^3? n → $(n)^3$ $(2)^3$ ← What do you cube to get 8? 2

So, $a = n$ and $b = 2$.

Step 2: Remember, $a^3 + b^3 = (a + b)(a^2 - ab + b^2)$.
Write the binomial factor, then write the trinomial.

$$n^3 + 8 = (n + 2)[(n)^2 - (n)(2) + (2)^2]$$

with *Same sign* under $(n + 2)$, *Opposite sign* under the minus, *Square a.* over $(n)^2$, *Product of a and b* over $(n)(2)$, *Square b.* over $(2)^2$.

Step 3: Simplify: $n^3 + 8 = (n + 2)(n^2 - 2n + 4)$

b) **Step 1:** Identify a and b.

$$c^3 - 64$$

What do you cube to get c^3? c → $(c)^3$ $(4)^3$ ← What do you cube to get 64? 4

So, $a = c$ and $b = 4$.

Step 2: Write the binomial factor, then write the trinomial. Remember, $a^3 - b^3 = (a - b)(a^2 + ab + b^2)$.

$$c^3 - 64 = (c - 4)[(c)^2 + (c)(4) + (4)^2]$$

with *Same sign* under $(c - 4)$, *Opposite sign* under the plus, *Square a.* over $(c)^2$, *Product of a and b* over $(c)(4)$, *Square b.* over $(4)^2$.

Step 3: Simplify: $c^3 - 64 = (c - 4)(c^2 + 4c + 16)$

c) $125r^3 + 27s^3$

Step 1: Identify a and b.

$$125r^3 + 27s^3$$

What do you cube to get $125r^3$? $5r$ → $(5r)^3$ $(3s)^3$ ← What do you cube to get $27s^3$? $3s$

So, $a = 5r$ and $b = 3s$.

> **W Hint**
> Write out the example as you are reading it!

Step 2: Write the binomial factor, then write the trinomial. Remember, $a^3 + b^3 = (a + b)(a^2 - ab + b^2)$.

$$125r^3 + 27s^3 = \underbrace{(5r + 3s)}_{\text{Same sign, Opposite sign}}[\underbrace{(5r)^2}_{\text{Square } a} - \underbrace{(5r)(3s)}_{\text{Product of } a \text{ and } b} + \underbrace{(3s)^2}_{\text{Square } b}]$$

Step 3: Simplify: $125r^3 + 27s^3 = (5r + 3s)(25r^2 - 15rs + 9s^2)$

[YOU TRY 4] Factor completely.

a) $r^3 + 1$ b) $p^3 - 1000$ c) $64x^3 - 125y^3$

Just as in the other factoring problems we've studied so far, the first step in factoring *any* polynomial should be to ask ourselves, "Can I factor out a GCF?"

EXAMPLE 6 Factor $3d^3 - 81$ completely.

Solution

"Can I factor out a GCF?" Yes. The GCF is 3.

$$3d^3 - 81 = 3(d^3 - 27)$$

Factor $d^3 - 27$. Use $a^3 - b^3 = (a - b)(a^2 + ab + b^2)$.

$$d^3 - 27 = (d - 3)[(d)^2 + (d)(3) + (3)^2]$$
$$(d)^3 - (3)^3 = (d - 3)(d^2 + 3d + 9)$$
$$3d^3 - 81 = 3(d^3 - 27)$$
$$= 3(d - 3)(d^2 + 3d + 9)$$

[YOU TRY 5] Factor completely.

a) $4t^3 + 4$ b) $72a^3 - 9b^6$

As always, the first thing you should do when factoring is ask yourself, "Can I factor out a GCF?" and the last thing you should do is ask yourself, "Can I factor again?" Now we will summarize the factoring methods discussed in this section.

Summary Special Factoring Rules

Perfect square trinomials: $a^2 + 2ab + b^2 = (a + b)^2$
$a^2 - 2ab + b^2 = (a - b)^2$

Difference of two squares: $a^2 - b^2 = (a + b)(a - b)$

Sum of two cubes: $a^3 + b^3 = (a + b)(a^2 - ab + b^2)$

Difference of two cubes: $a^3 - b^3 = (a - b)(a^2 + ab + b^2)$

ANSWERS TO YOU TRY EXERCISES

1) a) $(w + 4)^2$ b) $(a - 10)^2$ c) $(2d - 9)^2$ 2) a) $(r + 5)(r - 5)$ b) $(7p + 11q)(7p - 11q)$
c) $\left(x - \dfrac{5}{12}\right)\left(x + \dfrac{5}{12}\right)$ d) prime   3) a) $3p^2(2p + 3)(2p - 3)$ b) $(y^2 + 1)(y + 1)(y - 1)$
c) $2(n^2 + 36)$ 4) a) $(r + 1)(r^2 - r + 1)$ b) $(p - 10)(p^2 + 10p + 100)$
c) $(4x - 5y)(16x^2 + 20xy + 25y^2)$ 5) a) $4(t + 1)(t^2 - t + 1)$ b) $9(2a - b^2)(4a^2 + 2ab^2 + b^4)$

E Evaluate 7.3 Exercises

Do the exercises, and check your work.

Objective 1: Factor a Perfect Square Trinomial

1) Find the following.
   a) 6^2
   b) 10^2
   c) 4^2
   d) 11^2
   e) 3^2
   f) 8^2
   g) 12^2
   h) $\left(\dfrac{1}{2}\right)^2$
   i) $\left(\dfrac{3}{5}\right)^2$

2) What is a perfect square trinomial?

3) Fill in the blank with a term that has a positive coefficient.
   a) $(\_\_\_)^2 = n^4$
   b) $(\_\_\_)^2 = 25t^2$
   c) $(\_\_\_)^2 = 49k^2$
   d) $(\_\_\_)^2 = 16p^4$
   e) $(\_\_\_)^2 = \dfrac{1}{9}$
   f) $(\_\_\_)^2 = \dfrac{25}{4}$

4) If x^n is a perfect square, then n is divisible by what number?

5) What perfect square trinomial factors to $(z + 9)^2$?

6) What perfect square trinomial factors to $(2b - 7)^2$?

7) Why isn't $9c^2 - 12c + 16$ a perfect square trinomial?

8) Why isn't $k^2 + 6k + 8$ a perfect square trinomial?

Factor completely.

9) $t^2 + 16t + 64$
10) $x^2 + 12x + 36$
11) $g^2 - 18g + 81$
12) $q^2 - 22q + 121$
13) $4y^2 + 12y + 9$
14) $49r^2 + 14r + 1$
15) $9k^2 - 24k + 16$
16) $16b^2 - 24b + 9$
17) $a^2 + \dfrac{2}{3}a + \dfrac{1}{9}$
18) $m^2 + m + \dfrac{1}{4}$
19) $v^2 - 3v + \dfrac{9}{4}$
20) $h^2 - \dfrac{4}{5}h + \dfrac{4}{25}$
21) $x^2 + 6xy + 9y^2$
22) $36t^2 - 60tu + 25u^2$
23) $9a^2 - 12ab + 4b^2$
24) $81k^2 + 18km + m^2$
25) $4f^2 + 24f + 36$
26) $9j^2 - 18j + 9$
27) $2p^4 - 24p^3 + 72p^2$
28) $5r^3 + 40r^2 + 80r$
29) $-18d^2 - 60d - 50$
30) $-28z^2 + 28z - 7$
31) $12c^3 + 3c^2 + 27c$
32) $100n^4 - 8n^3 + 64n^2$

Objective 2: Factor the Difference of Two Squares

33) What binomial factors to
   a) $(x + 4)(x - 4)$?
   b) $(4 + x)(4 - x)$?

34) What binomial factors to
   a) $(y - 9)(y + 9)$?
   b) $(9 - y)(9 + y)$?

Factor completely.

35) $x^2 - 9$
36) $q^2 - 49$
37) $n^2 - 121$
38) $d^2 - 81$
39) $m^2 + 64$
40) $q^2 + 9$
41) $y^2 - \dfrac{1}{25}$
42) $t^2 - \dfrac{1}{100}$
43) $c^2 - \dfrac{9}{16}$
44) $m^2 - \dfrac{4}{25}$
45) $36 - h^2$
46) $4 - b^2$
47) $169 - a^2$
48) $121 - w^2$
49) $\dfrac{49}{64} - j^2$
50) $\dfrac{144}{49} - r^2$
51) $100m^2 - 49$
52) $36x^2 - 25$
53) $16p^2 - 81$
54) $9a^2 - 1$

386   CHAPTER 7   Factoring Polynomials

55) $4t^2 + 25$

56) $64z^2 + 9$

57) $\frac{1}{4}k^2 - \frac{4}{9}$

58) $\frac{1}{36}d^2 - \frac{4}{49}$

59) $b^4 - 64$

60) $u^4 - 49$

61) $144m^2 - n^4$

62) $64p^2 - 25q^4$

63) $r^4 - 1$

64) $k^4 - 81$

65) $16h^4 - g^4$

66) $b^4 - a^4$

67) $4a^2 - 100$

68) $3p^2 - 48$

69) $2m^2 - 128$

70) $6j^2 - 6$

71) $45r^4 - 5r^2$

72) $32n^5 - 200n^3$

Objective 3: Factor the Sum and Difference of Two Cubes

73) Find the following.

   a) 4^3
   b) 1^3
   c) 10^3
   d) 3^3
   e) 5^3
   f) 2^3

74) If x^n is a perfect cube, then n is divisible by what number?

75) Fill in the blank.

   a) $(\_\_)^3 = y^3$
   b) $(\_\_)^3 = 8c^3$
   c) $(\_\_)^3 = 125r^3$
   d) $(\_\_)^3 = x^6$

76) If x^n is a perfect square *and* a perfect cube, then n is divisible by what number?

Complete the factorization.

77) $x^3 + 27 = (x + 3)(\qquad)$

78) $t^3 - 125 = (t - 5)(\qquad)$

Factor completely.

79) $d^3 + 1$

80) $n^3 + 125$

81) $p^3 - 27$

82) $g^3 - 8$

83) $k^3 + 64$

84) $z^3 - 1000$

85) $27m^3 - 125$

86) $64c^3 + 1$

87) $125y^3 - 8$

88) $27a^3 + 64$

89) $1000c^3 - d^3$

90) $125v^3 + w^3$

91) $8j^3 + 27k^3$

92) $125m^3 - 27n^3$

93) $64x^3 + 125y^3$

94) $27a^3 - 1000b^3$

95) $6c^3 + 48$

96) $9k^3 - 9$

97) $7v^3 - 7000w^3$

98) $216a^3 + 64b^3$

99) $p^6 - 1$

100) $h^6 - 64$

Extend the concepts of this section to factor completely.

101) $(x + 5)^2 - (x - 2)^2$

102) $(r - 6)^2 - (r + 1)^2$

103) $(2p + 3)^2 - (p + 4)^2$

104) $(3d - 2)^2 - (d - 5)^2$

105) $(t + 5)^3 + 8$

106) $(c - 2)^3 + 27$

107) $(k - 9)^3 - 1$

108) $(y + 3)^3 - 125$

R Rethink

R1) How do you know if a trinomial is a *perfect square*?

R2) When does the product of two binomials result in the *difference of two squares*?

R3) Write down four examples of binomials that are differences of two squares.

R4) Write down four examples of binomials that are differences of two cubes.

R5) Why is it that you can factor a *sum of two cubes* but not a *sum of two squares*?

R6) Which exercises in this section do you find most challenging?

Putting It All Together

P Prepare

What are your objectives for Putting It All Together?	How can you accomplish each objective?
1 Learn Strategies for Factoring a Given Polynomial	<ul><li>Review the summary on **How To Factor a Polynomial**, and write it in your notes.</li><li>Know how to determine possible methods to choose from based on the given number of terms in the problem.</li><li>Be sure to check your answer.</li><li>Complete the given examples on your own.</li><li>Complete You Try 1.</li></ul>

O Organize

W Work

Read the explanations, follow the examples, take notes, and complete the You Trys.

1 Learn Strategies for Factoring a Given Polynomial

In this chapter, we have discussed several different types of factoring problems:

1) Factoring out a GCF (Section 7.1)
2) Factoring by grouping (Section 7.1)
3) Factoring a trinomial of the form $x^2 + bx + c$ (Section 7.2)
4) Factoring a trinomial of the form $ax^2 + bx + c$ (Section 7.2)
5) Factoring a perfect square trinomial (Section 7.3)
6) Factoring the difference of two squares (Section 7.3)
7) Factoring the sum and difference of two cubes (Section 7.3)

We have practiced the factoring methods separately in each section, but how do we know which factoring method to use given many different types of polynomials together? We will discuss some strategies in this section. First, recall the steps for factoring *any* polynomial:

> **Summary** To Factor a Polynomial
>
> 1) *Always* begin by asking yourself, *"Can I factor out a GCF?"* If so, factor it out.
> 2) Look at the expression to decide if it will factor further. Apply the appropriate method to factor. If there are
>   a) *two terms,* see if it is a difference of two squares or the sum or difference of two cubes as in Section 7.3.
>   b) *three terms,* see if it can be factored using the methods of Section 7.2 *or* determine if it is a perfect square trinomial (Section 7.3).
>   c) *four terms,* see if it can be factored by grouping as in Section 7.1.
> 3) After factoring *always* look carefully at the result and ask yourself, *"Can I factor it again?"* If so, factor again.

Next, we will discuss how to decide which factoring method should be used to factor a particular polynomial.

EXAMPLE 1

Factor completely.

a) $12a^2 - 27b^2$ b) $y^2 - y - 30$ c) $mn^2 - 4m + 5n^2 - 20$
d) $p^2 - 16p + 64$ e) $8x^2 + 26x + 20$ f) $27k^3 + 8$
g) $t^2 + 36$

Solution

a) *"Can I factor out a GCF?"* is the first thing you should ask yourself. *Yes.* Factor out 3.

$$12a^2 - 27b^2 = 3(4a^2 - 9b^2)$$

Ask yourself, *"Can I factor again?"* Examine $4a^2 - 9b^2$. It has two terms that are being subtracted, and each term is a perfect square. $4a^2 - 9b^2$ is the difference of squares.

$$4a^2 - 9b^2 = (2a + 3b)(2a - 3b)$$
$$(2a)^2 \quad (3b)^2$$

$$12a^2 - 27b^2 = 3(4a^2 - 9b^2)$$
$$= 3(2a + 3b)(2a - 3b)$$

"Can I factor again?" No. It is completely factored.

b) *"Can I factor out a GCF?"* No. To factor $y^2 - y - 30$, think of two numbers whose *product* is -30 and *sum* is -1. The numbers are -6 and 5.

$$y^2 - y - 30 = (y - 6)(y + 5)$$

"Can I factor again?" No. It is completely factored.

c) Look at $mn^2 - 4m + 5n^2 - 20$. *"Can I factor out a GCF?"* No. Notice that this polynomial has *four terms*. When a polynomial has *four terms*, think about *factoring by grouping*.

$$\underbrace{mn^2 - 4m}_{} + \underbrace{5n^2 - 20}_{}$$
$$= m(n^2 - 4) + 5(n^2 - 4) \quad \text{Take out the common factor from each pair of terms.}$$
$$= (n^2 - 4)(m + 5) \quad \text{Factor out } (n^2 - 4) \text{ using the distributive property.}$$

Examine $(n^2 - 4)(m + 5)$ and ask yourself, *"Can I factor again?"* Yes! $(n^2 - 4)$ is the difference of two squares. Factor again.

$$(n^2 - 4)(m + 5) = (n + 2)(n - 2)(m + 5)$$

"Can I factor again?" No. It is completely factored.

$$mn^2 - 4m + 5n^2 - 20 = (n + 2)(n - 2)(m + 5)$$

Note

Seeing four terms is a clue to try factoring by grouping.

Hint
Write out the examples in your notes, and include information about *why* a given procedure was used.

d) We cannot take out a GCF from $p^2 - 16p + 64$. It is a trinomial, and notice that the first and last terms are perfect squares. *Is this a perfect square trinomial?*

$$p^2 - 16p + 64$$
$$\downarrow \qquad \quad \downarrow$$
$$(p)^2 \qquad (8)^2$$

Does the middle term equal $2 \cdot p \cdot (8)$? Yes: $2 \cdot p \cdot (8) = 16p$.
Use $a^2 - 2ab + b^2 = (a - b)^2$ with $a = p$ and $b = 8$.
Then, $p^2 - 16p + 64 = (p - 8)^2$.
"*Can I factor again?*" No. It is completely factored.

e) It is tempting to jump right in and try to factor $8x^2 + 26x + 20$ as the product of two binomials, but ask yourself, "*Can I take out a GCF?*" Yes! Factor out 2.

$$8x^2 + 26x + 20 = 2(4x^2 + 13x + 10)$$

"*Can I factor again?*" Yes.

$$2(4x^2 + 13x + 10) = 2(4x + 5)(x + 2)$$

"*Can I factor again?*" No. So, $8x^2 + 26x + 20 = 2(4x + 5)(x + 2)$.

f) We cannot take out a GCF from $27k^3 + 8$. Notice that $27k^3 + 8$ has two terms, so think about squares and cubes. Neither term is a perfect square *and* the positive terms are being added, so this *cannot* be the difference of squares.

Is each term a perfect cube? Yes! $27k^3 + 8$ is the sum of two cubes. We will factor $27k^3 + 8$ using $a^3 + b^3 = (a + b)(a^2 - ab + b^2)$ with $a = 3k$ and $b = 2$.

$$27k^3 + 8 = (3k + 2)[(3k)^2 - (3k)(2) + (2)^2]$$
$$\downarrow \quad \downarrow$$
$$(3k)^3 \ (2)^3$$
$$= (3k + 2)(9k^2 - 6k + 4)$$

"*Can I factor again?*" No. It is completely factored.

g) Look at $t^2 + 36$ and ask yourself, "*Can I factor out a GCF?*" No. The binomial $t^2 + 36$ is the *sum* of two squares, so it does not factor. This polynomial is prime.

[YOU TRY 1] Factor completely.

a) $3p^2 + p - 10$ b) $2n^3 - n^2 + 12n - 6$ c) $4k^4 + 36k^3 + 32k^2$

d) $48 - 3y^4$ e) $8r^3 - 125$

ANSWERS TO [YOU TRY] EXERCISES

1) a) $(3p - 5)(p + 2)$ b) $(n^2 + 6)(2n - 1)$ c) $4k^2(k + 8)(k + 1)$ d) $3(4 + y^2)(2 + y)(2 - y)$
e) $(2r - 5)(4r^2 + 10r + 25)$

Putting It All Together Exercises

 **Evaluate** Do the exercises, and check your work.

Objective 1: Learn Strategies for Factoring a Given Polynomial

Factor completely.

1) $m^2 + 16m + 60$
2) $h^2 - 36$
3) $uv + 6u + 9v + 54$
4) $2y^2 + 5y - 18$
5) $3k^2 - 14k + 8$
6) $n^2 - 14n + 49$
7) $16d^6 + 8d^5 + 72d^4$
8) $b^2 - 3bc - 4c^2$
9) $60w^3 + 70w^2 - 50w$
10) $7c^3 - 7$
11) $t^3 + 1000$
12) $pq - 6p + 4q - 24$
13) $49 - p^2$
14) $h^2 - 15h + 56$
15) $4x^2 + 4xy + y^2$
16) $27c - 18$
17) $3z^4 - 21z^3 - 24z^2$
18) $9a^2 + 6a - 8$
19) $4b^2 + 1$
20) $5abc - 15ac + 10bc - 30c$
21) $40x^3 - 135$
22) $81z^2 + 36z + 4$
23) $c^2 - \dfrac{1}{4}$
24) $v^2 + 3v + 4$
25) $45s^2t + 4 - 36s^2 - 5t$
26) $12c^5d - 75cd^3$
27) $k^2 + 9km + 18m^2$
28) $64r^3 + 8$
29) $z^2 - 3z - 88$
30) $40f^4g^4 + 8f^3g^3 + 16fg^2$
31) $80y^2 - 40y + 5$
32) $4t^2 - t - 5$
33) $20c^2 + 26cd + 6d^2$
34) $x^2 - \dfrac{9}{49}$
35) $n^4 - 16m^4$
36) $k^2 - 21k + 108$
37) $2a^2 - 10a - 72$
38) $x^2y - 4y + 7x^2 - 28$
39) $r^2 - r + \dfrac{1}{4}$
40) $v^3 - 125$
41) $28gh + 16g - 63h - 36$
42) $-24x^3 + 30x^2 - 9x$
43) $8b^2 - 14b - 15$
44) $50u^2 + 60u + 18$
45) $55a^6b^3 + 35a^5b^3 - 10a^4b - 20a^2b$
46) $64 - u^2$
47) $2d^2 - 9d + 3$
48) $2v^4w + 14v^3w^2 + 12v^2w^3$
49) $9p^2 - 24pq + 16q^2$
50) $c^4 - 16$
51) $30y^2 + 37y - 7$
52) $g^2 + 49$
53) $80a^3 - 270b^3$
54) $26n^6 - 39n^4 + 13n^3$
55) $rt - r - t + 1$
56) $h^2 + 10h + 25$
57) $4g^2 - 4$
58) $25a^2 - 55ab + 24b^2$
59) $3c^2 - 24c + 48$
60) $9t^4 + 64u^2$
61) $144k^2 - 121$
62) $125p^3 - 64q^3$
63) $-48g^2 - 80g - 12$
64) $5d^2 + 60d + 55$
65) $q^3 + 1$
66) $9x^2 + 12x + 4$
67) $81u^4 - v^4$
68) $45v^2 + 9vw^2 + 30vw + 6w^3$
69) $11f^2 + 36f + 9$
70) $4y^3 - 4y^2 - 80y$
71) $2j^{11} - j^3$
72) $d^2 - \dfrac{169}{100}$
73) $w^2 - 2w - 48$
74) $16a^2 - 40a + 25$
75) $k^2 + 100$
76) $24y^3 + 375$
77) $m^2 + 4m + 4$
78) $r^2 - 15r + 54$
79) $100c^4 - 36c^2$
80) $9t^2 - 64$
81) $(2z + 1)y^2 + 6(2z + 1)y - 55(2z + 1)$
82) $(a + b)c^2 - 5(a + b)c - 24(a + b)$
83) $(r - 4)^2 + 11(r - 4) + 28$
84) $(n + 3)^2 - 2(n + 3) - 35$
85) $(3p - 4)^2 - 5(3p - 4) - 36$
86) $(5w - 2)^2 - 8(5w - 2) + 12$
87) $(4k + 1)^2 - (3k + 2)^2$
88) $(5z + 3)^2 - (3z - 1)^2$
89) $(x + y)^2 - (2x - y)^2$
90) $(3s - t)^2 - (2s + t)^2$
91) $n^2 + 12n + 36 - p^2$
92) $h^2 - 10h + 25 - k^2$
93) $x^2 - 2xy + y^2 - z^2$
94) $a^2 + 2ab + b^2 - c^2$

R Rethink

R1) Are you able to write down the factoring methods without using your notes?

R2) Do you need four terms to be able to *factor by grouping*?

R3) Would you be able to do these problems without looking back in the sections?

R4) Which exercises in this section do you find most challenging?

7.4 Solving Quadratic Equations by Factoring

P Prepare

What are your objectives for Section 7.4?

O Organize

How can you accomplish each objective?

1 Solve a Quadratic Equation of the Form $ab = 0$	• Learn the definition of a *quadratic equation*. • Learn the *zero product rule*. • Check your answers. • Complete the given example on your own. • Complete You Try 1.
2 Solve Quadratic Equations by Factoring	• Know how to write a quadratic equation in *standard form*. • Learn the procedure for **Solving a Quadratic Equation by Factoring**. • Complete the given examples on your own. • Complete You Trys 2 and 3.
3 Solve Higher Degree Equations by Factoring	• Be able to recognize a higher degree equation. • Follow the same procedure for **Solving a Quadratic Equation by Factoring**. • Complete the given example on your own. • Complete You Try 4.

W Work

Read the explanations, follow the examples, take notes, and complete the You Trys.

In Section 2.1 we began our study of linear equations in one variable. A *linear equation in one variable* is an equation that can be written in the form $ax + b = 0$, where a and b are real numbers and $a \neq 0$.

In this section, we will learn how to solve *quadratic equations*.

Definition

A **quadratic equation** can be written in the form $ax^2 + bx + c = 0$, where a, b, and c are real numbers and $a \neq 0$.

When a quadratic equation is written in the form $ax^2 + bx + c = 0$, we say that it is in **standard form.** But quadratic equations can be written in other forms too.

Some examples of quadratic equations are

$$x^2 + 12x + 27 = 0, \quad 2p(p - 5) = 0, \quad \text{and} \quad (c + 1)(c - 8) = 3.$$

Quadratic equations are also called *second-degree equations* because the highest power on the variable is 2.

There are many different ways to solve quadratic equations. In this section, we will learn how to solve them by factoring; other methods will be discussed later in this book.

Solving a quadratic equation by factoring is based on the *zero product rule:* if the product of two quantities is zero, then one or both of the quantities is zero.

For example, if $5y = 0$, then $y = 0$. If $p \cdot 4 = 0$, then $p = 0$. If $ab = 0$, then either $a = 0$, $b = 0$, or *both a and b* equal zero.

Definition
Zero product rule: If $ab = 0$, then $a = 0$ or $b = 0$.

We will use this idea to solve quadratic equations by factoring.

1 Solve a Quadratic Equation of the Form $ab = 0$

EXAMPLE 1

Solve each equation.

a) $x(x + 8) = 0$ b) $(4y - 3)(y + 9) = 0$

Solution

a) The zero product rule says that at least one of the factors on the left must equal zero in order for the *product* to equal zero.

$$x(x + 8) = 0$$

$x = 0 \quad \text{or} \quad x + 8 = 0 \qquad$ Set each factor equal to 0.
$\phantom{x = 0 \quad \text{or} \quad} x = -8 \qquad$ Solve.

Check the solutions in the original equation:

If $x = 0$,
$0(0 + 8) \stackrel{?}{=} 0$
$0(8) = 0$ ✓

If $x = -8$,
$-8(-8 + 8) \stackrel{?}{=} 0$
$-8(0) = 0$ ✓

The solution set is $\{-8, 0\}$.

W Hint
When x is isolated as a factor, one of the solutions will always be equal to zero.

Note
It is important to remember that the factor x gives us the solution 0.

b) At least one of the factors on the left must equal zero for the *product* to equal zero.

$$(4y - 3)(y + 9) = 0$$

$4y - 3 = 0$ or $y + 9 = 0$ Set each factor equal to 0.
$4y = 3$ $y = -9$ Solve each equation.
$y = \dfrac{3}{4}$

Check in the original equation:

If $y = \dfrac{3}{4}$,
$$\left[4\left(\dfrac{3}{4}\right) - 3\right]\left[\dfrac{3}{4} + 9\right] \stackrel{?}{=} 0$$
$$(3 - 3)\left(\dfrac{39}{4}\right) \stackrel{?}{=} 0$$
$$0\left(\dfrac{39}{4}\right) = 0 \checkmark$$

If $y = -9$,
$$[4(-9) - 3][-9 + 9] \stackrel{?}{=} 0$$
$$-39(0) = 0 \checkmark$$

The solution set is $\left\{-9, \dfrac{3}{4}\right\}$.

[**YOU TRY 1**] Solve each equation.

a) $c(c - 9) = 0$ b) $(5t + 2)(t - 7) = 0$

2 Solve Quadratic Equations by Factoring

If the equation is in standard form, $ax^2 + bx + c = 0$, begin by factoring the expression.

EXAMPLE 2 Solve $m^2 - 6m - 40 = 0$.

Solution

$$m^2 - 6m - 40 = 0$$
$$(m - 10)(m + 4) = 0 \quad \text{Factor.}$$

$m - 10 = 0$ or $m + 4 = 0$ Set each factor equal to zero.
$m = 10$ or $m = -4$ Solve.

Check in the original equation:

If $m = 10$,
$(10)^2 - 6(10) - 40 \stackrel{?}{=} 0$
$100 - 60 - 40 = 0 \checkmark$

If $m = -4$,
$(-4)^2 - 6(-4) - 40 \stackrel{?}{=} 0$
$16 + 24 - 40 = 0 \checkmark$

The solution set is $\{-4, 10\}$.

Here are the steps to use to solve a quadratic equation by factoring:

> **Procedure** Solving a Quadratic Equation by Factoring
>
> 1) Write the equation in the form $ax^2 + bx + c = 0$ (standard form) so that all terms are on one side of the equal sign and zero is on the other side.
> 2) Factor the expression.
> 3) Set each factor equal to zero, and solve for the variable. (Use the zero product rule.)
> 4) Check the answer(s).

[**YOU TRY 2**] Solve $h^2 + 9h + 18 = 0$.

EXAMPLE 3 Solve each equation by factoring.

a) $2r^2 + 3r = 20$ b) $6d^2 = -42d$ c) $k^2 = -12(k + 3)$

d) $2(x^2 + 5) + 5x = 6x(x - 1) + 16$ e) $(z - 8)(z - 4) = 5$

Solution

a) Begin by writing $2r^2 + 3r = 20$ in standard form, $ar^2 + br + c = 0$.

$$2r^2 + 3r - 20 = 0 \quad \text{Standard form}$$
$$(2r - 5)(r + 4) = 0 \quad \text{Factor.}$$
$$2r - 5 = 0 \quad \text{or} \quad r + 4 = 0 \quad \text{Set each factor equal to zero.}$$
$$2r = 5$$
$$r = \frac{5}{2} \quad \text{or} \quad r = -4 \quad \text{Solve.}$$

W Hint
Are you writing out the steps as you read the example?

Check in the original equation:

If $r = \frac{5}{2}$,

$$2\left(\frac{5}{2}\right)^2 + 3\left(\frac{5}{2}\right) \stackrel{?}{=} 20$$
$$2\left(\frac{25}{4}\right) + \frac{15}{2} \stackrel{?}{=} 20$$
$$\frac{25}{2} + \frac{15}{2} \stackrel{?}{=} 20$$
$$\frac{40}{2} = 20 \checkmark$$

If $r = -4$,

$$2(-4)^2 + 3(-4) \stackrel{?}{=} 20$$
$$2(16) - 12 \stackrel{?}{=} 20$$
$$32 - 12 = 20 \checkmark$$

The solution set is $\left\{-4, \frac{5}{2}\right\}$.

b) Write $6d^2 = -42d$ in standard form.

$$6d^2 + 42d = 0 \quad \text{Standard form}$$
$$6d(d + 7) = 0 \quad \text{Factor.}$$
$$6d = 0 \quad \text{or} \quad d + 7 = 0 \quad \text{Set each factor equal to zero.}$$
$$d = 0 \quad \text{or} \quad d = -7 \quad \text{Solve.}$$

Check. The solution set is $\{-7, 0\}$.

Because both terms in $6d^2 = -42d$ are divisible by 6, we could have started part b) by dividing by 6:

$$\frac{6d^2}{6} = \frac{-42d}{6}$$ Divide by 6.

$$d^2 = -7d$$

$$d^2 + 7d = 0$$ Write in standard form.

$$d(d + 7) = 0$$ Factor.

$d = 0$ or $d + 7 = 0$ Set each factor equal to zero.

$d = -7$ Solve.

The solution set is $\{-7, 0\}$. We get the same result.

BE CAREFUL We cannot divide by d even though each term contains a factor of d. Doing so would eliminate the solution of zero. *In general, we can divide an equation by a nonzero real number but we cannot divide an equation by a variable because we may eliminate a solution, and we may be dividing by zero.*

c) To solve $k^2 = -12(k + 3)$, begin by writing the equation in standard form.

$$k^2 = -12k - 36$$ Distribute.

$$k^2 + 12k + 36 = 0$$ Write in standard form.

$$(k + 6)^2 = 0$$ Factor.

Because $(k + 6)^2 = 0$ means $(k + 6)(k + 6) = 0$, setting each factor equal to zero will result in the same value for k.

$$k + 6 = 0$$ Set $k + 6 = 0$.

$$k = -6$$ Solve.

Check. The solution set is $\{-6\}$.

d) We will have to perform several steps to write the equation in standard form.

$$2(x^2 + 5) + 5x = 6x(x - 1) + 16$$

$$2x^2 + 10 + 5x = 6x^2 - 6x + 16$$ Distribute.

Move the terms on the left side of the equation to the right side so that the coefficient of x^2 is positive.

$$0 = 4x^2 - 11x + 6$$ Write in standard form.

$$0 = (4x - 3)(x - 2)$$ Factor.

$4x - 3 = 0$ or $x - 2 = 0$ Set each factor equal to zero.

$4x = 3$

$x = \dfrac{3}{4}$ or $x = 2$ Solve.

The check is left to the student. The solution set is $\left\{\dfrac{3}{4}, 2\right\}$.

e) It is tempting to solve $(z - 8)(z - 4) = 5$ like this:

$$(z - 8)(z - 4) = 5$$
$$z - 8 = 5 \quad \text{or} \quad z - 4 = 5 \qquad \text{This is incorrect!}$$

One side of the equation must equal zero in order to set each factor equal to zero. Begin by multiplying on the left.

$$\begin{aligned} (z - 8)(z - 4) &= 5 \\ z^2 - 12z + 32 &= 5 &&\text{Multiply using FOIL.} \\ z^2 - 12z + 27 &= 0 &&\text{Standard form} \\ (z - 9)(z - 3) &= 0 &&\text{Factor.} \end{aligned}$$

$$z - 9 = 0 \quad \text{or} \quad z - 3 = 0 \qquad \text{Set each factor equal to zero.}$$
$$z = 9 \quad \text{or} \quad z = 3 \qquad \text{Solve.}$$

The check is left to the student. The solution set is $\{3, 9\}$.

[YOU TRY 3] Solve.

a) $w^2 + 4w - 5 = 0$ b) $29b = 5(b^2 + 4)$ c) $(a + 6)(a + 4) = 3$

d) $t^2 = 8t$ e) $(2y + 1)^2 + 5 = y^2 + 2(y + 7)$

3 Solve Higher Degree Equations by Factoring

Sometimes, equations that are not quadratics can be solved by factoring as well.

EXAMPLE 4

Solve each equation.

a) $(2x - 1)(x^2 - 9x - 22) = 0$ b) $4w^3 - 100w = 0$

Solution

a) This is *not* a quadratic equation because if we multiplied the factors on the left we would get $2x^3 - 19x^2 - 35x + 22 = 0$. This is a *cubic* equation because the degree of the polynomial on the left is 3.

The original equation is the product of two factors so we can use the zero product rule.

W Hint
The degree of the polynomial is the same as the greatest number of solutions the equation *might* have.

$$(2x - 1)(x^2 - 9x - 22) = 0$$
$$(2x - 1)(x - 11)(x + 2) = 0 \qquad \text{Factor.}$$

$$2x - 1 = 0 \quad \text{or} \quad x - 11 = 0 \quad \text{or} \quad x + 2 = 0 \qquad \text{Set each factor equal to zero.}$$
$$2x = 1$$
$$x = \frac{1}{2} \quad \text{or} \quad x = 11 \quad \text{or} \quad x = -2 \qquad \text{Solve.}$$

The check is left to the student. The solution set is $\left\{-2, \dfrac{1}{2}, 11\right\}$.

b) The GCF of the terms in the equation is $4w$. Remember, however, that *we can divide an equation by a constant but we cannot divide an equation by a variable*. Dividing by a variable may eliminate a solution and may mean we are dividing by zero. So let's begin by dividing each term by 4.

$$\frac{4w^3}{4} - \frac{100w}{4} = \frac{0}{4}$$ Divide by 4.
$$w^3 - 25w = 0$$ Simplify.
$$w(w^2 - 25) = 0$$ Factor out w.
$$w(w + 5)(w - 5) = 0$$ Factor $w^2 - 25$.
$$w = 0 \quad \text{or} \quad w + 5 = 0 \quad \text{or} \quad w - 5 = 0$$ Set each factor equal to zero.
$$w = -5 \qquad\qquad w = 5$$ Solve.

Check. The solution set is $\{0, -5, 5\}$.

[**YOU TRY 4**] Solve.

a) $(c + 10)(2c^2 + 5c - 7) = 0$ b) $r^4 = 25r^2$

In this section, it was possible to solve all of the equations by factoring. Below we show the relationship between solving a quadratic equation by factoring and solving it using a graphing calculator. In Chapter 10 we will learn other methods for solving quadratic equations.

Using Technology

In this section, we learned how to solve a quadratic equation by factoring. We can use a graphing calculator to solve a quadratic equation as well. Let's see how the two are related by using the equation $x^2 - x - 6 = 0$.

$$x^2 - x - 6 = 0$$
$$(x + 2)(x - 3) = 0$$
$$x + 2 = 0 \quad \text{or} \quad x - 3 = 0$$
$$x = -2 \qquad\qquad x = 3$$

The solution set is $\{-2, 3\}$.

Next, solve $x^2 - x - 6 = 0$ using a graphing calculator. Recall from Chapter 4 that to find the x-intercepts of the graph of an equation we let $y = 0$ and solve the equation for x. If we let $y = x^2 - x - 6$, then solving $x^2 - x - 6 = 0$ is the same as finding the x-intercepts of the graph of $y = x^2 - x - 6$. X-intercepts are also called zeros of the equation since they are the values of x that make $y = 0$.

Use Y= to enter $y = x^2 - x - 6$ into the calculator, press ZOOM, and then enter 6 to display the graph using the standard viewing window as shown at right. We obtain a graph called a parabola, and we can see that it has two x-intercepts. If the scale for each tick mark on the graph is 1, then it appears that the x-intercepts are -2 and 3. To verify press TRACE, type -2, and press ENTER. Since $x = -2$ and $y = 0$, $x = -2$ is an x-intercept.

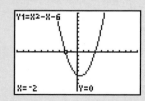

While still in "Trace mode," type 3 and press ENTER. Since $x = 3$ and $y = 0$, $x = 3$ is an x-intercept.

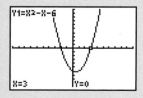

Sometimes an *x*-intercept is not an integer.

Solve $2x^2 + x - 15 = 0$ using a graphing calculator. Enter $2x^2 + x - 15$ into the calculator, and press GRAPH. The *x*-intercept on the right side of the graph is between two tick marks, so it is not an integer. To find the *x*-intercept, press 2nd TRACE and select 2:zero. Move the cursor to the left of one of the intercepts and press ENTER, then move the cursor again, so that it is to the right of the same intercept and press ENTER. Press ENTER one more time, and the calculator will reveal the intercept and, therefore one solution to the equation.

Press 2nd MODE to return to the home screen. Press X,T,Θ,n MATH ENTER ENTER to display the *x*-intercept in fraction form: $x = \frac{5}{2}$. Since the other *x*-intercept appears to be -3, press TRACE -3 ENTER to reveal that $x = -3$ and $y = 0$.

Solve using a graphing calculator.

1) $x^2 - 3x - 4 = 0$ 2) $2x^2 + 5x - 3 = 0$ 3) $x^2 - 5x - 14 = 0$
4) $3x^2 - 17x + 10 = 0$ 5) $x^2 - 7x + 12 = 0$ 6) $2x^2 - 9x + 10 = 0$

ANSWERS TO [YOU TRY] EXERCISES

1) a) $\{0, 9\}$ b) $\left\{-\frac{2}{5}, 7\right\}$ 2) $\{-6, -3\}$ 3) a) $\{-5, 1\}$ b) $\left\{\frac{4}{5}, 5\right\}$ c) $\{-7, -3\}$
   d) $\{0, 8\}$ e) $\left\{-2, \frac{4}{3}\right\}$ 4) a) $\left\{-10, -\frac{7}{2}, 1\right\}$ b) $\{0, 5, -5\}$

ANSWERS TO TECHNOLOGY EXERCISES

1) $\{-1, 4\}$ 2) $\left\{\frac{1}{2}, -3\right\}$ 3) $\{7, -2\}$ 4) $\left\{\frac{2}{3}, 5\right\}$ 5) $\{4, 3\}$ 6) $\left\{2, \frac{5}{2}\right\}$

E Evaluate 7.4 Exercises

Do the exercises, and check your work.

Objective 1: Solve a Quadratic Equation of the Form $ab = 0$

1) Explain the zero product rule.
2) When Ivan solves $c(c + 7) = 0$, he gets a solution set of $\{-7\}$. Is this correct? Why or why not?

Solve each equation.

3) $(m + 9)(m - 8) = 0$
4) $(a + 10)(a + 4) = 0$
5) $(q - 4)(q - 7) = 0$
6) $(x - 5)(x + 2) = 0$

7) $(4z + 3)(z - 9) = 0$

8) $(2n + 1)(n - 13) = 0$

9) $-5r(r - 8) = 0$

10) $11s(s + 15) = 0$

11) $(6x - 5)^2 = 0$

12) $(d + 7)^2 = 0$

13) $(4h + 7)(h + 3) = 0$

14) $(8p - 5)(3p - 11) = 0$

15) $\left(y + \dfrac{3}{2}\right)\left(y - \dfrac{1}{4}\right) = 0$

16) $\left(t - \dfrac{9}{8}\right)\left(t + \dfrac{5}{6}\right) = 0$

17) $q(q - 2.5) = 0$

18) $w(w + 0.8) = 0$

Objective 2: Solve Quadratic Equations by Factoring

19) Can we solve $(y + 6)(y - 11) = 8$ by setting each factor equal to 8 like this: $y + 6 = 8$ or $y - 11 = 8$? Why or why not?

20) Explain two ways you could begin to solve $5n^2 - 10n - 40 = 0$.

Solve each equation.

21) $v^2 + 15v + 56 = 0$

22) $y^2 + 2y - 35 = 0$

23) $k^2 + 12k - 45 = 0$

24) $z^2 - 12z + 11 = 0$

25) $3y^2 - y - 10 = 0$

26) $4f^2 - 15f + 14 = 0$

27) $14w^2 + 8w = 0$

28) $10a^2 + 20a = 0$

29) $d^2 - 15d = -54$

30) $j^2 + 11j = -28$

31) $t^2 - 49 = 0$

32) $k^2 - 100 = 0$

33) $36 = 25n^2$

34) $16 = 169p^2$

35) $m^2 = 60 - 7m$

36) $g^2 + 20 = 12g$

37) $55w = -20w^2 - 30$

38) $4v = 14v^2 - 48$

39) $p^2 = 11p$

40) $d^2 = d$

41) $45k + 27 = 18k^2$

42) $104r + 36 = 12r^2$

43) $b(b - 4) = 96$

44) $54 = w(15 - w)$

45) $-63 = 4j(j - 8)$

46) $g(3g + 11) = 70$

47) $10x(x + 1) - 6x = 9(x^2 + 5)$

48) $5r(3r + 7) = 2(4r^2 - 21)$

49) $3(h^2 - 4) = 5h(h - 1) - 9h$

50) $5(5 + u^2) + 10 = 3u(2u + 1) - u$

51) $\dfrac{1}{2}(m + 1)^2 = -\dfrac{3}{4}m(m + 5) - \dfrac{5}{2}$

52) $(2y - 3)^2 + y = (y - 5)^2 - 6$

53) $3t(t - 5) + 14 = 5 - t(t + 3)$

54) $\dfrac{1}{2}c(2 - c) - \dfrac{3}{2} = \dfrac{2}{5}c(c + 1) - \dfrac{7}{5}$

55) $33 = -m(14 + m)$

56) $-84 = s(s + 19)$

57) $(3w + 2)^2 - (w - 5)^2 = 0$

58) $(2j - 7)^2 - (j + 3)^2 = 0$

59) $(q + 3)^2 - (2q - 5)^2 = 0$

60) $(6n + 5)^2 - (3n + 4)^2 = 0$

Objective 3: Solve Higher Degree Equations by Factoring

The following equations are not quadratic but can be solved by factoring and applying the zero product rule. Solve each equation.

61) $8y(y + 4)(2y - 1) = 0$

62) $-13b(12b + 7)(b - 11) = 0$

63) $(9p - 2)(p^2 - 10p - 11) = 0$

64) $(4f + 5)(f^2 - 3f - 18) = 0$

65) $(2r - 5)(r^2 - 6r + 9) = 0$

66) $(3x - 1)(x^2 - 16x + 64) = 0$

67) $m^3 = 64m$

68) $r^3 = 81r$

69) $5w^2 + 36w = w^3$

70) $14a^2 - 49a = a^3$

71) $2g^3 = 120g - 14g^2$

72) $36z - 24z^2 = -3z^3$

73) $45h = 20h^3$

74) $64d^3 = 100d$

75) $2s^2(3s + 2) + 3s(3s + 2) - 35(3s + 2) = 0$

76) $10n^2(n - 8) + n(n - 8) - 2(n - 8) = 0$

77) $10a^2(4a + 3) + 2(4a + 3) = 9a(4a + 3)$

78) $12d^2(7d - 3) = 5d(7d - 3) + 2(7d - 3)$

79) $t^3 + 6t^2 - 4t - 24 = 0$

80) $k^3 - 8k^2 - 9k + 72 = 0$

Find the indicated values for the following polynomial functions.

81) $f(x) = x^2 + 10x + 21$. Find x so that $f(x) = 0$.

82) $h(t) = t^2 - 6t - 16$. Find t so that $h(t) = 0$.

83) $g(a) = 2a^2 - 13a + 24$. Find a so that $g(a) = 4$.

84) $Q(x) = 4x^2 - 4x + 9$. Find x so that $Q(x) = 8$.

85) $H(b) = b^2 + 3$. Find b so that $H(b) = 19$.

86) $f(z) = z^3 + 3z^2 - 54z + 5$. Find z so that $f(z) = 5$.

87) $h(k) = 5k^3 - 25k^2 + 20k$. Find k so that $h(k) = 0$.

88) $g(x) = 9x^2 - 10$. Find x so that $g(x) = -6$.

R Rethink

R1) Why do you factor a quadratic equation before solving it?

R2) Why do quadratic equations in this section have two solutions?

R3) Which types of equations in this section have more than two solutions?

R4) Can you just look at an equation and determine how many solutions it will have?

R5) Which exercises in this section do you find most challenging?

7.5 Applications of Quadratic Equations

▶ Prepare

What are your objectives for Section 7.5?

1. Solve Problems Involving Geometry

2. Solve Problems Involving Consecutive Integers

3. Solve Problems Using the Pythagorean Theorem

4. Solve an Applied Problem Using a Given Quadratic Equation

◯ Organize

How can you accomplish each objective?

- Review the procedure for **Solving Applied Problems.**
- Use the procedure for **Solving a Quadratic Equation by Factoring.**
- Complete the given example on your own.
- Complete You Try 1.

- Review the discussion of consecutive integer problems in Section 2.2 beginning on p. 55.
- Be aware that negative solutions are allowed.
- Complete the given example on your own.
- Complete You Try 2.

- Learn the *Pythagorean theorem*.
- Be able to recognize a right triangle.
- Complete the given examples on your own.
- Complete You Trys 3 and 4.

- Understand the meanings of the variables in the equation.
- Draw a diagram, when appropriate.
- Complete the given example on your own.
- Complete You Try 5.

Read the explanations, follow the examples, take notes, and complete the You Trys.

In Chapters 2 and 5 we explored applications of linear equations. In this section we will look at applications involving quadratic equations. Let's begin by restating the five steps for solving applied problems.

Procedure Steps for Solving Applied Problems

Step 1: **Read** the problem carefully, more than once if necessary, until you understand it. Draw a picture, if applicable. Identify what you are being asked to find.

Step 2: **Choose a variable** to represent an unknown quantity. If there are any other unknowns, define them in terms of the variable.

Step 3: **Translate** the problem from English into an equation using the chosen variable.

Step 4: **Solve** the equation.

Step 5: **Check** the answer in the original problem, and **interpret** the solution as it relates to the problem. Be sure your answer makes sense in the context of the problem.

1 Solve Problems Involving Geometry

EXAMPLE 1

Solve. A rectangular vegetable garden is 7 ft longer than it is wide. What are the dimensions of the garden if it covers 60 ft²?

Solution

Step 1: **Read** the problem carefully. Draw a picture.

Step 2: **Choose a variable** to represent the unknown.

$$\text{Let} \quad w = \text{the width}$$
$$w + 7 = \text{the length}$$

W Hint
Review basic geometry formulas, if necessary.

Step 3: **Translate** the information that appears in English into an algebraic equation. We must find the length and width of the garden. We are told that the *area* is 60 ft², so let's use the formula for the area of a rectangle. Then, substitute the expressions above for the length and width and 60 for the area.

$$(length)(width) = Area$$
$$(w + 7)(w) = 60 \qquad \text{length} = w + 7, \text{width} = w, \text{area} = 60$$

Step 4: **Solve** the equation.

$$w^2 + 7w = 60 \qquad \text{Distribute.}$$
$$w^2 + 7w - 60 = 0 \qquad \text{Write the equation in standard form.}$$
$$(w + 12)(w - 5) = 0 \qquad \text{Factor.}$$

$$w + 12 = 0 \quad \text{or} \quad w - 5 = 0 \qquad \text{Set each factor equal to zero.}$$
$$w = -12 \quad \text{or} \quad w = 5 \qquad \text{Solve.}$$

Step 5: **Check** the answer, and **interpret** the solution as it relates to the problem. Since w represents the width of the garden, it cannot be a negative number. So, $w = -12$ cannot be the solution. Therefore, the width is 5 ft, which will make the height $5 + 7 = 12$ ft. The area, then, is $(12 \text{ ft}) \cdot (5 \text{ ft}) = 60 \text{ ft}^2$.

[YOU TRY 1] *Solve.* The area of the surface of a desk is 8 ft². Find the dimensions of the desktop if the width is 2 ft less than the length.

2 Solve Problems Involving Consecutive Integers

In Chapter 2 we solved problems involving consecutive integers. Some applications involving consecutive integers lead to quadratic equations.

EXAMPLE 2

Solve. Twice the sum of three consecutive odd integers is 9 less than the product of the smaller two. Find the integers.

Solution

Step 1: **Read** the problem carefully, and identify what we are being asked to find.

We must find three consecutive odd integers.

Step 2: **Choose a variable** to represent an unknown, and define the other unknowns in terms of this variable.

$$x = \text{the first odd integer}$$
$$x + 2 = \text{the second odd integer}$$
$$x + 4 = \text{the third odd integer}$$

Step 3: **Translate** the information that appears in English into an algebraic equation. Read the problem slowly and carefully, breaking it into small parts.

Statement: Twice the sum of three consecutive odd integers | is | 9 less than | the product of the smaller two.

Equation: $2[x + (x + 2) + (x + 4)] = x(x + 2) - 9$

Step 4: **Solve** the equation.

$$2[x + (x + 2) + (x + 4)] = x(x + 2) - 9$$
$$2(3x + 6) = x^2 + 2x - 9 \quad \text{Combine like terms; distribute.}$$
$$6x + 12 = x^2 + 2x - 9 \quad \text{Distribute.}$$
$$0 = x^2 - 4x - 21 \quad \text{Write in standard form.}$$
$$0 = (x + 3)(x - 7) \quad \text{Factor.}$$

$$x + 3 = 0 \quad \text{or} \quad x - 7 = 0 \quad \text{Set each factor equal to zero.}$$
$$x = -3 \qquad\qquad x = 7 \quad \text{Solve.}$$

Step 5: **Check** the answer, and **interpret** the solution as it relates to the problem.

We get two sets of solutions. If $x = -3$, then the other odd integers are -1 and 1. If $x = 7$, the other odd integers are 9 and 11.

Check these numbers in the original statement of the problem.

$$2[-3 + (-1) + 1] = (-3)(-1) - 9 \qquad\qquad 2[7 + 9 + 11] = (7)(9) - 9$$
$$2(-3) = 3 - 9 \qquad\qquad\qquad\qquad 2(27) = 63 - 9$$
$$-6 = -6 \qquad\qquad\qquad\qquad\qquad 54 = 54$$

W Hint
In this application problem, one of the solutions is negative. Also note that this particular problem has two sets of solutions.

[YOU TRY 2] *Solve.* Find three consecutive even integers such that the product of the two smaller numbers is the same as twice the sum of the integers.

3 Solve Problems Using the Pythagorean Theorem

A **right triangle** is a triangle that contains a 90° **(right)** angle. We can label a right triangle as follows.

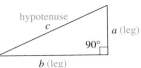

The side opposite the 90° angle is the longest side of the triangle and is called the **hypotenuse**. The other two sides are called the **legs**. The Pythagorean theorem states a relationship between the lengths of the sides of a right triangle. This is a very important relationship in mathematics and is one which is used in many different ways.

> **Definition** Pythagorean Theorem
>
> Given a right triangle with legs of length a and b and hypotenuse of length c,
>
> the Pythagorean theorem states that $a^2 + b^2 = c^2$ [or $(\text{leg})^2 + (\text{leg})^2 = (\text{hypotenuse})^2$].
>
> The Pythagorean theorem is true *only* for right triangles.

EXAMPLE 3

Find the length of the missing side.

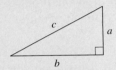

Solution

Since this is a right triangle, we can use the Pythagorean theorem to find the length of the side. Let a represent its length, and label the triangle.

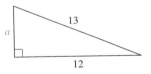

The length of the hypotenuse is 13, so $c = 13$. a and 12 are legs. Let $b = 12$.

$$a^2 + b^2 = c^2 \quad \text{Pythagorean theorem}$$
$$a^2 + (12)^2 = (13)^2 \quad \text{Substitute values.}$$
$$a^2 + 144 = 169$$
$$a^2 - 25 = 0 \quad \text{Write the equation in standard form.}$$
$$(a + 5)(a - 5) = 0 \quad \text{Factor.}$$
$$a + 5 = 0 \quad \text{or} \quad a - 5 = 0 \quad \text{Set each factor equal to 0.}$$
$$a = -5 \quad \text{or} \quad a = 5 \quad \text{Solve.}$$

W Hint
Notice that the negative solution does not make sense because it represents the side length of a triangle.

$a = -5$ does not make sense as an answer because the length of a side of a triangle cannot be negative. Therefore, $a = 5$.

Check: $5^2 + (12)^2 \stackrel{?}{=} (13)^2$
$25 + 144 = 169$ ✓

[YOU TRY 3] Find the length of the missing side.

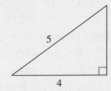

EXAMPLE 4

Solve. An animal holding pen situated between two buildings at a right angle with each other will have walls as two of its sides and a fence on the longest side. The side with the fence is 20 ft longer than the shortest side, while the third side is 10 ft longer than the shortest side. Find the length of the fence.

Solution

Step 1: **Read** the problem carefully, and identify what we are being asked to find. Draw a picture.

We must find the length of the fence.

Step 2: **Choose a variable** to represent an unknown, and define the other unknowns in terms of this variable. Draw and label the picture.

W **Hint**
Sketching a picture and labeling it is very useful in visualizing the application problem. This may make it easier to set up the equation.

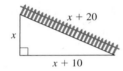

x = length of the shortest side (a leg)
$x + 10$ = length of the side along other building (a leg)
$x + 20$ = length of the fence (hypotenuse)

Step 3: **Translate** the information that appears in English into an algebraic equation. We will use the Pythagorean theorem.

$$a^2 + b^2 = c^2 \quad \text{Pythagorean theorem}$$
$$x^2 + (x + 10)^2 = (x + 20)^2 \quad \text{Substitute.}$$

Step 4: **Solve** the equation.

$$x^2 + (x + 10)^2 = (x + 20)^2$$
$$x^2 + x^2 + 20x + 100 = x^2 + 40x + 400 \quad \text{Multiply using FOIL.}$$
$$2x^2 + 20x + 100 = x^2 + 40x + 400$$
$$x^2 - 20x - 300 = 0 \quad \text{Write in standard form.}$$
$$(x - 30)(x + 10) = 0 \quad \text{Factor.}$$

$x - 30 = 0$ or $x + 10 = 0$ Set each factor equal to 0.
$x = 30$ or $x = -10$ Solve.

Step 5: **Check** the answer, and **interpret** the solution as it relates to the problem.

The length of the shortest side, x, cannot be a negative number, so x cannot equal -10. Therefore, the length of the shortest side must be 30 ft.

The length of the side along the other building is $x + 10$, so $30 + 10 = 40$ ft.

The length of the fence is $x + 20$, so $30 + 20 = 50$ ft.

Do these lengths satisfy the Pythagorean theorem? Yes.

$$a^2 + b^2 = c^2$$
$$(30)^2 + (40)^2 \stackrel{?}{=} (50)^2$$
$$900 + 1600 = 2500 \checkmark$$

Therefore, the length of the fence is 50 ft.

[YOU TRY 4] *Solve.* A wire is attached to the top of a pole. The wire is 4 ft longer than the pole, and the distance from the wire on the ground to the bottom of the pole is 4 ft less than the height of the pole. Find the length of the wire and the height of the pole.

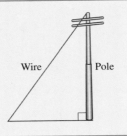

Next we will see how to use quadratic equations that model real-life situations.

4 Solve an Applied Problem Using a Given Quadratic Equation

EXAMPLE 5 A Little League baseball player throws a ball upward. The height h of the ball (in feet) t sec after the ball is released is given by the quadratic equation

$$h = -16t^2 + 30t + 4$$

a) What is the initial height of the ball?

b) How long does it take the ball to reach a height of 18 ft?

c) How long does it take for the ball to hit the ground?

Solution

W Hint
Try to sketch a picture for this problem.

a) We are asked to find the height at which the ball is released. Since t represents the number of seconds after the ball is thrown, $t = 0$ at the time of release.

Let $t = 0$, and solve for h.

$$h = -16(0)^2 + 30(0) + 4 \quad \text{Substitute 0 for } t.$$
$$= 0 + 0 + 4$$
$$= 4$$

The initial height of the ball is 4 ft.

b) We must find the *time* it takes for the ball to reach a height of 18 ft.

Find t when $h = 18$.

$$h = -16t^2 + 30t + 4$$
$$18 = -16t^2 + 30t + 4 \quad \text{Substitute 18 for } h.$$
$$0 = -16t^2 + 30t - 14 \quad \text{Write in standard form.}$$
$$0 = 8t^2 - 15t + 7 \quad \text{Divide by } -2.$$
$$0 = (8t - 7)(t - 1) \quad \text{Factor.}$$

$8t - 7 = 0 \quad \text{or} \quad t - 1 = 0 \quad$ Set each factor equal to 0.
$8t = 7$
$t = \dfrac{7}{8} \quad \text{or} \quad t = 1 \quad$ Solve.

How can two answers be possible? After $\frac{7}{8}$ sec the ball is 18 ft above the ground *on its way up,* and after 1 sec, the ball is 18 ft above the ground *on its way down.*

The ball reaches a height of 18 ft after $\frac{7}{8}$ sec *and* after 1 sec.

c) We must determine the amount of time it takes for the ball to hit the ground. When the ball hits the ground, how high off of the ground is it? *It is 0 ft high.* Find t when $h = 0$.

$$h = -16t^2 + 30t + 4$$
$$0 = -16t^2 + 30t + 4 \quad \text{Substitute 0 for } h.$$
$$0 = 8t^2 - 15t - 2 \quad \text{Divide by } -2.$$
$$0 = (8t + 1)(t - 2) \quad \text{Factor.}$$

$8t + 1 = 0$ or $t - 2 = 0$ Set each factor equal to 0.
$8t = -1$
$t = -\frac{1}{8}$ or $t = 2$ Solve.

Since t represents time, t cannot equal $-\frac{1}{8}$. We reject that as a solution. Therefore, $t = 2$. The ball will hit the ground after 2 sec.

Note

In Example 5, the equation can also be written using function notation $h(t) = -16t^2 + 30t + 4$ since the expression $-16t^2 + 30t + 4$ is a polynomial. Furthermore, $h(t) = -16t^2 + 30t + 4$ is a *quadratic function,* and we say that the height, h, is a function of the time, t. We will study quadratic functions in more detail in Chapter 10.

[YOU TRY 5] An object is thrown upward from a building. The height h of the object (in feet) t sec after the object is released is given by the quadratic equation

$$h = -16t^2 + 36t + 36$$

a) What is the initial height of the object?

b) How long does it take the object to reach a height of 44 ft?

c) How long does it take for the object to hit the ground?

ANSWERS TO [YOU TRY] EXERCISES

1) width = 2 ft; length = 4 ft   2) 6, 8, 10 or −2, 0, 2   3) 3
4) length of wire = 20 ft; height of pole = 16 ft
5) a) 36 ft   b) 0.25 sec and 2 sec   c) 3 sec

E Evaluate 7.5 Exercises

Do the exercises, and check your work.

Objective 1: Solve Problems Involving Geometry
Find the length and width of each rectangle.

1) Area = 36 in²

2) Area = 40 cm²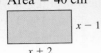

Find the base and height of each triangle.

3) Area = 12 cm²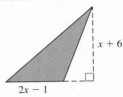

4) Area = 42 in²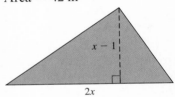

Find the base and height of each parallelogram.

5) Area = 18 in²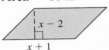

6) Area = 50 cm²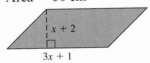

7) The volume of the box is 240 in³. Find its length and width.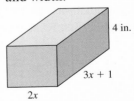

8) The volume of the box is 120 in³. Find its width and height.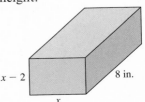

Write an equation, and solve.

9) A rectangular rug is 4 ft longer than it is wide. If its area is 45 ft², what is its length and width?

10) The surface of a rectangular bulletin board has an area of 300 in². Find its dimensions if it is 5 in. longer than it is wide.

11) Judy makes stained glass windows. She needs to cut a rectangular piece of glass with an area of 54 in² so that its width is 3 in. less than its length. Find the dimensions of the glass she must cut.

12) A rectangular painting is twice as long as it is wide. Find its dimensions if it has an area of 12.5 ft².

13) The volume of a rectangular storage box is 1440 in³. It is 20 in. long, and it is half as tall as it is wide. Find the width and height of the box.

14) A rectangular aquarium is 15 in. high, and its length is 8 in. more than its width. Find the length and width if the volume of the aquarium is 3600 in³.

15) The height of a triangle is 3 cm more than its base. Find the height and base if its area is 35 cm².

16) The area of a triangle is 16 cm². Find the height and base if its height is half the length of the base.

Objective 2: Solve Problems Involving Consecutive Integers
Write an equation, and solve.

17) The product of two consecutive integers is 19 more than their sum. Find the integers.

18) The product of two consecutive odd integers is 1 less than three times their sum. Find the integers.

19) Find three consecutive even integers such that the sum of the smaller two is one-fourth the product of the second and third integers.

20) Find three consecutive integers such that the square of the smallest is 29 less than the product of the larger two.

21) Find three consecutive integers such that the square of the largest is 22 more than the product of the smaller two.

22) Find three consecutive odd integers such that the product of the smaller two is 15 more than four times the sum of the three integers.

Objective 3: Solve Problems Using the Pythagorean Theorem

23) In your own words, explain the Pythagorean theorem.

24) Can the Pythagorean theorem be used to find a in this triangle? Why or why not?

Use the Pythagorean theorem to find the length of the missing side.

25)

26)

27)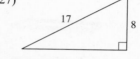

28)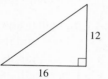

29)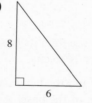

30)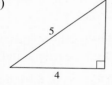

Find the lengths of the sides of each right triangle.

31)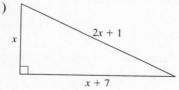

32)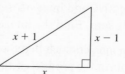

33)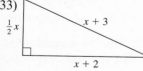

34)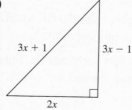

Write an equation, and solve.

35) The hypotenuse of a right triangle is 2 in. longer than the longer leg. The shorter leg measures 2 in. less than the longer leg. Find the measure of the longer leg of the triangle.

36) The longer leg of a right triangle is 7 cm more than the shorter leg. The length of the hypotenuse is 3 cm more than twice the length of the shorter leg. Find the length of the hypotenuse.

37) A 13-ft ladder is leaning against a wall. The distance from the top of the ladder to the bottom of the wall is 7 ft more than the distance from the bottom of the ladder to the wall. Find the distance from the bottom of the ladder to the wall.

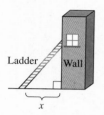

38) A wire is attached to the top of a pole. The pole is 2 ft shorter than the wire, and the distance from the wire on the ground to the bottom of the pole is 9 ft less than the length of the wire. Find the length of the wire and the height of the pole.

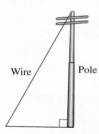

39) From a bike shop, Rana pedals due north while Yasmeen rides due west. When Yasmeen is 4 mi from the shop, the distance between her and Rana is two miles more than Rana's distance from the bike shop. Find the distance between Rana and Yasmeen.

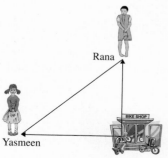

40) Henry and Allison leave home to go to work. Henry drives due west while his wife drives due south. At 8:30 am, Allison is 3 mi farther from home than Henry, and the distance between them is 6 mi more than Henry's distance from home. Find Henry's distance from his house.

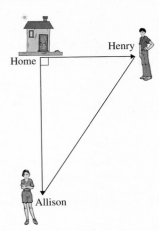

Objective 4: Solve Applied Problems Using Given Quadratic Equations
Solve.

41) A rock is dropped from a cliff and into the ocean. The height h (in feet) of the rock after t sec is given by $h = -16t^2 + 144$.

a) What is the initial height of the rock?

b) When is the rock 80 ft above the water?

c) How long does it take the rock to hit the water?

42) An object is launched from a platform with an initial velocity of 32 ft/sec. The height h (in feet) of the object after t sec is given by $h = -16t^2 + 32t + 20$.

a) What is the initial height of the object?

b) When is the object 32 ft above the ground?

c) How long does it take for the object to hit the ground?

Organizers of fireworks shows use quadratic and linear equations to help them design their programs. *Shells* contain the chemicals that produce the bursts we see in the sky. At a fireworks show the shells are shot from *mortars* and when the chemicals inside the shells ignite they explode, producing the brilliant bursts we see in the night sky.

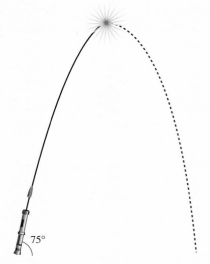

43) At a fireworks show, a 3-in. shell is shot from a mortar at an angle of 75°. The height, y (in feet), of the shell t sec after being shot from the mortar is given by the quadratic equation

$$y = -16t^2 + 144t$$

and the horizontal distance of the shell from the mortar, x (in feet), is given by the linear equation

$$x = 39t$$

(http://library.thinkquest.org/15384/physics/physics.html)

a) How high is the shell after 3 sec?

b) What is the shell's horizontal distance from the mortar after 3 sec?

c) The maximum height is reached when the shell explodes. How high is the shell when it bursts after 4.5 sec?

d) What is the shell's horizontal distance from its launching point when it explodes? (Round to the nearest foot.)

44) When a 10-in. shell is shot from a mortar at an angle of 75°, the height, y (in feet), of the shell t sec after being shot from the mortar is given by

$$y = -16t^2 + 264t$$

and the horizontal distance of the shell from the mortar, x (in feet), is given by

$$x = 71t$$

a) How high is the shell after 3 sec?

b) Find the shell's horizontal distance from the mortar after 3 sec.

c) The shell explodes after 8.25 sec. What is its height when it bursts?

d) What is the shell's horizontal distance from its launching point when it explodes? (Round to the nearest foot.)

e) Compare your answers to 43a) and 44a). What is the difference in their heights after 3 sec?

f) Compare your answers to 43c) and 44c). What is the difference in the shells' heights when they burst?

g) Assuming that the technicians timed the firings of the 3-in. shell and the 10-in. shell so that they exploded at the same time, how far apart would their respective mortars need to be so that the 10-in. shell would burst directly above the 3-in. shell?

45) The senior class at Richmond High School is selling t-shirts to raise money for its prom. The equation $R(p) = -25p^2 + 600p$ describes the revenue, R, in dollars, as a function of the price, p, in dollars, of a t-shirt. That is, the revenue is a function of price.

a) Determine the revenue if the group sells each shirt for $10.

b) Determine the revenue if the group sells each shirt for $15.

c) If the senior class hopes to have a revenue of $3600, how much should it charge for each t-shirt?

46) A famous comedian will appear at a comedy club for one performance. The equation $R(p) = -5p^2 + 300p$ describes the relationship between the price of a ticket, p, in dollars, and the revenue, R, in dollars, from ticket sales. That is, the revenue is a function of price.

a) Determine the club's revenue from ticket sales if the price of a ticket is $40.

b) Determine the club's revenue from ticket sales if the price of a ticket is $25.

c) If the club is expecting its revenue from ticket sales to be $4500, how much should it charge for each ticket?

47) An object is launched upward from the ground with an initial velocity of 200 ft/sec. The height h (in feet) of the object after t sec is given by $h(t) = -16t^2 + 200t$.

a) Find the height of the object after 1 sec.

b) Find the height of the object after 4 sec.

c) When is the object 400 ft above the ground?

d) How long does it take for the object to hit the ground?

48) The equation $R(p) = -7p^2 + 700p$ describes the revenue from ticket sales, R, in dollars, as a function of the price, p, in dollars, of a ticket to a fundraising dinner. That is, the revenue is a function of price.

a) Determine the revenue if the ticket price is $40.

b) Determine the revenue if the group sells each ticket for $70.

c) If the goal of the organizers is to have ticket revenue of $17,500, how much should it charge for each ticket?

R Rethink

R1) What is an application problem?

R2) Do you think a calculator can solve quadratic equations for you? How will a calculator help you solve the problems in this application section?

R3) Why do you sometimes eliminate negative solutions when you are solving application problems involving quadratic equations?

R4) What types of college courses do you think often use quadratic equations to solve application problems?

R5) Which exercises in this section do you find most challenging?

Group Activity — The Group Activity can be found online on Connect.

emPOWERme Switch "You" to "I"

When you are preparing for a final exam or another major academic challenge with a study group, you might find it hard to express yourself in a calm, constructive manner. As Carl Burton knows from his work during surgeries, stressful situations are often the most difficult ones in which to communicate calmly and effectively. Yet at the same time, these are the situations in which effective communication is probably most important.

To help improve your communications skills in times of stress, practice turning the following "you" statements into less aggressive "I" statements. For example, a possible "I" statement alternative to "You just don't get it, do you?" would be "I don't feel I'm making my feelings clear."

1. You just don't get it, do you?

2. You never listen to what I say.

3. You don't see where I'm coming from.

4. You don't really believe that, do you?

5. Do you even come to class?

6. Please stop interrupting me, and listen to what I'm saying for a change.

7. You're not making sense.

8. If you don't know what you're talking about, then just stop talking!

Chapter 7: Summary

Definition/Procedure	Example

7.1 The Greatest Common Factor and Factoring by Grouping

To **factor a polynomial** is to write it as a product of two or more polynomials:

To factor out a greatest common factor (GCF),

1) Identify the GCF of all of the terms of the polynomial.
2) Rewrite each term as the product of the GCF and another factor.
3) Use the distributive property to factor out the GCF from the terms of the polynomial.
4) Check the answer by multiplying the factors. (p. 358)

Factor out the greatest common factor.
$$16d^6 - 40d^5 + 72d^4$$
The GCF is $8d^4$.
$$16d^6 - 40d^5 + 72d^4 = (8d^4)(2d^2) - (8d^4)(5d) + (8d^4)(9)$$
$$= 8d^4(2d^2 - 5d + 9)$$
Check: $8d^4(2d^2 - 5d + 9) = 16d^6 - 40d^5 + 72d^4$ ✓

The first step in factoring any polynomial is to ask yourself, *"Can I factor out a GCF?"*

The last step in factoring any polynomial is to ask yourself, *"Can I factor again?"*

Try to **factor by grouping** when you are asked to factor a polynomial containing four terms.

1) Make two groups of two terms so that each group has a common factor.
2) Take out the common factor from each group of terms.
3) Factor out the common factor using the distributive property.
4) Check the answer by multiplying the factors. (p. 363)

Factor completely. $45tu + 27t + 20u + 12$

Since the four terms have a GCF of 1, we will not factor out a GCF. Begin by grouping two terms together so that each group has a common factor.

$$\underbrace{45tu + 27t}_{\downarrow} + \underbrace{20u + 12}_{\downarrow}$$
$$= 9t(5u + 3) + 4(5u + 3) \quad \text{Take out the common factor.}$$
$$= (5u + 3)(9t + 4) \quad \text{Factor out } (5u + 3).$$

Check: $(5u + 3)(9t + 4) = 45tu + 27t + 20u + 12$ ✓

7.2 Factoring Trinomials

Factoring $x^2 + bx + c$

If $x^2 + bx + c = (x + m)(x + n)$, then

1) if b and c are positive, then both m and n must be positive.
2) if c is positive and b is negative, then both m and n must be negative.
3) if c is negative, then one integer, m, must be positive and the other integer, n, must be negative. (p. 367)

Factor completely.

a) $y^2 + 7y + 12$

Think of two numbers whose *product* is 12 and whose *sum* is 7.   3 and 4. Then,
$$y^2 + 7y + 12 = (y + 3)(y + 4)$$

b) $2r^3 - 26r^2 + 60r$

Begin by factoring out the GCF of $2r$.
$$2r^3 - 26r^2 + 60r = 2r(r^2 - 13r + 30)$$
$$= 2r(r - 10)(r - 3)$$

Factoring $ax^2 + bx + c$ **by Grouping** (p. 370)

Factor completely. $5n^2 + 18n - 8$

$$\text{Sum is 18}$$
$$\downarrow$$
$$5n^2 + 18n - 8$$
$$\text{Product: } 5 \cdot (-8) = -40$$

Think of two integers whose *product* is -40 and whose *sum* is 18.   20 and -2

Definition/Procedure	Example
	Factor by grouping. $$5n^2 + 18n - 8 = \underbrace{5n^2 + 20n}_{} \underbrace{- 2n - 8}_{} \quad \text{Write } 18n \text{ as } 20 - 2n.$$ $$= 5n(n + 4) - 2(n + 4)$$ $$= (n + 4)(5n - 2)$$
Factoring $ax^2 + bx + c$ by Trial and Error When approaching a problem in this way, we must keep in mind that we are reversing the FOIL process. **(p. 371)**	Factor completely. $4x^2 - 16x + 15$ $$4x^2 - 16x + 15 = (2x - 3)(2x - 5)$$ $-6x$ $+ -10x$ $-16x$ $$4x^2 - 16x + 15 = (2x - 3)(2x - 5)$$

7.3 Special Factoring Techniques

A **perfect square trinomial** is a trinomial that results from squaring a binomial. **Factoring a Perfect Square Trinomial** $$a^2 + 2ab + b^2 = (a + b)^2$$ $$a^2 - 2ab + b^2 = (a - b)^2 \quad \textbf{(p. 378)}$$	Factor completely. a) $c^2 + 24c + 144 = (c + 12)^2$ $\quad a = c \quad b = 12$ b) $49p^2 - 56p + 16 = (7p - 4)^2$ $\quad a = 7p \quad b = 4$
Factoring the Difference of Two Squares $$a^2 - b^2 = (a + b)(a - b) \quad \textbf{(p. 380)}$$	Factor completely. $d^2 - 16 = (d + 4)(d - 4)$ $\downarrow \quad \downarrow$ $(d)^2 \quad (4)^2 \quad a = d, \quad b = 4$
Factoring the Sum and Difference of Two Cubes $$a^3 + b^3 = (a + b)(a^2 - ab + b^2)$$ $$a^3 - b^3 = (a - b)(a^2 + ab + b^2) \quad \textbf{(p. 383)}$$	Factor completely. $w^3 + 27 = (w + 3)[(w)^2 - (w)(3) + (3)^2]$ $\downarrow \quad \downarrow$ $(w)^3 \quad (3)^3 \quad a = w, \quad b = 3$ $w^3 + 27 = (w + 3)(w^2 - 3w + 9)$

7.4 Solving Quadratic Equations by Factoring

A **quadratic equation** can be written in the form $ax^2 + bx + c = 0$, where a, b, and c are real numbers and $a \neq 0$. **(p. 392)**	Some examples of quadratic equations are $3x^2 - 5x + 9 = 0, \quad t^2 = 4t + 21, \text{ and } 2(p - 3)^2 = 8 - 7p.$
To solve a quadratic equation by factoring, use the **zero product rule:** If $ab = 0$, then $a = 0$ or $b = 0$. **(p. 393)**	Solve $(y + 9)(y - 4) = 0$ $\downarrow \quad \downarrow$ $y + 9 = 0 \quad \text{or} \quad y - 4 = 0 \quad$ Set each factor equal to zero. $y = -9 \quad \text{or} \quad y = 4 \quad$ Solve. The solution set is $\{-9, 4\}$.

Definition/Procedure	Example
Steps for Solving a Quadratic Equation by Factoring 1) Write the equation in the form $ax^2 + bx + c = 0$. 2) Factor the expression. 3) Set each factor equal to zero, and solve for the variable. 4) Check the answer(s). **(p. 394)**	Solve $4m^2 - 11 = m^2 + 2m - 3$. $3m^2 - 2m - 8 = 0$ Standard form $(3m + 4)(m - 2) = 0$ Factor. $3m + 4 = 0$ or $m - 2 = 0$ $3m = -4$ $m = -\dfrac{4}{3}$ or $m = 2$ The solution set is $\left\{-\dfrac{4}{3}, 2\right\}$. Check the answers.
7.5 Applications of Quadratic Equations **Pythagorean Theorem** Given a right triangle with legs of length a and b and hypotenuse of length c, 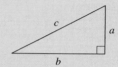 the Pythagorean theorem states that $a^2 + b^2 = c^2$ **(p. 405)**	Find the length of side a. 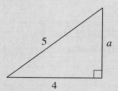 Let $b = 4$ and $c = 5$ in $a^2 + b^2 = c^2$. $a^2 + (4)^2 = (5)^2$ $a^2 + 16 = 25$ $a^2 - 9 = 0$ $(a + 3)(a - 3) = 0$ $a + 3 = 0$ or $a - 3 = 0$ $a = -3$ or $a = 3$ Reject -3 as a solution because the length of a side cannot be negative. Therefore, the length of side a is 3.

Chapter 7: Review Exercises

(7.1) Find the greatest common factor of each group of terms.

1) 18, 27
2) 56, 80, 24
3) $33p^5q^3$, $121p^4q^3$, $44p^7q^4$
4) $42r^4s^3$, $35r^2s^6$, $49r^2s^4$

Factor out the greatest common factor.

5) $48y + 84$
6) $30a^4 - 9a$
7) $7n^5 - 21n^4 + 7n^3$
8) $72u^3v^3 - 42u^3v^2 - 24uv$
9) $a(b + 6) - 2(b + 6)$
10) $u(13w - 9) + v(13w - 9)$

Factor by grouping.

11) $mn + 2m + 5n + 10$
12) $jk + 7j - 5k - 35$
13) $5qr - 10q - 6r + 12$
14) $cd^2 + 5c - d^2 - 5$
15) Factor out $-4x$ from $-8x^3 - 12x^2 + 4x$.
16) Factor out -1 from $-r^2 + 6r - 2$.

(7.2) Factor completely.

17) $p^2 + 13p + 40$
18) $f^2 - 17f + 60$
19) $x^2 + xy - 20y^2$
20) $t^2 - 2tu - 63u^2$
21) $3c^2 - 24c + 36$
22) $4m^3n + 8m^2n^2 - 60mn^3$
23) $5y^2 + 11y + 6$
24) $3g^2 + g - 44$
25) $4m^2 - 16m + 15$
26) $6t^2 - 49t + 8$
27) $56a^3 + 4a^2 - 16a$
28) $18n^2 + 98n + 40$
29) $3s^2 + 11st - 4t^2$
30) $8f^2(g - 11)^3 - 6f(g - 11)^3 - 35(g - 11)^3$
31) $(3c - 5)^2 + 10(3c - 5) + 24$
32) $2(k + 1)^2 - 15(k + 1) + 28$

(7.3) Factor completely.

33) $n^2 - 25$
34) $49a^2 - 4b^2$
35) $9t^2 + 16u^2$
36) $z^4 - 1$
37) $10q^2 - 810$
38) $48v - 27v^3$
39) $a^2 + 16a + 64$
40) $4x^2 - 20x + 25$
41) $h^3 + 8$
42) $q^3 - 1$
43) $27p^3 - 64q^3$
44) $16c^3 + 250d^3$

Mixed Exercises: Sections 7.1–7.3
Factor completely.

45) $7r^2 + 8r - 12$
46) $3y^2 + 60y + 300$
47) $\dfrac{9}{25} - x^2$
48) $81v^6 + 36v^5 - 9v^4$
49) $st - 5s - 8t + 40$
50) $n^2 - 11n + 30$
51) $w^5 - w^2$
52) $gh + 8g - 11h - 88$
53) $a^2 + 3a - 14$
54) $49k^2 - 144$
55) $(a - b)^2 - (a + b)^2$
56) $1000a^3 + 27b^3$
57) $6(y - 2)^2 - 13(y - 2) - 8$
58) $5a^2 + 22ab + 8b^2$

(7.4) Solve each equation.

59) $c(2c - 1) = 0$
60) $(4z + 7)^2 = 0$
61) $3x^2 + x = 2$
62) $f^2 - 1 = 0$
63) $n^2 = 12n + 45$
64) $10j^2 - 8 + 11j = 0$
65) $36 = 49d^2$
66) $-13w = w^2$
67) $8b + 64 = 2b^2$
68) $18 = a(9 - a)$
69) $y(5y - 9) = -4$
70) $(z + 2)^2 = -z(3z + 4) + 9$
71) $6a^3 - 3a^2 - 18a = 0$
72) $48 = 6r^2 + 12r$
73) $c(5c - 1) + 8 = 4(20 + c^2)$
74) $15t^3 + 40t = 70t^2$
75) $p^2(6p - 1) - 10p(6p - 1) + 21(6p - 1) = 0$
76) $k^2(4k - 3) - 3k(4k - 3) - 54(4k - 3) = 0$

(7.5)

77) Find the base and height if the area of the triangle is 15 in^2.

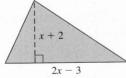

78) Find the length and width of the rectangle if its area is 28 cm^2.

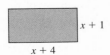

79) Find the height and length of the box if its volume is 96 in^3.

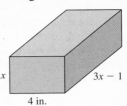

80) Find the length and width of the box if its volume is 360 in^3.

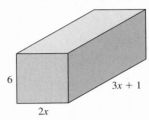

Use the Pythagorean theorem to find the length of the missing side.

81)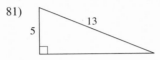

82)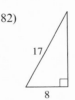

Write an equation, and solve.

83) A rectangular countertop has an area of 15 ft^2. If the width is 3.5 ft shorter than the length, what are the dimensions of the countertop?

84) Kelsey cuts a piece of fabric into a triangle to make a bandana for her dog. The base of the triangle is twice its height. Find the base and height if there is 144 in^2 of fabric.

85) The sum of three consecutive integers is one-third the square of the middle number. Find the integers.

86) Find two consecutive even integers such that their product is 6 more than 3 times their sum.

87) Seth builds a bike ramp in the shape of a right triangle. One leg is one inch shorter than the "ramp" while the other leg, the height of the ramp, is 8 in. shorter than the ramp. What is the height of the ramp?

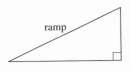

88) A car heads east from an intersection while a motorcycle travels south. After 20 min, the car is 2 mi farther from the intersection than the motorcycle. The distance between the two vehicles is 4 mi more than the motorcycle's distance from the intersection. What is the distance between the car and the motorcycle?

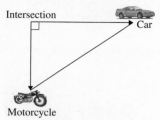

89) An object is launched with an initial velocity of 95 ft/sec. The height h (in feet) of the object after t sec is given by $h = -16t^2 + 96t$.

a) From what height is the object launched?

b) When does the object reach a height of 128 ft?

c) How high is the object after 3 sec?

d) When does the object hit the ground?

Chapter 7: Test

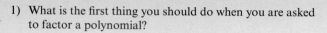

1) What is the first thing you should do when you are asked to factor a polynomial?

Factor completely.

2) $n^2 - 11n + 30$

3) $16 - b^2$

4) $5a^2 - 13a - 6$

5) $56p^6q^6 - 77p^4q^4 + 7p^2q^3$

6) $y^3 - 8z^3$

7) $2d^3 + 14d^2 - 36d$

8) $r^2 + 25$

9) $9h^2 + 24h + 16$

10) $24xy - 36x + 22y - 33$

11) $s^2 - 3st - 28t^2$

12) $16s^4 - 81t^4$

13) $4(3p + 2)^2 + 17(3p + 2) - 15$

14) $12b^2 - 44b + 35$

15) $m^{12} + m^9$

Solve each equation.

16) $b^2 + 7b + 12 = 0$

17) $25k = k^3$

18) $144m^2 = 25$

19) $(c - 5)(c + 2) = 18$

20) $4q(q - 5) + 14 = 11(2 + q)$

21) $24y^2 + 80 = 88y$

Write an equation, and solve.

22) Find the width and height of the storage locker pictured below if its volume is 120 ft³.

23) Find three consecutive odd integers such that the sum of the three numbers is 60 less than the square of the largest integer.

24) Cory and Isaac leave an intersection with Cory jogging north and Isaac jogging west. When Isaac is 1 mi farther from the intersection than Cory, the distance between them is 2 mi more than Cory's distance from the intersection. How far is Cory from the intersection?

25) The length of a rectangular dog run is 4 ft more than twice its width. Find the dimensions of the run if it covers 96 ft².

26) An object is thrown upward with an initial velocity of 68 ft/sec. The height h (in feet) of the object t sec after it is thrown is given by

$$h = -16t^2 + 68t + 60$$

a) How long does it take for the object to reach a height of 120 ft?

b) What is the initial height of the object?

c) What is the height of the object after 2 sec?

d) How long does it take the object to hit the ground?

Chapter 7: Cumulative Review for Chapters 1–7

Perform the indicated operation(s) and simplify.

1) $\dfrac{3}{8} - \dfrac{5}{6} + \dfrac{7}{12}$

2) $-\dfrac{15}{32} \cdot \dfrac{12}{25}$

Simplify. The answer should not contain any negative exponents.

3) $\dfrac{54t^5u^2}{36tu^8}$

4) $(8k^6)(-3k^4)$

5) Write 4.813×10^5 without exponents.

6) Solve $\dfrac{1}{3}(n - 2) + \dfrac{1}{4} = \dfrac{5}{12} + \dfrac{1}{6}n$

7) Solve for R.

$$A = P + PRT$$

8) *Write an equation, and solve.*
   A Twix candy bar is half the length of a Toblerone candy bar. Together, they are 12 in. long. Find the length of each candy bar.

9) Solve. Write the answer in interval notation.

$$2 + |9 - 5n| \geq 31$$

10) Graph $y = -\dfrac{3}{5}x + 7$.

11) Write the equation of the line perpendicular to $3x + y = 4$ containing the point $(-6, -1)$. Express the answer in slope-intercept form.

12) Use any method to solve this system of equations.

$$6(x + 2) + y = x - y - 2$$
$$5(2x - y + 1) = 2(x - y) - 5$$

Multiply and simplify.

13) $(6y + 5)(2y - 3)$

14) $(4p - 7)(2p^2 - 9p + 8)$

15) $(c + 8)^2$

16) Add $(4a^2b^2 - 17a^2b + 12ab - 11)$
    $+ (-a^2b + 10a^2b - 5ab^2 + 7ab + 3)$.

Divide.

17) $\dfrac{12x^4 - 30x^3 - 14x^2 + 27x + 20}{2x - 5}$

18) $\dfrac{12r^3 + 4r^2 - 10r + 3}{4r^2}$

Factor completely, if possible.

19) $bc + 8b - 7c - 56$

20) $54q^2 - 144q + 42$

21) $y^2 + 1$

22) $t^4 - 81$

23) $x^3 - 125$

Solve.

24) $z^2 + 3z = 40$

25) $-12j(1 - 2j) = 16(5 + j)$

CHAPTER 8
Rational Expressions, Equations, and Functions

OUTLINE

Study Strategies: The Writing Process

8.1 Simplifying, Multiplying, and Dividing Rational Expressions and Functions

8.2 Adding and Subtracting Rational Expressions

8.3 Simplifying Complex Fractions

8.4 Solving Rational Equations

Putting It All Together

8.5 Applications of Rational Equations

8.6 Variation

Group Activity

emPOWERme: Mad, Mad, Mad Math

Math at Work:
Chemical Engineer

Ramón Sanchez is a small businessperson with a background in chemistry and math. He has devised a new process for manufacturing an important industrial product, and now he makes his living pitching and selling this process to chemical manufacturers across the country. "The roots of my job are in math," Ramón says. "But for me to sell my product, I have to be able to explain mathematics to people who aren't necessarily as comfortable with it as I am."

Ramón has perfected his skills, then, both in the sophisticated mathematics of chemical engineering and in writing and speaking. His experiences in college gave him a solid foundation in both. "I didn't anticipate going into a career that would require me to use what I was learning in classes that weren't related to math and science," Ramón describes. "Now I realize that virtually everything you have the chance to learn in college can help you be successful, regardless of the specific field you choose."

In this chapter, we will discuss rational expressions, equations, and functions. We will also cover some strategies you can use to complete assignments that require you to do some writing.

# Study Strategies   The Writing Process

Many students enter college thinking that most people are born either good at math or good at communicating, but few people are "naturally" good at both. In truth, everyone needs to put in time and effort if they want to develop their skills in these areas. Further, anyone who works hard can excel both with numbers *and* with words.

Below are some strategies you can apply to completing writing assignments:

- Define your goals for the piece of writing. Are you going to share your opinion on a sensitive issue? Describe any research you've done.
- To get your brain focused and energized, do a freewriting exercise: Write continuously on your topic, without stopping, for 5 to 10 minutes.

- Define the main point, or thesis, of your piece of writing. Everything in the writing should relate in some way to this idea.
- Make an outline to ensure that your writing will have a solid structure.

- Create your first draft, giving yourself permission to be creative and make mistakes. Don't worry about spelling and punctuation at this stage—just try to get your thinking down on paper.
- Once you have completed your first draft, revise it, looking for ways to improve its effectiveness in meeting the goal you defined at the start.
- Now check your spelling and punctuation. Never show your writing, even in an early stage, to someone else without taking this step first.

- If you used anyone else's ideas in your writing, make sure you have given them proper credit.
- Take a step back from your writing. How might someone reading it respond to it?
- Meet with an instructor or a tutor in the writing lab to discuss ways to improve your writing further.

- Think over the whole process of writing. What worked for you? What might you do differently next time?

Chapter 8 POWER Plan

P Prepare | O Organize

What are your goals for Chapter 8?	How can you accomplish each goal?
1 Be prepared before and during class.	• _____ • _____ • _____ • _____
2 Understand the homework to the point where you could do it without needing any help or hints.	• _____ • _____ • _____
3 Use the P.O.W.E.R. framework to help you improve your writing skills: *Mad, Mad, Mad Math*.	• _____ • _____ • _____
4 Write your own goal. _____ _____	• _____ _____

What are your objectives for Chapter 8?	How can you accomplish each objective?
1 Be able to simplify, add, subtract, multiply, and divide rational expressions.	• Learn the different procedures for each of these operations on *rational expressions*. • Know how to find an LCD. • Review factoring, if needed.
2 Be able to simplify complex fractions.	• Learn the two methods for **Simplifying Complex Fractions**. • Be able to determine the LCD for the given fractions.
3 Be able to solve rational equations, including proportions.	• Know the difference between a *rational equation* and a *rational expression*. • Learn the procedure for **Solving Rational Equations**. • Follow the procedure for **Solving Proportions**.
4 Be able to solve applications of rational equations and variation problems.	• Review the procedure for **Solving Applied Problems in Section 2.1**. • Always draw a diagram or chart, when appropriate. • Review the formula for problems involving *distance, rate, and time*. • Be able to set up proportions and solve using cross multiplication. • Understand how to set up and solve *work problems*.
5 Write your own goal. _____ _____	• _____ _____

	W Work	Read Sections 8.1 to 8.6, and complete the exercises.	
E Evaluate	Complete the Chapter Review and Chapter Test. How did you do?	**R Rethink**	• Have you ever had to do any writing assignments in math? If so, what type of assignment was it? • Do writing assignments belong in a math class? Why or why not? • How are the operations on *rational expressions* different from the operations on fractions? How are they the same? • Name all the ways an LCD can be used to simplify or solve a problem.

8.1 Simplifying, Multiplying, and Dividing Rational Expressions and Functions

P Prepare	**O Organize**
What are your objectives for Section 8.1?	How can you accomplish each objective?
1 Define a Rational Function, and Determine the Domain	• Know the definition of a *rational expression*. • Understand the definition of the *domain* a rational function, and write it in your notes. • Review *interval notation*. • Complete the given examples on your own. • Complete You Trys 1 and 2.
2 Write a Rational Expression in Lowest Terms	• Learn the procedure for **Writing a Rational Expression in Lowest Terms.** • Review factoring. • Complete the given example on your own. • Complete You Try 3.
3 Simplify $\dfrac{a-b}{b-a}$	• Understand what it means to factor out -1. • Complete the given example on your own. • Complete You Try 4.
4 Write Equivalent Forms of a Rational Expression	• Understand the placement of a negative sign in a rational expression. • Complete the given example on your own. • Complete You Try 5.

(*continued*)

5 Multiply Rational Expressions	• Review factoring. • Learn the procedure for **Multiplying Rational Expressions.** • Complete the given example on your own. • Complete You Try 6.
6 Divide Rational Expressions	• Review how to divide fractions. • Review factoring. • Learn the procedure for **Dividing Rational Expressions.** • Complete the given example on your own. • Complete You Try 7.

 **Work** Read the explanations, follow the examples, take notes, and complete the You Trys.

In Section 1.1 we defined a **rational number** as the quotient of two integers provided that the denominator does not equal zero. We can define a rational expression in a similar way. A rational expression is a quotient of two polynomials provided that the denominator does not equal zero. We state the definition formally next.

> **Hint**
> In your own words, write the definition of a rational expression in your notes. Include two examples.

> **Definition**
> A **rational expression** is an expression of the form $\dfrac{P}{Q}$, where P and Q are polynomials and where $Q \neq 0$.

Some examples of rational expressions are

$$\frac{9n^4}{2}, \quad \frac{3x-5}{x+8}, \quad \frac{6}{c^2-2c-63}, \quad \text{and} \quad -\frac{3a+2b}{a^2+b^2}.$$

It is important to understand the following facts about rational expressions.

 Note

1) A fraction (rational expression) equals zero when its *numerator* equals zero.
2) A fraction (rational expression) is *undefined* when its denominator equals zero.

1 Define a Rational Function, and Determine the Domain

Hint
You do not need to look at the numerator of the *rational expression* when determining the *domain*.

Some functions are described by rational expressions. Such functions are called **rational functions**. $f(x) = \dfrac{x+3}{x-8}$ is an example of a **rational function** since $\dfrac{x+3}{x-8}$ is a rational expression and since each value that can be substituted for x will produce *only one* value for the expression.

Recall from Chapter 4 that the domain of a function $f(x)$ is the set of all real numbers that can be substituted for x. Since a rational expression is undefined when its denominator equals zero, we define the domain of a rational function as follows.

> The **domain of a rational function** consists of all real numbers except the value(s) of the variable that make(s) the denominator equal zero.

Therefore, to determine the domain of a rational function we set the denominator equal to zero and solve for the variable. Any value that makes the denominator equal to zero is *not* in the domain of the function.

To determine the domain of a rational function, sometimes it is helpful to ask yourself, "Is there any number that *cannot* be substituted for the variable?"

EXAMPLE 1

For $f(x) = \dfrac{x^2 - 9}{x + 7}$,

a) find $f(6)$.

b) find x so that $f(x) = 0$.

c) determine the domain of the function.

Solution

a) $f(6) = \dfrac{6^2 - 9}{6 + 7} = \dfrac{36 - 9}{13} = \dfrac{27}{13}$

b) To find the values of x that will make $f(x) = 0$, set the function equal to zero and solve for x.

$$\dfrac{x^2 - 9}{x + 7} = 0$$

Hint
You might need to factor when determining values that make an expression equal to zero or when determining values for the domain.

The expression $\dfrac{x^2 - 9}{x + 7} = 0$ when its *numerator* equals zero. Set the numerator equal to zero, and solve for x.

$$x^2 - 9 = 0$$
$$(x + 3)(x - 3) = 0 \quad \text{Factor.}$$
$$x + 3 = 0 \quad \text{or} \quad x - 3 = 0 \quad \text{Set each factor equal to 0.}$$
$$x = -3 \quad \text{or} \quad x = 3 \quad \text{Solve.}$$

Therefore, $f(x) = 0$ when $x = -3$ or $x = 3$.

c) To determine the domain of $f(x) = \dfrac{x^2 - 9}{x + 7}$ ask yourself, "Is there any number that *cannot* be substituted for *x*?" Yes. $f(x)$ is **undefined** when the denominator equals zero. Set the denominator equal to zero, and solve for *x*.

$$x + 7 = 0 \quad \text{Set the denominator} = 0.$$
$$x = -7 \quad \text{Solve.}$$

When $x = -7$, the denominator of $f(x) = \dfrac{x^2 - 9}{x + 7}$ equals zero. The domain contains all real numbers *except* -7. Write the domain in interval notation as $(-\infty, -7) \cup (-7, \infty)$.

[YOU TRY 1]

For $f(x) = \dfrac{x^2 - 25}{x - 2}$,

a) find $f(4)$.

b) find *x* so that $f(x) = 0$.

c) determine the domain of the function.

EXAMPLE 2

Determine the domain of each rational function.

a) $g(c) = \dfrac{6c + 5}{c^2 + 3c - 4}$ b) $h(n) = \dfrac{4n^2 - 9}{7}$

Solution

a) To determine the domain of $g(c) = \dfrac{6c + 5}{c^2 + 3c - 4}$, ask yourself, "Is there any number that *cannot* be substituted for *c*?" Yes. $g(c)$ is **undefined** when its *denominator* equals zero. Set the denominator equal to zero, and solve for *c*.

$$c^2 + 3c - 4 = 0 \quad \text{Set the denominator} = 0.$$
$$(c + 4)(c - 1) = 0 \quad \text{Factor.}$$
$$c + 4 = 0 \quad \text{or} \quad c - 1 = 0 \quad \text{Set each factor equal to 0.}$$
$$c = -4 \quad \text{or} \quad c = 1 \quad \text{Solve.}$$

When $c = -4$ or $c = 1$, the denominator of $g(c) = \dfrac{6c + 5}{c^2 + 3c - 4}$ equals zero. The domain contains all real numbers *except* -4 and 1. Write the domain in interval notation as $(-\infty, -4) \cup (-4, 1) \cup (1, \infty)$.

W Hint

In your notes and in your own words, explain how to find the domain of a rational function. Include an example.

b) Ask yourself, "Is there any number that *cannot* be substituted for *n*?" No! Looking at the denominator we see that it will never equal zero. Therefore, there *is no value* of *n* that makes $h(n) = \dfrac{4n^2 - 9}{7}$ undefined. Any real number may be substituted for *n*, and the function will be defined.

The domain of the function is the set of all real numbers. Write the domain in interval notation as $(-\infty, \infty)$.

[YOU TRY 2] Determine the domain of each rational function.

a) $f(x) = \dfrac{2x - 3}{x^2 - 8x + 12}$ b) $g(a) = \dfrac{a + 4}{10}$

All of the operations that can be performed with fractions can also be done with rational expressions. We begin our study of these operations with rational expressions by learning how to write a rational expression in lowest terms.

2 Write a Rational Expression in Lowest Terms

One way to think about writing a fraction such as $\dfrac{8}{12}$ in lowest terms is

$$\frac{8}{12} = \frac{2 \cdot 4}{3 \cdot 4} = \frac{2}{3} \cdot \frac{4}{4} = \frac{2}{3} \cdot 1 = \frac{2}{3}$$

Since $\dfrac{4}{4} = 1$, we can also think of reducing $\dfrac{8}{12}$ in the following way: $\dfrac{8}{12} = \dfrac{2 \cdot \cancel{4}}{3 \cdot \cancel{4}} = \dfrac{2}{3}$.

We can *factor* the numerator and denominator, then *divide* the numerator and denominator by the common factor, 4. This is the approach we use to write a rational expression in lowest terms.

Definition Fundamental Property of Rational Expressions
If P, Q, and C are polynomials such that $Q \neq 0$ and $C \neq 0$, then

$$\frac{PC}{QC} = \frac{P}{Q}$$

This property mirrors the example of writing $\dfrac{8}{12}$ in lowest terms because

$$\frac{PC}{QC} = \frac{P}{Q} \cdot \frac{C}{C} = \frac{P}{Q} \cdot 1 = \frac{P}{Q}.$$

Or, we can also think of the reducing procedure as dividing the numerator and denominator by the common factor, C.

$$\frac{P\cancel{C}}{Q\cancel{C}} = \frac{P}{Q}$$

Procedure Writing a Rational Expression in Lowest Terms
1) Completely **factor** the numerator and denominator.
2) **Divide** the numerator and denominator by the greatest common factor.

SECTION 8.1 Simplifying, Multiplying, and Dividing Rational Expressions and Functions

EXAMPLE 3 Write each rational expression in lowest terms.

a) $\dfrac{20c^6}{5c^4}$ b) $\dfrac{4m+12}{7m+21}$ c) $\dfrac{3x^2-3}{x^2+9x+8}$

Solution

a) We can simplify $\dfrac{20c^6}{5c^4}$ using the quotient rule presented in Chapter 6.

$$\dfrac{20c^6}{5c^4} = 4c^2$$ Divide 20 by 5, and use the quotient rule: $\dfrac{c^6}{c^4} = c^{6-4} = c^2$.

W Hint
Go back to Chapter 7 if you need to review factoring.

b) $\dfrac{4m+12}{7m+21} = \dfrac{4\cancel{(m+3)}}{7\cancel{(m+3)}}$ Factor.

$= \dfrac{4}{7}$ Divide out the common factor, $m+3$.

c) $\dfrac{3x^2-3}{x^2+9x+8} = \dfrac{3(x^2-1)}{(x+1)(x+8)}$ Factor.

$= \dfrac{3\cancel{(x+1)}(x-1)}{\cancel{(x+1)}(x+8)}$ Factor completely.

$= \dfrac{3(x-1)}{x+8}$ Divide out the common factor, $x+1$.

BE CAREFUL Notice that we divide by *factors* not *terms*.

$\dfrac{\cancel{x+5}}{2\cancel{(x+5)}} = \dfrac{1}{2}$ $\dfrac{x}{x+5} \ne \dfrac{1}{5}$

Divide by the *factor* $x+5$.   We cannot divide by x because the x in the denominator is a *term* in a sum.

[YOU TRY 3] Write each rational expression in lowest terms.

a) $\dfrac{6t-48}{t^2-8t}$ b) $\dfrac{b-2}{5b^2-6b-8}$ c) $\dfrac{v^3+27}{4v^4-12v^3+36v^2}$

3 Simplify $\dfrac{a-b}{b-a}$

W Hint
Don't forget that expressions such as $\dfrac{a+b}{b+a}$ are equal to 1 *not* -1.

Do you think that $\dfrac{x-4}{4-x}$ is in lowest terms? Let's look at it more closely to understand the answer.

$\dfrac{x-4}{4-x} = \dfrac{x-4}{-1(-4+x)}$ Factor -1 out of the denominator.

$= \dfrac{1\cancel{(x-4)}}{-1\cancel{(x-4)}}$ Rewrite $-4+x$ as $x-4$.

$= -1$

Therefore, $\dfrac{x-4}{4-x} = -1$.

We can generalize this result as

> **Note**
>
> 1) $b - a = -1(a - b)$ and   2) $\dfrac{a - b}{b - a} = -1$
>
> The terms in the numerator and denominator in 2) differ only in sign. They divide out to -1.

EXAMPLE 4   Write each rational expression in lowest terms.

a) $\dfrac{6 - d}{d - 6}$ b) $\dfrac{25z^2 - 9}{3 - 5z}$

Solution

a) $\dfrac{6 - d}{d - 6} = -1$ because $\dfrac{6 - d}{d - 6} = \dfrac{-1(d - 6)}{d - 6} = -1.$

b) $\dfrac{25z^2 - 9}{3 - 5z} = \dfrac{(5z + 3)(5z - 3)}{3 - 5z}$ Factor.

$\phantom{\dfrac{25z^2 - 9}{3 - 5z}} = -1(5z + 3)$ $\dfrac{5z - 3}{3 - 5z} = -1$

$\phantom{\dfrac{25z^2 - 9}{3 - 5z}} = -5z - 3$ Distribute.

YOU TRY 4   Write each rational expression in lowest terms.

a) $\dfrac{x - y}{y - x}$ b) $\dfrac{15n - 5m}{2m - 6n}$ c) $\dfrac{12 - 3y^2}{y^2 - 10y + 16}$

4 Write Equivalent Forms of a Rational Expression

Often, the same rational expression can be written in several different ways. You should be able to recognize equivalent forms of rational expressions because there isn't always just one way to write the correct answer.

EXAMPLE 5   Write $-\dfrac{8u(u - 1)}{1 + u}$ in three different ways.

Solution

The negative sign in front of a fraction can also be applied to the numerator or to the denominator. For example, $-\dfrac{4}{9} = \dfrac{-4}{9} = \dfrac{4}{-9}$. Applying this concept to rational expressions can result in expressions that look quite different but that are, actually, equivalent.

i) Apply the negative sign to the denominator.

$$-\frac{8u(u-1)}{1+u} = \frac{8u(u-1)}{-1(1+u)}$$
$$= \frac{8u(u-1)}{-1-u} \quad \text{Distribute.}$$

ii) Apply the negative sign to the numerator.

$$-\frac{8u(u-1)}{1+u} = \frac{-8u(u-1)}{1+u}$$

iii) Apply the negative sign to the numerator, then distribute the -1.

$$-\frac{8u(u-1)}{1+u} = \frac{(8u)(-1)(u-1)}{1+u}$$
$$= \frac{8u(-u+1)}{1+u} \quad \text{Distribute.}$$
$$= \frac{8u(1-u)}{1+u} \quad \text{Rewrite } -u+1 \text{ as } 1-u.$$

Therefore, $\frac{8u(u-1)}{-1-u}$, $\frac{-8u(u-1)}{1+u}$, and $\frac{8u(1-u)}{1+u}$ are *all* equivalent forms of $-\frac{8u(u-1)}{1+u}$.

Keep this idea of equivalent forms of rational expressions in mind when checking your answers against the answers in the back of the book. Sometimes students believe their answer is wrong because it "looks different" when, in fact, it is an *equivalent form* of the given answer!

[**YOU TRY 5**] Write $\frac{-(2-p)}{7p-9}$ in three different ways.

5 Multiply Rational Expressions

We multiply rational expressions the same way we multiply rational numbers. Multiply numerators, multiply denominators, and simplify.

> **Procedure** Multiplying Rational Expressions
>
> If $\frac{P}{Q}$ and $\frac{R}{T}$ are rational expressions, then $\frac{P}{Q} \cdot \frac{R}{T} = \frac{PR}{QT}$.
>
> To multiply two rational expressions, multiply their numerators, multiply their denominators, and simplify.

Let's begin by reviewing how we multiply two fractions. We can multiply numerators, multiply denominators, then simplify by dividing out common factors *or* we can divide out the common factors before multiplying. For example, let's find $\dfrac{8}{15} \cdot \dfrac{5}{6}$.

$$\dfrac{8}{15} \cdot \dfrac{5}{6} = \dfrac{\cancel{2} \cdot 4}{3 \cdot \cancel{5}} \cdot \dfrac{\cancel{5}}{\cancel{2} \cdot 3} \quad \text{Factor, and divide out common factors.}$$

$$= \dfrac{4}{3 \cdot 3} \quad \text{Multiply.}$$

$$= \dfrac{4}{9} \quad \text{Simplify.}$$

Multiplying rational expressions works the same way.

> **Procedure** Multiplying Rational Expressions
> 1) Factor.
> 2) Divide out common factors, and multiply.
>
> All products must be written in lowest terms.

EXAMPLE 6

Multiply.

a) $\dfrac{12a^4}{b^2} \cdot \dfrac{b^5}{6a^9}$ b) $\dfrac{8m + 48}{10m^6} \cdot \dfrac{m^2}{m^2 - 36}$

c) $\dfrac{3k^2 - 11k - 4}{k^2 + k - 20} \cdot \dfrac{k^2 + 10k + 25}{3k^2 + k}$

Solution

a) $\dfrac{12a^4}{b^2} \cdot \dfrac{b^5}{6a^9} = \dfrac{\overset{2}{\cancel{12a^4}}}{\cancel{b^2}} \cdot \dfrac{\cancel{b^2} \cdot b^3}{\cancel{6a^4} \cdot a^5} \quad$ Factor, and divide out common factors.

$= \dfrac{2b^3}{a^5} \quad$ Multiply.

b) $\dfrac{8m + 48}{10m^6} \cdot \dfrac{m^2}{m^2 - 36} = \dfrac{\overset{4}{\cancel{8(m+6)}}}{\underset{5}{\cancel{10m^2}} \cdot m^4} \cdot \dfrac{\cancel{m^2}}{\cancel{(m+6)}(m-6)} \quad$ Factor and simplify.

$= \dfrac{4}{5m^4(m-6)} \quad$ Multiply.

W Hint
Write out the example as you are reading it!

c) $\dfrac{3k^2 - 11k - 4}{k^2 + k - 20} \cdot \dfrac{k^2 + 10k + 25}{3k^2 + k} = \dfrac{\cancel{(3k+1)}\cancel{(k-4)}}{\cancel{(k+5)}\cancel{(k-4)}} \cdot \dfrac{\overset{(k+5)}{\cancel{(k+5)^2}}}{k\cancel{(3k+1)}} \quad$ Factor and simplify.

$= \dfrac{k+5}{k} \quad$ Multiply.

YOU TRY 6

Multiply.

a) $\dfrac{x^3}{14y^6} \cdot \dfrac{7y^2}{x^8}$ b) $\dfrac{r^2 - 25}{r^2 - 9r} \cdot \dfrac{r^2 + r - 90}{10 - 2r}$

6 Divide Rational Expressions

When we divide rational numbers we multiply by a reciprocal. For example, $\frac{7}{4} \div \frac{3}{8} = \frac{7}{\cancel{4}_1} \cdot \frac{\cancel{8}^2}{3} = \frac{14}{3}$. We divide rational expressions the same way. To divide rational expressions we multiply the first rational expression by the reciprocal of the second rational expression.

> **Procedure** Dividing Rational Expressions
>
> If $\frac{P}{Q}$ and $\frac{R}{T}$ are rational expressions with Q, R, and T not equal to zero, then
>
> $$\frac{P}{Q} \div \frac{R}{T} = \frac{P}{Q} \cdot \frac{T}{R} = \frac{PT}{QR}$$
>
> Multiply the first rational expression by the reciprocal of the second rational expression.

EXAMPLE 7 Divide.

a) $\dfrac{36p^5}{q^4} \div \dfrac{4p^2}{q^{10}}$ b) $\dfrac{s^2 + 8s + 16}{s^2 - 11s + 24} \div \dfrac{s^2 + 5s + 4}{15 - 5s}$

c) $\dfrac{2k^2 + k - 15}{k^3} \div (2k - 5)^2$

Solution

a) $\dfrac{36p^5}{q^4} \div \dfrac{4p^2}{q^{10}} = \dfrac{\cancel{36p^5}^{9p}}{\cancel{q^4}} \cdot \dfrac{\cancel{q^{10}}^{q^6}}{\cancel{4p^2}}$ Multiply by the reciprocal, and simplify.

$= 9p^3 q^6$ Multiply.

Notice that we used the *quotient rule* for exponents to simplify.

$$\frac{p^5}{p^2} = p^3, \quad \frac{q^{10}}{q^4} = q^6$$

b) $\dfrac{s^2 + 8s + 16}{s^2 - 11s + 24} \div \dfrac{s^2 + 5s + 4}{15 - 5s}$

$= \dfrac{s^2 + 8s + 16}{s^2 - 11s + 24} \cdot \dfrac{15 - 5s}{s^2 + 5s + 4}$ Multiply by the reciprocal.

$= \dfrac{\cancel{(s+4)}^{(s+4)}}{\cancel{(s-3)}(s-8)} \cdot \dfrac{5\cancel{(3-s)}^{-1}}{\cancel{(s+4)}(s+1)}$ Factor; $\dfrac{3-s}{s-3} = -1$.

$= \dfrac{-5(s+4)}{(s-8)(s+1)}$ Divide out common factors, and multiply.

c) $\dfrac{2k^2 + k - 15}{k^3} \div (2k - 5)^2 = \dfrac{(2k-5)(k + 3)}{k^3} \cdot \dfrac{1}{(2k-5)^2}$ Since $(2k - 5)^2$ can be written as $\dfrac{(2k - 5)^2}{1}$, its reciprocal is $\dfrac{1}{(2k - 5)^2}$. Simplify and multiply.

$= \dfrac{(k + 3)}{k^3(2k - 5)}$

[YOU TRY 7] Divide.

a) $\dfrac{r^6}{35t^9} \div \dfrac{r^4}{14t^3}$

b) $\dfrac{x^2 - 8x - 48}{7x^2 - 42x} \div \dfrac{x^2 - 16}{x^2 - 10x + 24}$

c) $\dfrac{5a^2 + 34a + 24}{a^2} \div (5a + 4)^2$

ANSWERS TO [YOU TRY] EXERCISES

1) a) $-\dfrac{9}{2}$ b) -5 or 5 c) $(-\infty, 2) \cup (2, \infty)$ 2) a) $(-\infty, 2) \cup (2, 6) \cup (6, \infty)$ b) $(-\infty, \infty)$

3) a) $\dfrac{6}{t}$ b) $\dfrac{1}{5b + 4}$ c) $\dfrac{v + 3}{4v^2}$ 4) a) -1 b) $-\dfrac{5}{2}$ c) $\dfrac{-3(y + 2)}{y - 8}$

5) Some possibilities are $\dfrac{p - 2}{7p - 9}, \dfrac{2 - p}{9 - 7p}, \dfrac{2 - p}{7p - 9}, \dfrac{2 - p}{-(7p - 9)}$ 6) a) $\dfrac{1}{2x^5y^4}$ b) $-\dfrac{(r + 5)(r + 10)}{2r}$

7) a) $\dfrac{2r^2}{5t^6}$ b) $\dfrac{x - 12}{7x}$ c) $\dfrac{a + 6}{a^2(5a + 4)}$

E Evaluate 8.1 Exercises

Do the exercises, and check your work.

Objective 1: Define a Rational Function, and Determine the Domain

1) When does a fraction or a rational expression equal 0?

2) When is a fraction or a rational expression undefined?

3) How do you determine the value of the variable for which a rational expression is undefined?

4) If $x^2 + 5$ is the numerator of a rational expression, can that expression equal zero? Give a reason.

For each rational function,

a) find $f(-2)$, if possible.

b) find x so that $f(x) = 0$.

c) determine the domain of the function.

5) $f(x) = \dfrac{x + 8}{x + 6}$

6) $f(x) = \dfrac{x}{3x - 1}$

7) $f(x) = \dfrac{5x - 3}{x + 2}$

8) $f(x) = \dfrac{9}{x - 1}$

9) $f(x) = \dfrac{6}{x^2 + 6x + 5}$

10) $f(x) = \dfrac{2x - 1}{x^2 + x - 12}$

Determine the domain of each rational function.

11) $f(p) = \dfrac{1}{p - 7}$

12) $h(z) = \dfrac{z + 8}{z + 3}$

13) $f(a) = \dfrac{6a}{7 - 2a}$

14) $k(r) = \dfrac{r}{5r + 2}$

15) $g(t) = \dfrac{3t - 4}{t^2 - 9t + 8}$

16) $r(c) = \dfrac{c + 9}{c^2 - c - 42}$

17) $h(w) = \dfrac{w + 7}{w^2 - 81}$

18) $k(t) = \dfrac{t}{t^2 - 14t + 33}$

19) $A(c) = \dfrac{8}{c^2 + 6}$

20) $C(n) = \dfrac{3n + 1}{2}$

21) Write your own example of a rational function, $f(x)$, that has a domain of $(-\infty, -8) \cup (-8, \infty)$.

22) Write your own example of a rational function, $g(x)$, that has a domain of $(-\infty, -5) \cup (-5, 6) \cup (6, \infty)$.

Objective 2: Write a Rational Expression in Lowest Terms

Write each rational expression in lowest terms.

23) $\dfrac{12d^5}{30d^8}$

24) $\dfrac{108g^4}{9g}$

25) $\dfrac{3c - 12}{5c - 20}$

26) $\dfrac{10d - 5}{12d - 6}$

27) $\dfrac{b^2 + b - 56}{b + 8}$

28) $\dfrac{g^2 + 9g + 20}{g^2 + 2g - 15}$

29) $\dfrac{r - 4}{r^2 - 16}$

30) $\dfrac{t + 2}{t^2 - 7t - 18}$

31) $\dfrac{3k^2 + 28k + 32}{k^2 + 10k + 16}$

32) $\dfrac{3c^2 - 36c + 96}{c - 8}$

33) $\dfrac{w^3 + 125}{5w^2 - 25w + 125}$

34) $\dfrac{4m^3 - 4}{m^2 + m + 1}$

35) $\dfrac{4m^2 - 20m + 4mn - 20n}{11m + 11n}$

36) $\dfrac{uv + 3u - 4v - 12}{v^2 - 9}$

37) $\dfrac{x^2 - y^2}{x^3 - y^3}$

38) $\dfrac{a^3 + b^3}{a^2 - b^2}$

Objective 3: Simplify $\dfrac{a - b}{b - a}$

39) Any rational expression of the form $\dfrac{a - b}{b - a}$ reduces to what?

40) Does $\dfrac{z + 9}{z - 9} = -1$?

Write each rational expression in lowest terms.

41) $\dfrac{12 - v}{v - 12}$

42) $\dfrac{q - 11}{11 - q}$

43) $\dfrac{a^2 - 8a - 33}{11 - a}$

44) $\dfrac{m - 10}{20 - 2m}$

45) $\dfrac{30 - 35x}{7x^2 + 8x - 12}$

46) $\dfrac{k^2 - 49}{7 - k}$

47) $\dfrac{16 - 4b^2}{b - 2}$

48) $\dfrac{16 - 2w}{w^2 - 64}$

49) $\dfrac{8t^3 - 27}{9 - 4t^2}$

50) $\dfrac{r^3 - 3r^2 + 2r - 6}{21 - 7r}$

Recall that the area of a rectangle is $A = lw$, where $w =$ width and $l =$ length. Solving for the width we get $w = \dfrac{A}{l}$ and solving for the length gives us $l = \dfrac{A}{w}$.

Find the missing side in each rectangle.

51) Area $= 5x^2 + 13x + 6$

$x + 2$

Find the length.

52) Area $= 2y^2 - y - 15$

$2y + 5$

Find the width.

53) Area $= c^3 - 2c^2 + 4c - 8$

$c^2 + 4$

Find the width.

54) Area $= 2n^3 - 8n^2 + n - 4$

$n - 4$

Find the length.

Objective 4: Write Equivalent Forms of a Rational Expression

Find three equivalent forms of each rational expression.

55) $-\dfrac{b + 7}{b - 2}$

56) $-\dfrac{8y - 1}{2y + 5}$

57) $-\dfrac{9 - 5t}{2t - 3}$

58) $\dfrac{-12m}{m^2 - 3}$

Objective 5: Multiply Rational Expressions

Multiply.

59) $\dfrac{9}{14} \cdot \dfrac{7}{6}$

60) $\dfrac{4}{15} \cdot \dfrac{25}{36}$

61) $\dfrac{14u^5}{15v^2} \cdot \dfrac{20v^6}{7u^8}$

62) $\dfrac{15s^3}{21t^2} \cdot \dfrac{42t^4}{5s^{12}}$

63) $\dfrac{5t^2}{(3t - 2)^2} \cdot \dfrac{3t - 2}{10t^3}$

64) $\dfrac{4u - 5}{9u^2} \cdot \dfrac{3u^6}{(4u - 5)^3}$

65) $\dfrac{8}{6p + 3} \cdot \dfrac{4p^2 - 1}{12}$

66) $\dfrac{n^2 + 7n + 12}{n + 3} \cdot \dfrac{4}{n + 4}$

67) $\dfrac{2v^2 + 15v + 18}{3v + 18} \cdot \dfrac{12v - 3}{8v + 12}$

68) $\dfrac{y^2 - 4y - 5}{3y^2 + y - 2} \cdot \dfrac{18y - 12}{4y^2}$

69) $(h^2 + 5h - 6) \cdot \dfrac{10h^2}{2h^2 + 12h}$

70) $(x - 8) \cdot \dfrac{4}{x^2 - 8x}$

71) $\dfrac{r^3 + 27}{4t + 20} \cdot \dfrac{rt + 5r - 2t - 10}{r^2 - 9}$

72) $\dfrac{36 - w^2}{wt + 6t - w - 6} \cdot \dfrac{8t - 8}{2w^2 - 11w - 6}$

Objective 6: Divide Rational Expressions
Divide.

73) $\dfrac{4}{5} \div \dfrac{8}{3}$

74) $\dfrac{16}{9} \div \dfrac{10}{3}$

75) $\dfrac{c^2}{6b} \div \dfrac{c^8}{b}$

76) $-\dfrac{15g^3}{14h} \div \dfrac{40g}{7h^3}$

77) $\dfrac{2a - 1}{8a^3} \div \dfrac{(2a - 1)^2}{24a^5}$

78) $\dfrac{2p^4}{(p + 7)^2} \div \dfrac{12p^5}{p + 7}$

79) $\dfrac{18y - 45}{18} \div \dfrac{4y^2 - 25}{10}$

80) $\dfrac{q^2 + q - 56}{5} \div \dfrac{q - 7}{q}$

81) $\dfrac{j^2 - 25}{5j + 25} \div \dfrac{7j - 35}{5}$

82) $\dfrac{n^2 + 3n - 18}{5n^2 + 30n} \div \dfrac{4n - 12}{8n}$

83) $\dfrac{z^2 + 18z + 80}{2z + 1} \div (z + 8)^2$

84) $\dfrac{6w^2 - 30w}{7} \div (w - 5)^2$

85) $\dfrac{36a - 12}{16} \div (9a^2 - 1)$

86) $\dfrac{h^2 - 21h + 108}{4h} \div (144 - h^2)$

87) $\dfrac{8d^2 - 8d + 8}{25 - 4d^2} \div \dfrac{d^3 + 1}{2d^2 - 3d - 5}$

88) $\dfrac{x^2 + 2xy + y^2}{3y - 12} \div \dfrac{7x + 7y}{xy - 4x + 3y - 12}$

89) In the division problem $\dfrac{12}{x} \div \dfrac{3y}{2}$, can $y = 0$? Explain your answer.

90) Find the polynomial in the second denominator so that $\dfrac{2a + 1}{a + 5} \cdot \dfrac{25 - a^2}{?} = -1$.

Mixed Exercises: Objectives 5 and 6
Perform the operations, and simplify.

91) $\dfrac{a^2 + 4a}{6a + 54} \cdot \dfrac{a^2 + 5a - 36}{16 - a^2}$

92) $\dfrac{3x + 2}{9x^2 - 4} \div \dfrac{4x}{15x^2 - 7x - 2}$

93) $\dfrac{r^3 + 8}{r + 2} \cdot \dfrac{7}{3r^2 - 6r + 12}$

94) $\dfrac{4a^3}{a^2 + a - 72} \cdot (a^2 - a - 56)$

95) $\dfrac{54x^8}{22x^3y^2} \div \dfrac{36xy^5}{11x^2y}$

96) $\dfrac{2t^2 - 6t + 18}{5t - 5} \cdot \dfrac{t^2 - 9}{t^3 + 27}$

97) $\dfrac{2a^2}{a^2 + a - 20} \cdot \dfrac{a^3 + 5a^2 + 4a + 20}{2a^2 + 8}$

98) $\dfrac{3m^2 + 8m + 4}{4} \div (12m + 8)$

99) $\dfrac{30}{4y^2 - 4x^2} \div \dfrac{10x^2 + 10xy + 10y^2}{x^3 - y^3}$

100) $\dfrac{28cd^9}{2c^3d} \cdot \dfrac{5d^2}{84c^{10}d^2}$

101) $\dfrac{4j^2 - 21j + 5}{j^3} \div \left(\dfrac{3j + 2}{j^3 - j^2} \cdot \dfrac{j^2 - 6j + 5}{j}\right)$

102) $\dfrac{t^3 - 8}{t - 2} \div \left(\dfrac{3t + 11}{5t + 15} \cdot \dfrac{t^2 + 2t + 4}{3t^2 + 11t}\right)$

103) If the area of a rectangle is $\dfrac{3}{2xy^6}$ and the width is $\dfrac{y^2}{12x^5}$, what is the length of the rectangle?

104) If the area of a triangle is $\dfrac{3m}{m^2 + 7m + 10}$ and the height is $\dfrac{m - 5}{m + 2}$, what is the length of the base of the triangle?

R Rethink

R1) Why is $\dfrac{x-6}{x+6}$ in lowest terms?

R2) Write a *rational expression* that is equal to 1 and write a *rational expression* that is equal to -1. What is the difference in the expressions?

R3) Why are you allowed to simplify $\dfrac{7x}{x}$ to 7 but not allowed to simplify $\dfrac{x+7}{x}$ to 7?

8.2 Adding and Subtracting Rational Expressions

P Prepare	O Organize
What are your objectives for Section 8.2?	How can you accomplish each objective?
1 Find the Least Common Denominator for a Group of Rational Expressions	• Review the procedure for **Finding the LCD for Fractions**. • Learn the procedure for **Finding the LCD for Rational Expressions**. • Know how to factor. • Complete the given examples on your own. • Complete You Trys 1 and 2.
2 Rewrite Rational Expressions with the LCD as Their Denominators	• Learn the procedure for **Writing Rational Expressions as Equivalent Expressions with the Least Common Denominator**. • Complete the given example on your own. • Complete You Try 3.
3 Add and Subtract Rational Expressions with a Common Denominator	• Know how to add and subtract fractions. • Learn the procedure for **Adding and Subtracting Rational Expressions**. • Complete the given examples on your own. • Complete You Trys 4 and 5.
4 Add and Subtract Rational Expressions with Different Denominators	• Know how to add and subtract fractions. • Learn the procedure for **Adding and Subtracting Rational Expressions with Different Denominators**. • Complete the given examples on your own. • Complete You Trys 6 and 7.
5 Add and Subtract Rational Expressions with Denominators Containing Factors $a-b$ and $b-a$	• Review the relationship between $a-b$ and $b-a$. • Complete the given example on your own. • Complete You Try 8.

1 Find the Least Common Denominator for a Group of Rational Expressions

Recall that to add or subtract fractions, they must have a common denominator. Similarly, rational expressions must have common denominators in order to be added or subtracted. In this section, we will discuss how to find the least common denominator (LCD) of rational expressions.

We begin by looking at the fractions $\frac{3}{8}$ and $\frac{5}{12}$. By inspection we can see that the LCD = 24. But, *why* is that true? Let's write each of the denominators, 8 and 12, as the product of their prime factors:

$$8 = 2 \cdot 2 \cdot 2 = 2^3$$
$$12 = 2 \cdot 2 \cdot 3 = 2^2 \cdot 3$$

The LCD will contain each factor the *greatest* number of times it appears in any single factorization.

2 appears as a factor *three* times in the factorization of 8, and it appears *twice* in the factorization of 12. *The LCD will contain* 2^3.

3 appears as a factor *one* time in the factorization of 12 but does not appear in the factorization of 8. *The LCD will contain* 3.

The LCD, then, is the product of the factors we have identified.

$$\text{LCD of } \frac{3}{8} \text{ and } \frac{5}{12} = 2^3 \cdot 3 = 8 \cdot 3 = 24$$

This is the same result as the one we obtained just by inspecting the two denominators.

The procedure we just illustrated is the one we use to find the least common denominator of rational expressions.

> **Procedure** Finding the Least Common Denominator (LCD)
>
> **Step 1:** Factor the denominators.
> **Step 2:** The LCD will contain each unique factor the *greatest* number of times it appears in any single factorization.
> **Step 3:** The LCD is the *product* of the factors identified in Step 2.

EXAMPLE 1 Find the LCD of each group of rational expressions.

a) $\dfrac{4}{9t^3}, \dfrac{5}{6t^2}$

b) $\dfrac{6}{x}, \dfrac{2}{x+5}$

c) $\dfrac{10}{c-8}, \dfrac{4c}{c^2-5c-24}$

d) $\dfrac{9}{w^2+2w+1}, \dfrac{1}{2w^2+2w}$

Solution

a) To find the LCD of $\dfrac{4}{9t^3}$ and $\dfrac{5}{6t^2}$,

Step 1: Factor the denominators.
$$9t^3 = 3 \cdot 3 \cdot t^3 = 3^2 \cdot t^3$$
$$6t^2 = 2 \cdot 3 \cdot t^2$$

Step 2: The LCD will contain each unique factor the *greatest* number of times it appears in any factorization. *It will contain* 2, 3^2, *and* t^3.

Step 3: The LCD is the *product* of the factors in Step 2.
$$\mathbf{LCD = 2 \cdot 3^2 \cdot t^3 = 18t^3}$$

b) The denominators of $\dfrac{6}{x}$ and $\dfrac{2}{x+5}$ are already in simplest form. It is important to recognize that x and $x+5$ are *different factors*.

The LCD will be the product of x and $x+5$: $\mathbf{LCD = x(x+5)}$

Usually, we leave the LCD in this form; we do not distribute.

> **W Hint**
> $x + 5$ is considered one unique factor even though it is two terms.

c) **Step 1:** Factor the denominators of $\dfrac{10}{c-8}$ and $\dfrac{4c}{c^2 - 5c - 24}$.

$c - 8$ cannot be factored.
$$c^2 - 5c - 24 = (c-8)(c+3)$$

Step 2: The LCD will contain each unique factor the *greatest* number of times it appears in any factorization. *It will contain* $c - 8$ *and* $c + 3$.

Step 3: The LCD is the *product* of the factors identified in Step 2.
$$\mathbf{LCD = (c-8)(c+3)}$$

d) **Step 1:** Factor the denominators of $\dfrac{9}{w^2 + 2w + 1}$ and $\dfrac{1}{2w^2 + 2w}$.
$$w^2 + 2w + 1 = (w+1)^2$$
$$2w^2 + 2w = 2w(w+1)$$

Step 2: The unique factors are 2, w, and $w + 1$ with $w + 1$ *appearing at most twice. The factors we will use in the LCD are* 2, w, *and* $(w+1)^2$.

Step 3: The LCD is the *product* of the factors identified in Step 2.
$$\mathbf{LCD = 2w(w+1)^2}$$

> **W Hint**
> In your notes and in your own words, explain how to find the LCD of two rational expressions. Include an example.

> **[YOU TRY 1]** Find the LCD of each group of rational expressions.
>
> a) $\dfrac{8}{9k^3}, \dfrac{1}{12k^5}$
> b) $\dfrac{5}{k}, \dfrac{8k}{k+2}$
>
> c) $\dfrac{12}{p^2 - 7p}, \dfrac{6}{p-7}$
> d) $\dfrac{m}{m^2 - 25}, \dfrac{8}{m^2 + 10m + 25}$

At first glance it may appear that the least common denominator of $\dfrac{2}{y-3}$ and $\dfrac{10}{3-y}$ is $(y-3)(3-y)$. This is *not* the case. Recall from Section 8.1 that $a-b = -1(b-a)$. We will use this idea to find the LCD of $\dfrac{2}{y-3}$ and $\dfrac{10}{3-y}$.

EXAMPLE 2

Find the LCD of $\dfrac{2}{y-3}$ and $\dfrac{10}{3-y}$.

Solution

Because $3-y = -(y-3)$, we can rewrite $\dfrac{10}{3-y}$ as $\dfrac{10}{-(y-3)} = -\dfrac{10}{y-3}$.

Therefore, we can now think of our task as finding the LCD of $\dfrac{2}{y-3}$ and $-\dfrac{10}{y-3}$. The least common denominator is $y-3$.

[YOU TRY 2]

Find the LCD of $\dfrac{9}{r-8}$ and $\dfrac{6}{8-r}$.

2 Rewrite Rational Expressions with the LCD as Their Denominators

In order to add or subtract fractions, they must have a common denominator. That means that sometimes we must rewrite one or both fractions so that their denominators contain the LCD. For example, to rewrite $\dfrac{5}{6}$ and $\dfrac{4}{9}$ as equivalent fractions with the least common denominator we begin by identifying the LCD as 18. Then, for each fraction we ask ourselves,

$\dfrac{5}{6}$: *By what number do we multiply 6 to get 18?*   3

Multiply the numerator and denominator of $\dfrac{5}{6}$ by 3 to obtain an equivalent fraction.

$$\dfrac{5}{6} \cdot \dfrac{3}{3} = \dfrac{15}{18}$$

$\dfrac{4}{9}$: *By what number do we multiply 9 to get 18?*   2

Multiply the numerator and denominator of $\dfrac{4}{9}$ by 2 to obtain an equivalent fraction.

$$\dfrac{4}{9} \cdot \dfrac{2}{2} = \dfrac{8}{18}$$

The procedure for rewriting rational expressions as equivalent expressions with the LCD is very similar to the process we use with fractions.

> **Procedure** Writing Rational Expressions as Equivalent Expressions with the Least Common Denominator
>
> **Step 1:** Identify and write down the LCD.
>
> **Step 2:** Look at each rational expression (with its denominator in factored form) and compare its denominator with the LCD. Ask yourself, *"What factors are missing?"*
>
> **Step 3:** Multiply the numerator and denominator by the "missing" factors to obtain an equivalent rational expression with the desired LCD. Multiply the terms in the numerator, but leave the denominator as the product of factors.

EXAMPLE 3   Identify the LCD of each pair of rational expressions, and rewrite each as an equivalent expression with the LCD as its denominator.

a) $\dfrac{7}{8n}, \dfrac{5}{6n^3}$ b) $\dfrac{t}{t-2}, \dfrac{4}{t+7}$ c) $\dfrac{r}{r-6}, \dfrac{2}{6-r}$

Solution

a) Follow the steps.

Step 1: Identify and write down the LCD of $\dfrac{7}{8n}$ and $\dfrac{5}{6n^3}$: **LCD = $24n^3$**

Step 2: Compare the denominators of $\dfrac{7}{8n}$ and $\dfrac{5}{6n^3}$ to the LCD and ask yourself, "What's missing?"

$\dfrac{7}{8n}$: $8n$ is "missing" the factors of 3 and n^2. | $\dfrac{5}{6n^3}$: $6n^3$ is "missing" the factor 4.

Step 3: Multiply the numerator and denominator by $3n^2$.

$\dfrac{7}{8n} \cdot \dfrac{3n^2}{3n^2} = \dfrac{21n^2}{24n^3}$

Multiply the numerator and denominator by 4.

$\dfrac{5}{6n^3} \cdot \dfrac{4}{4} = \dfrac{20}{24n^3}$

$\dfrac{7}{8n} = \dfrac{21n^2}{24n^3}$ and $\dfrac{5}{6n^3} = \dfrac{20}{24n^3}$

b) Follow the steps.

Step 1: Identify and write down the LCD of $\dfrac{t}{t-2}$ and $\dfrac{4}{t+7}$:
LCD = $(t-2)(t+7)$

Step 2: Compare the denominators of $\dfrac{t}{t-2}$ and $\dfrac{4}{t+7}$ to the LCD and ask yourself, "What's missing?"

$\dfrac{t}{t-2}$: $t-2$ is "missing" the factor $t+7$. | $\dfrac{4}{t+7}$: $t+7$ is "missing" the factor $t-2$.

Hint

After you finish Example 3, explain this procedure in your own words, and write it in your notes.

Step 3: Multiply the numerator and denominator by $t+7$.

$$\frac{t}{t-2} \cdot \frac{t+7}{t+7} = \frac{t(t+7)}{(t-2)(t+7)}$$
$$= \frac{t^2+7t}{(t-2)(t+7)}$$

Multiply the numerator and denominator by $t-2$.

$$\frac{4}{t+7} \cdot \frac{t-2}{t-2} = \frac{4(t-2)}{(t+7)(t-2)}$$
$$= \frac{4t-8}{(t-2)(t+7)}$$

Notice that we multiplied the factors in the numerator but left the denominator in factored form.

$$\frac{t}{t-2} = \frac{t^2+7t}{(t-2)(t+7)} \quad \text{and} \quad \frac{4}{t+7} = \frac{4t-8}{(t-2)(t+7)}$$

c) To find the LCD of $\dfrac{r}{r-6}$ and $\dfrac{2}{6-r}$ recall that $6-r$ can be rewritten as $-(r-6)$. So,

$$\frac{2}{6-r} = \frac{2}{-(r-6)} = -\frac{2}{r-6}$$

Therefore, the LCD of $\dfrac{r}{r-6}$ and $-\dfrac{2}{r-6}$ is $r-6$.

The expression $\dfrac{r}{r-6}$ already has the LCD, while $\dfrac{2}{6-r} = -\dfrac{2}{r-6}$.

[YOU TRY 3] Identify the least common denominator of each pair of rational expressions, and rewrite each as an equivalent expression with the LCD as its denominator.

a) $\dfrac{9}{7r^5}, \dfrac{4}{21r^2}$

b) $\dfrac{6}{y+4}, \dfrac{8}{3y-2}$

c) $\dfrac{d-1}{d^2+2d}, \dfrac{3}{d^2+12d+20}$

d) $\dfrac{k}{7-k}, \dfrac{4}{k-7}$

We know that in order to add or subtract fractions, they must have a common denominator. For example, $\dfrac{6}{7} - \dfrac{2}{7} = \dfrac{6-2}{7} = \dfrac{4}{7}$. The same is true for rational expressions.

3 Add and Subtract Rational Expressions with a Common Denominator

EXAMPLE 4

Add $\dfrac{3a}{2a-5} + \dfrac{4a+1}{2a-5}$.

Solution

Since $\dfrac{3a}{2a-5}$ and $\dfrac{4a+1}{2a-5}$ have the same denominator, add the terms in the numerator and keep the common denominator.

$$\frac{3a}{2a-5} + \frac{4a+1}{2a-5} = \frac{3a+(4a+1)}{2a-5} \quad \text{Add terms in the numerator.}$$
$$= \frac{7a+1}{2a-5} \quad \text{Combine like terms.}$$

SECTION 8.2 Adding and Subtracting Rational Expressions

We can generalize the procedure for adding and subtracting rational expressions that have a common denominator as follows. Write answers in lowest terms.

> **Procedure** Adding and Subtracting Rational Expressions
>
> If $\dfrac{P}{Q}$ and $\dfrac{R}{Q}$ are rational expressions with $Q \neq 0$, then
>
> 1) $\dfrac{P}{Q} + \dfrac{R}{Q} = \dfrac{P + R}{Q}$ and 2) $\dfrac{P}{Q} - \dfrac{R}{Q} = \dfrac{P - R}{Q}$

[YOU TRY 4] Add or subtract, as indicated.

a) $\dfrac{9}{11} - \dfrac{4}{11}$ b) $\dfrac{7c}{3c - 4} + \dfrac{2c + 5}{3c - 4}$

All answers to a sum or difference of rational expressions should be in lowest terms. Sometimes it is necessary to simplify our result to lowest terms by factoring the numerator and dividing the numerator and denominator by the greatest common factor.

EXAMPLE 5 Add or subtract, as indicated.

a) $\dfrac{11}{12t} + \dfrac{7}{12t}$ b) $\dfrac{n^2 - 8}{n(n + 5)} - \dfrac{7 - 2n}{n(n + 5)}$

Solution

a) $\dfrac{11}{12t} + \dfrac{7}{12t} = \dfrac{11 + 7}{12t}$ Add terms in the numerator.

$= \dfrac{18}{12t}$ Combine terms.

$= \dfrac{3}{2t}$ Write in lowest terms.

b) $\dfrac{n^2 - 8}{n(n + 5)} - \dfrac{7 - 2n}{n(n + 5)} = \dfrac{(n^2 - 8) - (7 - 2n)}{n(n + 5)}$ Subtract terms in the numerator.

$= \dfrac{n^2 - 8 - 7 + 2n}{n(n + 5)}$ Distribute.

$= \dfrac{n^2 + 2n - 15}{n(n + 5)}$ Combine like terms.

$= \dfrac{\cancel{(n + 5)}(n - 3)}{n\cancel{(n + 5)}}$ Factor the numerator.

$= \dfrac{n - 3}{n}$ Write in lowest terms.

[YOU TRY 5] Add or subtract, as indicated.

a) $\dfrac{3}{20c^2} - \dfrac{9}{20c^2}$ b) $\dfrac{k^2 + 2k + 5}{(k + 4)(k - 1)} + \dfrac{5k + 7}{(k + 4)(k + 1)}$

c) $\dfrac{20d - 9}{4d(3d + 1)} - \dfrac{5d - 14}{4d(3d + 1)}$

After combining like terms in the numerator, ask yourself, *"Can I factor the numerator?"* **If so, factor it. Sometimes, the expression can be simplified by dividing the numerator and denominator by the greatest common factor.**

4 Add and Subtract Rational Expressions with Different Denominators

If we are asked to add or subtract rational expressions with different denominators, we must begin by rewriting each expression with the least common denominator. Then, add or subtract. Simplify the result.

> **Procedure** Steps for Adding and Subtracting Rational Expressions with Different Denominators
>
> 1) Factor the denominators.
> 2) Write down the LCD.
> 3) Rewrite each rational expression as an equivalent rational expression with the LCD.
> 4) Add or subtract the numerators, and keep the common denominator in factored form.
> 5) After combining like terms in the numerator ask yourself, *"Can I factor it?"* If so, factor.
> 6) Simplify the rational expression, if possible.

EXAMPLE 6 Add or subtract, as indicated.

a) $\dfrac{m+8}{3} + \dfrac{m-1}{6}$ b) $\dfrac{3}{4x} - \dfrac{11}{10x^2}$ c) $\dfrac{4a-6}{a^2-9} + \dfrac{a}{a+3}$

Solution

a) The LCD is 6. $\dfrac{m-1}{6}$ already has the LCD.

Rewrite $\dfrac{m+8}{3}$ with the LCD: $\dfrac{m+8}{3} \cdot \dfrac{2}{2} = \dfrac{2(m+8)}{6}$.

$$\dfrac{m+8}{3} + \dfrac{m-1}{6} = \dfrac{2(m+8)}{6} + \dfrac{m-1}{6} \qquad \text{Write each expression with the LCD.}$$

$$= \dfrac{2(m+8)+(m-1)}{6} \qquad \text{Add the numerators.}$$

$$= \dfrac{2m+16+m-1}{6} \qquad \text{Distribute.}$$

$$= \dfrac{3m+15}{6} \qquad \text{Combine like terms.}$$

Ask yourself, "Can I factor the numerator?" Yes.

$$= \frac{\overset{1}{\cancel{3}}(m+5)}{\underset{2}{\cancel{6}}} \quad \text{Factor.}$$

$$= \frac{m+5}{2} \quad \text{Divide out the common factor.}$$

b) The LCD of $\frac{3}{4x}$ and $\frac{11}{10x^2}$ is $20x^2$. Rewrite each expression with the LCD.

$$\frac{3}{4x} \cdot \frac{5x}{5x} = \frac{15x}{20x^2} \quad \text{and} \quad \frac{11}{10x^2} \cdot \frac{2}{2} = \frac{22}{20x^2}$$

$$\frac{3}{4x} - \frac{11}{10x^2} = \frac{15x}{20x^2} - \frac{22}{20x^2} \quad \text{Write each expression with the LCD.}$$

$$= \frac{15x - 22}{20x^2} \quad \text{Subtract the numerators.}$$

"Can I factor the numerator?" No. The expression is in simplest form because the numerator and denominator have no common factors.

W Hint
Are you writing out the examples as you are reading them?

c) Begin by factoring the denominator of $\frac{4a-6}{a^2-9}$.

$$\frac{4a-6}{a^2-9} = \frac{4a-6}{(a+3)(a-3)}$$

The LCD of $\frac{4a-6}{(a+3)(a-3)}$ and $\frac{a}{a+3}$ is $(a+3)(a-3)$.

Rewrite $\frac{a}{a+3}$ with the LCD: $\frac{a}{a+3} \cdot \frac{a-3}{a-3} = \frac{a(a-3)}{(a+3)(a-3)}$.

$$\frac{4a-6}{a^2-9} + \frac{a}{a+3} = \frac{4a-6}{(a+3)(a-3)} + \frac{a}{a+3} \quad \text{Factor the denominator.}$$

$$= \frac{4a-6}{(a+3)(a-3)} + \frac{a(a-3)}{(a+3)(a-3)} \quad \text{Write each expression with the LCD.}$$

$$= \frac{4a-6+a(a-3)}{(a+3)(a-3)} \quad \text{Add the numerators.}$$

$$= \frac{4a-6+a^2-3a}{(a+3)(a-3)} \quad \text{Distribute.}$$

$$= \frac{a^2+a-6}{(a+3)(a-3)} \quad \text{Combine like terms.}$$

Ask yourself, "Can I factor the numerator?" Yes.

$$= \frac{\cancel{(a+3)}(a-2)}{\cancel{(a+3)}(a-3)} \quad \text{Factor.}$$

$$= \frac{a-2}{a-3} \quad \text{Write in lowest terms.}$$

[**YOU TRY 6**] Add or subtract, as indicated.

a) $\frac{7}{12t^3} + \frac{4}{9t}$ b) $\frac{k-3}{4} - \frac{k+3}{6}$ c) $\frac{6}{r-5} + \frac{r^2-17r}{r^2-25}$

EXAMPLE 7

Subtract $\dfrac{4w}{w^2 + 9w + 14} - \dfrac{2w + 5}{w^2 + 3w - 28}$.

Solution

Factor the denominators, then write down the LCD.

$$\dfrac{4w}{w^2 + 9w + 14} = \dfrac{4w}{(w + 7)(w + 2)}, \qquad \dfrac{2w + 5}{w^2 + 3w - 28} = \dfrac{2w + 5}{(w + 7)(w - 4)}$$

Rewrite each expression with the LCD, $(w + 7)(w + 2)(w - 4)$.

$$\dfrac{4w}{(w + 7)(w + 2)} \cdot \dfrac{w - 4}{w - 4} = \dfrac{4w(w - 4)}{(w + 7)(w + 2)(w - 4)}$$

$$\dfrac{2w + 5}{(w + 7)(w - 4)} \cdot \dfrac{w + 2}{w + 2} = \dfrac{(2w + 5)(w + 2)}{(w + 7)(w + 2)(w - 4)}$$

$$\dfrac{4w}{w^2 + 9w + 14} - \dfrac{2w + 5}{w^2 + 3w - 28}$$

$= \dfrac{4w}{(w + 7)(w + 2)} - \dfrac{2w + 5}{(w + 7)(w - 4)}$ Factor denominators.

$= \dfrac{4w(w - 4)}{(w + 7)(w + 2)(w - 4)} - \dfrac{(2w + 5)(w + 2)}{(w + 7)(w + 2)(w - 4)}$ Write each expression with the LCD.

$= \dfrac{4w(w - 4) - (2w + 5)(w + 2)}{(w + 7)(w + 2)(w - 4)}$ Subtract the numerators.

$= \dfrac{4w^2 - 16w - (2w^2 + 9w + 10)}{(w + 7)(w + 2)(w - 4)}$ Distribute. You must use parentheses.

$= \dfrac{4w^2 - 16w - 2w^2 - 9w - 10}{(w + 7)(w + 2)(w - 4)}$ Distribute.

$= \dfrac{2w^2 - 25w - 10}{(w + 7)(w + 2)(w - 4)}$ Combine like terms.

Ask yourself, *"Can I factor the numerator?"* No. The expression is in simplest form because the numerator and denominator have no common factors.

BE CAREFUL In Example 7, when you move from

$$\dfrac{4w(w - 4) - (2w + 5)(w + 2)}{(w + 7)(w + 2)(w - 4)} \quad \text{to} \quad \dfrac{4w^2 - 16w - (2w^2 + 9w + 10)}{(w + 7)(w + 2)(w - 4)}$$

you *must* use parentheses because the entire quantity $2w^2 + 9w + 10$ is being subtracted from $4w^2 - 16w$.

[YOU TRY 7] Subtract $\dfrac{3d}{d^2 + 13d + 40} - \dfrac{2d - 3}{d^2 + 7d - 8}$.

5 Add and Subtract Rational Expressions with Denominators Containing Factors $a - b$ and $b - a$

EXAMPLE 8

Add or subtract, as indicated.

a) $\dfrac{s}{s-3} - \dfrac{11}{3-s}$

b) $\dfrac{3}{4-h} + \dfrac{6}{h^2-16}$

Solution

a) Recall that $a - b = -(b - a)$. The least common denominator of $\dfrac{s}{s-3}$ and $\dfrac{11}{3-s}$ is $s - 3$ or $3 - s$. We will use LCD $= s - 3$.

Rewrite $\dfrac{11}{3-s}$ with the LCD: $\dfrac{11}{3-s} = \dfrac{11}{-(s-3)} = -\dfrac{11}{s-3}$.

$\dfrac{s}{s-3} - \dfrac{11}{3-s} = \dfrac{s}{s-3} - \left(-\dfrac{11}{s-3}\right)$ Write each expression with the LCD.

$= \dfrac{s}{s-3} + \dfrac{11}{s-3}$ Distribute.

$= \dfrac{s+11}{s-3}$ Add the numerators.

b) Factor the denominator of $\dfrac{6}{h^2-16}$: $\dfrac{6}{h^2-16} = \dfrac{6}{(h+4)(h-4)}$.

Rewrite $\dfrac{3}{4-h}$ with a denominator of $h - 4$: $\dfrac{3}{4-h} = \dfrac{3}{-(h-4)} = -\dfrac{3}{h-4}$.

Now we must find the LCD of $\dfrac{6}{(h+4)(h-4)}$ and $-\dfrac{3}{h-4}$.

$$\text{LCD} = (h+4)(h-4)$$

Rewrite $-\dfrac{3}{h-4}$ with the LCD.

$-\dfrac{3}{h-4} \cdot \dfrac{h+4}{h+4} = -\dfrac{3(h+4)}{(h+4)(h-4)} = \dfrac{-3(h+4)}{(h+4)(h-4)}$

$\dfrac{3}{4-h} + \dfrac{6}{h^2-16} = -\dfrac{3}{h-4} + \dfrac{6}{(h+4)(h-4)}$

$= \dfrac{-3(h+4)}{(h+4)(h-4)} + \dfrac{6}{(h+4)(h-4)}$ Write each expression with the LCD.

$= \dfrac{-3(h+4) + 6}{(h+4)(h-4)}$ Add the numerators.

$= \dfrac{-3h - 12 + 6}{(h+4)(h-4)}$ Distribute.

$= \dfrac{-3h - 6}{(h+4)(h-4)}$ Combine like terms.

Ask yourself, "Can I factor the numerator?" Yes.

$= \dfrac{-3(h+2)}{(h+4)(h-4)}$ Factor.

Although the numerator factors, the numerator and denominator do not contain any common factors. The result, $\dfrac{-3(h+2)}{(h+4)(h-4)}$, is in simplest form.

[YOU TRY 8] Add or subtract, as indicated.

a) $\dfrac{5}{8-y} + \dfrac{3}{y-8}$ b) $\dfrac{1}{x^2-36} - \dfrac{x+4}{6-x}$

ANSWERS TO [YOU TRY] EXERCISES

1) a) $36k^5$ b) $k(k+2)$ c) $p(p-7)$ d) $(m+5)^2(m-5)$ 2) $r-8$

3) a) LCD $= 21r^5$; $\dfrac{9}{7r^5} = \dfrac{27}{21r^5}$, $\dfrac{4}{21r^2} = \dfrac{4r^3}{21r^5}$

b) LCD $= (y+4)(3y-2)$; $\dfrac{6}{y+4} = \dfrac{18y-12}{(y+4)(3y-2)}$, $\dfrac{8}{3y-2} = \dfrac{8y+32}{(y+4)(3y-2)}$

c) LCD $= d(d+2)(d+10)$; $\dfrac{d-1}{d^2+2d} = \dfrac{d^2+9d-10}{d(d+2)(d+10)}$, $\dfrac{3}{d^2+12d+20} = \dfrac{3d}{d(d+2)(d+10)}$

d) LCD $= k-7$; $\dfrac{k}{7-k} = -\dfrac{k}{k-7}$, $\dfrac{4}{k-7} = \dfrac{4}{k-7}$ 4) a) $\dfrac{5}{11}$ b) $\dfrac{9c+5}{3c-4}$

5) a) $-\dfrac{3}{10c^2}$ b) $\dfrac{k+3}{k-1}$ c) $\dfrac{5}{4d}$ 6) a) $\dfrac{16t^2+21}{36t^3}$ b) $\dfrac{k-15}{12}$ c) $\dfrac{r-6}{r+5}$

7) $\dfrac{d^2-10d+15}{(d+8)(d+5)(d-1)}$ 8) a) $-\dfrac{2}{y-8}$ or $\dfrac{2}{8-y}$ b) $\dfrac{(x+5)^2}{(x+6)(x-6)}$

E Evaluate   **8.2 Exercises**   Do the exercises, and check your work.

Objective 1: Find the Least Common Denominator for a Group of Rational Expressions

Find the LCD of each group of rational expressions.

1) $\dfrac{5}{8}, \dfrac{9}{20}$

2) $\dfrac{11}{12}, \dfrac{5}{16}$

3) $\dfrac{6}{c^4}, \dfrac{7}{c^3}$

4) $\dfrac{4}{n^5}, \dfrac{1}{n^9}$

5) $\dfrac{8}{9p^3}, \dfrac{5}{12p^8}$

6) $-\dfrac{2}{9z^4}, \dfrac{16}{45z^6}$

7) $\dfrac{1}{8a^3b^3}, \dfrac{7}{12ab^4}$

8) $\dfrac{4}{27x^2y^2}, \dfrac{11}{3x^3y^2}$

9) $\dfrac{3}{n+4}, \dfrac{1}{2}$

10) $\dfrac{7}{9}, \dfrac{6}{z-8}$

11) $\dfrac{10}{w}, \dfrac{6}{2w+1}$

12) $\dfrac{2}{y}, -\dfrac{9}{3y+5}$

13) $\dfrac{1}{12a^2-4a}, \dfrac{15}{6a^4-2a^3}$

14) $\dfrac{1}{10p^3-15p^2}, \dfrac{6}{2p^5-3p^4}$

15) $\dfrac{8}{r+7}, \dfrac{4}{r-2}$

16) $\dfrac{m}{m-6}, \dfrac{3}{m-5}$

17) $\dfrac{w}{w^2-3w-10}, \dfrac{5}{w^2-2w-15}, \dfrac{9w}{w^2+5w+6}$

18) $\dfrac{10t}{t^2-5t-14}, -\dfrac{8}{t^2-49}, \dfrac{t}{t^2+9t+14}$

19) $\dfrac{6}{b-4}, \dfrac{5}{4-b}$

20) $\dfrac{u}{v-u}, \dfrac{2}{u-v}$

Objective 2: Rewrite Rational Expressions with the LCD as Their Denominators

21) Explain, in your own words, how to rewrite $\dfrac{5}{x+8}$ as an equivalent rational expression with a denominator of $(x+8)(x-2)$.

22) Explain, in your own words, how to rewrite $\dfrac{6}{3-n}$ as an equivalent rational expression with a denominator of $n-3$.

Identify the least common denominator of each group of rational expressions, and rewrite each as an equivalent rational expression with the LCD as its denominator.

23) $\dfrac{3}{t}, \dfrac{8}{t^3}$

24) $\dfrac{10}{p^5}, \dfrac{7}{p^2}$

25) $\dfrac{9}{8n^6}, \dfrac{2}{3n^2}$

26) $\dfrac{5}{6a}, \dfrac{7}{8a^5}$

27) $\dfrac{1}{x^3y}, \dfrac{6}{5xy^5}$

28) $\dfrac{5}{6a^2b^4}, \dfrac{5}{a^4b}$

29) $\dfrac{t}{5t-6}, \dfrac{10}{7}$

30) $\dfrac{8}{d}, \dfrac{2}{d-4}$

SECTION 8.2 Adding and Subtracting Rational Expressions

31) $\dfrac{7}{12x-4}, \dfrac{x}{18x-6}$ 32) $\dfrac{a}{24a+36}, \dfrac{1}{18a+27}$

33) $\dfrac{8}{a+9}, \dfrac{a}{2a+7}$ 34) $\dfrac{4}{h+5}, \dfrac{7h}{h-3}$

35) $\dfrac{9y}{y^2-y-42}, \dfrac{3}{2y^2+12y}$

36) $\dfrac{4q}{3q^2+24q}, \dfrac{5}{q^2+q-56}$

37) $\dfrac{z}{z^2-10z+25}, \dfrac{15z}{z^2-2z-15}$

38) $\dfrac{c}{c^2+11c+28}, \dfrac{6}{c^2+14c+49}$

39) $\dfrac{11}{g-3}, \dfrac{4}{9-g^2}$ 40) $\dfrac{10}{4k-1}, \dfrac{k}{1-16k^2}$

41) $\dfrac{4}{w^2-4w}, \dfrac{6}{7w^2-28w}, \dfrac{11}{w^2-8w+16}$

42) $\dfrac{t}{t^2-4t-21}, \dfrac{2}{t+3}, \dfrac{4}{t^2-49}$

Objective 3: Add and Subtract Rational Expressions with a Common Denominator

Add or subtract, as indicated.

43) $\dfrac{7}{20}+\dfrac{9}{20}$ 44) $\dfrac{11}{12}-\dfrac{5}{12}$

45) $\dfrac{8}{a}+\dfrac{2}{a}$ 46) $\dfrac{6}{5c}+\dfrac{14}{5c}$

47) $\dfrac{8}{x+4}+\dfrac{2x}{x+4}$ 48) $\dfrac{7m}{m+5}+\dfrac{35}{m+5}$

49) $\dfrac{7w-4}{w(3w-4)}-\dfrac{20-11w}{w(3w-4)}$

50) $\dfrac{10t+7}{t(2t+1)}-\dfrac{2t+3}{t(2t+1)}$

51) $\dfrac{d^2-12}{(d+4)(d+1)}+\dfrac{3-8d}{(d+4)(d+1)}$

52) $\dfrac{2r+15}{(r-5)(r+2)}+\dfrac{r^2-10r}{(r-5)(r+2)}$

Objective 4: Add and Subtract Rational Expressions with Different Denominators

53) For $\dfrac{8}{x-3}$ and $\dfrac{2}{x}$:

a) Find the LCD.

b) Explain, in your own words, how to rewrite each expression with the LCD.

c) Rewrite each expression with the LCD.

54) For $\dfrac{4}{9b^2}$ and $\dfrac{5}{6b^4}$:

a) Find the LCD.

b) Explain, in your own words, how to rewrite each expression with the LCD.

c) Rewrite each expression with the LCD.

55) How do you find the least common denominator of two rational expressions when their denominators have no common factors?

56) Explain, in your own words, how to add two rational expressions with different denominators.

Add or subtract, as indicated.

57) $\dfrac{5}{8}+\dfrac{1}{6}$ 58) $\dfrac{9}{10}-\dfrac{5}{6}$

59) $\dfrac{5x}{12}-\dfrac{4x}{15}$ 60) $\dfrac{9t}{5}+\dfrac{3}{4}$

61) $\dfrac{3}{2a}+\dfrac{6}{7a^2}$ 62) $\dfrac{3}{2f^2}-\dfrac{7}{f}$

63) $\dfrac{15}{d-8}-\dfrac{4}{d}$ 64) $\dfrac{8}{r-7}-\dfrac{4}{r}$

65) $\dfrac{1}{z+6}+\dfrac{4}{z+2}$ 66) $\dfrac{6}{c-5}+\dfrac{5}{c+3}$

67) $\dfrac{x}{2x+1}-\dfrac{3}{x+5}$ 68) $\dfrac{m}{3m+4}-\dfrac{2}{m-9}$

69) $\dfrac{t}{t+7}+\dfrac{11t-21}{t^2-49}$ 70) $\dfrac{-3u-5}{u^2-1}+\dfrac{u}{u+1}$

71) $\dfrac{b}{b^2-16}+\dfrac{10}{b^2-5b-36}$

72) $\dfrac{7g}{g^2-9g-10}+\dfrac{4}{g^2-100}$

73) $\dfrac{3c}{c^2+4c-12}-\dfrac{2c-5}{c^2+2c-24}$

74) $\dfrac{4a}{a^2-5a-24}-\dfrac{2a+3}{a^2-10a+16}$

75) $\dfrac{4b+1}{3b-12}+\dfrac{5b}{b^2-b-12}$

76) $\dfrac{k+9}{2k-24}+\dfrac{4k}{k^2-15k+36}$

Objective 5: Add and Subtract Rational Expressions with Denominators Containing Factors a − b and b − a

77) Is $(x-7)(7-x)$ the LCD for $\dfrac{5}{x-7} + \dfrac{2}{7-x}$? Why or why not?

78) What is the LCD of $\dfrac{n}{5-3n} - \dfrac{8}{3n-5}$?

Add or subtract, as indicated.

79) $\dfrac{15}{q-8} + \dfrac{9}{8-q}$

80) $\dfrac{9}{z-6} + \dfrac{2}{6-z}$

81) $\dfrac{2c}{12b-7c} - \dfrac{13}{7c-12b}$

82) $\dfrac{2}{4u-3v} - \dfrac{6u}{3v-4u}$

83) $\dfrac{5}{8-t} + \dfrac{10}{t^2-64}$

84) $\dfrac{8}{r^2-9} + \dfrac{2}{3-r}$

85) $\dfrac{a}{4a^2-9} - \dfrac{4}{3-2a}$

86) $\dfrac{2y}{9y^2-25} - \dfrac{2}{5-3y}$

Mixed Exercises: Objectives 4 and 5
Perform the indicated operations.

87) $\dfrac{2}{j^2+8j} + \dfrac{2j}{j+8} - \dfrac{1}{3j}$

88) $\dfrac{4}{w^2-3w} + \dfrac{9}{w} - \dfrac{10w}{w-3}$

89) $\dfrac{c}{c^2-8c+16} - \dfrac{5}{c^2-c-12}$

90) $\dfrac{n}{n^2+11n+30} - \dfrac{3}{n^2+10n+25}$

91) $\dfrac{1}{x+y} + \dfrac{x}{x^2-y^2} - \dfrac{4}{2x-2y}$

92) $\dfrac{8}{3a+3b} + \dfrac{3}{a-b} - \dfrac{3a}{a^2-b^2}$

93) $\dfrac{n+5}{4n^2+7n-2} - \dfrac{n-4}{3n^2+7n+2}$

94) $\dfrac{3v-4}{6v^2-v-5} - \dfrac{v-2}{3v^2+v-4}$

95) $\dfrac{y+6}{y^2-4y} + \dfrac{y}{2y^2-13y+20} - \dfrac{1}{2y^2-5y}$

96) $\dfrac{g-5}{5g^2-30g} + \dfrac{g}{2g^2-17g+30} - \dfrac{6}{2g^2-5g}$

For each rectangle, find a rational expression in simplest form to represent its a) area and b) perimeter.

97) dimensions: $\dfrac{x+1}{2}$ and $\dfrac{4}{x-3}$

98) dimensions: $\dfrac{10}{x+1}$ and $\dfrac{x-4}{6}$

99) dimensions: $\dfrac{w}{w+2}$ and $\dfrac{1}{w^2-4}$

100) dimensions: $\dfrac{t}{t+5}$ and $\dfrac{2}{t^2+9t+20}$

Recall that given functions $f(x)$ and $g(x)$, the domains of $(f+g)(x)$, $(f-g)(x)$, $(f \cdot g)(x)$, and $\left(\dfrac{f}{g}\right)(x)$ consist of all values, x, in the domain of f and in the domain of g as well as in the domain of the function obtained by adding, subtracting, multiplying, or dividing the two functions.

For Exercises 101–104, let $f(x) = \dfrac{6}{x}$ and $g(x) = \dfrac{3x+12}{x+5}$.

101) Find $(f+g)(x)$ and its domain.

102) Find $(f-g)(x)$ and its domain.

103) Find $(f \cdot g)(x)$ and its domain.

104) Find $\left(\dfrac{f}{g}\right)(x)$ and its domain.

R Rethink

R1) Can you use $(a-b)(b-a)$ for the LCD if one denominator is $a-b$ and the other denominator is $b-a$? Is there a better choice?

R2) Why do you need a common denominator when adding rational expressions?

R3) What do you find most difficult when finding a common denominator?

8.3 Simplifying Complex Fractions

P Prepare

What are your objectives for Section 8.3?	How can you accomplish each objective?
1 Simplify a Complex Fraction with One Term in the Numerator and One Term in the Denominator	• Learn the definition of a *complex fraction*. • Know the procedure for **Simplifying a Complex Fraction with One Term in the Numerator and Denominator.** • Complete the given example on your own. • Complete You Try 1.
2 Simplify a Complex Fraction with More Than One Term in the Numerator and/or Denominator by Rewriting It as a Division Problem	• Learn the procedure for **Simplifying by Rewriting as Division (Method 1).** • Complete the given example on your own. • Complete You Try 2.
3 Simplify a Complex Fraction with More Than One Term in the Numerator and/or Denominator by Multiplying by the LCD	• Learn the procedure for **Simplifying a Complex Fraction Using the LCD (Method 2).** • Know how to find the LCD. • Complete the given examples on your own. • Complete You Trys 3 and 4.
4 Simplify Rational Expressions Containing Negative Exponents	• Follow the rules for *making negative exponents positive*. • Be able to determine which method for **Simplifying Complex Fractions** should be used. • Complete the given example on your own. • Complete You Try 5.

W Work   Read the explanations, follow the examples, take notes, and complete the You Trys.

In algebra we sometimes encounter fractions that contain fractions in their numerators, denominators, or both. Such fractions are called *complex fractions*. Some examples of complex fractions are

$$\dfrac{\dfrac{5}{8}}{\dfrac{3}{4}}, \qquad \dfrac{\dfrac{1}{2}+\dfrac{1}{3}}{3-\dfrac{5}{4}}, \qquad \dfrac{\dfrac{2}{x^2y}}{\dfrac{1}{y}-\dfrac{y}{x}}, \qquad \dfrac{\dfrac{7k-28}{3}}{\dfrac{k-4}{k}}$$

Definition

A **complex fraction** is a rational expression that contains one or more fractions in its numerator, its denominator, or both.

A complex fraction is not considered to be an expression in simplest form. In this section, we will learn how to simplify complex fractions to lowest terms.

We begin by looking at two different types of complex fractions:

1) Complex fractions with *one term* in the numerator and *one term* in the denominator
2) Complex fractions that have *more than one term* in their numerators, their denominators, or both

1 Simplify a Complex Fraction with One Term in the Numerator and One Term in the Denominator

Remember that a fraction represents division. For example, $\dfrac{20}{4} = 20 \div 4 = 5$. We use this fact to simplify complex fractions that have one term in the numerator and one term in the denominator.

EXAMPLE 1

Simplify each complex fraction.

a) $\dfrac{\frac{7}{30}}{\frac{14}{25}}$ b) $\dfrac{\frac{7k-28}{3}}{\frac{k-4}{k}}$

Solution

a) There is one term in the numerator: $\dfrac{7}{30}$

There is one term in the denominator: $\dfrac{14}{25}$

$\dfrac{\frac{7}{30}}{\frac{14}{25}}$ means $\dfrac{7}{30} \div \dfrac{14}{25}$. Then,

$$\dfrac{7}{30} \div \dfrac{14}{25} = \dfrac{7}{30} \cdot \dfrac{25}{14} \quad \text{Multiply by the reciprocal.}$$

$$= \dfrac{\overset{1}{\cancel{7}}}{\underset{6}{\cancel{30}}} \cdot \dfrac{\overset{5}{\cancel{25}}}{\underset{2}{\cancel{14}}} \quad \text{Divide 7 and 14 by 7. Divide 25 and 30 by 5.}$$

$$= \dfrac{5}{12} \quad \text{Multiply.}$$

b) There is one term in the numerator: $\dfrac{7k-28}{3}$

There is one term in the denominator: $\dfrac{k-4}{k}$

W Hint
How could you check your answer quickly?

SECTION 8.3 **Simplifying Complex Fractions** 451

To simplify, rewrite as a division problem then carry out the division.

$$\frac{\frac{7k-28}{3}}{\frac{k-4}{k}} = \frac{7k-28}{3} \div \frac{k-4}{k} \qquad \text{Rewrite the complex fraction as a division problem.}$$

$$= \frac{7k-28}{3} \cdot \frac{k}{k-4} \qquad \text{Change division to multiplication by the reciprocal of } \frac{k-4}{k}.$$

$$= \frac{7\cancel{(k-4)}}{3} \cdot \frac{k}{\cancel{k-4}} \qquad \text{Factor. Divide the numerator and denominator by } k-4 \text{ to simplify.}$$

$$= \frac{7k}{3} \qquad \text{Multiply.}$$

[YOU TRY 1] Simplify.

a) $\dfrac{\frac{10}{27}}{\frac{20}{9}}$ \qquad b) $\dfrac{\frac{6n}{n^2-81}}{\frac{n^2}{2n+18}}$

> **Procedure** Simplify a Complex Fraction with One Term in the Numerator and Denominator
>
> To simplify a complex fraction containing one term in the numerator and one term in the denominator:
>
> 1) Rewrite the complex fraction as a division problem.
> 2) Perform the division by multiplying the first fraction by the reciprocal of the second.
>
> (We are multiplying the numerator of the complex fraction by the reciprocal of the denominator.)

2 Simplify a Complex Fraction with More Than One Term in the Numerator and/or Denominator by Rewriting It as a Division Problem

When a complex fraction has more than one term in the numerator and/or the denominator, we can use one of two methods to simplify.

> **Procedure** Simplify a Complex Fraction Using Method 1
>
> 1) Combine the terms in the numerator and combine the terms in the denominator so that each contains only one fraction.
> 2) Rewrite as a division problem.
> 3) Perform the division by multiplying the first fraction by the reciprocal of the second.

EXAMPLE 2

Simplify.

a) $\dfrac{\dfrac{1}{2} + \dfrac{1}{3}}{3 - \dfrac{5}{4}}$ b) $\dfrac{\dfrac{2}{x^2 y}}{\dfrac{1}{y} - \dfrac{y}{x}}$

Solution

a) The numerator is $\dfrac{1}{2} + \dfrac{1}{3}$, and it contains two terms. The denominator is $3 - \dfrac{5}{4}$, and it contains two terms.

We will add the terms in the numerator and subtract the terms in the denominator so that the numerator and denominator will each contain one fraction.

$$\dfrac{\dfrac{1}{2} + \dfrac{1}{3}}{3 - \dfrac{5}{4}} = \dfrac{\dfrac{3}{6} + \dfrac{2}{6}}{\dfrac{12}{4} - \dfrac{5}{4}} = \dfrac{\dfrac{5}{6}}{\dfrac{7}{4}}$$

Add the fractions in the numerator.

Subtract the terms in the denominator.

W Hint
Be sure you remember how to add and subtract fractions and rational expressions!

Rewrite as a division problem, multiply by the reciprocal, and simplify.

$$\dfrac{5}{6} \div \dfrac{7}{4} = \dfrac{5}{\underset{3}{\cancel{6}}} \cdot \dfrac{\overset{2}{\cancel{4}}}{7} = \dfrac{10}{21}$$

b) The numerator, $\dfrac{2}{x^2 y}$, contains one term; the denominator, $\dfrac{1}{y} - \dfrac{y}{x}$, contains two terms. We will subtract the terms in the denominator so that it, like the numerator, will contain only one term. The LCD of the expressions in the denominator is xy.

$$\dfrac{\dfrac{2}{x^2 y}}{\dfrac{1}{y} - \dfrac{y}{x}} = \dfrac{\dfrac{2}{x^2 y}}{\dfrac{x}{xy} - \dfrac{y^2}{xy}} = \dfrac{\dfrac{2}{x^2 y}}{\dfrac{x - y^2}{xy}}$$

Rewrite as a division problem, multiply by the reciprocal, and simplify.

$$\dfrac{2}{x^2 y} \div \dfrac{x - y^2}{xy} = \dfrac{2}{\underset{x}{\cancel{x^2 y}}} \cdot \dfrac{\cancel{xy}}{x - y^2} = \dfrac{2}{x(x - y^2)}$$

YOU TRY 2

Simplify.

a) $\dfrac{\dfrac{9}{8} - \dfrac{3}{4}}{\dfrac{2}{3} + \dfrac{1}{4}}$ b) $\dfrac{\dfrac{5}{a} + \dfrac{3}{ab}}{\dfrac{1}{ab} + 2}$

3 Simplify a Complex Fraction with More Than One Term in the Numerator and/or Denominator by Multiplying by the LCD

Another method we can use to simplify complex fractions involves multiplying the numerator and denominator of the complex fraction by the LCD of *all* of the fractions in the expression.

> **Procedure** Simplify a Complex Fraction Using Method 2
>
> 1) Identify and write down the LCD of *all* of the fractions in the complex fraction.
> 2) Multiply the numerator and denominator of the complex fraction by the LCD.
> 3) Simplify.

We will simplify the complex fractions we simplified in Example 2 using Method 2.

EXAMPLE 3   Simplify using Method 2.

a) $\dfrac{\dfrac{1}{2} + \dfrac{1}{3}}{3 - \dfrac{5}{4}}$ b) $\dfrac{\dfrac{2}{x^2 y}}{\dfrac{1}{y} - \dfrac{y}{x}}$

Solution

a) Look at *all* of the fractions in the complex fraction. Write down their LCD:

$$LCD = 12$$

Multiply the numerator and denominator of the complex fraction by the LCD, 12.

$$\frac{12\left(\dfrac{1}{2} + \dfrac{1}{3}\right)}{12\left(3 - \dfrac{5}{4}\right)}$$ We are multiplying the fraction by $\dfrac{12}{12}$, which equals 1.

$$= \frac{12 \cdot \dfrac{1}{2} + 12 \cdot \dfrac{1}{3}}{12 \cdot 3 - 12 \cdot \dfrac{5}{4}}$$ Distribute.

$$= \frac{6 + 4}{36 - 15} = \frac{10}{21}$$ Simplify.

This is the same result we obtained in Example 2 using Method 1.

> **Note**
>
> In the denominator we multiplied the 3 by 12 even though 3 is not a fraction. Remember, *all* terms, not just the fractions, must be multiplied by the LCD.

b) Look at *all* of the fractions in the complex fraction. Write down their LCD.

$$\text{LCD} = x^2y$$

Multiply the numerator and denominator of the complex fraction by the LCD, x^2y.

$$\frac{x^2y\left(\dfrac{2}{x^2y}\right)}{x^2y\left(\dfrac{1}{y} - \dfrac{y}{x}\right)} \qquad \text{We are multiplying the expression by } \frac{x^2y}{x^2y}, \text{ which equals 1.}$$

$$= \frac{x^2y \cdot \dfrac{2}{x^2y}}{x^2y \cdot \dfrac{1}{y} - x^2y \cdot \dfrac{y}{x}} \qquad \text{Distribute.}$$

$$= \frac{2}{x^2 - xy^2} = \frac{2}{x(x - y^2)} \qquad \text{Simplify.}$$

If the numerator and denominator factor, factor them. Sometimes, you can divide by a common factor to simplify.

Notice that the result is the same as what was obtained in Example 2 using Method 1.

Hint
Which method seems easier to use?

[**YOU TRY 3**] Simplify using Method 2.

a) $\dfrac{\dfrac{9}{8} - \dfrac{3}{4}}{\dfrac{2}{3} + \dfrac{1}{4}}$
b) $\dfrac{\dfrac{5}{a} + \dfrac{3}{ab}}{\dfrac{1}{ab} + 2}$

You should be familiar with both methods for simplifying complex fractions containing two terms in the numerator or denominator. After a lot of practice, you will be able to decide which method works best for a particular problem.

EXAMPLE 4 Determine which method to use to simplify each complex fraction, then simplify.

a) $\dfrac{\dfrac{2}{c} + \dfrac{1}{c+6}}{\dfrac{4}{c+6} - \dfrac{1}{c}}$
b) $\dfrac{\dfrac{a^2 - 9}{5a + 10}}{\dfrac{2a - 6}{a^2 - 4}}$

Solution

a) This complex fraction contains two terms in the numerator and two terms in the denominator. Let's use Method 2: Multiply the numerator and denominator by the LCD of all of the fractions.

List all of the fractions in the complex fraction: $\dfrac{2}{c}, \dfrac{1}{c+6}, \dfrac{4}{c+6}, \dfrac{1}{c}$.

Write down their LCD: $\text{LCD} = c(c + 6)$

Multiply the numerator and denominator of the complex fraction by the LCD, $c(c + 6)$, then simplify.

$$\frac{c(c+6)\left(\frac{2}{c} + \frac{1}{c+6}\right)}{c(c+6)\left(\frac{4}{c+6} - \frac{1}{c}\right)} = \frac{c(c+6) \cdot \frac{2}{c} + c(c+6) \cdot \frac{1}{c+6}}{c(c+6) \cdot \frac{4}{c+6} - c(c+6) \cdot \frac{1}{c}}$$
Multiply the numerator and denominator by $c(c+6)$ and distribute.

$$= \frac{2(c+6) + c}{4c - (c+6)}$$
Multiply.

$$= \frac{2c + 12 + c}{4c - c - 6}$$
Distribute.

$$= \frac{3c + 12}{3c - 6}$$
Combine like terms.

$$= \frac{3(c+4)}{3(c-2)} = \frac{c+4}{c-2}$$
Factor and simplify.

b) This complex fraction contains one term in the numerator, $\dfrac{a^2 - 9}{5a + 10}$, and one term in the denominator, $\dfrac{2a - 6}{a^2 - 4}$. To simplify, rewrite as a division problem, multiply by the reciprocal, and simplify.

$$\frac{\frac{a^2 - 9}{5a + 10}}{\frac{2a - 6}{a^2 - 4}} = \frac{a^2 - 9}{5a + 10} \div \frac{2a - 6}{a^2 - 4}$$
Rewrite as a division problem.

$$= \frac{a^2 - 9}{5a + 10} \cdot \frac{a^2 - 4}{2a - 6}$$
Multiply by the reciprocal.

$$= \frac{(a+3)\cancel{(a-3)}}{5\cancel{(a+2)}} \cdot \frac{\cancel{(a+2)}(a-2)}{2\cancel{(a-3)}}$$
Factor.

$$= \frac{(a+3)(a-2)}{10}$$
Divide out common factors, and multiply.

[YOU TRY 4] Determine which method to use to simplify each complex fraction, then simplify.

a) $\dfrac{\dfrac{8}{k} - \dfrac{1}{k+5}}{\dfrac{3}{k+5} + \dfrac{5}{k}}$ b) $\dfrac{\dfrac{c^2 - 9}{8c - 56}}{\dfrac{2c + 6}{c^2 - 49}}$

4 Simplify Rational Expressions Containing Negative Exponents

If a rational expression contains a negative exponent, rewrite it with positive exponents and simplify.

EXAMPLE 5

Simplify $\dfrac{x^{-2} - 3y^{-1}}{1 + x^{-1}}$.

Solution

First, rewrite the expression with positive exponents: $\dfrac{x^{-2} - 3y^{-1}}{1 + x^{-1}} = \dfrac{\dfrac{1}{x^2} - \dfrac{3}{y}}{1 + \dfrac{1}{x}}$

W Hint
Try doing this example using the other method.

Next we have to simplify. Identify the LCD of all the fractions: $x^2 y$

Multiply the numerator and denominator of the complex fraction by $x^2 y$, and simplify.

$$\dfrac{x^2 y \left(\dfrac{1}{x^2} - \dfrac{3}{y}\right)}{x^2 y \left(1 + \dfrac{1}{x}\right)} = \dfrac{x^2 y \cdot \dfrac{1}{x^2} - x^2 y \cdot \dfrac{3}{y}}{x^2 y \cdot 1 + x^2 y \cdot \dfrac{1}{x}}$$ Multiply and distribute.

$$= \dfrac{y - 3x^2}{x^2 y + xy}$$ Multiply.

$$= \dfrac{y - 3x^2}{xy(x + 1)}$$ Factor.

YOU TRY 5

Simplify $\dfrac{4a^{-2} + b^{-1}}{a^{-3} + b}$.

ANSWERS TO YOU TRY EXERCISES

1) a) $\dfrac{1}{6}$ b) $\dfrac{12}{n(n-9)}$ 2) a) $\dfrac{9}{22}$ b) $\dfrac{5b+3}{2ab+1}$ 3) a) $\dfrac{9}{22}$ b) $\dfrac{5b+3}{2ab+1}$
4) a) $\dfrac{7k+40}{8k+25}$ b) $\dfrac{(c-3)(c+7)}{16}$ 5) $\dfrac{a(4b+a^2)}{b(1+a^3 b)}$

E Evaluate 8.3 Exercises

Do the exercises, and check your work.

1) Explain, in your own words, two ways to simplify $\dfrac{\dfrac{2}{9}}{\dfrac{5}{18}}$. Then, simplify it both ways. Which method do you prefer and why?

2) Explain, in your own words, two ways to simplify $\dfrac{\dfrac{3}{2} - \dfrac{1}{5}}{\dfrac{1}{10} + \dfrac{3}{5}}$. Then, simplify it both ways. Which method do you prefer and why?

Objective 1: Simplify a Complex Fraction with One Term in the Numerator and One Term in the Denominator

Simplify completely.

3) $\dfrac{\dfrac{7}{10}}{\dfrac{5}{4}}$ 4) $\dfrac{\dfrac{3}{8}}{\dfrac{4}{3}}$ 5) $\dfrac{\dfrac{a^2}{b}}{\dfrac{a}{b^3}}$

6) $\dfrac{\dfrac{u^5}{v^2}}{\dfrac{u^2}{v}}$ 7) $\dfrac{\dfrac{s^3}{t^3}}{\dfrac{s^4}{t}}$ 8) $\dfrac{\dfrac{x^4}{y}}{\dfrac{x^2}{y^3}}$

SECTION 8.3 Simplifying Complex Fractions

9) $\dfrac{\dfrac{14m^5n^4}{9}}{\dfrac{35mn^6}{3}}$

10) $\dfrac{\dfrac{11b^4c^2}{4}}{\dfrac{55bc}{8}}$

11) $\dfrac{\dfrac{t-6}{5}}{\dfrac{t-6}{t}}$

12) $\dfrac{\dfrac{m-3}{m}}{\dfrac{m-3}{16}}$

13) $\dfrac{\dfrac{8}{y^2-64}}{\dfrac{6}{y+8}}$

14) $\dfrac{\dfrac{g^2-36}{15}}{\dfrac{g-6}{45}}$

15) $\dfrac{\dfrac{25w-35}{w^5}}{\dfrac{30w-42}{w}}$

16) $\dfrac{\dfrac{d^3}{16d-24}}{\dfrac{d}{40d-60}}$

17) $\dfrac{\dfrac{2x}{x+7}}{\dfrac{2}{x^2+4x-21}}$

18) $\dfrac{\dfrac{c^2-7c-8}{6c}}{\dfrac{c-8}{c}}$

29) $\dfrac{\dfrac{4}{3}+\dfrac{2}{5}}{\dfrac{1}{6}-\dfrac{2}{3}}$

30) $\dfrac{\dfrac{1}{4}-\dfrac{5}{6}}{\dfrac{3}{8}+\dfrac{1}{3}}$

31) $\dfrac{\dfrac{4}{x}-\dfrac{4}{y}}{\dfrac{3}{x^2}-\dfrac{3}{y^2}}$

32) $\dfrac{\dfrac{2}{a}-\dfrac{2}{b}}{\dfrac{1}{a^2}-\dfrac{1}{b^2}}$

33) $\dfrac{\dfrac{r}{s^2}+\dfrac{1}{rs}}{\dfrac{s}{r}+\dfrac{1}{r^2}}$

34) $\dfrac{\dfrac{n}{m^3}+\dfrac{m}{n}}{\dfrac{3}{n}-\dfrac{m}{n^4}}$

35) $\dfrac{1+\dfrac{4}{t-3}}{\dfrac{t}{t-3}+\dfrac{2}{t^2-9}}$

36) $\dfrac{1-\dfrac{4}{t+5}}{\dfrac{4}{t^2-25}+\dfrac{t}{t-5}}$

Mixed Exercises: Objectives 2 and 3

Simplify using Method 1 then by using Method 2. Think about which method you prefer and why.

19) $\dfrac{\dfrac{1}{4}+\dfrac{3}{2}}{\dfrac{2}{3}+\dfrac{1}{2}}$

20) $\dfrac{\dfrac{7}{9}-\dfrac{1}{3}}{2+\dfrac{1}{9}}$

21) $\dfrac{\dfrac{7}{c}+\dfrac{2}{d}}{1-\dfrac{5}{c}}$

22) $\dfrac{\dfrac{r}{s}-2}{\dfrac{1}{s}+\dfrac{3}{r}}$

23) $\dfrac{\dfrac{5}{z-2}-\dfrac{1}{z+1}}{\dfrac{1}{z-2}+\dfrac{4}{z+1}}$

24) $\dfrac{\dfrac{6}{w+4}+\dfrac{4}{w-1}}{\dfrac{5}{w-1}+\dfrac{3}{w+4}}$

Simplify using either Method 1 or Method 2.

25) $\dfrac{9+\dfrac{5}{y}}{\dfrac{9y+5}{8}}$

26) $\dfrac{4-\dfrac{12}{m}}{\dfrac{4m-12}{9}}$

27) $\dfrac{x-\dfrac{7}{x}}{x-\dfrac{11}{x}}$

28) $\dfrac{\dfrac{4}{c}-c}{3+\dfrac{8}{c}}$

Mixed Exercises: Objectives 1–3

Simplify completely.

37) $\dfrac{b+\dfrac{1}{b}}{b-\dfrac{3}{b}}$

38) $\dfrac{\dfrac{z+6}{4}}{\dfrac{z+6}{z}}$

39) $\dfrac{\dfrac{m}{n^2}}{\dfrac{m^4}{n}}$

40) $\dfrac{\dfrac{z^2+1}{5}}{z+\dfrac{1}{z}}$

41) $\dfrac{\dfrac{h^2-1}{4h-12}}{\dfrac{7h+7}{h^2-9}}$

42) $\dfrac{\dfrac{r^2+13r+40}{r^2-6r}}{\dfrac{r^2+2r-48}{3r}}$

43) $\dfrac{\dfrac{6}{x+3}-\dfrac{4}{x-1}}{\dfrac{2}{x-1}+\dfrac{1}{x+2}}$

44) $\dfrac{\dfrac{c^2}{d}+\dfrac{2}{c^2d}}{\dfrac{d}{c}-\dfrac{c}{d}}$

45) $\dfrac{\dfrac{r^2-6}{20}}{r-\dfrac{6}{r}}$

46) $\dfrac{\dfrac{1}{6}}{\dfrac{7}{8}}$

47) $\dfrac{\dfrac{a-4}{12}}{\dfrac{a-4}{a}}$

48) $\dfrac{\dfrac{8}{w}-w}{1+\dfrac{6}{w}}$

49) $\dfrac{\frac{5}{6}}{\frac{9}{15}}$

50) $\dfrac{\frac{5}{h+2} + \frac{7}{2h-3}}{\frac{1}{h-3} + \frac{3}{2h-3}}$

51) $\dfrac{\frac{5}{2n+1} + 1}{\frac{1}{n+3} + \frac{2}{2n+1}}$

52) $\dfrac{\frac{y^4}{z^3}}{\frac{y^6}{z^4}}$

Objective 4: Simplify Rational Expressions Containing Negative Exponents

Simplify.

53) $\dfrac{w^{-1} - v^{-1}}{2w^{-2} + v^{-1}}$

54) $\dfrac{4p^{-2} + q^{-1}}{p^{-1} + q^{-1}}$

55) $\dfrac{8x^{-2}}{x^{-1} - y^{-2}}$

56) $\dfrac{3d^{-1}}{2c^{-2} - d^{-1}}$

57) $\dfrac{a^{-3} + b^{-2}}{2b^{-2} - 7}$

58) $\dfrac{r^{-2} - t^{-2}}{5 + 7t^{-3}}$

59) $\dfrac{4m^{-1} - n^{-1}}{n^{-1} + m}$

60) $\dfrac{h^{-3} + 9}{k^{-2} - h}$

For Exercises 61 and 62, let $f(x) = \dfrac{1}{x}$.

61) Complete the table of values. As x gets larger, the value of $f(x)$ gets closer to what number?

x	f(x)
1	
2	
3	
10	
100	
1000	

62) Complete the table of values. As x gets smaller, what happens to the value of $f(x)$?

x	f(x)
1	
$\frac{1}{2}$	
$\frac{1}{3}$	
$\frac{1}{10}$	
$\frac{1}{100}$	
$\frac{1}{1000}$	

R Rethink

R1) Can all complex fractions be simplified using either Method 1 or Method 2?

R2) Can all complex fractions be simplified?

R3) Which method for simplifying complex fractions is easier for you?

R4) Do you need to factor at all when you use either of these methods?

8.4 Solving Rational Equations

Prepare / Organize

What are your objectives for Section 8.4?	How can you accomplish each objective?
1 Differentiate Between Rational Expressions and Rational Equations	• Be able to determine if the problem is an *expression* or an *equation*. • Review the procedures for **Adding and Subtracting Rational Expressions** and the procedures for **Solving Equations.** • Complete the given example on your own. • Complete You Try 1.
2 Solve Rational Equations	• Learn the procedure for **How to Solve a Rational Equation.** • Understand *why* we must check all proposed solutions. • Be sure you remember how to solve quadratic equations. • Complete the given examples on your own. • Complete You Trys 2–5.
3 Solve a Proportion	• Be able to recognize a *proportion*. • Learn the procedure for solving a proportion. • Complete the given example on your own. • Complete You Try 6.
4 Solve an Equation for a Specific Variable	• Follow same procedures as you would for solving equations. • Complete the given examples on your own. • Complete You Trys 7 and 8.

Work — Read the explanations, follow the examples, take notes, and complete the You Trys.

A **rational equation** is an equation that contains a rational expression. Some examples of rational equations are

$$\frac{1}{3}x + \frac{3}{4} = \frac{5}{6}x - 2, \quad \frac{5}{c-3} - \frac{c}{c+1} = 4, \quad \text{and} \quad \frac{2t}{t^2 + 6t + 8} + \frac{7}{t+2} = \frac{4}{t+4}.$$

1 Differentiate Between Rational Expressions and Rational Equations

In Chapter 2 we solved rational equations like the first one above, and we learned how to add and subtract rational expressions in Section 8.2. Let's summarize the difference between the two because this is often a point of confusion for students.

Summary Expressions vs. Equations

1) **The sum or difference of rational expressions does not contain an = sign.** To add or subtract, rewrite each expression with the LCD, and *keep the denominator* while performing the operations.

2) **An equation contains an = sign.** To solve an equation containing rational expressions, *multiply* the equation by the LCD of all fractions to *eliminate* the denominators, then solve.

EXAMPLE 1

Determine whether each is an equation or is a sum or difference of expressions. Then, solve the equation or find the sum or difference.

a) $\dfrac{k-2}{5} - \dfrac{k}{2} = \dfrac{7}{5}$ 　　b) $\dfrac{k-2}{5} - \dfrac{k}{2}$

Solution

a) This is an *equation* because it contains an = sign. We will *solve* for k using the method we learned in Chapter 2: Eliminate the denominators by multiplying by the LCD of all of the expressions.　**LCD = 10**

$$10\left(\dfrac{k-2}{5} - \dfrac{k}{2}\right) = 10 \cdot \dfrac{7}{5}$$　Multiply by the LCD of 10 to eliminate the denominators.

$$2(k-2) - 5k = 14$$　Distribute, and eliminate denominators.
$$2k - 4 - 5k = 14$$
$$-3k - 4 = 14$$
$$-3k = 18$$
$$k = -6$$

Check to verify that the solution set is $\{-6\}$.

b) $\dfrac{k-2}{5} - \dfrac{k}{2}$ is *not* an equation to be solved because it does *not* contain an = sign.

It is a difference of rational expressions. Rewrite each expression with the LCD, then subtract, *keeping the denominators* while performing the operations.

LCD = 10

$$\dfrac{(k-2)}{5} \cdot \dfrac{2}{2} = \dfrac{2(k-2)}{10}, \qquad \dfrac{k}{2} \cdot \dfrac{5}{5} = \dfrac{5k}{10}$$

$$\dfrac{k-2}{5} - \dfrac{k}{2} = \dfrac{2(k-2)}{10} - \dfrac{5k}{10}$$　Rewrite each expression with a denominator of 10.

$$= \dfrac{2(k-2) - 5k}{10}$$　Subtract the numerators.

$$= \dfrac{2k - 4 - 5k}{10}$$　Distribute.

$$= \dfrac{-3k - 4}{10}$$　Combine like terms.

W Hint
In your notes and in your own words, explain the difference between equations and the sum and difference of expressions. Give an example of each.

YOU TRY 1 Determine whether each is an equation or is a sum or difference of expressions. Then solve the equation or find the sum or difference.

a) $\dfrac{n}{9} + \dfrac{n-11}{6}$ b) $\dfrac{n}{9} + \dfrac{n-11}{6} = -1$

Let's look at more examples of how to solve equations containing rational expressions.

2 Solve Rational Equations

Procedure How to Solve a Rational Equation

1) If possible, factor all denominators.
2) Write down the LCD of all of the expressions.
3) Multiply both sides of the equation by the LCD to *eliminate* the denominators.
4) Solve the equation.
5) Check the solution(s) in the original equation. If a proposed solution makes a denominator equal 0, then it is rejected as a solution.

EXAMPLE 2 Solve $\dfrac{a}{3} + \dfrac{4}{a} = \dfrac{13}{3}$.

Solution

Because this is an equation, we will *eliminate* the denominators by multiplying the equation by the LCD of all of the expressions.

LCD = 3a

Hint
Write out the examples as you are reading them.

$$3a\left(\dfrac{a}{3} + \dfrac{4}{a}\right) = 3a\left(\dfrac{13}{3}\right)$$ Multiply both sides of the equation by the LCD, $3a$.

$$3a\left(\dfrac{a}{3}\right) + 3a\left(\dfrac{4}{a}\right) = 3a\left(\dfrac{13}{3}\right)$$ Distribute, and divide out common factors.

$$a^2 + 12 = 13a$$
$$a^2 - 13a + 12 = 0$$ Subtract $13a$.
$$(a - 12)(a - 1) = 0$$ Factor.

$a - 12 = 0$ or $a - 1 = 0$
$a = 12$ or $a = 1$

Check: $a = 12$ $a = 1$

$\dfrac{a}{3} + \dfrac{4}{a} \stackrel{?}{=} \dfrac{13}{3}$ $\dfrac{a}{3} + \dfrac{4}{a} \stackrel{?}{=} \dfrac{13}{3}$

$\dfrac{12}{3} + \dfrac{4}{12} \stackrel{?}{=} \dfrac{13}{3}$ $\dfrac{1}{3} + \dfrac{4}{1} \stackrel{?}{=} \dfrac{13}{3}$

$\dfrac{12}{3} + \dfrac{1}{3} = \dfrac{13}{3}$ ✓ $\dfrac{1}{3} + \dfrac{12}{3} = \dfrac{13}{3}$ ✓

The solution set is $\{1, 12\}$.

[YOU TRY 2] Solve $\dfrac{c}{2} + 1 = \dfrac{24}{c}$.

It is *very* important to check the proposed solution. Sometimes, what appears to be a solution actually is not.

EXAMPLE 3 Solve $3 - \dfrac{3}{x+3} = \dfrac{x}{x+3}$.

Solution

Because this is an equation, we will *eliminate* the denominators by multiplying the equation by the LCD of all of the expressions.

LCD $= x + 3$

$$(x+3)\left(3 - \dfrac{3}{x+3}\right) = (x+3)\left(\dfrac{x}{x+3}\right)$$ Multiply both sides of the equation by the LCD, $x + 3$.

$$(x+3)3 - \cancel{(x+3)} \cdot \dfrac{3}{\cancel{x+3}} = \cancel{(x+3)}\left(\dfrac{x}{\cancel{x+3}}\right)$$ Distribute, and divide out common factors.

$$3x + 9 - 3 = x$$ Multiply.
$$3x + 6 = x$$
$$6 = -2x$$ Subtract $3x$.
$$-3 = x$$ Divide by -2.

Check: $\quad 3 - \dfrac{3}{(-3)+3} \stackrel{?}{=} \dfrac{-3}{(-3)+3}$ Substitute -3 for x in the original equation.

$$3 - \dfrac{3}{0} \stackrel{?}{=} \dfrac{-3}{0}$$

Because $x = -3$ makes the denominator zero, -3 cannot be a solution to the equation. Therefore, this equation has no solution. The solution set is $\varnothing$.

BE CAREFUL Always check what *appears* to be the solution or solutions to an equation containing rational expressions. If one of these values makes a denominator zero, then it *cannot* be a solution of the equation.

[YOU TRY 3] Solve each equation.

a) $\dfrac{7}{s+1} + \dfrac{2s}{s+1} = 3$ b) $\dfrac{3m}{m-4} - 1 = \dfrac{12}{m-4}$

EXAMPLE 4

Solve $\dfrac{1}{3} - \dfrac{1}{t+2} = \dfrac{t+14}{3t^2 - 12}$.

Solution

This is an equation. *Eliminate* the denominators by multiplying by the LCD. Begin by factoring the denominator of $\dfrac{t+14}{3t^2 - 12}$.

$\dfrac{1}{3} - \dfrac{1}{t+2} = \dfrac{t+14}{3(t+2)(t-2)}$ Factor the denominator.

LCD = $3(t+2)(t-2)$ Write down the LCD of all of the expressions.

$3(t+2)(t-2)\left(\dfrac{1}{3} - \dfrac{1}{t+2}\right) = 3(t+2)(t-2)\left(\dfrac{t+14}{3(t+2)(t-2)}\right)$ Multiply both sides of the equation by the LCD.

$\cancel{3}(t+2)(t-2)\left(\dfrac{1}{\cancel{3}}\right) - 3\cancel{(t+2)}(t-2)\left(\dfrac{1}{\cancel{t+2}}\right) = \cancel{3}\cancel{(t+2)}\cancel{(t-2)}\left(\dfrac{t+14}{\cancel{3}\cancel{(t+2)}\cancel{(t-2)}}\right)$

Distribute, and divide out common factors.

$(t+2)(t-2) - 3(t-2) = t + 14$ Multiply.
$t^2 - 4 - 3t + 6 = t + 14$ Distribute.
$t^2 - 3t + 2 = t + 14$ Combine like terms.
$t^2 - 4t - 12 = 0$ Subtract t and subtract 14.
$(t-6)(t+2) = 0$ Factor.
$t - 6 = 0 \quad \text{or} \quad t + 2 = 0$ Set each factor equal to zero.
$t = 6 \quad \text{or} \quad t = -2$ Solve.

Hint
To save time, look for similar patterns when checking answers!

Look at the factored form of the equation. If $t = 6$, no denominator will equal zero. If $t = -2$, however, two of the denominators will equal zero. Therefore, we must reject $t = -2$ as a solution. Check only $t = 6$.

Check: $\dfrac{1}{3} - \dfrac{1}{6+2} \stackrel{?}{=} \dfrac{6+14}{3(6)^2 - 12}$ Substitute $t = 6$ into the original equation.

$\dfrac{1}{3} - \dfrac{1}{8} \stackrel{?}{=} \dfrac{20}{108 - 12}$ Simplify.

$\dfrac{1}{3} - \dfrac{1}{8} \stackrel{?}{=} \dfrac{20}{96}$ Simplify.

$\dfrac{8}{24} - \dfrac{3}{24} \stackrel{?}{=} \dfrac{5}{24}$ Get a common denominator and simplify $\dfrac{20}{96}$.

$\dfrac{5}{24} = \dfrac{5}{24}$ ✓   Subtract.

The solution set is $\{6\}$.

Examples 3 and 4 show why it is necessary to check all proposed solutions of equations containing rational expressions.

[YOU TRY 4] Solve $\dfrac{w}{w+6} - \dfrac{3}{w+2} = \dfrac{6-3w}{w^2 + 8w + 12}$.

EXAMPLE 5

Solve $\dfrac{5}{6n^2 + 18n + 12} = \dfrac{n}{2n + 2} + \dfrac{1}{3n + 6}$.

Solution

Because this is an equation, we will *eliminate* the denominators by multiplying by the LCD. Begin by factoring all denominators, then identify the LCD.

$$\dfrac{5}{6(n+2)(n+1)} = \dfrac{n}{2(n+1)} + \dfrac{1}{3(n+2)} \qquad \text{LCD} = 6(n+2)(n+1)$$

$$6(n+2)(n+1)\left(\dfrac{5}{6(n+2)(n+1)}\right) = 6(n+2)(n+1)\left(\dfrac{n}{2(n+1)} + \dfrac{1}{3(n+2)}\right) \qquad \text{Multiply by the LCD.}$$

$$\cancel{6(n+2)(n+1)}\left(\dfrac{5}{\cancel{6(n+2)(n+1)}}\right) = \overset{3}{\cancel{6}}(n+2)\cancel{(n+1)}\left(\dfrac{n}{\cancel{2(n+1)}}\right) + \overset{2}{\cancel{6}}\cancel{(n+2)}(n+1)\left(\dfrac{1}{\cancel{3(n+2)}}\right) \qquad \text{Distribute.}$$

$$
\begin{aligned}
5 &= 3n(n+2) + 2(n+1) & &\text{Multiply.} \\
5 &= 3n^2 + 6n + 2n + 2 & &\text{Distribute.} \\
5 &= 3n^2 + 8n + 2 & &\text{Combine like terms.} \\
0 &= 3n^2 + 8n - 3 & &\text{Subtract 5.} \\
0 &= (3n-1)(n+3) & &\text{Factor.} \\
3n - 1 &= 0 \quad \text{or} \quad n + 3 = 0 \\
n &= \dfrac{1}{3} \quad \text{or} \quad n = -3 & &\text{Solve.}
\end{aligned}
$$

If you look at the factored form of the equation you can see that neither $n = \dfrac{1}{3}$ nor $n = -3$ will make a denominator zero. Check the values in the original equation to verify that the solution set is $\left\{-3, \dfrac{1}{3}\right\}$.

YOU TRY 5

Solve $\dfrac{11}{6h^2 + 48h + 90} = \dfrac{h}{3h + 15} + \dfrac{1}{2h + 6}$.

3 Solve a Proportion

Recall that a **ratio** is a quotient of two quantities. A **proportion** is a statement that two ratios are equal. One way to determine if a proportion is true is to find the cross products. *If the cross products are equal, then the proportion is true.*

For example, we can show that the proportion $\dfrac{5}{9} = \dfrac{10}{18}$ is true by finding the cross products:

$$\dfrac{5}{9} \diagup\!\!\!\!\!\diagdown \dfrac{10}{18}$$

Multiply. Multiply.

$$5 \cdot 18 \overset{?}{=} 9 \cdot 10$$
$$90 = 90 \quad \text{True}$$

The cross products are equal, so $\dfrac{5}{9} = \dfrac{10}{18}$ is a true proportion. (If the cross products are not equal, then the proportion is not true.)

We can use cross products to solve equations that are proportions.

SECTION 8.4 Solving Rational Equations

EXAMPLE 6

Solve $\dfrac{20}{d+6} = \dfrac{8}{d}$.

Solution

This equation is a *proportion*. We can solve it like we have solved the other equations in this section, by multiplying both sides of the equation by the LCD. Or, *we can solve this proportion by setting the cross products equal to each other*.

> **W Hint**
> Multiplying the *cross products* eliminates the need to multiply both sides of the equations by the LCD.

$$\dfrac{20}{d+6} \diagdown\!\!\!\!\!= \dfrac{8}{d}$$
Multiply.   Multiply.

$20d = 8(d + 6)$	Set the cross products equal to each other.
$20d = 8d + 48$	Distribute.
$12d = 48$	Subtract $8d$.
$d = 4$	Solve.

Check: $\dfrac{20}{4+6} \stackrel{?}{=} \dfrac{8}{4}$ Substitute $d = 4$ into the original equation.

$\dfrac{20}{10} \stackrel{?}{=} 2$

$2 = 2$ ✓

The solution is $\{4\}$.

[YOU TRY 6]

Solve $\dfrac{9}{y} = \dfrac{5}{y-2}$.

4 Solve an Equation for a Specific Variable

In Section 2.3, we learned how to solve an equation for a specific variable. For example, to solve $2l + 2w = P$ for w, we do the following:

$2l + 2\boxed{w} = P$	Put a box around w, the variable for which we are solving.
$2\boxed{w} = P - 2l$	Subtract $2l$.
$w = \dfrac{P - 2l}{2}$	Divide by 2.

Next we discuss how to solve for a specific variable in a rational expression.

EXAMPLE 7

Solve $m = \dfrac{x}{a - A}$ for a.

Solution

Note that the equation contains a lowercase a and an uppercase A. These represent different quantities, so students should be sure to write them correctly. Put a in a box.

Since a is in the denominator of the rational expression, multiply both sides of the equation by $a - A$ to eliminate the denominator.

$$m = \frac{x}{\boxed{a} - A} \qquad \text{Put } a \text{ in a box.}$$

$$(\boxed{a} - A)m = (\boxed{a} - A)\left(\frac{x}{\boxed{a} - A}\right) \qquad \text{Multiply both sides by } a - A \text{ to eliminate the denominator.}$$

$$\boxed{a}m - Am = x \qquad \text{Distribute.}$$

$$\boxed{a}m = x + Am \qquad \text{Add } Am.$$

$$a = \frac{x + Am}{m} \qquad \text{Divide by } m.$$

[**YOU TRY 7**] Solve $y = \dfrac{c}{r + d}$ for d.

EXAMPLE 8

Solve $\dfrac{1}{a} + \dfrac{1}{b} = \dfrac{1}{c}$ for b.

Solution

Put the b in a box. The LCD of all of the fractions is abc. Multiply both sides of the equation by abc.

$$\frac{1}{a} + \frac{1}{\boxed{b}} = \frac{1}{c} \qquad \text{Put } b \text{ in a box.}$$

$$a\boxed{b}c\left(\frac{1}{a} + \frac{1}{\boxed{b}}\right) = a\boxed{b}c\left(\frac{1}{c}\right) \qquad \text{Multiply both sides by } abc \text{ to eliminate the denominator.}$$

$$a\boxed{b}c \cdot \frac{1}{a} + a\boxed{b}c \cdot \frac{1}{\boxed{b}} = a\boxed{b}c\left(\frac{1}{c}\right) \qquad \text{Distribute.}$$

$$\boxed{b}c + ac = a\boxed{b} \qquad \text{Divide out common factors.}$$

Because we are solving for b and there are terms containing b on each side of the equation, we must get bc and ab on one side of the equation and ac on the other side.

$$ac = a\boxed{b} - \boxed{b}c \qquad \text{Subtract } bc \text{ from each side.}$$

To isolate b, we will *factor* b out of each term on the right-hand side of the equation.

$$ac = \boxed{b}(a - c) \qquad \text{Factor out } b.$$

$$\frac{ac}{a - c} = b \qquad \text{Divide by } a - c.$$

[**YOU TRY 8**] Solve $\dfrac{1}{a} + \dfrac{1}{b} = \dfrac{1}{c}$ for a.

Using Technology

We can use a graphing calculator to solve a rational equation in one variable. First enter the left side of the equation in Y_1 and the right side of the equation in Y_2. Then enter $Y_1 - Y_2$ in Y_3, and graph the equations. The zero(s) [x-intercept(s)] of the graph are the solutions to the equation.

We will solve $\dfrac{2}{x+5} - \dfrac{3}{x-2} = \dfrac{4x}{x^2+3x-10}$ using a graphing calculator.

1) Enter $\dfrac{2}{x+5} - \dfrac{3}{x-2}$ in Y_1 by entering $2/(x+5) - 3/(x-2)$ in Y_1.

2) Enter $\dfrac{4x}{x^2+3x-10}$ in Y_2 by entering $4x/(x^2+3x-10)$ in Y_2.

3) Enter $Y_1 - Y_2$ in Y_3 as follows: press Y=, move the cursor to the right of \\Y$_3$ =, press VARS, select Y-VARS using the right arrow key, and press ENTER ENTER to select Y_1. Press −. Press VARS, select Y-VARS using the right arrow key, press ENTER 2 to select Y_2.

4) Move the cursor to the = right of \\Y$_1$ and press ENTER to deselect Y_1. Move the cursor to the = right of \\Y$_2$ and press ENTER to deselect Y_2. Press GRAPH to graph $Y_1 - Y_2$ as shown.

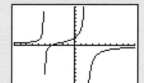

5) Press 2nd TRACE 2:zero, move the cursor to the left of the zero and press ENTER, move the cursor to the right of the zero and press ENTER, and move the cursor close to the zero and press ENTER to display the zero as shown on the graph.

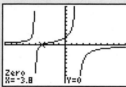

6) Press X,T,Θ,n MATH ENTER ENTER to display the zero $x = -\dfrac{19}{5}$, as shown on the image to the right.

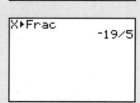

If there is more than one zero, repeat Steps 5 and 6 above for each zero.

Solve each equation using a graphing calculator.

1) $\dfrac{3x}{x-5} + \dfrac{3}{x+2} = \dfrac{5x}{x^2-3x-10}$

2) $\dfrac{5}{x-2} - \dfrac{3}{x+2} = \dfrac{2}{x^2-4}$

3) $\dfrac{3}{x+5} + \dfrac{2}{x-4} = \dfrac{6}{x^2+x-20}$

4) $\dfrac{5}{x-1} + \dfrac{2}{x+3} = \dfrac{4}{x^2+2x-3}$

5) $\dfrac{3}{x+5} + \dfrac{x}{x-2} = \dfrac{5x-2}{x^2+3x-10}$

6) $\dfrac{5}{x+1} - \dfrac{x}{x-4} = \dfrac{-25}{x^2-3x-4}$

ANSWERS TO YOU TRY EXERCISES

1) a) sum; $\dfrac{5n - 33}{9}$ b) equation; $\{3\}$ 2) $\{-8, 6\}$ 3) a) $\{4\}$ b) $\varnothing$ 4) $\{4\}$
5) $\left\{-4, -\dfrac{1}{2}\right\}$ 6) $\left\{\dfrac{9}{2}\right\}$ 7) $d = \dfrac{c - ry}{y}$ 8) $a = \dfrac{bc}{b - c}$

ANSWERS TO TECHNOLOGY EXERCISES

1) $\left\{-3, \dfrac{5}{3}\right\}$ 2) $\{-7\}$ 3) $\left\{\dfrac{8}{5}\right\}$ 4) $\left\{-\dfrac{9}{7}\right\}$ 5) $\{-4, 1\}$ 6) $\{5\}$

E Evaluate 8.4 Exercises

Do the exercises, and check your work.

Mixed Exercises: Objectives 1 and 2

1) When solving an equation containing rational expressions, do you keep the LCD throughout the problem or do you eliminate the denominators?

2) When adding or subtracting two rational expressions, do you keep the LCD throughout the problem or do you eliminate the denominators?

Determine whether each is an equation or is a sum or difference of expressions. Then, solve the equation or find the sum or difference.

3) $\dfrac{m}{8} + \dfrac{m - 7}{4}$

4) $\dfrac{2r + 9}{3} - \dfrac{r}{9}$

5) $\dfrac{2f - 19}{20} = \dfrac{f}{4} + \dfrac{2}{5}$

6) $\dfrac{2h}{5} + \dfrac{2}{3} = \dfrac{h + 3}{5}$

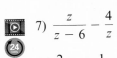

7) $\dfrac{z}{z - 6} - \dfrac{4}{z}$

8) $\dfrac{2}{a^2} + \dfrac{1}{a + 9}$

9) $1 + \dfrac{4}{c + 2} = \dfrac{9}{c + 2}$

10) $\dfrac{10}{b - 8} - 4 = \dfrac{2}{b - 8}$

Values that make the denominators equal zero cannot be solutions of an equation. Find *all* of the values that make the denominators zero and that, therefore, cannot be solutions of each equation. Do *not* solve the equation.

11) $\dfrac{t}{t + 10} - \dfrac{5}{t} = 3$

12) $\dfrac{k + 2}{k - 3} + 1 = \dfrac{7}{k}$

13) $\dfrac{2}{d^2 - 81} + \dfrac{8}{d} = \dfrac{6}{d + 9}$

14) $\dfrac{4}{p + 2} - \dfrac{5}{p} = \dfrac{p}{p^2 - 4}$

15) $\dfrac{v + 7}{v^2 - 13v + 36} - \dfrac{5}{3v - 12} = \dfrac{v}{v - 9}$

16) $\dfrac{3h}{h^2 + 3h - 40} + \dfrac{1}{h + 8} = \dfrac{h - 1}{4h - 20}$

Mixed Exercises: Objectives 2 and 3
Solve each equation.

17) $\dfrac{y}{3} - \dfrac{1}{2} = \dfrac{1}{6}$

18) $\dfrac{a}{2} + \dfrac{3}{10} = \dfrac{4}{5}$

19) $\dfrac{1}{2}h + h = -3$

20) $\dfrac{1}{6}j - j = -10$

21) $\dfrac{7u + 12}{15} = \dfrac{2u}{5} - \dfrac{3}{5}$

22) $\dfrac{8m - 7}{24} = \dfrac{m}{6} - \dfrac{7}{8}$

23) $\dfrac{4}{3t + 2} = \dfrac{2}{2t - 1}$

24) $\dfrac{9}{4x + 1} = \dfrac{3}{x + 2}$

25) $\dfrac{w}{3} = \dfrac{2w - 5}{12}$

26) $\dfrac{r - 2}{2} = \dfrac{3r - 5}{4}$

27) $\dfrac{12}{a} - 2 = \dfrac{6}{a}$

28) $\dfrac{16}{z} + 7 = -\dfrac{12}{z}$

29) $\dfrac{n}{n+2} + 3 = \dfrac{8}{n+2}$

30) $\dfrac{4q}{q+1} - 2 = \dfrac{3}{q+1}$

31) $\dfrac{2}{s+6} + 4 = \dfrac{2}{s+6}$

32) $\dfrac{u}{u-4} + 3 = \dfrac{4}{u-4}$

33) $\dfrac{c}{c-7} - 4 = \dfrac{10}{c-7}$

34) $\dfrac{2b}{b+9} - 5 = \dfrac{3}{b+9}$

35) $\dfrac{32}{g} + 10 = -\dfrac{8}{g}$

36) $\dfrac{9}{r} - 1 = \dfrac{6}{r}$

Solve each equation.

37) $\dfrac{1}{m-1} + \dfrac{2}{m+3} = \dfrac{4}{m+3}$

38) $\dfrac{4}{c+2} + \dfrac{2}{c-6} = \dfrac{5}{c+2}$

39) $\dfrac{4}{w-8} - \dfrac{10}{w+8} = \dfrac{40}{w^2 - 64}$

40) $\dfrac{4}{p-5} + \dfrac{7}{p+5} = \dfrac{18}{p^2 - 25}$

41) $\dfrac{3}{a+3} + \dfrac{14}{a^2 - 4a - 21} = \dfrac{5}{a-7}$

42) $\dfrac{5}{k+4} - \dfrac{3}{k+2} = \dfrac{8}{k^2 + 6k + 8}$

43) $\dfrac{9}{t+4} + \dfrac{8}{t^2 - 16} = \dfrac{1}{t-4}$

44) $\dfrac{12}{g^2 - 9} + \dfrac{2}{g+3} = \dfrac{7}{g-3}$

45) $\dfrac{4}{x^2 + 2x - 15} = \dfrac{8}{x-3} + \dfrac{2}{x+5}$

46) $\dfrac{4}{p-2} - \dfrac{9}{p^2 - 8p + 12} = \dfrac{9}{p-6}$

47) $\dfrac{k^2}{3} = \dfrac{k^2 + 2k}{4}$

48) $\dfrac{x^2}{2} = \dfrac{x^2 - 5x}{3}$

49) $\dfrac{5}{m^2 - 25} = \dfrac{4}{m^2 + 5m}$

50) $\dfrac{3}{t^2} = \dfrac{6}{t^2 + 5t}$

51) $\dfrac{10v}{3v - 12} - \dfrac{v+6}{v-4} = \dfrac{v}{3}$

52) $\dfrac{b-2}{2b-12} - \dfrac{b+2}{b-6} = \dfrac{b}{2}$

53) $\dfrac{w}{5} = \dfrac{w-3}{w+1} + \dfrac{12}{5w+5}$

54) $\dfrac{3y-2}{y+2} = \dfrac{y}{4} + \dfrac{1}{4y+8}$

55) $\dfrac{8}{p+2} + \dfrac{p}{p+1} = \dfrac{5p+2}{p^2 + 3p + 2}$

56) $\dfrac{6}{x+1} + \dfrac{x}{x-3} = \dfrac{3x+14}{x^2 - 2x - 3}$

57) $\dfrac{11}{c+9} = \dfrac{c}{c-4} - \dfrac{36-8c}{c^2 + 5c - 36}$

58) $\dfrac{3}{f+2} = \dfrac{f}{f+6} - \dfrac{2}{f^2 + 8f + 12}$

59) $\dfrac{8}{3g^2 - 7g - 6} + \dfrac{4}{g-3} = \dfrac{8}{3g+2}$

60) $\dfrac{1}{r-1} + \dfrac{2}{5r-3} = \dfrac{37}{5r^2 - 8r + 3}$

61) $\dfrac{h}{h^2 + 2h - 8} + \dfrac{4}{h^2 + 8h - 20} = \dfrac{4}{h^2 + 14h + 40}$

62) $\dfrac{b}{b^2 + b - 6} + \dfrac{3}{b^2 + 9b + 18} = \dfrac{8}{b^2 + 4b - 12}$

63) $\dfrac{u}{8} = \dfrac{2}{10 - u}$

64) $-\dfrac{a}{4} = \dfrac{3}{a+7}$

65) $\dfrac{5}{r+4} - \dfrac{2}{r} = -1$

66) $\dfrac{6}{c-5} - \dfrac{2}{c} = 1$

67) $\dfrac{q}{q^2 + 4q - 32} + \dfrac{2}{q^2 - 14q + 40} = \dfrac{6}{q^2 - 2q - 80}$

68) $\dfrac{r}{r^2 + 8r + 15} - \dfrac{2}{r^2 + r - 6} = \dfrac{2}{r^2 + 3r - 10}$

69) The average profit a king salmon fisherman receives is given by $A(x) = 6 - \dfrac{1500}{x}$, where x is the number of pounds of king salmon. How many pounds of salmon must he catch so that the profit he earns per pound (the average profit) is $4.00?

70) The average profit earned from the sale of x purses is given by $A(x) = 36 - \dfrac{8400}{x}$. How many purses must be sold so that the profit earned per purse (the average profit) is $20.00?

71) The formula $P = \dfrac{1}{f}$, where f is the focal length of a lens, in meters, and P is the power of the lens, in diopters, is used to determine an eyeglass prescription. If the power of a lens is 2.5 diopters, what is the focal length of the lens?

72) Magnetic stripe patterns on the ocean floor help geologists study the rates at which oceanic plates are spreading. Geologists use the formula

$$\text{Rate of spreading for stripes} = \frac{\text{width of stripe}}{\text{time duration}}$$

a) If a magnetic stripe on a plate on the ocean floor is 75 miles wide and formed over 2 million years, find the rate at which the plate is spreading.

b) If it continues to spread at this rate, how long will it have taken to become 90 miles wide?

Objective 4: Solve an Equation for a Specific Variable

Solve for the indicated variable.

73) $V = \dfrac{nRT}{P}$ for P

74) $W = \dfrac{CA}{m}$ for m

75) $a = \dfrac{rt}{2b}$ for b

76) $y = \dfrac{kx}{z}$ for z

77) $B = \dfrac{t + u}{3x}$ for x

78) $Q = \dfrac{n - k}{5r}$ for r

79) $z = \dfrac{a}{b + c}$ for b

80) $d = \dfrac{t}{l - n}$ for n

81) $A = \dfrac{4r}{q - t}$ for t

82) $h = \dfrac{3A}{r + s}$ for s

83) $w = \dfrac{na}{kc + b}$ for c

84) $r = \dfrac{kx}{y - az}$ for y

85) $\dfrac{1}{t} = \dfrac{1}{r} - \dfrac{1}{s}$ for r

86) $\dfrac{1}{R_1} + \dfrac{1}{R_2} = \dfrac{1}{R_3}$ for R_2

87) $\dfrac{2}{A} + \dfrac{1}{C} = \dfrac{3}{B}$ for C

88) $\dfrac{5}{x} = \dfrac{1}{y} - \dfrac{4}{z}$ for z

R Rethink

R1) What is the one main difference between *rational expressions* and *rational equations*?

R2) Why do you have to discard some solutions when solving a rational equation?

R3) Do all proportions have solutions? Why or why not?

Putting It All Together

P Prepare

What are your objectives for Putting It All Together?	How can you accomplish each objective?
1 Review the Concepts Presented in Sections 8.1–8.4	• Review the procedure for **Finding the Domain of a Rational Function**. • Know how to write rational expressions in lowest terms. • In your notes, summarize how to add, subtract, multiply, and divide rational expressions. • Be able to simplify complex fractions using more than one method. • Be able to solve rational equations. • Complete the given examples on your own. • Complete You Try 1.

O Organize

 **Work** — Read the explanations, follow the examples, take notes, and complete the You Trys.

1 Review the Concepts Presented in Sections 8.1–8.4

We began Section 8.1 with a discussion of rational functions. Let's review how to find the values of the variable that make a rational function equal zero and how to determine the domain of a rational function.

EXAMPLE 1

For $g(a) = \dfrac{a+6}{a^2 - 81}$,

a) find a so that $g(a) = 0$.

b) determine the domain of the function.

Solution

a) To find the value of a that will make $g(a) = 0$, set the function equal to zero and solve for a.

$\dfrac{a+6}{a^2 - 81} = 0$ when its *numerator* equals zero.

Let $a + 6 = 0$, and solve for a.

$$a + 6 = 0$$
$$a = -6$$

$g(a) = 0$ when $a = -6$.

b) Recall that the domain of a rational function consists of all real numbers except the values of the variable that make the denominator equal zero. Set the denominator equal to zero, and solve for a. All values that make the denominator equal zero are *not* in the domain of the function.

$$a^2 - 81 = 0$$
$$(a+9)(a-9) = 0 \quad \text{Factor.}$$
$$a + 9 = 0 \quad \text{or} \quad a - 9 = 0 \quad \text{Set each factor equal to zero.}$$
$$a = -9 \quad \text{or} \quad a = 9 \quad \text{Solve.}$$

When $a = 9$ or $a = -9$, the denominator of $g(a) = \dfrac{a+6}{a^2 - 81}$ equals zero. The domain contains all real numbers *except* -9 and 9. Write the domain in interval notation as $(-\infty, -9) \cup (-9, 9) \cup (9, \infty)$.

In Sections 8.1 and 8.2 we learned how to multiply, divide, add, and subtract rational expressions. Now we will practice these operations together so that we will learn to recognize the techniques needed to perform these operations. Write all answers in lowest terms.

EXAMPLE 2

Divide $\dfrac{y^2 - 6y - 27}{25y^2 - 49} \div \dfrac{y^2 - 9y}{14 - 10y}$.

Solution

Do we need a common denominator to divide? *No.* A common denominator is needed to add or subtract but not to multiply or divide.

To divide, multiply the first rational expression by the reciprocal of the second expression, then factor, divide out common factors, and multiply.

$$\dfrac{y^2 - 6y - 27}{25y^2 - 49} \div \dfrac{y^2 - 9y}{14 - 10y} = \dfrac{y^2 - 6y - 27}{25y^2 - 49} \cdot \dfrac{14 - 10y}{y^2 - 9y}$$ Multiply by the reciprocal.

$$= \dfrac{(y - 9)(y + 3)}{(5y + 7)(5y - 7)} \cdot \dfrac{2(7 - 5y)^{-1}}{y(y - 9)}$$ Factor.

$$= -\dfrac{2(y + 3)}{y(5y + 7)}$$ Divide out the common factors, and multiply.

Recall that $\dfrac{7 - 5y}{5y - 7} = -1$.

EXAMPLE 3

Add $\dfrac{n}{n - 6} + \dfrac{2}{n + 1}$.

Solution

To add or subtract rational expressions, we need a common denominator. We do not need to factor these denominators, so we are ready to identify the LCD.

$$\text{LCD} = (n - 6)(n + 1)$$

Rewrite each expression with the LCD.

$$\dfrac{n}{n - 6} \cdot \dfrac{n + 1}{n + 1} = \dfrac{n(n + 1)}{(n - 6)(n + 1)}, \qquad \dfrac{2}{n + 1} \cdot \dfrac{n - 6}{n - 6} = \dfrac{2(n - 6)}{(n - 6)(n + 1)}$$

$$\dfrac{n}{n - 6} + \dfrac{2}{n + 1} = \dfrac{n(n + 1)}{(n - 6)(n + 1)} + \dfrac{2(n - 6)}{(n - 6)(n + 1)}$$ Write each expression with the LCD.

$$= \dfrac{n(n + 1) + 2(n - 6)}{(n - 6)(n + 1)}$$ Add the numerators.

$$= \dfrac{n^2 + n + 2n - 12}{(n - 6)(n + 1)}$$ Distribute.

$$= \dfrac{n^2 + 3n - 12}{(n - 6)(n + 1)}$$ Combine like terms.

W Hint
In your notes, summarize how to add, subtract, multiply, and divide fractions.

Although this numerator will not factor, remember that sometimes it *is* possible to factor the numerator and simplify the result.

A complex fraction is not considered to be an expression in simplest form. Let's review how to simplify complex fractions.

EXAMPLE 4

Simplify each complex fraction.

a) $\dfrac{\dfrac{2a+12}{9}}{\dfrac{a+6}{5a}}$ b) $\dfrac{1-\dfrac{3}{4}}{\dfrac{1}{2}+\dfrac{1}{3}}$

Solution

a) This complex fraction contains one expression in the numerator and one in the denominator. Begin by rewriting it as a division problem.

$\dfrac{\dfrac{2a+12}{9}}{\dfrac{a+6}{5a}} = \dfrac{2a+12}{9} \div \dfrac{a+6}{5a}$ Rewrite the complex fraction as a division problem.

$= \dfrac{2a+12}{9} \cdot \dfrac{5a}{a+6}$ Change division to multiplication by the reciprocal of $\dfrac{a+6}{5a}$.

$= \dfrac{2\cancel{(a+6)}}{9} \cdot \dfrac{5a}{\cancel{a+6}}$ Factor, and divide numerator and denominator by $a+6$ to simplify.

$= \dfrac{10a}{9}$ Multiply.

W Hint
In your notes, summarize the different methods for simplifying complex fractions, and explain when you would use each method.

b) We can simplify the complex fraction $\dfrac{1-\dfrac{3}{4}}{\dfrac{1}{2}+\dfrac{1}{3}}$ in two different ways. We could combine the terms in the numerator, combine the terms in the denominator, and then proceed as in part a). Or, we can follow the steps below.

Step 1: Look at all of the fractions in the complex fraction. Write down their LCD.   LCD = 12

Step 2: Multiply the numerator and denominator of the complex fraction by the LCD, 12.

$$\dfrac{12\left(1-\dfrac{3}{4}\right)}{12\left(\dfrac{1}{2}+\dfrac{1}{3}\right)}$$

Step 3: Simplify.

$\dfrac{12\left(1-\dfrac{3}{4}\right)}{12\left(\dfrac{1}{2}+\dfrac{1}{3}\right)} = \dfrac{12\cdot 1 - 12\cdot \dfrac{3}{4}}{12\cdot \dfrac{1}{2} + 12\cdot \dfrac{1}{3}}$ Distribute.

$= \dfrac{12-9}{6+4}$ Multiply.

$= \dfrac{3}{10}$ Simplify.

Do you remember the difference between an equation and an expression? An equation contains an equal sign, and an expression does not. Remember that we can solve equations, and we can perform operations with rational expressions.

YOU TRY 1

a) Write in lowest terms: $\dfrac{2m^2 - 7m + 5}{1 - m^2}$

b) Subtract $\dfrac{a}{a + 3} - \dfrac{2}{a}$.

c) Multiply $\dfrac{x^3 - 8}{9} \cdot \dfrac{3x + 6}{x^2 - 4}$.

d) Solve $\dfrac{r}{r + 3} + 5 = \dfrac{12}{r + 3}$.

e) For $f(k) = \dfrac{3k - 4}{k^2 + 5k}$,
   i) find k so that $f(k) = 0$.
   ii) determine the domain of the function.

f) Simplify $\dfrac{\dfrac{w^2 - 25}{8w}}{\dfrac{3w + 15}{6w}}$.

ANSWERS TO YOU TRY EXERCISES

1) a) $\dfrac{5 - 2m}{m + 1}$ b) $\dfrac{a^2 - 2a - 6}{a(a + 3)}$ c) $\dfrac{x^2 + 2x + 4}{3}$ d) $\left\{-\dfrac{1}{2}\right\}$

e) i) $\dfrac{4}{3}$ ii) $(-\infty, -5) \cup (-5, 0) \cup (0, \infty)$ f) $\dfrac{w - 5}{4}$

Putting It All Together Exercises

 Do the exercises, and check your work.

Objective 1: Review the Concepts Presented in Sections 8.1–8.4

For each rational function,
a) find $f(2)$, if possible.
b) find x so that $f(x) = 0$.
c) determine the domain of the function.

1) $f(x) = \dfrac{4x}{x^2 - 9}$

2) $f(x) = \dfrac{3 - 5x}{x^2 + 2x - 8}$

3) $f(x) = \dfrac{12}{2x - 1}$

4) $f(x) = \dfrac{8}{5x}$

5) $f(x) = \dfrac{x - 1}{x^2 + 2}$

6) $f(x) = \dfrac{49 - x^2}{15}$

Write each rational expression in lowest terms.

7) $\dfrac{36n^9}{27n^{12}}$

8) $\dfrac{8w^{12}}{16w^7}$

9) $\dfrac{2j + 5}{2j^2 - 3j - 20}$

10) $\dfrac{m^2 + 5m - 24}{2m^2 + 2m - 24}$

Putting It All Together

11) $\dfrac{12 - 15n}{5n^2 + 6n - 8}$

12) $\dfrac{-x - y}{xy + y^2 + 3x + 3y}$

Perform the operations, and simplify.

13) $\dfrac{5}{f + 8} - \dfrac{2}{f}$

14) $\dfrac{2c^2 - 4c - 6}{c + 1} \div \dfrac{3c - 9}{7}$

15) $\dfrac{9a^3}{10b} \cdot \dfrac{40b^2}{81a}$

16) $\dfrac{4j}{j^2 - 81} + \dfrac{3}{j^2 - 3j - 54}$

17) $\dfrac{3}{q^2 - q - 20} + \dfrac{8q}{q^2 + 11q + 28}$

18) $\dfrac{12y^3}{4z} \cdot \dfrac{8z^4}{72y^6}$

19) $\dfrac{16 - m^2}{m + 4} \div \dfrac{8m - 32}{m + 7}$

20) $\dfrac{12p}{4p^2 + 11p + 6} - \dfrac{5}{p^2 - 4p - 12}$

21) $\dfrac{13}{r - 8} + \dfrac{4}{8 - r}$

22) $\dfrac{1}{4y} + \dfrac{7}{6y^2}$

23) $\dfrac{a^2 - 4}{a^3 + 8} \cdot \dfrac{5a^2 - 10a + 20}{3a - 6}$

24) $\dfrac{7}{d + 9} + \dfrac{6}{d^2}$

25) $\dfrac{10}{x - 8} + \dfrac{4}{x + 3}$

26) $\dfrac{xy - 4x + 3y - 12}{y^2 - 16} \div \dfrac{x^3 + 27}{6}$

27) $\dfrac{13}{5z} - \dfrac{1}{3z}$

28) $\dfrac{6}{k + 2} - \dfrac{1}{k + 5}$

29) $\dfrac{10q}{8p - 10q} + \dfrac{8p}{10q - 8p}$

30) $\dfrac{m}{7m - 4n} - \dfrac{20n}{4n - 7m}$

31) $\dfrac{6u + 1}{3u^2 - 2u} - \dfrac{u}{3u^2 + u - 2} + \dfrac{10}{u^2 + u}$

32) $\dfrac{2p + 3}{p^2 + 7p} - \dfrac{4p}{p^2 - p - 56} + \dfrac{5}{p^2 - 8p}$

33) $\dfrac{x}{2x^2 - 7x - 4} - \dfrac{x + 3}{4x^2 + 4x + 1}$

34) $\dfrac{f - 12}{f - 7} - \dfrac{5}{7 - f}$

35) $\left(\dfrac{2c}{c + 8} + \dfrac{4}{c - 2}\right) \div \dfrac{6}{5c + 40}$

36) $\left(\dfrac{3n}{3n - 1} - \dfrac{4}{n + 4}\right) \cdot \dfrac{9n^2 - 1}{21n^2 + 28}$

37) $\dfrac{3}{w^2 - w} + \dfrac{4}{5w} - \dfrac{3}{w - 1}$

38) $\dfrac{4}{k^2 + 4k} - \dfrac{3}{2k} + \dfrac{1}{k + 4}$

For each rectangle, find a rational expression in simplest form to represent its a) area and b) perimeter.

39) Rectangle with width $\dfrac{x - 3}{4}$ and height $\dfrac{x}{2}$.

40) Rectangle with width $\dfrac{z}{z + 5}$ and height $\dfrac{8}{z + 1}$.

Use $f(x) = \dfrac{x}{3x - 1}$ and $g(x) = \dfrac{2x^2}{9x^2 - 1}$ for Exercises 41 and 42.

41) If $h(x) = f(x) + g(x)$, find $h(x)$ in its simplified form, and determine its domain.

42) If $k(x) = f(x) - g(x)$, find $k(x)$ in its simplified form, and find x so that $k(x) = 0$.

Simplify each complex fraction.

43) $\dfrac{\dfrac{3c^3}{8c + 24}}{\dfrac{6c}{c + 3}}$

44) $\dfrac{\dfrac{3}{8} - \dfrac{1}{4}}{\dfrac{1}{6} + \dfrac{2}{3}}$

45) $\dfrac{\dfrac{5}{m} + \dfrac{2}{m-3}}{1 - \dfrac{4}{m}}$

46) $\dfrac{\dfrac{9k^2 - 1}{14k}}{\dfrac{3k+1}{21k^3}}$

47) $\dfrac{\dfrac{25t^2}{6u}}{\dfrac{10t}{9u^4}}$

48) $\dfrac{\dfrac{3}{xy} - \dfrac{y}{x}}{\dfrac{1}{y} + \dfrac{1}{x}}$

For Exercises 49 and 50, let $h(x) = \dfrac{f(x)}{g(x)}$. Find a simplified form for $h(x)$, and determine its domain.

49) $f(x) = 6x^2 + 43x - 40$, $g(x) = x + 8$

50) $f(x) = x^3 + 7x^2 - 4x - 28$, $g(x) = x^2 - 4$

Solve.

51) $\dfrac{3y}{y+7} - 6 = \dfrac{3}{y+7}$

52) $\dfrac{18}{t} - 2 = \dfrac{10}{t}$

53) $\dfrac{6}{m+5} + 1 = \dfrac{2}{m}$

54) $k - \dfrac{28}{k} = 3$

55) $\dfrac{3}{d+2} + \dfrac{10}{d^2 - 6d - 16} = \dfrac{5}{d-8}$

56) $\dfrac{a^2}{2} = \dfrac{a^2 - 6a}{3}$

57) $\dfrac{4}{3x-1} - 7 = \dfrac{4}{3x-1}$

58) $\dfrac{3}{n+4} = \dfrac{n}{n+6} - \dfrac{2}{n^2 + 10n + 24}$

R Rethink

R1) For which operation, addition or multiplication, do you need an LCD?

R2) For which operations, adding, subtracting, multiplying, or dividing, do you need to factor?

R3) Why do you need to find an LCD when solving rational equations?

8.5 Applications of Rational Equations

Prepare

What are your objectives for Section 8.5?	**Organize** How can you accomplish each objective?
1 Solve Problems Involving Proportions	• Use the **Steps for Solving Applied Problems.** • Understand how to write a *proportion*. • Complete the given examples on your own. • Complete You Trys 1 and 2.
2 Solve Problems Involving Distance, Rate, and Time	• Use the **Steps for Solving Applied Problems.** • Review how to use the *distance formula*, $d = rt$. • Make a chart to organize your information. • Complete the given example on your own. • Complete You Try 3.
3 Solve Problems Involving Work	• Use the **Steps for Solving Applied Problems.** • Learn the procedure for **Solving Work Problems.** • Complete the given example on your own. • Complete You Try 4.

Read the explanations, follow the examples, take notes, and complete the You Trys.

We have studied applications of linear and quadratic equations. Now we turn our attention to applications involving equations with rational expressions. We will continue to use the Steps for Solving Applied Problems outlined in Section 2.1.

1 Solve Problems Involving Proportions

Let's begin this section with two problems involving a proportion. A **proportion** is a statement that two ratios are equal. To solve a proportion, set the cross products equal and solve.

EXAMPLE 1

Write an equation, and solve.

If 3 lb of potatoes cost $1.77, how much would 5 lb of potatoes cost?

Solution

Step 1: **Read** the problem carefully, and identify what we are being asked to find.

We must find the cost of 5 lb of potatoes.

Step 2: **Choose a variable** to represent the unknown.

x = the cost of 5 lb of potatoes

> **W Hint**
> Notice that the numerators contain the same quantities and the denominators contain the same quantities.

Step 3: **Translate** the information that appears in English into an algebraic equation.

Write a proportion. We will write our ratios in the form of $\dfrac{\text{pounds of potatoes}}{\text{cost of potatoes}}$ so that the numerators contain the same quantities and the denominators contain the same quantities.

$$\text{pounds of potatoes} \to \dfrac{3}{1.77} = \dfrac{5}{x} \leftarrow \text{pounds of potatoes}$$
$$\text{cost of potatoes} \to \phantom{\dfrac{3}{1.77}} \phantom{\dfrac{5}{x}} \leftarrow \text{cost of potatoes}$$

The equation is $\dfrac{3}{1.77} = \dfrac{5}{x}$.

Step 4: Solve the equation.

$$\dfrac{3}{1.77} \underset{\text{Multiply.}}{\overset{\text{Multiply.}}{\times}} \dfrac{5}{x}$$

$3x = 1.77 \cdot 5$ Set the cross products equal.
$3x = 8.85$ Multiply.
$x = 2.95$ Divide by 3.

Step 5: **Check** the answer, and **interpret** the solution as it relates to the problem.

The cost of 5 lb of potatoes is $2.95. The check is left to the student.

[YOU TRY 1] How much would a customer pay for 16 gal of unleaded gasoline if another customer paid $23.76 for 12 gal at the same gas station?

EXAMPLE 2

Write an equation, and solve.

One morning at a coffee shop, the ratio of the number of customers who ordered regular coffee to the number who ordered decaffeinated coffee was 4 to 1. If the number of people who ordered regular coffee was 126 more than the number who ordered decaf, how many people ordered each type of coffee?

Solution

Step 1: **Read** the problem carefully, and identify what we are being asked to find.

We must find the number of customers who ordered regular coffee and the number who ordered decaffeinated coffee.

Step 2: **Choose a variable** to represent the unknown, and define the other unknown in terms of this variable.

$x =$ number of people who ordered decaffeinated coffee
$x + 126 =$ number of people who ordered regular coffee

Step 3: **Translate** the information that appears in English into an algebraic equation. Write a proportion. We will write our ratios so that the numerators contain the same quantities and the denominators contain the same quantities.

number who ordered regular coffee → $\dfrac{4}{1}$ = $\dfrac{x + 126}{x}$ ← number who ordered regular coffee
number who ordered decaf coffee →   ← number who ordered decaf coffee

The equation is $\dfrac{4}{1} = \dfrac{x + 126}{x}$.

Step 4: Solve the equation.

$$\dfrac{4}{1} \diagdown\!\!\!\!\!\diagup \dfrac{x+126}{x}$$

$4x = 1(x + 126)$ Set the cross products equal.
$4x = x + 126$
$3x = 126$ Subtract x.
$x = 42$ Divide by 3.

Step 5: **Check** the answer, and **interpret** the solution as it relates to the problem.

Therefore, 42 customers ordered decaffeinated coffee and $42 + 126 = 168$ people ordered regular coffee. The check is left to the student.

[**YOU TRY 2**] *Write an equation, and solve.*
During one week at a bookstore, the ratio of the number of romance novels sold to the number of travel books sold was 5 to 3. Determine the number of each type of book sold if customers bought 106 more romance novels than travel books.

2 Solve Problems Involving Distance, Rate, and Time

In Chapter 2 we solved problems involving distance (d), rate (r), and time (t).
The basic formula is $d = rt$. We can solve this formula for r and then for t to obtain

$$r = \dfrac{d}{t} \quad \text{and} \quad t = \dfrac{d}{r}.$$

In this section, we will encounter problems involving boats going with and against a current, and planes going with and against the wind. Both scenarios use the same idea.

Say a boat's speed is 18 mph in still water. If that same boat had a 4 mph current pushing *against* it, how fast would it be traveling? (The current will cause the boat to slow down.)

Speed *against* the current = 18 mph − 4 mph
= 14 mph

$$\dfrac{\text{Speed } against}{\text{the current}} = \dfrac{\text{Speed in}}{\text{still water}} - \dfrac{\text{Speed of}}{\text{the current}}$$

W Hint
Be sure you understand this concept before reading Example 3.

If the speed of the boat in still water is 18 mph and a 4 mph current is *pushing* the boat, how fast would the boat be traveling *with* the current? (The current will cause the boat to travel faster.)

$$\text{Speed } with \text{ the current} = 18 \text{ mph} + 4 \text{ mph}$$
$$= 22 \text{ mph}$$

$$\frac{\text{Speed } with}{\text{the current}} = \frac{\text{Speed in}}{\text{still water}} + \frac{\text{Speed of}}{\text{the current}}$$

A boat traveling *against* the current is said to be traveling *upstream*. A boat traveling *with* the current is said to be traveling *downstream*. We will use these ideas in Example 3.

EXAMPLE 3

Write an equation, and solve.
A boat can travel 8 mi downstream in the same amount of time it can travel 6 mi upstream. If the speed of the current is 2 mph, what is the speed of the boat in still water?

Solution

Step 1: **Read** the problem carefully, and identify what we are being asked to find.

First, we must understand that "8 mi downstream" means 8 mi *with the current*, and "6 mi upstream" means 6 mi *against the current*.

We must find the speed of the boat in still water.

Step 2: **Choose a variable** to represent the unknown, and define the other unknowns in terms of this variable.

x = the speed of the boat in still water
$x + 2$ = the speed of the boat *with* the current (downstream)
$x - 2$ = the speed of the boat *against* the current (upstream)

Step 3: **Translate** from English to an algebraic equation. Use a table to organize the information.

First, fill in the distances and the rates (or speeds).

	d	r	t
Downstream	8	$x + 2$	
Upstream	6	$x - 2$	

Next we must write expressions for the time it takes the boat to go downstream and upstream. We know that $d = rt$, so if we solve for t we get $t = \dfrac{d}{r}$.

Substitute the information from the table to get the expressions for the time.

$$\text{Downstream: } t = \frac{d}{r} = \frac{8}{x + 2} \qquad \text{Upstream: } t = \frac{d}{r} = \frac{6}{x - 2}$$

Put these expressions into the table.

	d	r	t
Downstream	8	$x + 2$	$\dfrac{8}{x + 2}$
Upstream	6	$x - 2$	$\dfrac{6}{x - 2}$

SECTION 8.5 **Applications of Rational Equations**

The problem states that it takes the boat the *same amount of time* to travel 8 mi downstream as it does to go 6 mi upstream. We can write an equation in English:

$$\begin{matrix} \text{Time for boat to go} \\ \text{8 mi downstream} \end{matrix} = \begin{matrix} \text{Time for boat to go} \\ \text{6 mi upstream} \end{matrix}$$

Looking at the table, we can write the algebraic equation using the expressions for time. The equation is $\dfrac{8}{x+2} = \dfrac{6}{x-2}$.

Step 4: **Solve** the equation.

$$\dfrac{8}{x+2} \diagdown\!\!\!\!\diagup \dfrac{6}{x-2}$$

$$\begin{aligned} 8(x-2) &= 6(x+2) & &\text{Set the cross products equal.} \\ 8x - 16 &= 6x + 12 & &\text{Distribute.} \\ 2x &= 28 \\ x &= 14 & &\text{Solve.} \end{aligned}$$

Step 5: **Check** the answer, and **interpret** the solution as it relates to the problem.

The speed of the boat in still water is 14 mph.

Check: The speed of the boat going downstream is $14 + 2 = 16$ mph, so the time to travel downstream is

$$t = \dfrac{d}{r} = \dfrac{8}{16} = \dfrac{1}{2} \text{ hr}$$

The speed of the boat going upstream is $14 - 2 = 12$ mph so the time to travel downstream is

$$t = \dfrac{d}{r} = \dfrac{6}{12} = \dfrac{1}{2} \text{ hr}$$

So, time upstream = time downstream. ✓

> **[YOU TRY 3]** *Write an equation, and solve.*
> It takes a boat the same amount of time to travel 10 mi upstream as it does to travel 15 mi downstream. Find the speed of the boat in still water if the speed of the current is 4 mph.

3 Solve Problems Involving Work

 Hint
Be sure you understand these concepts before you do Example 4.

Suppose it takes Brian 5 hr to paint his bedroom. What is the *rate* at which he does the job?

$$\text{rate} = \dfrac{1 \text{ room}}{5 \text{ hr}} = \dfrac{1}{5} \text{ room/hr}$$

Brian works at a rate of $\dfrac{1}{5}$ of a room per hour.

In general, we can say that if it takes t units of time to do a job, then the *rate* at which the job is done is $\dfrac{1}{t}$ job per unit of time.

This idea of *rate* is what we use to determine how long it can take for two or more people or things to do a job.

Let's assume, again, that Brian can paint his room in 5 hr. At this rate, how much of the job can he do in 2 hr?

$$\text{Fractional part of the job done} = \text{Rate of work} \cdot \text{Amount of time worked}$$

$$= \frac{1}{5} \cdot 2$$

$$= \frac{2}{5}$$

He can paint $\frac{2}{5}$ of the room in 2 hr.

Procedure Solving Work Problems

The basic equation used to solve work problems is:

$$\text{Fractional part of a job done by one person or thing} + \text{Fractional part of a job done by another person or thing} = 1 \text{ (whole job)}$$

EXAMPLE 4

Write an equation, and solve.

If Brian can paint his bedroom in 5 hr but his brother, Doug, could paint the room on his own in 4 hr, how long would it take for the two of them to paint the room together?

Solution

Step 1: Read the problem carefully, and identify what we are being asked to find.

We must determine how long it would take Brain and Doug to paint the room together.

Step 2: Choose a variable to represent the unknown.

t = the number of hours to paint the room together

Step 3: Translate the information that appears in English into an algebraic equation.

Let's write down their rates:

Brian's rate = $\frac{1}{5}$ room/hr (since the job takes him 5 hr)

Doug's rate = $\frac{1}{4}$ room/hr (since the job takes him 4 hr)

It takes them t hr to paint the room together. Recall that

$$\text{Fractional part of job done} = \text{Rate of work} \cdot \text{Amount of time worked}$$

Brian's fractional part = $\frac{1}{5} \cdot t = \frac{1}{5}t$

Doug's fractional part = $\frac{1}{4} \cdot t = \frac{1}{4}t$

The equation we can write comes from

$$\underset{\text{job done by Brian}}{\text{Fractional part of the}} + \underset{\text{job done by Doung}}{\text{Fractional part of the}} = 1 \text{ (whole job)}$$

$$\frac{1}{5}t + \frac{1}{4}t = 1$$

Step 4: Solve the equation.

$$20\left(\frac{1}{5}t + \frac{1}{4}t\right) = 20(1) \quad \text{Multiply by the LCD of 20 to eliminate the fractions.}$$
$$20\left(\frac{1}{5}t\right) + 20\left(\frac{1}{4}t\right) = 20(1) \quad \text{Distribute.}$$
$$4t + 5t = 20 \quad \text{Multiply.}$$
$$9t = 20 \quad \text{Combine like terms.}$$
$$t = \frac{20}{9} \quad \text{Divide by 9.}$$

Step 5: *Check* the answer, and **interpret** the solution as it relates to the problem.

Brian and Doug could paint the room together in $\frac{20}{9}$ hr or $2\frac{2}{9}$ hr.

Check: $\underset{\text{job done by Brian}}{\text{Fractional part of the}} + \underset{\text{job done by Doung}}{\text{Fractional part of the}} = 1 \text{ whole job}$

$$\frac{1}{5} \cdot \left(\frac{20}{9}\right) + \frac{1}{4} \cdot \left(\frac{20}{9}\right) \stackrel{?}{=} 1$$
$$\frac{4}{9} + \frac{5}{9} = 1 \checkmark$$

[YOU TRY 4] Write an equation, and solve.
Krutesh can mow a lawn in 2 hr while it takes Stefan 3 hr to mow the same lawn. How long would it take for them to mow the lawn if they worked together?

ANSWERS TO [YOU TRY] EXERCISES

1) $31.68   2) 159 travel books, 265 romance novels   3) 20 mph   4) $\frac{6}{5}$ hr or $1\frac{1}{5}$ hr

E Evaluate 8.5 Exercises

Do the exercises, and check your work.

Objective 1: Solve Problems Involving Proportions

Write an equation for each, and solve. See Examples 1 and 2.

1) Hector buys batteries for his Gameboy. He knows that 3 packs of batteries cost $4.26. How much should Hector expect to pay for 5 packs of batteries?

2) If 8 oz of granola costs $1.98, find the cost of 20 oz of granola.

3) A 12-oz serving of Mountain Dew contains 55 mg of caffeine. How much caffeine is in an 18-oz serving of Mountain Dew? (www.nsda.org)

4) An 8-oz serving of Coca-Cola Classic contains 23 mg of caffeine. How much caffeine is in a 12-oz serving of Coke? (www.nsda.org)

5) A nurse sets an intravenous fluid drip rate at 1000 ml every 8 hr. How much fluid would the patient receive in 3 hr?

6) A medication is to be given to a patient at a rate of 2.8 mg for every 40 lb of body weight. How much medication should be given to a patient who weighs 190 lb?

7) At a motocross race, the ratio of male spectators to female spectators was 10 to 3. If there were 370 male spectators, how many females were in the crowd?

8) The ratio of students in a history lecture who took notes in pen to those who took notes in pencil was 8 to 3. If 72 students took notes in pen, how many took notes in pencil?

9) In a gluten-free flour mixture, the ratio of potato-starch flour to tapioca flour is 2 to 1. If a mixture contains 3 more cups of potato-starch flour than tapioca flour, how much of each type of flour is in the mixture?

10) Rosa Cruz won an election over her opponent by a ratio of 6 to 5. If her opponent received 372 fewer votes than she did, how many votes did each candidate receive?

11) The ancient Greeks believed that the rectangle most pleasing to the eye, the golden rectangle, had sides in which the ratio of its  length to its width was approximately 8 to 5. They erected many buildings using this golden ratio, including the Parthenon. The marble floor of a museum foyer is to be designed as a golden rectangle. If its width is to be 18 ft less than its length, find the length and width of the foyer.

12) To obtain a particular color, a painter mixed two colors in a ratio of 7 parts blue to 3 parts yellow. If he used 8 fewer gallons of yellow than blue, how many gallons of blue paint did he use?

13) Ms. Hiramoto has invested her money so that the ratio of the amount in bonds to the amount in stocks is 3 to 2. If she has $4000 more invested in bonds than in stocks, how much does she have invested in each?

14) At a wildlife refuge, the ratio of deer to rabbits is 4 to 9. Determine the number of deer and rabbits at the refuge if there are 40 more rabbits than deer.

15) In a small town, the ratio of households with pets to those without pets is 5 to 4. If 271 more households have pets than do not, how many households have pets?

16) An industrial cleaning solution calls for 5 parts water to 2 parts concentrated cleaner. If a worker uses 15 more quarts of water than concentrated cleaner to make a solution,

   a) how much concentrated cleaner did she use?

   b) how much water did she use?

   c) how much solution did she make?

Objective 2: Solve Problems Involving Distance, Rate, and Time

17) If the speed of a boat in still water is 10 mph,

   a) what is its speed going *against* a 3 mph current?

   b) what is its speed *with* a 3 mph current?

18) If an airplane travels at a constant rate of 300 mph,

   a) what is its speed going *into* a 25 mph wind?

   b) what is its speed going *with* a 25 mph wind?

19) If an airplane travels at a constant rate of x mph,

   a) what is its speed going *with* a 30 mph wind?

   b) what is its speed going *against* a 30 mph wind?

20) If the speed of a boat in still water is 13 mph,

   a) what is its speed going *against* a current with a rate of x mph?

   b) what is its speed going *with* a current with a rate of x mph?

Write an equation for each, and solve. See Example 3.

21) A current flows at 5 mph. A boat can travel 20 mi downstream in the same amount of time it can go 12 mi upstream. What is the speed of the boat in still water?

22) With a current flowing at 4 mph, a boat can travel 32 mi with the current in the same amount of time it can go 24 mi against the current. Find the speed of the boat in still water.

23) A boat travels at 16 mph in still water. It takes the same amount of time for the boat to travel 15 mi downstream as to go 9 mi upstream. Find the speed of the current.

24) A boat can travel 12 mi downstream in the time it can go 6 mi upstream. If the speed of the boat in still water is 9 mph, what is the speed of the current?

25) An airplane flying at constant speed can fly 350 mi with the wind in the same amount of time it can fly 300 mi against the wind. What is the speed of the plane if the wind blows at 20 mph?

26) When the wind is blowing at 25 mph, a plane flying at a constant speed can travel 500 mi with the wind in the same amount of time it can fly 400 mi against the wind. Find the speed of the plane.

27) In still water the speed of a boat is 10 mph. Against the current it can travel 4 mi in the same amount of time it can travel 6 mi with the current. What is the speed of the current?

28) Flying at a constant speed, a plane can travel 800 mi with the wind in the same amount of time it can fly 650 mi against the wind. If the wind blows at 30 mph, what is the speed of the plane?

Objective 3: Solve Problems Involving Work

29) Toby can finish a computer programming job in 4 hr. What is the rate at which he does the job?

30) It takes Crystal 3 hr to paint her backyard fence. What is the rate at which she works?

31) Eloise can fertilize her lawn in t hr. What is the rate at which she does this job?

32) It takes Manu twice as long to clean a pool as it takes Anders. If it takes Anders t hr to clean the pool, at what rate does Manu do the job?

Write an equation for each, and solve. See Example 4.

33) It takes Arlene 2 hr to trim the bushes at a city park while the same job takes Andre 3 hr. How long would it take for them to do the job together?

34) A hot water faucet can fill a sink in 8 min while it takes the cold water faucet only 6 min. How long would it take to fill the sink if both faucets were on?

35) Jermaine and Sue must put together notebooks for each person attending a conference. Working alone it would take Jermaine 5 hr while it would take Sue 8 hr. How long would it take for them to assemble the notebooks together?

36) The Williams family has two printers on which they can print out their vacation pictures. The larger printer can print all of the pictures in 3 hr, while it would take 5 hr on the smaller printer. How long would it take to print the vacation pictures using both printers?

37) A faucet can fill a tub in 12 min. The leaky drain can empty the tub in 30 min. If the faucet is on and the drain is leaking, how long would it take to fill the tub?

38) It takes Deepak 50 min to shovel snow from his sidewalk and driveway. When he works with his brother, Kamal, it takes only 30 min. How long would it take Kamal to do the shoveling himself?

39) Fatima and Antonio must cut out shapes for an art project at a day-care center. Fatima can do the job twice as fast as Antonio. Together, it takes 2 hr to cut out all of the shapes. How long would it take Fatima to cut out the shapes herself?

40) It takes Burt three times as long as Phong to set up a new alarm system. Together they can set it up in 90 min. How long would it take Phong to set up the alarm system by himself?

41) Working together it takes 2 hr for a new worker and an experienced worker to paint a billboard. If the new employee worked alone, it would take him 6 hr. How long would it take the experienced worker to paint the billboard by himself?

42) Audrey can address party invitations in 40 min, while it would take her mom 1 hr. How long would it take for them to address the invitations together?

43) Homer uses the moving walkway to get to his gate at the airport. He can travel 126 ft when he is walking on the moving walkway in the same amount of time it would take for him to walk only 66 ft on the floor next to it. If the walkway is moving at 2 ft/sec, how fast does Homer walk?

44) Another walkway at the airport moves at $2\frac{1}{2}$ ft/sec. If Bart can travel 140 ft when he is walking on the moving walkway in the same amount of time he can walk 80 ft on the floor next to it, how fast does Bart walk?

Extension
Another application of proportions is for solving similar triangles.

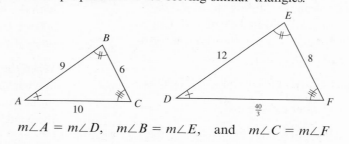

$m\angle A = m\angle D$, $m\angle B = m\angle E$, and $m\angle C = m\angle F$

△ABC and △DEF are *similar triangles*. Two triangles are **similar** if they have the same shape. The corresponding angles have the same measure, and the corresponding sides are *proportional*.

The ratio of each pair of corresponding sides is $\frac{3}{4}: \frac{9}{12} = \frac{3}{4}; \frac{6}{8} = \frac{3}{4}; \frac{10}{\frac{40}{3}} = 10 \cdot \frac{3}{40} = \frac{3}{4}.$

We can set up and solve a proportion to find the length of an unknown side in two similar triangles.

Given the following similar triangles, find *x*.

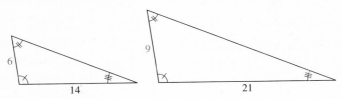

$\frac{6}{9} = \frac{x}{24}$ Set the ratios of the two corresponding sides equal to each other. (Set up a proportion.)

$9x = 6 \cdot 24$ Solve the proportion.

$9x = 144$ Multiply.

$x = 16$ Divide by 9.

Find the length of the indicated side, *x*, by setting up a proportion.

45)

46)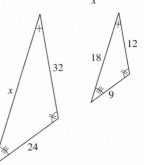

47)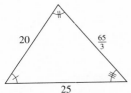

48)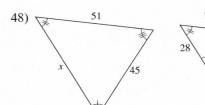

R Rethink

R1) Which of the application problems are hardest for you to understand?

R2) Why do the units for time have to be the same within a work problem?

R3) Can you think of an example of a *work problem* that you encountered this past week?

8.6 Variation

P Prepare

What are your objectives for Section 8.6?

1. Solve Direct Variation Problems
2. Solve Inverse Variation Problems
3. Solve Joint Variation Problems
4. Solve Combined Variation Problems

O Organize

How can you accomplish each objective?

- Learn the definition of *direct variation*.
- Learn the procedure for **Solving a Variation Problem.**
- Complete the given examples on your own.
- Complete You Trys 1–3.

- Learn the definition of *inverse variation*.
- Follow the procedure for **Solving a Variation Problem.**
- Complete the given examples on your own.
- Complete You Trys 4 and 5.

- Learn the definition of *joint variation*.
- Follow the procedure for **Solving a Variation Problem.**
- Complete the given example on your own.
- Complete You Try 6.

- Follow the procedure for **Solving a Variation Problem.**
- Complete the given example on your own.
- Complete You Try 7.

Read the explanations, follow the examples, take notes, and complete the You Trys.

1 Solve Direct Variation Problems

In Section 4.5 we discussed the following situation:

If you are driving on a highway at a constant speed of 60 mph, the distance you travel depends on the amount of time spent driving.

Let y = the distance traveled, in miles, and let x = the number of hours spent driving. An equation relating x and y is $y = 60x$, and y is a function of x.

We can make a table of values relating x and y. We can say that the distance traveled, y, is *directly proportional to* the time spent traveling, x. Or y *varies directly as* x.

x	y
1	60
1.5	90
2	120
3	180

Definition

Direct Variation: y varies directly as x (or y is directly proportional to x) means

$$y = kx$$

where k is a nonzero real number. k **is called the constant of variation.**

488 CHAPTER 8 Rational Expressions, Equations, and Functions

If two quantities vary directly and $k > 0$, then as one quantity increases the other increases as well. And, as one quantity decreases, the other decreases.

In our example of driving distance, $y = 60x$, 60 is the *constant of variation*. Given information about how variables are related, we can write an equation and solve a variation problem.

EXAMPLE 1

Suppose y varies directly as x. If $y = 18$ when $x = 3$,

a) find the constant of variation, k.

b) write a variation equation relating x and y using the value of k found in a).

c) find y when $x = 11$.

Solution

a) To find the constant of variation, write a *general* variation equation relating x and y. y varies directly as x means $y = kx$.

We are told that $y = 18$ when $x = 3$. Substitute these values into the equation, and solve for k.

$$y = kx$$
$$18 = k(3) \quad \text{Substitute 3 for } x \text{ and 18 for } y.$$
$$6 = k \quad \text{Divide by 3.}$$

b) The *specific* variation equation is the equation obtained when we substitute 6 for k in $y = kx$: Therefore, $y = 6x$.

c) To find y when $x = 11$, substitute 11 for x in $y = 6x$ and evaluate.

$$y = 6x$$
$$= 6(11) \quad \text{Substitute 11 for } x.$$
$$= 66 \quad \text{Multiply.}$$

W Hint
Learn this procedure, and write it in your notes.

Procedure Solving a Variation Problem

Step 1: Write the *general* variation equation.

Step 2: Find k by substituting the known values into the equation and solving for k.

Step 3: Write the *specific* variation equation by substituting the value of k into the *general* variation equation.

Step 4: Use the specific variation equation to solve the problem.

[YOU TRY 1]

Suppose y varies directly as x. If $y = 40$ when $x = 5$,

a) find the constant of variation.

b) write the specific variation equation relating x and y.

c) find y when $x = 3$.

EXAMPLE 2   Suppose p varies directly as the square of z. If $p = 12$ when $z = 2$, find p when $z = 5$.

Solution

Step 1: Write the *general* variation equation.

p varies directly as the *square* of z means $p = kz^2$.

Step 2: Find k using the known values: $p = 12$ when $z = 2$.

$$p = kz^2$$
$$12 = k(2)^2 \quad \text{Substitute 2 for } z \text{ and 12 for } p.$$
$$12 = k(4)$$
$$3 = k$$

W Hint
Write out these examples, using the *steps*, as you are reading them.

Step 3: Substitute $k = 3$ into $p = kz^2$ to get the *specific* variation equation, $p = 3z^2$.

Step 4: We are asked to find p when $z = 5$. Substitute $z = 5$ into $p = 3z^2$ to get p.

$$p = 3z^2$$
$$= 3(5)^2 \quad \text{Substitute 5 for } z.$$
$$= 3(25)$$
$$= 75$$

[YOU TRY 2]   Suppose w varies directly as the cube of n. If $w = 135$ when $n = 3$, find w when $n = 2$.

EXAMPLE 3   A theater's nightly revenue varies directly as the number of tickets sold. If the revenue from the sale of 80 tickets is $3360, find the revenue from the sale of 95 tickets.

Solution

Let n = the number of tickets sold, and let R = revenue.

We will follow the four steps for solving a variation problem.

Step 1: Write the *general* variation equation: $R = kn$.

Step 2: Find k using the known values: $R = 3360$ when $n = 80$.

$$R = kn$$
$$3360 = k(80) \quad \text{Substitute 80 for } n \text{ and 3360 for } R.$$
$$42 = k \quad \text{Divide by 80.}$$

Step 3: Substitute $k = 42$ into $R = 42n$ to get the *specific* variation equation, $R = 42n$.

Step 4: We must find the revenue from the sale of 95 tickets. Substitute $n = 95$ into $R = 42n$ to find R.

$$R = 42n$$
$$R = 42(95)$$
$$R = 3990$$

The revenue is $3990.

[YOU TRY 3]   The cost to carpet a room varies directly as the area of the room. If it costs $525.00 to carpet a room of area 210 ft^2, how much would it cost to carpet a room of area 288 ft^2?

2 Solve Inverse Variation Problems

If two quantities vary *inversely* (are *inversely* proportional) then as one value increases, the other decreases. Likewise, as one value decreases, the other increases.

> **Definition**
>
> **Inverse Variation: *y* varies inversely as *x*** (or ***y* is inversely proportional to *x***) means
>
> $$y = \frac{k}{x}$$
>
> where *k* is a nonzero real number. ***k* is the constant of variation.**

A good example of inverse variation is the relationship between the time, *t*, it takes to travel a given distance, *d*, as a function of the rate (or speed), *r*. We can define this relationship as $t = \frac{d}{r}$. As the rate, *r*, increases, the time, *t*, that it takes to travel *d* mi decreases. Likewise, as *r* decreases, the time, *t*, that it takes to travel *d* mi increases. Therefore, *t* varies *inversely* as *r*.

EXAMPLE 4

Suppose *q* varies inversely as *h*. If $q = 4$ when $h = 15$, find *q* when $h = 10$.

Solution

Step 1: Write the *general* variation equation, $q = \frac{k}{h}$.

Step 2: Find *k* using the known values: $q = 4$ when $h = 15$.

$$q = \frac{k}{h}$$
$$4 = \frac{k}{15} \quad \text{Substitute 15 for } h \text{ and 4 for } q.$$
$$60 = k \quad \text{Multiply by 15.}$$

Step 3: Substitute $k = 60$ into $q = \frac{k}{h}$ to get the *specific* variation equation, $q = \frac{60}{h}$.

Step 4: Substitute 10 for *h* in $q = \frac{60}{h}$ to find *q*.

$$q = \frac{60}{10} = 6$$

[YOU TRY 4] Suppose *m* varies inversely as the square of *v*. If $m = 1.5$ when $v = 4$, find *m* when $v = 2$.

EXAMPLE 5

The intensity of light (in lumens) varies inversely as the square of the distance from the source. If the intensity of the light is 40 lumens 5 ft from the source, what is the intensity of the light 4 ft from the source?

Solution

Let d = distance from the source (in feet), and let I = intensity of the light (in lumens).

Step 1: Write the *general* variation equation, $I = \dfrac{k}{d^2}$.

Step 2: Find k using the known values: $I = 40$ when $d = 5$.

$$I = \frac{k}{d^2}$$
$$40 = \frac{k}{(5)^2} \quad \text{Substitute 5 for } d \text{ and 40 for } I.$$
$$40 = \frac{k}{25}$$
$$1000 = k \quad \text{Multiply by 25.}$$

Step 3: Substitute $k = 1000$ into $I = \dfrac{k}{d^2}$ to get the *specific* variation equation, $I = \dfrac{1000}{d^2}$.

Step 4: Find the intensity, I, of the light 4 ft from the source. Substitute $d = 4$ into $I = \dfrac{1000}{d^2}$ to find I.

$$I = \frac{1000}{(4)^2} = \frac{1000}{16} = 62.5$$

The intensity of the light is 62.5 lumens.

[YOU TRY 5] If the voltage in an electrical circuit is held constant (stays the same), then the current in the circuit varies inversely as the resistance. If the current is 40 amps when the resistance is 3 ohms, find the current when the resistance is 8 ohms.

3 Solve Joint Variation Problems

If a variable varies directly as the *product* of two or more other variables, the first variable *varies jointly* as the other variables.

Definition

Joint Variation: *y* **varies jointly as** *x* **and** *z* **means** $y = kxz$ where k is a nonzero real number.

EXAMPLE 6

For a given amount invested in a bank account (called the principal), the interest earned varies jointly as the interest rate (expressed as a decimal) and the time the principal is in the account. If Graham earns $80 in interest when he invests his money for 1 yr at 4%, how much interest would the same principal earn if he invested it at 5% for 2 yr?

Solution

Let r = interest rate (as a decimal)
t = the number of years the principal is invested
I = interest earned

Step 1: Write the *general* variation equation, $I = krt$.

Step 2: Find k using the known values: $I = 80$ when $t = 1$ and $r = 0.04$.

$$I = krt$$
$$80 = k(0.04)(1) \quad \text{Substitute the values into } I = krt.$$
$$80 = 0.04k$$
$$2000 = k \quad \text{Divide by 0.04.}$$

(The amount he invested, the principal, is $2000.)

Step 3: Substitute $k = 2000$ into $I = krt$ to get the *specific* variation equation, $I = 2000rt$.

Step 4: Find the interest Graham would earn if he invested $2000 at 5% interest for 2 yr. Let $r = 0.05$ and $t = 2$. Solve for I.

$$I = 2000(0.05)(2) \quad \text{Substitute 0.05 for } r \text{ and 2 for } t.$$
$$= 200 \quad \text{Multiply.}$$

Graham would earn $200.

[YOU TRY 6]

The volume of a box of constant height varies jointly as its length and width. A box with a volume of 9 ft³ has a length of 3 ft and a width of 2 ft. Find the volume of a box with the same height, if its length is 4 ft and its width is 3 ft.

4 Solve Combined Variation Problems

A combined variation problem involves both direct and inverse variation.

EXAMPLE 7

Suppose y varies directly as the square root of x and inversely as z. If $y = 12$ when $x = 36$ and $z = 5$, find y when $x = 81$ and $z = 15$.

Solution

Step 1: Write the *general* variation equation.

$$y = \frac{k\sqrt{x}}{z} \quad \begin{array}{l} \leftarrow y \text{ varies directly as the square root of } x. \\ \leftarrow y \text{ varies inversely as } z. \end{array}$$

Step 2: Find k using the known values: $y = 12$ when $x = 36$ and $z = 5$.

$$12 = \frac{k\sqrt{36}}{5} \quad \text{Substitute the values.}$$
$$60 = 6k \quad \text{Multiply by 5; } \sqrt{36} = 6$$
$$10 = k \quad \text{Divide by 6.}$$

Step 3: Substitute $k = 10$ into $y = \dfrac{k\sqrt{x}}{z}$ to get the specific variation equation, $y = \dfrac{10\sqrt{x}}{z}$.

Step 4: Find y when $x = 81$ and $z = 15$.

$$y = \frac{10\sqrt{81}}{15}$$
Substitute 81 for x and 15 for z.

$$y = \frac{10 \cdot 9}{15} = \frac{90}{15} = 6$$

[YOU TRY 7] Suppose a varies directly as b and inversely as the square of c. If $a = 28$ when $b = 12$ and $c = 3$, find a when $b = 36$ and $c = 4$.

ANSWERS TO [YOU TRY] **EXERCISES**

1) a) 8 b) $y = 8x$ c) 24   2) 40   3) $720.00   4) 6   5) 15 amps   6) 18 ft^3 7) 47.25

8.6 Exercises

Do the exercises, and check your work.

Mixed Exercises: Objectives 1–4

1) If z varies directly as y, then as y increases, the value of z _____.

2) If a varies inversely as b, then as b increases, the value of a _____.

Decide whether each equation represents direct, inverse, joint, or combined variation.

3) $y = 6x$

4) $c = 4ab$

5) $f = \dfrac{15}{t}$

6) $z = 3\sqrt{x}$

7) $p = \dfrac{8q^2}{r}$

8) $w = \dfrac{11}{v^2}$

Write a general variation equation using k as the constant of variation.

9) M varies directly as n.

10) q varies directly as r.

11) h varies inversely as j.

12) R varies inversely as B.

13) T varies inversely as the square of c.

14) b varies directly as the cube of w.

15) s varies jointly as r and t.

16) C varies jointly as A and D.

17) Q varies directly as the square root of z and inversely as m.

18) r varies directly as d and inversely as the square of L.

19) Suppose z varies directly as x. If $z = 63$ when $x = 7$,
   a) find the constant of variation.
   b) write the specific variation equation relating z and x.
   c) find z when $x = 6$.

20) Suppose A varies directly as D. If $A = 12$ when $D = 3$,
   a) find the constant of variation.
   b) write the specific variation equation relating A and D.
   c) find A when $D = 11$.

21) Suppose N varies inversely as y. If $N = 4$ when $y = 12$,
   a) find the constant of variation.
   b) write the specific variation equation relating N and y.
   c) find N when $y = 3$.

22) Suppose j varies inversely as m. If $j = 7$ when $m = 9$,
   a) find the constant of variation.
   b) write the specific variation equation relating j and m.
   c) find j when $m = 21$.

23) Suppose Q varies directly as the square of r and inversely as w. If $Q = 25$ when $r = 10$ and $w = 20$,

   a) find the constant of variation.

   b) write the specific variation equation relating Q, r, and w.

   c) find Q when $r = 6$ and $w = 4$.

24) Suppose y varies jointly as a and the square root of b. If $y = 42$ when $a = 3$ and $b = 49$,

   a) find the constant of variation.

   b) write the specific variation equation relating y, a, and b.

   c) find y when $a = 4$ and $b = 9$.

Solve.

25) If B varies directly as R, and $B = 35$ when $R = 5$, find B when $R = 8$.

26) If q varies directly as p, and $q = 10$ when $p = 4$, find q when $p = 10$.

27) If L varies inversely as the square of h, and $L = 8$ when $h = 3$, find L when $h = 2$.

28) If w varies inversely as d, and $w = 3$ when $d = 10$, find w when $d = 5$.

29) If y varies jointly as x and z, and $y = 60$ when $x = 4$ and $z = 3$, find y when $x = 7$ and $z = 2$.

30) If R varies directly as P and inversely as the square of Q, and $R = 5$ when $P = 10$ and $Q = 4$, find R when $P = 18$ and $Q = 3$.

Solve each problem by writing a variation equation.

31) Kosta is paid hourly at his job. His weekly earnings vary directly as the number of hours worked. If Kosta earned $437.50 when he worked 35 hr, how much would he earn if he worked 40 hr?

32) If distance is held constant, the time it takes to travel that distance is inversely proportional to the speed at which one travels. If it takes 14 hr to travel the given distance at 60 mph, how long would it take to travel the same distance at 70 mph?

33) The cost of manufacturing a certain brand of spiral notebook is inversely proportional to the number produced. When 16,000 notebooks are produced, the cost per notebook is $0.60. What is the cost of each notebook when 12,000 are produced?

34) The surface area of a cube varies directly as the square of the length of one of its sides. A cube has a surface area of 54 in^2 when the length of each side is 3 in. What is the surface area of a cube with a side of length 6 in.?

35) The power in an electrical system varies jointly as the current and the square of the resistance. If the power is 100 watts when the current is 4 amps and the resistance is 5 ohms, what is the power when the current is 5 amps and the resistance is 6 ohms?

36) The force exerted on an object varies jointly as the mass and acceleration of the object. If a 20-newton force is exerted on an object of mass 10 kg and an acceleration of 2 m/sec^2, how much force is exerted on a 50 kg object with an acceleration of 8 m/sec^2?

37) The volume of a cylinder varies jointly as its height and the square of its radius. The volume of a cylindrical can is 108π cm^3 when its radius is 3 cm and it is 12 cm high. Find the volume of a cylindrical can with a radius of 4 cm and a height of 3 cm.

38) The kinetic energy of an object varies jointly as its mass and the square of its speed. When a roller coaster car with a mass of 1000 kg is traveling at 15 m/sec, its kinetic energy is 112,500 J (joules). What is the kinetic energy of the same car when it travels at 18 m/sec?

39) The frequency of a vibrating string varies inversely as its length. If a 5-ft-long piano string vibrates at 100 cycles/sec, what is the frequency of a piano string that is 2.5 ft long?

40) The amount of pollution produced varies directly as the population. If a city of 500,000 people produces 800,000 tons of pollutants, how many tons of pollutants would be produced by a city of 1,000,000 people?

41) The resistance of a wire varies directly as its length and inversely as its cross-sectional area. A wire of length 40 cm and cross-sectional area 0.05 cm^2 has a resistance of 2 ohms. Find the resistance of 60 cm of the same type of wire.

42) When a rectangular beam is positioned horizontally, the maximum weight that it can support varies jointly as its width and the square of its thickness and inversely as its length. A beam is $\frac{3}{4}$ ft wide, $\frac{1}{3}$ ft thick, and 8 ft long, and it can support 17.5 tons. How much weight can a similar beam support if it is 1 ft wide, $\frac{1}{2}$ ft thick and 12 ft long?

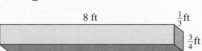

43) Hooke's law states that the force required to stretch a spring is proportional to the distance that the spring is stretched from its original length. A force of 200 lb is required to stretch a spring 5 in. from its natural length. How much force is needed to stretch the spring 8 in. beyond its natural length?

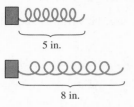

44) The weight of an object on Earth varies inversely as the square of its distance from the center of the Earth. If an object weighs 300 lb on the surface of the Earth (4000 mi from the center), what is the weight of the object if it is 800 mi above the Earth? (Round to the nearest pound.)

 Rethink

R1) Can you think of an example of a direct variation problem that you encountered this week?

R2) What is the purpose of the constant of variation, *k*?

R3) Which variation problem is hardest for you to remember?

Group Activity — The Group Activity can be found online on Connect.

emPOWERme Mad, Mad, Mad Math

Our society has a stereotype of mathematicians as rather dull figures. Yet the history of mathematics is a fascinating one, filled with colorful and often controversial personalities. Select a figure from math's history—you might choose John Nash, Werner Heisenberg, Galileo Galilei, or Hypatia—and write a 200-word essay detailing your thoughts on him or her, and whether he or she fits your initial ideas of what a mathematician was like.

When you are finished, share your piece of writing with a classmate. Ask him or her to be honest in evaluating it. Did your classmate feel he or she learned something from your writing? What about the essay was successful? What might be improved? Also, discuss with your classmate the entire process of writing. Compare thoughts on what was enjoyable and what proved challenging.

Chapter 8: Summary

Definition/Procedure	Example

8.1 Simplifying, Multiplying, and Dividing Rational Expressions and Functions

A **rational expression** is an expression of the form $\frac{P}{Q}$, where P and Q are polynomials and where $Q \neq 0$. (p. 424)

Some examples of rational expressions are
$$\frac{4a-9}{a+2}, \frac{7w^3}{8}, \frac{12}{k^2-9k+18}, \text{ and } \frac{5x+3y}{11xy^2}.$$

Rational Functions

$f(x) = \frac{x-9}{x+2}$ is a rational function because $\frac{x-9}{x+2}$ is a rational expression and because each value that can be substituted for x will produce only one value for the expression.

To determine the value of the variable that makes the function equal zero, set the numerator equal to zero and solve for the variable.

The **domain** of a rational function consists of all real numbers except the value(s) of the variable which make the denominator equal zero. (p. 425)

$f(x) = \frac{x-9}{x+2}$

a) Find x so that $f(x) = 0$.

$\frac{x-9}{x+2} = 0$ when $x - 9 = 0$.

$x - 9 = 0$
$x = 9$

When $x = 9$, $f(x) = 0$.

b) Determine the domain of $f(x)$.

$f(x) = \frac{x-9}{x+2}$

$x + 2 = 0$ Set the denominator $= 0$.
$x = -2$ Solve.

When $x = -2$, the denominator of $f(x) = \frac{x-9}{x+2}$ equals zero. The domain contains all real numbers *except* -2. Write the domain in interval notation as $(-\infty, -2) \cup (-2, \infty)$.

Writing a Rational Expression in Lowest Terms

To write an expression in lowest terms,
1) completely **factor** the numerator and denominator.
2) **divide** the numerator and denominator by the greatest common factor. (p. 427)

Simplify $\frac{2r^2 - 11r + 15}{4r^2 - 36}$.

$\frac{2r^2 - 11r + 15}{4r^2 - 36} = \frac{(2r-5)(r-3)}{4(r+3)(r-3)}$

$= \frac{2r-5}{4(r+3)}$

Simplifying $\frac{a-b}{b-a}$

A rational expression of the form $\frac{a-b}{b-a}$ simplifies to -1. (p. 428)

Simplify $\frac{4-w}{w^2 - 16}$.

$\frac{4-w}{w^2-16} = \frac{\overset{-1}{\cancel{4-w}}}{(w+4)\cancel{(w-4)}}$

$= -\frac{1}{w+4}$

Multiplying Rational Expressions

1) Factor the numerators and denominators.
2) Divide out the common factors, and multiply. (p. 431)

Multiply $\frac{15v^3}{v^2 + 8v + 12} \cdot \frac{2v+12}{5v}$.

$\frac{15v^3}{v^2+8v+12} \cdot \frac{2v+12}{5v} = \frac{\overset{3}{\cancel{15}}v^2 \cdot \cancel{v}}{(v+2)\cancel{(v+6)}} \cdot \frac{2\cancel{(v+6)}}{\cancel{5v}}$

$= \frac{6v^2}{v+2}$

Definition/Procedure	Example
Dividing Rational Expressions To divide rational expressions, multiply the first expression by the reciprocal of the second. (p. 432)	Divide $\dfrac{3x^2 + 4x}{x + 1} \div \dfrac{9x^2 - 16}{21x - 28}$. $\dfrac{3x^2 + 4x}{x + 1} \div \dfrac{9x^2 - 16}{21x - 28} = \dfrac{3x^2 + 4x}{x + 1} \cdot \dfrac{21x - 28}{9x^2 - 16}$ $= \dfrac{x(3x+4)}{x + 1} \cdot \dfrac{7(3x-4)}{(3x+4)(3x-4)}$ $= \dfrac{7x}{x + 1}$

8.2 Adding and Subtracting Rational Expressions

To Find the Least Common Denominator (LCD) **Step 1:** Factor the denominators. **Step 2:** The LCD will contain each unique factor the greatest number of times it appears in any single factorization. **Step 3:** The LCD is the *product* of the factors identified in Step 2. (p. 437)	Find the LCD of $\dfrac{5a}{a^2 + 7a}$ and $\dfrac{2}{a^2 + 14a + 49}$. **Step 1:** $a^2 + 7a = a(a + 7)$ $a^2 + 14a + 49 = (a + 7)^2$ **Step 2:** The factors we will use in the LCD are a and $(a + 7)^2$. **Step 3:** LCD $= a(a + 7)^2$
Adding and Subtracting Rational Expressions 1) Factor the denominators. 2) Write down the LCD. 3) Rewrite each rational expression as an equivalent expression with the LCD. 4) Add or subtract the numerators, and keep the common denominator in factored form. 5) After combining like terms in the numerator ask yourself, "*Can I factor it?*" If so, factor. 6) Simplify the rational expression, if possible. (p. 443)	Add $\dfrac{y}{y + 5} + \dfrac{4y - 30}{y^2 - 25}$. Factor the denominator of $\dfrac{4y - 30}{y^2 - 25}$: $\dfrac{4y - 30}{y^2 - 25} = \dfrac{4y - 30}{(y + 5)(y - 5)}$ The LCD is $(y + 5)(y - 5)$. Rewrite $\dfrac{y}{y + 5}$ with the LCD. $\dfrac{y}{y + 5} \cdot \dfrac{y - 5}{y - 5} = \dfrac{y(y - 5)}{(y + 5)(y - 5)}$ $\dfrac{y}{y + 5} + \dfrac{4y - 30}{y^2 - 25} = \dfrac{y(y - 5)}{(y + 5)(y - 5)} + \dfrac{4y - 30}{(y + 5)(y - 5)}$ $= \dfrac{y(y - 5) + 4y - 30}{(y + 5)(y - 5)}$ $= \dfrac{y^2 - 5y + 4y - 30}{(y + 5)(y - 5)}$ $= \dfrac{y^2 - y - 30}{(y + 5)(y - 5)}$ $= \dfrac{(y+5)(y - 6)}{(y+5)(y - 5)}$ Factor. $= \dfrac{y - 6}{y - 5}$ Simplify.

8.3 Simplifying Complex Fractions

A **complex fraction** is a rational expression that contains one or more fractions in its numerator, its denominator, or both. (p. 450)	Some examples of complex fractions are $\dfrac{\frac{9}{10}}{\frac{3}{2}}$, $\dfrac{\frac{b + 5}{2}}{\frac{4b + 20}{7}}$, and $\dfrac{\frac{1}{x} - \frac{1}{y}}{1 - \frac{x}{y}}$.

Definition/Procedure	Example
To simplify a complex fraction containing one term in the numerator and one term in the denominator, 1) Rewrite the complex fraction as a division problem. 2) Perform the division by multiplying the first fraction by the reciprocal of the second. **(p. 452)**	Simplify $\dfrac{\frac{b+5}{2}}{\frac{4b+20}{7}}$. $\dfrac{\frac{b+5}{2}}{\frac{4b+20}{7}} = \dfrac{b+5}{2} \div \dfrac{4b+20}{7}$ $= \dfrac{b+5}{2} \cdot \dfrac{7}{4(b+5)}$ $= \dfrac{\cancel{b+5}}{2} \cdot \dfrac{7}{4\cancel{(b+5)}}$ $= \dfrac{7}{8}$
To simplify complex fractions containing more than one term in the numerator and/or the denominator, **Method 1** 1) Combine the terms in the numerator, and combine the terms in the denominator so that each contains only one fraction. 2) Rewrite as a division problem. 3) Perform the division. **(p. 452)**	**Method 1** Simplify $\dfrac{\frac{1}{x}-\frac{1}{y}}{1-\frac{x}{y}}$. $\dfrac{\frac{1}{x}-\frac{1}{y}}{1-\frac{x}{y}} = \dfrac{\frac{y}{xy}-\frac{x}{xy}}{\frac{y}{y}-\frac{x}{y}} = \dfrac{\frac{y-x}{xy}}{\frac{y-x}{y}}$ $= \dfrac{y-x}{xy} \div \dfrac{y-x}{y}$ $= \dfrac{\cancel{y-x}}{x\cancel{y}} \cdot \dfrac{\cancel{y}}{\cancel{y-x}} = \dfrac{1}{x}$
Method 2 1) Identify and write down the LCD of *all* of the fractions in the complex fraction. 2) Multiply the numerator and denominator of the complex fraction by the LCD. 3) Simplify. **(p. 454)**	**Method 2** Simplify $\dfrac{\frac{1}{x}-\frac{1}{y}}{1-\frac{x}{y}}$. Identify the LCD: **LCD = xy** Multiply the numerator and denominator by the LCD. $\dfrac{xy\left(\frac{1}{x}-\frac{1}{y}\right)}{xy\left(1-\frac{x}{y}\right)} = \dfrac{xy \cdot \frac{1}{x} - xy \cdot \frac{1}{y}}{xy \cdot 1 - xy \cdot \frac{x}{y}}$ Distribute. $= \dfrac{y-x}{xy-x^2}$ Simplify. $= \dfrac{\cancel{y-x}}{x\cancel{(y-x)}}$ Factor and simplify. $= \dfrac{1}{x}$

Definition/Procedure	Example

8.4 Solving Rational Equations

An equation contains an = sign. To solve a rational equation, **multiply** the equation by the LCD to **eliminate** the denominators, then solve.

Always check the answer to be sure the proposed solution does not make a denominator equal zero. (p. 461)

Solve $\dfrac{n}{n+4} + 1 = \dfrac{20}{n+4}$.

This is an equation because it contains an = sign. We must eliminate the denominators. Identify the LCD of all of the expressions in the equation.

$$\text{LCD} = (n+4)$$

Multiply both sides of the equation by $(n+4)$.

$$(n+4)\left(\dfrac{n}{n+4} + 1\right) = (n+4)\left(\dfrac{20}{n+4}\right)$$

$$\cancel{(n+4)} \cdot \dfrac{n}{\cancel{(n+4)}} + (n+4) \cdot 1 = \cancel{(n+4)} \cdot \dfrac{20}{\cancel{n+4}}$$

$$n + n + 4 = 20$$
$$2n + 4 = 20$$
$$2n = 16$$
$$n = 8$$

The solution set is $\{8\}$. The check is left to the student.

Solving an Equation for a Specific Variable (p. 466)

Solve for n: $x = \dfrac{2a}{n+m}$.

Because we are solving for n, put it in a box.

$$x = \dfrac{2a}{\boxed{n}+m}$$

$$(\boxed{n}+m)x = (\boxed{n}+m) \cdot \dfrac{2a}{\boxed{n}+m}$$

$$(\boxed{n}+m)x = 2a \quad \text{Simplify.}$$
$$\boxed{n}\,x + mx = 2a \quad \text{Distribute.}$$
$$\boxed{n}\,x = 2a - mx$$
$$n = \dfrac{2a - mx}{x}$$

8.5 Applications of Rational Equations

Use the Steps for Solving Applied Problems outlined in Section 2.1. (p. 482)

Write an equation, and solve.

Dimos can put up the backyard pool in 6 hr, but it takes his father only 4 hr to put up the pool. How long would it take the two of them to put up the pool together?

Step 1: **Read** the problem carefully.
Step 2: **Choose** a variable to represent the unknown.
   t = number of hours to put up the pool together.
Step 3: **Translate** from English to an algebraic equation.

$$\text{Dimos' rate} = \dfrac{1}{6} \text{ pool/hr} \quad \text{Father's rate} = \dfrac{1}{4} \text{ pool/hr}$$

Fractional part = rate · time

$$\text{Dimos' part} = \dfrac{1}{6} \cdot t = \dfrac{1}{6}t$$

$$\text{Father's part} = \dfrac{1}{4} \cdot t = \dfrac{1}{4}t$$

Definition/Procedure	Example
	Fractional job by Dimos + Fractional job by his father = 1 whole job $$\frac{1}{6}t + \frac{1}{4}t = 1$$ Equation: $\frac{1}{6}t + \frac{1}{4}t = 1$. **Step 4:** Solve the equation. $$12\left(\frac{1}{6}t + \frac{1}{4}t\right) = 12(1) \quad \text{Multiply by 12, the LCD.}$$ $$12 \cdot \frac{1}{6}t + 12 \cdot \frac{1}{4}t = 12(1) \quad \text{Distribute.}$$ $$2t + 3t = 12 \quad \text{Multiply.}$$ $$5t = 12$$ $$t = \frac{12}{5}$$ **Step 5:** Check the answer, and interpret the solution as it relates to the problem. Dimos and his father could put up the pool together in $\frac{12}{5}$ hr or $2\frac{2}{5}$ hr. The check is left to the student.

8.6 Variation

Definition/Procedure	Example
Direct Variation *y* **varies directly as** *x* (or *y* **is directly proportional to** *x*) means $$y = kx$$ where *k* is a nonzero real number. *k* is called the **constant of variation.** (p. 488)	The circumference, *C*, of a circle is given by $C = 2\pi r$. *C* varies directly as *r*, where $k = 2\pi$.
Inverse Variation *y* **varies inversely as** *x* (or *y* **is inversely proportional to** *x*) means $$y = \frac{k}{x}$$ where *k* is a nonzero real number. (p. 491)	The time, *t* (in hours), it takes to drive 600 mi is inversely proportional to the rate, *r*, at which you drive. $$t = \frac{600}{r}$$ where $k = 600$.
Joint Variation *y* **varies jointly as** *x* **and** *z* means $y = kxz$, where *k* is a nonzero real number. (p. 492)	For a given amount, called the principal, deposited in a bank account, the interest earned, *I*, varies jointly as the interest rate, *r*, and the time, *t*, the principal is in the account. $$I = 1000rt$$ $k = 1000$, the principal.
Combined Variation A **combined variation** problem involves both direct and inverse variation. (p. 493)	The resistance of a wire, *R*, varies directly as its length, *L*, and inversely as its cross-sectional area, *A*. $$R = \frac{0.002L}{A}$$ The constant of variation, *k*, is 0.002. This is the resistivity of the material from which the wire was made.

Definition/Procedure	Example
Solving a Variation Problem **Step 1:** Write the *general* variation equation. **Step 2:** Find k by substituting the known values into the equation and solving for k. **Step 3:** Write the *specific* variation equation by substituting the value of k into the *general* variation equation. **Step 4:** Use the specific variation equation to solve the problem. (p. 489)	The cost of manufacturing a certain soccer ball is inversely proportional to the number produced. When 15,000 are made, the cost per ball is $4.00. What is the cost to manufacture each soccer ball when 25,000 are produced? Let n = number of soccer balls produced and let C = cost of producing each ball **Step 1:** Write the *general* variation equation: $C = \dfrac{k}{n}$ **Step 2:** Find k using $C = 4$ when $n = 15{,}000$. $$4 = \dfrac{k}{15{,}000}$$ $$60{,}000 = k$$ **Step 3:** Write the *specific* variation equation: $C = \dfrac{60{,}000}{n}$ **Step 4:** Find the cost, C, per ball when $n = 25{,}000$. $$C = \dfrac{60{,}000}{25{,}000} \quad \text{Substitute 25,000 for } n.$$ $$C = 2.4$$ The cost per ball is $2.40.

Chapter 8: Review Exercises

(8.1) Let $f(x) = \dfrac{P(x)}{Q(x)}$ be a rational function.

1) How do we determine the domain of $f(x)$?

2) How do we determine where $f(x) = 0$?

For each rational function,
a) find $f(5)$.
b) find x so that $f(x) = 0$.
c) determine the domain of the function.

3) $f(x) = \dfrac{x + 9}{5x - 1}$

4) $f(x) = \dfrac{8}{x^2 - 100}$

Determine the domain of each rational function.

5) $h(a) = \dfrac{9a}{a^2 - 2a - 24}$

6) $k(t) = \dfrac{6t - 1}{t^2 + 7}$

Write each rational expression in lowest terms.

7) $\dfrac{63a^2}{9a^{11}}$

8) $\dfrac{15c - 55}{33c - 121}$

9) $\dfrac{2z - 7}{6z^2 - 19z - 7}$

10) $\dfrac{10 - x}{x^2 - 100}$

11) $\dfrac{y^2 + 9y - yz - 9z}{yz - 12y - z^2 + 12z}$

12) Find three equivalent forms of $-\dfrac{u - 6}{u + 2}$.

Find the missing side in each rectangle.

13) Area $= 2l^2 - 5l - 3$

   $l - 3$

Find the length.

14) Area $= 3b^2 + 17b + 20$

$3b + 5$

Find the width.

Perform the operations, and simplify.

15) $\dfrac{16k^4}{3m^2} \div \dfrac{4k^2}{27m}$

16) $\dfrac{t + 4}{6} \cdot \dfrac{3(t - 2)}{(t + 4)^2}$

17) $\dfrac{6w - 1}{6w^2 + 5w - 1} \cdot \dfrac{3w + 3}{12w}$

18) $\dfrac{3x^2 + 14x + 16}{15x + 40} \div \dfrac{11x + 22}{x - 5}$

19) $\dfrac{25 - a^2}{4a^2 + 12a} \div \dfrac{a^3 - 125}{a^2 + 3a}$

20) $\dfrac{3p^5}{20q^2} \cdot \dfrac{4q^3}{21p^7}$

(8.2) Find the LCD of each group of fractions.

21) $\dfrac{2}{k^2}, \dfrac{9}{k}$

22) $\dfrac{4}{9x^2y}, \dfrac{13}{4xy^4}$

23) $\dfrac{3}{m}, \dfrac{4}{m + 5}$

24) $\dfrac{8}{2d^2 - d}, \dfrac{11}{6d - 3}$

25) $\dfrac{1}{3x + 7}, \dfrac{6x}{x - 9}$

26) $\dfrac{9}{2 - b}, \dfrac{4b}{b - 2}$

27) $\dfrac{5c - 1}{c^2 + 10c + 24}, \dfrac{9c}{c^2 - 3c - 28}$

28) $\dfrac{6}{x^2 + 8x}, \dfrac{1}{3x^2 + 24x}, \dfrac{13}{x^2 + 16x + 64}$

29) $\dfrac{3}{a^2 - 13a + 40}, \dfrac{a + 12}{a^2 - 7a - 8}, \dfrac{1}{a^2 - 4a - 5}$

30) $\dfrac{8c}{c^2 - d^2}, \dfrac{d}{d - c}$

Rewrite each rational expression with the indicated denominator.

31) $\dfrac{6}{5r} = \dfrac{}{20r^3}$

32) $\dfrac{8}{3z + 4} = \dfrac{}{z(3z + 4)}$

33) $\dfrac{t - 3}{2t + 1} = \dfrac{}{(2t + 1)(t + 5)}$

34) $\dfrac{n}{4 - n} = \dfrac{}{n - 4}$

Identify the LCD of each group of fractions, and rewrite each as an equivalent fraction with the LCD as its denominator.

35) $\dfrac{3}{8a^3b}, \dfrac{6}{5ab^5}$

36) $\dfrac{8}{p + 7}, \dfrac{2}{p}$

37) $\dfrac{9c}{c^2 + 6c - 16}, \dfrac{4}{c^2 - 4c + 4}$

38) $\dfrac{7}{2r^2 - 12r}, \dfrac{3r}{36 - r^2}, \dfrac{r - 5}{2r^2 + 12r}$

Add or subtract, as indicated.

39) $\dfrac{3}{8c} + \dfrac{7}{8c}$

40) $\dfrac{4m}{m - 3} - \dfrac{5}{m - 3}$

41) $\dfrac{2}{5z^2} + \dfrac{9}{10z}$

42) $\dfrac{n}{2n - 5} - \dfrac{4}{n}$

43) $\dfrac{5}{y - 2} - \dfrac{6}{y + 3}$

44) $\dfrac{8d - 3}{d^2 - 3d - 28} + \dfrac{2d}{5d - 35}$

45) $\dfrac{10p + 3}{4p + 4} - \dfrac{8}{p^2 - 6p - 7}$

46) $\dfrac{k - 3}{k^2 + 14k + 49} - \dfrac{2}{k^2 + 7k}$

47) $\dfrac{2}{m - 11} + \dfrac{19}{11 - m}$

48) $\dfrac{1}{8 - r} + \dfrac{16}{r^2 - 64}$

49) $\dfrac{x^2}{x^2 - y^2} + \dfrac{x}{y - x}$

50) $\dfrac{8}{w^2 + 7w} + \dfrac{3w}{w + 7} + \dfrac{2}{5w}$

51) $\dfrac{3}{g^2 - 7g} + \dfrac{2g}{5g - 35} - \dfrac{6}{5g}$

52) $\dfrac{d + 4}{d^2 + 2d} + \dfrac{d}{5d^2 + 7d - 6} - \dfrac{10}{5d^2 - 3d}$

For each rectangle, find a rational expression in simplest form to represent its a) area and b) perimeter.

53)

width: $\dfrac{x}{8}$, length: $\dfrac{12}{x - 4}$

54)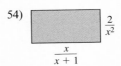

width: $\dfrac{2}{x^2}$, length: $\dfrac{x}{x + 1}$

For Exercises 55 and 56, let $f(x) = \dfrac{5x + 3}{x - 2}$ and $g(x) = \dfrac{4}{x}$.

55) Find $(f + g)(x)$ and its domain.

56) Find $\left(\dfrac{g}{f}\right)(x)$ and its domain.

(8.3) Simplify completely.

57) $\dfrac{\tfrac{2}{5}}{\tfrac{7}{15}}$

58) $\dfrac{\tfrac{f}{g}}{\tfrac{f^2}{g}}$

59) $\dfrac{p + \tfrac{6}{p}}{\tfrac{8}{p} + p}$

60) $\dfrac{\tfrac{a}{b} - \tfrac{2a}{b^2}}{\tfrac{4}{ab} - \tfrac{a}{b}}$

61) $\dfrac{\tfrac{6n + 48}{n^2}}{4n + 32}$

62) $\dfrac{\tfrac{2}{3} - \tfrac{4}{5}}{\tfrac{1}{6} + \tfrac{1}{2}}$

63) $\dfrac{1 - \tfrac{1}{y - 9}}{\tfrac{2}{y + 3} + 1}$

64) $\dfrac{\tfrac{4q}{7q + 63}}{\tfrac{q^2}{8q + 72}}$

65) $\dfrac{\dfrac{c}{c+2}+\dfrac{1}{c^2-4}}{1-\dfrac{3}{c+2}}$

66) $\dfrac{1+\dfrac{b}{a-b}}{\dfrac{b}{a^2-b^2}+\dfrac{1}{a+b}}$

67) $\dfrac{2x^{-2}+y^{-1}}{x^{-1}-y^{-2}}$

68) $\dfrac{12a^{-1}}{4a+b^{-2}}$

(8.4) Solve each equation.

69) $\dfrac{5w}{6}-\dfrac{1}{2}=-\dfrac{1}{6}$

70) $\dfrac{r}{r+6}+3=\dfrac{10}{r+6}$

71) $\dfrac{4}{y-6}=\dfrac{12}{y+2}$

72) $\dfrac{3}{x-5}+\dfrac{2}{2x+1}=\dfrac{1}{2x^2-9x-5}$

73) $\dfrac{16}{9t-27}+\dfrac{2t-4}{t-3}=\dfrac{t}{9}$

74) $\dfrac{5}{j+4}+\dfrac{j}{j-2}=\dfrac{2j^2-2j}{j^2+2j-8}$

75) $\dfrac{3}{b+2}=\dfrac{16}{b^2-4}-\dfrac{4}{b-2}$

76) $\dfrac{c}{c^2+3c-28}-\dfrac{5}{c^2+15c+56}=\dfrac{5}{c^2+4c-32}$

77) $\dfrac{3k}{k+9}=\dfrac{3}{k+1}$

78) $\dfrac{a}{a^2-1}+\dfrac{4}{a^2+9a+8}=\dfrac{8}{a^2+7a-8}$

Solve for the indicated variable.

79) $A=\dfrac{2p}{c}$ for c

80) $R=\dfrac{s+T}{D}$ for D

81) $n=\dfrac{t}{a+b}$ for a

82) $w=\dfrac{N}{c-ak}$ for k

83) $\dfrac{1}{r}=\dfrac{1}{s}+\dfrac{1}{t}$ for s

84) $\dfrac{1}{R_1}+\dfrac{1}{R_2}=\dfrac{1}{R_3}$ for R_1

(8.5) Write an equation, and solve.

85) The ratio of saturated fat to total fat in a Starbucks tall Caramel Frappuccino is 2 to 3. If there are 4 more grams of total fat in the drink than there are grams of saturated fat, how much total fat is in a Caramel Frappuccino? (Starbucks brochure)

86) A boat can travel 9 mi downstream in the same amount of time it can travel 6 mi upstream. If the speed of the boat in still water is 10 mph, find the speed of the current.

87) When the wind is blowing at 40 mph, a plane flying at a constant speed can travel 800 mi with the wind in the same amount of time it can fly 600 mi against the wind. Find the speed of the plane.

88) Wayne can clean the carpets in his house in 4 hr, but it would take his son, Garth, 6 hr to clean them on his own. How long would it take both of them to clean the carpets together?

(8.6)

89) Suppose c varies directly as m. If $c=56$ when $m=8$, find c when $m=3$.

90) Suppose A varies jointly as t and r. If $A=15$ when $t=\dfrac{1}{2}$ and $r=5$, find A when $t=3$ and $r=4$.

91) Suppose z varies inversely as the cube of w. If $z=16$ when $w=2$, find z when $w=4$.

92) Suppose p varies directly as n and inversely as the square of d. If $p=42$ when $n=7$ and $d=2$, find p when $n=12$ and $d=3$.

Solve each problem by writing a variation equation.

93) The weight of a ball varies directly as the cube of its radius. If a ball with a radius of 2 in. weighs 0.96 lb, how much would a ball made out of the same material weigh if it had a radius of 3 in.?

94) If the temperature remains the same, the volume of a gas is inversely proportional to the pressure. If the volume of a gas is 10L (liters) at a pressure of 1.25 atm (atmospheres), what is the volume of the gas at 2 atm?

Chapter 8: Test

1) $f(x) = \dfrac{x^2 + 4}{x^2 - 2x - 48}$

   a) Find $f(-2)$.

   b) Find x so that $f(x) = 0$.

   c) Determine the domain of the function.

2) Determine the domain of $g(x) = \dfrac{x + 9}{2x + 3}$.

Write each rational expression in lowest terms.

3) $\dfrac{54w^3}{24w^8}$

4) $\dfrac{7v^2 + 55v - 8}{v^2 - 64}$

5) Write three equivalent forms of $\dfrac{9 - h}{2h - 3}$.

6) If three rational expressions have denominators of k, $k^2 + 4k + 4$, and $2k^2 + k - 6$, find their least common denominator.

Perform the operations, and simplify.

7) $\dfrac{7}{12z} + \dfrac{5}{12z}$

8) $\dfrac{21m^4}{n} \div \dfrac{12m^8}{n^3}$

9) $\dfrac{r}{2r + 1} + \dfrac{3}{r + 5}$

10) $\dfrac{a^3 - 8}{6a - 66} \cdot \dfrac{a^2 - 9a - 22}{4 - a^2}$

11) $\dfrac{c - 3}{c - 15} + \dfrac{c + 8}{15 - c}$

12) $\dfrac{x}{x^2 - 49} - \dfrac{3}{x^2 - 2x - 63}$

13) Let $f(x) = \dfrac{2}{x}$ and $g(x) = \dfrac{x - 5}{x + 7}$. If $h(x) = f(x) - g(x)$, find $h(x)$ in its simplest form and determine its domain.

Simplify completely.

14) $\dfrac{1 + \dfrac{2}{d - 3}}{\dfrac{-2d}{d - 3} - d}$

15) $\dfrac{\dfrac{15}{7}}{\dfrac{20}{21}}$

16) $\dfrac{\dfrac{1}{x} - \dfrac{1}{y}}{\dfrac{1}{y^2} - \dfrac{1}{x^2}}$

17) Write an expression for the base of the triangle if the area is given by $12k^2 + 28k$.

18) Find all values that cannot be solutions to the equation $\dfrac{3}{5} - \dfrac{x + 2}{4x - 1} = \dfrac{7}{x}$. Do not solve the equation.

Solve each equation.

19) $\dfrac{7t}{12} + \dfrac{t - 4}{6} = \dfrac{7}{3}$

20) $\dfrac{30}{x^2 - 9} = \dfrac{5}{x - 3} - \dfrac{2}{x + 3}$

21) $\dfrac{w + 2}{6} = \dfrac{2w - 3}{4}$

22) $\dfrac{5}{n^2 + 10n + 24} + \dfrac{5}{n^2 + 3n - 18} = \dfrac{n}{n^2 + n - 12}$

23) Solve $y = \dfrac{kxz}{c}$ for c.

24) Solve $\dfrac{1}{p} + \dfrac{1}{q} = \dfrac{1}{r}$ for p.

Write an equation for each, and solve.

25) The ratio of Khloe's Facebook friends to Twitter followers is 19 to 4. If she has 540 more Facebook friends than Twitter followers, how many of each does Khloe have?

26) A river flows at 3 mph. If a boat can travel 16 mi downstream in the same amount of time it can go 10 mi upstream, find the speed of the boat in still water.

27) Suppose n varies jointly as r and the square of s. If $n = 72$ when $r = 2$ and $s = 3$, find n when $r = 3$ and $s = 5$.

28) The loudness of a sound is inversely proportional to the square of the distance between the source of the sound and the listener. If the sound level measures 112.5 dB 4 ft from a speaker, how loud is the sound 10 ft from the speaker?

Chapter 8: Cumulative Review for Chapters 1–8

Simplify. Assume all variables represent nonzero real numbers. The answers should not contain negative exponents.

1) $-5(3w^4)^2$

2) $\left(\dfrac{2n^{-10}}{n^{-4}}\right)^3$

Solve.

3) $-\dfrac{12}{5}c - 7 = 20$

4) *Write an equation in one variable, and solve.*
How many milliliters of a 12% alcohol solution and how many milliliters of a 4% alcohol solution must be mixed to obtain 60 ml of a 10% alcohol solution?

5) Find the x- and y-intercepts of $5x - 2y = 10$ and sketch a graph of the equation.

6) Graph $f(x) = -\dfrac{1}{3}x + 2$.

7) Write an equation of the line containing the points $(-5, 8)$ and $(1, 2)$. Express the answer in standard form.

8) Solve this system using the elimination method.
$-7x + 2y = 6$
$9x - y = 8$

9) *Write a system of two equations in two variables, and solve.*
The length of a rectangle is 7 cm less than twice its width. The perimeter of the rectangle is 76 cm. What are the dimensions of the figure?

10) Solve the compound inequality, and write the answer in interval notation. $3y + 16 < 4$ or $8 - y \geq 7$

11) Solve $|6p + 13| = 8$.

12) Graph $2x - y > 4$.

13) Graph $x \geq -3$ and $y \geq -\dfrac{1}{2}x - 1$.

14) $h(t) = 2t^2 - 11t + 4$

   a) Find $h(3)$.

   b) Find t so that $h(t) = -8$.

Perform the indicated operation(s).

15) $(4q^2 + 11q - 2) - 3(6q^2 - 5q + 4) + 2(-10q - 3)$

16) $(3d^2 - 7)(4d^2 + 6d - 1)$

17) $\dfrac{8a^4b^4 - 20a^3b^2 + 56ab + 8b}{8a^3b^3}$

18) $\dfrac{17v^3 - 22v + 7v^4 + 24 - 47v^2}{7v - 4}$

19) $\dfrac{7}{c - 6} - \dfrac{4}{c}$

Factor completely.

20) $25n^2 - 81$

21) $3xy^2 + 15xy - 72x$

22) $r^2 + 8rt + 16t^2$

23) Determine the domain of $g(a) = \dfrac{a + 3}{8 - a}$.

24) Simplify completely.

   a) $\dfrac{\dfrac{12k^4}{8k^3 - 27}}{\dfrac{9k^5}{3 - 2k}}$

   b) $\dfrac{1 + \dfrac{2}{r}}{\dfrac{1}{r} + \dfrac{r}{r + 3}}$

Write an equation, and solve.

25) Leticia can assemble a swing set in 3 hr while it takes Betty 5 hr. How long would it take for them to assemble the swing set together?

CHAPTER 9

Radicals and Rational Exponents

Math at Work:
Computer Support Specialist

OUTLINE

Study Strategies: Working with Technology
9.1 Radical Expressions and Functions
9.2 Rational Exponents
9.3 Simplifying Expressions Containing Square Roots
9.4 Simplifying Expressions Containing Higher Roots
9.5 Adding, Subtracting, and Multiplying Radicals
9.6 Dividing Radicals
Putting It All Together
9.7 Solving Radical Equations
9.8 Complex Numbers
Group Activity
emPOWERme: Information, Please!

Leslie Burns is proud to have the sort of job that didn't even exist a generation ago. She has always loved technology, and her career lets her pursue this passion at work. As a computer support specialist for a major American software company, she has to maintain a firm grasp on everything her company does and understand how it fits with changes in the larger tech industry. More importantly, she has to be able to explain all of this to customers. "Our customers are small businesspeople and individuals," Leslie describes. "When they have a problem with one of our products, I'm the person who helps them resolve it."

Leslie says that a lot of her customers are frightened by technology. She argues that they don't need to be. "A lot of times I compare technology to a math problem you've never seen before," Leslie describes. "At first, the unfamiliar symbols can be scary. But once you understand what they mean and how they connect to what you already know, you quickly become more comfortable."

In this chapter, we will discuss working with radicals and rational exponents and offer some strategies for using new technology effectively.

 ## Study Strategies Working with Technology

Today, technology develops at breathtaking speed. Just about every week there is a new phone, a new social media platform, or a new app that is splashed across the news and that seemingly everyone is talking about—until the next one comes along, at least. The pace of change can be daunting, but all this new technology can benefit you in your education and in your career. Here are some strategies for approaching a piece of new technology.

P Prepare
- Before starting to use the technology, much less spending money on it, try to get a sense of what it does. You can do this by talking to friends who use it or by seeking out reports in the media.
- Consider what your goals are for using the technology. These goals can be general (staying in better touch with friends, for example) or very specific (not getting lost anymore as you drive to campus, for instance).

O Organize
- Take the steps to acquire the new technology. This might involve downloading software from the Internet or making a purchase at an electronics store. (Remember to save your receipt!)
- Set aside a specific amount of time to familiarize yourself with the technology.

W Work
- Start trying to use the technology to accomplish your goal.
- Don't give up if you are initially confused—as with anything, you need to put in effort to master a new technology.
- Ask a tech savvy friend or family member for help if you become frustrated.

E Evaluate
- After a few days with the technology, ask yourself if it is helping you meet your goal. If not, is this because you are using it incorrectly, or because it simply does not fit your needs?
- Decide whether the technology is really worth your time and money.

R Rethink
- Technology is meant to make our lives easier, but sometimes it seems it only makes it more hectic. Consider what role technology plays in your life.

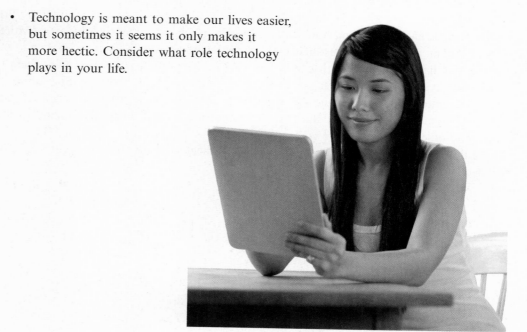

Chapter 9 POWER Plan

P Prepare	**O Organize**
What are your goals for Chapter 9?	How can you accomplish each goal?
1 Be prepared before and during class.	• _____ • _____ • _____ • _____
2 Understand the homework to the point where you could do it without needing any help or hints.	• _____ • _____ • _____
3 Use the P.O.W.E.R. framework to help you learn how to gather information: *Information, Please!*	• _____ • _____ • _____
4 Write your own goal. _____	• _____

What are your objectives for Chapter 9?	How can you accomplish each objective?
1 Be able to simplify square roots and higher roots. Be able to graph radical functions and determine the domain.	• Learn the different definitions for *square roots* and higher roots. • Be able to make a table of values for *radical functions* by choosing a proper domain.
2 Be able to perform operations with rational exponents.	• Learn the definition of $a^{1/n}$, $a^{m/n}$, and $a^{-m/n}$. • Review the rules of exponents found in Sections 6.1 and 6.2. • Know how to convert from radical form to exponential form.
3 Be able to simplify expressions containing square roots and higher roots including expressions with variables.	• Know the *product rules* and the *quotient rules*. • Learn how to tell when a *radical expression* is simplified. • Know how to simplify exponents with even and odd exponents.
4 Be able to add, subtract, and multiply radicals.	• Review the procedure for **Simplifying Radicals**. • Know the definition of *like radicals*. • Review the rules for multiplying polynomials so you can apply them to multiplying radicals. • Learn the procedure for **Multiplying Radicals**.
5 Be able to divide radical expressions and rationalize denominators.	• Learn the procedure for **Rationalizing Denominators,** including denominators with two terms. • Be able to find the *conjugate* of a radical expression, when appropriate. • Know how to determine when radical expressions are simplified.

www.mhhe.com/messersmith

6 Be able to solve radical equations, and check your solutions.	• Know the procedure for **Solving Radical Equations Containing Square Roots**. • Be able to determine when a solution is an *extraneous* solution. • Check all solutions.
7 Be able to simplify, add, subtract, multiply, and divide with complex numbers.	• Know the definition of *i*. • Know the procedure for **Adding and Subtracting Complex Numbers**. • Know the procedure for **Multiplying and Dividing Complex Numbers**.
8 Write your own goal. _____ _____	• _____ _____

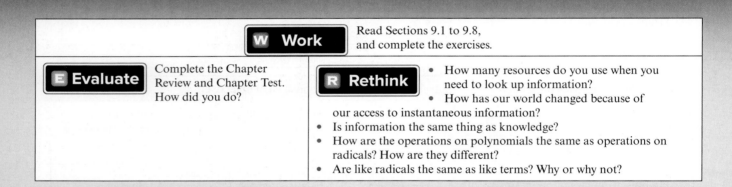

9.1 Radical Expressions and Functions

Prepare | Organize

What are your objectives for Section 9.1?	How can you accomplish each objective?
1 Find Square Roots and Principal Square Roots	• Write the definitions of *square root, principal square root, negative square root, square root symbol* or *radical sign, radicand,* and *radical* in your notes with an example next to each. • Write the properties for **Radicands and Square Roots** in your notes, and learn them. • Complete the given examples on your own. • Complete You Trys 1 and 2.
2 Find Higher Roots	• Know the definition of an *index*. • Understand the definitions of the *nth root*, and write examples in your notes that match the definitions. • Complete the given examples on your own. • Complete You Trys 3 and 4.
3 Evaluate $\sqrt[n]{a^n}$	• Write the procedure for **Evaluating** $\sqrt[n]{a^n}$ in your own words. • Complete the given example on your own. • Complete You Try 5.
4 Determine the Domains of Square Root and Cube Root Functions	• Know how to identify a *radical expression, radical function,* and a *square root function*. • Write the definition of the *domain of a square root function* and *domain of a cube root function* in your own words. • Complete the given examples on your own. • Complete You Trys 6–8.
5 Graph a Square Root Function	• Follow Example 9 to create your own procedure for **Graphing a Square Root Function**. • Complete You Try 9.
6 Graph a Cube Root Function	• Follow Example 10 to create your own procedure for **Graphing a Cube Root Function**. • Complete You Try 10.

Read the explanations, follow the examples, take notes, and complete the You Trys.

Recall that exponential notation represents repeated multiplication. For example,

3^2 means $3 \cdot 3$, so $3^2 = 9$.

2^4 means $2 \cdot 2 \cdot 2 \cdot 2$, so $2^4 = 16$.

In this chapter we will study the opposite, or inverse, procedure, finding **roots** of numbers.

1 Find Square Roots and Principal Square Roots

EXAMPLE 1 Find all square roots of 25.

Solution

To find a *square* root of 25 ask yourself, "What number do I *square* to get 25?" Or, "What number *multiplied by itself* equals 25?" One number is 5 since $5^2 = 25$. Another number is -5 since $(-5)^2 = 25$. So, 5 and -5 are square roots of 25.

YOU TRY 1 Find all square roots of 64.

The $\sqrt{}$ symbol represents the *positive* square root, or the **principal square root,** of a nonnegative number. For example, $\sqrt{25} = 5$.

BE CAREFUL Notice that $\sqrt{25} = 5$, but $\sqrt{25} \neq -5$. The $\sqrt{}$ symbol represents *only* the principal square root (positive square root).

To find the **negative square root** of a nonnegative number we must put a negative sign in front of the $\sqrt{}$. For example, $-\sqrt{25} = -5$.

Next we will define some terms associated with the $\sqrt{}$ symbol.

The symbol $\sqrt{}$ is the **square root symbol** or the **radical sign.** The number under the radical sign is the **radicand.** The entire expression, $\sqrt{25}$, is called a **radical.**

Radical sign → $\underbrace{\sqrt{25}}_{\text{Radical}}$ ← Radicand = 25

EXAMPLE 2 Find each square root, if possible.

a) $\sqrt{100}$ b) $-\sqrt{16}$ c) $\sqrt{\dfrac{4}{25}}$ d) $-\sqrt{\dfrac{81}{49}}$ e) $\sqrt{-9}$

Solution

a) $\sqrt{100} = 10$ since $(10)^2 = 100$.

b) $-\sqrt{16}$ means $-1 \cdot \sqrt{16}$. Therefore, $-\sqrt{16} = -1 \cdot \sqrt{16} = -1 \cdot 4 = -4$.

c) $\sqrt{\dfrac{4}{25}} = \dfrac{2}{5}$ since $\left(\dfrac{2}{5}\right)^2 = \dfrac{4}{25}$.

d) $-\sqrt{\dfrac{81}{49}}$ means $-1 \cdot \sqrt{\dfrac{81}{49}}$. So, $-\sqrt{\dfrac{81}{49}} = -1 \cdot \sqrt{\dfrac{81}{49}} = -1 \cdot \dfrac{9}{7} = -\dfrac{9}{7}$.

e) To find $\sqrt{-9}$, ask yourself, "What number do I *square* to get -9?" There is no such real number since $3^2 = 9$ and $(-3)^2 = 9$. Therefore, $\sqrt{-9}$ is not a real number.

W Hint
In Example 2, we are finding the **principal square roots** of 100 and $\dfrac{4}{25}$ and the **negative square roots** of 16 and $\dfrac{81}{49}$.

[**YOU TRY 2**] Find each square root.

a) $\sqrt{9}$ b) $-\sqrt{144}$ c) $\sqrt{\dfrac{25}{36}}$ d) $-\sqrt{\dfrac{1}{64}}$ e) $\sqrt{-49}$

Let's review what we know about radicands and add a third fact.

Property Radicands and Square Roots

1) If the radicand is a *perfect square,* the square root is a *rational* number.

   Example: $\sqrt{16} = 4$ 16 is a perfect square. $\sqrt{\dfrac{100}{49}} = \dfrac{10}{7}$ $\dfrac{100}{49}$ is a perfect square.

2) If the radicand is a *negative number,* the square root is *not* a real number.

   Example: $\sqrt{-25}$ is *not* a real number.

3) If the radicand is *positive and not a perfect square,* then the square root is an *irrational* number.

   Example: $\sqrt{13}$ is irrational.   13 is not a perfect square.

The square root of such a number is a real number that is a nonrepeating, nonterminating decimal.

Sometimes, we must plot points containing radicals. For the purposes of graphing, approximating a radical to the nearest tenth is sufficient. A calculator with a $\sqrt{}$ key will give a better approximation of the radical.

W Hint
Do you think you could estimate $\sqrt{13}$? How could you do that?

2 Find Higher Roots

We saw in Example 2a) that $\sqrt{100} = 10$ since $(10)^2 = 100$. We can also find higher roots of numbers like $\sqrt[3]{a}$ (read as "the cube root of a"), $\sqrt[4]{a}$ (read as "the fourth root of a"), $\sqrt[5]{a}$, etc. We will look at a few roots of numbers before learning important rules.

EXAMPLE 3   Find each root.

a) $\sqrt[3]{125}$ b) $\sqrt[5]{32}$

Solution

a) To find $\sqrt[3]{125}$ (the cube root of 125) ask yourself, "What number do I *cube* to get 125?" That number is 5.

$$\sqrt[3]{125} = 5 \text{ since } 5^3 = 125$$

Finding the cube root of a number is the *opposite,* or *inverse* procedure, of cubing a number.

Hint
Could you write an equation that would help you find the root?

b) To find $\sqrt[5]{32}$ (the fifth root of 32) ask yourself, "What number do I raise to the *fifth power* to get 32?" That number is 2.

$$\sqrt[5]{32} = 2 \text{ since } 2^5 = 32$$

Finding the fifth root of a number and raising a number to the fifth power are *opposite,* or *inverse,* procedures.

[YOU TRY 3] Find each root.

a) $\sqrt[3]{27}$ b) $\sqrt[3]{8}$

The symbol $\sqrt[n]{a}$ is read as "the *n*th root of *a*." If $\sqrt[n]{a} = b$, then $b^n = a$.

Index → $\sqrt[n]{a}$ ← Radicand = *a*

Radical

We call *n* the **index** of the radical.

Note
When finding *square* roots we do not write $\sqrt[2]{a}$. The square root of *a* is written as $\sqrt{a}$, and the index is understood to be 2.

We know that a positive number, say 36, has a principal square root ($\sqrt{36}$, or 6) and a negative square root ($-\sqrt{36}$, or -6). This is true for all even roots of positive numbers: square roots, fourth roots, sixth roots, and so on. For example, 81 has a principal fourth root ($\sqrt[4]{81}$, or 3) and a negative fourth root ($-\sqrt[4]{81}$, or -3).

Definition *n*th Root

For any *positive* number *a* and any *even* index *n*,
  the **principal *n*th root** of *a* is $\sqrt[n]{a}$. $\sqrt[even]{positive}$ = principal (positive) root
  the **negative *n*th root** of *a* is $-\sqrt[n]{a}$. $-\sqrt[even]{positive}$ = negative root

For any *negative* number *a* and any *even* index *n*,
  there is **no** real *n*th root of *a*. $\sqrt[even]{negative}$ = no real root

For any number *a* and any *odd* index *n*,
  there is **one** real *n*th root of *a*, $\sqrt[n]{a}$. $\sqrt[odd]{any number}$ = exactly one root

Hint
Write examples in your notes to help you understand the different situations.

BE CAREFUL The definition means that $\sqrt[4]{81}$ cannot be -3 because $\sqrt[4]{81}$ is *defined* as the principal fourth root of 81, which must be positive. $\sqrt[4]{81} = 3$

EXAMPLE 4

Find each root, if possible.

a) $\sqrt[4]{16}$ b) $-\sqrt[4]{16}$ c) $\sqrt[4]{-16}$ d) $\sqrt[3]{64}$ e) $\sqrt[3]{-64}$

Solution

a) To find $\sqrt[4]{16}$ ask yourself, "What *positive* number do I raise to the *fourth power* to get 16?" Since $2^4 = 16$ and 2 is positive, $\sqrt[4]{16} = \mathbf{2}$.

b) In part a) we found that $\sqrt[4]{16} = 2$, so $-\sqrt[4]{16} = -(\sqrt[4]{16}) = \mathbf{-2}$.

c) To find $\sqrt[4]{-16}$ ask yourself, "What number do I raise to the *fourth power* to get -16?" There is no such real number since $2^4 = 16$ and $(-2)^4 = 16$. Therefore, $\sqrt[4]{-16}$ **has *no real root*.** (Recall from the definition that $\sqrt[\text{even}]{\text{negative}}$ has no real root.)

d) To find $\sqrt[3]{64}$ ask yourself, "What number do I *cube* to get 64?" Since $4^3 = 64$ and since we know that $\sqrt[\text{odd}]{\text{any number}}$ gives exactly one root, $\sqrt[3]{64} = \mathbf{4}$.

e) To find $\sqrt[3]{-64}$ ask yourself, "What number do I *cube* to get -64?" Since $(-4)^3 = -64$ and since we know that $\sqrt[\text{odd}]{\text{any number}}$ gives exactly one root, $\sqrt[3]{-64} = \mathbf{-4}$.

[YOU TRY 4]

Find each root, if possible.

a) $\sqrt[6]{-64}$ b) $\sqrt[3]{-125}$ c) $-\sqrt[4]{81}$ d) $\sqrt[3]{1}$ e) $\sqrt[4]{81}$

3 Evaluate $\sqrt[n]{a^n}$

Earlier we said that the $\sqrt{\ }$ symbol represents only the *positive* square root of a number. For example, $\sqrt{9} = 3$. It is also true that $\sqrt{(-3)^2} = \sqrt{9} = 3$.

If a variable is in the radicand and we do not know whether the variable represents a positive number, then we must use the absolute value symbol to evaluate the radical. Then we know that the result will be a positive number. For example, $\sqrt{a^2} = |a|$.

What if the index is greater than 2? Let's look at how to find the following roots:

$$\sqrt[4]{(-2)^4} = \sqrt[4]{16} = 2 \qquad \sqrt[3]{(-4)^3} = \sqrt[3]{-64} = -4$$

When the index on the radical is any positive, even integer and we do not know whether the variable in the radicand represents a positive number, we must use the absolute value symbol to write the root. However, when the index is a positive, odd integer, we do not need to use the absolute value symbol.

> **Procedure** Evaluating $\sqrt[n]{a^n}$
>
> 1) If n is a positive, *even* integer, then $\sqrt[n]{a^n} = |a|$.
> 2) If n is a positive, *odd* integer, then $\sqrt[n]{a^n} = a$.

EXAMPLE 5 Simplify.

a) $\sqrt{(-7)^2}$ b) $\sqrt{k^2}$ c) $\sqrt[3]{(-5)^3}$ d) $\sqrt[7]{n^7}$

e) $\sqrt[4]{(y-9)^4}$ f) $\sqrt[5]{(8p+1)^5}$

Solution

a) $\sqrt{(-7)^2} = |-7| = 7$ — When the index is even, use the absolute value symbol to be certain that the result is not negative.

b) $\sqrt{k^2} = |k|$ — When the index is even, use the absolute value symbol to be certain that the result is not negative.

c) $\sqrt[3]{(-5)^3} = -5$ — The index is odd, so the absolute value symbol is not necessary.

d) $\sqrt[7]{n^7} = n$ — The index is odd, so the absolute value symbol is not necessary.

e) $\sqrt[4]{(y-9)^4} = |y-9|$ — Even index: use the absolute value symbol to be certain that the result is not negative.

f) $\sqrt[5]{(8p+1)^5} = 8p+1$ — Odd index: the absolute value symbol is not necessary.

W Hint
In your notes, summarize when you need the absolute value symbol and when you do not.

[YOU TRY 5] Simplify.

a) $\sqrt{(-12)^2}$ b) $\sqrt{w^2}$ c) $\sqrt[3]{(-3)^3}$ d) $\sqrt[5]{r^5}$

e) $\sqrt[6]{(t+4)^6}$ f) $\sqrt[7]{(4h-3)^7}$

4 Determine the Domains of Square Root and Cube Root Functions

An algebraic expression containing a radical is called a **radical expression.** When real numbers are substituted for the variable in radical expressions like $\sqrt{x}$, $\sqrt{4t+1}$, and $\sqrt[3]{p}$ so that the expression is defined, each value that is substituted will produce *only one* value for the expression. Therefore, function notation can be used to represent radical expressions.

Radical functions are functions of the form $f(x) = \sqrt[n]{x}$. Let's look at some square root and cube root functions.

Two examples of **square root functions** are $f(x) = \sqrt{x}$ and $g(r) = \sqrt{2r-9}$.

EXAMPLE 6 Let $f(x) = \sqrt{x}$ and $g(r) = \sqrt{2r-9}$. Find the function values, if possible.

a) $f(64)$ b) $g(7)$ c) $f(-25)$ d) $g(3)$

Solution

a) $f(64) = \sqrt{64} = 8$

b) $g(7) = \sqrt{2(7)-9} = \sqrt{14-9} = \sqrt{5}$

c) $f(-25) = \sqrt{-25}$; not a real number

d) $g(3) = \sqrt{2(3)-9} = \sqrt{-3}$; not a real number

[YOU TRY 6] Let $f(x) = \sqrt{x}$ and $h(t) = \sqrt{3t-10}$. Find the function values, if possible.

a) $f(25)$ b) $h(9)$ c) $f(-11)$ d) $h(2)$

Parts c) and d) of Example 6 illustrate that when the radicand of a square root function is negative, the function is undefined. Therefore, *any value that makes the radicand negative is not in the domain of a square root function.*

> ### Definition
> The **domain of a square root function** consists of all of the real numbers that can be substituted for the variable so that radicand is nonnegative.
> To determine the domain of a square root function, set up an inequality so that the radicand ≥ 0. Solve for the variable. These are the real numbers in the domain of the function.

EXAMPLE 7   Determine the domain of each square root function.

a) $f(x) = \sqrt{x}$ b) $g(r) = \sqrt{2r - 9}$

Solution

a) The radicand, x, must be greater than or equal to zero. We write that as the inequality $x \geq 0$. In interval notation, we write the domain as $[0, \infty)$.

b) In the square root function $g(r) = \sqrt{2r - 9}$, the radicand, $2r - 9$, must be nonnegative. We write this as $2r - 9 \geq 0$. To determine the domain of the function, solve the inequality $2r - 9 \geq 0$.

$$2r - 9 \geq 0 \qquad \text{The value of the radicand must be} \geq 0.$$
$$2r \geq 9$$
$$r \geq \frac{9}{2} \qquad \text{Solve.}$$

Any value of r that satisfies $r \geq \frac{9}{2}$ will make the radicand greater than or equal to zero. The domain of $g(r) = \sqrt{2r - 9}$ is $\left[\frac{9}{2}, \infty\right)$.

[YOU TRY 7]   Determine the domain of each square root function.

a) $h(x) = \sqrt{x - 9}$ b) $k(t) = \sqrt{7t + 2}$

Two examples of cube root functions are $f(x) = \sqrt[3]{x}$ and $h(a) = \sqrt[3]{a - 5}$. Let's look at these next.

EXAMPLE 8   Let $f(x) = \sqrt[3]{x}$ and $h(a) = \sqrt[3]{a - 5}$. Find the function values, if possible.

a) $f(125)$ b) $f(-7)$ c) $h(10)$ d) $h(-3)$

Solution

a) $f(125) = \sqrt[3]{125} = 5$

b) $f(-7) = \sqrt[3]{-7}$

c) $h(10) = \sqrt[3]{10 - 5} = \sqrt[3]{5}$

d) $h(-3) = \sqrt[3]{-3 - 5} = \sqrt[3]{-8} = -2$

[YOU TRY 8] Let $f(x) = \sqrt[3]{x}$ and $g(c) = \sqrt[3]{2c + 3}$. Find the function values, if possible.

a) $f(25)$ b) $f(-27)$ c) $g(-2)$ d) $g(2)$

Unlike square root functions, it is possible to evaluate cube root functions when the radicand is negative. Therefore, *any real number may be substituted into a cube root function and the function will be defined.*

Definition

The **domain of a cube root function** is the set of all real numbers. We can write this in interval notation as $(-\infty, \infty)$.

In fact we can say that when n is an odd, positive number, the domain of $f(x) = \sqrt[n]{x}$ is all real numbers, or $(-\infty, \infty)$. This is because the odd root of any real number is, itself, a real number.

5 Graph a Square Root Function

We need to know the domain of a square root function in order to sketch its graph.

EXAMPLE 9 Graph each function.

a) $f(x) = \sqrt{x}$ b) $g(x) = \sqrt{x + 4}$

Solution

a) In Example 7 we found that the domain of $f(x) = \sqrt{x}$ is $[0, \infty)$. When we make a table of values, we will start by letting $x = 0$, the smallest number in the domain, and then choose real numbers greater than 0. Usually it is easiest to choose values for x that are perfect squares so that it will be easier to plot the points. We will also plot the point $(6, \sqrt{6})$ so that you can see where it lies on the graph. Connect the points with a smooth curve.

$f(x) = \sqrt{x}$	
x	$f(x)$
0	0
1	1
4	2
6	$\sqrt{6} \approx 2.4$
9	3

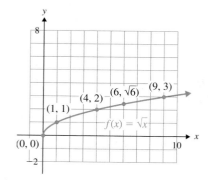

The graph reinforces the fact that this is a function. It passes the vertical line test.

> **Hint**
> Create a procedure for graphing a square root function.

b) To graph $g(x) = \sqrt{x + 4}$ we will begin by determining its domain. Solve $x + 4 \geq 0$.

$$x + 4 \geq 0 \quad \text{The value of the radicand must be} \geq 0.$$
$$x \geq -4 \quad \text{Solve.}$$

The domain of $g(x)$ is $[-4, \infty)$. When we make a table of values, we will start by letting $x = -4$, the smallest number in the domain, and then choose real numbers greater than -4. We will choose values for x so that the radicand will be a perfect square. This will make it easier to plot the points. We will also plot the point $(1, \sqrt{5})$ so that you can see where it lies on the graph. Connect the points with a smooth curve.

$g(x) = \sqrt{x + 4}$	
x	$g(x)$
-4	0
-3	1
0	2
1	$\sqrt{5} \approx 2.2$
5	3

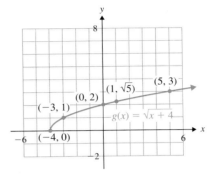

Since this graph represents a function, it passes the vertical line test.

[YOU TRY 9] Graph $f(x) = \sqrt{x + 2}$.

6 Graph a Cube Root Function

The domain of a cube root function consists of all real numbers. Therefore, we can substitute any real number into the function and it will be defined. However, we want to choose our numbers carefully. *To make the table of values, pick values in the domain so that the radicand will be a perfect cube, and choose values for the variable that will give us positive numbers, negative numbers, and zero for the value of the radicand.* This will help us to graph the function correctly.

EXAMPLE 10   Graph each function.

a) $f(x) = \sqrt[3]{x}$ b) $g(x) = \sqrt[3]{x - 1}$

Solution

a) Make a table of values. Choose x-values that are perfect cubes. Also, remember to choose x-values that are positive, negative, and zero. Plot the points, and connect them with a smooth curve.

$f(x) = \sqrt[3]{x}$	
x	$f(x)$
0	0
1	1
8	2
-1	-1
-8	-2

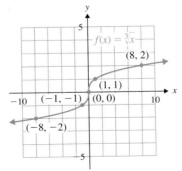

The graph passes the vertical line test for functions.

b) Remember, for the table of values we want to choose values for x that will give us positive numbers, negative numbers, and zero *in the radicand*. First we will determine what value of x will make the radicand in $g(x) = \sqrt[3]{x-1}$ equal to zero.

$$x - 1 = 0$$
$$x = 1$$

If $x = 1$, the radicand equals zero. Therefore, the first value we will put in the table of values is $x = 1$. Then, choose a couple of numbers *greater than* 1 and a couple that are *less than* 1 so that we get positive and negative numbers in the radicand. Also, we will choose our x-values so that the radicand will be a perfect cube. Plot the points, and connect them with a smooth curve.

x	$g(x) = \sqrt[3]{x-1}$
1	$\sqrt[3]{1-1} = \sqrt[3]{0} = 0$
2	$\sqrt[3]{2-1} = \sqrt[3]{1} = 1$
9	$\sqrt[3]{9-1} = \sqrt[3]{8} = 2$
0	$\sqrt[3]{0-1} = \sqrt[3]{-1} = -1$
-7	$\sqrt[3]{-7-1} = \sqrt[3]{-8} = -2$

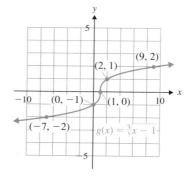

Since this graph represents a function, it passes the vertical line test.

[YOU TRY 10]   Graph $f(x) = \sqrt[3]{x-3}$.

Using Technology

We can evaluate square roots, cube roots, or even higher roots using a graphing calculator. A radical sometimes evaluates to an integer and sometimes must be approximated using a decimal.

To evaluate a square root:

For example, to evaluate $\sqrt{9}$ press [2nd] [x^2], enter the radicand [9], and then press [)] [ENTER]. The result is 3 as shown on the screen on the left below. When the radicand is a perfect square such as 9, 16, or 25, then the square root evaluates to a whole number. For example $\sqrt{16}$ evaluates to 4 and $\sqrt{25}$ evaluates to 5 as shown.

If the radicand of a square root is not a perfect square, then the result is a decimal approximation. For example, to evaluate $\sqrt{19}$ press [2nd] [x^2], enter the radicand [1] [9], and then press [)] [ENTER]. The result is approximately 4.3589, rounded to four decimal places.

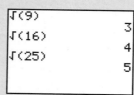

To evaluate a cube root:

For example, to evaluate $\sqrt[3]{27}$ press **MATH** $\boxed{4}$, enter the radicand $\boxed{2}$ $\boxed{7}$, and then press $\boxed{)}$ **ENTER**. The result is 3 as shown.

If the radicand is a perfect cube such as 27, then the cube root evaluates to an integer. Since 28 is not a perfect cube, the cube root evaluates to approximately 3.0366.

To evaluate radicals with an index greater than 3:

For example, to evaluate $\sqrt[4]{16}$ enter the index $\boxed{4}$, press **MATH** $\boxed{5}$, enter the radicand $\boxed{1}$ $\boxed{6}$, and press **ENTER**. The result is 2.

Since the fifth root of 18 evaluates to a decimal, the result is an approximation of 1.7826, rounded to four decimal places as shown.

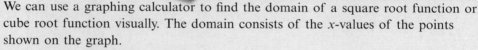

Evaluate each root using a graphing calculator. If necessary, approximate to the nearest tenth.

1) $\sqrt{25}$ 2) $\sqrt[3]{216}$ 3) $\sqrt{29}$ 4) $\sqrt{324}$ 5) $\sqrt[5]{1024}$ 6) $\sqrt[3]{343}$

Using Technology

We can use a graphing calculator to find the domain of a square root function or cube root function visually. The domain consists of the *x*-values of the points shown on the graph.

We first consider the basic shape of a square root function. To graph the equation $f(x) = \sqrt{x}$, press $\boxed{2^{nd}}$ $\boxed{x^2}$ $\boxed{X,T,\Theta,n}$ $\boxed{)}$ to the right of \Y$_1$ =. Press **ZOOM** and select 6:ZStandard to graph the equation.

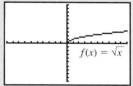

The left side of the graph begins at the point (0, 0), and the right side of the graph continues up and to the right forever. The *x*-values of the graph consist of all *x*-values greater than or equal to 0. In interval notation, the domain is [0, ∞).

The domain of any square root function can be found using a similar approach. First graph the function, and then look at the *x*-values of the points on the graph. The graph of a square root function will always start at a number and extend to positive or negative infinity.

For example, consider the graph of the function $g(x) = \sqrt{3 - x}$ as shown.

The largest *x*-value on the graph is 3. The *x*-values of the graph consist of all *x*-values less than or equal to 3. In interval notation, the domain is (−∞, 3].

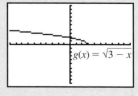

Next consider the basic shape of a cube root function.

To graph the equation $f(x) = \sqrt[3]{x}$ press **MATH**, select 4: $\sqrt[3]{}$ (, and press $\boxed{X,T,\Theta,n}$ $\boxed{)}$ to the right of \Y$_1$=. Press **ZOOM** and select 6:ZStandard to graph the equation as shown on the graph at right.

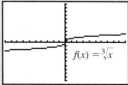

The left side of the graph extends down and to the left forever, and the right side of the graph extends up and to the right forever. In interval notation, the domain is $(-\infty, \infty)$. This is true for any cube root function, so the domain is always $(-\infty, \infty)$.

Determine the domain using a graphing calculator. Use interval notation in your answer.

7) $f(x) = \sqrt{x-2}$ 8) $g(x) = \sqrt{x+3}$ 9) $h(x) = \sqrt{2-x}$

10) $f(x) = -\sqrt{x+1}$ 11) $f(x) = \sqrt[3]{x+5}$ 12) $g(x) = \sqrt[3]{4-x}$

ANSWERS TO [YOU TRY] EXERCISES

1) 8, −8   2) a) 3 b) −12 c) $\frac{5}{6}$ d) $-\frac{1}{8}$ e) not a real number   3) a) 3 b) 2
4) a) not a real number b) −5 c) −3 d) 1 e) 3   5) a) 12 b) $|w|$ c) −3 d) r
e) $|t+4|$ f) $4h-3$ 6) a) 5 b) $\sqrt{17}$ c) not a real number d) not a real number
7) a) $[9, \infty)$ b) $\left[-\frac{2}{7}, \infty\right)$ 8) a) $\sqrt[3]{25}$ b) −3 c) −1 d) $\sqrt[3]{7}$

9) 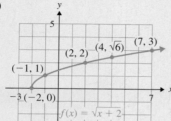   10)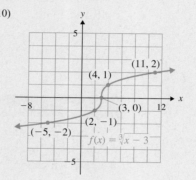

ANSWERS TO TECHNOLOGY EXERCISES

1) 5   2) 6   3) 5.4   4) 18   5) 4   6) 7   7) $[2, \infty)$ 8) $[-3, \infty)$
9) $(-\infty, 2]$ 10) $[-1, \infty)$ 11) $(-\infty, \infty)$ 12) $(-\infty, \infty)$

E Evaluate 9.1 Exercises

Do the exercises, and check your work.

Objective 1: Find Square Roots and Principal Square Roots

Decide whether each statement is true or false. If it is false, explain why.

1) $\sqrt{121} = 11$ and -11.

2) $\sqrt{81} = 9$

3) The square root of a negative number is a negative number.

4) The even root of a negative number is a negative number.

Find all square roots of each number.

5) 144

6) 2500

7) $\dfrac{36}{25}$

8) 0.01

Find each square root, if possible.

9) $\sqrt{49}$

10) $\sqrt{169}$

11) $\sqrt{-4}$

12) $\sqrt{-100}$

13) $\sqrt{\dfrac{81}{25}}$

14) $\sqrt{\dfrac{121}{4}}$

15) $-\sqrt{36}$

16) $-\sqrt{0.04}$

Objective 2: Find Higher Roots

Decide if each statement is true or false. If it is false, explain why.

17) The cube root of a negative number is a negative number.

18) The odd root of a negative number is not a real number.

19) Every nonnegative real number has two real, even roots.

20) $-\sqrt[4]{10{,}000} = -10$

21) Explain how to find $\sqrt[3]{64}$.

22) Explain how to find $\sqrt[4]{16}$.

23) Does $\sqrt[4]{-81} = -3$? Why or why not?

24) Does $\sqrt[3]{-8} = -2$? Why or why not?

Find each root, if possible.

25) $\sqrt[3]{125}$ 26) $\sqrt[3]{27}$

27) $\sqrt[3]{-1}$ 28) $\sqrt[3]{-8}$

29) $\sqrt[4]{81}$ 30) $\sqrt[4]{16}$

31) $\sqrt[4]{-1}$ 32) $\sqrt[4]{-81}$

33) $-\sqrt[4]{16}$ 34) $-\sqrt[4]{1}$

35) $\sqrt[5]{-32}$ 36) $-\sqrt[6]{64}$

37) $-\sqrt[3]{-1000}$ 38) $-\sqrt[3]{-27}$

39) $\sqrt[6]{-64}$ 40) $\sqrt[4]{-16}$

41) $\sqrt[3]{\dfrac{8}{125}}$ 42) $\sqrt[4]{\dfrac{81}{16}}$

43) $\sqrt{60 - 11}$ 44) $\sqrt{100 + 21}$

45) $\sqrt[3]{9 - 36}$ 46) $\sqrt{1 - 9}$

47) $\sqrt{5^2 + 12^2}$ 48) $\sqrt{3^2 + 4^2}$

Objective 3: Evaluate $\sqrt[n]{a^n}$

49) If n is a positive, even integer and we are not certain that $a \geq 0$, then we must use the absolute value symbol to evaluate $\sqrt[n]{a^n}$. That is, $\sqrt[n]{a^n} = |a|$. Why must we use the absolute value symbol?

50) If n is a positive, odd integer then $\sqrt[n]{a^n} = a$ for any value of a. Why don't we need to use the absolute value symbol?

Simplify.

51) $\sqrt{8^2}$ 52) $\sqrt{5^2}$

53) $\sqrt{(-6)^2}$ 54) $\sqrt{(-11)^2}$

55) $\sqrt{y^2}$ 56) $\sqrt{d^2}$

57) $\sqrt[3]{5^3}$ 58) $\sqrt[3]{(-4)^3}$

59) $\sqrt[3]{z^3}$ 60) $\sqrt[7]{t^7}$

61) $\sqrt[4]{h^4}$ 62) $\sqrt[6]{m^6}$

63) $\sqrt{(x+7)^2}$ 64) $\sqrt{(a-9)^2}$

65) $\sqrt[3]{(2t-1)^3}$ 66) $\sqrt[5]{(6r+7)^5}$

67) $\sqrt[4]{(3n+2)^4}$ 68) $\sqrt[3]{(x-6)^3}$

69) $\sqrt[7]{(d-8)^7}$ 70) $\sqrt[6]{(4y+3)^6}$

Objective 4: Determine the Domains of Square Root and Cube Root Functions

71) Is -1 in the domain of $f(x) = \sqrt{x}$? Explain your answer.

72) Is -1 in the domain of $f(x) = \sqrt[3]{x}$? Explain your answer.

73) How do you find the domain of a square root function?

74) What is the domain of a cube root function?

Let $f(x) = \sqrt{x}$ and $g(t) = \sqrt{3t + 4}$. Find each of the following, if possible, and simplify.

75) $f(100)$ 76) $f(9)$

77) $f(-49)$ 78) $f(-64)$

79) $g(-1)$ 80) $g(3)$

81) $g(-5)$ 82) $g(-2)$

83) $f(a)$ 84) $g(w)$

85) $f(t + 4)$ 86) $f(5p - 9)$

87) $g(2n - 1)$ 88) $g(m + 10)$

Let $f(a) = \sqrt[3]{a}$ and $g(x) = \sqrt[3]{4x - 1}$. Find each of the following, and simplify.

89) $f(64)$ 90) $f(-125)$

91) $f(-27)$ 92) $g(3)$

93) $g(-4)$ 94) $g(0)$

95) $g(r)$ 96) $f(z)$

97) $f(c + 8)$ 98) $f(5k - 2)$

99) $g(2a - 3)$ 100) $g(6 - w)$

Determine the domain of each function.

101) $h(n) = \sqrt{n + 2}$ 102) $g(c) = \sqrt{c + 10}$

103) $p(a) = \sqrt{a - 8}$ 104) $f(a) = \sqrt{a - 1}$

105) $f(a) = \sqrt[3]{a - 7}$ 106) $h(t) = \sqrt[3]{t}$

107) $r(k) = \sqrt{3k + 7}$

108) $k(x) = \sqrt{2x - 5}$

109) $g(x) = \sqrt[3]{2x - 5}$

110) $h(c) = \sqrt[3]{-c}$

111) $g(t) = \sqrt{-t}$

112) $h(x) = \sqrt{3 - x}$

113) $r(a) = \sqrt{9 - 7a}$

114) $g(c) = \sqrt{8 - 5c}$

Objective 5: Graph a Square Root Function
Determine the domain, and then graph each function.

115) $f(x) = \sqrt{x - 1}$

116) $g(x) = \sqrt{x - 4}$

117) $g(x) = \sqrt{x + 3}$

118) $h(x) = \sqrt{x + 1}$

119) $h(x) = \sqrt{x} - 2$

120) $f(x) = \sqrt{x} + 2$

121) $f(x) = \sqrt{-x}$

122) $g(x) = \sqrt{-x} - 3$

Objective 6: Graph a Cube Root Function
Determine the domain, and then graph each function.

123) $f(x) = \sqrt[3]{x + 1}$

124) $g(x) = \sqrt[3]{x + 2}$

125) $h(x) = \sqrt[3]{x - 2}$

126) $g(x) = \sqrt[3]{-x}$

127) $g(x) = \sqrt[3]{x} + 1$

128) $h(x) = \sqrt[3]{x} - 1$

129) A stackable storage cube has a storage capacity of 8 ft³. The length of a side of the cube, s, is given by $s = \sqrt[3]{V}$, where V is the volume of the cube. How long is each side of the cube?

130) A die (singular of dice) has a surface area of 13.5 cm³. Since the die is in the shape of a cube the length of a side of the die, s, is given by $s = \sqrt{\dfrac{A}{6}}$, where A is the surface area of the cube. How long is each side of the cube?

131) If an object is dropped from the top of a building 160 ft tall, then the formula $t = \sqrt{\dfrac{160 - h}{16}}$ describes how many seconds, t, it takes for the object to reach a height of h ft above the ground. If a piece of ice falls off the top of this building, how long would it take to reach the ground? Give an exact answer and an approximation to two decimal places.

132) A circular flower garden has an area of 51 ft². The radius, r, of a circle in terms of its area is given by $r = \sqrt{\dfrac{A}{\pi}}$, where A is the area of the circle. What is the radius of the garden? Give an exact answer and an approximation to two decimal places.

133) The speed limit on a street in a residential neighborhood is 25 mph. A car involved in an accident on this road left skid marks 40 ft long. Accident investigators determine that the speed of the car can be described by the function $S(d) = \sqrt{22.5d}$, where $S(d)$ is the speed of the car, in miles per hour, at the time of impact and d is the length of the skid marks, in feet. Was the driver speeding at the time of the accident? Show all work to support your answer.

134) A car involved in an accident on a wet highway leaves skid marks 170 ft long. Accident investigators determine that the speed of the car, $S(d)$ in miles per hour, can be described by the function $S(d) = \sqrt{19.8d}$, where d is the length of the skid marks, in feet. The speed limit on the highway is 65 mph. Was the car speeding when the accident occurred? Show all work to support your answer.

Use the following information for Exercises 135–140.

The period of a pendulum is the time it takes for the pendulum to make one complete swing back and forth. The period, $T(L)$ in seconds, can be described by the function $T(L) = 2\pi\sqrt{\dfrac{L}{32}}$, where L is the length of the pendulum, in feet. For each exercise give an *exact answer and an answer rounded to two decimal places*. Use 3.14 for π.

135) Find the period of the pendulum whose length is 8 ft.

136) Find the period of the pendulum whose length is 2 ft.

137) Find $T\left(\dfrac{1}{2}\right)$, and explain what it means in the context of the problem.

138) Find $T(1.5)$, and explain what it means in the context of the problem.

139) Find the period of a 30-inch-long pendulum.

140) Find the period of a 54-inch-long pendulum.

R Rethink

R1) Would you be able to complete exercises similar to these on a quiz?

R2) Do you have a firm grasp on graphing?

9.2 Rational Exponents

Prepare

What are your objectives for Section 9.2?	How can you accomplish each objective?
1 Evaluate Expressions of the Form $a^{1/n}$	• Write the definition of $a^{1/n}$ in your own words. • Complete the given example on your own. • Complete You Try 1.
2 Evaluate Expressions of the Form $a^{m/n}$	• Write the definition of $a^{m/n}$ in your own words. • Complete the given example on your own. • Complete You Try 2.
3 Evaluate Expression of the Form $a^{-m/n}$	• Write the definition of $a^{-m/n}$ in your own words. • Complete the given example on your own. • Complete You Try 3.
4 Combine the Rules of Exponents	• Review the rules of exponents found in Sections 6.1 and 6.2 to help follow the examples. • Complete the given examples on your own. • Complete You Trys 4 and 5.
5 Convert a Radical Expression to Exponential Form and Simplify	• Use the same rules of exponents to convert between radical and exponential forms. • Complete the given examples on your own. • Complete You Trys 6 and 7.

Work Read the explanations, follow the examples, take notes, and complete the You Trys.

1 Evaluate Expressions of the Form $a^{1/n}$

In this section, we will explain the relationship between radicals and rational exponents (fractional exponents). Sometimes, converting between these two forms makes it easier to simplify expressions.

Definition

If n is a positive integer greater than 1 and $\sqrt[n]{a}$ is a real number, then

$$a^{1/n} = \sqrt[n]{a}$$

(The denominator of the fractional exponent is the index of the radical.)

EXAMPLE 1 Write in radical form, and evaluate.

a) $8^{1/3}$ b) $49^{1/2}$ c) $81^{1/4}$
d) $-64^{1/6}$ e) $(-16)^{1/4}$ f) $(-125)^{1/3}$

Solution

a) The denominator of the fractional exponent is the index of the radical. Therefore, $8^{1/3} = \sqrt[3]{8} = 2$.

b) The denominator in the exponent of $49^{1/2}$ is 2, so the index on the radical is 2, meaning *square* root.
$$49^{1/2} = \sqrt{49} = 7$$

c) $81^{1/4} = \sqrt[4]{81} = 3$

d) $-64^{1/6} = -(64^{1/6}) = -\sqrt[6]{64} = -2$

e) $(-16)^{1/4} = \sqrt[4]{-16}$, which is not a real number. Remember, the even root of a negative number is not a real number.

f) $(-125)^{1/3} = \sqrt[3]{-125} = -5$ The odd root of a negative number is a negative number.

[YOU TRY 1] Write in radical form, and evaluate.

a) $16^{1/4}$ b) $121^{1/2}$ c) $(-121)^{1/2}$ d) $(-27)^{1/3}$ e) $-81^{1/4}$

2 Evaluate Expressions of the Form $a^{m/n}$

We can add another relationship between rational exponents and radicals.

Definition

If m and n are positive integers and $\dfrac{m}{n}$ is in lowest terms, then
$$a^{m/n} = (a^{1/n})^m = (\sqrt[n]{a})^m$$
provided that $a^{1/n}$ is a real number.

(The denominator of the fractional exponent is the index of the radical, and the numerator is the power to which we raise the radical expression.) We can also think of $a^{m/n}$ this way: $a^{m/n} = (a^m)^{1/n} = \sqrt[n]{a^m}$.

EXAMPLE 2 Write in radical form, and evaluate.

a) $25^{3/2}$ b) $-64^{2/3}$ c) $(-81)^{3/2}$ d) $-81^{3/2}$ e) $(-1000)^{2/3}$

Solution

a) The *denominator* of the fractional exponent is the *index* of the radical, and the *numerator* is the *power* to which we raise the radical expression.

$25^{3/2} = (25^{1/2})^3$ Use the definition to rewrite the exponent.
$\phantom{25^{3/2}} = (\sqrt{25})^3$ Rewrite as a radical.
$\phantom{25^{3/2}} = 5^3$ $\sqrt{25} = 5$
$\phantom{25^{3/2}} = 125$

> **W Hint**
> Write a procedure in your notes for converting from exponential form to radical form.

b) To evaluate $-64^{2/3}$, *first* evaluate $64^{2/3}$, *then* take the negative of that result.

$$-64^{2/3} = -(64^{2/3}) = -(64^{1/3})^2 \quad \text{Use the definition to rewrite the exponent.}$$
$$= -(\sqrt[3]{64})^2 \quad \text{Rewrite as a radical.}$$
$$= -(4)^2 \quad \sqrt[3]{64} = 4$$
$$= -16$$

c) $(-81)^{3/2} = [(-81)^{1/2}]^3$
$= (\sqrt{-81})^3$ Not a real number.   The even root of a negative number is not a real number.

d) $-81^{3/2} = -(81^{1/2})^3 = -(\sqrt{81})^3 = -(9)^3 = -729$

e) $(-1000)^{2/3} = [(-1000)^{1/3}]^2 = (\sqrt[3]{-1000})^2 = (-10)^2 = 100$

[**YOU TRY 2**] Write in radical form, and evaluate.

a) $32^{2/5}$ b) $-100^{3/2}$ c) $(-100)^{3/2}$ d) $(-1)^{4/5}$ e) $-1^{5/3}$

 BE CAREFUL In Example 2, notice how the parentheses affect how we evaluate an expression. The base of the expression $(-81)^{3/2}$ is -81, while the base of $-81^{3/2}$ is 81.

3 Evaluate Expressions of the Form $a^{-m/n}$

Recall that if n is any integer and $a \neq 0$, then $a^{-n} = \left(\dfrac{1}{a}\right)^n = \dfrac{1}{a^n}$.

That is, to rewrite the expression with a *positive* exponent, take the reciprocal of the base. For example,

$$2^{-4} = \left(\frac{1}{2}\right)^4 = \frac{1}{16}$$

We can extend this idea to rational exponents.

> ### Definition
> If $a^{m/n}$ is a nonzero real number, then
> $$a^{-m/n} = \left(\frac{1}{a}\right)^{m/n} = \frac{1}{a^{m/n}}$$
> (To rewrite the expression with a *positive* exponent, take the reciprocal of the base.)

EXAMPLE 3 Rewrite with a positive exponent, and evaluate.

a) $36^{-1/2}$ b) $32^{-2/5}$ c) $\left(\dfrac{125}{64}\right)^{-2/3}$

Solution

W Hint
Write out the steps as you are reading the example!

a) To write $36^{-1/2}$ with a positive exponent, take the reciprocal of the base.

$$36^{-1/2} = \left(\frac{1}{36}\right)^{1/2}$$ The reciprocal of 36 is $\frac{1}{36}$.

$$= \sqrt{\frac{1}{36}}$$ The denominator of the fractional exponent is the index of the radical.

$$= \frac{1}{6}$$

b) $$32^{-2/5} = \left(\frac{1}{32}\right)^{2/5}$$ The reciprocal of 32 is $\frac{1}{32}$.

$$= \left(\sqrt[5]{\frac{1}{32}}\right)^2$$ The denominator of the fractional exponent is the index of the radical.

$$= \left(\frac{1}{2}\right)^2 \quad \sqrt[5]{\frac{1}{32}} = \frac{1}{2}$$

$$= \frac{1}{4}$$

W Hint
The negative exponent does not make the expression negative!

c) $$\left(\frac{125}{64}\right)^{-2/3} = \left(\frac{64}{125}\right)^{2/3}$$ The reciprocal of $\frac{125}{64}$ is $\frac{64}{125}$.

$$= \left(\sqrt[3]{\frac{64}{125}}\right)^2$$ The denominator of the fractional exponent is the index of the radical.

$$= \left(\frac{4}{5}\right)^2 \quad \sqrt[3]{\frac{64}{125}} = \frac{4}{5}$$

$$= \frac{16}{25}$$

[**YOU TRY 3**] Rewrite with a positive exponent, and evaluate.

a) $144^{-1/2}$ b) $16^{-3/4}$ c) $\left(\frac{8}{27}\right)^{-2/3}$

4 Combine the Rules of Exponents

We can combine the rules presented in this section with the rules of exponents we learned in Chapter 6 to simplify expressions containing numbers or variables.

EXAMPLE 4

Simplify completely. The answer should contain only positive exponents.

a) $(6^{1/5})^2$ b) $25^{3/4} \cdot 25^{-1/4}$ c) $\dfrac{8^{2/9}}{8^{11/9}}$

Solution

a) $(6^{1/5})^2 = 6^{2/5}$ Multiply exponents.

b) $25^{3/4} \cdot 25^{-1/4} = 25^{\frac{3}{4}+\left(-\frac{1}{4}\right)}$ Add exponents.

$\qquad\qquad\qquad = 25^{2/4} = 25^{1/2} = 5$

c) $\dfrac{8^{2/9}}{8^{11/9}} = 8^{\frac{2}{9} - \frac{11}{9}}$ Subtract exponents.

$= 8^{-9/9}$ Subtract $\dfrac{2}{9} - \dfrac{11}{9}$.

$= 8^{-1}$ Reduce $-\dfrac{9}{9}$.

$= \left(\dfrac{1}{8}\right)^1 = \dfrac{1}{8}$

[YOU TRY 4] Simplify completely. The answer should contain only positive exponents.

a) $49^{3/8} \cdot 49^{1/8}$ b) $(16^{1/12})^3$ c) $\dfrac{7^{2/5}}{7^{4/5}}$

EXAMPLE 5

Simplify completely. Assume the variables represent positive real numbers. The answer should contain only positive exponents.

a) $r^{1/8} \cdot r^{3/8}$ b) $\left(\dfrac{x^{2/3}}{y^{1/4}}\right)^6$ c) $\dfrac{n^{-5/6} \cdot n^{1/3}}{n^{-1/6}}$ d) $\left(\dfrac{a^{-7}b^{1/2}}{a^5 c^{1/3}}\right)^{-3/4}$

Solution

a) $r^{1/8} \cdot r^{3/8} = r^{\frac{1}{8} + \frac{3}{8}}$ Add exponents.

$= r^{4/8} = r^{1/2}$

b) $\left(\dfrac{x^{2/3}}{y^{1/4}}\right)^6 = \dfrac{x^{\frac{2}{3} \cdot 6}}{y^{\frac{1}{4} \cdot 6}}$ Multiply exponents.

$= \dfrac{x^4}{y^{3/2}}$ Reduce.

c) $\dfrac{n^{-5/6} \cdot n^{1/3}}{n^{-1/6}} = \dfrac{n^{-\frac{5}{6} + \frac{1}{3}}}{n^{-1/6}} = \dfrac{n^{-\frac{5}{6} + \frac{2}{6}}}{n^{-1/6}} = \dfrac{n^{-3/6}}{n^{-1/6}}$ Add exponents.

$= n^{-\frac{3}{6} - \left(-\frac{1}{6}\right)} = n^{-2/6} = n^{-1/3} = \dfrac{1}{n^{1/3}}$ Subtract exponents.

d) $\left(\dfrac{a^{-7}b^{1/2}}{a^5 b^{1/3}}\right)^{-3/4} = \left(\dfrac{a^5 b^{1/3}}{a^{-7}b^{1/2}}\right)^{3/4}$ Eliminate the negative from the outermost exponent by taking the reciprocal of the base.

Simplify the expression inside the parentheses by subtracting the exponents.

$= (a^{5-(-7)}b^{1/3 - 1/2})^{3/4} = (a^{5+7}b^{2/6 - 3/6})^{3/4} = (a^{12}b^{-1/6})^{3/4}$

Apply the power rule, and simplify.

$= (a^{12})^{3/4}(b^{-1/6})^{3/4} = a^9 b^{-1/8} = \dfrac{a^9}{b^{1/8}}$

[YOU TRY 5] Simplify completely. Assume the variables represent positive real numbers. The answer should contain only positive exponents.

a) $(a^3 b^{1/5})^{10}$ b) $\dfrac{t^{3/10}}{t^{7/10}}$ c) $\dfrac{s^{3/4}}{s^{1/2} \cdot s^{-5/4}}$ d) $\left(\dfrac{x^4 y^{3/8}}{x^9 y^{1/4}}\right)^{-2/5}$

5 Convert a Radical Expression to Exponential Form and Simplify

Some radicals can be simplified by first putting them into rational exponent form and then converting them back to radicals.

EXAMPLE 6

Rewrite each radical in exponential form, then simplify. Write the answer in simplest (or radical) form. Assume the variable represents a nonnegative real number.

a) $\sqrt[8]{9^4}$ b) $\sqrt[6]{s^4}$

Solution

a) Because the index of the radical is the denominator of the exponent and the power is the numerator, we can write

$\sqrt[8]{9^4} = 9^{4/8}$ Write with a rational exponent.
$= 9^{1/2} = 3$

b) $\sqrt[6]{s^4} = s^{4/6}$ Write with a rational exponent.
$= s^{2/3} = \sqrt[3]{s^2}$

The expression $\sqrt[6]{s^4}$ is not in simplest form because the 4 and the 6 contain a common factor of 2, but $\sqrt[3]{s^2}$ is in simplest form because 2 and 3 do not have any common factors besides 1.

YOU TRY 6

Rewrite each radical in exponential form, then simplify. Write the answer in simplest (or radical) form. Assume the variable represents a nonnegative real number.

a) $\sqrt[6]{125^2}$ b) $\sqrt[10]{p^4}$

In Section 9.1 we said that if a is negative and n is a positive, even number, then $\sqrt[n]{a^n} = |a|$. For example, if we are *not* told that k is positive, then $\sqrt{k^2} = |k|$. However, if we assume that k is positive, then $\sqrt{k^2} = k$. **In the rest of this chapter, we will assume that all variables represent positive, real numbers unless otherwise stated.** When we make this assumption, we do not need to use absolute values when simplifying even roots. And if we consider this together with the relationship between radicals and rational exponents we have another way to explain why $\sqrt[n]{a^n} = a$.

EXAMPLE 7

Simplify.

a) $\sqrt[3]{5^3}$ b) $(\sqrt[4]{9})^4$ c) $\sqrt{k^2}$

Solution

a) $\sqrt[3]{5^3} = (5^3)^{1/3} = 5^{3 \cdot \frac{1}{3}} = 5^1 = 5$

b) $(\sqrt[4]{9})^4 = (9^{1/4})^4 = 9^{\frac{1}{4} \cdot 4} = 9^1 = 9$

c) $\sqrt{k^2} = (k^2)^{1/2} = k^{2 \cdot \frac{1}{2}} = k^1 = k$

YOU TRY 7

Simplify.

a) $(\sqrt{10})^2$ b) $\sqrt[3]{7^3}$ c) $\sqrt[4]{t^4}$

Using Technology

We can evaluate square roots, cube roots, or even higher roots by first rewriting the radical in exponential form and then using a graphing calculator.

For example, to evaluate $\sqrt{49}$, first rewrite the radical as $49^{1/2}$, then enter [4] [9], press [^] [(], enter [1] [÷] [2], and press [)] [ENTER]. The result is 7, as shown on the screen on the left below.

To approximate $\sqrt[3]{12^2}$ rounded to the nearest tenth, first rewrite the radical as $12^{2/3}$, then enter [1] [2], press [^] [(], enter [2] [÷] [3], and press [)] [ENTER]. The result is 5.241482788 as shown on the screen on the right below. The result rounded to the nearest tenth is then 5.2.

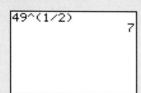

To evaluate radicals with an index greater than 3, follow the same procedure explained above. Evaluate by rewriting in exponential form if necessary and then using a graphing calculator. If necessary, approximate to the nearest tenth.

1) $16^{1/2}$ 2) $\sqrt[3]{512}$ 3) $\sqrt{37}$ 4) $361^{1/2}$ 5) $4096^{2/3}$ 6) $2401^{1/4}$

ANSWERS TO [YOU TRY] EXERCISES

1) a) 2   b) 11   c) not a real number   d) −3   e) −3   2) a) 4   b) −1000
c) not a real number   d) 1   e) −1   3) a) $\frac{1}{12}$ b) $\frac{1}{8}$ c) $\frac{9}{4}$ 4) a) 7   b) 2   c) $\frac{1}{7^{2/5}}$
5) a) $a^{30}b^2$ b) $\frac{1}{t^{2/5}}$ c) $s^{3/2}$ d) $\frac{x^2}{y^{1/20}}$ 6) a) 5   b) $\sqrt[5]{p^2}$ 7) a) 10   b) 7   c) t

ANSWERS TO TECHNOLOGY EXERCISES

1) 4   2) 8   3) 6.1   4) 19   5) 256   6) 7

E Evaluate 9.2 Exercises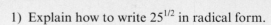

Do the exercises, and check your work.

Objective 1: Evaluate Expressions of the Form $a^{1/n}$

1) Explain how to write $25^{1/2}$ in radical form.
2) Explain how to write $1^{1/3}$ in radical form.

Write in radical form, and evaluate.

3) $9^{1/2}$
4) $64^{1/2}$
5) $1000^{1/3}$
6) $27^{1/3}$
7) $32^{1/5}$
8) $81^{1/4}$
9) $-64^{1/6}$
10) $-125^{1/3}$
11) $\left(\frac{4}{121}\right)^{1/2}$
12) $\left(\frac{4}{9}\right)^{1/2}$
13) $\left(\frac{125}{64}\right)^{1/3}$
14) $\left(\frac{16}{81}\right)^{1/4}$
15) $-\left(\frac{36}{169}\right)^{1/2}$
16) $-\left(\frac{1000}{27}\right)^{1/3}$
17) $(-81)^{1/4}$
18) $(-169)^{1/2}$
19) $(-1)^{1/7}$
20) $(-8)^{1/3}$

Objective 2: Evaluate Expressions of the Form $a^{m/n}$

21) Explain how to write $16^{3/4}$ in radical form.

22) Explain how to write $100^{3/2}$ in radical form.

Write in radical form, and evaluate.

23) $8^{4/3}$
24) $81^{3/4}$
25) $64^{5/6}$
26) $32^{3/5}$
27) $(-125)^{2/3}$
28) $(-1000)^{2/3}$
29) $-36^{3/2}$
30) $-27^{4/3}$
31) $(-81)^{3/4}$
32) $(-25)^{3/2}$
33) $\left(\dfrac{16}{81}\right)^{3/4}$
34) $-16^{5/4}$
35) $-\left(\dfrac{1000}{27}\right)^{2/3}$
36) $-\left(\dfrac{8}{27}\right)^{4/3}$

Objective 3: Evaluate Expressions of the Form $a^{-m/n}$

Decide whether each statement is true or false. Explain your answer.

37) $81^{-1/2} = -9$
38) $\left(\dfrac{1}{100}\right)^{-3/2} = \left(\dfrac{1}{100}\right)^{2/3}$

Rewrite with a positive exponent, and evaluate.

Fill It In

Fill in the blanks with either the missing mathematical step or reason for the given step.

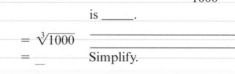

39) $64^{-1/2} = \left(\right)^{1/2}$ The reciprocal of 64 is ___.

$= \sqrt{\dfrac{1}{64}}$ _____

$= \underline{}$ Simplify.

40) $\left(\dfrac{1}{1000}\right)^{-1/3} = ()^{1/3}$ The reciprocal of $\dfrac{1}{1000}$ is ___.

$= \sqrt[3]{1000}$ _____

$= \underline{}$ Simplify.

41) $49^{-1/2}$
42) $100^{-1/2}$
43) $1000^{-1/3}$
44) $27^{-1/3}$
45) $\left(\dfrac{1}{81}\right)^{-1/4}$
46) $\left(\dfrac{1}{32}\right)^{-1/5}$
47) $-\left(\dfrac{1}{64}\right)^{-1/3}$
48) $-\left(\dfrac{1}{125}\right)^{-1/3}$
49) $64^{-5/6}$
50) $81^{-3/4}$

51) $125^{-2/3}$
52) $64^{-2/3}$
53) $\left(\dfrac{25}{4}\right)^{-3/2}$
54) $\left(\dfrac{9}{100}\right)^{-3/2}$
55) $\left(\dfrac{64}{125}\right)^{-2/3}$
56) $\left(\dfrac{81}{16}\right)^{-3/4}$

Objective 4: Combine the Rules of Exponents

Simplify completely. The answer should contain only positive exponents.

57) $2^{2/3} \cdot 2^{7/3}$
58) $5^{3/4} \cdot 5^{5/4}$
59) $(9^{1/4})^2$
60) $(7^{2/3})^3$
61) $8^{7/5} \cdot 8^{-3/5}$
62) $6^{-4/3} \cdot 6^{5/3}$
63) $\dfrac{2^{23/4}}{2^{3/4}}$
64) $\dfrac{5^{3/2}}{5^{9/2}}$
65) $\dfrac{4^{2/5}}{4^{6/5} \cdot 4^{3/5}}$
66) $\dfrac{6^{-1}}{6^{1/2} \cdot 6^{-5/2}}$

Simplify completely. The answer should contain only positive exponents.

67) $z^{1/6} \cdot z^{5/6}$
68) $h^{1/6} \cdot h^{-3/4}$
69) $(-9v^{5/8})(8v^{3/4})$
70) $(-3x^{-1/3})(8x^{4/9})$
71) $\dfrac{a^{5/9}}{a^{4/9}}$
72) $\dfrac{x^{1/6}}{x^{5/6}}$
73) $\dfrac{48w^{3/10}}{10w^{2/5}}$
74) $\dfrac{20c^{-2/3}}{72c^{5/6}}$
75) $(x^{-2/9})^3$
76) $(n^{-2/7})^3$
77) $(z^{1/5})^{2/3}$
78) $(r^{4/3})^{5/2}$
79) $(81u^{8/3}v^4)^{3/4}$
80) $(64x^6 y^{12/5})^{5/6}$
81) $(32r^{1/3}s^{4/9})^{3/5}$
82) $(125a^9 b^{1/4})^{2/3}$
83) $\left(\dfrac{f^{6/7}}{27g^{-5/3}}\right)^{1/3}$
84) $\left(\dfrac{16c^{-8}}{b^{-11/3}}\right)^{3/4}$
85) $\left(\dfrac{x^{-5/3}}{w^{3/2}}\right)^{-6}$
86) $\left(\dfrac{t^{-3/2}}{u^{1/4}}\right)^{-4}$
87) $\dfrac{y^{1/2} \cdot y^{-1/3}}{y^{5/6}}$
88) $\dfrac{t^5}{t^{1/2} \cdot t^{3/4}}$
89) $\left(\dfrac{a^4 b^3}{32a^{-2}b^4}\right)^{2/5}$
90) $\left(\dfrac{16c^{-8}d^3}{c^4 d^5}\right)^{3/2}$
91) $\left(\dfrac{r^{4/5}t^{-2}}{r^{2/3}t^5}\right)^{-3/2}$
92) $\left(\dfrac{x^{10}y^{1/6}}{x^{-8}y^{2/3}}\right)^{-2/3}$
93) $\left(\dfrac{h^{-2}k^{5/2}}{h^{-8}k^{5/6}}\right)^{-5/6}$
94) $\left(\dfrac{c^{1/8}d^{-4}}{c^{3/4}d}\right)^{-8/5}$
95) $p^{1/2}(p^{2/3} + p^{1/2})$
96) $w^{4/3}(w^{1/2} - w^3)$

Objective 5: Convert a Radical Expression to Exponential Form and Simplify

Rewrite each radical in exponential form, then simplify. Write the answer in simplest (or radical) form.

Fill It In
Fill in the blanks with either the missing mathematical step or reason for the given step.

97) $\sqrt[12]{25^6}$ = _____ Write with a rational exponent.
    = _____ Reduce the exponent.
    = 5   _____

98) $\sqrt[10]{c^4}$ = _____ Write with a rational exponent.
    = $c^{2/5}$ _____
    = _____ Write in radical form.

99) $\sqrt[6]{49^3}$ 100) $\sqrt[9]{8^3}$
101) $\sqrt[4]{81^2}$ 102) $\sqrt{3^2}$
103) $(\sqrt{5})^2$ 104) $(\sqrt[3]{10})^3$
105) $(\sqrt[3]{12})^3$ 106) $(\sqrt[4]{15})^4$
107) $\sqrt[3]{x^{12}}$ 108) $\sqrt[4]{t^8}$
109) $\sqrt[6]{k^2}$ 110) $\sqrt[9]{w^6}$
111) $\sqrt[4]{z^2}$ 112) $\sqrt[8]{m^4}$
113) $\sqrt{d^4}$ 114) $\sqrt{s^6}$

The wind chill temperature, WC, measures how cold it feels outside (for temperatures under 50 degrees F) when the velocity of the wind, V, is considered along with the air temperature, T. The stronger the wind at a given air temperature, the colder it feels.

The formula for calculating wind chill is

$$WC = 35.74 + 0.6215T - 35.75V^{4/25} + 0.4275TV^{4/25}$$

where WC and T are in degrees Fahrenheit and V is in miles per hour. (http://www.nws.noaa.gov/om/windchill/windchillglossary.shtml)

Use this information for Exercises 115 and 116, and round all answers to the nearest degree.

115) Determine the wind chill when the air temperature is 20 degrees and the wind is blowing at the given speed.

   a) 5 mph
   b) 15 mph

116) Determine the wind chill when the air temperature is 10 degrees and the wind is blowing at the given speed. Round your answer to the nearest degree.

   a) 12 mph
   b) 20 mph

R Rethink

R1) Describe how having a good understanding of the rules of exponents has helped you move quickly through the exercises.

R2) Where could you spend more time mastering an objective?

9.3 Simplifying Expressions Containing Square Roots

P Prepare

O Organize

What are your objectives for Section 9.3?	How can you accomplish each objective?
1 Multiply Square Roots	• Write the definition of the *product rule for square roots* in your own words. • Complete the given example on your own. • Complete You Try 1.
2 Simplify the Square Root of a Whole Number	• Write the property for **When Is a Square Root Simplified?** • Complete the given example on your own. • Complete You Try 2.
3 Use the Quotient Rule for Square Roots	• Write the definition of the *quotient rule for square roots* in your own words. • Complete the given example on your own. • Complete You Try 3.
4 Simplify Square Root Expressions Containing Variables with Even Exponents	• Write the property for $\sqrt{a^m}$ in your notes. • Complete the given example on your own. • Complete You Try 4.
5 Simplify Square Root Expressions Containing Variables with Odd Exponents	• Write the procedure for **Simplifying a Radical Containing Variables** in your own words. • Complete the given examples on your own. • Complete You Trys 5–7.
6 Simplify More Square Root Expressions Containing Variables	• Apply the rules from previous objectives, and follow the example. • Complete You Try 8.

Read the explanations, follow the examples, take notes, and complete the You Trys.

In this section, we will introduce rules for finding the product and quotient of square roots as well as for simplifying expressions containing square roots.

1 Multiply Square Roots

Let's begin with the product $\sqrt{4} \cdot \sqrt{9}$. We can find the product like this: $\sqrt{4} \cdot \sqrt{9} = 2 \cdot 3 = 6$. Also notice that $\sqrt{4} \cdot \sqrt{9} = \sqrt{4 \cdot 9} = \sqrt{36} = 6$.

We obtain the same result. This leads us to the product rule for multiplying expressions containing square roots.

> **Definition** Product Rule for Square Roots
>
> Let a and b be nonnegative real numbers. Then,
>
> $$\sqrt{a} \cdot \sqrt{b} = \sqrt{a \cdot b}$$
>
> In other words, the product of two square roots equals the square root of the product.

EXAMPLE 1 Multiply. a) $\sqrt{5} \cdot \sqrt{2}$ b) $\sqrt{3} \cdot \sqrt{x}$

Solution

a) $\sqrt{5} \cdot \sqrt{2} = \sqrt{5 \cdot 2} = \sqrt{10}$ b) $\sqrt{3} \cdot \sqrt{x} = \sqrt{3 \cdot x} = \sqrt{3x}$

 BE CAREFUL We can multiply radicals this way *only if* the indices are the same. We will see later how to multiply radicals with different indices such as $\sqrt{5} \cdot \sqrt[3]{t}$.

[YOU TRY 1] Multiply.

a) $\sqrt{6} \cdot \sqrt{5}$ b) $\sqrt{10} \cdot \sqrt{r}$

2 Simplify the Square Root of a Whole Number

Knowing how to simplify radicals is very important in the study of algebra. We begin by discussing how to simplify expressions containing square roots.

How do we know when a square root is simplified?

> **Property** When Is a Square Root Simplified?
>
> An expression containing a square root is simplified when all of the following conditions are met:
>
> 1) The radicand does not contain any factors (other than 1) that are perfect squares.
> 2) The radicand does not contain any fractions.
> 3) There are no radicals in the denominator of a fraction.
>
> Note: Condition 1) implies that the radical cannot contain variables with exponents greater than or equal to 2, the index of the square root.

We will discuss higher roots in Section 9.4.

To simplify expressions containing square roots we reverse the process of multiplying. That is, we use the product rule that says $\sqrt{a \cdot b} = \sqrt{a} \cdot \sqrt{b}$ where a or b are perfect squares.

EXAMPLE 2 Simplify completely.

a) $\sqrt{18}$ b) $\sqrt{500}$ c) $\sqrt{21}$ d) $\sqrt{48}$

Solution

a) The radical $\sqrt{18}$ is not in simplest form since 18 contains a factor (other than 1) that is a perfect square. Think of two numbers that multiply to 18 so that at least one of the numbers is a perfect square: $18 = 9 \cdot 2$.

(While it is true that $18 = 6 \cdot 3$, neither 6 nor 3 is a perfect square.)
Rewrite $\sqrt{18}$:

$$\begin{aligned}\sqrt{18} &= \sqrt{9 \cdot 2} &&\text{9 is a perfect square.}\\ &= \sqrt{9} \cdot \sqrt{2} &&\text{Product rule}\\ &= 3\sqrt{2} &&\sqrt{9} = 3\end{aligned}$$

$3\sqrt{2}$ is completely simplified because 2 does not have any factors that are perfect squares.

b) Does 500 have a factor that is a perfect square? *Yes!* $500 = 100 \cdot 5$. To simplify $\sqrt{500}$, rewrite it as

$$\begin{aligned}\sqrt{500} &= \sqrt{100 \cdot 5} &&\text{100 is a perfect square.}\\ &= \sqrt{100} \cdot \sqrt{5} &&\text{Product rule}\\ &= 10\sqrt{5} &&\sqrt{100} = 10\end{aligned}$$

$10\sqrt{5}$ is completely simplified because 5 does not have any factors that are perfect squares.

W Hint
What is the first question you should ask when simplifying?

c) $21 = 3 \cdot 7$ Neither 3 nor 7 is a perfect square.
   $21 = 1 \cdot 21$ Although 1 is a perfect square, it will not help us simplify $\sqrt{21}$.

$\sqrt{21}$ is in simplest form.

d) There are different ways to simplify $\sqrt{48}$. We will look at two of them.

  i) Two numbers that multiply to 48 are 16 and 3 with 16 being a perfect square. We can write

  $$\sqrt{48} = \sqrt{16 \cdot 3} = \sqrt{16} \cdot \sqrt{3} = 4\sqrt{3}$$

  ii) We can also think of 48 as $4 \cdot 12$ since 4 is a perfect square. We can write

  $$\sqrt{48} = \sqrt{4 \cdot 12} = \sqrt{4} \cdot \sqrt{12} = 2\sqrt{12}$$

  Therefore, $\sqrt{48} = 2\sqrt{12}$. Is $\sqrt{12}$ in simplest form? *No, because $12 = 4 \cdot 3$ and 4 is a perfect square.* We must continue to simplify.

  $$\begin{aligned}\sqrt{48} &= 2\sqrt{12}\\ &= 2\sqrt{4 \cdot 3} = 2\sqrt{4} \cdot \sqrt{3} = 2 \cdot 2 \cdot \sqrt{3} = 4\sqrt{3}\end{aligned}$$

  $4\sqrt{3}$ is completely simplified because 3 does not have any factors that are perfect squares.

Example 2(d) shows that using either $\sqrt{48} = \sqrt{16 \cdot 3}$ or $\sqrt{48} = \sqrt{4 \cdot 12}$ leads us to the same result. Furthermore, this example illustrates that a radical is not always *completely* simplified after just one iteration of the simplification process. It is necessary to always examine the radical to determine whether or not it can be simplified more.

> **Note**
> After simplifying a radical, look at the result and ask yourself, "Is the radical in simplest form?" If it is not, simplify again. Asking yourself this question will help you to be sure that the radical *is* completely simplified.

[YOU TRY 2] Simplify completely.

a) $\sqrt{28}$ b) $\sqrt{75}$ c) $\sqrt{72}$

3 Use the Quotient Rule for Square Roots

Let's simplify $\dfrac{\sqrt{36}}{\sqrt{9}}$. We can say $\dfrac{\sqrt{36}}{\sqrt{9}} = \dfrac{6}{3} = 2$. It is also true that $\dfrac{\sqrt{36}}{\sqrt{9}} = \sqrt{\dfrac{36}{9}} = \sqrt{4} = 2$.

This leads us to the quotient rule for dividing expressions containing square roots.

> **Definition** Quotient Rule for Square Roots
>
> Let a and b be nonnegative real numbers such that $b \neq 0$. Then,
>
> $$\sqrt{\dfrac{a}{b}} = \dfrac{\sqrt{a}}{\sqrt{b}}$$
>
> The square root of a quotient equals the quotient of the square roots.

EXAMPLE 3 Simplify completely.

a) $\sqrt{\dfrac{9}{49}}$ b) $\sqrt{\dfrac{200}{2}}$ c) $\dfrac{\sqrt{72}}{\sqrt{6}}$ d) $\sqrt{\dfrac{5}{81}}$

Solution

a) Because 9 and 49 are each perfect squares, find the square root of each separately.

$$\sqrt{\dfrac{9}{49}} = \dfrac{\sqrt{9}}{\sqrt{49}} \quad \text{Quotient rule}$$

$$= \dfrac{3}{7} \quad \sqrt{9} = 3 \text{ and } \sqrt{49} = 7$$

b) Neither 200 nor 2 is a perfect square, but if we simplify $\dfrac{200}{2}$ we get 100, which *is* a perfect square.

$$\sqrt{\dfrac{200}{2}} = \sqrt{100} \quad \text{Simplify } \dfrac{200}{2}.$$

$$= 10$$

Hint
Which method do you prefer?

c) We can simplify $\dfrac{\sqrt{72}}{\sqrt{6}}$ using two different methods.

i) Begin by applying the quotient rule to obtain a fraction under *one* radical and simplify the fraction.

$$\dfrac{\sqrt{72}}{\sqrt{6}} = \sqrt{\dfrac{72}{6}} \quad \text{Quotient rule}$$
$$= \sqrt{12} = \sqrt{4 \cdot 3} = \sqrt{4} \cdot \sqrt{3} = 2\sqrt{3}$$

ii) We can apply the product rule to rewrite $\sqrt{72}$ then simplify the fraction.

$$\dfrac{\sqrt{72}}{\sqrt{6}} = \dfrac{\sqrt{6} \cdot \sqrt{12}}{\sqrt{6}} \quad \text{Product rule}$$
$$= \dfrac{\cancel{\sqrt{6}}^{1} \cdot \sqrt{12}}{\cancel{\sqrt{6}}_{1}} \quad \text{Divide out the common factor.}$$
$$= \sqrt{12} = \sqrt{4 \cdot 3} = \sqrt{4} \cdot \sqrt{3} = 2\sqrt{3}$$

Either method will produce the same result.

d) The fraction $\dfrac{5}{81}$ is in simplest form, and 81 *is* a perfect square. Begin by applying the quotient rule.

$$\sqrt{\dfrac{5}{81}} = \dfrac{\sqrt{5}}{\sqrt{81}} \quad \text{Quotient rule}$$
$$= \dfrac{\sqrt{5}}{9} \quad \sqrt{81} = 9$$

[YOU TRY 3] Simplify completely.

a) $\sqrt{\dfrac{100}{169}}$ b) $\sqrt{\dfrac{27}{3}}$ c) $\dfrac{\sqrt{250}}{\sqrt{5}}$ d) $\sqrt{\dfrac{11}{36}}$

4 Simplify Square Root Expressions Containing Variables with Even Exponents

Recall that a square root is not simplified if it contains any factors that are perfect squares.
This means that a square root containing variables is simplified if the power on each variable is less than 2. For example, $\sqrt{r^6}$ is not in simplified form. If r represents a nonnegative real number, then we can use rational exponents to simplify $\sqrt{r^6}$.

$$\sqrt{r^6} = (r^6)^{1/2} = r^{6 \cdot \frac{1}{2}} = r^{6/2} = r^3$$

Multiplying $6 \cdot \frac{1}{2}$ is the same as dividing 6 by 2. We can generalize this result with the following statement.

Property $\sqrt{a^m}$

If a is a nonnegative real number and m is an integer, then

$$\sqrt{a^m} = a^{m/2}$$

We can combine this property with the product and quotient rules to simplify radical expressions.

EXAMPLE 4

Simplify completely.

a) $\sqrt{z^2}$ b) $\sqrt{49t^2}$ c) $\sqrt{18b^{14}}$ d) $\sqrt{\dfrac{32}{n^{20}}}$

Solution

a) $\sqrt{z^2} = z^{2/2} = z^1 = z$

b) $\sqrt{49t^2} = \sqrt{49} \cdot \sqrt{t^2} = 7 \cdot t^{2/2} = 7t$

c) $\sqrt{18b^{14}} = \sqrt{18} \cdot \sqrt{b^{14}}$ Product rule
$= \sqrt{9} \cdot \sqrt{2} \cdot b^{14/2}$ 9 is a perfect square.
$= 3\sqrt{2} \cdot b^7$ Simplify.
$= 3b^7\sqrt{2}$ Rewrite using the commutative property.

W Hint
Are you writing out the examples as you are reading them?

d) We begin by using the quotient rule.

$$\sqrt{\dfrac{32}{n^{20}}} = \dfrac{\sqrt{32}}{\sqrt{n^{20}}} = \dfrac{\sqrt{16} \cdot \sqrt{2}}{n^{20/2}} = \dfrac{4\sqrt{2}}{n^{10}}$$

[YOU TRY 4]

Simplify completely.

a) $\sqrt{y^{10}}$ b) $\sqrt{144p^{16}}$ c) $\sqrt{\dfrac{45}{w^4}}$

5 Simplify Square Root Expressions Containing Variables with Odd Exponents

How do we simplify an expression containing a square root if the power under the square root is odd? We can use the product rule for radicals and fractional exponents to help us understand how to simplify such expressions.

EXAMPLE 5

Simplify completely.

a) $\sqrt{x^7}$ b) $\sqrt{c^{11}}$

Solution

a) To simplify $\sqrt{x^7}$, write x^7 as the product of two factors so that the exponent of one of the factors is the *largest* number less than 7 that is divisible by 2 (the index of the radical).

$\sqrt{x^7} = \sqrt{x^6 \cdot x^1}$ 6 is the largest number less than 7 that is divisible by 2.
$= \sqrt{x^6} \cdot \sqrt{x}$ Product rule
$= x^{6/2} \cdot \sqrt{x}$ Use a fractional exponent to simplify.
$= x^3\sqrt{x}$ $6 \div 2 = 3$

b) To simplify $\sqrt{c^{11}}$, write c^{11} as the product of two factors so that the exponent of one of the factors is the *largest* number less than 11 that is divisible by 2 (the index of the radical).

$\sqrt{c^{11}} = \sqrt{c^{10} \cdot c^1}$ 10 is the largest number less than 11 that is divisible by 2.
$= \sqrt{c^{10}} \cdot \sqrt{c}$ Product rule
$= c^{10/2} \cdot \sqrt{c}$ Use a fractional exponent to simplify.
$= c^5\sqrt{c}$ $10 \div 2 = 5$

[YOU TRY 5] Simplify completely.

a) $\sqrt{m^5}$ b) $\sqrt{z^{19}}$

We used the product rule to simplify each radical in Example 5. During the simplification, however, we always divided an exponent by 2. This idea of division gives us another way to simplify radical expressions. Once again, let's look at the radicals and their simplified forms in Example 5 to see how we can simplify radical expressions using division.

$$\sqrt{x^7} = x^3\sqrt{x^1} = x^3\sqrt{x}$$

Index of radical → $2\overline{)7}$ → Quotient 3, -6, Remainder 1

$$\sqrt{c^{11}} = c^5\sqrt{c^1} = c^5\sqrt{c}$$

Index of radical → $2\overline{)11}$ → Quotient 5, -10, Remainder 1

W Hint
Write a sample expression that follows these steps in your notes.

Procedure Simplifying a Radical Containing Variables

To simplify a radical expression containing variables:

1) Divide the original exponent in the radicand by the index of the radical.
2) The exponent on the variable *outside* of the radical will be the *quotient* of the division problem.
3) The exponent on the variable *inside* of the radical will be the *remainder* of the division problem.

EXAMPLE 6

Simplify completely.

a) $\sqrt{t^9}$ b) $\sqrt{16b^5}$ c) $\sqrt{45y^{21}}$

Solution

a) To simplify $\sqrt{t^9}$, divide: $2\overline{)9}$ → Quotient 4, -8, Remainder 1

$\sqrt{t^9} = t^4\sqrt{t^1} = t^4\sqrt{t}$

b) $\sqrt{16b^5} = \sqrt{16} \cdot \sqrt{b^5}$ Product rule
$= 4 \cdot b^2\sqrt{b^1}$ $5 \div 2$ gives a quotient of 2 and a remainder of 1.
$= 4b^2\sqrt{b}$

c) $\sqrt{45y^{21}} = \sqrt{45} \cdot \sqrt{y^{21}}$ Product rule
$= \sqrt{9} \cdot \sqrt{5} \cdot y^{10}\sqrt{y^1}$
↑ Product rule   ↑ $21 \div 2$ gives a quotient of 10 and a remainder of 1.
$= 3\sqrt{5} \cdot y^{10}\sqrt{y}$ $\sqrt{9} = 3$
$= 3y^{10} \cdot \sqrt{5} \cdot \sqrt{y}$ Use the commutative property to rewrite the expression.
$= 3y^{10}\sqrt{5y}$ Use the product rule to write the expression with one radical.

[YOU TRY 6] Simplify completely.

a) $\sqrt{m^{13}}$ b) $\sqrt{100v^7}$ c) $\sqrt{32a^3}$

If a radical contains more than one variable, apply the product or quotient rule.

EXAMPLE 7

Simplify completely.

a) $\sqrt{8a^{15}b^3}$ b) $\sqrt{\dfrac{5r^{27}}{s^8}}$

Solution

a) $\sqrt{8a^{15}b^3} = \sqrt{8} \cdot \sqrt{a^{15}} \cdot \sqrt{b^3}$
$= \sqrt{4} \cdot \sqrt{2} \cdot a^7\sqrt{a^1} \cdot b^1\sqrt{b^1}$

    Product rule   15 ÷ 2 gives a quotient   3 ÷ 2 gives a quotient
                      of 7 and a remainder of 1. of 1 and a remainder of 1.

$= 2\sqrt{2} \cdot a^7\sqrt{a} \cdot b\sqrt{b}$ $\sqrt{4} = 2$
$= 2a^7b \cdot \sqrt{2} \cdot \sqrt{a} \cdot \sqrt{b}$ Use the commutative property to rewrite the expression.
$= 2a^7b\sqrt{2ab}$ Use the product rule to write the expression with one radical.

b) $\sqrt{\dfrac{5r^{27}}{s^8}} = \dfrac{\sqrt{5r^{27}}}{\sqrt{s^8}}$ Quotient rule

$= \dfrac{\sqrt{5} \cdot \sqrt{r^{27}}}{s^4}$ Product rule
                        8 ÷ 2 = 4

$= \dfrac{\sqrt{5} \cdot r^{13}\sqrt{r^1}}{s^4}$ 27 ÷ 2 gives a quotient of 13 and a remainder of 1.

$= \dfrac{r^{13} \cdot \sqrt{5} \cdot \sqrt{r}}{s^4}$ Use the commutative property to rewrite the expression.

$= \dfrac{r^{13}\sqrt{5r}}{s^4}$ Use the product rule to write the expression with one radical.

[YOU TRY 7]

Simplify completely.

a) $\sqrt{c^5d^{12}}$ b) $\sqrt{27x^{10}y^9}$ c) $\sqrt{\dfrac{40u^{13}}{v^{20}}}$

6 Simplify More Square Root Expressions Containing Variables

Next we will look at some examples of multiplying and dividing radical expressions that contain variables. Remember to always look at the result and ask yourself, "*Is the radical in simplest form?*" If it is not, simplify completely.

EXAMPLE 8

Perform the indicated operation, and simplify completely.

a) $\sqrt{6t} \cdot \sqrt{3t}$ b) $\sqrt{2a^3b} \cdot \sqrt{8a^2b^5}$ c) $\dfrac{\sqrt{20x^5}}{\sqrt{5x}}$

Solution

a) $\sqrt{6t} \cdot \sqrt{3t} = \sqrt{6t \cdot 3t}$ Product rule
$= \sqrt{18t^2}$
$= \sqrt{18} \cdot \sqrt{t^2}$ Product rule
$= \sqrt{9 \cdot 2} \cdot t = \sqrt{9} \cdot \sqrt{2} \cdot t = 3\sqrt{2} \cdot t = 3t\sqrt{2}$

Hint
In parts b) and c), think about which method you prefer and why.

b) $\sqrt{2a^3b} \cdot \sqrt{8a^2b^5}$

There are two good methods for multiplying these radicals.

i) Multiply the radicands to obtain one radical.

$\sqrt{2a^3b} \cdot \sqrt{8a^2b^5} = \sqrt{2a^3b \cdot 8a^2b^5}$ Product rule
$= \sqrt{16a^5b^6}$ Multiply.

Is the radical in simplest form? *No.*

$= \sqrt{16} \cdot \sqrt{a^5} \cdot \sqrt{b^6}$ Product rule
$= 4 \cdot a^2\sqrt{a} \cdot b^3$ Evaluate.
$= 4a^2b^3\sqrt{a}$ Commutative property

ii) Simplify each radical, then multiply.

$\sqrt{2a^3b} = \sqrt{2} \cdot \sqrt{a^3} \cdot \sqrt{b}$ $\sqrt{8a^2b^5} = \sqrt{8} \cdot \sqrt{a^2} \cdot \sqrt{b^5}$
$\phantom{\sqrt{2a^3b}} = \sqrt{2} \cdot a\sqrt{a} \cdot \sqrt{b}$ $\phantom{\sqrt{8a^2b^5}} = 2\sqrt{2} \cdot a \cdot b^2\sqrt{b}$
$\phantom{\sqrt{2a^3b}} = a\sqrt{2ab}$ $\phantom{\sqrt{8a^2b^5}} = 2ab^2\sqrt{2b}$

Then, $\sqrt{2a^3b} \cdot \sqrt{8a^2b^5} = a\sqrt{2ab} \cdot 2ab^2\sqrt{2b}$
$= a \cdot 2ab^2 \cdot \sqrt{2ab} \cdot \sqrt{2b}$ Commutative property
$= 2a^2b^2\sqrt{4ab^2}$ Multiply.
$= 2a^2b^2 \cdot 2 \cdot b \cdot \sqrt{a}$ $\sqrt{4ab^2} = 2b\sqrt{a}$
$= 4a^2b^3\sqrt{a}$ Multiply.

Both methods give the same result.

c) We can use the quotient rule first or simplify first.

i) $\dfrac{\sqrt{20x^5}}{\sqrt{5x}} = \sqrt{\dfrac{20x^5}{5x}}$ Use the quotient rule first.
$= \sqrt{4x^4} = \sqrt{4} \cdot \sqrt{x^4} = 2x^2$

ii) $\dfrac{\sqrt{20x^5}}{\sqrt{5x}} = \dfrac{\sqrt{20} \cdot \sqrt{x^5}}{\sqrt{5x}}$ Simplify first by using the product rule.
$= \dfrac{\sqrt{4} \cdot \sqrt{5} \cdot x^2\sqrt{x}}{\sqrt{5x}}$ Product rule; simplify $\sqrt{x^5}$.
$= \dfrac{2\sqrt{5} \cdot x^2\sqrt{x}}{\sqrt{5x}}$ $\sqrt{4} = 2$
$= \dfrac{2x^2\sqrt{5x}}{\sqrt{5x}}$ Product rule
$= 2x^2$ Divide out the common factor.

Both methods give the same result. In this case, the second method was longer. Sometimes, however, this method *can* be more efficient.

YOU TRY 8 Perform the indicated operation, and simplify completely.

a) $\sqrt{2n^3} \cdot \sqrt{6n}$ b) $\sqrt{15cd^5} \cdot \sqrt{3c^2d}$ c) $\dfrac{\sqrt{128k^9}}{\sqrt{2k}}$

ANSWERS TO YOU TRY EXERCISES

1) a) $\sqrt{30}$ b) $\sqrt{10r}$ 2) a) $2\sqrt{7}$ b) $5\sqrt{3}$ c) $6\sqrt{2}$ 3) a) $\frac{10}{13}$ b) 3 c) $5\sqrt{2}$ d) $\frac{\sqrt{11}}{6}$
4) a) y^5 b) $12p^8$ c) $\frac{3\sqrt{5}}{w^2}$ 5) a) $m^2\sqrt{m}$ b) $z^9\sqrt{z}$ 6) a) $m^6\sqrt{m}$ b) $10v^3\sqrt{v}$ c) $4a\sqrt{2a}$
7) a) $c^2d^6\sqrt{c}$ b) $3x^5y^4\sqrt{3y}$ c) $\frac{2u^6\sqrt{10u}}{v^{10}}$ 8) a) $2n^2\sqrt{3}$ b) $3cd^3\sqrt{5c}$ c) $8k^4$

E Evaluate 9.3 Exercises

Do the exercises, and check your work.

Unless otherwise stated, assume all variables represent nonnegative real numbers.

Objective 1: Multiply Square Roots
Multiply and simplify.

1) $\sqrt{3} \cdot \sqrt{7}$ 2) $\sqrt{11} \cdot \sqrt{5}$
3) $\sqrt{10} \cdot \sqrt{3}$ 4) $\sqrt{7} \cdot \sqrt{2}$
5) $\sqrt{6} \cdot \sqrt{y}$ 6) $\sqrt{5} \cdot \sqrt{p}$

Objective 2: Simplify the Square Root of a Whole Number
Label each statement as true or false. Give a reason for your answer.

7) $\sqrt{20}$ is in simplest form.
8) $\sqrt{35}$ is in simplest form.
9) $\sqrt{42}$ is in simplest form.
10) $\sqrt{63}$ is in simplest form.

Simplify completely.

Fill It In
Fill in the blanks with either the missing mathematical step or reason for the given step.

11) $\sqrt{60} = \sqrt{4 \cdot 15}$ _____
    = _____   Product rule
    = _____   Simplify.

12) $\sqrt{200} =$ _____   Factor.
    $= \sqrt{100} \cdot \sqrt{2}$ _____
    = _____   Simplify.

Simplify completely. If the radical is already simplified, then say so.

13) $\sqrt{20}$ 14) $\sqrt{12}$
15) $\sqrt{54}$ 16) $\sqrt{63}$
17) $\sqrt{33}$ 18) $\sqrt{15}$

19) $\sqrt{108}$ 20) $\sqrt{80}$
21) $\sqrt{98}$ 22) $\sqrt{96}$
23) $\sqrt{38}$ 24) $\sqrt{46}$
25) $\sqrt{400}$ 26) $\sqrt{900}$
27) $\sqrt{750}$ 28) $\sqrt{420}$

Objective 3: Use the Quotient Rule for Square Roots
Simplify completely.

29) $\sqrt{\frac{144}{25}}$ 30) $\sqrt{\frac{16}{81}}$
31) $\frac{\sqrt{4}}{\sqrt{49}}$ 32) $\frac{\sqrt{64}}{\sqrt{121}}$
33) $\frac{\sqrt{54}}{\sqrt{6}}$ 34) $\frac{\sqrt{48}}{\sqrt{3}}$
35) $\sqrt{\frac{60}{5}}$ 36) $\sqrt{\frac{40}{5}}$
37) $\frac{\sqrt{120}}{\sqrt{6}}$ 38) $\frac{\sqrt{54}}{\sqrt{3}}$
39) $\frac{\sqrt{35}}{\sqrt{5}}$ 40) $\frac{\sqrt{30}}{\sqrt{2}}$
41) $\sqrt{\frac{6}{49}}$ 42) $\sqrt{\frac{2}{81}}$
43) $\sqrt{\frac{45}{16}}$ 44) $\sqrt{\frac{60}{49}}$

Objective 4: Simplify Square Root Expressions Containing Variables with Even Exponents
Simplify completely.

45) $\sqrt{x^8}$ 46) $\sqrt{q^6}$
47) $\sqrt{w^{14}}$ 48) $\sqrt{t^{16}}$
49) $\sqrt{100c^2}$ 50) $\sqrt{9z^8}$

SECTION 9.3 Simplifying Expressions Containing Square Roots

51) $\sqrt{64k^6m^{10}}$
52) $\sqrt{25p^{20}q^{14}}$
53) $\sqrt{28r^4}$
54) $\sqrt{27z^{12}}$
55) $\sqrt{300q^{22}t^{16}}$
56) $\sqrt{50n^4y^4}$
57) $\sqrt{\dfrac{81}{c^6}}$
58) $\sqrt{\dfrac{h^2}{169}}$
59) $\dfrac{\sqrt{40}}{\sqrt{t^8}}$
60) $\dfrac{\sqrt{18}}{\sqrt{m^{30}}}$
61) $\sqrt{\dfrac{75x^2}{y^{12}}}$
62) $\sqrt{\dfrac{44}{w^2z^{18}}}$

Objective 5: Simplify Square Root Expressions Containing Variables with Odd Exponents
Simplify completely.

Fill It In
Fill in the blanks with either the missing mathematical step or reason for the given step.

63) $\sqrt{w^9} = \sqrt{w^8 \cdot w^1}$
   = _____ Product rule
   = $w^4\sqrt{w}$ _____

64) $\sqrt{z^{19}} = \sqrt{z^{18} \cdot z^1}$ _____
   = $\sqrt{z^{18}} \cdot \sqrt{z^1}$ _____
   = _____ Simplify.

65) $\sqrt{a^5}$
66) $\sqrt{c^7}$
67) $\sqrt{g^{13}}$
68) $\sqrt{k^{15}}$
69) $\sqrt{h^{31}}$
70) $\sqrt{b^{25}}$
71) $\sqrt{72x^3}$
72) $\sqrt{100a^5}$
73) $\sqrt{13q^7}$
74) $\sqrt{20c^9}$
75) $\sqrt{75t^{11}}$
76) $\sqrt{45p^{17}}$
77) $\sqrt{c^8d^2}$
78) $\sqrt{r^4s^{12}}$
79) $\sqrt{a^4b^3}$
80) $\sqrt{x^2y^9}$
81) $\sqrt{u^5v^7}$
82) $\sqrt{f^3g^9}$
83) $\sqrt{36m^9n^4}$
84) $\sqrt{4t^6u^5}$
85) $\sqrt{44x^{12}y^5}$
86) $\sqrt{63c^7d^4}$

87) $\sqrt{32t^5u^7}$
88) $\sqrt{125k^3l^9}$
89) $\sqrt{\dfrac{a^7}{81b^6}}$
90) $\sqrt{\dfrac{x^5}{49y^6}}$
91) $\sqrt{\dfrac{3r^9}{s^2}}$
92) $\sqrt{\dfrac{17h^{11}}{k^8}}$

Objective 6: Simplify More Square Root Expressions Containing Variables
Perform the indicated operation, and simplify. Assume all variables represent positive real numbers.

93) $\sqrt{5} \cdot \sqrt{10}$
94) $\sqrt{8} \cdot \sqrt{6}$
95) $\sqrt{21} \cdot \sqrt{3}$
96) $\sqrt{2} \cdot \sqrt{14}$
97) $\sqrt{w} \cdot \sqrt{w^5}$
98) $\sqrt{d^3} \cdot \sqrt{d^{11}}$
99) $\sqrt{n^3} \cdot \sqrt{n^4}$
100) $\sqrt{a^{10}} \cdot \sqrt{a^3}$
101) $\sqrt{2k} \cdot \sqrt{8k^5}$
102) $\sqrt{5z^9} \cdot \sqrt{5z^3}$
103) $\sqrt{5a^6b^5} \cdot \sqrt{10ab^4}$
104) $\sqrt{6x^4y^3} \cdot \sqrt{2x^5y^2}$
105) $\sqrt{8c^9d^2} \cdot \sqrt{5cd^7}$
106) $\sqrt{6t^3u^3} \cdot \sqrt{3t^7u^4}$
107) $\dfrac{\sqrt{18k^{11}}}{\sqrt{2k^3}}$
108) $\dfrac{\sqrt{48m^{15}}}{\sqrt{3m^9}}$
109) $\dfrac{\sqrt{120h^8}}{\sqrt{3h^2}}$
110) $\dfrac{\sqrt{72c^{10}}}{\sqrt{6c^2}}$
111) $\dfrac{\sqrt{50a^{16}b^9}}{\sqrt{5a^7b^4}}$
112) $\dfrac{\sqrt{21y^8z^{18}}}{\sqrt{3yz^{13}}}$

113) The velocity v of a moving object can be determined from its mass m and its kinetic energy KE using the formula $v = \sqrt{\dfrac{2\text{KE}}{m}}$, where the velocity is in meters/second, the mass is in kilograms, and the KE is measured in joules. A 600-kg roller coaster car is moving along a track and has kinetic energy of 120,000 joules. What is the velocity of the car?

114) The length of a side s of an equilateral triangle is a function of its area A and can be described by $s(A) = \sqrt{\dfrac{4\sqrt{3}A}{3}}$. If an equilateral triangle has an area of $6\sqrt{3}$ cm^2, how long is each side of the triangle?

R Rethink

R1) Describe the steps you take to manually check your answers.

R2) After completing the exercises, do you feel that you have memorized the definitions, properties, and procedures by practicing them? Explain.

9.4 Simplifying Expressions Containing Higher Roots

P Prepare

What are your objectives for Section 9.4?	**O Organize** How can you accomplish each objective?
1 Multiply Higher Roots	• Write the *product rule for higher roots* in your own words. • Complete the given example on your own. • Complete You Try 1.
2 Simplify Higher Roots of Integers	• Write the property for **When Is a Radical Simplified?** in your own words. • Know the two different methods for simplifying higher roots. • Complete the given example on your own. • Complete You Try 2.
3 Use the Quotient Rule for Higher Roots	• Write the *quotient rule for higher roots* in your own words. • Complete the given example on your own. • Complete You Try 3.
4 Simplify Radicals Containing Variables	• Write the property for $\sqrt[n]{a^m}$ in your notes. • Complete the given examples on your own. • Complete You Trys 4–6.
5 Multiply and Divide Radicals with Different Indices	• Create a procedure by following Example 7, and write it in your notes. • Complete You Try 7.

W Work Read the explanations, follow the examples, take notes, and complete the You Trys.

In Section 9.1 we first discussed finding higher roots like $\sqrt[4]{16} = 2$ and $\sqrt[3]{-27} = -3$. In this section, we will extend what we learned about multiplying, dividing, and simplifying *square* roots to doing the same with higher roots.

1 Multiply Higher Roots

Definition Product Rule for Higher Roots

If $\sqrt[n]{a}$ and $\sqrt[n]{b}$ are real numbers, then

$$\sqrt[n]{a} \cdot \sqrt[n]{b} = \sqrt[n]{a \cdot b}$$

This rule enables us to multiply and simplify radicals with any index in a way that is similar to multiplying and simplifying square roots.

EXAMPLE 1

Multiply.

a) $\sqrt[3]{2} \cdot \sqrt[3]{7}$ b) $\sqrt[4]{t} \cdot \sqrt[4]{10}$

Solution

a) $\sqrt[3]{2} \cdot \sqrt[3]{7} = \sqrt[3]{2 \cdot 7} = \sqrt[3]{14}$ b) $\sqrt[4]{t} \cdot \sqrt[4]{10} = \sqrt[4]{t \cdot 10} = \sqrt[4]{10t}$

[YOU TRY 1] Multiply.

a) $\sqrt[4]{6} \cdot \sqrt[4]{5}$ b) $\sqrt[5]{8} \cdot \sqrt[5]{k^2}$

BE CAREFUL Remember that we can apply the product rule in this way *only* if the indices of the radicals are the same. Later in this section we will discuss how to multiply radicals with different indices.

2 Simplify Higher Roots of Integers

In Section 9.3 we said that a simplified *square root* cannot contain any *perfect squares*. Next we list the conditions that determine when a radical with *any* index is in simplest form.

Property When Is a Radical Simplified?

Let P be an expression and let n be an integer greater than 1. Then $\sqrt[n]{P}$ is completely simplified when all of the following conditions are met:

1) The radicand does not contain any factors (other than 1) that are perfect nth powers.
2) The exponents in the radicand and the index of the radical do not have any common factors (other than 1).
3) The radicand does not contain any fractions.
4) There are no radicals in the denominator of a fraction.

Note
Condition 1) implies that the radical cannot contain variables with exponents greater than or equal to n, the index of the radical.

To simplify radicals with any index, use the product rule $\sqrt[n]{a \cdot b} = \sqrt[n]{a} \cdot \sqrt[n]{b}$, where a or b is an nth power.

Remember, to be certain that a radical is simplified completely, always look at the radical carefully and ask yourself, "*Is the radical in simplest form?*"

EXAMPLE 2

Simplify completely.

a) $\sqrt[3]{250}$ b) $\sqrt[4]{48}$

Solution

a) We will look at two methods for simplifying $\sqrt[3]{250}$.

  i) Since we must simplify the *cube* root of 250, think of two numbers that multiply to 250 so that at least one of the numbers is a *perfect cube*.

  $$250 = 125 \cdot 2$$
  $$\sqrt[3]{250} = \sqrt[3]{125 \cdot 2} \quad \text{125 is a perfect cube.}$$
  $$= \sqrt[3]{125} \cdot \sqrt[3]{2} \quad \text{Product rule}$$
  $$= 5\sqrt[3]{2} \quad \sqrt[3]{125} = 5$$

  Is $5\sqrt[3]{2}$ in simplest form? Yes, because 2 does not have any factors that are perfect cubes.

  ii) Use a factor tree to find the prime factorization of 250: $250 = 2 \cdot 5^3$.

  $$\sqrt[3]{250} = \sqrt[3]{2 \cdot 5^3} \quad 2 \cdot 5^3 \text{ is the prime factorization of 250.}$$
  $$= \sqrt[3]{2} \cdot \sqrt[3]{5^3} \quad \text{Product rule}$$
  $$= \sqrt[3]{2} \cdot 5 \quad \sqrt[3]{5^3} = 5$$
  $$= 5\sqrt[3]{2} \quad \text{Commutative property}$$

  We obtain the same result using either method.

b) We will use two methods for simplifying $\sqrt[4]{48}$.

  i) Since we must simplify the *fourth* root of 48, think of two numbers that multiply to 48 so that at least one of the numbers is a *perfect fourth power*.

  $$48 = 16 \cdot 3$$
  $$\sqrt[4]{48} = \sqrt[4]{16 \cdot 3} \quad \text{16 is a perfect fourth power.}$$
  $$= \sqrt[4]{16} \cdot \sqrt[4]{3} \quad \text{Product rule}$$
  $$= 2\sqrt[4]{3} \quad \sqrt[4]{16} = 2$$

  Is $2\sqrt[4]{3}$ in simplest form? Yes, because 3 does not have any factors that are perfect fourth powers.

  ii) Use a factor tree to find the prime factorization of 48: $48 = 2^4 \cdot 3$.

  $$\sqrt[4]{48} = \sqrt[4]{2^4 \cdot 3} \quad 2^4 \cdot 3 \text{ is the prime factorization of 48.}$$
  $$= \sqrt[4]{2^4} \cdot \sqrt[4]{3} \quad \text{Product rule}$$
  $$= 2\sqrt[4]{3} \quad \sqrt[4]{2^4} = 2$$

  Once again, both methods give us the same result.

> **W Hint**
> In your notes and in your own words, explain how you know that a radical with any index is in simplest form.

[YOU TRY 2]   Simplify completely.

a) $\sqrt[3]{40}$ b) $\sqrt[5]{63}$

3 Use the Quotient Rule for Higher Roots

Definition Quotient Rule for Higher Roots

If $\sqrt[n]{a}$ and $\sqrt[n]{b}$ are real numbers, $b \neq 0$, and n is a natural number then

$$\sqrt[n]{\frac{a}{b}} = \frac{\sqrt[n]{a}}{\sqrt[n]{b}}$$

We apply the quotient rule when working with *n*th roots the same way we apply it when working with square roots.

EXAMPLE 3

Simplify completely.

a) $\sqrt[3]{-\dfrac{81}{3}}$ b) $\dfrac{\sqrt[3]{96}}{\sqrt[3]{2}}$

Solution

a) We can think of $-\dfrac{81}{3}$ as $\dfrac{-81}{3}$ or $\dfrac{81}{-3}$. Let's think of it as $\dfrac{-81}{3}$.

Neither -81 nor 3 is a perfect cube, but if we simplify $\dfrac{-81}{3}$ we get -27, which *is* a perfect cube.

$$\sqrt[3]{-\dfrac{81}{3}} = \sqrt[3]{-27} = -3$$

W Hint
Make a list of perfect cubes and numbers raised to fourth and fifth powers.

b) Let's begin by applying the quotient rule to obtain a fraction under *one* radical, then simplify the fraction.

$$\begin{aligned}\dfrac{\sqrt[3]{96}}{\sqrt[3]{2}} &= \sqrt[3]{\dfrac{96}{2}} &&\text{Quotient rule}\\ &= \sqrt[3]{48} &&\text{Simplify } \dfrac{96}{2}.\\ &= \sqrt[3]{8 \cdot 6} &&\text{8 is a perfect cube.}\\ &= \sqrt[3]{8} \cdot \sqrt[3]{6} &&\text{Product rule}\\ &= 2\sqrt[3]{6} &&\sqrt[3]{8} = 2\end{aligned}$$

Is $2\sqrt[3]{6}$ in simplest form? Yes, because 6 does not have any factors that are perfect cubes.

[YOU TRY 3]

Simplify completely.

a) $\sqrt[4]{\dfrac{1}{81}}$ b) $\dfrac{\sqrt[3]{162}}{\sqrt[3]{3}}$

4 Simplify Radicals Containing Variables

In Section 9.2 we discussed the relationship between radical notation and fractional exponents. Recall that

Property $\sqrt[n]{a^m}$

If a is a nonnegative number and m and n are integers such that $n > 1$, then
$$\sqrt[n]{a^m} = a^{m/n}.$$

That is, the index of the radical becomes the denominator of the fractional exponent, and the power in the radicand becomes the numerator of the fractional exponent.

This is the principle we use to simplify radicals with indices greater than 2.

EXAMPLE 4

Simplify completely.

a) $\sqrt[3]{y^{15}}$ b) $\sqrt[4]{16t^{24}u^8}$ c) $\sqrt[5]{\dfrac{c^{10}}{d^{30}}}$

Solution

a) $\sqrt[3]{y^{15}} = y^{15/3} = y^5$

b) $\sqrt[4]{16t^{24}u^8} = \sqrt[4]{16} \cdot \sqrt[4]{t^{24}} \cdot \sqrt[4]{u^8}$ Product rule
$= 2 \cdot t^{24/4} \cdot u^{8/4}$ Write with rational exponents.
$= 2t^6 u^2$ Simplify exponents.

c) $\sqrt[5]{\dfrac{c^{10}}{d^{30}}} = \dfrac{\sqrt[5]{c^{10}}}{\sqrt[5]{d^{30}}} = \dfrac{c^{10/5}}{d^{30/5}} = \dfrac{c^2}{d^6}$ Quotient rule

[YOU TRY 4]

Simplify completely.

a) $\sqrt[3]{a^3 b^{21}}$ b) $\sqrt[4]{\dfrac{m^{12}}{16n^{20}}}$

To simplify a radical expression if the power in the radicand does not divide evenly by the index, we use the same methods we used in Section 9.3 for simplifying similar expressions with square roots. We can use the product rule or we can use the idea of quotient and remainder in a division problem.

EXAMPLE 5

Simplify $\sqrt[4]{x^{23}}$ completely in two ways: i) use the product rule and ii) divide the exponent by the index and use the quotient and remainder.

Solution

i) Using the product rule:
To simplify $\sqrt[4]{x^{23}}$, write x^{23} as the product of two factors so that the exponent of one of the factors is the *largest* number less than 23 that is divisible by 4 (the index).

$\sqrt[4]{x^{23}} = \sqrt[4]{x^{20} \cdot x^3}$ 20 is the largest number less than 23 that is divisible by 4.
$= \sqrt[4]{x^{20}} \cdot \sqrt[4]{x^3}$ Product rule
$= x^{20/4} \cdot \sqrt[4]{x^3}$ Use a fractional exponent to simplify.
$= x^5 \sqrt[4]{x^3}$ $20 \div 4 = 5$

> **W Hint**
> Be sure to do this example both ways. Which method do you prefer and why?

ii) Using the quotient and remainder:

To simplify $\sqrt[4]{x^{23}}$, divide $4\overline{)23}$ with quotient 5 and remainder 3.

Recall from our work with square roots in Section 9.3 that
i) the exponent on the variable *outside* of the radical will be the *quotient* of the division problem,

and

ii) the exponent on the variable *inside* of the radical will be the *remainder* of the division problem.

$$\sqrt[4]{x^{23}} = x^5 \sqrt[4]{x^3}$$

Is $x^5 \sqrt[4]{x^3}$ in *simplest form*? Yes, because the exponent inside of the radical is less than the index, and they contain no common factors other than 1.

[YOU TRY 5] Simplify $\sqrt[5]{r^{32}}$ completely using both methods shown in Example 5.

We can apply the product and quotient rules together with the methods in Example 5 to simplify certain radical expressions.

EXAMPLE 6 Completely simplify $\sqrt[3]{56a^{16}b^8}$.

Solution

$$\sqrt[3]{56a^{16}b^8} = \sqrt[3]{56} \cdot \sqrt[3]{a^{16}} \cdot \sqrt[3]{b^8} \qquad \text{Product rule}$$
$$= \sqrt[3]{8} \cdot \sqrt[3]{7} \cdot a^5\sqrt[3]{a^1} \cdot b^2\sqrt[3]{b^2}$$

Product rule ↗   16 ÷ 3 gives a quotient ↑ of 5 and a remainder of 1.   8 ÷ 3 gives a quotient ↖ of 2 and a remainder of 2.

$$= 2\sqrt[3]{7} \cdot a^5\sqrt[3]{a} \cdot b^2\sqrt[3]{b^2} \qquad \text{Simplify } \sqrt[3]{8}.$$
$$= 2a^5b^2 \cdot \sqrt[3]{7} \cdot \sqrt[3]{a} \cdot \sqrt[3]{b^2} \qquad \text{Use the commutative property to rewrite the expression.}$$
$$= 2a^5b^2\sqrt[3]{7ab^2} \qquad \text{Product rule}$$

[YOU TRY 6] Simplify completely.

a) $\sqrt[4]{48x^{15}y^{22}}$ b) $\sqrt[3]{\dfrac{r^{19}}{27s^{12}}}$

5 Multiply and Divide Radicals with Different Indices

The product and quotient rules for radicals apply only when the radicals have the *same* indices. To multiply or divide radicals with *different* indices, we first change the radical expressions to rational exponent form.

EXAMPLE 7 Multiply the expressions, and write the answer in simplest radical form.

$$\sqrt[3]{x^2} \cdot \sqrt{x}$$

Solution

The indices of $\sqrt[3]{x^2}$ and $\sqrt{x}$ are different, so we *cannot* use the product rule right now. Rewrite each radical as a fractional exponent, use the product rule for *exponents*, then convert the answer back to radical form.

$$\sqrt[3]{x^2} \cdot \sqrt{x} = x^{2/3} \cdot x^{1/2} \qquad \text{Change radicals to fractional exponents.}$$
$$= x^{4/6} \cdot x^{3/6} \qquad \text{Get a common denominator to add exponents.}$$
$$= x^{\frac{4}{6}+\frac{3}{6}} = x^{7/6} \qquad \text{Add exponents.}$$
$$= \sqrt[6]{x^7} = x\sqrt[6]{x} \qquad \text{Rewrite in radical form, and simplify.}$$

W Hint Write your own procedure in your notes.

[YOU TRY 7] Perform the indicated operation, and write the answer in simplest radical form.

a) $\sqrt[4]{y} \cdot \sqrt[6]{y}$ b) $\dfrac{\sqrt[3]{c^2}}{\sqrt{c}}$

ANSWERS TO [YOU TRY] EXERCISES

1) a) $\sqrt[4]{30}$ b) $\sqrt[5]{8k^2}$ 2) a) $2\sqrt[3]{5}$ b) simplified   3) a) $\dfrac{1}{3}$ b) $3\sqrt[3]{2}$ 4) a) ab^7 b) $\dfrac{m^3}{2n^5}$

5) $r^6\sqrt[5]{r^2}$ 6) a) $2x^3y^5\sqrt[4]{3x^3y^2}$ b) $\dfrac{r^6\sqrt[3]{r}}{3s^4}$ 7) a) $\sqrt[12]{y^5}$ b) $\sqrt[6]{c}$

E Evaluate 9.4 Exercises
Do the exercises, and check your work.

Mixed Exercises: Objectives 1–3

1) In your own words, explain the product rule for radicals.

2) In your own words, explain the quotient rule for radicals.

3) How do you know that a radical expression containing a cube root is completely simplified?

4) How do you know that a radical expression containing a fourth root is completely simplified?

Assume all variables represent positive real numbers.

Objective 1: Multiply Higher Roots
Multiply.

5) $\sqrt[5]{6}\cdot\sqrt[5]{2}$ 6) $\sqrt[3]{5}\cdot\sqrt[3]{4}$

7) $\sqrt[5]{9}\cdot\sqrt[5]{m^2}$ 8) $\sqrt[4]{11}\cdot\sqrt[4]{h^3}$

9) $\sqrt[3]{a^2}\cdot\sqrt[3]{b}$ 10) $\sqrt[5]{t^2}\cdot\sqrt[5]{u^4}$

Mixed Exercises: Objectives 2 and 3
Simplify completely.

Fill It In
Fill in the blanks with either the missing mathematical step or reason for the given step.

11) $\sqrt[3]{56} = \sqrt[3]{8\cdot 7}$ _____
    = _____ Product rule
    = _____ Simplify.

12) $\sqrt[4]{80} = \sqrt[4]{16\cdot 5}$ _____
    $= \sqrt[4]{16}\cdot\sqrt[4]{5}$ _____
    = _____ Simplify.

13) $\sqrt[3]{24}$ 14) $\sqrt[3]{48}$

15) $\sqrt[4]{64}$ 16) $\sqrt[4]{32}$

17) $\sqrt[3]{54}$ 18) $\sqrt[3]{88}$

19) $\sqrt[3]{2000}$ 20) $\sqrt[3]{108}$

21) $\sqrt[5]{64}$ 22) $\sqrt[4]{162}$

23) $\sqrt[3]{\dfrac{1}{125}}$ 24) $\sqrt[4]{\dfrac{1}{16}}$

25) $\sqrt[3]{-\dfrac{54}{2}}$ 26) $\sqrt[4]{\dfrac{48}{3}}$

27) $\dfrac{\sqrt[3]{48}}{\sqrt[3]{2}}$ 28) $\dfrac{\sqrt[3]{500}}{\sqrt[3]{2}}$

29) $\dfrac{\sqrt[4]{240}}{\sqrt[4]{3}}$ 30) $\dfrac{\sqrt[3]{8000}}{\sqrt[3]{4}}$

Objective 4: Simplify Radicals Containing Variables
Simplify completely.

31) $\sqrt[3]{d^6}$ 32) $\sqrt[3]{g^9}$

33) $\sqrt[4]{n^{20}}$ 34) $\sqrt[4]{t^{36}}$

35) $\sqrt[5]{x^5 y^{15}}$ 36) $\sqrt[6]{a^{12} b^6}$

37) $\sqrt[3]{w^{14}}$ 38) $\sqrt[3]{b^{19}}$

39) $\sqrt[4]{y^9}$ 40) $\sqrt[4]{m^7}$

41) $\sqrt[3]{d^5}$ 42) $\sqrt[3]{c^{29}}$

43) $\sqrt[3]{u^{10} v^{15}}$ 44) $\sqrt[3]{x^9 y^{16}}$

45) $\sqrt[3]{b^{16} c^5}$ 46) $\sqrt[4]{r^{15} s^9}$

47) $\sqrt[4]{m^3 n^{18}}$ 48) $\sqrt[3]{a^{11} b}$

49) $\sqrt[3]{24 x^{10} y^{12}}$ 50) $\sqrt[3]{54 y^{10} z^{24}}$

51) $\sqrt[3]{72 t^{17} u^7}$ 52) $\sqrt[3]{250 w^4 x^{16}}$

53) $\sqrt[4]{\dfrac{m^8}{81}}$ 54) $\sqrt[4]{\dfrac{16}{x^{12}}}$

55) $\sqrt[5]{\dfrac{32 a^{23}}{b^{15}}}$ 56) $\sqrt[3]{\dfrac{h^{17}}{125 k^{21}}}$

57) $\sqrt[4]{\dfrac{t^9}{81 s^{24}}}$ 58) $\sqrt[5]{\dfrac{32 c^9}{d^{20}}}$

59) $\sqrt[3]{\dfrac{u^{28}}{v^3}}$ 60) $\sqrt[4]{\dfrac{m^{13}}{n^8}}$

Perform the indicated operation, and simplify.

61) $\sqrt[3]{6} \cdot \sqrt[3]{4}$
62) $\sqrt[3]{4} \cdot \sqrt[3]{10}$
63) $\sqrt[3]{9} \cdot \sqrt[3]{12}$
64) $\sqrt[3]{9} \cdot \sqrt[3]{6}$
65) $\sqrt[3]{20} \cdot \sqrt[3]{4}$
66) $\sqrt[3]{28} \cdot \sqrt[3]{2}$
67) $\sqrt[3]{m^4} \cdot \sqrt[3]{m^5}$
68) $\sqrt[3]{t^5} \cdot \sqrt[3]{t}$
69) $\sqrt[4]{k^7} \cdot \sqrt[4]{k^9}$
70) $\sqrt[4]{a^9} \cdot \sqrt[4]{a^{11}}$
71) $\sqrt[3]{r^7} \cdot \sqrt[3]{r^4}$
72) $\sqrt[3]{y^2} \cdot \sqrt[3]{y^{17}}$
73) $\sqrt[5]{p^{14}} \cdot \sqrt[5]{p^9}$
74) $\sqrt[5]{c^{17}} \cdot \sqrt[5]{c^9}$
75) $\sqrt[3]{9z^{11}} \cdot \sqrt[3]{3z^8}$
76) $\sqrt[3]{2h^4} \cdot \sqrt[3]{4h^{16}}$
77) $\sqrt[3]{\dfrac{h^{14}}{h^2}}$
78) $\sqrt[3]{\dfrac{a^{20}}{a^{14}}}$
79) $\sqrt[3]{\dfrac{c^{11}}{c^4}}$
80) $\sqrt[3]{\dfrac{z^{16}}{z^5}}$
81) $\sqrt[4]{\dfrac{162d^{21}}{2d^2}}$
82) $\sqrt[4]{\dfrac{48t^{11}}{3t^6}}$

Objective 5: Multiply and Divide Radicals with Different Indices

The following radical expressions do not have the same indices. Perform the indicated operation, and write the answer in simplest radical form.

Fill It In

Fill in the blanks with either the missing mathematical step or reason for the given step.

83) $\sqrt{a} \cdot \sqrt[4]{a^3} = a^{1/2} \cdot a^{3/4}$ _____

$= a^{2/4} \cdot a^{3/4}$ _____

$=$ _____ Add exponents.

$= \sqrt[4]{a^5}$ _____

$=$ _____ Simplify.

84) $\sqrt[5]{r^4} \cdot \sqrt[3]{r^2} =$ _____ Change radicals to fractional exponents.

$=$ _____ Rewrite exponents with a common denominator.

$= r^{22/15}$ _____

$=$ _____ Rewrite in radical form.

$=$ _____ Simplify.

85) $\sqrt{p} \cdot \sqrt[3]{p}$
86) $\sqrt[3]{y^2} \cdot \sqrt[4]{y}$
87) $\sqrt[4]{n^3} \cdot \sqrt{n}$
88) $\sqrt[5]{k^4} \cdot \sqrt{k}$
89) $\sqrt[5]{c^3} \cdot \sqrt[3]{c^2}$
90) $\sqrt[3]{a^2} \cdot \sqrt[5]{a^2}$
91) $\dfrac{\sqrt{w}}{\sqrt[4]{w}}$
92) $\dfrac{\sqrt[3]{m^3}}{\sqrt{m}}$
93) $\dfrac{\sqrt[4]{h^3}}{\sqrt[3]{h^2}}$
94) $\dfrac{\sqrt[5]{t^4}}{\sqrt[3]{t^2}}$

95) A block of candle wax in the shape of a cube has a volume of 64 in^3. The length of a side of the block, s, is given by $s = \sqrt[3]{V}$, where V is the volume of the block of candle wax. How long is each side of the block?

96) The radius $r(V)$ of a sphere is a function of its volume V and can be described by the function $r(V) = \sqrt[3]{\dfrac{3V}{4\pi}}$. If a spherical water tank has a volume of $\dfrac{256\pi}{3}$ ft^3, what is the radius of the tank?

R Rethink

R1) Which method did you use to simplify in Objective 2? Why?

R2) How did it help you to learn these same rules, but only apply them to square roots before applying them to higher roots in this section?

9.5 Adding, Subtracting, and Multiplying Radicals

P Prepare
What are your objectives for Section 9.5?

O Organize
How can you accomplish each objective?

1	Add and Subtract Radical Expressions	• Write the definition of *like radicals* in your notes. • Write the property for **Adding and Subtracting Radicals** in your notes. • Complete the given examples on your own. • Complete You Trys 1 and 2.
2	Simplify Before Adding and Subtracting	• Write the procedure for **Adding and Subtracting Radicals** in your own words. • Complete the given example on your own. • Complete You Try 3.
3	Multiply a Binomial Containing Radical Expressions by a Monomial	• Write a procedure that outlines how to multiply in this objective. • Complete the given example on your own. • Complete You Try 4.
4	Multiply Radical Expressions Using FOIL	• Compare using FOIL in Chapter 6 to using it for binomials containing radicals, and note what is similar and what is different. • Complete the given example on your own. • Complete You Try 5.
5	Square a Binomial Containing Radical Expressions	• Review the formulas for squaring a binomial in Chapter 6, if necessary. • Complete the given example on your own. • Complete You Try 6.
6	Multiply Radical Expressions of the Form $(a + b)(a - b)$	• Use the same formula derived in Chapter 6. • Complete the given example on your own. • Complete You Try 7.

Read the explanations, follow the examples, take notes, and complete the You Trys.

Just as we can add and subtract like terms such as $4x + 6x = 10x$, we can add and subtract *like radicals* such as $4\sqrt{3} + 6\sqrt{3}$.

Definition

Like radicals have the same index and the same radicand.

Some examples of like radicals are

$$4\sqrt{3} \text{ and } 6\sqrt{3}, \quad -\sqrt[3]{5} \text{ and } 8\sqrt[3]{5}, \quad \sqrt{x} \text{ and } 7\sqrt{x}, \quad 2\sqrt[3]{a^2b} \text{ and } \sqrt[3]{a^2b}$$

In this section, assume all variables represent nonnegative real numbers.

1 Add and Subtract Radical Expressions

Property Adding and Subtracting Radicals

In order to add or subtract radicals, they must be *like* radicals.

We add and subtract like radicals in the same way we add and subtract like terms—add or subtract the "coefficients" of the radicals and multiply that result by the radical. We are using the distributive property when we combine like terms in this way.

W Hint Where have you seen similar definitions and procedures before?

EXAMPLE 1 Perform the operations, and simplify.

a) $4x + 6x$
b) $4\sqrt{3} + 6\sqrt{3}$
c) $\sqrt[4]{5} - 9\sqrt[4]{5}$
d) $7\sqrt{2} + 4\sqrt{3}$

Solution

a) First notice that $4x$ and $6x$ are like terms. Therefore, they can be added.

$$4x + 6x = (4 + 6)x \quad \text{Distributive property}$$
$$= 10x \quad \text{Simplify.}$$

Or, we can say that by just adding the coefficients, $4x + 6x = 10x$.

W Hint When adding or subtracting "like radicals," the radical does not change.

b) Before attempting to add $4\sqrt{3}$ and $6\sqrt{3}$, we must be certain that they are like radicals. Since they *are* like, they can be added.

$$4\sqrt{3} + 6\sqrt{3} = (4 + 6)\sqrt{3} \quad \text{Distributive property}$$
$$= 10\sqrt{3} \quad \text{Simplify.}$$

Or, we can say that by just adding the coefficients of $\sqrt{3}$, we get $4\sqrt{3} + 6\sqrt{3} = 10\sqrt{3}$.

c) $\sqrt[4]{5} - 9\sqrt[4]{5} = 1\sqrt[4]{5} - 9\sqrt[4]{5} = (1 - 9)\sqrt[4]{5} = -8\sqrt[4]{5}$

d) The radicands in $7\sqrt{2} + 4\sqrt{3}$ are different, so these expressions cannot be combined.

YOU TRY 1 Perform the operations, and simplify.

a) $9c + 7c$
b) $9\sqrt{10} + 7\sqrt{10}$
c) $2\sqrt[3]{4} - 8\sqrt[3]{4}$
d) $5\sqrt{6} - 2\sqrt{3}$

EXAMPLE 2 Perform the operations, and simplify. $6\sqrt{x} + 11\sqrt[3]{x} + 2\sqrt{x} - 6\sqrt[3]{x}$

Solution

Begin by noticing that there are *two* different types of radicals: $\sqrt{x}$ and $\sqrt[3]{x}$. Write the like radicals together.

$$6\sqrt{x} + 11\sqrt[3]{x} + 2\sqrt{x} - 6\sqrt[3]{x} = 6\sqrt{x} + 2\sqrt{x} + 11\sqrt[3]{x} - 6\sqrt[3]{x} \quad \text{Commutative property}$$
$$= (6 + 2)\sqrt{x} + (11 - 6)\sqrt[3]{x} \quad \text{Distributive property}$$
$$= 8\sqrt{x} + 5\sqrt[3]{x}$$

Is $8\sqrt{x} + 5\sqrt[3]{x}$ in simplest form? Yes. The radicals are not like (they have different indices) so they cannot be combined further. Also, each radical, $\sqrt{x}$ and $\sqrt[3]{x}$, is in simplest form.

YOU TRY 2 Perform the operations, and simplify. $8\sqrt[3]{2n} - 3\sqrt{2n} + 5\sqrt{2n} + 5\sqrt[3]{2n}$

2 Simplify Before Adding and Subtracting

Sometimes it looks like two radicals cannot be added or subtracted. But if the radicals can be *simplified* and they turn out to be *like* radicals, then we can add or subtract them.

> **Procedure** Adding and Subtracting Radicals
> 1) Write each radical expression in simplest form.
> 2) Combine like radicals.

EXAMPLE 3

Perform the operations, and simplify.

a) $8\sqrt{2} + 3\sqrt{50} - \sqrt{45}$ b) $-7\sqrt[3]{40} + \sqrt[3]{5}$

c) $10\sqrt{8t} - 9\sqrt{2t}$ d) $\sqrt[3]{xy^6} + \sqrt[3]{x^7}$

Solution

a) The radicals $8\sqrt{2}$, $3\sqrt{50}$, and $\sqrt{45}$ are not like. The first radical is in simplest form, but $3\sqrt{50}$ and $\sqrt{45}$ should be simplified to determine if any of the radicals can be combined.

$$\begin{aligned} 8\sqrt{2} + 3\sqrt{50} - \sqrt{45} &= 8\sqrt{2} + 3\sqrt{25 \cdot 2} - \sqrt{9 \cdot 5} & \text{Factor.} \\ &= 8\sqrt{2} + 3\sqrt{25} \cdot \sqrt{2} - \sqrt{9} \cdot \sqrt{5} & \text{Product rule} \\ &= 8\sqrt{2} + 3 \cdot 5 \cdot \sqrt{2} - 3\sqrt{5} & \text{Simplify radicals.} \\ &= 8\sqrt{2} + 15\sqrt{2} - 3\sqrt{5} & \text{Multiply.} \\ &= 23\sqrt{2} - 3\sqrt{5} & \text{Add like radicals.} \end{aligned}$$

Hint
Write out each step very carefully to avoid making mistakes.

b) $\begin{aligned} -7\sqrt[3]{40} + \sqrt[3]{5} &= -7\sqrt[3]{8 \cdot 5} + \sqrt[3]{5} & \text{8 is a perfect cube.} \\ &= -7\sqrt[3]{8} \cdot \sqrt[3]{5} + \sqrt[3]{5} & \text{Product rule} \\ &= -7 \cdot 2 \cdot \sqrt[3]{5} + \sqrt[3]{5} & \sqrt[3]{8} = 2 \\ &= -14\sqrt[3]{5} + \sqrt[3]{5} & \text{Multiply.} \\ &= -13\sqrt[3]{5} & \text{Add like radicals.} \end{aligned}$

c) The radical $\sqrt{2t}$ is simplified, but $\sqrt{8t}$ is not. We must simplify $\sqrt{8t}$:

$$\sqrt{8t} = \sqrt{8} \cdot \sqrt{t} = \sqrt{4} \cdot \sqrt{2} \cdot \sqrt{t} = 2\sqrt{2} \cdot \sqrt{t} = 2\sqrt{2t}$$

Substitute $2\sqrt{2t}$ for $\sqrt{8t}$ in the original expression.

$$\begin{aligned} 10\sqrt{8t} - 9\sqrt{2t} &= 10(2\sqrt{2t}) - 9\sqrt{2t} & \text{Substitute } 2\sqrt{2t} \text{ for } \sqrt{8t}. \\ &= 20\sqrt{2t} - 9\sqrt{2t} & \text{Multiply.} \\ &= 11\sqrt{2t} & \text{Subtract.} \end{aligned}$$

d) Each radical in the expression $\sqrt[3]{xy^6} + \sqrt[3]{x^7}$ must be simplified.

$$\sqrt[3]{xy^6} = \sqrt[3]{x} \cdot \sqrt[3]{y^6} = \sqrt[3]{x} \cdot y^2 = y^2\sqrt[3]{x} \qquad \sqrt[3]{x^7} = x^2\sqrt[3]{x^1} \quad \text{7 ÷ 3 gives a quotient of 2 and a remainder of 1.}$$

$$\begin{aligned} \sqrt[3]{xy^6} + \sqrt[3]{x^7} &= y^2\sqrt[3]{x} + x^2\sqrt[3]{x} & \text{Substitute the simplified radicals in the original expression.} \\ &= (y^2 + x^2)\sqrt[3]{x} & \text{Factor out } \sqrt[3]{x} \text{ from each term.} \end{aligned}$$

In this problem we cannot *add* $y^2\sqrt[3]{x} + x^2\sqrt[3]{x}$ like we added radicals in previous examples, but we *can* factor out $\sqrt[3]{x}$.

$(y^2 + x^2)\sqrt[3]{x}$ is the completely simplified form of the sum.

[YOU TRY 3] Perform the operations, and simplify.

a) $7\sqrt{3} - \sqrt{12}$
b) $2\sqrt{63} - 11\sqrt{28} + 2\sqrt{21}$
c) $\sqrt[3]{54} + 5\sqrt[3]{16}$
d) $2\sqrt{6k} + 4\sqrt{54k}$
e) $\sqrt[4]{mn^{11}} + \sqrt[4]{81mn^3}$

In the rest of this section, we will learn how to simplify expressions that combine multiplication, addition, and subtraction of radicals.

3 Multiply a Binomial Containing Radical Expressions by a Monomial

EXAMPLE 4 Multiply and simplify.

a) $4(\sqrt{5} - \sqrt{20})$
b) $\sqrt{2}(\sqrt{10} + \sqrt{15})$
c) $\sqrt{x}(\sqrt{x} + \sqrt{32y})$

Solution

a) Because $\sqrt{20}$ can be simplified, we will do that first.

$$\sqrt{20} = \sqrt{4 \cdot 5} = \sqrt{4} \cdot \sqrt{5} = 2\sqrt{5}$$

Substitute $2\sqrt{5}$ for $\sqrt{20}$ in the original expression.

$$4(\sqrt{5} - \sqrt{20}) = 4(\sqrt{5} - 2\sqrt{5}) \quad \text{Substitute } 2\sqrt{5} \text{ for } \sqrt{20}.$$
$$= 4(-\sqrt{5}) \quad \text{Subtract.}$$
$$= -4\sqrt{5} \quad \text{Multiply.}$$

W Hint
Use a pen with multiple colors to perform the steps in different colors as it is done in the examples.

b) Neither $\sqrt{10}$ nor $\sqrt{15}$ can be simplified. Begin by applying the distributive property.

$$\sqrt{2}(\sqrt{10} + \sqrt{15}) = \sqrt{2} \cdot \sqrt{10} + \sqrt{2} \cdot \sqrt{15} \quad \text{Distribute.}$$
$$= \sqrt{20} + \sqrt{30} \quad \text{Product rule}$$

Is $\sqrt{20} + \sqrt{30}$ in simplest form? No. $\sqrt{20}$ can be simplified.

$$= \sqrt{4 \cdot 5} + \sqrt{30} = \sqrt{4} \cdot \sqrt{5} + \sqrt{30} = 2\sqrt{5} + \sqrt{30}$$

c) Since $\sqrt{32y}$ can be simplified, we will do that first.

$$\sqrt{32y} = \sqrt{32} \cdot \sqrt{y} = \sqrt{16 \cdot 2} \cdot \sqrt{y} = \sqrt{16} \cdot \sqrt{2} \cdot \sqrt{y} = 4\sqrt{2y}$$

Substitute $4\sqrt{2y}$ for $\sqrt{32y}$ in the original expression.

$$\sqrt{x}(\sqrt{x} + \sqrt{32y}) = \sqrt{x}(\sqrt{x} + 4\sqrt{2y}) \quad \text{Substitute } 4\sqrt{2y} \text{ for } \sqrt{32y}.$$
$$= \sqrt{x} \cdot \sqrt{x} + \sqrt{x} \cdot 4\sqrt{2y} \quad \text{Distribute.}$$
$$= x + 4\sqrt{2xy} \quad \text{Multiply.}$$

[YOU TRY 4] Multiply and simplify.

a) $6(\sqrt{75} + 2\sqrt{3})$
b) $\sqrt{3}(\sqrt{3} + \sqrt{21})$
c) $\sqrt{c}(\sqrt{c^3} - \sqrt{100d})$

4 Multiply Radical Expressions Using FOIL

In Chapter 6, we first multiplied binomials using **FOIL** (**F**irst **O**uter **I**nner **L**ast).

$$(2x + 3)(x + 4) = \underset{F}{2x \cdot x} + \underset{O}{2x \cdot 4} + \underset{I}{3 \cdot x} + \underset{L}{3 \cdot 4}$$
$$= 2x^2 + 8x + 3x + 12$$
$$= 2x^2 + 11x + 12$$

We can multiply binomials containing radicals the same way.

EXAMPLE 5 Multiply and simplify.

a) $(2 + \sqrt{5})(4 + \sqrt{5})$ b) $(2\sqrt{3} + \sqrt{2})(\sqrt{3} - 5\sqrt{2})$

c) $(\sqrt{r} + \sqrt{3s})(\sqrt{r} + 8\sqrt{3s})$

Solution

a) Since we must multiply two binomials, we will use FOIL.

$$(2 + \sqrt{5})(4 + \sqrt{5}) = \underset{F}{2 \cdot 4} + \underset{O}{2 \cdot \sqrt{5}} + \underset{I}{4 \cdot \sqrt{5}} + \underset{L}{\sqrt{5} \cdot \sqrt{5}}$$
$$= 8 + 2\sqrt{5} + 4\sqrt{5} + 5 \qquad \text{Multiply.}$$
$$= 13 + 6\sqrt{5} \qquad \text{Combine like terms.}$$

W Hint
Multiplication of binomials containing radicals uses the same procedure as multiplying binomials in Chapter 6!

b) $(2\sqrt{3} + \sqrt{2})(\sqrt{3} - 5\sqrt{2})$

$$= \underset{F}{2\sqrt{3} \cdot \sqrt{3}} + \underset{O}{2\sqrt{3} \cdot (-5\sqrt{2})} + \underset{I}{\sqrt{2} \cdot \sqrt{3}} + \underset{L}{\sqrt{2} \cdot (-5\sqrt{2})}$$
$$= 2 \cdot 3 + (-10\sqrt{6}) + \sqrt{6} + (-5 \cdot 2) \qquad \text{Multiply.}$$
$$= 6 - 10\sqrt{6} + \sqrt{6} - 10 \qquad \text{Multiply.}$$
$$= -4 - 9\sqrt{6} \qquad \text{Combine like terms.}$$

c) $(\sqrt{r} + \sqrt{3s})(\sqrt{r} + 8\sqrt{3s})$

$$= \underset{F}{\sqrt{r} \cdot \sqrt{r}} + \underset{O}{\sqrt{r} \cdot 8\sqrt{3s}} + \underset{I}{\sqrt{3s} \cdot \sqrt{r}} + \underset{L}{\sqrt{3s} \cdot 8\sqrt{3s}}$$
$$= r + 8\sqrt{3rs} + \sqrt{3rs} + 8 \cdot 3s \qquad \text{Multiply.}$$
$$= r + 8\sqrt{3rs} + \sqrt{3rs} + 24s \qquad \text{Multiply.}$$
$$= r + 9\sqrt{3rs} + 24s \qquad \text{Combine like terms.}$$

[YOU TRY 5] Multiply and simplify.

a) $(6 - \sqrt{7})(5 + \sqrt{7})$ b) $(\sqrt{2} + 4\sqrt{5})(3\sqrt{2} + \sqrt{5})$

c) $(\sqrt{6p} - \sqrt{2q})(\sqrt{6p} - 3\sqrt{2q})$

5 Square a Binomial Containing Radical Expressions

Recall, again, from Chapter 6, that we can use FOIL to square a binomial or we can use these special formulas:

$$(a + b)^2 = a^2 + 2ab + b^2 \qquad (a - b)^2 = a^2 - 2ab + b^2$$

For example,

$$(k+7)^2 = (k)^2 + 2(k)(7) + (7)^2 \quad \text{and} \quad (2p-5)^2 = (2p)^2 - 2(2p)(5) + (5)^2$$
$$= k^2 + 14k + 49 \qquad\qquad\qquad\qquad = 4p^2 - 20p + 25$$

To square a binomial containing radicals, we can either use FOIL or we can use the formulas above. Understanding how to use the formulas to square a binomial will make it easier to solve radical equations in Section 9.7.

EXAMPLE 6

Multiply and simplify.

a) $(\sqrt{10} + 3)^2$ b) $(2\sqrt{x} - 6)^2$

Solution

a) Use $(a+b)^2 = a^2 + 2ab + b^2$.

$$(\sqrt{10} + 3)^2 = (\sqrt{10})^2 + 2(\sqrt{10})(3) + (3)^2 \quad \text{Substitute } \sqrt{10} \text{ for } a \text{ and } 3 \text{ for } b.$$
$$= 10 + 6\sqrt{10} + 9 \qquad \text{Multiply.}$$
$$= 19 + 6\sqrt{10} \qquad \text{Combine like terms.}$$

b) Use $(a-b)^2 = a^2 - 2ab + b^2$.

$$(2\sqrt{x} - 6)^2 = (2\sqrt{x})^2 - 2(2\sqrt{x})(6) + (6)^2 \quad \text{Substitute } 2\sqrt{x} \text{ for } a \text{ and } 6 \text{ for } b.$$
$$= (4 \cdot x) - (4\sqrt{x})(6) + 36 \qquad \text{Multiply.}$$
$$= 4x - 24\sqrt{x} + 36 \qquad \text{Multiply.}$$

[YOU TRY 6]

Multiply and simplify.

a) $(\sqrt{6} + 5)^2$ b) $(3\sqrt{2} - 4)^2$ c) $(\sqrt{w} + \sqrt{11})^2$

6 Multiply Radical Expressions of the Form $(a+b)(a-b)$

We will review one last rule from Chapter 6 on multiplying binomials. We will use this in Section 9.6 when we divide radicals.

$$(a+b)(a-b) = a^2 - b^2$$

For example, $(t+8)(t-8) = (t)^2 - (8)^2 = t^2 - 64$.

The same rule applies when we multiply binomials containing radicals.

EXAMPLE 7

Multiply and simplify $(2\sqrt{x} + \sqrt{y})(2\sqrt{x} - \sqrt{y})$.

Solution

Use $(a+b)(a-b) = a^2 - b^2$.

$$(2\sqrt{x} + \sqrt{y})(2\sqrt{x} - \sqrt{y}) = (2\sqrt{x})^2 - (\sqrt{y})^2 \quad \text{Substitute } 2\sqrt{x} \text{ for } a \text{ and } \sqrt{y} \text{ for } b.$$
$$= 4(x) - y \qquad \text{Square each term.}$$
$$= 4x - y \qquad \text{Simplify.}$$

> **Note**
> When we multiply expressions of the form $(a + b)(a - b)$ containing square roots, the radicals are eliminated. *This will always be true.*

[YOU TRY 7] Multiply and simplify.

a) $(4 + \sqrt{10})(4 - \sqrt{10})$ b) $(\sqrt{5h} + \sqrt{k})(\sqrt{5h} - \sqrt{k})$

ANSWERS TO [YOU TRY] EXERCISES

1) a) $16c$ b) $16\sqrt{10}$ c) $-6\sqrt[3]{4}$ d) $5\sqrt{6} - 2\sqrt{3}$ 2) $13\sqrt[3]{2n} + 2\sqrt{2n}$ 3) a) $5\sqrt{3}$
b) $-16\sqrt{7} + 2\sqrt{21}$ c) $13\sqrt[3]{2}$ d) $14\sqrt[4]{6k}$ e) $(n^2 + 3)\sqrt[4]{mn^3}$ 4) a) $42\sqrt{3}$ b) $3 + 3\sqrt{7}$
c) $c^2 - 10\sqrt{cd}$ 5) a) $23 + \sqrt{7}$ b) $26 + 13\sqrt{10}$ c) $6p - 8\sqrt{3pq} + 6q$
6) a) $31 + 10\sqrt{6}$ b) $34 - 24\sqrt{2}$ c) $w + 2\sqrt{11w} + 11$ 7) a) 6 b) $5h - k$

E Evaluate 9.5 Exercises

Do the exercises, and check your work.

Assume all variables represent nonnegative real numbers.

Objective 1: Add and Subtract Radical Expressions

1) How do you know if two radicals are *like* radicals?

2) Are $5\sqrt{3}$ and $7\sqrt[3]{3}$ like radicals? Why or why not?

Perform the operations, and simplify.

3) $5\sqrt{2} + 9\sqrt{2}$ 4) $11\sqrt{7} + 7\sqrt{7}$

5) $7\sqrt[3]{4} + 8\sqrt[3]{4}$ 6) $10\sqrt[3]{5} - 2\sqrt[3]{5}$

 7) $6 - \sqrt{13} + 5 - 2\sqrt{13}$

8) $-8 + 3\sqrt{6} - 4\sqrt{6} + 9$

9) $15\sqrt[3]{z^2} - 20\sqrt[3]{z^2}$

10) $7\sqrt[3]{p} - 4\sqrt[3]{p}$

11) $2\sqrt[3]{n^2} + 9\sqrt[5]{n^2} - 11\sqrt[3]{n^2} + \sqrt[5]{n^2}$

12) $5\sqrt[4]{s} - 3\sqrt[3]{s} + 2\sqrt[3]{s} + 4\sqrt[4]{s}$

13) $\sqrt{5c} - 8\sqrt{6c} + \sqrt{5c} + 6\sqrt{6c}$

14) $10\sqrt{2m} + 6\sqrt{3m} - \sqrt{2m} + 8\sqrt{3m}$

Objective 2: Simplify Before Adding and Subtracting

15) What are the steps for adding or subtracting radicals?

16) Is $6\sqrt{2} + \sqrt{10}$ in simplified form? Explain.

Perform the operations, and simplify.

Fill It In
Fill in the blanks with either the missing mathematical step or reason for the given step.

17) $\sqrt{48} + \sqrt{3}$
 $= \sqrt{16 \cdot 3} + \sqrt{3}$
 $= \underline{}$ Product rule
 $= 4\sqrt{3} + \sqrt{3}$
 $= \underline{}$ Add like radicals.

18) $\sqrt{44} - 8\sqrt{11}$
 $= \sqrt{4 \cdot 11} - 8\sqrt{11}$
 $= \sqrt{4} \cdot \sqrt{11} - 8\sqrt{11}$
 $= \underline{}$ Simplify.
 $= \underline{}$ Subtract like radicals.

19) $6\sqrt{3} - \sqrt{12}$ 20) $\sqrt{45} + 4\sqrt{5}$

21) $\sqrt{32} - 3\sqrt{8}$ 22) $3\sqrt{24} + \sqrt{96}$

23) $\sqrt{12} + \sqrt{75} - \sqrt{3}$ 24) $\sqrt{96} + \sqrt{24} - 5\sqrt{54}$

25) $8\sqrt[3]{9} + \sqrt[3]{72}$ 26) $5\sqrt[3]{88} + 2\sqrt[3]{11}$

27) $\sqrt[3]{6} - \sqrt[3]{48}$ 28) $11\sqrt[3]{16} + 7\sqrt[3]{2}$

29) $6q\sqrt{q} + 7\sqrt{q^3}$ 30) $11\sqrt{m^3} + 8m\sqrt{m}$

31) $4d^2\sqrt{d} - 24\sqrt{d^5}$ 32) $16k^4\sqrt{k} - 13\sqrt{k^9}$

33) $9t^3\sqrt[3]{t} - 5\sqrt[3]{t^{10}}$ 34) $8r^4\sqrt[3]{r} - 16\sqrt[3]{r^{13}}$

35) $5a\sqrt[4]{a^7} + \sqrt[4]{a^{11}}$ 36) $-3\sqrt[4]{c^{11}} + 6c^2\sqrt[4]{c^3}$

37) $2\sqrt{8p} - 6\sqrt{2p}$ 38) $4\sqrt{63t} + 6\sqrt{7t}$

39) $7\sqrt[3]{81a^5} + 4a\sqrt[3]{3a^2}$

40) $3\sqrt[3]{40x} - 12\sqrt[3]{5x}$

41) $\sqrt{xy^3} + 3y\sqrt{xy}$

42) $5a\sqrt{ab} + 2\sqrt{a^3b}$

43) $6c^2\sqrt{8d^3} - 9d\sqrt{2c^4d}$

44) $11v\sqrt{5u^3} - 2u\sqrt{45uv^2}$

45) $8p^2q\sqrt[3]{11pq^2} + 3p^2\sqrt[3]{88pq^5}$

46) $18a^5\sqrt[3]{7a^2b} + 2a^3\sqrt[3]{7a^8b}$

47) $15cd\sqrt[4]{9cd} - \sqrt[4]{9c^5d^5}$

48) $7yz^2\sqrt[4]{11y^4z} + 3z\sqrt[4]{11y^8z^5}$

49) $\sqrt[3]{a^9b} - \sqrt[3]{b^7}$

50) $\sqrt[3]{c^8} + \sqrt[3]{c^2d^3}$

Objective 3: Multiply a Binomial Containing Radical Expressions by a Monomial
Multiply and simplify.

51) $3(x + 5)$
52) $8(k + 3)$
53) $7(\sqrt{6} + 2)$
54) $5(4 - \sqrt{7})$
55) $\sqrt{10}(\sqrt{3} - 1)$
56) $\sqrt{2}(9 + \sqrt{11})$
57) $-6(\sqrt{32} + \sqrt{2})$
58) $10(\sqrt{12} - \sqrt{3})$
59) $4(\sqrt{45} - \sqrt{20})$
60) $-3(\sqrt{18} + \sqrt{50})$
61) $\sqrt{5}(\sqrt{24} - \sqrt{54})$
62) $\sqrt{2}(\sqrt{20} + \sqrt{45})$
63) $\sqrt[4]{3}(5 - \sqrt[4]{27})$
64) $\sqrt[3]{4}(2\sqrt[3]{5} + 7\sqrt[3]{4})$
65) $\sqrt{t}(\sqrt{t} - \sqrt{81u})$
66) $\sqrt{s}(\sqrt{12r} + \sqrt{7s})$
67) $\sqrt{2xy}(\sqrt{2y} - y\sqrt{x})$
68) $\sqrt{ab}(\sqrt{5a} + \sqrt{27b})$
69) $\sqrt[3]{c^2}(\sqrt[3]{c^2} + \sqrt[3]{125cd})$
70) $\sqrt[5]{mn^3}(\sqrt[5]{2m^2n} - n\sqrt[5]{mn^2})$

Mixed Exercises: Objectives 4–6

71) How are the problems *Multiply* $(x + 8)(x + 3)$ and *Multiply* $(3 + \sqrt{2})(1 + \sqrt{2})$ similar? What method can be used to multiply each of them?

72) How are the problems *Multiply* $(y - 5)^2$ and *Multiply* $(\sqrt{7} - 2)^2$ similar? What method can be used to multiply each of them?

73) What formula can be used to multiply $(5 + \sqrt{6})(5 - \sqrt{6})$?

74) What happens to the radical terms whenever we multiply $(a + b)(a - b)$ where the binomials contain square roots?

Objective 4: Multiply Radical Expressions Using FOIL
Multiply and simplify.

75) $(p + 7)(p + 6)$
76) $(z - 8)(z + 2)$

Fill It In
Fill in the blanks with either the missing mathematical step or reason for the given step.

77) $(6 + \sqrt{7})(2 + \sqrt{7})$
= _____ Use FOIL.
= $12 + 6\sqrt{7} + 2\sqrt{7} + 7$
= _____ Combine like terms.

78) $(3 + \sqrt{5})(1 + \sqrt{5})$
= $3 \cdot 1 + 3\sqrt{5} + 1\sqrt{5} + \sqrt{5} \cdot \sqrt{5}$ _____
= _____ Multiply.
= _____ Combine like terms.

79) $(\sqrt{2} + 8)(\sqrt{2} - 3)$
80) $(\sqrt{6} - 7)(\sqrt{6} + 2)$
81) $(5\sqrt{2} - \sqrt{3})(2\sqrt{3} - \sqrt{2})$
82) $(\sqrt{5} - 4\sqrt{3})(2\sqrt{5} - \sqrt{3})$
83) $(5 + 2\sqrt{3})(\sqrt{7} + \sqrt{2})$
84) $(\sqrt{5} + 4)(\sqrt{3} - 6\sqrt{2})$
85) $(\sqrt[3]{25} - 3)(\sqrt[3]{5} - \sqrt[3]{6})$
86) $(\sqrt[4]{8} - \sqrt[4]{3})(\sqrt[4]{6} + \sqrt[4]{2})$
87) $(\sqrt{6p} - 2\sqrt{q})(8\sqrt{q} + 5\sqrt{6p})$
88) $(4\sqrt{3r} + \sqrt{s})(3\sqrt{s} - 2\sqrt{3r})$

Objective 5: Square a Binomial Containing Radical Expressions

89) $(\sqrt{3} + 1)^2$
90) $(2 + \sqrt{5})^2$
91) $(\sqrt{11} - \sqrt{5})^2$
92) $(\sqrt{3} + \sqrt{13})^2$
93) $(\sqrt{h} + \sqrt{7})^2$
94) $(\sqrt{m} + \sqrt{3})^2$
95) $(\sqrt{x} - \sqrt{y})^2$
96) $(\sqrt{b} - \sqrt{a})^2$

Objective 6: Multiply Radical Expressions of the Form $(a + b)(a - b)$

97) $(c + 9)(c - 9)$
98) $(g - 7)(g + 7)$
99) $(6 - \sqrt{5})(6 + \sqrt{5})$
100) $(4 - \sqrt{7})(4 + \sqrt{7})$

101) $(4\sqrt{3} + \sqrt{2})(4\sqrt{3} - \sqrt{2})$
102) $(2\sqrt{2} - 2\sqrt{7})(2\sqrt{2} + 2\sqrt{7})$
103) $(\sqrt[3]{2} - 3)(\sqrt[3]{2} + 3)$
104) $(1 + \sqrt[3]{6})(1 - \sqrt[3]{6})$
105) $(\sqrt{c} + \sqrt{d})(\sqrt{c} - \sqrt{d})$
106) $(\sqrt{2y} + \sqrt{z})(\sqrt{2y} - \sqrt{z})$
107) $(8\sqrt{f} - \sqrt{g})(8\sqrt{f} + \sqrt{g})$
108) $(\sqrt{a} + 3\sqrt{4b})(\sqrt{a} - 3\sqrt{4b})$

Extension

Multiply and simplify.

109) $(1 + 2\sqrt[3]{5})(1 - 2\sqrt[3]{5} + 4\sqrt[3]{25})$
110) $(3 + \sqrt[3]{2})(9 - 3\sqrt[3]{2} + \sqrt[3]{4})$

Let $f(x) = x^2$. Find each function value.

111) $f(\sqrt{7} + 2)$
112) $f(5 - \sqrt{6})$
113) $f(1 - 2\sqrt{3})$
114) $f(3\sqrt{2} + 4)$

R Rethink

R1) Explain the difference between mastering the concepts from Chapter 6 to completing these exercises. How did it compare?

R2) Explain how you could substitute "x" for the radical $\sqrt{3}$ and "y" for the radical $\sqrt{2}$ in Exercise 101 to help you multiply.

9.6 Dividing Radicals

P Prepare

What are your objectives for Section 9.6?	How can you accomplish each objective?
1 Rationalize a Denominator: One Square Root	• Understand what *rationalizing a denominator* means. • Complete the given examples on your own, and develop a procedure for rationalizing a denominator that contains one square root. • Complete You Trys 1–3.
2 Rationalize a Denominator: One Higher Root	• Follow the explanation to understand the logic behind rationalizing a higher root. • Complete the given examples on your own, and develop a procedure for rationalizing a denominator that contains one higher root. • Complete You Trys 4 and 5.
3 Rationalize a Denominator Containing Two Terms	• Understand the definition of a *conjugate*, and write a few examples in your notes. • Recall the formula for $(a + b)(a - b)$, and know how it can help you multiply conjugates. • Learn the procedure for **Rationalizing a Denominator That Contains Two Terms.** • Complete the given examples on your own. • Complete You Trys 6 and 7.

(*continued*)

What are your objectives for Section 9.6?	How can you accomplish each objective?
4 Rationalize a Numerator	• Follow the same procedures developed for rationalizing the denominator, but focus instead on the numerator. • Complete the given example on your own. • Complete You Try 8.
5 Divide Out Common Factors from the Numerator and Denominator	• Know how to apply the properties of real numbers, especially factoring and distributing, to simplify. • Complete the given example on your own. • Complete You Try 9.

W Work   Read the explanations, follow the examples, take notes, and complete the You Trys.

It is generally agreed that a radical expression is *not* in simplest form if its denominator contains a radical. For example, $\frac{1}{\sqrt{3}}$ is not simplified, but an equivalent form, $\frac{\sqrt{3}}{3}$, is simplified.

Later we will show that $\frac{1}{\sqrt{3}} = \frac{\sqrt{3}}{3}$. The process of eliminating radicals from the denominator of an expression is called **rationalizing the denominator**. We will look at two types of rationalizing problems.

    1) Rationalizing a denominator containing one term
    2) Rationalizing a denominator containing two terms

To rationalize a denominator, we will use the fact that multiplying the numerator and denominator of a fraction by the same quantity results in an equivalent fraction:

$$\frac{2}{3} \cdot \frac{4}{4} = \frac{8}{12} \qquad \frac{2}{3} \text{ and } \frac{8}{12} \text{ are equivalent because } \frac{4}{4} = 1$$

We use the same idea to rationalize the denominator of a radical expression.

1 Rationalize a Denominator: One Square Root

The goal of rationalizing a denominator is to eliminate the radical from the denominator. With regard to square roots, recall that $\sqrt{a} \cdot \sqrt{a} = \sqrt{a^2} = a$ for $a \geq 0$. For example,

$$\sqrt{19} \cdot \sqrt{19} = \sqrt{(19)^2} = 19 \text{ and } \sqrt{t} \cdot \sqrt{t} = \sqrt{t^2} = t \ (t \geq 0)$$

We will use this property to rationalize the denominators of the following expressions.

EXAMPLE 1   Rationalize the denominator of each expression.

     a) $\frac{1}{\sqrt{3}}$ b) $\frac{36}{\sqrt{18}}$ c) $\frac{5\sqrt{3}}{\sqrt{2}}$

Solution

a) To eliminate the square root from the denominator of $\dfrac{1}{\sqrt{3}}$, ask yourself, "By what do I multiply $\sqrt{3}$ to get a *perfect square* under the square root?" The answer is $\sqrt{3}$ because $\sqrt{3} \cdot \sqrt{3} = \sqrt{3^2} = \sqrt{9} = 3$. Multiply by $\sqrt{3}$ in the numerator *and* denominator. (We are actually multiplying by 1.)

$$\dfrac{1}{\sqrt{3}} = \dfrac{1}{\sqrt{3}} \cdot \underbrace{\dfrac{\sqrt{3}}{\sqrt{3}}}_{\text{Rationalize the denominator.}} = \dfrac{\sqrt{3}}{\sqrt{3^2}} = \dfrac{\sqrt{3}}{\sqrt{9}} = \dfrac{\sqrt{3}}{3}$$

BE CAREFUL $\dfrac{\sqrt{3}}{3}$ is in simplest form. We cannot "simplify" terms inside and outside of the radical.

$$\dfrac{\sqrt{3}}{3} = \dfrac{\sqrt{3}^{1}}{3_{1}} = \sqrt{1} = 1 \quad \text{Incorrect!}$$

b) First, simplify the denominator of $\dfrac{36}{\sqrt{18}}$.

W Hint
Be sure you understand *why* you multiply by a particular radical to rationalize the denominator!

$$\dfrac{36}{\sqrt{18}} = \dfrac{36}{3\sqrt{2}} = \dfrac{12}{\sqrt{2}} = \dfrac{12}{\sqrt{2}} \cdot \dfrac{\sqrt{2}}{\sqrt{2}} = \dfrac{12\sqrt{2}}{2} = 6\sqrt{2}$$

Simplify $\sqrt{18}$. Simplify. Rationalize the denominator.

c) To rationalize $\dfrac{5\sqrt{3}}{\sqrt{2}}$, multiply the numerator and denominator by $\sqrt{2}$.

$$\dfrac{5\sqrt{3}}{\sqrt{2}} = \dfrac{5\sqrt{3}}{\sqrt{2}} \cdot \dfrac{\sqrt{2}}{\sqrt{2}} = \dfrac{5\sqrt{6}}{2}$$

[**YOU TRY 1**] Rationalize the denominator of each expression.

a) $\dfrac{1}{\sqrt{7}}$ b) $\dfrac{15}{\sqrt{27}}$ c) $\dfrac{9\sqrt{6}}{\sqrt{5}}$

Sometimes we will apply the quotient or product rule before rationalizing.

EXAMPLE 2 Simplify completely.

a) $\sqrt{\dfrac{3}{24}}$ b) $\sqrt{\dfrac{5}{14}} \cdot \sqrt{\dfrac{7}{3}}$

Solution

a) Begin by simplifying the fraction $\dfrac{3}{24}$ under the radical.

$$\sqrt{\dfrac{3}{24}} = \sqrt{\dfrac{1}{8}} \quad \text{Simplify.}$$

$$= \dfrac{\sqrt{1}}{\sqrt{8}} = \dfrac{1}{\sqrt{4} \cdot \sqrt{2}} = \dfrac{1}{2\sqrt{2}} = \dfrac{1}{2\sqrt{2}} \cdot \dfrac{\sqrt{2}}{\sqrt{2}} = \dfrac{\sqrt{2}}{2 \cdot 2} = \dfrac{\sqrt{2}}{4}$$

b) Begin by using the product rule to multiply the radicands.

$$\sqrt{\frac{5}{14}} \cdot \sqrt{\frac{7}{3}} = \sqrt{\frac{5}{14} \cdot \frac{7}{3}} \quad \text{Product rule}$$

Multiply the fractions under the radical.

$$= \sqrt{\frac{5}{\underset{2}{\cancel{14}}} \cdot \frac{\cancel{7}^1}{3}} = \sqrt{\frac{5}{6}} \quad \text{Multiply.}$$

$$= \frac{\sqrt{5}}{\sqrt{6}} = \frac{\sqrt{5}}{\sqrt{6}} \cdot \frac{\sqrt{6}}{\sqrt{6}} = \frac{\sqrt{30}}{6}$$

[**YOU TRY 2**] Simplify completely.

a) $\sqrt{\dfrac{10}{35}}$ b) $\sqrt{\dfrac{21}{10}} \cdot \sqrt{\dfrac{2}{7}}$

We work with radical expressions containing variables the same way. **In the rest of this section, we will assume that all variables represent positive real numbers.**

EXAMPLE 3

Simplify completely.

a) $\dfrac{2}{\sqrt{x}}$ b) $\sqrt{\dfrac{12m^3}{7n}}$ c) $\sqrt{\dfrac{6cd^2}{cd^3}}$

Solution

a) Ask yourself, "By what do I multiply $\sqrt{x}$ to get a *perfect square* under the square root?" The perfect square we want to get is $\sqrt{x^2}$.

$$\sqrt{x} \cdot \sqrt{?} = \sqrt{x^2} = x$$
$$\sqrt{x} \cdot \sqrt{x} = \sqrt{x^2} = x$$

$$\frac{2}{\sqrt{x}} = \frac{2}{\sqrt{x}} \cdot \frac{\sqrt{x}}{\sqrt{x}} = \frac{2\sqrt{x}}{\sqrt{x^2}} = \frac{2\sqrt{x}}{x}$$
↑
Rationalize the denominator.

W Hint
Ask yourself the questions found in parts a) and b). This will help you understand the next objective too.

b) Before rationalizing, apply the quotient rule and simplify the numerator.

$$\sqrt{\frac{12m^3}{7n}} = \frac{\sqrt{12m^3}}{\sqrt{7n}} = \frac{2m\sqrt{3m}}{\sqrt{7n}}$$

Rationalize the denominator. "By what do I multiply $\sqrt{7n}$ to get a *perfect square* under the square root?" The perfect square we want to get is $\sqrt{7^2 n^2}$ or $\sqrt{49n^2}$.

$$\sqrt{7n} \cdot \sqrt{?} = \sqrt{7^2 n^2} = 7n$$
$$\sqrt{7n} \cdot \sqrt{7n} = \sqrt{7^2 n^2} = 7n$$

$$\sqrt{\frac{12m^3}{7n}} = \frac{2m\sqrt{3m}}{\sqrt{7n}}$$
$$= \frac{2m\sqrt{3m}}{\sqrt{7n}} \cdot \frac{\sqrt{7n}}{\sqrt{7n}} = \frac{2m\sqrt{21mn}}{7n}$$
↑
Rationalize the denominator.

c) $\sqrt{\dfrac{6cd^2}{cd^3}} = \sqrt{\dfrac{6}{d}}$ Simplify the radicand using the quotient rule for exponents.

$= \dfrac{\sqrt{6}}{\sqrt{d}} = \dfrac{\sqrt{6}}{\sqrt{d}} \cdot \dfrac{\sqrt{d}}{\sqrt{d}} = \dfrac{\sqrt{6d}}{d}$

[YOU TRY 3] Simplify completely.

a) $\dfrac{5}{\sqrt{p}}$ b) $\sqrt{\dfrac{18k^5}{10m}}$ c) $\sqrt{\dfrac{20r^3 s}{s^2}}$

2 Rationalize a Denominator: One Higher Root

Many students assume that to rationalize *all* denominators we simply multiply the numerator and denominator of the expression by the denominator as in $\dfrac{4}{\sqrt{3}} = \dfrac{4}{\sqrt{3}} \cdot \dfrac{\sqrt{3}}{\sqrt{3}} = \dfrac{4\sqrt{3}}{3}$. We will see, however, why this reasoning is incorrect.

To rationalize an expression like $\dfrac{4}{\sqrt{3}}$ we asked ourselves, "By what do I multiply $\sqrt{3}$ to get a *perfect square* under the *square root*?"

To rationalize an expression like $\dfrac{5}{\sqrt[3]{2}}$ we must ask ourselves, "By what do I multiply $\sqrt[3]{2}$ to get a *perfect cube* under the *cube root*?" The perfect cube we want is 2^3 (since we began with 2) so that $\sqrt[3]{2} \cdot \sqrt[3]{2^2} = \sqrt[3]{2^3} = 2$.

We will practice some fill-in-the-blank problems to eliminate radicals before we move on to rationalizing.

EXAMPLE 4

Fill in the blank.

a) $\sqrt[3]{5} \cdot \sqrt[3]{?} = \sqrt[3]{5^3} = 5$ b) $\sqrt[3]{3} \cdot \sqrt[3]{?} = \sqrt[3]{3^3} = 3$

c) $\sqrt[3]{x^2} \cdot \sqrt[3]{?} = \sqrt[3]{x^3} = x$ d) $\sqrt[5]{8} \cdot \sqrt[5]{?} = \sqrt[5]{2^5} = 2$

e) $\sqrt[4]{27} \cdot \sqrt[4]{?} = \sqrt[4]{3^4} = 3$

Solution

a) Ask yourself, "By what do I multiply $\sqrt[3]{5}$ to get $\sqrt[3]{5^3}$?" The answer is $\sqrt[3]{5^2}$.

$\sqrt[3]{5} \cdot \sqrt[3]{?} = \sqrt[3]{5^3} = 5$
$\sqrt[3]{5} \cdot \sqrt[3]{5^2} = \sqrt[3]{5^3} = 5$

W Hint
Get in the habit of asking yourself these questions!

b) "By what do I multiply $\sqrt[3]{3}$ to get $\sqrt[3]{3^3}$?" $\sqrt[3]{3^2}$

$\sqrt[3]{3} \cdot \sqrt[3]{?} = \sqrt[3]{3^3} = 3$
$\sqrt[3]{3} \cdot \sqrt[3]{3^2} = \sqrt[3]{3^3} = 3$

c) "By what do I multiply $\sqrt[3]{x^2}$ to get $\sqrt[3]{x^3}$?" $\sqrt[3]{x}$

$\sqrt[3]{x^2} \cdot \sqrt[3]{?} = \sqrt[3]{x^3} = x$
$\sqrt[3]{x^2} \cdot \sqrt[3]{x} = \sqrt[3]{x^3} = x$

d) In this example, $\sqrt[5]{8} \cdot \sqrt[5]{?} = \sqrt[5]{2^5} = 2$, why are we trying to obtain $\sqrt[5]{2^5}$ instead of $\sqrt[5]{8^5}$? Because in the first radical, $\sqrt[5]{8}$, 8 *is a power of* 2. Before attempting to fill in the blank, rewrite 8 as 2^3.

$$\sqrt[5]{8} \cdot \sqrt[5]{?} = \sqrt[5]{2^5} = 2$$
$$\sqrt[5]{2^3} \cdot \sqrt[5]{?} = \sqrt[5]{2^5} = 2$$
$$\sqrt[5]{2^3} \cdot \sqrt[5]{2^2} = \sqrt[5]{2^5} = 2$$

> **W Hint**
> Don't move to the next example unless you have fully grasped this concept first!

e) $\sqrt[4]{27} \cdot \sqrt[4]{?} = \sqrt[4]{3^4} = 3$
$\sqrt[4]{3^3} \cdot \sqrt[4]{?} = \sqrt[4]{3^4} = 3$ Since 27 is a power of 3, rewrite $\sqrt[4]{27}$ as $\sqrt[4]{3^3}$.
$\sqrt[4]{3^3} \cdot \sqrt[4]{3} = \sqrt[4]{3^4} = 3$

[**YOU TRY 4**] Fill in the blank.

a) $\sqrt[3]{2} \cdot \sqrt[3]{?} = \sqrt[3]{2^3} = 2$ b) $\sqrt[5]{t^2} \cdot \sqrt[5]{?} = \sqrt[5]{t^5} = t$

c) $\sqrt[4]{125} \cdot \sqrt[4]{?} = \sqrt[4]{5^4} = 5$

We will use the technique presented in Example 4 to rationalize denominators with indices higher than 2.

EXAMPLE 5   Rationalize the denominator.

a) $\dfrac{7}{\sqrt[3]{3}}$ b) $\sqrt[5]{\dfrac{3}{4}}$ c) $\dfrac{7}{\sqrt[4]{n}}$

Solution

a) *First* identify what we want the denominator to be *after* multiplying. **We want to obtain $\sqrt[3]{3^3}$ since $\sqrt[3]{3^3} = 3$.**

$$\dfrac{7}{\sqrt[3]{3}} \cdot \dfrac{}{} = \dfrac{}{\sqrt[3]{3^3}} \quad \longleftarrow \text{This is what we want to get.}$$
$\uparrow$
What is needed here?

Ask yourself, "By what do I multiply $\sqrt[3]{3}$ to get $\sqrt[3]{3^3}$?" $\sqrt[3]{3^2}$

> **W Hint**
> Be sure to ask yourself this question so that you multiply by the correct radical!

$$\dfrac{7}{\sqrt[3]{3}} \cdot \dfrac{\sqrt[3]{3^2}}{\sqrt[3]{3^2}} = \dfrac{7\sqrt[3]{3^2}}{\sqrt[3]{3^3}} \quad \text{Multiply.}$$
$$= \dfrac{7\sqrt[3]{9}}{3} \quad \text{Simplify.}$$

b) Use the quotient rule for radicals to rewrite $\sqrt[5]{\dfrac{3}{4}}$ as $\dfrac{\sqrt[5]{3}}{\sqrt[5]{4}}$. Then, write 4 as 2^2 to get

$$\dfrac{\sqrt[5]{3}}{\sqrt[5]{4}} = \dfrac{\sqrt[5]{3}}{\sqrt[5]{2^2}}$$

What denominator do we want to get *after* multiplying? **We want to obtain $\sqrt[5]{2^5}$ since $\sqrt[5]{2^5} = 2$.**

$$\dfrac{\sqrt[5]{3}}{\sqrt[5]{2^2}} \cdot \dfrac{}{} = \dfrac{}{\sqrt[5]{2^5}} \quad \longleftarrow \text{This is what we want to get.}$$
$\uparrow$
What is needed here?

"By what do I multiply $\sqrt[5]{2^2}$ to get $\sqrt[5]{2^5}$?" $\sqrt[5]{2^3}$

$$\frac{\sqrt[5]{3}}{\sqrt[5]{2^2}} \cdot \frac{\sqrt[5]{2^3}}{\sqrt[5]{2^3}} = \frac{\sqrt[5]{3} \cdot \sqrt[5]{2^3}}{\sqrt[5]{2^5}} \qquad \text{Multiply.}$$

$$= \frac{\sqrt[5]{3} \cdot \sqrt[5]{8}}{2} = \frac{\sqrt[5]{24}}{2} \qquad \text{Multiply.}$$

c) What denominator do we want to get *after* multiplying? **We want to obtain $\sqrt[4]{n^4}$ since $\sqrt[4]{n^4} = n$.**

 Hint
In your own words, summarize how to rationalize denominators in Objective 2.

$$\frac{7}{\sqrt[4]{n}} \cdot \frac{}{} = \frac{}{\sqrt[4]{n^4}} \quad \longleftarrow \text{ This is what we want to get.}$$
↑
What is needed here?

Ask yourself, "By what do I multiply $\sqrt[4]{n}$ to get $\sqrt[4]{n^4}$?" $\sqrt[4]{n^3}$

$$\frac{7}{\sqrt[4]{n}} \cdot \frac{\sqrt[4]{n^3}}{\sqrt[4]{n^3}} = \frac{7\sqrt[4]{n^3}}{\sqrt[4]{n^4}} \qquad \text{Multiply.}$$

$$= \frac{7\sqrt[4]{n^3}}{n} \qquad \text{Simplify.}$$

[YOU TRY 5] Rationalize the denominator.

a) $\dfrac{4}{\sqrt[3]{7}}$ b) $\sqrt[4]{\dfrac{2}{27}}$ c) $\sqrt[5]{\dfrac{8}{w^3}}$

3 Rationalize a Denominator Containing Two Terms

To rationalize the denominator of an expression like $\dfrac{1}{5 + \sqrt{3}}$, we multiply the numerator and the denominator of the expression by the *conjugate* of $5 + \sqrt{3}$.

Definition
The **conjugate** of a binomial is the binomial obtained by changing the sign between the two terms.

Expression	Conjugate
$\sqrt{7} - 2\sqrt{5}$	$\sqrt{7} + 2\sqrt{5}$
$\sqrt{a} + \sqrt{b}$	$\sqrt{a} - \sqrt{b}$

In Section 9.5 we applied the formula $(a + b)(a - b) = a^2 - b^2$ to multiply binomials containing square roots. Recall that the terms containing the square roots were eliminated.

EXAMPLE 6

Multiply $8 - \sqrt{6}$ by its conjugate.

Solution

The conjugate of $8 - \sqrt{6}$ is $8 + \sqrt{6}$. We will first multiply using FOIL to show *why* the radical drops out, then we will multiply using the formula

$$(a + b)(a - b) = a^2 - b^2$$

i) Use FOIL to multiply.

$$(8 - \sqrt{6})(8 + \sqrt{6}) = \underset{F}{8 \cdot 8} + \underset{O}{8 \cdot \sqrt{6}} - \underset{I}{8 \cdot \sqrt{6}} - \underset{L}{\sqrt{6} \cdot \sqrt{6}}$$
$$= 64 - 6$$
$$= 58$$

ii) Use $(a + b)(a - b) = a^2 - b^2$.

$$(8 - \sqrt{6})(8 + \sqrt{6}) = (8)^2 - (\sqrt{6})^2 \qquad \text{Substitute 8 for } a \text{ and } \sqrt{6} \text{ for } b.$$
$$= 64 - 6 = 58$$

Each method gives the same result.

[YOU TRY 6]

Multiply $2 + \sqrt{11}$ by its conjugate.

Procedure Rationalize a Denominator Containing Two Terms

If the denominator of an expression contains two terms, including one or two square roots, then to rationalize the denominator we multiply the numerator and denominator of the expression by the conjugate of the denominator.

EXAMPLE 7

Rationalize the denominator, and simplify completely.

a) $\dfrac{3}{5 + \sqrt{3}}$ b) $\dfrac{\sqrt{a} + b}{\sqrt{b} - a}$

Solution

a) The denominator of $\dfrac{3}{5 + \sqrt{3}}$ has two terms, so we multiply the numerator and denominator by $5 - \sqrt{3}$, the conjugate of the denominator.

$$\dfrac{3}{5 + \sqrt{3}} \cdot \dfrac{5 - \sqrt{3}}{5 - \sqrt{3}} \qquad \text{Multiply by the conjugate.}$$
$$= \dfrac{3(5 - \sqrt{3})}{(5)^2 - (\sqrt{3})^2} \qquad (a + b)(a - b) = a^2 - b^2$$
$$= \dfrac{15 - 3\sqrt{3}}{25 - 3} \qquad \text{Simplify.}$$
$$= \dfrac{15 - 3\sqrt{3}}{22} \qquad \text{Subtract.}$$

b) $\dfrac{\sqrt{a}+b}{\sqrt{b}-a} \cdot \dfrac{\sqrt{b}+a}{\sqrt{b}+a}$ Multiply by the conjugate.

In the numerator we must multiply $(\sqrt{a}+b)(\sqrt{b}+a)$. We will use FOIL.

$$\dfrac{\sqrt{a}+b}{\sqrt{b}-a} \cdot \dfrac{\sqrt{b}+a}{\sqrt{b}+a} = \dfrac{\sqrt{ab}+a\sqrt{a}+b\sqrt{b}+ab}{(\sqrt{b})^2-(a)^2}$$ ← Use FOIL in the numerator.
← $(a+b)(a-b) = a^2 - b^2$

$$= \dfrac{\sqrt{ab}+a\sqrt{a}+b\sqrt{b}+ab}{b-a^2}$$ Square the terms.

[YOU TRY 7] Rationalize the denominator, and simplify completely.

a) $\dfrac{1}{\sqrt{7}-2}$ b) $\dfrac{c+\sqrt{d}}{c-\sqrt{d}}$

4 Rationalize a Numerator

In higher-level math courses, sometimes it is necessary to rationlize the *numerator* of a radical expression so that the numerator does not contain a radical.

EXAMPLE 8 Rationalize the numerator, and simplify completely.

a) $\dfrac{\sqrt{7}}{\sqrt{2}}$ b) $\dfrac{8-\sqrt{5}}{3}$

Solution

a) Rationalizing the numerator of $\dfrac{\sqrt{7}}{\sqrt{2}}$ means eliminating the square root from the *numerator*. Multiply the numerator and denominator by $\sqrt{7}$.

$$\dfrac{\sqrt{7}}{\sqrt{2}} = \dfrac{\sqrt{7}}{\sqrt{2}} \cdot \dfrac{\sqrt{7}}{\sqrt{7}} = \dfrac{7}{\sqrt{14}}$$

W Hint
Follow the same process you did for rationalizing the denominator, but focus on the numerator.

b) To rationalize the numerator we must multiply the numerator and denominator by $8+\sqrt{5}$, the conjugate of the numerator.

$$\dfrac{8-\sqrt{5}}{3} \cdot \dfrac{8+\sqrt{5}}{8+\sqrt{5}}$$ Multiply by the conjugate.

$$= \dfrac{8^2 - (\sqrt{5})^2}{3(8+\sqrt{5})}$$ ← $(a+b)(a-b) = a^2 - b^2$
← Multiply.

$$= \dfrac{64-5}{24+3\sqrt{5}} = \dfrac{59}{24+3\sqrt{5}}$$

[YOU TRY 8] Rationalize the numerator, and simplify completely.

a) $\dfrac{\sqrt{3}}{\sqrt{5}}$ b) $\dfrac{6+\sqrt{7}}{4}$

5 Divide Out Common Factors from the Numerator and Denominator

Sometimes it is necessary to simplify a radical expression by dividing out common factors from the numerator and denominator. This is a skill we will need in Chapter 10 to solve quadratic equations, so we will look at an example here.

EXAMPLE 9

Simplify completely: $\dfrac{4\sqrt{5} + 12}{4}$.

Solution

 It is tempting to do one of the following:

$$\dfrac{\cancel{4}\sqrt{5} + 12}{\cancel{4}} = \sqrt{5} + 12 \qquad \text{Incorrect!}$$

or

$$\dfrac{4\sqrt{5} + \overset{3}{\cancel{12}}}{\cancel{4}} = 4\sqrt{5} + 3 \qquad \text{Incorrect!}$$

Each is incorrect because $4\sqrt{5}$ is a *term* in a sum and 12 is a *term* in a sum.

The correct way to simplify $\dfrac{4\sqrt{5} + 12}{4}$ is to begin by factoring out a 4 in the numerator and *then* divide the numerator and denominator by any common factors.

$$\dfrac{4\sqrt{5} + 12}{4} = \dfrac{4(\sqrt{5} + 3)}{4} \qquad \text{Factor out 4 from the numerator.}$$

$$= \dfrac{\overset{1}{\cancel{4}}(\sqrt{5} + 3)}{\underset{1}{\cancel{4}}} \qquad \text{Divide by 4.}$$

$$= \sqrt{5} + 3 \qquad \text{Simplify.}$$

We can divide the numerator and denominator by 4 in $\dfrac{4(\sqrt{5} + 3)}{4}$ because the 4 in the numerator is part of a *product* not a sum or difference.

[YOU TRY 9] Simplify completely.

a) $\dfrac{5\sqrt{7} - 40}{5}$ b) $\dfrac{20 + 6\sqrt{2}}{4}$

ANSWERS TO [YOU TRY] EXERCISES

1) a) $\dfrac{\sqrt{7}}{7}$ b) $\dfrac{5\sqrt{3}}{3}$ c) $\dfrac{9\sqrt{30}}{5}$ 2) a) $\dfrac{\sqrt{14}}{7}$ b) $\dfrac{\sqrt{15}}{5}$ 3) a) $\dfrac{5\sqrt{p}}{p}$ b) $\dfrac{3k^2\sqrt{5km}}{5m}$
c) $\dfrac{2r\sqrt{5rs}}{s}$ 4) a) 2^2 or 4 b) t^3 c) 5   5) a) $\dfrac{4\sqrt[3]{49}}{7}$ b) $\dfrac{\sqrt[4]{6}}{3}$ c) $\dfrac{\sqrt[5]{8w^2}}{w}$ 6) -7
7) a) $\dfrac{\sqrt{7} + 2}{3}$ b) $\dfrac{c^2 + 2c\sqrt{d} + d}{c^2 - d}$ 8) a) $\dfrac{3}{\sqrt{15}}$ b) $\dfrac{29}{24 - 4\sqrt{7}}$ 9) a) $\sqrt{7} - 8$ b) $\dfrac{10 + 3\sqrt{2}}{2}$

9.6 Exercises

Do the exercises, and check your work.

Assume all variables represent positive real numbers.

Objective 1: Rationalize a Denominator: One Square Root

1) What does it mean to rationalize the denominator of a radical expression?

2) In your own words, explain how to rationalize the denominator of an expression containing one term in the denominator.

Rationalize the denominator of each expression.

3) $\dfrac{1}{\sqrt{5}}$

4) $\dfrac{1}{\sqrt{6}}$

5) $\dfrac{9}{\sqrt{6}}$

6) $\dfrac{25}{\sqrt{10}}$

7) $-\dfrac{20}{\sqrt{8}}$

8) $-\dfrac{18}{\sqrt{45}}$

9) $\dfrac{\sqrt{3}}{\sqrt{28}}$

10) $\dfrac{\sqrt{8}}{\sqrt{27}}$

11) $\sqrt{\dfrac{20}{60}}$

12) $\sqrt{\dfrac{12}{80}}$

13) $\dfrac{\sqrt{56}}{\sqrt{48}}$

14) $\dfrac{\sqrt{66}}{\sqrt{12}}$

Multiply and simplify.

15) $\sqrt{\dfrac{10}{7}} \cdot \sqrt{\dfrac{7}{3}}$

16) $\sqrt{\dfrac{11}{5}} \cdot \sqrt{\dfrac{5}{2}}$

17) $\sqrt{\dfrac{6}{5}} \cdot \sqrt{\dfrac{1}{8}}$

18) $\sqrt{\dfrac{11}{10}} \cdot \sqrt{\dfrac{8}{11}}$

Simplify completely.

19) $\dfrac{8}{\sqrt{y}}$

20) $\dfrac{4}{\sqrt{w}}$

21) $\dfrac{\sqrt{5}}{\sqrt{t}}$

22) $\dfrac{\sqrt{2}}{\sqrt{m}}$

23) $\sqrt{\dfrac{64v^7}{5w}}$

24) $\sqrt{\dfrac{81c^5}{2d}}$

25) $\sqrt{\dfrac{a^3b^3}{3ab^4}}$

26) $\sqrt{\dfrac{m^2n^5}{7m^3n}}$

27) $-\dfrac{\sqrt{75}}{\sqrt{b^3}}$

28) $-\dfrac{\sqrt{24}}{\sqrt{v^3}}$

29) $\dfrac{\sqrt{13}}{\sqrt{j^5}}$

30) $\dfrac{\sqrt{22}}{\sqrt{w^7}}$

Objective 2: Rationalize a Denominator: One Higher Root

Fill in the blank.

31) $\sqrt[3]{2} \cdot \sqrt[3]{?} = \sqrt[3]{2^3} = 2$

32) $\sqrt[3]{5} \cdot \sqrt[3]{?} = \sqrt[3]{5^3} = 5$

33) $\sqrt[3]{9} \cdot \sqrt[3]{?} = \sqrt[3]{3^3} = 3$

34) $\sqrt[3]{4} \cdot \sqrt[3]{?} = \sqrt[3]{2^3} = 2$

35) $\sqrt[3]{c} \cdot \sqrt[3]{?} = \sqrt[3]{c^3} = c$

36) $\sqrt[3]{p} \cdot \sqrt[3]{?} = \sqrt[3]{p^3} = p$

37) $\sqrt[5]{4} \cdot \sqrt[5]{?} = \sqrt[5]{2^5} = 2$

38) $\sqrt[5]{16} \cdot \sqrt[5]{?} = \sqrt[5]{2^5} = 2$

39) $\sqrt[4]{m^3} \cdot \sqrt[4]{?} = \sqrt[4]{m^4} = m$

40) $\sqrt[4]{k} \cdot \sqrt[4]{?} = \sqrt[4]{k^4} = k$

Rationalize the denominator of each expression.

41) $\dfrac{4}{\sqrt[3]{3}}$

42) $\dfrac{26}{\sqrt[3]{5}}$

43) $\dfrac{12}{\sqrt[3]{2}}$

44) $\dfrac{21}{\sqrt[3]{3}}$

45) $\dfrac{9}{\sqrt[3]{25}}$

46) $\dfrac{6}{\sqrt[3]{4}}$

47) $\sqrt[4]{\dfrac{5}{9}}$

48) $\sqrt[4]{\dfrac{2}{25}}$

49) $\sqrt[5]{\dfrac{3}{8}}$

50) $\sqrt[5]{\dfrac{7}{4}}$

51) $\dfrac{10}{\sqrt[3]{z}}$

52) $\dfrac{6}{\sqrt[3]{u}}$

53) $\sqrt[3]{\dfrac{3}{n^2}}$

54) $\sqrt[3]{\dfrac{5}{x^2}}$

55) $\dfrac{\sqrt[3]{7}}{\sqrt[3]{2k^2}}$

56) $\dfrac{\sqrt[3]{2}}{\sqrt[3]{25t}}$

57) $\dfrac{9}{\sqrt[5]{a^3}}$

58) $\dfrac{8}{\sqrt[5]{h^2}}$

59) $\sqrt[4]{\dfrac{5}{2m}}$

60) $\sqrt[4]{\dfrac{2}{3t^2}}$

Objective 3: Rationalize a Denominator Containing Two Terms

61) How do you find the conjugate of an expression with two radical terms?

62) When you multiply a binomial containing a square root by its conjugate, what happens to the radical?

Find the conjugate of each expression. Then, multiply the expression by its conjugate.

63) $(5 + \sqrt{2})$
64) $(\sqrt{5} - 4)$
65) $(\sqrt{2} + \sqrt{6})$
66) $(\sqrt{3} - \sqrt{10})$
67) $(\sqrt{t} - 8)$
68) $(\sqrt{p} + 5)$

Rationalize the denominator, and simplify completely.

Fill It In
Fill in the blanks with either the missing mathematical step or reason for the given step.

69) $\dfrac{6}{4 - \sqrt{5}} = \dfrac{6}{4 - \sqrt{5}} \cdot \dfrac{4 + \sqrt{5}}{4 + \sqrt{5}}$ _____

$= \dfrac{6(4 + \sqrt{5})}{(4)^2 - (\sqrt{5})^2}$ _____

$=$ _____ Multiply terms in numerator; square terms in denominator.

$=$ _____ Simplify.

70) $\dfrac{\sqrt{6}}{\sqrt{7} + \sqrt{2}} = \dfrac{\sqrt{6}}{\sqrt{7} + \sqrt{2}} \cdot \dfrac{\sqrt{7} - \sqrt{2}}{\sqrt{7} - \sqrt{2}}$ _____

$= \dfrac{\sqrt{6}(\sqrt{7} - \sqrt{2})}{(\sqrt{7})^2 - (\sqrt{2})^2}$ _____

$=$ _____ Multiply terms in numerator; square terms in denominator.

$=$ _____ Simplify.

71) $\dfrac{3}{2 + \sqrt{3}}$
72) $\dfrac{8}{6 - \sqrt{5}}$
73) $\dfrac{10}{9 - \sqrt{2}}$
74) $\dfrac{5}{4 + \sqrt{6}}$
75) $\dfrac{\sqrt{8}}{\sqrt{3} + \sqrt{2}}$
76) $\dfrac{\sqrt{32}}{\sqrt{5} - \sqrt{7}}$
77) $\dfrac{\sqrt{3} - \sqrt{5}}{\sqrt{10} - \sqrt{3}}$
78) $\dfrac{\sqrt{3} + \sqrt{6}}{\sqrt{2} + \sqrt{5}}$
79) $\dfrac{\sqrt{m}}{\sqrt{m} + \sqrt{n}}$
80) $\dfrac{\sqrt{u}}{\sqrt{u} - \sqrt{v}}$
81) $\dfrac{b - 25}{\sqrt{b} - 5}$
82) $\dfrac{d - 9}{\sqrt{d} + 3}$
83) $\dfrac{\sqrt{x} + \sqrt{y}}{\sqrt{x} - \sqrt{y}}$
84) $\dfrac{\sqrt{f} - \sqrt{g}}{\sqrt{f} + \sqrt{g}}$

Objective 4: Rationalize a Numerator
Rationalize the numerator of each expression, and simplify.

85) $\dfrac{\sqrt{5}}{3}$
86) $\dfrac{\sqrt{2}}{9}$
87) $\dfrac{\sqrt{x}}{\sqrt{7}}$
88) $\dfrac{\sqrt{8a}}{\sqrt{b}}$
89) $\dfrac{2 + \sqrt{3}}{6}$
90) $\dfrac{1 + \sqrt{7}}{3}$
91) $\dfrac{\sqrt{x} - 2}{x - 4}$
92) $\dfrac{3 - \sqrt{n}}{n - 9}$
93) $\dfrac{4 - \sqrt{c + 11}}{c - 5}$
94) $\dfrac{\sqrt{x + h} - \sqrt{x}}{h}$

95) Does rationalizing the denominator of an expression change the value of the original expression? Explain your answer.

96) Does rationalizing the numerator of an expression change the value of the original expression? Explain your answer.

Objective 5: Divide Out Common Factors from the Numerator and Denominator
Simplify completely.

97) $\dfrac{5 + 10\sqrt{3}}{5}$
98) $\dfrac{18 - 6\sqrt{7}}{6}$
99) $\dfrac{30 - 18\sqrt{5}}{4}$
100) $\dfrac{36 + 20\sqrt{2}}{12}$
101) $\dfrac{\sqrt{45} + 6}{9}$
102) $\dfrac{\sqrt{48} + 28}{4}$
103) $\dfrac{-10 - \sqrt{50}}{5}$
104) $\dfrac{-35 + \sqrt{200}}{15}$

105) The function $r(A) = \sqrt{\dfrac{A}{\pi}}$ describes the radius of a circle, $r(A)$, in terms of its area, A.

a) If the area of a circle is measured in square inches, find $r(8\pi)$ and explain what it means in the context of the problem.

b) If the area of a circle is measured in square inches, find $r(7)$ and rationalize the denominator. Explain the meaning of $r(7)$ in the context of the problem.

c) Obtain an equivalent form of the function by rationalizing the denominator.

106) The function $r(V) = \sqrt[3]{\dfrac{3V}{4\pi}}$ describes the radius of a sphere, $r(V)$, in terms of its volume, V.

a) If the volume of a sphere is measured in cubic centimeters, find $r(36\pi)$ and explain what it means in the context of the problem.

b) If the volume of a sphere is measured in cubic centimeters, find $r(11)$ and rationalize the denominator. Explain the meaning of $r(11)$ in the context of the problem.

c) Obtain an equivalent form of the function by rationalizing the denominator.

R Rethink

R1) Why is it important to always read the directions closely?

R2) Which section of the book could you go to for help or review if you struggled with this section?

R3) Did you circle any problems you struggled with? Ask about them in class!

Putting It All Together

P Prepare

What are your objectives?
1 Review the concepts of Sections 9.1–9.6.

O Organize

How can you accomplish each objective?
• Complete the given examples on your own. • Complete You Try 1.

W Work

Read the explanations, follow the examples, take notes, and complete the You Trys.

1 Review the Concepts Presented in 9.1–9.6

In Section 9.1, we learned how to find roots of numbers. For example, $\sqrt[3]{-64} = -4$ because $(-4)^3 = -64$. We also, learned about square root and cube root functions.

In Section 9.2, we learned about the relationship between rational exponents and radicals. Recall that if m and n are positive integers and $\dfrac{m}{n}$ is in lowest terms, then $a^{m/n} = (a^{1/n})^m = (\sqrt[n]{a})^m$ provided that $a^{1/n}$ is a real number. **For the rest of this section we will assume that all variables represent positive real numbers.**

EXAMPLE 1

Simplify completely. The answer should contain only positive exponents.

a) $(32)^{4/5}$ b) $\left(\dfrac{a^7 b^{9/8}}{25 a^9 b^{3/4}}\right)^{-3/2}$

Solution

a) The denominator of the fractional exponent is the index of the radical, and the numerator is the power to which we raise the radical expression.

$$32^{4/5} = (\sqrt[5]{32})^4 \quad \text{Write in radical form.}$$
$$= (2)^4 \quad \sqrt[5]{32} = 2$$
$$= 16$$

b) $\left(\dfrac{a^7 b^{9/8}}{25 a^9 b^{3/4}}\right)^{-3/2} = \left(\dfrac{25 a^9 b^{3/4}}{a^7 b^{9/8}}\right)^{3/2}$ Eliminate the negative from the outermost exponent by taking the reciprocal of the base.

Simplify the expression inside the parentheses by subtracting the exponents.

$$= (25 a^{9-7} b^{3/4 - 9/8})^{3/2} = (25 a^2 b^{6/8 - 9/8})^{3/2} = (25 a^2 b^{-3/8})^{3/2}$$

Apply the power rule, and simplify.

$$= (25)^{3/2} (a^2)^{3/2} (b^{-3/8})^{3/2} = (\sqrt{25})^3 a^3 b^{-9/16} = 5^3 a^3 b^{-9/16} = \dfrac{125 a^3}{b^{9/16}}$$

In Sections 9.3–9.6 we learned how to simplify, multiply, divide, add, and subtract radicals. Let's look at these operations together so that we will learn to recognize the techniques needed to perform these operations.

EXAMPLE 2

Perform the operations, and simplify.

a) $\sqrt{3} + 10\sqrt{6} - 4\sqrt{3}$ b) $\sqrt{3}(10\sqrt{6} - 4\sqrt{3})$

Solution

a) This is the *sum and difference* of radicals. Remember that we can only add and subtract radicals that are like radicals.

$$\sqrt{3} + 10\sqrt{6} - 4\sqrt{3} = \sqrt{3} - 4\sqrt{3} + 10\sqrt{6} \quad \text{Write like radicals together.}$$
$$= -3\sqrt{3} + 10\sqrt{6} \quad \text{Subtract.}$$

b) This is the *product* of radical expressions. We must multiply the binomial $10\sqrt{6} - 4\sqrt{3}$ by $\sqrt{3}$ using the distributive property.

$$\sqrt{3}(10\sqrt{6} - 4\sqrt{3}) = \sqrt{3} \cdot 10\sqrt{6} - \sqrt{3} \cdot 4\sqrt{3} \quad \text{Distribute.}$$
$$= 10\sqrt{18} - 4 \cdot 3 \quad \text{Product rule; } \sqrt{3} \cdot \sqrt{3} = 3.$$
$$= 10\sqrt{18} - 12 \quad \text{Multiply.}$$

Ask yourself, "Is $10\sqrt{18} - 12$ in simplest form?" *No.* $\sqrt{18}$ can be simplified.

$$= 10\sqrt{9 \cdot 2} - 12 \quad \text{9 is a perfect square.}$$
$$= 10\sqrt{9} \cdot \sqrt{2} - 12 \quad \text{Product rule}$$
$$= 10 \cdot 3\sqrt{2} - 12 \quad \sqrt{9} = 3$$
$$= 30\sqrt{2} - 12 \quad \text{Multiply.}$$

The expression is now in simplest form.

Next we will look at multiplication problems involving binomials. Remember that the rules we used to multiply binomials like $(x + 4)(x - 9)$ are the same rules we use to multiply binomials containing radicals.

EXAMPLE 3

Multiply and simplify.

a) $(8 + \sqrt{2})(9 - \sqrt{11})$ b) $(\sqrt{n} + \sqrt{7})(\sqrt{n} - \sqrt{7})$

c) $(2\sqrt{5} - 3)^2$

Solution

a) Since we must multiply two binomials, we will use FOIL.

$$(8 + \sqrt{2})(9 - \sqrt{11}) = \overset{F}{8 \cdot 9} - \overset{O}{8 \cdot \sqrt{11}} + \overset{I}{9 \cdot \sqrt{2}} - \overset{L}{\sqrt{2} \cdot \sqrt{11}} \quad \text{Use FOIL.}$$
$$= 72 - 8\sqrt{11} + 9\sqrt{2} - \sqrt{22} \quad \text{Multiply.}$$

All radicals are simplified and none of them are like radicals, so this expression is in simplest form.

W Hint
Notice the connection between what you learned in Chapter 6 and multiplying binomials containing radicals.

b) We can multiply $(\sqrt{n} + \sqrt{7})(\sqrt{n} - \sqrt{7})$ using FOIL or, if we notice that this product is in the form $(a + b)(a - b)$ we can apply the rule $(a + b)(a - b) = a^2 - b^2$. Either method will give us the correct answer. We will use the second method.

$$(a + b)(a - b) = a^2 - b^2$$
$$(\sqrt{n} + \sqrt{7})(\sqrt{n} - \sqrt{7}) = (\sqrt{n})^2 - (\sqrt{7})^2 = n - 7 \quad \text{Substitute } \sqrt{n} \text{ for } a \text{ and } \sqrt{7} \text{ for } b.$$

c) Once again we can either use FOIL to expand $(2\sqrt{5} - 3)^2$, or we can use the special formula we learned for squaring a binomial.

We will use $(a - b)^2 = a^2 - 2ab + b^2$.

$$(2\sqrt{5} - 3)^2 = (2\sqrt{5})^2 - 2(2\sqrt{5})(3) + (3)^2 \quad \text{Substitute } 2\sqrt{5} \text{ for } a \text{ and } 3 \text{ for } b.$$
$$= (4 \cdot 5) - 4\sqrt{5}(3) + 9 \quad \text{Multiply.}$$
$$= 20 - 12\sqrt{5} + 9 \quad \text{Multiply.}$$
$$= 29 - 12\sqrt{5} \quad \text{Combine like terms.}$$

Remember that an expression is not considered to be in simplest form if it contains a radical in its denominator. To rationalize the denominator of a radical expression, we must keep in mind the index on the radical and the number of terms in the denominator.

EXAMPLE 4

Rationalize the denominator of each expression.

a) $\dfrac{10}{\sqrt{2x}}$ b) $\dfrac{10}{\sqrt[3]{2x}}$ c) $\dfrac{\sqrt{10}}{\sqrt{2} - 1}$

Solution

a) First, notice that the denominator of $\dfrac{10}{\sqrt{2x}}$ contains only one term and it is a *square* root. Ask yourself, "By what do I multiply $\sqrt{2x}$ to get a perfect *square* under the radical?" The answer is $\sqrt{2x}$ since $\sqrt{2x} \cdot \sqrt{2x} = \sqrt{4x^2} = 2x$. Multiply the numerator and denominator by $\sqrt{2x}$, and simplify.

$$\dfrac{10}{\sqrt{2x}} = \dfrac{10}{\sqrt{2x}} \cdot \dfrac{\sqrt{2x}}{\sqrt{2x}} \quad \text{Rationalize the denominator.}$$
$$= \dfrac{10\sqrt{2x}}{\sqrt{4x^2}} = \dfrac{10\sqrt{2x}}{2x} = \dfrac{5\sqrt{2x}}{x}$$

Hint
In your notes, summarize the similarities and differences in rationalizing the denominators in each expression in this example.

b) The denominator of $\dfrac{10}{\sqrt[3]{2x}}$ contains only one term, but it is a *cube* root. Ask yourself, "By what do I multiply $\sqrt[3]{2x}$ to get a radicand that is a perfect *cube*?" The answer is $\sqrt[3]{4x^2}$ since $\sqrt[3]{2x} \cdot \sqrt[3]{4x^2} = \sqrt[3]{8x^3} = 2x$. Multiply the numerator and denominator by $\sqrt[3]{4x^2}$, and simplify.

$$\dfrac{10}{\sqrt[3]{2x}} = \dfrac{10}{\sqrt[3]{2x}} \cdot \dfrac{\sqrt[3]{4x^2}}{\sqrt[3]{4x^2}} \qquad \text{Rationalize the denominator.}$$

$$= \dfrac{10\sqrt[3]{4x^2}}{\sqrt[3]{8x^3}} = \dfrac{10\sqrt[3]{4x^2}}{2x} = \dfrac{5\sqrt[3]{4x^2}}{x}$$

c) The denominator of $\dfrac{\sqrt{10}}{\sqrt{2}-1}$ contains two terms, so how do we rationalize the denominator of this expression? We multiply the numerator and denominator by the *conjugate* of the denominator.

$$\dfrac{\sqrt{10}}{\sqrt{2}-1} = \dfrac{\sqrt{10}}{\sqrt{2}-1} \cdot \dfrac{\sqrt{2}+1}{\sqrt{2}+1} \qquad \text{Multiply by the conjugate.}$$

$$= \dfrac{\sqrt{10}(\sqrt{2}+1)}{(\sqrt{2})^2 - (1)^2} \qquad \begin{array}{l}\text{Multiply.}\\ (a+b)(a-b) = a^2 - b^2\end{array}$$

$$= \dfrac{\sqrt{20} + \sqrt{10}}{1} \qquad \begin{array}{l}\text{Distribute.}\\ \text{Simplify.}\end{array}$$

$$= 2\sqrt{5} + \sqrt{10} \qquad \sqrt{20} = 2\sqrt{5};\ \text{simplify.}$$

YOU TRY 1

a) Perform the operations, and simplify.
   i) $(\sqrt{w} + 8)^2$ ii) $(3 - \sqrt{5a})(4 + \sqrt{5a})$ iii) $\sqrt{2}(9\sqrt{10} - \sqrt{2})$
   iv) $\sqrt{2} + 9\sqrt{10} - 5\sqrt{2}$ v) $(2\sqrt{3} + y)(2\sqrt{3} - y)$

b) Find each root.
   i) $-\sqrt{\dfrac{121}{16}}$ ii) $\sqrt[3]{-1000}$ iii) $\sqrt{0.09}$ iv) $\sqrt{-49}$

Hint
Read all directions first.

c) Simplify completely. The answer should contain only positive exponents.
   i) $(-64)^{2/3}$ ii) $\left(\dfrac{81x^3 y^{1/2}}{x^{-5} y^6}\right)^{-3/4}$

d) Rationalize the denominator of each expression.
   i) $\dfrac{24}{\sqrt[3]{9h}}$ ii) $\dfrac{7 + \sqrt{6}}{4 + \sqrt{6}}$ iii) $\dfrac{56}{\sqrt{7}}$

ANSWERS TO YOU TRY EXERCISES

1) a) i) $w + 16\sqrt{w} + 64$ ii) $12 - \sqrt{5a} - 5a$ iii) $18\sqrt{5} - 2$ iv) $-4\sqrt{2} + 9\sqrt{10}$ v) $12 - y^2$
   b) i) $-\dfrac{11}{4}$ ii) -10 iii) 0.3 iv) not a real number
   c) i) 16 ii) $\dfrac{y^{33/8}}{27x^6}$ d) i) $\dfrac{8\sqrt[3]{3h^2}}{h}$ ii) $\dfrac{22 - 3\sqrt{6}}{10}$ iii) $8\sqrt{7}$

Putting It All Together Exercises

E Evaluate Do the exercises, and check your work.

Objective 1: Review the Concepts Presented in 9.1–9.6
Assume all variables represent positive real numbers.

Find each root, if possible.

1) $\sqrt[4]{81}$
2) $\sqrt[3]{-1000}$
3) $-\sqrt[6]{64}$
4) $\sqrt{121}$
5) $\sqrt{-169}$
6) $\sqrt{\dfrac{144}{49}}$

Simplify completely. The answer should contain only positive exponents.

7) $(144)^{1/2}$
8) $(-32)^{4/5}$
9) $-1000^{2/3}$
10) $\left(-\dfrac{16}{81}\right)^{3/4}$
11) $125^{-1/3}$
12) $\left(\dfrac{100}{9}\right)^{-3/2}$
13) $k^{-3/5} \cdot k^{3/10}$
14) $(t^{3/8})^{16}$
15) $\left(\dfrac{27a^{-8}}{b^9}\right)^{2/3}$
16) $\left(\dfrac{18x^{-9}y^{4/3}}{2x^3y}\right)^{-5/2}$

Simplify completely.

17) $\sqrt{24}$
18) $\sqrt[4]{32}$
19) $\sqrt[3]{72}$
20) $\sqrt[3]{\dfrac{500}{2}}$
21) $\sqrt[4]{243}$
22) $\sqrt{45c^{11}}$
23) $\sqrt[3]{96m^7n^{15}}$
24) $\sqrt[5]{\dfrac{64x^{19}}{y^{20}}}$

Perform the operations, and simplify.

25) $\sqrt[3]{12} \cdot \sqrt[3]{2}$
26) $\sqrt[4]{\dfrac{96k^{11}}{2k^3}}$
27) $(6 + \sqrt{7})(2 + \sqrt{7})$
28) $4c^2\sqrt[3]{108c} - 15\sqrt[3]{32c^7}$
29) $\dfrac{18}{\sqrt{6}}$
30) $\dfrac{5}{\sqrt{3} - \sqrt{2}}$
31) $3\sqrt{75m^3n} + m\sqrt{12mn}$
32) $\sqrt{6p^7q^3} \cdot \sqrt{15pq^2}$
33) $\dfrac{\sqrt{60t^8u^3}}{\sqrt{5t^2u}}$
34) $\dfrac{9}{\sqrt[3]{2}}$
35) $(2\sqrt{3} + 10)^2$
36) $(\sqrt{2} + 3)(\sqrt{2} - 3)$
37) $\dfrac{\sqrt{2}}{4 + \sqrt{10}}$
38) $\sqrt[3]{r^2} \cdot \sqrt{r}$
39) $\sqrt[3]{\dfrac{b^2}{9c}}$
40) $\dfrac{\sqrt[4]{32}}{\sqrt[4]{w^{11}}}$

For each function, find the domain and graph the function.

41) $f(x) = \sqrt{x - 2}$
42) $g(x) = \sqrt[3]{x}$
43) $h(x) = \sqrt[3]{x} + 1$
44) $k(x) = \sqrt{x + 1}$
45) $g(x) = \sqrt{-x}$
46) $f(x) = \sqrt[3]{x} - 2$

R Rethink

R1) Which types of problems did you struggle with the most?

R2) Would you be able to complete exercises like these without any help?

9.7 Solving Radical Equations

P Prepare

What are your objectives for Section 9.7?

O Organize

How can you accomplish each objective?

What are your objectives for Section 9.7?	How can you accomplish each objective?
1 Understand the Steps for Solving a Radical Equation	• Write the procedure for **Solving Radical Equations Containing Square Roots** in your own words, and be aware of *extraneous solutions*.
2 Solve an Equation Containing One Square Root	• Use the steps for **Solving Radical Equations Containing Square Roots**. • Be aware that you will, often, have to square a binomial. • Complete the given examples on your own. • Complete You Trys 1 and 2.
3 Solve an Equation Containing Two Square Roots	• Use the steps for **Solving Radical Equations Containing Square Roots**. • Recognize when you need to square both sides of the equation twice. • Complete the given examples on your own. • Complete You Trys 3–5.
4 Solve an Equation Containing a Cube Root	• Follow the explanation to create your own procedure for solving radical equations containing a cube root. • Complete the given example on your own. • Complete You Try 6.

W Work — Read the explanations, follow the examples, take notes, and complete the You Trys.

In this section, we will learn how to solve *radical equations*.

An equation containing a variable in the radicand is a **radical equation**. Some examples of radical equations are

$$\sqrt{p} = 7, \quad \sqrt[3]{n} = 2, \quad \sqrt{2x+1} + 1 = x, \quad \sqrt{5w+6} - \sqrt{4w+1} = 1$$

1 Understand the Steps for Solving a Radical Equation

Let's review what happens when we square a square root expression: If $x \geq 0$, then $(\sqrt{x})^2 = x$. That is, to eliminate the radical from $\sqrt{x}$, we *square* the expression. Therefore to solve equations like those above containing *square roots*, we *square* both sides of the equation to obtain new equations. The solutions of the new equations contain all of the solutions of the original equation and may also contain *extraneous solutions*.

An **extraneous solution** is a value that satisfies one of the new equations but does not satisfy the original equation. Extraneous solutions occur frequently when solving radical equations, so we *must* check all possible solutions in the original equation and discard any that are extraneous.

> **Procedure** Solving Radical Equations Containing Square Roots
>
> **Step 1:** Get a radical on a side by itself.
> **Step 2:** Square both sides of the equation to eliminate a radical.
> **Step 3:** Combine like terms on each side of the equation.
> **Step 4:** If the equation still contains a radical, repeat Steps 1–3.
> **Step 5:** Solve the equation.
> **Step 6:** Check the proposed solutions *in the original equation,* and discard extraneous solutions.

2 Solve an Equation Containing One Square Root

EXAMPLE 1

Solve.

a) $\sqrt{c-2} = 3$ b) $\sqrt{t+5} + 6 = 0$

Solution

a) **Step 1:** The radical *is* on a side by itself: $\sqrt{c-2} = 3$

Step 2: *Square* both sides to eliminate the *square root.*

$$(\sqrt{c-2})^2 = 3^2 \quad \text{Square both sides.}$$
$$c - 2 = 9$$

W Hint
Write out the steps as you are reading them. Be sure to write neatly and in a very orderly way.

Steps 3 and 4 do not apply because there are no like terms to combine and no radicals remain.

Step 5: Solve the equation.

$$c = 11 \quad \text{Add 2 to each side.}$$

Step 6: Check $c = 11$ in the *original* equation.

$$\sqrt{c-2} = 3$$
$$\sqrt{11-2} \stackrel{?}{=} 3$$
$$\sqrt{9} = 3 \checkmark$$

The solution set is $\{11\}$.

b) The first step is to get the radical on a side by itself.

$$\sqrt{t+5} + 6 = 0$$
$$\sqrt{t+5} = -6 \quad \text{Subtract 6 from each side.}$$
$$(\sqrt{t+5})^2 = (-6)^2 \quad \text{Square both sides to eliminate the radical.}$$
$$t + 5 = 36 \quad \text{The square root has been eliminated.}$$
$$t = 31 \quad \text{Solve the equation.}$$

Check $t = 31$ in the *original* equation.

$$\sqrt{t+5} + 6 = 0$$
$$\sqrt{31+5} + 6 \stackrel{?}{=} 0$$
$$6 + 6 \stackrel{?}{=} 0 \quad \text{FALSE}$$

Because $t = 31$ gives us a false statement, it is an *extraneous solution.* The equation has no real solution. The solution set is $\emptyset$.

[YOU TRY 1] Solve.

a) $\sqrt{a+4} = 7$ b) $\sqrt{m-7} + 12 = 9$

Sometimes, we have to square a binomial in order to solve a radical equation. Don't forget that when we square a binomial we can either use FOIL or one of the following formulas: $(a+b)^2 = a^2 + 2ab + b^2$ or $(a-b)^2 = a^2 - 2ab + b^2$.

EXAMPLE 2   Solve $\sqrt{2x+1} + 1 = x$.

Solution

Start by getting the radical on a side by itself.

$\sqrt{2x+1} = x - 1$ Subtract 1 from each side.
$(\sqrt{2x+1})^2 = (x-1)^2$ Square both sides to eliminate the radical.
$2x + 1 = x^2 - 2x + 1$ Simplify; square the binomial.
$0 = x^2 - 4x$ Subtract $2x$; subtract 1.
$0 = x(x-4)$ Factor.

$x = 0$ or $x - 4 = 0$ Set each factor equal to zero.
$x = 0$ or $x = 4$ Solve.

W Hint
Review Section 6.4 if you need help squaring a binomial.

Check $x = 0$ and $x = 4$ in the *original* equation.

$x = 0$: $\sqrt{2x+1} + 1 = x$ $x = 4$: $\sqrt{2x+1} + 1 = x$
   $\sqrt{2(0)+1} + 1 \stackrel{?}{=} 0$ $\sqrt{2(4)+1} + 1 \stackrel{?}{=} 4$
   $\sqrt{1} + 1 \stackrel{?}{=} 0$ $\sqrt{9} + 1 \stackrel{?}{=} 4$
   $2 \stackrel{?}{=} 1$ FALSE $3 + 1 = 4$ TRUE

$x = 4$ is a solution but $x = 0$ is **not** because $x = 0$ does not satisfy the original equation. The solution set is $\{4\}$.

[YOU TRY 2] Solve.

a) $\sqrt{3p+10} - 4 = p$ b) $\sqrt{4h-3} - h = -2$

3 Solve an Equation Containing Two Square Roots

Next, we will take our first look at solving an equation containing two square roots.

EXAMPLE 3   Solve $\sqrt{2a+4} - 3\sqrt{a-5} = 0$.

Solution

Begin by getting a radical on a side by itself.

$\sqrt{2a+4} = 3\sqrt{a-5}$ Add $3\sqrt{a-5}$ to each side.
$(\sqrt{2a+4})^2 = (3\sqrt{a-5})^2$ Square both sides to eliminate the radicals.
$2a + 4 = 9(a-5)$ $3^2 = 9$
$2a + 4 = 9a - 45$ Distribute.
$-7a = -49$
$a = 7$ Solve.

Check $a = 7$ in the original equation.

$$\sqrt{2a+4} - 3\sqrt{a-5} = 0$$
$$\sqrt{2(7)+4} - 3\sqrt{7-5} = 0$$
$$\sqrt{14+4} - 3\sqrt{2} \stackrel{?}{=} 0$$
$$\sqrt{18} - 3\sqrt{2} \stackrel{?}{=} 0$$
$$3\sqrt{2} - 3\sqrt{2} = 0 \checkmark$$

The solution set is $\{7\}$.

[**YOU TRY 3**] Solve $4\sqrt{r-3} - \sqrt{6r+2} = 0$.

Recall from Section 9.5 that we can square binomials containing radical expressions just like we squared $(x-1)^2$ in Example 2. We can use FOIL or the formulas

$$(a+b)^2 = a^2 + 2ab + b^2 \quad \text{or} \quad (a-b)^2 = a^2 - 2ab + b^2$$

EXAMPLE 4 Square and simplify $(3 - \sqrt{m+2})^2$.

Solution

Use the formula $(a-b)^2 = a^2 - 2ab + b^2$.

$$(3 - \sqrt{m+2})^2 = (3)^2 - 2(3)(\sqrt{m+2}) + (\sqrt{m+2})^2 \quad \text{Substitute 3 for } a \text{ and } \sqrt{m+2} \text{ for } b.$$
$$= 9 - 6\sqrt{m+2} + (m+2)$$
$$= m + 11 - 6\sqrt{m+2} \quad \text{Combine like terms.}$$

W Hint
Make sure you understand this example before going on to the next.

[**YOU TRY 4**] Square and simplify each expression.

a) $(\sqrt{z} - 4)^2$ b) $(5 + \sqrt{3d-1})^2$

To solve the next two equations, we will have to square both sides of the equation twice to eliminate the radicals. Be very careful when you are squaring the binomials that contain a radical.

EXAMPLE 5 Solve each equation.

a) $\sqrt{x+5} + \sqrt{x} = 5$ b) $\sqrt{5w+6} - \sqrt{4w+1} = 1$

Solution

a) This equation contains two radicals *and* a constant. Get one of the radicals on a side by itself, then square both sides.

$$\sqrt{x+5} = 5 - \sqrt{x} \quad \text{Subtract } \sqrt{x} \text{ from each side.}$$
$$(\sqrt{x+5})^2 = (5 - \sqrt{x})^2 \quad \text{Square both sides.}$$
$$x + 5 = (5)^2 - 2(5)(\sqrt{x}) + (\sqrt{x})^2 \quad \text{Use the formula } (a-b)^2 = a^2 - 2ab + b^2.$$
$$x + 5 = 25 - 10\sqrt{x} + x \quad \text{Simplify.}$$

W Hint
Always use parentheses to organize your work when squaring both sides.

The equation still contains a radical. Therefore, repeat Steps 1–3. Begin by getting the radical on a side by itself.

$$5 = 25 - 10\sqrt{x}$$ Subtract x from each side.
$$-20 = -10\sqrt{x}$$ Subtract 25 from each side.
$$2 = \sqrt{x}$$ Divide by -10.
$$2^2 = (\sqrt{x})^2$$ Square both sides.
$$4 = x$$ Solve.

The check is left to the student. The solution set is $\{4\}$.

b) **Step 1:** Get a radical on a side by itself.

$$\sqrt{5w + 6} - \sqrt{4w + 1} = 1$$
$$\sqrt{5w + 6} = 1 + \sqrt{4w + 1}$$ Add $\sqrt{4w+1}$ to each side.

> **W Hint**
> These problems can be very long, so write out each step *carefully* and *neatly* to minimize mistakes.

Step 2: Square both sides of the equation to eliminate a radical.

$$(\sqrt{5w+6})^2 = (1 + \sqrt{4w+1})^2$$ Square both sides.
$$5w + 6 = (1)^2 + 2(1)(\sqrt{4w+1}) + (\sqrt{4w+1})^2$$ Use the formula $(a+b)^2 = a^2 + 2ab + b^2$.
$$5w + 6 = 1 + 2\sqrt{4w+1} + 4w + 1$$

Step 3: Combine like terms on the right side.

$$5w + 6 = 4w + 2 + 2\sqrt{4w+1}$$ Combine like terms.

Step 4: The equation still contains a radical, so repeat Steps 1–3.

Step 1: Get the radical on a side by itself.

$$5w + 6 = 4w + 2 + 2\sqrt{4w+1}$$
$$w + 4 = 2\sqrt{4w+1}$$ Subtract $4w$ and subtract 2.

We do not need to eliminate the 2 from in front of the radical before squaring both sides. The radical must not be a part of a *sum* or *difference* when we square.

Step 2: Square both sides of the equation to eliminate the radical.

$$(w + 4)^2 = (2\sqrt{4w+1})^2$$ Square both sides.
$$w^2 + 8w + 16 = 4(4w + 1)$$ Square the binomial; $2^2 = 4$.

Steps 3 and 4 no longer apply.

Step 5: Solve the equation.

$$w^2 + 8w + 16 = 16w + 4$$ Distribute.
$$w^2 - 8w + 12 = 0$$ Subtract $16w$ and subtract 4.
$$(w - 2)(w - 6) = 0$$ Factor.
$$w - 2 = 0 \quad \text{or} \quad w - 6 = 0$$ Set each factor equal to zero.
$$w = 2 \quad \text{or} \quad w = 6$$ Solve.

Step 6: The check is left to the student. Verify that $w = 2$ and $w = 6$ each satisfy the original equation. The solution set is $\{2, 6\}$.

[YOU TRY 5] Solve each equation.

a) $\sqrt{2y + 1} - \sqrt{y} = 1$ b) $\sqrt{3t + 4} + \sqrt{t + 2} = 2$

BE CAREFUL Watch out for two common mistakes that students make when solving an equation like the one in Example 5b.

1) Do not square both sides before getting a radical on a side by itself.

   This is incorrect: $(\sqrt{5w+6} - \sqrt{4w+1})^2 = 1^2$
   $$5w + 6 - (4w + 1) = 1$$

2) The *second* time we perform Step 2, watch out for this common error.

   This is incorrect: $(w + 4)^2 = (2\sqrt{4w+1})^2$
   $$w^2 + 16 = 2(4w+1)$$

   On the left we must multiply using FOIL or the formula $(a+b)^2 = a^2 + 2ab + b^2$, and on the right we must remember to square the 2.

4 Solve an Equation Containing a Cube Root

We can solve many equations containing cube roots the same way we solve equations containing square roots except, to eliminate a *cube root*, we *cube* both sides of the equation.

EXAMPLE 6 Solve $\sqrt[3]{7a+1} - 2\sqrt[3]{a-1} = 0$.

Solution

Begin by getting a radical on a side by itself.

$$\sqrt[3]{7a+1} = 2\sqrt[3]{a-1} \quad \text{Add } 2\sqrt[3]{a-1} \text{ to each side.}$$
$$(\sqrt[3]{7a+1})^3 = (2\sqrt[3]{a-1})^3 \quad \text{Cube both sides to eliminate the radicals.}$$
$$7a + 1 = 8(a-1) \quad \text{Simplify; } 2^3 = 8.$$
$$7a + 1 = 8a - 8 \quad \text{Distribute.}$$
$$9 = a \quad \text{Subtract } 7a; \text{ add } 8.$$

Check $a = 9$ in the original equation.

$$\sqrt[3]{7a+1} - 2\sqrt[3]{a-1} = 0$$
$$\sqrt[3]{7(9)+1} - 2\sqrt[3]{9-1} \stackrel{?}{=} 0$$
$$\sqrt[3]{64} - 2\sqrt[3]{8} \stackrel{?}{=} 0$$
$$4 - 2(2) \stackrel{?}{=} 0$$
$$4 - 4 = 0 \checkmark$$

The solution set is $\{9\}$.

W Hint Write a procedure to help you solve equations containing a cube root.

[**YOU TRY 6**] Solve $3\sqrt[3]{r-4} - \sqrt[3]{5r+2} = 0$.

Using Technology

We can use a graphing calculator to solve a radical equation in one variable. First subtract every term on the right side of the equation from both sides of the equation, and enter the result in Y_1. Graph the equation in Y_1. The zeros or *x*-intercepts of the graph are the solutions to the equation.

We will solve $\sqrt{x+3} = 2$ using a graphing calculator.

1) Enter $\sqrt{x+3} - 2$ in Y$_1$.
2) Press ZOOM 6 to graph the function in Y$_1$ as shown.
3) Press 2nd TRACE 2:zero, move the cursor to the left of the zero and press ENTER, move the cursor to the right of the zero and press ENTER, and move the cursor close to the zero and press ENTER to display the zero. The solution to the equation is $x = 1$.

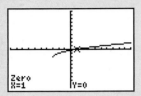

Solve each equation using a graphing calculator.

1) $\sqrt{x-2} = 1$ 2) $\sqrt{3x-2} = 5$ 3) $\sqrt{3x-2} = \sqrt{x}+2$
4) $\sqrt{4x-5} = \sqrt{x+4}$ 5) $\sqrt{2x-7} = \sqrt{x}-1$ 6) $\sqrt{\sqrt{x}-1} = 1$

ANSWERS TO YOU TRY EXERCISES

1) a) {45} b) ∅   2) a) {−3, −2} b) {7}   3) {5}   4) a) $z - 8\sqrt{z} + 16$
b) $3d + 24 + 10\sqrt{3d-1}$ 5) a) {0, 4} b) {−1}   6) {5}

ANSWERS TO TECHNOLOGY EXERCISES

1) {3}   2) {9}   3) {9}   4) {3}   5) {4}   6) {4}

E Evaluate 9.7 Exercises

Do the exercises, and check your work.

Objective 1: Understand the Steps for Solving a Radical Equation

1) Why is it necessary to check the proposed solutions to a radical equation in the original equation?

2) How do you know, without actually solving and checking the solution, that $\sqrt{y} = -3$ has no solution?

Objective 2: Solve an Equation Containing One Square Root
Solve.

3) $\sqrt{q} = 7$
4) $\sqrt{z} = 10$
5) $\sqrt{w} - \dfrac{2}{3} = 0$
6) $\sqrt{r} - \dfrac{3}{5} = 0$

7) $\sqrt{a} + 5 = 3$
8) $\sqrt{k} + 8 = 2$
9) $\sqrt{b-11} - 3 = 0$
10) $\sqrt{d+3} - 5 = 0$
11) $\sqrt{4g-1} + 7 = 1$
12) $\sqrt{3v+4} + 10 = 6$
13) $\sqrt{3f+2} + 9 = 11$
14) $\sqrt{5u-4} + 12 = 17$
15) $m = \sqrt{m^2 - 3m + 6}$
16) $b = \sqrt{b^2 + 4b - 24}$
17) $\sqrt{9r^2 - 2r + 10} = 3r$
18) $\sqrt{4p^2 - 3p + 6} = 2p$

Square each binomial, and simplify.

19) $(n+5)^2$
20) $(z-3)^2$
21) $(c-6)^2$
22) $(2k+1)^2$

Solve.

23) $c - 7 = \sqrt{2c + 1}$

24) $p + 6 = \sqrt{12 + p}$

25) $6 + \sqrt{c^2 + 3c - 9} = c$

26) $-4 + \sqrt{z^2 + 5z - 8} = z$

27) $w - \sqrt{10w + 6} = -3$

28) $3 - \sqrt{8t + 9} = -t$

29) $3v = 8 + \sqrt{3v + 4}$

30) $4k = 3 + \sqrt{10k + 5}$

31) $m + 4 = 5\sqrt{m}$

32) $b + 5 = 6\sqrt{b}$

33) $y + 2\sqrt{6 - y} = 3$

34) $r - 3\sqrt{r + 2} = 2$

35) $\sqrt{r^2 - 8r - 19} = r - 9$

36) $\sqrt{x^2 + x + 4} = x + 8$

Objective 3: Solve an Equation Containing Two Square Roots

Solve.

37) $5\sqrt{1 - 5h} = 4\sqrt{1 - 8h}$

38) $3\sqrt{6a - 2} - 4\sqrt{3a + 3} = 0$

39) $3\sqrt{3x + 6} - 2\sqrt{9x - 9} = 0$

40) $5\sqrt{q + 11} = 2\sqrt{8q + 25}$

41) $\sqrt{m} = 3\sqrt{7}$

42) $4\sqrt{3} = \sqrt{p}$

43) $2\sqrt{3t + 4} + \sqrt{t - 6} = 0$

44) $\sqrt{2w - 1} + 2\sqrt{w + 4} = 0$

Square each expression, and simplify.

45) $(\sqrt{x} + 5)^2$

46) $(\sqrt{y} - 8)^2$

47) $(9 - \sqrt{a + 4})^2$

48) $(4 + \sqrt{p + 5})^2$

49) $(2\sqrt{3n - 1} + 7)^2$

50) $(5 - 3\sqrt{2k - 3})^2$

Solve.

51) $\sqrt{2y - 1} = 2 + \sqrt{y - 4}$

52) $\sqrt{3n + 4} = \sqrt{2n + 1} + 1$

53) $1 + \sqrt{3s - 2} = \sqrt{2s + 5}$

54) $\sqrt{4p + 12} - 1 = \sqrt{6p - 11}$

55) $\sqrt{5a + 19} - \sqrt{a + 12} = 1$

56) $\sqrt{2u + 3} - \sqrt{5u + 1} = -1$

57) $\sqrt{3k + 1} - \sqrt{k - 1} = 2$

58) $\sqrt{4z - 3} - \sqrt{5z + 1} = -1$

59) $\sqrt{3x + 4} - 5 = \sqrt{3x - 11}$

60) $\sqrt{4c - 7} = \sqrt{4c + 1} - 4$

61) $\sqrt{3v + 3} - \sqrt{v - 2} = 3$

62) $\sqrt{2y + 1} - \sqrt{y} = 1$

Objective 4: Solve an Equation Containing a Cube Root

63) How do you eliminate the radical from an equation like $\sqrt[3]{x} = 2$?

64) Give a reason why $\sqrt[3]{h} = -3$ has no extraneous solutions.

Solve.

65) $\sqrt[3]{y} = 5$

66) $\sqrt[3]{c} = 3$

67) $\sqrt[3]{m} = -4$

68) $\sqrt[3]{t} = -2$

69) $\sqrt[3]{2x - 5} + 3 = 1$

70) $\sqrt[3]{4a + 1} + 7 = 4$

71) $\sqrt[3]{6j - 2} = \sqrt[3]{j - 7}$

72) $\sqrt[3]{w + 3} = \sqrt[3]{2w - 11}$

73) $\sqrt[3]{3y - 1} - \sqrt[3]{2y - 3} = 0$

74) $\sqrt[3]{2 - 2b} + \sqrt[3]{b - 5} = 0$

75) $\sqrt[3]{2n^2} = \sqrt[3]{7n + 4}$

76) $\sqrt[3]{4c^2 - 5c + 11} = \sqrt[3]{c^2 + 9}$

Extension

Solve.

77) $p^{1/2} = 6$

78) $\dfrac{2}{3} = t^{1/2}$

79) $7 = (2z - 3)^{1/2}$

80) $(3k + 1)^{1/2} = 4$

81) $(y + 4)^{1/3} = 3$

82) $-5 = (a - 2)^{1/3}$

83) $\sqrt[4]{n + 7} = 2$

84) $\sqrt[4]{x - 3} = -1$

85) $\sqrt{13 + \sqrt{r}} = \sqrt{r + 7}$

86) $\sqrt{m - 1} = \sqrt{m} - \sqrt{m - 4}$

87) $\sqrt{y + \sqrt{y + 5}} = \sqrt{y + 2}$

88) $\sqrt{2d - \sqrt{d + 6}} = \sqrt{d + 6}$

Mixed Exercises: Objectives 2 and 4

Solve for the indicated variable.

89) $v = \sqrt{\dfrac{2E}{m}}$ for E

90) $V = \sqrt{\dfrac{300VP}{m}}$ for P

91) $c = \sqrt{a^2 + b^2}$ for b^2

92) $r = \sqrt{\dfrac{A}{\pi}}$ for A

93) $T = \sqrt[4]{\dfrac{E}{\sigma}}$ for σ

94) $r = \sqrt[3]{\dfrac{3V}{4\pi}}$ for V

95) The speed of sound is proportional to the square root of the air temperature in still air. The speed of sound is given by the formula.

$$V_S = 20\sqrt{T + 273}$$

where V_S is the speed of sound in meters/second and T is the temperature of the air in °Celsius.

a) What is the speed of sound when the temperature is $-17°C$ (about $1°F$)?

b) What is the speed of sound when the temperature is $16°C$ (about $61°F$)?

c) What happens to the speed of sound as the temperature increases?

d) Solve the equation for T.

96) If the area of a square is A and each side has length l, then the length of a side is given by

$$l = \sqrt{A}$$

A square rug has an area of 25 ft².

a) Find the dimensions of the rug.

b) Solve the equation for A.

97) Let V represent the volume of a cylinder, h represent its height, and r represent its radius. V, h, and r are related according to the formula

$$r = \sqrt{\dfrac{V}{\pi h}}$$

a) A cylindrical soup can has a volume of 28π in³. It is 7 in. high. What is the radius of the can?

b) Solve the equation for V.

98) For shallow water waves, the wave velocity is given by

$$c = \sqrt{gH}$$

where g is the acceleration due to gravity (32 ft/sec²) and H is the depth of the water (in feet).

a) Find the velocity of a wave in 8 ft of water.

b) Solve the equation for H.

99) Refer to the formula given in Problem 98.

The catastrophic Indian Ocean tsunami that hit Banda Aceh, Sumatra, Indonesia, on December 26, 2004 was caused by an earthquake whose epicenter was off the coast of northern Sumatra. The tsunami originated in about 14,400 ft of water.

a) Find the velocity of the wave near the epicenter, in miles per hour. Round the answer to the nearest unit. (Hint: 1 mile = 5280 ft.)

b) Banda Aceh, the area hardest hit by the tsunami, was about 60 mi from the tsunami's origin. Approximately how many minutes after the earthquake occurred did the tsunami hit Banda Aceh? (*Exploring Geology*, McGraw-Hill, 2008.)

100) The radius r of a cone with height h and volume V is given by $r = \sqrt{\dfrac{3V}{\pi h}}$.

A hanging glass vase in the shape of a cone is 8 in. tall, and the radius of the top of the cone is 2 in. How much water will the vase hold? Give an exact answer and an approximation to the tenths place.

Use the following information for Exercises 101 and 102.

The distance a person can see to the horizon is approximated by the function $D(h) = 1.2\sqrt{h}$, where D is the number of miles a person can see to the horizon from a height of h ft.

101) Sig is the captain of an Alaskan crab fishing boat and can see 4.8 mi to the horizon when he is sailing his ship. Find his height above the sea.

102) Phil is standing on the deck of a boat and can see 3.6 mi to the horizon. What is his height above the water?

Use the following information for Problems 103 and 104.

When the air temperature is 0°F, the wind chill temperature, W, in degrees Fahrenheit is a function of the velocity of the wind, V, in miles per hour and is given by the formula

$$W(V) = 35.74 - 35.75V^{4/25}$$

103) Calculate the wind speed when the wind chill temperature is $-10°F$. Round to the nearest whole number

104) Find V so that $W(V) = -20$. Round to the nearest whole number. Explain your result in the context of the problem. (http://www.nws.noaa.gov/om/windchill/windchillglossary.shtml)

R Rethink

R1) How has a good understanding of binomials helped you complete these exercises?

R2) Could you explain how to solve these problems to a friend?

9.8 Complex Numbers

P Prepare

What are your objectives for Section 9.8?

O Organize

How can you accomplish each objective?

1	Find the Square Root of a Negative Number	• Understand the definition of the *imaginary number, i.* • Learn the definition of a *complex number.* • Understand the property of **Real Numbers and Complex Numbers.** • Complete the given example on your own. • Complete You Try 1.
2	Multiply and Divide Square Roots Containing Negative Numbers	• Follow the explanation and example to create a procedure for this objective. • Complete the given example on your own. • Complete You Try 2.
3	Add and Subtract Complex Numbers	• Write the procedure for **Adding and Subtracting Complex Numbers** in your own words. • Complete the given example on your own. • Complete You Try 3.
4	Multiply Complex Numbers	• Compare the steps for multiplying complex numbers to the notes you took on multiplying polynomials. • Complete the given example on your own. • Complete You Try 4.
5	Multiply a Complex Number by Its Conjugate	• Understand the definition of a *complex conjugate.* • Follow the explanation, and write the summary of **Complex Conjugates** in your notes. • Complete the given example on your own. • Complete You Try 5.
6	Divide Complex Numbers	• Write the procedure for **Dividing Complex Numbers** in your own words. • Complete the given example on your own. • Complete You Try 6.
7	Simplify Powers of *i*	• Follow the explanation, and summarize how to simplify powers of *i* in your notes. • Complete the given example on you own. • Complete You Try 7.

Read the explanations, follow the examples, take notes, and complete the You Trys.

1 Find the Square Root of a Negative Number

We have seen throughout this chapter that the square root of a negative number does not exist in the real number system because there is no real number that, when squared, will result in a negative number. For example, $\sqrt{-4}$ is not a real number because there is no real number whose square is -4.

The square roots of negative numbers do exist, however, under another system of numbers called *complex numbers*. Before we define a complex number, we must define the number i. The number i is called an *imaginary number*.

Definition

The **imaginary number** i is defined as
$$i = \sqrt{-1}.$$
Therefore, squaring both sides gives us
$$i^2 = -1.$$

Note

$i = \sqrt{-1}$ and $i^2 = -1$ are two *very* important facts to remember. We will be using them often!

Definition

A **complex number** is a number of the form $a + bi$, where a and b are real numbers; a is called the **real part**, and b is called the **imaginary part**.

The following table lists some examples of complex numbers and their real and imaginary parts.

Complex Number	Real Part	Imaginary Part
$-5 + 2i$	-5	2
$\dfrac{1}{3} - 7i$	$\dfrac{1}{3}$	-7
$8i$	0	8
4	4	0

Note

The complex number $8i$ can be written in the form $a + bi$ as $0 + 8i$. Likewise, besides being a real number, 4 is a complex number because it can be written as $4 + 0i$.

Because all real numbers, a, can be written in the form $a + 0i$, all real numbers are also complex numbers.

> **Property** Real Numbers and Complex Numbers
>
> The set of real numbers is a subset of the set of complex numbers.

Since we defined i as $i = \sqrt{-1}$, we can now evaluate square roots of negative numbers.

EXAMPLE 1

Simplify.

a) $\sqrt{-9}$ b) $\sqrt{-7}$ c) $\sqrt{-12}$

Solution

a) $\sqrt{-9} = \sqrt{-1 \cdot 9} = \sqrt{-1} \cdot \sqrt{9} = i \cdot 3 = 3i$

b) $\sqrt{-7} = \sqrt{-1 \cdot 7} = \sqrt{-1} \cdot \sqrt{7} = i\sqrt{7}$

c) $\sqrt{-12} = \sqrt{-1 \cdot 12} = \sqrt{-1} \cdot \sqrt{12} = i\sqrt{4}\sqrt{3} = i \cdot 2\sqrt{3} = 2i\sqrt{3}$

W Hint
Create a procedure for simplifying square roots of negative numbers.

Note
In Example 1b) we wrote $i\sqrt{7}$ instead of $\sqrt{7}i$, and in Example 1c) we wrote $2i\sqrt{3}$ instead of $2\sqrt{3}i$. We do this to be clear that the i is not under the radical. It is good practice to write the i before the radical.

[YOU TRY 1]

Simplify.

a) $\sqrt{-36}$ b) $\sqrt{-13}$ c) $\sqrt{-20}$

2 Multiply and Divide Square Roots Containing Negative Numbers

When multiplying or dividing square roots with negative radicands, write each radical in terms of i first. Remember, also, that since $i = \sqrt{-1}$ it follows that $i^2 = -1$. We must keep this in mind when simplifying expressions.

Note
Whenever an i^2 appears in an expression, replace it with -1.

EXAMPLE 2

Multiply and simplify. $\sqrt{-8} \cdot \sqrt{-2}$

Solution

$\sqrt{-8} \cdot \sqrt{-2} = i\sqrt{8} \cdot i\sqrt{2}$ Write each radical in terms of i *before* multiplying.
$= i^2\sqrt{16}$ Multiply.
$= (-1)(4)$ Replace i^2 with -1.
$= -4$

[YOU TRY 2] Perform the operation, and simplify.

a) $\sqrt{-6} \cdot \sqrt{-3}$ b) $\dfrac{\sqrt{-72}}{\sqrt{-2}}$

3 Add and Subtract Complex Numbers

Just as we can add, subtract, multiply, and divide real numbers, we can perform all of these operations with complex numbers.

> **Procedure** Adding and Subtracting Complex Numbers
>
> 1) To add complex numbers, add the real parts and add the imaginary parts.
>
> 2) To subtract complex numbers, apply the distributive property and combine the real parts and combine the imaginary parts.

EXAMPLE 3   Add or subtract.

a) $(8 + 3i) + (4 + 2i)$ b) $(7 + i) - (3 - 4i)$

Solution

a) $(8 + 3i) + (4 + 2i) = (8 + 4) + (3 + 2)i$ Add real parts; add imaginary parts.
$= 12 + 5i$

b) $(7 + i) - (3 - 4i) = 7 + i - 3 + 4i$ Distributive property
$= (7 - 3) + (1 + 4)i$ Add real parts; add imaginary parts.
$= 4 + 5i$

[YOU TRY 3] Add or subtract.

a) $(-10 + 6i) + (1 + 8i)$ b) $(2 - 5i) - (-1 + 6i)$

4 Multiply Complex Numbers

We multiply complex numbers just like we would multiply polynomials. There may be an additional step, however. Remember to replace i^2 with -1.

EXAMPLE 4   Multiply and simplify.

a) $5(-2 + 3i)$ b) $(8 + 3i)(-1 + 4i)$ c) $(6 + 2i)(6 - 2i)$

Solution

a) $5(-2 + 3i) = -10 + 15i$ Distributive property

W Hint
How does this compare to the techniques you learned in Section 6.4?

b) Look carefully at $(8 + 3i)(-1 + 4i)$. Each complex number has two terms, similar to, say, $(x + 3)(x + 4)$. How can we multiply these two binomials? We can use FOIL.

$$
\begin{aligned}
(8 + 3i)(-1 + 4i) &= \overset{F}{(8)(-1)} + \overset{O}{(8)(4i)} + \overset{I}{(3i)(-1)} + \overset{L}{(3i)(4i)} \\
&= -8 + 32i - 3i + 12i^2 \\
&= -8 + 29i + 12(-1) \quad \text{Replace } i^2 \text{ with } -1. \\
&= -8 + 29i - 12 \\
&= -20 + 29i
\end{aligned}
$$

c) Use FOIL to find the product $(6 + 2i)(6 - 2i)$.

$$
\begin{aligned}
(6 + 2i)(6 - 2i) &= \overset{F}{(6)(6)} + \overset{O}{(6)(-2i)} + \overset{I}{(2i)(6)} + \overset{L}{(2i)(-2i)} \\
&= 36 - 12i + 12i - 4i^2 \\
&= 36 - 4(-1) \quad \text{Replace } i^2 \text{ with } -1. \\
&= 36 + 4 \\
&= 40
\end{aligned}
$$

[YOU TRY 4] Multiply and simplify.

a) $-3(6 - 7i)$ b) $(5 - i)(4 + 8i)$ c) $(-2 - 9i)(-2 + 9i)$

5 Multiply a Complex Number by Its Conjugate

In Section 9.6, we learned about conjugates of radical expressions. For example, the conjugate of $3 + \sqrt{5}$ is $3 - \sqrt{5}$.

The complex numbers in Example 4c, $6 + 2i$ and $6 - 2i$, are **complex conjugates.**

Definition

The **conjugate** of $a + bi$ is $a - bi$.

We found that $(6 + 2i)(6 - 2i) = 40$ which is a real number. **The product of a complex number and its conjugate is *always* a real number,** as illustrated next.

$$
\begin{aligned}
(a + bi)(a - bi) &= \overset{F}{(a)(a)} + \overset{O}{(a)(-bi)} + \overset{I}{(bi)(a)} + \overset{L}{(bi)(-bi)} \\
&= a^2 - abi + abi - b^2i^2 \\
&= a^2 - b^2(-1) \quad \text{Replace } i^2 \text{ with } -1. \\
&= a^2 + b^2
\end{aligned}
$$

We can summarize these facts about complex numbers and their conjugates as follows:

Summary Complex Conjugates

1) The conjugate of $a + bi$ is $a - bi$.
2) The product of $a + bi$ and $a - bi$ is a real number.
3) We can find the product $(a + bi)(a - bi)$ by using FOIL or by using $(a + bi)(a - bi) = a^2 + b^2$.

EXAMPLE 5

Multiply $-3 + 4i$ by its conjugate using the formula $(a + bi)(a - bi) = a^2 + b^2$.

Solution

The conjugate of $-3 + 4i$ is $-3 - 4i$.

$$(-3 + 4i)(-3 - 4i) = (-3)^2 + (4)^2 \quad a = -3, b = 4$$
$$= 9 + 16$$
$$= 25$$

 Hint
Always use parentheses!

YOU TRY 5 Multiply $2 - 9i$ by its conjugate using the formula $(a + bi)(a - bi) = a^2 + b^2$.

6 Divide Complex Numbers

To rationalize the denominator of a radical expression like $\dfrac{2}{3 + \sqrt{5}}$, we multiply the numerator and denominator by $3 - \sqrt{5}$, the conjugate of the denominator. We divide complex numbers in the same way.

Procedure Dividing Complex Numbers

To divide complex numbers, multiply the numerator and denominator by the *conjugate of the denominator*. Write the quotient in the form $a + bi$.

EXAMPLE 6

Divide. Write the quotient in the form $a + bi$.

a) $\dfrac{3}{4 - 5i}$ b) $\dfrac{6 - 2i}{-7 + i}$

Solution

a) $\dfrac{3}{4 - 5i} = \dfrac{3}{(4 - 5i)} \cdot \dfrac{(4 + 5i)}{(4 + 5i)}$ Multiply the numerator and denominator by the conjugate of the denominator.

$= \dfrac{12 + 15i}{(4)^2 + (5)^2}$ Multiply numerators. $(a + bi)(a - bi) = a^2 + b^2$

$= \dfrac{12 + 15i}{16 + 25}$

$= \dfrac{12 + 15i}{41}$

$= \dfrac{12}{41} + \dfrac{15}{41}i$ Write the quotient in the form $a + bi$.

Recall that we can find the product $(4 - 5i)(4 + 5i)$ using FOIL *or* by using the formula $(a + bi)(a - bi) = a^2 + b^2$.

b) $\dfrac{6 - 2i}{-7 + i} = \dfrac{(6 - 2i)}{(-7 + i)} \cdot \dfrac{(-7 - i)}{(-7 - i)}$ Multiply the numerator and denominator by the conjugate of the denominator.

$= \dfrac{-42 - 6i + 14i + 2i^2}{(-7)^2 + (1)^2}$ Multiply using FOIL. $(a + bi)(a - bi) = a^2 + b^2$

$= \dfrac{-42 + 8i - 2}{49 + 1} = \dfrac{-44 + 8i}{50} = -\dfrac{44}{50} + \dfrac{8}{50}i = -\dfrac{22}{25} + \dfrac{4}{25}i$

> [YOU TRY 6] Divide. Write the result in the form $a + bi$.
>
> a) $\dfrac{6}{-2 + i}$ b) $\dfrac{5 + 3i}{-6 - 4i}$

7 Simplify Powers of i

All powers of i larger than i^1 (or just i) can be simplified. We use the fact that $i^2 = -1$ to simplify powers of i.

Let's write i through i^4 in their simplest forms.

i is in simplest form.
$i^2 = -1$
$i^3 = i^2 \cdot i = -1 \cdot i = -i$
$i^4 = (i^2)^2 = (-1)^2 = 1$

Let's continue by simplifying i^5 and i^6.

$$i^5 = i^4 \cdot i \qquad\qquad i^6 = (i^2)^3$$
$$= (i^2)^2 \cdot i \qquad\qquad\; = (-1)^3$$
$$= (-1)^2 \cdot i \qquad\qquad = -1$$
$$= 1i$$
$$= i$$

The pattern repeats so that all powers of i can be simplified to i, -1, $-i$, or 1.

EXAMPLE 7

Simplify each power of i.

a) i^8 b) i^{14} c) i^{11} d) i^{37}

Solution

a) Use the power rule for exponents to simplify i^8. Since the exponent is even, we can rewrite it in terms of i^2.

$i^8 = (i^2)^4$ Power rule
$\;\;\; = (-1)^4$ $i^2 = -1$
$\;\;\; = 1$ Simplify.

b) As in Example 7a), the exponent is even. Rewrite i^{14} in terms of i^2.

$i^{14} = (i^2)^7$ Power rule
$\;\;\;\;\; = (-1)^7$ $i^2 = -1$
$\;\;\;\;\; = -1$ Simplify.

c) The exponent of i^{11} is odd, so first use the product rule to write i^{11} as a product of i and i^{11-1} or i^{10}.

$i^{11} = i^{10} \cdot i$ Product rule
$\;\;\;\;\; = (i^2)^5 \cdot i$ 10 is even; write i^{10} in terms of i^2.
$\;\;\;\;\; = (-1)^5 \cdot i$ $i^2 = -1$
$\;\;\;\;\; = -1 \cdot i$ Simplify.
$\;\;\;\;\; = -i$ Multiply.

d) The exponent of i^{37} is odd. Use the product rule to write i^{37} as a product of i and i^{37-1} or i^{36}.

$i^{37} = i^{36} \cdot i$ Product rule
$\;\;\;\;\; = (i^2)^{18} \cdot i$ 36 is even; write i^{36} in terms of i^2.
$\;\;\;\;\; = (-1)^{18} \cdot i$ $i^2 = -1$
$\;\;\;\;\; = 1 \cdot i$ Simplify.
$\;\;\;\;\; = i$ Multiply.

> [YOU TRY 7] Simplify each power of i.
>
> a) i^{18} b) i^{32} c) i^7 d) i^{25}

Using Technology

We can use a graphing calculator to perform operations on complex numbers or to evaluate square roots of negative numbers.

If the calculator is in the default REAL mode the result is an error message "ERR: NONREAL ANS," which indicates that $\sqrt{-4}$ is a complex number rather than a real number. Before evaluating $\sqrt{-4}$ on the home screen of your calculator, check the mode by pressing **MODE** and looking at row 7. Change the mode to complex numbers by selecting $a + bi$, as shown at the left below.

Now evaluating $\sqrt{-4}$ on the home screen results in the correct answer $2i$, as shown on the right below.

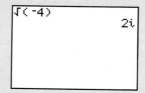

Operations can be performed on complex numbers with the calculator in either REAL or $a + bi$ mode. Simply use the arithmetic operators on the right column on your calculator. To enter the imaginary number i, press **2nd** **.**. To add $2 - 5i$ and $4 + 3i$, enter $(2 - 5i) + (4 + 3i)$ on the home screen and press **ENTER** as shown on the left screen below. To subtract $8 + 6i$ from $7 - 2i$, enter $(7 - 2i) - (8 + 6i)$ on the home screen and press **ENTER** as shown.

To multiply $3 - 5i$ and $7 + 4i$, enter $(3 - 5i) \cdot (7 + 4i)$ on the home screen and press **ENTER** as shown on the middle screen below. To divide $2 + 9i$ by $2 - i$, enter $(2 + 9i)/(2 - i)$ on the home screen and press **ENTER** as shown.

To raise $3 - 4i$ to the fifth power, enter $(3 - 4i)\wedge 5$ on the home screen and press **ENTER** as shown.

Consider the quotient $(5 + 3i)/(4 - 7i)$. The exact answer is $-\dfrac{1}{65} + \dfrac{47}{65}i$. The calculator automatically displays the decimal result. Press **MATH** **1** **ENTER** to convert the decimal result to the exact fractional result, as shown on the right screen below.

Perform the indicated operation using a graphing calculator.

1) Simplify $\sqrt{-36}$
2) $(3 + 7i) + (5 - 8i)$
3) $(10 - 3i) - (4 + 8i)$
4) $(3 + 2i)(6 - 3i)$
5) $(4 + 3i) \div (1 - i)$
6) $(5 - 3i)^3$

ANSWERS TO [YOU TRY] EXERCISES

1) a) $6i$ b) $i\sqrt{13}$ c) $2i\sqrt{5}$ 2) a) $-3\sqrt{2}$ b) 6 3) a) $-9 + 14i$ b) $3 - 11i$
4) a) $-18 + 21i$ b) $28 + 36i$ c) 85 5) 85 6) a) $-\dfrac{12}{5} - \dfrac{6}{5}i$ b) $-\dfrac{21}{26} + \dfrac{1}{26}i$
7) a) -1 b) 1 c) $-i$ d) i

ANSWERS TO TECHNOLOGY EXERCISES

1) $6i$ 2) $8 - i$ 3) $6 - 11i$ 4) $24 + 3i$ 5) $\frac{1}{2} + \frac{7}{2}i$ 6) $-10 - 198i$

E Evaluate 9.8 Exercises

Do the exercises, and check your work.

Objective 1: Find the Square Root of a Negative Number

Determine if each statement is true or false.

1) Every complex number is a real number.
2) Every real number is a complex number.
3) Since $i = \sqrt{-1}$, it follows that $i^2 = -1$.
4) In the complex number $-6 + 5i$, -6 is the real part and $5i$ is the imaginary part.

Simplify.

5) $\sqrt{-81}$
6) $\sqrt{-16}$
7) $\sqrt{-25}$
8) $\sqrt{-169}$
9) $\sqrt{-6}$
10) $\sqrt{-30}$
11) $\sqrt{-27}$
12) $\sqrt{-75}$
13) $\sqrt{-60}$
14) $\sqrt{-28}$

Objective 2: Multiply and Divide Square Roots Containing Negative Numbers

Find the error in each of the following exercises, then find the correct answer.

15) $\sqrt{-5} \cdot \sqrt{-10} = \sqrt{-5 \cdot (-10)}$
$= \sqrt{50}$
$= \sqrt{25} \cdot \sqrt{2}$
$= 5\sqrt{2}$

16) $(\sqrt{-7})^2 = \sqrt{(-7)^2}$
$= \sqrt{49}$
$= 7$

Perform the indicated operation, and simplify.

17) $\sqrt{-1} \cdot \sqrt{-5}$
18) $\sqrt{-5} \cdot \sqrt{-15}$
19) $\sqrt{-20} \cdot \sqrt{-5}$
20) $\sqrt{-12} \cdot \sqrt{-3}$
21) $\dfrac{\sqrt{-60}}{\sqrt{-15}}$
22) $\dfrac{\sqrt{-2}}{\sqrt{-128}}$
23) $(\sqrt{-13})^2$
24) $(\sqrt{-1})^2$

Mixed Exercises: Objectives 3–6

25) Explain how to add complex numbers.
26) How is multiplying $(1 + 3i)(2 - 7i)$ similar to multiplying $(x + 3)(2x - 7)$?
27) When i^2 appears in an expression, it should be replaced with what?
28) Explain how to divide complex numbers.

Objective 3: Add and Subtract Complex Numbers

Perform the indicated operations.

29) $(-4 + 9i) + (7 + 2i)$
30) $(6 + i) + (8 - 5i)$
31) $(13 - 8i) - (9 + i)$
32) $(-12 + 3i) - (-7 - 6i)$
33) $\left(-\dfrac{3}{4} - \dfrac{1}{6}i\right) - \left(-\dfrac{1}{2} + \dfrac{2}{3}i\right)$
34) $\left(\dfrac{1}{2} + \dfrac{7}{9}i\right) - \left(\dfrac{7}{8} - \dfrac{1}{6}i\right)$
35) $16i - (3 + 10i) + (3 + i)$
36) $(-6 - 5i) + (2 + 6i) - (-4 + i)$

Objective 4: Multiply Complex Numbers

Multiply and simplify.

37) $3(8 - 5i)$
38) $-6(8 - i)$
39) $\dfrac{2}{3}(-9 + 2i)$
40) $\dfrac{1}{2}(18 + 7i)$
41) $-4i(6 + 11i)$
42) $6i(5 + 6i)$
43) $(2 + 5i)(1 + 6i)$
44) $(2 + i)(10 + 5i)$
45) $(-1 + 3i)(4 - 6i)$
46) $(-4 - 9i)(3 - i)$
47) $(5 - 3i)(9 - 3i)$
48) $(3 - 4i)(6 + 7i)$
49) $\left(\dfrac{3}{4} + \dfrac{3}{4}i\right)\left(\dfrac{2}{5} + \dfrac{1}{5}i\right)$
50) $\left(\dfrac{1}{3} - \dfrac{4}{3}i\right)\left(\dfrac{3}{4} + \dfrac{2}{3}i\right)$

Objective 5: Multiply a Complex Number by Its Conjugate

Identify the conjugate of each complex number, then multiply the number and its conjugate.

51) $11 + 4i$
52) $-1 - 2i$
53) $-3 - 7i$
54) $4 + 9i$
55) $-6 + 4i$
56) $6 - 5i$

57) How are conjugates of complex numbers like conjugates of expressions containing real numbers and radicals?

58) Is the product of two complex numbers always a complex number? Explain your answer.

Objective 6: Divide Complex Numbers
Divide. Write the result in the form $a + bi$.

59) $\dfrac{4}{2 - 3i}$

60) $\dfrac{-10}{8 - 9i}$

61) $\dfrac{8i}{4 + i}$

62) $\dfrac{i}{6 - 5i}$

63) $\dfrac{2i}{-3 + 7i}$

64) $\dfrac{9i}{-4 + 10i}$

65) $\dfrac{3 - 8i}{-6 + 7i}$

66) $\dfrac{-5 + 2i}{4 - i}$

67) $\dfrac{2 + 3i}{5 - 6i}$

68) $\dfrac{1 + 6i}{5 + 2i}$

69) $\dfrac{9}{i}$

70) $\dfrac{16 + 3i}{-i}$

Objective 7: Simplify Powers of i
Simplify each power of i.

Fill It In
Fill in the blanks with either the missing mathematical step or reason for the given step.

71) $i^{24} = $ ___   Rewrite i^{24} in terms of i^2 using the power rule.
$= (-1)^{12}$ ___
$= $ ___   Simplify.

72) $i^{31} = i^{30} \cdot i$ ___
$= (i^2)^{15} \cdot i$
$= $ ___ $i^2 = -1$
$= $ ___   Simplify.
$= $ ___   Multiply.

73) i^{24}
74) i^{16}
75) i^{28}
76) i^{30}
77) i^9
78) i^{19}
79) i^{35}
80) i^{29}
81) i^{23}
82) i^{40}
83) i^{42}
84) i^{33}
85) $(2i)^5$
86) $(2i)^6$
87) $(-i)^{14}$
88) $(-i)^{15}$

Expand.

89) $(-2 + 5i)^3$
90) $(3 - 4i)^3$

Simplify each expression. Write the result in the form $a + bi$.

91) $1 + \sqrt{-8}$
92) $-7 - \sqrt{-48}$
93) $8 - \sqrt{-45}$
94) $3 + \sqrt{-20}$
95) $\dfrac{-12 + \sqrt{-32}}{4}$
96) $\dfrac{21 - \sqrt{-18}}{3}$

Used in the field of electronics, the **impedance**, Z, is the total opposition to the current flow of an alternating current (AC) within an electronic component, circuit, or system. It is expressed as a complex number $Z = R + Xj$, where the i used to represent an imaginary number in most areas of mathematics is replaced by j in electronics. R represents the resistance of a substance, and X represents the reactance.

The **total impedance, Z,** of components connected in series is the *sum* of the individual impedances of each component.

Each exercise contains the impedance of individual circuits. Find the total impedance of a system formed by connecting the circuits in series by finding the sum of the individual impedances.

97) $Z_1 = 3 + 2j$
    $Z_2 = 7 + 4j$

98) $Z_1 = 5 + 3j$
    $Z_2 = 9 + 6j$

99) $Z_1 = 5 - 2j$
    $Z_2 = 11 + 6j$

100) $Z_1 = 4 - 1.5j$
     $Z_2 = 3 + 0.5j$

R Rethink

R1) Did you check your answers manually before looking up the answers?

R2) Which learning objectives from previous chapters helped you master the concepts of this exercise set?

Group Activity — The Group Activity can be found online on Connect.

emPOWERme Information, Please!

One key to effective use of the Internet is being able to identify accurate, reliable sources of information. To practice your skills in this area, go online to find answers to the following questions. However, you have to find the information on *five different websites*. So, if you use Wikipedia to answer the first question, you can't use it to answer any of the remaining four. Along with your answer, note which site you used to find the information.

1. What is the origin of the word "algebra"?

2. Who was Euclid?

3. What is the number e?

4. Who was the first winner of the Nobel Prize in Mathematics?

5. What is the ECU?

How easy was it for you to find the answers to the questions? How easy was it to find multiple, reliable sites?

Chapter 9: Summary

Definition/Procedure	Example
9.1 Radical Expressions and Functions	
If the radicand is a perfect square, then the square root is a *rational* number. (p. 513)	$\sqrt{49} = 7$ since $7^2 = 49$.
If the radicand is a negative number, then the square root is *not* a real number. (p. 513)	$\sqrt{-36}$ is not a real number.
If the radicand is positive and not a perfect square, then the square root is an *irrational* number. (p. 513)	$\sqrt{7}$ is irrational because 7 is not a perfect square.
The symbol $\sqrt[n]{a}$ is read as "the *n*th root of *a*." If $\sqrt[n]{a} = b$, then $b^n = a$. We call *n* the **index** of the radical.	$\sqrt[5]{32} = 2$ since $2^5 = 32$.
For any *positive* number *a* and any *even* index *n*, the **principal *n*th root** of *a* is $\sqrt[n]{a}$, and the **negative *n*th root** of *a* is $-\sqrt[n]{a}$. (p. 514)	$\sqrt[4]{16} = 2$ $-\sqrt[4]{16} = -2$
The *odd root* of a negative number is a negative number. (p. 514)	$\sqrt[3]{-125} = -5$ since $(-5)^3 = -125$.
The *even root* of a negative number is not a real number. (p. 514)	$\sqrt[4]{-16}$ is not a real number.
If *n* is a positive, *even* integer, then $\sqrt[n]{a^n} = |a|$. If *n* is a positive, *odd* integer, then $\sqrt[n]{a^n} = a$. (p. 515)	$\sqrt[4]{(-2)^4} = |-2| = 2$ $\sqrt[3]{(-2)^3} = -2$
The **domain of a square root function** consists of all of the real numbers that can be substituted for the variable so that the radicand is nonnegative. (p. 517)	Determine the domain of the square root function. $f(x) = \sqrt{6x - 7}$ $6x - 7 \geq 0$ The value of the radicand must be ≥ 0. $6x \geq 7$ $x \geq \dfrac{7}{6}$ Solve. The domain of $f(x) = \sqrt{6x-7}$ is $\left[\dfrac{7}{6}, \infty\right)$.
The **domain of a cube root function** is the set of all real numbers. We can write this in interval notation as $(-\infty, \infty)$. (p. 518)	The domain of $g(x) = \sqrt[3]{x}$ is $(-\infty, \infty)$.
To graph a square root function, find the domain, make a table of values, and graph. (p. 519)	Graph $g(x) = \sqrt{x}$. 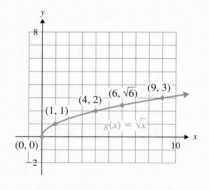 	x	g(x)	 |---|---| 	0	0	 	1	1	 	4	2	 	6	$\sqrt{6}$	 	9	3	
9.2 Rational Exponents	
If *n* is a positive integer greater than 1 and $\sqrt[n]{a}$ is a real number, then $a^{1/n} = \sqrt[n]{a}$. (p. 525)	$8^{1/3} = \sqrt[3]{8} = 2$
If *m* and *n* are positive integers and $\dfrac{m}{n}$ is in lowest terms, then $a^{m/n} = (a^{1/n})^m = (\sqrt[n]{a})^m$ provided that $a^{1/n}$ is a real number. (p. 526)	$16^{3/4} = (\sqrt[4]{16})^3 = 2^3 = 8$

Definition/Procedure	Example
If $a^{m/n}$ is a nonzero real number, then $$a^{-m/n} = \left(\frac{1}{a}\right)^{m/n} = \frac{1}{a^{m/n}}.$$ (p. 527)	$$25^{-3/2} = \left(\frac{1}{25}\right)^{3/2} = \left(\sqrt{\frac{1}{25}}\right)^3 = \left(\frac{1}{5}\right)^3 = \frac{1}{125}$$ **BE CAREFUL** The negative exponent does not make the expression negative.

9.3 Simplifying Expressions Containing Square Roots

Product Rule for Square Roots Let a and b be nonnegative real numbers. Then, $\sqrt{a} \cdot \sqrt{b} = \sqrt{a \cdot b}$. (p. 535)	$\sqrt{5} \cdot \sqrt{7} = \sqrt{5 \cdot 7} = \sqrt{35}$
An expression containing a square root is simplified when all of the following conditions are met: 1) The radicand does not contain any factors (other than 1) that are perfect squares. 2) The radicand does not contain any fractions. 3) There are no radicals in the denominator of a fraction. To *simplify square roots*, rewrite using the product rule as $\sqrt{a \cdot b} = \sqrt{a} \cdot \sqrt{b}$, where a or b is a perfect square. After simplifying a radical, look at the result and ask yourself, "*Is the radical in simplest form?*" If it is not, simplify again. (p. 535)	Simplify $\sqrt{24}$. $\sqrt{24} = \sqrt{4 \cdot 6}$ 4 is a perfect square. $\phantom{\sqrt{24}} = \sqrt{4} \cdot \sqrt{6}$ Product rule $\phantom{\sqrt{24}} = 2\sqrt{6}$ $\sqrt{4} = 2$
Quotient Rule for Square Roots Let a and b be nonnegative real numbers such that $b \neq 0$. Then, $\sqrt{\dfrac{a}{b}} = \dfrac{\sqrt{a}}{\sqrt{b}}$. (p. 537)	$\sqrt{\dfrac{72}{25}} = \dfrac{\sqrt{72}}{\sqrt{25}}$ Quotient rule $\phantom{\sqrt{\dfrac{72}{25}}} = \dfrac{\sqrt{36} \cdot \sqrt{2}}{5}$ Product rule; $\sqrt{25} = 5$ $\phantom{\sqrt{\dfrac{72}{25}}} = \dfrac{6\sqrt{2}}{5}$ $\sqrt{36} = 6$
If a is a nonnegative real number and m is an integer, then $\sqrt{a^m} = a^{m/2}$. (p. 538)	$\sqrt{k^{18}} = k^{18/2} = k^9$ (provided k represents a nonnegative real number)
Two Approaches to Simplifying Radical Expressions Containing Variables Let a represent a nonnegative real number. To simplify $\sqrt{a^n}$ where n is odd and positive, **i) Method 1:** Write a^n as the product of two factors so that the exponent of one of the factors is the *largest* number less than n that is divisible by 2 (the index of the radical). (p. 539) **ii) Method 2:** 1) Divide the exponent in the radicand by the index of the radical. 2) The exponent on the variable *outside* of the radical will be the *quotient* of the division problem. 3) The exponent on the variable *inside* of the radical will be the *remainder* of the division problem. (p. 540)	i) Simplify $\sqrt{x^9}$. $\sqrt{x^9} = \sqrt{x^8 \cdot x^1}$ 8 is the largest number less than 9 that is divisible by 2. $\phantom{\sqrt{x^9}} = \sqrt{x^8} \cdot \sqrt{x}$ Product rule $\phantom{\sqrt{x^9}} = x^{8/2}\sqrt{x}$ $\phantom{\sqrt{x^9}} = x^4\sqrt{x}$ $8 \div 2 = 4$ ii) Simplify $\sqrt{p^{15}}$. $\sqrt{p^{15}} = p^7\sqrt{p^1}$ $15 \div 2$ gives a quotient of $\phantom{\sqrt{p^{15}}} = p^7\sqrt{p}$ 7 and a remainder of 1.

Definition/Procedure	Example

9.4 Simplifying Expressions Containing Higher Roots

Definition/Procedure	Example
Product Rule for Higher Roots If $\sqrt[n]{a}$ and $\sqrt[n]{b}$ are real numbers such that the roots exist, then $\sqrt[n]{a} \cdot \sqrt[n]{b} = \sqrt[n]{a \cdot b}$. (p. 545)	$\sqrt[3]{3} \cdot \sqrt[3]{5} = \sqrt[3]{15}$
Let P be an expression and let n be a positive integer greater than 1. Then $\sqrt[n]{P}$ **is completely simplified when all of the following conditions are met:** 1) The radicand does not contain any factors (other than 1) that are perfect nth powers. 2) The exponents in the radicand and the index of the radical do not have any common factors (other than 1). 3) The radicand does not contain any fractions. 4) There are no radicals in the denominator of a fraction. To *simplify radicals with any index,* reverse the process of multiplying radicals, where a or b is an nth power. $\sqrt[n]{a \cdot b} = \sqrt[n]{a} \cdot \sqrt[n]{b}$ (p. 546)	Simplify $\sqrt[3]{40}$. **Method 1:** Think of two numbers that multiply to 40 so that one of them is a *perfect cube*. $40 = 8 \cdot 5$ 8 is a perfect cube. Then, $\sqrt[3]{40} = \sqrt[3]{8 \cdot 5}$ $= \sqrt[3]{8} \cdot \sqrt[3]{5}$ Product rule $= 2\sqrt[3]{5}$ $\sqrt[3]{8} = 2$ **Method 2:** Begin by using a factor tree to find the prime factorization of 40. $40 = 2^3 \cdot 5$ $\sqrt[3]{40} = \sqrt[3]{2^3 \cdot 5}$ $= \sqrt[3]{2^3} \cdot \sqrt[3]{5}$ Product rule $= 2\sqrt[3]{5}$ $\sqrt[3]{2^3} = 2$
Quotient Rule for Higher Roots If $\sqrt[n]{a}$ and $\sqrt[n]{b}$ are real numbers, $b \neq 0$ and n is a natural number, then $\sqrt[n]{\dfrac{a}{b}} = \dfrac{\sqrt[n]{a}}{\sqrt[n]{b}}$. (p. 547)	$\sqrt[4]{\dfrac{32}{81}} = \dfrac{\sqrt[4]{32}}{\sqrt[4]{81}} = \dfrac{\sqrt[4]{16} \cdot \sqrt[4]{2}}{3} = \dfrac{2\sqrt[4]{2}}{3}$
Simplifying Higher Roots with Variables in the Radicand If a is a nonnegative number and m and n are integers such that $n > 1$, then $\sqrt[n]{a^m} = a^{m/n}$. (p. 548)	Simplify $\sqrt[4]{a^{12}}$. $\sqrt[4]{a^{12}} = a^{12/4} = a^3$
If the exponent does not divide evenly by the index, we can use two methods for simplifying the radical expression. If a is a nonnegative number and m and n are integers such that $n > 1$, then i) **Method 1:** Use the product rule. To simplify $\sqrt[n]{a^m}$, write a^m as the product of two factors so that the exponent of one of the factors is the *largest* number less than m that is divisible by n (the index). ii) **Method 2:** Use the quotient and remainder (presented in Section 9.3). (p. 549)	i) Simplify $\sqrt[5]{c^{17}}$. $\sqrt[5]{c^{17}} = \sqrt[5]{c^{15} \cdot c^2}$ 15 is the largest number less than 17 that is divisible by 5. $= \sqrt[5]{c^{15}} \cdot \sqrt[5]{c^2}$ Product rule $= c^{15/5} \cdot \sqrt[5]{c^2}$ $= c^3 \sqrt[5]{c^2}$ $15 \div 5 = 3$ ii) Simplify $\sqrt[4]{m^{11}}$. $\sqrt[4]{m^{11}} = m^2 \sqrt[4]{m^3}$ $11 \div 4$ gives a quotient of 2 and a remainder of 3.

9.5 Adding, Subtracting, and Multiplying Radicals

Definition/Procedure	Example
Like radicals have the same index and the same radicand. In order to add or subtract radicals, they must be like radicals. **Steps for Adding and Subtracting Radicals** 1) Write each radical expression in simplest form. 2) Combine like radicals. (p. 555)	Perform the operations, and simplify. a) $5\sqrt{2} + 9\sqrt{7} - 3\sqrt{2} + 4\sqrt{7}$ $= 2\sqrt{2} + 13\sqrt{7}$ b) $\sqrt{72} + \sqrt{18} - \sqrt{45}$ $= \sqrt{36} \cdot \sqrt{2} + \sqrt{9} \cdot \sqrt{2} - \sqrt{9} \cdot \sqrt{5}$ $= 6\sqrt{2} + 3\sqrt{2} - 3\sqrt{5}$ $= 9\sqrt{2} - 3\sqrt{5}$

Definition/Procedure	Example
Combining Multiplication, Addition, and Subtraction of Radicals Multiply expressions containing radicals using the same techniques that are used for multiplying polynomials. (p. 556)	Multiply and simplify. a) $\sqrt{m}(\sqrt{2m} + \sqrt{n})$ $= \sqrt{m} \cdot \sqrt{2m} + \sqrt{m} \cdot \sqrt{n}$ Distribute. $= \sqrt{2m^2} + \sqrt{mn}$ Multiply. $= m\sqrt{2} + \sqrt{mn}$ Simplify. b) $(\sqrt{k} + \sqrt{6})(\sqrt{k} - \sqrt{2})$ Since we are multiplying two binomials, multiply using FOIL. $(\sqrt{k} + \sqrt{6})(\sqrt{k} - \sqrt{2})$ FOIL $= \sqrt{k} \cdot \sqrt{k} - \sqrt{2} \cdot \sqrt{k} + \sqrt{6} \cdot \sqrt{k} - \sqrt{6} \cdot \sqrt{2}$ $= k^2 - \sqrt{2k} + \sqrt{6k} - \sqrt{12}$ Product rule $= k^2 - \sqrt{2k} + \sqrt{6k} - 2\sqrt{3}$ $\sqrt{12} = 2\sqrt{3}$
Squaring a Radical Expression with Two Terms To square a binomial we can either use FOIL or one of the special formulas from Chapter 6: $(a + b)^2 = a^2 + 2ab + b^2$ $(a - b)^2 = a^2 - 2ab + b^2$ (p. 557)	$(\sqrt{7} + 5)^2 = (\sqrt{7})^2 + 2(\sqrt{7})(5) + (5)^2$ $= 7 + 10\sqrt{7} + 25$ $= 32 + 10\sqrt{7}$
Multiply $(a + b)(a - b)$ To multiply binomials of the form $(a + b)(a - b)$ use the formula $(a + b)(a - b) = a^2 - b^2$. (p. 558)	$(3 + \sqrt{10})(3 - \sqrt{10}) = (3)^2 - (\sqrt{10})^2$ $= 9 - 10$ $= -1$

9.6 Dividing Radicals

The process of eliminating radicals from the denominator of an expression is called **rationalizing the denominator.** First, we give examples of rationalizing denominators containing one term. (p. 562)	Rationalize the denominator of each expression. a) $\dfrac{9}{\sqrt{2}} = \dfrac{9}{\sqrt{2}} \cdot \dfrac{\sqrt{2}}{\sqrt{2}} = \dfrac{9\sqrt{2}}{2}$ b) $\dfrac{5}{\sqrt[3]{2}} = \dfrac{5}{\sqrt[3]{2}} \cdot \dfrac{\sqrt[3]{2^2}}{\sqrt[3]{2^2}} = \dfrac{5\sqrt[3]{2^2}}{\sqrt[3]{2^3}} = \dfrac{5\sqrt[3]{4}}{2}$
The **conjugate** of an expression of the form $a + b$ is $a - b$. (p. 567)	$\sqrt{11} - 4$ conjugate: $\sqrt{11} + 4$ $-8 + \sqrt{5}$ conjugate: $-8 - \sqrt{5}$
Rationalizing a Denominator with Two Terms If the denominator of an expression contains two terms, including one or two square roots, then to rationalize the denominator, we multiply the numerator and denominator of the expression by the conjugate of the denominator. (p. 568)	Rationalize the denominator of $\dfrac{4}{\sqrt{2} - 3}$. $\dfrac{4}{\sqrt{2} - 3} = \dfrac{4}{\sqrt{2} - 3} \cdot \dfrac{\sqrt{2} + 3}{\sqrt{2} + 3}$ Multiply by the conjugate of the denominator. $= \dfrac{4(\sqrt{2} + 3)}{(\sqrt{2})^2 - (3)^2}$ $(a + b)(a - b) = a^2 - b^2$ $= \dfrac{4(\sqrt{2} + 3)}{2 - 9}$ Square the terms. $= \dfrac{4(\sqrt{2} + 3)}{-7} = -\dfrac{4\sqrt{2} + 12}{7}$

Definition/Procedure	Example
9.7 Solving Radical Equations	
Solving Radical Equations Containing Square Roots	Solve $t = 2 + \sqrt{2t - 1}$.
Step 1: Get a radical on a side by itself. *Step 2:* Square both sides of the equation to eliminate a radical. *Step 3:* Combine like terms on each side of the equation. *Step 4:* If the equation still contains a radical, repeat Steps 1–3. *Step 5:* Solve the equation. *Step 6:* Check the proposed solutions *in the original equation,* and discard extraneous solutions. **(p. 579)**	$t - 2 = \sqrt{2t - 1}$ Get the radical by itself. $(t - 2)^2 = (\sqrt{2t - 1})^2$ Square both sides. $t^2 - 4t + 4 = 2t - 1$ $t^2 - 6t + 5 = 0$ Get all terms on the same side. $(t - 5)(t - 1) = 0$ Factor. $t - 5 = 0$ or $t - 1 = 0$ $t = 5$ or $t = 1$ Check $t = 5$ and $t = 1$ in the *original* equation. $t = 5: t = 2 + \sqrt{2t - 1}$ $t = 1: t = 2 + \sqrt{2t - 1}$ $5 \stackrel{?}{=} 2 + \sqrt{2(5) - 1}$ $1 \stackrel{?}{=} 2 + \sqrt{2(1) - 1}$ $5 \stackrel{?}{=} 2 + \sqrt{9}$ $1 \stackrel{?}{=} 2 + 1$ $5 = 2 + 3$ $1 = 3$ True False $t = 5$ *is* a solution, but $t = 1$ is *not* because $t = 1$ does not satisfy the original equation. The solution set is $\{5\}$.
9.8 Complex Numbers	
Definition of *i*: $i = \sqrt{-1}$ Therefore, $i^2 = -1$. A **complex number** is a number of the form $a + bi$, where a and b are real numbers. a is called the **real part** and b is called the **imaginary part**. The set of real numbers is a subset of the set of complex numbers. **(p. 588)**	*Examples of complex numbers:* $-2 + 7i$ 5 (since it can be written $5 + 0i$) $4i$ (since it can be written $0 + 4i$)
Simplifying Square Roots with Negative Radicands Use the product rule and $i = \sqrt{-1}$. **(p. 589)**	Simplify $\sqrt{-25}$. $\sqrt{-25} = \sqrt{-1} \cdot \sqrt{25}$ $= i \cdot 5$ $= 5i$
When multiplying or dividing square roots with negative radicands, write each radical in terms of *i* first. **(p. 589)**	Multiply $\sqrt{-12} \cdot \sqrt{-3}$. $\sqrt{-12} \cdot \sqrt{-3} = i\sqrt{12} \cdot i\sqrt{3} = i^2\sqrt{36}$ $= -1 \cdot 6 = -6$
Adding and Subtracting Complex Numbers To add and subtract complex numbers, combine the real parts and combine the imaginary parts. **(p. 590)**	Subtract $(10 + 7i) - (-2 + 4i)$. $(10 + 7i) - (-2 + 4i) = 10 + 7i + 2 - 4i$ $= 12 + 3i$
Multiply complex numbers like we multiply polynomials. Remember to replace i^2 with -1. **(p. 590)**	Multiply and simplify. a) $4(9 + 5i) = 36 + 20i$ F O I L b) $(-3 + i)(2 - 7i) = -6 + 21i + 2i - 7i^2$ $= -6 + 23i - 7(-1)$ $= -6 + 23i + 7$ $= 1 + 23i$

Definition/Procedure	Example
Complex Conjugates 1) The conjugate of $a + bi$ is $a - bi$. 2) The product of $a + bi$ and $a - bi$ is a real number. 3) Find the product $(a + bi)(a - bi)$ using FOIL or recall that $(a + bi)(a - bi) = a^2 + b^2$. (p. 591)	Multiply $-5 - 3i$ by its conjugate. The conjugate of $-5 - 3i$ is $-5 + 3i$. Use $(a + bi)(a - bi) = a^2 + b^2$. $$(-5 - 3i)(-5 + 3i) = (-5)^2 + (3)^2$$ $$= 25 + 9$$ $$= 34$$
Dividing Complex Numbers To divide complex numbers, multiply the numerator and denominator by the *conjugate of the denominator*. Write the quotient in the form $a + bi$. (p. 592)	Divide $\dfrac{6i}{2 + 5i}$. Write the result in the form $a + bi$. $$\dfrac{6i}{2 + 5i} = \dfrac{6i}{2 + 5i} \cdot \dfrac{(2 - 5i)}{(2 - 5i)}$$ $$= \dfrac{12i - 30i^2}{(2)^2 + (5)^2}$$ $$= \dfrac{12i - 30(-1)}{29}$$ $$= \dfrac{30}{29} + \dfrac{12}{29}i$$
Simplify Powers of i We can simplify powers of i using $i^2 = -1$. (p. 593)	Simplify i^{14}. $i^{14} = (i^2)^7$ Power rule $= (-1)^7$ $i^2 = -1$ $= -1$ Simplify.

Chapter 9: Review Exercises

(9.1) Find each root, if possible.

1) $\sqrt{\dfrac{169}{4}}$
2) $\sqrt{-16}$
3) $-\sqrt{81}$
4) $\sqrt[5]{32}$
5) $\sqrt[3]{-1}$
6) $-\sqrt[4]{81}$
7) $\sqrt[6]{-64}$
8) $\sqrt{9 - 16}$

Simplify. Use absolute values when necessary.

9) $\sqrt{(-13)^2}$
10) $\sqrt[5]{(-8)^5}$
11) $\sqrt{p^2}$
12) $\sqrt[6]{c^6}$
13) $\sqrt[3]{h^3}$
14) $\sqrt[4]{(y + 7)^4}$

15) $f(x) = \sqrt{5x + 3}$

   a) Find $f(4)$.
   b) Find $f(p)$.
   c) Find the domain of f.

16) $g(x) = \sqrt[3]{x - 12}$

   a) Find $g(4)$.
   b) Find $g(t + 7)$.
   c) Find the domain of g.

17) Graph $k(x) = \sqrt{x + 4}$.
18) Graph $h(x) = -\sqrt[3]{x}$.

(9.2)

19) Explain how to write $8^{2/3}$ in radical form.

20) Explain how to eliminate the negative from the exponent in an expression like $9^{-1/2}$.

Evaluate.

21) $36^{1/2}$
22) $32^{1/5}$
23) $\left(\dfrac{27}{125}\right)^{1/3}$
24) $-16^{1/4}$
25) $32^{3/5}$
26) $\left(\dfrac{64}{27}\right)^{2/3}$
27) $81^{-1/2}$
28) $\left(\dfrac{1}{27}\right)^{-1/3}$
29) $81^{-3/4}$
30) $1000^{-2/3}$
31) $\left(\dfrac{27}{1000}\right)^{-2/3}$
32) $\left(\dfrac{25}{16}\right)^{-3/2}$

From this point forward, assume all variables represent positive real numbers.

Simplify completely. The answer should contain only positive exponents.

33) $3^{6/7} \cdot 3^{8/7}$
34) $(169^4)^{1/8}$
35) $(8^{1/5})^{10}$
36) $\dfrac{8^2}{8^{11/3}}$

37) $\dfrac{7^2}{7^{5/3} \cdot 7^{1/3}}$

38) $(2k^{-5/6})(3k^{1/2})$

39) $(64a^4 b^{12})^{5/6}$

40) $\left(\dfrac{t^4 u^3}{7t^7 u^5}\right)^{-2}$

41) $\left(\dfrac{81c^{-5} d^9}{16c^{-1} d^2}\right)^{-1/4}$

Rewrite each radical in exponential form, then simplify. Write the answer in simplest (or radical) form.

42) $\sqrt[4]{36^2}$

43) $\sqrt[12]{27^4}$

44) $(\sqrt{17})^2$

45) $\sqrt[3]{7^3}$

46) $\sqrt[5]{t^{20}}$

47) $\sqrt[4]{k^{28}}$

48) $\sqrt{x^{10}}$

49) $\sqrt{w^6}$

(9.3) Simplify completely.

50) $\sqrt{28}$

51) $\sqrt{1000}$

52) $\dfrac{\sqrt{63}}{\sqrt{7}}$

53) $\sqrt{\dfrac{18}{49}}$

54) $\dfrac{\sqrt{48}}{\sqrt{121}}$

55) $\sqrt{k^{12}}$

56) $\sqrt{\dfrac{40}{m^4}}$

57) $\sqrt{x^9}$

58) $\sqrt{y^5}$

59) $\sqrt{45t^2}$

60) $\sqrt{80n^{21}}$

61) $\sqrt{72x^7 y^{13}}$

62) $\sqrt{\dfrac{m^{11}}{36n^2}}$

Perform the indicated operation, and simplify.

63) $\sqrt{5} \cdot \sqrt{3}$

64) $\sqrt{6} \cdot \sqrt{15}$

65) $\sqrt{2} \cdot \sqrt{12}$

66) $\sqrt{b^7} \cdot \sqrt{b^3}$

67) $\sqrt{11x^5} \cdot \sqrt{11x^8}$

68) $\sqrt{5a^2 b} \cdot \sqrt{15a^6 b^4}$

69) $\dfrac{\sqrt{200k^{21}}}{\sqrt{2k^5}}$

70) $\dfrac{\sqrt{63c^{17}}}{\sqrt{7c^9}}$

(9.4) Simplify completely.

71) $\sqrt[3]{16}$

72) $\sqrt[3]{250}$

73) $\sqrt[4]{48}$

74) $\sqrt[3]{\dfrac{81}{3}}$

75) $\sqrt[4]{z^{24}}$

76) $\sqrt[5]{p^{40}}$

77) $\sqrt[3]{a^{20}}$

78) $\sqrt[5]{x^{14} y^7}$

79) $\sqrt[3]{16z^{15}}$

80) $\sqrt[3]{80m^{17} n^{10}}$

81) $\sqrt[4]{\dfrac{h^{12}}{81}}$

82) $\sqrt[5]{\dfrac{c^{22}}{32d^{10}}}$

Perform the indicated operation, and simplify.

83) $\sqrt[3]{3} \cdot \sqrt[3]{7}$

84) $\sqrt[3]{25} \cdot \sqrt[3]{10}$

85) $\sqrt[4]{4t^7} \cdot \sqrt[4]{8t^{10}}$

86) $\sqrt[5]{\dfrac{x^{21}}{x^{16}}}$

87) $\sqrt[3]{n} \cdot \sqrt{n}$

88) $\dfrac{\sqrt[4]{a^3}}{\sqrt[3]{a}}$

(9.5) Perform the operations, and simplify.

89) $8\sqrt{5} + 3\sqrt{5}$

90) $\sqrt{125} + \sqrt{80}$

91) $\sqrt{80} - \sqrt{48} + \sqrt{20}$

92) $9\sqrt[3]{72} - 8\sqrt[3]{9}$

93) $3p\sqrt{p} - 7\sqrt{p^3}$

94) $9n\sqrt{n} - 4\sqrt{n^3}$

95) $10d^2 \sqrt{8d} - 32d\sqrt{2d^3}$

96) $\sqrt{6}(\sqrt{7} - \sqrt{6})$

97) $3\sqrt{k}(\sqrt{20k} + \sqrt{2})$

98) $(5 - \sqrt{3})(2 + \sqrt{3})$

99) $(\sqrt{2r} + 5\sqrt{s})(3\sqrt{s} + 4\sqrt{2r})$

100) $(2\sqrt{5} - 4)^2$

101) $(1 + \sqrt{y+1})^2$

102) $(\sqrt{6} - \sqrt{5})(\sqrt{6} + \sqrt{5})$

(9.6) Rationalize the denominator of each expression.

103) $\dfrac{14}{\sqrt{3}}$

104) $\dfrac{20}{\sqrt{6}}$

105) $\dfrac{\sqrt{18k}}{\sqrt{n}}$

106) $\dfrac{\sqrt{45}}{\sqrt{m^5}}$

107) $\dfrac{7}{\sqrt[3]{2}}$

108) $-\dfrac{15}{\sqrt[3]{9}}$

109) $\dfrac{\sqrt[3]{x^2}}{\sqrt[3]{y}}$

110) $\sqrt[4]{\dfrac{3}{4k^2}}$

111) $\dfrac{2}{3 + \sqrt{3}}$

112) $\dfrac{z-4}{\sqrt{z}+2}$

Simplify completely.

113) $\dfrac{8 - 24\sqrt{2}}{8}$

114) $\dfrac{-\sqrt{48} - 6}{10}$

(9.7) Solve.

115) $\sqrt{x+8} = 3$

116) $10 - \sqrt{3r-5} = 2$

117) $\sqrt{3j+4} = -\sqrt{4j-1}$

118) $\sqrt[3]{6d-14} = -2$

119) $a = \sqrt{a+8} - 6$

120) $1 + \sqrt{6m+7} = 2m$

121) $\sqrt{4a+1} - \sqrt{a-2} = 3$

122) $\sqrt{6x+9} - \sqrt{2x+1} = 4$

123) Solve for V: $r = \sqrt{\dfrac{3V}{\pi h}}$

124) The velocity of a wave in shallow water is given by $c = \sqrt{gH}$, where g is the acceleration due to gravity (32 ft/sec^2), and H is the depth of the water (in feet). Find the velocity of a wave in 10 ft of water.

(9.8) Simplify.

125) $\sqrt{-49}$

126) $\sqrt{-8}$

127) $\sqrt{-2} \cdot \sqrt{-8}$

128) $\sqrt{-6} \cdot \sqrt{-3}$

Perform the indicated operations.

129) $(2 + i) + (10 - 4i)$

130) $(4 + 3i) - (11 - 4i)$

131) $\left(\dfrac{4}{5} - \dfrac{1}{3}i\right) - \left(\dfrac{1}{2} + i\right)$

132) $\left(-\dfrac{3}{8} - 2i\right) + \left(\dfrac{5}{8} + \dfrac{3}{2}i\right) - \left(\dfrac{1}{4} - \dfrac{1}{2}i\right)$

Multiply and simplify.

133) $5(-6 + 7i)$

134) $-8i(4 + 3i)$

135) $3i(-7 + 12i)$

136) $(3 - 4i)(2 + i)$

137) $(4 - 6i)(3 - 6i)$

138) $\left(\dfrac{1}{5} - \dfrac{2}{3}i\right)\left(\dfrac{3}{2} - \dfrac{2}{3}i\right)$

Identify the conjugate of each complex number, then multiply the number and its conjugate.

139) $2 - 7i$

140) $-2 + 3i$

Divide. Write the quotient in the form $a + bi$.

141) $\dfrac{6}{2 + 5i}$

142) $\dfrac{-12}{4 - 3i}$

143) $\dfrac{8}{i}$

144) $\dfrac{4i}{1 - 3i}$

145) $\dfrac{9 - 4i}{6 - i}$

146) $\dfrac{5 - i}{-2 + 6i}$

Simplify.

147) i^{10}

148) i^{51}

149) i^{33}

150) i^{24}

Chapter 9: Test

Find each real root, if possible.

1) $\sqrt{144}$

2) $\sqrt[3]{-27}$

3) $\sqrt{-16}$

Simplify. Use absolute values when necessary.

4) $\sqrt[4]{w^4}$

5) $\sqrt[5]{(-19)^5}$

6) Let $h(c) = \sqrt{3c + 7}$.

  a) Find $h(-2)$.

  b) Find $h(a - 4)$.

  c) Determine the domain of h.

7) Determine the domain of $f(x) = \sqrt{x - 2}$, and graph the function.

Evaluate.

8) $16^{1/4}$

9) $27^{4/3}$

10) $(49)^{-1/2}$

11) $\left(\dfrac{8}{125}\right)^{-2/3}$

From this point forward, assume all variables represent positive real numbers.

Simplify completely. The answer should contain only positive exponents.

12) $m^{3/8} \cdot m^{1/4}$

13) $\dfrac{35a^{1/6}}{14a^{5/6}}$

14) $(2x^{3/10}y^{-2/5})^{-5}$

Simplify completely.

15) $\sqrt{75}$

16) $\sqrt[3]{48}$

17) $\sqrt{\dfrac{24}{2}}$

Simplify completely.

18) $\sqrt{y^6}$

19) $\sqrt[4]{p^{24}}$

20) $\sqrt{t^9}$

21) $\sqrt{63m^5n^8}$

22) $\sqrt[3]{c^{23}}$

23) $\sqrt[3]{\dfrac{a^{14}b^7}{27}}$

Perform the operations, and simplify.

24) $\sqrt{3} \cdot \sqrt{12}$

25) $\sqrt[3]{z^4} \cdot \sqrt[3]{z^6}$

26) $\dfrac{\sqrt{120w^{15}}}{\sqrt{2w^4}}$

27) $9\sqrt{7} - 3\sqrt{7}$

28) $\sqrt{12} - \sqrt{108} + \sqrt{18}$

29) $2h^3\sqrt[4]{h} - 16\sqrt[4]{h^{13}}$

Multiply and simplify.

30) $\sqrt{6}(\sqrt{2} - 5)$

31) $(3 - 2\sqrt{5})(\sqrt{2} + 1)$

32) $(\sqrt{7} + \sqrt{3})(\sqrt{7} - \sqrt{3})$

33) $(\sqrt{2p + 1} + 2)^2$

34) $2\sqrt{t}(\sqrt{t} - \sqrt{3u})$

Rationalize the denominator of each expression.

35) $\dfrac{2}{\sqrt{5}}$

36) $\dfrac{8}{\sqrt{7} + 3}$

37) $\dfrac{\sqrt{6}}{\sqrt{a}}$

38) $\dfrac{5}{\sqrt[3]{9}}$

39) Simplify completely. $\dfrac{2 - \sqrt{48}}{2}$

Solve.

40) $\sqrt{5h + 4} = 3$

41) $z = \sqrt{1 - 4z} - 5$

42) $\sqrt[3]{n - 5} - \sqrt[3]{2n - 18} = 0$

43) $\sqrt{3k+1} - \sqrt{2k-1} = 1$

44) In the formula $r = \sqrt{\dfrac{V}{\pi h}}$, V represents the volume of a cylinder, h represents the height of the cylinder, and r represents the radius.

   a) A cylindrical container has a volume of 72π in^3. It is 8 in. high. What is the radius of the container?

   b) Solve the formula for V.

Simplify.

45) $\sqrt{-64}$

46) $\sqrt{-45}$

47) i^{19}

Perform the indicated operation, and simplify. Write the answer in the form $a + bi$.

48) $(-10 + 3i) - (6 + i)$

49) $(2 - 7i)(-1 + 3i)$

50) $\dfrac{8 + i}{2 - 3i}$

Chapter 9: Cumulative Review for Chapters 1–9

1) Combine like terms.

   $4x - 3y + 9 - \dfrac{2}{3}x + y - 1$

2) Write in scientific notation.

   8,723,000

3) Solve $3(2c - 1) + 7 = 9c + 5(c + 2)$.

4) Graph $3x + 2y = 12$.

5) Write the equation of the line containing the points $(5, 3)$ and $(1, -2)$. Write the equation in slope-intercept form.

6) Solve by substitution.

   $2x + 7y = -12$
   $x - 4y = -6$

7) Multiply.

   $(5p^2 - 2)(3p^2 - 4p - 1)$

8) Divide.

   $\dfrac{8n^3 - 1}{2n - 1}$

Factor completely.

9) $4w^2 + 5w - 6$

10) $8 - 18t^2$

11) Solve $6y^2 - 4 = 5y$.

12) Solve $3(k^2 + 20) - 4k = 2k^2 + 11k + 6$.

13) *Write an equation, and solve.* The width of a rectangle is 5 in. less than its length. The area is 84 in^2. Find the dimensions of the rectangle.

Perform the operations, and simplify.

14) $\dfrac{5a^2 + 3}{a^2 + 4a} - \dfrac{3a - 2}{a + 4}$

15) $\dfrac{10m^2}{9n} \cdot \dfrac{6n^2}{35m^5}$

16) Solve $\dfrac{3}{r^2 + 8r + 15} - \dfrac{4}{r + 3} = 1$.

17) Solve $|6g + 1| \geq 11$. Write the answer in interval notation.

18) Solve using Gaussian elimination.

   $x + 3y + z = 3$
   $2x - y - 5z = -1$
   $-x + 2y + 3z = 0$

19) Simplify. Assume all variables represent nonnegative real numbers.

   a) $\sqrt{500}$

   b) $\sqrt[3]{56}$

   c) $\sqrt{p^{10}q^7}$

   d) $\sqrt[4]{32a^{15}}$

20) Evaluate.

   a) $81^{1/2}$

   b) $8^{4/3}$

   c) $(27)^{-1/3}$

21) Multiply and simplify $2\sqrt{3}(5 - \sqrt{3})$.

22) Rationalize the denominator. Assume the variables represent positive real numbers.

   a) $\sqrt{\dfrac{20}{50}}$

   b) $\dfrac{6}{\sqrt[3]{2}}$

   c) $\dfrac{x}{\sqrt[3]{y^2}}$

   d) $\dfrac{\sqrt{a} - 2}{1 - \sqrt{a}}$

23) Solve.

   a) $\sqrt{2b - 1} + 7 = 6$

   b) $\sqrt{3z + 10} = 2 - \sqrt{z + 4}$

24) Simplify.

   a) $\sqrt{-49}$

   b) $\sqrt{-56}$

   c) i^8

25) Perform the indicated operation, and simplify. Write the answer in the form $a + bi$.

   a) $(-3 + 4i) + (5 + 3i)$

   b) $(3 + 6i)(-2 + 7i)$

   c) $\dfrac{2 - i}{-4 + 3i}$

CHAPTER 10

Quadratic Equations and Functions

OUTLINE

Study Strategies: Developing Financial Literacy

10.1 The Square Root Property and Completing the Square

10.2 The Quadratic Formula

Putting It All Together

10.3 Equations in Quadratic Form

10.4 Formulas and Applications

10.5 Quadratic Functions and Their Graphs

10.6 Applications of Quadratic Functions and Graphing Other Parabolas

10.7 Quadratic and Rational Inequalities

Group Activity

emPOWERme: Determine Your Saving Style

Math at Work:
Economist

Anyone who has tried to keep track of their personal finances—managing checking accounts, credit cards, car loans, rent, utilities, and often more—knows how complicated money matters can be even for just one person. Jeremy Norton's career accomplishments, then, seem even more impressive. As an economist writing for newspapers, magazines, and blogs, Jeremy needs to understand how financial forces across the country, and even around the world, interact and affect each other.

"Economics connects what happens at the local grocery store to what is happening at oil fields on the other side of the globe," Jeremy explains. "Without mathematical models and sophisticated equations tracking economic forces, we wouldn't be able to understand or predict how prices are going to rise and fall, whether companies will start to hire, and so on."

Jeremy says that while some elements of his job are very complicated, the fundamentals of economics are the same ones individuals use to manage their personal bottom lines. "The truth is," Jeremy says, "that the most valuable economic lessons I learned were the ones I picked up outside the classroom, working to keep my expenses under control so I could pay my tuition."

We will introduce quadratic equations and functions in this chapter and also introduce strategies you can use to keep your personal finances in order.

 **Study Strategies** Developing Financial Literacy

Would you start a road trip without looking at a map? Would you start cooking dinner by putting ingredients in a bowl without any idea what you were trying to make? The point is that for any project to succeed, you need a plan. When it comes to your personal finances, that plan is a budget. A budget will help you keep your spending under control and remain on track to meet your larger financial goals. Use the strategies below to create an effective budget.

- Write down your financial goals. These might include paying off a car loan, saving money for a big purchase, or avoiding falling into credit card debt.

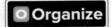
- Figure out how you spend your money. For a week, write down or use a phone app to keep track of *every* purchase you make, large and small. This will help you understand where your money goes on a daily basis.
- Make a list of everything you know you will need to spend money on in the coming year. This list should include things like tuition, insurance payments, your rent or mortgage, and so forth.
- Make a list of all your income for the coming year, such as the salary you earn at your job and any financial aid you might receive.

- Create your budget by adding up all your sources of income and everything you spend money on, including small, daily purchases and your larger expenses.
- If your spending is greater than your income, find ways to reduce your costs. The emPOWERme on page 691 can help you save money.
- You could also look for ways to increase your income, such as by finding a part-time job, assuming your schedule allows.

- Review your finances every month, and adjust your budget to ensure that it is accurate. For example, if your rent goes up, you want to be sure your budget accounts for that.

- Revisit your financial goals. Are you getting closer to achieving them? If not, how can you change your budget to improve this situation?
- Be on the lookout for issues that might create budget chaos, such as credit cards with interest rates that spike after a certain amount of time.

Chapter 10 POWER Plan

P Prepare | O Organize

What are your goals for Chapter 10?	How can you accomplish each goal?
1 Be prepared before and during class.	• _____ • _____ • _____ • _____
2 Understand the homework to the point where you could do it without needing any help or hints.	• _____ • _____ • _____
3 Use the P.O.W.E.R. framework to help you analyze how you save money: *Determine Your Saving Style*.	• _____ • _____ • _____
4 Write your own goal. _____	• _____

What are your objectives for Chapter 10?	How can you accomplish each objective?
1 Be able to solve quadratic equations of different forms using several different methods. Be able to choose the most efficient method.	• Learn the different methods for **Solving Quadratic Equations**. • Know when to use *factoring, the square root property, completing the square,* and the *quadratic formula*.
2 Be able to use the quadratic formula to solve an applied problem, and use the discriminant to determine the type and number of solutions.	• Learn the different types of solutions you can get for a quadratic equation by calculating the value of the *discriminant*. • Review the procedure for **Solving Applied Problems**.
3 Be able to solve equations that are in quadratic form, including equations with radicals.	• Review the procedures for **Eliminating Fractions in Equations**. • Review factoring methods. • Learn how to use *substitution* to solve quadratic equations.
4 Be able to solve several different types of applied problems using any method in this chapter.	• Learn the procedure for **Solving Applied Problems**. • Always draw a diagram or chart, when appropriate. • Learn the formula for problems involving volume and area.
5 Be able to graph parabolas given different forms of an equation. Be able to identify several things about the graph, including the vertex and max/min values if it is a quadratic function.	• Learn how to graph quadratic functions and understand how the graph shifts based on the information given in the equation. • Be able to identify the *vertex* of a *parabola* from its equation. • Recognize which equations represent parabolas that open vertically and which represent parabolas that open horizontally.

6 Be able to solve quadratic inequalities by graphing or using test points. Also, be able to solve inequalities of higher degree.	• Learn the procedure for **Graphing a Quadratic Inequality**. • Know how to pick test points to determine the solution to an inequality. • Review interval notation.
7 Write your own goal. _____ _____	• _____

	W Work	Read Sections 10.1 to 10.7, and complete the exercises.	
E Evaluate	Complete the Chapter Review and Chapter Test. How did you do?	**R Rethink**	• How are saving money and math related? • Does budgeting money require good math skills? • Can you make a list of all the methods used to factor a quadratic equation? Do you know when each method can be used? • Which type of quadratic equation is hardest for you to graph? • Which type of applied problems use the vertex to find the answer?

10.1 The Square Root Property and Completing the Square

P Prepare | O Organize

What are your objectives for Section 10.1?	How can you accomplish each objective?
1 Solve an Equation of the Form $x^2 = k$	• Write the definition of a *quadratic equation* in your own words. • Understand the *square root property*. • Complete the given example on your own. • Complete You Try 1.
2 Solve an Equation of the Form $(ax + b)^2 = k$	• Be able to recognize a binomial that is being squared. • Understand the *square root property*. • Review the procedure for solving equations. • Complete the given examples on your own. • Complete You Trys 2 and 3.

What are your objectives for Section 10.1?	How can you accomplish each objective?
3 Use the Distance Formula	• Understand the definition of *the distance formula*. • Review the **Pythagorean theorem.** • Complete the given example on your own. • Complete You Try 4.
4 Complete the Square for an Expression of the Form $x^2 + bx$	• Learn the procedure for **Completing the Square for $x^2 + bx$.** • Review factoring of *perfect square trinomials*. • Complete the given examples on your own. • Complete You Trys 5 and 6.
5 Solve a Quadratic Equation by Completing the Square	• Learn the procedure for **Solving a Quadratic Equation by Completing the Square.** • Review the *square root property*. • Complete the given example on your own. • Complete You Try 7.

W Work Read the explanations, follow the examples, take notes, and complete the You Trys.

We defined a quadratic equation in Chapter 7. Let's restate the definition:

Definition

A **quadratic equation** can be written in the form $ax^2 + bx + c = 0$, where a, b, and c are real numbers and $a \neq 0$.

W Hint
Look at Section 7.4 if you need a detailed review of solving quadratic equations by factoring.

In Section 7.4 we learned how to solve quadratic equations by factoring. For example, we can use the zero product rule to solve $x^2 - 3x - 40 = 0$.

$$x^2 - 3x - 40 = 0$$
$$(x - 8)(x + 5) = 0 \quad \text{Factor.}$$
$$x - 8 = 0 \quad \text{or} \quad x + 5 = 0 \quad \text{Set each factor equal to zero.}$$
$$x = 8 \quad \text{or} \quad x = -5 \quad \text{Solve.}$$

The solution set is $\{-5, 8\}$.

It is not easy to solve all quadratic equations by factoring, however. Therefore, we need to learn other methods. In this chapter, we will discuss three more methods for solving quadratic equations. Let's begin with the square root property.

1 Solve an Equation of the Form $x^2 = k$

Look at the equation $x^2 = 9$. We can solve this equation by factoring, like this:

$$x^2 = 9$$
$$x^2 - 9 = 0 \quad \text{Get all terms on the same side.}$$
$$(x + 3)(x - 3) = 0 \quad \text{Factor.}$$
$$x + 3 = 0 \quad \text{or} \quad x - 3 = 0 \quad \text{Set each factor equal to zero.}$$
$$x = -3 \quad \text{or} \quad x = 3 \quad \text{Solve.}$$

The solution set is $\{-3, 3\}$.

Or, we can solve an equation like $x^2 = 9$ using the **square root property** as we will see in Example 1a).

> **Definition** The Square Root Property
> Let k be a constant. If $x^2 = k$, then $x = \sqrt{k}$ or $x = -\sqrt{k}$.
>
> (The solution is often written as $x = \pm\sqrt{k}$, read as "x equals plus or minus the square root of k.")

> **Note**
> We can use the square root property to solve an equation containing a squared quantity and a constant. To do so we will get the squared quantity containing the variable on one side of the equal sign and the constant on the other side.

EXAMPLE 1 Solve using the square root property.

a) $x^2 = 9$ b) $t^2 - 20 = 0$ c) $2a^2 + 21 = 3$

Solution

a)
$$x^2 = 9$$
$$x = \sqrt{9} \quad \text{or} \quad x = -\sqrt{9} \qquad \text{Square root property}$$
$$x = 3 \quad \text{or} \quad x = -3$$

The solution set is $\{-3, 3\}$. The check is left to the student.

An equivalent way to solve $x^2 = 9$ is to write it as

$$x^2 = 9$$
$$x = \pm\sqrt{9} \qquad \text{Square root property}$$
$$x = \pm 3$$

The solution set is $\{-3, 3\}$. We will use this approach when solving equations using the square root property.

W Hint
Be sure you are writing out each step as you are reading the example.

b) To solve $t^2 - 20 = 0$, begin by getting t^2 on a side by itself.

$$t^2 - 20 = 0$$
$$t^2 = 20 \qquad \text{Add 20 to each side.}$$
$$t = \pm\sqrt{20} \qquad \text{Square root property}$$
$$t = \pm\sqrt{4} \cdot \sqrt{5} \qquad \text{Product rule for radicals}$$
$$t = \pm 2\sqrt{5} \qquad \sqrt{4} = 2$$

Check:

$t = 2\sqrt{5}$:
$$t^2 - 20 = 0$$
$$(2\sqrt{5})^2 - 20 \stackrel{?}{=} 0$$
$$(4 \cdot 5) - 20 \stackrel{?}{=} 0$$
$$20 - 20 = 0 \checkmark$$

$t = -2\sqrt{5}$:
$$t^2 - 20 = 0$$
$$(-2\sqrt{5})^2 - 20 \stackrel{?}{=} 0$$
$$(4 \cdot 5) - 20 \stackrel{?}{=} 0$$
$$20 - 20 = 0 \checkmark$$

The solution set is $\{-2\sqrt{5}, 2\sqrt{5}\}$.

c) $2a^2 + 21 = 3$
$2a^2 = -18$ Subtract 21.
$a^2 = -9$ Divide by 2.
$a = \pm\sqrt{-9}$ Square root property
$a = \pm 3i$

Check:

W Hint
Remember: $i^2 = -1$.

$a = 3i$: $2a^2 + 21 = 3$
$2(3i)^2 + 21 \stackrel{?}{=} 3$
$2(9i^2) + 21 \stackrel{?}{=} 3$
$2(9)(-1) + 21 \stackrel{?}{=} 3$
$-18 + 21 = 3$ ✓

$a = -3i$: $2a^2 + 21 = 3$
$2(-3i)^2 + 21 \stackrel{?}{=} 3$
$2(9i^2) + 21 \stackrel{?}{=} 3$
$2(9)(-1) + 21 \stackrel{?}{=} 3$
$-18 + 21 = 3$ ✓

The solution set is $\{-3i, 3i\}$.

[**YOU TRY 1**] Solve using the square root property.

a) $p^2 = 100$ b) $w^2 - 32 = 0$ c) $3m^2 + 19 = 7$

Can we solve $(w - 4)^2 = 25$ using the square root property? Yes. The equation has a *squared quantity* and a *constant*.

2   Solve an Equation of the Form $(ax + b)^2 = k$

EXAMPLE 2   Solve $x^2 = 25$ and $(w - 4)^2 = 25$ using the square root property.

Solution

While the equation $(w - 4)^2 = 25$ has a *binomial* that is being squared, the two equations are actually in the same form.

$x^2 = 25$ $(w - 4)^2 = 25$
↑   ↑   ↑   ↑
x squared = constant $(w - 4)$ squared = constant

W Hint
The squared quantity must be isolated before taking the square root of each side.

Solve $x^2 = 25$:

$x^2 = 25$
$x = \pm\sqrt{25}$ Square root property
$x = \pm 5$

The solution set is $\{-5, 5\}$.

We solve $(w - 4)^2 = 25$ in the same way with some additional steps.

$(w - 4)^2 = 25$
$w - 4 = \pm\sqrt{25}$ Square root property
$w - 4 = \pm 5$

This means $w - 4 = 5$ or $w - 4 = -5$. Solve both equations.

$w - 4 = 5$ or $w - 4 = -5$
$w = 9$ or $w = -1$ Add 4 to each side.

Check:

$w = 9$: $(w - 4)^2 = 25$
$(9 - 4)^2 \stackrel{?}{=} 25$
$5^2 = 25$ ✓

$w = -1$: $(w - 4)^2 = 25$
$(-1 - 4)^2 \stackrel{?}{=} 25$
$(-5)^2 = 25$ ✓

The solution set is $\{-1, 9\}$.

[YOU TRY 2] Solve $(c + 6)^2 = 81$ using the square root property.

EXAMPLE 3 Solve.

a) $(3t + 4)^2 = 9$ b) $(2m - 5)^2 = 12$ c) $(z + 8)^2 + 11 = 7$

d) $(6k - 5)^2 + 20 = 0$

Solution

a) $(3t + 4)^2 = 9$
$3t + 4 = \pm\sqrt{9}$ Square root property
$3t + 4 = \pm 3$

This means $3t + 4 = 3$ or $3t + 4 = -3$. Solve both equations.

$3t + 4 = 3$ or $3t + 4 = -3$
$3t = -1$ $3t = -7$ Subtract 4 from each side.
$t = -\dfrac{1}{3}$ or $t = -\dfrac{7}{3}$ Divide by 3.

The solution set is $\left\{-\dfrac{7}{3}, -\dfrac{1}{3}\right\}$.

W Hint
Do you need to review how to simplify square roots?

b) $(2m - 5)^2 = 12$
$2m - 5 = \pm\sqrt{12}$ Square root property
$2m - 5 = \pm 2\sqrt{3}$ Simplify $\sqrt{12}$.
$2m = 5 \pm 2\sqrt{3}$ Add 5 to each side.
$m = \dfrac{5 \pm 2\sqrt{3}}{2}$ Divide by 2.

One solution is $\dfrac{5 + 2\sqrt{3}}{2}$, and the other is $\dfrac{5 - 2\sqrt{3}}{2}$.

The solution set, $\left\{\dfrac{5 - 2\sqrt{3}}{2}, \dfrac{5 + 2\sqrt{3}}{2}\right\}$, can also be written as $\left\{\dfrac{5 \pm 2\sqrt{3}}{2}\right\}$.

c) $(z + 8)^2 + 11 = 7$
$(z + 8)^2 = -4$ Subtract 11 from each side.
$z + 8 = \pm\sqrt{-4}$ Square root property
$z + 8 = \pm 2i$ Simplify $\sqrt{-4}$.
$z = -8 \pm 2i$ Subtract 8 from each side.

The check is left to the student. The solution set is $\{-8 - 2i, -8 + 2i\}$.

d) $(6k - 5)^2 + 20 = 0$
  $(6k - 5)^2 = -20$ Subtract 20 from each side.
  $6k - 5 = \pm\sqrt{-20}$ Square root property
  $6k - 5 = \pm 2i\sqrt{5}$ Simplify $\sqrt{-20}$.
  $6k = 5 \pm 2i\sqrt{5}$ Add 5 to each side.
  $k = \dfrac{5 \pm 2i\sqrt{5}}{6}$ Divide by 6.

The check is left to the student. The solution set is $\left\{\dfrac{5 - 2i\sqrt{5}}{6}, \dfrac{5 + 2i\sqrt{5}}{6}\right\}$.

[YOU TRY 3] Solve.

a) $(7q + 1)^2 = 36$
b) $(5a - 3)^2 = 24$
c) $(c - 7)^2 + 100 = 0$
d) $(2y + 3)^2 - 5 = -23$

Did you notice in Examples 1c), 3c), and 3d) that a complex number *and* its conjugate were the solutions to the equations? This will always be true provided that the variables in the equation have real number coefficients.

Note
If $a + bi$ is a solution of a quadratic equation having only real coefficients, then $a - bi$ is also a solution.

3 Use the Distance Formula

In mathematics, we sometimes need to find the distance between two points in a plane. The **distance formula** enables us to do that. We can use the Pythagorean theorem and the square root property to develop the distance formula.

Suppose we want to find the distance between any two points with coordinates (x_1, y_1) and (x_2, y_2) as pictured here. [We also include the point (x_2, y_1) in our drawing so that we get a right triangle.]

The lengths of the legs are a and b. The length of the hypotenuse is c. Our goal is to find the *distance* between (x_1, y_1) and (x_2, y_2), which is the same as finding the length of c.

How long is side a? $|x_2 - x_1|$
How long is side b? $|y_2 - y_1|$

The Pythagorean theorem states that $a^2 + b^2 = c^2$. Substitute $|x_2 - x_1|$ for a and $|y_2 - y_1|$ for b, then solve for c.

$a^2 + b^2 = c^2$ Pythagorean theorem
$|x_2 - x_1|^2 + |y_2 - y_1|^2 = c^2$ Substitute values.
$\pm\sqrt{(x_2 - x_1)^2 + (y_2 - y_1)^2} = c$ Solve for c using the square root property.

W Hint
Distances will always be a positive number or zero.

The distance between the points (x_1, y_1) and (x_2, y_2) is $c = \sqrt{(x_2 - x_1)^2 + (y_2 - y_1)^2}$. We want only the positive square root since c is a length.

Because this formula represents the *distance* between two points, we usually use the letter *d* instead of *c*.

> **Hint**
> Learn this formula! It is **very** important.

> **Definition** The Distance Formula
> The distance, d, between two points with coordinates (x_1, y_1) and (x_2, y_2) is given by
> $$d = \sqrt{(x_2 - x_1)^2 + (y_2 - y_1)^2}.$$

EXAMPLE 4 Find the distance between the points $(-4, 1)$ and $(2, 5)$.

Solution
Begin by labeling the points: $(\overset{x_1, y_1}{-4, 1})$, $(\overset{x_2, y_2}{2, 5})$.

Substitute the values into the distance formula.

$$\begin{aligned}
d &= \sqrt{(x_2 - x_1)^2 + (y_2 - y_1)^2} \\
&= \sqrt{[2 - (-4)]^2 + (5 - 1)^2} \quad \text{Substitute values.} \\
&= \sqrt{(2 + 4)^2 + (4)^2} \\
&= \sqrt{(6)^2 + (4)^2} = \sqrt{36 + 16} = \sqrt{52} = 2\sqrt{13}
\end{aligned}$$

[YOU TRY 4] Find the distance between the points $(1, 2)$ and $(7, -3)$.

The next method we will learn for solving a quadratic equation is *completing the square*. We need to review an idea first presented in Section 7.3.

A **perfect square trinomial** is a trinomial whose factored form is the square of a binomial. Some examples of perfect square trinomials are

Perfect Square Trinomials	Factored Form
$x^2 + 10x + 25$	$(x + 5)^2$
$d^2 - 8d + 16$	$(d - 4)^2$

In the trinomial $x^2 + 10x + 25$, x^2 is called the *quadratic term*, $10x$ is called the *linear term*, and 25 is called the *constant*.

4 Complete the Square for an Expression of the Form $x^2 + bx$

In a perfect square trinomial where the coefficient of the quadratic term is 1, the constant term is related to the coefficient of the linear term in the following way: *If you find half of the linear coefficient and square the result, you will get the constant term.*

$x^2 + 10x + 25$: The constant, 25, is obtained by

1) finding half of the coefficient of x;   then   2) squaring the result.

$\frac{1}{2}(10) = 5$ $5^2 = 25$ (the constant)

616 CHAPTER 10 **Quadratic Equations and Functions**   www.mhhe.com/messersmith

$d^2 - 8d + 16$: The constant, 16, is obtained by

1) finding half of the coefficient of d; then 2) squaring the result.

$$\frac{1}{2}(-8) = -4 \qquad\qquad (-4)^2 = 16 \text{ (the constant)}$$

We can generalize this procedure so that we can find the constant needed to obtain the perfect square trinomial for any quadratic expression of the form $x^2 + bx$. Finding this perfect square trinomial is called **completing the square** because the trinomial will factor to the square of a binomial.

Procedure Completing the Square for $x^2 + bx$

To find the constant needed to complete the square for $x^2 + bx$:

Step 1: Find half of the coefficient of x: $\frac{1}{2}b$.

Step 2: Square the result: $\left(\frac{1}{2}b\right)^2$.

Step 3: Then add it to $x^2 + bx$ to get $x^2 + bx + \left(\frac{1}{2}b\right)^2$. The factored form is $\left(x + \frac{1}{2}b\right)^2$.

BE CAREFUL The coefficient of the squared term *must* be 1 before you complete the square!

EXAMPLE 5

Complete the square for each expression to obtain a perfect square trinomial. Then, factor.

a) $y^2 + 6y$ b) $t^2 - 14t$

Solution

a) Find the constant needed to complete the square for $y^2 + 6y$.

Step 1: Find half of the coefficient of y:

$$\frac{1}{2}(6) = 3$$

Step 2: Square the result:

$$3^2 = 9$$

Step 3: Add 9 to $y^2 + 6y$:

$$y^2 + 6y + 9$$

The perfect square trinomial is $y^2 + 6y + 9$. The factored form is $(y + 3)^2$.

b) Find the constant needed to complete the square for $t^2 - 14t$.

Step 1: Find half of the coefficient of t:

$$\frac{1}{2}(-14) = -7$$

Step 2: Square the result:

$$(-7)^2 = 49$$

Step 3: Add 49 to $t^2 - 14t$:

$$t^2 - 14t + 49$$

The perfect square trinomial is $t^2 - 14t + 49$. The factored form is $(t - 7)^2$.

[YOU TRY 5]   Complete the square for each expression to obtain a perfect square trinomial. Then, factor.
a) $w^2 + 2w$ b) $z^2 - 16z$

We've seen the following perfect square trinomials and their factored forms. Let's look at the relationship between the constant in the factored form and the coefficient of the linear term.

Perfect Square Trinomial	Factored Form
$x^2 + 10x + 25$ 5 is $\frac{1}{2}(10)$.	$(x + 5)^2$
$d^2 - 8d + 16$ -4 is $\frac{1}{2}(-8)$.	$(d - 4)^2$
$y^2 + 6y + 9$ 3 is $\frac{1}{2}(6)$.	$(d + 3)^2$
$t^2 - 14t + 49$ -7 is $\frac{1}{2}(-14)$.	$(t - 7)^2$

W Hint
Be sure you understand these relationships before reading Example 6.

This pattern will always hold true and can be helpful in factoring some perfect square trinomials.

EXAMPLE 6   Complete the square for $n^2 + 5n$ to obtain a perfect square trinomial. Then, factor.

Solution

Find the constant needed to complete the square for $n^2 + 5n$.

Step 1: Find half of the coefficient of n: $\frac{1}{2}(5) = \frac{5}{2}$

Step 2: Square the result: $\left(\frac{5}{2}\right)^2 = \frac{25}{4}$

Step 3: Add $\frac{25}{4}$ to $n^2 + 5n$. The perfect square trinomial is $n^2 + 5n + \frac{25}{4}$.

The factored form is $\left(n + \frac{5}{2}\right)^2$.

$\frac{5}{2}$ is $\frac{1}{2}(5)$, the coefficient of n.

Check: $\left(n + \frac{5}{2}\right)^2 = n^2 + 2n\left(\frac{5}{2}\right) + \left(\frac{5}{2}\right)^2 = n^2 + 5n + \frac{25}{4}$

[YOU TRY 6]   Complete the square for $p^2 - 3p$ to obtain a perfect square trinomial. Then, factor.

5 Solve a Quadratic Equation by Completing the Square

Any quadratic equation of the form $ax^2 + bx + c = 0$ ($a \neq 0$) can be written in the form $(x - h)^2 = k$ by completing the square. Once an equation is in this form, we can use the square root property to solve for the variable.

> **Procedure** Solve a Quadratic Equation ($ax^2 + bx + c = 0$) by Completing the Square
>
> **Step 1:** The coefficient of the squared term must be 1. If it is not 1, divide both sides of the equation by a to obtain a leading coefficient of 1.
>
> **Step 2:** Get the variables on one side of the equal sign and the constant on the other side.
>
> **Step 3:** Complete the square. Find half of the linear coefficient, then square the result. Add that quantity to *both* sides of the equation.
>
> **Step 4:** Factor.
>
> **Step 5:** Solve using the square root property.

EXAMPLE 7

Solve by completing the square.

a) $x^2 + 6x + 8 = 0$ b) $12h + 4h^2 = -24$

Solution

a) $x^2 + 6x + 8 = 0$

   Step 1: The coefficient of x^2 is already 1.

   Step 2: Get the variables on one side of the equal sign and the constant on the other side: $x^2 + 6x = -8$

   Step 3: Complete the square: $\frac{1}{2}(6) = 3$
   $$3^2 = 9$$

   Add 9 to both sides of the equation: $x^2 + 6x + 9 = -8 + 9$
   $$x^2 + 6x + 9 = 1$$

   Step 4: Factor: $(x + 3)^2 = 1$

   Step 5: Solve using the square root property.

   $$(x + 3)^2 = 1$$
   $$x + 3 = \pm\sqrt{1}$$
   $$x + 3 = \pm 1$$

   $x + 3 = 1$ or $x + 3 = -1$
   $x = -2$ or $x = -4$

   The check is left to the student. The solution set is $\{-4, -2\}$.

W Hint
Are you writing out the steps as you are reading them?

> **Note**
> We would have obtained the same result if we had solved the equation by factoring.
>
> $$x^2 + 6x + 8 = 0$$
> $$(x + 4)(x + 2) = 0$$
> $$x + 4 = 0 \quad \text{or} \quad x + 2 = 0$$
> $$x = -4 \quad \text{or} \quad x = -2$$

b) $12h + 4h^2 = -24$

Step 1: Because the coefficient of h^2 is *not* 1, divide the whole equation by 4.

$$\frac{12h}{4} + \frac{4h^2}{4} = \frac{-24}{4}$$
$$3h + h^2 = -6$$

Step 2: The constant is on a side by itself. Rewrite the left side of the equation.

$$h^2 + 3h = -6$$

Step 3: Complete the square: $\frac{1}{2}(3) = \frac{3}{2}$

$$\left(\frac{3}{2}\right)^2 = \frac{9}{4}$$

Add $\frac{9}{4}$ to both sides of the equation.

$$h^2 + 3h + \frac{9}{4} = -6 + \frac{9}{4}$$
$$h^2 + 3h + \frac{9}{4} = -\frac{24}{4} + \frac{9}{4} \qquad \text{Get a common denominator.}$$
$$h^2 + 3h + \frac{9}{4} = -\frac{15}{4}$$

Step 4: Factor.

$$\left(h + \frac{3}{2}\right)^2 = -\frac{15}{4}$$
$$\uparrow$$
$$\frac{3}{2} \text{ is } \frac{1}{2}(3), \text{ the coefficient of } h.$$

Step 5: Solve using the square root property.

$$\left(h + \frac{3}{2}\right)^2 = -\frac{15}{4}$$
$$h + \frac{3}{2} = \pm\sqrt{-\frac{15}{4}}$$
$$h + \frac{3}{2} = \pm\frac{\sqrt{15}}{2}i \qquad \text{Simplify the radical.}$$
$$h = -\frac{3}{2} \pm \frac{\sqrt{15}}{2}i \qquad \text{Subtract } \frac{3}{2}.$$

W Hint
This quadratic equation produced nonreal, complex solutions.

The check is left to the student. The solution set is
$\left\{-\frac{3}{2} - \frac{\sqrt{15}}{2}i, -\frac{3}{2} + \frac{\sqrt{15}}{2}i\right\}$.

[YOU TRY 7] Solve by completing the square.

a) $q^2 + 10q - 24 = 0$ b) $2m^2 + 16 = 10m$

ANSWERS TO [YOU TRY] EXERCISES

1) a) $\{-10, 10\}$ b) $\{-4\sqrt{2}, 4\sqrt{2}\}$ c) $\{-2i, 2i\}$ 2) $\{-15, 3\}$
3) a) $\left\{-1, \dfrac{5}{7}\right\}$ b) $\left\{\dfrac{3 - 2\sqrt{6}}{5}, \dfrac{3 + 2\sqrt{6}}{5}\right\}$ c) $\{7 - 10i, 7 + 10i\}$
d) $\left\{-\dfrac{3}{2} - \dfrac{3\sqrt{2}}{2}i, \dfrac{3}{2} + \dfrac{3\sqrt{2}}{2}i\right\}$ 4) $\sqrt{61}$
5) a) $w^2 + 2w + 1$; $(w + 1)^2$ b) $z^2 - 16z + 64$; $(z - 8)^2$
6) $p^2 - 3p + \dfrac{9}{4}$; $\left(p - \dfrac{3}{2}\right)^2$ 7) a) $\{-12, 2\}$ b) $\left\{\dfrac{5}{2} - \dfrac{\sqrt{7}}{2}i, \dfrac{5}{2} + \dfrac{\sqrt{7}}{2}i\right\}$

E Evaluate 10.1 Exercises

Do the exercises, and check your work.

Objective 1: Solve an Equation of the Form $x^2 = k$

1) Choose two methods to solve $y^2 - 16 = 0$. Solve the equation using both methods.

2) If k is a negative number and $x^2 = k$, what can you conclude about the solution to the equation?

Solve using the square root property.

3) $b^2 = 36$
4) $h^2 = 64$
5) $r^2 - 27 = 0$
6) $a^2 - 30 = 0$
7) $n^2 = \dfrac{4}{9}$
8) $v^2 = \dfrac{121}{16}$
9) $q^2 = -4$
10) $w^2 = -121$
11) $z^2 + 3 = 0$
12) $h^2 + 14 = -23$
13) $z^2 + 5 = 19$
14) $q^2 - 3 = 15$
15) $4m^2 + 1 = 37$
16) $2d^2 + 5 = 55$
17) $5f^2 + 39 = -21$
18) $2y^2 + 56 = 0$

Objective 2: Solve an Equation of the Form $(ax + b)^2 = k$

Solve using the square root property.

19) $(r + 10)^2 = 4$
20) $(x - 5)^2 = 81$
21) $(q - 7)^2 = 1$
22) $(c + 12)^2 = 25$
23) $(p + 4)^2 - 18 = 0$
24) $(d + 2)^2 - 7 = 13$
25) $(c + 3)^2 - 4 = -29$
26) $(u - 15)^2 - 4 = -8$
27) $1 = 15 + (k - 2)^2$
28) $2 = 14 + (g + 4)^2$
29) $20 = (2w + 1)^2$
30) $(5b - 6)^2 = 11$
31) $8 = (3q - 10)^2 - 6$
32) $22 = (6x + 11)^2 + 4$
33) $36 + (4p - 5)^2 = 6$
34) $(3k - 1)^2 + 20 = 4$
35) $(6g + 11)^2 + 50 = 1$
36) $9 = 38 + (9s - 4)^2$
37) $\left(\dfrac{3}{4}n - 8\right)^2 = 4$
38) $\left(\dfrac{2}{3}j + 10\right)^2 = 16$
39) $(5y - 2)^2 + 6 = 22$
40) $-6 = 3 - (2q - 9)^2$

Objective 3: Use the Distance Formula

Find the distance between the given points.

41) $(7, -1)$ and $(3, 2)$
42) $(3, 10)$ and $(12, 6)$
43) $(-5, -6)$ and $(-2, -8)$
44) $(5, -2)$ and $(-3, 4)$
45) $(0, 3)$ and $(3, -1)$
46) $(-8, 3)$ and $(2, 1)$
47) $(-4, 11)$ and $(2, 6)$
48) $(0, 13)$ and $(0, 7)$
49) $(3, -3)$ and $(5, -7)$
50) $(-5, -6)$ and $(-1, 2)$

Objective 4: Complete the Square for an Expression of the Form $x^2 + bx$

51) What is a perfect square trinomial? Give an example.

52) Can you complete the square on $3y^2 + 15y$ as it is given? Why or why not?

Complete the square for each expression to obtain a perfect square trinomial. Then, factor.

Fill It In
Fill in the blanks with either the missing mathematical step or reason for the given step.

53) $w^2 + 8w$

_____ Find half of the coefficient of w.

_____ Square the result.

_____ Add the constant to the expression.

The perfect square trinomial is _____

The factored form of the trinomial is _____

54) $n^2 - n$

$\frac{1}{2}(-1) = -\frac{1}{2}$ _____

$\left(-\frac{1}{2}\right)^2 = \frac{1}{4}$ _____

$n^2 - n + \frac{1}{4}$ _____

The perfect square trinomial is _____

The factored form of the trinomial is _____

55) $a^2 + 12a$ 　　 56) $g^2 + 4g$

57) $c^2 - 18c$ 　　 58) $k^2 - 16k$

59) $t^2 + 5t$ 　　 60) $z^2 - 7z$

61) $b^2 - 9b$ 　　 62) $r^2 + 3r$

63) $x^2 + \frac{1}{3}x$ 　　 64) $y^2 - \frac{3}{5}y$

Objective 5: Solve a Quadratic Equation by Completing the Square

65) What is the first thing you should do if you want to solve $2p^2 - 7p = 8$ by completing the square?

66) Can $x^3 + 10x - 3 = 0$ be solved by completing the square? Give a reason for your answer.

Solve by completing the square.

67) $x^2 + 6x + 8 = 0$ 　　 68) $t^2 + 12t - 13 = 0$

69) $k^2 - 8k + 15 = 0$ 　　 70) $v^2 - 6v - 27 = 0$

71) $u^2 - 9 = 2u$ 　　 72) $s^2 + 10 = -10s$

73) $p^2 = -10p - 26$ 　　 74) $t^2 = 2t - 9$

75) $a^2 + 19 = 8a$ 　　 76) $v^2 + 4v + 8 = 0$

77) $m^2 + 3m - 40 = 0$ 　　 78) $p^2 + 5p + 4 = 0$

79) $x^2 - 7x + 12 = 0$ 　　 80) $d^2 + d - 72 = 0$

81) $r^2 - r = 3$ 　　 82) $y^2 - 3y = 7$

83) $c^2 + 5c + 7 = 0$ 　　 84) $b^2 + 14 = 7b$

85) $3k^2 - 6k + 12 = 0$ 　　 86) $4f^2 + 16f + 48 = 0$

87) $4r^2 + 24r = 8$ 　　 88) $3h^2 + 6h = 15$

89) $10d = 2d^2 + 12$ 　　 90) $54x - 6x^2 = 48$

91) $2n^2 + 8 = 5n$ 　　 92) $2t^2 + 3t + 4 = 0$

93) $4a^2 - 7a + 3 = 0$ 　　 94) $n + 2 = 3n^2$

95) $(y + 5)(y - 3) = 5$ 　　 96) $(b - 4)(b + 10) = -17$

97) $(2m + 1)(m - 3) = -7$ 　　 98) $(3c + 4)(c + 2) = 3$

Use the Pythagorean theorem and the square root property to find the length of the missing side.

99) 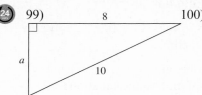　　100)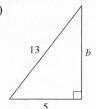

101) 　　102)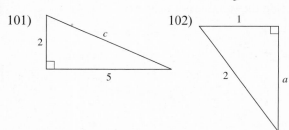

Write an equation, and solve. (Hint: Draw a picture.)

103) The width of a rectangle is 4 in., and its diagonal is $2\sqrt{13}$ in. long. What is the length of the rectangle?

104) Find the length of the diagonal of a rectangle if it has a width of 5 cm and a length of $4\sqrt{2}$ cm.

Write an equation, and solve.

105) A 13-ft ladder is leaning against a wall so that the base of the ladder is 5 ft away from the wall. How high on the wall does the ladder reach?

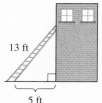

106) Salma is flying a kite. It is 30 ft from her horizontally, and it is 40 ft above her hand. How long is the kite string?

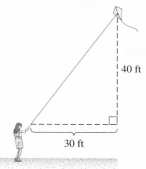

107) Let $f(x) = (x + 3)^2$. Find x so that $f(x) = 49$.

108) Let $g(t) = (t - 5)^2$. Find t so that $g(t) = 12$.

Solve each problem by writing an equation and solving it by completing the square.

109) The length of a rectangular garden is 8 ft more than its width. Find the dimensions of the garden if it has an area of 153 ft².

110) The rectangular screen on a laptop has an area of 375 cm². Its width is 10 cm less than its length. What are the dimensions of the screen?

R Rethink

R1) Why must the coefficient be "1" before you use the square root property?

R2) Can you think of a real-life situation where you may need to use the distance formula?

R3) What do you find most difficult about completing the square? After you complete the square, which method do you use next?

10.2 The Quadratic Formula

P Prepare	O Organize
What are your objectives for Section 10.2?	How can you accomplish each objective?
1 Derive the Quadratic Formula	• Understand where the *quadratic formula* comes from. • Learn the *quadratic formula*. Practice writing it several times in your notes.
2 Solve a Quadratic Equation Using the Quadratic Formula	• Know how to write a quadratic equation in standard form. • Be able to identify a, b, and c in the quadratic formula. • Complete the given examples on your own. • Complete You Trys 1 and 2.
3 Determine the Number and Type of Solutions of a Quadratic Equation Using the Discriminant	• Write out all the possible types/number of solutions for a *discriminant* in your notes. • Complete the given example on your own. • Complete You Try 3.
4 Solve an Applied Problem Using the Quadratic Formula	• Read the problem carefully. • In a motion equation, be able to determine if you are solving for height (h) or solving for time (t), if applicable. • Review the **Pythagorean theorem.** • Complete the given example on your own. • Complete You Try 4.

W Work — Read the explanations, follow the examples, take notes, and complete the You Trys.

1 Derive the Quadratic Formula

In Section 10.1, we saw that any quadratic equation of the form $ax^2 + bx + c = 0$ ($a \neq 0$) can be solved by completing the square. Therefore, we can solve equations like $x^2 - 8x + 5 = 0$ and $2x^2 + 3x - 1 = 0$ using this method.

We can develop another method for solving quadratic equations by completing the square on the general quadratic equation $ax^2 + bx + c = 0$ ($a \neq 0$). This will let us derive the *quadratic formula*.

The steps we use to complete the square on $ax^2 + bx + c = 0$ are *exactly* the same steps we use to solve an equation like $2x^2 + 3x - 1 = 0$. We will do these steps side by side so that you can more easily understand how we are solving $ax^2 + bx + c = 0$ for x by completing the square.

Solve for x by Completing the Square.

$$2x^2 + 3x - 1 = 0 \qquad\qquad ax^2 + bx + c = 0$$

Step 1: The coefficient of the squared term must be 1.

$$2x^2 + 3x - 1 = 0 \qquad\qquad ax^2 + bx + c = 0$$
$$\frac{2x^2}{2} + \frac{3x}{2} - \frac{1}{2} = \frac{0}{2} \quad \text{Divide by 2.} \qquad \frac{ax^2}{a} + \frac{bx}{a} + \frac{c}{a} = \frac{0}{a} \quad \text{Divide by } a.$$
$$x^2 + \frac{3}{2}x - \frac{1}{2} = 0 \quad \text{Simplify.} \qquad x^2 + \frac{b}{a}x + \frac{c}{a} = 0 \quad \text{Simplify.}$$

W Hint
Notice that we are using the same steps that we used in Section 10.1 to solve *both* equations.

Step 2: Get the constant on the other side of the equal sign.

$$x^2 + \frac{3}{2}x = \frac{1}{2} \quad \text{Add } \tfrac{1}{2}. \qquad\qquad x^2 + \frac{b}{a}x = -\frac{c}{a} \quad \text{Subtract } \tfrac{c}{a}.$$

Step 3: Complete the square.

$$\frac{1}{2}\left(\frac{3}{2}\right) = \frac{3}{4} \quad \tfrac{1}{2} \text{ of } x\text{-coefficient} \qquad \frac{1}{2}\left(\frac{b}{a}\right) = \frac{b}{2a} \quad \tfrac{1}{2} \text{ of } x\text{-coefficient}$$

$$\left(\frac{3}{4}\right)^2 = \frac{9}{16} \quad \text{Square the result.} \qquad \left(\frac{b}{2a}\right)^2 = \frac{b^2}{4a^2} \quad \text{Square the result.}$$

Add $\dfrac{9}{16}$ to both sides of the equation. Add $\dfrac{b^2}{4a^2}$ to both sides of the equation.

$$x^2 + \frac{3}{2}x + \frac{9}{16} = \frac{1}{2} + \frac{9}{16} \qquad\qquad x^2 + \frac{b}{a}x + \frac{b^2}{4a^2} = -\frac{c}{a} + \frac{b^2}{4a^2}$$

$$x^2 + \frac{3}{2}x + \frac{9}{16} = \frac{8}{16} + \frac{9}{16} \quad \begin{array}{l}\text{Get a}\\\text{common}\\\text{denominator.}\end{array} \qquad x^2 + \frac{b}{a}x + \frac{b^2}{4a^2} = -\frac{4ac}{4a^2} + \frac{b^2}{4a^2} \quad \begin{array}{l}\text{Get a}\\\text{common}\\\text{denominator.}\end{array}$$

$$x^2 + \frac{3}{2}x + \frac{9}{16} = \frac{17}{16} \quad \text{Add.} \qquad x^2 + \frac{b}{a}x + \frac{b^2}{4a^2} = \frac{b^2 - 4ac}{4a^2} \quad \text{Add.}$$

Step 4: Factor.

$$\left(x + \frac{3}{4}\right)^2 = \frac{17}{16}$$

$\frac{3}{4}$ is $\frac{1}{2}\left(\frac{3}{2}\right)$, the coefficient of x.

$$\left(x + \frac{b}{2a}\right)^2 = \frac{b^2 - 4ac}{4a^2}$$

$\frac{b}{2a}$ is $\frac{1}{2}\left(\frac{b}{a}\right)$, the coefficient of x.

Step 5: Solve using the square root property.

$$\left(x + \frac{3}{4}\right)^2 = \frac{17}{16}$$

$$x + \frac{3}{4} = \pm\sqrt{\frac{17}{16}}$$

$$x + \frac{3}{4} = \frac{\pm\sqrt{17}}{4} \quad \sqrt{16} = 4$$

$$x = -\frac{3}{4} \pm \frac{\sqrt{17}}{4} \quad \text{Subtract } \frac{3}{4}.$$

$$x = \frac{-3 \pm \sqrt{17}}{4} \quad \text{Same denominators, add numerators.}$$

$$\left(x + \frac{b}{2a}\right)^2 = \frac{b^2 - 4ac}{4a^2}$$

$$x + \frac{b}{2a} = \pm\sqrt{\frac{b^2 - 4ac}{4a^2}}$$

$$x + \frac{b}{2a} = \frac{\pm\sqrt{b^2 - 4ac}}{2a} \quad \sqrt{4a^2} = 2a$$

$$x = -\frac{b}{2a} \pm \frac{\sqrt{b^2 - 4ac}}{2a} \quad \text{Subtract } \frac{b}{2a}.$$

$$x = \frac{-b \pm \sqrt{b^2 - 4ac}}{2a} \quad \text{Same denominators, add numerators.}$$

The result on the right is called the *quadratic formula*.

Hint

Memorize this formula!

Definition The Quadratic Formula

The solutions of any quadratic equation of the form $ax^2 + bx + c = 0$ $(a \neq 0)$ are

$$x = \frac{-b \pm \sqrt{b^2 - 4ac}}{2a}$$

This formula is called the **quadratic formula.**

Note

1) To use the quadratic formula, write the equation to be solved in the form $ax^2 + bx + c = 0$ so that a, b, and c can be identified correctly.

2) $x = \dfrac{-b \pm \sqrt{b^2 - 4ac}}{2a}$ represents the two solutions

   $x = \dfrac{-b + \sqrt{b^2 - 4ac}}{2a}$ and $x = \dfrac{-b - \sqrt{b^2 - 4ac}}{2a}$.

3) Notice that the fraction bar runs under $-b$ and under the radical.

   $x = \dfrac{-b \pm \sqrt{b^2 - 4ac}}{2a}$ $x = -b \pm \dfrac{\sqrt{b^2 - 4ac}}{2a}$

   Correct   Incorrect

4) When deriving the quadratic formula, using the $\pm$ allows us to say that $\sqrt{4a^2} = 2a$.

5) The quadratic formula is a *very* important result, and we will use it often. *It should be memorized!*

2 Solve a Quadratic Equation Using the Quadratic Formula

EXAMPLE 1

Solve using the quadratic formula.

a) $2x^2 + 3x - 1 = 0$ b) $k^2 = 10k - 29$

Solution

a) Is $2x^2 + 3x - 1 = 0$ in the form $ax^2 + bx + c = 0$? Yes. Identify the values of a, b, and c, and substitute them into the quadratic formula.

$$a = 2 \quad b = 3 \quad c = -1$$

$$\begin{aligned}
x &= \frac{-b \pm \sqrt{b^2 - 4ac}}{2a} & &\text{Quadratic formula} \\
&= \frac{-(3) \pm \sqrt{(3)^2 - 4(2)(-1)}}{2(2)} & &\text{Substitute } a = 2, b = 3, \text{ and } c = -1. \\
&= \frac{-3 \pm \sqrt{9 - (-8)}}{4} & &\text{Perform the operations.} \\
&= \frac{-3 \pm \sqrt{17}}{4}
\end{aligned}$$

W Hint
Write down the quadratic formula and the values of a, b, and c on your paper.

The solution set is $\left\{\dfrac{-3 - \sqrt{17}}{4}, \dfrac{-3 + \sqrt{17}}{4}\right\}$. This is the same result we obtained when we solved this equation by completing the square at the beginning of the section.

b) Is $k^2 = 10k - 29$ in the form $ax^2 + bx + c = 0$? No. Begin by writing the equation in the correct form.

$$k^2 - 10k + 29 = 0 \quad \text{Subtract } 10k, \text{ and add 29 to both sides.}$$

$$a = 1 \quad b = -10 \quad c = 29 \quad \text{Identify } a, b, \text{ and } c.$$

$$\begin{aligned}
k &= \frac{-b \pm \sqrt{b^2 - 4ac}}{2a} & &\text{Quadratic formula} \\
&= \frac{-(-10) \pm \sqrt{(-10)^2 - 4(1)(29)}}{2(1)} & &\text{Substitute } a = 1, b = -10, \text{ and } c = 29. \\
&= \frac{10 \pm \sqrt{100 - 116}}{2} & &\text{Perform the operations.} \\
&= \frac{10 \pm \sqrt{-16}}{2} & &100 - 116 = -16 \\
&= \frac{10 \pm 4i}{2} & &\sqrt{-16} = 4i \\
&= \frac{10}{2} \pm \frac{4}{2}i = 5 \pm 2i
\end{aligned}$$

The solution set is $\{5 - 2i, 5 + 2i\}$.

[YOU TRY 1]

Solve using the quadratic formula.

a) $n^2 + 9n + 18 = 0$ b) $5t^2 + t - 2 = 0$

Equations in various forms may be solved using the quadratic formula.

EXAMPLE 2

Solve using the quadratic formula.
$$(3p - 1)(3p + 4) = 3p - 5$$

Solution

Is $(3p - 1)(3p + 4) = 3p - 5$ in the form $ax^2 + bx + c = 0$? No. Before we can apply the quadratic formula, we must write it in that form.

$(3p - 1)(3p + 4) = 3p - 5$
$9p^2 + 9p - 4 = 3p - 5$ Multiply using FOIL.
$9p^2 + 6p + 1 = 0$ Subtract $3p$, and add 5 to both sides.

The equation is in the correct form. Identify a, b, and c: $a = 9$ $b = 6$ $c = 1$

$p = \dfrac{-b \pm \sqrt{b^2 - 4ac}}{2a}$ Quadratic formula

$= \dfrac{-(6) \pm \sqrt{(6)^2 - 4(9)(1)}}{2(9)}$ Substitute $a = 9$, $b = 6$, and $c = 1$.

$= \dfrac{-6 \pm \sqrt{36 - 36}}{18}$ Perform the operations.

$= \dfrac{-6 \pm \sqrt{0}}{18}$

$= \dfrac{-6 \pm 0}{18} = \dfrac{-6}{18} = -\dfrac{1}{3}$

The solution set is $\left\{-\dfrac{1}{3}\right\}$.

YOU TRY 2

Solve using the quadratic formula.

a) $3 - 2z = -2z^2$ b) $(d + 6)(d - 2) = -10$

To solve a quadratic equation containing fractions, first multiply by the LCD to eliminate the fractions. Then, solve using the quadratic formula.

3   Determine the Number and Type of Solutions of a Quadratic Equation Using the Discriminant

We can find the solutions of any quadratic equation of the form $ax^2 + bx + c = 0$ ($a \neq 0$) using the quadratic formula.

$$x = \dfrac{-b \pm \sqrt{b^2 - 4ac}}{2a}$$

The radicand in the quadratic formula determines the type of solution a quadratic equation has.

Property The Discriminant and Solutions

The expression under the radical, $b^2 - 4ac$, is called the **discriminant**. The discriminant tells us what kind of solution a quadratic equation has. If a, b, and c are integers, then

1) if $b^2 - 4ac$ is *positive and the square of an integer*, the equation has *two rational solutions*.
2) if $b^2 - 4ac$ is *positive but not a perfect square*, the equation has *two irrational solutions*.
3) if $b^2 - 4ac$ is *negative*, the equation has *two nonreal, complex solutions of the form $a + bi$ and $a - bi$*.
4) if $b^2 - 4ac = 0$, the equation has *one rational solution*.

EXAMPLE 3

Find the value of the discriminant. Then, determine the number and type of solutions of each equation.

a) $z^2 + 6z - 4 = 0$ b) $5h^2 = 6h - 2$

Solution

a) Is $z^2 + 6z - 4 = 0$ in the form $ax^2 + bx + c = 0$? Yes. Identify a, b, and c.

$$a = 1 \qquad b = 6 \qquad c = -4$$

$$\text{Discriminant} = b^2 - 4ac = (6)^2 - 4(1)(-4) = 36 + 16 = 52$$

Because 52 is positive but *not* a perfect square, the equation will have *two irrational solutions*. ($\sqrt{52}$, or $2\sqrt{13}$, will appear in the solution, and $2\sqrt{13}$ is irrational.)

b) Is $5h^2 = 6h - 2$ in the form $ax^2 + bx + c = 0$? No. Rewrite the equation in that form, and identify a, b, and c.

$$5h^2 - 6h + 2 = 0$$

$$a = 5 \qquad b = -6 \qquad c = 2$$

$$\text{Discriminant} = b^2 - 4ac = (-6)^2 - 4(5)(2) = 36 - 40 = -4$$

The discriminant is -4, so the equation will have *two nonreal, complex solutions of the form $a + bi$ and $a - bi$*, where $b \neq 0$.

The discriminant is $b^2 - 4ac$ not $\sqrt{b^2 - 4ac}$.

[YOU TRY 3] Find the value of the discriminant. Then, determine the number and type of solutions of each equation.

a) $2x^2 + x + 5 = 0$ b) $m^2 + 5m = 24$ c) $-3v^2 = 4v - 1$

d) $4r(2r - 3) = -1 - 6r - r^2$

4 Solve an Applied Problem Using the Quadratic Formula

EXAMPLE 4

A ball is thrown upward from a height of 20 ft. The height h of the ball (in feet) t sec after the ball is released is given by

$$h = -16t^2 + 16t + 20$$

a) How long does it take the ball to reach a height of 8 ft?

b) How long does it take the ball to hit the ground?

Solution

a) Find the *time* it takes for the ball to reach a height of 8 ft.

Find t when $h = 8$.

$$h = -16t^2 + 16t + 20$$
$$8 = -16t^2 + 16t + 20 \quad \text{Substitute 8 for } h.$$
$$0 = -16t^2 + 16t + 12 \quad \text{Write in standard form.}$$
$$0 = 4t^2 - 4t - 3 \quad \text{Divide by } -4.$$

$$t = \frac{-b \pm \sqrt{b^2 - 4ac}}{2a} \quad \text{Quadratic formula}$$

$$= \frac{-(-4) \pm \sqrt{(-4)^2 - 4(4)(-3)}}{2(4)} \quad \text{Substitute } a = 4, b = -4, \text{ and } c = -3.$$

$$= \frac{4 \pm \sqrt{16 + 48}}{8} \quad \text{Perform the operations.}$$

$$= \frac{4 \pm \sqrt{64}}{8} = \frac{4 \pm 8}{8}$$

$$t = \frac{4 + 8}{8} \quad \text{or} \quad t = \frac{4 - 8}{8} \quad \text{The equation has two rational solutions.}$$

$$t = \frac{12}{8} = \frac{3}{2} \quad \text{or} \quad t = \frac{-4}{8} = -\frac{1}{2}$$

Because t represents time, t cannot equal $-\frac{1}{2}$. We reject that as a solution.

Therefore, $t = \frac{3}{2}$ sec or 1.5 sec. The ball will be 8 ft above the ground after 1.5 sec.

W Hint
Think *carefully* about the problem you are asked to solve. What does it mean in terms of the formula?

b) When the ball hits the ground, it is 0 ft above the ground.

Find t when $h = 0$.

$$h = -16t^2 + 16t + 20$$
$$0 = -16t^2 + 16t + 20 \quad \text{Substitute 0 for } h.$$
$$0 = 4t^2 - 4t - 5 \quad \text{Divide by } -4.$$

$$t = \frac{-(-4) \pm \sqrt{(-4)^2 - 4(4)(-5)}}{2(4)} \quad \text{Substitute } a = 4, b = -4, \text{ and } c = -5.$$

$$= \frac{4 \pm \sqrt{16 + 80}}{8} \quad \text{Perform the operations.}$$

$$= \frac{4 \pm \sqrt{96}}{8}$$

$$= \frac{4 \pm 4\sqrt{6}}{8} \quad \sqrt{96} = \sqrt{16} \cdot \sqrt{6} = 4\sqrt{6}$$

$$= \frac{4(1 \pm \sqrt{6})}{8} \quad \text{Factor out 4 in the numerator.}$$

$$t = \frac{1 \pm \sqrt{6}}{2} \quad \text{Divide numerator and denominator by 4 to simplify.}$$

$$t = \frac{1 + \sqrt{6}}{2} \quad \text{or} \quad t = \frac{1 - \sqrt{6}}{2} \quad \text{The equation has two irrational solutions.}$$

$$t \approx \frac{1 + 2.4}{2} \quad \text{or} \quad t \approx \frac{1 - 2.4}{2} \quad \sqrt{6} \approx 2.4$$

$$t \approx \frac{3.4}{2} = 1.7 \quad \text{or} \quad t \approx -0.7$$

Because t represents time, t cannot equal $\frac{1 - \sqrt{6}}{2}$. We reject this as a solution.

Therefore, $t = \frac{1 + \sqrt{6}}{2}$ sec or $t \approx 1.7$ sec. The ball will hit the ground after about 1.7 sec.

[YOU TRY 4] An object is thrown upward from a height of 12 ft. The height h of the object (in feet) t sec after the object is thrown is given by

$$h = -16t^2 + 56t + 12$$

a) How long does it take the object to reach a height of 36 ft?

b) How long does it take the object to hit the ground?

ANSWERS TO [YOU TRY] EXERCISES

1) a) $\{-6, -3\}$ b) $\left\{\frac{-1 - \sqrt{41}}{10}, \frac{-1 + \sqrt{41}}{10}\right\}$ 2) a) $\left\{\frac{1}{2} - \frac{\sqrt{5}}{2}i, \frac{1}{2} + \frac{\sqrt{5}}{2}i\right\}$
b) $\{-2 - \sqrt{6}, -2 + \sqrt{6}\}$ 3) a) -39; two nonreal, complex solutions
b) 121; two rational solutions   c) 28; two irrational solutions   d) 0; one rational solution
4) a) It takes $\frac{1}{2}$ sec to reach 36 ft on its way up and 3 sec to reach 36 ft on its way down.
b) $\frac{7 + \sqrt{61}}{4}$ sec or approximately 3.7 sec

10.2 Exercises

Do the exercises, and check your work.

Mixed Exercises: Objectives 2 and 3
Find the error in each, and correct the mistake.

1) The solution to $ax^2 + bx + c = 0$ ($a \neq 0$) can be found using the quadratic formula.
$$x = -b \pm \frac{\sqrt{b^2 - 4ac}}{2a}$$

2) In order to solve $5n^2 - 3n = 1$ using the quadratic formula, a student substitutes a, b, and c into the formula in this way: $a = 5$, $b = -3$, $c = 1$.
$$n = \frac{-(-3) \pm \sqrt{(-3)^2 - 4(5)(1)}}{2(5)}$$

3) $\dfrac{-2 \pm 6\sqrt{11}}{2} = -1 \pm 6\sqrt{11}$

4) The discriminant of $3z^2 - 4z + 1 = 0$ is
$$\sqrt{b^2 - 4ac} = \sqrt{(-4)^2 - 4(3)(1)}$$
$$= \sqrt{16 - 12}$$
$$= \sqrt{4}$$
$$= 2.$$

Objective 2: Solve a Quadratic Equation Using the Quadratic Formula
Solve using the quadratic formula.

5) $x^2 + 4x + 3 = 0$
6) $v^2 - 8v + 7 = 0$
7) $3t^2 + t - 10 = 0$
8) $6q^2 + 11q + 3 = 0$
9) $k^2 + 2 = 5k$
10) $n^2 = 5 - 3n$
11) $y^2 = 8y - 25$
12) $-4x + 5 = -x^2$
13) $3 - 2w = -5w^2$
14) $2d^2 = -4 - 5d$
15) $r^2 + 7r = 0$
16) $p^2 - 10p = 0$
17) $2k(k - 3) = -3$
18) $3v(v + 3) = 7v + 4$
19) $(2c - 5)(c - 5) = -3$
20) $-11 = (3z - 1)(z - 5)$
21) $\dfrac{1}{6}u^2 + \dfrac{4}{3}u = \dfrac{5}{2}$
22) $\dfrac{1}{6}h + \dfrac{1}{2} = \dfrac{3}{4}h^2$
23) $2(p + 10) = (p + 10)(p - 2)$
24) $(t - 8)(t - 3) = 3(3 - t)$
25) $4g^2 + 9 = 0$
26) $25q^2 - 1 = 0$
27) $x(x + 6) = -34$
28) $c(c - 4) = -22$
29) $(2s + 3)(s - 1) = s^2 - s + 6$
30) $(3m + 1)(m - 2) = (2m - 3)(m + 2)$
31) $3(3 - 4y) = -4y^2$
32) $5a(5a + 2) = -1$
33) $-\dfrac{1}{6} = \dfrac{2}{3}p^2 + \dfrac{1}{2}p$
34) $\dfrac{1}{2}n = \dfrac{3}{4}n^2 + 2$

35) $4q^2 + 6 = 20q$
36) $4w^2 = 6w + 16$
37) Let $f(x) = x^2 + 6x - 2$. Find x so that $f(x) = 0$.
38) Let $g(x) = 3x^2 - 4x - 1$. Find x so that $g(x) = 0$.
39) Let $h(t) = 2t^2 - t + 7$. Find t so that $h(t) = 12$.
40) Let $P(a) = a^2 + 8a + 9$. Find a so that $P(a) = -3$.
41) Let $f(x) = 5x^2 + 21x - 1$ and $g(x) = 2x + 3$. Find all values of x such that $f(x) = g(x)$.
42) Let $F(x) = -x^2 + 3x - 2$ and $G(x) = x^2 + 12x + 6$. Find all values of x such that $F(x) = G(x)$.

Objective 3: Determine the Number and Type of Solutions of a Quadratic Equation Using the Discriminant

43) If the discriminant of a quadratic equation is zero, what do you know about the solutions of the equation?

44) If the discriminant of a quadratic equation is negative, what do you know about the solutions of the equation?

Find the value of the discriminant. Then, determine the number and type of solutions of each equation. *Do not solve.*

45) $10d^2 - 9d + 3 = 0$
46) $3j^2 + 8j + 2 = 0$
47) $4y^2 + 49 = -28y$
48) $3q = 1 + 5q^2$
49) $-5 = u(u + 6)$
50) $g^2 + 4 = 4g$
51) $2w^2 - 4w - 5 = 0$
52) $3 + 2p^2 - 7p = 0$

Find the value of a, b, or c so that each equation has only one rational solution.

53) $z^2 + bz + 16 = 0$
54) $k^2 + bk + 49 = 0$
55) $4y^2 - 12y + c = 0$
56) $25t^2 - 20t + c = 0$
57) $ap^2 + 12p + 9 = 0$
58) $ax^2 - 6x + 1 = 0$

Objective 4: Solve an Applied Problem Using the Quadratic Formula
Write an equation, and solve.

59) One leg of a right triangle is 1 in. more than twice the other leg. The hypotenuse is $\sqrt{29}$ in. long. Find the lengths of the legs.

60) The hypotenuse of a right triangle is $\sqrt{34}$ in. long. The length of one leg is 1 in. less than twice the other leg. Find the lengths of the legs.

Solve.

61) An object is thrown upward from a height of 24 ft. The height h of the object (in feet) t sec after the object is released is given by $h = -16t^2 + 24t + 24$.

   a) How long does it take the object to reach a height of 8 ft?

   b) How long does it take the object to hit the ground?

62) A ball is thrown upward from a height of 6 ft. The height h of the ball (in feet) t sec after the ball is released is given by $h = -16t^2 + 44t + 6$.

   a) How long does it take the ball to reach a height of 16 ft?

   b) How long does it take the object to hit the ground?

R Rethink

R1) How many different types of solutions are possible for a quadratic equation?

R2) Do you know any songs that might help you remember the quadratic formula?

R3) What do you find most difficult about solving applied problems?

Putting It All Together

P Prepare

What are your objectives for Putting It All Together?	How can you accomplish each objective?
1 Decide Which Method to Use to Solve a Quadratic Equation	• Be able to write out all the different methods in your notes. • Review characteristics of each method, and be able to identify the most efficient method for each problem. • Try solving some problems using more than one method, if time permits, and check your answers. • Complete the given example on your own. • Complete You Try 1.

W Work

Read the explanations, follow the examples, take notes, and complete the You Try.

We have learned four methods for solving quadratic equations.

Methods for Solving Quadratic Equations

1) Factoring
2) Square root property
3) Completing the square
4) Quadratic formula

While it is true that the quadratic formula can be used to solve *every* quadratic equation of the form $ax^2 + bx + c = 0$ ($a \neq 0$), it is not always the most *efficient* method. In this section we will discuss how to decide which method to use to solve a quadratic equation.

1 Decide Which Method to Use to Solve a Quadratic Equation

EXAMPLE 1

Solve.

a) $p^2 - 6p = 16$
b) $m^2 - 8m + 13 = 0$
c) $3t^2 + 8t + 7 = 0$
d) $(2z - 7)^2 - 6 = 0$

Solution

a) Write $p^2 - 6p = 16$ in standard form: $p^2 - 6p - 16 = 0$

Does $p^2 - 6p - 16$ factor? Yes. *Solve by factoring.*

$$(p - 8)(p + 2) = 0$$

$p - 8 = 0$ or $p + 2 = 0$ Set each factor equal to 0.
$p = 8$ or $p = -2$ Solve.

The solution set is $\{-2, 8\}$.

W Hint

Write the solutions to the equations in this example in your notes. In your own words, explain why each method is chosen to solve each equation.

b) To solve $m^2 - 8m + 13 = 0$ ask yourself, "Can I factor $m^2 - 8m + 13$?" No, it does not factor. We could solve this using the quadratic formula, but *completing the square* is also a good method for solving this equation. Why?

Note

Completing the square is a good method for solving a quadratic equation when the coefficient of the squared term is 1 or -1 and when the coefficient of the linear term is even.

We will solve $m^2 - 8m + 13 = 0$ by completing the square.

Step 1: The coefficient of m^2 is 1.

Step 2: Get the variables on one side of the equal sign and the constant on the other side.

$$m^2 - 8m = -13$$

Step 3: Complete the square: $\frac{1}{2}(-8) = -4$

$$(-4)^2 = 16$$

Add 16 to both sides of the equation.

$$m^2 - 8m + 16 = -13 + 16$$
$$m^2 - 8m + 16 = 3$$

Step 4: Factor: $(m - 4)^2 = 3$

Step 5: Solve using the square root property:

$$(m - 4)^2 = 3$$
$$m - 4 = \pm\sqrt{3}$$
$$m = 4 \pm \sqrt{3}$$

The solution set is $\{4 - \sqrt{3}, 4 + \sqrt{3}\}$.

c) Ask yourself, "Can I solve $3t^2 + 8t + 7 = 0$ by factoring?" No, $3t^2 + 8t + 7$ does not factor. Completing the square would not be a very efficient way to solve the equation because the coefficient of t^2 is 3, and dividing the equation by 3 would give us $t^2 + \frac{8}{3}t + \frac{7}{3} = 0$.

We will solve $3t^2 + 8t + 7 = 0$ using the quadratic formula.

Identify a, b, and c: $a = 3$ $b = 8$ $c = 7$

$$t = \frac{-b \pm \sqrt{b^2 - 4ac}}{2a}$$ Quadratic formula

$$= \frac{-(8) \pm \sqrt{(8)^2 - 4(3)(7)}}{2(3)}$$ Substitute $a = 3$, $b = 8$, and $c = 7$.

$$= \frac{-8 \pm \sqrt{64 - 84}}{6}$$ Perform the operations.

$$= \frac{-8 \pm \sqrt{-20}}{6}$$

$$= \frac{-8 \pm 2i\sqrt{5}}{6}$$ $\sqrt{-20} = i\sqrt{4}\sqrt{5} = 2i\sqrt{5}$

$$= \frac{2(-4 \pm i\sqrt{5})}{6}$$ Factor out 2 in the numerator.

$$= \frac{-4 \pm i\sqrt{5}}{3}$$ Divide numerator and denominator by 2 to simplify.

$$= -\frac{4}{3} \pm \frac{\sqrt{5}}{3}i$$ Write in the form $a + bi$.

The solution set is $\left\{ -\frac{4}{3} - \frac{\sqrt{5}}{3}i, -\frac{4}{3} + \frac{\sqrt{5}}{3}i \right\}$.

d) Which method should we use to solve $(2z - 7)^2 - 6 = 0$?

We *could* square the binomial, combine like terms, then solve, possibly, by factoring or using the quadratic formula. However, this would be very inefficient. The equation contains a squared quantity and a constant.

We will solve $(2z - 7)^2 - 6 = 0$ using the square root property.

$$(2z - 7)^2 - 6 = 0$$
$$(2z - 7)^2 = 6$$ Add 6 to each side.
$$2z - 7 = \pm\sqrt{6}$$ Square root property
$$2z = 7 \pm \sqrt{6}$$ Add 7 to each side.
$$z = \frac{7 \pm \sqrt{6}}{2}$$ Divide by 2.

The solution set is $\left\{ \frac{7 - \sqrt{6}}{2}, \frac{7 + \sqrt{6}}{2} \right\}$.

> **YOU TRY 1** Solve.
> a) $2k^2 + 3 = 9k$
> b) $2r^2 + 3r - 2 = 0$
> c) $(n - 8)^2 + 9 = 0$
> d) $y^2 + 4y = -10$

ANSWERS TO YOU TRY EXERCISES

1) a) $\left\{\dfrac{9 - \sqrt{57}}{4}, \dfrac{9 + \sqrt{57}}{4}\right\}$ b) $\left\{-2, \dfrac{1}{2}\right\}$ c) $\{8 - 3i, 8 + 3i\}$ d) $\{-2 - i\sqrt{6}, -2 + i\sqrt{6}\}$

Putting It All Together Exercises

E Evaluate Do the exercises, and check your work.

Objective 1: Decide Which Method to Use to Solve a Quadratic Equation

Keep in mind the four methods we have learned for solving quadratic equations: *factoring, the square root property, completing the square, and the quadratic formula.* Solve the equations using one of these methods.

1) $z^2 - 50 = 0$
2) $j^2 - 6j = 8$
3) $a(a + 1) = 20$
4) $2x^2 + 6 = 3x$
5) $u^2 + 7u + 9 = 0$
6) $3p^2 - p - 4 = 0$
7) $2k(2k + 7) = 3(k + 1)$
8) $2 = (w + 3)^2 + 8$
9) $m^2 + 14m + 60 = 0$
10) $\dfrac{1}{2}y^2 = \dfrac{3}{4} - \dfrac{1}{2}y$
11) $10 + (3b - 1)^2 = 4$
12) $c^2 + 8c + 25 = 0$
13) $\dfrac{9}{2a^2} = \dfrac{1}{6} + \dfrac{1}{a}$
14) $100 = 4d^2$
15) $r^2 - 4r = 3$
16) $2t^3 + 108t = -30t^2$
17) $p(p + 8) = 3(p^2 + 2) + p$
18) $h^2 = h$
19) $\dfrac{10}{z} = 1 + \dfrac{21}{z^2}$
20) $2s(2s + 3) = 4s + 5$
21) $(3v + 4)(v - 2) = -9$
22) $34 = 6y - y^2$
23) $(c - 5)^2 + 16 = 0$
24) $(2b + 1)(b + 5) = -7$
25) $3g = g^2$
26) $5z^2 + 15z + 30 = 0$
27) $4m^3 = 9m$
28) $1 = \dfrac{x^2}{12} - \dfrac{x}{3}$
29) $\dfrac{1}{3}q^2 + \dfrac{5}{6}q + \dfrac{4}{3} = 0$
30) $-3 = (12d + 5)^2 + 6$

R Rethink

R1) Which method is easiest to use? Which one is the most difficult?

R2) Why would you need to complete the square? Why must the coefficient be equal to 1 before you complete the square?

10.3 Equations in Quadratic Form

P Prepare / O Organize

What are your objectives for Section 10.3?	How can you accomplish each objective?
1 Solve Quadratic Equations Resulting from Equations Containing Fractions or Radicals	• Review the procedure for eliminating fractions in equations by using the LCD. • Review the procedures for solving equations containing radicals. • Complete the given examples on your own. • Complete You Trys 1 and 2.
2 Solve an Equation in Quadratic Form by Factoring	• Be able to recognize when an equation is in *quadratic form*. • Apply the same methods for solving equations in *quadratic form* as you would use with *quadratic equations*. • Check all solutions in the original equation. • Complete the given example on your own. • Complete You Try 3.
3 Solve an Equation in Quadratic Form Using Substitution	• Understand why and how we can use substitution. • Be familiar with *all* methods for solving quadratic equations. • Substitute the original variables back into the equation to finish solving. • Complete the given example on your own. • Complete You Trys 4 and 5.
4 Use Substitution for a Binomial to Solve a Quadratic Equation	• Choose a variable for the binomial quantity and substitute this into the equation. • Be familiar with *all* methods for solving quadratic equations. • Substitute the original binomial back into the equation to finish solving. • Complete the given example on your own. • Complete You Try 6.

 Read the explanations, follow the examples, take notes, and complete the You Trys.

In Chapters 8 and 9, we solved some equations that were *not* quadratic but could be rewritten in the form of a quadratic equation, $ax^2 + bx + c = 0$. Two such examples are:

$$\frac{10}{x} - \frac{7}{x+1} = \frac{2}{3} \qquad \text{and} \qquad r + \sqrt{r} = 12$$

Rational equation (Ch. 8)   Radical equation (Ch. 9)

We will review how to solve each type of equation.

1 Solve Quadratic Equations Resulting from Equations Containing Fractions or Radicals

EXAMPLE 1

Solve $\dfrac{10}{x} - \dfrac{7}{x+1} = \dfrac{2}{3}$.

Solution

To solve an equation containing rational expressions, *multiply the equation by the LCD of all of the fractions to eliminate the denominators,* then solve.

$$\text{LCD} = 3x(x+1)$$

> **W Hint**
> Why do you have to check your solutions in the original equations?

$$3x(x+1)\left(\dfrac{10}{x} - \dfrac{7}{x+1}\right) = 3x(x+1)\left(\dfrac{2}{3}\right)$$ Multiply both sides of the equation by the LCD of the fractions.

$$3\cancel{x}(x+1) \cdot \dfrac{10}{\cancel{x}} - 3x\cancel{(x+1)} \cdot \dfrac{7}{\cancel{x+1}} = \cancel{3}x(x+1) \cdot \left(\dfrac{2}{\cancel{3}}\right)$$ Distribute, and divide out common factors.

$$30(x+1) - 3x(7) = 2x(x+1)$$
$$30x + 30 - 21x = 2x^2 + 2x \quad \text{Distribute.}$$
$$9x + 30 = 2x^2 + 2x \quad \text{Combine like terms.}$$
$$0 = 2x^2 - 7x - 30 \quad \text{Write in the form } ax^2 + bx + c = 0.$$
$$0 = (2x+5)(x-6) \quad \text{Factor.}$$

$$2x + 5 = 0 \quad \text{or} \quad x - 6 = 0 \quad \text{Set each factor equal to zero.}$$
$$2x = -5$$
$$x = -\dfrac{5}{2} \quad \text{or} \quad x = 6 \quad \text{Solve.}$$

Recall that you *must* check the proposed solutions in the original equation to be certain they do not make a denominator equal zero. The solution set is $\left\{-\dfrac{5}{2}, 6\right\}$.

[YOU TRY 1] Solve $\dfrac{1}{m} = \dfrac{1}{2} + \dfrac{m}{m+4}$.

EXAMPLE 2

Solve $r + \sqrt{r} = 12$.

Solution

The first step in solving a radical equation is getting a radical on a side by itself.

$$r + \sqrt{r} = 12$$
$$\sqrt{r} = 12 - r \quad \text{Subtract } r \text{ from each side.}$$
$$(\sqrt{r})^2 = (12 - r)^2 \quad \text{Square both sides.}$$
$$r = 144 - 24r + r^2$$
$$0 = r^2 - 25r + 144 \quad \text{Write in the form } ax^2 + bx + c = 0.$$
$$0 = (r - 16)(r - 9) \quad \text{Factor.}$$

$$r - 16 = 0 \quad \text{or} \quad r - 9 = 0 \quad \text{Set each factor equal to zero.}$$
$$r = 16 \quad \text{or} \quad r = 9 \quad \text{Solve.}$$

SECTION 10.3   **Equations in Quadratic Form**

Recall that you *must* check the proposed solutions in the *original* equation.

Check $r = 16$:
$$r + \sqrt{r} = 12$$
$$16 + \sqrt{16} \stackrel{?}{=} 12$$
$$16 + 4 = 12 \quad \text{False}$$

Check $r = 9$:
$$r + \sqrt{r} = 12$$
$$9 + \sqrt{9} \stackrel{?}{=} 12$$
$$9 + 3 = 12 \quad \text{True}$$

16 is an extraneous solution. The solution set is {9}.

[**YOU TRY 2**]   Solve $y + 3\sqrt{y} = 10$.

2 Solve an Equation in Quadratic Form by Factoring

Some equations that are not quadratic can be solved using the same methods that can be used to solve quadratic equations. These are called **equations in quadratic form.** Some examples of equations in quadratic form are:

$$x^4 - 10x^2 + 9 = 0, \qquad t^{2/3} + t^{1/3} - 6 = 0, \qquad 2n^4 - 5n^2 = -1$$

Let's compare the equations above to *quadratic equations* to understand why they are said to be in quadratic form.

> **Note**
>
> **COMPARE**
>
> | An Equation in Quadratic Form | to | A Quadratic Equation |
> |---|---|---|
> | This exponent is *twice* this exponent. | | This exponent is *twice* this exponent. |
> | $x^4 - 10x^2 + 9 = 0$ | | $x^2 - 10x^1 + 9 = 0$ |
> | This exponent is *twice* this exponent. | | This exponent is *twice* this exponent. |
> | $t^{2/3} + t^{1/3} - 6 = 0$ | | $t^2 + t^1 - 6 = 0$ |
> | This exponent is *twice* this exponent. | | This exponent is *twice* this exponent. |
> | $2n^4 - 5n^2 = -1$ | | $2n^2 - 5n^1 = -1$ |

This pattern enables us to work with equations in quadratic form like we can work with quadratic equations.

EXAMPLE 3

Solve.

a) $x^4 - 10x^2 + 9 = 0$ b) $t^{2/3} + t^{1/3} - 6 = 0$

Solution

a) Let's compare $x^4 - 10x^2 + 9 = 0$ to $x^2 - 10x + 9 = 0$.

We can factor $x^2 - 10x + 9$:

$$(x - 9)(x - 1)$$

Confirm by multiplying using FOIL:

$$(x - 9)(x - 1) = x^2 - x - 9x + 9$$
$$= x^2 - 10x + 9$$

Factor $x^4 - 10x^2 + 9$ in a similar way since the exponent, 4, of the first term is twice the exponent, 2, of the second term:

$$x^4 - 10x^2 + 9 = (x^2 - 9)(x^2 - 1)$$

Confirm by multiplying using FOIL:

$$(x^2 - 9)(x^2 - 1) = x^4 - x^2 - 9x^2 + 9$$
$$= x^4 - 10x^2 + 9$$

We can solve $x^4 - 10x^2 + 9 = 0$ by factoring.

$$x^4 - 10x^2 + 9 = 0$$
$$(x^2 - 9)(x^2 - 1) = 0 \quad \text{Factor.}$$

$x^2 - 9 = 0$ or $x^2 - 1 = 0$ Set each factor equal to 0.
$x^2 = 9$ $x^2 = 1$ Square root property
$x = \pm 3$ $x = \pm 1$

The check is left to the student. Check the answers in the *original* equation. The solution set is $\{-3, -1, 1, 3\}$.

> **W Hint**
> Be sure you understand *why* you are performing each step as you are writing out the examples.

b) Compare $t^{2/3} + t^{1/3} - 6 = 0$ to $t^2 + t - 6 = 0$.

We can factor $t^2 + t - 6$:

$$(t + 3)(t - 2)$$

Confirm by multiplying using FOIL:

$$(t + 3)(t - 2) = t^2 - 2t + 3t - 6$$
$$= t^2 + t - 6$$

Factor $t^{2/3} + t^{1/3} - 6$ in a similar way since the exponent, $\frac{2}{3}$, of the first term is twice the exponent, $\frac{1}{3}$, of the second term:

$$t^{2/3} + t^{1/3} - 6 = (t^{1/3} + 3)(t^{1/3} - 2)$$

Confirm by multiplying using FOIL:

$$(t^{1/3} + 3)(t^{1/3} - 2) = t^{2/3} - 2t^{1/3} + 3t^{1/3} - 6$$
$$= t^{2/3} + t^{1/3} - 6$$

We can solve $t^{2/3} + t^{1/3} - 6 = 0$ by factoring.

$$t^{2/3} + t^{1/3} - 6 = 0$$
$$(t^{1/3} + 3)(t^{1/3} - 2) = 0 \quad \text{Factor.}$$

$t^{1/3} + 3 = 0$ or $t^{1/3} - 2 = 0$ Set each factor equal to 0.
$t^{1/3} = -3$ $t^{1/3} = 2$ Isolate the constant.
$\sqrt[3]{t} = -3$ $\sqrt[3]{t} = 2$ $t^{1/3} = \sqrt[3]{t}$
$(\sqrt[3]{t})^3 = (-3)^3$ $(\sqrt[3]{t})^3 = 2^3$ Cube both sides.
$t = -27$ or $t = 8$ Solve.

The check is left to the student. The solution set is $\{-27, 8\}$.

[YOU TRY 3] Solve.

a) $r^4 - 13r^2 + 36 = 0$ b) $c^{2/3} + 4c^{1/3} - 5 = 0$

3 Solve an Equation in Quadratic Form Using Substitution

The equations in Example 3 can also be solved using a method called **substitution**. We will illustrate the method in Example 4.

EXAMPLE 4 Solve $x^4 - 10x^2 + 9 = 0$ using substitution.

Solution

$$x^4 - 10x^2 + 9 = 0$$
$$\downarrow$$
$$x^4 = (x^2)^2$$

To rewrite $x^4 - 10x^2 + 9 = 0$ in quadratic form, let $u = x^2$.

If $u = x^2$, then $u^2 = x^4$.

W Hint
When using substitution, make special notations so that you do not forget to substitute back to the original variables.

$$x^4 - 10x^2 + 9 = 0$$
$$u^2 - 10u + 9 = 0 \qquad \text{Substitute } u^2 \text{ for } x^4 \text{ and } u \text{ for } x^2.$$
$$(u - 9)(u - 1) = 0 \qquad \text{Solve by factoring.}$$
$$u - 9 = 0 \quad \text{or} \quad u - 1 = 0 \qquad \text{Set each factor equal to 0.}$$
$$u = 9 \quad \text{or} \quad u = 1 \qquad \text{Solve for } u.$$

Be careful! $u = 9$ and $u = 1$ are *not* the solutions to $x^4 - 10x^2 + 9 = 0$. **We still need to solve for x.** Above we let $u = x^2$. To solve for x, substitute 9 for u and solve for x, and then substitute 1 for u and solve for x.

	$u = x^2$	$u = x^2$	
Substitute 9 for u.	$9 = x^2$	$1 = x^2$	Substitute 1 for u.
Square root property	$\pm 3 = x$	$\pm 1 = x$	Square root property

The solution set is $\{-3, -1, 1, 3\}$. This is the same as the result we obtained in Example 3a).

[YOU TRY 4] Solve by substitution.

a) $r^4 - 13r^2 + 36 = 0$ b) $c^{2/3} + 4c^{1/3} - 5 = 0$

If, after substitution, an equation cannot be solved by factoring, we can use the quadratic formula.

EXAMPLE 5 Solve $2n^4 - 5n^2 = -1$.

Solution

Write the equation in standard form: $2n^4 - 5n^2 + 1 = 0$.

Can we solve the equation by factoring? *No.*

We will solve $2n^4 - 5n^2 + 1 = 0$ *using the quadratic formula.* Begin with substitution.

If $u = n^2$, then $u^2 = n^4$.

$$2n^4 - 5n^2 + 1 = 0$$
$$2u^2 - 5u + 1 = 0 \quad \text{Substitute } u^2 \text{ for } n^4 \text{ and } u \text{ for } n^2.$$

$$u = \frac{-(-5) \pm \sqrt{(-5)^2 - 4(2)(1)}}{2(2)} \quad a = 2, b = -5, c = 1$$

$$u = \frac{5 \pm \sqrt{25 - 8}}{4} = \frac{5 \pm \sqrt{17}}{4}$$

Note that $u = \frac{5 \pm \sqrt{17}}{4}$ does not solve the *original* equation. We must solve for x using the fact that $u = x^2$. Since $u = \frac{5 \pm \sqrt{17}}{4}$ means $u = \frac{5 + \sqrt{17}}{4}$ or $u = \frac{5 - \sqrt{17}}{4}$, we get

$$u = x^2 \qquad\qquad u = x^2$$
$$\frac{5 + \sqrt{17}}{4} = x^2 \qquad \frac{5 - \sqrt{17}}{4} = x^2$$
$$\pm\sqrt{\frac{5 + \sqrt{17}}{4}} = x \qquad \pm\sqrt{\frac{5 - \sqrt{17}}{4}} = x \quad \text{Square root property}$$
$$\frac{\pm\sqrt{5 + \sqrt{17}}}{2} = x \qquad \frac{\pm\sqrt{5 - \sqrt{17}}}{2} = x \quad \sqrt{4} = 2$$

The solution set is $\left\{ \frac{\sqrt{5 + \sqrt{17}}}{2}, -\frac{\sqrt{5 + \sqrt{17}}}{2}, \frac{\sqrt{5 - \sqrt{17}}}{2}, -\frac{\sqrt{5 - \sqrt{17}}}{2} \right\}$.

[YOU TRY 5] Solve $2k^4 + 3 = 9k^2$.

4 Use Substitution for a Binomial to Solve a Quadratic Equation

We can use substitution to solve an equation like the one in Example 6.

EXAMPLE 6 Solve $2(3a + 1)^2 - 7(3a + 1) - 4 = 0$.

Solution

The binomial $3a + 1$ appears as a *squared quantity* and as a *linear quantity*. Begin by using substitution.

$$\text{Let } u = 3a + 1. \quad \text{Then, } u^2 = (3a + 1)^2.$$

Substitute: $2(3a + 1)^2 - 7(3a + 1) - 4 = 0$
$2u^2 - 7u - 4 = 0$

Does $2u^2 - 7u - 4 = 0$ factor? *Yes.* Solve by factoring.

$$(2u + 1)(u - 4) = 0 \qquad \text{Factor } 2u^2 - 7u - 4 = 0.$$

$2u + 1 = 0 \quad \text{or} \quad u - 4 = 0 \qquad$ Set each factor equal to 0.

$u = -\dfrac{1}{2} \quad \text{or} \quad u = 4 \qquad$ Solve for u.

Solve for a using $u = 3a + 1$.

W Hint
Don't forget to solve for the variable in the *original* equation.

When $u = -\dfrac{1}{2}$:
$u = 3a + 1$
$-\dfrac{1}{2} = 3a + 1$
Subtract 1. $-\dfrac{3}{2} = 3a$
Multiply by $\dfrac{1}{3}$. $-\dfrac{1}{2} = a$

When $u = 4$:
$u = 3a + 1$
$4 = 3a + 1$
$3 = 3a$ Subtract 1.
$1 = a$ Divide by 3.

The solution set is $\left\{-\dfrac{1}{2}, 1\right\}$. Check these values in the original equation.

[YOU TRY 6] Solve $3(2p - 1)^2 - 11(2p - 1) + 10 = 0$.

ANSWERS TO [YOU TRY] EXERCISES

1) $\left\{-2, \dfrac{4}{3}\right\}$ 2) $\{4\}$ 3) a) $\{-3, -2, 2, 3\}$ b) $\{-125, 1\}$ 4) a) $\{-3, -2, 2, 3\}$
b) $\{-125, 1\}$ 5) $\left\{\dfrac{\sqrt{9 + \sqrt{57}}}{2}, -\dfrac{\sqrt{9 + \sqrt{57}}}{2}, \dfrac{\sqrt{9 - \sqrt{57}}}{2}, -\dfrac{\sqrt{9 - \sqrt{57}}}{2}\right\}$ 6) $\left\{\dfrac{4}{3}, \dfrac{3}{2}\right\}$

Evaluate 10.3 Exercises

Do the exercises, and check your work.

Objective 1: Solve Quadratic Equations Resulting from Equations Containing Fractions or Radicals

Solve.

1) $t - \dfrac{48}{t} = 8$

2) $z + 11 = -\dfrac{24}{z}$

3) $\dfrac{2}{x} + \dfrac{6}{x-2} = -\dfrac{5}{2}$

4) $\dfrac{3}{y} - \dfrac{6}{y-1} = \dfrac{1}{2}$

5) $1 = \dfrac{2}{c} + \dfrac{1}{c-5}$

6) $\dfrac{2}{g} = 1 + \dfrac{g}{g+5}$

7) $\dfrac{3}{2v+2} + \dfrac{1}{v} = \dfrac{3}{2}$

8) $\dfrac{1}{b+3} + \dfrac{1}{b} = \dfrac{1}{3}$

9) $\dfrac{9}{n^2} = 5 + \dfrac{4}{n}$

10) $3 - \dfrac{16}{a^2} = \dfrac{8}{a}$

11) $\dfrac{5}{6r} = 1 - \dfrac{r}{6r-6}$

12) $\dfrac{7}{4} - \dfrac{x}{4x+4} = \dfrac{1}{x}$

13) $g = \sqrt{g + 20}$

14) $c = \sqrt{7c - 6}$

15) $a = \sqrt{\dfrac{14a - 8}{5}}$

16) $k = \sqrt{\dfrac{6 - 11k}{2}}$

17) $v + \sqrt{v} = 2$

18) $p - \sqrt{p} = 6$

19) $x = 5\sqrt{x} - 4$

20) $10 = m - 3\sqrt{m}$

21) $2 + \sqrt{2y - 1} = y$

22) $1 - \sqrt{5t + 1} = -t$

23) $2 = \sqrt{6k + 4} - k$

24) $\sqrt{10 - 3q} - 6 = q$

Mixed Exercises: Objectives 2–3

Determine whether each is an equation in quadratic form. Do *not* solve.

25) $n^4 - 12n^2 + 32 = 0$

26) $p^6 + 8p^3 - 9 = 0$

27) $2t^6 + 3t^3 - 5 = 0$

28) $a^4 - 4a - 3 = 0$

29) $c^{2/3} - 4c - 6 = 0$

30) $3z^{2/3} + 2z^{1/3} + 1 = 0$

31) $m + 9m^{1/2} = 4$

32) $2x^{1/2} - 5x^{1/4} = 2$

33) $5k^4 + 6k - 7 = 0$

34) $r^{-2} = 10 - 4r^{-1}$

Solve.

35) $x^4 - 10x^2 + 9 = 0$

36) $d^4 - 29d^2 + 100 = 0$

37) $p^4 - 11p^2 + 28 = 0$

38) $k^4 - 9k^2 + 8 = 0$

39) $c^4 + 9c^2 = -18$

40) $a^4 + 12a^2 = -35$

41) $b^{2/3} + 3b^{1/3} + 2 = 0$

42) $z^{2/3} + z^{1/3} - 12 = 0$

43) $p^{2/3} - p^{1/3} = 6$

44) $t^{2/3} - 6t^{1/3} = 40$

45) $4h^{1/2} + 21 = h$

46) $s + 12 = -7s^{1/2}$

47) $2a - 5a^{1/2} - 12 = 0$

48) $2w = 9w^{1/2} + 18$

49) $9n^4 = -15n^2 - 4$

50) $4h^4 + 19h^2 + 12 = 0$

51) $z^4 - 2z^2 = 15$

52) $a^4 + 2a^2 = 24$

53) $w^4 - 6w^2 + 2 = 0$

54) $p^4 - 8p^2 + 3 = 0$

55) $2m^4 + 1 = 7m^2$

56) $8x^4 + 2 = 9x^2$

57) $t^{-2} - 4t^{-1} - 12 = 0$

58) $d^{-2} + d^{-1} - 6 = 0$

59) $4 = 13y^{-1} - 3y^{-2}$

60) $14h^{-1} + 3 = 5h^{-2}$

Objective 4: Use Substitution for a Binomial to Solve a Quadratic Equation

Solve.

61) $(x - 2)^2 + 11(x - 2) + 24 = 0$

62) $(r + 1)^2 - 3(r + 1) - 10 = 0$

63) $2(3q + 4)^2 - 13(3q + 4) + 20 = 0$

64) $4(2b - 3)^2 - 9(2b - 3) - 9 = 0$

65) $(5a - 3)^2 + 6(5a - 3) = -5$

66) $(3z - 2)^2 - 8(3z - 2) = 20$

67) $3(k + 8)^2 + 5(k + 8) = 12$

68) $5(t + 9)^2 + 37(t + 9) + 14 = 0$

69) $1 - \dfrac{8}{2w + 1} = -\dfrac{16}{(2w + 1)^2}$

70) $1 - \dfrac{8}{4p + 3} = -\dfrac{12}{(4p + 3)^2}$

71) $1 + \dfrac{2}{h - 3} = \dfrac{1}{(h - 3)^2}$

72) $\dfrac{2}{(c + 6)^2} + \dfrac{2}{(c + 6)} = 1$

Write an equation and solve.

73) It takes Kevin 3 hr longer than Walter to build a tree house. Together they can do the job in 2 hr. How long would it take each man to build the tree house on his own?

74) It takes one pipe 4 hours more to empty a pool than it takes another pipe to fill a pool. If both pipes are accidentally left open, it takes 24 hr to fill the pool. How long does it take the single pipe to fill the pool?

75) A boat can travel 9 mi downstream and then 6 mi back upstream in 1 hr. If the speed of the current is 3 mph, what is the speed of the boat in still water?

76) A plane can travel 800 mi with the wind and then 650 mi back against the wind in 5 hr. If the wind blows at 30 mph, what is the speed of the plane?

77) A large fish tank at an aquarium needs to be emptied so that it can be cleaned. When its large and small drains are opened together, the tank can be emptied in 2 hr. By itself, it takes the small drain 3 hr longer to empty the tank than it takes the large drain to empty the tank on its own. How much time would it take for each drain to empty the pool on its own?

78) Working together, a professor and her teaching assistant can grade a set of exams in 1.2 hr. On her own, the professor can grade the tests 1 hr faster than the teaching assistant can grade them on her own. How long would it take for each person to grade the test by herself?

79) Miguel took his son to college in Boulder, Colorado, 600 mi from their hometown. On his way home, he was slowed by a snowstorm so that his speed was 10 mph less than when he was driving to Boulder. His total driving time was 22 hr. How fast did Miguel drive on each leg of the trip?

80) Nariko was training for a race and went out for a run. Her speed was 2 mph faster during the first 6 mi than it was for the last 3 mi. If her total running time was $1\frac{3}{4}$ hr, what was her speed on each part of the run?

 Rethink

R1) What does it mean for an equation to be in quadratic form?

R2) What is the hardest part of using substitution?

R3) Can you solve any of the problems in this section by using two different methods?

10.4 Formulas and Applications

P Prepare

What are your objectives for Section 10.4?	How can you accomplish each objective?
1 Solve a Formula for a Variable	• Use the procedures for solving quadratic equations or equations containing radicals. • Complete the given examples on your own. • Complete You Trys 1 and 2.
2 Solve an Applied Problem Involving Volume	• Use the **Steps for Solving Applied Problems**. • Complete the given example on your own. • Complete You Try 3.
3 Solve an Applied Problem Involving Area	• Use the **Steps for Solving Applied Problems**. • Complete the given example on your own. • Complete You Try 4.
4 Solve an Applied Problem Using a Quadratic Equation	• Read the problem carefully, and understand how the given information relates to the given equation. • Complete the given example on your own.

W Work Read the explanations, follow the examples, take notes, and complete the You Trys.

Sometimes, solving a formula for a variable involves using one of the techniques we've learned for solving a quadratic equation or for solving an equation containing a radical.

1 Solve a Formula for a Variable

EXAMPLE 1

Solve $v = \sqrt{\dfrac{300VP}{m}}$ for m.

Solution

Put a box around the m. The goal is to get m on a side by itself.

$$v = \sqrt{\dfrac{300VP}{\boxed{m}}}$$

$$v^2 = \dfrac{300VP}{\boxed{m}} \qquad \text{Square both sides.}$$

Since we are solving for m and it is in the denominator, multiply both sides by m to eliminate the denominator.

$$\boxed{m}v^2 = 300VP \qquad \text{Multiply both sides by } m.$$

$$m = \frac{300VP}{v^2} \qquad \text{Divide both sides by } v^2.$$

[YOU TRY 1] Solve $v = \sqrt{\dfrac{2E}{m}}$ for m.

We may need to use the quadratic formula to solve a formula for a variable. Compare the following equations. Each equation is *quadratic in x* because each is written in the form $ax^2 + bx + c = 0$.

$$8x^2 + 3x - 2 = 0 \qquad \text{and} \qquad 8x^2 + tx - z = 0$$

$$a = 8 \quad b = 3 \quad c = -2 \qquad\qquad a = 8 \quad b = t \quad c = -z$$

To solve the equations for x, we can use the quadratic formula.

EXAMPLE 2 Solve for x.

a) $8x^2 + 3x - 2 = 0$ b) $8x^2 + tx - z = 0$

Solution

a) $8x^2 + 3x - 2$ does not factor, so we will solve using the quadratic formula.

$$8x^2 + 3x - 2 = 0$$
$$a = 8 \quad b = 3 \quad c = -2$$
$$x = \frac{-3 \pm \sqrt{(3)^2 - 4(8)(-2)}}{2(8)} = \frac{-3 \pm \sqrt{9 + 64}}{16} = \frac{-3 \pm \sqrt{73}}{16}$$

The solution set is $\left\{ \dfrac{-3 - \sqrt{73}}{16}, \dfrac{-3 + \sqrt{73}}{16} \right\}$.

b) Solve $8x^2 + tx - z = 0$ for x using the quadratic formula.

$$a = 8 \quad b = t \quad c = -z$$

$$x = \frac{-t \pm \sqrt{t^2 - 4(8)(-z)}}{2(8)} \qquad x = \frac{-b \pm \sqrt{b^2 - 4ac}}{2a}$$

$$= \frac{-t \pm \sqrt{t^2 + 32z}}{16} \qquad \text{Perform the operations.}$$

The solution set is $\left\{ \dfrac{-t - \sqrt{t^2 + 32z}}{16}, \dfrac{-t + \sqrt{t^2 + 32z}}{16} \right\}$.

[YOU TRY 2] Solve for n.

a) $3n^2 + 5n - 1 = 0$ b) $3n^2 + pn - r = 0$

2 Solve an Applied Problem Involving Volume

EXAMPLE 3

A rectangular piece of cardboard is 5 in. longer than it is wide. A square piece that measures 2 in. on each side is cut from each corner, then the sides are turned up to make an uncovered box with volume 252 in^3. Find the length and width of the original piece of cardboard.

Solution

Step 1: **Read** the problem carefully. Draw a picture.

Step 2: **Choose a variable** to represent the unknown, and define the other unknown in terms of this variable.

$$\text{Let} \quad x = \text{the width of the cardboard}$$
$$x + 5 = \text{the length of the cardboard}$$

> **W Hint**
> Draw a picture to help you understand the problem.

Step 3: **Translate** the information that appears in English into an algebraic equation.

The volume of a box is (length)(width)(height). We will use the formula (length)(width)(height) = 252.

Original Cardboard

Box

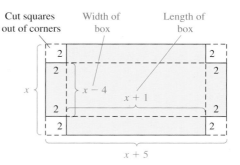

The figure on the left shows the original piece of cardboard with the sides labeled. The figure on the right illustrates how to label the box when the squares are cut out of the corners. When the sides are folded along the dotted lines, we must label the length, width, and height of the box.

Length of box = Length of original cardboard − Length of side cut out on the left − Length of side cut out on the right

= $x + 5$ − 2 − 2
= $x + 1$

Width of box = Width of original cardboard − Length of side cut out on top − Length of side cut out on bottom

= x − 2 − 2
= $x - 4$

Height of box = Length of side cut out
= 2

Statement: Volume of box = (length)(width)(height)
Equation: $252 = (x + 1)(x - 4)(2)$

SECTION 10.4 **Formulas and Applications** 647

Step 4: **Solve** the equation.

$$252 = (x + 1)(x - 4)(2)$$
$$126 = (x + 1)(x - 4) \quad \text{Divide both sides by 2.}$$
$$126 = x^2 - 3x - 4 \quad \text{Multiply.}$$
$$0 = x^2 - 3x - 130 \quad \text{Write in standard form.}$$
$$0 = (x + 10)(x - 13) \quad \text{Factor.}$$
$$x + 10 = 0 \quad \text{or} \quad x - 13 = 0 \quad \text{Set each factor equal to zero.}$$
$$x = -10 \quad \text{or} \quad x = 13 \quad \text{Solve.}$$

Step 5: **Check** the answer, and **interpret** the solution as it relates to the problem.

Because x represents the width, it cannot be negative. Therefore, the width of the original piece of cardboard is 13 in.

The length of the cardboard is $x + 5$, so $13 + 5 = 18$ in.

Width of cardboard = 13 in.   Length of cardboard = 18 in.

Check:

Width of box = $13 - 4 = 9$ in.; Length of box = $13 + 1 = 14$ in.;
Height of box = 2 in.
Volume of box = $9(14)(2) = 252$ in^3.

[YOU TRY 3]   The width of a rectangular piece of cardboard is 2 in. less than its length. A square piece that measures 3 in. on each side is cut from each corner, then the sides are turned up to make a box with volume 504 in^3. Find the length and width of the original piece of cardboard.

3 Solve an Applied Problem Involving Area

EXAMPLE 4   A rectangular pond is 20 ft long and 12 ft wide. The pond is bordered by a strip of grass of uniform (the same) width. The area of the grass is 320 ft^2. How wide is the border of grass around the pond?

Solution

Step 1: **Read** the problem carefully. Draw a picture.

Step 2: **Choose a variable** to represent the unknown, and define the other unknowns in terms of this variable.

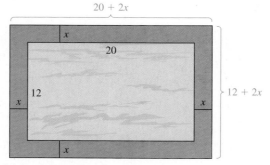

x = width of the strip of grass
$20 + 2x$ = length of pond plus two strips of grass
$12 + 2x$ = width of pond plus two strips of grass

Step 3: **Translate** from English into an algebraic equation.

We know that the area of the grass border is 320 ft². We can calculate the area of the pond since we know its length and width. The pond plus grass border forms a large rectangle of length $20 + 2x$ and width $12 + 2x$. The equation will come from the following relationship:

$$\text{Statement:} \quad \begin{array}{c}\text{Area of pond}\\ \text{plus grass}\end{array} - \begin{array}{c}\text{Area of}\\ \text{pond}\end{array} = \begin{array}{c}\text{Area of}\\ \text{grass border}\end{array}$$

$$\text{Equation:} \quad (20 + 2x)(12 + 2x) \ - \ 20(12) \ = \ 320$$

Step 4: **Solve** the equation.

$$\begin{aligned} (20 + 2x)(12 + 2x) - 20(12) &= 320 \\ 240 + 64x + 4x^2 - 240 &= 320 &&\text{Multiply.} \\ 4x^2 + 64x &= 320 &&\text{Combine like terms.} \\ x^2 + 16x &= 80 &&\text{Divide by 4.} \\ x^2 + 16x - 80 &= 0 &&\text{Write in standard form.} \\ (x + 20)(x - 4) &= 0 &&\text{Factor.} \\ x + 20 = 0 \quad \text{or} \quad x - 4 &= 0 &&\text{Set each factor equal to 0.} \\ x = -20 \quad \text{or} \quad x &= 4 &&\text{Solve.} \end{aligned}$$

Step 5: **Check** the answer, and **interpret** the solution as it relates to the problem.

x represents the width of the strip of grass, so x cannot equal -20.

The width of the strip of grass is 4 ft.

Check: Substitute $x = 4$ into the equation written in Step 3.

$$\begin{aligned} [20 + 2(4)][12 + 2(4)] - 20(12) &\stackrel{?}{=} 320 \\ (28)(20) - 240 &\stackrel{?}{=} 320 \\ 560 - 240 &= 320 \ \checkmark \end{aligned}$$

[**YOU TRY 4**] A rectangular pond is 6 ft wide and 10 ft long and is surrounded by a concrete border of uniform width. The area of the border is 80 ft². Find the width of the border.

4 Solve an Applied Problem Using a Quadratic Equation

EXAMPLE 5

The total tourism-related output in the United States from 2000 to 2004 can be modeled by

$$y = 16.4x^2 - 50.6x + 896$$

where x is the number of years since 2000 and y is the total tourism-related output in billions of dollars. (www.bea.gov)

a) According to the model, how much money was generated in 2002 due to tourism-related output?

b) In what year was the total tourism-related output about $955 billion?

Solution

a) Because x is the number of years *after* 2000, the year 2002 corresponds to $x = 2$.

$$y = 16.4x^2 - 50.6x + 896$$
$$y = 16.4(2)^2 - 50.6(2) + 896 \quad \text{Substitute 2 for } x.$$
$$y = 860.4$$

The total tourism-related output in 2002 was approximately \$860.4 billion.

b) Because y represents the total tourism-related output (in billions), substitute 955 for y and solve for x.

$$y = 16.4x^2 - 50.6x + 896$$
$$955 = 16.4x^2 - 50.6x + 896 \quad \text{Substitute 955 for } y.$$
$$0 = 16.4x^2 - 50.6x - 59 \quad \text{Write in standard form.}$$

Use the quadratic formula to solve for x.

$$a = 16.4 \quad b = -50.6 \quad c = -59$$
$$x = \frac{50.6 \pm \sqrt{(-50.6)^2 - 4(16.4)(-59)}}{2(16.4)} \quad \text{Substitute the values into the quadratic formula.}$$
$$x \approx 3.99 \approx 4 \text{ or } x \approx -0.90$$

The negative value of x does not make sense in the context of the problem. Use $x \approx 4$, which corresponds to the year 2004. The total tourism-related output was about \$955 billion in 2004.

W Hint
Think carefully about how to relate the information given to the variables in the equation.

ANSWERS TO [YOU TRY] EXERCISES

1) $m = \dfrac{2E}{v^2}$ 2) a) $\left\{\dfrac{-5 - \sqrt{37}}{6}, \dfrac{-5 + \sqrt{37}}{6}\right\}$ b) $\left\{\dfrac{-p - \sqrt{p^2 + 12r}}{6}, \dfrac{-p + \sqrt{p^2 + 12r}}{6}\right\}$
3) length = 20 in., width = 18 in. 4) 2 ft

E Evaluate 10.4 Exercises

Do the exercises, and check your work.

Objective 1: Solve a Formula for a Variable
Solve for the indicated variable.

1) $A = \pi r^2$ for r

2) $V = \dfrac{1}{3}\pi r^2 h$ for r

3) $a = \dfrac{v^2}{r}$ for v

4) $K = \dfrac{1}{2}Iw^2$ for w

5) $E = \dfrac{I}{d^2}$ for d

6) $L = \dfrac{2U}{I^2}$ for I

7) $F = \dfrac{kq_1q_2}{r^2}$ for r

8) $E = \dfrac{kq}{r^2}$ for r

9) $d = \sqrt{\dfrac{4A}{\pi}}$ for A

10) $d = \sqrt{\dfrac{12V}{\pi h}}$ for V

11) $T_p = 2\pi\sqrt{\dfrac{l}{g}}$ for l

12) $V = \sqrt{\dfrac{3RT}{M}}$ for T

13) $T_p = 2\pi\sqrt{\dfrac{l}{g}}$ for g

14) $V = \sqrt{\dfrac{3RT}{M}}$ for M

15) Compare the equations $3x^2 - 5x + 4 = 0$ and $rx^2 + 5x + s = 0$.

  a) How are the equations alike?

  b) How can both equations be solved for x?

16) What method could be used to solve $2t^2 + 7t + 1 = 0$ and $kt^2 + mt + n = 0$ for t? Why?

Solve for the indicated variable.

17) $rx^2 - 5x + s = 0$ for x

18) $cx^2 + dx - 3 = 0$ for x

19) $pz^2 + rz - q = 0$ for z

20) $hr^2 - kr + j = 0$ for r

21) $da^2 - ha = k$ for a

22) $kt^2 + mt = -n$ for t

23) $s = \frac{1}{2}gt^2 + vt$ for t

24) $s = 2\pi rh + \pi r^2$ for r

Mixed Exercises: Objectives 2 and 3
Write an equation, and solve.

25) The length of a rectangular piece of sheet metal is 3 in. longer than its width. A square piece that measures 1 in. on each side is cut from each corner, then the sides are turned up to make a box with volume 70 in^3. Find the length and width of the original piece of sheet metal.

26) The width of a rectangular piece of cardboard is 8 in. less than its length. A square piece that measures 2 in. on each side is cut from each corner, then the sides are turned up to make a box with volume 480 in^3. Find the length and width of the original piece of cardboard.

27) A rectangular swimming pool is 60 ft wide and 80 ft long. A nonskid surface of uniform width is to be installed around the pool. If there is 576 ft^2 of the nonskid material, how wide can the strip of the nonskid surface be?

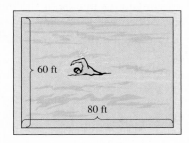

28) A picture measures 10 in. by 12 in. Emilio will get it framed with a border around it so that the total area of the picture plus the frame of uniform width is 168 in^2. How wide is the border?

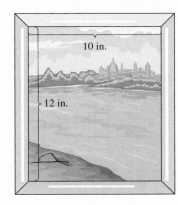

29) The height of a triangular sail is 1 ft less than twice the base of the sail. Find its height and the length of its base if the area of the sail is 60 ft^2.

30) Chandra cuts fabric into isosceles triangles for a quilt. The height of each triangle is 1 in. less than the length of the base. The area of each triangle is 15 in^2. Find the height and base of each triangle.

31) Valerie makes a bike ramp in the shape of a right triangle. The base of the ramp is 4 in. more than twice its height, and the length of the incline is 4 in. less than three times its height. How high is the ramp?

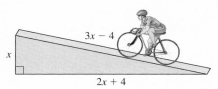

32) The width of a widescreen TV is 10 in. less than its length. The diagonal of the rectangular screen is 10 in. more than the length. Find the length and width of the screen.

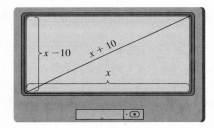

Objective 4: Solve an Applied Problem Using a Quadratic Equation
Solve.

33) An object is propelled upward from a height of 4 ft. The height h of the object (in feet) t sec after the object is released is given by

$$h = -16t^2 + 60t + 4$$

a) How long does it take the object to reach a height of 40 ft?

b) How long does it take the object to hit the ground?

34) An object is launched from the ground. The height h of the object (in feet) t sec after the object is released is given by

$$h = -16t^2 + 64t$$

When will the object be 48 ft in the air?

35) Attendance at Broadway plays from 1996 to 2000 can be modeled by

$$y = -0.25x^2 + 1.5x + 9.5$$

where x represents the number of years after 1996 and y represents the number of people who attended a Broadway play (in millions).
(*Statistical Abstracts of the United States*)

a) Approximately how many people saw a Broadway play in 1996?

b) In what year did approximately 11.75 million people see a Broadway play?

36) The illuminance E (measure of the light emitted, in lux) of a light source is given by

$$E = \frac{I}{d^2}$$

where I is the luminous intensity (measured in candela) and d is the distance, in meters, from the light source. The luminous intensity of a lamp is 2700 candela at a distance of 3 m from the lamp. Find the illuminance, E, in lux.

37) A sandwich shop has determined that the demand for its turkey sandwich is $\frac{65}{P}$ per day, where P is the price of the sandwich in dollars. The daily supply is given by $10P + 3$. Find the price at which the demand for the sandwich equals the supply.

38) A hardware store determined that the demand for shovels one winter was $\frac{2800}{P}$, where P is the price of the shovel in dollars. The supply was given by $12P + 32$. Find the price at which demand for the shovels equals the supply.

Use the following formula for Exercises 39 and 40.

A wire is stretched between two poles separated by a distance d, and a weight is in the center of the wire of length L so that the wire is pulled taut as pictured here. The vertical distance, D, between the weight on the wire and the top of the poles is given by $D = \frac{\sqrt{L^2 - d^2}}{2}$.

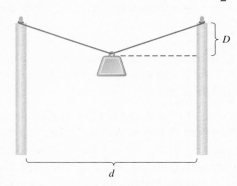

39) A 12.5-ft clothesline is attached to the top of two poles that are 12 ft apart. A shirt is hanging in the middle of the clothesline. Find the distance, D, that the shirt is hanging down.

40) An 11-ft wire is attached to a ceiling in a loft apartment by hooks that are 10 ft apart. A light fixture is hanging in the middle of the wire. Find the distance, D, between the ceiling and the top of the light fixture. Round the answer to the nearest tenths place.

R Rethink

R1) Have you ever had to calculate volume or area before studying this section?

R2) Are you allowed to have negative answers for volume or area?

R3) What do you find most difficult about doing applied problems?

10.5 Quadratic Functions and Their Graphs

P Prepare

What are your objectives for Section 10.5?

O Organize

How can you accomplish each objective?

1 Graph a Quadratic Function by Shifting the Graph of $f(x) = x^2$	• Be able to write the definition of a *quadratic function* in your own words. • Be familiar with the graph of $f(x) = x^2$. • Know the definition of a *parabola* and a *vertex*. • Be able to write the properties of **Vertical Shifts** and **Horizontal Shifts** in your own words, with examples. • Understand the property of **Reflection about the x-axis** and how that occurs. • Complete the given examples on your own. • Complete You Trys 1–3.
2 Graph $f(x) = a(x - h)^2 + k$ Using Characteristics of a Parabola	• Follow the procedure for **Graphing a Quadratic Function of the Form** $f(x) = a(x - h)^2 + k$. • Review procedures for finding intercepts. • Follow the procedure for **Graphing Parabolas from the Form** $f(x) = ax^2 + bx + c$. • Complete the given example on your own. • Complete You Try 4.
3 Graph $f(x) = ax^2 + bx + c$ by Completing the Square	• Follow the procedure for **Rewriting** $f(x) = ax^2 + bx + c$ in the Form $f(x) = a(x - h)^2 + k$ by Completing the Square. • Complete the given example on your own. • Complete You Try 5.
4 Graph $f(x) = ax^2 + bx + c$ Using $\left(-\dfrac{b}{2a}, f\left(-\dfrac{b}{2b}\right)\right)$	• Learn the **Vertex Formula**. • Complete the given example on your own. • Complete You Try 6.

W Work

Read the explanations, follow the examples, take notes, and complete the You Trys.

1 Graph a Quadratic Function by Shifting the Graph of $f(x) = x^2$

We were introduced to quadratic functions in Chapter 7, and in this chapter we have learned different methods for solving quadratic equations. In this section, we will learn how to graph quadratic functions. Let's begin with the definition of a quadratic function.

> **Definition**
>
> A **quadratic function** is a function that can be written in the form
>
> $$f(x) = ax^2 + bx + c$$
>
> where a, b, and c are real numbers and $a \neq 0$. An example is $f(x) = x^2 + 6x + 10$. The domain of a quadratic function is $(-\infty, \infty)$.

The simplest form of a quadratic function is $f(x) = x^2$. Let's graph this function as well as a similar function, $g(x) = x^2 + 2$.

EXAMPLE 1 Graph $f(x) = x^2$ and $g(x) = x^2 + 2$ on the same axes.

Solution

We will make a table of values for each function, then plot the points.

$f(x) = x^2$	
x	f(x)
0	0
1	1
2	4
−1	1
−2	4

$g(x) = x^2 + 2$	
x	g(x)
0	2
1	3
2	6
−1	3
−2	6

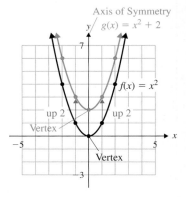

The domain of $f(x)$ is $(-\infty, \infty)$, and the range is $[0, \infty)$. The domain of $g(x)$ is $(-\infty, \infty)$, and the range is $[2, \infty)$.

> **Definition**
>
> The graph of a quadratic function is called a **parabola.** The lowest point on a parabola that opens upward or the highest point on a parabola that opens downward is called the **vertex.**

The vertex of the graph of $f(x)$ in Example 1 is $(0, 0)$, and the vertex of the graph of $g(x)$ is $(0, 2)$.

Every parabola has symmetry. Let's look at the graph of $f(x) = x^2$. If we were to fold the paper along the y-axis, one half of the graph of $f(x) = x^2$ would fall exactly on the other half. The y-axis, or the line $x = 0$, is the **axis of symmetry** of $f(x) = x^2$. (It is also true that the line $x = 0$ is the axis of symmetry of $g(x) = x^2 + 2$.)

We can see from the tables of values in Example 1 that although the x-values are the same in each table, the corresponding y-values in the table for $g(x)$ are *2 more than* the y-values in the first table. In other words, if $f(x) = x^2$, then $g(x) = x^2 + 2$, so $g(x) = f(x) + 2$.

The y-coordinates of the ordered pairs of $g(x)$ are *2 more than* the y-coordinates of the ordered pairs of $f(x)$ when the ordered pairs of f and g have the same x-coordinates. This means that **the graph of g is the same shape as the graph of f, but g is shifted up 2 units.**

We can make the following general statement about shifting the graph of a function vertically.

W Hint
In your own words, define the **vertex** of a parabola, the **axis of symmetry,** and **vertical shift.**

> **Property** Vertical Shifts
>
> Given the graph of $f(x)$, if $g(x) = f(x) + k$, where k is a constant, then the graph of $g(x)$ is the same shape as the graph of $f(x)$ but g is shifted **vertically** k units.

[YOU TRY 1] Graph $g(x) = x^2 + 1$.

Now let's look at how we can shift the parabola $f(x) = x^2$ horizontally.

EXAMPLE 2

Graph $f(x) = x^2$ and $g(x) = (x + 3)^2$ on the same axes.

Solution

$f(x) = x^2$	
x	f(x)
0	0
1	1
2	4
−1	1
−2	4

$g(x) = (x + 3)^2$	
x	g(x)
−3	0
−2	1
−1	4
−4	1
−5	4

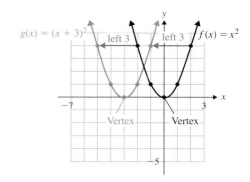

The functions $f(x)$ and $g(x)$ each have a domain of $(-\infty, \infty)$. Each has a range of $[0, \infty)$.

Notice that the y-values are the same in each table. The corresponding x-values in the table for $g(x)$, however, are *3 less than* the x-values in the first table.

The x-coordinates of the ordered pairs of $g(x)$ are *3 less than* the x-coordinates of the ordered pairs of $f(x)$ when the ordered pairs of f and g have the same y-coordinates. This means that **the graph of g is the same shape as the graph of f, but g is shifted left 3 units.**

W Hint
In your own words, describe how to shift a graph horizontally.

Property Horizontal Shifts

Given the graph of $f(x)$, if $g(x) = f(x - h)$, where h is a constant, then the graph of $g(x)$ is the same shape as the graph of $f(x)$ but g is shifted **horizontally** h units.

We can think of Example 2 in terms of this horizontal shift. Since $f(x) = x^2$ and $g(x) = (x + 3)^2$, $h = -3$ in $g(x)$ because we can think of $g(x)$ as $g(x) = (x - (-3))^2$.

The graph of g is the same shape as the graph of f but g is shifted -3 units horizontally or 3 units to the **left.**

Note
This vertical and horizontal shifting works for any function, not just quadratic functions.

[YOU TRY 2] Graph $g(x) = (x + 4)^2$.

Next we will learn about reflecting the graph of $f(x) = x^2$ about the x-axis.

EXAMPLE 3

Graph $f(x) = x^2$ and $g(x) = -x^2$ on the same axes.

Solution

$f(x) = x^2$	
x	$f(x)$
0	0
1	1
2	4
-1	1
-2	4

$g(x) = -x^2$	
x	$g(x)$
0	0
1	-1
2	-4
-1	-1
-2	-4

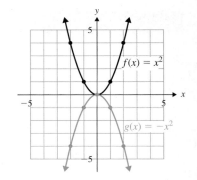

The tables of values show us that although the x-values are the same in each table, the corresponding y-values in the table for $g(x)$ are the *negatives* of the y-values in the first table. With the exception of the vertex, all of the y-coordinates of the points on the graph of g are negative. That is why the graph of $g(x) = -x^2$ is below the x-axis.

Each function has a domain of $(-\infty, \infty)$. The range of $f(x)$ is $[0, \infty)$, and the range of $g(x)$ is $(-\infty, 0]$.

We say that *the graph of g is the reflection of the graph of f about the x-axis.* (The graph of g is the mirror image of the graph of f.)

W Hint

In your own words, explain how to reflect the graph of $f(x)$ about the x-axis.

Property Reflection about the x-axis

Given the graph of any function $f(x)$, if $g(x) = -f(x)$ then the graph of $g(x)$ will be the **reflection of the graph of f about the x-axis.** That is, obtain the graph of g by keeping the x-coordinate of each point on f the same but take the negative of the y-coordinate.

[**YOU TRY 3**] Graph $g(x) = -(x + 2)^2$.

EXAMPLE 4

Graph $g(x) = (x - 2)^2 - 1$.

Solution

If we compare $g(x)$ to $f(x) = x^2$, what do the constants in $g(x)$ tell us about transforming the graph of $f(x)$?

$$g(x) = (x - 2)^2 - 1$$
         ↑   ↑
    Shift $f(x)$ Shift $f(x)$
    right 2.   down 1.

Sketch the graph of $f(x) = x^2$, then move every point on the graph of f right 2 and down 1 to obtain the graph of $g(x)$. This moves the vertex from (0, 0) to (2, −1). Notice that the axis of symmetry of $g(x)$ moves 2 units to the right also. Its equation is $x = 2$. The domain of $g(x)$ is $(-\infty, \infty)$; the range is $[-1, \infty)$.

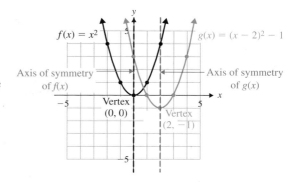

2 Graph $f(x) = a(x - h)^2 + k$ Using Characteristics of a Parabola

When a quadratic function is in the form $f(x) = a(x - h)^2 + k$, we can read the vertex directly from the equation. Furthermore, the value of a tells us if the parabola opens upward or downward and whether the graph is narrower, wider, or the same width as $y = x^2$.

> **Procedure** Graphing a Quadratic Function of the Form $f(x) = a(x - h)^2 + k$
>
> 1) The vertex of the parabola is (h, k).
> 2) The axis of symmetry is the vertical line with equation $x = h$.
> 3) If a is positive, the parabola opens upward.
>   If a is negative, the parabola opens downward.
> 4) If $|a| < 1$, then the graph of $f(x) = a(x - h)^2 + k$ is *wider* than the graph of $y = x^2$.
>   If $|a| > 1$, then the graph of $f(x) = a(x - h)^2 + k$ is *narrower* than the graph of $y = x^2$.
>   If $a = 1$ or $a = -1$, the graph is the *same* width as $y = x^2$.

EXAMPLE 5 Graph $f(x) = 2(x + 1)^2 - 4$. Also find the *x*- and *y*-intercepts.

Solution

Here is the information we can get from the equation:

1) $h = -1$ and $k = -4$. The vertex is $(-1, -4)$.
2) The axis of symmetry is $x = -1$.
3) $a = 2$. Because a is positive, the parabola opens upward.
4) Since $|a| > 1$, the graph of $f(x) = 2(x + 1)^2 - 4$ is *narrower* than the graph of $f(x) = x^2$.

To graph the function, start by putting the vertex on the axes. Then, choose a couple of values of x to the left or right of the vertex to plot more points. Use the axis of symmetry to find the points $(-2, -2)$ and $(-3, 4)$ on the graph of $f(x) = 2(x + 1)^2 - 4$.

x	y
0	-2
1	4

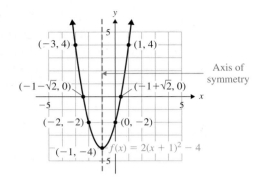

We can read the *y-intercept* from the graph: $(0, -2)$. To find the *x-intercepts*, let $f(x) = 0$ and solve for x.

$$f(x) = 2(x + 1)^2 - 4$$
$$0 = 2(x + 1)^2 - 4 \quad \text{Substitute 0 for } f(x).$$
$$4 = 2(x + 1)^2 \quad \text{Add 4.}$$
$$2 = (x + 1)^2 \quad \text{Divide by 2.}$$
$$\pm\sqrt{2} = x + 1 \quad \text{Square root property}$$
$$-1 \pm \sqrt{2} = x \quad \text{Add } -1.$$

The *x*-intercepts are $(-1 - \sqrt{2}, 0)$ and $(-1 + \sqrt{2}, 0)$. The domain is $(-\infty, \infty)$; the range is $[-4, \infty)$.

[YOU TRY 4] Graph $f(x) = 2(x - 1)^2 - 2$. Also find the *x*- and *y*-intercepts.

When a quadratic function is written in the form $f(x) = ax^2 + bx + c$, there are two methods we can use to graph the function.

Procedure Graphing Parabolas from the Form $f(x) = ax^2 + bx + c$

There are two methods we can use to graph the function $f(x) = ax^2 + bx + c$.

Method 1: Rewrite $f(x) = ax^2 + bx + c$ in the form $f(x) = a(x - h)^2 + k$ by *completing the square*.

Method 2: Use the formula $x = -\dfrac{b}{2a}$ to find the *x*-coordinate of the vertex. Then, the vertex has coordinates $\left(-\dfrac{b}{2a}, f\left(-\dfrac{b}{2a}\right)\right)$.

We will begin with Method 1. We will modify the steps we used in Section 10.1 to solve quadratic equations by completing the square.

3 Graph $f(x) = ax^2 + bx + c$ by Completing the Square

> **Procedure** Rewriting $f(x) = ax^2 + bx + c$ in the Form $f(x) = a(x - h)^2 + k$ by Completing the Square
>
> **Step 1:** The coefficient of the square term must be 1. If it is not 1, multiply or divide both sides of the equation (*including* $f(x)$) by the appropriate value to obtain a leading coefficient of 1.
>
> **Step 2:** Separate the constant from the terms containing the variables by grouping the variable terms with parentheses.
>
> **Step 3:** Complete the square for the quantity in the parentheses. Find half of the linear coefficient, then square the result. *Add* that quantity inside the parentheses, and *subtract* the quantity from the constant. (Adding and subtracting the same number on the same side of an equation is like adding 0 to the equation.)
>
> **Step 4:** Factor the expression inside the parentheses.
>
> **Step 5:** Solve for $f(x)$.

EXAMPLE 6 Graph each function. Begin by completing the square to rewrite each function in the form $f(x) = a(x - h)^2 + k$. Include the intercepts.

a) $f(x) = x^2 + 6x + 10$

b) $g(x) = -\dfrac{1}{2}x^2 + 4x - 6$

Solution

a) **Step 1:** The coefficient of x^2 is 1.

Step 2: Separate the constant from the variable terms using parentheses.
$$f(x) = (x^2 + 6x) + 10$$

Step 3: Complete the square for the quantity in the parentheses.
$$\frac{1}{2}(6) = 3$$
$$3^2 = 9$$

Add 9 inside the parentheses, and subtract 9 from the 10. This is like adding 0 to the equation.
$$f(x) = (x^2 + 6x + 9) + 10 - 9$$
$$f(x) = (x^2 + 6x + 9) + 1$$

Step 4: Factor the expression inside the parentheses.
$$f(x) = (x + 3)^2 + 1$$

Step 5: The equation *is* solved for $f(x)$.

From the equation $f(x) = (x + 3)^2 + 1$ we can see that

  i) The vertex is $(-3, 1)$.
  ii) The axis of symmetry is $x = -3$.
  iii) $a = 1$ so the parabola opens upward.
  iv) Since $a = 1$, the graph is the same width as $y = x^2$.

Find some other points on the parabola. Use the axis of symmetry.

To find the x-intercepts, let $f(x) = 0$ and solve for x. Use *either* form of the equation. We will use $f(x) = (x + 3)^2 + 1$.

$$0 = (x + 3)^2 + 1 \quad \text{Let } f(x) = 0.$$
$$-1 = (x + 3)^2 \quad \text{Subtract 1.}$$
$$\pm\sqrt{-1} = x + 3 \quad \text{Square root property}$$
$$-3 \pm i = x \quad \sqrt{-1} = i; \text{ subtract 3.}$$

x	f(x)
−2	2
−1	5

Because the solutions to $f(x) = 0$ are *not* real numbers, *there are no x-intercepts*. To find the y-intercept, let $x = 0$ and solve for $f(0)$.

$$f(x) = (x + 3)^2 + 1$$
$$f(0) = (0 + 3)^2 + 1$$
$$f(0) = 9 + 1 = 10$$

The y-intercept is $(0, 10)$. The domain is $(-\infty, \infty)$, and the range is $[1, \infty)$.

b) **Step 1:** The coefficient of x^2 is $-\dfrac{1}{2}$. Multiply both sides of the equation (including the $g(x)$) by -2 so that the coefficient of x^2 will be 1.

$$g(x) = -\frac{1}{2}x^2 + 4x - 6$$
$$-2g(x) = -2\left(-\frac{1}{2}x^2 + 4x - 6\right) \quad \text{Multiply by } -2.$$
$$-2g(x) = x^2 - 8x + 12 \quad \text{Distribute.}$$

Step 2: Separate the constant from the variable terms using parentheses.
$$-2g(x) = (x^2 - 8x) + 12$$

Step 3: Complete the square for the quantity in parentheses.
$$\frac{1}{2}(-8) = -4$$
$$(-4)^2 = 16$$

Add 16 inside the parentheses, and subtract 16 from the 12.
$$-2g(x) = (x^2 - 8x + 16) + 12 - 16$$
$$-2g(x) = (x^2 - 8x + 16) - 4$$

Step 4: Factor the expression inside the parentheses.
$$-2g(x) = (x - 4)^2 - 4$$

> **W Hint**
> Write out each step as you are reading the examples, and be sure you *understand* each step!

Step 5: Solve the equation for g(x) by dividing by −2.

$$\frac{-2g(x)}{-2} = \frac{(x-4)^2}{-2} - \frac{4}{-2}$$

$$g(x) = -\frac{1}{2}(x-4)^2 + 2$$

From $g(x) = -\frac{1}{2}(x-4)^2 + 2$ we can see that

i) The vertex is (4, 2).
ii) The axis of symmetry is $x = 4$.
iii) $a = -\frac{1}{2}$ (the same as in the form $g(x) = -\frac{1}{2}x^2 + 4x - 6$) so the parabola opens downward.
iv) Since $a = -\frac{1}{2}$, the graph of g(x) will be wider than $y = x^2$.

Find some other points on the parabola. Use the axis of symmetry.

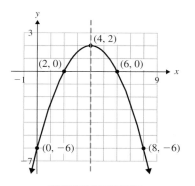

x	g(x)
6	0
8	−6

Using the axis of symmetry, we can see that the x-intercepts are (6, 0) and (2, 0) and that the y-intercept is (0, −6). The domain is (−∞, ∞); the range is (−∞, 2].

[YOU TRY 5] Graph each function. Begin by completing the square to rewrite each function in the form $f(x) = a(x-h)^2 + k$. Include the intercepts.

a) $f(x) = x^2 + 4x + 3$ b) $g(x) = -2x^2 + 12x - 8$

4 Graph $f(x) = ax^2 + bx + c$ Using $\left(-\frac{b}{2a}, f\left(-\frac{b}{2a}\right)\right)$

We can also graph quadratic functions of the form $f(x) = ax^2 + bx + c$ by using the formula $h = -\frac{b}{2a}$ to find the x-coordinate of the vertex. This formula comes from completing the square on $f(x) = ax^2 + bx + c$.

Although there is a formula for k, it is only necessary to remember the formula for h. The y-coordinate of the vertex, then, is $k = f\left(-\frac{b}{2a}\right)$. The axis of symmetry is $x = h$.

Property The Vertex Formula

The **vertex** of the graph of $f(x) = ax^2 + bx + c$ $(a \neq 0)$ has coordinates $\left(-\frac{b}{2a}, f\left(-\frac{b}{2a}\right)\right)$.

EXAMPLE 7

Graph $f(x) = x^2 - 6x + 3$ using the vertex formula. Include the intercepts.

Solution

$a = 1$, $b = -6$, $c = 3$. Since $a = +1$, the graph opens upward. The x-coordinate, h, of the vertex is

$$h = -\frac{b}{2a} = -\frac{(-6)}{2(1)} = \frac{6}{2} = 3$$

$h = 3$. Then, the y-coordinate, k, of the vertex is $k = f(3)$.

$$f(x) = x^2 - 6x + 3$$
$$f(3) = (3)^2 - 6(3) + 3$$
$$= 9 - 18 + 3 = -6$$

The vertex is $(3, -6)$. The axis of symmetry is $x = 3$.

Find more points on the graph of $f(x) = x^2 - 6x + 3$, then use the axis of symmetry to find other points on the parabola.

x	f(x)
4	-5
5	-2
6	3

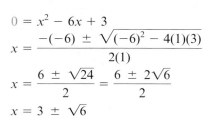

To find the x-intercepts, let $f(x) = 0$ and solve for x.

$$0 = x^2 - 6x + 3$$
$$x = \frac{-(-6) \pm \sqrt{(-6)^2 - 4(1)(3)}}{2(1)} \quad \text{Solve using the quadratic formula.}$$
$$x = \frac{6 \pm \sqrt{24}}{2} = \frac{6 \pm 2\sqrt{6}}{2} \quad \text{Simplify.}$$
$$x = 3 \pm \sqrt{6}$$

The x-intercepts are $(3 + \sqrt{6}, 0)$ and $(3 - \sqrt{6}, 0)$.

We can see from the graph that the y-intercept is $(0, 3)$. The domain is $(-\infty, \infty)$; the range is $[-6, \infty)$.

YOU TRY 6

Graph $f(x) = -x^2 - 8x - 13$ using the vertex formula. Include the intercepts.

Using Technology

In Section 7.4 we said that the solutions of the equation $x^2 - x - 6 = 0$ are the x-intercepts of the graph of $y = x^2 - x - 6$.

The x-intercepts are also called the zeros of the equation since they are the values of x that make $y = 0$. Enter $x^2 - x - 6$ in Y_1 then find the x-intercepts shown on the graph by pressing 2nd TRACE and then selecting 2:zero. Move the cursor to the left of an x-intercept using the right arrow key and press ENTER. Move the cursor to the right of the x-intercept using the right arrow key

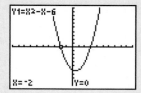

and press ENTER. Move the cursor close to the x-intercept using the left arrow key and press ENTER. Repeat these steps for each x-intercept. The x-intercepts are (−2, 0) and (3, 0) as shown in the graphs to the right.

The y-intercept is found by graphing the function and pressing TRACE 0 ENTER. As shown on the graph, the y-intercept for $y = x^2 - x - 6$ is (0, −6).

The x-value of the vertex can be found using the vertex formula. In this case, $a = 1$ and $b = -1$, so $-\frac{b}{2a} = \frac{1}{2}$.

To find the vertex on the graph, press TRACE, type 1/2, and press ENTER. The vertex is shown as (0.5, −6.25) on the graph.

Remember, you can convert the coordinates of the vertex to fractions. Go to the home screen by pressing 2nd MODE. To display the x-value of the vertex, press X, T, Θ, n MATH ENTER ENTER. To display the y-value of the vertex, press ALPHA 1 MATH ENTER ENTER. The vertex is then $\left(\frac{1}{2}, -\frac{25}{4}\right)$.

Find the x-intercepts, y-intercept, and vertex using a graphing calculator.

1) $y = x^2 - 2x + 2$
2) $y = x^2 - 4x - 5$
3) $y = -(x + 1)^2 + 4$
4) $y = x^2 - 4$
5) $y = x^2 - 6x + 9$
6) $y = -x^2 - 8x - 19$

ANSWERS TO [YOU TRY] EXERCISES

1)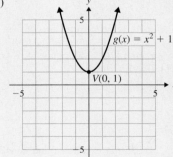

2)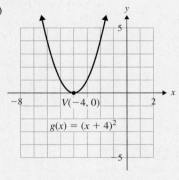

3)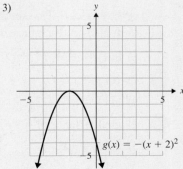

4) x-ints: (0, 0), (2, 0); y-int: (0, 0)
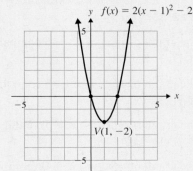

5) a) x-ints: (−3, 0), (−1, 0); y-int: (0, 3)   b) x-ints: (3 + √5, 0), (3 − √5, 0); y-int: (0, −8)

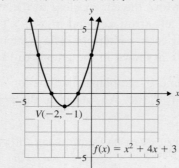
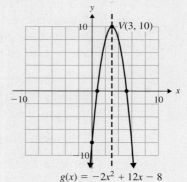

6) x-ints: (−4 + √3, 0), (−4 − √3, 0); y-int: (0, −13)

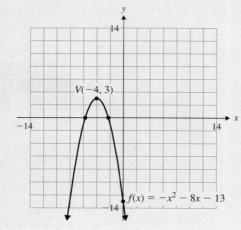

ANSWERS TO TECHNOLOGY EXERCISES

1) no x-intercepts; y-intercept: (0, 2); vertex: (1, 1)
2) x-intercepts: (−1, 0), (5, 0); y-intercept: (0, −5); vertex: (2, −9)
3) x-intercepts: (−3, 0), (1, 0); y-intercept: (0, 3); vertex: (−1, 4)
4) x-intercepts: (−2, 0), (2, 0); y-intercept: (0, −4); vertex: (0, −4)
5) x-intercept: (3, 0); y-intercept: (0, 9); vertex: (3, 0)
6) no x-intercepts; y-intercept: (0, −19); vertex: (−4, −3)

Evaluate 10.5 Exercises

Do the exercises, and check your work.

Objective 1: Graph a Quadratic Function by Shifting the Graph of $f(x) = x^2$

1) How does the graph of $g(x) = x^2 + 6$ compare to the graph of $f(x) = x^2$?

2) How does the graph of $h(x) = x^2 - 5$ compare to the graph of $f(x) = x^2$?

3) How does the graph of $h(x) = (x + 5)^2$ compare to the graph of $f(x) = x^2$?

4) How does the graph of $g(x) = (x - 4)^2$ compare to the graph of $f(x) = x^2$?

For Exercises 5–18, sketch the graph of $f(x) = x^2$. Then graph $g(x)$ on the same axes by shifting the graph of $f(x)$.

5) $g(x) = x^2 + 3$
6) $g(x) = x^2 + 5$
7) $g(x) = x^2 - 4$
8) $g(x) = x^2 - 1$
9) $g(x) = (x + 2)^2$
10) $g(x) = (x + 1)^2$
11) $g(x) = (x - 3)^2$
12) $g(x) = (x - 4)^2$
13) $g(x) = -x^2$
14) $g(x) = -(x - 1)^2$
15) $g(x) = \frac{1}{2}x^2$
16) $g(x) = 2x^2$
17) $g(x) = (x - 1)^2 - 3$
18) $g(x) = (x + 2)^2 + 1$

Objective 2: Graph $f(x) = a(x - h)^2 + k$ Using Characteristics of a Parabola

19) Given a quadratic function of the form $f(x) = a(x - h)^2 + k$,

   a) what is the vertex?

   b) what is the equation of the axis of symmetry?

   c) how do you know if the parabola opens upward?

   d) how do you know if the parabola opens downward?

   e) how do you know if the parabola is narrower than the graph of $y = x^2$?

   f) how do you know if the parabola is wider than the graph of $y = x^2$?

For each quadratic function, identify the vertex, axis of symmetry, and x- and y-intercepts. Then, graph the function. Determine the domain and range.

20) $g(x) = (x - 3)^2 - 1$ 21) $f(x) = (x + 1)^2 - 4$

22) $h(x) = (x + 2)^2 + 7$ 23) $g(x) = (x - 2)^2 + 3$

24) $y = (x + 1)^2 - 5$ 25) $y = (x - 4)^2 - 2$

26) $g(x) = -(x - 3)^2 + 2$ 27) $f(x) = -(x + 3)^2 + 6$

28) $f(x) = -(x - 2)^2 - 4$ 29) $y = -(x + 1)^2 - 5$

30) $y = 2(x + 1)^2 - 2$ 31) $f(x) = 2(x - 1)^2 - 8$

32) $h(x) = \frac{1}{2}(x + 4)^2$ 33) $g(x) = \frac{1}{4}x^2 - 1$

34) $y = -x^2 + 5$

35) $f(x) = -\frac{1}{3}(x + 4)^2 + 3$ 36) $y = -\frac{1}{2}(x - 4)^2 + 2$

37) $g(x) = 3(x + 2)^2 + 5$ 38) $f(x) = 2(x - 3)^2 + 3$

In Exercises 39 and 40, match each function to its graph.

39) $f(x) = x^2 - 3$, $g(x) = (x - 3)^2$,
    $h(x) = -(x + 3)^2$, $k(x) = -x^2 + 3$

a)

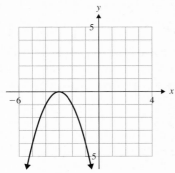

b)

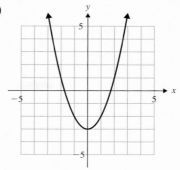

c)

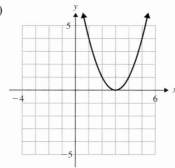

d)

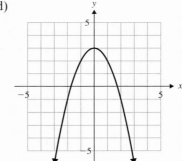

40) $f(x) = (x + 1)^2$, $g(x) = x^2 + 1$,
    $h(x) = -(x + 1)^2$, $k(x) = -x^2 + 1$

a)

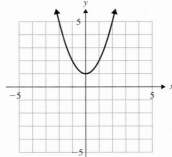

b)

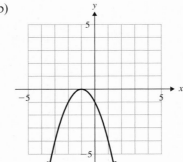

c)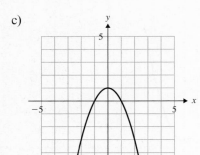

d)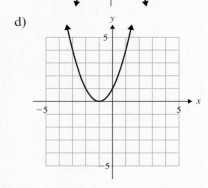

Exercises 41–46 each present a shift that is performed on the graph of $f(x) = x^2$ to obtain the graph of $g(x)$. Write the equation of $g(x)$.

41) $f(x)$ is shifted 8 units to the right.

42) $f(x)$ is shifted down 9 units.

43) $f(x)$ is shifted up 3.5 units.

44) $f(x)$ is shifted left 1.2 units.

45) $f(x)$ is shifted left 4 units and down 7 units.

46) $f(x)$ is shifted up 2 units and right 5 units.

Objective 3: Graph $f(x) = ax^2 + bx + c$ by Completing the Square

Rewrite each function in the form $f(x) = a(x - h)^2 + k$.

Fill It In

Fill in the blanks with either the missing mathematical step or the reason for the given step.

47) $f(x) = x^2 + 8x + 11$

_____ Group the variable terms together using parentheses.

_____ Find the number that completes the square in the parentheses.

$f(x) = (x^2 + 8x + 16) + 11 - 16$

_____

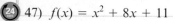

Factor and simplify.

48) $f(x) = x^2 - 4x - 7$
$f(x) = (x^2 - 4x) - 7$

_____ Find the number that completes the square in the parentheses.

_____ Add and subtract the number above to the same side of the equation.

$f(x) = (x - 2)^2 - 11$ _____

Rewrite each function in the form $f(x) = a(x - h)^2 + k$ by completing the square. Then, graph the function. Include the intercepts. Determine the domain and range.

49) $f(x) = x^2 - 2x - 3$ 50) $g(x) = x^2 + 6x + 8$

51) $y = x^2 + 6x + 7$ 52) $h(x) = x^2 - 4x + 1$

53) $g(x) = x^2 + 4x$ 54) $y = x^2 - 8x + 18$

55) $h(x) = -x^2 - 4x + 5$ 56) $f(x) = -x^2 - 2x + 3$

57) $y = -x^2 + 6x - 10$ 58) $g(x) = -x^2 - 4x - 6$

59) $y = 2x^2 - 8x + 2$ 60) $f(x) = 2x^2 - 8x + 4$

61) $g(x) = -\dfrac{1}{3}x^2 - 2x - 9$

62) $h(x) = -\dfrac{1}{2}x^2 - 3x - \dfrac{19}{2}$

63) $y = x^2 - 3x + 2$ 64) $f(x) = x^2 + 5x + \dfrac{21}{4}$

Objective 4: Graph $f(x) = ax^2 + bx + c$ Using $\left(-\dfrac{b}{2a}, f\left(-\dfrac{b}{2a}\right)\right)$

Graph each function using the vertex formula. Include the intercepts. Determine the domain and range.

65) $y = x^2 + 2x - 3$ 66) $g(x) = x^2 - 6x + 8$

67) $f(x) = -x^2 - 8x - 13$ 68) $y = -x^2 + 2x + 2$

69) $g(x) = 2x^2 - 4x + 4$

70) $f(x) = -4x^2 - 8x - 6$ 71) $y = -3x^2 + 6x + 1$

72) $h(x) = 2x^2 - 12x + 9$ 73) $f(x) = \dfrac{1}{2}x^2 - 4x + 5$

74) $y = \dfrac{1}{2}x^2 + 2x - 3$

75) $h(x) = -\dfrac{1}{3}x^2 - 2x - 5$ 76) $g(x) = \dfrac{1}{5}x^2 - 2x + 8$

R Rethink

R1) What shape is a quadratic function? What causes the graph to open up?

R2) What are the different ways to find the vertex of a parabola? Which do you prefer?

R3) What do you find most difficult about completing the square?

10.6 Applications of Quadratic Functions and Graphing Other Parabolas

P Prepare | O Organize

What are your objectives for Section 10.6?	How can you accomplish each objective?
1 Find the Maximum or Minimum Value of a Quadratic Function	• Learn how to find the **Maximum or Minimum Value of a Quadratic Function.** • Complete the given example on your own. • Complete You Try 1.
2 Given a Quadratic Function, Solve an Applied Problem Involving a Maximum or Minimum Value	• Be able to determine whether the problem involves a maximum or minimum. • Complete the given example on your own. • Complete You Try 2.
3 Write a Quadratic Function to Solve an Applied Problem Involving a Maximum or Minimum Value	• Follow the procedure for **Solving a Max/Min Problem.** Write this procedure in your notes. • Review methods for solving quadratic equations. • Complete the given example on your own. • Complete You Try 3.
4 Graph Parabolas of the Form $x = a(y - k)^2 + h$	• Follow the procedure for **Graphing an Equation of the Form** $x = a(y - k)^2 + h$. • Complete the given example on your own. • Complete You Try 4.
5 Rewrite $x = ay^2 + by + c$ as $x = a(y - k)^2 + h$ by Completing the Square	• Review the procedure for **Completing the Square.** Write the procedure in your own words. • Complete the given example on your own. • Complete You Try 5.
6 Find the Vertex of the Graph of $x = ay^2 + by + c$ Using $y = -\dfrac{b}{2a}$, and Graph the Equation	• Review the vertex formula in Section 10.5, and notice the slight differences when using the formula for these graphs. • Know the procedures for **Graphing Parabolas from the Form** $x = ay^2 + by + c$. • Complete the given example on your own. • Complete You Try 6.

 Read the explanations, follow the examples, take notes, and complete the You Trys.

1 Find the Maximum or Minimum Value of a Quadratic Function

From our work with quadratic functions, we have seen that the vertex is either the lowest point or the highest point on the graph depending on whether the parabola opens upward or downward.

If the parabola opens upward, the vertex is the *lowest* point on the parabola.

If the parabola opens downward, the vertex is the *highest* point on the parabola.

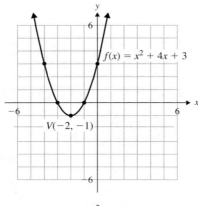

$f(x) = x^2 + 4x + 3$

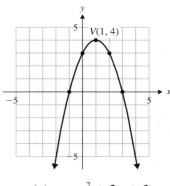

$g(x) = -x^2 + 2x + 3$

W Hint
In your notes and in your own words, summarize this information about the minimum and maximum values of a quadratic function.

The y-coordinate of the vertex, -1, is the *smallest* y-value the function will have. We say that **-1 is the minimum value of the function.** $f(x)$ has no maximum because the graph continues upward indefinitely—the y-values get larger without bound.

The y-coordinate of the vertex, 4, is the *largest* y-value the function will have. We say that **4 is the maximum value of the function.** $g(x)$ has no minimum because the graph continues downward indefinitely—the y-values get smaller without bound.

> **Property** Maximum and Minimum Values of a Quadratic Function
>
> Let $f(x) = ax^2 + bx + c$.
>
> 1) If a is **positive,** the graph of $f(x)$ opens upward, so the vertex is the lowest point on the parabola. The y-coordinate of the vertex is the **minimum** value of the function $f(x)$.
>
> 2) If a is **negative,** the graph of $f(x)$ opens downward, so the vertex is the highest point on the parabola. The y-coordinate of the vertex is the **maximum** value of the function $f(x)$.

We can use this information about the vertex to help us solve problems.

EXAMPLE 1

Let $f(x) = -x^2 + 4x + 2$.

a) Does the function attain a minimum or maximum value at its vertex?
b) Find the vertex of the graph of $f(x)$.
c) What is the minimum or maximum value of the function?
d) Graph the function to verify parts a)–c).

Solution

a) Because $a = -1$, the graph of $f(x)$ will open downward. Therefore, the vertex will be the *highest* point on the parabola. The function will attain its *maximum* value at the vertex.

b) Use $x = -\dfrac{b}{2a}$ to find the x-coordinate of the vertex. For $f(x) = -x^2 + 4x + 2$,

$$x = -\frac{b}{2a} = -\frac{(4)}{2(-1)} = 2$$

The y-coordinate of the vertex is $f(2)$.

$$f(2) = -(2)^2 + 4(2) + 2 = -4 + 8 + 2 = 6$$

The vertex is $(2, 6)$.

c) $f(x)$ has no minimum value. The *maximum* value of the function is 6, the y-coordinate of the vertex. (The largest y-value of the function is 6.)

We say that the maximum value of the function is 6 and that it occurs at $x = 2$ (the x-coordinate of the vertex).

d) From the graph of $f(x)$, we can see that our conclusions in parts a)–c) make sense.

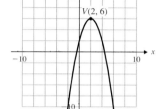

[YOU TRY 1]   Let $f(x) = x^2 + 6x + 7$. Repeat parts a)–d) from Example 1.

2 Given a Quadratic Function, Solve an Applied Problem Involving a Maximum or Minimum Value

EXAMPLE 2

A ball is thrown upward from a height of 24 ft. The height h of the ball (in feet) t sec after the ball is released is given by

$$h(t) = -16t^2 + 16t + 24.$$

a) How long does it take the ball to reach its maximum height?
b) What is the maximum height attained by the ball?

Solution

a) Begin by understanding what the function $h(t)$ tells us: $a = -16$, so the graph of h would open downward. Therefore, the vertex is the highest point on the parabola. The maximum value of the function occurs at the vertex. The ordered pairs that satisfy $h(t)$ are of the form $(t, h(t))$.

To determine how long it takes the ball to reach its maximum height, we must find the t-coordinate of the vertex.

$$t = -\frac{b}{2a} = -\frac{16}{2(-16)} = \frac{1}{2}$$

The ball will reach its maximum height after $\dfrac{1}{2}$ sec.

b) The maximum height the ball reaches is the y-coordinate (or h(t)-coordinate) of the vertex. Since the ball attains its maximum height when $t = \frac{1}{2}$, find $h\left(\frac{1}{2}\right)$.

$$h\left(\frac{1}{2}\right) = -16\left(\frac{1}{2}\right)^2 + 16\left(\frac{1}{2}\right) + 24$$
$$= -16\left(\frac{1}{4}\right) + 8 + 24$$
$$= -4 + 32 = 28$$

The ball reaches a maximum height of 28 ft.

[**YOU TRY 2**] An object is propelled upward from a height of 10 ft. The height h of the object (in feet) t sec after the ball is released is given by

$$h(t) = -16t^2 + 32t + 10$$

a) How long does it take the object to reach its maximum height?
b) What is the maximum height attained by the object?

3 Write a Quadratic Function to Solve an Applied Problem Involving a Maximum or Minimum Value

EXAMPLE 3 Ayesha plans to put a fence around her rectangular garden. If she has 32 ft of fencing, what is the maximum area she can enclose?

Solution

Begin by drawing a picture.

Let x = the width of the garden
Let y = the length of the garden
Label the picture.

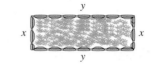

We will write two equations for a problem like this:

W Hint
Read this example **slowly** and **carefully**. Be sure you understand everything being done to solve this problem.

1) *The maximize or minimize equation;* this equation describes what we are trying to maximize or minimize.
2) *The constraint equation;* this equation describes the restrictions on the variables or the conditions the variables must meet.

Here is how we will get the equations.

1) We will write a *maximize* equation because we are trying to find the *maximum area* of the garden.

Let A = area of the garden

The area of the rectangle above is xy. Our equation is

Maximize: $A = xy$

2) To write the *constraint* equation, think about the restriction put on the variables. We cannot choose *any* two numbers for x and y. Since Ayesha has 32 ft of fencing, the distance around the garden is 32 ft. This is the *perimeter* of the rectangular garden. The perimeter of the rectangle drawn above is $2x + 2y$, and it must equal 32 ft.

The constraint equation is

$$\text{Constraint:} \quad 2x + 2y = 32$$

Set up this maximization problem as

$$\text{Maximize:} \quad A = xy$$
$$\text{Constraint:} \quad 2x + 2y = 32$$

Solve the constraint for a variable, and then substitute the expression into the maximize equation.

$$2x + 2y = 32$$
$$2y = 32 - 2x$$
$$y = 16 - x \quad \text{Solve the constraint for } y.$$

Substitute $y = 16 - x$ into $A = xy$.

$$A = x(16 - x)$$
$$A = 16x - x^2 \quad \text{Distribute.}$$
$$A = -x^2 + 16x \quad \text{Write in descending powers.}$$

Look carefully at $A = -x^2 + 16x$. This is a quadratic function! Its graph is a parabola that opens downward (since $a = -1$). At the vertex, the function attains its maximum. The ordered pairs that satisfy this function are of the form $(x, A(x))$, where x represents the width and $A(x)$ represents the area of the rectangular garden. *The second coordinate of the vertex is the maximum area we are looking for.*

$$A = -x^2 + 16x$$

> **W Hint**
> Do you understand how the solution of this problem is related to the graph of a quadratic equation?

Use $x = -\dfrac{b}{2a}$ with $a = -1$ and $b = 16$ to find the x-coordinate of the vertex (the width of the rectangle that produces the maximum area).

$$x = -\dfrac{16}{2(-1)} = 8$$

Substitute $x = 8$ into $A = -x^2 + 16x$ to find the maximum area.

$$A = -(8)^2 + 16(8)$$
$$A = -64 + 128$$
$$A = 64$$

The graph of $A = -x^2 + 16x$ is a parabola that opens downward with vertex (8, 64).

The maximum area of the garden is 64 ft², and this will occur when the width of the garden is 8 ft. (The length will be 8 ft as well.)

Let's summarize the steps we can use to solve a max/min problem.

SECTION 10.6 **Applications of Quadratic Functions and Graphing Other Parabolas**

Procedure Steps for Solving a Max/Min Problem Like Example 3

1) Draw a picture, if applicable.
2) Define the unknowns. Label the picture.
3) Write the max/min equation.
4) Write the constraint equation.
5) Solve the constraint for a variable. Substitute the expression into the max/min equation to obtain a quadratic function.
6) Find the vertex of the parabola using the vertex formula, $x = -\dfrac{b}{2a}$.
7) Answer the question being asked.

[YOU TRY 3] Find the maximum area of a rectangle that has a perimeter of 28 in.

4 Graph Parabolas of the Form $x = a(y - k)^2 + h$

Not all parabolas are functions. Parabolas can open in the *x*-direction as illustrated below. Clearly, these fail the vertical line test for functions.

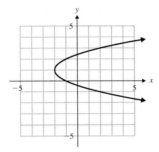
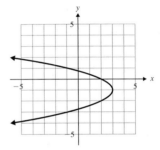

W Hint
When graphing equations that begin with $x =$, many earlier processes are switched when graphing.

Parabolas that open in the *y*-direction, or vertically, result from the functions

$$y = a(x - h)^2 + k \quad \text{or} \quad y = ax^2 + bx + c.$$

If we interchange the *x* and *y*, we obtain the equations

$$x = a(y - k)^2 + h \quad \text{or} \quad x = ay^2 + by + c.$$

The graphs of these equations are parabolas that open in the *x*-direction, or horizontally.

Procedure Graphing an Equation of the Form $x = a(y - k)^2 + h$

1) The vertex of the parabola is (h, k). (Notice, however, that *h* and *k* have changed their positions in the equation when compared to a quadratic function.)
2) The axis of symmetry is the horizontal line $y = k$.
3) If *a* is positive, the graph opens to the right.
   If *a* is negative, the graph opens to the left.

EXAMPLE 4

Graph each equation. Find the x- and y-intercepts and the domain and range.

a) $x = (y + 2)^2 - 1$ b) $x = -2(y - 2)^2 + 4$

Solution

a) 1) $h = -1$ and $k = -2$. The vertex is $(-1, -2)$.
   2) The axis of symmetry is $y = -2$.
   3) $a = +1$, so the parabola opens to the right. It is the same width as $y = x^2$.

To find the x-intercept, let $y = 0$ and solve for x.

$$x = (y + 2)^2 - 1$$
$$x = (0 + 2)^2 - 1$$
$$x = 4 - 1 = 3$$

The x-intercept is $(3, 0)$.

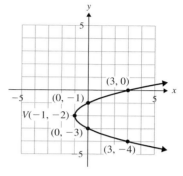

Find the y-intercepts by substituting 0 for x and solving for y.

$$x = (y + 2)^2 - 1$$
$$0 = (y + 2)^2 - 1 \quad \text{Substitute 0 for } x.$$
$$1 = (y + 2)^2 \quad \text{Add 1.}$$
$$\pm 1 = y + 2 \quad \text{Square root property}$$

$1 = y + 2$ or $-1 = y + 2$
$-1 = y$ $-3 = y$ Solve.

The y-intercepts are $(0, -3)$ and $(0, -1)$. Use the axis of symmetry to locate the point $(3, -4)$ on the graph. The domain is $[-1, \infty)$, and the range is $(-\infty, \infty)$.

b) $x = -2(y - 2)^2 + 4$

1) $h = 4$ and $k = 2$. The vertex is $(4, 2)$.
2) The axis of symmetry is $y = 2$.
3) $a = -2$, so the parabola opens to the left. It is narrower than $y = x^2$.

To find the x-intercept, let $y = 0$ and solve for x.

$$x = -2(y - 2)^2 + 4$$
$$x = -2(0 - 2)^2 + 4$$
$$x = -2(4) + 4 = -4$$

The x-intercept is $(-4, 0)$.

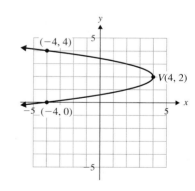

W Hint
Do you notice the similarities and differences between graphing these equations and graphing quadratic functions?

Find the *y*-intercepts by substituting 0 for *x* and solving for *y*.

$$x = -2(y-2)^2 + 4$$
$$0 = -2(y-2)^2 + 4 \quad \text{Substitute 0 for } x.$$
$$-4 = -2(y-2)^2 \quad \text{Subtract 4.}$$
$$2 = (y-2)^2 \quad \text{Divide by } -2.$$
$$\pm\sqrt{2} = y - 2 \quad \text{Square root property}$$
$$2 \pm \sqrt{2} = y \quad \text{Add 2.}$$

The *y*-intercepts are $(0, 2 - \sqrt{2})$ and $(0, 2 + \sqrt{2})$. Use the axis of symmetry to locate the point $(-4, 4)$ on the graph. The domain is $(-\infty, 4]$; the range is $(-\infty, \infty)$.

[**YOU TRY 4**] Graph $x = -(y+1)^2 - 3$. Find the *x*- and *y*-intercepts and the domain and range.

Procedure Graphing Parabolas from the Form $x = ay^2 + by + c$

We can use two methods to graph $x = ay^2 + by + c$.

Method 1: Rewrite $x = ay^2 + by + c$ in the form $x = a(y-k)^2 + h$ by completing the square.

> **W Hint**
> Notice how this compares to the procedure for graphing $f(x) = ax^2 + bx + c$.

Method 2: Use the formula $y = -\dfrac{b}{2a}$ to find the *y-coordinate* of the vertex. Find the *x*-coordinate by substituting the *y*-value into the equation $x = ay^2 + by + c$.

5 Rewrite $x = ay^2 + by + c$ as $x = a(y - k)^2 + h$ by Completing the Square

EXAMPLE 5 Rewrite $x = 2y^2 - 4y + 8$ in the form $x = a(y-k)^2 + h$ by completing the square.

Solution

To complete the square, follow the same procedure used for quadratic functions. (This is outlined on p. 659 in Section 10.5.)

Step 1: Divide the equation by 2 so that the coefficient of y^2 is 1.

$$\frac{x}{2} = y^2 - 2y + 4$$

Step 2: Separate the constant from the variable terms using parentheses.

$$\frac{x}{2} = (y^2 - 2y) + 4$$

Step 3: Complete the square for the quantity in parentheses. Add 1 *inside* the parentheses, and *subtract* 1 from the 4.

$$\frac{x}{2} = (y^2 - 2y + 1) + 4 - 1$$
$$\frac{x}{2} = (y^2 - 2y + 1) + 3$$

Step 4: Factor the expression inside the parentheses.

$$\frac{x}{2} = (y - 1)^2 + 3$$

Step 5: Solve the equation for x by multiplying by 2.

$$2\left(\frac{x}{2}\right) = 2[(y - 1)^2 + 3]$$
$$x = 2(y - 1)^2 + 6$$

[**YOU TRY 5**] Rewrite $x = -y^2 - 6y - 1$ in the form $x = a(y - k)^2 + h$ by completing the square.

6 Find the Vertex of the Graph of $x = ay^2 + by + c$ Using $y = -\frac{b}{2a}$, and Graph the Equation

EXAMPLE 6

Graph $x = y^2 - 2y + 5$. Find the vertex using the vertex formula. Find the x- and y-intercepts and the domain and range.

Solution

Since this equation is solved for x and is quadratic in y, it opens in the x-direction. $a = 1$, so it opens to the right. Use the vertex formula to find the y-coordinate of the vertex.

$$y = -\frac{b}{2a}$$
$$y = -\frac{-2}{2(1)} = 1 \qquad a = 1, b = -2$$

Substitute $y = 1$ into $x = y^2 - 2y + 5$ to find the x-coordinate of the vertex.

$$x = (1)^2 - 2(1) + 5$$
$$x = 1 - 2 + 5 = 4$$

The vertex is (4, 1). Because the vertex is (4, 1) and the parabola opens to the right, the graph has *no y-intercepts*.

To find the x-intercept, let $y = 0$ and solve for x.

$$x = y^2 - 2y + 5$$
$$x = (0)^2 - 2(0) + 5$$
$$x = 5$$

The x-intercept is (5, 0).

Find another point on the parabola by choosing a value for y that is close to the y-coordinate of the vertex. Let $y = -1$. Find x.

$$x = (-1)^2 - 2(-1) + 5$$
$$x = 1 + 2 + 5 = 8$$

Another point on the parabola is (8, −1). Use the axis of symmetry to locate the additional points (5, 2) and (8, 3). The domain is [4, ∞), and the range is (−∞, ∞)

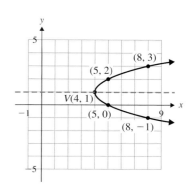

[YOU TRY 6] Graph $x = y^2 + 6y + 3$. Find the vertex using the vertex formula. Find the x- and y-intercepts and the domain and range.

Using Technology

To graph a parabola that is a function, just enter the equation and press GRAPH.

Example 1: Graph $f(x) = -x^2 + 2$.

Enter $Y_1 = -x^2 + 2$ to graph the function on a calculator.

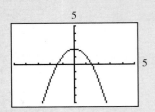

To graph an equation on a calculator, it must be entered so that y is a function of x. Since a parabola that opens horizontally is not a function, we must solve for y in terms of x so that the equation is represented by two different functions.

Example 2: Graph $x = y^2 - 4$ on a calculator.

Solve for y.

$$x = y^2 - 4$$
$$x + 4 = y^2$$
$$\pm\sqrt{x + 4} = y$$

Now the equation $x = y^2 - 4$ is rewritten so that y is in terms of x. In the graphing calculator, enter $y = \sqrt{x + 4}$ as Y_1. This represents the top half of the parabola since the y-values are positive above the x-axis. Enter $y = -\sqrt{x + 4}$ as Y_2. This represents the bottom half of the parabola since the y-values are negative below the x-axis. Set an appropriate window and press GRAPH.

 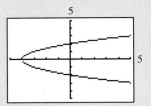

Graph each parabola on a graphing calculator. Where appropriate, rewrite the equation for y in terms of x. These problems come from the homework exercises so that the graphs can be found in the Answers to Exercises appendix.

1) $f(x) = x^2 + 6x + 9$; Exercise 9
2) $x = y^2 + 2$; Exercise 33
3) $x = \frac{1}{4}(y + 2)^2$; Exercise 39
4) $f(x) = -\frac{1}{2}x^2 + 4x - 6$; Exercise 11
5) $x = -(y - 4)^2 + 5$; Exercise 35
6) $x = y^2 - 4y + 5$; Exercise 41

ANSWERS TO [YOU TRY] EXERCISES

1) a) minimum value   b) vertex $(-3, -2)$
   c) The minimum value of the function is -2.
   d)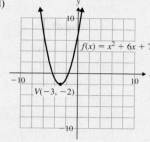

2) a) 1 sec   b) 26 ft   3) 49 in^2
4) $V(-3, -1)$; x-int: $(-4, 0)$; y-int: none; domain: $(-\infty, -1]$; range: $(-\infty, \infty)$

5) $x = -(y + 3)^2 + 8$
6) $V(-6, -3)$; x-int: $(3, 0)$; y-int: $(0, -3 - \sqrt{6})$, $(0, -3 + \sqrt{6})$; domain: $[-6, \infty)$; range: $(-\infty, \infty)$

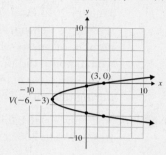

ANSWERS TO TECHNOLOGY EXERCISES

1) The equation can be entered as it is.
2) $Y_1 = \sqrt{x - 2}$; $Y_2 = -\sqrt{x - 2}$
3) $Y_1 = -2 + \sqrt{4x}$; $Y_2 = -2 - \sqrt{4x}$
4) The equation can be entered as it is.
5) $Y_1 = 4 + \sqrt{5 - x}$; $Y_2 = 4 - \sqrt{5 - x}$
6) $Y_1 = 2 + \sqrt{x - 1}$; $Y_2 = 2 - \sqrt{x - 1}$

E Evaluate 10.6 Exercises

Do the exercises, and check your work.

Objective 1: Find the Maximum or Minimum Value of a Quadratic Function

For Exercises 1–6, determine whether the function has a maximum value, minimum value, or neither.

1)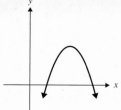

2)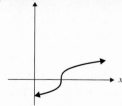

3)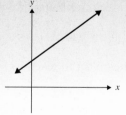

4)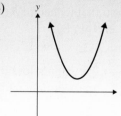

5)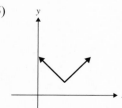

6)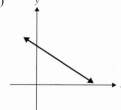

SECTION 10.6   Applications of Quadratic Functions and Graphing Other Parabolas

7) Let $f(x) = ax^2 + bx + c$. How do you know whether the function has a maximum or minimum value at the vertex?

8) Is there a maximum value of the function $y = 2x^2 + 12x + 11$? Explain your answer.

For Problems 9–12, answer parts a)–d) for each function, $f(x)$.

a) Does the function attain a minimum or maximum value at its vertex?

b) Find the vertex of the graph of $f(x)$.

c) What is the minimum or maximum value of the function?

d) Graph the function to verify parts a)–c).

9) $f(x) = x^2 + 6x + 9$ 10) $f(x) = -x^2 + 2x + 4$

11) $f(x) = -\dfrac{1}{2}x^2 + 4x - 6$ 12) $f(x) = 2x^2 + 4x$

Objective 2: Given a Quadratic Function, Solve an Applied Problem Involving a Maximum or Minimum Value

Solve.

13) An object is fired upward from the ground so that its height h (in feet) t sec after being fired is given by

$$h(t) = -16t^2 + 320t$$

a) How long does it take the object to reach its maximum height?

b) What is the maximum height attained by the object?

c) How long does it take the object to hit the ground?

14) An object is thrown upward from a height of 64 ft so that its height h (in feet) t sec after being thrown is given by

$$h(t) = -16t^2 + 48t + 64$$

a) How long does it take the object to reach its maximum height?

b) What is the maximum height attained by the object?

c) How long does it take the object to hit the ground?

15) The average number of traffic tickets issued in a city on any given day Sunday–Saturday can be approximated by

$$T(x) = -7x^2 + 70x + 43$$

where x represents the number of days after Sunday ($x = 0$ represents Sunday, $x = 1$ represents Monday, etc.), and $T(x)$ represents the number of traffic tickets issued. On which day are the most tickets written? How many tickets are issued on that day?

16) The number of guests staying at the Toasty Inn from January to December 2013 can be approximated by

$$N(x) = -10x^2 + 120x + 120$$

where x represents the number of months after January 2013 ($x = 0$ represents January, $x = 1$ represents February, etc.), and $N(x)$ represents the number of guests who stayed at the inn. During which month did the inn have the greatest number of guests? How many people stayed at the inn during that month?

17) The number of babies born to teenage mothers from 1989 to 2002 can be approximated by

$$N(t) = -0.721t^2 + 2.75t + 528$$

where t represents the number of years after 1989 and $N(t)$ represents the number of babies born (in thousands). According to this model, in what year was the number of babies born to teen mothers the greatest? How many babies were born that year? (U.S. Census Bureau)

18) The number of violent crimes in the United States from 1985 to 1999 can be modeled by

$$C(x) = -49.2x^2 + 636x + 12{,}468$$

where x represents the number of years after 1985 and $C(x)$ represents the number of violent crimes (in thousands). During what year did the greatest number of violent crimes occur, and how many were there? (U.S. Census Bureau)

Objective 3: Write a Quadratic Function to Solve an Applied Problem Involving a Maximum or Minimum Value

Solve.

19) Every winter Rich makes a rectangular ice rink in his backyard. He has 100 ft of material to use as the border. What is the maximum area of the ice rink?

20) Find the dimensions of the rectangular garden of greatest area that can be enclosed with 40 ft of fencing.

21) The Soo family wants to fence in a rectangular area to hold their dogs. One side of the pen will be their barn. Find the dimensions of the pen of greatest area that can be enclosed with 48 ft of fencing.

22) A farmer wants to enclose a rectangular area with 120 ft of fencing. One side is a river and will not require a fence. What is the maximum area that can be enclosed?

23) Find two integers whose sum is 18 and whose product is a maximum.

24) Find two integers whose sum is 26 and whose product is a maximum.

25) Find two integers whose difference is 12 and whose product is a minimum.

26) Find two integers whose difference is 30 and whose product is a minimum.

Objective 4: Graph Parabolas of the Form $x = a(y - k)^2 + h$

Given a quadratic equation of the form $x = a(y - k)^2 + h$, answer the following.

27) What is the vertex?

28) What is the equation of the axis of symmetry?

29) If a is negative, which way does the parabola open?

30) If a is positive, which way does the parabola open?

For each equation, identify the vertex, axis of symmetry, and x- and y-intercepts. Then, graph the equation. Determine the domain and range.

31) $x = (y - 1)^2 - 4$ 32) $x = (y + 3)^2 - 1$

33) $x = y^2 + 2$ 34) $x = (y - 4)^2$

35) $x = -(y - 4)^2 + 5$ 36) $x = -(y + 1)^2 - 7$

37) $x = -2(y - 2)^2 - 9$ 38) $x = -\dfrac{1}{2}(y - 4)^2 + 7$

39) $x = \dfrac{1}{4}(y + 2)^2$ 40) $x = 2y^2 + 3$

Objective 5: Rewrite $x = ay^2 + by + c$ as $x = a(y - k)^2 + h$ by Completing the Square

Rewrite each equation in the form $x = a(y - k)^2 + h$ by completing the square and graph it. Determine the domain and range.

41) $x = y^2 + 4y - 6$ 42) $x = y^2 - 4y + 5$

43) $x = -y^2 - 2y - 5$ 44) $x = -y^2 + 6y + 6$

45) $x = \dfrac{1}{3}y^2 + \dfrac{8}{3}y - \dfrac{5}{3}$ 46) $x = 2y^2 - 4y + 5$

47) $x = -4y^2 - 8y - 10$ 48) $x = \dfrac{1}{2}y^2 + 4y - 1$

Objective 6: Find the Vertex of $x = ay^2 + by + c$ Using $y = -\dfrac{b}{2a}$, and Graph the Equation

Graph each equation using the vertex formula. Find the x- and y-intercepts. Determine the domain and range.

49) $x = y^2 - 4y + 3$ 50) $x = -y^2 + 2y + 2$

51) $x = -y^2 + 4y$ 52) $x = y^2 + 6y - 4$

53) $x = -2y^2 + 4y - 6$ 54) $x = 3y^2 + 6y - 1$

55) $x = 4y^2 - 16y + 13$ 56) $x = 2y^2 + 4y + 8$

57) $x = \dfrac{1}{4}y^2 - \dfrac{1}{2}y + \dfrac{25}{4}$ 58) $x = -\dfrac{3}{4}y^2 + \dfrac{3}{2}y - \dfrac{11}{4}$

Mixed Exercises

Exercises 59–68 contain parabolas that open either horizontally or vertically. Graph each equation. Determine the domain and range.

59) $h(x) = -x^2 + 6$ 60) $y = x^2 - 6x - 1$

61) $x = y^2$ 62) $f(x) = -3x^2 + 12x - 8$

63) $x = -\dfrac{1}{2}y^2 - 4y - 5$ 64) $x = (y - 4)^2 + 3$

65) $y = x^2 + 2x - 3$ 66) $x = -3(y + 2)^2 + 11$

67) $f(x) = -2(x - 4)^2 + 3$

68) $g(x) = \dfrac{3}{2}x^2 - 12x + 20$

R Rethink

R1) What does it mean to have a maximum or a minimum?

R2) How does a parabola open left or right?

R3) What do you find most difficult about solving applied problems? Which type is the hardest for you?

10.7 Quadratic and Rational Inequalities

P Prepare

What are your objectives for Section 10.7?

O Organize

How can you accomplish each objective?

1 Solve a Quadratic Inequality by Graphing	• Know the definition of a *quadratic inequality*. • Review the procedure for **Graphing Quadratic Functions Using the Vertex Formula.** • Review interval notation, and understand the meanings of $<$, $\leq$, $>$, and $\geq$ on a graph. • Complete the given example on your own. • Complete You Try 1.
2 Solve a Quadratic Inequality Using Test Points	• Learn the procedure for **How to Solve a Quadratic Inequality.** • Review methods for solving quadratic equations. • Write solutions in interval notation and review that procedure, if needed. • Complete the given example on your own. • Complete You Try 2.
3 Solve Quadratic Inequalities with Special Solutions	• Analyze the inequality carefully to determine if picking test points or graphing is necessary. • Complete the given example on your own. • Complete You Try 3.
4 Solve an Inequality of Higher Degree	• Use the procedure for **How to Solve a Quadratic Inequality.** You may use more intervals and more test points. • Complete the given example on your own. • Complete You Try 4.
5 Solve a Rational Inequality	• Know the definition of a *rational inequality*. • Learn the procedure for **How to Solve a Rational Inequality.** • Write the answer in interval notation. • Complete the given examples on your own. • Complete You Trys 5 and 6.

Read the explanations, follow the examples, take notes, and complete the You Trys.

In Chapter 3, we learned how to solve *linear* inequalities such as $3x - 5 \leq 16$. In this section, we will discuss how to solve *quadratic* and *rational* inequalities.

> **Definition**
>
> A **quadratic inequality** can be written in the form
>
> $$ax^2 + bx + c \leq 0 \quad \text{or} \quad ax^2 + bx + c \geq 0$$
>
> where a, b, and c are real numbers and $a \neq 0$. ($<$ and $>$ may be substituted for $\leq$ and $\geq$.)

1 Solve a Quadratic Inequality by Graphing

To understand how to solve a quadratic inequality, let's look at the graph of a quadratic function.

EXAMPLE 1

a) Graph $y = x^2 - 2x - 3$.

b) Solve $x^2 - 2x - 3 < 0$.

c) Solve $x^2 - 2x - 3 \geq 0$.

Solution

a) The graph of the quadratic function $y = x^2 - 2x - 3$ is a parabola that opens upward. Use the vertex formula to confirm that the vertex is $(1, -4)$.

To find the y-intercept, let $x = 0$ and solve for y.

$$y = 0^2 - 2(0) - 3$$
$$y = -3$$

The y-intercept is $(0, -3)$.

To find the x-intercepts, let $y = 0$ and solve for x.

$$0 = x^2 - 2x - 3$$
$$0 = (x-3)(x+1) \quad \text{Factor.}$$
$$x - 3 = 0 \quad \text{or} \quad x + 1 = 0 \quad \text{Set each factor equal to 0.}$$
$$x = 3 \quad \text{or} \quad x = -1 \quad \text{Solve.}$$

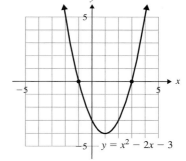

 Hint
Be sure to notice the difference between $<$ and $\leq$ as well as $>$ and $\geq$.

b) We will use the graph of $y = x^2 - 2x - 3$ to solve the inequality $x^2 - 2x - 3 < 0$. That is, to solve $x^2 - 2x - 3 < 0$ we must ask ourselves, "Where are the y-values of the function *less than* zero?"

The y-values of the function are less than zero when the x-values are greater than -1 and less than 3, as shown to the right.

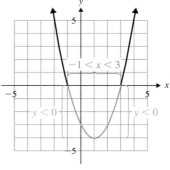

The solution set of $x^2 - 2x - 3 < 0$ (in interval notation) is $(-1, 3)$.

c) To solve $x^2 - 2x - 3 \geq 0$ means to find the x-values for which the y-values of the function $y = x^2 - 2x - 3$ are *greater than or equal to* zero. (Recall that the x-intercepts are where the function equals zero.)

The y-values of the function are greater than or equal to zero when $x \leq -1$ or when $x \geq 3$. The solution set of $x^2 - 2x - 3 \geq 0$ is $(-\infty, -1] \cup [3, \infty)$.

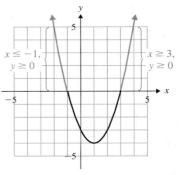

When $x \leq -1$ or $x \geq 3$, the y-values are greater than or equal to 0.

[YOU TRY 1] a) Graph $y = x^2 + 6x + 5$. b) Solve $x^2 + 6x + 5 \leq 0$.
c) Solve $x^2 + 6x + 5 > 0$.

2 Solve a Quadratic Inequality Using Test Points

Example 1 illustrates how the x-intercepts of $y = x^2 - 2x - 3$ break up the x-axis into the three separate intervals: $x < -1$, $-1 < x < 3$, and $x > 3$. We can use this idea of intervals to solve a quadratic inequality without graphing.

EXAMPLE 2

Solve $x^2 - 2x - 3 < 0$.

Solution

Begin by solving the equation $x^2 - 2x - 3 = 0$.

$$x^2 - 2x - 3 = 0$$
$$(x - 3)(x + 1) = 0 \qquad \text{Factor.}$$
$$x - 3 = 0 \quad \text{or} \quad x + 1 = 0 \qquad \text{Set each factor equal to 0.}$$
$$x = 3 \quad \text{or} \quad x = -1 \qquad \text{Solve.}$$

(These are the x-intercepts of $y = x^2 - 2x - 3$.)

W Hint
Be sure that you *understand* what is being done in each step.

Note

The $<$ indicates that we want to find the values of x that will make $x^2 - 2x - 3 < 0$; that is, find the values of x that make $x^2 - 2x - 3$ a *negative* number.

Put $x = 3$ and $x = -1$ on a number line with the smaller number on the left. This breaks up the number line into three intervals: $x < -1$, $-1 < x < 3$, and $x > 3$.

Choose a test number in each interval and substitute it into $x^2 - 2x - 3$ to determine whether that value makes $x^2 - 2x - 3$ positive or negative. (If one number in the interval makes $x^2 - 2x - 3$ positive, then *all* numbers in that interval will make $x^2 - 2x - 3$ positive.) Indicate the result on the number line.

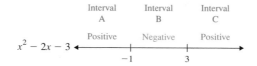

Hint
Read this example *very* carefully.

Interval A: $(x < -1)$ As a test number, choose any number less than -1. We will choose -2. Evaluate $x^2 - 2x - 3$ for $x = -2$.

$$x^2 - 2x - 3 = (-2)^2 - 2(-2) - 3 \quad \text{Substitute } -2 \text{ for } x.$$
$$= 4 + 4 - 3$$
$$= 8 - 3 = 5$$

When $x = -2$, $x^2 - 2x - 3$ is *positive*. Therefore, $x^2 - 2x - 3$ will be positive for all values of x in this interval. Indicate this on the number line as seen above.

Interval B: $(-1 < x < 3)$ As a test number, choose any number between -1 and 3. We will choose 0. Evaluate $x^2 - 2x - 3$ for $x = 0$.

$$x^2 - 2x - 3 = (0)^2 - 2(0) - 3 \quad \text{Substitute } 0 \text{ for } x.$$
$$= 0 - 0 - 3 = -3$$

When $x = 0$, $x^2 - 2x - 3$ is *negative*. Therefore, $x^2 - 2x - 3$ will be negative for all values of x in this interval. Indicate this on the number line above.

Interval C: $(x > 3)$ As a test number, choose any number greater than 3. We will choose 4. Evaluate $x^2 - 2x - 3$ for $x = 4$.

$$x^2 - 2x - 3 = (4)^2 - 2(4) - 3 \quad \text{Substitute } 4 \text{ for } x.$$
$$= 16 - 8 - 3$$
$$= 8 - 3 = 5$$

When $x = 4$, $x^2 - 2x - 3$ is *positive*. Therefore, $x^2 - 2x - 3$ will be positive for all values of x in this interval. Indicate this on the number line.

Look at the number line above. The solution set of $x^2 - 2x - 3 < 0$ consists of the interval(s) where $x^2 - 2x - 3$ is *negative*. This is in **Interval B,** $(-1, 3)$.

The graph of the solution set is

The solution set is $(-1, 3)$. This is the same as the result we obtained in Example 1 by graphing.

[**YOU TRY 2**] Solve $x^2 + 5x + 4 \leq 0$. Graph the solution set, and write the solution in interval notation.

Next we will summarize how to solve a quadratic inequality.

W Hint
In your own words, summarize this procedure.

> **Procedure** How to Solve a Quadratic Inequality
>
> **Step 1:** Write the inequality in the form $ax^2 + bx + c \leq 0$ or $ax^2 + bx + c \geq 0$. ($<$ and $>$ may be substituted for $\leq$ and ≥ 0.) If the inequality symbol is $<$ or $\leq$, we are looking for a *negative* quantity in the interval on the number line. If the inequality symbol is $>$ or $\geq$, we are looking for a *positive* quantity in the interval.
>
> **Step 2:** Solve the equation $ax^2 + bx + c = 0$.
>
> **Step 3:** Put the solutions of $ax^2 + bx + c = 0$ on a number line. These values break up the number line into intervals.
>
> **Step 4:** Choose a test number in each interval to determine whether $ax^2 + bx + c$ is positive or negative in each interval. Indicate this on the number line.
>
> **Step 5:** If the inequality is in the form $ax^2 + bx + c \leq 0$ or $ax^2 + bx + c < 0$, then the solution set contains the numbers in the interval where $ax^2 + bx + c$ is *negative*.
>
> If the inequality is in the form $ax^2 + bx + c \geq 0$ or $ax^2 + bx + c > 0$, then the solution set contains the numbers in the interval where $ax^2 + bx + c$ is *positive*.
>
> **Step 6:** If the inequality symbol is $\leq$ or $\geq$, then the endpoints of the interval(s) (the numbers found in Step 3) are included in the solution set. Indicate this with brackets in the interval notation.
>
> If the inequality symbol is $<$ or $>$, then the endpoints of the interval(s) are not included in the solution set. Indicate this with parentheses in interval notation.

3 Solve Quadratic Inequalities with Special Solutions

We should look carefully at the inequality before trying to solve it. Sometimes, it is not necessary to go through all of the steps.

EXAMPLE 3

Solve.

a) $(y + 4)^2 \geq -5$ b) $(t - 8)^2 < -3$

Solution

a) The inequality $(y + 4)^2 \geq -5$ says that a squared quantity, $(y + 4)^2$, is greater than or equal to a *negative* number, -5. *This is always true.* (A squared quantity will *always* be greater than or equal to zero.) Any real number, y, will satisfy the inequality.

The solution set is (∞, ∞).

b) The inequality $(t - 8)^2 < -3$ says that a squared quantity, $(t - 8)^2$, is less than a *negative* number, -3. There is no real number value for t so that $(t - 8)^2 < -3$.

The solution set is $\varnothing$.

[YOU TRY 3]

Solve.

a) $(k + 2)^2 \leq -4$ b) $(z - 9)^2 > -1$

4 Solve an Inequality of Higher Degree

Other polynomial inequalities in factored form can be solved in the same way that we solve quadratic inequalities.

EXAMPLE 4 Solve $(c - 2)(c + 5)(c - 4) < 0$.

Solution

This is the factored form of a third-degree polynomial. Since the inequality is $<$, the solution set will contain the intervals where $(c - 2)(c + 5)(c - 4)$ is *negative*.

Solve $(c - 2)(c + 5)(c - 4) = 0$.

$c - 2 = 0$ or $c + 5 = 0$ or $c - 4 = 0$ Set each factor equal to 0.
$c = 2$ or $c = -5$ or $c = 4$ Solve.

Put $c = 2$, $c = -5$, and $c = 4$ on a number line, and test a number in each interval.

Interval	$c < -5$	$-5 < c < 2$	$2 < c < 4$	$c > 4$
Test number	$c = -6$	$c = 0$	$c = 3$	$c = 5$
Evaluate $(c - 2)(c + 5)(c - 4)$	$(-6 - 2)(-6 + 5)(-6 - 4)$ $= (-8)(-1)(-10)$ $= -80$	$(0 - 2)(0 + 5)(0 - 4)$ $= (-2)(5)(-4)$ $= 40$	$(3 - 2)(3 + 5)(3 - 4)$ $= (1)(8)(-1)$ $= -8$	$(5 - 2)(5 + 5)(5 - 4)$ $= (3)(10)(1)$ $= 30$
Sign	Negative	Positive	Negative	Positive

$(c - 2)(c + 5)(c - 4)$ ← Negative | Positive | Negative | Positive → at -5, 2, 4

We can see that the intervals where $(c - 2)(c + 5)(c - 4)$ is negative are $(-\infty, -5)$ and $(2, 4)$. The endpoints are not included because the inequality is $<$.

The graph of the solution set is ←++++◇++++++◇+++◇+++→
                                   $-8\,-7\,-6\,-5\,-4\,-3\,-2\,-1\ 0\ 1\ 2\ 3\ 4\ 5\ 6\ 7\ 8$

The solution set of $(c - 2)(c + 5)(c - 4) < 0$ is $(-\infty, -5) \cup (2, 4)$.

[YOU TRY 4] Solve $(y + 3)(y - 1)(y + 1) \geq 0$. Graph the solution set, and write the solution in interval notation.

5 Solve a Rational Inequality

An inequality containing a rational expression, $\dfrac{p}{q}$, where p and q are polynomials, is called a **rational inequality**. The way we solve rational inequalities is very similar to the way we solve quadratic inequalities.

W Hint
Summarize this procedure in your notes, in your own words.

> **Procedure** How to Solve a Rational Inequality
>
> **Step 1:** Write the inequality so that there is a 0 on one side and only one rational expression on the other side. If the inequality symbol is $<$ or $\leq$, we are looking for a *negative* quantity in the interval on the number line. If the inequality symbol is $>$ or $\geq$, we are looking for a *positive* quantity in the interval.
>
> **Step 2:** Find the numbers that make the numerator equal 0 and any numbers that make the denominator equal 0.
>
> **Step 3:** Put the numbers found in Step 2 on a number line. These values break up the number line into intervals.
>
> **Step 4:** Choose a test number in each interval to determine whether the rational inequality is positive or negative in each interval. Indicate this on the number line.
>
> **Step 5:** If the inequality is in the form $\dfrac{p}{q} \leq 0$ or $\dfrac{p}{q} < 0$, then the solution set contains the numbers in the interval where $\dfrac{p}{q}$ is *negative*.
>
> If the inequality is in the form $\dfrac{p}{q} \geq 0$ or $\dfrac{p}{q} > 0$, then the solution set contains the numbers in the interval where $\dfrac{p}{q}$ is *positive*.
>
> **Step 6:** Determine whether the endpoints of the intervals are included in or excluded from the solution set. Do not include any values that make the denominator equal 0.

EXAMPLE 5

Solve $\dfrac{5}{x+3} > 0$. Graph the solution set, and write the solution in interval notation.

Solution

Step 1: The inequality is in the correct form—zero on one side and only one rational expression on the other side. Since the inequality symbol is > 0, the solution set will contain the interval(s) where $\dfrac{5}{x+3}$ is *positive*.

Step 2: Find the numbers that make the numerator equal 0 and any numbers that make the denominator equal 0.

Numerator: 5	Denominator: $x + 3$
The numerator is a constant, 5, so it cannot equal 0.	Set $x + 3 = 0$ and solve for x. $x + 3 = 0$ $x = -3$

Step 3: Put -3 on a number line to break it up into intervals.

$$\dfrac{5}{x+3} \qquad \xleftarrow{\hspace{1cm}|\hspace{2cm}}\rightarrow$$
$$-3$$

Step 4: Choose a test number in each interval to determine whether $\dfrac{5}{x+3}$ is positive or negative in each interval.

Interval	$x < -3$	$x > -3$
Test number	$x = -4$	$x = 0$
Evaluate $\dfrac{5}{x+3}$	$\dfrac{5}{-4+3} = \dfrac{5}{-1} = -5$	$\dfrac{5}{0+3} = \dfrac{5}{3}$
Sign	Negative	Positive

686 CHAPTER 10 **Quadratic Equations and Functions**

Step 5: The solution set of $\dfrac{5}{x+3} > 0$ contains the numbers in the interval where $\dfrac{5}{x+3}$ is *positive*. This interval is $(-3, \infty)$.

$$\dfrac{5}{x+3} \quad \underset{-3}{\longleftarrow \text{Negative} \quad | \quad \text{Positive} \longrightarrow}$$

Step 6: Since the inequality symbol is $>$, the endpoint of the interval, -3, is not included in the solution set.

The graph of the solution set is $\underset{-5\,-4\,-3\,-2\,-1\ 0\ 1\ 2\ 3\ 4\ 5}{\longleftarrow\!\!\!+\!\!+\!\!\diamond\!\!+\!\!+\!\!+\!\!+\!\!+\!\!+\!\!+\!\!+\!\!+\!\!\longrightarrow}$

The solution set is $(-3, \infty)$.

[**YOU TRY 5**] Solve $\dfrac{2}{y-6} < 0$. Graph the solution set, and write the solution in interval notation.

EXAMPLE 6

Solve $\dfrac{7}{a+2} \leq 3$. Graph the solution set, and write the solution in interval notation.

Solution

Step 1: Get a zero on one side of the inequality symbol and only one rational expression on the other side.

$$\dfrac{7}{a+2} \leq 3$$

$$\dfrac{7}{a+2} - 3 \leq 0 \qquad \text{Subtract 3.}$$

$$\dfrac{7}{a+2} - \dfrac{3(a+2)}{a+2} \leq 0 \qquad \text{Get a common denominator.}$$

$$\dfrac{7}{a+2} - \dfrac{3a+6}{a+2} \leq 0 \qquad \text{Distribute.}$$

$$\dfrac{1-3a}{a+2} \leq 0 \qquad \text{Combine numerators, and combine like terms.}$$

From this point forward, we will work with the inequality $\dfrac{1-3a}{a+2} \leq 0$. It is equivalent to the original inequality. Because the inequality symbol is $\leq$, the solution set contains the interval(s) where $\dfrac{1-3a}{a+2}$ is *negative*.

Step 2: Find the numbers that make the numerator equal 0 and any numbers that make the denominator equal 0.

Numerator	Denominator
$1 - 3a = 0$	$a + 2 = 0$
$-3a = -1$	$a = -2$
$a = \dfrac{1}{3}$	

Step 3: Put $\dfrac{1}{3}$ and -2 on a number line to break it up into intervals.

$$\dfrac{1-3a}{a+2} \quad \underset{-2 \qquad \frac{1}{3}}{\longleftarrow \quad | \quad\quad | \quad \longrightarrow}$$

SECTION 10.7 **Quadratic and Rational Inequalities**

Step 4: Choose a test number in each interval.

Interval	$a < -2$	$-2 < a < \dfrac{1}{3}$	$a > \dfrac{1}{3}$
Test number	$a = -3$	$a = 0$	$a = 1$
Evaluate $\dfrac{1-3a}{a+2}$	$\dfrac{1-3(-3)}{-3+2} = \dfrac{10}{-1} = -10$	$\dfrac{1-3(0)}{0+2} = \dfrac{1}{2}$	$\dfrac{1-3(1)}{1+2} = -\dfrac{2}{3}$
Sign	Negative	Positive	Negative

Step 5: The solution set of $\dfrac{1-3a}{a+2} \leq 0$ $\left(\text{and therefore } \dfrac{7}{a+2} \leq 3\right)$ will contain the numbers in the intervals where $\dfrac{1-3a}{a+2}$ is *negative*. These are the first and last intervals.

$\dfrac{1-3a}{a+2}$ ← Negative | Positive | Negative →
                        -2 $\dfrac{1}{3}$

Step 6: Determine whether the endpoints of the intervals, -2 and $\dfrac{1}{3}$, are included in the solution set. The endpoint $\dfrac{1}{3}$ is included because it does not make the denominator equal 0. *But* -2 *is not included because it makes the denominator equal 0.*

The graph of the solution set of $\dfrac{7}{a+2} \leq 3$ is

The solution set is $(-\infty, -2) \cup \left[\dfrac{1}{3}, \infty\right)$.

 Although an inequality symbol may be $\leq$ or $\geq$, an endpoint cannot be included in the solution set if it makes the denominator equal 0.

[YOU TRY 6] Solve $\dfrac{3}{z+4} \geq 2$. Graph the solution set, and write the solution in interval notation.

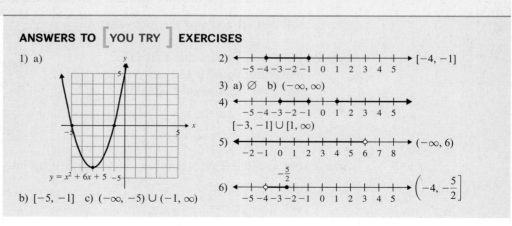

ANSWERS TO [YOU TRY] EXERCISES

1) a) [graph of $y = x^2 + 6x + 5$] b) $[-5, -1]$ c) $(-\infty, -5) \cup (-1, \infty)$

2) $[-4, -1]$

3) a) $\varnothing$ b) $(-\infty, \infty)$

4) $[-3, -1] \cup [1, \infty)$

5) $(-\infty, 6)$

6) $\left(-4, -\dfrac{5}{2}\right]$

Evaluate 10.7 Exercises

Do the exercises, and check your work.

1) When solving a quadratic inequality, how do you know when to include and when to exclude the endpoints in the solution set?

2) If a rational inequality contains a $\leq$ or $\geq$ symbol, will the endpoints of the solution set always be included? Explain your answer.

Objective 1: Solve a Quadratic Inequality by Graphing

For Exercises 3–6, use the graph of the function to solve each inequality.

3) $y = x^2 + 4x - 5$

4) $y = x^2 - 6x + 8$

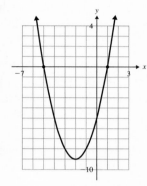

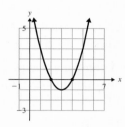

a) $x^2 - 6x + 8 > 0$
b) $x^2 - 6x + 8 \leq 0$

a) $x^2 + 4x - 5 \leq 0$
b) $x^2 + 4x - 5 > 0$

5) $y = -\frac{1}{2}x^2 + x + \frac{3}{2}$

6) $y = -x^2 - 8x - 12$

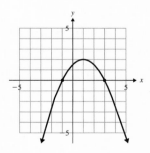

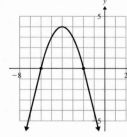

a) $-\frac{1}{2}x^2 + x + \frac{3}{2} \geq 0$
b) $-\frac{1}{2}x^2 + x + \frac{3}{2} < 0$

a) $-x^2 - 8x - 12 < 0$
b) $-x^2 - 8x - 12 \geq 0$

Objective 2: Solve a Quadratic Inequality Using Test Points

Solve each quadratic inequality. Graph the solution set, and write the solution in interval notation.

7) $x^2 + 6x - 7 \geq 0$
8) $m^2 - 2m - 24 > 0$
9) $c^2 + 5c < 36$
10) $t^2 + 36 \leq 15t$
11) $3z^2 + 14z - 24 \leq 0$
12) $5k^2 + 36k + 7 \geq 0$
13) $7p^2 - 4 > 12p$
14) $4w^2 - 19w < 30$
15) $b^2 - 9b > 0$
16) $c^2 + 12c \leq 0$
17) $m^2 - 64 < 0$
18) $p^2 - 144 > 0$
19) $121 - h^2 \leq 0$
20) $1 - d^2 > 0$

Objective 3: Solve Quadratic Inequalities with Special Solutions

Solve each inequality.

21) $(h + 5)^2 \geq -2$
22) $(3v - 11)^2 > -20$
23) $(2y - 1)^2 < -8$
24) $(r + 4)^2 < -3$
25) $(4d - 3)^2 > -1$
26) $(5s - 2)^2 \leq -9$

Objective 4: Solve an Inequality of Higher Degree

Solve each inequality. Graph the solution set, and write the solution in interval notation.

27) $(r + 2)(r - 5)(r - 1) \leq 0$
28) $(b + 2)(b - 3)(b - 12) > 0$
29) $(6c + 1)(c + 7)(4c - 3) < 0$
30) $(t + 2)(4t - 7)(5t - 1) \geq 0$

Objective 5: Solve a Rational Inequality

Solve each rational inequality. Graph the solution set, and write the solution in interval notation.

31) $\dfrac{7}{p + 6} > 0$
32) $\dfrac{3}{v - 2} < 0$
33) $\dfrac{5}{z + 3} \leq 0$
34) $\dfrac{9}{m - 4} \geq 0$
35) $\dfrac{x - 4}{x - 3} > 0$
36) $\dfrac{a - 2}{a + 1} < 0$

37) $\dfrac{h-9}{3h+1} \leq 0$

38) $\dfrac{2c+1}{c+4} \geq 0$

39) $\dfrac{k}{k+3} \leq 0$

40) $\dfrac{r}{r-7} \geq 0$

41) $\dfrac{7}{t+6} < 3$

42) $\dfrac{3}{x+7} < -2$

43) $\dfrac{3}{a+7} \geq 1$

44) $\dfrac{5}{w-3} \leq 1$

45) $\dfrac{2y}{y-6} \leq -3$

46) $\dfrac{3z}{z+4} \geq 2$

47) $\dfrac{3w}{w+2} > -4$

48) $\dfrac{4h}{h+3} < 1$

49) $\dfrac{(4t-3)^2}{t-5} > 0$

50) $\dfrac{(2y+3)^2}{y+3} < 0$

51) $\dfrac{m+1}{m^2+3} \geq 0$

52) $\dfrac{w-7}{w^2+8} \leq 0$

53) $\dfrac{s^2+2}{s-4} \leq 0$

54) $\dfrac{z^2+10}{z+6} \leq 0$

Mixed Exercises: Objectives 2 and 5
Write an inequality, and solve.

55) Compu Corp. estimates that its total profit function, $P(x)$, for producing x thousand units is given by $P(x) = -2x^2 + 32x - 96$.

a) At what level of production does the company make a profit?

b) At what level of production does the company lose money?

56) A model rocket is launched from the ground with an initial velocity of 128 ft/s. The height $s(t)$, in ft, of the rocket t seconds after liftoff is given by the function $s(t) = -16t^2 + 128t$.

a) When is the rocket more than 192 ft above the ground?

b) When does the rocket hit the ground?

57) A designer purse company has found that the average cost, $\overline{C}(x)$, of producing x purses per month can be described by the function $\overline{C}(x) = \dfrac{10x + 100{,}000}{x}$. How many purses must the company produce each month so that the average cost of producing each purse is no more than $20?

58) A company that produces clay pigeons for target shooting has determined that the average cost, $\overline{C}(x)$, of producing x cases of clay pigeons per month can be described by the function $\overline{C}(x) = \dfrac{2x + 15{,}000}{x}$. How many cases of clay pigeons must the company produce each month so that the average cost of producing each case is no more than $3?

R Rethink

R1) What is a rational inequality? How is this different from a rational equation?

R2) Are there any similarities between a quadratic inequality and a rational inequality?

R3) Which type of inequality do you find easiest to solve?

R4) Why do you need to pick test points?

Group Activity — The Group Activity can be found online on Connect.

emPOWERme Determine Your Saving Style

There are as many ways to save money as there are people looking to save it. To help you save money and keep your budget in order, identify your saving style. Read each of the following statements, and rate how well it describes you, using this scale:

1 = That's me
2 = Sometimes
3 = That's not me

	1	2	3
1. I count the change I'm given by cashiers in stores and restaurants.			
2. I always pick up all the change I receive from a transaction in a store, even if it's only a few cents.			
3. I don't buy something right away if I'm pretty sure it will go on sale soon.			
4. I feel a real sense of accomplishment if I buy something on sale.			
5. I always remember how much I paid for something.			
6. If something goes on sale soon after I've bought it, I feel cheated.			
7. I have money in at least one interest-bearing account.			
8. I rarely lend people money.			
9. If I lend money to someone repeatedly without getting it back, I stop lending it to that person.			
10. I share resources (e.g., CDs, books, magazines) with other people to save money.			
11. I'm good at denying myself small purchases when I know I am low on cash.			
12. I believe most generic or off-brand items are just as good as name brands.			

Add up your ratings. Interpret your total score according to this informal guide:

12–15: Very aggressive saving style
16–20: Careful saving style
21–27: Fairly loose saving style
28–32: Loose saving style
33–36: Nonexistent saving style

What are the advantages and disadvantages of your saving style? How do you think your saving style affects your ability to keep a healthy budget? If you are dissatisfied with your saving style, how might you be able to change it?

Chapter 10: Summary

Definition/Procedure	Example

10.1 The Square Root Property and Completing the Square

The Square Root Property

Let k be a constant. If $x^2 = k$, then $x = \sqrt{k}$ or $x = -\sqrt{k}$. (p. 612)

Solve $6p^2 = 54$.

$p^2 = 9$ Divide by 6.
$p = \pm\sqrt{9}$ Square root property
$p = \pm 3$ $\sqrt{9} = 3$

The solution set is $\{-3, 3\}$.

The Distance Formula

The **distance**, d, between two points with coordinates (x_1, y_1) and (x_2, y_2) is given by $d = \sqrt{(x_2 - x_1)^2 + (y_2 - y_1)^2}$. (p. 616)

Find the distance between the points $(6, -2)$ and $(0, 2)$.

Label the points: $(\overset{x_1}{6}, \overset{y_1}{-2})\ (\overset{x_2}{0}, \overset{y_2}{2})$

Substitute the values into the distance formula.

$$d = \sqrt{(0 - 6)^2 + (2 - (-2))^2}$$
$$= \sqrt{(-6)^2 + (4)^2}$$
$$= \sqrt{36 + 16} = \sqrt{52} = 2\sqrt{13}$$

A **perfect square trinomial** is a trinomial whose factored form is the square of a binomial. (p. 616)

Perfect Square Trinomial	Factored Form
$y^2 + 8y + 16$	$(y + 4)^2$
$9t^2 - 30t + 25$	$(3t - 5)^2$

Complete the Square for $x^2 + bx$

To find the constant needed to complete the square for $x^2 + bx$,

Step 1: Find half of the coefficient of x: $\dfrac{1}{2}b$

Step 2: Square the result: $\left(\dfrac{1}{2}b\right)^2$

Step 3: Add it to $x^2 + bx$: $x^2 + bx + \left(\dfrac{1}{2}b\right)^2$. The factored form is $\left(x + \dfrac{1}{2}b\right)^2$. (p. 616)

Complete the square for $x^2 + 12x$ to obtain a perfect square trinomial. Then, factor.

Step 1: Find half of the coefficient of x: $\dfrac{1}{2}(12) = 6$

Step 2: Square the result: $6^2 = 36$

Step 3: Add 36 to $x^2 + 12x$: $x^2 + 12x + 36$

The perfect square trinomial is $x^2 + 12x + 36$.
The factored form is $(x + 6)^2$.

Solve a Quadratic Equation ($ax^2 + bx + c = 0$) by Completing the Square

Step 1: The coefficient of the squared term must be 1. If it is not 1, divide both sides of the equation by a to obtain a leading coefficient of 1.
Step 2: Get the variables on one side of the equal sign and the constant on the other side.
Step 3: Complete the square. Find half of the linear coefficient, then square the result. Add that quantity to both sides of the equation.
Step 4: Factor.
Step 5: Solve using the square root property. (p. 619)

Solve $x^2 + 6x + 7 = 0$ by completing the square.

$x^2 + 6x + 7 = 0$ The coefficient of x^2 is 1.
$x^2 + 6x = -7$ Get the constant on the other side of the equal sign.

Complete the square: $\dfrac{1}{2}(6) = 3$
$(3)^2 = 9$

Add 9 to both sides of the equation.

$x^2 + 6x + 9 = -7 + 9$
$(x + 3)^2 = 2$ Factor.
$x + 3 = \pm\sqrt{2}$ Square root property
$x = -3 \pm \sqrt{2}$

The solution set is $\{-3 - \sqrt{2}, -3 + \sqrt{2}\}$.

Definition/Procedure	Example

10.2 The Quadratic Formula

The Quadratic Formula

The solutions of any quadratic equation of the form $ax^2 + bx + c = 0$ ($a \neq 0$) are

$$x = \frac{-b \pm \sqrt{b^2 - 4ac}}{2a}$$

This formula is called the **quadratic formula**. (p. 625)

Solve $2x^2 - 5x - 2 = 0$ using the quadratic formula.

$$a = 2 \quad b = -5 \quad c = -2$$

Substitute the values into the quadratic formula, and simplify.

$$x = \frac{-(-5) \pm \sqrt{(-5)^2 - 4(2)(-2)}}{2(2)}$$

$$x = \frac{5 \pm \sqrt{25 + 16}}{4} = \frac{5 \pm \sqrt{41}}{4}$$

The solution set is $\left\{\dfrac{5 - \sqrt{41}}{4}, \dfrac{5 + \sqrt{41}}{4}\right\}$.

The expression under the radical, $b^2 - 4ac$ is called the **discriminant**.

1) If $b^2 - 4ac$ is **positive and the square of an integer**, the equation has **two rational solutions**.
2) If $b^2 - 4ac$ is **positive but not a perfect square**, the equation has **two irrational solutions**.
3) If $b^2 - 4ac$ is **negative**, the equation has **two nonreal, complex solutions of the form** $a + bi$ **and** $a - bi$.
4) If $b^2 - 4ac = 0$, the equation has **one rational solution**. (p. 628)

Find the value of the discriminant for $3m^2 + 4m + 5 = 0$, and determine the number and type of solutions of the equation.

$$a = 3 \quad b = 4 \quad c = 5$$

$$b^2 - 4ac = (4)^2 - 4(3)(5) = 16 - 60 = -44$$

Discriminant $= -44$. The equation has two nonreal, complex solutions of the form $a + bi$ and $a - bi$.

10.3 Equations in Quadratic Form

Some equations that are not quadratic can be solved using the same methods that can be used to solve quadratic equations. These are called **equations in quadratic form**. (p. 638)

Solve $r^4 + 2r^2 - 24 = 0$.

$$(r^2 - 4)(r^2 + 6) = 0 \quad \text{Factor.}$$

$$r^2 - 4 = 0 \quad \text{or} \quad r^2 + 6 = 0$$
$$r^2 = 4 \quad\quad\quad\quad r^2 = -6$$
$$r = \pm\sqrt{4} \quad\quad\quad r = \pm\sqrt{-6}$$
$$r = \pm 2 \quad\quad\quad\quad r = \pm i\sqrt{6}$$

The solution set is $\{-i\sqrt{6}, i\sqrt{6}, -2, 2\}$.

10.4 Formulas and Applications

Solve a Formula for a Variable. (p. 645)

Solve for s: $g = \dfrac{10}{s^2}$

$s^2 g = 10$ Multiply both sides by s^2.

$s^2 = \dfrac{10}{g}$ Divide both sides by g.

$s = \pm\sqrt{\dfrac{10}{g}}$ Square root property

$s = \dfrac{\pm\sqrt{10}}{\sqrt{g}} \cdot \dfrac{\sqrt{g}}{\sqrt{g}}$ Rationalize the denominator.

$s = \dfrac{\pm\sqrt{10g}}{g}$

Definition/Procedure	Example
Solving Application Problems Using a Quadratic Equation. (p. 649)	A woman dives off of a cliff 49 m above the ocean. Her height, $h(t)$, in meters, above the water is given by $$h(t) = -9.8t^2 + 49$$ where t is the time, in seconds, after she leaves the cliff. When will she hit the water? Let $h(t) = 0$, and solve for t. $h(t) = -9.8t^2 + 49$ $0 = -9.8t^2 + 49$ Substitute 0 for h. $9.8t^2 = 49$ Add $9.8t^2$ to each side. $t^2 = 5$ Divide by 9.8. $t = \pm\sqrt{5}$ Square root property Since t represents time, we discard $-\sqrt{5}$. She will hit the water in $\sqrt{5}$, or about 2.2, sec.

10.5 Quadratic Functions and Their Graphs

A **quadratic function** is a function that can be written in the form $f(x) = ax^2 + bx + c$, where a, b, and c are real numbers and $a \neq 0$. The graph of a quadratic function is called a **parabola**. The lowest point on an upward-opening parabola or the highest point on a downward-opening parabola is called the **vertex.** (p. 653)	$f(x) = 5x^2 + 7x - 9$ is a quadratic function.
A quadratic function can also be written in the form $f(x) = a(x - h)^2 + k$: 1) The vertex of the parabola is (h, k). 2) The axis of symmetry is the vertical line with equation $x = h$. 3) If a is positive, the parabola opens upward. If a is negative, the parabola opens downward. 4) If $|a| < 1$, then the graph of $f(x) = a(x - h)^2 + k$ is *wider* than the graph of $y = x^2$. If $|a| > 1$, then the graph of $f(x) = a(x - h)^2 + k$ is *narrower* than the graph of $y = x^2$. (p. 657)	Graph $f(x) = -(x + 3)^2 + 4$. 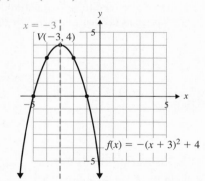 Vertex: $(-3, 4)$ Axis of symmetry: $x = -3$ $a = -1$, so the graph opens downward. The domain is $(-\infty, \infty)$; the range is $(-\infty, 4]$.

Definition/Procedure	Example

When a quadratic function is written in the form $f(x) = ax^2 + bx + c$, there are two methods we can use to graph the function.

Method 1: Rewrite $f(x) = ax^2 + bx + c$ in the form $f(x) = a(x - h)^2 + k$ by *completing the square*.

Method 2: Use the formula $h = -\dfrac{b}{2a}$ to find the *x*-coordinate of the vertex. The vertex has coordinates $\left(-\dfrac{b}{2a}, f\left(-\dfrac{b}{2a}\right)\right)$. **(p. 658)**

Graph $f(x) = x^2 + 4x + 5$.

Method 1: Complete the square.

$$f(x) = x^2 + 4x + 5$$
$$f(x) = (x^2 + 4x + 2^2) + 5 - 2^2$$
$$f(x) = (x^2 + 4x + 4) + 5 - 4$$
$$f(x) = (x + 2)^2 + 1$$

The vertex of the parabola is $(-2, 1)$. The axis of symmetry is $x = -2$. The parabola opens upward and has the same shape as $f(x) = x^2$. The graph is shown.

Method 2: Use the formula $h = -\dfrac{b}{2a}$.

$$h = -\dfrac{4}{2(1)} = -2. \text{ Then, } f(-2) = 1.$$

The vertex of the parabola is $(-2, 1)$. The axis of symmetry is $x = -2$. The domain is $(-\infty, \infty)$; the range is $[-2, \infty)$.

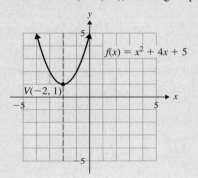

10.6 Applications of Quadratic Functions and Graphing Other Parabolas

Maximum and Minimum Values of a Quadratic Function

Let $f(x) = ax^2 + bx + c$.

1) If a is *positive*, the *y*-coordinate of the vertex is the **minimum** value of the function $f(x)$.
2) If a is *negative*, the *y*-coordinate of the vertex is the **maximum** value of the function $f(x)$. **(p. 668)**

Find the minimum value of the function $f(x) = 2x^2 + 12x + 7$.

Because a is positive ($a = 2$), the function's *minimum value* is at the vertex.

The *x*-coordinate of the vertex is $h = -\dfrac{b}{2a} = -\dfrac{12}{2(2)} = -3$.

The *y*-coordinate of the vertex is

$$f(-3) = 2(-3)^2 + 12(-3) + 7$$
$$= 18 - 36 + 7 = -11$$

The minimum value of the function is -11.

Definition/Procedure	Example
The graph of the quadratic equation $x = ay^2 + by + c$ is a parabola that opens in the x-direction, or horizontally. The quadratic equation $x = ay^2 + by + c$ can also be written in the form $x = a(y - k)^2 + h$. When it is written in this form we can find the following. 1) The vertex of the parabola is (h, k). 2) The axis of symmetry is the horizontal line $y = k$. 3) If a is positive, the graph opens to the right. If a is negative, the graph opens to the left. **(p. 672)**	Graph $x = \dfrac{1}{2}(y + 4)^2 - 2$. Vertex: $(-2, -4)$ Axis of symmetry: $y = -4$ $a = \dfrac{1}{2}$, so the graph opens to the right. The domain is $[-2, \infty)$; the range is $(-\infty, \infty)$. 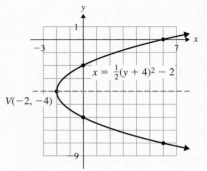

10.7 Quadratic and Rational Inequalities

A **quadratic inequality** can be written in the form $\quad ax^2 + bx + c \leq 0 \quad$ or $\quad ax^2 + bx + c \geq 0$ where a, b, and c are real numbers and $a \neq 0$. ($<$ and $>$ may be substituted for $\leq$ and $\geq$.) **(p. 681)** An inequality containing a rational expression, like $\dfrac{c - 5}{c + 1} \leq 0$, is called a **rational inequality**. **(p. 685)**	Solve $r^2 - 4r \geq 12$. **Step 1:** $r^2 - 4r - 12 \geq 0 \quad$ Subtract 12. Because the inequality symbol is $\geq$, the solution set contains the interval(s) where the quantity $r^2 - 4r - 12$ is *positive*. **Step 2:** Solve $r^2 - 4r - 12 = 0$. $\quad (r - 6)(r + 2) = 0 \quad$ Factor. $\quad r - 6 = 0 \quad$ or $\quad r + 2 = 0$ $\quad r = 6 \quad$ or $\quad r = -2$ **Step 3:** Put $r = 6$ and $r = -2$ on a number line. $r^2 - 4r - 12$: Positive — Negative — Positive, with endpoints at -2 and 6. **Step 4:** Choose a test number in each interval to determine the sign of $r^2 - 4r - 12$. **Step 5:** The solution set will contain the numbers in the intervals where $r^2 - 4r - 12$ is *positive*. **Step 6:** The endpoints of the intervals are included because the inequality is $\geq$. The graph of the solution set is (number line from -5 to 10 with closed circles at -2 and 6) The solution set of $r^2 - 4r - 12$ is $(-\infty, -2] \cup [6, \infty)$.

Chapter 10: Review Exercises

(10.1) Solve using the square root property.

1) $d^2 = 144$
2) $m^2 = 75$
3) $v^2 + 4 = 0$
4) $2c^2 - 11 = 25$
5) $(b - 3)^2 = 49$
6) $(6y + 7)^2 - 15 = 0$
7) $27k^2 - 30 = 0$
8) $(j - 14)^2 + 5 = 0$
9) Find the distance between the points $(-8, 3)$ and $(-12, 5)$.
10) A rectangle has a length of $5\sqrt{2}$ in. and a width of 4 in. How long is its diagonal?

Complete the square for each expression to obtain a perfect square trinomial. Then, factor.

11) $r^2 + 10r$

12) $z^2 - 12z$

13) $c^2 - 5c$

14) $x^2 + x$

15) $a^2 + \dfrac{2}{3}a$

16) $d^2 - \dfrac{5}{2}d$

Solve by completing the square.

17) $p^2 - 6p - 16 = 0$

18) $w^2 - 2w - 35 = 0$

19) $n^2 + 10n = 6$

20) $t^2 + 9 = -4t$

21) $f^2 + 3f + 1 = 0$

22) $j^2 - 7j = 4$

23) $-3q^2 + 7q = 12$

24) $6v^2 - 15v + 3 = 0$

(10.2) Solve using the quadratic formula.

25) $m^2 + 4m - 12 = 0$

26) $3y^2 = 10y - 8$

27) $10g - 5 = 2g^2$

28) $20 = 4x - 5x^2$

29) $\dfrac{1}{6}t^2 - \dfrac{1}{3}t + \dfrac{2}{3} = 0$

30) $(s - 3)(s - 5) = 9$

31) $(6r + 1)(r - 4) = -2(12r + 1)$

32) $z^2 - \dfrac{3}{2}z + \dfrac{13}{16} = 0$

Find the value of the discriminant. Then, determine the number and type of solutions of each equation. *Do not solve*.

33) $3n^2 - 2n - 5 = 0$

34) $t^2 = -3(t + 2)$

35) Find the value of b so that $4k^2 + bk + 9 = 0$ has only one rational solution.

36) A ball is thrown upward from a height of 4 ft. The height, $h(t)$, of the ball (in feet) t sec after the ball is released is given by $h(t) = -16t^2 + 52t + 4$.

   a) How long does it take the ball to reach a height of 16 ft?

   b) How long does it take the ball to hit the ground?

(10.1–10.2) Keep in mind the four methods we have learned for solving quadratic equations: *factoring, the square root property, completing the square, and the quadratic formula*. Solve the equations using one of these methods.

37) $3k^2 + 4 = 7k$

38) $n^2 - 6n + 11 = 0$

39) $15 = 3 + (y + 8)^2$

40) $(2a + 1)(a + 2) = 14$

41) $\dfrac{1}{3}w^2 + w = -\dfrac{5}{6}$

42) $4t^2 + 5 = 7$

43) $6 + p(p - 10) = 2(4p - 15)$

44) $6 = 2m - 3m^2$

45) $x^3 = x$

46) $\dfrac{1}{12}b^2 - \dfrac{9}{2} = \dfrac{1}{4}b$

47) Let $f(x) = (2x - 1)^2$. Find all values of x so that $f(x) = 25$.

48) Let $f(x) = \dfrac{1}{10}x^2 + 3x$ and $g(x) = 4x - \dfrac{11}{5}$. Find all values of x such that $f(x) = g(x)$.

(10.3) Solve.

49) $\dfrac{5k}{k + 1} = 3k - 4$

50) $\dfrac{10}{m} = 3 + \dfrac{8}{m^2}$

51) $f = \sqrt{7f - 12}$

52) $x - 4\sqrt{x} = 5$

53) $n^4 - 17n^2 + 16 = 0$

54) $b^4 + 5b^2 - 14 = 0$

55) $q^{2/3} + 2q^{1/3} - 3 = 0$

56) $y + 2 = 3y^{1/2}$

57) $2r^4 = 7r^2 - 2$

58) $2(v + 2)^2 + (v + 2) - 3 = 0$

59) $(2k - 5)^2 - 5(2k - 5) - 6 = 0$

Write an equation, and solve.

60) At the end of the day, the employees at Forever Young have to put all clothes left in the dressing room back to their proper places. Working together, Lorena and Erica can put away the clothes in 1 hr 12 min. On her own, it takes Lorena 1 hr longer to put away the clothes than it takes Erica to do it by herself. How long does it take each girl to put away the clothes by herself?

(10.4) Solve for the indicated variable.

61) $F = \dfrac{mv^2}{r}$ for v

62) $U = \dfrac{1}{2}kx^2$ for x

63) $r = \sqrt{\dfrac{A}{\pi}}$ for A

64) $r = \sqrt{\dfrac{V}{\pi l}}$ for V

65) $kn^2 - ln - m = 0$ for n

66) $2p^2 + t = rp$ for p

Write an equation, and solve.

67) Ayesha is making a pillow sham by sewing a border onto an old pillow case. The rectangular pillow case measures 18 in. by 27 in. When she sews a border of uniform width around the pillowcase, the total area of the surface of the pillow sham will be 792 in². How wide is the border?

68) The width of a rectangular piece of cardboard is 4 in. less than its length. A square piece that measures 2 in. on each side is cut from each corner, then the sides are turned up to make a box with volume 280 in³. Find the length and width of the original piece of cardboard.

69) A flower shop determined that the demand, $D(P)$, for its tulip bouquet is $D(P) = \dfrac{240}{P}$ per week, where P is the price of the bouquet in dollars. The weekly supply, $S(P)$, is given by $S(P) = 4p - 2$. Find the price at which demand for the tulips equals the supply.

70) U.S. sales of a certain brand of wine can be modeled by
$$y = -0.20x^2 + 4.0x + 8.4$$
for the years 1995–2010. x is the number of years after 1995, and y is the number of bottles sold, in millions.

   a) How many bottles were sold in 1995?

   b) How many bottles were sold in 2008?

   c) In what year did sales reach 28.4 million bottles?

(10.5)

71) Given a quadratic function in the form
$$f(x) = a(x - h)^2 + k,$$

   a) what is the vertex?

   b) what is the equation of the axis of symmetry?

   c) what does the sign of a tell us about the graph of f?

72) Match each function to its graph.
$f(x) = (x - 2)^2$, $g(x) = x^2 - 2$, $h(x) = -x^2 + 2$, $k(x) = (x + 2)^2$

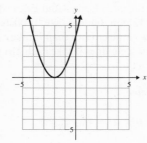

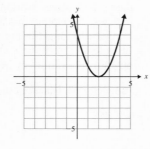

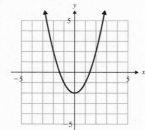

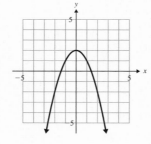

For each quadratic function, identify the vertex, axis of symmetry, and x- and y-intercepts. Then, graph the function. Determine the domain and range.

73) $f(x) = x^2 - 4$

74) $h(x) = -(x + 1)^2$

75) $f(x) = (x + 2)^2 - 1$

76) $y = 2x^2$

77) $y = (x - 4)^2 + 2$

78) $g(x) = -\dfrac{1}{2}(x - 3)^2 - 2$

79) If the graph of $f(x) = x^2$ is shifted 6 units to the right to obtain the graph of $g(x)$, what is the equation of $g(x)$?

80) What are two ways to find the vertex of the graph of $f(x) = ax^2 + bx + c$?

Rewrite each function in the form $f(x) = a(x - h)^2 + k$ by completing the square. Then, graph the function. Include the intercepts. Determine the domain and range.

81) $f(x) = x^2 - 2x + 3$

82) $y = x^2 + 4x - 1$

83) $y = \dfrac{1}{2}x^2 - 4x + 9$

84) $f(x) = -2x^2 - 8x + 2$

Graph each equation using the vertex formula. Include the intercepts. Determine the domain and range.

85) $y = -x^2 - 6x - 10$

86) $f(x) = x^2 - 2x - 4$

(10.6) Solve.

87) An object is thrown upward from a height of 240 ft so that its height h (in feet) t sec after being thrown is given by
$$h(t) = -16t^2 + 32t + 240.$$

   a) How long does it take the object to reach its maximum height?

   b) What is the maximum height attained by the object?

   c) How long does it take the object to hit the ground?

88) A restaurant wants to add outdoor seating to its inside service. It has 56 ft of fencing to enclose a rectangular, outdoor café. Find the dimensions of the outdoor café of maximum area if the building will serve as one side of the café.

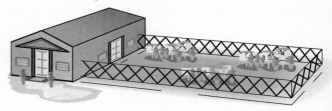

For each quadratic equation, identify the vertex, axis of symmetry, and x- and y-intercepts. Then, graph the equation. Determine the domain and range.

89) $x = -(y - 3)^2 + 11$

90) $x = (y + 1)^2 - 5$

Rewrite each equation in the form $x = a(y - k)^2 + h$ by completing the square. Then, graph the equation. Include the intercepts. Determine the domain and range.

91) $x = y^2 + 8y + 7$

92) $x = -y^2 + 4y - 4$

Graph each equation using the vertex formula. Include the intercepts. Determine the domain and range.

93) $x = -\dfrac{1}{2}y^2 - 3y - \dfrac{5}{2}$

94) $x = 3y^2 - 12y$

(10.7) Solve each inequality. Graph the solution set, and write the solution in interval notation.

95) $a^2 + 2a - 3 < 0$

96) $4m^2 + 8m \geq 21$

97) $64v^2 \geq 25$

98) $36 - r^2 > 0$

99) $(5c + 2)(c - 4)(3c + 1) < 0$

100) $(p - 6)^2 \leq -5$

101) $\dfrac{t + 7}{2t - 3} > 0$

102) $\dfrac{6}{g - 7} \leq 0$

103) $\dfrac{z}{z - 2} \leq 3$

104) $\dfrac{1}{n - 4} > -3$

105) $\dfrac{r^2 + 4}{r - 7} \geq 0$

Solve.

106) Custom Bikes, Inc., estimates that its total profit function, $P(x)$, for producing x thousand units is given by $P(x) = -2x^2 + 32x - 110$. At what level of production does the company make a profit?

Chapter 10: Test

1) Solve $b^2 + 4b - 7 = 0$ by completing the square.

2) Solve $x^2 - 8x + 17 = 0$ using the quadratic formula.

Solve using any method.

3) $(c + 5)^2 + 8 = 2$

4) $3q^2 + 2q = 8$

5) $(4n + 1)^2 + 9(4n + 1) + 18 = 0$

6) $(2t - 3)(t - 2) = 2$

7) $p^4 + p^2 - 72 = 0$

8) $\dfrac{3}{10x} = \dfrac{x}{x - 1} - \dfrac{4}{5}$

9) Find the value of the discriminant. Then, determine the number and type of solutions of the equation. *Do not solve.*
$$5z^2 - 6z - 1 = 0$$

10) Find the length of the missing side.

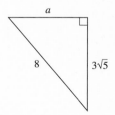

11) Let $P(x) = 5x^2$ and $Q(x) = 2x$. Find all values of x so that $P(x) = Q(x)$.

12) Find the distance between the points $(7, -4)$ and $(5, 6)$.

Write an equation, and solve.

13) A rectangular piece of sheet metal is 6 in. longer than it is wide. A square piece that measures 3 in. on each side is cut from each corner, then the sides are turned up to make a box with volume 273 in³. Find the length and width of the original piece of sheet metal.

14) Solve for V. $r = \sqrt{\dfrac{3V}{\pi h}}$

15) Solve for t. $rt^2 - st = 6$

16) Graph $f(x) = x^2$ and $g(x) = x^2 + 2$.

17) If the graph of $f(x) = x^2$ is shifted 3 units to the left to obtain the graph of $g(x)$, what is the equation of $g(x)$?

Graph each equation. Identify the vertex, axis of symmetry, and intercepts. Determine the domain and range.

18) $f(x) = -(x + 2)^2 + 4$

19) $x = y^2 - 3$

20) $x = 3y^2 - 6y + 5$

21) $g(x) = x^2 - 6x + 8$

22) A ball is projected upward from the top of a 200-ft tall building. The height $h(t)$ of the ball above the ground (in feet) t sec after the ball is released is given by
$$h(t) = -16t^2 + 24t + 200.$$

a) What is the maximum height attained by the ball?

b) When will the ball be 40 ft above the ground?

c) When will the ball hit the ground?

Solve each inequality. Graph the solution set, and write the solution in interval notation.

23) $y^2 + 4y - 45 \geq 0$

24) $\dfrac{m - 5}{m + 3} \geq 0$

25) A company has determined that the average cost, $\overline{C}(x)$, of producing x backpacks per month can be described by the function $\overline{C}(x) = \dfrac{5x + 80{,}000}{x}$. How many backpacks must the company produce each month so that the average cost of producing each backpack is no more than \$15?

Chapter 10: Cumulative Review for Chapters 1–10

1) Simplify $\dfrac{\frac{12}{35}}{\frac{24}{49}}$.

Simplify. The final answer should contain only positive exponents.

2) $(5x^4y^{-10})(3xy^3)^2$

3) *Write an equation, and solve.*
   In December 2010, an electronics store sold 108 digital cameras. This is a 20% increase over their sales in December 2009. How many digital cameras did they sell in December 2009?

4) Solve for m. $y = mx + b$

5) Given the relation $\{(4, 0), (3, 1), (3, -1), (0, 2)\}$,
   a) what is the domain?
   b) what is the range?
   c) is the relation a function?

6) Let $f(x) = \sqrt{x + 3}$.
   a) Find $f(1)$.
   b) Find the domain of f.
   c) Graph the function.

7) *Write a system of two equations in two variables, and solve.*
   Two bags of chips and three cans of soda cost $3.85, while one bag of chips and two cans of soda cost $2.30. Find the cost of a bag of chips and a can of soda.

8) Subtract
   $(4x^2y^2 - 11x^2y + xy + 2) - (x^2y^2 - 6x^2y + 3xy^2 + 10xy - 6)$

9) Multiply and simplify $3(r - 5)^2$.

Factor completely.

10) $4p^3 + 14p^2 - 8p$

11) $a^3 + 125$

12) Add $\dfrac{z - 8}{z + 4} + \dfrac{3}{z}$

13) Simplify $\dfrac{2 + \frac{6}{c}}{\frac{2}{c^2} - \frac{8}{c}}$

14) Solve this system: $\begin{array}{l} 4x - 2y + z = -7 \\ -3x + y - 2z = 5 \\ 2x + 3y + 5z = 4 \end{array}$

Simplify. Assume all variables represent nonnegative real numbers.

15) $\sqrt{75}$

16) $\sqrt[3]{40}$

17) $\sqrt{63x^7y^4}$

18) Simplify $64^{2/3}$.

19) Rationalize the denominator: $\dfrac{5}{2 + \sqrt{3}}$.

20) Multiply and simplify $(10 + 3i)(1 - 8i)$.

Solve.

21) $1 - \dfrac{1}{3h - 2} = \dfrac{20}{(3h - 2)^2}$

22) $p^2 + 6p = 27$

23) Solve for V: $r = \sqrt{\dfrac{V}{\pi h}}$.

24) Graph $f(x) = (x - 1)^2 - 4$.

25) Solve $25p^2 \le 144$.

Exponential and Logarithmic Functions

CHAPTER 11

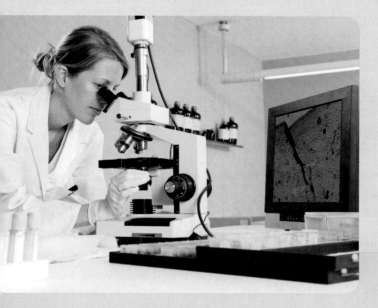

OUTLINE

Study Strategies: Coping with Stress

11.1 Composite and Inverse Functions

11.2 Exponential Functions

11.3 Logarithmic Functions

11.4 Properties of Logarithms

11.5 Common and Natural Logarithms and Change of Base

11.6 Solving Exponential and Logarithmic Equations

Group Activity

emPOWERme: Progressive Relaxation

Math at Work:
Epidemiologist

Hannah Novak makes her living taking a very close look at something most of us try to avoid: bacteria. As an epidemiologist at a hospital, Hannah is responsible for studying, tracking, and trying to prevent outbreaks of the flu and other illnesses. "The first thing you have to understand about bacteria is that they can spread exponentially," Hannah says. "One week you might see two cases of a certain type of flu, and if you're not careful, the next week you could see two hundred. My job is to stop that from happening."

Hannah says that the best way for people to fight the spread of swarms of bacteria is to practice basic hygiene. "Washing your hands is one of the simplest and yet most effective ways to stay healthy," Hannah advises. She also notes that while many college students worry a lot about getting sick during flu season, they often ignore a more immediate health risk: stress.

"Stress may not be caused by bacteria," Hannah jokes, "but I think there's an epidemic of it on many college campuses."

In this chapter, we will discuss exponential and logarithmic equations. We will also present strategies you can use to reduce the stress in your life.

Study Strategies — Coping with Stress

The truth is that for most college students, at least some stress is probably unavoidable. There will always be tests to take and difficult problem sets to complete. The key is being able to cope with stress so that it does not take an unnecessary toll on your health and mental well-being. The strategies below will help you manage stress effectively.

Prepare
- Make time to exercise. Jogging, biking, and other physical activities help you stay in shape and prepare you to deal with stress.
- Caffeinated beverages like coffee and soda can make you feel anxious and jittery even when you're not stressed. Try to replace them with noncaffeinated alternatives.

Organize
- Figure out the causes of your stress in your life. Simply knowing why you are stressed can help you manage the feeling better.

Work
- Think of specific actions you can take to address sources of stress. Even small steps will help you feel more in control.
- Talk about your feelings of stress with friends and family. Stress only grows when it is bottled up.
- Keep perspective. Some things can be very stressful as they happen, but if we take a step back, we realize they don't really matter very much at all.
- Relax! Do yoga, go to the movies, play video games—give yourself permission to do the things that make you feel less stressed.

Evaluate
- Be open to new ways to reduce your stress, such as taking up meditation or learning to play a musical instrument. You can also try using the progressive relaxation exercise described in the emPOWERme on page 777.

Rethink
- Make peace with stress. Try to think of it as an unavoidable but manageable outcome of working toward your goals!

Chapter 11 POWER Plan

P Prepare

What are your goals for Chapter 11?

1. Be prepared before and during class.
2. Understand the homework to the point where you could do it without needing any help or hints.
3. Use the P.O.W.E.R. framework to help you deal with stress: *Progressive Relaxation*.
4. Write your own goal.

O Organize

How can you accomplish each goal?

1.
   -
   -
   -
   -
2.
   -
   -
   -
3.
   -
   -
   -
4.
   -

What are your objectives for Chapter 11?

1. Be able to find the composition of functions and inverse of functions, and be able to graph functions and their inverses.
2. Be able to solve exponential and logarithmic equations.
3. Be able to simplify logarithmic expressions.
4. Be able to graph exponential and logarithmic functions.
5. Be able to evaluate and solve equations containing a natural or common logarithm. Be able to graph a natural logarithm function.
6. Be able to solve applied problems involving exponential functions.
7. Write your own goal.

How can you accomplish each objective?

1.
   - Learn the different definitions associated with functions, including *composition of functions* and *inverse of functions*.
   - Know the procedure for **Graphing a Function** and the procedure for **Graphing Its Inverse**.
2.
   - Review powers of numbers so that you can solve exponential equations.
   - Learn how to convert a logarithmic equation to an exponential equation.
   - Learn the **Properties of Logarithms**.
3.
   - Learn the **Product Rule, Quotient Rule, and Power Rule for Logarithms**.
4.
   - Learn how to make an appropriate table of values.
   - Be familiar with graphs of exponential and logarithmic functions and the domain and range of these functions.
5.
   - Know how to convert from the logarithmic form of an equation to the exponential form of the equation.
   - Be familiar with using a calculator for entering log expressions or for making a table of values.
6.
   - Learn formulas for compound interest and continuous compounding.
   - Learn formulas for exponential growth and decay.
7.
   -

	W Work	Read Sections 11.1 to 11.6, and complete the exercises.
E Evaluate Complete the Chapter Review and Chapter Test. How did you do?	**R Rethink**	• How does dealing with stress affect your success in life? • Are there any relaxation techniques you currently use before you take a test? • How are exponential and logarithmic graphs the same? How are they different? • What does the letter *e* stand for when used in math?

11.1 Composite and Inverse Functions

P Prepare	**O Organize**
What are your objectives for Section 11.1?	How can you accomplish each objective?
1 Find the Composition of Functions	• Write the definition of a *composition function* in your own words. • Understand how to evaluate the *composition of functions* for a given value. • Understand what it means to find the *decomposition of functions*. • Complete the given examples on your own. • Complete You Trys 1–3.
2 Use Function Composition	• Be able to use the composition of functions to solve an applied problem. • Complete the given example on your own. • Complete You Try 4.
3 Determine Whether a Function Is One-to-One	• Write the definition of a *one-to-one function* in your own words. • Use the *horizontal line test* to determine if a function is one-to-one. • Complete the given examples on your own. • Complete You Trys 5 and 6.
4 Find the Inverse of a Function	• Write the definition of an *inverse function* in your own words. • Know how to write the inverse of a function. • Know how to graph a function and its inverse on the same axes. • Complete the given examples on your own. • Complete You Trys 7–9.
5 Given the Graph of $f(x)$, Graph $f^{-1}(x)$	• Review how to determine the domain and range of a function. • Complete the given example on your own. • Complete You Try 10.

W Work — Read the explanations, follow the examples, take notes, and complete the You Trys.

Later in this chapter we will learn about exponential and logarithmic functions. These functions are important not only in mathematics but also in areas such as economics, finance, chemistry, and biology. Before we can begin our study of these functions, we must learn about composite, one-to-one, and inverse functions. This is because exponential and logarithmic functions are related in a special way: They are inverses of each other.

1 Find the Composition of Functions

W Hint — Read this explanation carefully so that you understand *how* the composition of functions works.

Earlier we learned how to add, subtract, multiply, and divide functions. Now we will combine functions in a new way, using *function composition*. We use these *composite functions* when we are given certain two-step processes and want to combine them into a single step.

For example, if you work x hours per week earning \$8 per hour, your earnings before taxes and other deductions can be described by the function $f(x) = 8x$. Your take-home pay is different, however, because of taxes and other deductions. So, if your take-home pay is 75% of your earnings before taxes, then $g(x) = 0.75x$ can be used to compute your take-home pay when x is your earnings before taxes.

We can describe what is happening with two tables of values.

$f(x) = 8x$		$g(x) = 0.75x$	
Hours Worked x	Earnings Before Deductions $f(x)$	Earnings Before Deductions x	Take-Home Pay $g(x)$
6	48	48	36
10	80	80	60
20	160	160	120
40	320	320	240

$x \longrightarrow f(x) \qquad x \longrightarrow g(x)$

One function, $f(x)$, describes total earnings before deductions in terms of the number of hours worked. Another function, $g(x)$, describes take-home pay in terms of the total earnings before deductions. It would be convenient to have a function that would allow us to compute, directly, the take-home pay in terms of the number of hours worked.

	$f(x) = 8x$	$g(x) = 0.75x$
Hours Worked	Earnings Before Deductions	Take-Home Pay
6	48	36
10	80	60
20	160	120
40	320	240
x	$f(x)$	$h(x) = g(f(x))$

$$h(x) = (g \circ f)(x) = g(f(x))$$

If we substitute the function $f(x)$ for x in the function $g(x)$, we will get a new function, $h(x)$, where $h(x) = g(f(x))$. The take-home pay in terms of the number of hours worked, $h(x)$, is given by the composition function $g(f(x))$, read as "g of f of x" and is given by

$$h(x) = g(f(x)) = g(8x)$$
$$= 0.75(8x)$$
$$= 6x$$

Therefore, $h(x) = 6x$ allows us to directly compute the take-home pay from the number of hours worked. To find out your take-home pay when you work 20 hr in a week, find $h(20)$.

$$h(x) = 6x$$
$$h(20) = 6(20) = 120$$

Working 20 hr will result in take-home pay of \$120. Notice that this is the same as the take-home pay computed in the tables.

Another way to write $g(f(x))$ is $(g \circ f)(x)$, and both can be read as "g of f of x," or "g composed with f," or "the composition of g and f." Likewise, $f(g(x)) = (f \circ g)(x)$, and these can be read as "f of g of x," or "f composed with g," or "the composition of f and g."

Definition

Given the function $f(x)$ and $g(x)$, the **composition function** $f \circ g$ (read "f of g") is defined as

$$(f \circ g)(x) = f(g(x))$$

where $g(x)$ is in the domain of f.

EXAMPLE 1

Let $f(x) = 2x - 5$ and $g(x) = x + 8$. Find

a) $g(3)$ b) $(f \circ g)(3)$ c) $(f \circ g)(x)$

Solution

a) $g(x) = x + 8$
$g(3) = 3 + 8 = 11$

b) $(f \circ g)(3) = f(g(3))$ In part a) we found $g(3) = 11$.
$= f(11)$
$= 2(11) - 5$ Substitute 11 for x in $f(x) = 2x - 5$.
$= 17$

c) $(f \circ g)(x) = f(g(x))$
$= f(x + 8)$ Substitute $x + 8$ for $g(x)$.
$= 2(x + 8) - 5$ Substitute $x + 8$ for x in $f(x)$.
$= 2x + 11$

We can also find $(f \circ g)(3)$, the question for part b), by substituting 3 for x in $(f \circ g)(x)$ found in part c).

$$(f \circ g)(x) = 2x + 11$$
$$(f \circ g)(3) = 2(3) + 11 = 17$$

Notice that this is the same as the result we obtained in b).

[YOU TRY 1] Let $f(x) = 3x + 4$ and $g(x) = x - 10$. Find

a) $g(-2)$ b) $(f \circ g)(-2)$ c) $(f \circ g)(x)$

 BE CAREFUL The notation $(f \circ g)(x)$ represents the *composition* of functions, $f(g(x))$; the notation $(f \cdot g)(x)$ represents the *product* of functions, $f(x) \cdot g(x)$.

EXAMPLE 2 Let $f(x) = 4x - 1$, $g(x) = x^2$, and $h(x) = x^2 + 5x - 2$. Find

a) $(f \circ g)(x)$ b) $(g \circ f)(x)$ c) $(h \circ f)(x)$

Solution

a) $(f \circ g)(x) = f(g(x))$
$= f(x^2)$ Substitute x^2 for $g(x)$.
$= 4(x^2) - 1$ Substitute x^2 for x in $f(x)$.
$= 4x^2 - 1$

W Hint
In general,
$(f \circ g)(x) \neq (g \circ f)(x)$.

b) $(g \circ f)(x) = g(f(x))$
$= g(4x - 1)$ Substitute $4x - 1$ for $f(x)$.
$= (4x - 1)^2$ Substitute $4x - 1$ for x in $g(x)$.
$= 16x^2 - 8x + 1$ Expand the binomial.

c) $(h \circ f)(x) = h(f(x))$
$= h(4x - 1)$ Substitute $4x - 1$ for $f(x)$.
$= (4x - 1)^2 + 5(4x - 1) - 2$ Substitute $4x - 1$ for x in $h(x)$.
$= 16x^2 - 8x + 1 + 20x - 5 - 2$ Distribute.
$= 16x^2 + 12x - 6$ Combine like terms.

[YOU TRY 2] Let $f(x) = x^2 + 6$, $g(x) = 2x - 3$, and $h(x) = x^2 - 4x + 9$. Find

a) $(g \circ f)(x)$ b) $(f \circ g)(x)$ c) $(h \circ g)(x)$

Sometimes, it is necessary to rewrite a single function in terms of the composition of two other functions. This is called the **decomposition** of functions.

EXAMPLE 3 Let $h(x) = \sqrt{x^2 + 5}$. Find f and g such that $h(x) = (f \circ g)(x)$.

Solution

Think about what is happening in the function $h(x)$. We can "build" $h(x)$ in the following way: first find $x^2 + 5$, then take the square root of that quantity. So if we let $g(x) = x^2 + 5$ and $f(x) = \sqrt{x}$, we will get $h(x) = (f \circ g)(x)$. Let's check by finding the composition function, $(f \circ g)(x)$.

$$g(x) = x^2 + 5 \quad f(x) = \sqrt{x}$$

$(f \circ g)(x) = f(g(x))$
$= f(x^2 + 5)$ Substitute $x^2 + 5$ for $g(x)$.
$= \sqrt{x^2 + 5}$ Substitute $x^2 + 5$ for x in $f(x) = \sqrt{x}$.

Our result is $h(x) = \sqrt{x^2 + 5}$.

W Hint
In your notes and in your own words, explain how to decompose a function. Give an example.

In Example 3, forming $h(x)$ using $f(x) = \sqrt{x}$ and $g(x) = x^2 + 5$ is probably the easiest decomposition to "see." However, there is more than one way to decompose a function, $h(x)$, into two functions $f(x)$ and $g(x)$ so that $h(x) = (f \circ g)(x)$.

For example, if $f(x) = \sqrt{x + 5}$ and $g(x) = x^2$, we get

$$(f \circ g)(x) = f(g(x))$$
$$= f(x^2) \quad \text{Substitute } x^2 \text{ for } g(x).$$
$$= \sqrt{x^2 + 5} \quad \text{Substitute } x^2 \text{ for } x \text{ in } f(x) = \sqrt{x+5}.$$

This is another way to obtain $h(x) = \sqrt{x^2 + 5}$.

[YOU TRY 3] Let $h(x) = \sqrt{2x^2 + 1}$. Find f and g such that $h(x) = (f \circ g)(x)$.

2 Use Function Composition

EXAMPLE 4

The area, A, of a square expressed in terms of its perimeter, P, is defined by the function

$$A(P) = \frac{1}{16}P^2$$

The perimeter of a square with a side of length x, is defined by the function

$$P(x) = 4x$$

a) Find $(A \circ P)(x)$, and explain what it represents.
b) Find $(A \circ P)(3)$, and explain what it represents.

Solution

a) $(A \circ P)(x) = A(P(x))$
$ = A(4x) \quad$ Substitute $4x$ for $P(x)$.
$ = \dfrac{1}{16}(4x)^2 \quad$ Substitute $4x$ for P in $A(P) = \dfrac{1}{16}P^2$.
$ = \dfrac{1}{16}(16x^2)$
$ = x^2$

$(A \circ P)(x) = x^2$. This is the formula for the area of a square in terms of the length of a side, x.

b) To find $(A \circ P)(3)$, use the result obtained in a).

$$(A \circ P)(x) = x^2$$
$$(A \circ P)(3) = 3^2 = 9$$

A square that has a side of length 3 units has an area of 9 square units.

[YOU TRY 4] Let $f(x) = 100x$ represent the number of centimeters in x meters. Let $g(y) = 1000y$ represent the number of meters in y kilometers.

a) Find $(f \circ g)(y)$, and explain what it represents.
b) Find $(f \circ g)(4)$, and explain what it represents.

3 Determine Whether a Function Is One-to-One

Recall from Section 4.5 that a relation is a *function* if each x-value corresponds to exactly one y-value. Let's look at two functions, f and g.

$$f = \{(1, -3), (2, -1), (4, 3), (7, 9)\} \quad g = \{(0, 3), (1, 4), (-1, 4), (2, 7)\}$$

In functions f and g, each x-value corresponds to exactly one y-value. That is why they are functions. In function f, each *y-value also corresponds to exactly one x-value*. Therefore, f is a *one-to-one function*. In function g, however, each y-value does *not* correspond to exactly one x-value. (The y-value of 4 corresponds to $x = 1$ and $x = -1$.) Therefore, g is *not* a one-to-one function.

Definition

In order for a function to be a **one-to-one function,** each x-value corresponds to exactly one y-value, and each y-value corresponds to exactly one x-value.

Alternatively we can say that a function is one-to-one if each value in its domain corresponds to exactly one value in its range *and* if each value in its range corresponds to exactly one value in its domain.

EXAMPLE 5

Determine whether each function is one-to-one.

a) $f = \{(-1, 9), (1, -3), (2, -6), (4, -6)\}$

b) $g = \{(-3, 13), (-1, 5), (5, -19), (8, -31)\}$

c)

State	Number of Representatives in U.S. House of Representatives (2012)
Alaska	1
California	53
Connecticut	5
Delaware	1
Ohio	18

d)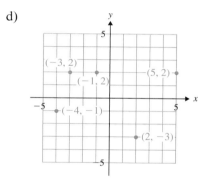

Solution

W Hint
In your notes, describe the difference between a function that is one-to-one and a function that is not.

a) f is *not* a one-to-one function because the y-value -6 corresponds to two different x-values: $(2, -6)$ and $(4, -6)$.

b) g is a one-to-one function because each y-value corresponds to exactly one x-value.

c) The information in the table does *not* represent a one-to-one function because the value 1 in the range corresponds to two different values in the domain, Alaska and Delaware.

d) The graph does *not* represent a one-to-one function because three points have the same y-value: $(-3, 2)$, $(-1, 2)$, and $(5, 2)$.

[YOU TRY 5] Determine whether each function is one-to-one.

a) $f = \{(-2, -13), (0, -7), (4, 5), (5, 8)\}$ b) $g = \{(-4, 2), (-1, 1), (0, 2), (3, 5)\}$

c)
Element	Atomic Mass (in amu)
Hydrogen	1.00794
Lithium	6.941
Sulfur	32.066
Lead	207.2

d)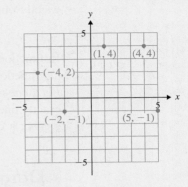

Just as we can use the vertical line test to determine if a graph represents a function, we can use the *horizontal line test* to determine if a function is one-to-one.

Definition

Horizontal Line Test: If every horizontal line that could be drawn through a function would intersect the graph at most once, then the function is one-to-one.

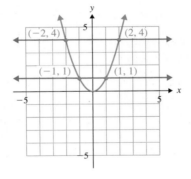

Look at the graph of the given function, in blue. We can see that if a horizontal line intersects the graph more than once, then one y-value corresponds to more than one x-value. This means that the function is not one-to-one. For example, the y-value of 1 corresponds to $x = 1$ and $x = -1$.

EXAMPLE 6

Determine whether each graph represents a one-to-one function.

a) 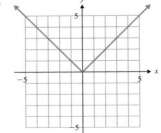   b)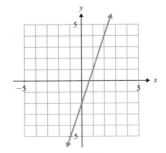

Solution

a) *Not* one-to-one. It is possible to draw a horizontal line through the graph so that it intersects the graph more than once.

b) *Is* one-to-one. Every horizontal line that could be drawn through the graph would intersect the graph at most once.

[YOU TRY 6] Determine whether each graph represents a one-to-one function.

a) 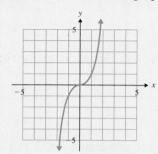   b)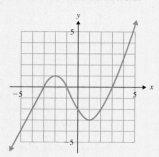

4 Find the Inverse of a Function

One-to-one functions lead to other special functions—inverse functions. *A one-to-one function has an inverse function.*

To find the inverse of a one-to-one function, we interchange the coordinates of the ordered pairs.

EXAMPLE 7   Find the inverse function of $f = \{(4, 2), (9, 3), (36, 6)\}$.

Solution

To find the inverse of f, switch the x- and y-coordinates of each ordered pair. The inverse of f is $\{(2, 4), (3, 9), (6, 36)\}$.

[YOU TRY 7] Find the inverse function of $f = \{(-5, -1), (-3, 2), (0, 7), (4, 13)\}$.

We use special notation to represent the inverse of a function. If f is a one-to-one function, then f^{-1} (read "f inverse") represents the inverse of f. For Example 7, we can write the inverse as $f^{-1} = \{(2, 4), (3, 9), (6, 36)\}$.

Definition

Inverse Function: Let f be a one-to-one function. The **inverse** of f, denoted by f^{-1}, is a one-to-one function that contains the set of all ordered pairs (y, x), where (x, y) belongs to f.

BE CAREFUL

1) f^{-1} is read "f inverse" *not* "f to the negative one."

2) f^{-1} does *not* mean $\dfrac{1}{f}$.

3) If a function is not one-to-one, it does not have an inverse.

We said that if (x, y) belongs to the one-to-one function $f(x)$, then (y, x) belongs to its inverse, $f^{-1}(x)$ (read as *f inverse of x*). We use this idea to find the equation for the inverse of $f(x)$.

Procedure How to Find an Equation of the Inverse of $y = f(x)$

Step 1: Replace $f(x)$ with y.
Step 2: Interchange x and y.
Step 3: Solve for y.
Step 4: Replace y with the inverse notation, $f^{-1}(x)$.

EXAMPLE 8   Find an equation of the inverse of $f(x) = 3x + 4$.

Solution

$$f(x) = 3x + 4$$

Step 1: $y = 3x + 4$ Replace $f(x)$ with y.

Step 2: $x = 3y + 4$ Interchange x and y.

Step 3: Solve for y.

$$x - 4 = 3y \quad \text{Subtract 4.}$$
$$\frac{x - 4}{3} = y \quad \text{Divide by 3.}$$
$$\frac{1}{3}x - \frac{4}{3} = y \quad \text{Simplify.}$$

Step 4: $f^{-1}(x) = \dfrac{1}{3}x - \dfrac{4}{3}$ Replace y with $f^{-1}(x)$.

[YOU TRY 8]   Find an equation of the inverse of $f(x) = -5x + 10$.

In Example 9, we look more closely at the relationship between a function and its inverse.

EXAMPLE 9   Find the equation of the inverse of $f(x) = 2x - 4$. Then, graph $f(x)$ and $f^{-1}(x)$ on the same axes.

Solution

$$f(x) = 2x - 4$$

Step 1: $y = 2x - 4$ Replace $f(x)$ with y.

Step 2: $x = 2y - 4$ Interchange x and y.

Step 3: Solve for y.

$$x + 4 = 2y \qquad \text{Add 4.}$$
$$\frac{x+4}{2} = y \qquad \text{Divide by 2.}$$
$$\frac{1}{2}x + 2 = y \qquad \text{Simplify.}$$

Step 4: $\qquad f^{-1}(x) = \frac{1}{2}x + 2 \qquad$ Replace y with $f^{-1}(x)$.

We will graph $f(x)$ and $f^{-1}(x)$ by making a table of values for each. Then we can see another relationship between the two functions.

$f(x) = 2x - 4$	
x	$y = f(x)$
0	−4
1	−2
2	0
5	6

$f^{-1}(x) = \frac{1}{2}x + 2$	
x	$y = f^{-1}(x)$
−4	0
−2	1
0	2
6	5

Notice that the x- and y-coordinates have switched when we compare the tables of values. Graph $f(x)$ and $f^{-1}(x)$.

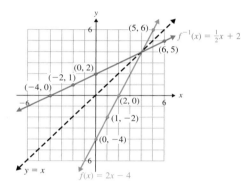

[YOU TRY 9] Find the equation of the inverse of $f(x) = -3x + 1$. Then, graph $f(x)$ and $f^{-1}(x)$ on the same axes.

5 Given the Graph of f(x), Graph $f^{-1}(x)$

Look again at the tables in Example 9. The x-values for $f(x)$ become the y-values of $f^{-1}(x)$, and the y-values of $f(x)$ become the x-values of $f^{-1}(x)$. This is true not only for the values in the tables but for *all* values of x and y. That is, for all ordered pairs (x, y) that belong to $f(x)$, (y, x) belongs to $f^{-1}(x)$. Another way to say this is *the domain of f becomes the range of f^{-1}, and the range of f becomes the domain of f^{-1}.*

Let's turn our attention to the graph in Example 9. The graphs of $f(x)$ and $f^{-1}(x)$ are mirror images of one another with respect to the line $y = x$. We say that *the graphs of $f(x)$ and $f^{-1}(x)$ are symmetric with respect to the line $y = x$.* This is true for every function $f(x)$ and its inverse, $f^{-1}(x)$.

 Hint
Write down the relationships between a function and its inverse that you have learned so far.

Note
The graphs of $f(x)$ and $f^{-1}(x)$ are symmetric with respect to the line $y = x$.

EXAMPLE 10 Given the graph of $f(x)$, graph $f^{-1}(x)$.

Solution

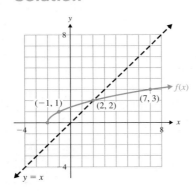 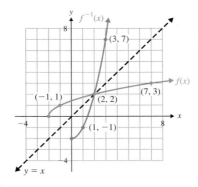

Some points on the graph of $f(x)$ are $(-2, 0)$, $(-1, 1)$, $(2, 2)$, and $(7, 3)$. We can obtain points on the graph of $f^{-1}(x)$ by interchanging the x- and y-values.

Some points on the graph of $f^{-1}(x)$ are $(0, -2)$, $(1, -1)$, $(2, 2)$, and $(3, 7)$. Plot these points to get the graph of $f^{-1}(x)$. Notice that the graphs are symmetric with respect to the line $y = x$.

YOU TRY 10 Given the graph of $f(x)$, graph $f^{-1}(x)$.

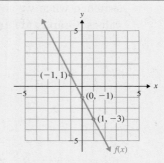

Using Technology

The composition of two functions can be evaluated analytically, numerically, and graphically using a graphing calculator.

Consider the composition $h(x) = f(g(x))$ given the functions $f(x) = x^2$ and $g(x) = 3x - 2$. The function $h(x)$ is determined analytically by substituting $g(x)$ into the function f.

We can evaluate $h(2) = f(g(2))$ by substituting 2 in for x.

$$\begin{aligned} h(x) &= f(g(x)) \\ &= f(3x - 2) \\ &= (3x - 2)^2 \\ &= 9x^2 - 12x + 4 \end{aligned}$$

$$9(2)^2 - 12(2) + 4 = 16$$

Using a graphing calculator, enter $f(x) = x^2$ into Y_1, enter $g(x) = 3x - 2$ into Y_2, and enter $Y_1(Y_2)$ into Y_3 to represent the composition h, as shown. Recall that Y_1 and Y_2 are found by pressing , pressing the right arrow key, and pressing ENTER.

To evaluate $h(2) = f(g(2))$ using the calculator, press 2nd MODE to return to the home screen and $Y_3(2)$ as shown.

 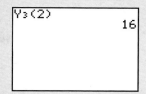

Start with the function Y_1, Y_2 as shown above. $h(2) = f(g(2))$ can be evaluated numerically by setting up a table showing x, Y_1, and Y_2 near $x = 2$. Press 2nd WINDOW and enter 0 after TblStart =. Then press 2nd ENTER to display the table. First evaluate $g(2) = Y_2(2)$ by moving the cursor down to $x = 2$ and then across to the column under Y_2 as shown.

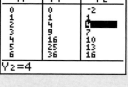

Then substitute the result $g(2) = Y_2(2) = 4$ into the function f to evaluate $f(4) = Y_1(4) = 16$ by moving the cursor down to $x = 4$ and then across to the column under Y_1 as shown. The result is 16 as desired.

To illustrate this composition graphically, change the window by increasing Ymax to 20. $h(2) = f(g(2))$ can be evaluated graphically using the following approach.

First evaluate $g(2) = Y_2(2)$ by pressing TRACE, pressing the down arrow key to switch to Y_2, and pressing 2 ENTER resulting in the point $(2, 4)$ on the graph of g, as shown.

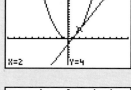

Next evaluate $f(4) = Y_1(4)$ by pressing the down arrow key to switch to Y_1, and pressing 4 ENTER resulting in the point $(4, 16)$ on the graph of f as shown on the graph. The result is 16 as desired.

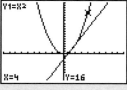

Given the functions $f(x) = x^2 - 5x$ and $g(x) = 2x + 3$, evaluate the following function values using a graphing calculator.

1) $(f \circ g)(3)$ 2) $(f \circ g)(-2)$ 3) $(g \circ f)(2)$
4) $(g \circ f)(-2)$ 5) $(f \circ f)(1)$ 6) $(g \circ g)(2)$

ANSWERS TO YOU TRY EXERCISES

1) a) -12 b) -32 c) $3x - 26$ 2) a) $2x^2 + 9$ b) $4x^2 - 12x + 15$ c) $4x^2 - 20x + 30$
3) $f(x) = \sqrt{x}, g(x) = 2x^2 + 1$; answers may vary.
4) a) $(f \circ g)(y) = 100,000y$. This tells us the number of centimeters in y kilometers.
   b) $(f \circ g)(4) = 400,000$. There are 400,000 cm in 4 kilometers.
5) a) yes   b) no   c) yes   d) no   6) a) yes   b) no
7) $\{(-1, -5), (2, -3), (7, 0), (13, 4)\}$ 8) $f^{-1}(x) = -\dfrac{1}{5}x + 2$

9) $f^{-1}(x) = -\dfrac{1}{3}x + \dfrac{1}{3}$

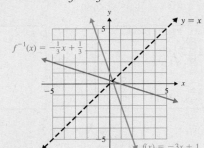

10)

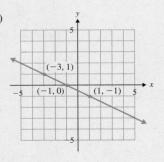

ANSWERS TO TECHNOLOGY EXERCISES

1) 36   2) 6   3) −9   4) 31   5) 36   6) 17

E Evaluate   11.1 Exercises

Do the exercises, and check your work.

Objective 1: Find the Composition of Functions

1) Given two functions $f(x)$ and $g(x)$, explain how to find $(f \circ g)(x)$.

2) Given two functions $f(x)$ and $g(x)$, explain the difference between $(f \circ g)(x)$ and $(f \cdot g)(x)$.

For Exercises 3–6, find

  a) $g(4)$
  b) $(f \circ g)(4)$ using the result in part a)
  c) $(f \circ g)(x)$
  d) $(f \circ g)(4)$ using the result in part c)

3) $f(x) = 3x + 1$, $g(x) = 2x - 9$

4) $f(x) = -x + 5$, $g(x) = x + 7$

5) $f(x) = x^2 - 5$, $g(x) = x + 3$

6) $f(x) = x^2 + 2$, $g(x) = x - 1$

7) Let $f(x) = 5x - 4$ and $g(x) = x + 7$. Find
  a) $(f \circ g)(x)$ b) $(g \circ f)(x)$
  c) $(f \circ g)(3)$

8) Let $f(x) = x - 10$ and $g(x) = 4x + 3$. Find
  a) $(f \circ g)(x)$ b) $(g \circ f)(x)$
  c) $(f \circ g)(-6)$

9) Let $h(x) = -2x + 9$ and $k(x) = 3x - 1$. Find
  a) $(k \circ h)(x)$ b) $(h \circ k)(x)$
  c) $(k \circ h)(-1)$

10) Let $r(x) = 6x + 2$ and $v(x) = -7x - 5$. Find
  a) $(v \circ r)(x)$ b) $(r \circ v)(x)$
  c) $(r \circ v)(2)$

11) Let $g(x) = x^2 - 6x + 11$ and $h(x) = x - 4$. Find
  a) $(h \circ g)(x)$ b) $(g \circ h)(x)$
  c) $(g \circ h)(4)$

12) Let $f(x) = x^2 + 7x - 9$ and $g(x) = x + 2$. Find
  a) $(g \circ f)(x)$ b) $(f \circ g)(x)$
  c) $(g \circ f)(3)$

13) Let $m(x) = x + 8$ and $n(x) = -x^2 + 3x - 8$. Find
  a) $(n \circ m)(x)$ b) $(m \circ n)(x)$
  c) $(m \circ n)(0)$

14) Let $f(x) = -x^2 + 10x + 4$ and $g(x) = x + 1$. Find
  a) $(g \circ f)(x)$ b) $(f \circ g)(x)$
  c) $(f \circ g)(-2)$

15) Let $f(x) = \sqrt{x + 10}$, $g(x) = x^2 - 6$. Find
  a) $(f \circ g)(x)$ b) $(g \circ f)(x)$
  c) $(f \circ g)(-3)$

16) Let $h(x) = x^2 + 7$, $k(x) = \sqrt{x - 1}$. Find
  a) $(h \circ k)(x)$ b) $(k \circ h)(x)$
  c) $(k \circ h)(0)$

17) Let $P(t) = \dfrac{1}{t+8}$, $Q(t) = t^2$. Find

   a) $(P \circ Q)(t)$ b) $(Q \circ P)(t)$

   c) $(Q \circ P)(-5)$

18) Let $F(a) = \dfrac{1}{5a}$, $G(a) = a^2$. Find

   a) $(G \circ F)(a)$ b) $(F \circ G)(a)$

   c) $(G \circ F)(-2)$

For Exercises 19–24, find $f(x)$ and $g(x)$ such that $h(x) = (f \circ g)(x)$.

19) $h(x) = \sqrt{x^2 + 13}$

20) $h(x) = (8x - 3)^2$

21) $h(x) = \dfrac{1}{6x + 5}$

Objective 2: Use Function Composition

22) The sales tax on goods in a major metropolitan area is 7% so that the final cost of an item, $f(x)$, is given by $f(x) = 1.07x$, where x is the cost of the item. A women's clothing store is having a sale so that all of its merchandise is 20% off. If the regular price of an item is x dollars then the sale price, $s(x)$, is given by $s(x) = 0.80x$. Find each of the following and explain their meanings.

   a) $s(40)$

   b) $f(32)$

   c) $(f \circ s)(x)$

   d) $(f \circ s)(40)$

23) Oil spilled from a ship off the coast of Alaska with the oil spreading out in a circle across the surface of the water. The radius of the oil spill is given by $r(t) = 4t$ where t is the number of minutes after the leak began and $r(t)$ is in feet. The area of the spill is given by $A(r) = \pi r^2$ where r represents the radius of the oil slick. Find each of the following, and explain their meanings.

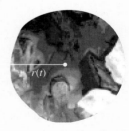

   a) $r(5)$ b) $A(20)$ c) $A(r(t))$ d) $A(r(5))$

24) The radius of a circle is half its diameter. We can express this with the function $r(d) = \dfrac{1}{2}d$, where d is the diameter of a circle and r is the radius. The area of a circle in terms of its radius is $A(r) = \pi r^2$. Find each of the following and explain their meanings.

   a) $r(6)$ b) $A(3)$ c) $A(r(d))$ d) $A(r(6))$

Objective 3: Determine Whether a Function Is One-to-One

Determine whether each function is one-to-one. If it is one-to-one, find its inverse.

25) $f = \{(-6, 3), (-1, 8), (4, 3)\}$

26) $g = \{(0, -7), (1, -6), (4, -5), (25, -2)\}$

27) $h = \{(-5, -16), (-1, -4), (3, 8)\}$

28) $f = \{(-4, 3), (-2, -3), (2, -3), (6, 13)\}$

Determine whether each function is one-to-one.

29) The table shows the average temperature during selected months in Tulsa, Oklahoma. The function matches each month with the average temperature, in °F. Is it one-to-one? (www.noaa.gov)

Month	Average Temp. (°F)
Jan.	36.4
Apr.	60.8
July	83.5
Oct.	62.6

30) The table shows some NCAA conferences and the number of schools in the conference in 2009. The function matches each conference with the number of schools it contains. Is it one-to-one?

Conference	Number of Member Schools
ACC	12
Big 10	11
Big 12	12
MVC	10
Pac10	10

Mixed Exercises: Objectives 3–5

31) Do all functions have inverses? Explain your answer.

32) What test can be used to determine whether the graph of a function has an inverse?

Determine whether each statement is true or false. If it is false, rewrite the statement so that it is true.

33) $f^{-1}(x)$ is read as "f to the negative one of x."

34) If f^{-1} is the inverse of f, then $(f^{-1} \circ f)(x) = x$ and $(f \circ f^{-1})(x) = x$.

35) The domain of f is the range of f^{-1}.

36) If f is one-to-one and $(5, 9)$ is on the graph of f, then $(-5, -9)$ is on the graph of f^{-1}.

37) The graphs of $f(x)$ and $f^{-1}(x)$ are symmetric with respect to the x-axis.

38) Let $f(x)$ be one-to-one. If $f(7) = 2$, then $f^{-1}(2) = 7$.

For each function graphed here, answer the following.
   a) Determine whether it is one-to-one.
   b) If it is one-to-one, graph its inverse.

 39) 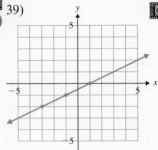   40)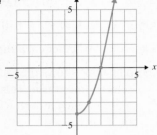

41) 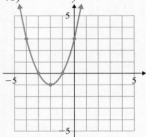   42)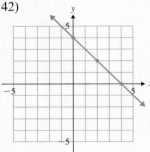

43) 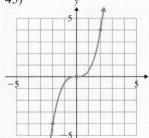   44)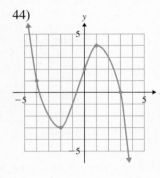

Objective 4: Find the Inverse of a Function
Find the inverse of each one-to-one function.

Fill It In
Fill in the blanks with either the missing mathematical step or the reason for the given step.

45) $f(x) = 2x - 10$
   $y = 2x - 10$ _____
   _____   Interchange x and y.
   Solve for y.
   $x + 10 = 2y$ _____
   _____   Divide by 2, and simplify.
   $f^{-1}(x) = \dfrac{1}{2}x + 5$ _____

46) $g(x) = \dfrac{1}{3}x + 4$
   _____   Replace $g(x)$ with y.
   $x = \dfrac{1}{3}y + 4$ _____
   Solve for y.
   _____   Subtract 4.
   $3x - 12 = y$ _____
   _____   Replace y with $g^{-1}(x)$.

Find the inverse of each one-to-one function. Then, graph the function and its inverse on the same axes.

47) $g(x) = x - 6$ 48) $h(x) = x + 3$
49) $f(x) = -2x + 5$ 50) $g(x) = 4x - 9$
51) $g(x) = \dfrac{1}{2}x$ 52) $h(x) = -\dfrac{1}{3}x$
53) $f(x) = x^3$ 54) $g(x) = \sqrt[3]{x} + 4$

Find the inverse of each one-to-one function.

55) $f(x) = 2x - 6$ 56) $g(x) = -4x + 8$
57) $h(x) = -\dfrac{3}{2}x + 4$ 58) $f(x) = \dfrac{2}{5}x + 1$
59) $h(x) = \sqrt[3]{x - 7}$ 60) $g(x) = \sqrt[3]{x + 2}$
61) $f(x) = \sqrt{x},\ x \geq 0$ 62) $g(x) = \sqrt{x + 3},\ x \geq -3$

R Rethink

R1) What is a composition of functions?

R2) Is a composition of functions commutative? If not, can you think of an example that is not commutative?

R3) Could you explain to a classmate how to tell if a function is one-to-one?

R4) How can you tell whether the graphs of two functions are inverses of each other?

11.2 Exponential Functions

P Prepare	**O Organize**
What are your objectives for Section 11.2?	How can you accomplish each objective?
1 Define an Exponential Function	• Learn the definition of an *exponential function*. • Understand why $a > 0$ and $a \neq 1$ in the exponential function $f(x) = a^x$.
2 Graph $f(x) = a^x$	• Understand why we choose values of x that are negative, positive, and zero. • Know what the graph should look like based on the value of a. • Complete the given examples on your own. • Complete You Trys 1 and 2.
3 Graph $f(x) = a^{x+c}$	• Understand why we choose values of x that will make the exponent negative, positive, and zero. • Complete the given example on your own. • Complete You Try 3.
4 Define the Number e and Graph $f(x) = e^x$	• Learn the definition of e and its numerical approximation. • Be able to use a calculator to generate a table of values for $f(x) = e^x$. • Complete the given example on your own.
5 Solve an Exponential Equation	• Review the rules of exponents. • Learn the procedure for **Solving an Exponential Equation.** • Complete the given example on your own. • Complete You Try 4.
6 Solve an Applied Problem Using a Given Exponential Function	• Read the given example carefully. • Complete the given example on your own. • Complete You Try 5.

Read the explanations, follow the examples, take notes, and complete the You Trys.

W Hint
Do you remember what the graph of each of these functions looks like?

We have already studied the following types of functions:

Linear functions like $f(x) = 2x + 5$
Quadratic functions like $g(x) = x^2 - 6x + 8$
Square root functions like $k(x) = \sqrt{x - 3}$

1 Define an Exponential Function

In this section, we will learn about *exponential functions*.

> **Definition**
>
> An **exponential function** is a function of the form
> $$f(x) = a^x$$
> where $a > 0$, $a \neq 1$, and x is a real number.

Note

1) We stipulate that *a* is a positive number ($a > 0$) because if *a* were a negative number, some expressions would not be real numbers.

   Example: If $a = -2$ and $x = \dfrac{1}{2}$, we get $f(x) = (-2)^{1/2} = \sqrt{-2}$ (not real).

   Therefore, *a* must be a positive number.

2) We add the condition that $a \neq 1$ because if $a = 1$, the function would be linear, not exponential.

   Example: If $a = 1$, then $f(x) = 1^x$. This is equivalent to $f(x) = 1$, which is a linear function.

2 Graph $f(x) = a^x$

We can graph exponential functions by plotting points. *It is important to choose many values for the variable so that we obtain positive numbers, negative numbers, and zero in the exponent.*

EXAMPLE 1

Graph $f(x) = 2^x$ and $g(x) = 3^x$ on the same axes. Determine the domain and range.

Solution

Make a table of values for each function. Be sure to choose values for *x* that will give us *positive numbers, negative numbers, and zero* in the exponent.

Hint
Be sure to choose values of *x* that will make the exponent *positive*, *negative*, and *zero* so that we can see the complete graph.

$f(x) = 2^x$	
x	f(x)
0	1
1	2
2	4
3	8
−1	$\dfrac{1}{2}$
−2	$\dfrac{1}{4}$

$g(x) = 3^x$	
x	g(x)
0	1
1	3
2	9
3	27
−1	$\dfrac{1}{3}$
−2	$\dfrac{1}{9}$

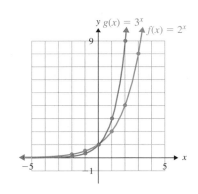

Plot each set of points, and connect them with a smooth curve. Note that the larger the value of *a*, the more rapidly the *y*-values increase. Additionally, as *x* increases, the value of *y* also increases. Here are some other interesting facts to note about the graphs of these functions.

1) Each graph passes the vertical line test so the graphs *do* represent functions.
2) Each graph passes the horizontal line test, so the functions are one-to-one.
3) The *y*-intercept of each function is (0, 1).
4) The domain of each function is $(-\infty, \infty)$, and the range is $(0, \infty)$.

[**YOU TRY 1**] Graph $f(x) = 4^x$. Determine the domain and range.

EXAMPLE 2

Graph $f(x) = \left(\dfrac{1}{2}\right)^x$. Determine the domain and range.

Solution

Make a table of values and be sure to choose values for *x* that will give us *positive numbers, negative numbers, and zero* in the exponent.

$f(x) = \left(\dfrac{1}{2}\right)^x$

x	f(x)
0	1
1	$\dfrac{1}{2}$
2	$\dfrac{1}{4}$
−1	2
−2	4
−3	8

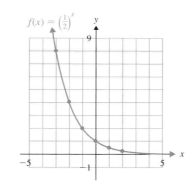

Like the graphs of $f(x) = 2^x$ and $g(x) = 3^x$ in Example 1, the graph of $f(x) = \left(\dfrac{1}{2}\right)^x$ passes both the vertical and horizontal line tests, making it a one-to-one function. The *y*-intercept is (0, 1). The domain is $(-\infty, \infty)$, and the range is $(0, \infty)$.

In the case of $f(x) = \left(\dfrac{1}{2}\right)^x$, however, as the value of *x* increases, the value of *y* decreases. This is because $0 < a < 1$.

[**YOU TRY 2**] Graph $g(x) = \left(\dfrac{1}{3}\right)^x$. Determine the domain and range.

Hint

Remember that the domain contains the x-values, and the range contains the y-values.

We can summarize what we have learned so far about exponential functions:

Summary Characteristics of $f(x) = a^x$, where $a > 0$ and $a \neq 1$

1) If $f(x) = a^x$ where $a > 1$, the value of y increases as the value of x increases.

2) If $f(x) = a^x$, where $0 < a < 1$, the value of y decreases as the value of x increases.

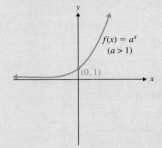
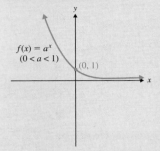

3) The function is one-to-one.
4) The y-intercept is $(0, 1)$.
5) The domain is $(-\infty, \infty)$, and the range is $(0, \infty)$.

3 Graph $f(x) = a^{x+c}$

Next we will graph an exponential function with an expression other than x as its exponent.

EXAMPLE 3

Graph $f(x) = 3^{x-2}$. Determine the domain and range.

Solution

Remember, for the table of values we want to choose values of x that will give us positive numbers, negative numbers, and zero *in the exponent*. First we will determine which value of x will make the exponent equal zero.

$$x - 2 = 0$$
$$x = 2$$

If $x = 2$, the exponent equals zero. Choose a couple of numbers *greater than* 2 and a couple that are *less than* 2 to get positive and negative numbers in the exponent.

	x	$x - 2$	$f(x) = 3^{x-2}$	Plot
	2	0	$3^0 = 1$	$(2, 1)$
Values greater than 2	3	1	$3^1 = 3$	$(3, 3)$
	4	2	$3^2 = 9$	$(4, 9)$
Values less than 2	1	-1	$3^{-1} = \dfrac{1}{3}$	$\left(1, \dfrac{1}{3}\right)$
	0	-2	$3^{-2} = \dfrac{1}{9}$	$\left(0, \dfrac{1}{9}\right)$

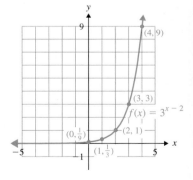

Note that the y-intercept is not $(0, 1)$ because the exponent is $x - 2$, not x, as in $f(x) = a^x$. The graph of $f(x) = 3^{x-2}$ is the same shape as the graph of $g(x) = 3^x$ except that the graph of f is shifted 2 units to the right. This is because $f(x) = g(x - 2)$. The domain of f is $(-\infty, \infty)$, and the range is $(0, \infty)$.

[**YOU TRY 3**] Graph $f(x) = 2^{x+4}$. Determine the domain and range.

4 Define the Number e and Graph $f(x) = e^x$

Next we will introduce a special exponential function, one with a base of e.

Like the number π, e is an irrational number that has many uses in mathematics. In the 1700s, the work of Swiss mathematician Leonhard Euler led him to the approximation of e.

> **Definition**
>
> **Approximation of e**
>
> $$e \approx 2.718281828459045235$$

One of the questions Euler set out to answer was, what happens to the value of $\left(1 + \dfrac{1}{n}\right)^n$ as n gets larger and larger? He found that as n gets larger, $\left(1 + \dfrac{1}{n}\right)^n$ gets closer to a fixed number. This number is e. Euler approximated e to the 18 decimal places in the definition, and the letter e was chosen to represent this number in his honor. It should be noted that there are other ways to generate e. Finding the value that $\left(1 + \dfrac{1}{n}\right)^n$ approaches as n gets larger and larger is just one way. Also, since e is irrational, it is a nonterminating, nonrepeating decimal.

EXAMPLE 4 Graph $f(x) = e^x$. Determine the domain and range.

Solution

A calculator is needed to generate a table of values. We will use either the $\boxed{e^x}$ key or the two keys $\boxed{\text{INV}}$ (or $\boxed{\text{2ND}}$) and $\boxed{\ln x}$ to find powers of e. (Calculators will approximate powers of e to a few decimal places.)

For example, if a calculator has an $\boxed{e^x}$ key, find e^2 by pressing the following keys:

$\boxed{2}\ \boxed{e^x}$ or $\boxed{e^x}\ \boxed{2}\ \boxed{\text{ENTER}}$

To four decimal places, $e^2 \approx 7.3891$.

If a calculator has an $\boxed{\ln x}$ key with e^x written above it, find e^2 by pressing the following keys:

$\boxed{2}\ \boxed{\text{INV}}\ \boxed{\ln x}$ or $\boxed{\text{INV}}\ \boxed{\ln x}\ \boxed{2}\ \boxed{\text{ENTER}}$

The same approximation for e^2 is obtained.

Remember to choose positive numbers, negative numbers, and zero for x when making the table of values. We will approximate the values of e^x to four decimal places.

$f(x) = e^x$	
x	f(x)
0	1
1	2.7183
2	7.3891
3	20.0855
−1	0.3679
−2	0.1353

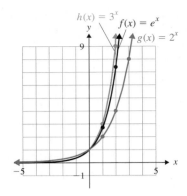

Notice that the graph of $f(x) = e^x$ is between the graphs of $g(x) = 2^x$ and $h(x) = 3^x$. This is because $2 < e < 3$, so e^x grows more quickly than 2^x, but e^x grows more slowly than 3^x. The domain of $f(x) = e^x$ is $(-\infty, \infty)$, and the range is $(0, \infty)$.

We will study e^x and its special properties in more detail later in the chapter.

5 Solve an Exponential Equation

An **exponential equation** is an equation that has a variable in the exponent. Some examples of exponential equations are

$$2^x = 8, \quad 3^{a-5} = \frac{1}{9}, \quad e^t = 14, \quad 5^{2y-1} = 6^{y+4}$$

In this section, we will learn how to solve exponential equations like the first two examples. We can solve those equations by getting the same base.

We know that the exponential function $f(x) = a^x$ ($a > 0, a \neq 1$) is one-to-one. This leads to the following property that enables us to solve many exponential equations.

$$\text{If } a^x = a^y, \text{ then } x = y. \quad (a > 0, a \neq 1)$$

This property says that if two sides of an equation have the same base, set the exponents equal and solve for the unknown variable.

> **Procedure** Solving an Exponential Equation
>
> **Step 1:** If possible, express each side of the equation with the same base. If it is *not* possible to get the same base, a different method must be used. (This is presented in Section 11.6.)
>
> **Step 2:** Use the rules of exponents to simplify the exponents.
>
> **Step 3:** Set the exponents equal, and solve for the variable.

EXAMPLE 5 Solve each equation.

a) $2^x = 8$ b) $49^{c+3} = 7^{3c}$ c) $9^{6n} = 27^{n-4}$ d) $3^{a-5} = \frac{1}{9}$

Solution

a) **Step 1:** Express each side of the equation with the same base.

$$2^x = 8$$
$$2^x = 2^3 \quad \text{Rewrite 8 with a base of 2: } 8 = 2^3.$$

> **W Hint**
> Don't forget to check your answers in the original equation.

Step 2: The exponents are simplified.

Step 3: Since the bases are the same, set the exponents equal and solve.

$$x = 3$$

The solution set is $\{3\}$.

b) *Step 1:* Express each side of the equation with the same base.

$$49^{c+3} = 7^{3c}$$
$$(7^2)^{c+3} = 7^{3c} \quad \text{Both sides are powers of 7; } 49 = 7^2.$$

Step 2: Use the rules of exponents to simplify the exponents.

$$7^{2(c+3)} = 7^{3c} \quad \text{Power rule for exponents}$$
$$7^{2c+6} = 7^{3c} \quad \text{Distribute.}$$

Step 3: Since the bases are the same, set the exponents equal and solve.

$$2c + 6 = 3c \quad \text{Set the exponents equal.}$$
$$6 = c \quad \text{Subtract } 2c.$$

The solution set is $\{6\}$. Check the answer in the original equation.

> **W Hint**
> Do you see why it is important to know the powers of numbers? Review them, if necessary.

c) *Step 1:* Express each side of the equation with the same base. 9 *and* 27 *are each powers of* 3.

$$9^{6n} = 27^{n-4}$$
$$(3^2)^{6n} = (3^3)^{n-4} \quad 9 = 3^2; 27 = 3^3$$

Step 2: Use the rules of exponents to simplify the exponents.

$$3^{2(6n)} = 3^{3(n-4)} \quad \text{Power rule for exponents}$$
$$3^{12n} = 3^{3n-12} \quad \text{Multiply.}$$

Step 3: Since the bases are the same, set the exponents equal and solve.

$$12n = 3n - 12 \quad \text{Set the exponents equal.}$$
$$9n = -12 \quad \text{Subtract } 3n.$$
$$n = -\frac{12}{9} = -\frac{4}{3} \quad \text{Divide by 9; simplify.}$$

The solution set is $\left\{-\dfrac{4}{3}\right\}$.

d) *Step 1:* Express each side of the equation $3^{a-5} = \dfrac{1}{9}$ with the same base. $\dfrac{1}{9}$ *can be expressed with a base of* 3: $\dfrac{1}{9} = \left(\dfrac{1}{3}\right)^2 = 3^{-2}$.

$$3^{a-5} = \frac{1}{9}$$
$$3^{a-5} = 3^{-2} \quad \text{Rewrite } \frac{1}{9} \text{ with a base of 3.}$$

Step 2: The exponents are simplified.

Step 3: Set the exponents equal and solve.

$$a - 5 = -2 \quad \text{Set the exponents equal.}$$
$$a = 3 \quad \text{Add 5.}$$

The solution set is $\{3\}$.

[YOU TRY 4] Solve each equation.

a) $(12)^x = 144$ b) $6^{t-5} = 36^{t+4}$ c) $32^{2w} = 8^{4w-1}$ d) $8^k = \dfrac{1}{64}$

6 Solve an Applied Problem Using a Given Exponential Function

EXAMPLE 6 The value of a car depreciates (decreases) over time. The value, $V(t)$, in dollars, of a sedan t yr after it is purchased is given by

$$V(t) = 18{,}200(0.794)^t$$

a) What was the purchase price of the car?

b) What will the car be worth 5 yr after purchase?

Solution

a) To find the purchase price of the car, let $t = 0$. Evaluate $V(0)$ given that $V(t) = 18{,}200(0.794)^t$.

$$\begin{aligned} V(0) &= 18{,}200(0.794)^0 \\ &= 18{,}200(1) \\ &= 18{,}200 \end{aligned}$$

The purchase price of the car was $18,200.

b) To find the value of the car after 5 yr, let $t = 5$. Use a calculator to find $V(5)$.

$$\begin{aligned} V(5) &= 18{,}200(0.794)^5 \\ &= 5743.46 \end{aligned}$$

The car will be worth about $5743.46.

[YOU TRY 5] The value, $V(t)$, in dollars, of a pickup truck t yr after it is purchased is given by

$$V(t) = 23{,}500(0.785)^t$$

a) What was the purchase price of the pickup?

b) What will the pickup truck be worth 4 yr after purchase?

ANSWERS TO [YOU TRY] EXERCISES

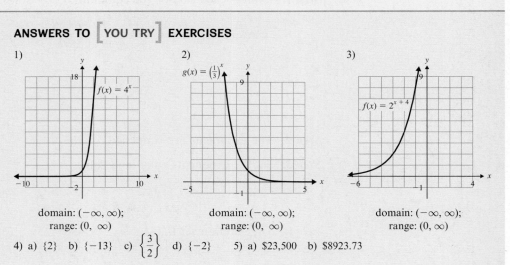

1) domain: $(-\infty, \infty)$; range: $(0, \infty)$ 2) domain: $(-\infty, \infty)$; range: $(0, \infty)$ 3) domain: $(-\infty, \infty)$; range: $(0, \infty)$

4) a) $\{2\}$ b) $\{-13\}$ c) $\left\{\dfrac{3}{2}\right\}$ d) $\{-2\}$ 5) a) $23,500 b) $8923.73

Evaluate 11.2 Exercises

Do the exercises, and check your work.

Mixed Exercises: Objectives 1 and 2

1) When making a table of values to graph an exponential function, what kind of values should be chosen for the variable?

2) What is the y-intercept of the graph of $f(x) = a^x$ where $a > 0$ and $a \neq 1$?

Graph each exponential function. Determine the domain and range.

3) $y = 2^x$

4) $g(x) = 4^x$

5) $h(x) = \left(\dfrac{1}{3}\right)^x$

6) $y = \left(\dfrac{1}{4}\right)^x$

For an exponential function of the form $f(x) = a^x$ ($a > 0$, $a \neq 1$), answer the following.

7) What is the domain?

8) What is the range?

Objective 3: Graph $f(x) = a^{x+c}$

Graph each exponential function. State the domain and range.

9) $g(x) = 2^{x+1}$

10) $y = 3^{x+2}$

11) $f(x) = 3^{x-4}$

12) $h(x) = 2^{x-3}$

13) $f(x) = 2^{2x}$

14) $h(x) = 3^{\frac{1}{2}x}$

15) $y = 2^x + 1$

16) $f(x) = 2^x - 3$

17) $g(x) = 3^x - 2$

18) $h(x) = 3^x + 1$

19) $y = -2^x$

20) $f(x) = -\left(\dfrac{1}{3}\right)^x$

21) As the value of x gets larger, would you expect $f(x) = 2x$ or $g(x) = 2^x$ to grow faster? Why?

22) Let $f(x) = \left(\dfrac{1}{5}\right)^x$. The graph of $f(x)$ gets very close to the line $y = 0$ (the x-axis) as the value of x gets larger. Why?

23) If you are given the graph of $f(x) = a^x$, where $a > 0$ and $a \neq 1$, how would you obtain the graph of $g(x) = a^x - 2$?

24) If you are given the graph of $f(x) = a^x$, where $a > 0$ and $a \neq 1$, how would you obtain the graph of $g(x) = a^{x-3}$?

Objective 4: Define the Number e and Graph $f(x) = e^x$

25) What is the approximate value of e to four decimal places?

26) Is e a rational or an irrational number? Explain your answer.

For Exercises 27–30, match each exponential function with its graph.

A)
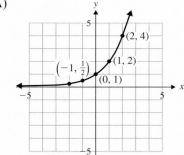

B)
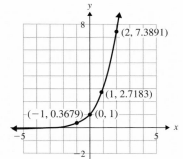

C)
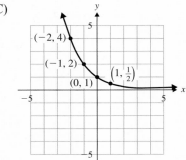

D)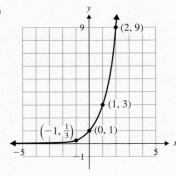

27) $f(x) = e^x$
28) $g(x) = 2^x$
29) $h(x) = 3^x$
30) $k(x) = \left(\dfrac{1}{2}\right)^x$

Graph each function. State the domain and range.

31) $f(x) = e^x - 2$
32) $g(x) = e^x + 1$
33) $y = e^{x+1}$
34) $h(x) = e^{x-3}$
35) $g(x) = \dfrac{1}{2}e^x$
36) $y = 2e^x$
37) $h(x) = -e^x$
38) $f(x) = e^{-x}$

39) Graph $y = e^x$, and compare it with the graph of $h(x) = -e^x$ in Exercise 41. What can you say about these graphs?

40) Graph $y = e^x$, and compare it with the graph of $f(x) = e^{-x}$ in Exercise 38. What can you say about these graphs?

Objective 5: Solve an Exponential Equation
Solve each exponential equation.

Fill It In
Fill in the blanks with either the missing mathematical step or the reason for the given step.

41) $6^{3n} = 36^{n-4}$
_____ Express each side with the same base.
$6^{3n} = 6^{2(n-4)}$ _____
$6^{3n} = 6^{2n-8}$ _____
_____ Set the exponents equal.
$n = -8$ Solve for n.
The solution set is ____.

42) $125^{2w} = 5^{w+2}$
$(5^3)^{2w} = 5^{w+2}$ _____
_____ Power rule for exponents
$5^{6w} = 5^{w+2}$ _____
$6w = w + 2$ _____
_____ Solve for w.
The solution set is ____.

43) $9^x = 81$
44) $4^y = 16$
45) $5^{4d} = 125$
46) $4^{3a} = 64$
47) $3^{5t} = 9^{t+4}$
48) $16^{m-2} = 2^{3m}$
49) $(1000)^{2p-3} = 10^{4p+1}$
50) $7^{2k-6} = 49^{3k+1}$
51) $6^x = \dfrac{1}{36}$
52) $11^t = \dfrac{1}{121}$
53) $9^r = \dfrac{1}{27}$
54) $16^c = \dfrac{1}{8}$
55) $\left(\dfrac{5}{6}\right)^{3x+7} = \left(\dfrac{36}{25}\right)^{2x}$
56) $\left(\dfrac{7}{2}\right)^{5w} = \left(\dfrac{4}{49}\right)^{4w+3}$

Objective 6: Solve an Applied Problem Using a Given Exponential Function
Solve each application.

57) The value of a car depreciates (decreases) over time. The value, $V(t)$, in dollars, of an SUV t yr after it is purchased is given by
$$V(t) = 32{,}700(0.812)^t$$
a) What was the purchase price of the SUV?
b) What will the SUV be worth 3 yr after purchase?

58) The value, $V(t)$, in dollars, of a sports car t yr after it is purchased is given by
$$V(t) = 48{,}600(0.820)^t$$
a) What was the purchase price of the sports car?
b) What will the sports car be worth 4 yr after purchase?

59) From 1995 to 2005, the value of homes in a suburb increased by 3% per year. The value, $V(t)$, in dollars, of a particular house t yr after 1995 is given by
$$V(t) = 185{,}200(1.03)^t$$
a) How much was the house worth in 1995?
b) How much was the house worth in 2002?

60) From 2000 to 2010, the value of condominiums in a big city high-rise building increased by 2% per year. The value, $V(t)$, in dollars, of a particular condo t yr after 2000 is given by

$$V(t) = 420{,}000(1.02)^t$$

a) How much was the condominium worth in 2000?

b) How much was the condominium worth in 2010?

An *annuity* is an account into which money is deposited every year. The amount of money, A in dollars, in the account after t yr of depositing c dollars at the beginning of every year earning an interest rate r (as a decimal) is

$$A = c\left[\frac{(1+r)^t - 1}{r}\right](1+r)$$

Use the formula for Exercises 61 and 62.

61) After Fernando's daughter is born, he decides to begin saving for her college education. He will deposit $2000 every year in an annuity for 18 yr at a rate of 9%. How much will be in the account after 18 yr?

62) To save for retirement, Susan plans to deposit $6000 per year in an annuity for 30 yr at a rate of 8.5%. How much will be in the account after 30 yr?

63) After taking a certain antibiotic, the amount of amoxicillin $A(t)$, in milligrams, remaining in the patient's system t hr after taking 1000 mg of amoxicillin is

$$A(t) = 1000e^{-0.5332t}$$

How much amoxicillin is in the patient's system 6 hr after taking the medication?

64) Some cockroaches can reproduce according to the formula

$$y = 2(1.65)^t$$

where y is the number of cockroaches resulting from the mating of two cockroaches and their offspring t months after the first two cockroaches mate.

If Morris finds two cockroaches in his kitchen (assuming one is male and one is female) how large can the cockroach population become after 12 months?

R Rethink

R1) What is a composition of functions?

R2) Is a composition of functions commutative? If not, can you think of an example that is not commutative?

R3) Could you explain to a classmate how to tell if a function is one-to-one?

R4) How can you tell if two functions on a graph are inverses of each other?

11.3 Logarithmic Functions

P Prepare	**O Organize**
What are your objectives for Section 11.3?	How can you accomplish each objective?
1 Define a Logarithm	• Learn the definition of a *logarithm*. • Know that $a > 0$, $x > 0$, and $a \neq 1$. • In your notes, write the relationship between the logarithmic form of an equation and the exponential form of an equation.
2 Convert from Logarithmic Form to Exponential Form	• Review the conversion from log form to exponential form. • Complete the given example on your own. • Complete You Try 1.
3 Convert from Exponential Form to Logarithmic Form	• Review the conversion from exponential form to log form. • Complete the given example on your own. • Complete You Try 2.
4 Solve an Equation of the Form $\log_a b = c$	• Be able to identify a *logarithmic equation*. • Learn the procedure for **Solving an Equation of the Form $\log_a b = c$.** • Complete the given example on your own. • Complete You Try 3.
5 Evaluate a Logarithm	• Know what it means to evaluate a logarithmic expression. • Complete the given example on your own. • Complete You Try 4.
6 Evaluate Common Logarithms, and Solve Equations of the Form $\log b = c$	• Learn the definition of a *common logarithm*. • Complete the given example on your own. • Complete You Try 5.
7 Use the Properties $\log_a a = 1$ and $\log_a 1 = 0$	• Learn the **Properties of Logarithms** in this section. • Complete the given example on your own. • Complete You Try 6.
8 Define and Graph a Logarithmic Function	• Learn the definition for a *logarithmic function*. • When graphing a logarithmic function, choose values for y in the table of values. • Review the **Summary of Characteristics of Logarithmic Functions.** • Complete the given examples on your own. • Complete You Trys 7–9.
9 Solve an Applied Problem Using a Logarithmic Equation	• Read the given example carefully. • Choose the correct value for t, and be sure you understand what this value represents. • Complete the given example on your own. • Complete You Try 10.

W Work Read the explanations, follow the examples, take notes, and complete the You Trys.

1 Define a Logarithm

In Section 11.2, we graphed $f(x) = 2^x$ by making a table of values and plotting the points. The graph passes the horizontal line test, making the function one-to-one. Recall that if (x, y) is on the graph of a function, then (y, x) is on the graph of its inverse. We can graph the inverse of $f(x) = 2^x$, $f^{-1}(x)$, by switching the x- and y-coordinates in the table of values and plotting the points.

$f(x) = 2^x$	
x	y = f(x)
0	1
1	2
2	4
3	8
−1	$\frac{1}{2}$
−2	$\frac{1}{4}$

$f^{-1}(x)$	
x	y = $f^{-1}(x)$
1	0
2	1
4	2
8	3
$\frac{1}{2}$	−1
$\frac{1}{4}$	−2

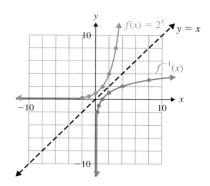

Above is the graph of $f(x) = 2^x$ and its inverse. Notice that, like the graphs of all functions and their inverses, they are symmetric with respect to the line $y = x$.

What is the equation of $f^{-1}(x)$ if $f(x) = 2^x$? We will use the procedure outlined in Section 11.1 to find the equation of $f^{-1}(x)$.

If $f(x) = 2^x$, then find the equation of $f^{-1}(x)$ as follows.

Step 1: Replace $f(x)$ with y.

$$y = 2^x$$

Step 2: Interchange x and y.

$$x = 2^y$$

Step 3: Solve for y.

How do we solve $x = 2^y$ for y? To answer this question, we must introduce another concept called *logarithms*.

Definition

Definition of Logarithm: If $a > 0$, $a \neq 1$, and $x > 0$, then for every real number y,

$$y = \log_a x \text{ means } x = a^y$$

The word **log** is an abbreviation for **logarithm**. We read $\log_a x$ as "log of x to the base a" or "log to the base a of x." *This definition of a logarithm should be memorized!*

Note

It is very important to note that the base of the logarithm must be positive and not equal to 1, and that x must be positive as well.

The relationship between the logarithmic form of an equation ($y = \log_a x$) and the exponential form of an equation ($x = a^y$) is one that has many uses. Notice the relationship between the two forms.

Logarithmic Form	Exponential Form
Value of the logarithm ↓	Exponent ↓
$y = \log_a x$	$x = a^y$
↑ Base	↑ Base

From the above, you can see that *a logarithm is an exponent*. $\log_a x$ is the power to which we raise a to get x.

2 Convert from Logarithmic Form to Exponential Form

Much of our work with logarithms involves converting between logarithmic and exponential notation. After working with logs and exponential form, we will come back to the question of how to solve $x = 2^y$ for y.

EXAMPLE 1

Write in exponential form.

a) $\log_6 36 = 2$ b) $\log_4 \dfrac{1}{64} = -3$ c) $\log_7 1 = 0$

Solution

a) $\log_6 36 = 2$ means that 2 is the power to which we raise 6 to get 36. The exponential form is $6^2 = 36$.

$$\log_6 36 = 2 \text{ means } 6^2 = 36.$$

W Hint
You will need to understand this example before you can do Example 3.

b) $\log_4 \dfrac{1}{64} = -3$ means $4^{-3} = \dfrac{1}{64}$.

c) $\log_7 1 = 0$ means $7^0 = 1$.

[YOU TRY 1] Write in exponential form.

a) $\log_3 81 = 4$ b) $\log_5 \dfrac{1}{25} = -2$ c) $\log_{64} 8 = \dfrac{1}{2}$ d) $\log_{13} 13 = 1$

3 Convert from Exponential Form to Logarithmic Form

EXAMPLE 2

Write in logarithmic form.

a) $10^4 = 10{,}000$ b) $9^{-2} = \dfrac{1}{81}$ c) $8^1 = 8$

d) $\sqrt{25} = 5$

Solution

a) $10^4 = 10{,}000$ means $\log_{10} 10{,}000 = 4$.

b) $9^{-2} = \dfrac{1}{81}$ means $\log_9 \dfrac{1}{81} = -2$.

c) $8^1 = 8$ means $\log_8 8 = 1$.

d) To write $\sqrt{25} = 5$ in logarithmic form, rewrite $\sqrt{25}$ as $25^{1/2}$.

$25^{1/2} = 5$ means $\log_{25} 5 = \dfrac{1}{2}$.

Note

When working with logarithms, we will often change radical notation to the equivalent fractional exponent. This is because a logarithm *is* an exponent.

[YOU TRY 2] Write in logarithmic form.

a) $7^2 = 49$ b) $5^{-4} = \dfrac{1}{625}$ c) $19^0 = 1$ d) $\sqrt{144} = 12$

---

4 Solve an Equation of the Form $\log_a b = c$

A **logarithmic equation** is an equation in which at least one term contains a logarithm. In this section, we will learn how to solve a logarithmic equation of the form $\log_a b = c$. We will learn how to solve other types of logarithmic equations in Sections 11.5 and 11.6.

Procedure Solve an Equation of the Form $\log_a b = c$

To solve a logarithmic equation of the form $\log_a b = c$, write the equation in exponential form ($a^c = b$) and solve for the variable. Check the answer in the original equation.

EXAMPLE 3 Solve each logarithmic equation.

a) $\log_{10} r = 3$ b) $\log_w 25 = 2$ c) $\log_2 16 = c$ d) $\log_{36} \sqrt[4]{6} = x$

Solution

a) Write the equation in exponential form, and solve for r.

$$\log_{10} r = 3 \quad \text{means} \quad 10^3 = r$$
$$1000 = r$$

The solution set is $\{1000\}$.

b) Write $\log_w 25 = 2$ in exponential form and solve for w.

$$\log_w 25 = 2 \quad \text{means} \quad w^2 = 25$$
$$w = \pm 5 \quad \text{Square root property}$$

> **W Hint**
> Part b) is a good example of why you should check your answer.

Although we get $w = 5$ or $w = -5$ when we solve $w^2 = 25$, recall that the base of a logarithm must be a positive number. Therefore, $w = -5$ is *not* a solution of the original equation.

The solution set is $\{5\}$.

c) Write $\log_2 16 = c$ in exponential form, and solve for c.

$$\log_2 16 = c \quad \text{means} \quad 2^c = 16$$
$$c = 4$$

Verify that the solution set is $\{4\}$.

d) $\log_{36} \sqrt[4]{6} = x \quad \text{means} \quad 36^x = \sqrt[4]{6}$

$(6^2)^x = 6^{1/4}$ Express each side with the same base; rewrite the radical as a fractional exponent.

$6^{2x} = 6^{1/4}$ Power rule for exponents

$2x = \dfrac{1}{4}$ Set the exponents equal.

$x = \dfrac{1}{8}$ Divide by 2.

The solution set is $\left\{\dfrac{1}{8}\right\}$. The check is left to the student.

[YOU TRY 3] Solve each logarithmic equation.

a) $\log_2 y = 5$ b) $\log_x 169 = 2$ c) $\log_6 36 = n$

d) $\log_{64} \sqrt[5]{8} = k$ e) $\log_5(3p + 11) = 3$

5 Evaluate a Logarithm

Often when working with logarithms, we are asked to *evaluate* them or to find the value of a log.

EXAMPLE 4   Evaluate.

a) $\log_3 9$ b) $\log_{10} \dfrac{1}{10}$ c) $\log_{25} 5$

Solution

a) To *evaluate* (or *find the value of*) $\log_3 9$ means to find the power to which we raise 3 to get 9. That power is **2**.

$$\log_3 9 = 2 \quad \text{because} \quad 3^2 = 9$$

b) To evaluate $\log_{10} \frac{1}{10}$ means to find the power to which we raise 10 to get $\frac{1}{10}$.

That power is **−1.**

If you don't see that this is the answer, set the expression $\log_{10} \frac{1}{10}$ equal to x, write the equation in exponential form, and solve for x as in Example 3.

$$\log_{10} \frac{1}{10} = x \quad \text{means} \quad 10^x = \frac{1}{10}$$
$$10^x = 10^{-1} \qquad \frac{1}{10} = 10^{-1}$$
$$x = -1$$

Then, $\log_{10} \frac{1}{10} = -1$.

c) To evaluate $\log_{25} 5$ means to find the power to which we raise 25 to get 5. That power is $\frac{1}{2}$.

Once again, we can also find the value of $\log_{25} 5$ by setting it equal to x, writing the equation in exponential form, and solving for x.

$$\log_{25} 5 = x \quad \text{means} \quad 25^x = 5$$
$$(5^2)^x = 5 \qquad \text{Express each side with the same base.}$$
$$5^{2x} = 5^1 \qquad \text{Power rule; } 5 = 5^1$$
$$2x = 1 \qquad \text{Set the exponents equal.}$$
$$x = \frac{1}{2} \qquad \text{Divide by 2.}$$

Therefore, $\log_{25} 5 = \frac{1}{2}$.

[**YOU TRY 4**] Evaluate.

a) $\log_{10} 100$ b) $\log_8 \frac{1}{8}$ c) $\log_{144} 12$

---

6 Evaluate Common Logarithms, and Solve Equations of the Form log b = c

Logarithms have many applications not only in mathematics but also in other areas such as chemistry, biology, engineering, and economics.

Since our number system is a base 10 system, logarithms to the base 10 are very widely used and are called **common logarithms** or **common logs**. A base 10 log has a special notation—$\log_{10} x$ is written as $\log x$. When a log is written in this way, the base is assumed to be 10.

W Hint
Write the definition of a common logarithm in your notes.

$$\log x \text{ means } \log_{10} x$$

This also means that $\log 100 = 2$ and $\log 1000 = 3$. We must keep this in mind when evaluating logarithms and when solving logarithmic equations.

EXAMPLE 5

Solve $\log(3x - 8) = 1$.

Solution

$\log(3x - 8) = 1$ is equivalent to $\log_{10}(3x - 8) = 1$. Write the equation in exponential form, and solve for x.

$$\log(3x - 8) = 1 \quad \text{means} \quad 10^1 = 3x - 8$$
$$10 = 3x - 8$$
$$18 = 3x \qquad \text{Add 8.}$$
$$6 = x \qquad \text{Divide by 3.}$$

Check $x = 6$ in the original equation. The solution set is $\{6\}$.

[YOU TRY 5] Solve $\log(12q + 16) = 2$.

We will study common logs in more depth in Section 11.5.

7 Use the Properties $\log_a a = 1$ and $\log_a 1 = 0$

There are a couple of properties of logarithms that can simplify our work.

If a is any real number, then $a^1 = a$. Furthermore, if $a \neq 0$, then $a^0 = 1$. Write $a^1 = a$ and $a^0 = 1$ in logarithmic form to obtain these two properties of logarithms:

> **Properties of Logarithms**
>
> If $a > 0$ and $a \neq 1$,
>
> 1) $\log_a a = 1$
> 2) $\log_a 1 = 0$

EXAMPLE 6

Use the properties of logarithms to evaluate each.

a) $\log_{12} 1$ b) $\log 10$

Solution

a) By Property 2, $\log_{12} 1 = 0$.

b) The base of $\log 10$ is 10. Therefore, $\log 10 = \log_{10} 10$. By Property 1, $\log 10 = 1$.

[YOU TRY 6] Use the properties of logarithms to evaluate each.

a) $\log_{1/3} 1$ b) $\log_{\sqrt{11}} \sqrt{11}$

8 Define and Graph a Logarithmic Function

Next we define a logarithmic function.

Definition
For $a > 0$, $a \neq 1$, and $x > 0$, $f(x) = \log_a x$ is the **logarithmic function with base a**.

Note
$f(x) = \log_a x$ can also be written as $y = \log_a x$. Changing $y = \log_a x$ to exponential form, we get $a^y = x$. Remembering that a is a positive number not equal to 1, it follows that

1) any real number may be substituted for y. Therefore, **the range of $y = \log_a x$ is $(-\infty, \infty)$**.

2) x must be a positive number. So, **the domain of $y = \log_a x$ is $(0, \infty)$**.

Let's return to the problem of finding the equation of the inverse of $f(x) = 2^x$ that was first introduced on p. 731.

EXAMPLE 7 Find the equation of the inverse of $f(x) = 2^x$.

Solution

Step 1: Replace $f(x)$ with y: $y = 2^x$

Step 2: Interchange x and y: $x = 2^y$

Step 3: Solve for y.

To solve $x = 2^y$ for y, write the equation in logarithmic form.

$$x = 2^y \quad \text{means} \quad y = \log_2 x$$

Step 4: Replace y with $f^{-1}(x)$.

$$f^{-1}(x) = \log_2 x$$

The inverse of the exponential function $f(x) = 2^x$ is $f^{-1}(x) = \log_2 x$.

[YOU TRY 7] Find the equation of the inverse of $f(x) = 6^x$.

Hint
Summarize this Note box in your notes.

Note
The inverse of the exponential function $f(x) = a^x$ (where $a > 0$, $a \neq 1$, and x is any real number) is $f^{-1}(x) = \log_a x$. Furthermore,

1) the domain of $f(x)$ is the range of $f^{-1}(x)$.
2) the range of $f(x)$ is the domain of $f^{-1}(x)$.

Their graphs are symmetric with respect to $y = x$.

To graph a logarithmic function, write it in exponential form first. Then make a table of values, plot the points, and draw the curve through the points.

EXAMPLE 8

Graph $f(x) = \log_2 x$.

Solution

Substitute y for $f(x)$ and write the equation in exponential form.

$$y = \log_2 x \quad \text{means} \quad 2^y = x$$

To make a table of values, it will be easier to *choose values for y* and compute the corresponding values of x. Remember to choose values of y that will give positive numbers, negative numbers, and zero in the exponent.

W Hint
This is the result from Example 7!

$2^y = x$	
x	y
1	0
2	1
4	2
8	3
$\frac{1}{2}$	-1
$\frac{1}{4}$	-2

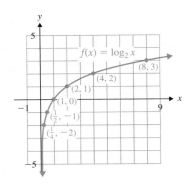

From the graph, we can see that the domain of f is $(0, \infty)$, and the range of f is $(-\infty, \infty)$.

YOU TRY 8

Graph $f(x) = \log_4 x$.

EXAMPLE 9

Graph $f(x) = \log_{1/3} x$.

Solution

Substitute y for $f(x)$ and write the equation in exponential form.

$$y = \log_{1/3} x \quad \text{means} \quad \left(\frac{1}{3}\right)^y = x$$

For the table of values, *choose values for y* and compute the corresponding values of x.

$\left(\frac{1}{3}\right)^y = x$	
x	y
1	0
$\frac{1}{3}$	1
$\frac{1}{9}$	2
3	-1
9	-2

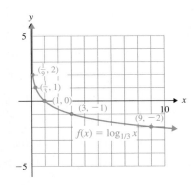

The domain of f is $(0, \infty)$, and the range is $(-\infty, \infty)$.

[YOU TRY 9] Graph $f(x) = \log_{1/4} x$.

The graphs in Examples 8 and 9 are typical of the graphs of logarithmic functions—Example 8 for functions where $a > 1$ and Example 9 for functions where $0 < a < 1$. Next is a summary of some characteristics of logarithmic functions.

Hint
Learn these characteristics, and summarize them in your notes.

Summary Characteristics of a Logarithmic Function $f(x) = \log_a x$, where $a > 0$ and $a \neq 1$

1) If $f(x) = \log_a x$ where $a > 1$, the value of y increases as the value of x increases.

2) If $f(x) = \log_a x$ where $0 < a < 1$, the value of y decreases as the value of x increases.

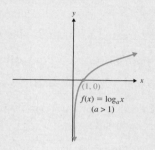
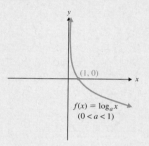

3) The function is one-to-one.
4) The x-intercept is $(1, 0)$.
5) The domain is $(0, \infty)$, and the range is $(-\infty, \infty)$.
6) The inverse of $f(x) = \log_a x$ is $f^{-1}(x) = a^x$.

Compare these characteristics of logarithmic functions to the characteristics of exponential functions on p. 722 in Section 11.2. The domain and range of logarithmic and exponential functions are interchanged because they are inverse functions.

9 Solve an Applied Problem Using a Logarithmic Equation

EXAMPLE 10 A hospital has found that the function $A(t) = 50 + 8 \log_2(t + 2)$ approximates the number of people treated each year since 1995 for severe allergic reactions to peanuts. If $t = 0$ represents the year 1995, answer the following.

a) How many people were treated in 1995?

b) How many people were treated in 2001?

c) In what year were approximately 82 people treated for allergic reactions to peanuts?

Solution

a) The year 1995 corresponds to $t = 0$. Let $t = 0$, and find $A(0)$.

$$A(0) = 50 + 8\log_2(0 + 2) \quad \text{Substitute 0 for } t.$$
$$= 50 + 8\log_2 2$$
$$= 50 + 8(1) \quad \log_2 2 = 1$$
$$= 58$$

In 1995, 58 people were treated for peanut allergies.

b) The year 2001 corresponds to $t = 6$. Let $t = 6$, and find $A(6)$.

$$A(6) = 50 + 8\log_2(6 + 2) \quad \text{Substitute 6 for } t.$$
$$= 50 + 8\log_2 8$$
$$= 50 + 8(3) \quad \log_2 8 = 3$$
$$= 50 + 24$$
$$= 74$$

In 2001, 74 people were treated for peanut allergies.

c) To determine in what year 82 people were treated, let $A(t) = 82$ and solve for t.

$$82 = 50 + 8\log_2(t + 2) \quad \text{Substitute 82 for } A(t).$$

To solve for t, we first need to get the term containing the logarithm on a side by itself. Subtract 50 from each side.

$$32 = 8\log_2(t + 2) \quad \text{Subtract 50.}$$
$$4 = \log_2(t + 2) \quad \text{Divide by 8.}$$
$$2^4 = t + 2 \quad \text{Write in exponential form.}$$
$$16 = t + 2$$
$$14 = t$$

$t = 14$ corresponds to the year 2009. (Add 14 to the year 1995.)
82 people were treated for peanut allergies in 2009.

[YOU TRY 10] The amount of garbage (in millions of pounds) collected in a certain town each year since 1990 can be approximated by $G(t) = 6 + \log_2(t + 1)$, where $t = 0$ represents the year 1990.

a) How much garbage was collected in 1990?

b) How much garbage was collected in 1997?

c) In what year would it be expected that 11,000,000 pounds of garbage will be collected? [Hint: Let $G(t) = 11$.]

ANSWERS TO [YOU TRY] EXERCISES

1) a) $3^4 = 81$ b) $5^{-2} = \dfrac{1}{25}$ c) $64^{1/2} = 8$ d) $13^1 = 13$

2) a) $\log_7 49 = 2$ b) $\log_5 \dfrac{1}{625} = -4$ c) $\log_{19} 1 = 0$ d) $\log_{144} 12 = \dfrac{1}{2}$

3) a) $\{32\}$ b) $\{13\}$ c) $\{2\}$ d) $\left\{\dfrac{1}{10}\right\}$ e) $\{38\}$

4) a) 2 b) -1 c) $\dfrac{1}{2}$ 5) $\{7\}$ 6) a) 0 b) 1   7) $f^{-1}(x) = \log_6 x$

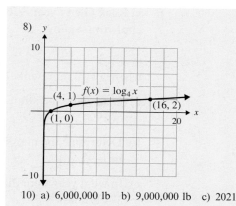
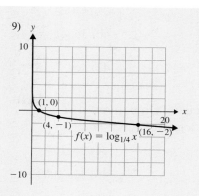

10) a) 6,000,000 lb   b) 9,000,000 lb   c) 2021

E Evaluate   11.3 Exercises

Do the exercises, and check your work.

Mixed Exercises: Objectives 1 and 2

1) In the equation $y = \log_a x$, a must be what kind of number?

2) In the equation $y = \log_a x$, x must be what kind of number?

3) What is the base of $y = \log x$?

4) A base 10 logarithm is called a _____ logarithm.

Write in exponential form.

5) $\log_7 49 = 2$

6) $\log_{11} 121 = 2$

7) $\log_9 \dfrac{1}{81} = -2$

8) $\log_8 \dfrac{1}{64} = -2$

9) $\log_{25} 5 = \dfrac{1}{2}$

10) $\log_{64} 4 = \dfrac{1}{3}$

11) $\log_9 1 = 0$

12) $\log_{13} 13 = 1$

Objective 3: Convert from Exponential Form to Logarithmic Form

Write in logarithmic form.

13) $9^2 = 81$

14) $12^2 = 144$

15) $2^{-5} = \dfrac{1}{32}$

16) $3^{-4} = \dfrac{1}{81}$

17) $10^1 = 10$

18) $10^0 = 1$

19) $169^{1/2} = 13$

20) $27^{1/3} = 3$

21) $\sqrt[3]{64} = 4$

22) $\sqrt[4]{81} = 3$

Mixed Exercises: Objectives 4 and 6

23) Explain how to solve a logarithmic equation of the form $\log_a b = c$.

24) A student solves $\log_x 9 = 2$ and gets the solution set $\{-3, 3\}$. Is this correct? Why or why not?

Solve each logarithmic equation.

Fill It In

Fill in the blanks with either the missing mathematical step or the reason for the given step.

25) $\log_2 x = 6$
    $2^6 = x$ _____
    _____ Solve for x.
    The solution set is ____.

26) $\log_5 t = -3$
    _____ Rewrite in exponential form.
    _____ Solve for t.
    The solution set is _____.

Solve each logarithmic equation.

27) $\log_5 k = 3$

28) $\log_{11} x = 2$

29) $\log w = 2$

30) $\log p = 5$

31) $\log_m 49 = 2$

32) $\log_x 4 = 2$

33) $\log_6 h = -2$

34) $\log_4 b = -3$

35) $\log_2(a + 2) = 4$

36) $\log_6(5y + 1) = 2$

37) $\log_{125} \sqrt{5} = c$

38) $\log_{16} \sqrt[5]{4} = k$

39) $\log_8 x = \dfrac{2}{3}$

40) $\log_{16} t = \dfrac{3}{4}$

41) $\log_{(3m-1)} 25 = 2$

42) $\log_{(y-1)} 4 = 2$

Mixed Exercises: Objectives 5–7
Evaluate each logarithm.

43) $\log_2 32$

44) $\log_4 64$

45) $\log 100$

46) $\log 1000$

47) $\log_{49} 7$

48) $\log_{36} 6$

49) $\log_8 \dfrac{1}{8}$

50) $\log_3 \dfrac{1}{3}$

51) $\log_5 5$

52) $\log_2 1$

53) $\log_{1/4} 16$

54) $\log_{1/3} 27$

Objective 8: Define and Graph a Logarithmic Function

55) Explain how to graph a logarithmic function of the form $f(x) = \log_a x$.

56) What are the domain and range of $f(x) = \log_a x$?

Graph each logarithmic function.

57) $f(x) = \log_2 x$

58) $f(x) = \log_5 x$

59) $f(x) = \log_{1/2} x$

60) $f(x) = \log_{1/3} x$

Find the inverse of each function.

61) $f(x) = 3^x$

62) $f(x) = 4^x$

63) $f(x) = \log_2 x$

64) $f(x) = \log_5 x$

Objective 9: Solve an Applied Problem Using a Logarithmic Equation
Solve each problem.

65) The function $L(t) = 1800 + 68 \log_3(t + 3)$ approximates the number of dog licenses issued by a city each year since 1980. If $t = 0$ represents the year 1980, answer the following.

   a) How many dog licenses were issued in 1980?

   b) How many were issued in 2004?

   c) In what year would it be expected that 2072 dog licenses will be issued?

66) Until the 1990s, Rock Glen was a rural community outside of a large city. In 1994, subdivisions of homes began to be built. The number of houses in Rock Glen t years after 1994 can be approximated by

$$H(t) = 142 + 58 \log_2(t + 1)$$

where $t = 0$ represents 1994.

   a) Determine the number of homes in Rock Glen in 1994.

   b) Determine the number of homes in Rock Glen in 1997.

   c) In what year were there approximately 374 homes?

67) A company plans to introduce a new type of cookie to the market. The company predicts that its sales over the next 24 months can be approximated by

$$S(t) = 14 \log_3(2t + 1)$$

where t is the number of months after the product is introduced, and $S(t)$ is in thousands of boxes of cookies.

   a) How many boxes of cookies were sold after they were on the market for 1 month?

   b) How many boxes were sold after they were on the market for 4 months?

   c) After 13 months, sales were approximately 43,000. Does this number fall short of, meet, or exceed the number of sales predicted by the formula?

68) Based on previous data, city planners have calculated that the number of tourists (in millions) to their city each year can be approximated by

$$N(t) = 10 + 1.2 \log_2(t + 2)$$

where t is the number of years after 1995.

   a) How many tourists visited the city in 1995?

   b) How many tourists visited the city in 2001?

   c) In 2009, actual data put the number of tourists at 14,720,000. How does this number compare to the number predicted by the formula?

R Rethink

R1) Why do you need logarithms?

R2) Could you explain to a classmate how to convert from logarithmic form to exponential form?

R3) Are log functions and exponential functions inverses of each other? How do you know?

R4) What is the most difficult part of learning logarithmic functions?

11.4 Properties of Logarithms

P Prepare / O Organize

What are your objectives for Section 11.4?	How can you accomplish each objective?
1 Use the Product Rule for Logarithms	• Learn the **Product Rule for Logarithms.** • Review the rules for exponents and understand how they relate to logarithms. • Complete the given examples on your own. • Complete You Trys 1 and 2.
2 Use the Quotient Rule for Logarithms	• Learn the **Quotient Rule for Logarithms.** • Complete the given examples on your own. • Complete You Trys 3 and 4.
3 Use the Power Rule for Logarithms	• Learn the **Power Rule for Logarithms.** • Review the rules for rewriting radicals as fractional exponents. • Complete the given example on your own. • Complete You Try 5.
4 Use the Properties $\log_a a^x = x$ and $a^{\log_a x} = x$	• Learn the **Other Properties of Logarithms.** • Complete the given example on your own. • Complete You Try 6.
5 Combine the Properties of Logarithms	• Review the **Summary of Properties of Logarithms,** and write the summary in your notes. • Review the list of common errors. • Write an explanation next to each step to help you remember the rules. • Complete the given examples on your own. • Complete You Trys 7–9.

W Work — Read the explanations, follow the examples, take notes, and complete the You Trys.

Logarithms have properties that are very useful in applications and in higher mathematics.

In this section, we will learn more properties of logarithms, and we will practice using them because they can make some very difficult mathematical calculations much easier. *The properties of logarithms come from the properties of exponents.*

1 Use the Product Rule for Logarithms

The product rule for logarithms can be derived from the product rule for exponents.

> **Property** The Product Rule for Logarithms
>
> Let x, y, and a be positive real numbers where $a \neq 1$. Then,
>
> $$\log_a xy = \log_a x + \log_a y$$
>
> The logarithm of a product, xy, is the same as the sum of the logarithms of each factor, x and y.

W Hint
Write this in your notes, and include an example.

 BE CAREFUL $\log_a xy \neq (\log_a x)(\log_a y)$

EXAMPLE 1

Rewrite as the sum of logarithms and simplify, if possible. Assume the variables represent positive real numbers.

a) $\log_6(4 \cdot 7)$ b) $\log_4 16t$ c) $\log_8 y^3$ d) $\log 10pq$

Solution

a) The logarithm of a product equals the *sum* of the logs of the factors. Therefore,

$$\log_6(4 \cdot 7) = \log_6 4 + \log_6 7 \quad \text{Product rule}$$

b) $\log_4 16t = \log_4 16 + \log_4 t$ Product rule
   $ = 2 + \log_4 t$ $\log_4 16 = 2$

Evaluate logarithms, like $\log_4 16$, when possible.

W Hint
Think about exponent rules when doing logarithm rules.

c) $\log_8 y^3 = \log_8(y \cdot y \cdot y)$ Write y^3 as $y \cdot y \cdot y$.
   $ = \log_8 y + \log_8 y + \log_8 y$ Product rule
   $ = 3 \log_8 y$

d) Recall that if no base is written, then it is assumed to be 10.

$$\log 10pq = \log 10 + \log p + \log q \quad \text{Product rule}$$
$$ = 1 + \log p + \log q \quad \log 10 = 1$$

[YOU TRY 1]

Rewrite as the sum of logarithms and simplify, if possible. Assume the variables represent positive real numbers.

a) $\log_9(2 \cdot 5)$ b) $\log_2 32k$ c) $\log_6 c^4$ d) $\log 100yz$

We can use the product rule for exponents in the "opposite" direction, too. That is, given the sum of logarithms we can write a single logarithm.

EXAMPLE 2

Write as a single logarithm. Assume the variables represent positive real numbers.

a) $\log_8 5 + \log_8 3$ b) $\log 7 + \log r$ c) $\log_3 x + \log_3(x + 4)$

Solution

a) $\log_8 5 + \log_8 3 = \log_8(5 \cdot 3)$ Product rule
$= \log_8 15$ $5 \cdot 3 = 15$

b) $\log 7 + \log r = \log 7r$ Product rule

c) $\log_3 x + \log_3(x + 4) = \log_3 x(x + 4)$ Product rule
$= \log_3(x^2 + 4x)$ Distribute.

BE CAREFUL $\log_a(x + y) \neq \log_a x + \log_a y$. Therefore, $\log_3(x^2 + 4x)$ does *not* equal $\log_3 x^2 + \log_3 4x$.

YOU TRY 2

Write as a single logarithm. Assume the variables represent positive real numbers.

a) $\log_5 9 + \log_5 4$ b) $\log_6 13 + \log_6 c$ c) $\log y + \log(y - 6)$

2 Use the Quotient Rule for Logarithms

The quotient rule for logarithms can be derived from the quotient rule for exponents.

Hint
Write this in your notes, and include an example.

Property The Quotient Rule for Logarithms

Let x, y, and a be positive real numbers where $a \neq 1$. Then,

$$\log_a \frac{x}{y} = \log_a x - \log_a y$$

The logarithm of a quotient, $\frac{x}{y}$, is the same as the logarithm of the numerator *minus* the logarithm of the denominator.

BE CAREFUL $\log_a \frac{x}{y} \neq \frac{\log_a x}{\log_a y}$.

EXAMPLE 3

Write as the difference of logarithms and simplify, if possible. Assume $w > 0$.

a) $\log_7 \dfrac{3}{10}$ b) $\log_3 \dfrac{81}{w}$

Solution

a) $\log_7 \dfrac{3}{10} = \log_7 3 - \log_7 10$ Quotient rule

b) $\log_3 \dfrac{81}{w} = \log_3 81 - \log_3 w$ Quotient rule

$\phantom{\log_3 \dfrac{81}{w}} = 4 - \log_3 w$ $\log_3 81 = 4$

[YOU TRY 3]

Write as the difference of logarithms and simplify, if possible. Assume $n > 0$.

a) $\log_6 \dfrac{2}{9}$ b) $\log_5 \dfrac{n}{25}$

EXAMPLE 4

Write as a single logarithm. Assume the variable is defined so that the expressions are positive.

a) $\log_2 18 - \log_2 6$ b) $\log_4(z - 5) - \log_4(z^2 + 9)$

Solution

a) $\log_2 18 - \log_2 6 = \log_2 \dfrac{18}{6}$ Quotient rule

$ = \log_2 3$ $\dfrac{18}{6} = 3$

b) $\log_4(z - 5) - \log_4(z^2 + 9) = \log_4 \dfrac{z - 5}{z^2 + 9}$ Quotient rule

$\log_a(x - y) \neq \log_a x - \log_a y$

[YOU TRY 4]

Write as a single logarithm. Assume the variable is defined so that the expressions are positive.

a) $\log_4 36 - \log_4 3$ b) $\log_5(c^2 - 2) - \log_5(c + 1)$

3 Use the Power Rule for Logarithms

In Example 1c), we saw that $\log_8 y^3 = 3 \log_8 y$ because

$$\log_8 y^3 = \log_8(y \cdot y \cdot y)$$
$$= \log_8 y + \log_8 y + \log_8 y$$
$$= 3 \log_8 y$$

This result can be generalized as the next property and comes from the power rule for exponents.

Hint
Write this in your notes, and include an example.

Property The Power Rule for Logarithms

Let x and a be positive real numbers, where $a \neq 1$, and let r be any real number. Then,

$$\log_a x^r = r \log_a x$$

 The rule applies to $\log_a x^r$ not $(\log_a x)^r$. Be sure you can distinguish between the two expressions.

EXAMPLE 5 Rewrite each expression using the power rule and simplify, if possible. Assume the variables represent positive real numbers and that the variable bases are positive real numbers not equal to 1.

a) $\log_9 y^4$ b) $\log_2 8^5$ c) $\log_a \sqrt{3}$ d) $\log_w \dfrac{1}{w}$

Solution

a) $\log_9 y^4 = 4 \log_9 y$ Power rule

b) $\log_2 8^5 = 5 \log_2 8$ Power rule
   $= 5(3)$ $\log_2 8 = 3$
   $= 15$ Multiply.

Hint
Are you writing out each step as you read the example?

c) *It is common practice to rewrite radicals as fractional exponents when applying the properties of logarithms. This will be our first step.*

$\log_a \sqrt{3} = \log_a 3^{1/2}$ Rewrite as a fractional exponent.
$= \dfrac{1}{2} \log_a 3$ Power rule

d) Rewrite $\dfrac{1}{w}$ as w^{-1}: $\log_w \dfrac{1}{w} = \log_w w^{-1}$ $\dfrac{1}{w} = w^{-1}$
$= -1 \log_w w$ Power rule
$= -1(1)$ $\log_w w = 1$
$= -1$ Multiply.

YOU TRY 5 Rewrite each expression using the power rule and simplify, if possible. Assume the variables represent positive real numbers and that the variable bases are positive real numbers not equal to 1.

a) $\log_8 t^9$ b) $\log_3 9^7$ c) $\log_a \sqrt[3]{5}$ d) $\log_m \dfrac{1}{m^8}$

The next properties we will look at can be derived from the power rule and from the fact that $f(x) = a^x$ and $g(x) = \log_a x$ are inverse functions.

4 Use the Properties $\log_a a^x = x$ and $a^{\log_a x} = x$

> **W Hint**
> Write this in your notes, and include an example.

Other Properties of Logarithms

Let a be a positive real number such that $a \neq 1$. Then,

1) $\log_a a^x = x$ for any real number x.
2) $a^{\log_a x} = x$ for $x > 0$.

EXAMPLE 6

Evaluate each expression.

a) $\log_6 6^7$ b) $\log 10^8$ c) $5^{\log_5 3}$

Solution

a) $\log_6 6^7 = 7$ $\log_a a^x = x$
b) $\log 10^8 = 8$ The base of the log is 10.
c) $5^{\log_5 3} = 3$ $a^{\log_a x} = x$

[YOU TRY 6]

Evaluate each expression.

a) $\log_3 3^{10}$ b) $\log 10^{-6}$ c) $7^{\log_7 9}$

Next is a summary of the properties of logarithms. The properties presented in Section 11.3 are included as well.

Summary Properties of Logarithms

Let x, y, and a be positive real numbers where $a \neq 1$, and let r be any real number. Then,

> **W Hint**
> Learn these properties! We will use them, again, in later sections.

1) $\log_a a = 1$
2) $\log_a 1 = 0$
3) $\log_a xy = \log_a x + \log_a y$ Product rule
4) $\log_a \dfrac{x}{y} = \log_a x - \log_a y$ Quotient rule
5) $\log_a x^r = r \log_a x$ Power rule
6) $\log_a a^x = x$ for any real number x
7) $a^{\log_a x} = x$

Many students make the same mistakes when working with logarithms. Keep in mind the following to avoid these common errors.

1) $\log_a xy \neq (\log_a x)(\log_a y)$
2) $\log_a(x + y) \neq \log_a x + \log_a y$
3) $\log_a \dfrac{x}{y} \neq \dfrac{\log_a x}{\log_a y}$
4) $\log_a(x - y) \neq \log_a x - \log_a y$
5) $(\log_a x)^r \neq r \log_a x$

5 Combine the Properties of Logarithms

Not only can the properties of logarithms simplify some very complicated computations, they are also needed for solving some types of logarithmic equations. The properties of logarithms are also used in calculus and many areas of science.

Next, we will see how to use different properties of logarithms together to rewrite logarithmic expressions.

EXAMPLE 7

Write each expression as the sum or difference of logarithms in simplest form. Assume all variables represent positive real numbers and that the variable bases are positive real numbers not equal to 1.

a) $\log_8 r^5 t$ b) $\log_3 \dfrac{27}{ab^2}$ c) $\log_7 \sqrt{7p}$ d) $\log_a(4a + 5)$

Solution

a) $\log_8 r^5 t = \log_8 r^5 + \log_8 t$ Product rule
$ = 5 \log_8 r + \log_8 t$ Power rule

b) $\log_3 \dfrac{27}{ab^2} = \log_3 27 - \log_3 ab^2$ Quotient rule
$\phantom{\log_3 \dfrac{27}{ab^2}} = 3 - (\log_3 a + \log_3 b^2)$ $\log_3 27 = 3$; product rule
$\phantom{\log_3 \dfrac{27}{ab^2}} = 3 - (\log_3 a + 2 \log_3 b)$ Power rule
$\phantom{\log_3 \dfrac{27}{ab^2}} = 3 - \log_3 a - 2 \log_3 b$ Distribute.

c) $\log_7 \sqrt{7p} = \log_7 (7p)^{1/2}$ Rewrite the radical as a fractional exponent.
$\phantom{\log_7 \sqrt{7p}} = \dfrac{1}{2} \log_7 (7p)$ Power rule
$\phantom{\log_7 \sqrt{7p}} = \dfrac{1}{2}(\log_7 7 + \log_7 p)$ Product rule
$\phantom{\log_7 \sqrt{7p}} = \dfrac{1}{2}(1 + \log_7 p)$ $\log_7 7 = 1$
$\phantom{\log_7 \sqrt{7p}} = \dfrac{1}{2} + \dfrac{1}{2} \log_7 p$ Distribute.

d) $\log_a(4a + 5)$ is in simplest form and cannot be rewritten using any properties of logarithms. [Recall that $\log_a(x + y) \neq \log_a x + \log_a y$.]

[YOU TRY 7]

Write each expression as the sum or difference of logarithms in simplest form. Assume all variables represent positive real numbers and that the variable bases are positive real numbers not equal to 1.

a) $\log_2 8s^2 t^5$ b) $\log_a \dfrac{4c^2}{b^3}$ c) $\log_5 \sqrt[3]{\dfrac{25}{n}}$ d) $\dfrac{\log_4 k}{\log_4 m}$

EXAMPLE 8

Write each as a single logarithm in simplest form. Assume the variable represents a positive real number.

a) $2 \log_7 5 + 3 \log_7 2$ b) $\dfrac{1}{2} \log_6 s - 3 \log_6(s^2 + 1)$

Solution

a) $\begin{aligned} 2 \log_7 5 + 3 \log_7 2 &= \log_7 5^2 + \log_7 2^3 && \text{Power rule} \\ &= \log_7 25 + \log_7 8 && 5^2 = 25;\ 2^3 = 8 \\ &= \log_7 (25 \cdot 8) && \text{Product rule} \\ &= \log_7 200 && \text{Multiply.} \end{aligned}$

b) $\begin{aligned} \dfrac{1}{2} \log_6 s - 3 \log_6(s^2 + 1) &= \log_6 s^{1/2} - \log_6 (s^2 + 1)^3 && \text{Power rule} \\ &= \log_6 \sqrt{s} - \log_6 (s^2 + 1)^3 && \text{Write in radical form.} \\ &= \log_6 \dfrac{\sqrt{s}}{(s^2 + 1)^3} && \text{Quotient rule} \end{aligned}$

YOU TRY 8

Write each as a single logarithm in simplest form. Assume the variables are defined so that the expressions are positive.

a) $2 \log 4 + \log 5$ b) $\dfrac{2}{3} \log_5 c + \dfrac{1}{3} \log_5 d - 2 \log_5(c - 6)$

Given the values of logarithms, we can compute the values of other logarithms using the properties we have learned in this section.

EXAMPLE 9

Given that $\log 6 \approx 0.7782$ and $\log 4 \approx 0.6021$, use the properties of logarithms to approximate the following.

a) $\log 24$ b) $\log \sqrt{6}$

Solution

a) To find the value of $\log 24$, we must determine how to write 24 in terms of 6 or 4 or some combination of the two. Because $24 = 6 \cdot 4$, we can write

$\begin{aligned} \log 24 &= \log(6 \cdot 4) && 24 = 6 \cdot 4 \\ &= \log 6 + \log 4 && \text{Product rule} \\ &\approx 0.7782 + 0.6021 && \text{Substitute.} \\ &= 1.3803 && \text{Add.} \end{aligned}$

b) We can write $\sqrt{6}$ as $6^{1/2}$.

$\begin{aligned} \log \sqrt{6} &= \log 6^{1/2} && \sqrt{6} = 6^{1/2} \\ &= \dfrac{1}{2} \log 6 && \text{Power rule} \\ &\approx \dfrac{1}{2}(0.7782) && \log 6 \approx 0.7782 \\ &= 0.3891 && \text{Multiply.} \end{aligned}$

YOU TRY 9 Using the values given in Example 9, use the properties of logarithms to approximate the following.

a) $\log 16$ b) $\log \dfrac{6}{4}$ c) $\log \sqrt[3]{4}$ d) $\log \dfrac{1}{6}$

ANSWERS TO YOU TRY EXERCISES

1) a) $\log_9 2 + \log_9 5$ b) $5 + \log_2 k$ c) $4\log_6 c$ d) $2 + \log y + \log z$ 2) a) $\log_5 36$
b) $\log_6 13c$ c) $\log(y^2 - 6y)$ 3) a) $\log_6 2 - \log_6 9$ b) $\log_5 n - 2$ 4) a) $\log_4 12$
b) $\log_5 \dfrac{c^2 - 2}{c + 1}$ 5) a) $9\log_8 t$ b) 14 c) $\dfrac{1}{3}\log_a 5$ d) -8 6) a) 10 b) -6 c) 9
7) a) $3 + 2\log_2 s + 5\log_2 t$ b) $\log_a 4 + 2\log_a c - 3\log_a b$ c) $\dfrac{2}{3} - \dfrac{1}{3}\log_5 n$
d) cannot be simplified   8) a) $\log 80$ b) $\log_5 \dfrac{\sqrt[3]{c^2 d}}{(c-6)^2}$ 9) a) 1.2042 b) 0.1761
c) 0.2007 d) -0.7782

E Evaluate 11.4 Exercises

Do the exercises, and check your work.

Mixed Exercises: Objectives 1–5

Decide whether each statement is true or false.

1) $\log_6 8c = \log_6 8 + \log_6 c$

2) $\log_5 \dfrac{m}{3} = \log_5 m - \log_5 3$

3) $\log_9 \dfrac{7}{2} = \dfrac{\log_9 7}{\log_9 2}$

4) $\log 1000 = 3$

5) $(\log_4 k)^2 = 2\log_4 k$

6) $\log_2(x^2 + 8) = \log_2 x^2 + \log_2 8$

7) $5^{\log_5 4} = 4$

8) $\log_3 4^5 = 5\log_3 4$

Write as the sum or difference of logarithms and simplify, if possible. Assume all variables represent positive real numbers.

Fill It In

Fill in the blanks with either the missing mathematical step or the reason for the given step.

9) $\log_5 25y$
   $\log_5 25y = \log_5 25 + \log_5 y$ _____
   $= $ _____   Evaluate $\log_5 25$.

10) $\log_3 \dfrac{81}{n^2}$

$\log_3 \dfrac{81}{n^2} = \log_3 81 - \log_3 n^2$ _____

$=$ _____   Evaluate $\log_3 81$; use power rule.

11) $\log_8(3 \cdot 10)$

12) $\log_2(6 \cdot 5)$

13) $\log_7 5d$

14) $\log_4 6w$

15) $\log_5 \dfrac{20}{17}$

16) $\log_9 \dfrac{4}{7}$

17) $\log_8 10^4$

18) $\log_5 2^3$

19) $\log p^8$

20) $\log_3 z^5$

21) $\log_3 \sqrt{7}$

22) $\log_7 \sqrt[3]{4}$

23) $\log_2 16p$

24) $\log_5 25t$

25) $\log_2 \dfrac{8}{k}$

26) $\log_3 \dfrac{x}{9}$

27) $\log_7 49^3$

28) $\log_8 64^{12}$

29) $\log 1000b$

30) $\log_3 27m$

31) $\log_2 2^9$

32) $\log_2 32^7$

33) $\log_5 \sqrt{5}$

34) $\log \sqrt[3]{10}$

35) $\log \sqrt[3]{100}$

36) $\log_2 \sqrt{8}$

37) $\log_6 w^4 z^3$

38) $\log_5 x^2 y$

39) $\log_7 \dfrac{a^2}{b^5}$

40) $\log_4 \dfrac{s^4}{t^6}$

41) $\log \dfrac{\sqrt[5]{11}}{y^2}$

42) $\log_3 \dfrac{\sqrt{x}}{y^4}$

43) $\log_2 \dfrac{4\sqrt{n}}{m^3}$

44) $\log_9 \dfrac{gf^2}{h^3}$

45) $\log_4 \dfrac{x^3}{yz^2}$

46) $\log \dfrac{3}{ab^2}$

47) $\log_5 \sqrt{5c}$

48) $\log_8 \sqrt[3]{\dfrac{z}{8}}$

49) $\log k(k-6)$

50) $\log_2 \dfrac{m^5}{m^2+3}$

Write as a single logarithm. Assume the variables are defined so that the variable expressions are positive and so that the bases are positive real numbers not equal to 1.

Fill It In
Fill in the blanks with either the missing mathematical step or the reason for the given step.

51) $2 \log_6 x + \log_6 y$

$2 \log_6 x + \log_6 y = \log_6 x^2 + \log_6 y$ _____

$= $ _____ Product rule

52) $5 \log 2 + \log c - 3 \log d$

$5 \log 2 + \log c - 3 \log d$

$= $ _____ Power rule

$= \log 32 + \log c - \log d^3$ _____

$= $ _____ Product rule

$= \log \dfrac{32c}{d^3}$ _____

53) $\log_a m + \log_a n$

54) $\log_4 7 + \log_4 x$

55) $\log_7 d - \log_7 3$

56) $\log_p r - \log_p s$

57) $4 \log_3 f + \log_3 g$

58) $5 \log_y m + 2 \log_y n$

59) $\log_8 t + 2 \log_8 u - 3 \log_8 v$

60) $3 \log a + 4 \log c - 6 \log b$

61) $\log(r^2 + 3) - 2 \log(r^2 - 3)$

62) $2 \log_2 t - 3 \log_2(5t + 1)$

63) $3 \log_n 2 + \dfrac{1}{2} \log_n k$

64) $2 \log_z 9 + \dfrac{1}{3} \log_z w$

65) $\dfrac{1}{3} \log_d 5 - 2 \log_d z$

66) $\dfrac{1}{2} \log_5 a - 4 \log_5 b$

67) $\log_6 y - \log_6 3 - 3 \log_6 z$

68) $\log_7 8 - 4 \log_7 x - \log_7 y$

69) $4 \log_3 t - 2 \log_3 6 - 2 \log_3 u$

70) $2 \log_9 m - 4 \log_9 2 - 4 \log_9 n$

71) $\dfrac{1}{2} \log_b (c + 4) - 2 \log_b (c + 3)$

72) $\dfrac{1}{2} \log_a r + \dfrac{1}{2} \log_a (r - 2) - \log_a (r + 2)$

73) $\log(a^2 + b^2) - \log(a^4 - b^4)$

74) $\log_n(x^3 - y^3) - \log_n(x - y)$

Given that $\log 5 \approx 0.6990$ and $\log 9 \approx 0.9542$, use the properties of logarithms to approximate the following. **Do not use a calculator.**

75) $\log 45$

76) $\log 25$

77) $\log 81$

78) $\log \dfrac{9}{5}$

79) $\log \dfrac{5}{9}$

80) $\log \sqrt{5}$

81) $\log 3$

82) $\log \dfrac{1}{9}$

83) $\log \dfrac{1}{5}$

84) $\log 5^8$

85) $\log \dfrac{1}{81}$

86) $\log 90$

87) $\log 50$

88) $\log \dfrac{25}{9}$

89) Since $8 = (-4)(-2)$, can we use the properties of logarithms in the following way? Explain.

$\log_2 8 = \log_2(-4)(-2)$
$\qquad = \log_2(-4) + \log_2(-2)$

90) Derive the product rule for logarithms from the product rule for exponents. Assume a, x, and y are positive real numbers with $a \neq 1$. Let $a^m = x$ so that $\log_a x = m$, and let $a^n = y$ so that $\log_a y = n$. Since $a^m \cdot a^n = xy$, show that $\log_a xy = \log_a x + \log_a y$.

R Rethink

R1) How are the rules of exponents and the properties of logarithms similar? How are they different?

R2) Could you write all the rules without looking at your notes?

R3) Which rule is the hardest to remember?

11.5 Common and Natural Logarithms and Change of Base

P Prepare

What are your objectives for Section 11.5?	How can you accomplish each objective?
1 Evaluate Common Logarithms Using a Calculator	• Know the definition of a *common logarithm*. • Complete the given example on your own. • Complete You Try 1.
2 Solve an Equation Containing a Common Logarithm	• Review converting from the logarithmic form of an equation to the exponential form of an equation. • Know what it means to find an exact solution and an approximate solution. • Complete the given example on your own. • Complete You Try 2.
3 Solve an Applied Problem Given an Equation Containing a Common Logarithm	• Read the problem carefully. • Complete the given example on your own. • Complete You Try 3.
4 Define and Evaluate a Natural Logarithm	• Learn the definition of a *natural logarithm,* and write it in your notes. • Complete the given example on your own. • Complete You Try 4.
5 Graph a Natural Logarithm Function	• Use a table of values to find ordered pairs. • Review how to find the domain and range of a function. • Complete the given example on your own. • Complete You Try 5.

(continued)

What are your objectives for Section 11.5?	How can you accomplish each objective?
6 Solve an Equation Containing a Natural Logarithm	• Review converting from the logarithmic form of an equation to the exponential form of an equation. • Know what it means to find an exact solution and an approximate solution. • Complete the given example on your own. • Complete You Try 6.
7 Solve Applied Problems Using Exponential Functions	• Understand how to use the *compound interest* formula. • Understand *continuous compounding,* and learn the formula. • Complete the given examples on your own. • Complete You Trys 7 and 8.
8 Use the Change-of-Base Formula	• Learn the **Change-of-Base Formula,** and write it in your own words. • Practice using either base 10 or base *e*. • Complete the given example on your own. • Complete You Try 9.

 Read the explanations, follow the examples, take notes, and complete the You Trys.

In this section, we will focus our attention on two widely used logarithmic bases—base 10 and base *e*.

1 Evaluate Common Logarithms Using a Calculator

In Section 11.3, we said that a base 10 logarithm is called a **common logarithm.** It is often written as log *x*.

$$\log x \text{ means } \log_{10} x$$

We can evaluate many logarithms without the use of a calculator because we can write them in terms of a base of 10.

For example, $\log 1000 = 3$ and $\log \dfrac{1}{100} = -2$.

Common logarithms are used throughout mathematics and other fields to make calculations easier to solve in applications. Often, however, we need a calculator to evaluate the logarithms. Next we will learn how to use a calculator to find the value of a base 10 logarithm. **We will approximate the value to four decimal places.**

W Hint
Expressions are entered into different calculators in different ways. Be sure you know how to use yours.

EXAMPLE 1

Find log 12.

Solution

Enter 12 [LOG] or [LOG] 12 [ENTER] into your calculator.

$$\log 12 \approx 1.0792$$

(Note that $10^{1.0792} \approx 12$. Press 10 $[y^x]$ 1.0792 $[=]$ to evaluate $10^{1.0792}$.)

[YOU TRY 1] Find log 3.

We can solve logarithmic equations with or without the use of a calculator.

2 Solve an Equation Containing a Common Logarithm

For the equation in Example 2, we will give an exact solution *and* a solution that is approximated to four decimal places. This will give us an idea of the size of the exact solution.

EXAMPLE 2

Solve log x = 2.4. Give an exact solution and a solution that is approximated to four decimal places.

Solution

Change to exponential form, and solve for x.

$$\log x = 2.4 \quad \text{means} \quad \log_{10} x = 2.4$$
$$10^{2.4} = x \quad \text{Exponential form}$$
$$251.1886 \approx x \quad \text{Approximation}$$

The exact solution is $\{10^{2.4}\}$. This is approximately $\{251.1886\}$.

[YOU TRY 2] Solve log x = 0.7. Give an exact solution and a solution that is approximated to four decimal places.

3 Solve an Applied Problem Given an Equation Containing a Common Logarithm

EXAMPLE 3

The loudness of sound, $L(I)$ in decibels (dB), is given by

$$L(I) = 10 \log \frac{I}{10^{-12}}$$

where I is the intensity of sound in watts per square meter (W/m²). Fifty meters from the stage at a concert, the intensity of sound is 0.01 W/m². Find the loudness of the music at the concert 50 m from the stage.

Solution

Substitute 0.01 for I and find $L(0.01)$.

$$L(0.01) = 10 \log \frac{0.01}{10^{-12}}$$
$$= 10 \log \frac{10^{-2}}{10^{-12}} \quad 0.01 = 10^{-2}$$
$$= 10 \log 10^{10} \quad \text{Quotient rule for exponents}$$
$$= 10(10) \quad \log 10^{10} = 10$$
$$= 100$$

The sound level of the music 50 m from the stage is 100 dB. (To put this in perspective, a normal conversation has a loudness of about 50 dB.)

[**YOU TRY 3**] The intensity of sound from a thunderstorm is about 0.001 W/m². Find the loudness of the storm, in decibels.

4 Define and Evaluate a Natural Logarithm

Another base that is often used for logarithms is the number e. In Section 11.2, we said that e, like π, is an irrational number. To four decimal places, $e \approx 2.7183$.

A base e logarithm is called a **natural logarithm** or **natural log.** The notation used for a base e logarithm is ln x (read as "*the natural log of x*" or "*ln of x*"). Since it is a base e logarithm, it is important to remember that

$$\ln x \text{ means } \log_e x$$

Using the properties $\log_a a^x = x$ and $\log_a a = 1$, we can find the value of some natural logarithms without using a calculator. For some natural logs, we will use a calculator to approximate the value to four decimal places.

W Hint
In your notes, summarize this information about the natural logarithm, ln x.

EXAMPLE 4

Evaluate.

a) ln e b) ln e^2 c) ln 5

Solution

a) To evaluate ln e, remember that $\ln e = \log_e e = 1$ since $\log_a a = 1$.

$$\ln e = 1$$

This is a value you should remember. We will use this in Section 11.6 to solve exponential equations with base e.

b) $\ln e^2 = \log_e e^2 = 2 \quad \log_a a^x = x$

c) We can use a calculator to *approximate natural logarithms to four decimal places* if the properties do not give us an exact value.

Enter 5 LN or LN 5 ENTER into your calculator.

$$\ln 5 \approx 1.6094$$

W Hint
Remember that ln $e = 1$. We will use it when solving some equations.

[**YOU TRY 4**] Evaluate.

a) 5 ln e b) ln e^8 c) ln 9

5 Graph a Natural Logarithm Function

We can graph $y = \ln x$ by substituting values for x and using a calculator to approximate the values of y.

EXAMPLE 5

Graph $y = \ln x$. Determine the domain and range.

Solution

Choose values for x, and use a calculator to approximate the corresponding values of y. Remember that $\ln e = 1$, so e is a good choice for x.

x	y
1	0
$e \approx 2.72$	1
6	1.79
0.5	−0.69
0.25	−1.39

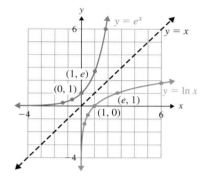

The domain of $y = \ln x$ is $(0, \infty)$, and the range is $(-\infty, \infty)$.

The graph of the inverse of $y = \ln x$ is also shown. We can obtain the graph of the inverse of $y = \ln x$ by reflecting the graph about the line $y = x$. The inverse of $y = \ln x$ is $y = e^x$.

Notice that the domain of $y = e^x$ is $(-\infty, \infty)$, while the range is $(0, \infty)$, the opposite of the domain and range of $y = \ln x$. This is a direct result of the relationship between a function and its inverse.

[YOU TRY 5] Graph $y = \ln(x + 4)$. Determine the domain and range.

W Hint
Remember that $y = \ln x$ is equivalent to $y = \log_e x$.

It is important to remember that $y = \ln x$ means $y = \log_e x$. Understanding this relationship allows us to make the following connections:

$$y = \ln x \text{ is equivalent to } y = \log_e x$$

and

$$y = \log_e x \text{ can be written in exponential form as } e^y = x.$$

Therefore, $y = \ln x$ can be written in exponential form as $e^y = x$.

We can use this relationship to show that the inverse of $y = \ln x$ is $y = e^x$. (You will be asked to verify this relationship in Exercise 83.) Also, in Example 5, notice that the graphs of $y = \ln x$ and $y = e^x$ are symmetric with respect to $y = x$.

6 Solve an Equation Containing a Natural Logarithm

Note

To solve an equation containing a natural logarithm, like ln x = 4, we change to exponential form and solve for the variable. We can give an exact solution and a solution that is approximated to four decimal places.

EXAMPLE 6

Solve each equation. Give an exact solution and a solution that is approximated to four decimal places.

a) $\ln x = 4$ b) $\ln(2x + 5) = 3.8$

Solution

a) $\ln x = 4$ means $\log_e x = 4$

$$e^4 = x \quad \text{Exponential form}$$
$$54.5982 \approx x \quad \text{Approximation}$$

The exact solution is $\{e^4\}$. This is approximately $\{54.5982\}$.

b) $\ln(2x + 5) = 3.8$ means $\log_e(2x + 5) = 3.8$

$$e^{3.8} = 2x + 5 \quad \text{Exponential form}$$
$$e^{3.8} - 5 = 2x \quad \text{Subtract 5.}$$
$$\frac{e^{3.8} - 5}{2} = x \quad \text{Divide by 2.}$$
$$19.8506 \approx x \quad \text{Approximation}$$

The exact solution is $\left\{\dfrac{e^{3.8} - 5}{2}\right\}$. This is approximately $\{19.8506\}$.

YOU TRY 6

Solve each equation. Give an exact solution and a solution that is approximated to four decimal places.

a) $\ln y = 2.7$ b) $\ln(3a - 1) = 0.5$

7 Solve Applied Problems Using Exponential Functions

One of the most practical applications of exponential functions is for compound interest.

Definition

Compound Interest: The amount of money, A, in dollars, in an account after t years is given by

$$A = P\left(1 + \frac{r}{n}\right)^{nt}$$

where P (the principal) is the amount of money (in dollars) deposited in the account, r is the annual interest rate, and n is the number of times the interest is compounded (paid) per year.

> **Note**
> We can also think of this formula in terms of the amount of money owed, A, after t yr when P is the amount of money loaned.

EXAMPLE 7

If $2000 is deposited in an account paying 4% per year, find the total amount in the account after 5 yr if the interest is compounded

a) quarterly.   b) monthly.

(We assume no withdrawals or additional deposits are made.)

Solution

a) If interest compounds quarterly, then interest is paid four times per year. Use

$$A = P\left(1 + \frac{r}{n}\right)^{nt}$$

W Hint
To use the interest rate in the formula, you must change the percent to a decimal.

with $P = 2000$, $r = 0.04$, $t = 5$, $n = 4$.

$$A = 2000\left(1 + \frac{0.04}{4}\right)^{4(5)}$$
$$= 2000(1.01)^{20}$$
$$\approx 2440.3801$$

Because A is an amount of money, round to the nearest cent. The account will contain $2440.38 after 5 yr.

b) If interest is compounded monthly, then interest is paid 12 times per year. Use

$$A = P\left(1 + \frac{r}{n}\right)^{nt}$$

with $P = 2000$, $r = 0.04$, $t = 5$, $n = 12$.

$$A = 2000\left(1 + \frac{0.04}{12}\right)^{12(5)}$$
$$\approx 2441.9932$$

Round A to the nearest cent. The account will contain $2441.99 after 5 yr.

[YOU TRY 7]

If $1500 is deposited in an account paying 5% per year, find the total amount in the account after 8 yr if the interest is compounded

a) monthly.   b) weekly.

In Example 7 we saw that the account contained more money after 5 yr when the interest compounded monthly (12 times per year) versus quarterly (four times per year). This will always be true. The more often interest is compounded each year, the more money that accumulates in the account.

If interest *compounds continuously*, we obtain the formula for *continuous compounding*, $A = Pe^{rt}$.

Definition

Continuous Compounding: If P dollars is deposited in an account earning interest rate r compounded continuously, then the amount of money, A (in dollars), in the account after t years is given by

$$A = Pe^{rt}$$

EXAMPLE 8 Determine the amount of money in an account after 5 yr if $2000 was initially invested at 4% compounded continuously.

Solution

Use $A = Pe^{rt}$ with $P = 2000$, $r = 0.04$, and $t = 5$.

$$\begin{aligned} A &= 2000e^{0.04(5)} &&\text{Substitute values.} \\ &= 2000e^{0.20} &&\text{Multiply } (0.04)(5). \\ &\approx 2442.8055 &&\text{Evaluate using a calculator.} \end{aligned}$$

Round A to the nearest cent.

The account will contain $2442.81 after 5 yr. Note that, as expected, this is more than the amounts obtained in Example 7 when the same amount was deposited for 5 yr at 4% but the interest was compounded quarterly and monthly.

[YOU TRY 8] Determine the amount of money in an account after 8 yr if $1500 was initially invested at 5% compounded continuously.

8 Use the Change-of-Base Formula

Sometimes, we need to find the value of a logarithm with a base other than 10 or e—like $\log_3 7$. Some calculators, however, do not calculate logarithms other than common logs (base 10) and natural logs (base e). In such cases, we can use the change-of-base formula to evaluate logarithms with bases other than 10 or e.

Definition

Change-of-Base Formula: If a, b, and x are positive real numbers and $a \neq 1$ and $b \neq 1$, then

$$\log_a x = \frac{\log_b x}{\log_b a}$$

Note

We can choose any positive real number not equal to 1 for b, but it is most convenient to choose 10 or e since these will give us common logarithms and natural logarithms, respectively.

EXAMPLE 9

Find the value of $\log_3 7$ to four decimal places using

a) common logarithms. b) natural logarithms.

Solution

a) The base we will use to evaluate $\log_3 7$ is 10; this is the base of a common logarithm. Then

$$\log_3 7 = \frac{\log_{10} 7}{\log_{10} 3} \quad \text{Change-of-base formula}$$
$$\approx 1.7712 \quad \text{Use a calculator.}$$

b) The base of a natural logarithm is e. Then

$$\log_3 7 = \frac{\log_e 7}{\log_e 3}$$
$$= \frac{\ln 7}{\ln 3}$$
$$\approx 1.7712 \quad \text{Use a calculator.}$$

Using either base 10 or base e gives us the same result.

W Hint
Decide whether you prefer to use a common log or a natural log in the change-of-base formula.

YOU TRY 9

Find the value of $\log_5 38$ to four decimal places using

a) common logarithms. b) natural logarithms.

Using Technology

Graphing calculators will graph common logarithmic functions and natural logarithmic functions directly using the log or LN keys.

For example, let's graph $f(x) = \ln x$.

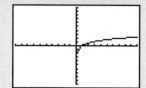

To graph a logarithmic function with a base other than 10 or e, it is necessary to use the change-of-base formula. For example, to graph the function $f(x) = \log_2 x$, first rewrite the function as a quotient of natural logarithms or common logarithms: $f(x) = \log_2 x = \dfrac{\ln x}{\ln 2}$ or $\dfrac{\log x}{\log 2}$. Enter one of these quotients in Y_1 and press GRAPH to graph as shown below. To illustrate that the same graph results in either case, trace to the point where $x = 3$.

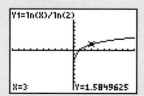

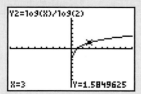

SECTION 11.5 Common and Natural Logarithms and Change of Base

Graph the following functions using a graphing calculator.

1) $f(x) = \log_3 x$ 2) $f(x) = \log_5 x$ 3) $f(x) = 4\log_2 x + 1$

4) $f(x) = \log_2(x - 3)$ 5) $f(x) = 2 - \log_4 x$ 6) $f(x) = 3 - \log_2(x + 1)$

ANSWERS TO [YOU TRY] EXERCISES

1) 0.4771   2) $\{10^{0.7}\}$; $\{5.0119\}$ 3) 90 dB   4) a) 5 b) 8 c) 2.1972
5) domain: $(-4, \infty)$; range: $(-\infty, \infty)$

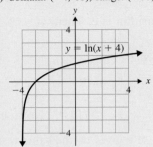

6) a) $\{e^{2.7}\}$; $\{14.8797\}$ b) $\left\{\dfrac{e^{0.5} + 1}{3}\right\}$; $\{0.8829\}$
7) a) $2235.88 b) $2237.31   8) $2237.74
9) a) 2.2602 b) 2.2602

ANSWERS TO TECHNOLOGY EXERCISES

1) 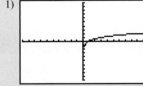 2)  3)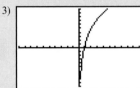

4) 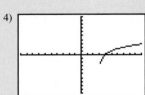 5) 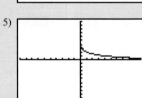 6)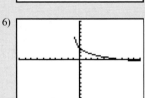

E Evaluate 11.5 Exercises

Do the exercises, and check your work.

Mixed Exercises: Objectives 1 and 4

1) What is the base of $\ln x$?
2) What is the base of $\log x$?

Evaluate each logarithm. Do *not* use a calculator.

 3) $\log 100$ 4) $\log 10{,}000$
5) $\log \dfrac{1}{1000}$ 6) $\log \dfrac{1}{100{,}000}$
7) $\log 0.1$ 8) $\log 0.01$
9) $\log 10^9$ 10) $\log 10^7$
11) $\log \sqrt[4]{10}$ 12) $\log \sqrt[5]{10}$
13) $\ln e^{10}$ 14) $\ln e^6$
15) $\ln \sqrt{e}$ 16) $\ln \sqrt[3]{e}$
17) $\ln \dfrac{1}{e^5}$ 18) $\ln \dfrac{1}{e^2}$
19) $\ln 1$ 20) $\log 1$

Mixed Exercises: Objectives 1 and 4
Use a calculator to find the approximate value of each logarithm to four decimal places.

21) log 16
22) log 23
23) log 0.5
24) log 627
25) ln 3
26) ln 6
27) ln 1.31
28) ln 0.218

Objective 2: Solve an Equation Containing a Common Logarithm
Solve each equation. Do *not* use a calculator.

29) log x = 3
30) log z = 5
31) log k = −1
32) log c = −2
33) log(4a) = 2
34) log(5w) = 1
35) log(3t + 4) = 1
36) log(2p + 12) = 2

Mixed Exercises: Objectives 2 and 6
Solve each equation. Give an exact solution and a solution that is approximated to four decimal places.

37) log a = 1.5
38) log y = 1.8
39) log r = 0.8
40) log c = 0.3
41) ln x = 1.6
42) ln p = 1.1
43) ln t = −2
44) ln z = 0.25
45) ln(3q) = 2.1
46) $\ln\left(\frac{1}{4}m\right) = 3$
47) $\log\left(\frac{1}{2}c\right) = 0.47$
48) log(6k) = −1
49) log(5y − 3) = 3.8
50) log(8x + 15) = 2.7
51) ln(10w + 19) = 1.85
52) ln(7a − 4) = 0.6
53) ln(2d − 5) = 0
54) log(3t + 14) = 2.4

Objective 5: Graph a Natural Logarithm Function
Graph each function. State the domain and range.

55) f(x) = ln x − 3
56) y = ln x + 2
57) g(x) = ln(x + 3)
58) f(x) = ln(x − 2)
59) y = −ln x
60) f(x) = ln(−x)
61) h(x) = log x
62) k(x) = log(x + 4)

63) If you are given the graph of f(x) = ln x, how could you obtain the graph of g(x) = ln(x + 5) without making a table of values and plotting points?

64) If you are given the graph of f(x) = ln x, how could you obtain the graph of h(x) = ln x + 4 without making a table of values and plotting points?

Objective 8: Use the Change-of-Base Formula
Use the change-of-base formula with either base 10 or base *e* to approximate each logarithm to four decimal places.

65) $\log_2 13$
66) $\log_6 25$
67) $\log_{1/3} 16$
68) $\log_5 3$

Mixed Exercises: Objectives 3 and 7
For Exercises 69 and 70, use the formula

$$L(I) = 10 \log \frac{I}{10^{-12}}$$

where *I* is the intensity of sound, in watts per square meter, and L(I) is the loudness of sound in decibels. Do *not* use a calculator.

69) The intensity of sound from a dishwasher is about 0.000001 W/m². Find the loudness of the dishwasher, in decibels.

70) The intensity of sound from the takeoff of a space shuttle is 1,000,000 W/m². Find the loudness of the sound made by the space shuttle at takeoff, in decibels.

Use the formula $A = P\left(1 + \frac{r}{n}\right)^{nt}$ to solve each problem. See Example 7.

71) Isabel deposits $3000 in an account earning 5% per year compounded monthly. How much will be in the account after 3 yr?

72) How much will Anna owe at the end of 4 yr if she borrows $5000 at a rate of 7.2% compounded weekly?

Use the formula $A = Pe^{rt}$ to solve each problem. See Example 8.

73) If $3000 is deposited in an account earning 5% compounded continuously, how much will be in the account after 3 yr?

74) Find the amount Nadia owes at the end of 5 yr if she borrows $4500 at a rate of 6.8% compounded continuously.

75) The number of bacteria, N(t), in a culture *t* hr after the bacteria are placed in a dish is given by

$$N(t) = 5000e^{0.0617t}$$

a) How many bacteria were originally in the culture?

b) How many bacteria are present after 8 hr?

76) The number of bacteria, $N(t)$, in a culture t hr after the bacteria are placed in a dish is given by

$$N(t) = 8000e^{0.0342t}$$

a) How many bacteria were originally in the culture?

b) How many bacteria are present after 10 hr?

77) The function $N(t) = 10{,}000e^{0.0492t}$ describes the number of bacteria in a culture t hr after 10,000 bacteria were placed in the culture. How many bacteria are in the culture after 1 day?

78) How many bacteria are present 2 days after 6000 bacteria are placed in a culture if the number of bacteria in the culture is

$$N(t) = 6000e^{0.0285t}$$

t hr after the bacteria are placed in a dish?

In chemistry, the pH of a solution is given by

$$\text{pH} = -\log[\text{H}^+]$$

where $[\text{H}^+]$ is the molar concentration of the hydronium ion. A neutral solution has pH = 7. *Acidic solutions* have pH < 7, and *basic solutions* have pH > 7.

For Exercises 79–82, the hydronium ion concentrations, $[\text{H}^+]$, are given for some common substances. Find the pH of each substance (to the tenths place), and determine whether each substance is acidic or basic.

79) Cola: $[\text{H}^+] = 2 \times 10^{-3}$

80) Tomatoes: $[\text{H}^+] = 1 \times 10^{-4}$

81) Ammonia: $[\text{H}^+] = 6 \times 10^{-12}$

82) Egg white: $[\text{H}^+] = 2 \times 10^{-8}$

Extension

83) Show that the inverse of $y = \ln x$ is $y = e^x$.

R Rethink

R1) What is the difference between a common logarithm and a natural logarithm?

R2) When using the change-of-base formula, can you use either base 10 or base e?

R3) Which part of using the calculator is hardest for you when evaluating logarithms?

11.6 Solving Exponential and Logarithmic Equations

P Prepare
What are your objectives for Section 11.6?

1 Solve an Exponential Equation

2 Solve Logarithmic Equations Using the Properties of Logarithms

3 Solve Applied Problems Involving Exponential Functions Using a Calculator

4 Solve an Applied Problem Involving Exponential Growth or Decay

O Organize
How can you accomplish each objective?

- Review the properties of logarithms.
- Learn the procedure for **Solving an Exponential Equation,** and summarize it in your notes.
- Complete the given examples on your own.
- Complete You Trys 1 and 2.

- Learn the procedure for **How to Solve an Equation Where Each Term Contains a Logarithm.**
- Learn the procedure for **How to Solve an Equation Where One Term Does Not Contain a Logarithm.**
- Complete the given examples on your own.
- Complete You Trys 3 and 4.

- Review the formula for **Continuous Compounding.**
- Complete the given examples on your own.
- Complete You Trys 5 and 6.

- Understand and learn the formula for **Exponential Growth or Decay.**
- Understand the meaning of *half-life*.
- Complete the given example on your own.
- Complete You Try 7.

Read the explanations, follow the examples, take notes, and complete the You Trys.

In this section, we will learn another property of logarithms that will allow us to solve additional types of exponential and logarithmic equations.

Properties for Solving Exponential and Logarithmic Equations

Let a, x, and y be positive, real numbers, where $a \neq 1$.
1) If $x = y$, then $\log_a x = \log_a y$.
2) If $\log_a x = \log_a y$, then $x = y$.

For example, 1) tells us that if $x = 3$, then $\log_a x = \log_a 3$. Likewise, 2) tells us that if $\log_a 5 = \log_a y$, then $5 = y$. We can use the properties above to solve exponential and logarithmic equations that we could not solve previously.

1 Solve an Exponential Equation

We will look at two types of exponential equations—equations where both sides *can* be expressed with the same base and equations where both sides *cannot* be expressed with the same base. If the two sides of an exponential equation *cannot* be expressed with the same base, we will use logarithms to solve the equation.

EXAMPLE 1

Solve.

a) $2^x = 8$ b) $2^x = 12$

Solution

a) Because 8 is a power of 2, we can solve $2^x = 8$ by expressing each side of the equation with the same base and setting the exponents equal to each other.

$$2^x = 8$$
$$2^x = 2^3 \quad 8 = 2^3$$
$$x = 3 \quad \text{Set the exponents equal.}$$

W Hint
Write the solutions to these equations in your notes. Explain, in your own words, why they are solved differently.

The solution set is $\{3\}$.

b) Can we express both sides of $2^x = 12$ with the same base? *No. We will use property* 1) *to solve* $2^x = 12$ *by taking the logarithm of each side.*

We can use a logarithm of *any* base. It is most convenient to use base 10 (common logarithm) or base e (natural logarithm) because this is what we can find most easily on our calculators. *We will take the natural log of both sides.*

$$2^x = 12$$
$$\ln 2^x = \ln 12 \quad \text{Take the natural log of each side.}$$
$$x \ln 2 = \ln 12 \quad \log_a x^r = r \log_a x$$
$$x = \frac{\ln 12}{\ln 2} \quad \text{Divide by } \ln 2.$$

The exact solution is $\left\{\dfrac{\ln 12}{\ln 2}\right\}$. Use a calculator to get an approximation to four decimal places: $x \approx 3.5850$.

The approximation is $\{3.5850\}$. We can verify the solution by substituting it for x in $2^x = 12$: $2^{3.5850} \approx 12$.

BE CAREFUL The exact solution is written as $\left\{\dfrac{\ln 12}{\ln 2}\right\}$. Recall that $\dfrac{\ln 12}{\ln 2} \neq \ln 6$.

Procedure Solving an Exponential Equation

Begin by asking yourself, "Can I express each side with the same base?"

1) If the answer is **yes**, then write each side of the equation with the same base, set the exponents equal, and solve for the variable.

2) If the answer is **no**, then take the natural logarithm of each side, use the properties of logarithms, and solve for the variable.

W Hint
Get in the habit of asking yourself this question.

[YOU TRY 1] Solve.

a) $3^{a-5} = 9$ b) $3^t = 24$

EXAMPLE 2 Solve.

a) $5^{x-2} = 16$ b) $e^{5n} = 4$

Solution

a) Ask yourself, *"Can I express each side with the same base?"* **No.** Therefore, take the natural log of each side.

$$5^{x-2} = 16$$
$$\ln 5^{x-2} = \ln 16 \qquad \text{Take the natural log of each side.}$$
$$(x - 2)\ln 5 = \ln 16 \qquad \log_a x^r = r \log_a x$$

$(x - 2)$ *must* be in parentheses because it contains two terms.

$$x \ln 5 - 2 \ln 5 = \ln 16 \qquad \text{Distribute.}$$
$$x \ln 5 = \ln 16 + 2 \ln 5 \qquad \text{Add 2 ln 5 to get the } x\text{-term by itself.}$$
$$x = \frac{\ln 16 + 2 \ln 5}{\ln 5} \qquad \text{Divide by ln 5.}$$

The exact solution is $\left\{\dfrac{\ln 16 + 2 \ln 5}{\ln 5}\right\}$. This is approximately $\{3.7227\}$.

b) Begin by taking the natural log of each side.

$$e^{5n} = 4$$
$$\ln e^{5n} = \ln 4 \qquad \text{Take the natural log of each side.}$$
$$5n \ln e = \ln 4 \qquad \log_a x^r = r \log_a x$$
$$5n(1) = \ln 4 \qquad \ln e = 1$$
$$5n = \ln 4$$
$$n = \frac{\ln 4}{5} \qquad \text{Divide by 5.}$$

W Hint
Remember that ln e = 1.

The exact solution is $\left\{\dfrac{\ln 4}{5}\right\}$. The approximation is $\{0.2773\}$.

[YOU TRY 2] Solve.

a) $9^{k+4} = 2$ b) $e^{6c} = 2$

2 Solve Logarithmic Equations Using the Properties of Logarithms

We learned earlier that to solve a logarithmic equation like $\log_2(t + 5) = 4$, we write the equation in exponential form and solve for the variable.

$$\log_2(t + 5) = 4$$
$$2^4 = t + 5 \qquad \text{Write in exponential form.}$$
$$16 = t + 5 \qquad 2^4 = 16$$
$$11 = t \qquad \text{Subtract 5.}$$

In this section, we will learn how to solve other types of logarithmic equations as well. We will look at equations where

1) each term in the equation contains a logarithm.
2) one term in the equation does *not* contain a logarithm.

Let's begin with the first case.

> **Procedure** How to Solve an Equation Where Each Term Contains a Logarithm
>
> 1) Use the properties of logarithms to write the equation in the form $\log_a x = \log_a y$.
> 2) Set $x = y$, and solve for the variable.
> 3) Check the proposed solution(s) in the original equation to be sure the values satisfy the equation.

EXAMPLE 3

Solve.

a) $\log_5(m - 4) = \log_5 9$ b) $\log x + \log(x + 6) = \log 16$

Solution

a) To solve $\log_5(m - 4) = \log_5 9$, use the property that states if $\log_a x = \log_a y$, then $x = y$.

$$\log_5(m - 4) = \log_5 9$$
$$m - 4 = 9$$
$$m = 13 \quad \text{Add 4.}$$

Check to be sure that $m = 13$ satisfies the original equation.

$$\log_5(13 - 4) \stackrel{?}{=} \log_5 9$$
$$\log_5 9 = \log_5 9 \checkmark$$

The solution set is $\{13\}$.

 Hint
Review the properties of logarithms on p. 748, if necessary.

b) To solve $\log x + \log(x + 6) = \log 16$, we must begin by using the product rule for logarithms to obtain one logarithm on the left side.

$$\log x + \log(x + 6) = \log 16$$
$$\log x(x + 6) = \log 16 \quad \text{Product rule}$$
$$x(x + 6) = 16 \quad \text{If } \log_a x = \log_a y, \text{ then } x = y.$$
$$x^2 + 6x = 16 \quad \text{Distribute.}$$
$$x^2 + 6x - 16 = 0 \quad \text{Subtract 16.}$$
$$(x + 8)(x - 2) = 0 \quad \text{Factor.}$$
$$x + 8 = 0 \quad \text{or} \quad x - 2 = 0 \quad \text{Set each factor equal to 0.}$$
$$x = -8 \quad \text{or} \quad x = 2 \quad \text{Solve.}$$

Check to be sure that $x = -8$ and $x = 2$ satisfy the original equation.

Check $x = -8$:

$\log x + \log(x + 6) = \log 16$
$\log(-8) + \log(-8 + 6) \stackrel{?}{=} \log 16$
FALSE

We reject $x = -8$ as a solution because it leads to $\log(-8)$, which is undefined.

The solution set is $\{2\}$.

Check $x = 2$:

$\log x + \log(x + 6) = \log 16$
$\log 2 + \log(2 + 6) \stackrel{?}{=} \log 16$
$\log 2 + \log 8 \stackrel{?}{=} \log 16$
$\log(2 \cdot 8) \stackrel{?}{=} \log 16$
$\log 16 = \log 16$ ✓

$x = 2$ satisfies the original equation.

BE CAREFUL Just because a proposed solution is a negative number does *not* mean it should be rejected. You *must* check it in the original equation; it may satisfy the equation.

[YOU TRY 3] Solve.

a) $\log_8(z + 3) = \log_8 5$ b) $\log_3 c + \log_3(c - 1) = \log_3 12$

Procedure How to Solve an Equation Where One Term Does *Not* Contain a Logarithm

1) Use the properties of logarithms to get one logarithm on one side of the equation and a constant on the other side. That is, write the equation in the form $\log_a x = y$.
2) Write $\log_a x = y$ in exponential form, $a^y = x$, and solve for the variable.
3) Check the proposed solution(s) in the original equation to be sure the values satisfy the equation.

EXAMPLE 4   Solve $\log_2 3w - \log_2(w - 5) = 3$.

Solution

Notice that one term in the equation $\log_2 3w - \log_2(w - 5) = 3$ does *not* contain a logarithm. Therefore, we want to use the properties of logarithms to get *one* logarithm on the left. Then, write the equation in exponential form, and solve.

Hint
Are you writing down the example as you are reading it?

$\log_2 3w - \log_2(w - 5) = 3$

$\log_2 \dfrac{3w}{w - 5} = 3$ Quotient rule

$2^3 = \dfrac{3w}{w - 5}$ Write in exponential form.

$8 = \dfrac{3w}{w - 5}$ $2^3 = 8$

$8(w - 5) = 3w$ Multiply by $w - 5$.
$8w - 40 = 3w$ Distribute.
$-40 = -5w$ Subtract $8w$.
$8 = w$ Divide by -5.

Verify that $w = 8$ satisfies the original equation. The solution set is $\{8\}$.

[**YOU TRY 4**] Solve.

a) $\log_4(7p + 1) = 3$ b) $\log_3 2x - \log_3(x - 14) = 2$

W Hint
Write these equations and their solutions in your notes. In your own words, explain the difference between them.

Let's look at the two types of equations we have discussed side by side. Notice the difference between them.

Solve each equation

1) $\log_3 x + \log_3(2x + 5) = \log_3 12$

Use the properties of logarithms to get one log on the left.

$$\log_3 x(2x + 5) = \log_3 12$$

Since *both terms contain logarithms*, use the property that states if $\log_a x = \log_a y$, then $x = y$.

$$x(2x + 5) = 12$$
$$2x^2 + 5x = 12$$
$$2x^2 + 5x - 12 = 0$$
$$(2x - 3)(x + 4) = 0$$
$$2x - 3 = 0 \quad \text{or} \quad x + 4 = 0$$
$$x = \frac{3}{2} \quad \text{or} \quad x = -4$$

Check. Reject -4 as a solution. The solution set is $\left\{\dfrac{3}{2}\right\}$.

2) $\log_3 x + \log_3(2x + 5) = 1$

Use the properties of logarithms to get one log on the left.

$$\log_3 x(2x + 5) = 1$$

The term on the right does *not* contain a logarithm. Write the equation in exponential form, and solve.

$$3^1 = x(2x + 5)$$
$$3 = 2x^2 + 5x$$
$$0 = 2x^2 + 5x - 3$$
$$0 = (2x - 1)(x + 3)$$
$$2x - 1 = 0 \quad \text{or} \quad x + 3 = 0$$
$$x = \frac{1}{2} \quad \text{or} \quad x = -3$$

Check. Reject $x = -3$ as a solution. The solution set is $\left\{\dfrac{1}{2}\right\}$.

3 Solve Applied Problems Involving Exponential Functions Using a Calculator

Recall that $A = Pe^{rt}$ is the formula for continuous compound interest where P (the principal) is the amount invested, r is the interest rate, and A is the amount (in dollars) in the account after t yr. Here we will look at how we can use the formula to solve a different problem from the type we solved in Section 11.5.

EXAMPLE 5

If $3000 is invested at 5% interest compounded continuously, how long would it take for the investment to grow to $4000?

Solution

In this problem, we are asked to find t, the amount of *time* it will take for $3000 to grow to $4000 when invested at 5% compounded continuously.

Use $A = Pe^{rt}$ with $P = 3000$, $A = 4000$, and $r = 0.05$.

$$A = Pe^{rt}$$
$$4000 = 3000e^{0.05t} \quad \text{Substitute the values.}$$
$$\frac{4}{3} = e^{0.05t} \quad \text{Divide by 3000.}$$
$$\ln\frac{4}{3} = \ln e^{0.05t} \quad \text{Take the natural log of both sides.}$$
$$\ln\frac{4}{3} = 0.05t \ln e \quad \log_a x^r = r \log_a x$$
$$\ln\frac{4}{3} = 0.05t(1) \quad \ln e = 1$$
$$\ln\frac{4}{3} = 0.05t$$
$$\frac{\ln\frac{4}{3}}{0.05} = t \quad \text{Divide by 0.05.}$$
$$5.75 \approx t \quad \text{Use a calculator to get the approximation.}$$

It would take about 5.75 yr for $3000 to grow to $4000.

[**YOU TRY 5**] If $4500 is invested at 6% interest compounded continuously, how long would it take for the investment to grow to $5000?

The amount of time it takes for a quantity to double in size is called the *doubling time*. We can use this in many types of applications.

EXAMPLE 6

The number of bacteria, $N(t)$, in a culture t hr after the bacteria are placed in a dish is given by

$$N(t) = 5000e^{0.0462t}$$

where 5000 bacteria are initially present. How long will it take for the number of bacteria to double?

Solution

If there are 5000 bacteria present initially, there will be $2(5000) = 10{,}000$ bacteria when the number doubles. This is $N(t)$.

Find t when $N(t) = 10{,}000$.

$$N(t) = 5000e^{0.0462t}$$
$$10{,}000 = 5000e^{0.0462t} \quad \text{Substitute 10,000 for } N(t).$$
$$2 = e^{0.0462t} \quad \text{Divide by 5000.}$$
$$\ln 2 = \ln e^{0.0462t} \quad \text{Take the natural log of both sides.}$$
$$\ln 2 = 0.0462t \ln e \quad \log_a x^r = r \log_a x$$
$$\ln 2 = 0.0462t(1) \quad \ln e = 1$$
$$\ln 2 = 0.0462t$$
$$\frac{\ln 2}{0.0462} = t \quad \text{Divide by 0.0462.}$$
$$15 \approx t$$

W Hint
Don't forget: $\ln e = 1$.

It will take about 15 hr for the number of bacteria to double.

> **[YOU TRY 6]** The number of bacteria, $N(t)$, in a culture t hr after the bacteria are placed in a dish is given by
> $$N(t) = 12{,}000e^{0.0385t}$$
> where 12,000 bacteria are initially present. How long will it take for the number of bacteria to double?

4 Solve an Applied Problem Involving Exponential Growth or Decay

W Hint
Learn this formula, and understand what each variable represents.

We can generalize the formulas used in Examples 5 and 6 with a formula widely used to model situations that grow or decay exponentially. That formula is

$$y = y_0 e^{kt}$$

where y_0 is the initial amount or quantity at time $t = 0$, y is the amount present after time t, and k is a constant. If k is positive, it is called a *growth constant* because the quantity will *increase* over time. If k is negative, it is called a *decay constant* because the quantity will *decrease* over time.

EXAMPLE 7

In April 1986, an accident at the Chernobyl nuclear power plant released many radioactive substances into the environment. One such substance was cesium-137. Cesium-137 decays according to the equation

$$y = y_0 e^{-0.0230t}$$

where y_0 is the initial amount present at time $t = 0$ and y is the amount present after t yr. If a sample of soil contains 10 g of cesium-137 immediately after the accident,

a) how many grams will remain after 15 yr?

b) how long would it take for the initial amount of cesium-137 to decay to 2 g?

c) the **half-life** of a substance is the amount of time it takes for a substance to decay to half its original amount. What is the half-life of cesium-137?

Solution

a) The initial amount of cesium-137 is 10 g, so $y_0 = 10$. We must find y when $y_0 = 10$ and $t = 15$.

$$\begin{aligned} y &= y_0 e^{-0.0230t} \\ &= 10 e^{-0.0230(15)} && \text{Substitute the values.} \\ &\approx 7.08 && \text{Use a calculator to get the approximation.} \end{aligned}$$

There will be about 7.08 g of cesium-137 remaining after 15 yr.

b) The initial amount of cesium-137 is $y_0 = 10$. To determine how long it will take to decay to 2 g, let $y = 2$ and solve for t.

$$\begin{aligned} y &= y_0 e^{-0.0230t} \\ 2 &= 10 e^{-0.0230t} && \text{Substitute 2 for } y \text{ and 10 for } y. \\ 0.2 &= e^{-0.0230t} && \text{Divide by 10.} \\ \ln 0.2 &= \ln e^{-0.0230t} && \text{Take the natural log of both sides.} \\ \ln 0.2 &= -0.0230t \ln e && \log_a x^r = r \log_a x \\ \ln 0.2 &= -0.0230t && \ln e = 1 \\ \frac{\ln 0.2}{-0.0230} &= t && \text{Divide by } -0.0230. \\ 69.98 &\approx t && \text{Use a calculator to get the approximation.} \end{aligned}$$

It will take about 69.98 yr for 10 g of cesium-137 to decay to 2 g.

> **Hint**
> Write a definition of **half-life** in your notes, in your own words.

c) Since there are 10 g of cesium-137 in the original sample, to determine the half-life we will determine how long it will take for the 10 g to decay to 5 g because $\frac{1}{2}(10) = 5$.

Let $y_0 = 10$, $y = 5$, and solve for t.

$$y = y_0 e^{-0.0230t}$$
$$5 = 10e^{-0.0230t} \quad \text{Substitute the values.}$$
$$0.5 = e^{-0.0230t} \quad \text{Divide by 10.}$$
$$\ln 0.5 = \ln e^{-0.0230t} \quad \text{Take the natural log of both sides.}$$
$$\ln 0.5 = -0.0230t \ln e \quad \log_a x^r = r \log_a x$$
$$\ln 0.5 = -0.0230t \quad \ln e = 1$$
$$\frac{\ln 0.5}{-0.0230} = t \quad \text{Divide by } -0.0230.$$
$$30.14 \approx t \quad \text{Use a calculator to get the approximation.}$$

The half-life of cesium-137 is about 30.14 yr. This means that it would take about 30.14 yr for any quantity of cesium-137 to decay to half of its original amount.

[YOU TRY 7] Radioactive strontium-90 decays according to the equation

$$y = y_0 e^{-0.0244t}$$

where t is in years. If a sample contains 40 g of strontium-90,

a) how many grams will remain after 8 yr?
b) how long would it take for the initial amount of strontium-90 to decay to 30 g?
c) what is the half-life of strontium-90?

Using Technology

We can solve exponential and logarithmic equations in the same way that we solved other equations—by graphing both sides of the equation and finding where the graphs intersect.

In Example 2 of this section, we learned how to solve $5^{x-2} = 16$. Because the right side of the equation is 16, the graph will have to go at least as high as 16. So set the $Y_{\max}$ to be 20, enter the left side of the equation as Y_1 and the right side as Y_2, and press GRAPH:

 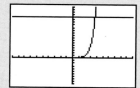

Recall that the x-coordinate of the point of intersection is the solution to the equation. To find the point of intersection, press 2nd TRACE and then highlight 5:intersect and press ENTER. Press ENTER three more times to see that the x-coordinate of the point of intersection is approximately 3.723.

Remember, while the calculator can sometimes save you time, it will often give an approximate answer and not an exact solution.

Use a graphing calculator to solve each equation. Round your answer to the nearest thousandth.

1) $7^x = 49$
2) $6^{2b+1} = 13$
3) $5^{4a+7} = 8^{2a}$
4) $\ln x = 1.2$
5) $\log(k + 9) = \log 11$
6) $\ln(x + 3) = \ln(x - 2)$

ANSWERS TO [YOU TRY] EXERCISES

1) a) $\{7\}$ b) $\left\{\dfrac{\ln 24}{\ln 3}\right\}; \{2.8928\}$ 2) a) $\left\{\dfrac{\ln 2 - 4\ln 9}{\ln 9}\right\}; \{-3.6845\}$ b) $\left\{\dfrac{\ln 2}{6}\right\}; \{0.1155\}$
3) a) $\{2\}$ b) $\{4\}$ 4) a) $\{9\}$ b) $\{18\}$ 5) 1.76 yr   6) 18 hr
7) a) 32.91 g   b) 11.79 yr   c) 28.41 yr

ANSWERS TO TECHNOLOGY EXERCISES

1) $\{2\}$ 2) $\{0.216\}$ 3) $\{-4.944\}$ 4) $\{3.320\}$ 5) $\{2\}$ 6) $\varnothing$

E Evaluate 11.6 Exercises

Do the exercises, and check your work.

Objective 1: Solve an Exponential Equation

Solve each equation. Give the exact solution. If the answer contains a logarithm, approximate the solution to four decimal places.

1) $7^x = 49$
2) $5^c = 125$
3) $7^n = 15$
4) $5^a = 38$
5) $8^z = 3$
6) $4^y = 9$
7) $6^{5p} = 36$
8) $2^{3t} = 32$
9) $4^{6k} = 2.7$
10) $3^{2x} = 7.8$
11) $2^{4n+1} = 5$
12) $6^{2b+1} = 13$
13) $5^{3a-2} = 8$
14) $3^{2x-3} = 14$
15) $4^{2c+7} = 64^{3c-1}$
16) $27^{5m-2} = 3^{m+6}$
17) $9^{5d-2} = 4^{3d}$
18) $5^{4a+7} = 8^{2a}$

Solve each equation. Give the exact solution and the approximation to four decimal places.

19) $e^y = 12.5$
20) $e^t = 0.36$
21) $e^{-4x} = 9$
22) $e^{3p} = 4$
23) $e^{0.01r} = 2$
24) $e^{-0.08k} = 10$
25) $e^{0.006t} = 3$
26) $e^{0.04a} = 12$
27) $e^{-0.4y} = 5$
28) $e^{-0.005c} = 16$

Objective 2: Solve Logarithmic Equations Using the Properties of Logarithms

Solve each equation.

29) $\log_6(k + 9) = \log_6 11$
30) $\log_5(d - 4) = \log_5 2$
31) $\log_7(3p - 1) = \log_7 9$
32) $\log_4(5y + 2) = \log_4 10$
33) $\log x + \log(x - 2) = \log 15$
34) $\log_9 r + \log_9(r + 7) = \log_9 18$
35) $\log_3 n + \log_3(12 - n) = \log_3 20$
36) $\log m + \log(11 - m) = \log 24$
37) $\log_2(-z) + \log_2(z - 8) = \log_2 15$
38) $\log_5 8y - \log_5(3y - 4) = \log_5 2$

39) $\log_3(4c + 5) = 3$ 40) $\log_6(5b - 4) = 2$

41) $\log(3p + 4) = 1$ 42) $\log(7n - 11) = 1$

43) $\log_3 y + \log_3(y - 8) = 2$

44) $\log_4 k + \log_4(k - 6) = 2$

45) $\log_2 r + \log_2(r + 2) = 3$

46) $\log_9(z + 8) + \log_9 z = 1$

47) $\log_4 20c - \log_4(c + 1) = 2$

48) $\log_6 40x - \log_6(1 + x) = 2$

49) $\log_2 8d - \log_2(2d - 1) = 4$

50) $\log_6(13 - x) + \log_6 x = 2$

Mixed Exercises: Objectives 3 and 4

Use the formula $A = Pe^{rt}$ to solve Exercises 51–58.

51) If $2000 is invested at 6% interest compounded continuously, how long would it take

   a) for the investment to grow to $2500?

   b) for the initial investment to double?

52) If $5000 is invested at 7% interest compounded continuously, how long would it take

   a) for the investment to grow to $6000?

   b) for the initial investment to double?

53) How long would it take for an investment of $7000 to earn $800 in interest if it is invested at 7.5% compounded continuously?

54) How long would it take for an investment of $4000 to earn $600 in interest if it is invested at 6.8% compounded continuously?

55) Cynthia wants to invest some money now so that she will have $5000 in the account in 10 yr. How much should she invest in an account earning 8% compounded continuously?

56) How much should Leroy invest now at 7.2% compounded continuously so that the account contains $8000 in 12 yr?

57) Raj wants to invest $3000 now so that it grows to $4000 in 4 yr. What interest rate should he look for? (Round to the nearest tenth of a percent.)

58) Marisol wants to invest $12,000 now so that it grows to $20,000 in 7 yr. What interest rate should she look for? (Round to the nearest tenth of a percent.)

59) The number of bacteria, $N(t)$, in a culture t hr after the bacteria are placed in a dish is given by

$$N(t) = 4000e^{0.0374t}$$

where 4000 bacteria are initially present.

   a) After how many hours will there be 5000 bacteria in the culture?

   b) How long will it take for the number of bacteria to double?

60) The number of bacteria, $N(t)$, in a culture t hr after the bacteria are placed in a dish is given by

$$N(t) = 10{,}000e^{0.0418t}$$

where 10,000 bacteria are initially present.

   a) After how many hours will there be 15,000 bacteria in the culture?

   b) How long will it take for the number of bacteria to double?

61) The population of an Atlanta suburb is growing at a rate of 3.6% per year. If 21,000 people lived in the suburb in 2004, determine how many people will live in the town in 2012. Use $y = y_0 e^{0.036t}$.

62) The population of a Seattle suburb is growing at a rate of 3.2% per year. If 30,000 people lived in the suburb in 2008, determine how many people will live in the town in 2015. Use $y = y_0 e^{0.032t}$.

63) A rural town in South Dakota is losing residents at a rate of 1.3% per year. The population of the town was 2470 in 1990. Use $y = y_0 e^{-0.013t}$ to answer the following questions.

   a) What was the population of the town in 2005?

   b) In what year would it be expected that the population of the town is 1600?

64) In 1995, the population of a rural town in Kansas was 1682. The population is decreasing at a rate of 0.8% per year. Use $y = y_0 e^{-0.008t}$ to answer the following questions.

   a) What was the population of the town in 2000?

   b) In what year would it be expected that the population of the town is 1000?

65) Radioactive carbon-14 is a substance found in all living organisms. After the organism dies, the carbon-14 decays according to the equation

$$y = y_0 e^{-0.000121t}$$

where t is in years, y_0 is the initial amount present at time $t = 0$, and y is the amount present after t yr.

a) If a sample initially contains 15 g of carbon-14, how many grams will be present after 2000 yr?

b) How long would it take for the initial amount to decay to 10 g?

c) What is the half-life of carbon-14?

66) Plutonium-239 decays according to the equation
$$y = y_0 e^{-0.0000287t}$$
where t is in years, y_0 is the initial amount present at time $t = 0$, and y is the amount present after t yr.

a) If a sample initially contains 8 g of plutonium-239, how many grams will be present after 5000 yr?

b) How long would it take for the initial amount to decay to 5 g?

c) What is the half-life of plutonium-239?

67) Radioactive iodine-131 is used in the diagnosis and treatment of some thyroid-related illnesses. The concentration of the iodine in a patient's system is given by
$$y = 0.4e^{-0.086t}$$
where t is in days, and y is in the appropriate units.

a) How much iodine-131 is given to the patient?

b) How much iodine-131 remains in the patient's system 7 days after treatment?

68) The amount of cobalt-60 in a sample is given by
$$y = 30e^{-0.131t}$$
where t is in years, and y is in grams.

a) How much cobalt-60 is originally in the sample?

b) How long would it take for the initial amount to decay to 10 g?

Extension

Solve. Where appropriate, give the exact solution and the approximation to four decimal places.

69) $\log_2 (\log_2 x) = 2$

70) $\log_3 (\log y) = 1$

71) $\log_3 \sqrt{n^2 + 5} = 1$

72) $\log(p - 7)^2 = 4$

73) $e^{|t|} = 13$

74) $e^{r^2 - 25} = 1$

75) $e^{2y} + 3e^y - 4 = 0$

76) $e^{2x} - 9e^x + 8 = 0$

77) $5^{2c} - 4 \cdot 5^c - 21 = 0$

78) $9^{2a} + 5 \cdot 9^a - 24 = 0$

79) $(\log x)^2 = \log x^3$

80) $\log 6^y = y^2$

R Rethink

R1) How are exponential equations and logarithmic equations the same? How are they different?

R2) Can you think of a real-life situation that you have encountered that involves decay or growth?

R3) What does continuous compounding mean?

Group Activity — The Group Activity can be found online on Connect.

emPOWERme Progressive Relaxation

If you are feeling stressed, progressive relaxation can help. Progressive relaxation does some of the same things that meditation does, but in a more direct way. To use progressive relaxation, you systematically tense and then relax different groups of muscles. You can undertake progressive relaxation almost anywhere, including the library, a sports field, or a classroom, since tensing and relaxing muscles is quiet and unobtrusive. Although the following exercise suggests you lie down, you can use parts of it no matter where you are.

1. Lie flat on your back, get comfortable, and focus on your toes.

2. Become aware of your left toes. Bunch them up into a tight ball, then let them go. Then let them relax even further.

3. Now work on your left foot, from the toes to the heel. Without tensing your toes, tighten up the rest of your foot and then let it relax. Then relax it more.

4. Work your way up your left leg, first tensing and then relaxing each part. You may move up as slowly or as quickly as you wish, using big leaps (e.g., the entire lower leg) or small steps (e.g., the ankle, the calf, the front of the lower leg, the knee, etc.).

5. Repeat the process for the right leg.

6. Now tense and relax progressively your groin, buttocks, abdomen, lower back, ribs, upper back, and shoulders.

7. Work your way down each arm, one at a time, until you reach the fingers.

8. Return to the neck, then the jaw, cheeks, nose, eyes, ears, forehead, and skull.

By now you should be completely relaxed. In fact, you may even be asleep—this technique works well as a sleep-induction strategy. To vary the routine, play with it. Try going from top to bottom, or starting from your extremities and working inward.

Chapter 11: Summary

Definition/Procedure	Example

11.1 Composite and Inverse Functions

Composition of Functions

Given the functions $f(x)$ and $g(x)$, the **composition function** $f \circ g$ $f \circ$ (read "f of g") is defined as

$$(f \circ g)(x) = f(g(x))$$

where $g(x)$ is in the domain of f. (p. 706)

$f(x) = 4x - 10$ and $g(x) = -3x + 2$.

$$\begin{aligned}(f \circ g)(x) &= f(g(x)) \\ &= f(-3x + 2) \\ &= 4(-3x + 2) - 10 \quad \text{Substitute } -3x + 2 \text{ for } x \text{ in } f(x). \\ &= -12x + 8 - 10 \\ &= -12x - 2\end{aligned}$$

One-to-One Function

In order for a function to be a **one-to-one function**, each x-value corresponds to exactly one y-value, and each y-value corresponds to exactly one x-value. (p. 709)

The **horizontal line test** tells us how we can determine whether a graph represents a one-to-one function:

If every horizontal line that could be drawn through a function would intersect the graph at most once, then the function is one-to-one. (p. 710)

Determine whether each function is one-to-one.

a) $f = \{(-2, 9), (1, 3), (3, -1), (7, -9)\}$ *is* one-to-one.
b) $g = \{(0, 9), (2, 1), (4, 1), (5, 4)\}$ is *not* one-to-one since the y-value 1 corresponds to two different x-values.

c)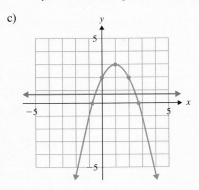

No. It fails the horizontal line test.

Inverse Function

Let f be a one-to-one function. The **inverse** of f, denoted by f^{-1}, is a one-to-one function that contains the set of all ordered pairs (y, x) where (x, y) belongs to f.

How to Find an Equation of the Inverse of $y = f(x)$

Step 1: Replace $f(x)$ with y.

Step 2: Interchange x and y.

Step 3: Solve for y.

Step 4: Replace y with the inverse notation, $f^{-1}(x)$.

The graphs of $f(x)$ and $f^{-1}(x)$ are symmetric with respect to the line $y = x$. (p. 711)

Find an equation of the inverse of $f(x) = 2x - 4$.

Step 1: $y = 2x - 4$ Replace $f(x)$ with y.
Step 2: $x = 2y - 4$ Interchange x and y.
Step 3: Solve for y.

$$\begin{aligned} x + 4 &= 2y \quad \text{Add 4.} \\ \frac{x + 4}{2} &= y \quad \text{Divide by 2.} \\ \frac{1}{2}x + 2 &= y \quad \text{Simplify.} \end{aligned}$$

Step 4: $f^{-1}(x) = \dfrac{1}{2}x + 2$ Replace y with $f^{-1}(x)$.

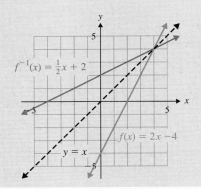

Definition/Procedure	Example

11.2 Exponential Functions

An **exponential function** is a function of the form

$$f(x) = a^x$$

where $a > 0$, $a \neq 1$, and x is a real number. (p. 720)

$f(x) = 3^x$

Characteristics of an Exponential Function

$$f(x) = a^x$$

1) If $f(x) = a^x$, where $a > 1$, the value of y increases as the value of x increases.
2) If $f(x) = a^x$ where $0 < a < 1$, the value of y decreases as the value of x increases.
3) The function is one-to-one.
4) The y-intercept is $(0, 1)$.
5) The domain is $(-\infty, \infty)$, and the range is $(0, \infty)$. (p. 722)

$f(x) = e^x$ is a special exponential function that has many uses in mathematics. Like the number π, e is an irrational number. (p. 722)

$$e \approx 2.7183$$

1)
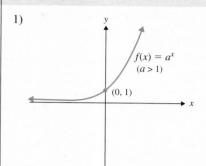

2)
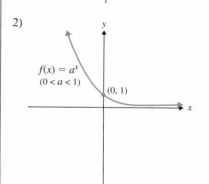

Solving an Exponential Equation

Step 1: If possible, express each side of the equation with the same base. If it is not possible to get the same base, a method in Section 11.6 can be used.
Step 2: Use the rules of exponents to simplify the exponents.
Step 3: Set the exponents equal, and solve for the variable. (p. 724)

Solve $5^{4x-1} = 25^{3x+4}$.

Step 1: $5^{4x-1} = (5^2)^{(3x+4)}$ Both sides are powers of 5.
Step 2: $5^{4x-1} = 5^{2(3x+4)}$ Power rule for exponents
        $5^{4x-1} = 5^{6x+8}$ Distribute.
Step 3: $4x - 1 = 6x + 8$ The bases are the same. Set the exponents equal.
        $-2x = 9$ Subtract $6x$; add 1.
        $x = -\dfrac{9}{2}$ Divide by -2.

The solution set is $\left\{-\dfrac{9}{2}\right\}$.

11.3 Logarithmic Functions

Definition of Logarithm
If $a > 0$, $a \neq 1$, and $x > 0$, then for every real number y, $y = \log_a x$ means $x = a^y$. (p. 731)

Write $\log_5 125 = 3$ in exponential form.

$$\log_5 125 = 3 \text{ means } 5^3 = 125$$

A **logarithmic equation** is an equation in which at least one term contains a logarithm.
To solve a logarithmic equation of the form

$$\log_a b = c$$

write the equation in exponential form ($a^c = b$) and solve for the variable. (p. 733)

Solve $\log_2 k = 3$.
Write the equation in exponential form, and solve for k.

$$\log_2 k = 3 \text{ means } 2^3 = k.$$
$$8 = k$$

The solution set is $\{8\}$.

Definition/Procedure	Example
To evaluate $\log_a b$ means *to find the power to which we raise a to get b.* (p. 734)	Evaluate $\log_7 49$. $\log_7 49 = 2$ because $7^2 = 49$
A base 10 logarithm is called a **common logarithm**. A base 10 logarithm is often written without the base. (p. 735)	$\log x$ means $\log_{10} x$.
Characteristics of a Logarithmic Function $f(x) = \log_a x$, where $a > 0$ and $a \neq 1$ 1) If $f(x) = \log_a x$, where $a > 1$, the value of y increases as the value of x increases. 2) If $f(x) = \log_a x$, where $0 < a < 1$, the value of y decreases as the value of x increases. 3) The function is one-to-one. 4) The x-intercept is $(1, 0)$. 5) The domain is $(0, \infty)$, and the range is $(-\infty, \infty)$. 6) The inverse of $f(x) = \log_a x$ is $f^{-1}(x) = a^x$. (p. 739)	1) 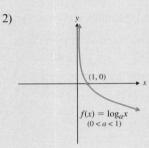 2)

11.4 Properties of Logarithms

Let x, y, and a be positive real numbers where $a \neq 1$, and let r be any real number. Then, 1) $\log_a a = 1$ 2) $\log_a 1 = 0$ 3) $\log_a xy = \log_a x + \log_a y$ Product rule 4) $\log_a \dfrac{x}{y} = \log_a x - \log_a y$ Quotient rule 5) $\log_a x^r = r \log_a x$ Power rule 6) $\log_a a^x = x$ for any real number x 7) $a^{\log_a x} = x$ (p. 748)	Write $\log_4 \dfrac{c^5}{d^2}$ as the sum or difference of logarithms in simplest form. Assume c and d represent positive real numbers. $\log_4 \dfrac{c^5}{d^2} = \log_4 c^5 - \log_4 d^2$ Quotient rule $\phantom{\log_4 \dfrac{c^5}{d^2}} = 5 \log_4 c - 2 \log_4 d$ Power rule

11.5 Common and Natural Logarithms and Change of Base

We can evaluate common logarithms with or without a calculator. (p. 754)	Find the value of each. a) $\log 100$ b) $\log 53$ a) $\log 100 = \log_{10} 100 = \log_{10} 10^2 = 2$ b) Using a calculator, we get $\log 53 \approx 1.7243$.

Definition/Procedure	Example
The number e is approximately equal to 2.7183. A base e logarithm is called a **natural logarithm**. The notation used for a natural logarithm is $\ln x$. $\quad f(x) = \ln x \quad$ means $\quad f(x) = \log_e x$ The domain of $f(x) = \ln x$ is $(0, \infty)$, and the range is $(-\infty, \infty)$. **(p. 756)**	The graph of $f(x) = \ln x$ looks like this: 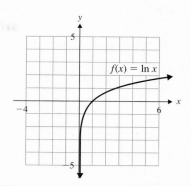
$\ln e = 1$ because $\ln e = 1$ means $\log_e e = 1$. We can find the values of some natural logarithms using the properties of logarithms. We can approximate the values of other natural logarithms using a calculator. **(p. 756)**	Find the value of each. a) $\ln e^{12}$ b) $\ln 18$ a) $\ln e^{12} = 12 \ln e \quad$ Power rule $\qquad\quad\; = 12(1) \quad\;\; \ln e = 1$ $\qquad\quad\; = 12$ b) Using a calculator, we get $\ln 18 \approx 2.8904$.
To solve an equation such as $\ln x = 1.6$, change to exponential form and solve for the variable. **(p. 758)**	Solve $\ln x = 1.6$. $\ln x = 1.6$ means $\log_e x = 1.6$. $\quad \log_e x = 1.6$ $\quad\;\; e^{1.6} = x \quad$ Exponential form $\;\; 4.9530 \approx x \quad$ Approximation The exact solution is $\{e^{1.6}\}$. The approximation is $\{4.9530\}$.
Applications of Exponential Functions **Continuous Compounding** If P dollars are deposited in an account earning interest rate r compounded continuously, then the amount of money, A (in dollars), in the account after t years is given by $A = Pe^{rt}$. **(p. 760)**	Determine the amount of money in an account after 6 yr if \$3000 was initially invested at 5% compounded continuously. $\quad A = Pe^{rt}$ $\quad\;\;\; = 3000e^{0.05(6)} \quad$ Substitute the values. $\quad\;\;\; = 3000e^{0.30} \quad\;\;$ Multiply $(0.05)(6)$. $\quad\;\;\; \approx 4049.5764 \quad\;\;$ Evaluate using a calculator. $\quad\;\;\; \approx \$4049.58 \quad\;\;\;$ Round to the nearest cent.
Change-of-Base Formula If a, b, and x are positive real numbers and $a \neq 1$ and $b \neq 1$, then $\qquad \log_a x = \dfrac{\log_b x}{\log_b a} \qquad$ **(p. 760)**	Find $\log_2 75$ to four decimal places. $\qquad \log_2 75 = \dfrac{\log_{10} 75}{\log_{10} 2} \approx 6.2288$

Definition/Procedure	Example
11.6 Solving Exponential and Logarithmic Equations	
Let a, x, and y be positive real numbers, where $a \neq 1$. 1) If $x = y$, then $\log_a x = \log_a y$. 2) If $\log_a x = \log_a y$, then $x = y$. (p. 765) **How to Solve an Exponential Equation** Begin by asking yourself, "*Can I express each side with the same base?*" 1) If the answer is **yes**, then write each side of the equation with the same base, set the exponents equal, and solve for the variable. 2) If the answer is **no**, then take the natural logarithm of each side, use the properties of logarithms, and solve for the variable. (p. 766)	Solve each equation. a) $4^x = 64$ Ask yourself, "*Can I express both sides with the same base?*" Yes. $$4^x = 64$$ $$4^x = 4^3$$ $$x = 3 \quad \text{Set the exponents equal.}$$ The solution set is $\{3\}$. b) $4^x = 9$ Ask yourself, "*Can I express both sides with the same base?*" No. Take the natural logarithm of each side. $$4^x = 9$$ $$\ln 4^x = \ln 9 \quad \text{Take the natural log of each side.}$$ $$x \ln 4 = \ln 9 \quad \log_a x^r = r \log_a x$$ $$x = \frac{\ln 9}{\ln 4} \quad \text{Divide by } \ln 4.$$ $$x \approx 1.5850 \quad \text{Use a calculator to get the approximation.}$$ The exact solution is $\left\{\dfrac{\ln 9}{\ln 4}\right\}$. The approximation is $\{1.5850\}$.
Solve an exponential equation with base e by taking the natural logarithm of each side. (p. 767)	Solve $e^y = 35.8$. $$\ln e^y = \ln 35.8 \quad \text{Take the natural log of each side.}$$ $$y \ln e = \ln 35.8 \quad \log_a x^r = r \log_a x$$ $$y(1) = \ln 35.8 \quad \ln e = 1$$ $$y = \ln 35.8$$ $$y \approx 3.5779 \quad \text{Approximation}$$ The exact solution is $\{\ln 35.8\}$. The approximation is $\{3.5779\}$.
Solving Logarithmic Equations Sometimes we must use the properties of logarithms to solve logarithmic equations. (p. 768)	Solve $\log x + \log(x - 3) = \log 28$. $$\log x + \log(x - 3) = \log 28$$ $$\log x(x - 3) = \log 28 \quad \text{Product rule}$$ $$x(x - 3) = 28 \quad \text{If } \log_a x = \log_a y, \text{ then } x = y.$$ $$x^2 - 3x = 28 \quad \text{Distribute.}$$ $$x^2 - 3x - 28 = 0 \quad \text{Subtract 28.}$$ $$(x - 7)(x + 4) = 0 \quad \text{Factor.}$$ $$x - 7 = 0 \text{ or } x + 4 = 0 \quad \text{Set each factor equal to 0.}$$ $$x = 7 \text{ or } x = -4 \quad \text{Solve.}$$ Verify that only 7 satisfies the original equation. The solution set is $\{7\}$.

Chapter 11: Review Exercises

(11.1)

1) Let $f(x) = x + 6$ and $g(x) = 2x - 9$. Find
   a) $(g \circ f)(x)$
   b) $(f \circ g)(x)$
   c) $(f \circ g)(5)$

2) Let $h(x) = 2x - 1$ and $k(x) = x^2 + 5x - 4$. Find
   a) $(k \circ h)(x)$
   b) $(h \circ k)(x)$
   c) $(h \circ k)(-3)$

3) Antoine's gross weekly pay, G, in terms of the number of hours, h, he worked is given by $G(h) = 12h$. His net weekly pay, N, in terms of his gross pay is given by $N(G) = 0.8G$.
   a) Find $(N \circ G)(h)$, and explain what it represents.
   b) Find $(N \circ G)(30)$, and explain what it represents.
   c) What is his net pay if he works 40 hr in 1 week?

4) Given $h(x)$, find f and g such that $h(x) = (f \circ g)(x)$.
   a) $h(x) = (3x + 10)^2$
   b) $h(x) = (8x - 7)^3$
   c) $h(x) = \sqrt{x^2 + 6}$
   d) $h(x) = \dfrac{1}{2 - 5x}$

Determine whether each function is one-to-one. If it is one-to-one, find its inverse.

5) $f = \{(-7, -4), (-2, 1), (1, 5), (6, 11)\}$
6) $g = \{(1, 4), (3, 7), (6, 4), (10, 9)\}$

Determine whether each function is one-to-one. If it is one-to-one, graph its inverse.

7)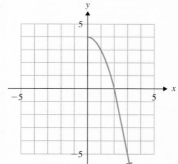

8)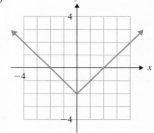

Find the inverse of each one-to-one function. Graph each function and its inverse on the same axes.

9) $f(x) = x + 4$
10) $g(x) = 2x - 10$
11) $h(x) = \dfrac{1}{3}x - 1$
12) $f(x) = \sqrt[3]{x} + 2$

In Exercises 13 and 14, determine whether each statement is *true* or *false*. If it is false, explain why.

13) Every function has an inverse.
14) The domain of $f(x)$ is the range of $f^{-1}(x)$.

(11.2) Graph each exponential function. State the domain and range.

15) $f(x) = 2x$
16) $h(x) = \left(\dfrac{1}{3}\right)^x$
17) $y = 2^x - 4$
18) $f(x) = 3^{x-2}$
19) $f(x) = e^x$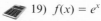
20) $g(x) = e^x + 2$

Solve each exponential equation.

21) $2^c = 64$
22) $7^{m+5} = 49$
23) $16^{3z} = 32^{2z-1}$
24) $9^y = \dfrac{1}{81}$
25) $\left(\dfrac{3}{2}\right)^{x+4} = \left(\dfrac{4}{9}\right)^{x-3}$

26) The value, $V(t)$, in dollars, of a luxury car t yr after it is purchased is given by $V(t) = 38{,}200(0.816)^t$.
   a) What was the purchase price of the car?
   b) What will the car be worth 4 yr after purchase?

(11.3)

27) What is the domain of $y = \log_a x$?
28) In the equation $y = \log_a x$, a must be what kind of number?

Write in exponential form.

29) $\log_5 125 = 3$
30) $\log_{16} \dfrac{1}{4} = -\dfrac{1}{2}$
31) $\log 100 = 2$
32) $\log 1 = 0$

Write in logarithmic form.

33) $3^4 = 81$
34) $\left(\dfrac{2}{3}\right)^{-2} = \dfrac{9}{4}$
35) $10^3 = 1000$
36) $\sqrt{121} = 11$

Solve.

37) $\log_2 x = 3$
38) $\log_9(4x + 1) = 2$
39) $\log_{32} 16 = x$
40) $\log(2x + 5) = 1$

Evaluate.

41) $\log_8 64$ 42) $\log_3 27$

43) $\log 1000$ 44) $\log 1$

45) $\log_{1/2} 16$ 46) $\log_{1/5} \dfrac{1}{25}$

Graph each logarithmic function.

47) $f(x) = \log_2 x$ 48) $f(x) = \log_{1/4} x$

Find the inverse of each function.

49) $f(x) = 5^x$ 50) $h(x) = \log_6 x$

Solve.

51) A company plans to test market its new dog food in a large metropolitan area before taking it nationwide. The company predicts that its sales over the next 12 months can be approximated by

$$S(t) = 10 \log_3(2t + 1)$$

where t is the number of months after the dog food is introduced, and $S(t)$ is in thousands of bags of dog food.

a) How many bags of dog food were sold after 1 month on the market?

b) How many bags of dog food were sold after 4 months on the market?

(11.4) Decide whether each statement is true or false.

52) $\log_5(x + 4) = \log_5 x + \log_5 4$

53) $\log_2 \dfrac{k}{6} = \log_2 k - \log_2 6$

Write as the sum or difference of logarithms and simplify, if possible. Assume all variables represent positive real numbers.

54) $\log_8 3z$ 55) $\log_7 \dfrac{49}{t}$

56) $\log_4 \sqrt{64}$ 57) $\log \dfrac{1}{100}$

58) $\log_5 c^4 d^3$ 59) $\log_4 m\sqrt{n}$

60) $\log_a \dfrac{xy}{z^3}$ 61) $\log_4 \dfrac{a^2}{bc^4}$

62) $\log p(p + 8)$ 63) $\log_6 \dfrac{r^3}{r^2 - 5}$

Write as a single logarithm. Assume the variables are defined so that the variable expressions are positive and so that the bases are positive real numbers not equal to 1.

64) $\log c + \log d$

65) $9 \log_2 a + 3 \log_2 b$

66) $\log_5 r - 2 \log_5 t$

67) $\log_3 5 + 4 \log_3 m - 2 \log_3 n$

68) $\dfrac{1}{2} \log_z a - \log_z b$

Given that $\log 7 \approx 0.8451$ and $\log 9 \approx 0.9542$, use the properties of logarithms to approximate the following. Do NOT use a calculator.

69) $\log 49$ 70) $\log \dfrac{1}{7}$

(11.5)

71) What is the base of $\ln x$?   72) Evaluate $\ln e$.

Evaluate each logarithm. Do not use a calculator.

73) $\log 100$ 74) $\log \sqrt{10}$

75) $\log \dfrac{1}{100}$ 76) $\log 0.001$

77) $\ln 1$ 78) $\ln \sqrt[3]{e}$

Use a calculator to find the approximate value of each logarithm to four decimal places.

79) $\log 8$ 80) $\log 0.3$

81) $\ln 1.75$ 82) $\ln 0.924$

Solve each equation. Do not use a calculator.

83) $\log p = 2$ 84) $\log(5n) = 3$

85) $\log\left(\dfrac{1}{2}c\right) = -1$ 86) $\log(6z - 5) = 1$

Solve each equation. Give an exact solution and a solution that is approximated to four decimal places.

87) $\log x = 2.1$ 88) $\log k = -1.4$

89) $\ln y = 2$ 90) $\ln c = -0.5$

91) $\log(4t) = 1.75$ 92) $\ln(2a - 3) = 1$

Graph each function. State the domain and range.

93) $f(x) = \ln(x - 3)$ 94) $g(x) = \ln x - 2$

Use the change-of-base formula with either base 10 or base e to approximate each logarithm to four decimal places.

95) $\log_4 19$ 96) $\log_9 42$

97) $\log_{1/2} 38$ 98) $\log_6 0.82$

For Exercises 99 and 100, use the formula $L(I) = 10 \log \dfrac{1}{10^{-12}}$, where I is the intensity of sound, in watts per square meter, and $L(I)$ is the loudness of sound in decibels. Do *not* use a calculator.

99) The intensity of sound from the crowd at a college basketball game reached 0.1 W/m². Find the loudness of the crowd, in decibels.

100) Find the intensity of the sound of a jet taking off if the noise level can reach 140 dB 25 m from the jet.

Use the formula $A = P\left(1 + \dfrac{r}{n}\right)^{nt}$ and a calculator to solve.

101) Pedro deposits $2500 in an account earning 6% interest compounded quarterly. How much will be in the account after 5 yr?

Use the formula $A = Pe^{rt}$ and a calculator to solve.

102) Find the amount Liang will owe at the end of 4 yr if he borrows $9000 at a rate of 6.2% compounded continuously.

103) The number of bacteria, $N(t)$, in a culture t hr after the bacteria are placed in a dish is given by

$$N(t) = 6000e^{0.0514t}$$

a) How many bacteria were originally in the culture?

b) How many bacteria are present after 12 hr?

104) The pH of a solution is given by pH $= -\log[\text{H}^+]$, where $[\text{H}^+]$ is the molar concentration of the hydronium ion. Find the ideal pH of blood if $[\text{H}^+] = 3.98 \times 10^{-8}$.

(11.6) Solve each equation. Give the exact solution. If the answer contains a logarithm, approximate the solution to four decimal places. *Some of these exercises require the use of a calculator to obtain a decimal approximation.*

105) $2^y = 16$

106) $3^n = 7$

107) $125^{m-4} = 25^{1-m}$

108) $e^z = 22$

109) $e^{0.03t} = 19$

Solve each logarithmic equation.

110) $\log(3n - 5) = 3$

111) $\log_2 x + \log_2(x + 2) = \log_2 24$

112) $\log_7 10p - \log_7(p - 8) = \log_7 6$

113) $\log_4 k + \log_4(k - 12) = 3$

114) $\log_3 12m - \log_3(1 + m) = 2$

Use the formula $A = Pe^{rt}$ to solve Exercises 131 and 132.

115) Jamar wants to invest some money now so that he will have $10,000 in the account in 6 yr. How much should he invest in an account earning 6.5% compounded continuously?

116) Samira wants to invest $6000 now so that it grows to $9000 in 5 yr. What interest rate (compounded continuously) should she look for? (Round to the nearest tenth of a percent.)

117) The population of a suburb is growing at a rate of 1.6% per year. The population of the suburb was 16,410 in 1990. Use $y = y_0 e^{0.016t}$ to answer the following questions.

a) What was the population of the town in 1995?

b) In what year would it be expected that the population of the town is 23,000?

118) Radium-226 decays according to the equation

$$y = y_0 e^{-0.000436t}$$

where t is in years, y_0 is the initial amount present at time $t = 0$, and y is the amount present after t yr.

a) If a sample initially contains 80 g of radium-226, how many grams will be present after 500 yr?

b) How long would it take for the initial amount to decay to 20 g?

c) What is the half-life of radium-226?

Chapter 11: Test

Use a calculator only where indicated.

1) Let $h(x) = 2x + 7$ and $k(x) = x^2 + 5x - 3$. Find
   a) $(h \circ k)(x)$
   b) $(k \circ h)(x)$
   c) $(k \circ h)(-3)$

2) Let $h(x) = (9x - 7)^3$. Find f and g such that $h(x) = (f \circ g)(x)$.

Determine whether each function is one-to-one. If it is one-to-one, find its inverse.

3) $f = \{(-4, 5), (-2, 7), (0, 3), (6, 5)\}$

4) $g = \left\{(2, 4), (6, 6), \left(9, \dfrac{15}{2}\right), (14, 10)\right\}$

5) Is this function one-to-one? If it is one-to-one, graph its inverse.

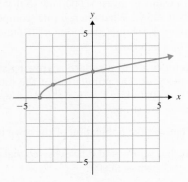

6) Find an equation of the inverse of $f(x) = -3x + 12$.

Use $f(x) = 2x$ and $g(x) = \log_2 x$ for Exercises 7–10.

7) Graph $f(x)$.

8) Graph $g(x)$.

9) a) What is the domain of $g(x)$?
   b) What is the range of $g(x)$?

10) How are the functions $f(x)$ and $g(x)$ related?

11) Write $3^{-2} = \dfrac{1}{9}$ in logarithmic form.

Solve each equation.

12) $9^{4x} = 81$

13) $125^{2c} = 25^{c-4}$

14) $\log_5 y = 3$

15) $\log(3r + 13) = 2$

16) $\log_6(2m) + \log_6(2m - 3) = \log_6 40$

17) Evaluate.
    a) $\log_2 16$
    b) $\log_7 \sqrt{7}$

18) Find $\ln e$.

Write as the sum or difference of logarithms and simplify, if possible. Assume all variables represent positive real numbers.

19) $\log_8 5n$

20) $\log_3 \dfrac{9a^4}{b^5 c}$

21) Write as a single logarithm.
$$2 \log x - 3 \log (x + 1)$$

Use a calculator for the rest of the problems.

Solve each equation. Give an exact solution and a solution that is approximated to four decimal places.

22) $\log w = 0.8$

23) $e^{0.3t} = 5$

24) $\ln x = -0.25$

25) $4^{4a+3} = 9$

Graph each function. State the domain and range.

26) $y = e^x - 4$

27) $f(x) = \ln(x + 1)$

28) Approximate $\log_5 17$ to four decimal places.

29) If $6000 is deposited in an account earning 7.4% interest compounded continuously, how much will be in the account after 5 yr? Use $A = Pe^{rt}$.

30) Polonium-210 decays according to the equation
$$y = y_0 e^{-0.00495t}$$
where t is in days, y_0 is the initial amount present at time $t = 0$, and y is the amount present after t days.

   a) If a sample initially contains 100 g of polonium-210, how many grams will be present after 30 days?
   b) How long would it take for the initial amount to decay to 20 g?
   c) What is the half-life of polonium-210?

Chapter 11: Cumulative Review for Chapters 1–11

Simplify. The answer should not contain any negative exponents.

1) $(-5a^2)(3a^4)$

2) $\left(\dfrac{2c^{10}}{d^3}\right)^{-3}$

3) Write 0.00009231 in scientific notation.

4) *Write an equation, and solve.*
   A watch is on sale for $38.40. This is 20% off of the regular price. What was the regular price of the watch?

5) Solve $-4x + 7 < 13$. Graph the solution set, and write the answer in interval notation.

6) Solve the system.
$$6x + 5y = -8$$
$$3x - y = 3$$

7) Divide $(6c^3 - 7c^2 - 22c + 5) \div (2c - 5)$.

8) Factor $4w^2 + w - 18$.

9) Solve $x^2 + 14x = -48$.

10) Solve $\dfrac{9}{y+6} + \dfrac{4}{y-6} = \dfrac{-4}{y^2 - 36}$.

11) Graph the compound inequality $x + 2y \geq 6$ and $y - x \leq -2$.

Simplify. Assume all variables represent positive real numbers.

12) $\sqrt{45t^9}$

13) $\sqrt{\dfrac{36a^5}{a^3}}$

14) Solve $\sqrt{h^2 + 2h - 7} = h - 3$.

15) Solve $k^2 - 8k + 4 = 0$ by completing the square.

Solve each equation in Exercises 16–18.

16) $r^2 + 5r = -2$

17) $t^2 = 10t - 41$

18) $4m^4 + 4 = 17m^2$

19) Evaluate $\left(\dfrac{81}{16}\right)^{-5/4}$.

20) Let $f(x) = x^2 - 6x + 2$ and $g(x) = x - 3$.
    a) Find $f(-1)$.
    b) Find $(f \circ g)(x)$.
    c) Find x so that $g(x) = -7$.

21) Graph $f(x) = 2^x - 3$. State the domain and range.

22) Solve $\log_4(5x + 1) = 2$.

23) Write as a single logarithm.
$$\log a + 2\log b - 5\log c$$

24) Solve $\log 5r - \log(r + 6) = \log 2$.

25) Solve $e^{-0.04t} = 6$. Give an exact solution and an approximation to four decimal places.

CHAPTER

12 Nonlinear Functions, Conic Sections, and Nonlinear Systems

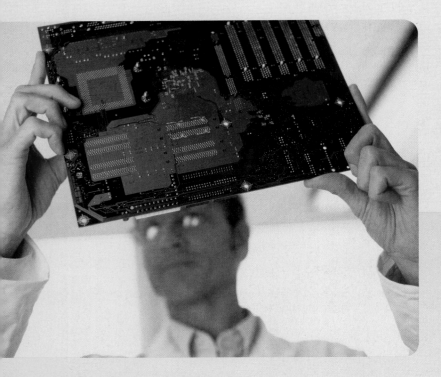

OUTLINE

Study Strategies: Improving Your Memory
12.1 Graphs of Other Useful Functions
12.2 The Circle
12.3 The Ellipse
12.4 The Hyperbola
Putting It All Together
12.5 Nonlinear Systems of Equations
12.6 Second-Degree Inequalities and Systems of Inequalities
Group Activity
emPOWERme: Memory Devices

Math at Work:
Electrical Engineer

Before you ever plug in a toaster or a hair dryer, someone like Gerald Cohen makes sure that it is going to work properly. As an electrical engineer at a small consumer device maker, Gerald's job is to make sure the circuits inside common household appliances function properly. "I've always been fascinated by how things work," Gerald says. "As a little kid, I used to drive my parents crazy taking apart things around the house and then trying to put them back together again."

Gerald's job requires more than a taste for opening things up and looking at their wiring, of course. "In college, I learned the math that allows me to do my job today," says Gerald. "For example, nonlinear functions and systems are essential to understanding power flow." Gerald also has developed an excellent memory that allows him to recall the information he needs to do his job at the moment he needs it.

We'll introduce nonlinear functions, conic sections, and nonlinear systems in this chapter, and cover ways you can enhance your memory.

Study Strategies: Improving Your Memory

Whatever field you choose to pursue, you will need to recall a great deal of information. You will need to remember concepts you learned in college as well as new ideas and practices related to your specific job. The good news is that all of us are equipped with an astonishing ability to remember enormous amounts of information. The strategies below will help you get the most out of your memory.

- Identify the specific information you need to memorize: the essential ideas, equations, or principles you will want to use over and over.

- Relate new information to what you already know. For example, try to think of new math skills in terms of real-world contexts that are familiar to you.

- Rehearse new material, repeating it again and again, until you can do so automatically.
- Try to create acronyms and acrostics to help you remember lists or orders of steps. The emPOWERme on page 845 will help you.
- Engage your senses as you try to memorize new material: Read key passages from your textbook out loud, draw diagrams, trace numbers with your fingers, and so forth.
- Think positively! If you connect new material with positive emotions, you have a better chance of remembering it.

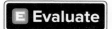

- Test your recall of the material you've tried to memorize. Flashcards are one great method of testing yourself.
- Form a study group and work together to assess your memories.

- Your mind needs time to form long-lasting memories. Don't leave memorization to the last minute.
- Return to the material you've tried to memorize a few days later. This will both further fix the information in your mind and help you identify the areas where you still need more work.

Chapter 12 POWER Plan

P Prepare | O Organize

What are your goals for Chapter 12?	How can you accomplish each goal?
1 Be prepared before and during class.	• _____ • _____ • _____ • _____
2 Understand the homework to the point where you could do it without needing any help or hints.	• _____ • _____ • _____
3 Use the P.O.W.E.R. framework to help you improve your memory: *Memory Devices*.	• _____ • _____ • _____
4 Write your own goal. _____ _____	• _____

What are your objectives for Chapter 12?	How can you accomplish each objective?
1 Be able to graph absolute value, piecewise, and greatest integer functions.	• Learn the shapes of each of these functions and the standard forms of these equations.
2 Be able to identify the standard form of the equations of the different conic sections and graph them.	• Learn definitions of the *circle, ellipse,* and *hyperbola*. • Know the standard form for an equation of each of these conics.
3 Be able to identify and solve a nonlinear system of equations.	• Know the substitution and elimination methods and when to use them. • Be able to sketch the graphs to determine the number of possible solutions for the system.
4 Be able to graph second-degree inequalities.	• Review graphing of conic sections and other functions. • Be able to pick test points to determine the solution set of the system.
5 Write your own goal. _____ _____	• _____

W Work	Read Sections 12.1 to 12.6, and complete the exercises.
E Evaluate Complete the Chapter Review and Chapter Test. How did you do?	**R Rethink** <ul><li>Have you used any acronyms in any of your classes to help with memorizing concepts? If yes, what was the acronym and did it help?</li><li>Are there any acronyms you currently use for your math classes?</li><li>Can you list all the conic sections and draw a graph of each?</li><li>What are examples of ellipses that you have encountered in your daily life? What about hyperbolas?</li></ul>

12.1 Graphs of Other Useful Functions

P Prepare	O Organize
What are your objectives for Section 12.1?	How can you accomplish each objective?
1 Review the Rules for Translating the Graphs of Functions	<ul><li>Review the property for **Translating and Reflecting the Graphs of Functions.** Write each rule in your own words.</li><li>Write a master sheet of all the functions you have learned so far, and add the ones in this chapter to that list.</li><li>Review how to determine the domain and range of a function from its graph.</li><li>Complete the given examples on your own.</li><li>Complete You Trys 1 and 2.</li></ul>
2 Graph Absolute Value Functions	<ul><li>Use the techniques of the transformation of graphs to graph absolute value functions.</li><li>Complete the given examples on your own.</li><li>Complete You Trys 3 and 4.</li></ul>
3 Graph a Piecewise Function	<ul><li>Learn the definition of a *piecewise function*.</li><li>Be careful when choosing values for the domain.</li><li>Complete the given example on your own.</li><li>Complete You Try 5.</li></ul>
4 Define, Graph, and Apply the Greatest Integer Function	<ul><li>Learn the definition of the *greatest integer function*.</li><li>Understand how to make a table of values.</li><li>For applied problems, include the appropriate values for the domain.</li><li>Complete the given examples on your own.</li><li>Complete You Trys 6–8.</li></ul>

Work Read the explanations, follow the examples, take notes, and complete the You Trys.

1 Review the Rules for Translating the Graphs of Functions

In Section 10.5 we discussed shifting the graph of $f(x) = x^2$ both horizontally and vertically. We also learned how to reflect the graph about the x-axis. Let's review the rules for translating the graph of $f(x) = x^2$.

EXAMPLE 1 Graph $g(x) = (x + 3)^2 - 1$ by shifting the graph of $f(x) = x^2$.

Solution

If we compare $g(x)$ to $f(x) = x^2$, what do the constants that have been added to $g(x)$ tell us about translating the graph of $f(x)$?

$$g(x) = (x + 3)^2 - 1$$
$$\uparrow \qquad \uparrow$$
Shift $f(x)$ Shift $f(x)$
left 3.   down 1.

Sketch the graph of $f(x) = x^2$, then move every point on f left 3 and down 1 to obtain the graph of $g(x)$. This moves the vertex from $(0, 0)$ to $(-3, -1)$.

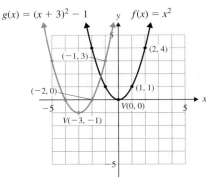

The rules we learned for shifting and reflecting $f(x) = x^2$ works for any function, so we will restate them here and apply them to other functions.

[YOU TRY 1] Graph $g(x) = (x - 1)^2 + 2$ by shifting the graph of $f(x) = x^2$.

 Hint
Summarize these properties in your notes and in your own words.

Property Translating and Reflecting the Graphs of Functions

1) **Vertical Shifts:** Given the graph of any function, $f(x)$, if $g(x) = f(x) + k$ where k is a constant, then the graph of $g(x)$ will be the same shape as the graph of $f(x)$ but g will be shifted **vertically** k units.

2) **Horizontal Shifts:** Given the graph of any function $f(x)$, if $g(x) = f(x - h)$ where h is a constant, then the graph of $g(x)$ will be the same shape as the graph of $f(x)$ but g will be shifted **horizontally** h units.

3) **Reflection about the x-axis:** Given the graph of any function $f(x)$, if $g(x) = -f(x)$, then the graph of $g(x)$ will be the **reflection of the graph of f about the x-axis.** That is, obtain the graph of g by keeping the x-coordinate of each point on f the same but take the negative of the y-coordinate.

Next we will review how to graph the square root functions we learned about in Chapter 9.

EXAMPLE 2

Graph $f(x) = \sqrt{x}$ and $g(x) = -\sqrt{x}$ on the same axes. Identify the domain and range.

Solution

Compare $f(x)$ and $g(x)$.

$$f(x) = \sqrt{x} \qquad g(x) = -\sqrt{x}$$
$$g(x) = -f(x) \qquad \text{Substitute } f(x) \text{ for } \sqrt{x}.$$

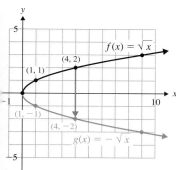

Since $g(x) = -f(x)$, the graph of g is the reflection of the graph of f about the x-axis. This means that when the x-coordinates of their ordered pairs are the same, the y-coordinates of the ordered pairs of g are the *negatives* of the y-coordinates in the ordered pairs of f.

The domain of each function is $[0, \infty)$. The range of $f(x)$ is $[0, \infty)$, and the range of $g(x)$ is $(-\infty, 0]$.

[YOU TRY 2]

Graph $f(x) = \sqrt{x}$ and $g(x) = -\sqrt{x+2}$ on the same axes. Identify the domain and range.

2 Graph Absolute Value Functions

EXAMPLE 3

Graph $f(x) = |x|$ and $g(x) = |x| + 2$ on the same axes. Identify the domain and range.

Solution

Compare $f(x)$ and $g(x)$.

$$f(x) = |x| \qquad g(x) = |x| + 2$$
$$g(x) = f(x) + 2 \qquad \text{Substitute } f(x) \text{ for } |x|.$$

The graph of $g(x) = |x| + 2$ is the same shape as the graph of $f(x)$ except it will be *shifted up* 2 units. Let's make a table of values for each function so that you can see why this is true.

$f(x) = \|x\|$	
x	f(x)
0	0
1	1
2	2
−1	1
−2	2

$g(x) = \|x\| + 2$	
x	g(x)
0	2
1	3
2	4
−1	3
−2	4

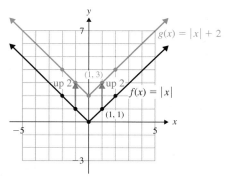

Absolute value functions have V-shaped graphs. The domain of $f(x) = |x|$ is $(-\infty, \infty)$, and the range is $[0, \infty)$. The domain of $g(x) = |x| + 2$ is $(-\infty, \infty)$, and the range is $[2, \infty)$.

The absolute value functions we will study have V-shaped graphs. The graph of a quadratic function is *not* shaped like a V. It is a parabola.

[**YOU TRY 3**] Graph $g(x) = |x| - 1$. Identify the domain and range.

We can combine the techniques used in the transformation of the graphs of functions to help us graph more complicated absolute value functions.

EXAMPLE 4

Graph $h(x) = |x + 2| - 3$.

Solution

The graph of $h(x)$ will be the same shape as the graph of $f(x) = |x|$. So, let's see what the constants in $h(x)$ tell us about transforming the graph of $f(x) = |x|$.

$$h(x) = |x + 2| - 3$$
$$\quad\quad\quad\uparrow \quad\quad\quad \uparrow$$
$$\text{Shift } f(x) \quad \text{Shift } f(x)$$
$$\text{left 2.} \quad\quad \text{down 3.}$$

Sketch the graph of $f(x) = |x|$, including some key points, then *move every point on the graph of f left 2 and down 3 to obtain the graph of h.*

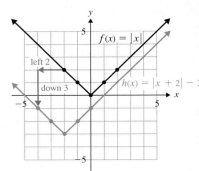

[**YOU TRY 4**] Graph $h(x) = |x - 2| - 4$.

3 Graph a Piecewise Function

Definition

A **piecewise function** is a single function defined by two or more different rules.

EXAMPLE 5

Graph the piecewise function

$$f(x) = \begin{cases} 2x - 4, & x \geq 3 \\ -x + 2, & x < 3 \end{cases}$$

Solution

This is a piecewise function because $f(x)$ is defined by two different rules. *The rule we use to find $f(x)$ depends on which value is substituted for x.*

Graph $f(x)$ by making two separate tables of values, one for each rule.

When $x \geq 3$, use the rule

$$f(x) = 2x - 4$$

The first x-value we will put in the table of values is 3 because it is the smallest number (lower bound) of the domain of $f(x) = 2x - 4$. *The other values we choose for x must be greater than 3 because this is when we use the rule $f(x) = 2x - 4$.* **This part of the graph will not extend to the left of (3, 2).**

$$f(x) = 2x - 4$$
$$(x \geq 3)$$

x	f(x) = 2x − 4, x ≥ 3
3	2
4	4
5	6
6	8

When $x < 3$, use the rule

$$f(x) = -x + 2$$

The first x-value we will put in the table of values is 3 because it is the upper bound of the domain. *Notice that 3 is not included in the domain (the inequality is <, not ≤) so that the point (3, $f(3)$) will be represented as an open circle on the graph.* The other values we choose for x must be less than 3 because this is when we use the rule $f(x) = -x + 2$. **This part of the graph will not extend to the right of (3, −1).**

$$f(x) = -x + 2$$
$$(x < 3)$$

x	f(x) = −x + 2, x < 3
3	−1
2	0
1	1
0	2

(3, −1) is an open circle.

The graph of $f(x)$ is at the left.

[YOU TRY 5]

Graph the piecewise function

$$f(x) = \begin{cases} -2x + 3, & x \leq -2 \\ \dfrac{3}{2}x - 1, & x > -2 \end{cases}$$

4 Define, Graph, and Apply the Greatest Integer Function

Another function that has many practical applications is the greatest integer function.

Hint
You always "round down" when evaluating greatest integer functions.

Definition

The **greatest integer function**

$$f(x) = [\![x]\!]$$

represents the largest integer less than or equal to x.

EXAMPLE 6

Let $f(x) = [\![x]\!]$. Find the following function values.

a) $f\left(9\dfrac{1}{2}\right)$ b) $f(6)$ c) $f(-2.3)$

Solution

a) $f\left(9\dfrac{1}{2}\right) = \left[\!\left[9\dfrac{1}{2}\right]\!\right]$. This is the largest integer *less than or equal to* $9\dfrac{1}{2}$. That number is 9. So $f\left(9\dfrac{1}{2}\right) = \left[\!\left[9\dfrac{1}{2}\right]\!\right] = 9$.

b) $f(x) = [\![6]\!] = 6$. The largest integer *less than or equal to* 6 is 6.

c) To help us understand how to find this function value we will locate -2.3 on a number line.

The largest integer *less than or equal to* -2.3 is -3, so $f(-2.3) = [\![-2.3]\!] = -3$.

W Hint
Be sure you *really* understand how to find function values before doing Example 7.

YOU TRY 6

Let $f(x) = [\![x]\!]$. Find the following function values.

a) $f(5.1)$ b) $f(0)$ c) $f\left(-5\dfrac{1}{4}\right)$

EXAMPLE 7

Graph $f(x) = [\![x]\!]$.

Solution

First, let's look at the part of this function between $x = 0$ and $x = 1$ (when $0 \le x \le 1$).

x	$f(x) = [\![x]\!]$
0	0
$\frac{1}{4}$	0
$\frac{1}{2}$	0
$\frac{3}{4}$	0
$\vdots$	0
1	1

For all values of x greater than or equal to 0 and less than 1, the function value, $f(x)$, equals zero.

⟶ When $x = 1$ the function value changes to 1.

The graph has an open circle at $(1, 0)$ because if $x < 1, f(x) = 0$. That means that x can get *very close to* 1 and the function value will be zero, but $f(1) \ne 0$.

This pattern continues so that for the x-values in the interval $[1, 2)$, the function values are 1. The graph has an open circle at $(2, 1)$.

For the x-values in the interval $[2, 3)$, $f(x) = 2$. The graph has an open circle at $(3, 2)$.

Continuing in this way we get the graph to the right.

The domain of the function is $(-\infty, \infty)$.
The range is the set of all integers
$\{\ldots, -3, -2, -1, 0, 1, 2, 3, \ldots\}$.

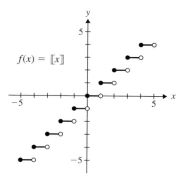

Because of the appearance of the graph, $f(x) = [\![x]\!]$ is also called a **step function**.

[YOU TRY 7] Graph $f(x) = [\![x]\!] - 3$.

EXAMPLE 8

To mail a large envelope within the United States in 2009, the U.S. Postal Service charged $0.88 for the first ounce and $0.17 for each additional ounce or fraction of an ounce. Let $C(x)$ represent the cost of mailing a large envelope within the United States, and let x represent the weight of the envelope, in ounces. Graph $C(x)$ for any large envelope weighing up to (and including) 5 oz. (www.usps.com)

Solution

If a large envelope weighs between 0 and 1 oz ($0 < x \leq 1$), the cost, $C(x)$, is $0.88.

If it weighs more than 1 oz but less than or equal to 2 oz ($1 < x \leq 2$), the cost, $C(x)$, is $0.88 + $0.17 = $1.05.

The pattern continues, and we get the graph to the right.

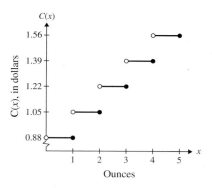

[YOU TRY 8] To mail a package within the United States at *library rate* in 2009, the U.S. Postal Service charged $2.26 for the first pound and $0.37 for each additional pound or fraction of a pound. Let $C(x)$ represent the cost of mailing a package at library rate, and let x represent the weight of the package, in pounds. Graph $C(x)$ for any package weighing up to (and including) 5 lb. (www.usps.com)

Using Technology

We can graph piecewise functions using a graphing calculator by entering each piece separately. Suppose that we wish to graph the piecewise function
$f(x) = \begin{cases} 2x - 4, & x \geq 3 \\ -x + 2, & x < 3 \end{cases}$. Enter $2x - 4$ in Y_1 and $-x + 2$ in Y_2. First put parentheses around each function, and then put parentheses around the interval of x-values for which that part of the function is defined. In order to enter the inequality symbols, press [2nd] [MATH] and use the arrow keys to scroll to the desired symbol before pressing [ENTER].

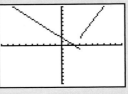

Display the piecewise function using the standard view.

Graph the following functions using a graphing calculator.

1) $f(x) = \begin{cases} 3x - 2, & x \geq 1 \\ -x - 5, & x < 1 \end{cases}$
2) $f(x) = \begin{cases} x + 3, & x \geq 2 \\ -2x + 1, & x < 2 \end{cases}$

3) $f(x) = \begin{cases} -\frac{1}{2}x - 1, & x \leq -3 \\ x - 3, & x > -3 \end{cases}$
4) $f(x) = \begin{cases} 4, & x < 4 \\ -x + 2, & x \geq 4 \end{cases}$

5) $f(x) = \begin{cases} -\frac{2}{3}x + 1, & x \geq -1 \\ x + 3, & x < -1 \end{cases}$
6) $f(x) = \begin{cases} x, & x \geq 0 \\ 5x - 2, & x < 0 \end{cases}$

ANSWERS TO [YOU TRY] EXERCISES

1)
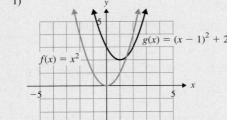

2) $f(x) = \sqrt{x}$; domain: $[0, \infty)$; range: $[0, \infty)$
$g(x) = -\sqrt{x + 2}$; domain: $[-2, \infty)$; range: $(-\infty, 0]$

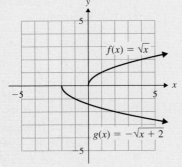

3) domain: $(-\infty, \infty)$, range: $[-1, \infty)$

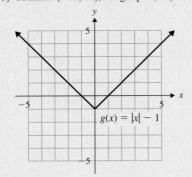

4)
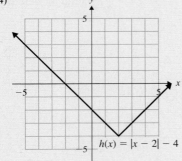

5)

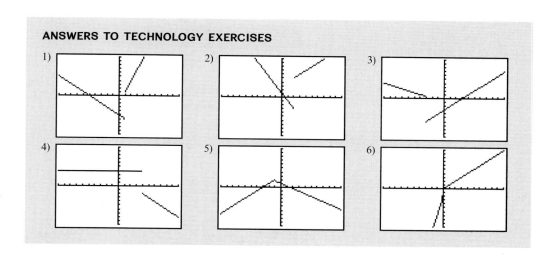

$f(x) = \begin{cases} -2x + 3, & x \leq -2 \\ \frac{3}{2}x - 1, & x > -2 \end{cases}$

6) a) 5 b) 0 c) −6

7) $f(x) = [\![x]\!] - 3$

8)

ANSWERS TO TECHNOLOGY EXERCISES

1)   2)   3)

4)   5)   6)

E Evaluate 12.1 Exercises

Do the exercises, and check your work.

Mixed Exercises: Objectives 1 and 2

Sketch the graph of $f(x)$. Then, graph $g(x)$ on the same axes using the transformation techniques reviewed in this section.

1) $f(x) = |x|$
   $g(x) = |x| - 2$

2) $f(x) = |x|$
   $g(x) = |x| + 1$

3) $f(x) = |x|$
   $g(x) = |x| + 3$

4) $f(x) = |x|$
   $g(x) = |x| - 4$

5) $f(x) = |x|$
   $g(x) = |x - 1|$

6) $f(x) = |x|$
   $g(x) = |x + 3|$

7) $f(x) = |x|$
   $g(x) = |x + 2| + 3$

8) $f(x) = |x|$
   $g(x) = |x - 3| - 4$

9) $f(x) = |x|$
   $g(x) = -|x|$

10) $f(x) = |x|$
    $g(x) = -|x| + 4$

11) $f(x) = |x - 3|$
    $g(x) = -|x - 3|$

12) $f(x) = |x + 4|$
    $g(x) = -|x + 4|$

SECTION 12.1 **Graphs of Other Useful Functions**

Match each function to its graph.

13) $f(x) = -|x - 2|$, $g(x) = |x + 2|$, $h(x) = -|x| - 2$, $k(x) = |x| + 2$

a)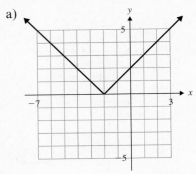

b)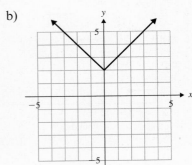

c)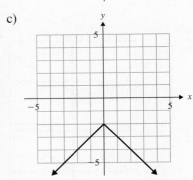

d)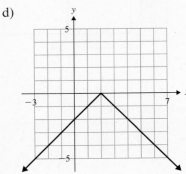

If the following transformations are performed on the graph of $f(x)$ to obtain the graph of $g(x)$, write the equation of $g(x)$.

14) $f(x) = |x|$ is shifted right 1 unit and up 4 units.

15) $f(x) = |x|$ is shifted left 2 units and down 1 unit.

16) $f(x) = |x|$ is reflected about x-axis.

Objective 3: Graph a Piecewise Function

Graph the following piecewise functions.

17) $f(x) = \begin{cases} -x - 3, & x \leq -1 \\ 2x + 2, & x > -1 \end{cases}$

18) $g(x) = \begin{cases} x - 1, & x \geq 2 \\ -3x + 3, & x < 2 \end{cases}$

19) $h(x) = \begin{cases} -x + 5, & x \geq 3 \\ \frac{1}{2}x + 1, & x < 3 \end{cases}$

20) $f(x) = \begin{cases} 2x + 13, & x \leq -4 \\ -\frac{1}{2}x + 1, & x > -4 \end{cases}$

21) $g(x) = \begin{cases} -\frac{3}{2}x - 3, & x < 0 \\ 1, & x \geq 0 \end{cases}$

22) $h(x) = \begin{cases} -\frac{2}{3}x - \frac{7}{3}, & x \geq -1 \\ 2, & x < -1 \end{cases}$

23) $k(x) = \begin{cases} x + 1, & x \geq -2 \\ 2x + 8, & x < -2 \end{cases}$

24) $g(x) = \begin{cases} x, & x \leq 0 \\ 2x + 3, & x > 0 \end{cases}$

25) $f(x) = \begin{cases} 2x - 4, & x > 1 \\ -\frac{1}{3}x - \frac{5}{3}, & x \leq 1 \end{cases}$

26) $k(x) = \begin{cases} \frac{1}{2}x + \frac{5}{2}, & x < 3 \\ -x + 7, & x \geq 3 \end{cases}$

Objective 4: Define, Graph, and Apply the Greatest Integer Function

Let $f(x) = [\![x]\!]$. Find the following function values.

27) $f\left(3\frac{1}{4}\right)$

28) $f\left(10\frac{3}{8}\right)$

29) $f(7.8)$

30) $f(9.2)$

31) $f(8)$

32) $f\left(\frac{4}{5}\right)$

33) $f\left(-6\frac{2}{5}\right)$

34) $f\left(-1\frac{3}{4}\right)$

35) $f(-8.1)$

36) $f(-3.6)$

Graph the following greatest integer functions.

37) $f(x) = [\![x]\!] + 1$
38) $g(x) = [\![x]\!] - 2$
39) $h(x) = [\![x]\!] - 4$
40) $k(x) = [\![x]\!] + 3$
41) $g(x) = [\![x + 2]\!]$
42) $h(x) = [\![x - 1]\!]$
43) $k(x) = \left[\!\left[\dfrac{1}{2}x\right]\!\right]$
44) $f(x) = [\![2x]\!]$

45) To ship small packages within the United States, a shipping company charges $3.75 for the first pound and $1.10 for each additional pound or fraction of a pound. Let $C(x)$ represent the cost of shipping a package, and let x represent the weight of the package. Graph $C(x)$ for any package weighing up to (and including) 6 lb.

46) To deliver small packages overnight, an express delivery service charges $15.40 for the first pound and $4.50 for each additional pound or fraction of a pound. Let $C(x)$ represent the cost of shipping a package overnight, and let x represent the weight of the package. Graph $C(x)$ for any package weighing up to (and including) 6 lb.

47) Visitors to downtown Hinsdale must pay the parking meters to park their cars. The cost of parking is 5¢ for the first 12 min and 5¢ for each additional 12 min or fraction of this time. Let $P(t)$ represent the cost of parking, and let t represent the number of minutes the car is parked at the meter. Graph $P(t)$ for parking a car for up to (and including) 1 hr.

48) To consult with an attorney costs $35 for every 10 min or fraction of this time. Let $C(t)$ represent the cost of meeting an attorney, and let t represent the length of the meeting, in minutes. Graph $C(t)$ for meeting with the attorney for up to (and including) 1 hr.

Extension

Use the translation techniques reviewed in this section to graph each function in Exercises 49–56. State the domain and range of each function.

49) a) $f(x) = x^3$
    b) $f(x) = \sqrt{x}$
    c) $f(x) = \sqrt[3]{x}$

50) $g(x) = (x + 2)^3$
51) $y = \sqrt{x - 1}$
52) $g(x) = \sqrt[3]{x} + 4$
53) $h(x) = x^3 - 3$
54) $h(x) = \sqrt{x - 3}$
55) $k(x) = \sqrt[3]{x - 2}$
56) $g(x) = x^3 + 1$

Mixed Exercises: Objectives 1–3

Graph each function. State the domain and range.

57) $g(x) = |x + 2| + 3$
58) $h(x) = |x + 1| - 5$
59) $k(x) = \dfrac{1}{2}|x|$
60) $g(x) = 2|x|$
61) $f(x) = \sqrt{x + 4} - 2$
62) $y = (x - 3)^3 + 1$
63) $h(x) = -x^3$
64) $g(x) = -\sqrt[3]{x}$

If the following transformations are performed on the graph of $f(x)$ to obtain the graph of $g(x)$, write the equation of $g(x)$.

65) $f(x) = \sqrt{x}$ is shifted 5 units to the left.
66) $f(x) = \sqrt[3]{x}$ is shifted down 6 units.
67) $f(x) = x^3$ is shifted left 2 units and down 1 unit.
68) $f(x) = \sqrt{x}$ is shifted right 1 unit and up 4 units.
69) $f(x) = \sqrt[3]{x}$ is reflected about the x-axis.
70) $f(x) = x^3$ is reflected about the x-axis.

R Rethink

R1) Can you think of an applied problem that would produce a piecewise function?

R2) What makes the graph of a function open down?

R3) Which of the graphs is the hardest to graph?

12.2 The Circle

Prepare

What are your objectives for Section 12.2?

1 Graph a Circle Given in the Form $(x - h)^2 + (y - k)^2 = r^2$

2 Graph a Circle of the Form $Ax^2 + Ay^2 + Cx + Dy + E = 0$

Organize

How can you accomplish each objective?

- Understand what a *conic section* is and how it is formed. Write all the different conic sections in your notes with an example of each.
- Learn the definitions of a *circle*, the *center* of a circle, and the *radius* of a circle.
- Draw a circle in your notes and label the parts of the circle.
- Know the definition of the *standard form for the equation of a circle*. Understand what the variables represent in the equation.
- Complete the given examples on your own.
- Complete You Trys 1–3.

- Know the definition of the *general form for the equation of a circle*.
- Review the procedure for **Completing the Square**.
- Complete the given example on your own.
- Complete You Try 4.

Read the explanations, follow the examples, take notes, and complete the You Trys.

In this chapter, we will study the *conic sections*. When a right circular cone is intersected by a plane, the result is a **conic section**. The conic sections are parabolas, circles, ellipses, and hyperbolas. The following figures show how each conic section is obtained from the intersection of a cone and a plane.

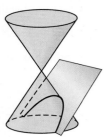
Parabola

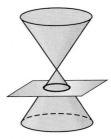
Circle

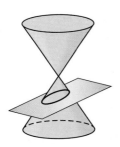
Ellipse

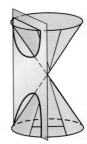
Hyperbola

In Chapter 10 we learned how to graph parabolas. The graph of a quadratic function, $f(x) = ax^2 + bx + c$, is a *parabola* that opens vertically. Another form this function may take is $f(x) = a(x - h)^2 + k$. The graph of a quadratic equation of the form $x = ay^2 + by + c$, or $x = a(y - k)^2 + h$, is a *parabola* that opens horizontally. The next conic section we will discuss is the circle.

We will use the distance formula, presented in Section 10.1, to derive the equation of a circle.

1 Graph a Circle Given in the Form $(x - h)^2 + (y - k)^2 = r^2$

A **circle** is defined as the set of all points in a plane equidistant (the same distance) from a fixed point. The fixed point is the **center** of the circle. The distance from the center to a point on the circle is the **radius** of the circle.

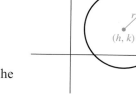

Let the center of a circle have coordinates (h, k) and let (x, y) represent any point on the circle. Let r represent the distance between these two points. r is the radius of the circle.

We will use the distance formula to find the distance between the center, (h, k), and the point (x, y) on the circle.

$$d = \sqrt{(x_2 - x_1)^2 + (y_2 - y_1)^2} \quad \text{Distance formula}$$

Substitute (x, y) for (x_2, y_2), (h, k) for (x_1, y_1), and r for d.

$$r = \sqrt{(x - h)^2 + (y - k)^2}$$
$$r^2 = (x - h)^2 + (y - k)^2 \quad \text{Square both sides.}$$

This is the **standard form** for the equation of a circle.

W Hint
Write this formula in your notes. Include an example.

Definition

Standard Form for the Equation of a Circle: The standard form for the equation of a circle with center (h, k) and radius r is

$$(x - h)^2 + (y - k)^2 = r^2$$

EXAMPLE 1

Graph $(x - 2)^2 + (y + 1)^2 = 9$.

Solution

Standard form is $(x - h)^2 + (y - k)^2 = r^2$.
Our equation is $(x - 2)^2 + (y + 1)^2 = 9$.

$$h = 2 \quad k = -1 \quad r = \sqrt{9} = 3$$

The center is $(2, -1)$. The radius is 3.
To graph the circle, first plot the center $(2, -1)$. Use the radius to locate four points on the circle. From the center, move 3 units up, down, left, and right. Draw a circle through the four points.

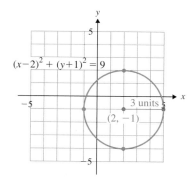

[YOU TRY 1] Graph $(x + 3)^2 + (y - 1)^2 = 16$.

EXAMPLE 2

Graph $x^2 + y^2 = 1$.

Solution

Standard form is $(x - h)^2 + (y - k)^2 = r^2$.
Our equation is $\quad x^2 \;\; + \;\; y^2 \;\; = 1$.

$$h = 0 \quad k = 0 \quad r = \sqrt{1} = 1$$

The center is (0, 0). The radius is 1. Plot (0, 0), then use the radius to locate four points on the circle. From the center, move 1 unit up, down, left, and right. Draw a circle through the four points.

The circle $x^2 + y^2 = 1$ is used often in other areas of mathematics such as trigonometry. $x^2 + y^2 = 1$ is called the **unit circle**.

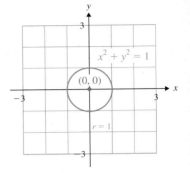

[YOU TRY 2]

Graph $x^2 + y^2 = 25$.

If we are told the center and radius of a circle, we can write its equation.

EXAMPLE 3

Find an equation of the circle with center $(-5, 0)$ and radius $\sqrt{7}$.

Solution

The x-coordinate of the center is h: $h = -5$
The y-coordinate of the center is k: $k = 0$

$$r = \sqrt{7}$$

Substitute these values into $(x - h)^2 + (y - k)^2 = r^2$.

$$[x - (-5)]^2 + (y - 0)^2 = (\sqrt{7})^2 \quad \text{Substitute } -5 \text{ for } x, 0 \text{ for } k, \text{ and } \sqrt{7} \text{ for } r.$$
$$(x + 5)^2 + y^2 = 7$$

[YOU TRY 3]

Find an equation of the circle with center (4, 7) and radius 5.

2 Graph a Circle of the Form $Ax^2 + Ay^2 + Cx + Dy + E = 0$

The equation of a circle can take another form—general form.

 Hint

Write this formula in your notes along with an example.

Definition

General Form for the Equation of a Circle: An equation of the form $Ax^2 + Ay^2 + Cx + Dy + E = 0$, where A, C, D, and E are real numbers, is the **general form** for the equation of a circle.

The coefficients of x^2 and y^2 must be the same in order for this to be the equation of a circle.

To graph a circle given in this form, we complete the square on x and on y to put it into standard form.

After we learn *all* of the conic sections, it is very important that we understand how to identify each one. To do this we will usually look at the coefficients of the square terms.

EXAMPLE 4

Graph $x^2 + y^2 + 6x + 2y + 6 = 0$.

Solution

The coefficients of x^2 and y^2 are each 1. Therefore, this is the equation of a circle.

Our goal is to write the given equation in standard form, $(x - h)^2 + (y - k)^2 = r^2$, so that we can identify its center and radius. To do this we will group x^2 and $6x$ together, group y^2 and $2y$ together, then complete the square on each group of terms.

$$x^2 + y^2 + 6x + 2y + 6 = 0 \quad \text{Group } x^2 \text{ and } 6x \text{ together.}$$
$$(x^2 + 6x) + (y^2 + 2y) = -6 \quad \begin{array}{l}\text{Group } y^2 \text{ and } 2y \text{ together.}\\ \text{Move the constant to the other side.}\end{array}$$

Complete the square for each group of terms.

$$(x^2 + 6x + 9) + (y^2 + 2y + 1) = -6 + 9 + 1 \quad \begin{array}{l}\text{Because 9 and 1 are added on the left,}\\ \text{they must also be added on the right.}\end{array}$$
$$(x + 3)^2 + (y + 1)^2 = 4 \quad \text{Factor; add.}$$

The center of the circle is $(-3, -1)$. The radius is 2.

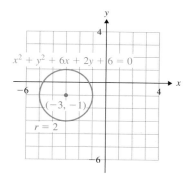

 Hint
See Section 10.1 if you need to review how to complete the square.

YOU TRY 4

Graph $x^2 + y^2 + 10x - 4y + 20 = 0$.

Note

If we rewrite $Ax^2 + Ay^2 + Cx + Dy + E = 0$ in standard form and get $(x - h)^2 + (y - k)^2 = 0$, then the graph is just the point (h, k). If the constant on the right side of the standard form equation is a negative number then the equation has no graph.

Using Technology

Recall that the equation of a circle is not a function. However, if we want to graph an equation on a graphing calculator, it must be entered as a function or a pair of functions. Therefore, to graph a circle we must solve the equation for y in terms of x.

Let's discuss how to graph $x^2 + y^2 = 4$ on a graphing calculator.

We must solve the equation for y.

$$x^2 + y^2 = 4$$
$$y^2 = 4 - x^2$$
$$y = \pm\sqrt{4 - x^2}$$

Now the equation of the circle $x^2 + y^2 = 4$ is rewritten so that y is in terms of x. In the graphing calculator, enter $y = \sqrt{4 - x^2}$ as Y_1. This represents the top half of the circle since the y-values are positive above the x-axis. Enter $y = -\sqrt{4 - x^2}$ as Y_2. This represents the bottom half of the circle since the y-values are negative below the x-axis. Here we have the window set from -3 to 3 in both the x- and y-directions. Press GRAPH.

The graph is distorted and does not actually look like a circle! This is because the screen is rectangular, and the graph is longer in the x-direction. We can "fix" this by squaring the window.

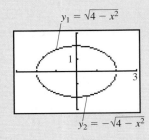

To square the window and get a better representation of the graph of $x^2 + y^2 = 4$, press ZOOM and choose 5:ZSquare. The graph reappears on a "squared" window and now looks like a circle.

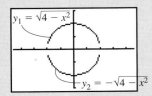

Identify the center and radius of each circle. Then, rewrite each equation for y in terms of x, and graph each circle on a graphing calculator. These problems come from the homework exercises.

1) $x^2 + y^2 = 36$; Exercise 11
2) $x^2 + y^2 = 9$; Exercise 13
3) $(x + 3)^2 + y^2 = 4$; Exercise 7
4) $x^2 + (y - 1)^2 = 25$; Exercise 15

ANSWERS TO [YOU TRY] EXERCISES

1)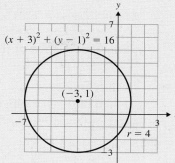

2)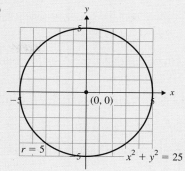

3) $(x-4)^2 + (y-7)^2 = 25$

4)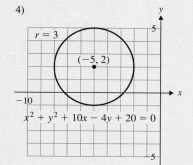

ANSWERS TO TECHNOLOGY EXERCISES

1) Center (0, 0); radius = 6; $Y_1 = \sqrt{36 - x^2}$, $Y_2 = -\sqrt{36 - x^2}$

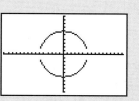

2) Center (0, 0); radius = 3; $Y_1 = \sqrt{9 - x^2}$, $Y_2 = -\sqrt{9 - x^2}$

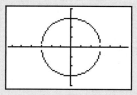

3) Center −3, 0; radius = 2; $Y_1 = \sqrt{4 - (x + 3)^2}$, $Y_2 = -\sqrt{4 - (x + 3)^2}$

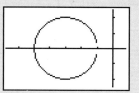

4) Center (0, 1); radius = 5; $Y_1 = 1 + \sqrt{25 - x^2}$, $Y_2 = 1 - \sqrt{25 - x^2}$

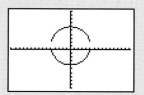

Evaluate 12.2 Exercises

Do the exercises, and check your work.

Objective 1: Graph a Circle Given in the Form $(x - h)^2 + (y - k)^2 = r^2$

1) Is the equation of a circle a function? Explain your answer.

2) The standard form for the equation of a circle is
$$(x - h)^2 + (y - k)^2 = r^2$$
Identify the center and the radius.

Identify the center and radius of each circle and graph.

3) $(x + 2)^2 + (y - 4)^2 = 9$

4) $(x + 1)^2 + (y + 3)^2 = 25$

5) $(x - 5)^2 + (y - 3)^2 = 1$

6) $x^2 + (y - 5)^2 = 9$

7) $(x + 3)^2 + y^2 = 4$

8) $(x - 2)^2 + (y - 2)^2 = 36$

9) $(x - 6)^2 + (y + 3)^2 = 16$

10) $(x + 8)^2 + (y - 4)^2 = 4$

11) $x^2 + y^2 = 36$ 12) $x^2 + y^2 = 16$

13) $x^2 + y^2 = 9$ 14) $x^2 + y^2 = 25$

15) $x^2 + (y - 1)^2 = 25$ 16) $(x + 3)^2 + y^2 = 1$

Find an equation of the circle with the given center and radius.

17) Center (4, 1); radius = 5

18) Center (3, 5); radius = 2

19) Center (−3, 2); radius = 1

20) Center (4, −6); radius = 3

21) Center (−1, −5); radius = $\sqrt{3}$

22) Center (−2, −1); radius = $\sqrt{5}$

23) Center (0, 0); radius = $\sqrt{10}$

24) Center (0, 0); radius = $\sqrt{6}$

25) Center (6, 0); radius = 4

26) Center (0, −3); radius = 5

27) Center (0, −4); radius = $2\sqrt{2}$

28) Center (1, 0); radius = $3\sqrt{2}$

Objective 2: Graph a Circle of the Form $Ax^2 + Ay^2 + Cx + Dy + E = 0$

Write the equation of the circle in standard form.

Fill It In
Fill in the blanks with either the missing mathematical step or the reason for the given step.

29) $x^2 + y^2 - 8x + 2y + 8 = 0$
$(x^2 - 8x) + (y^2 + 2y) = -8$ _____

_____ Complete the square.
_____ Factor.

30) $x^2 + y^2 + 2x + 10y + 10 = 0$
$(x^2 + 2x) + (y^2 + 10y) = -10$ _____

_____ Complete the square.
_____ Factor.

Put the equation of each circle in the form $(x - h)^2 + (y - k)^2 = r^2$, identify the center and the radius, and graph.

31) $x^2 + y^2 + 2x + 10y + 17 = 0$

32) $x^2 + y^2 - 4x - 6y + 9 = 0$

33) $x^2 + y^2 + 8x - 2y - 8 = 0$

34) $x^2 + y^2 - 6x + 8y + 24 = 0$

35) $x^2 + y^2 - 10x - 14y + 73 = 0$

36) $x^2 + y^2 + 12x + 12y + 63 = 0$

37) $x^2 + y^2 + 6y + 5 = 0$

38) $x^2 + y^2 + 2x - 24 = 0$

39) $x^2 + y^2 - 4x - 1 = 0$

40) $x^2 + y^2 - 10y + 22 = 0$

41) $x^2 + y^2 - 8x + 8y - 4 = 0$

42) $x^2 + y^2 - 6x + 2y - 6 = 0$

43) $4x^2 + 4y^2 - 12x - 4y - 6 = 0$
(Hint: Begin by dividing the equation by 4.)

44) $16x^2 + 16y^2 + 16x - 24y - 3 = 0$
(Hint: Begin by dividing the equation by 16.)

Mixed Exercises: Objectives 1 and 2

45) The London Eye is a Ferris wheel that opened in London in March 2000. It is 135 m high, and the bottom of the wheel is 7 m off the ground.

a) What is the diameter of the wheel?

b) What is the radius of the wheel?

c) Using the axes in the illustration, what are the coordinates of the center of the wheel?

d) Write the equation of the wheel.

(www.aviewoncities.com/london/londoneye.htm)

46) The first Ferris wheel was designed and built by George W. Ferris in 1893 for the Chicago World's Fair. It was 264 ft tall, and the wheel had a diameter of 250 ft.

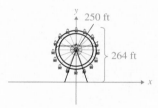

a) What is the radius of the wheel?

b) Using the axes in the illustration, what are the coordinates of the center of the wheel?

c) Write the equation of the wheel.

47) A CD is placed on axes as shown in the figure where the units of measurement for x and y are millimeters. Using $\pi = 3.14$, what is the surface area of a CD (to the nearest square millimeter)?

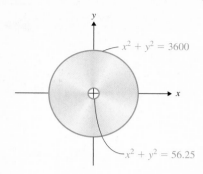

48) A storage container is in the shape of a right circular cylinder. The top of the container may be described by the equation $x^2 + y^2 = 5.76$, as shown in the figure (x and y are in feet). If the container is 3.2 ft tall, what is the storage capacity of the container (to the nearest ft^3)?

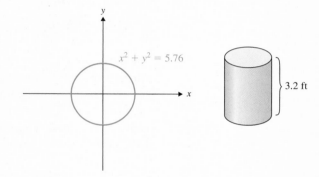

Use a graphing calculator to graph each equation. Square the viewing window.

49) $(x - 5)^2 + (y - 3)^2 = 1$

50) $x^2 + y^2 = 16$

51) $(x + 2)^2 + (y - 4)^2 = 9$

52) $(x + 3)^2 + y^2 = 1$

R Rethink

R1) Can you write down the equation for a circle without your notes?

R2) Can you explain how to graph a circle to a classmate?

R3) What profession might deal with circles on a daily basis?

12.3 The Ellipse

P Prepare
What are your objectives for Section 12.3?

O Organize
How can you accomplish each objective?

1 Graph an Ellipse	• Know the definition of the *standard form for the equation of an ellipse.* Understand what the variables represent in the equation. • Know how to identify the *center* of the ellipse from the equation. • Understand the properties of an ellipse with center at origin. • Review the procedure for **Completing the Square.** • Complete the given examples on your own. • Complete You Trys 1–4.

W Work Read the explanations, follow the examples, take notes, and complete the You Trys.

1 Graph an Ellipse

The next conic section we will study is the *ellipse*. An **ellipse** is the set of all points in a plane such that the *sum* of the distances from a point on the ellipse to two fixed points is constant. Each fixed point is called a **focus** (plural: **foci**). The point halfway between the foci is the **center** of the ellipse.

The orbits of planets around the sun as well as satellites around the Earth are elliptical. Statuary Hall in the U.S. Capitol building is an ellipse. If a person stands at one focus of this ellipse and whispers, a person standing across the room on the other focus can clearly hear what was said. Properties of the ellipse are used in medicine as well. One procedure for treating kidney stones involves immersing the patient in an elliptical tub of water. The kidney stone is at one focus, while at the other focus, high energy shock waves are produced, which destroy the kidney stone.

W Hint
Write this equation in your notes along with the information above, in bold. Include an example.

Definition

Standard Form for the Equation of an Ellipse: The standard form for the equation of an ellipse is

$$\frac{(x-h)^2}{a^2} + \frac{(y-k)^2}{b^2} = 1$$

The center of the ellipse is (h, k).

It is important to remember that the terms on the left are *both* positive quantities.

EXAMPLE 1

Graph $\dfrac{(x-3)^2}{16} + \dfrac{(y-1)^2}{4} = 1$.

Solution

Standard form is $\dfrac{(x-h)^2}{a^2} + \dfrac{(y-k)^2}{b^2} = 1$.

Our equation is $\dfrac{(x-3)^2}{16} + \dfrac{(y-1)^2}{4} = 1$.

$h = 3 \quad k = 1$
$a = \sqrt{16} = 4 \quad b = \sqrt{4} = 2$

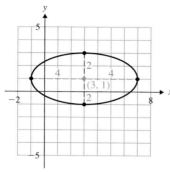

The center is (3, 1).

To graph the ellipse, first plot the center (3, 1). Since $a = 4$ and a^2 is under the squared quantity containing the x, move 4 units each way in the x-direction from the center. These are two points on the ellipse.

Since $b = 2$ and b^2 is under the squared quantity containing the y, move 2 units each way in the y-direction from the center. These are two more points on the ellipse. Sketch the ellipse through the four points.

[YOU TRY 1]

Graph $\dfrac{(x+2)^2}{25} + \dfrac{(y-3)^2}{16} = 1$.

EXAMPLE 2

Graph $\dfrac{x^2}{9} + \dfrac{y^2}{25} = 1$.

Solution

Standard form is $\dfrac{(x-h)^2}{a^2} + \dfrac{(y-k)^2}{b^2} = 1$.

Our equation is $\dfrac{x^2}{9} + \dfrac{y^2}{25} = 1$.

$h = 0 \quad k = 0$
$a = \sqrt{9} = 3 \quad b = \sqrt{25} = 5$

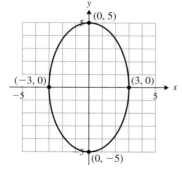

The center is (0, 0).

Plot the center (0, 0). Since $a = 3$ and a^2 is under the x^2, move 3 units each way in the x-direction from the center. These are two points on the ellipse.

Since $b = 5$ and b^2 is under the y^2, move 5 units each way in the y-direction from the center. These are two more points on the ellipse. Sketch the ellipse through the four points.

[YOU TRY 2]

Graph $\dfrac{x^2}{36} + \dfrac{y^2}{9} = 1$.

In Example 2, note that the *origin*, (0, 0), is the center of the ellipse. Notice that $a = 3$ and the *x*-intercepts are (3, 0) and (−3, 0); $b = 5$ and the *y*-intercepts are (0, 5) and (0, −5). We can generalize these relationships as follows.

> **Property** Equation of an Ellipse with Center at Origin
>
> The graph of $\dfrac{x^2}{a^2} + \dfrac{y^2}{b^2} = 1$ is an ellipse with center at the origin, *x*-intercepts $(a, 0)$ and $(-a, 0)$, and *y*-intercepts $(0, b)$ and $(0, -b)$.
>
>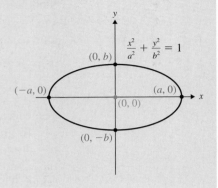

Looking at Examples 1 and 2 we can make another interesting observation.

Example 1	**Example 2**
$\dfrac{(x-3)^2}{16} + \dfrac{(y-1)^2}{4} = 1$	$\dfrac{x^2}{9} + \dfrac{y^2}{25} = 1$
$a^2 = 16 \quad b^2 = 4$	$a^2 = 9 \quad b^2 = 25$
$a^2 > b^2$	$b^2 > a^2$
The number under $(x-3)^2$ is greater than the number under $(y-1)^2$. *The ellipse is longer in the x-direction.*	The number under y^2 is greater than the number under x^2. *The ellipse is longer in the y-direction.*

This relationship between a^2 and b^2 will always produce the same result. The equation of an ellipse can take other forms.

EXAMPLE 3

Graph $4x^2 + 25y^2 = 100$.

Solution

How can we tell if this is a circle or an ellipse? We look at the coefficients of x^2 and y^2. Both of the coefficients are positive, *and* they are different. *This is an ellipse.* (If this were a circle, the coefficients would be the same.)

Since the standard form for the equation of an ellipse has a 1 on one side of the = sign, divide both sides of $4x^2 + 25y^2 = 100$ by 100 to obtain a 1 on the right.

$$4x^2 + 25y^2 = 100$$

$$\frac{4x^2}{100} + \frac{25y^2}{100} = \frac{100}{100} \qquad \text{Divide both sides by 100.}$$

$$\frac{x^2}{25} + \frac{y^2}{4} = 1 \qquad \text{Simplify.}$$

> **W Hint**
> Use the coefficients of x^2 and y^2 to help you identify the type of conic section you are given.

The center is (0, 0). $a = \sqrt{25} = 5$ and $b = \sqrt{4} = 2$. Plot (0, 0). Move 5 units each way from the center in the x-direction. Move 2 units each way from the center in the y-direction.

Notice that the x-intercepts are (5, 0) and (−5, 0). The y-intercepts are (2, 0) and (−2, 0).

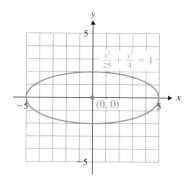

[**YOU TRY 3**] Graph $x^2 + 4y^2 = 4$.

Just like with the equation of a circle, sometimes we must complete the square to put the equation of an ellipse into standard form.

EXAMPLE 4 Graph $9x^2 + 4y^2 + 18x - 16y - 11 = 0$.

Solution

Our goal is to write the given equation in standard form, $\dfrac{(x-h)^2}{a^2} + \dfrac{(y-k)^2}{b^2} = 1$. To do this we will group the x-terms together, group the y-terms together, then complete the square on each group of terms.

$9x^2 + 4y^2 + 18x - 16y - 11 = 0$ Group the x-terms together and the y-terms together.
$(9x^2 + 18x) + (4y^2 - 16y) = 11$ Move the constant to the other side.
$9(x^2 + 2x) + 4(y^2 - 4y) = 0$ Factor out the coefficients of the squared terms.

Complete the square, inside the parentheses, for each group of terms.

$9 \cdot 1 = 9 \qquad 4 \cdot 4 = 16$
$9(x^2 + 2x + 1) + 4(y^2 - 4y + 4) = 11 + 9 + 16$ Since 9 and 16 are added on the left, they must also be added on the right.
$9(x+1)^2 + 4(y-2)^2 = 36$ Factor; add.
$\dfrac{9(x+1)^2}{36} + \dfrac{4(y-2)^2}{36} = \dfrac{36}{36}$ Divide by 36 to get 1 on the right.
$\dfrac{(x+1)^2}{4} + \dfrac{(y-2)^2}{9} = 1$ Standard form

The center of the ellipse is (−1, 2). The graph is at right.

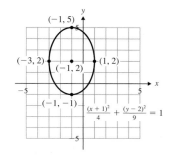

[**YOU TRY 4**] Graph $4x^2 + 25y^2 - 24x - 50y - 39 = 0$.

Using Technology

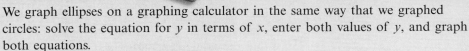

We graph ellipses on a graphing calculator in the same way that we graphed circles: solve the equation for y in terms of x, enter both values of y, and graph both equations.

Let's graph the ellipse given in Example 3: $4x^2 + 25y^2 = 100$.
Solve for y.
$$25y^2 = 100 - 4x^2$$
$$y^2 = 4 - \frac{4x^2}{25}$$
$$y = \pm\sqrt{4 - \frac{4x^2}{25}}$$

Enter $Y_1 = \sqrt{4 - \frac{4x^2}{25}}$, the top half of the ellipse (the y-values are positive above the x-axis). Enter $Y_2 = -\sqrt{4 - \frac{4x^2}{25}}$, the bottom half of the ellipse (the y-values are negative below the x-axis). Set an appropriate window, and press GRAPH.

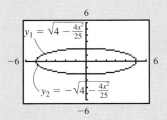

Identify the center of each ellipse, rewrite each equation for y in terms of x, and graph each equation on a graphing calculator.

1) $4x^2 + 9y^2 = 36$
2) $x^2 + \frac{y^2}{4} = 1$
3) $\frac{x^2}{25} + (y+4)^2 = 1$
4) $25x^2 + y^2 = 25$

ANSWERS TO [YOU TRY] EXERCISES

1)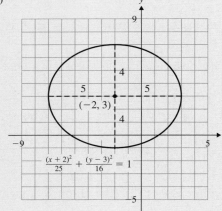

2)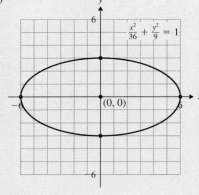

3)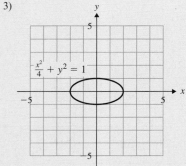

4)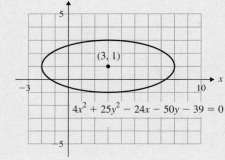

ANSWERS TO TECHNOLOGY EXERCISES

1) Center $(0, 0)$; $Y_1 = \sqrt{4 - \dfrac{4x^2}{9}}$; $Y_2 = -\sqrt{4 - \dfrac{4x^2}{9}}$

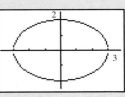

2) Center $(0, 0)$; $Y_1 = \sqrt{4 - 4x^2}$; $Y_2 = -\sqrt{4 - 4x^2}$

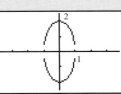

3) Center $(0, -4)$; $Y_1 = -4 + \sqrt{1 - \dfrac{x^2}{25}}$; $Y_2 = -4 - \sqrt{1 - \dfrac{x^2}{25}}$

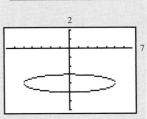

4) Center $(0, 0)$; $Y_1 = \sqrt{25 - 25x^2}$; $Y_2 = -\sqrt{25 - 25x^2}$

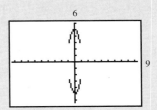

E Evaluate 12.3 Exercises

Do the exercises, and check your work.

Objective 1: Graph an Ellipse

Determine whether each statement is *true* or *false*.

1) The graph of an ellipse is a function.

2) The center of the ellipse with equation
$$\dfrac{(x-1)^2}{16} + \dfrac{(y-5)^2}{9} = 1 \text{ is } (-1, -5).$$

3) The center of the ellipse with equation
$$\dfrac{(x-8)^2}{9} + \dfrac{(y+3)^2}{25} = 1 \text{ is } (8, -3).$$

4) The center of the ellipse with equation
$$\dfrac{x^2}{7} + \dfrac{y^2}{4} = 1 \text{ is } (0, 0).$$

5) The graph of $4x^2 + 25y^2 = 100$ is an ellipse.

6) The graph of $\dfrac{(x+3)^2}{4} - \dfrac{(y+2)^2}{9} = 1$ is an ellipse.

7) The graph of $9y^2 - x^2 = 9$ is an ellipse.

8) The equation of $4x^2 + y^2 + 8x - 10y + 8 = 0$ can be put into the standard form for the equation of an ellipse by completing the square.

Identify the center of each ellipse, and graph the equation.

9) $\dfrac{(x+2)^2}{9} + \dfrac{(y-1)^2}{4} = 1$

10) $\dfrac{(x-4)^2}{4} + \dfrac{(y-3)^2}{16} = 1$

11) $\dfrac{(x-3)^2}{9} + \dfrac{(y+2)^2}{16} = 1$

12) $\dfrac{(x+4)^2}{25} + \dfrac{(y-5)^2}{16} = 1$ 13) $\dfrac{x^2}{36} + \dfrac{y^2}{16} = 1$

14) $\dfrac{x^2}{36} + \dfrac{y^2}{4} = 1$ 15) $x^2 + \dfrac{y^2}{4} = 1$

16) $\dfrac{x^2}{9} + y^2 = 1$

17) $\dfrac{x^2}{25} + (y + 4)^2 = 1$

18) $(x + 3)^2 + \dfrac{(y + 4)^2}{9} = 1$

19) $\dfrac{(x + 1)^2}{4} + \dfrac{(y + 3)^2}{9} = 1$

20) $\dfrac{(x - 2)^2}{16} + \dfrac{y^2}{25} = 1$

21) $4x^2 + 9y^2 = 36$

22) $x^2 + 4y^2 = 16$

23) $25x^2 + y^2 = 25$

24) $9x^2 + y^2 = 36$

Write the equation of the ellipse in standard form.

Fill It In

Fill in the blanks with either the missing mathematical step or the reason for the given step.

25) $3x^2 + 2y^2 - 6x + 4y - 7 = 0$
$(3x^2 - 6x) + (2y^2 + 4y) = 7$

_____ Factor out the coefficients of the squared terms.

$3(x^2 - 2x + 1) + 2(y^2 + 2y + 1)$
$= 7 + 3(1) + 2(1)$

_____ Factor.

_____ Divide both sides by 12.

26) $4x^2 + 9y^2 + 16x + 54y + 61 = 0$
$(4x^2 + 16x) + (9y^2 + 54y) = -61$

_____ Factor out the coefficients of the squared terms.

_____ Complete the square.

$4(x + 2)^2 + 9(y + 3)^2 = 36$

_____ Divide both sides by 36.

Put each equation into the standard form for the equation of an ellipse, and graph.

27) $x^2 + 4y^2 - 2x - 24y + 21 = 0$

28) $9x^2 + 4y^2 + 36x - 8y + 4 = 0$

29) $9x^2 + y^2 + 72x + 2y + 136 = 0$

30) $x^2 + 4y^2 - 6x - 40y + 105 = 0$

31) $4x^2 + 9y^2 - 16x - 54y + 61 = 0$

32) $4x^2 + 25y^2 + 8x + 200y + 304 = 0$

33) $25x^2 + 4y^2 + 150x + 125 = 0$

34) $4x^2 + y^2 - 2y - 15 = 0$

Extension

Write an equation of the ellipse containing the following points.

35) $(-3, 0), (3, 0), (0, -5),$ and $(0, 5)$

36) $(-6, 0), (6, 0), (0, -2),$ and $(0, 2)$

37) $(-7, 0), (7, 0), (0, -1),$ and $(0, 1)$

38) $(-1, 0), (1, 0), (0, -9),$ and $(0, 9)$

39) $(3, 5), (3, -3), (1, 1),$ and $(5, 1)$

40) $(-1, 1), (-1, -5), (4, -2),$ and $(-6, -2)$

41) Is a circle a special type of ellipse? Explain your answer.

42) The Oval Office in the White House is an ellipse about 36 ft long and 29 ft wide. If the center of the room is at the origin of a Cartesian coordinate system and the length of the room is along the x-axis, write an equation of the elliptical room. (www.whitehousehistory.org)

43) The fuselage of a Boeing 767 jet has an elliptical cross section that is 198 in. wide and 213 in. tall. If the center of this cross section is at the origin and the width is along the x-axis, write an equation of this ellipse. (Jan Roskam, *Airplane Design*, p. 89).

44) The arch of a bridge over a canal in Amsterdam is half of an ellipse. At water level the arch is 14 ft wide, and it is 6 ft tall at its highest point.

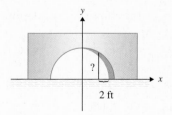

a) Write an equation of the arch.

b) What is the height of the arch (to the nearest foot) 2 ft from the bottom edge?

Use a graphing calculator to graph the following. Square the viewing window.

45) $\dfrac{x^2}{36} + \dfrac{y^2}{16} = 1$

46) $9x^2 + y^2 = 36$

R Rethink

R1) What is the difference between a circle and an ellipse?

R2) What is the difference in the equation between a circle and an ellipse?

R3) Which of the graphs is the hardest to graph, a circle or an ellipse?

12.4 The Hyperbola

P Prepare

What are your objectives for Section 12.4?	How can you accomplish each objective?
1 Graph a Hyperbola in Standard Form	• Learn the definition of a *hyperbola*. Draw an example in your notes, and label the diagram. • Learn the definitions of the *focus* and the *center* of a hyperbola. • Learn the definition of the *standard form for the equation of a hyperbola*. • Understand the properties of an equation of a hyperbola with center at the origin. • Complete the given examples on your own. • Complete You Trys 1–3.
2 Graph a Hyperbola in Nonstandard Form	• Learn the definition of a *nonstandard form for the equation of a hyperbola*. • Complete the given example on your own. • Complete You Try 4.
3 Graph Other Square Root Functions	• Know the standard forms of equations of conic sections. • Review the squaring procedure when rearranging equations. • Complete the given example on your own. • Complete You Try 5.

W Work

Read the explanations, follow the examples, take notes, and complete the You Trys.

1 Graph a Hyperbola in Standard Form

The last of the conic sections is the *hyperbola*. A **hyperbola** is the set of all points, *P*, in a plane such that the absolute value of the *difference* of the distances $|d_1 - d_2|$, from two fixed points is constant. Each fixed point is called a **focus**. The point halfway between the foci is the **center** of the hyperbola.

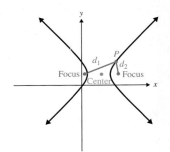

Some navigation systems used by ships are based on the properties of hyperbolas. A lamp casts a hyperbolic shadow on a wall, and many telescopes use hyperbolic lenses.

A hyperbola is a graph consisting of two branches.

> **Hint**
> Write this definition in your notes, and include an example.

Definition Standard Form for the Equation of a Hyperbola

1) A hyperbola with center (h, k) and branches that open in the *x-direction* has equation

$$\frac{(x - h)^2}{a^2} - \frac{(y - k)^2}{b^2} = 1.$$

Its graph is to the right.

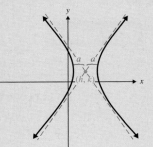

2) A hyperbola with center (h, k) and branches that open in the *y-direction* has equation

$$\frac{(y - k)^2}{b^2} - \frac{(x - h)^2}{a^2} = 1.$$

Its graph is to the right.

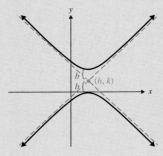

Notice in 1) that $\dfrac{(x - h)^2}{a^2}$ is the positive quantity, and the branches open in the *x*-direction. In 2), the positive quantity is $\dfrac{(y - k)^2}{b^2}$, and the branches open in the *y*-direction.

In 1) and 2) notice how the branches of the hyperbola get closer to the dotted lines as the branches continue indefinitely. These dotted lines are called **asymptotes**. They are not an actual part of the graph of the hyperbola, but we can use them to help us obtain the hyperbola.

EXAMPLE 1

Graph $\dfrac{(x+2)^2}{9} - \dfrac{(y-1)^2}{4} = 1$.

Solution

How do we know that this is a hyperbola and not an ellipse? *It is a hyperbola because there is a subtraction sign between the two quantities on the left.* If it was addition, it would be an ellipse.

Standard form is $\dfrac{(x-h)^2}{a^2} - \dfrac{(y-k)^2}{b^2} = 1$.

Our equation is $\dfrac{(x+2)^2}{9} - \dfrac{(y-1)^2}{4} = 1$.

$$h = -2 \quad k = 1$$
$$a = \sqrt{9} = 3 \quad b = \sqrt{4} = 2$$

The center is $(-2, 1)$. Because the quantity $\dfrac{(x-h)^2}{a^2}$ is the *positive* quantity, the branches of the hyperbola will open in the *x*-direction.

We will use the center, $a = 3$, and $b = 2$ to draw a *reference rectangle*. The diagonals of this rectangle are the asymptotes of the hyperbola.

First, plot the center $(-2, 1)$. Since $a = 3$ and a^2 is under the squared quantity containing the *x*, move 3 units each way in the *x*-direction from the center. These are two points on the rectangle.

Since $b = 2$ and b^2 is under the squared quantity containing the *y*, move 2 units each way in the *y*-direction from the center. These are two more points on the rectangle.

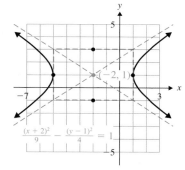

Draw the rectangle containing these four points, then draw the diagonals of the rectangle as dotted lines. These are the asymptotes of the hyperbola.

Sketch the branches of the hyperbola opening in the *x*-direction with the branches approaching the asymptotes.

YOU TRY 1

Graph $\dfrac{(x+1)^2}{9} - \dfrac{(y+1)^2}{16} = 1$.

EXAMPLE 2

Graph $\dfrac{y^2}{4} - \dfrac{x^2}{25} = 1$.

Solution

Standard form is $\dfrac{(y-k)^2}{b^2} - \dfrac{(x-h)^2}{a^2} = 1$. Our equation is $\dfrac{y^2}{4} - \dfrac{x^2}{25} = 1$.

$$k = 0 \quad h = 0$$
$$b = \sqrt{4} = 2 \quad a = \sqrt{25} = 5$$

W Hint
After reading Examples 1 and 2, create a procedure for graphing a hyperbola.

The center is (0, 0). *Because the quantity $\frac{y^2}{4}$ is the positive quantity, the branches of the hyperbola will open in the y-direction.*

Use the center, $a = 5$, and $b = 2$ to draw the reference rectangle and its diagonals.

Plot the center (0, 0). Because $a = 5$ and a^2 is under the x^2, move 5 units each way in the x-direction from the center to get two points on the rectangle.

Because $b = 2$ and b^2 is under the y^2, move 2 units each way in the y-direction from the center to get two more points on the rectangle.

Draw the rectangle and its diagonals as dotted lines. These are the asymptotes of the hyperbola.

Sketch the branches of the hyperbola opening in the y-direction approaching the asymptotes.

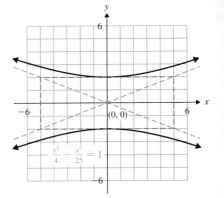

[**YOU TRY 2**] Graph $\frac{y^2}{16} - \frac{x^2}{16} = 1$.

W Hint
Summarize these properties in your notes. Include examples.

> **Property** Equation of a Hyperbola with Center at the Origin
>
> 1) The graph of $\frac{x^2}{a^2} - \frac{y^2}{b^2} = 1$ is a hyperbola with center (0, 0) and x-intercepts $(a, 0)$ and $(-a, 0)$ as shown below.
>
> 2) The graph of $\frac{y^2}{b^2} - \frac{x^2}{a^2} = 1$ is a hyperbola with center (0, 0) and y-intercepts $(0, b)$ and $(0, -b)$ as shown below.
>
>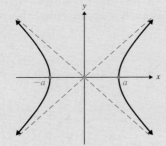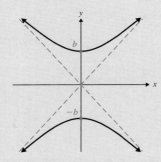
>
> The equations of the asymptotes are $y = \frac{b}{a}x$ and $y = -\frac{b}{a}x$.

Let's look at another example.

EXAMPLE 3

Graph $y^2 - 9x^2 = 9$.

Solution

This is a hyperbola because there is a subtraction sign between the two terms.

Since the standard form for the equation of the hyperbola has a 1 on one side of the = sign, divide both sides of $y^2 - 9x^2 = 9$ by 9 to obtain a 1 on the right.

$$y^2 - 9x^2 = 9$$
$$\frac{y^2}{9} - \frac{9x^2}{9} = \frac{9}{9} \quad \text{Divide both sides by 9.}$$
$$\frac{y^2}{9} - x^2 = 1 \quad \text{Simplify.}$$

The center is (0, 0). *The branches of the hyperbola will open in the y-direction since $\frac{y^2}{9}$ is a positive quantity.*

x^2 is the same as $\frac{x^2}{1}$, so $a = \sqrt{1} = 1$ and $b = \sqrt{9} = 3$

Plot the center at the origin. Move 1 unit each way in the *x*-direction from the center and 3 units each way in the *y*-direction. Draw the rectangle and the asymptotes.

We can find the equations of the asymptotes using $a = 1$ and $b = 3$:

$$y = \frac{3}{1}x \text{ and } y = -\frac{3}{1}x \qquad y = \frac{b}{a}x \text{ and } y = -\frac{b}{a}x.$$

The equations of the asymptotes are $y = 3x$ and $y = -3x$.

Sketch the branches of the hyperbola opening in the *y*-direction approaching the asymptotes. The *y*-intercepts are (0, 3) and (0, −3). There are no *x*-intercepts.

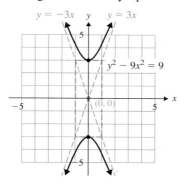

[YOU TRY 3]

Graph $4x^2 - 9y^2 = 36$.

2 Graph a Hyperbola in Nonstandard Form

Equations of hyperbolas can take other forms. We look at one here.

W Hint
Write this definition in your notes. Include an example.

Definition A Nonstandard Form for the Equation of a Hyperbola

The graph of the equation $xy = c$, where c is a nonzero constant, is a hyperbola whose asymptotes are the *x*- and *y*-axes.

EXAMPLE 4

Graph $xy = 4$.

Solution

Solve for y.

$$y = \frac{4}{x} \quad \text{Divide by } x.$$

Notice that we cannot substitute 0 for x in the equation $y = \dfrac{4}{x}$ because then the denominator would equal zero. Also notice that as $|x|$ gets larger, the value of y gets closer to 0. Likewise, we cannot substitute 0 for y. The x-axis is a horizontal asymptote, and the y-axis is a vertical asymptote.

Make a table of values, plot the points, and sketch the branches of the hyperbola so that they approach the asymptotes.

x	y
1	4
-1	-4
2	2
-2	-2
4	1
-4	-1
8	$\frac{1}{2}$
-8	$-\frac{1}{2}$

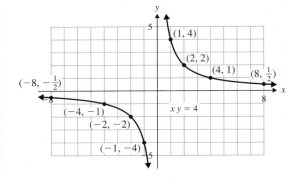

[YOU TRY 4]

Graph $xy = 8$.

3 Graph Other Square Root Functions

We have already learned how to graph square root functions like $f(x) = \sqrt{x}$ and $g(x) = \sqrt{x - 3}$. Next, we will learn how to graph other square root functions by relating them to the graphs of conic sections.

The vertical line test shows that horizontal parabolas, circles, ellipses, and some hyperbolas are not the graphs of functions. What happens, however, if we look at a *portion* of the graph of a conic section? Let's start with a circle.

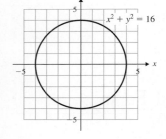

The graph of $x^2 + y^2 = 16$, at the left, is a circle with center $(0, 0)$ and radius 4. If we solve this equation for y we get $y = \pm\sqrt{16 - x^2}$. This represents two equations, $y = \sqrt{16 - x^2}$ and $y = -\sqrt{16 - x^2}$.

The graph of $y = \sqrt{16 - x^2}$ is the **top half of the circle** since the y-coordinates of all points on the graph will be non-negative. The domain is $[-4, 4]$, and the range is $[0, 4]$.

Because of the negative sign out front of the radical, the graph of $y = -\sqrt{16 - x^2}$ is the **bottom half of the circle** since the y-coordinates of all points on the graph will be non-positive. The domain is $[-4, 4]$, and the range is $[-4, 0]$.

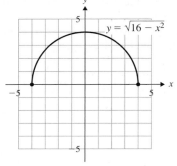
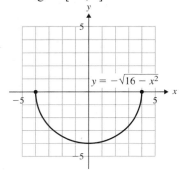

Therefore, to graph $y = \sqrt{16 - x^2}$ it is helpful if we recognize that it is the top half of the circle with equation $x^2 + y^2 = 16$. Likewise, if we are asked to graph $y = -\sqrt{16 - x^2}$, we should recognize that it is the bottom half of the graph of $x^2 + y^2 = 16$.

Let's graph another square root function by first relating it to a conic section.

EXAMPLE 5

Graph $f(x) = -5\sqrt{1 - \dfrac{x^2}{9}}$. Identify the domain and range.

Solution

The graph of this function is half of the graph of a conic section. First, notice that $f(x)$ is always a nonpositive quantity. Since all nonpositive values of y are on or below the x-axis, the graph of this function will only be on or below the x-axis.

Replace $f(x)$ with y, and rearrange the equation into a form we recognize as a conic section.

$$y = -5\sqrt{1 - \dfrac{x^2}{9}} \quad \text{Replace } f(x) \text{ with } y.$$

$$-\dfrac{y}{5} = \sqrt{1 - \dfrac{x^2}{9}} \quad \text{Divide by } -5.$$

$$\dfrac{y^2}{25} = 1 - \dfrac{x^2}{9} \quad \text{Square both sides.}$$

$$\dfrac{x^2}{9} + \dfrac{y^2}{25} = 1 \quad \text{Add } \dfrac{x^2}{9}.$$

The equation $\dfrac{x^2}{9} + \dfrac{y^2}{25} = 1$ represents an ellipse centered at the origin. Graph it. The domain is $[-3, 3]$, and the range is $[-5, 5]$. It is not a function.

The graph of $f(x) = -5\sqrt{1 - \dfrac{x^2}{9}}$ is the *bottom half* of the ellipse, shown below, and it is a function. Its domain is $[-3, 3]$, and its range is $[-5, 0]$.

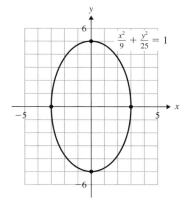

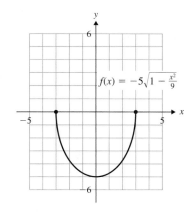

YOU TRY 5

Graph $f(x) = \sqrt{\dfrac{x^2}{4} - 1}$. Identify the domain and range.

More about the conic sections and their characteristics are studied in later mathematics courses.

Using Technology

To sketch the graph of a hyperbola on a graphing calculator, we use the same technique we used for graphing an ellipse: Solve the equation for y in terms of x, enter both values of y, and graph both equations. If we are graphing a hyperbola such as $xy = 4$, we will solve for y and enter the single equation.

Graph $9x^2 - y^2 = 16$. First, solve for y.

$$9x^2 - y^2 = 16$$
$$-y^2 = 16 - 9x^2$$
$$y^2 = 9x^2 - 16$$
$$y = \pm\sqrt{9x^2 - 16}$$

Enter $y = \sqrt{9x^2 - 16}$ as Y_1. This represents the portions of the branches of the hyperbola that are on and above the x-axis since the y-values are all non-negative. Enter $y = -\sqrt{9x^2 - 16}$ as Y_2. This represents the portions of the branches of the hyperbola that are on and below the x-axis since the y-values are nonpositive. Set an appropriate window, and press GRAPH.

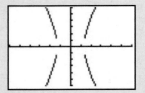

Although it appears that there is a "break" in the graph near the x-axis, the graph does actually continue across the x-axis with the x-intercepts at $\left(\frac{4}{3}, 0\right)$ and $\left(-\frac{4}{3}, 0\right)$.

Graph each equation on a graphing calculator. These problems come from the homework exercises.

1) $9x^2 - y^2 = 36$; Exercise 17
2) $y^2 - x^2 = 1$; Exercise 19
3) $\dfrac{y^2}{16} - \dfrac{x^2}{4} = 1$; Exercise 7
4) $\dfrac{x^2}{9} - \dfrac{y^2}{25} = 1$; Exercise 5
5) $y^2 - \dfrac{(x-1)^2}{9} = 1$; Exercise 13
6) $xy = 1$; Exercise 29

ANSWERS TO [YOU TRY] EXERCISES

1)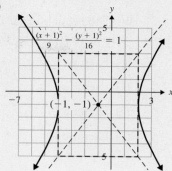

2)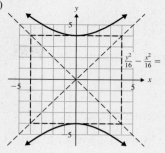

3)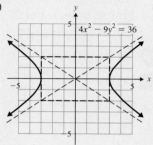

4)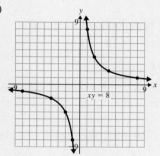

5) domain: $(-\infty, -2] \cup [2, \infty)$; range: $[0, \infty)$

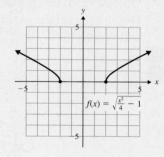

ANSWERS TO TECHNOLOGY EXERCISES

1) $9x^2 - y^2 = 36$

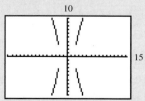

2) $y^2 - x^2 = 1$

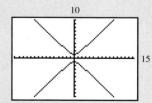

3) $\dfrac{y^2}{16} - \dfrac{x^2}{4} = 1$

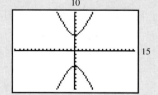

4) $\dfrac{x^2}{9} - \dfrac{y^2}{25} = 1$

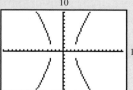

5) $y^2 - \dfrac{(x-1)^2}{9} = 1$

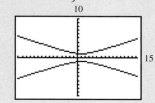

6) $xy = 1$

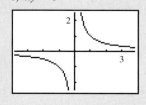

E Evaluate 12.4 Exercises

Do the exercises, and check your work.

Objective 1: Graph a Hyperbola in Standard Form

Determine whether each statement is *true* or *false*.

1) The graph of $\dfrac{(x-7)^2}{9} + \dfrac{(y-2)^2}{16} = 1$ is a hyperbola with center $(7, 2)$.

2) The center of the hyperbola with equation $\dfrac{(x-5)^2}{16} - \dfrac{(y+3)^2}{25} = 1$ is $(5, -3)$.

3) The center of the hyperbola with equation $\dfrac{(y-6)^2}{25} - \dfrac{(x+4)^2}{4} = 1$ is $(-4, 6)$.

4) The graph of $9y^2 - x^2 = 9$ is a hyperbola with center at the origin.

Graph each hyperbola. Identify the center and sketch the asymptotes.

5) $\dfrac{x^2}{9} - \dfrac{y^2}{25} = 1$

6) $\dfrac{x^2}{9} - \dfrac{y^2}{4} = 1$

7) $\dfrac{y^2}{16} - \dfrac{x^2}{4} = 1$

8) $\dfrac{y^2}{4} - \dfrac{x^2}{4} = 1$

9) $\dfrac{(x-2)^2}{9} - \dfrac{(y+3)^2}{16} = 1$

10) $\dfrac{(x+3)^2}{4} - \dfrac{(y+1)^2}{16} = 1$

11) $\dfrac{(y+1)^2}{25} - \dfrac{(x+4)^2}{4} = 1$

12) $\dfrac{(y-1)^2}{36} - \dfrac{(x+1)^2}{9} = 1$

13) $y^2 - \dfrac{(x-1)^2}{9} = 1$

14) $\dfrac{(y+4)^2}{4} - x^2 = 1$

15) $\dfrac{(x-1)^2}{25} - \dfrac{(y-2)^2}{25} = 1$

16) $\dfrac{(x-2)^2}{16} - \dfrac{(y-3)^2}{9} = 1$

17) $9x^2 - y^2 = 36$

18) $4y^2 - x^2 = 16$

19) $y^2 - x^2 = 1$

20) $x^2 - y^2 = 25$

Write the equations of the asymptotes of the graph in each of the following exercises. *Do not graph.*

21) Exercise 5   22) Exercise 6
23) Exercise 7   24) Exercise 8
25) Exercise 17 26) Exercise 18
27) Exercise 19 28) Exercise 20

Objective 2: Graph a Hyperbola in Nonstandard Form

Graph each equation.

29) $xy = 1$ 30) $xy = 6$
31) $xy = 2$ 32) $xy = 10$
33) $xy = -4$ 34) $xy = -8$
35) $xy = -6$ 36) $xy = -1$

Objective 3: Graph Other Square Root Functions

Graph each square root function. Identify the domain and range.

37) $g(x) = \sqrt{25 - x^2}$

38) $f(x) = \sqrt{9 - x^2}$

39) $h(x) = -\sqrt{1 - x^2}$

40) $k(x) = -\sqrt{9 - x^2}$

41) $g(x) = -2\sqrt{1 - \dfrac{x^2}{9}}$

42) $f(x) = 3\sqrt{1 - \dfrac{x^2}{16}}$

43) $h(x) = -3\sqrt{\dfrac{x^2}{4} - 1}$

44) $k(x) = 2\sqrt{\dfrac{x^2}{16} - 1}$

Sketch the graph of each equation.

45) $x = \sqrt{16 - y^2}$

46) $x = -\sqrt{4 - y^2}$

47) $x = -3\sqrt{1 - \dfrac{y^2}{4}}$

48) $x = \sqrt{1 - \dfrac{y^2}{9}}$

Extension

We have learned that sometimes it is necessary to complete the square to put the equations of circles and ellipses into standard form. The same is true for hyperbolas. In Exercises 49 and 50, practice going through the steps of putting the equation of a hyperbola into standard form.

Fill It In

Fill in the blanks with either the missing mathematical step or the reason for the given step.

49) $4x^2 - 9y^2 - 8x - 18y - 41 = 0$
$4x^2 - 8x - 9y^2 - 18y = 41$

_____ | Factor out the coefficients of the squared terms.

$4(x^2 - 2x + 1) - 9(y^2 + 2y + 1)$
$= 41 + 4(1) - 9(1)$ | _____

_____ | Factor.

_____ | Divide both sides by 36.

50) $-x^2 + 4y^2 + 6x - 16y + 3 = 0$
$-x^2 + 6x + 4y^2 - 16y = -3$

_____ | Factor out the coefficients of the squared terms. Complete the square.

_____ |

$-(x - 3)^2 + 4(y - 2)^2 = 4$ |

_____ | Divide both sides by 4.

For Exercises 51–54, put each equation into the standard form for the equation of a hyperbola and graph.

51) $x^2 - 4y^2 - 2x - 24y - 51 = 0$

52) $9x^2 - 4y^2 + 90x - 16y + 173 = 0$

53) $16y^2 - 9x^2 + 18x - 64y - 89 = 0$

54) $y^2 - 4x^2 - 16x - 6y - 23 = 0$

We know that the standard form for the equation of a hyperbola is

$$\frac{(x-h)^2}{a^2} - \frac{(y-k)^2}{b^2} = 1 \text{ or } \frac{(y-k)^2}{b^2} - \frac{(x-h)^2}{a^2} = 1.$$

The equations for the asymptotes are

$$y - k = \frac{b}{a}(x - h) \text{ and } y - k = -\frac{b}{a}(x - h)$$

Use these formulas to write the equations of the asymptotes of the graph in each of the following exercises. *Do not graph.*

55) Exercise 9
56) Exercise 10
57) Exercise 11
58) Exercise 12
59) Exercise 13
60) Exercise 14
61) Exercise 15
62) Exercise 16

63) A hyperbola centered at the origin opens in the y-direction and has asymptotes with equations $y = \frac{1}{2}x$ and $y = -\frac{1}{2}x$. Write an equation of the hyperbola.

64) A hyperbola centered at the origin opens in the x-direction and has asymptotes with equations $y = \frac{3}{2}x$ and $y = -\frac{3}{2}x$. Write an equation of the hyperbola.

Use a graphing calculator to graph the following. Square the viewing window.

65) $\dfrac{(x-2)^2}{9} - \dfrac{(y+3)^2}{16} = 1$

66) $\dfrac{(x+3)^2}{4} - \dfrac{(y+1)^2}{16} = 1$

67) $\dfrac{(y+1)^2}{25} - \dfrac{(x+4)^2}{4} = 1$

68) $\dfrac{(y-1)^2}{36} - \dfrac{(x+1)^2}{9} = 1$

R Rethink

R1) Can you explain when the branches of a hyperbola would open in the x direction? What about opening in the y direction?

R2) Can you draw a graph of a hyperbola and explain the asymptotes to a classmate?

R3) What is the hardest part of graphing a hyperbola?

Putting It All Together

P Prepare

What are your objectives for Putting It All Together?

1. Identify and Graph Different Types of Conic Sections

O Organize

How can you accomplish each objective?

- Review the equations for all the different conic sections, and summarize them in your notes.
- Draw a graph of all the conic sections in your notes, and label the graphs.
- Complete the given examples on your own.
- Complete You Try 1.

W Work

Read the explanations, follow the examples, take notes, and complete the You Try.

1 Identify and Graph Different Types of Conic Sections

Sometimes the most difficult part of graphing a conic section is identifying which type of graph will result from the given equation. In this section, we will discuss how to look at an equation and determine which type of conic section it represents.

EXAMPLE 1 Graph $x^2 + y^2 + 4x - 6y + 9 = 0$.

Solution

First, notice that this equation has two squared terms. Therefore, its graph cannot be a parabola because the equation of a parabola contains only one squared term. Next, observe that the coefficients of x^2 and y^2 are each 1. Since the coefficients are the same, *this is the equation of a circle.*

Write the equation in the form $(x - h)^2 + (y - k)^2 = r^2$ by completing the square on the x-terms and on the y-terms.

$$x^2 + y^2 + 4x - 6y + 9 = 0$$
$$(x^2 + 4x) + (y^2 - 6y) = -9$$

Group the x-terms together, and group the y-terms together. Move the constant to the other side.

$$(x^2 + 4x + 4) + (y^2 - 6y + 9) = -9 + 4 + 9$$
$$(x + 2)^2 + (y - 3)^2 = 4$$

Complete the square for each group of terms. Factor; add.

The center of the circle is $(-2, 3)$. The radius is 2.

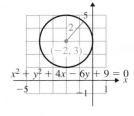

W Hint

Pay close attention to how you identify the different types of conic sections in Examples 1–4.

EXAMPLE 2

Graph $x = y^2 + 4y + 3$.

Solution

This equation contains only one squared term. Therefore, *this is the equation of a parabola.* Since the squared term is y^2 and $a = 1$, the parabola will open to the right.

Use the formula $y = -\dfrac{b}{2a}$ to find the y-coordinate of the vertex.

$$a = 1 \quad b = 4 \quad c = 3$$
$$y = -\dfrac{4}{2(1)} = -2$$
$$x = (-2)^2 + 4(-2) + 3 = -1$$

The vertex is $(-1, -2)$. Make a table of values to find other points on the parabola, and use the axis of symmetry to find more points.

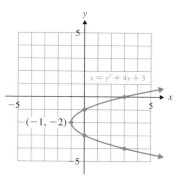

x	y
0	−1
3	0

Plot the points in the table. Locate the points $(0, -3)$ and $(3, -4)$ using the axis of symmetry, $y = -2$.

EXAMPLE 3

Graph $\dfrac{(y-1)^2}{9} - \dfrac{(x-3)^2}{4} = 1$.

Solution

In this equation we see the *difference* of two squares. *The graph of this equation is a hyperbola.* The branches of the hyperbola will open in the y-direction because the quantity containing the variable y, $\dfrac{(y-1)^2}{9}$, is the positive, squared quantity.

The center is $(3, 1)$; $a = \sqrt{4} = 2$ and $b = \sqrt{9} = 3$.

Draw the reference rectangle and its diagonals, the asymptotes of the graph.
   The branches of the hyperbola approach the asymptotes.

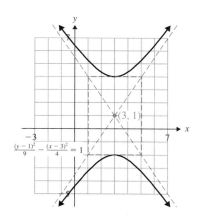

EXAMPLE 4

Graph $x^2 + 9y^2 = 36$.

Solution

This equation contains the *sum* of two squares with *different* coefficients. *This is the equation of an ellipse.* (If the coefficients were the same, the graph would be a circle.)

Divide both sides of the equation by 36 to get 1 on the right side of the = sign.

$$x^2 + 9y^2 = 36$$
$$\frac{x^2}{36} + \frac{9y^2}{36} = \frac{36}{36} \quad \text{Divide both sides by 36.}$$
$$\frac{x^2}{36} + \frac{y^2}{4} = 1 \quad \text{Simplify.}$$

The center is $(0, 0)$, $a = 6$ and $b = 2$.

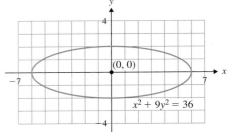

YOU TRY 1

Determine whether the graph of each equation is a parabola, circle, ellipse, or hyperbola. Then, graph each equation.

a) $4x^2 - 25y^2 = 100$ b) $x^2 + y^2 - 6x - 12y + 9 = 0$

c) $y = -x^2 - 2x + 4$ d) $x^2 + \frac{(y+4)^2}{9} = 1$

W Hint

After completing Examples 1–4, summarize what you've learned about each of the conic sections. Explain, in your own words, how to determine which type of conic section an equation represents. Include examples.

ANSWERS TO [YOU TRY] EXERCISES

1) a) hyperbola

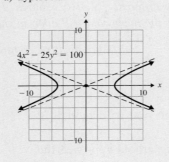

b) circle

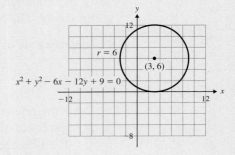

c) parabola

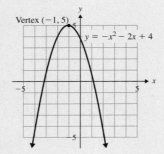

d) ellipse

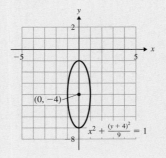

Putting It All Together Exercises

E Evaluate — Do the exercises, and check your work.

Objective 1: Identify and Graph Different Types of Conic Sections

Determine whether the graph of each equation is a parabola, circle, ellipse, or hyperbola. Then, graph each equation.

1) $y = x^2 + 4x + 8$

2) $(x + 5)^2 + (y - 3)^2 = 25$

3) $\dfrac{(y + 4)^2}{9} - \dfrac{(x + 1)^2}{4} = 1$

4) $x = (y - 1)^2 + 8$

5) $16x^2 + 9y^2 = 144$

6) $x^2 - 4y^2 = 36$

7) $x^2 + y^2 + 8x - 6y - 11 = 0$

8) $\dfrac{(x - 2)^2}{25} + \dfrac{(y - 2)^2}{36} = 1$

9) $(x - 1)^2 + \dfrac{y^2}{16} = 1$

10) $x^2 + y^2 + 8y + 7 = 0$

11) $x = -(y + 4)^2 - 3$

12) $\dfrac{(y + 4)^2}{4} - (x + 2)^2 = 1$

13) $25x^2 - 4y^2 = 100$

14) $4x^2 + y^2 = 16$

15) $(x - 3)^2 + y^2 = 16$

16) $y = -x^2 + 6x - 7$

17) $x = \dfrac{1}{2}y^2 + 2y + 3$

18) $\dfrac{(x - 3)^2}{16} + y^2 = 1$

19) $(x - 2)^2 - (y + 1)^2 = 9$

20) $x^2 + y^2 + 6x - 8y + 9 = 0$

Where appropriate, write the equation in standard form. Then, graph each equation.

21) $xy = 5$

22) $25x^2 - 4y^2 + 150x + 125 = 0$

23) $9x^2 + y^2 - 54x + 4y + 76 = 0$

24) $xy = -2$

25) $9y^2 - 4x^2 - 18y + 16x - 43 = 0$

26) $4x^2 + 9y^2 - 8x - 54y + 49 = 0$

Use a graphing calculator to graph Exercises 27–30. Square the viewing window.

27) $x^2 + y^2 + 8x - 6y - 11 = 0$

28) $(x - 1)^2 + \dfrac{y^2}{16} = 1$

29) $x = -(y + 4)^2 - 3$

30) $25x^2 - 4y^2 = 100$

R Rethink

R1) Can you write down, without using your notes, the equations for the different conic sections?

R2) Are all conic sections functions? Why or why not?

R3) Which of the conic sections is the hardest to graph?

12.5 Nonlinear Systems of Equations

Prepare

What are your objectives for Section 12.5?	How can you accomplish each objective?
1 Define a Nonlinear System of Equations	• Know the definition of a *nonlinear system of equations*.
2 Solve a Nonlinear System by Substitution	• Know what the graphs of each equation look like in order to know how many possible intersection points exist. Sketch a rough graph in your notes. • Review the substitution method for solving a system of equations. • Check the proposed solutions in the original equations. • Complete the given examples on your own. • Complete You Trys 1 and 2.
3 Solve a Nonlinear System Using the Elimination Method	• Be able to recognize when both equations are second-degree equations. • Review the elimination method for solving a system of equations. • Check the proposed solutions in the original equations. • Complete the given examples on your own. • Complete You Trys 3 and 4.

Organize

Work Read the explanations, follow the examples, take notes, and complete the You Trys.

1 Define a Nonlinear System of Equations

In Chapter 5, we learned to solve systems of linear equations by graphing, substitution, and the elimination method. We can use these same techniques for solving a *nonlinear system of equations* in two variables. A **nonlinear system of equations** is a system in which at least one of the equations is not linear.

Solving a nonlinear system by graphing is not practical since it would be very difficult (if not impossible) to accurately read the points of intersection. Therefore, we will solve the systems using substitution and the elimination method. We will graph the equations, however, so that we can visualize the solution(s) as the point(s) of intersection of the graphs.

We are interested only in real-number solutions. If a system has imaginary solutions, then the graphs of the equations do not intersect in the real-number plane.

2 Solve a Nonlinear System by Substitution

When one of the equations in a system is linear, it is often best to use the substitution method to solve the system.

EXAMPLE 1

Solve the system $x^2 - 2y = 2$ (1)
$-x + y = 3$ (2)

Solution

The graph of equation (1) is a parabola, and the graph of equation (2) is a line. Let's begin by thinking about the number of possible points of intersection the graphs can have.

Hint
Make a rough sketch for yourself so that you know how many solutions the system *could* have.

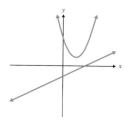
No points of intersection
The system has no solution.

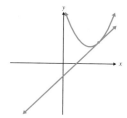
One point of intersection
The system has one solution.

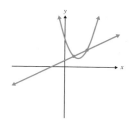
Two points of intersection
The system has two solutions.

Solve the linear equation for one of the variables.

$$-x + y = 3$$
$$y = x + 3 \quad (3) \qquad \text{Solve for } y.$$

Substitute $x + 3$ for y in equation (1).

$$x^2 - 2y = 2 \qquad \text{Equation (1)}.$$
$$x^2 - 2(x + 3) = 2 \qquad \text{Substitute}.$$
$$x^2 - 2x - 6 = 2 \qquad \text{Distribute}.$$
$$x^2 - 2x - 8 = 0 \qquad \text{Subtract 2}.$$
$$(x - 4)(x + 2) = 0 \qquad \text{Factor}.$$
$$x = 4 \text{ or } x = -2 \qquad \text{Solve}.$$

To find the corresponding value of y for each value of x, we can substitute $x = 4$ and then $x = -2$ into *either* equation (1), (2), or (3). No matter which equation you choose, you should always check the solutions in *both* of the original equations. We will substitute the values into equation (3) because this is just an alternative form of equation (2), and it is already solved for y.

Substitute each value into equation (3) to find y.

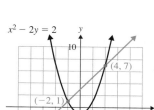

$x = 4$: $y = x + 3$ $x = -2$: $y = x + 3$
         $y = 4 + 3$ $y = -2 + 3$
         $y = 7$ $y = 1$

The proposed solutions are $(4, 7)$ and $(-2, 1)$. Verify that they solve the system by checking them in equation (1). (Remember, the ordered pair must satisfy *both* equations in the system.) The solution set is $\{(4, 7), (-2, 1)\}$. We can see on the graph to the left that these are the points of intersection of the graphs.

[YOU TRY 1]   Solve the system $x^2 + 3y = 6$
$x + y = 2$

EXAMPLE 2

Solve the system $x^2 + y^2 = 1$ (1)
$x + 2y = -1$ (2)

Solution

The graph of equation (1) is a circle, and the graph of equation (2) is a line. These graphs can intersect at zero, one, or two points. Therefore, this system will have zero, one, or two solutions.

> **W Hint**
> Write out each step as you are reading the example.

We will not solve equation (1) for a variable because doing so would give us a radical in the expression. It will be easiest to solve equation (2) for x because its coefficient is 1.

$$x + 2y = -1 \quad (2)$$
$$x = -2y - 1 \quad (3) \quad \text{Solve for } x.$$

Substitute $-2y - 1$ for x in equation (1).

$$x^2 + y^2 = 1 \quad (1)$$
$$(-2y - 1)^2 + y^2 = 1 \quad \text{Substitute.}$$
$$4y^2 + 4y + 1 + y^2 = 1 \quad \text{Expand } (-2y - 1)^2.$$
$$5y^2 + 4y = 0 \quad \text{Combine like terms; subtract 1.}$$
$$y(5y + 4) = 0 \quad \text{Factor.}$$
$$y = 0 \quad \text{or} \quad 5y + 4 = 0 \quad \text{Set each factor equal to zero.}$$
$$y = -\frac{4}{5} \quad \text{Solve for } y.$$

Substitute $y = 0$ and then $y = -\dfrac{4}{5}$ into equation (3) to find their corresponding values of x.

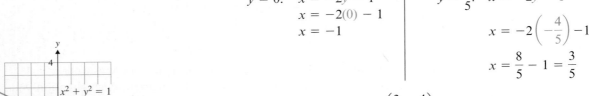

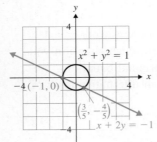

The proposed solutions are $(-1, 0)$ and $\left(\dfrac{3}{5}, -\dfrac{4}{5}\right)$. Check them in equations (1) and (2).

The solution set is $\left\{(-1, 0), \left(\dfrac{3}{5}, -\dfrac{4}{5}\right)\right\}$. The graph at left shows that these are the points of intersection of the graphs.

[YOU TRY 2] Solve the system $x^2 + y^2 = 25$
$x - y = 7$

Note

We must always check the proposed solutions in *each* equation in the system.

3 Solve a Nonlinear System Using the Elimination Method

The elimination method can be used to solve a system when both equations are second-degree equations.

EXAMPLE 3

Solve the system $5x^2 + 3y^2 = 21$ (1)
$4x^2 - y^2 = 10$ (2)

Solution

Each equation is a second-degree equation. The first is an ellipse, and the second is a hyperbola. They can have zero, one, two, three, or four points of intersection. Multiply equation (2) by 3. Then adding the two equations will eliminate the y^2-terms.

W Hint
Make a rough sketch of the possible ways an ellipse and a hyperbola can (or might not) intersect.

Original System
$5x^2 + 3y^2 = 21$
$4x^2 - y^2 = 10$

$\longrightarrow$

Rewrite the System
$5x^2 + 3y^2 = 21$
$12x^2 - 3y^2 = 30$

$5x^2 + 3y^2 = 21$
$+\ 12x^2 - 3y^2 = 30$
$17x^2 = 51$ Add the equations to eliminate y^2.
$x^2 = 3$
$x = \pm\sqrt{3}$

Find the corresponding values of y for $x = \sqrt{3}$ and $x = -\sqrt{3}$.

$x = \sqrt{3}$:
$4x^2 - y^2 = 10$ (2)
$4(\sqrt{3})^2 - y^2 = 10$
$12 - y^2 = 10$
$-y^2 = -2$
$y^2 = 2$
$y = \pm\sqrt{2}$

This gives us $(\sqrt{3}, \sqrt{2})$ and $(\sqrt{3}, -\sqrt{2})$.

$x = -\sqrt{3}$:
$4x^2 - y^2 = 10$ (2)
$4(-\sqrt{3})^2 - y^2 = 10$
$12 - y^2 = 10$
$-y^2 = -2$
$y^2 = 2$
$y = \pm\sqrt{2}$

This gives us $(-\sqrt{3}, \sqrt{2})$ and $(-\sqrt{3}, -\sqrt{2})$.

Check the proposed solutions in equation (1) to verify that they satisfy that equation as well.

The solution set is $\{(\sqrt{3}, \sqrt{2}), (\sqrt{3}, -\sqrt{2}), (-\sqrt{3}, \sqrt{2}), (-\sqrt{3}, -\sqrt{2})\}$.

[YOU TRY 3] Solve the system $2x^2 - 13y^2 = 20$
$-x^2 + 10y^2 = 4$

For solving some systems, using *either* substitution or the elimination method works well. Look carefully at each system to decide which method to use.

We will see in Example 4 that not all systems have solutions.

EXAMPLE 4

Solve the system
$$y = \sqrt{x} \quad (1)$$
$$y^2 - 4x^2 = 4 \quad (2)$$

Solution

The graph of the square root function $y = \sqrt{x}$ is half of a parabola. The graph of equation (2) is a hyperbola. Solve this system by substitution. Replace y in equation (2) with $\sqrt{x}$ from equation (1).

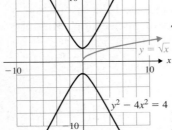

$$y^2 - 4x^2 = 4 \quad (2)$$
$$(\sqrt{x})^2 - 4x^2 = 4 \qquad \text{Substitute } y = \sqrt{x} \text{ into equation (2).}$$
$$x - 4x^2 = 4$$
$$0 = 4x^2 - x + 4$$

The right-hand side does not factor, so solve it using the quadratic formula.

$$4x^2 - x + 4 = 0 \qquad a = 4 \quad b = -1 \quad c = 4$$
$$x = \frac{-(-1) \pm \sqrt{(-1)^2 - 4(4)(4)}}{2(4)} = \frac{1 \pm \sqrt{1 - 64}}{8} = \frac{1 \pm \sqrt{-63}}{8}$$

Because $\sqrt{-63}$ is not a real number, there are no real-number values for x. The system has no solution, so the solution set is $\varnothing$. The graph is shown on the left.

[YOU TRY 4] Solve the system $4x^2 + y^2 = 4$
$\qquad\qquad\qquad\qquad\;\; x - y = 3$

Using Technology

We can solve systems of nonlinear equations on the graphing calculator just like we solved systems of linear equations in Chapter 5—graph the equations and find their points of intersection.

Let's look at Example 3:

$$5x^2 + 3y^2 = 21$$
$$4x^2 - y^2 = 10$$

Solve each equation for y and enter them into the calculator.

Solve $5x^2 + 3y^2 = 21$ for y: Solve $4x^2 - y^2 = 10$ for y:

$$y = \pm\sqrt{7 - \frac{5}{3}x^2}$$
$$y = \pm\sqrt{4x^2 - 10}$$

Enter $\sqrt{7 - \frac{5}{3}x^2}$ as Y_1.   Enter $\sqrt{4x^2 - 10}$ as Y_3.

Enter $-\sqrt{7 - \frac{5}{3}x^2}$ as Y_2.   Enter $-\sqrt{4x^2 - 10}$ as Y_4.

After entering the equations, press GRAPH.

 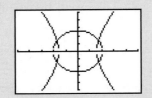

The system has four real solutions since the graphs have four points of intersection. We can use the INTERSECT option to find the solutions. Since we graphed four functions, we must tell the calculator which point of intersection we want to find. Note that the point where the graphs intersect in the first quadrant comes from the intersection of equations Y_1 and Y_3. Press 2nd TRACE and choose 5:intersect and you will see the screen to the right.

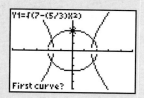

Notice that the top left of the screen to the right displays the function Y_1. Since we want to find the intersection of Y_1 and Y_3, press ENTER when Y_1 is displayed. Now Y_2 appears at the top left, but we do not need this function. Press the down arrow to see the equation for Y_3 and be sure that the cursor is close to the intersection point in quadrant I. Press ENTER twice. You will see the approximate solution (1.732, 1.414), as shown to the right.

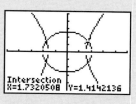

In Example 3 we found the exact solutions algebraically. The calculator solution, (1.732, 1.414), is an approximation of the exact solution ($\sqrt{3}$, $\sqrt{2}$).

The other solutions of the system can be found in the same way.

Use the graphing calculator to find all real-number solutions of each system. These are taken from the examples in the section and from the Chapter Summary.

1) $x^2 - 2y = 2$
   $-x + y = 3$

2) $x^2 + y^2 = 1$
   $x + 2y = -1$

3) $x - y^2 = 3$
   $x - 2y = 6$

4) $y = \sqrt{x}$
   $y^2 - 4x^2 = 4$

ANSWERS TO [YOU TRY] EXERCISES

1) $\{(0, 2), (3, -1)\}$ 2) $\{(4, -3), (3, -4)\}$ 3) $\{(6, 2), (6, -2), (-6, 2), (-6, -2)\}$ 4) $\varnothing$

ANSWERS TO TECHNOLOGY EXERCISES

1) $\{(4, 7), (-2, 1)\}$ 2) $(-1, 0), (0.6, -0.8)$ 3) $\{(12, 3), (4, -1)\}$ 4) $\varnothing$

E Evaluate 12.5 Exercises

Do the exercises, and check your work.

Objective 1: Define a Nonlinear System of Equations

If a nonlinear system consists of equations with the following graphs,

a) sketch the different ways in which the graphs can intersect.

b) make a sketch in which the graphs do not intersect.

c) how many possible solutions can each system have?

1) circle and line
2) parabola and line
3) parabola and ellipse
4) ellipse and hyperbola
5) parabola and hyperbola
6) circle and ellipse

Mixed Exercises: Objectives 2 and 3

Solve each system using either substitution or the elimination method.

7) $x^2 + 4y = 8$
   $x + 2y = -8$

8) $x^2 + y = 1$
   $-x + y = -5$

9) $x + 2y = 5$
   $x^2 + y^2 = 10$

10) $y = 2$
    $x^2 + y^2 = 8$

11) $y = x^2 - 6x + 10$
    $y = 2x - 6$

12) $y = x^2 - 10x + 22$
    $y = 4x - 27$

13) $x^2 + 2y^2 = 11$
    $x^2 - y^2 = 8$

14) $2x^2 - y^2 = 7$
    $2y^2 - 3x^2 = 2$

15) $x^2 + y^2 = 6$
    $2x^2 + 5y^2 = 18$

16) $5x^2 - y^2 = 16$
    $x^2 + y^2 = 14$

17) $3x^2 + 4y = -1$
    $x^2 + 3y = -12$

18) $2x^2 + y = 9$
    $y = 3x^2 + 4$

19) $y = 6x^2 - 1$
    $2x^2 + 5y = -5$

20) $x^2 + 2y = 5$
    $-3x^2 + 2y = 5$

21) $x^2 + y^2 = 4$
    $-2x^2 + 3y = 6$

22) $x^2 + y^2 = 49$
    $x - 2y^2 = 7$

23) $x^2 + y^2 = 3$
    $x + y = 4$

24) $y - x = 1$
    $4y^2 - 16x^2 = 64$

25) $x = \sqrt{y}$
    $x^2 - 9y^2 = 9$

26) $x = \sqrt{y}$
    $x^2 - y^2 = 4$

27) $9x^2 + y^2 = 9$
    $x^2 + y^2 = 5$

28) $x^2 + y^2 = 6$
    $5x^2 + y^2 = 10$

29) $x^2 + y^2 = 1$
    $y = x^2 + 1$

30) $y = -x^2 - 2$
    $x^2 + y^2 = 4$

33) The perimeter of a rectangular computer screen is 38 in. Its area is 88 in². Find the dimensions of the screen.

34) The area of a rectangular bulletin board is 180 in², and its perimeter is 54 in. Find the dimensions of the bulletin board.

35) A sporting goods company estimates that the cost y, in dollars, to manufacture x thousands of basketballs is given by

$$y = 6x^2 + 33x + 12$$

The revenue y, in dollars, from the sale of x thousands of basketballs is given by

$$y = 15x^2$$

The company breaks even on the sale of basketballs when revenue equals cost. The point, (x, y), at which this occurs is called the *break-even point*. Find the break-even point for the manufacture and sale of the basketballs.

36) A backpack manufacturer estimates that the cost y, in dollars, to make x thousands of backpacks is given by

$$y = 9x^2 + 30x + 18$$

The revenue y, in dollars, from the sale of x thousands of backpacks is given by

$$y = 21x^2$$

Find the break-even point for the manufacture and sale of the backpacks. (See Exercise 35 for an explanation.)

Write a system of equations, and solve.

31) Find two numbers whose product is 40 and whose sum is 13.

32) Find two numbers whose product is 28 and whose sum is 11.

R Rethink

R1) Do all nonlinear systems of equations have solutions? Why or why not?

R2) Why do you need to check the solutions in the original equations?

R3) Why is it helpful to sketch a graph of the equations before solving them?

12.6 Second-Degree Inequalities and Systems of Inequalities

P Prepare

What are your objectives for Section 12.6?

1 Graph Second-Degree Inequalities

2 Graph Systems of Nonlinear Inequalities

O Organize

How can you accomplish each objective?

- Learn the definition of a *second-degree inequality*.
- Know how to graph conic sections.
- Recognize when to graph a conic section as a solid curve or as a dotted curve, depending on the sign of the inequality.
- Review how to choose a test point to determine which region to shade.
- Complete the given examples on your own.
- Complete You Trys 1 and 2.

- Understand the definition of a *solution set of a system of inequalities*. Know that the solution set is the intersection of the shaded regions.
- Know how to graph the conic sections and other functions we have learned so far.
- Complete the given examples on your own.
- Complete You Trys 3 and 4.

Read the explanations, follow the examples, take notes, and complete the You Trys.

1 Graph Second-Degree Inequalities

In Section 4.4, we learned how to graph linear inequalities in two variables such as $2x + y \leq 3$. To graph this inequality, first graph the boundary line $2x + y = 3$. Then we can choose a test point on one side of the line, say $(0, 0)$. Since $(0, 0)$ satisfies the inequality $2x + y \leq 3$, we shade the side of the line containing $(0, 0)$. All points in the shaded region satisfy $2x + y \leq 3$. (If the test point had *not* satisfied the inequality, we would have shaded the other side of the line.)

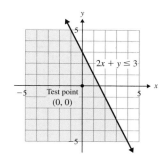

The graph of $2x + y \leq 3$ is to the left.

A **second-degree inequality** contains at least one squared term and no variable with degree greater than 2. We graph second-degree inequalities the same way we graph linear inequalities in two variables.

EXAMPLE 1

Graph $x^2 + y^2 < 25$.

Solution

Begin by graphing the *circle*, $x^2 + y^2 = 25$, as a dotted curve since the inequality is $<$. (Points *on* the circle will not satisfy the inequality.)

Next, choose a test point not on the boundary curve: $(0, 0)$. Does the test point satisfy $x^2 + y^2 < 25$? Yes: $0^2 + 0^2 < 25$ is true. Shade the region inside the circle. All points in the shaded region satisfy $x^2 + y^2 < 25$.

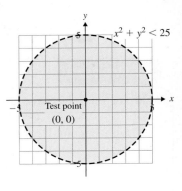

YOU TRY 1

Graph $\dfrac{x^2}{25} + \dfrac{y^2}{9} < 1$.

EXAMPLE 2

Graph $4x^2 - 9y^2 \geq 36$.

Solution

First, graph the *hyperbola*, $4x^2 - 9y^2 = 36$, as a solid curve because the inequality is $\geq$.

$$4x^2 - 9y^2 = 36$$
$$\frac{4x^2}{36} - \frac{9y^2}{36} = \frac{36}{36} \qquad \text{Divide by 36.}$$
$$\frac{x^2}{9} - \frac{y^2}{4} = 1 \qquad \text{Simplify.}$$

The center is $(0, 0)$, $a = 3$, and $b = 2$.

Next, choose a test point not on the boundary curve: $(0, 0)$. Does $(0, 0)$ satisfy $4x^2 - 9y^2 \geq 36$? No: $4(0)^2 - 9(0)^2 \geq 36$ is false.

Since $(0, 0)$ does not satisfy the inequality, we do not shade the region containing $(0, 0)$. Shade the other side of the branches of the hyperbola.

All points in the shaded region, as well as *on* the hyperbola, satisfy $4x^2 - 9y^2 \geq 36$.

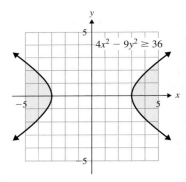

YOU TRY 2

Graph $y \geq -x^2 + 3$.

2 Graph Systems of Nonlinear Inequalities

The **solution set of a system of inequalities** consists of the set of points that satisfy *all* the inequalities in the system.

We first discussed this in Section 4.4 when we graphed the solution set of a system like

$$x \geq -2 \text{ and } y \leq x - 3$$

The solution set of such a system of linear inequalities is the intersection of their graphs. The solution set of a system of non-linear inequalities is also the intersection of the graphs of the individual inequalities.

EXAMPLE 3

Graph the solution set of the system.

$$4x^2 + y^2 < 16$$
$$-x + 2y > 2$$

Solution

First, graph the *ellipse*, $4x^2 + y^2 = 16$, as a dotted curve because the inequality is $<$.

$$4x^2 + y^2 = 16$$
$$\frac{4x^2}{16} + \frac{y^2}{16} = \frac{16}{16} \quad \text{Divide by 16.}$$
$$\frac{x^2}{4} + \frac{y^2}{16} = 1 \quad \text{Simplify.}$$

The test point $(0, 0)$ satisfies the inequality $4x^2 + y^2 < 16$, so shade inside the dotted curve of the ellipse, as shown to the right.

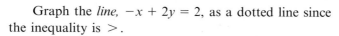

Graph the *line*, $-x + 2y = 2$, as a dotted line since the inequality is $>$.

$$-x + 2y > 2$$
$$2y > x + 2$$
$$y > \frac{1}{2}x + 1 \quad \text{Solve for } y.$$

Shade above the line since the test point $(0, 0)$ does *not* satisfy $y > \frac{1}{2}x + 1$.

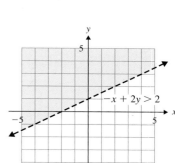

W Hint
Notice the similarities between this system and linear systems of inequalities.

The solution set of the system is the *intersection* of the two graphs. The shaded regions overlap as shown to the right. This is the solution set of the system.

All points in the shaded region satisfy *both* inequalities.

$$4x^2 + y^2 < 16$$
$$-x + 2y > 2$$

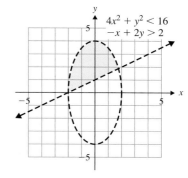

YOU TRY 3

Graph the solution set of the system.

$$y < x - 4$$
$$x^2 + y > 0$$

EXAMPLE 4

Graph the solution set of the system.

$$x \geq 0$$
$$y - x^2 \leq 1$$
$$x^2 + y^2 \leq 4$$

Solution

First, the graph of $x = 0$ is the y-axis. Therefore, the graph of $x \geq 0$ consists of quadrants I and IV. See the figure below left.

Graph the *parabola*, $y - x^2 = 1$, as a solid curve since the inequality is $\leq$. Rewrite the inequality as $y \leq x^2 + 1$ to determine that the vertex is $(0, 1)$. The test point $(0, 0)$ satisfies the inequality $y \leq x^2 + 1$, so shade outside of the parabola. See the figure below center.

The graph of $x^2 + y^2 \leq 4$ is the inside of the circle $x^2 + y^2 = 4$. See the figure below right.

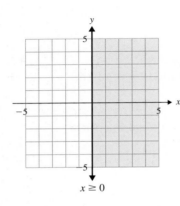
$x \geq 0$

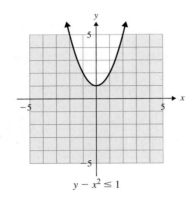
$y - x^2 \leq 1$

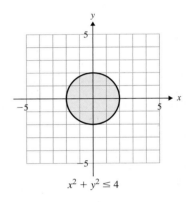
$x^2 + y^2 \leq 4$

Finally, the graph of the solution set of the system is the intersection, or overlap, of these three regions, as shown to the right.

All points in the shaded region satisfy each of the inequalities in the system

$$x \geq 0$$
$$y - x^2 \leq 1$$
$$x^2 + y^2 \leq 4$$

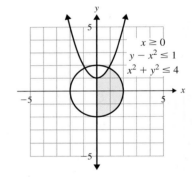

[YOU TRY 4]

Graph the solution set of the system.

$$y \geq 0$$
$$x^2 + y^2 \leq 9$$
$$y - x^2 \geq -2$$

ANSWERS TO [YOU TRY] EXERCISES

1)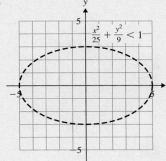

2)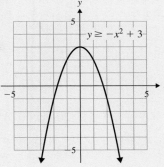

3)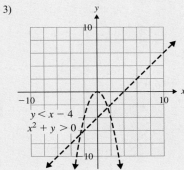

4)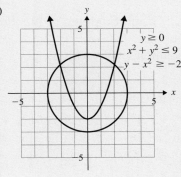

E Evaluate 12.6 Exercises
Do the exercises, and check your work.

Objective 1: Graph Second-Degree Inequalities

The graphs of second-degree inequalities are given below. For each, find three points which satisfy the inequality and three points that are not in the solution set.

1) $x^2 + y^2 \geq 36$

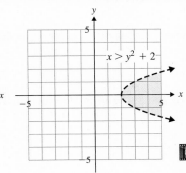

2) $x > y^2 + 2$

3) $4y^2 - x^2 < 4$

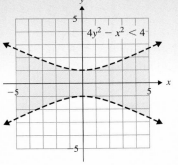

4) $25x^2 + 4y^2 \leq 100$

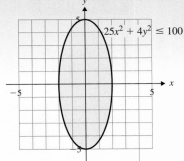

Graph each inequality.

5) $\dfrac{x^2}{9} + \dfrac{y^2}{16} < 1$

6) $y \geq (x - 2)^2 - 1$

7) $y < (x + 1)^2 + 3$

8) $x^2 + y^2 < 16$

9) $25y^2 - 4x^2 \geq 100$
10) $x^2 - 4y^2 \leq 4$
11) $x^2 + y^2 \geq 3$
12) $y \geq -x^2 + 5$
13) $x > y^2 - 2$
14) $x^2 + 4y^2 \leq 4$
15) $\dfrac{x^2}{4} - \dfrac{y^2}{9} \leq 1$
16) $x \geq -y^2 - 2y + 1$
17) $x^2 + (y+4)^2 < 9$
18) $y^2 - \dfrac{x^2}{9} > 1$
19) $x^2 + 9y^2 \geq 9$
20) $(x+3)^2 + y^2 \geq 4$
21) $y \leq -x^2 - 2x + 3$
22) $\dfrac{x^2}{36} + \dfrac{y^2}{25} > 1$

35) $\dfrac{x^2}{16} + \dfrac{y^2}{9} \leq 1$
     $4x^2 - y^2 \geq 16$
36) $\dfrac{x^2}{25} + \dfrac{y^2}{9} > 1$
     $y > x^2$
37) $y \leq -x$
     $y - 2x^2 \leq -2$
38) $x^2 + y^2 \geq 1$
     $x^2 + 25y^2 \leq 25$
39) $y \leq 0$
     $x^2 + 4y^2 \leq 36$
40) $x \geq 0$
     $x^2 + y^2 \leq 25$
41) $x > 0$
     $y^2 - 4x^2 < 4$
42) $y < 0$
     $4x^2 - 9y^2 < 36$
43) $x \geq 0$
     $y \geq x^2 + 4$
     $x + 2y \leq 12$
44) $y \geq 0$
     $4x^2 + 25y^2 \leq 100$
     $2x + 5y \leq 10$

Objective 2: Graph Systems of Nonlinear Inequalities
Graph the solution set of each system.

23) $y \geq x - 2$
     $x^2 + y \leq 1$
24) $y + x < 3$
     $x^2 + y^2 < 9$
25) $2y - x > 4$
     $4x^2 + 9y^2 > 36$
26) $y < x - 3$
     $y - x^2 < -5$
27) $x^2 + y^2 \geq 16$
     $25x^2 - 4y^2 \leq 100$
28) $x^2 - y^2 \leq 1$
     $4x^2 + 9y^2 \leq 36$
29) $x^2 + y^2 \leq 9$
     $9x^2 + 4y^2 \geq 36$
30) $x^2 + y^2 \leq 16$
     $y^2 - 4x^2 \geq 4$
31) $y^2 - x^2 < 1$
     $4x^2 + y^2 > 16$
32) $x - y^2 > -4$
     $2y - 3x > 4$
33) $x + y^2 < 0$
     $x^2 + y^2 < 16$
34) $x^2 + y^2 \geq 4$
     $x + y \geq 1$

45) $y < 0$
     $x^2 + y^2 < 9$
     $y > x + 1$
46) $y < 0$
     $y < -x^2 + 4$
     $y < \dfrac{1}{2}x - 1$
47) $y \geq 0$
     $y \leq x^2$
     $\dfrac{x^2}{4} + \dfrac{y^2}{9} \leq 1$
48) $x \geq 0$
     $y \leq x^2 - 1$
     $x^2 + y^2 \leq 16$
49) $y < 0$
     $4x^2 + 9y^2 < 36$
     $x^2 - y^2 > 1$
50) $x < 0$
     $x^2 + y^2 < 9$
     $y < -x^2$
51) $x \geq 0$
     $y \geq 0$
     $x^2 + y^2 \geq 4$
     $x \geq y^2$
52) $x \geq 0$
     $y \geq 0$
     $y \geq x^2$
     $x^2 + 4y^2 \geq 16$

R Rethink

R1) When do you use a dotted line to graph, and when do you use a solid line to graph?

R2) What do the solutions of a nonlinear inequality represent?

Group Activity — The Group Activity can be found online on Connect.

emPOWERme Memory Devices

Sometimes it's easier for us to remember information if we "repackage it." That's the idea behind two of the most common memory devices, acronyms and acrostics. **Acronyms** are words or phrases formed by the first letters of a list of terms. P.O.W.E.R. is one acronym you're probably familiar with by now! **Acrostics** are sentences in which the first letters spell out something that needs to be recalled. "**P**lease **E**xcuse **M**y **D**ear **A**unt **S**ally" is a common acrostic that tells you the order of operations in doing a math problem: **P**arentheses, **E**xponents, **M**ultiplication, **D**ivision, **A**ddition, **S**ubtraction.

Practice creating acronyms and acrostics by completing the following exercise:

1. Figure out a Great Lakes acronym using the first letters of their names (which are Erie, Huron, Michigan, Ontario, and Superior.)

2. Devise an acrostic to help remember the names of the nine planets in order of their average distance from the sun. Their names, in order, are Mercury, Venus, Earth, Mars, Jupiter, Saturn, Uranus, Neptune, and Pluto. (There is still disagreement about whether or not Pluto is a planet. We include it here.)

After you've tried to create the acronym and acrostic, compare your answers with others in your class, and answer the following questions: How successful were you in devising effective acronyms and acrostics? Do other people's creations seem to be more effective than others? Why? Do you think the act of creating an acronym or an acrostic is an important component of helping to remember what they represent?

For your information, a common acronym for the Great Lakes is **HOMES** (**H**uron, **O**ntario, **M**ichigan, **E**rie, **S**uperior), and a popular acrostic for the order of the planets is **M**y **V**ery **E**ducated **M**other **J**ust **S**erved **U**s **N**ine **P**izzas.

Chapter 12: Summary

Definition/Procedure	Example

12.1 Graphs of Other Useful Functions

Absolute value functions have V-shaped graphs. (p. 793)

Graph $f(x) = |x|$.

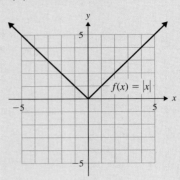

A **piecewise function** is a single function defined by two or more different rules. (p. 794)

$$f(x) = \begin{cases} x + 3, & x > -2 \\ -\dfrac{1}{2}x + 2, & x \leq -2 \end{cases}$$

The **greatest integer function,** $f(x) = [\![x]\!]$, represents the largest integer less than or equal to x. (p. 795)

$[\![8.3]\!] = 8$, $\left[\!\left[-4\dfrac{3}{8}\right]\!\right] = -5$

12.2 The Circle

Parabolas, circles, ellipses, and hyperbolas are called **conic sections**.

The **standard form for the equation of a circle** with center (h, k) and radius r is

$$(x - h)^2 + (y - k)^2 = r^2 \quad \text{(p. 803)}$$

Graph $(x + 3)^2 + y^2 = 4$.

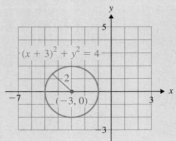

The center is $(-3, 0)$. The radius is $\sqrt{4} = 2$.

The **general form for the equation of a circle** is

$$Ax^2 + Ay^2 + Cx + Dy + E = 0$$

where A, C, D, and E are real numbers.

To rewrite the equation in the form $(x - h)^2 + (y - k)^2 = r^2$, divide the equation by A so that the coefficient of each squared term is 1, then complete the square on x and on y to put it into standard form. (p. 804)

Write $x^2 + y^2 - 16x + 4y + 67 = 0$ in the form $(x - h)^2 + (y - k)^2 = r^2$.

Group the x-terms together, and group the y-terms together.

$$(x^2 - 16x) + (y^2 + 4y) = -67$$

Complete the square for each group of terms.

$$(x^2 - 16x + 64) + (y^2 + 4y + 4) = -67 + 64 + 4$$
$$(x - 8)^2 + (y + 2)^2 = 1$$

Definition/Procedure	Example

12.3 The Ellipse

The **standard form for the equation of an ellipse** is

$$\frac{(x-h)^2}{a^2} + \frac{(y-k)^2}{b^2} = 1$$

The center of the ellipse is (h, k). **(p. 810)**

Graph $\dfrac{(x-1)^2}{9} + \dfrac{(y-2)^2}{4} = 1$.

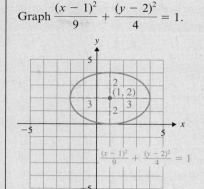

The center is $(1, 2)$.
$a = \sqrt{9} = 3$
$b = \sqrt{4} = 2$

12.4 The Hyperbola

Standard Form for the Equation of a Hyperbola

1) A hyperbola with center (h, k) with branches that open in the *x-direction* has equation

$$\frac{(x-h)^2}{a^2} - \frac{(y-k)^2}{b^2} = 1$$

2) A hyperbola with center (h, k) with branches that open in the *y-direction* has equation

$$\frac{(y-k)^2}{b^2} - \frac{(x-h)^2}{a^2} = 1$$

Notice in 1) that $\dfrac{(x-h)^2}{a^2}$ is the positive quantity, and the branches open in the *x*-direction.

In 2), the positive quantity is $\dfrac{(y-k)^2}{b^2}$, and the branches open in the *y*-direction. **(p. 818)**

Graph $\dfrac{(y-1)^2}{9} - \dfrac{(x-4)^2}{4} = 1$.

The center is $(4, 1)$, $a = \sqrt{4} = 2$, and $b = \sqrt{9} = 3$.

Use the center, $a = 2$, and $b = 3$ to draw the reference rectangle. The diagonals of the rectangle are the asymptotes of the hyperbola.

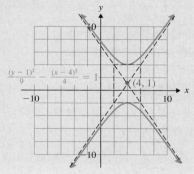

A Nonstandard Form for the Equation of a Hyperbola

The graph of the equation $xy = c$, where c is a nonzero constant, is a hyperbola whose asymptotes are the *x*- and *y*-axes. **(p. 821)**

Graph $xy = 4$.

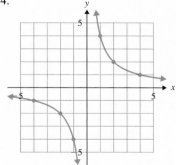

12.5 Nonlinear Systems of Equations

A **nonlinear system of equations** is a system in which at least one of the equations is not linear. We can solve nonlinear systems by substitution or the elimination method. **(p. 832)**

Solve $x - y^2 = 3$ (1)
$\quad\quad\;\; x - 2y = 6$ (2)

$x - y^2 = 3$ (1) Solve equation (1) for x.
$x = y^2 + 3$ (3)

Definition/Procedure	Example
	Substitute $x = y^2 + 3$ into equation (2). $$(y^2 + 3) - 2y = 6$$ $$y^2 - 2y - 3 = 0 \quad \text{Subtract 6.}$$ $$(y - 3)(y + 1) = 0 \quad \text{Factor.}$$ $y - 3 = 0 \quad \text{or} \quad y + 1 = 0 \quad$ Set each factor equal to 0. $\quad y = 3 \quad \text{or} \quad \quad y = -1 \quad$ Solve. Substitute each value into equation (3). $y = 3:\quad x = y^2 + 3 \qquad\qquad y = -1:\quad x = y^2 + 3$ $\qquad\qquad x = (3)^2 + 3 \qquad\qquad\qquad\qquad x = (-1)^2 + 3$ $\qquad\qquad x = 12 \qquad\qquad\qquad\qquad\qquad\quad\; x = 4$ The proposed solutions are $(12, 3)$ and $(4, -1)$. Verify that they also satisfy (2). The solution set is $\{(12, 3), (4, -1)\}$.

12.6 Second-Degree Inequalities and Systems of Inequalities

To graph a **second-degree inequality,** graph the boundary curve, choose a test point, and shade the appropriate region. **(p. 839)**	Graph $x^2 + y^2 < 9$. Graph the *circle* $x^2 + y^2 = 9$ as a dotted curve since the inequality is $<$. Choose a test point not on the boundary curve: $(0, 0)$. Since the test point $(0, 0)$ satisfies the inequality, shade the region inside the circle. 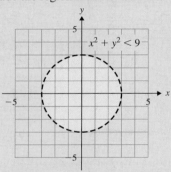 All points in the shaded region satisfy $x^2 + y^2 < 9$.
The **solution set of a system of inequalities** consists of the set of points that satisfy all inequalities in the system. The solution set is the intersection of the graphs of the individual inequalities. **(p. 840)**	Graph the solution set of the system $$x^2 + y \leq 3$$ $$x + y \geq 1$$ First, graph the *parabola* $x^2 + y = 3$ as a solid curve since the inequality is $\leq$. The test point $(0, 0)$ satisfies the inequality $x^2 + y \leq 3$, so shade inside the parabola. Graph the *line* $x + y = 1$ as a solid line since the inequality is $\geq$. Shade above the line since the test point $(0, 0)$ does not satisfy $x + y \geq 1$. The solution set of the system is the *intersection* of the shaded regions of the two graphs. All points in the shaded region satisfy both inequalities. 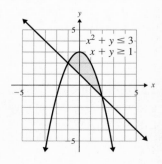

Chapter 12: Review Exercises

(12.1) Graph each function, and identify the domain and range.

1) $g(x) = |x|$
2) $f(x) = |x + 1|$
3) $h(x) = |x| - 4$
4) $h(x) = |x - 1| + 2$
5) $k(x) = -|x| + 5$
6) If the graph of $f(x) = |x|$ is shifted right 5 units to obtain the graph of $g(x)$, write the equation of $g(x)$.

Graph each piecewise function.

7) $f(x) = \begin{cases} -\frac{1}{2}x - 2, & x \le 2 \\ x - 3, & x > 2 \end{cases}$

8) $g(x) = \begin{cases} 1, & x < -3 \\ x + 4, & x \ge -3 \end{cases}$

Let $f(x) = [\![x]\!]$. Find the following function values.

9) $f\left(7\frac{2}{3}\right)$
10) $f(2.1)$
11) $f\left(-8\frac{1}{2}\right)$
12) $f(-5.8)$
13) $f\left(\frac{3}{8}\right)$

Graph each greatest integer function.

14) $f(x) = [\![x]\!]$
15) $g(x) = \left[\!\left[\frac{1}{2}x\right]\!\right]$

16) To mail a letter from the United States to Mexico in 2009 cost $0.60 for the first ounce, $0.85 over 1 oz but less than or equal to 2 oz, then $0.40 for each additional ounce or fraction of an ounce. Let $C(x)$ represent the cost of mailing a letter from the United States to Mexico, and let x represent the weight of the letter, in ounces. Graph $C(x)$ for any letter weighing up to (and including) 5 oz. (www.usps.com)

(12.2) Identify the center and radius of each circle and graph.

17) $(x + 3)^2 + (y - 5)^2 = 36$
18) $x^2 + (y + 4)^2 = 9$
19) $x^2 + y^2 - 10x - 4y + 13 = 0$
20) $x^2 + y^2 + 4x + 16y + 52 = 0$

Find an equation of the circle with the given center and radius.

21) Center $(3, 0)$; radius $= 4$
22) Center $(1, -5)$; radius $= \sqrt{7}$

(12.3)

23) When is an ellipse also a circle?

Identify the center of the ellipse, and graph the equation.

24) $\frac{x^2}{25} + \frac{y^2}{36} = 1$
25) $\frac{(x + 3)^2}{9} + \frac{(y - 3)^2}{4} = 1$
26) $(x - 4)^2 + \frac{(y - 2)^2}{16} = 1$
27) $25x^2 + 4y^2 = 100$

Write each equation in standard form, and graph.

28) $4x^2 + 9y^2 + 8x + 36y + 4 = 0$
29) $25x^2 + 4y^2 - 100x = 0$

(12.4)

30) How can you distinguish between the equation of an ellipse and the equation of a hyperbola?

Identify the center of the hyperbola, and graph the equation.

31) $\frac{y^2}{9} - \frac{x^2}{25} = 1$
32) $\frac{(y - 3)^2}{4} - \frac{(x + 2)^2}{9} = 1$
33) $\frac{(x + 1)^2}{4} - \frac{(y + 2)^2}{4} = 1$
34) $16x^2 - y^2 = 16$
35) Graph $xy = 6$.
36) Write the equations of the asymptotes of the graph in Exercise 31.

Write each equation in standard form, and graph.

37) $16y^2 - x^2 + 2x + 96y + 127 = 0$
38) $x^2 - y^2 - 4x + 6y - 9 = 0$

Graph each function. Identify the domain and range.

39) $h(x) = 2\sqrt{1 - \frac{x^2}{9}}$
40) $f(x) = -\sqrt{4 - x^2}$

(12.2–12.4) Determine whether the graph of each equation is a parabola, circle, ellipse, or hyperbola, then graph each equation.

41) $x^2 + 9y^2 = 9$
42) $x^2 + y^2 = 25$
43) $x = -y^2 + 6y - 5$
44) $x^2 - y = 3$
45) $\frac{(x - 3)^2}{16} - \frac{(y - 4)^2}{25} = 1$
46) $\frac{(x + 3)^2}{16} + \frac{(y + 1)^2}{25} = 1$
47) $x^2 + y^2 - 2x + 2y - 2 = 0$
48) $4y^2 - 9x^2 = 36$
49) $y = \frac{1}{2}(x + 2)^2 + 1$
50) $x^2 + y^2 - 6x - 8y + 16 = 0$

(12.5)

51) If a nonlinear system of equations consists of an ellipse and a hyperbola, how many possible solutions can the system have?

52) If a nonlinear system of equations consists of a line and a circle, how many possible solutions can the system have?

Solve each system.

53) $\begin{aligned} -4x^2 + 3y^2 &= 3 \\ 7x^2 - 5y^2 &= 7 \end{aligned}$

54) $\begin{aligned} y - x^2 &= 7 \\ 3x^2 + 4y &= 28 \end{aligned}$

55) $y = 3 - x^2$
$x - y = -1$

56) $x^2 + y^2 = 9$
$8x + y^2 = 21$

57) $4x^2 + 9y^2 = 36$
$y = \frac{1}{3}x - 5$

58) $4x + 3y = 0$
$4x^2 + 4y^2 = 25$

Write a system of equations, and solve.

59) Find two numbers whose product is 36 and whose sum is 13.

60) The perimeter of a rectangular window is 78 in., and its area is 378 in². Find the dimensions of the window.

(12.6) Graph each inequality.

61) $x^2 + y^2 \leq 4$

62) $y > -(x - 3)^2 + 2$

63) $\frac{x^2}{9} + \frac{y^2}{4} > 1$

64) $\frac{x^2}{9} - \frac{y^2}{4} \geq 1$

65) $x < -y^2 + 2y - 12$

66) $4x^2 + 25y^2 < 100$

Graph the solution set of each system.

67) $y - x^2 > 2$
$x + y > 5$

68) $x^2 + y^2 \leq 25$
$y - x \leq 2$

69) $\frac{x^2}{16} + \frac{y^2}{25} < 1$
$y + 3 > x^2$

70) $x^2 - y^2 \leq 4$
$x^2 + y^2 \geq 16$

71) $y \geq 0$
$x^2 + y^2 \leq 36$
$x + y \leq 0$

72) $x \leq 0$
$4x^2 + 25y^2 \geq 100$
$-x + y \leq 1$

Chapter 12: Test

Graph each function.

1) $g(x) = |x| + 2$

2) $h(x) = -|x - 1|$

3) $f(x) = \begin{cases} x + 3, & x > -1 \\ -2x - 5, & x \leq -1 \end{cases}$

Determine whether the graph of each equation is a parabola, circle, ellipse, or hyperbola, then graph each equation.

4) $\frac{(x - 2)^2}{25} + \frac{(y + 3)^2}{4} = 1$

5) $y = -2x^2 + 6$

6) $y^2 - 4x^2 = 16$

7) $x^2 + (y - 1)^2 = 9$

8) $xy = 2$

9) Write $x^2 + y^2 + 2x - 6y - 6 = 0$ in the form $(x - h)^2 + (y - k)^2 = r^2$. Identify the center and radius, and graph the equation.

10) Write an equation of the circle with center (5, 2) and radius $\sqrt{11}$.

11) The Colosseum in Rome is an ellipse measuring 188 m long and 156 m wide. If the Colosseum is represented on a Cartesian coordinate system with the center of the ellipse at the origin and the longer axis along the x-axis, write an equation of this elliptical structure.
(www.romaviva.com/Colosseo/colosseum.htm)

12) Graph $f(x) = -\sqrt{25 - x^2}$. State the domain and range.

13) Suppose a nonlinear system consists of the equation of a parabola and a circle.
   a) Sketch the different ways in which the graphs can intersect.
   b) Make a sketch in which the graphs do not intersect.
   c) How many possible solutions can the system have?

Solve each system.

14) $x - 2y^2 = -1$
$x + 4y = -1$

15) $2x^2 + 3y^2 = 21$
$-x^2 + 12y^2 = 3$

16) The perimeter of a rectangular picture frame is 44 in. The area is 112 in². Find the dimensions of the frame.

Graph each inequality.

17) $y \geq x^2 - 2$

18) $x^2 - 4y^2 < 36$

Graph the solution set of each system.

19) $x^2 + y^2 > 9$
$y < 2x - 1$

20) $x \geq 0$
$y \leq x^2$
$4x^2 + 9y^2 \leq 36$

Chapter 12: Cumulative Review for Chapters 1–12

Perform the indicated operations, and simplify.

1) $\dfrac{1}{6} - \dfrac{11}{12}$

2) $16 + 20 \div 4 - (5 - 2)^2$

Find the area and perimeter of each figure.

3)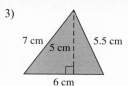

4)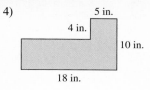

Evaluate.

5) $(-1)^5$

6) 2^4

7) Simplify $\left(\dfrac{2a^8 b}{a^2 b^{-4}}\right)^{-3}$.

8) Solve $\dfrac{3}{8}k + 11 = -4$.

9) Solve for n. $an + z = c$

10) Solve $8 - 5p \le 28$.

11) Write an equation, and solve.
The sum of three consecutive odd integers is 13 more than twice the largest integer. Find the numbers.

12) Find the slope of the line containing the points $(-6, 4)$ and $(2, -4)$.

13) What is the slope of the line with equation $y = 3$?

14) Graph $y = -2x + 5$.

15) Write the slope-intercept form of the line containing the points $(-4, 7)$ and $(4, 1)$.

16) Solve the system. $3x + 4y = 3$
$5x + 6y = 4$

17) Write a system of equations, and solve.
How many milliliters of an 8% alcohol solution and how many milliliters of a 16% alcohol solution must be mixed to make 20 mL of a 14% alcohol solution?

18) Subtract $5p^2 - 8p + 4$ from $2p^2 - p + 10$.

19) Multiply and simplify. $(4w - 3)(2w^2 + 9w - 5)$

20) Divide. $(x^3 - 7x - 36) \div (x - 4)$

Factor completely.

21) $6c^2 - 14c + 8$

22) $m^3 - 8$

23) Solve $(x + 1)(x + 2) = 2(x + 7) + 5x$.

24) Multiply and simplify. $\dfrac{a^2 + 3a - 54}{4a + 36} \cdot \dfrac{10}{36 - a^2}$

25) Simplify $\dfrac{\dfrac{t^2 - 9}{4}}{\dfrac{t - 3}{24}}$.

26) Solve $|3n + 11| = 7$.

27) Solve $|5r + 3| > 12$.

Simplify. Assume all variables represent nonnegative real numbers.

28) $\sqrt{75}$

29) $\sqrt[3]{48}$

30) $\sqrt[3]{27 a^5 b^{13}}$

31) $(16)^{-3/4}$

32) $\dfrac{18}{\sqrt{12}}$

33) Rationalize the denominator of $\dfrac{5}{\sqrt{3} + 4}$.

Solve.

34) $(2p - 1)^2 + 16 = 0$

35) $y^2 = -7y - 3$

36) Solve $25p^2 \le 144$.

37) Solve $\dfrac{t - 3}{2t + 5} > 0$.

38) Given the relation $\{(4, 0), (3, 1), (3, -1), (0, 2)\}$,
  a) what is the domain?
  b) what is the range?
  c) is the relation a function?

39) Graph $f(x) = \sqrt{x}$ and $g(x) = \sqrt{x + 3}$ on the same axes.

40) $f(x) = 3x - 7$ and $g(x) = 2x + 5$.
  a) Find $g(-4)$.
  b) Find $(g \circ f)(x)$.
  c) Find x so that $f(x) = 11$.

41) Given the function $f = \{(-5, 9), (-2, 11), (3, 14), (7, 9)\}$,
  a) is f one-to-one?
  b) does f have an inverse?

42) Find an equation of the inverse of $f(x) = \dfrac{1}{3}x + 4$.

Solve the equations in Exercises 43 and 44.

43) $8^{5t} = 4^{t-3}$

44) $\log_3(4n - 11) = 2$

45) Evaluate $\log 100$.

46) Graph $f(x) = \log_2 x$.

47) Solve $e^{3k} = 8$. Give an exact solution and an approximation to four decimal places.

48) Graph $\dfrac{y^2}{4} - \dfrac{x^2}{9} = 1$.

49) Graph $x^2 + y^2 - 2x + 6y - 6 = 0$.

50) Solve the system. $y - 5x^2 = 3$
$x^2 + 2y = 6$

Answers to Selected Exercises

Chapter 1

Section 1.1

1) a) 6, 0   b) −14, 6, 0   c) $\sqrt{19}$ d) 6
   e) −14, 6, $\frac{2}{5}$, 0, $3.\overline{28}$, $-1\frac{3}{7}$, 0.95
   f) −14, 6, $\frac{2}{5}$, $\sqrt{19}$, 0, $3.\overline{28}$, $-1\frac{3}{7}$, 0.95

3) true   5) false   7) true

9) [number line showing −4, $-1\frac{1}{2}$, 0, $\frac{3}{4}$, 6]

11) [number line showing −5, $-2\frac{1}{2}$, $-\frac{4}{5}$, 1.7, $3\frac{1}{5}$]

13) [number line showing −6.8, $-4\frac{1}{3}$, $-\frac{3}{8}$, 0.2, $1\frac{8}{9}$]

15) 23   17) $\frac{3}{2}$ 19) −10   21) −19   23) −11

25) 7   27) 4.2   29) −10, −2, 0, $\frac{9}{10}$, 3.8, 7

31) −9, $-4\frac{1}{2}$, −0.3, $\frac{1}{4}$, $\frac{5}{8}$, 1   33) true   35) true

37) false   39) false   41) true   43) −27

45) −0.5%   47) −1827   49) 30,391

Section 1.2

1) Answers may vary.   3) Answers may vary.

5) −4   7) −14   9) 13   11) 27   13) −1451

15) $\frac{1}{2}$ 17) $-\frac{31}{24}$ or $-1\frac{7}{24}$ 19) $-\frac{13}{36}$ 21) −3.9

23) 7.31   25) −5.2   27) 6   29) 22   31) 0.4

33) true   35) false   37) true

39) 29,035 − (−36,201) = 65,236. There is a 65,236-ft difference between Mt. Everest and the Mariana Trench.

41) 54,091 − 57,278 = −3187. The median income for a male with a bachelor's degree decreased by $3187 from 2008 to 2009.

43) 12 − 73 = −61. The coldest temperature recorded in Colorado was −61°F.

45) 7 + 4 + 1 + 6 − 10 = 8. The Patriots' net yardage on this offensive drive was 8 yd.

47) a) −5000   b) 10,000   c) 18,000   d) −21,000

49) positive   51) −45   53) 42   55) −42   57) $-\frac{14}{15}$

59) −0.3   61) −64   63) 0   65) negative   67) 7

69) 8   71) −24   73) $\frac{10}{13}$ 75) 0   77) $-\frac{9}{7}$ or $-1\frac{2}{7}$

79) $\frac{4}{3}$ or $1\frac{1}{3}$ 81) −0.05

Section 1.3

1) 9^4 3) negative   5) positive   7) 11^4 9) 3.2^3

11) $\left(\frac{1}{5}\right)^4$ 13) $(-7)^6$ 15) 32   17) 121   19) 16

21) −49   23) −8   25) $\frac{1}{125}$ 27) 0.25 or $\frac{1}{4}$

29) False; the $\sqrt{\ }$ symbol means to find only the positive square root of 49.

31) true   33) 8 and −8   35) 20 and −20

37) $\frac{5}{4}$ and $-\frac{5}{4}$ 39) 6   41) −30   43) not real

45) $\frac{10}{11}$ 47) $-\frac{1}{8}$ 49) Answers may vary.   51) 43

53) 48   55) 5   57) 10   59) 23   61) $\frac{7}{25}$

63) 3   65) −27   67) −20   69) 8   71) $\frac{5}{6}$

Section 1.4

1) a) 37   b) 28   3) −9   5) −70   7) −1   9) $-\frac{9}{5}$

11) 0   13) $\frac{1}{6}$ 15) associative   17) commutative

19) associative   21) distributive   23) identity

25) distributive   27) $7u + 7v$ 29) $4 + k$ 31) $-4z$

33) No. Subtraction is not commutative.

35) $5 \cdot 4 + 5 \cdot 3 = 20 + 15 = 35$

37) $(-2) \cdot 5 + (-2) \cdot 7 = -10 + (-14) = -24$

39) $(-7) \cdot 2 + (-7) \cdot (-6) = -14 + 42 = 28$

41) $-6 - 1 = -7$

43) $(-10) \cdot 5 + 3 \cdot 5 = -50 + 15 = -35$

SA-1

45) $9g + 9 \cdot 6 = 9g + 54$

47) $-5z + (-5) \cdot 3 = -5z - 15$

49) $-8u + (-8) \cdot (-4) = -8u + 32$ 51) $-v + 6$

53) $10m + 10 \cdot 5n + 10 \cdot (-3) = 10m + 50n - 30$

55) $8c - 9d + 14$

Chapter 1: Review Exercises

1) a) 0, 2   b) 2   c) $-6, 0, 2$ d) $-6, 14.38, \dfrac{3}{11}, 2, 5.\overline{7}, 0$
   e) $\sqrt{23}, 9.21743819\ldots$

3) 10   5) 75   7) -5.1 9) 70   11) -7.77

13) 60   15) 2   17) $-\dfrac{25}{18}$ or $-1\dfrac{7}{18}$ 19) -25

21) 81   23) -64 25) 7   27) -6 29) -2

31) -40 33) $\dfrac{15}{28}$ 35) $\dfrac{1}{6}$ 37) 9

39)

Term	Coeff.
c^4	1
$12c^3$	12
$-c^2$	-1
$-3.8c$	-3.8
11	11

41) $-\dfrac{29}{9}$ or $-3\dfrac{2}{9}$ 43) associative   45) commutative

47) $3 \cdot 10 - 3 \cdot 6 = 30 - 18 = 12$ 49) $-12 - 5 = -17$

Chapter 1: Test

1) a) 41, -8, 0   b) $\sqrt{75}, 6.37528861\ldots$ c) 41
   d) 41, -8, 0, $2.\overline{83}$, 6.5, $4\dfrac{5}{8}$ e) 41, 0

2) [number line from -7 to 7 with points at $-4, -1.2, -\tfrac{2}{3}, \tfrac{7}{8}, 4\tfrac{3}{4}, 6$]

3) $\dfrac{3}{16}$ 4) $\dfrac{3}{5}$ 5) $3\dfrac{1}{12}$ 6) $\dfrac{2}{13}$ 7) $-\dfrac{11}{42}$

8) -30 9) 28   10) $-\dfrac{9}{25}$ 11) 48   12) -6.5

13) 0   14) 1   15) 32   16) -81 17) 92

18) -10 19) -24 20) 23   21) $-\dfrac{5}{8}$

22) true   23) false   24) false   25) false

26) 14,787 ft   27) 0

28) a) distributive   b) inverse   c) commutative

29) a) $(-2) \cdot 5 + (-2) \cdot 3 = -10 + (-6) = -16$
   b) $5t + 5 \cdot 9u + 5 \cdot 1 = 5t + 45u + 5$

30) No. Subtraction is not commutative.

Chapter 2

Section 2.1

1) equation   3) expression   5) No, it is an expression.

7) b, d   9) yes   11) no   13) $\{17\}$ 15) $\{-4\}$

17) $\left\{-\dfrac{1}{8}\right\}$ 19) $\{10\}$ 21) $\{-48\}$ 23) $\{-15\}$

25) $\{18\}$ 27) $\{12\}$ 29) $\{8\}$ 31) $\{0\}$ 33) $\{27\}$

35) $\left\{-\dfrac{7}{5}\right\}$ 37) $\{6\}$ 39) $\{-3\}$ 41) $\left\{-\dfrac{5}{4}\right\}$

43) $\{0\}$ 45) $\{3\}$ 47) $\{-3\}$ 49) $\left\{\dfrac{7}{3}\right\}$ 51) $\{4\}$

53) $\varnothing$ 55) {all real numbers}   57) {all real numbers}

59) $\{0\}$ 61) $\left\{\dfrac{11}{2}\right\}$ 63) $\{5\}$ 65) $\{3\}$

67) $\left\{-\dfrac{15}{2}\right\}$ 69) $\{-8\}$ 71) $\{5\}$ 73) $\{24\}$

75) $\{300\}$ 77) $x + 4 = 15; 11$ 79) $x - 7 = 22; 29$

81) $2x = -16; -8$ 83) $2x + 7 = 35; 14$

85) $3x - 8 = 40; 16$ 87) $\dfrac{1}{2}x + 10 = 3; -14$

89) $5x - 12 = x + 16; 7$ 91) $\dfrac{1}{3}x + 10 = x - 2; 18$

93) $2(x + 5) = 16; 3$ 95) $3x = 15 + \dfrac{1}{2}x; 6$

97) $x - 6 = 5 + 2x; -11$

Section 2.2

1) $c + 5$ 3) $p - 31$ 5) $3w$ 7) $14 - x$

9) Pepsi = 9.8 tsp, Gatorade = 3.3 tsp

11) China: 88, Germany: 44

13) Columbia: 1240 mi, Ohio: 1310 mi

15) 11 in. and 25 in.   17) 12 ft and 6 ft

19) 38, 39   21) 12, 14   23) $-15, -13, -11$

25) 172, 173   27) $63.75   29) $11.60   31) $140.00

33) $10.95   35) $32.50   37) 80   39) 425

41) $32   43) $380   45) $9000 at 6% and $6000 at 7%

47) $1400 at 6% and $1600 at 5%

49) $2800 at 9.5% and $4200 at 7%

51) ride: 4 mi, walk: 3 mi   53) 2500

55) 12 in., 17 in., and 24 in.   57) 72, 74, 76

59) *Harry Potter:* $381 million, *Twilight:* $281 million

61) CD: $1500, IRA: $3000, mutual fund: $2500

63) $38,600

Section 2.3

1) 25 ft   3) 6 in.   5) 415 ft^2 7) 12 ft
9) 6 in.   11) $m\angle A = 26°, m\angle B = 52°$
13) $m\angle A = 37°, m\angle B = 55°, m\angle C = 88°$ 15) 123°, 123°
17) 150°, 150°   19) 133°, 47°   21) 79°, 101°
23) $180 - x$ 25) 17°
27) angle: 20°, comp.: 70°, supp.: 160°
29) 35°   31) 40°   33) 2   35) 4   37) 18
39) 2   41) 7   43) 20
45) a) $x = 23$ b) $x = p - n$ c) $x = v - q$
47) a) $y = 9$ b) $y = \dfrac{x}{a}$ c) $y = \dfrac{r}{p}$
49) a) $c = 21$ b) $c = ur$ c) $c = xt$
51) a) $d = 3$ b) $d = \dfrac{z + a}{k}$
53) a) $z = -\dfrac{5}{2}$ b) $z = \dfrac{w - t}{y}$ 55) $m = \dfrac{F}{a}$
57) $c = nv$ 59) $\sigma = \dfrac{E}{T^4}$ 61) $h = \dfrac{3V}{\pi r^2}$
63) $E = IR$ 65) $R = \dfrac{I}{PT}$
67) $I = \dfrac{P - 2w}{2}$ or $I = \dfrac{P}{2} - w$
69) $N = \dfrac{2.5H}{D^2}$ 71) $b_2 = \dfrac{2A}{h} - b_1$ or $b_2 = \dfrac{2A - hb_1}{h}$
73) a) $w = \dfrac{P - 2l}{2}$ or $w = \dfrac{P}{2} - l$ b) 3 cm
75) a) $F = \dfrac{9}{5}C + 32$ b) 77°F

Section 2.4

1) a) $0.80   b) 80¢   3) a) $2.17   b) 217¢
5) a) $2.95   b) 295¢   7) a) $0.25q$ dollars   b) $25q$ cents
9) a) $0.10d$ dollars   b) $10d$ cents
11) a) $0.01p + 0.25q$ dollars   b) $p + 25q$ cents
13) 9 nickels, 17 quarters   15) 11 $5 bills, 14 $1 bills
17) 38 adult tickets, 19 children's tickets
19) 2 oz   21) 7.6 mL
23) 16 oz of the 4% acid solution, 8 oz of the 10% acid solution
25) $2\dfrac{1}{2}$ L   27) 2 lb of Aztec and 3 lb of Cinnamon
29) eastbound: 65 mph; westbound: 73 mph   31) $\dfrac{5}{6}$ hr
33) passenger train: 50 mph; freight train: 30 mph
35) 36 min   37) 4.30 P.M.   39) 48 mph
41) 23 dimes, 16 quarters
43) jet: 400 mph, small plane: 200 mph
45) 8 cc of the 0.08% solution and 12 cc of the 0.03% solution

Chapter 2: Review Exercises

1) yes   3) no
5) The variables are eliminated and you get a false statement like 5 = 13.
7) $\left\{-\dfrac{10}{3}\right\}$ 9) $\{19\}$ 11) $\left\{\dfrac{45}{14}\right\}$ 13) $\{35\}$
15) $\{-2\}$ 17) $\{2\}$ 19) $\{-2\}$
21) {all real numbers}   23) $\{3\}$ 25) $\{1.3\}$
27) $\{10\}$ 29) $\left\{-\dfrac{20}{3}\right\}$ 31) 17   33) $26 - c$
35) Kelly Clarkson: 297,000 copies, Clay Aiken: 613,000 copies
37) 125 ft   39) $10.1 billion   41) 12 cm
43) $m\angle A = 55°, m\angle B = 55°, m\angle C = 70°$
45) 46°, 46°   47) $p = z + n$ 49) $R = \dfrac{pV}{nT}$
51) 58 dimes, 33 quarters   53) 45 min

Chapter 2: Test

1) $\{2\}$ 2) $\{-3\}$ 3) $\{7\}$ 4) $\varnothing$ 5) $\left\{\dfrac{39}{2}\right\}$
6) $\{15\}$ 7) 21   8) 36, 38, 40   9) 10 in. × 15 in.
10) 3.5 qt of regular oil, 1.5 qt of synthetic oil
11) 50   12) 70 ft
13) eastbound: 66 mph; westbound: 72 mph
14) $t = \dfrac{5R}{k}$ 15) $h = \dfrac{S - 2\pi r^2}{2\pi r}$ 16) 32°, 148°
17) $m\angle A = 26°, m\angle B = 115°, m\angle C = 39°$ 18) $1\dfrac{1}{5}$ gal

Chapter 2: Cumulative Review for Chapters 1–2

1) $-\dfrac{13}{36}$ 2) $\dfrac{2}{45}$ 3) 64   4) 56   5) -81
6) 9   7) $\left\{-13.7, \dfrac{19}{7}, 0, 8, 0.\overline{61}, \sqrt{81}, -2\right\}$
8) $\{0, 8, \sqrt{81}, -2\}$ 9) $\{0, 8, \sqrt{81}\}$ 10) $\{8, \sqrt{81}\}$
11) associative property   12) distributive property
13) inverse property   14) $5 + 8$ 15) $\{-70\}$
16) $\left\{\dfrac{19}{4}\right\}$ 17) {all real numbers}

18) $R = \dfrac{A - P}{PT}$ 19) 121°   20) 57°

21) area = 105 cm²; perimeter = 44 cm

22) 123°; obtuse   23) generic: 48; name brand: 24

24) $\dfrac{1}{2}$ hour   25) $40

Chapter 3
Section 3.1

1) You use parentheses when there is a < or > symbol or when you use ∞ or −∞.

3)
a) $\{x | x \geq 3\}$ b) $[3, \infty)$

5)
a) $\{c | c < -1\}$ b) $(-\infty, -1)$

7)
a) $\left\{w \left| w > -\dfrac{11}{3}\right.\right\}$ b) $\left(-\dfrac{11}{3}, \infty\right)$

9)
a) $\{r | r \leq 4\}$ b) $(-\infty, 4]$

11)
a) $\{y | y \geq -4\}$ b) $[-4, \infty)$

13)
a) $\{c | c > 4\}$ b) $(4, \infty)$

15)
a) $\left\{k \left| k < -\dfrac{11}{3}\right.\right\}$ b) $\left(-\infty, -\dfrac{11}{3}\right)$

17)
a) $\{b | b \geq -8\}$ b) $[-8, \infty)$

19)
a) $\{w | w < 3\}$ b) $(-\infty, 3)$

21)
a) $\{z | z \geq -15\}$ b) $[-15, \infty)$

23)
a) $\{y | y > 8\}$ b) $(8, \infty)$

25)
$(-1, \infty)$

27)
$\left(-\infty, -\dfrac{3}{7}\right]$

29)
$(-3, \infty)$

31)
$\left(\dfrac{6}{11}, \infty\right)$

33)
$(-\infty, 4)$

35)
$(0, \infty)$

37)
$(-\infty, 5]$

39)
a) $\{n | 1 \leq n \leq 4\}$ b) $[1, 4]$

41)
a) $\{a | -2 < a < 1\}$ b) $(-2, 1)$

43)
a) $\left\{z \left| \dfrac{1}{2} < z \leq 3\right.\right\}$ b) $\left(\dfrac{1}{2}, 3\right]$

45)
$[-3, 1]$

47)
$\left(\dfrac{3}{2}, 3\right)$

49)
$[-4, -1]$

51)
$\left[\dfrac{7}{4}, 3\right)$

53)
$(-8, 4)$

55)
$\left[-1, -\dfrac{2}{5}\right]$

57)
$[5, 8)$

59)
$\left[-1, \dfrac{4}{3}\right]$

61) $(-7, \infty)$ 63) $\left(-\infty, \dfrac{4}{3}\right]$ 65) $(-\infty, -12)$

67) $\left(-15, -\dfrac{15}{4}\right]$ 69) $[-9, \infty)$ 71) $[-2, 0)$

73) at most $5\dfrac{1}{2}$ hr   75) at most 8 mi   77) 89 or higher

Section 3.2

1) $A \cap B$ means "A intersect B." $A \cap B$ is the set of all numbers which are in set A and in set B.

3) $\{8, 10\}$ 5) $\{2, 4, 5, 6, 7, 8, 9, 10\}$ 7) $\emptyset$

9) $\{1, 2, 3, 4, 5, 6, 8, 10\}$

11) $[-3, 2]$

13) $(-1, 3)$

15) $[3, \infty)$

17) $\emptyset$

19) $[2, 5]$

21) $(-2, 3)$

23) $(-3, 4]$

25) $\emptyset$

27) $(3, \infty)$

29) $[-4, 1]$

31) $(-\infty, -1) \cup (5, \infty)$

33) $\left(-\infty, \dfrac{5}{3}\right] \cup (4, \infty)$

35) $(1, \infty)$

37) $(-\infty, \infty)$

39) $(-\infty, -1) \cup (3, \infty)$

41) $\left(-\infty, \dfrac{7}{2}\right] \cup (6, \infty)$

43) $(-5, \infty)$

45) $(-\infty, -6) \cup [-3, \infty)$

47) $(-\infty, \infty)$

49) $(-\infty, -2]$

51) $\left[-5, \dfrac{1}{2}\right]$ 53) $\left(-\infty, -\dfrac{9}{4}\right) \cup [5, \infty)$

55) $(-\infty, \infty)$ 57) $(-\infty, 0)$ 59) $[-8, -4]$

61) {Liliane Bettancourt, Alice Walton}

63) {Liliane Bettancourt, J. K. Rowling, Oprah Winfrey}

Section 3.3

1) Answers may vary.   3) $\{-6, 6\}$ 5) $\{2, 8\}$

7) $\left\{-\dfrac{1}{2}, 3\right\}$ 9) $\left\{-\dfrac{1}{2}, -\dfrac{1}{3}\right\}$ 11) $\{-24, 15\}$

13) $\left\{-\dfrac{10}{3}, \dfrac{50}{3}\right\}$ 15) $\emptyset$ 17) $\{-10, 22\}$

19) $\{-5, 0\}$ 21) $\{-14\}$ 23) $\emptyset$

25) $\left\{-\dfrac{16}{5}, 2\right\}$ 27) $\emptyset$ 29) $\left\{-\dfrac{14}{3}, 4\right\}$ 31) $\left\{\dfrac{1}{2}, 4\right\}$

33) $\left\{\dfrac{2}{5}, 2\right\}$ 35) $\{10\}$ 37) $|x| = 9$, may vary

39) $|x| = \dfrac{1}{2}$, may vary

41) $[-1, 5]$

43) $(-\infty, 2) \cup (9, \infty)$

45) $\left(-\infty, -\dfrac{9}{2}\right] \cup \left[\dfrac{3}{5}, \infty\right)$

47) $[-7, 7]$

49) $(-4, 4)$

51) $(-2, 6)$

53) $\left[-\dfrac{14}{3}, -2\right]$

55) $\left[\dfrac{2}{3}, \dfrac{5}{3}\right]$

57) $\emptyset$

59) $\emptyset$

61) $(-8, 3)$

63) $[-12, 4]$

65) $(-\infty, -7] \cup [7, \infty)$

67) $(-\infty, -14] \cup [-6, \infty)$

69) $\left(-\infty, -\dfrac{3}{2}\right] \cup [3, \infty)$

71) $(-\infty, 2) \cup \left(\dfrac{11}{3}, \infty\right)$

73) $(-\infty, \infty)$

75) $(-\infty, \infty)$

77) $(-\infty, -12] \cup [0, \infty)$

79) $\left(-\infty, -\dfrac{27}{5}\right] \cup \left[\dfrac{21}{5}, \infty\right)$

81) The absolute value of a quantity is always 0 or positive; it cannot be less than 0.

83) The absolute value of a quantity is always 0 or positive, so for any real number, x, the quantity $|2x + 1|$ will be greater than -3.

85) $(-\infty, -6) \cup (-3, \infty)$

87) $\left\{-2, -\dfrac{1}{2}\right\}$ 89) $\left(-\infty, -\dfrac{1}{4}\right] \cup [2, \infty)$

91) $(-3, \infty)$ 93) $\emptyset$ 95) $\{-21, -3\}$

97) $\emptyset$ 99) $\left(-\infty, -\dfrac{1}{25}\right]$ 101) $(-\infty, \infty)$

103) $[-15, -1]$ 105) $\left(-\infty, \dfrac{1}{5}\right) \cup (3, \infty)$

107) $|a - 128| \leq 0.75$; $127.25 \leq a \leq 128.75$; there is between 127.25 oz and 128.75 oz of milk in the container.

109) $|b - 38| \leq 5$; $33 \leq b \leq 43$; he will spend between \$33 and \$43 on his daughter's gift.

Chapter 3: Review Exercises

1) $[8, \infty)$

3) $(-3, \infty)$

5) $(-\infty, 4]$

7) $(-2, 3]$

9) $(-\infty, 4)$

11) $(4, 8]$

15) $[1, 3]$

17) $(-1, \infty)$

19) {Toyota}   21) $\{-9, 9\}$

23) $\left\{-1, \dfrac{1}{7}\right\}$ 25) $\left\{-\dfrac{15}{8}, -\dfrac{7}{8}\right\}$ 27) $\left\{\dfrac{11}{5}, \dfrac{13}{5}\right\}$

29) $\left\{-8, \dfrac{4}{15}\right\}$ 31) $\emptyset$ 33) $\left\{-\dfrac{4}{9}\right\}$

35) $|a| = 4$, may vary

37) $[-3, 3]$

39) $(-\infty, -2) \cup (2, \infty)$

41) $(-\infty, -1] \cup \left[\dfrac{1}{6}, \infty\right)$

43) $(-5, 13)$

45) $\left[-\dfrac{15}{4}, -\dfrac{3}{4}\right]$

47) $\left(-\infty, -\dfrac{19}{5}\right] \cup [-1, \infty)$

49) $(-\infty, \infty)$

51) $\left\{-\dfrac{1}{12}\right\}$

Chapter 3: Test

1) $(-\infty, -5]$

2) $(-\infty, -2]$

3) $\left(\dfrac{2}{3}, \infty\right)$

4) $\left[\dfrac{3}{4}, \infty\right)$

5) $(1, 7]$

6) $(-3, 0)$

7) at most 6 hr   8) a) $\{1, 2, 3, 6, 9, 12\}$ b) $\{1, 2, 12\}$

9) $(-\infty, -8) \cup \left(\dfrac{7}{3}, \infty\right)$ 10) $[0, 3]$

11) $(-\infty, \infty)$ 12) $\left\{-\dfrac{1}{2}, 5\right\}$ 13) $\{-16, 4\}$

14) $\left\{-8, \dfrac{3}{2}\right\}$ 15) $\varnothing$ 16) $|x| = 8$, may vary

17) $[-1, 8]$

18) $\left(-\infty, -\dfrac{11}{2}\right] \cup [1, \infty)$ 19) $\varnothing$

20) $|w - 168| \le 0.75$; $167.25 \le w \le 168.75$; Thanh's weight is between 167.25 lb and 168.75 lb.

Chapter 3: Cumulative Review for Chapters 1–3

1) $-\dfrac{11}{24}$ 2) $\dfrac{15}{2}$ 3) 54   4) -87 5) -47

6) 27 cm²   7) $\{-5, 0, 9\}$ 8) $\left\{\dfrac{3}{4}, -5, 2.5, 0, 0.\overline{4}, 9\right\}$

9) $\{0, 9\}$ 10) distributive

11) No. For example, $10 - 3 \ne 3 - 10$.   12) $17y^2 - 18y$

13) $\left\{-\dfrac{23}{2}\right\}$ 14) $\{4\}$ 15) {all real numbers}

16) $\left\{\dfrac{11}{8}\right\}$ 17) $\{18\}$ 18) car: 60 mph, train: 70 mph

19) $(-\infty, 6)$ 20) $\left[-\dfrac{2}{3}, \infty\right)$ 21) $\left(-4, -\dfrac{7}{6}\right)$

22) $(-\infty, -3] \cup \left[\dfrac{11}{4}, \infty\right)$ 23) $[-36, 8]$

24) $(-\infty, -1) \cup (7, \infty)$ 25) at least 77

Chapter 4

Section 4.1

1) a) x represents the season and y represents the number of people who watched the finale of that season.
b) 28.8 million people watched the Season 3 finale.
c) 36.4 million   d) Season 6   e) (1, 22.8)

3) A: (5, 1); quadrant I; B: (2, −3); quadrant IV
C: (−2, 4); quadrant II; D: (−3, −4); quadrant III
E: (3, 0); no quadrant; F: (0, −2); no quadrant

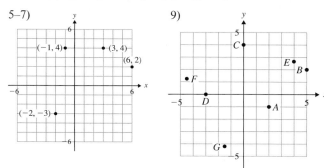

11) positive   13) negative   15) zero   17) yes

19) yes   21) no   23) yes   25) no

27) Answers may vary.

29) A line: The line represents all solutions to the equation. Every point on the line is a solution of the equation.

31) It is the point where the graph intersects the y-axis. To find the y-intercept, let $x = 0$ and solve for y.

33)

x	y
0	5
−1	7
2	1
3	−1

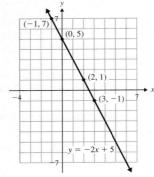

35)

x	y
0	6
2	11
−2	1
−4	−4

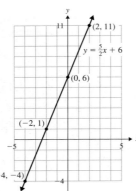

37)

x	y
0	$\frac{3}{2}$
−3	0
5	4
−1	1

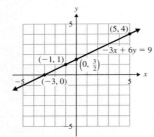

39)

x	y
0	−4
−3	−4
−1	−4
2	−4

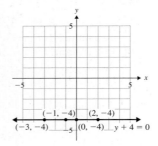

41) a) $(3, -5), (6, -3), (-3, -9)$

b) $\left(1, -\frac{19}{3}\right), \left(5, -\frac{11}{3}\right), \left(-2, -\frac{25}{3}\right)$

c) The x-values in part a) are multiples of the denominator of $\frac{2}{3}$. So, when you multiply $\frac{2}{3}$ by a multiple of 3 the fraction is eliminated.

43) $(3, 0), (0, 6), (1, 4)$

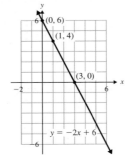

45) $(4, 0), (0, -3), \left(2, -\frac{3}{2}\right)$

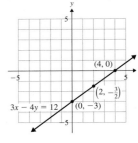

47) $(-8, 0), (0, -12), (-4, -6)$

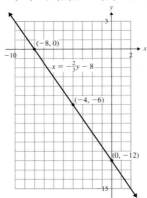

49) $(6, 0), \left(0, -\frac{3}{2}\right), (2, -1)$

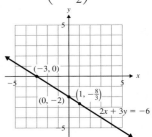

51) $(0, 0), (1, -1), (-1, 1)$

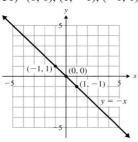

53) $(0, 0), (5, 2), (-5, -2)$

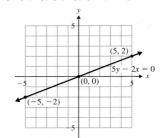

55) $(5, 0), (5, 2), (5, -1)$
Answers may vary.

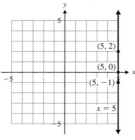

57) $(0, 0), (1, 0), (-2, 0)$
Answers may vary.

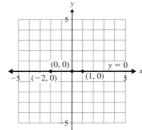

59) $(0, -3), (1, -3), (-3, -3)$
Answers may vary.

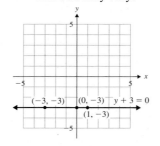

SA-8 Answers to Selected Exercises

61) $(8, 0)$, $\left(0, \dfrac{8}{3}\right)$, $(2, 2)$

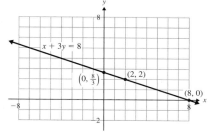

63) a) $y = 0$ b) $x = 0$ 65) $(4, 6)$ 67) $(-3, -3)$

69) $\left(-1, -\dfrac{9}{2}\right)$ 71) $\left(\dfrac{1}{2}, -\dfrac{5}{2}\right)$ 73) $\left(2, \dfrac{5}{4}\right)$

75) $(-0.7, 3.6)$

77) a) $(16, 800)$, $(17, 1300)$, $(18, 1800)$, $(19, 1900)$

b)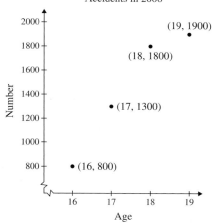

c) There were 1800 18-year old drivers in fatal motor vehicle accidents in 2006.

79) a)

x	y
1	120
3	160
4	180
6	220

$(1, 120)$, $(3, 160)$, $(4, 180)$, $(6, 220)$

b)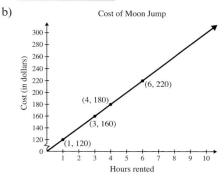

c) The cost of renting the moon jump for 4 hours is $180.
d) 9 hours

81) a)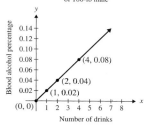

$(0, 0)$, $(1, 0.02)$, $(2, 0.04)$, $(4, 0.08)$

x	y
0	0
1	0.02
2	0.04
4	0.08

b) $(0, 0)$: If no drinks are consumed, the blood alcohol percentage is 0.
$(1, 0.02)$: After 1 drink, the blood alcohol percentage is 0.02.
$(2, 0.04)$: After 2 drinks, the blood alcohol percentage is 0.04.
$(4, 0.08)$: After 4 drinks, the blood alcohol percentage is 0.08.

c)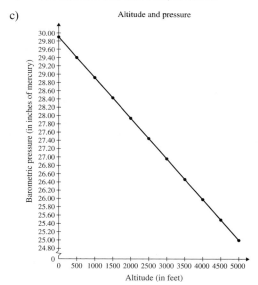

83) a) Answers will vary.
b) 29.86 in.; 28.86 in.; 26.36 in.; 24.86 in.

c)

d) No, because the problem states that the equation applies to altitudes 0 ft–5000 ft.

Section 4.2

1) The slope of a line is the ratio of vertical change to horizontal change. It is $\dfrac{\text{change in } y}{\text{change in } x}$ or $\dfrac{\text{rise}}{\text{run}}$ or $\dfrac{y_2 - y_1}{x_2 - x_1}$ where (x_1, y_1) and (x_2, y_2) are points on the line.

3) It slants downward from left to right.   5) undefined

7) $m = \dfrac{3}{4}$ 9) $m = -\dfrac{1}{3}$ 11) $m = -5$

13) Slope is undefined.

15)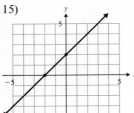

17) $\frac{1}{2}$ 19) -1

21) $-\frac{2}{9}$ 23) 0   25) undefined   27) $\frac{14}{3}$

29) 2.5   31) a) $\frac{5}{6}$ b) $\frac{2}{3}$ c) $m = \frac{1}{3}$; 4-12 pitch

33) Yes. The slope of the driveway will be 10%.

35) a) $22,000   b) negative   c) The value of the car is decreasing over time.   d) $m = -2000$; the value of the car is decreasing $2000 per year.

37) 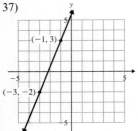   39)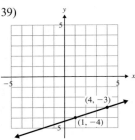

41)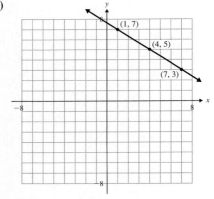

43)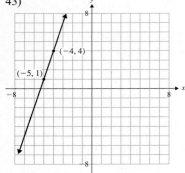

45)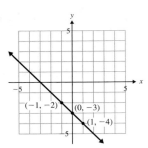

47)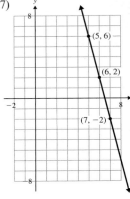

49)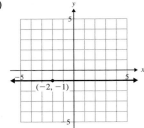

51) 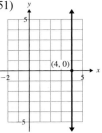   53)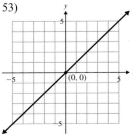

55) The slope is m, and the y-intercept is $(0, b)$.

57) $m = \frac{2}{5}$, y-int: $(0, -6)$ 59) $m = -\frac{5}{3}$, y-int: $(0, 4)$

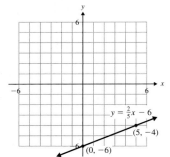   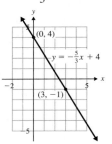

61) $m = \frac{3}{4}$, y-int: $(0, 1)$ 63) $m = 4$, y-int: $(0, -2)$

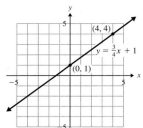   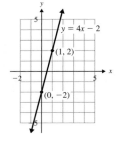

65) $m = -1$, y-int: $(0, 5)$ 67) $m = \dfrac{3}{2}$, y-int: $\left(0, \dfrac{1}{2}\right)$

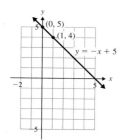
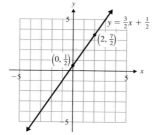

69) $m = 0$, y-int: $(0, -2)$ 71) $y = -\dfrac{1}{3}x - 2$

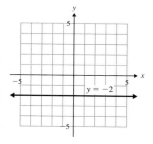
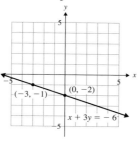

73) $y = \dfrac{3}{2}x - 4$ 75) This cannot be written in slope-intercept form.

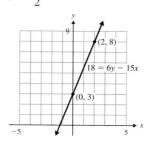
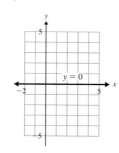

77) $y = \dfrac{5}{2}x + 3$ 79) $y = 0$

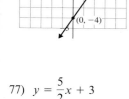
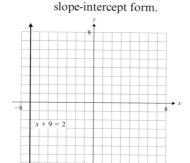

81) a) $(0, 34{,}000)$; if Dave has $0 in sales, his income is $34,000.
  b) $m = 0.05$; Dave earns $0.05 for every $1 in sales.
  c) $38,000

83) a) In 1945, 40.53 gal of whole milk were consumed per person, per year.
  b) $m = -0.59$; since 1945, Americans have been consuming 0.59 fewer gallons of whole milk each year.
  c) estimate from the graph: 7.5 gal; consumption from the equation: 8.08 gal.

85) a) $(0, 68{,}613)$; in 2000, the average annual salary of a pharmacist was $68,613.
  b) $m = 3986$; the average salary of a pharmacist is increasing by $3986 per year.
  c) $84,557 d) $124,417
  e) Answers may vary.

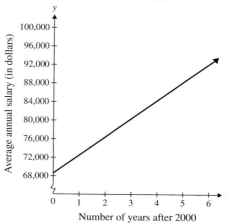

87) $y = -4x + 7$ 89) $y = \dfrac{8}{5}x - 6$ 91) $y = \dfrac{1}{3}x + 5$
93) $y = -x$ 95) $y = -2$

Section 4.3

1) Substitute the slope and y-intercept into $y = mx + b$.
3) $y = -7x + 2$ 5) $x - y = 3$
7) $x + 3y = -12$ 9) $y = x$
11) a) $y - y_1 = m(x - x_1)$ b) Substitute the slope and point into the point-slope formula.
13) $y = 5x + 1$ 15) $y = -x - 5$ 17) $4x - y = -7$
19) $y = \dfrac{1}{6}x - \dfrac{13}{3}$ 21) $5x + 9y = 30$
23) Find the slope, and use it and one of the points in the point-slope formula.
25) $y = x + 1$ 27) $y = -\dfrac{1}{3}x + \dfrac{10}{3}$ 29) $3x - 4y = 17$
31) $x + 2y = -3$ 33) $y = 5.0x - 8.3$ 35) $y = \dfrac{3}{2}x - 4$
37) $y = -x - 2$ 39) $y = 5$ 41) $y = 4x + 15$
43) $y = \dfrac{8}{3}x - 9$ 45) $y = -\dfrac{3}{4}x - \dfrac{11}{4}$ 47) $x = 3$
49) $y = 4$ 51) $y = -\dfrac{1}{3}x + 2$ 53) $y = x$
55) Answers may vary.   57) perpendicular
59) parallel   61) neither   63) perpendicular
65) parallel   67) parallel   69) parallel
71) perpendicular   73) neither   75) $y = 4x + 2$
77) $x - 2y = -6$ 79) $x + 4y = 12$

81) $y = -\dfrac{1}{5}x + 10$ 83) $y = -\dfrac{3}{2}x + 6$

85) $x - 5y = 10$ 87) $y = x$ 89) $5x + 8y = 24$

91) $y = -3x + 8$ 93) $y = 2x - 5$ 95) $x = -1$

97) $x = 2$ 99) $y = -\dfrac{2}{7}x + \dfrac{1}{7}$ 101) $y = -\dfrac{5}{2}$

103) a) $L = \dfrac{1}{3}S + \dfrac{22}{3}$ b) 12.5

105) a) $y = 8700x + 1{,}257{,}900$
b) The population of Maine is increasing by 8700 people per year.
c) 1,257,900; 1,292,700   d) 2018

107) a) $y = -6.4x + 124$
b) The number of farms with milk cows is decreasing by 6.4 thousand (6400) per year.
c) 79.2 thousand (79,200)

109) a) $y = 12{,}318.7x + 6479$
b) The number of registered hybrid vehicles is increasing by 12,318.7 per year.
c) 31,116.4; this is slightly lower than the actual value.
d) 129,666

Section 4.4

1) Answers may vary.   3) Answers may vary.

5) Answers may vary.   7) dotted

9)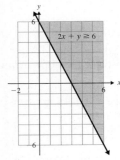

11)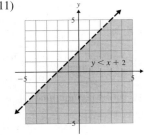

13)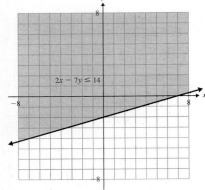

15)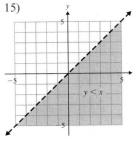

17)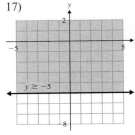

19) below

21)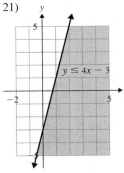

23)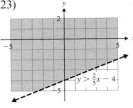

25)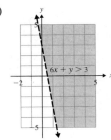

27)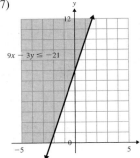

29)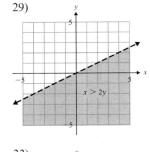

31) Answers may vary.

33)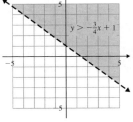

35)

SA-12 Answers to Selected Exercises   www.mhhe.com/messersmith

37)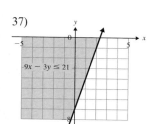

39)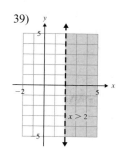

41)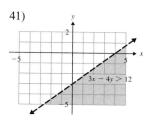

43) No; it does not satisfy $2x + y < 7$.

45) 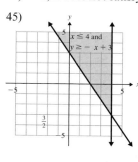 47)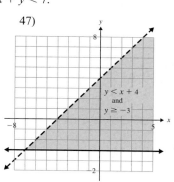

49)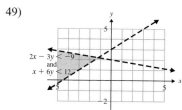

51)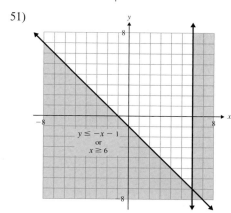

53)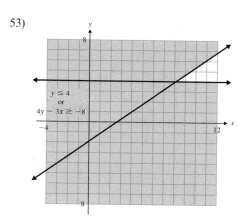

55)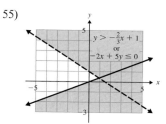

57)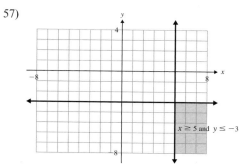

59) 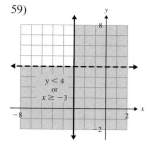 61)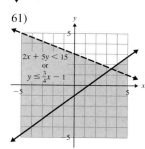

63)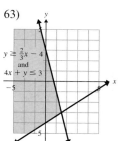

Section 4.5

1) a) any set of ordered pairs
   b) Answers may vary.
   c) Answers may vary.

3) domain: $\{-8, -2, 1, 5\}$; range: $\{-3, 4, 6, 13\}$; function

5) domain: $\{1, 9, 25\}$; range: $\{-3, -1, 1, 5, 7\}$; not a function

7) domain: $\{-1, 2, 5, 8\}$; range: $\{-7, -3, 12, 19\}$; not a function

9) domain: $(-\infty, \infty)$; range: $(-\infty, \infty)$; function

11) domain: $[-5, 1]$; range: $[-6, 0]$; not a function

13) domain: $(-\infty, \infty)$; range: $(-\infty, 6]$; function

15) yes   17) yes   19) no   21) no

23) $(-\infty, \infty)$; function   25) $(-\infty, \infty)$; function

27) $[0, \infty)$; not a function   29) $(-\infty, 0) \cup (0, \infty)$; function

31) $(-\infty, -4) \cup (-4, \infty)$; function

33) $(-\infty, 5) \cup (5, \infty)$; function

35) $\left(-\infty, \dfrac{3}{5}\right) \cup \left(\dfrac{3}{5}, \infty\right)$; function

37) $\left(-\infty, -\dfrac{4}{3}\right) \cup \left(-\dfrac{4}{3}, \infty\right)$; function

39) $(-\infty, 3) \cup (3, \infty)$; function   41) $(-\infty, \infty)$; function

43) y is a function, and y is a function of x.

45) a) $y = 7$ b) $f(3) = 7$ 47) -13 49) 7

51) 50   53) -10 55) $-\dfrac{25}{4}$ 57) -105

59) $f(-1) = \dfrac{5}{2}, f(4) = 5$ 61) $f(-1) = 6, f(4) = 2$

63) $f(-1) = 7, f(4) = 3$ 65) -4 67) 6

69) $f(n - 3) = -9(n - 3) + 2$; Distribute.; $= -9n + 29$

71) a) $f(c) = -7c + 2$ b) $f(t) = -7t + 2$
    c) $f(a + 4) = -7a - 26$ d) $f(z - 9) = -7z + 65$
    e) $g(k) = k^2 - 5k + 12$ f) $g(m) = m^2 - 5m + 12$
    g) $-7x - 7h + 2$ h) $-7h$

73)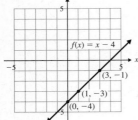

75)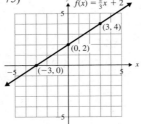

77)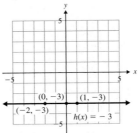

79) x-int: $(-1, 0)$; y-int: $(0, 3)$
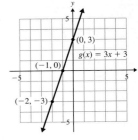

81) x-int: $(4, 0)$; y-int: $(0, 2)$
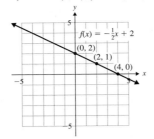

83) intercept: $(0, 0)$
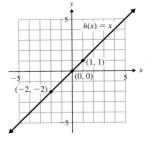

85) $m = -4$; y-int: $(0, -1)$
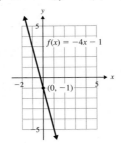

87) $m = -\dfrac{1}{4}$; y-int: $(0, -2)$
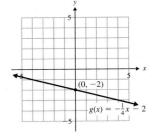

89) $m = 2$; y-int: $\left(0, \dfrac{1}{2}\right)$
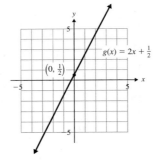

91)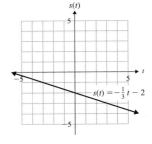

93)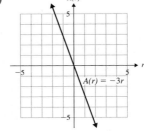

95) a) 108 mi b) 216 mi c) 2.5 hr
d)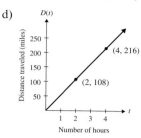

97) a) $C(8) = 28$; 8 gal of gas cost $28.00.
b) $C(15) = 52.5$; 15 gal of gas cost $52.50.
c) $g = 12$; 12 gal of gas can be purchased for $42.00.

99) a) 253.56 MB b) 1267.80 MB c) 20 sec
d)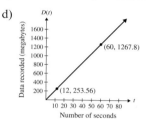

101) a) 60,000 b) 3.5 sec

103) a) $S(50) = 2205$; after 50 sec, the CD player reads 2,205,000 samples of sound.
b) $S(180) = 7938$; after 180 sec (or 3 min), the CD player reads 7,938,000 samples of sound.
c) $t = 60$; the CD player reads 2,646,000 samples of sound in 60 seconds (or 1 min).

105) a) 2 hr; 400 mg b) after about 30 min and after 6 hr
c) 200 mg d) $A(8) = 50$. After 8 hr there are 50 mg of ibuprofen in Sasha's bloodstream.

Chapter 4: Review Exercises

1) yes   3) yes   5) 28   7) -8

9)

x	y
0	-11
3	-8
-1	-12
-5	-16

11)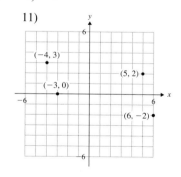

13) a)

x	y
1	0.10
2	0.20
7	0.70
10	1.00

(1, 0.10), (2, 0.20), (7, 0.70), (10, 1.00)

b)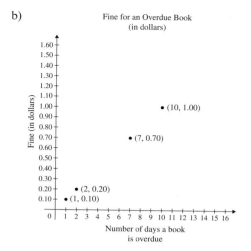

c) If a book is 14 days overdue, the fine is $1.40.

15)

x	y
0	3
1	1
2	-1
-2	7

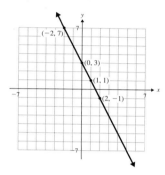

17) (6, 0), (0, -3); (2, -2) may vary.
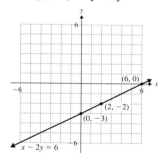

19) (24, 0), (0, 4); (12, 2) may vary.
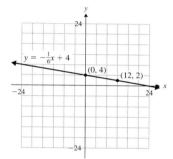

21) (5, 0), (5, 1) may vary, (5, 2) may vary.
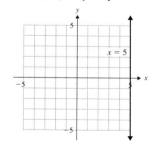

23) (4, 5)   25) $\left(\dfrac{13}{2}, -\dfrac{7}{2}\right)$

27) $-\dfrac{2}{5}$ 29) 1   31) $-\dfrac{13}{5}$ 33) -2 35) undefined

37) a) In 2008, one share of the stock was worth $32.
b) The slope is positive, so the value of the stock is increasing over time.
c) $m = 3$; the value of one share of stock is increasing by $3.00 per year.

39) 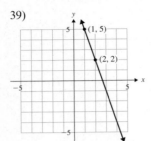   41)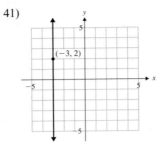

43) $m = 1$, y-int: $(0, -3)$ 45) $m = -\dfrac{3}{4}$, y-int: $(0, 1)$

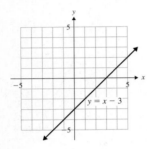   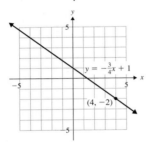

47) $m = \dfrac{1}{3}$, y-int: $(0, 2)$ 49) $m = -1$, y-int: $(0, 0)$

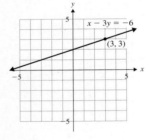   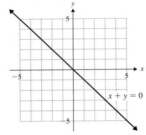

51) a) $(0, 5920.1)$; in 1998 the amount of money spent for personal consumption was $5920.1 billion.
b) It has been increasing by $371.5 billion per year.
c) estimate from the graph: $7400 billion; number from the equation: $7406.1 billion

53) $y = 7x - 9$ 55) $y = -\dfrac{4}{9}x + 2$ 57) $y = -\dfrac{1}{3}x - 5$

59) $y = 9$ 61) $x - 2y = -6$ 63) $3x + y = 5$

65) $6x - y = 0$ 67) $3x - 4y = -1$

69) a) $y = 186.2x + 944.2$
b) The number of worldwide wireless subscribers is increasing by 186.2 million per year.
c) 1316.6 million; this is slightly less than the number given on the chart.

71) parallel   73) perpendicular   75) neither

77) $y = 5x + 6$ 79) $x - 4y = -7$ 81) $y = 2x - 7$

83) $y = -\dfrac{11}{2}x + 4$ 85) $x = 2$ 87) $x = -1$

89) 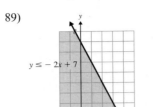   91)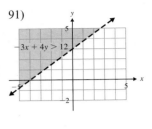

93) 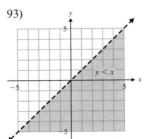   95)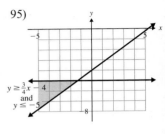

97)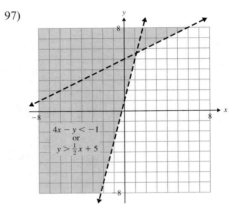

99)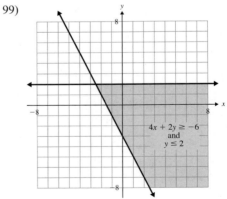

101) domain: $\{-3, 5, 12\}$; range: $\{1, 3, -3, 4\}$; not a function

103) domain: {Beagle, Siamese, Parrot}; range: {Dog, Cat, Bird}; function

105) domain: $[0, 4]$; range: $(0, 2)$; not a function

107) $(-\infty, -3) \cup (-3, \infty)$; function

109) $[0, \infty)$; not a function

111) $\left(-\infty, \frac{2}{7}\right) \cup \left(\frac{2}{7}, \infty\right)$; function

113) $f(3) = 27, f(-2) = -8$

115) a) 8 b) -27 c) 32 d) 5 e) $5a - 12$
f) $t^2 + 6t + 5$ g) $5k + 28$ h) $5c - 22$
i) $5x + 5h - 12$ j) $5h$

117) $\frac{1}{3}$

119) a) 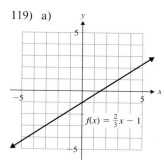 b)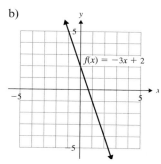

121)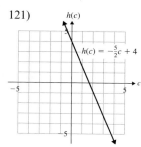

123) a) 960 MB; 2880 MB
b) 2.5 sec

Chapter 4: Test

1) yes

2)
x	y
0	4
3	-2
-1	6
2	0

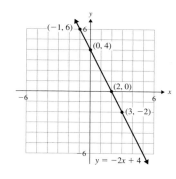

3) negative; positive

4) a) $(6, 0)$ b) $(0, -4)$ c) Answers may vary.
d)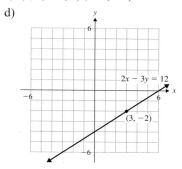

5)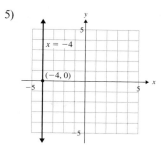

6) $\left(6, -\frac{5}{2}\right)$ 7) a) $-\frac{3}{4}$ b) 0

8) 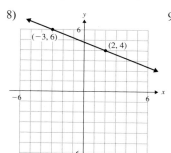 9)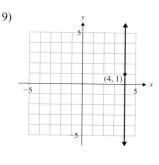

10) $y = -4x + 5$ 11) $x - 2y = -13$

12) perpendicular 13) a) $y = \frac{1}{3}x - 9$ b) $y = \frac{5}{2}x - 6$

14) a) $55{,}200 b) $y = 5.8x + 32$ c) $m = 5.8$; the profit is increasing by $5.8 thousand or $5800 per year.
d) y-int: $(0, 32)$; the profit in 2007 was $32,000. e) 2017

15) 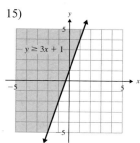 16)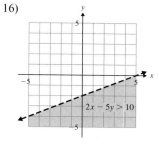

17) 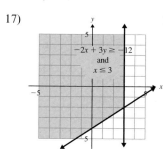 18)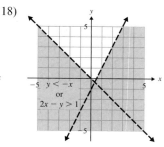

19) domain: $\{-2, 1, 3, 8\}$; range: $\{-5, -1, 1, 4\}$; function

20) domain: $[-3, \infty)$; range: $(-\infty, \infty)$; not a function

21) a) $(-\infty, \infty)$ b) yes

22) a) $\left(-\infty, \frac{5}{2}\right) \cup \left(\frac{5}{2}, \infty\right)$ b) yes

23) -3 24) 5 25) -22 26) 5

27) $t^2 - 3t + 7$ 28) $-4h + 30$

29)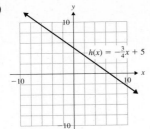

30) a) 36 MB   b) 11 sec

Chapter 4: Cumulative Review for Chapters 1–4

1) $\dfrac{3}{10}$ 2) 37 in.   3) -64 4) $\dfrac{5}{24}$ 5) $\dfrac{13}{5}$

6) $2(11) - 53; -31$ 7) $\{-3\}$ 8) $\varnothing$

9) $w = \dfrac{r - t}{z}$ 10) $\left\{-\dfrac{18}{7}, 2\right\}$ 11) $\left(-\infty, -\dfrac{3}{2}\right]$

12) $(-2, 4)$ 13) 300 calories

14) $m\angle A = 30°, m\angle B = 122°$

15) Lynette's age = 41; daughter's age = 16

16) $m = 1$

17)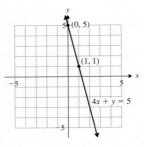

18) $5x + 4y = -36$ 19) $y = -3x$

20) a) yes   b) $(-\infty, -7) \cup (-7, \infty)$ 21) -37

22) $8a + 3$ 23) $8t + 19$

24) 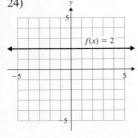   25)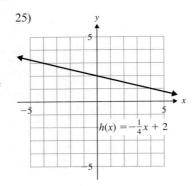

Chapter 5
Section 5.1

1) no   3) no   5) The lines are parallel.

7) dependent

9) $(3, 1)$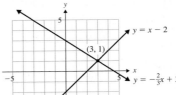

11) $(1, -2)$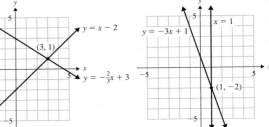

13) $\varnothing$; inconsistent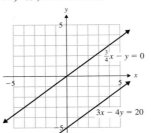

15) $(1, -1)$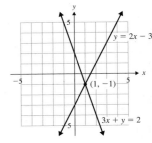

17) $\{(x, y) | y = -3x + 1\}$; dependent equations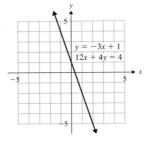

19) Answers may vary.

21) Answers may vary.   23) $B; (-2, 1)$ is in quadrant II.

25) The slopes are different.   27) infinite number of solutions

29) one solution   31) no solution

33) a) after 2001   b) 2001; 5.3 million
c) 1999–2001   d) 1999–2001

35) $(3, -2)$ 37) $(-5, -1)$ 39) $(-0.5, -1.25)$

41) It is the only variable with a coefficient of 1.

43) $(-1, -7)$ 45) $(4, 0)$ 47) $\varnothing$; inconsistent

49) $(2, 2)$ 51) $\left(-\dfrac{4}{5}, 3\right)$

53) $\{(x, y) | 6x + y = -6\}$; dependent   55) $(0, -6)$

57) Multiply the equation by the LCD of the fractions to eliminate the fractions.

59) $(6, 1)$ 61) $\varnothing$; inconsistent   63) $(8, 0)$ 65) $(3, 5)$

67) $(1, 3)$ 69) $\left(\dfrac{3}{4}, 0\right)$ 71) $\varnothing$; inconsistent

73) $\{(x, y) | x - 6y = -5\}$; dependent   75) $(0, 2)$

77) $\{(x, y) | 7x + 2y = 12\}$; dependent   79) $(-6, 1)$

81) $(1, 2)$ 83) $(12, -1)$ 85) $\varnothing$; inconsistent

87) a) Rent-for-Less: $24; Frugal: $30
    b) Rent-for-Less: $64; Frugal: $60
    c) (120, 48); if the car is driven 120 mi, the cost would be the same from each company: $48.
    d) If a car is driven less than 120 mi, it is cheaper to rent from Rent-for-Less. If a car is driven more than 120 mi, it is cheaper to rent from Frugal Rentals. If a car is driven exactly 120 mi, the cost is the same from each company.

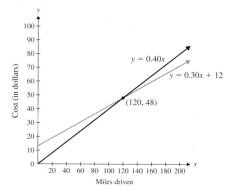

89) $(-3, 7)$ 91) $\left(5, -\frac{3}{2}\right)$ 93) $(8, 0)$

95) $\varnothing$; inconsistent   97) $(-6, 2)$

99) $(-3, 3)$ 101) $\left(-\frac{3}{2}, 4\right)$

103) a) The variables are eliminated, and you get a false statement.
     b) The variables are eliminated, and you get a true statement.

105) a) 3   b) c can be any real number except 3.

107) 5   109) $\left(\frac{2}{a}, \frac{3}{b}\right)$ 111) $\left(-1, -\frac{1}{b}\right)$

113) $\left(0, \frac{c}{b}\right)$ 115) $(8, -2)$ 117) $\left(\frac{1}{3}, 3\right)$

119) 4   121) -1

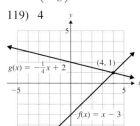
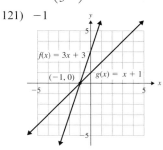

Section 5.2

1) yes   3) no   5) Answers may vary.   7) $(-2, 0, 5)$

9) $(1, -1, 4)$ 11) $\left(2, -\frac{1}{2}, \frac{5}{2}\right)$ 13) $\varnothing$; inconsistent

15) $\{(x, y, z) | 5x + y - 3z = -1\}$; dependent equations

17) $\{(a, b, c) | -a + 4b - 3c = -1\}$; dependent equations

19) $(2, 5, -5)$ 21) $\left(-4, \frac{3}{5}, 4\right)$ 23) $(0, -7, 6)$

25) $(1, 5, 2)$ 27) $\left(-\frac{1}{4}, -5, 3\right)$ 29) $\varnothing$; inconsistent

31) $\left(4, -\frac{3}{2}, 0\right)$ 33) $(4, 4, 4)$

35) $\{(x, y, z) | -4x + 6y + 3z = 3\}$; dependent equations

37) $\left(1, -7, \frac{1}{3}\right)$ 39) $(0, -1, 2)$ 41) $(-3, -1, 1)$

43) Answers may vary.   45) $(3, 1, 2, 1)$

47) $(0, -3, 1, -4)$

Section 5.3

1) 17 and 19   3) USC: 33; Michigan: 20

5) IHOP: 1156; Waffle House: 1470

7) 1939: 5500; 2011: 70,376

9) beef: 63.4 lb; chicken: 57.1 lb

11) length: 26 in.; width: 13 in.

13) width: 30 in.; height: 80 in.

15) length: 9 cm; width: 5 cm

17) $m\angle x = 92°$; $m\angle y = 46°$

19) Marc Anthony: $86; Santana: $66.50

21) two-item: $5.19; three-item: $6.39

23) deluxe: $7; regular: $4

25) cantaloupe: $1.50; watermelon: $3.00

27) hamburger: $0.61; small fries: $1.39

29) 9%: 3 oz; 17%: 9 oz   31) $0.44: 12; $0.28: 8

33) 3%: $2800; 5%: $1200   35) 52 quarters; 58 dimes

37) 4 L of pure acid; 8 L of 10% solution

39) passenger train: 50 mph; freight train: 30 mph

41) walking: 4 mph; biking: 11 mph

43) Nick: 14 mph; Scott: 12 mph

45) hot dog: $2.00; fries: $1.50; soda: $2.00

47) Clif Bar: 11 g; Balance Bar: 15 g; PowerBar: 24 g

49) Knicks: $160 million; Lakers: $149 million; Bulls: $119 million

51) bronze: $24; silver: $45; gold: $55   53) 104°, 52°, 24°

55) 80°, 64°, 36°   57) 12 cm, 10 cm, 7 cm

Section 5.4

1) $\begin{bmatrix} 1 & -7 & | & 15 \\ 4 & 3 & | & -1 \end{bmatrix}$ 3) $\begin{bmatrix} 1 & 6 & -1 & | & -2 \\ 3 & 1 & 4 & | & 7 \\ -1 & -2 & 3 & | & 8 \end{bmatrix}$

5) $3x + 10y = -4$
   $x - 2y = 5$

7) $x - 6y = 8$
   $y = -2$

9) $x - 3y + 2z = 7$
$4x - y + 3z = 0$
$-2x + 2y - 3z = -9$

11) $x + 5y + 2z = 14$
$y - 8z = 2$
$z = -3$

13) $(3, -1)$ 15) $(10, -4)$ 17) $(0, -2)$
19) $(-1, 4, 8)$ 21) $(10, 1, -4)$ 23) $(0, 1, 8)$
25) $\emptyset$; inconsistent   27) $(-5, 2, 1, -1)$ 29) $(3, 0, -2, 1)$

Chapter 5: Review Exercises

1) no

3) $(-2, -3)$ 5) $\emptyset$; inconsistent

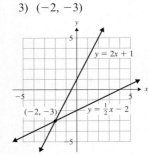   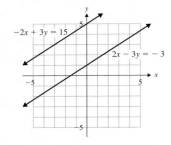

7) $\{(x, y) | 4x + y = -4\}$; dependent equations

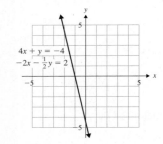

9) no solution   11) no solution   13) $(-10, 1)$
15) $(9, 0)$ 17) $(2, -1)$ 19) $(-5, 3)$ 21) $(1, 4)$
23) $(1, -6)$ 25) $(-3, -3)$
27) $\{(x, y) | 6x - 4y = 12\}$; dependent   29) $\emptyset$; inconsistent
31) no   33) $(3, -1, 4)$ 35) $\left(-1, 2, \dfrac{1}{2}\right)$
37) $\left(3, \dfrac{2}{3}, -\dfrac{1}{2}\right)$ 39) $\emptyset$; inconsistent
41) $\{(a, b, c) | 3a - 2b + c = 2\}$; dependent
43) $(1, 0, 3)$ 45) $\left(\dfrac{3}{4}, -2, 1\right)$ 47) 34 dogs, 17 cats
49) hot dog: $5.00; soda: $3.25
51) $m\angle x = 38°$; $m\angle y = 19°$
53) pure alcohol: 20 ml; 4% solution: 460 ml
55) Propel: 35 mg; Powerade: 52 mg; Gatorade: 110 mg
57) Blair: 65; Serena: 50; Chuck: 25
59) ice cream cone: $1.50; shake: $2.50; sundae: $3.00
61) 92°, 66°, 22°   63) $(-9, 2)$ 65) $(1, 0)$ 67) $(5, -2, 6)$

Chapter 5: Test

1) yes

2) $(2, 1)$

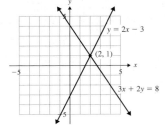

3) $\emptyset$; inconsistent

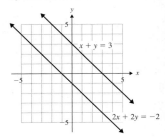

4) $(-3, 2)$ 5) $\left\{(x, y) \big| y = \dfrac{3}{4}x + \dfrac{7}{4}\right\}$; dependent equations
6) $(1, 6)$ 7) $(4, 0)$ 8) $(0, 3, -2)$
9) $(9, 1)$ 10) $\left(-4, \dfrac{1}{3}\right)$ 11) $(-2, 7)$
12) Subway: 280 cal; Whopper: 670 cal
13) screws: $4; nails: $3   14) 105°, 42°, 33°
15) $(6, -2)$ 16) $(1, -1, 1)$

Chapter 5: Cumulative Review for Chapters 1–5

1) $-\dfrac{1}{6}$ 2) $3\dfrac{1}{8}$ 3) -29 4) 30 in^2

5) $-24x^2 + 8x + 56$ 6) $\left\{\dfrac{1}{2}\right\}$ 7) $\emptyset$

8) $\left[-6, \dfrac{11}{7}\right]$ 9) $9.0 billion

10) $\left(-\infty, -\dfrac{1}{2}\right) \cup \left(\dfrac{5}{4}, \infty\right)$ 11) a) $h = \dfrac{2A}{b_1 + b_2}$ b) 6 cm

12) 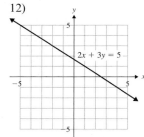   13)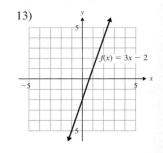

14) x-int: $\left(\dfrac{5}{2}, 0\right)$; y-int: $(0, -2)$ 15) $y = -\dfrac{7}{8}x - \dfrac{17}{8}$

16) perpendicular   17) $(-\infty, \infty)$

18) $\left(-\infty, \dfrac{7}{2}\right) \cup \left(\dfrac{7}{2}, \infty\right)$

19) a) -11 b) $4p + 9$ c) $4n + 17$

20) $(7, -8)$ 21) $(3, 1)$

22) $\emptyset$; inconsistent   23) $(-2, -1, 0)$

24) 4-ft boards: 16; 6-ft boards: 32

25) kitty: 3.8 million; puppy: 2.2 million; pig: 1.1 million

Chapter 6
Section 6.1

1) base: $-7r$; exponent: 5   3) 9^6

5) $(3+4)^2 = 49$, $3^2 + 4^2 = 25$. They are not equivalent because when evaluating $(3+4)^2$, first add $3+4$ to get 7, then square the 7.

7) No. $-3^4 = -1 \cdot 3^4 = -1 \cdot 81 = -81$; $(-3)^4 = (-3)\cdot(-3)\cdot(-3)\cdot(-3) = 81$

9) 64   11) 675   13) $(-4)^7$ 15) a^5 17) k^6

19) $-24p^7$ 21) $-60n^{12}$ 23) $14t^{13}$ 25) a^{36}

27) 64   29) $\dfrac{1}{32}$ 31) $\dfrac{64}{y^3}$ 33) $10{,}000r^4$

35) $-64a^3b^3$ 37) k^{24} 39) 64   41) $288k^{14}t^4$

43) $125m^{22}$ 45) $\dfrac{d^{18}}{4c^{30}}$ 47) $\dfrac{3a^{40}b^{15}}{4c^3}$ 49) $676u^8v^{26}$

51) $\dfrac{64r^{18}s^6}{t^9}$ 53) $\dfrac{81n^8}{m^{16}p^6}$ 55) false   57) true

59) 1   61) -1 63) 2   65) -6 67) $\dfrac{1}{36}$

69) 49   71) $\dfrac{27}{64}$ 73) -32 75) $-\dfrac{1}{64}$ 77) $\dfrac{1}{16}$

79) $\dfrac{83}{81}$ 81) $\dfrac{1}{y^4}$ 83) $\dfrac{b^3}{a^{10}}$ 85) $\dfrac{x^4y^5}{10}$ 87) $\dfrac{8x^3}{y^7}$

89) $\dfrac{8a^6c^{10}}{5bd}$ 91) $\dfrac{36}{a^2}$ 93) $\dfrac{c^2d^2}{144b^2}$ 95) $-\dfrac{6}{r^2}$

97) $-\dfrac{1}{p^8}$ 99) x 101) n^5 103) 64   105) $\dfrac{1}{125}$

107) $\dfrac{5}{9}w^5$ 109) $\dfrac{1}{t^2}$ 111) $\dfrac{1}{x^9}$ 113) $-\dfrac{6}{k^3}$ 115) $\dfrac{w^6}{9v^3}$

117) $\dfrac{5}{m^{11}n^2}$ 119) $(x+y)^5$

Section 6.2

1) $270g^{12}$ 3) $\dfrac{23}{t^{10}}$ 5) $\dfrac{8}{27}x^{30}y^{18}$ 7) $\dfrac{s^{16}}{49r^6}$

9) $-k^{12}$ 11) $-32m^{25}n^{10}$ 13) $-12w^8$ 15) $\dfrac{b^{24}}{a^{30}}$

17) $a^{14}b^3c^7$ 19) $\dfrac{27u^{30}}{64v^{21}}$ 21) $81t^8u^{20}$ 23) $\dfrac{1}{h^{21}}$

25) $\dfrac{h^5}{32}$ 27) $56c^{14}$ 29) $\dfrac{3}{a^5}$ 31) $\dfrac{6}{55}d^{11}$

33) $\dfrac{9}{64n^6}$ 35) $-\dfrac{32r^{40}}{t^{15}}$ 37) $\dfrac{c}{24a^{29}b^{32}}$

39) $\dfrac{1}{(x+y)^5}$ 41) p^{10n} 43) y^{11m}

45) $\dfrac{1}{x^{3a}}$ 47) $\dfrac{3}{5c^{6x}}$ 49) $\dfrac{3}{8}x^2$ sq units

51) yes   53) no   55) yes   57) no

59) Answers may vary.   61) -0.000068 63) $34{,}502.9$

65) -0.00005 67) 0.00081 69) $26{,}450.67$

71) -0.03921 73) 2.1105×10^3 75) 4.8×10^{-3}

77) -4×10^5 79) 1.1×10^4 81) 8×10^{-4}

83) -5.4×10^{-2} 85) 6.5×10^3 87) $\$1.5 \times 10^7$

89) 1×10^{-8} cm   91) $20{,}000$ 93) $690{,}000$

95) $400{,}000$ 97) -0.0009 99) $8{,}000{,}000$

101) 6.7392×10^8 m   103) 20 metric tons

105) $24{,}000{,}000{,}000$ kW-hr

Section 6.3

1) Yes; the coefficients are real numbers and the exponents are whole numbers.

3) No; one of the exponents is a negative number.

5) No; two of the exponents are fractions.

7) binomial   9) trinomial   11) monomial

13) It is the same as the degree of the term in the polynomial with the highest degree.

15) Add the exponents on the variables.

17)

Term	Coeff.	Degree
$4d^2$	4	2
$12d$	12	1
-9	-9	0

Degree is 2.

19)

Term	Coeff.	Degree
$-9r^3s^2$	-9	5
$-r^2s^2$	-1	4
$\dfrac{1}{2}rs$	$\dfrac{1}{2}$	2
$6s$	6	1

Degree is 5.

21) $5w^2 + 7w - 6$ 23) $-8p + 7$

25) $-5a^4 + 9a^3 - a^2 - \dfrac{3}{2}a + \dfrac{5}{8}$

27) $\dfrac{25}{8}x^3 + \dfrac{1}{12}$ 29) $2.6k^3 + 8.7k^2 - 7.3k - 4.4$

31) $17x - 8$ 33) $12r^2 + 6r + 11$

35) $1.5q^3 - q^2 + 4.6q + 15$ 37) $-7a^4 - 12a^2 + 14$

39) $8j^2 + 12j$ 41) $10h^5 - h^4 + 6$

43) $-b^4 - 10b^3 + b + 20$ 45) $-\dfrac{5}{14}r^2 + r - \dfrac{7}{6}$

47) $15v - 6$ 49) $-b^2 - 12b + 7$

51) $3a^4 - 11a^3 + 7a^2 - 7a + 8$ 53) Answers may vary.

55) No. If the coefficients of the like terms are opposite in sign, their sum will be zero. Example:
$(3x^2 + 4x + 5) + (2x^2 - 4x + 1) = 5x^2 + 6$

57) $-9b^2 - 4$ 59) $\frac{5}{4}n^3 - \frac{3}{2}n^2 + 3n - \frac{1}{8}$

61) $5u^3 - 10u^2 - u - 6$ 63) $-\frac{5}{8}k^2 - 9k + \frac{1}{2}$

65) $3t^3 - 7t^2 - t - 3$ 67) $9a^3 - 8a + 13$

69) $3a + 8b$ 71) $-m + \frac{11}{6}n - \frac{1}{4}$

73) $5y^2z^2 + 7y^2z - 25yz^2 + 1$

75) $10x^3y^2 - 11x^2y^2 + 6x^2y - 12$

77) $5v^2 + 3v - 8$ 79) $4g^2 + 10g - 10$

81) $-8n^2 + 11n$ 83) $8x + 14$ units

85) $6w^2 - 2w - 6$ units   87) a) 16 b) 4

89) a) 29 b) 5   91) 6   93) 40

95) a) $-x - 10$ b) -15 c) $-5x + 12$ d) 2

97) a) $5x^2 - 4x - 7$ b) 98 c) $3x^2 - 10x + 5$ d) -3

99) $-4t^2 + 4t + 6$ 101) -74 103) 0   105) $-\frac{9}{4}$

107) Answers may vary.

109) a) $P(x) = 4x - 2000$ b) $4000

111) a) $P(x) = 3x - 2400$ b) $0

113) a) $P(x) = -0.2x^2 + 19x - 9$ b) $291,000

Section 6.4

1) Answers may vary   3) $14k^4$ 5) $\frac{7}{4}d^{11}$

7) $28y^2 - 63y$ 9) $6v^5 - 24v^4 - 12v^3$

11) $-27t^5 + 18t^4 + 12t^3 + 21t^2$

13) $2x^4y^3 + 16x^4y^2 - 22x^3y^2 + 4x^3y$

15) $-15t^7 - 6t^6 + \frac{15}{4}t^5$ 17) $18g^3 + 10g^2 + 14g + 28$

19) $-47a^3b^3 - 39a^3b^2 + 108a^2b - 51$

21) $9m^3 + 85m^2 + 29m - 63$ 23) $2p^3 - 9p^2 - 23p + 30$

25) $15y^4 - 23y^3 + 28y^2 - 29y - 4$

27) $6k^4 + \frac{5}{2}k^3 + 31k^2 + 15k - 30$

29) $a^4 + 3a^3 - 3a^2 + 14a - 6$

31) $-24v^5 + 26v^4 - 22v^3 + 27v^2 - 5v + 10$

33) $8x^3 - 22x^2 + 19x - 6$

35) First, Outer, Inner, Last

37) $w^2 + 15w + 56$ 39) $n^2 - 15n + 44$

41) $4p^2 - 7p - 15$ 43) $24n^2 + 41n + 12$

45) $0.04g^2 + 0.35g - 0.99$ 47) $12a^2 + ab - 20b^2$

49) $60p^2 + 68pq + 15q^2$ 51) $2a^2 + \frac{15}{4}ab - \frac{1}{2}b^2$

53) a) $4y + 8$ units b) $y^2 + 4y - 12$ sq units

55) a) $2a^2 + 4a + 16$ units b) $3a^3 - 3a^2 + 24a$ sq units

57) Both are correct.   59) $15y^2 + 54y - 24$

61) $-24g^4 - 36g^3 + 60g^2$ 63) $c^3 + 6c^2 + 5c - 12$

65) $5n^5 + 55n^3 + 150n$ 67) $2r^3 - 3r^2t - 3rt^2 + 2t^3$

69) $9m^2 - 4$ 71) $49a^2 - 64$ 73) $4p^2 - 49q^2$

75) $n^2 - \frac{1}{4}$ 77) $\frac{4}{9} - k^2$ 79) $0.09x^2 - 0.16y^2$

81) $25x^4 - 16$ 83) $y^2 + 16y + 64$

85) $t^2 - 22t + 121$ 87) $16w^2 + 8w + 1$

89) $4d^2 - 20d + 25$ 91) $36a^2 - 60ab + 25b^2$

93) No. The order of operations tells us to perform exponents, $(t + 3)^2$, before multiplying by 4.

95) $6x^2 + 12x + 6$ 97) $2a^3 + 12a^2 + 18a$

99) $9m^2 + 6mn + n^2 + 12m + 4n + 4$

101) $x^2 - 8x + 16 - 2xy + 8y + y^2$

103) $r^3 + 15r^2 + 75r + 125$ 105) $s^3 - 6s^2 + 12s - 8$

107) $c^4 - 18c^2 + 81$ 109) $y^4 + 8y^3 + 24y^2 + 32y + 16$

111) $v^2 - 10vw + 25w^2 - 16$ 113) $4a^2 + 4ab + b^2 - c^2$

115) No; $(x + 5)^2 = x^2 + 10x + 25$

117) $h^3 + 6h^2 + 12h + 8$ cubic units

119) $9x^2 + 33x + 14$ sq units   121) $3x^3 - 5x^2$

123) 52   125) $6x^3 - 19x^2 + 12x + 5$ 127) $-\frac{4}{3}$

129) It is the same because multiplication of functions is defined as $(fg)(x) = f(x) \cdot g(x)$.

Section 6.5

1) dividend: $12c^3 + 20c^2 - 4c$; divisor: $4c$; quotient: $3c^2 + 5c - 1$

3) Answers may vary.   5) $7k^2 + 2k - 10$

7) $2u^5 + 2u^3 + 5u^2 - 8$ 9) $-5d^3 + 1$

11) $\frac{3}{2}w^2 + 7w - 1 + \frac{1}{2w}$

13) $\frac{5}{2}v^3 - 9v - \frac{11}{2} - \frac{5}{4v^2} + \frac{1}{4v^4}$

15) $9a^3b + 6a^2b - 4a^2 + 10a$

17) $-t^4u^2 + 7t^3u^2 + 12t^2u^2 - \frac{1}{9}t^2$

19) The answer is incorrect. When you divide $4t$ by $4t$, you get 1. The quotient should be $4t^2 - 9t + 1$.

21) $g + 4$ 23) $p + 6$ 25) $k - 5$ 27) $h + 8$

29) $2a^2 - 7a - 3$ 31) $3p^2 + 7p - 1$

33) $6t + 23 + \dfrac{119}{t - 5}$ 35) $4z^2 + 8z + 7 - \dfrac{72}{3z + 5}$

37) $w^2 - 4w + 16$ 39) $2r^2 + 8r + 3$

41) $x^2y^2 + 5x^2y - \dfrac{1}{6} + \dfrac{1}{2xy}$

43) $-2g^3 - 5g^2 + g - 4$

45) $6t + 5 + \dfrac{20}{t - 8}$ 47) $4n^2 + 10n + 25$

49) $5x^2 - 7x + 3$ 51) $4a^2 - 3a + 7$

53) $5h^2 - 3h - 2 + \dfrac{1}{2h^2 - 9}$ 55) $3d^2 - d - 8$

57) $9c^3 + 8c + 3 + \dfrac{7c + 4}{c^2 - 10c + 4}$ 59) $k^2 - 9$

61) $-\dfrac{5}{7}a^3 - 7a + 2 + \dfrac{15}{7a}$ 63) $\dfrac{1}{2}x + 3$

65) $\dfrac{2}{3}w + 2$ 67) $4y + 1$ 69) $2a^2 - 5a + 1$

71) $12h^2 + 6h + 2$

73) $\left(\dfrac{f}{g}\right)(x) = 5x - 9$, where $x \neq -3$

75) $\left(\dfrac{f}{g}\right)(x) = 6x^2 - 9x + 1$, where $x \neq 0$

77) $\left(\dfrac{f}{g}\right)(x) = 3x^3 - 7x^2 + 2x + 4$, where $x \neq 1$

79) $2x - 1$, where $x \neq -\dfrac{1}{2}$ 81) 9

83) $\dfrac{2x + 1}{3x}$, where $x \neq 0$ 85) $\dfrac{1}{6}$

Chapter 6: Review Exercises

1) 81   3) $\dfrac{64}{125}$ 5) z^{18} 7) $7r^5$

9) $-54t^7$ 11) $\dfrac{1}{k^8}$ 13) $-\dfrac{40b^4}{a^6}$ 15) $\dfrac{4q^{30}}{9p^6}$

17) 14   19) x^{8t} 21) y^{6p}

23) False. $-x^2 = -1 \cdot x \cdot x$. If $x \neq 0$, then $-x^2$ is a negative number. But $(-x)^2 = -x \cdot (-x)$, which is a positive number when $x \neq 0$.

25) 938,000   27) 0.00000105   29) 5.75×10^{-5}

31) 3.2×10^7 33) 0.0000004   35) 7500

37) 30,000 quills

39)

Term	Coeff.	Degree
$4r^3$	4	3
$-7r^2$	-7	2
r	1	1
5	5	0

Degree is 3.

41) 5   43) $7.9p^3 + 5.1p^2 + 4.8p - 3.6$

45) $-2a^2b^2 + 17a^2b - 4ab + 5$ 47) Answers may vary

49) a) -41 b) -6

51) a) 123   b) 151
c) $N(1) = 135$. In 2003, there were approximately 135 cruise ships operating in North America.

53) $-54m^5 + 18m^4 - 42m^3$

55) $-24w^4 - 48w^3 + 26w^2 - 4w + 15$

57) $y^2 + y - 56$ 59) $a^2b^2 + 11ab + 30$

61) $-24d^2 - 46d - 21$ 63) $p^3 - p^2 - 24p - 36$

65) $\dfrac{1}{5}m^2 - \dfrac{26}{15}m - 8$ 67) $z^2 - 81$ 69) $\dfrac{49}{64} - r^4$

71) $b^2 + 14b + 49$ 73) $25q^2 - 20q + 4$

75) $x^3 - 6x^2 + 12x - 8$ 77) $-18c^2 + 48c - 32$

79) $m^2 - 10m + 25 + 2mn - 10n + n^2$

81) a) $-10x^2 + 53x - 36$ b) 30

83) a) $4m^2 - 7m - 15$ sq units   b) $10m + 4$ units

85) $4t^2 - 7t - 10$ 87) $c + 10$ 89) $4r^2 - 7r + 3$

91) $-\dfrac{5}{2}x^2y^3 + 7xy^3 + 1 - \dfrac{5}{3x^2}$ 93) $2q - 4 - \dfrac{7}{3q + 7}$

95) $3a + 7 + \dfrac{21}{5a - 4}$ 97) $t^2 + 4t - 4 + \dfrac{4}{8t^2 - 11}$

99) $f^2 - 5f + 25$ 101) $5k^2 + 6k + 4$

103) $3c^2 - c - 2 + \dfrac{-3c + 2}{2c^2 + 5c - 4}$ 105) $2x^2 + x + 5$

Chapter 6: Test

1) a) -81 b) $\dfrac{1}{32}$ c) -2 d) $\dfrac{1000}{27}$ e) $\dfrac{1}{64}$

2) $-30p^{12}$ 3) $\dfrac{1}{a^4b^6}$ 4) $\dfrac{8}{y^9}$ 5) $\dfrac{9x^4}{4y^{18}}$ 6) t^{13k}

7) 728,300   8) 1.65×10^{-4} 9) $-50,000$

10) 9.1×10^{-18} g   11) a) -1 b) 3   12) -1

13) $14r^3s^2 - 2r^2s^2 - 5rs + 8$ 14) $6j^2 - 8j - 6$

15) $c^2 - 15c + 56$ 16) $6y^2 + 13y + 5$ 17) $u^2 - \dfrac{9}{16}$

18) $6a^2 - 13ab - 5b^2$ 19) $49m^2 - 70m + 25$

20) $-8n^3 - 43n^2 + 12n - 9$

21) $-16m^3 - 26m^2 + 68m - 21$ 22) $3x^3 + 24x^2 + 48x$

23) $25a^2 - 10ab + b^2 - 30a + 6b + 9$

24) $s^3 - 12s^2 + 48s - 64$ 25) $r + 3$

26) $2t^2 - 5t + 1 - \dfrac{2}{3t}$ 27) $6v^2 - 3v + 4 - \dfrac{7}{5v - 6}$

28) $3k + 7$ 29) a) $P(x) = 2x + 1$ b) $17,000

30) a) $x^2 - 4x - 32$ b) -20 c) $x^2 - 10x + 16$ d) -8
    e) $x + 3$ where $x \neq 8$ f) 10

Chapter 6: Cumulative Review for Chapters 1–6

1) a) 43, 0   b) $-14, 43, 0$ c) $\dfrac{6}{11}, -14, 2.7, 43, 0.\overline{65}, 0$

2) -28 3) $\dfrac{12}{5}$ or $2\dfrac{2}{5}$ 4) $\left\{-\dfrac{45}{4}\right\}$ 5) $\varnothing$

6) $\{-2, 11\}$ 7) $b_2 = \dfrac{2A}{h} - b_1$ 8) $(-\infty, 1]$

9) 45 ml of 12% solution, 15 ml of 4% solution

10) x-int: $(2, 0)$; y-int: $(0, -5)$ 11)

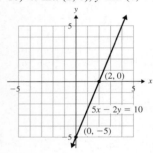
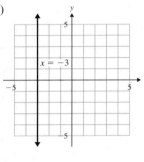

12) $x + y = 3$ 13) $(-\infty, \infty)$ 14) $(2, 10)$

15) width: 15 cm; length: 23 cm   16) $-45w^8$

17) $\dfrac{8}{n^{18}}$ 18) p^{14k} 19) $-14q^2 + 6q - 20$

20) $16g^2 - 81$ 21) $ab - \dfrac{5}{2b} + \dfrac{7}{a^2b^2} + \dfrac{1}{a^3b^2}$

22) $v^3 + 3v^2 - 5v - 6$ 23) $2p^2 - 3p + 1 + \dfrac{5}{p + 4}$

24) $2c^3 - 17c^2 + 48c - 45$

25) a) $x^4 + 5x^3 - 2x^2 - 30x - 24$ b) -36

Chapter 7
Section 7.1

1) $5m^2$ 3) $6k^5$ 5) $8r^2s^5$ 7) $4uv^3$ 9) s^2t

11) $(n - 7)$ 13) Answers may vary.   15) $6(5s + 3)$

17) $4(6z - 1)$ 19) $3d(d - 2)$ 21) $5b(6b^2 - 1)$

23) $r^2(r^7 + 1)$ 25) $\dfrac{1}{2}c(c + 5)$ 27) $5n^3(2n^2 - n + 8)$

29) $2v^5(v^3 - 9v^2 - 12v + 1)$ 31) does not factor

33) $a^3b^2(a + 4b)$ 35) $10x^2y(5xy^2 - 7xy + 4)$

37) $(n - 12)(m + 8)$ 39) $(9c + 4)(a - b)$

41) $(z + 11)(y + 1)$ 43) $(3r + 4)(2k^2 - 1)$

45) $-8(8m + 5)$ 47) $-5t^2(t - 2)$ 49) $-a(3a^2 - 7a + 1)$

51) $-1(b - 8)$ 53) $(t + 3)(k + 8)$ 55) $(g - 7)(f + 4)$

57) $(s - 3)(2r + 5)$ 59) $(3x - 2)(y + 9)$

61) $(2b + 5c)(4b + c^2)$ 63) $(a^2 - 3b)(4a + b)$

65) $(k + 7)(t - 5)$ 67) $(n - 8)(m - 10)$

69) $(g - 1)(d + 1)$ 71) $(5u + 6)(t - 1)$

73) $(12g^3 + h)(3g - 8h)$ 75) Answers may vary.

77) $5(mn + 3m + 2n + 6)$; Group the terms and factor out the GCF from each group; $5(n + 3)(m + 2)$

79) $7(p + 4)(q + 2)$ 81) $8s(s - 5)(t + 2)$

83) $(d + 4)(7c + 3)$ 85) $(7k - 3d)(6k^2 - 5d)$

87) $9fj(f + 1)(j + 5)$ 89) $2x^3(2x - 1)(y + 7)$

91) $(q - 8)(p + 3)$ 93) $a^2b^2(a - 2b)(a + 4b)$

95) $(3h^2 - 2k)(h + 4k^2)$ 97) $2c(c^2 + 7)(c + 6)$

99) $-8v(2v^2 + 7v - 1)$

Section 7.2

1) They are negative.   3) Can I factor out a GCF?

5) Can I factor again?   7) $(g + 6)(g + 2)$

9) $(w + 7)(w + 6)$ 11) $(c - 4)(c - 9)$

13) $(m - 11)(m + 10)$ 15) $(u + 12)(u - 11)$

17) $(q - 5)(q - 3)$ 19) prime

21) $(p - 10)(p - 10)$ or $(p - 10)^2$ 23) $3(p - 9)(p + 1)$

25) $2k(k - 8)(k - 5)$ 27) $ab(a + 12)(a - 3)$

29) $-(a + 8)(a + 2)$ 31) $-(h - 5)(h + 3)$

33) $-(k - 7)(k - 4)$ 35) $-(n + 7)(n + 7)$ or $-(n + 7)^2$

37) $(a + 5b)(a + b)$ 39) $(m - 3n)(m + 7n)$

41) $(p - 8q)(p - 9q)$ 43) $(f + g)(f - 11g)$

45) $(c - 5d)(c + 11d)$ 47) $(2r + 5)(r + 3)$

49) $(5p - 1)(p - 4)$ 51) $(11m + 4)(m - 2)$

53) $(3v + 7)(2v - 1)$ 55) $(5c + 2)(2c + 3)$

57) $(6a - 5b)(a + b)$

59) because 2 can be factored out of $2x - 4$, but 2 cannot be factored out of $2x^2 + 13x - 24$

61) $(5w + 6)(w + 1)$ 63) $(3u - 5)(u - 6)$

65) $(7k - 6)(k + 3)$ 67) $(4r + 3)(2r + 5)$

69) $(6v - 7)(v - 2)$ 71) $(5a - 4b)(2a - b)$

73) $(3c + 2d)(2c + 9d)$ 75) $(n + 14)(n + 2)$

77) $(k - 4)(k - 11)$ 79) $(2w - 1)(w - 4)$

81) $(6y - 7)(4y - 1)$ 83) $4q(q - 3)(q - 4)$

85) $(t + 6)(t + 1)$ 87) $3(4c - 3)(c + 2)$

89) $(3b + 25)(b + 3)$ 91) $(7s - 3t)(s - 2t)$
93) $-(5z - 2)(2z - 3)$ 95) prime   97) $(r - 9)(r - 2)$
99) $(q - 1)^2(3p - 7)(4p - 7)$

Section 7.3

1) a) 36 b) 100 c) 16 d) 121 e) 9
   f) 64 g) 144 h) $\dfrac{1}{4}$ i) $\dfrac{9}{25}$

3) a) n^2 b) $5t$ c) $7k$ d) $4p^2$ e) $\dfrac{1}{3}$ f) $\dfrac{5}{2}$

5) $z^2 + 18z + 81$

7) The middle term does not equal $2(3c)(-4)$. It would have to equal $-24c$ to be a perfect square trinomial.

9) $(t + 8)^2$ 11) $(g - 9)^2$ 13) $(2y + 3)^2$

15) $(3k - 4)^2$ 17) $\left(a + \dfrac{1}{3}\right)^2$ 19) $\left(v - \dfrac{3}{2}\right)^2$

21) $(x + 3y)^2$ 23) $(3a - 2b)^2$ 25) $4(f + 3)^2$

27) $2p^2(p - 6)^2$ 29) $-2(3d + 5)^2$ 31) $3c(4c^2 + c + 9)$

33) a) $x^2 - 16$ b) $16 - x^2$ 35) $(x + 3)(x - 3)$

37) $(n + 11)(n - 11)$ 39) prime

41) $\left(y + \dfrac{1}{5}\right)\left(y - \dfrac{1}{5}\right)$ 43) $\left(c + \dfrac{3}{4}\right)\left(c - \dfrac{3}{4}\right)$

45) $(6 + h)(6 - h)$ 47) $(13 + a)(13 - a)$

49) $\left(\dfrac{7}{8} + j\right)\left(\dfrac{7}{8} - j\right)$ 51) $(10m + 7)(10m - 7)$

53) $(4p + 9)(4p - 9)$ 55) prime

57) $\left(\dfrac{1}{2}k + \dfrac{2}{3}\right)\left(\dfrac{1}{2}k - \dfrac{2}{3}\right)$ 59) $(b^2 + 8)(b^2 - 8)$

61) $(12m + n^2)(12m - n^2)$ 63) $(r^2 + 1)(r + 1)(r - 1)$

65) $(4h^2 + g^2)(2h + g)(2h - g)$ 67) $4(a + 5)(a - 5)$

69) $2(m + 8)(m - 8)$ 71) $5r^2(3r + 1)(3r - 1)$

73) a) 64 b) 1 c) 1000 d) 27 e) 125 f) 8

75) a) y b) $2c$ c) $5r$ d) x^2 77) $x^2 - 3x + 9$

79) $(d + 1)(d^2 - d + 1)$ 81) $(p - 3)(p^2 + 3p + 9)$

83) $(k + 4)(k^2 - 4k + 16)$

85) $(3m - 5)(9m^2 + 15m + 25)$

87) $(5y - 2)(25y^2 + 10y + 4)$

89) $(10c - d)(100c^2 + 10cd + d^2)$

91) $(2j + 3k)(4j^2 - 6jk + 9k^2)$

93) $(4x + 5y)(16x^2 - 20xy + 25y^2)$

95) $6(c + 2)(c^2 - 2c + 4)$

97) $7(v - 10w)(v^2 + 10vw + 100w^2)$

99) $(p + 1)(p - 1)(p^2 - p + 1)(p^2 + p + 1)$

101) $7(2x + 3)$ 103) $(3p + 7)(p - 1)$

105) $(t + 7)(t^2 + 8t + 19)$ 107) $(k - 10)(k^2 - 17k + 73)$

Chapter 7: Putting It All Together

1) $(m + 10)(m + 6)$ 2) $(h + 6)(h - 6)$
3) $(u + 9)(v + 6)$ 4) $(2y + 9)(y - 2)$
5) $(3k - 2)(k - 4)$ 6) $(n - 7)^2$ 7) $8d^4(2d^2 + d + 9)$
8) $(b + c)(b - 4c)$ 9) $10w(3w + 5)(2w - 1)$
10) $7(c - 1)(c^2 + c + 1)$ 11) $(t + 10)(t^2 - 10t + 100)$
12) $(p + 4)(q - 6)$ 13) $(7 + p)(7 - p)$
14) $(h - 7)(h - 8)$ 15) $(2x + y)^2$ 16) $9(3c - 2)$
17) $3z^2(z - 8)(z + 1)$ 18) $(3a - 2)(3a + 4)$
19) prime   20) $5c(a + 2)(b - 3)$
21) $5(2x - 3)(4x^2 + 6x + 9)$ 22) $(9z + 2)^2$
23) $\left(c + \dfrac{1}{2}\right)\left(c - \dfrac{1}{2}\right)$ 24) prime
25) $(3s + 1)(3s - 1)(5t - 4)$ 26) $3cd(2c^2 + 5d)(2c^2 - 5d)$
27) $(k + 3m)(k + 6m)$ 28) $8(2r + 1)(4r^2 - 2r + 1)$
29) $(z - 11)(z + 8)$ 30) $8fg^2(5f^3g^2 + f^2g + 2)$
31) $5(4y - 1)^2$ 32) $(4t - 5)(t + 1)$
33) $2(10c + 3d)(c + d)$ 34) $\left(x + \dfrac{3}{7}\right)\left(x - \dfrac{3}{7}\right)$
35) $(n^2 + 4m^2)(n + 2m)(n - 2m)$ 36) $(k - 12)(k - 9)$
37) $2(a - 9)(a + 4)$ 38) $(x + 2)(x - 2)(y + 7)$
39) $\left(r - \dfrac{1}{2}\right)^2$ 40) $(v - 5)(v^2 + 5v + 25)$
41) $(4g - 9)(7h + 4)$ 42) $-3x(2x - 1)(4x - 3)$
43) $(4b + 3)(2b - 5)$ 44) $2(5u + 3)^2$
45) $5a^2b(11a^4b^2 + 7a^3b^2 - 2a^2 - 4)$ 46) $(8 + u)(8 - u)$
47) prime   48) $2v^2w(v + w)(v + 6w)$ 49) $(3p - 4q)^2$
50) $(c^2 + 4)(c + 2)(c - 2)$ 51) $(6y - 1)(5y + 7)$
52) prime   53) $10(2a - 3b)(4a^2 + 6ab + 9b^2)$
54) $13n^3(2n^3 - 3n + 1)$ 55) $(r - 1)(t - 1)$
56) $(h + 5)^2$ 57) $4(g + 1)(g - 1)$
58) $(5a - 3b)(5a - 8b)$ 59) $3(c - 4)^2$
60) prime   61) $(12k + 11)(12k - 11)$
62) $(5p - 4q)(25p^2 + 20pq + 16q^2)$
63) $-4(6g + 1)(2g + 3)$ 64) $5(d + 11)(d + 1)$
65) $(q + 1)(q^2 - q + 1)$ 66) $(3x + 2)^2$
67) $(9u^2 + v^2)(3u + v)(3u - v)$ 68) $3(5v + w^2)(3v + 2w)$
69) $(11f + 3)(f + 3)$ 70) $4y(y - 5)(y + 4)$
71) $j^3(2j^8 - 1)$ 72) $\left(d + \dfrac{13}{10}\right)\left(d - \dfrac{13}{10}\right)$

73) $(w - 8)(w + 6)$ 74) $(4a - 5)^2$ 75) prime
76) $3(2y + 5)(4y^2 - 10y + 25)$ 77) $(m + 2)^2$
78) $(r - 9)(r - 6)$ 79) $4c^2(5c + 3)(5c - 3)$
80) $(3t + 8)(3t - 8)$ 81) $(2z + 1)(y + 11)(y - 5)$
82) $(a + b)(c - 8)(c + 3)$ 83) $r(r + 3)$
84) $(n - 4)(n + 8)$ 85) $3p(3p - 13)$
86) $(5w - 8)(5w - 4)$ 87) $(7k + 3)(k - 1)$
88) $4(4z + 1)(z + 2)$ 89) $-3x(x - 2y)$
90) $5s(s - 2t)$ 91) $(n + p + 6)(n - p + 6)$
92) $(h + k - 5)(h - k - 5)$ 93) $(x - y + z)(x - y - z)$
94) $(a + b + c)(a + b - c)$

Section 7.4

1) It says that if the product of two quantities equals 0, then one or both of the quantities must be zero.
3) $\{-9, 8\}$ 5) $\{4, 7\}$ 7) $\left\{-\dfrac{3}{4}, 9\right\}$
9) $\{0, 8\}$ 11) $\left\{\dfrac{5}{6}\right\}$ 13) $\left\{-3, -\dfrac{7}{4}\right\}$
15) $\left\{-\dfrac{3}{2}, \dfrac{1}{4}\right\}$ 17) $\{0, 2.5\}$
19) No; the product of the factors must equal zero.
21) $\{-8, -7\}$ 23) $\{-15, 3\}$ 25) $\left\{-\dfrac{5}{3}, 2\right\}$
27) $\left\{-\dfrac{4}{7}, 0\right\}$ 29) $\{6, 9\}$ 31) $\{-7, 7\}$
33) $\left\{-\dfrac{6}{5}, \dfrac{6}{5}\right\}$ 35) $\{-12, 5\}$ 37) $\left\{-2, -\dfrac{3}{4}\right\}$
39) $\{0, 11\}$ 41) $\left\{-\dfrac{1}{2}, 3\right\}$ 43) $\{-8, 12\}$
45) $\left\{\dfrac{7}{2}, \dfrac{9}{2}\right\}$ 47) $\{-9, 5\}$ 49) $\{1, 6\}$
51) $\left\{-3, -\dfrac{4}{5}\right\}$ 53) $\left\{\dfrac{3}{2}\right\}$ 55) $\{-11, -3\}$
57) $\left\{-\dfrac{7}{2}, \dfrac{3}{4}\right\}$ 59) $\left\{\dfrac{2}{3}, 8\right\}$ 61) $\left\{-4, 0, \dfrac{1}{2}\right\}$
63) $\left\{-1, \dfrac{2}{9}, 11\right\}$ 65) $\left\{\dfrac{5}{2}, 3\right\}$ 67) $\{0, -8, 8\}$
69) $\{-4, 0, 9\}$ 71) $\{-12, 0, 5\}$ 73) $\left\{0, -\dfrac{3}{2}, \dfrac{3}{2}\right\}$
75) $\left\{-5, -\dfrac{2}{3}, \dfrac{7}{2}\right\}$ 77) $\left\{-\dfrac{3}{4}, \dfrac{2}{5}, \dfrac{1}{2}\right\}$
79) $\{-6, -2, 2\}$ 81) $-7, -3$
83) $\dfrac{5}{2}, 4$ 85) $-4, 4$ 87) $0, 1, 4$

Section 7.5

1) length = 12 in.; width = 3 in.
3) base = 3 cm; height = 8 cm
5) base = 6 in.; height = 3 in.
7) length = 10 in.; width = 6 in.
9) length = 9 ft; width = 5 ft
11) length = 9 in.; width = 6 in.
13) width = 12 in.; height = 6 in.
15) height = 10 cm; base = 7 cm
17) 5 and 6 or -4 and -3 19) 0, 2, 4 or 2, 4, 6
21) 6, 7, 8   23) Answers may vary.   25) 9
27) 15   29) 10   31) 8, 15, 17   33) 5, 12, 13
35) 8 in.   37) 5 ft   39) 5 mi
41) a) 144 ft   b) after 2 sec   c) 3 sec
43) a) 288 ft   b) 117 ft   c) 324 ft   d) 176 ft
45) a) $3500   b) $3375   c) $12
47) a) 184 ft   b) 544 ft
   c) when $t = 2\dfrac{1}{2}$ sec and when $t = 10$ sec   d) $12\dfrac{1}{2}$ sec

Chapter 7: Review Exercises

1) 9   3) $11p^4q^3$ 5) $12(4y + 7)$
7) $7n^3(n^2 - 3n + 1)$ 9) $(b + 6)(a - 2)$
11) $(n + 2)(m + 5)$ 13) $(r - 2)(5q - 6)$
15) $-4x(2x^2 + 3x - 1)$ 17) $(p + 8)(p + 5)$
19) $(x + 5y)(x - 4y)$ 21) $3(c - 6)(c - 2)$
23) $(5y + 6)(y + 1)$ 25) $(2m - 5)(2m - 3)$
27) $4a(7a + 4)(2a - 1)$ 29) $(3s - t)(s + 4t)$
31) $(3c + 1)(3c - 1)$ 33) $(n + 5)(n - 5)$
35) prime   37) $10(q + 9)(q - 9)$ 39) $(a + 8)^2$
41) $(h + 2)(h^2 - 2h + 4)$
43) $(3p - 4q)(9p^2 + 12pq + 16q^2)$
45) $(7r - 6)(r + 2)$ 47) $\left(\dfrac{3}{5} + x\right)\left(\dfrac{3}{5} - x\right)$
49) $(s - 8)(t - 5)$ 51) $w^2(w - 1)(w^2 + w + 1)$
53) prime   55) $-4ab$ 57) $(3y - 14)(2y - 3)$
59) $\left\{0, \dfrac{1}{2}\right\}$ 61) $\left\{-1, \dfrac{2}{3}\right\}$ 63) $\{-3, 15\}$
65) $\left\{-\dfrac{6}{7}, \dfrac{6}{7}\right\}$ 67) $\{-4, 8\}$ 69) $\left\{\dfrac{4}{5}, 1\right\}$
71) $\left\{0, -\dfrac{3}{2}, 2\right\}$ 73) $\{-8, 9\}$ 75) $\left\{\dfrac{1}{6}, 3, 7\right\}$
77) base = 5 in.; height = 6 in.

79) height = 3 in.; length = 8 in.
81) 12   83) length = 6 ft; width = 2.5 ft
85) −1, 0, 1 or 8, 9, 10   87) 5 in.
89) a) 0 ft   b) after 2 sec and after 4 sec
    c) 144 ft   d) after 6 sec

Chapter 7: Test

1) Determine whether you can factor out a GCF.
2) $(n - 6)(n - 5)$ 3) $(4 + b)(4 - b)$
4) $(5a + 2)(a - 3)$ 5) $7p^2q^3(8p^4q^3 - 11p^2q + 1)$
6) $(y - 2z)(y^2 + 2yz + 4z^2)$ 7) $2d(d + 9)(d - 2)$
8) prime   9) $(3h + 4)^2$ 10) $(2y - 3)(12x + 11)$
11) $(s - 7t)(s + 4t)$ 12) $(4s^2 + 9t^2)(2s + 3t)(2s - 3t)$
13) $(12p + 5)(3p + 7)$ 14) $(2b - 5)(6b - 7)$
15) $m^9(m + 1)(m^2 - m + 1)$ 16) $\{-4, -3\}$
17) $\{0, -5, 5\}$ 18) $\left\{-\dfrac{5}{12}, \dfrac{5}{12}\right\}$
19) $\{-4, 7\}$ 20) $\left\{-\dfrac{1}{4}, 8\right\}$ 21) $\left\{\dfrac{5}{3}, 2\right\}$
22) height = 10 ft; width = 4 ft   23) 5, 7, 9
24) 3 mi   25) height = 16 ft; width = 6 ft
26) a) $\dfrac{5}{4}$ sec and 3 sec   b) 60 ft   c) 132 ft   d) 5 sec

Chapter 7: Cumulative Review for Chapters 1–7

1) $\dfrac{1}{8}$ 2) $-\dfrac{9}{40}$ 3) $\dfrac{3t^4}{2u^6}$ 4) $-24k^{10}$
5) 481,300   6) $\{5\}$
7) $R = \dfrac{A - P}{PT}$ 8) Twix: 4 in.; Toblerone: 8 in.
9) $(-\infty, -4] \cup \left[\dfrac{38}{5}, \infty\right)$
10)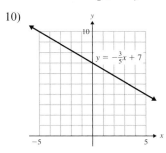
11) $y = \dfrac{1}{3}x + 1$ 12) $(-2, -2)$ 13) $12y^2 - 8y - 15$
14) $8p^3 - 50p^2 + 95p - 56$ 15) $c^2 + 16c + 64$
16) $3a^2b^2 - 7a^2b - 5ab^2 + 19ab - 8$ 17) $6x^3 - 7x - 4$
18) $3r + 1 - \dfrac{5}{2r} + \dfrac{3}{4r^2}$ 19) $(b - 7)(c + 8)$

20) $6(3q - 7)(3q - 1)$ 21) prime
22) $(t^2 + 9)(t + 3)(t - 3)$ 23) $(x - 5)(x^2 + 5x + 25)$
24) $\{-8, 5\}$ 25) $\left\{-\dfrac{4}{3}, \dfrac{5}{2}\right\}$

Chapter 8

Section 8.1

1) when its numerator equals zero
3) Set the denominator equal to zero, and solve for the variable. That value cannot be substituted into the expression because it will make the denominator equal to zero.
5) a) $\dfrac{3}{2}$ b) -8 c) $(-\infty, -6) \cup (-6, \infty)$
7) a) undefined   b) $\dfrac{3}{5}$ c) $(-\infty, -2) \cup (-2, \infty)$
9) a) -2 b) never equals zero
   c) $(-\infty, -5) \cup (-5, -1) \cup (-1, \infty)$
11) $(-\infty, 7) \cup (7, \infty)$ 13) $\left(-\infty, \dfrac{7}{2}\right) \cup \left(\dfrac{7}{2}, \infty\right)$
15) $(-\infty, 1) \cup (1, 8) \cup (8, \infty)$
17) $(-\infty, -9) \cup (-9, 9) \cup (9, \infty)$
19) $(-\infty, \infty)$ 21) Answers may vary.
23) $\dfrac{2}{5d^3}$ 25) $\dfrac{3}{5}$ 27) $b - 7$ 29) $\dfrac{1}{r + 4}$
31) $\dfrac{3k + 4}{k + 2}$ 33) $\dfrac{w + 5}{5}$ 35) $\dfrac{4(m - 5)}{11}$
37) $\dfrac{x + y}{x^2 + xy + y^2}$ 39) -1 41) -1 43) $-a - 3$
45) $-\dfrac{5}{x + 2}$ 47) $-4(b + 2)$ 49) $-\dfrac{4t^2 + 6t + 9}{2t + 3}$
51) $5x + 3$ 53) $c - 2$
55) Possible answers:
    $\dfrac{-b - 7}{b - 2}, \dfrac{-(b + 7)}{b - 2}, \dfrac{b + 7}{2 - b}, \dfrac{b + 7}{-(b - 2)}, \dfrac{b + 7}{-b + 2}$
57) Possible answers:
    $\dfrac{-9 + 5t}{2t - 3}, \dfrac{5t - 9}{2t - 3}, \dfrac{-(9 - 5t)}{2t - 3}, \dfrac{9 - 5t}{-2t + 3}, \dfrac{9 - 5t}{3 - 2t}, \dfrac{9 - 5t}{-(2t - 3)}$
59) $\dfrac{3}{4}$ 61) $\dfrac{8v^4}{3u^3}$ 63) $\dfrac{1}{2t(3t - 2)}$ 65) $\dfrac{2(2p - 1)}{9}$
67) $\dfrac{4v - 1}{4}$ 69) $5h(h - 1)$ 71) $\dfrac{(r - 2)(r^2 - 3r + 9)}{4(r - 3)}$
73) $\dfrac{3}{10}$ 75) $\dfrac{1}{6c^6}$ 77) $\dfrac{3a^2}{2a - 1}$
79) $\dfrac{5}{2y + 5}$ 81) $\dfrac{1}{7}$ 83) $\dfrac{z + 10}{(2z + 1)(z + 8)}$
85) $\dfrac{3}{4(3a + 1)}$ 87) $-\dfrac{8}{2d + 5}$

89) No. To divide the rational expressions we rewrite the problem as $\dfrac{12}{x} \cdot \dfrac{2}{3y}$. If $y = 0$, then the denominator of $\dfrac{2}{3y}$ will equal zero and the expression will be undefined.

91) $-\dfrac{a}{6}$ 93) $\dfrac{7}{3}$ 95) $\dfrac{3x^6}{4y^6}$ 97) $\dfrac{a^2}{a-4}$

99) $-\dfrac{3}{4(x+y)}$ 101) $\dfrac{4j-1}{3j+2}$ 103) $\dfrac{18x^4}{y^8}$

Section 8.2

1) 40 3) c^4 5) $36p^8$ 7) $24a^3b^4$

9) $2(n+4)$ 11) $w(2w+1)$ 13) $4a^3(3a-1)$

15) $(r+7)(r-2)$ 17) $(w-5)(w+2)(w+3)$

19) $b-4$ or $4-b$ 21) Answers may vary.

23) $\dfrac{3}{t} = \dfrac{3t^2}{t^3}; \dfrac{8}{t^3} = \dfrac{8}{t^3}$ 25) $\dfrac{9}{8n^6} = \dfrac{27}{24n^6}; \dfrac{2}{3n^2} = \dfrac{16n^4}{24n^6}$

27) $\dfrac{1}{x^3 y} = \dfrac{5y^4}{5x^3y^5}; \dfrac{6}{5xy^5} = \dfrac{6x^2}{5x^3y^5}$

29) $\dfrac{t}{5t-6} = \dfrac{7t}{7(5t-6)}; \dfrac{10}{7} = \dfrac{50t-60}{7(5t-6)}$

31) $\dfrac{7}{12x-4} = \dfrac{21}{12(3x-1)}; \dfrac{x}{18x-6} = \dfrac{2x}{12(3x-1)}$

33) $\dfrac{8}{a+9} = \dfrac{16a+56}{(a+9)(2a+7)}; \dfrac{a}{2a+7} = \dfrac{a^2+9a}{(a+9)(2a+7)}$

35) $\dfrac{9y}{y^2-y-42} = \dfrac{18y^2}{2y(y+6)(y-7)}; \dfrac{3}{2y^2+12y} = \dfrac{3y-21}{2y(y+6)(y-7)}$

37) $\dfrac{z}{z^2-10z+25} = \dfrac{z^2+3z}{(z-5)^2(z+3)}; \dfrac{15z}{z^2-2z-15} = \dfrac{15z^2-75z}{(z-5)^2(z+3)}$

39) $\dfrac{11}{g-3} = \dfrac{11g+33}{(g+3)(g-3)}; \dfrac{4}{9-g^2} = -\dfrac{4}{(g+3)(g-3)}$

41) $\dfrac{4}{w^2-4w} = \dfrac{28w-112}{7w(w-4)^2}; \dfrac{6}{7w^2-28w} = \dfrac{6w-24}{7w(w-4)^2}; \dfrac{11}{w^2-8w+16} = \dfrac{77w}{7w(w-4)^2}$

43) $\dfrac{4}{5}$ 45) $\dfrac{10}{a}$ 47) 2 49) $\dfrac{6}{w}$ 51) $\dfrac{d-9}{d+4}$

53) a) $x(x-3)$

b) Multiply the numerator and denominator of $\dfrac{8}{x-3}$ by x and the numerator and denominator of $\dfrac{2}{x}$ by $x-3$.

c) $\dfrac{8}{x-3} = \dfrac{8x}{x(x-3)}; \dfrac{2}{x} = \dfrac{2x-6}{x(x-3)}$

55) Find the product of the denominators.

57) $\dfrac{19}{24}$ 59) $\dfrac{3x}{20}$ 61) $\dfrac{3(7a+4)}{14a^2}$ 63) $\dfrac{11d+32}{d(d-8)}$

65) $\dfrac{5z+26}{(z+6)(z+2)}$ 67) $\dfrac{x^2-x-3}{(2x+1)(x+5)}$ 69) $\dfrac{t-3}{t-7}$

71) $\dfrac{b^2+b-40}{(b+4)(b-4)(b-9)}$ 73) $\dfrac{(c-5)(c+2)}{(c+6)(c-2)(c-4)}$

75) $\dfrac{4b^2+28b+3}{3(b-4)(b+3)}$

77) No; if the sum is rewritten as $\dfrac{5}{x-7} - \dfrac{2}{x-7}$ then the LCD = $x - 7$. If the sum is rewritten as $\dfrac{-5}{7-x} + \dfrac{2}{7-x}$, then the LCD is $7 - x$.

79) $\dfrac{6}{q-8}$ or $-\dfrac{6}{8-q}$ 81) $\dfrac{2c+13}{12b-7c}$ or $-\dfrac{2c+13}{7c-12b}$

83) $-\dfrac{5(t+6)}{(t+8)(t-8)}$ 85) $\dfrac{3(3a+4)}{(2a+3)(2a-3)}$

87) $\dfrac{6j^2-j-2}{3j(j+8)}$ 89) $\dfrac{c^2-2c+20}{(c-4)^2(c+3)}$

91) $-\dfrac{3y}{(x+y)(x-y)}$ 93) $\dfrac{-n^2+33n+1}{(4n-1)(3n+1)(n+2)}$

95) $\dfrac{3y^2+6y+26}{y(y-4)(2y-5)}$

97) a) $\dfrac{2(x+1)}{x-3}$ b) $\dfrac{x^2-2x+5}{x-3}$

99) a) $\dfrac{w}{(w+2)^2(w-2)}$ b) $\dfrac{2(w-1)^2}{(w+2)(w-2)}$

101) $\dfrac{3(x^2+6x+10)}{x(x+5)}$; Domain: $(-\infty, -5) \cup (-5, 0) \cup (0, \infty)$

103) $\dfrac{18(x+12)}{x(x+5)}$; Domain: $(-\infty, -5) \cup (-5, 0) \cup (0, \infty)$

Section 8.3

1) Method 1: Rewrite it as a division problem, then simplify.

$$\dfrac{2}{9} \div \dfrac{5}{18} = \dfrac{2}{9} \cdot \dfrac{\overset{2}{\cancel{18}}}{5} = \dfrac{4}{5}$$

Method 2: Multiply the numerator and denominator by 18, the LCD of $\dfrac{2}{9}$ and $\dfrac{5}{18}$. Then, simplify.

$$\dfrac{\overset{2}{\cancel{18}}\left(\dfrac{2}{9}\right)}{\underset{1}{\cancel{18}}\left(\dfrac{5}{18}\right)} = \dfrac{4}{5}$$

3) $\dfrac{14}{25}$ 5) ab^2 7) $\dfrac{1}{st^2}$ 9) $\dfrac{2m^4}{15n^2}$

11) $\dfrac{t}{5}$ 13) $\dfrac{4}{3(y-8)}$

15) $\dfrac{5}{6w^4}$ 17) $x(x-3)$ 19) $\dfrac{3}{2}$

21) $\dfrac{7d + 2c}{d(c - 5)}$ 23) $\dfrac{4z + 7}{5z - 7}$

25) $\dfrac{8}{y}$ 27) $\dfrac{x^2 - 7}{x^2 - 11}$ 29) $-\dfrac{52}{15}$

31) $\dfrac{4xy}{3(x + y)}$ 33) $\dfrac{r(r^2 + s)}{s^2(sr + 1)}$ 35) $\dfrac{t + 3}{t + 2}$

37) $\dfrac{b^2 + 1}{b^2 - 3}$ 39) $\dfrac{1}{m^3 n}$ 41) $\dfrac{(h - 1)(h + 3)}{28}$

43) $\dfrac{2(x - 9)(x + 2)}{3(x + 3)(x + 1)}$ 45) $\dfrac{r}{20}$ 47) $\dfrac{a}{12}$ 49) $\dfrac{25}{18}$

51) $\dfrac{2(n + 3)^2}{4n + 7}$ 53) $\dfrac{w(v - w)}{2v + w^2}$ 55) $\dfrac{8y^2}{x(y^2 - x)}$

57) $\dfrac{a^3 + b^2}{a^3(2 - 7b^2)}$ 59) $\dfrac{4n - m}{m(1 + mn)}$

61) 0

x	f(x)
1	1
2	$\dfrac{1}{2}$
3	$\dfrac{1}{3}$
10	$\dfrac{1}{10}$
100	$\dfrac{1}{100}$
1000	$\dfrac{1}{1000}$

Section 8.4

1) Eliminate the denominators.

3) sum; $\dfrac{3m - 14}{8}$ 5) equation; $\{-9\}$

7) difference; $\dfrac{z^2 - 4z + 24}{z(z - 6)}$ 9) equation; $\{3\}$

11) $0, -10$ 13) $0, 9, -9$ 15) $4, 9$ 17) $\{2\}$

19) $\{-2\}$ 21) $\{-21\}$ 23) $\{4\}$ 25) $\left\{-\dfrac{5}{2}\right\}$

27) $\{3\}$ 29) $\left\{\dfrac{1}{2}\right\}$ 31) $\emptyset$ 33) $\{6\}$ 35) $\{-4\}$

37) $\{5\}$ 39) $\{12\}$ 41) $\{-11\}$ 43) $\emptyset$ 45) $\{-3\}$

47) $\{0, 6\}$ 49) $\{-20\}$ 51) $\{2, 9\}$ 53) $\{1, 3\}$

55) $\{-3\}$ 57) $\{-4, -2\}$ 59) $\{-10\}$ 61) $\{-6\}$

63) $\{2, 8\}$ 65) $\{-8, 1\}$ 67) $\emptyset$ 69) 750 lb

71) 0.4 m   73) $P = \dfrac{nRT}{V}$ 75) $b = \dfrac{rt}{2a}$

77) $x = \dfrac{t + u}{3B}$ 79) $b = \dfrac{a - zc}{z}$ 81) $t = \dfrac{Aq - 4r}{A}$

83) $c = \dfrac{na - wb}{wk}$ 85) $r = \dfrac{st}{s + t}$ 87) $C = \dfrac{AB}{3A - 2B}$

Chapter 8: Putting It All Together

1) a) $-\dfrac{8}{5}$ b) 0   c) $(-\infty, -3) \cup (-3, 3) \cup (3, \infty)$

2) a) undefined   b) $\dfrac{3}{5}$ c) $(-\infty, -4) \cup (-4, 2) \cup (2, \infty)$

3) a) 4   b) never equals 0   c) $\left(-\infty, \dfrac{1}{2}\right) \cup \left(\dfrac{1}{2}, \infty\right)$

4) a) $\dfrac{4}{5}$ b) never equals 0   c) $(-\infty, 0) \cup (0, \infty)$

5) a) $\dfrac{1}{6}$ b) 1   c) $(-\infty, \infty)$

6) a) 3   b) -7 and 7 c) $(-\infty, \infty)$

7) $\dfrac{4}{3n^3}$ 8) $\dfrac{w^5}{2}$ 9) $\dfrac{1}{j - 4}$ 10) $\dfrac{m + 8}{2(m + 4)}$

11) $-\dfrac{3}{n + 2}$ 12) $-\dfrac{1}{y + 3}$ 13) $\dfrac{3f - 16}{f(f + 8)}$

14) $\dfrac{14}{3}$ 15) $\dfrac{4a^2 b}{9}$ 16) $\dfrac{4j^2 + 27j + 27}{(j + 9)(j - 9)(j + 6)}$

17) $\dfrac{8q^2 - 37q + 21}{(q - 5)(q + 4)(q + 7)}$ 18) $\dfrac{z^3}{3y^3}$ 19) $-\dfrac{m + 7}{8}$

20) $\dfrac{12p^2 - 92p - 15}{(4p + 3)(p + 2)(p - 6)}$ 21) $\dfrac{9}{r - 8}$ 22) $\dfrac{3y + 14}{12y^2}$

23) $\dfrac{5}{3}$ 24) $\dfrac{7d^2 + 6d + 54}{d^2(d + 9)}$ 25) $\dfrac{2(7x - 1)}{(x - 8)(x + 3)}$

26) $\dfrac{6}{(y + 4)(x^2 - 3x + 9)}$ 27) $\dfrac{34}{15z}$

28) $\dfrac{5k + 28}{(k + 2)(k + 5)}$ 29) -1 30) $\dfrac{m + 20n}{7m - 4n}$

31) $\dfrac{5u^2 + 37u - 19}{u(3u - 2)(u + 1)}$ 32) $\dfrac{-2p^2 - 8p + 11}{p(p + 7)(p - 8)}$

33) $\dfrac{x^2 + 2x + 12}{(2x + 1)^2 (x - 4)}$ 34) 1   35) $\dfrac{5(c^2 + 16)}{3(c - 2)}$

36) $\dfrac{3n + 1}{7(n + 4)}$ 37) $-\dfrac{11}{5w}$ 38) $-\dfrac{1}{2k}$

39) a) $\dfrac{x(x - 3)}{8}$ b) $\dfrac{3(x - 1)}{2}$

40) a) $\dfrac{8z}{(z + 5)(z + 1)}$ b) $\dfrac{2(z^2 + 9z + 40)}{(z + 5)(z + 1)}$

41) $h(x) = \dfrac{x(5x + 1)}{(3x + 1)(3x - 1)}$;

Domain: $\left(-\infty, -\dfrac{1}{3}\right) \cup \left(-\dfrac{1}{3}, \dfrac{1}{3}\right) \cup \left(\dfrac{1}{3}, \infty\right)$

42) $k(x) = \dfrac{x(x+1)}{(3x+1)(3x-1)}$; $k(x) = 0$ when $x = 0$ or $x = -1$

43) $\dfrac{c^2}{16}$ 44) $\dfrac{3}{20}$ 45) $\dfrac{7m-15}{(m-3)(m-4)}$

46) $\dfrac{3k^2(3k-1)}{2}$ 47) $\dfrac{15tu^3}{4}$ 48) $\dfrac{3-y^2}{x+y}$

49) $h(x) = 6x - 5$; Domain: $(-\infty, -8) \cup (-8, \infty)$

50) $h(x) = x + 7$; Domain: $(-\infty, -2) \cup (-2, 2) \cup (2, \infty)$

51) $\{-15\}$ 52) $\{4\}$ 53) $\{-10, 1\}$ 54) $\{-10, 1\}$

55) $\{-12\}$ 56) $\{0, -12\}$ 57) $\varnothing$ 58) $\{-5, 4\}$

Section 8.5

1) $7.10 3) 82.5 mg 5) 375 ml 7) 111

9) 3 cups of tapioca flour and 6 cups of potato-starch flour

11) length: 48 ft; width: 30 ft

13) stocks: $8000; bonds: $12,000

15) 1355 17) a) 7 mph b) 13 mph

19) a) $x + 30$ mph b) $x - 30$ mph

21) 20 mph 23) 4 mph 25) 260 mph 27) 2 mph

29) $\dfrac{1}{4}$ job/hr 31) $\dfrac{1}{t}$ job/hr 33) $1\dfrac{1}{5}$ hr 35) $3\dfrac{1}{13}$ hr

37) 20 min 39) 3 hr 41) 3 hr 43) $2\dfrac{1}{5}$ ft/sec

45) $x = 10$ 47) $x = 13$

Section 8.6

1) increases 3) direct 5) inverse 7) combined

9) $M = kn$ 11) $h = \dfrac{k}{j}$ 13) $T = \dfrac{k}{c^2}$ 15) $s = krt$

17) $Q = \dfrac{k\sqrt{z}}{m}$ 19) a) 9 b) $z = 9x$ c) 54

21) a) 48 b) $N = \dfrac{48}{y}$ c) 16

23) a) 5 b) $Q = \dfrac{5r^2}{w}$ c) 45

25) 56 27) 18 29) 70 31) $500.00

33) $0.80 35) 180 watts 37) 48π cm^3

39) 200 cycles/sec 41) 3 ohms 43) 320 lb

Chapter 8: Review Exercises

1) Set $Q(x) = 0$, and solve for x. The domain of $f(x)$ contains all real numbers except the values of x that make $Q(x) = 0$.

3) a) $\dfrac{7}{12}$ b) -9 c) $\left(-\infty, \dfrac{1}{5}\right) \cup \left(\dfrac{1}{5}, \infty\right)$

5) $(-\infty, -4) \cup (-4, 6) \cup (6, \infty)$ 7) $\dfrac{7}{a^9}$ 9) $\dfrac{1}{3z+1}$

11) $\dfrac{y+9}{z-12}$ 13) $2l + 1$ 15) $\dfrac{36k^2}{m}$ 17) $\dfrac{1}{4w}$

19) $-\dfrac{a+5}{4(a^2+5a+25)}$ 21) k^2 23) $m(m+5)$

25) $(3x+7)(x-9)$ 27) $(c+4)(c+6)(c-7)$

29) $(a-5)(a-8)(a+1)$ 31) $\dfrac{24r^2}{20r^3}$

33) $\dfrac{t^2 + 2t - 15}{(2t+1)(t+5)}$ 35) $\dfrac{3}{8a^3b} = \dfrac{15b^4}{40a^3b^5}$; $\dfrac{6}{5ab^5} = \dfrac{48a^2}{40a^3b^5}$

37) $\dfrac{9c}{c^2+6c-16} = \dfrac{9c^2-18c}{(c-2)^2(c+8)}$; 6) $\dfrac{4}{c^2-4c+4} = \dfrac{4c+32}{(c-2)^2(c+8)}$

39) $\dfrac{5}{4c}$ 41) $\dfrac{4+9z}{10z^2}$ 43) $\dfrac{27-y}{(y-2)(y+3)}$

45) $\dfrac{10p^2 - 67p - 53}{4(p+1)(p-7)}$ 47) $\dfrac{17}{11-m}$ or $-\dfrac{17}{m-11}$

49) $-\dfrac{xy}{(x+y)(x-y)}$ 51) $\dfrac{2g^2 - 6g + 57}{5g(g-7)}$

53) a) $\dfrac{3x}{2(x-4)}$ sq units b) $\dfrac{x^2 - 4x + 96}{4(x-4)}$ units

55) $\dfrac{5x^2 + 7x - 8}{x(x-2)}$; Domain: $(-\infty, 0) \cup (0, 2) \cup (2, \infty)$

57) $\dfrac{6}{7}$ 59) $\dfrac{p^2+6}{p^2+8}$ 61) $\dfrac{2}{3n}$ 63) $\dfrac{(y+3)(y-10)}{(y-9)(y+5)}$

65) $\dfrac{c-1}{c-2}$ 67) $\dfrac{y(x^2+2y)}{x(y^2-x)}$ 69) $\left\{\dfrac{2}{5}\right\}$ 71) $\{10\}$

73) $\{1, 20\}$ 75) $\varnothing$ 77) $\{-3, 5\}$ 79) $c = \dfrac{2p}{A}$

81) $a = \dfrac{t-nb}{n}$ 83) $s = \dfrac{rt}{t-r}$ 85) 12 g

87) 280 mph 89) 21 91) 2 93) 3.24 lb

Chapter 8: Test

1) a) $-\dfrac{1}{5}$ b) $f(x)$ never equals zero

c) $(-\infty, -8) \cup (-8, 6) \cup (6, \infty)$

2) $\left(-\infty, -\dfrac{3}{2}\right) \cup \left(-\dfrac{3}{2}, \infty\right)$ 3) $\dfrac{9}{4w^5}$ 4) $\dfrac{7v-1}{v-8}$

5) Possible answers: $-\dfrac{h-9}{2h-3}$, $\dfrac{h-9}{3-2h}$, $\dfrac{-h+9}{2h-3}$, $\dfrac{h-9}{-2h+3}$

6) $k(2k-3)(k+2)^2$ 7) $\dfrac{1}{z}$ 8) $\dfrac{7n^2}{4m^4}$

9) $\dfrac{r^2+11r+3}{(2r+1)(r+5)}$ 10) $-\dfrac{a^2+2a+4}{6}$

11) $\dfrac{11}{15-c}$ or $-\dfrac{11}{c-15}$ 12) $\dfrac{x^2-12x+21}{(x+7)(x-7)(x-9)}$

13) $h(x)=\dfrac{-x^2+7x+14}{x(x+7)}$;
Domain: $(-\infty,-7)\cup(-7,0)\cup(0,\infty)$

14) $-\dfrac{1}{d}$ 15) $\dfrac{9}{4}$ 16) $-\dfrac{xy}{x+y}$ 17) $3k+7$

18) $0,\dfrac{1}{4}$ 19) $\{4\}$ 20) $\varnothing$ 21) $\left\{\dfrac{13}{4}\right\}$

22) $\{-1,5\}$ 23) $c=\dfrac{kxz}{y}$ 24) $p=\dfrac{qr}{q-r}$

25) Facebook friends: 684; Twitter followers: 144

26) 13 mph   27) 300   28) 18 dB

Chapter 8: Cumulative Review for Chapters 1–8

1) $-45w^8$ 2) $\dfrac{8}{n^{18}}$ 3) $\left\{-\dfrac{45}{4}\right\}$

4) 45 ml of 12% solution; 15 ml of 4% solution

5) x-int: $(2, 0)$; y-int: $(0, -5)$

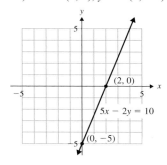
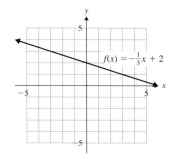

7) $x+y=3$ 8) $(2,10)$ 9) width: 15 cm; length: 23 cm

10) $(-\infty,1]$ 11) $\left\{-\dfrac{7}{2},-\dfrac{5}{6}\right\}$

12)
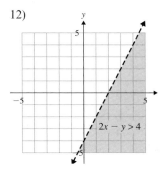

13)
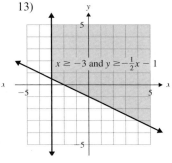

14) a) -11 b) $\dfrac{3}{2}$ or 4 15) $-14q^2+6q-20$

16) $12d^4+18d^3-31d^2-42d+7$

17) $ab-\dfrac{5}{2b}+\dfrac{7}{a^2b^2}+\dfrac{1}{a^3b^2}$ 18) v^3+3v^2-5v-6

19) $\dfrac{3(c+8)}{c(c-6)}$ 20) $(5n+9)(5n-9)$

21) $3x(y+8)(y-3)$ 22) $(r+4t)^2$

23) $(-\infty,8)\cup(8,\infty)$

24) a) $-\dfrac{4}{3k(4k^2+6k+9)}$ b) $\dfrac{(r+3)(r+2)}{r^2+r+3}$

25) $1\dfrac{7}{8}$ hr

Chapter 9
Section 9.1

1) False; $\sqrt{121}=11$ because the $\sqrt{}$ symbol means principal square root.

3) False; the square root of a negative number is not a real number.

5) 12 and -12 7) $\dfrac{6}{5}$ and $-\dfrac{6}{5}$ 9) 7

11) not real   13) $\dfrac{9}{5}$ 15) -6 17) True

19) False; the only even root of zero is zero.

21) $\sqrt[3]{64}$ is the number you cube to get 64. $\sqrt[3]{64}=4$

23) No; the even root of a negative number is not a real number.

25) 5   27) -1 29) 3   31) not real   33) -2

35) -2 37) 10   39) not real   41) $\dfrac{2}{5}$ 43) 7

45) -3 47) 13

49) If a is negative and we didn't use the absolute values, the result would be negative. This is incorrect because if a is negative and n is even, then $a^n>0$ so that $\sqrt[n]{a^n}>0$. Using absolute values ensures a positive result.

51) 8   53) 6   55) $|y|$ 57) 5

59) z 61) $|h|$ 63) $|x+7|$ 65) $2t-1$

67) $|3n+2|$ 69) $d-8$

71) No, because $\sqrt{-1}$ is not a real number.

73) Set up an inequality so that the radicand is greater than or equal to 0. Solve for the variable. These are the real numbers in the domain of the function.

75) 10   77) not a real number

79) 1   81) not a real number   83) $\sqrt{a}$

85) $\sqrt{t+4}$ 87) $\sqrt{6n+1}$ 89) 4   91) -3

93) $\sqrt[3]{-17}$ 95) $\sqrt[3]{4r-1}$ 97) $\sqrt[3]{c+8}$

99) $\sqrt[3]{8a-13}$ 101) $[-2,\infty)$ 103) $[8,\infty)$

105) $(-\infty,\infty)$ 107) $\left[-\dfrac{7}{3},\infty\right)$ 109) $(-\infty,\infty)$

111) $(-\infty,0]$ 113) $\left(-\infty,\dfrac{9}{7}\right]$

115) Domain: $[1, \infty)$ 117) Domain: $[-3, \infty)$

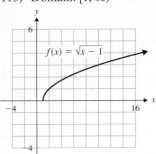 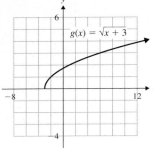

119) Domain: $[0, \infty)$ 121) Domain: $(-\infty, 0]$

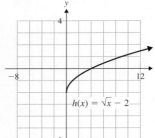 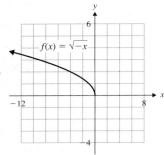

123) Domain: $(-\infty, \infty)$ 125) Domain: $(-\infty, \infty)$

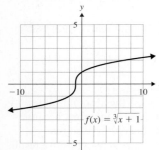 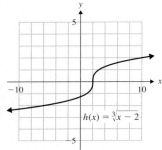

127) Domain: $(-\infty, \infty)$

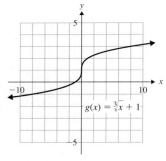

129) 2 ft   131) $\sqrt{10}$ sec; 3.16 sec

133) Yes. The car was traveling at 30 mph.

135) π sec; 3.14 sec

137) $T\left(\dfrac{1}{2}\right) = \dfrac{\pi}{4}$; A $\dfrac{1}{2}$-ft-long pendulum has a period of $\dfrac{\pi}{4}$ sec. This is approximately 0.79 sec.

139) $\dfrac{\sqrt{5}}{4}\pi$ sec; 1.76 sec

Section 9.2

1) The denominator of 2 becomes the index of the radical. $25^{1/2} = \sqrt{25}$

3) 3   5) 10   7) 2   9) -2 11) $\dfrac{2}{11}$ 13) $\dfrac{5}{4}$

15) $-\dfrac{6}{13}$ 17) not a real number   19) -1

21) The denominator of 4 becomes the index of the radical. The numerator of 3 is the power to which we raise the radical expression. $16^{3/4} = (\sqrt[4]{16})^3$

23) 16   25) 32   27) 25   29) -216

31) not a real number   33) $\dfrac{8}{27}$ 35) $-\dfrac{100}{9}$

37) False; the negative exponent does not make the result negative. $81^{-1/2} = \dfrac{1}{9}$

39) $\dfrac{1}{64}$; $\dfrac{1}{64}$; The denominator of the fractional exponent is the index of the radical.; $\dfrac{1}{8}$

41) $\dfrac{1}{7}$ 43) $\dfrac{1}{10}$ 45) 3   47) -4 49) $\dfrac{1}{32}$

51) $\dfrac{1}{25}$ 53) $\dfrac{8}{125}$ 55) $\dfrac{25}{16}$ 57) 8   59) 3

61) $8^{4/5}$ 63) 32   65) $\dfrac{1}{4^{7/5}}$ 67) z 69) $-72v^{11/8}$

71) $a^{1/9}$ 73) $\dfrac{24}{5w^{1/10}}$ 75) $\dfrac{1}{x^{2/3}}$ 77) $z^{2/15}$

79) $27u^2v^3$ 81) $8r^{1/5}s^{4/15}$ 83) $\dfrac{f^{2/7}g^{5/9}}{3}$ 85) $x^{10}w^9$

87) $\dfrac{1}{y^{2/3}}$ 89) $\dfrac{a^{12/5}}{4b^{2/5}}$ 91) $\dfrac{t^{21/2}}{r^{1/5}}$ 93) $\dfrac{1}{h^5 k^{25/18}}$

95) $p^{7/6} + p$ 97) $25^{6/12}$; $25^{1/2}$; Evaluate.   99) 7

101) 9   103) 5   105) 12   107) x^4 109) $\sqrt[3]{k}$

111) $\sqrt{z}$ 113) d^2 115) a) 13 degrees F   b) 6 degrees F

Section 9.3

1) $\sqrt{21}$ 3) $\sqrt{30}$ 5) $\sqrt{6y}$

7) False; 20 contains a factor of 4 which is a perfect square.

9) True; 42 does not have any factors (other than 1) that are perfect squares.

11) Factor; $\sqrt{4} \cdot \sqrt{15}$; $2\sqrt{15}$

13) $2\sqrt{5}$ 15) $3\sqrt{6}$ 17) simplified   19) $6\sqrt{3}$

21) $7\sqrt{2}$ 23) simplified   25) 20   27) $5\sqrt{30}$

29) $\dfrac{12}{5}$ 31) $\dfrac{2}{7}$ 33) 3   35) $2\sqrt{3}$

37) $2\sqrt{5}$ 39) $\sqrt{7}$ 41) $\dfrac{\sqrt{6}}{7}$ 43) $\dfrac{3\sqrt{5}}{4}$

45) x^4 47) w^7 49) $10c$ 51) $8k^3m^5$

53) $2r^2\sqrt{7}$ 55) $10q^{11}t^8\sqrt{3}$ 57) $\dfrac{9}{c^3}$ 59) $\dfrac{2\sqrt{10}}{t^4}$

61) $\dfrac{5x\sqrt{3}}{y^6}$ 63) Factor; $\sqrt{w^8} \cdot \sqrt{w^1}$; Simplify.

65) $a^2\sqrt{a}$ 67) $g^6\sqrt{g}$ 69) $h^{15}\sqrt{h}$ 71) $6x\sqrt{2x}$

73) $q^3\sqrt{13q}$ 75) $5t^5\sqrt{3t}$ 77) c^4d 79) $a^2b\sqrt{b}$

81) $u^3v^3\sqrt{uv}$ 83) $6m^4n^2\sqrt{m}$ 85) $2x^6y^2\sqrt{11y}$

87) $4t^2u^3\sqrt{2tu}$ 89) $\dfrac{a^3\sqrt{a}}{9b^3}$ 91) $\dfrac{r^4\sqrt{3r}}{s}$ 93) $5\sqrt{2}$

95) $3\sqrt{7}$ 97) w^3 99) $n^3\sqrt{n}$ 101) $4k^3$

103) $5a^3b^4\sqrt{2ab}$ 105) $2c^5d^4\sqrt{10d}$ 107) $3k^4$

109) $2h^3\sqrt{10}$ 111) $a^4b^2\sqrt{10ab}$ 113) 20 m/s

Section 9.4

1) Answers may vary.

3) i) Its radicand will not contain any factors that are perfect cubes.
   ii) The radicand will not contain fractions.
   iii) There will be no radical in the denominator of a fraction.

5) $\sqrt[5]{12}$ 7) $\sqrt[5]{9m^2}$ 9) $\sqrt[3]{a^2b}$

11) Factor; $\sqrt[3]{8} \cdot \sqrt[3]{7}$; $2\sqrt[3]{7}$

13) $2\sqrt[3]{3}$ 15) $2\sqrt[4]{4}$ 17) $3\sqrt[3]{2}$ 19) $10\sqrt[3]{2}$

21) $2\sqrt[5]{2}$ 23) $\dfrac{1}{5}$ 25) -3 27) $2\sqrt[3]{3}$ 29) $2\sqrt[4]{5}$

31) d^2 33) n^5 35) xy^3 37) $w^4\sqrt[3]{w^2}$ 39) $y^2\sqrt[4]{y}$

41) $d\sqrt[3]{d^2}$ 43) $u^3v^5\sqrt[3]{u}$ 45) $b^5c\sqrt[3]{bc^2}$

47) $n^4\sqrt[4]{m^3n^2}$ 49) $2x^3y^4\sqrt[3]{3x}$ 51) $2t^5u^2\sqrt[3]{9t^2u}$

53) $\dfrac{m^2}{3}$ 55) $\dfrac{2a^4\sqrt[4]{a^3}}{b^3}$ 57) $\dfrac{t^2\sqrt[4]{t}}{3s^6}$ 59) $\dfrac{u^9\sqrt[3]{u}}{v}$

61) $2\sqrt[3]{3}$ 63) $3\sqrt[3]{4}$ 65) $2\sqrt[3]{10}$ 67) m^3

69) k^4 71) $r^3\sqrt[3]{r^2}$ 73) $p^4\sqrt[5]{p^3}$ 75) $3z^6\sqrt[3]{z}$

77) h^4 79) $c^2\sqrt[3]{c}$ 81) $3d^4\sqrt[4]{d^3}$

83) Change radicals to fractional exponents.; Rewrite exponents with a common denominator.; $a^{5/4}$; Rewrite in radical form. $a\sqrt[4]{a}$

85) $\sqrt[6]{p^5}$ 87) $n\sqrt[4]{n}$ 89) $c\sqrt[15]{c^4}$ 91) $\sqrt[4]{w}$

93) $\sqrt[12]{h}$ 95) 4 in.

Section 9.5

1) They have the same index and the same radicand.

3) $14\sqrt{2}$ 5) $15\sqrt[3]{4}$ 7) $11 - 3\sqrt{13}$ 9) $-5\sqrt[3]{z^2}$

11) $-9\sqrt[3]{n^2} + 10\sqrt[5]{n^2}$ 13) $2\sqrt{5c} - 2\sqrt{6c}$

15) i) Write each radical expression in simplest form.
    ii) Combine like radicals.

17) Factor.; $\sqrt{16} \cdot \sqrt{3} + \sqrt{3}$; Simplify.; $5\sqrt{3}$

19) $4\sqrt{3}$ 21) $-2\sqrt{2}$ 23) $6\sqrt{3}$ 25) $10\sqrt[3]{9}$

27) $-\sqrt[3]{6}$ 29) $13q\sqrt{q}$ 31) $-20d^2\sqrt{d}$

33) $4t^3\sqrt[3]{t}$ 35) $6a^2\sqrt[4]{a^3}$ 37) $-2\sqrt{2p}$

39) $25a\sqrt[3]{3a^2}$ 41) $4y\sqrt{xy}$ 43) $3c^2d\sqrt{2d}$

45) $14p^2q\sqrt[3]{11pq^2}$ 47) $14cd\sqrt[4]{9cd}$ 49) $\sqrt[3]{b}(a^3 - b^2)$

51) $3x + 15$ 53) $7\sqrt{6} + 14$ 55) $\sqrt{30} - \sqrt{10}$

57) $-30\sqrt{2}$ 59) $4\sqrt{5}$ 61) $-\sqrt{30}$ 63) $5\sqrt[4]{3} - 3$

65) $t - 9\sqrt{tu}$ 67) $2y\sqrt{x} - xy\sqrt{2y}$

69) $c\sqrt[3]{c} + 5c\sqrt[3]{d}$

71) Both are examples of multiplication of two binomials. They can be multiplied using FOIL.

73) $(a+b)(a-b) = a^2 - b^2$ 75) $p^2 + 13p + 42$

77) $6 \cdot 2 + 6\sqrt{7} + 2\sqrt{7} + \sqrt{7} \cdot \sqrt{7}$; Multiply.; $19 + 8\sqrt{7}$

79) $-22 + 5\sqrt{2}$ 81) $-16 + 11\sqrt{6}$

83) $5\sqrt{7} + 5\sqrt{2} + 2\sqrt{21} + 2\sqrt{6}$

85) $5 - \sqrt[3]{150} - 3\sqrt[3]{5} + 3\sqrt[3]{6}$

87) $-2\sqrt{6pq} + 30p - 16q$ 89) $4 + 2\sqrt{3}$

91) $16 - 2\sqrt{55}$ 93) $h + 2\sqrt{7h} + 7$

95) $x - 2\sqrt{xy} + y$ 97) $c^2 - 81$ 99) 31   101) 46

103) $\sqrt[3]{4} - 9$ 105) $c - d$ 107) $64f - g$ 109) 41

111) $11 + 4\sqrt{7}$ 113) $13 - 4\sqrt{3}$

Section 9.6

1) Eliminate the radical from the denominator.

3) $\dfrac{\sqrt{5}}{5}$ 5) $\dfrac{3\sqrt{6}}{2}$ 7) $-5\sqrt{2}$ 9) $\dfrac{\sqrt{21}}{14}$

11) $\dfrac{\sqrt{3}}{3}$ 13) $\dfrac{\sqrt{42}}{6}$ 15) $\dfrac{\sqrt{30}}{3}$ 17) $\dfrac{\sqrt{15}}{10}$

19) $\dfrac{8\sqrt{y}}{y}$ 21) $\dfrac{\sqrt{5t}}{t}$ 23) $\dfrac{8v^3\sqrt{5vw}}{5w}$ 25) $\dfrac{a\sqrt{3b}}{3b}$

27) $-\dfrac{5\sqrt{3b}}{b^2}$ 29) $\dfrac{\sqrt{13j}}{j^3}$ 31) 2^2 or 4   33) 3

35) c^2 37) 2^3 or 8   39) m 41) $\dfrac{4\sqrt[3]{9}}{3}$ 43) $6\sqrt[3]{4}$

45) $\dfrac{9\sqrt[3]{5}}{5}$ 47) $\dfrac{\sqrt[4]{45}}{3}$ 49) $\dfrac{\sqrt[5]{12}}{2}$ 51) $\dfrac{10\sqrt[3]{z^2}}{z}$

53) $\dfrac{\sqrt[3]{3n}}{n}$ 55) $\dfrac{\sqrt[3]{28k}}{2k}$ 57) $\dfrac{9\sqrt[5]{a^2}}{a}$ 59) $\dfrac{\sqrt[4]{40m^3}}{2m}$

61) Change the sign between the two terms.

63) $(5 - \sqrt{2})$; 23   65) $(\sqrt{2} - \sqrt{6})$; -4

67) $(\sqrt{t} + 8)$; $t - 64$

69) Multiply by the conjugate.; $(a+b)(a-b) = a^2 - b^2$; $\dfrac{24 + 6\sqrt{5}}{16 - 5}$; $\dfrac{24 + 6\sqrt{5}}{11}$

71) $6 - 3\sqrt{3}$ 73) $\dfrac{90 + 10\sqrt{2}}{79}$ 75) $2\sqrt{6} - 4$

77) $\dfrac{\sqrt{30} - 5\sqrt{2} + 3 - \sqrt{15}}{7}$ 79) $\dfrac{m - \sqrt{mn}}{m - n}$

81) $\sqrt{b} + 5$ 83) $\dfrac{x + 2\sqrt{xy} + y}{x - y}$ 85) $\dfrac{5}{3\sqrt{5}}$

87) $\dfrac{x}{\sqrt{7x}}$ 89) $\dfrac{1}{12 - 6\sqrt{3}}$

91) $\dfrac{1}{\sqrt{x} + 2}$ 93) $-\dfrac{1}{4 + \sqrt{c + 11}}$

95) No, because when we multiply the numerator and denominator by the conjugate of the denominator, we are multiplying the original expression by 1.

97) $1 + 2\sqrt{3}$ 99) $\dfrac{15 - 9\sqrt{5}}{2}$

101) $\dfrac{\sqrt{5} + 2}{3}$ 103) $-2 - \sqrt{2}$

105) a) $r(8\pi) = 2\sqrt{2}$; When the area of a circle is 8π in^2, its radius is $2\sqrt{2}$ in.

b) $r(7) = \dfrac{\sqrt{7\pi}}{\pi}$; When the area of a circle is 7 in^2, its radius is $\dfrac{\sqrt{7\pi}}{\pi}$ in. (This is approximately 1.5 in.)

c) $r(A) = \dfrac{\sqrt{A\pi}}{\pi}$

Chapter 9: Putting It All Together

1) 3   2) -10 3) -2 4) 11   5) not a real number

6) $\dfrac{12}{7}$ 7) 12   8) 16   9) -100

10) not a real number   11) $\dfrac{1}{5}$ 12) $\dfrac{27}{1000}$ 13) $\dfrac{1}{k^{3/10}}$

14) t^6 15) $\dfrac{9}{a^{16/3}b^6}$ 16) $\dfrac{x^{30}}{243y^{5/6}}$ 17) $2\sqrt{6}$

18) $2\sqrt[4]{2}$ 19) $2\sqrt[3]{9}$ 20) $5\sqrt[3]{2}$ 21) $3\sqrt[4]{3}$

22) $3c^5\sqrt{5c}$ 23) $2m^2n^5\sqrt[3]{12m}$ 24) $\dfrac{2x^3\sqrt[5]{2x^4}}{y^4}$

25) $2\sqrt[3]{3}$ 26) $2k^2\sqrt[4]{3}$ 27) $19 + 8\sqrt{7}$

28) $-18c^2\sqrt[3]{4c}$ 29) $3\sqrt{6}$ 30) $5\sqrt{3} + 5\sqrt{2}$

31) $17m\sqrt{3mn}$ 32) $3p^4q^2\sqrt{10q}$ 33) $2t^3u\sqrt{3}$

34) $\dfrac{9\sqrt[3]{4}}{2}$ 35) $112 + 40\sqrt{3}$ 36) -7

37) $\dfrac{2\sqrt{2} - \sqrt{5}}{3}$ 38) $r\sqrt[6]{r}$ 39) $\dfrac{\sqrt[3]{3b^2c^2}}{3c}$

40) $\dfrac{2\sqrt[4]{2w}}{w^3}$

41) Domain: $[2, \infty)$
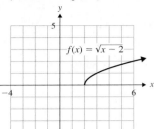

42) Domain: $(-\infty, \infty)$
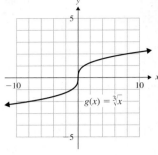

43) Domain: $(-\infty, \infty)$
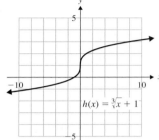

44) Domain: $[-1, \infty)$
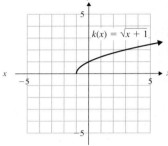

45) Domain: $(-\infty, 0]$
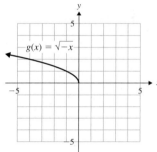

46) Domain: $[0, \infty)$
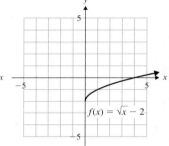

Section 9.7

1) Sometimes there are extraneous solutions.

3) $\{49\}$ 5) $\left\{\dfrac{4}{9}\right\}$ 7) $\varnothing$ 9) $\{20\}$ 11) $\varnothing$

13) $\left\{\dfrac{2}{3}\right\}$ 15) $\{2\}$ 17) $\{5\}$ 19) $n^2 + 10n + 25$

21) $c^2 - 12c + 36$ 23) $\{12\}$ 25) $\varnothing$ 27) $\{1, 3\}$

29) $\{4\}$ 31) $\{1, 16\}$ 33) $\{-3\}$ 35) $\{10\}$

37) $\{-3\}$ 39) $\{10\}$ 41) $\{63\}$ 43) $\varnothing$

45) $x + 10\sqrt{x} + 25$ 47) $85 - 18\sqrt{a + 4} + a$

49) $12n + 28\sqrt{3n - 1} + 45$ 51) $\{5, 13\}$ 53) $\{2\}$

55) $\left\{\dfrac{1}{4}\right\}$ 57) $\{1, 5\}$ 59) $\varnothing$ 61) $\{2, 11\}$

63) Raise both sides of the equation to the third power.

65) $\{125\}$ 67) $\{-64\}$ 69) $\left\{-\dfrac{3}{2}\right\}$ 71) $\{-1\}$

73) $\{-2\}$ 75) $\left\{-\dfrac{1}{2}, 4\right\}$ 77) $\{36\}$ 79) $\{26\}$

81) $\{23\}$ 83) $\{9\}$ 85) $\{9\}$ 87) $\{-1\}$

89) $E = \dfrac{mv^2}{2}$ 91) $b^2 = c^2 - a^2$ 93) $\sigma = \dfrac{E}{T^4}$

95) a) 320 m/sec   b) 340 m/sec
    c) The speed of sound increases.   d) $T = \dfrac{V_s^2}{400} - 273$

97) a) 2 in.   b) $V = \pi r^2 h$

99) a) 463 mph   b) about 8 min.

101) 16 ft   103) 5 mph

Section 9.8

1) False   3) True   5) $9i$ 7) $5i$

9) $i\sqrt{6}$ 11) $3i\sqrt{3}$ 13) $2i\sqrt{15}$

15) Write each radical in terms of i before multiplying.
$$\begin{aligned}\sqrt{-5} \cdot \sqrt{-10} &= i\sqrt{5} \cdot i\sqrt{10}\\ &= i^2\sqrt{50}\\ &= (-1)\sqrt{25} \cdot \sqrt{2}\\ &= -5\sqrt{2}\end{aligned}$$

17) $-\sqrt{5}$ 19) -10 21) 2   23) -13

25) Add the real parts, and add the imaginary parts.

27) -1 29) $3 + 11i$ 31) $4 - 9i$ 33) $-\dfrac{1}{4} - \dfrac{5}{6}i$

35) $7i$ 37) $24 - 15i$ 39) $-6 + \dfrac{4}{3}i$

41) $44 - 24i$ 43) $-28 + 17i$ 45) $14 + 18i$

47) $36 - 42i$ 49) $\dfrac{3}{20} + \dfrac{9}{20}i$

51) conjugate: $11 - 4i$;   53) conjugate: $-3 + 7i$;
    product: 137   product: 58

55) conjugate: $-6 - 4i$;   57) Answers may vary.
    product: 52

59) $\dfrac{8}{13} + \dfrac{12}{13}i$ 61) $\dfrac{8}{17} + \dfrac{32}{17}i$ 63) $\dfrac{7}{29} - \dfrac{3}{29}i$

65) $-\dfrac{74}{85} + \dfrac{27}{85}i$ 67) $-\dfrac{8}{61} + \dfrac{27}{61}i$ 69) $-9i$

71) $(i^2)^{12}$; $i^2 = -1$; 1   73) 1   75) 1   77) i

79) $-i$ 81) $-i$ 83) -1 85) $32i$ 87) -1

89) $142 - 65i$ 91) $1 + 2i\sqrt{2}$ 93) $8 - 3i\sqrt{5}$

95) $-3 + i\sqrt{2}$ 97) $Z = 10 + 6j$ 99) $Z = 16 + 4j$

Chapter 9: Review Exercises

1) $\dfrac{13}{2}$ 3) -9 5) -1

7) not real   9) 13   11) $|p|$ 13) h

15) a) $\sqrt{23}$ b) $\sqrt{5p + 3}$ c) $\left[-\dfrac{3}{5}, \infty\right)$

17)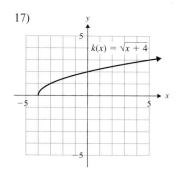

19) The denominator of the fractional exponent becomes the index on the radical. The numerator is the power to which we raise the radical expression. $8^{2/3} = (\sqrt[3]{8})^2$

21) 6   23) $\dfrac{3}{5}$ 25) 8   27) $\dfrac{1}{9}$ 29) $\dfrac{1}{27}$

31) $\dfrac{100}{9}$ 33) 9   35) 64   37) 1

39) $32a^{10/3}b^{10}$ 41) $\dfrac{2c}{3d^{7/4}}$ 43) 3

45) 7   47) k^7 49) w^3 51) $10\sqrt{10}$

53) $\dfrac{3\sqrt{2}}{7}$ 55) k^6 57) $x^4\sqrt{x}$ 59) $3t\sqrt{5}$

61) $6x^3y^6\sqrt{2xy}$ 63) $\sqrt{15}$ 65) $2\sqrt{6}$ 67) $11x^6\sqrt{x}$

69) $10k^8$ 71) $2\sqrt[3]{2}$ 73) $2\sqrt[4]{3}$ 75) z^6

77) $a^6\sqrt[3]{a^2}$ 79) $2z^5\sqrt[3]{2}$ 81) $\dfrac{h^3}{3}$ 83) $\sqrt[3]{21}$

85) $2t^4\sqrt[4]{2t}$ 87) $\sqrt[6]{n^5}$ 89) $11\sqrt{5}$

91) $6\sqrt{5} - 4\sqrt{3}$ 93) $-4p\sqrt{p}$ 95) $-12d^2\sqrt{2d}$

97) $6k\sqrt{5} + 3\sqrt{2k}$ 99) $23\sqrt{2rs} + 8r + 15s$

101) $2 + 2\sqrt{y+1} + y$ 103) $\dfrac{14\sqrt{3}}{3}$ 105) $\dfrac{3\sqrt{2kn}}{n}$

107) $\dfrac{7\sqrt[3]{4}}{2}$ 109) $\dfrac{\sqrt[3]{x^2y^2}}{y}$ 111) $\dfrac{3 - \sqrt{3}}{3}$

113) $1 - 3\sqrt{2}$ 115) $\{1\}$ 117) $\varnothing$ 119) $\{-4\}$

121) $\{2, 6\}$ 123) $V = \dfrac{1}{3}\pi r^2 h$ 125) $7i$ 127) -4

129) $12 - 3i$ 131) $\dfrac{3}{10} - \dfrac{4}{3}i$ 133) $-30 + 35i$

135) $-36 - 21i$ 137) $-24 - 42i$

139) conjugate: $2 + 7i$; product: 53   141) $\dfrac{12}{29} - \dfrac{30}{29}i$

143) $-8i$ 145) $\dfrac{58}{37} - \dfrac{15}{37}i$ 147) -1 149) i

Chapter 9: Test

1) 12   2) -3 3) not real   4) $|w|$ 5) -19

6) a) 1   b) $\sqrt{3a - 5}$ c) $\left[-\dfrac{7}{3}, \infty\right)$

7) Domain: $[2, \infty)$

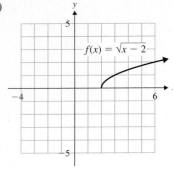

8) 2 9) 81 10) $\dfrac{1}{7}$ 11) $\dfrac{25}{4}$ 12) $m^{5/8}$

13) $\dfrac{5}{2a^{2/3}}$ 14) $\dfrac{y^2}{32x^{3/2}}$ 15) $5\sqrt{3}$ 16) $2\sqrt[3]{6}$

17) $2\sqrt{3}$ 18) y^3 19) p^6 20) $t^4\sqrt{t}$

21) $3m^2n^4\sqrt{7m}$ 22) $c^7\sqrt[3]{c^2}$ 23) $\dfrac{a^4b^2\sqrt[3]{a^2b}}{3}$

24) 6 25) $z^3\sqrt[3]{z}$ 26) $2w^5\sqrt{15w}$ 27) $6\sqrt{7}$

28) $3\sqrt{2} - 4\sqrt{3}$ 29) $-14h^3\sqrt[4]{h}$ 30) $2\sqrt{3} - 5\sqrt{6}$

31) $3\sqrt{2} + 3 - 2\sqrt{10} - 2\sqrt{5}$ 32) 4

33) $2p + 5 + 4\sqrt{2p+1}$ 34) $2t - 2\sqrt{3tu}$ 35) $\dfrac{2\sqrt{5}}{5}$

36) $12 - 4\sqrt{7}$ 37) $\dfrac{\sqrt{6a}}{a}$ 38) $\dfrac{5\sqrt[3]{3}}{3}$ 39) $1 - 2\sqrt{3}$

40) $\{1\}$ 41) $\{-2\}$ 42) $\{13\}$ 43) $\{1, 5\}$

44) a) 3 in. b) $V = \pi r^2 h$ 45) $8i$ 46) $3i\sqrt{5}$

47) $-i$ 48) $-16 + 2i$ 49) $19 + 13i$ 50) $1 + 2i$

Chapter 9: Cumulative Review for Chapters 1–9

1) $\dfrac{10}{3}x - 2y + 8$ 2) 8.723×10^6 3) $\left\{-\dfrac{3}{4}\right\}$

4)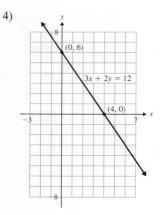

5) $y = \dfrac{5}{4}x - \dfrac{13}{4}$ 6) $(-6, 0)$

7) $15p^4 - 20p^3 - 11p^2 + 8p + 2$ 8) $4n^2 + 2n + 1$

9) $(4w - 3)(w + 2)$ 10) $2(2 + 3t)(2 - 3t)$

11) $\left\{-\dfrac{1}{2}, \dfrac{4}{3}\right\}$ 12) $\{6, 9\}$

13) length = 12 in., width = 7 in.

14) $\dfrac{2a^2 + 2a + 3}{a(a+4)}$ 15) $\dfrac{4n}{21m^3}$ 16) $\{-8, -4\}$

17) $(-\infty, -2] \cup \left[\dfrac{5}{3}, \infty\right)$ 18) $(4, -1, 2)$

19) a) $10\sqrt{5}$ b) $2\sqrt[3]{7}$ c) $p^5q^3\sqrt{q}$ d) $2a^3\sqrt[4]{2a^3}$

20) a) 9 b) 16 c) $\dfrac{1}{3}$ 21) $10\sqrt{3} - 6$

22) a) $\dfrac{\sqrt{10}}{5}$ b) $3\sqrt[3]{4}$ c) $\dfrac{x\sqrt[3]{y}}{y}$ d) $\dfrac{a - 2 - \sqrt{a}}{1-a}$

23) a) $\varnothing$ b) $\{-3\}$ 24) a) $7i$ b) $2i\sqrt{14}$ c) 1

25) a) $2 + 7i$ b) $-48 + 9i$ c) $-\dfrac{11}{25} - \dfrac{2}{25}i$

Chapter 10

Section 10.1

1) Methods may vary; $\{-4, 4\}$ 3) $\{-6, 6\}$

5) $\{-3\sqrt{3}, 3\sqrt{3}\}$ 7) $\left\{-\dfrac{2}{3}, \dfrac{2}{3}\right\}$ 9) $\{-2i, 2i\}$

11) $\{-i\sqrt{3}, i\sqrt{3}\}$ 13) $\{-\sqrt{14}, \sqrt{14}\}$ 15) $\{-3, 3\}$

17) $\{-2i\sqrt{3}, 2i\sqrt{3}\}$ 19) $\{-12, -8\}$ 21) $\{6, 8\}$

23) $\{-4 - 3\sqrt{2}, -4 + 3\sqrt{2}\}$ 25) $\{-3 - 5i, -3 + 5i\}$

27) $\{2 - i\sqrt{14}, 2 + i\sqrt{14}\}$

29) $\left\{\dfrac{-1 - 2\sqrt{5}}{2}, \dfrac{-1 + 2\sqrt{5}}{2}\right\}$

31) $\left\{\dfrac{10 - \sqrt{14}}{3}, \dfrac{10 + \sqrt{14}}{3}\right\}$

33) $\left\{\dfrac{5}{4} - \dfrac{\sqrt{30}}{4}i, \dfrac{5}{4} + \dfrac{\sqrt{30}}{4}i\right\}$

35) $\left\{-\dfrac{11}{6} - \dfrac{7}{6}i, -\dfrac{11}{6} + \dfrac{7}{6}i\right\}$

37) $\left\{8, \dfrac{40}{3}\right\}$ 39) $\left\{-\dfrac{2}{5}, \dfrac{6}{5}\right\}$

41) 5 43) $\sqrt{13}$ 45) 5

47) $\sqrt{61}$ 49) $2\sqrt{5}$

51) A trinomial whose factored form is the square of a binomial; examples will vary.

53) $\dfrac{1}{2}(8) = 4; 4^2 = 16; w^2 + 8w + 16;$ $w^2 + 8w + 16; (w + 4)^2$

55) $a^2 + 12a + 36; (a + 6)^2$ 57) $c^2 - 18c + 81; (c - 9)^2$

59) $t^2 + 5t + \dfrac{25}{4}; \left(t + \dfrac{5}{2}\right)^2$ 61) $b^2 - 9b + \dfrac{81}{4}; \left(b - \dfrac{9}{2}\right)^2$

63) $x^2 + \dfrac{1}{3}x + \dfrac{1}{36}; \left(x + \dfrac{1}{6}\right)^2$

65) Divide both sides of the equation by 2.
67) $\{-4, -2\}$ 69) $\{3, 5\}$
71) $\{1 - \sqrt{10}, 1 + \sqrt{10}\}$ 73) $\{-5 - i, -5 + i\}$
75) $\{4 - i\sqrt{3}, 4 + i\sqrt{3}\}$ 77) $\{-8, 5\}$ 79) $\{3, 4\}$
81) $\left\{\dfrac{1}{2} - \dfrac{\sqrt{13}}{2}, \dfrac{1}{2} + \dfrac{\sqrt{13}}{2}\right\}$
83) $\left\{-\dfrac{5}{2} - \dfrac{\sqrt{3}}{2}i, -\dfrac{5}{2} + \dfrac{\sqrt{3}}{2}i\right\}$
85) $\{1 - i\sqrt{3}, 1 + i\sqrt{3}\}$ 87) $\{-3 - \sqrt{11}, -3 + \sqrt{11}\}$
89) $\{2, 3\}$ 91) $\left\{\dfrac{5}{4} - \dfrac{\sqrt{39}}{4}i, \dfrac{5}{4} + \dfrac{\sqrt{39}}{4}i\right\}$
93) $\left\{\dfrac{3}{4}, 1\right\}$ 95) $\{-1 - \sqrt{21}, -1 + \sqrt{21}\}$
97) $\left\{\dfrac{5}{4} - \dfrac{\sqrt{7}}{4}i, \dfrac{5}{4} + \dfrac{\sqrt{7}}{4}i\right\}$ 99) 6   101) $\sqrt{29}$
103) 6 in.   105) 12 ft   107) $-10, 4$
109) width = 9 ft, length = 17 ft

Section 10.2

1) The fraction bar should also be under $-b$:
$$x = \dfrac{-b \pm \sqrt{b^2 - 4ac}}{2a}$$

3) You cannot divide only the -2 by 2.
$$\dfrac{-2 \pm 6\sqrt{11}}{2} = \dfrac{2(-1 \pm 3\sqrt{11})}{2} = -1 \pm 3\sqrt{11}$$

5) $\{-3, -1\}$ 7) $\left\{-2, \dfrac{5}{3}\right\}$
9) $\left\{\dfrac{5 - \sqrt{17}}{2}, \dfrac{5 + \sqrt{17}}{2}\right\}$ 11) $\{4 - 3i, 4 + 3i\}$
13) $\left\{\dfrac{1}{5} - \dfrac{\sqrt{14}}{5}i, \dfrac{1}{5} + \dfrac{\sqrt{14}}{5}i\right\}$ 15) $\{-7, 0\}$
17) $\left\{\dfrac{3 - \sqrt{3}}{2}, \dfrac{3 + \sqrt{3}}{2}\right\}$ 19) $\left\{\dfrac{7}{2}, 4\right\}$
21) $\{-4 - \sqrt{31}, -4 + \sqrt{31}\}$ 23) $\{-10, 4\}$
25) $\left\{-\dfrac{3}{2}i, \dfrac{3}{2}i\right\}$ 27) $\{-3 - 5i, -3 + 5i\}$
29) $\{-1 - \sqrt{10}, -1 + \sqrt{10}\}$ 31) $\left\{\dfrac{3}{2}\right\}$
33) $\left\{-\dfrac{3}{8} - \dfrac{\sqrt{7}}{8}i, -\dfrac{3}{8} + \dfrac{\sqrt{7}}{8}i\right\}$
35) $\left\{\dfrac{5 - \sqrt{19}}{2}, \dfrac{5 + \sqrt{19}}{2}\right\}$
37) $-3 - \sqrt{11}, -3 + \sqrt{11}$
39) $\dfrac{1 - \sqrt{41}}{4}, \dfrac{1 + \sqrt{41}}{4}$ 41) $-4, \dfrac{1}{5}$

43) There is one rational solution.
45) -39; two nonreal, complex solutions
47) 0; one rational solution
49) 16; two rational solutions
51) 56; two irrational solutions
53) -8 or 8   55) 9   57) 4   59) 2 in., 5 in.
61) a) 2 sec   b) $\dfrac{3 + \sqrt{33}}{4}$ sec or about 2.2 sec

Chapter 10: Putting It All Together

1) $\{-5\sqrt{2}, 5\sqrt{2}\}$ 2) $\{3 - \sqrt{17}, 3 + \sqrt{17}\}$
3) $\{-5, 4\}$ 4) $\left\{\dfrac{3}{4} - \dfrac{\sqrt{39}}{4}i, \dfrac{3}{4} + \dfrac{\sqrt{39}}{4}i\right\}$
5) $\left\{\dfrac{-7 - \sqrt{13}}{2}, \dfrac{-7 + \sqrt{13}}{2}\right\}$ 6) $\left\{-1, \dfrac{4}{3}\right\}$
7) $\left\{-3, \dfrac{1}{4}\right\}$ 8) $\{-3 - i\sqrt{6}, -3 + i\sqrt{6}\}$
9) $\{-7 - i\sqrt{11}, -7 + i\sqrt{11}\}$
10) $\left\{\dfrac{-1 - \sqrt{7}}{2}, \dfrac{-1 + \sqrt{7}}{2}\right\}$
11) $\left\{\dfrac{1}{3} - \dfrac{\sqrt{6}}{3}i, \dfrac{1}{3} + \dfrac{\sqrt{6}}{3}i\right\}$ 12) $\{-4 - 3i, -4 + 3i\}$
13) $\{-9, 3\}$ 14) $\{-5, 5\}$ 15) $\{2 - \sqrt{7}, 2 + \sqrt{7}\}$
16) $\{-9, -6, 0\}$ 17) $\left\{\dfrac{3}{2}, 2\right\}$ 18) $\{0, 1\}$
19) $\{3, 7\}$ 20) $\left\{\dfrac{-1 - \sqrt{21}}{4}, \dfrac{-1 + \sqrt{21}}{4}\right\}$
21) $\left\{\dfrac{1}{3} - \dfrac{\sqrt{2}}{3}i, \dfrac{1}{3} + \dfrac{\sqrt{2}}{3}i\right\}$ 22) $\{3 - 5i, 3 + 5i\}$
23) $\{5 - 4i, 5 + 4i\}$ 24) $\left\{-4, -\dfrac{3}{2}\right\}$ 25) $\{0, 3\}$
26) $\left\{-\dfrac{3}{2} - \dfrac{\sqrt{15}}{2}i, -\dfrac{3}{2} + \dfrac{\sqrt{15}}{2}i\right\}$ 27) $\left\{-\dfrac{3}{2}, 0, \dfrac{3}{2}\right\}$
28) $\{-2, 6\}$ 29) $\left\{-\dfrac{5}{4} - \dfrac{\sqrt{39}}{4}i, -\dfrac{5}{4} + \dfrac{\sqrt{39}}{4}i\right\}$
30) $\left\{-\dfrac{5}{12} - \dfrac{1}{4}i, -\dfrac{5}{12} + \dfrac{1}{4}i\right\}$

Section 10.3

1) $\{-4, 12\}$ 3) $\left\{-2, \dfrac{4}{5}\right\}$
5) $\{4 - \sqrt{6}, 4 + \sqrt{6}\}$ 7) $\left\{\dfrac{1 - \sqrt{7}}{3}, \dfrac{1 + \sqrt{7}}{3}\right\}$
9) $\left\{-\dfrac{9}{5}, 1\right\}$ 11) $\left\{\dfrac{11 - \sqrt{21}}{10}, \dfrac{11 + \sqrt{21}}{10}\right\}$ 13) $\{5\}$
15) $\left\{\dfrac{4}{5}, 2\right\}$ 17) $\{1\}$ 19) $\{1, 16\}$ 21) $\{5\}$

23) $\{0, 2\}$ 25) yes   27) yes   29) no   31) yes

33) no   35) $\{-3, -1, 1, 3\}$ 37) $\{-\sqrt{7}, -2, 2, \sqrt{7}\}$

39) $\{-i\sqrt{6}, -i\sqrt{3}, i\sqrt{3}, i\sqrt{6}\}$ 41) $\{-8, -1\}$

43) $\{-8, 27\}$ 45) $\{49\}$ 47) $\{16\}$

49) $\left\{-\dfrac{2\sqrt{3}}{3}i, -\dfrac{\sqrt{3}}{3}i, \dfrac{\sqrt{3}}{3}i, \dfrac{2\sqrt{3}}{3}i\right\}$

51) $\{-\sqrt{5}, \sqrt{5}, -i\sqrt{3}, i\sqrt{3}\}$

53) $\{-\sqrt{3}+\sqrt{7}, \sqrt{3}+\sqrt{7}, -\sqrt{3}-\sqrt{7}, \sqrt{3}-\sqrt{7}\}$

55) $\left\{-\dfrac{\sqrt{7}+\sqrt{41}}{2}, \dfrac{\sqrt{7}+\sqrt{41}}{2}, -\dfrac{\sqrt{7}-\sqrt{41}}{2}, \dfrac{\sqrt{7}-\sqrt{41}}{2}\right\}$

57) $\left\{-\dfrac{1}{2}, \dfrac{1}{6}\right\}$ 59) $\left\{\dfrac{1}{4}, 3\right\}$ 61) $\{-6, -1\}$

63) $\left\{-\dfrac{1}{2}, 0\right\}$ 65) $\left\{-\dfrac{2}{5}, \dfrac{2}{5}\right\}$ 67) $\left\{-11, -\dfrac{20}{3}\right\}$

69) $\left\{\dfrac{3}{2}\right\}$ 71) $\{2 - \sqrt{2}, 2 + \sqrt{2}\}$

73) Walter: 3 hr; Kevin: 6 hr   75) 15 mph

77) large drain: 3 hr; small drain: 6 hr

79) to Boulder: 60 mph; going home: 50 mph

Section 10.4

1) $r = \dfrac{\pm\sqrt{A\pi}}{\pi}$ 3) $v = \pm\sqrt{ar}$

5) $d = \dfrac{\pm\sqrt{IE}}{E}$ 7) $r = \dfrac{\pm\sqrt{kq_1q_2F}}{F}$

9) $A = \dfrac{1}{4}\pi d^2$ 11) $l = \dfrac{gT_P^2}{4\pi^2}$ 13) $g = \dfrac{4\pi^2 l}{T_P^2}$

15) a) Both are written in the standard form for a quadratic equation, $ax^2 + bx + c = 0$.
    b) Use the quadratic formula.

17) $x = \dfrac{5 \pm \sqrt{25 - 4rs}}{2r}$ 19) $z = \dfrac{-r \pm \sqrt{r^2 + 4pq}}{2p}$

21) $a = \dfrac{h \pm \sqrt{h^2 + 4dk}}{2d}$ 23) $t = \dfrac{-v \pm \sqrt{v^2 + 2gs}}{g}$

25) length = 12 in., width = 9 in.   27) 2 ft

29) base = 8 ft, height = 15 ft   31) 10 in.

33) a) 0.75 sec on the way up, 3 sec on the way down
    b) $\dfrac{15 + \sqrt{241}}{8}$ sec or about 3.8 sec

35) a) 9.5 million   b) 1999   37) $2.40   39) 1.75 ft

Section 10.5

1) The graph of $g(x)$ is the same shape as the graph of $f(x)$, but $g(x)$ is shifted up 6 units.

3) The graph of $h(x)$ is the same shape as the graph of $f(x)$, but $h(x)$ is shifted left 5 units.

5) 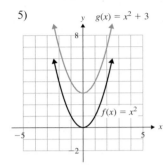   7)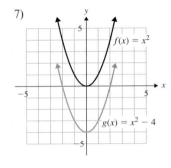

9) 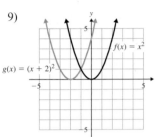   11)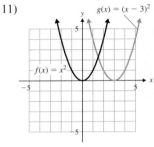

13) 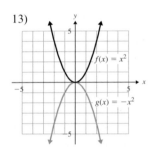   15)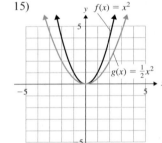

17)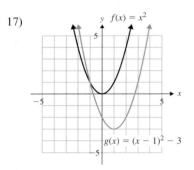

19) a) (h, k) b) $x = h$ c) a is positive.
    d) a is negative.
    e) $a > 1$ or $a < -1$
    f) $0 < a < 1$ or $-1 < a < 0$

21) $V(-1, -4)$; $x = -1$; x-ints: $(-3, 0)$, $(1, 0)$; y-int: $(0, -3)$; domain: $(-\infty, \infty)$; range: $[-4, \infty)$

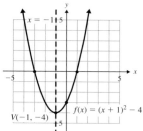

23) $V(2, 3)$; $x = 2$; x-ints: none; y-int: $(0, 7)$; domain: $(-\infty, \infty)$; range: $[3, \infty)$

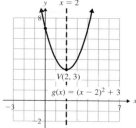

35) $V(-4, 3)$; $x = -4$; x-ints: $(-7, 0)$, $(-1, 0)$; y-int: $\left(0, -\dfrac{7}{3}\right)$; domain: $(-\infty, \infty)$; range: $(-\infty, 3]$

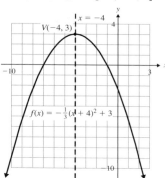

37) $V(-2, 5)$; $x = -2$; x-int: none; y-int: $(0, 17)$; domain: $(-\infty, \infty)$; range: $[5, \infty)$

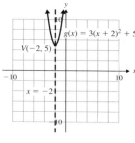

25) $V(4, -2)$; $x = 4$; x-ints: $(4 - \sqrt{2}, 0)$, $(4 + \sqrt{2}, 0)$; y-int: $(0, 14)$; domain: $(-\infty, \infty)$; range: $[-2, \infty)$

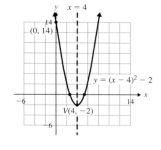

27) $V(-3, 6)$; $x = -3$; x-ints: $(-3 - \sqrt{6}, 0)$, $(-3 + \sqrt{6}, 0)$; y-int: $(0, -3)$; domain: $(-\infty, \infty)$; range: $(-\infty, 6]$

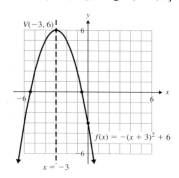

39) a) $h(x)$ b) $f(x)$ c) $g(x)$ d) $k(x)$

41) $g(x) = (x - 8)^2$ 43) $g(x) = x^2 + 3.5$

45) $g(x) = (x + 4)^2 - 7$

47) $f(x) = (x^2 + 8x) + 11$; $\left[\dfrac{1}{2}(8)\right]^2 = (4)^2 = 16$; Add and subtract the number above to the same side of the equation.; $f(x) = (x + 4)^2 - 5$

49) $f(x) = (x - 1)^2 - 4$; x-ints: $(-1, 0)$, $(3, 0)$; y-int: $(0, -3)$; domain: $(-\infty, \infty)$; range: $[-4, \infty)$

51) $y = (x + 3)^2 - 2$; x-ints: $(-3 - \sqrt{2}, 0)$, $(-3 + \sqrt{2}, 0)$; y-int: $(0, 7)$; domain: $(-\infty, \infty)$; range: $[-2, \infty)$

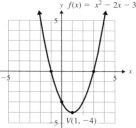

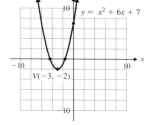

29) $V(-1, -5)$; $x = -1$; x-ints: none; y-int: $(0, -6)$; domain: $(-\infty, \infty)$; range: $(-\infty, -5]$

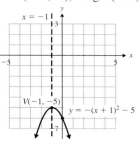

31) $V(1, -8)$; $x = 1$; x-ints: $(-1, 0)$, $(3, 0)$; y-int: $(0, -6)$; domain: $(-\infty, \infty)$; range: $[-8, \infty)$

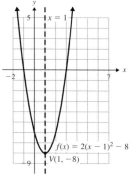

53) $g(x) = (x + 2)^2 - 4$; x-ints: $(-4, 0)$, $(0, 0)$; y-int: $(0, 0)$; domain: $(-\infty, \infty)$; range: $[-4, \infty)$

55) $h(x) = -(x + 2)^2 + 9$; x-ints: $(-5, 0)$, $(1, 0)$; y-int: $(0, 5)$; domain: $(-\infty, \infty)$; range: $(-\infty, 9]$

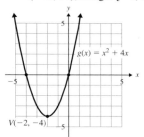

33) $V(0, -1)$; $x = 0$; x-ints: $(-2, 0)$, $(2, 0)$; y-int: $(0, -1)$; domain: $(-\infty, \infty)$; range: $[-1, \infty)$

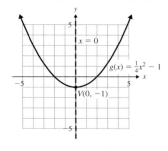

57) $y = -(x-3)^2 - 1$; x-ints: none; y-int: (0, −10); domain: $(-\infty, \infty)$; range: $(-\infty, -1]$

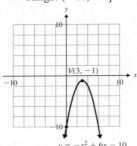

59) $y = 2(x-2)^2 - 6$; x-ints: $(2 - \sqrt{3}, 0)$, $(2 + \sqrt{3}, 0)$; y-int: (0, 2); domain: $(-\infty, \infty)$; range: $[-6, \infty)$

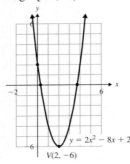

61) $g(x) = -\dfrac{1}{3}(x+3)^2 - 6$; x-ints: none; y-int: (0, −9); domain: $(-\infty, \infty)$; range: $(-\infty, -6]$

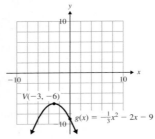

63) $y = \left(x - \dfrac{3}{2}\right)^2 - \dfrac{1}{4}$; x-ints: (1, 0), (2, 0); y-int: (0, 2); domain: $(-\infty, \infty)$; range: $\left[-\dfrac{1}{4}, \infty\right)$

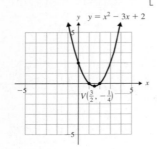

65) $V(-1, -4)$; x-ints: $(-3, 0)$, $(1, 0)$; y-int: $(0, -3)$; domain: $(-\infty, \infty)$; range: $[-4, \infty)$

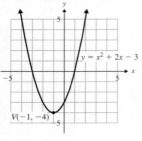

67) $V(-4, 3)$; x-ints: $(-4 - \sqrt{3}, 0)$, $(-4 + \sqrt{3}, 0)$; y-int: $(0, -13)$; domain: $(-\infty, \infty)$; range: $(-\infty, 3]$

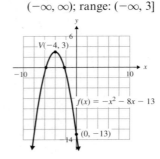

69) $V(1, 2)$; x-int: none; y-int: (0, 4); domain: $(-\infty, \infty)$; range: $[2, \infty)$

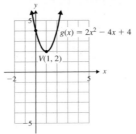

71) $V(1, 4)$; x-ints: $\left(1 + \dfrac{2\sqrt{3}}{3}, 0\right)$, $\left(1 - \dfrac{2\sqrt{3}}{3}, 0\right)$; y-int: (0, 1); domain: $(-\infty, \infty)$; range: $(-\infty, 4]$

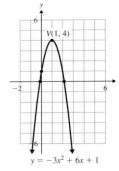

73) $V(4, -3)$; x-ints: $(4 - \sqrt{6}, 0)$, $(4 + \sqrt{6}, 0)$; y-int: (0, 5); domain: $(-\infty, \infty)$; range: $[-3, \infty)$

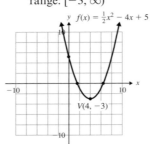

75) $V(-3, -2)$; x-int: none; y-int: $(0, -5)$; domain: $(-\infty, \infty)$; range: $(-\infty, -2]$

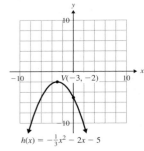

Section 10.6

1) maximum 3) neither 5) minimum

7) If a is positive the graph opens upward, so the y-coordinate of the vertex is the minimum value of the function. If a is negative the graph opens downward, so the y-coordinate of the vertex is the maximum value of the function.

9) a) minimum b) $(-3, 0)$ c) 0 d)

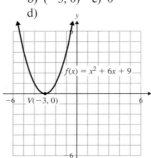

11) a) maximum b) (4, 2) c) 2 d)

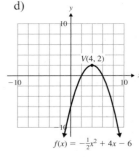

13) a) 10 sec b) 1600 ft c) 20 sec

15) Friday; 218 tickets   17) 1991; 531,000

19) 625 ft²   21) 12 ft × 24 ft   23) 9 and 9

25) 6 and −6   27) (h, k) 29) to the left

31) $V(-4, 1)$; $y = 1$;
x-int: $(-3, 0)$;
y-ints: $(0, -1)$, $(0, 3)$;
domain: $[-4, \infty)$;
range: $(-\infty, \infty)$

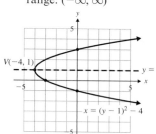

33) $V(2, 0)$; $y = 0$;
x-int: $(2, 0)$; y-int: none;
domain: $[2, \infty)$;
range: $(-\infty, \infty)$

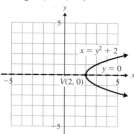

45) $x = \frac{1}{3}(y + 4)^2 - 7$;
domain: $[-7, \infty)$;
range: $(-\infty, \infty)$

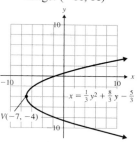

47) $x = -4(y + 1)^2 - 6$;
domain: $(-\infty, -6]$;
range: $(-\infty, \infty)$

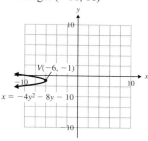

35) $V(5, 4)$; $y = 4$; x-int: $(-11, 0)$; y-ints: $(0, 4 - \sqrt{5})$, $(0, 4 + \sqrt{5})$; domain: $(-\infty, 5]$; range: $(-\infty, \infty)$

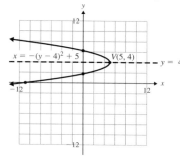

49) $V(-1, 2)$; x-int: $(3, 0)$;
y-ints: $(0, 1)$, $(0, 3)$;
domain: $[-1, \infty)$;
range: $(-\infty, \infty)$

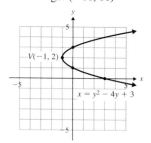

51) $V(4, 2)$; x-int: $(0, 0)$;
y-ints: $(0, 0)$, $(0, 4)$;
domain: $(-\infty, 4]$;
range: $(-\infty, \infty)$

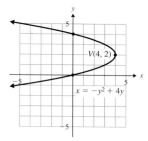

37) $V(-9, 2)$; $y = 2$; x-int: $(-17, 0)$; y-int: none;
domain: $(-\infty, -9]$;
range: $(-\infty, \infty)$

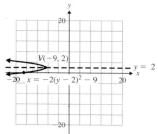

39) $V(0, -2)$; $y = -2$;
x-int: $(1, 0)$;
y-int: $(0, -2)$;
domain: $[0, \infty)$;
range: $(-\infty, \infty)$

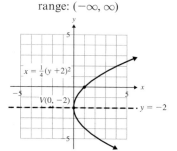

53) $V(-4, 1)$;
x-int: $(-6, 0)$;
y-int: none;
domain: $(-\infty, -4]$;
range: $(-\infty, \infty)$

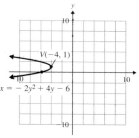

55) $V(-3, 2)$; x-int: $(13, 0)$;
y-ints: $\left(0, 2 - \frac{\sqrt{3}}{2}\right)$,
$\left(0, 2 + \frac{\sqrt{3}}{2}\right)$;
domain: $[-3, \infty)$;
range: $(-\infty, \infty)$

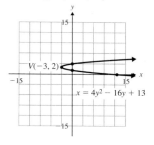

41) $x = (y + 2)^2 - 10$;
domain: $[-10, \infty)$;
range: $(-\infty, \infty)$

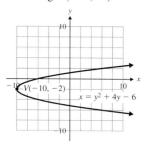

43) $x = -(y + 1)^2 - 4$;
domain: $(-\infty, -4]$;
range: $(-\infty, \infty)$

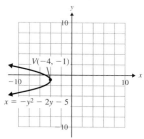

57) $V(6, 1)$; x-int: $\left(\frac{25}{4}, 0\right)$;
y-int: none;
domain: $[6, \infty)$;
range: $(-\infty, \infty)$

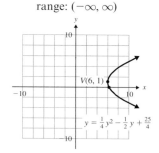

59) domain: $(-\infty, \infty)$;
range: $(-\infty, 6]$

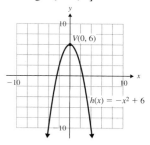

61) domain: $[0, \infty)$; range: $(-\infty, \infty)$

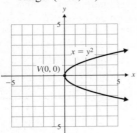

63) domain: $(-\infty, 3]$; range: $(-\infty, \infty)$

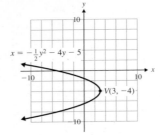

65) domain: $(-\infty, \infty)$; range: $[-4, \infty)$

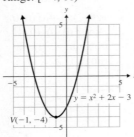

67) domain: $(-\infty, \infty)$; range: $(-\infty, 3]$

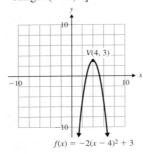

Section 10.7

1) The endpoints are included when the inequality symbol is $\leq$ or $\geq$. The endpoints are not included when the symbol is $<$ or $>$.

3) a) $[-5, 1]$ b) $(-\infty, -5) \cup (1, \infty)$

5) a) $[-1, 3]$ b) $(-\infty, -1) \cup (3, \infty)$

7) $(-\infty, -7] \cup [1, \infty)$

9) $(-9, 4)$

11) $\left[-6, \dfrac{4}{3}\right]$

13) $\left(-\infty, -\dfrac{2}{7}\right) \cup (2, \infty)$

15) $(-\infty, 0) \cup (9, \infty)$

17) $(-8, 8)$

19) $(-\infty, -11] \cup [11, \infty)$

21) $(-\infty, \infty)$ 23) $\varnothing$ 25) $(-\infty, \infty)$

27) $(-\infty, -2] \cup [1, 5]$

29) $(-\infty, -7) \cup \left(-\dfrac{1}{6}, \dfrac{3}{4}\right)$

31) $(-6, \infty)$

33) $(-\infty, -3)$

35) $(-\infty, 3) \cup (4, \infty)$

37) $\left(-\dfrac{1}{3}, 9\right]$

39) $(-3, 0]$

41) $(-\infty, -6) \cup \left(-\dfrac{11}{3}, \infty\right)$

43) $(-7, -4]$

45) $\left[\dfrac{18}{5}, 6\right)$

47) $(-\infty, -2) \cup \left(-\dfrac{8}{7}, \infty\right)$

49) $(5, \infty)$

51) $[-1, \infty)$

53) $(-\infty, 4)$

55) a) between 4000 and 12,000 units
b) when it produces less than 4000 units or more than 12,000 units

57) 10,000 or more

Chapter 10: Review Exercises

1) $\{-12, 12\}$ 3) $\{-2i, 2i\}$ 5) $\{-4, 10\}$

7) $\left\{-\dfrac{\sqrt{10}}{3}, \dfrac{\sqrt{10}}{3}\right\}$ 9) $2\sqrt{5}$

11) $r^2 + 10r + 25; (r + 5)^2$

13) $c^2 - 5c + \dfrac{25}{4}; \left(c - \dfrac{5}{2}\right)^2$

15) $a^2 + \dfrac{2}{3}a + \dfrac{1}{9}; \left(a + \dfrac{1}{3}\right)^2$ 17) $\{-2, 8\}$

19) $\{-5 - \sqrt{31}, -5 + \sqrt{31}\}$

21) $\left\{-\dfrac{3}{2} - \dfrac{\sqrt{5}}{2}, -\dfrac{3}{2} + \dfrac{\sqrt{5}}{2}\right\}$

23) $\left\{\dfrac{7}{6} - \dfrac{\sqrt{95}}{6}i, \dfrac{7}{6} + \dfrac{\sqrt{95}}{6}i\right\}$ 25) $\{-6, 2\}$

27) $\left\{\dfrac{5 - \sqrt{15}}{2}, \dfrac{5 + \sqrt{15}}{2}\right\}$ 29) $\{1 - i\sqrt{3}, 1 + i\sqrt{3}\}$

31) $\left\{-\dfrac{2}{3}, \dfrac{1}{2}\right\}$ 33) 64; two rational solutions

35) -12 or 12 37) $\left\{1, \dfrac{4}{3}\right\}$

39) $\{-8 - 2\sqrt{3}, -8 + 2\sqrt{3}\}$

41) $\left\{-\dfrac{3}{2} - \dfrac{1}{2}i, -\dfrac{3}{2} + \dfrac{1}{2}i\right\}$

43) $\{9 - 3\sqrt{5}, 9 + 3\sqrt{5}\}$ 45) $\{-1, 0, 1\}$

47) $-2, 3$ 49) $\left\{\dfrac{3 - \sqrt{21}}{3}, \dfrac{3 + \sqrt{21}}{3}\right\}$

51) $\{3, 4\}$ 53) $\{-4, -1, 1, 4\}$ 55) $\{-27, 1\}$

57) $\left\{-\dfrac{\sqrt{7 + \sqrt{33}}}{2}, \dfrac{\sqrt{7 - \sqrt{33}}}{2}, -\dfrac{\sqrt{7 - \sqrt{33}}}{2}, \dfrac{\sqrt{7 + \sqrt{33}}}{2}\right\}$

59) $\left\{2, \dfrac{11}{2}\right\}$ 61) $v = \dfrac{\pm\sqrt{Frm}}{m}$ 63) $A = \pi r^2$

65) $n = \dfrac{l \pm \sqrt{l^2 + 4km}}{2k}$ 67) 3 in. 69) \$8.00

71) a) (h, k) b) $x = h$
c) If a is positive, the parabola opens upward. If a is negative, the parabola opens downward.

73) $V(0, -4)$; $x = 0$; x-ints: $(-2, 0), (2, 0)$; y-int: $(0, -4)$; domain: $(-\infty, \infty)$; range: $[-4, \infty)$

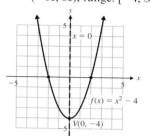

75) $V(-2, -1)$; $x = -2$; x-ints: $(-3, 0), (-1, 0)$; y-int: $(0, 3)$; domain: $(-\infty, \infty)$; range: $[-1, \infty)$

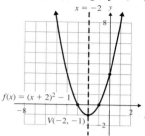

77) $V(4, 2)$; $x = 4$; x-ints: none; y-int: $(0, 18)$; domain: $(-\infty, \infty)$; range: $[2, \infty)$

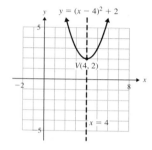

79) $g(x) = (x - 6)^2$

81) $f(x) = (x - 1)^2 + 2$; x-ints: none; y-int: $(0, 3)$; domain: $(-\infty, \infty)$; range: $[2, \infty)$

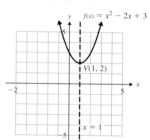

83) $y = \dfrac{1}{2}(x - 4)^2 + 1$; x-ints: none; y-int: $(0, 9)$; domain: $(-\infty, \infty)$; range: $[1, \infty)$

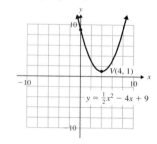

85) $V(-3, -1)$; x-ints: none; y-int: $(0, -10)$; domain: $(-\infty, \infty)$; range: $(-\infty, -1]$

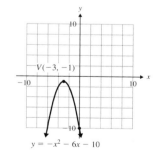

87) a) 1 sec b) 256 ft c) 5 sec

89) $V(11, 3)$; $y = 3$; x-int: $(2, 0)$;
    y-ints: $(0, 3 - \sqrt{11})$, $(0, 3 + \sqrt{11})$;
    domain: $(-\infty, 11]$; range: $(-\infty, \infty)$

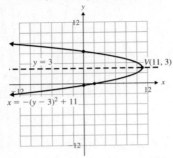

91) $x = (y + 4)^2 - 9$; x-int: $(7, 0)$; y-ints: $(0, -1)$, $(0, -7)$;
    domain: $[-9, \infty)$; range: $(-\infty, \infty)$

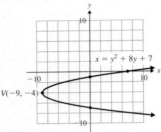

93) $V(2, -3)$; x-int: $\left(-\dfrac{5}{2}, 0\right)$; y-ints: $(0, -5)$, $(0, -1)$;
    domain: $(-\infty, 2]$; range: $(-\infty, \infty)$

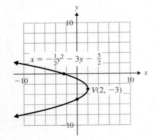

95) $(-3, 1)$

97) $\left(-\infty, -\dfrac{5}{8}\right] \cup \left[\dfrac{5}{8}, \infty\right)$

99) $\left(-\infty, -\dfrac{2}{5}\right) \cup \left(-\dfrac{1}{3}, 4\right)$

101) $(-\infty, -7) \cup \left(\dfrac{3}{2}, \infty\right)$

103) $(-\infty, 2) \cup [3, \infty)$

105) $(7, \infty)$

Chapter 10: Test

1) $\{-2 - \sqrt{11}, -2 + \sqrt{11}\}$ 2) $\{4 - i, 4 + i\}$

3) $\{-5 - i\sqrt{6}, -5 + i\sqrt{6}\}$ 4) $\left\{-2, \dfrac{4}{3}\right\}$

5) $\left\{-\dfrac{7}{4}, -1\right\}$ 6) $\left\{\dfrac{7 - \sqrt{17}}{4}, \dfrac{7 + \sqrt{17}}{4}\right\}$

7) $\{-2\sqrt{2}, 2\sqrt{2}, -3i, 3i\}$ 8) $\left\{-\dfrac{3}{2}, -1\right\}$

9) 56; two irrational solutions   10) $\sqrt{19}$

11) $0, \dfrac{2}{5}$ 12) $2\sqrt{26}$ 13) width = 13 in., length = 19 in.

14) $V = \dfrac{1}{3}\pi r^2 h$ 15) $t = \dfrac{s \pm \sqrt{s^2 + 24r}}{2r}$

16) 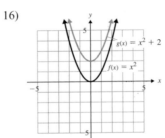   17) $g(x) = (x + 3)^2$

18) domain: $(-\infty, \infty)$;   19) domain: $[-3, \infty)$;
    range: $(-\infty, 4]$ range: $(-\infty, \infty)$

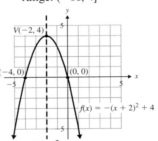 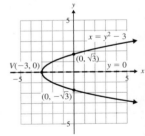

20) domain: $[2, \infty)$;   21) domain: $(-\infty, \infty)$;
    range: $(-\infty, \infty)$ range: $[-1, \infty)$

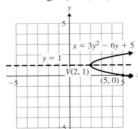 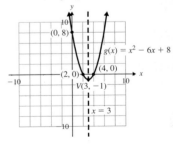

22) a) 209 ft b) after 4 sec
    c) after $\dfrac{3 + \sqrt{209}}{4}$ sec or about 4.4 sec

23) $(-\infty, -9] \cup [5, \infty)$

24) $(-\infty, -3) \cup [5, \infty)$

25) 8000 or more

Chapter 10: Cumulative Review for Chapters 1–10

1) $\dfrac{7}{10}$ 2) $\dfrac{45x^6}{y^4}$ 3) 90 4) $m = \dfrac{y-b}{x}$

5) a) $\{0, 3, 4\}$ b) $\{-1, 0, 1, 2\}$ c) no

6) a) 2 b) $[-3, \infty)$
   c)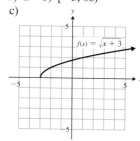

7) chips: $0.80, soda: $0.75

8) $3x^2y^2 - 5x^2y - 3xy^2 - 9xy + 8$ 9) $3r^2 - 30r + 75$

10) $2p(2p-1)(p+4)$ 11) $(a+5)(a^2 - 5a + 25)$

12) $\dfrac{z^2 - 5z + 12}{z(z+4)}$ 13) $\dfrac{c(c+3)}{1-4c}$ 14) $(0, 3, -1)$

15) $5\sqrt{3}$ 16) $2\sqrt[3]{5}$ 17) $3x^3y^2\sqrt{7x}$ 18) 16

19) $10 - 5\sqrt{3}$ 20) $34 - 77i$ 21) $\left\{-\dfrac{2}{3}, \dfrac{7}{3}\right\}$

22) $\{-9, 3\}$ 23) $V = \pi r^2 h$

24)   25) $\left[-\dfrac{12}{5}, \dfrac{12}{5}\right]$

Chapter 11
Section 11.1

1) $(f \circ g)(x) = f(g(x))$, so substitute the function $g(x)$ into the function $f(x)$ and simplify.

3) a) -1 b) -2 c) $6x - 26$ d) -2

5) a) 7 b) 44 c) $x^2 + 6x + 4$ d) 44

7) a) $5x + 31$ b) $5x + 3$ c) 46

9) a) $-6x + 26$ b) $-6x + 11$ c) 32

11) a) $x^2 - 6x + 7$ b) $x^2 - 14x + 51$ c) 11

13) a) $-x^2 - 13x - 48$ b) $-x^2 + 3x$ c) 0

15) a) $\sqrt{x^2 + 4}$ b) $x + 4$ c) $\sqrt{13}$

17) a) $\dfrac{1}{t^2 + 8}$ b) $\dfrac{1}{(t+8)^2}$ c) $\dfrac{1}{9}$

19) $f(x) = \sqrt{x}, g(x) = x^2 + 13$; answers may vary.

21) $f(x) = \dfrac{1}{x}, g(x) = 6x + 5$; answers may vary.

23) a) $r(5) = 20$. The radius of the spill 5 min after the ship started leaking was 20 ft.
    b) $A(20) = 400\pi$. The area of the oil slick is 400π ft^2 when its radius is 20 ft.
    c) $A(r(t)) = 16\pi t^2$. This is the area of the oil slick in terms of t, the number of minutes after the leak began.
    d) $A(r(5)) = 400\pi$. The area of the oil slick 5 min after the ship began leaking was 400π ft^2.

25) no   27) yes; $h^{-1} = \{(-16, -5), (-4, -1), (8, 3)\}$

29) yes   31) No; only one-to-one functions have inverses.

33) False; it is read "f inverse of x."   35) true

37) False; they are symmetric with respect to $y = x$.

39) a) yes b) 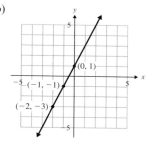   41) no

43) a) yes b)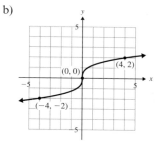

45) Replace $f(x)$ with y.; $x = 2y - 10$; Add 10.; $\dfrac{1}{2}x + 5 = y$; Replace y with $f^{-1}(x)$.

47) $g^{-1}(x) = x + 6$

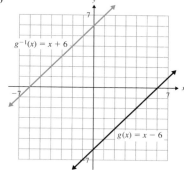

49) $f^{-1}(x) = -\dfrac{1}{2}x + \dfrac{5}{2}$

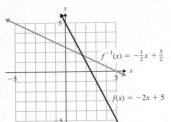

51) $g^{-1}(x) = 2x$

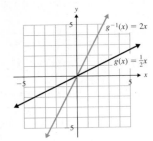

53) $f^{-1}(x) = \sqrt[3]{x}$

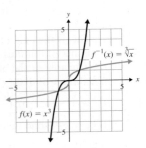

55) $f^{-1}(x) = \dfrac{1}{2}x + 3$ 57) $h^{-1}(x) = -\dfrac{2}{3}x + \dfrac{8}{3}$

59) $h^{-1}(x) = x^3 + 7$ 61) $f^{-1}(x) = x^2,\ x \geq 0$

Section 11.2

1) Choose values for the variable that will give positive numbers, negative numbers, and zero in the exponent.

3)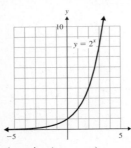
domain: $(-\infty, \infty)$;
range: $(0, \infty)$

5)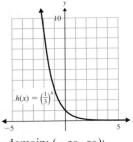
domain: $(-\infty, \infty)$;
range: $(0, \infty)$

7) $(-\infty, \infty)$

9)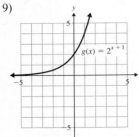
domain: $(-\infty, \infty)$;
range: $(0, \infty)$

11)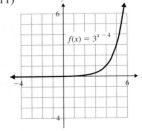
domain: $(-\infty, \infty)$;
range: $(0, \infty)$

13)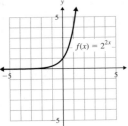
domain: $(-\infty, \infty)$;
range: $(0, \infty)$

15)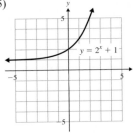
domain: $(-\infty, \infty)$;
range: $(1, \infty)$

17)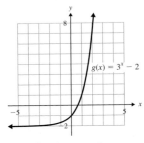
domain: $(-\infty, \infty)$;
range: $(-2, \infty)$

19)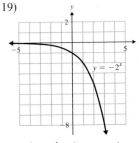
domain: $(-\infty, \infty)$;
range: $(-\infty, 0)$

21) $g(x) = 2^x$ would grow faster because for values of $x > 2$, $2^x > 2x$.

23) Shift the graph of $f(x)$ down 2 units.

25) 2.7183   27) B   29) D

31)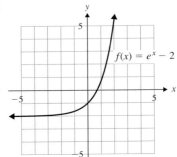
domain: $(-\infty, \infty)$; range: $(-2, \infty)$

33)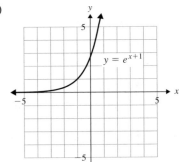
domain: $(-\infty, \infty)$; range: $(0, \infty)$

35)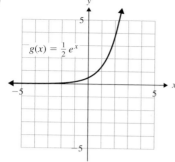
domain: $(-\infty, \infty)$; range: $(0, \infty)$

37)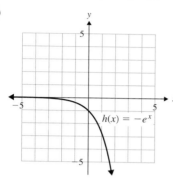
domain: $(-\infty, \infty)$; range: $(-\infty, 0)$

39) They are symmetric with respect to the x-axis.

41) $6^{3n} = (6^2)^{n-4}$; Power rule for exponents; Distribute.; $3n = 2n - 8$; $n = -8$; $\{-8\}$

43) $\{2\}$ 45) $\left\{\dfrac{3}{4}\right\}$ 47) $\left\{\dfrac{8}{3}\right\}$ 49) $\{5\}$ 51) $\{-2\}$

53) $\left\{-\dfrac{3}{2}\right\}$ 55) $\{-1\}$ 57) a) $32,700 b) $17,507.17

59) a) $185,200 b) $227,772.64

61) $90,036.92   63) 40.8 mg

Section 11.3

1) a must be a positive real number that is not equal to 1.

3) 10   5) $7^2 = 49$ 7) $9^{-2} = \dfrac{1}{81}$ 9) $25^{1/2} = 5$

11) $9^0 = 1$ 13) $\log_9 81 = 2$ 15) $\log_2 \dfrac{1}{32} = -5$

17) $\log_{10} 10 = 1$ 19) $\log_{169} 13 = \dfrac{1}{2}$ 21) $\log_{64} 4 = \dfrac{1}{3}$

23) Write the equation in exponential form, then solve for the variable.

25) Rewrite in exponential form.; $64 = x$; $\{64\}$

27) $\{125\}$ 29) $\{100\}$ 31) $\{7\}$ 33) $\left\{\dfrac{1}{36}\right\}$

35) $\{14\}$ 37) $\left\{\dfrac{1}{6}\right\}$ 39) $\{4\}$ 41) $\{2\}$ 43) 5

45) 2   47) $\dfrac{1}{2}$ 49) -1 51) 1   53) -2

55) Replace $f(x)$ with y, write $y = \log_a x$ in exponential form, make a table of values, then plot the points and draw the curve.

57) 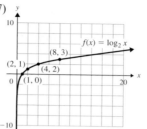   59)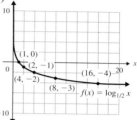

61) $f^{-1}(x) = \log_3 x$ 63) $f^{-1}(x) = 2^x$

65) a) 1868   b) 2004   c) 2058

67) a) 14,000   b) 28,000
c) It is 1000 more than what was predicted by the formula.

Section 11.4

1) true   3) false   5) false   7) true

9) Product rule; $2 + \log_5 y$ 11) $\log_8 3 + \log_8 10$

13) $\log_7 5 + \log_7 d$ 15) $\log_5 20 - \log_5 17$ 17) $4 \log_8 10$

19) $8 \log p$ 21) $\dfrac{1}{2}\log_3 7$ 23) $4 + \log_2 p$

25) $3 - \log_2 k$ 27) 6   29) $3 + \log b$ 31) 9

33) $\dfrac{1}{2}$ 35) $\dfrac{2}{3}$ 37) $4 \log_6 w + 3 \log_6 z$

39) $2 \log_7 a - 5 \log_7 b$ 41) $\dfrac{1}{5}\log 11 - 2 \log y$

43) $2 + \dfrac{1}{2}\log_2 n - 3 \log_2 m$

45) $3 \log_4 x - \log_4 y - 2 \log_4 z$ 47) $\dfrac{1}{2} + \dfrac{1}{2}\log_5 c$

49) $\log k + \log(k - 6)$ 51) Power rule; $\log_6 x^2 y$

53) $\log_a mn$ 55) $\log_7 \dfrac{d}{3}$ 57) $\log_3 f^4 g$ 59) $\log_8 \dfrac{tu^2}{v^3}$

61) $\log \dfrac{r^2 + 3}{(r^2 - 3)^2}$ 63) $\log_n 8\sqrt{k}$ 65) $\log_d \dfrac{\sqrt[3]{5}}{z^2}$

67) $\log_6 \dfrac{y}{3z^3}$ 69) $\log_3 \dfrac{t^4}{36u^2}$ 71) $\log_b \dfrac{\sqrt{c+4}}{(c+3)^2}$

73) $-\log(a^2 - b^2)$ 75) 1.6532   77) 1.9084

79) -0.2552 81) 0.4771   83) -0.6990

85) -1.9084 87) 1.6990

89) No. $\log_a xy$ is defined only if x and y are positive.

Section 11.5

1) e 3) 2   5) -3 7) -1 9) 9   11) $\dfrac{1}{4}$

13) 10   15) $\dfrac{1}{2}$ 17) -5 19) 0   21) 1.2041

23) -0.3010 25) 1.0986 27) 0.2700 29) $\{1000\}$

31) $\left\{\dfrac{1}{10}\right\}$ 33) $\{25\}$ 35) $\{2\}$ 37) $\{10^{1.5}\}$; $\{31.6228\}$

39) $\{10^{0.8}\}$; $\{6.3096\}$ 41) $\{e^{1.6}\}$; $\{4.9530\}$

43) $\left\{\dfrac{1}{e^2}\right\}$; $\{0.1353\}$ 45) $\left\{\dfrac{e^{2.1}}{3}\right\}$; $\{2.7221\}$

47) $\{2 \cdot 10^{0.47}\}$; $\{5.9024\}$ 49) $\left\{\dfrac{3 + 10^{3.8}}{5}\right\}$; $\{1262.5147\}$

51) $\left\{\dfrac{e^{1.85} - 19}{10}\right\}$; $\{-1.2640\}$ 53) $\{3\}$

55)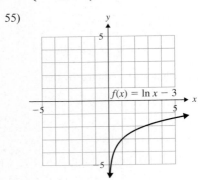

domain: $(0, \infty)$; range: $(-\infty, \infty)$

57)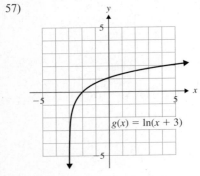

domain: $(-3, \infty)$; range: $(-\infty, \infty)$

59)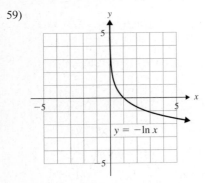

domain: $(0, \infty)$; range: $(-\infty, \infty)$

61)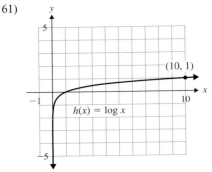

domain: $(0, \infty)$; range: $(-\infty, \infty)$

63) Shift the graph of $f(x)$ left 5 units.

65) 3.7004 67) -2.5237 69) 60 dB   71) $\$3484.42$

73) $\$3485.50$ 75) a) 5000 b) 8191 77) $32,570$

79) 2.7; acidic   81) 11.2; basic   83) Answers may vary.

Section 11.6

1) $\{2\}$ 3) $\left\{\dfrac{\ln 15}{\ln 7}\right\}$; $\{1.3917\}$ 5) $\left\{\dfrac{\ln 3}{\ln 8}\right\}$; $\{0.5283\}$

7) $\left\{\dfrac{2}{5}\right\}$ 9) $\left\{\dfrac{\ln 2.7}{6 \ln 4}\right\}$; $\{0.1194\}$

11) $\left\{\dfrac{\ln 5 - \ln 2}{4 \ln 2}\right\}$; $\{0.3305\}$

13) $\left\{\dfrac{\ln 8 + 2 \ln 5}{3 \ln 5}\right\}$; $\{1.0973\}$

15) $\left\{\dfrac{10}{7}\right\}$ 17) $\left\{\dfrac{2 \ln 9}{5 \ln 9 - 3 \ln 4}\right\}$; $\{0.6437\}$

19) $\{\ln 12.5\}$; $\{2.5257\}$ 21) $\left\{-\dfrac{\ln 9}{4}\right\}$; $\{-0.5493\}$

23) $\left\{\dfrac{\ln 2}{0.01}\right\}$; $\{69.3147\}$ 25) $\left\{\dfrac{\ln 3}{0.006}\right\}$; $\{183.1021\}$

27) $\left\{-\dfrac{\ln 5}{0.4}\right\}$; $\{-4.0236\}$ 29) $\{2\}$ 31) $\left\{\dfrac{10}{3}\right\}$

33) $\{5\}$ 35) $\{2, 10\}$ 37) $\varnothing$ 39) $\left\{\dfrac{11}{2}\right\}$ 41) $\{2\}$

43) $\{9\}$ 45) $\{2\}$ 47) $\{4\}$ 49) $\left\{\dfrac{2}{3}\right\}$

51) a) 3.72 yr b) 11.55 yr   53) 1.44 yr

55) $\$2246.64$ 57) 7.2% 59) a) 6 hr b) 18.5 hr

61) $28,009$ 63) a) 2032 b) 2023

65) a) 11.78 g b) 3351 yr c) 5728 yr

67) a) 0.4 units b) 0.22 units   69) $\{16\}$ 71) $\{-2, 2\}$

73) $\{-\ln 13, \ln 13\}$; $\{-2.5649, 2.5649\}$ 75) $\{0\}$

77) $\left\{\dfrac{\ln 7}{\ln 5}\right\}$; $\{1.2091\}$ 79) $\{1, 1000\}$

Chapter 11: Review Exercises

1) a) $2x + 3$ b) $2x - 3$ c) 7

3) a) $(N \circ G)(h) = 9.6h$. This is Antoine's net pay in terms of how many hours he has worked.
   b) $(N \circ G)(30) = 288$. When Antoine works 30 hr, his net pay is $288.
   c) $384

5) yes; $\{(-4, -7), (1, -2), (5, 1), (11, 6)\}$

7) yes

9) $f^{-1}(x) = x - 4$

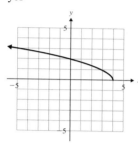

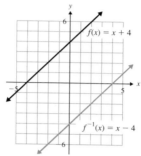

11) $h^{-1}(x) = 3x + 3$

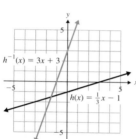

13) False. Only one-to-one functions have inverses.

15)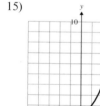
domain: $(-\infty, \infty)$; range: $(0, \infty)$

17) domain: $(-\infty, \infty)$; range: $(-4, \infty)$

19)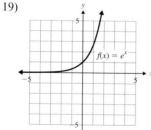
domain: $(-\infty, \infty)$; range: $(0, \infty)$

21) $\{6\}$ 23) $\left\{-\dfrac{5}{2}\right\}$ 25) $\left\{\dfrac{2}{3}\right\}$

27) $(0, \infty)$ 29) $5^3 = 125$ 31) $10^2 = 100$

33) $\log_3 81 = 4$ 35) $\log 1000 = 3$ 37) $\{8\}$

39) $\left\{\dfrac{4}{5}\right\}$ 41) 2 43) 3 45) -4

47)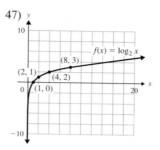

49) $f^{-1}(x) = \log_5 x$ 51) a) 10,000 b) 20,000

53) true 55) $2 - \log_7 t$ 57) -2

59) $\log_4 m + \dfrac{1}{2}\log_4 n$ 61) $2\log_4 a - \log_4 b - 4\log_4 c$

63) $3\log_6 r - \log_6(r^2 - 5)$ 65) $\log_2 a^9 b^3$ 67) $\log_3 \dfrac{5m^4}{n^2}$

69) 1.6902 71) e 73) 2 75) -2 77) 0

79) 0.9031 81) 0.5596 83) $\{100\}$ 85) $\left\{\dfrac{1}{5}\right\}$

87) $\{10^{2.1}\}$; $\{125.8925\}$ 89) $\{e^2\}$; $\{7.3891\}$

91) $\left\{\dfrac{10^{1.75}}{4}\right\}$; $\{14.0585\}$

93) domain: $(3, \infty)$; range: $(-\infty, \infty)$

95) 2.1240 97) -5.2479 99) 110 dB

101) $3367.14 103) a) 6000 b) 11,118 105) $\{4\}$

107) $\left\{\dfrac{14}{5}\right\}$ 109) $\left\{\dfrac{\ln 19}{0.03}\right\}$; $\{98.1480\}$ 111) $\{4\}$

113) $\{16\}$ 115) $6770.57 117) a) 17,777 b) 2011

Chapter 11: Test

1) a) $2x^2 + 10x + 1$ b) $4x^2 + 38x + 81$ c) 3

2) $f(x) = x^3$, $g(x) = 9x - 7$; answers may vary.

3) no 4) yes; $g^{-1} = \left\{(4, 2), (6, 6), \left(\dfrac{15}{2}, 9\right), (10, 14)\right\}$

5) yes

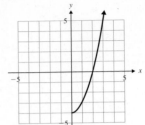

6) $f^{-1}(x) = -\dfrac{1}{3}x + 4$

7)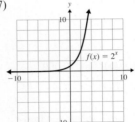

8)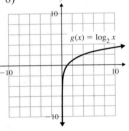

9) a) $(0, \infty)$ b) $(-\infty, \infty)$ 10) They are inverses.

11) $\log_3 \dfrac{1}{9} = -2$ 12) $\left\{\dfrac{1}{2}\right\}$ 13) $\{-2\}$ 14) $\{125\}$

15) $\{29\}$ 16) $\{4\}$ 17) a) 4   b) $\dfrac{1}{2}$

18) $\{1\}$ 19) $\log_8 5 + \log_8 n$

20) $2 + 4\log_3 a - 5\log_3 b - \log_3 c$

21) $\log \dfrac{x^2}{(x+1)^3}$ 22) $\{10^{0.08}\}$; $\{6.3096\}$

23) $\left\{\dfrac{\ln 5}{0.3}\right\}$; $\{5.3648\}$ 24) $\{e^{-0.25}\}$; $\{0.7788\}$

25) $\left\{\dfrac{\ln 9 - 3\ln 4}{4\ln 4}\right\}$; $\{-0.3538\}$

26)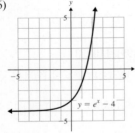

27)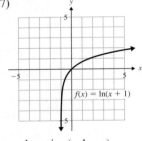

domain: $(-\infty, \infty)$; range: $(-4, \infty)$

domain: $(-1, \infty)$; range: $(-\infty, \infty)$

28) 1.7604   29) $8686.41

30) a) 86.2 g   b) 325.1 days   c) 140 days

Chapter 11: Cumulative Review for Chapters 1–11

1) $-15a^6$ 2) $\dfrac{d^9}{8c^{30}}$ 3) 9.231×10^{-5} 4) $48.00

5) 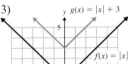; $\left(-\dfrac{3}{2}, \infty\right)$

6) $\left(\dfrac{1}{3}, -2\right)$ 7) $3c^2 + 4c - 1$ 8) $(4w + 9)(w - 2)$

9) $\{-8, -6\}$ 10) $\{2\}$

11)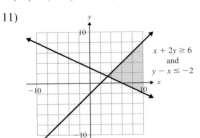

12) $3t^4\sqrt{5t}$ 13) $6a$ 14) $\varnothing$

15) $\{4 - 2\sqrt{3}, 4 + 2\sqrt{3}\}$

16) $\left\{-\dfrac{5}{2} - \dfrac{\sqrt{17}}{2}, -\dfrac{5}{2} + \dfrac{\sqrt{17}}{2}\right\}$

17) $\{5 + 4i, 5 - 4i\}$

18) $\left\{-2, -\dfrac{1}{2}, \dfrac{1}{2}, 2\right\}$ 19) $\dfrac{32}{243}$

20) a) 9   b) $x^2 - 12x + 29$ c) -4

21)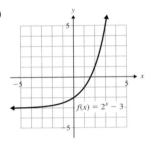

domain: $(-\infty, \infty)$; range: $(-3, \infty)$

22) $\{3\}$ 23) $\log\dfrac{ab^2}{c^5}$ 24) $\{4\}$

25) $\left\{-\dfrac{\ln 6}{0.04}\right\}$; $\{-44.7940\}$

Chapter 12

Section 12.1

1)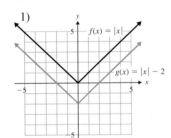

3)

5)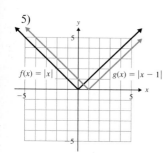
7)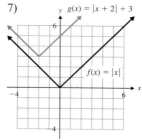
37)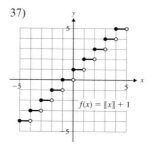
39)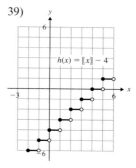

9)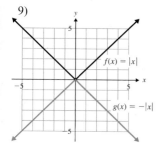
11)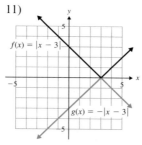
41)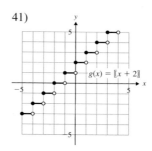
43)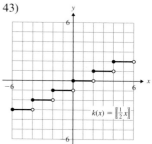

13) a) $g(x)$ b) $k(x)$ c) $h(x)$ d) $f(x)$

15) $g(x) = |x + 2| - 1$

17)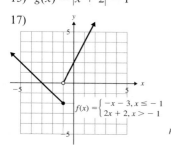
19)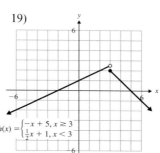

45)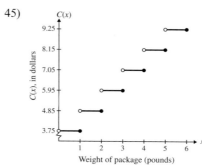

21)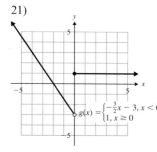
23)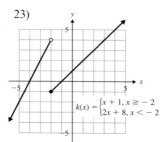

47)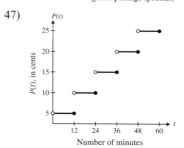

25)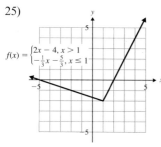

49) a) domain: $(-\infty, \infty)$; b) domain: $[0, \infty)$;
   range: $(-\infty, \infty)$ range: $[0, \infty)$

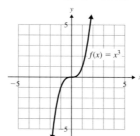
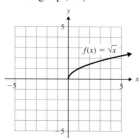

27) 3   29) 7   31) 8   33) -7 35) -9

c) domain: $(-\infty, \infty)$;
range: $(-\infty, \infty)$

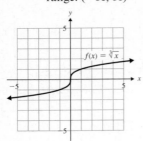

51) domain: $[1, \infty)$;
range: $[0, \infty)$

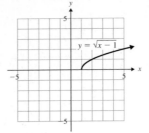

3) center: $(-2, 4)$; $r = 3$

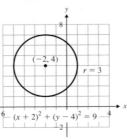

5) center: $(5, 3)$; $r = 1$

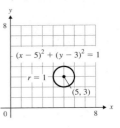

53) domain: $(-\infty, \infty)$;
range: $(-\infty, \infty)$

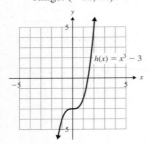

55) domain: $(-\infty, \infty)$;
range: $(-\infty, \infty)$

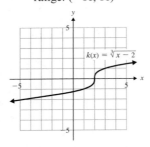

7) center: $(-3, 0)$; $r = 2$

9) center: $(6, -3)$; $r = 4$

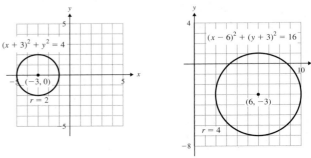

57) domain: $(-\infty, \infty)$;
range: $[3, \infty)$

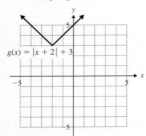

59) domain: $(-\infty, \infty)$;
range: $[0, \infty)$

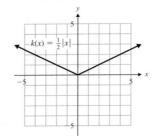

11) center: $(0, 0)$; $r = 6$

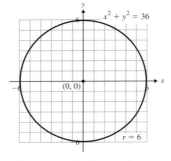

13) center: $(0, 0)$; $r = 3$

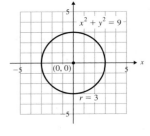

61) domain: $[-4, \infty)$;
range: $[-2, \infty)$

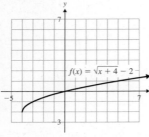

63) domain: $(-\infty, \infty)$;
range: $(-\infty, \infty)$

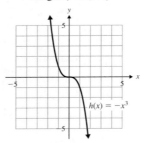

15) center: $(0, 1)$; $r = 5$

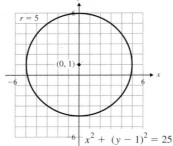

65) $g(x) = \sqrt{x + 5}$ 67) $g(x) = (x + 2)^3 - 1$

69) $g(x) = -\sqrt[3]{x}$

Section 12.2

1) No; there are values in the domain that give more than one value in the range. The graph fails the vertical line test.

17) $(x - 4)^2 + (y - 1)^2 = 25$

19) $(x + 3)^2 + (y - 2)^2 = 1$

21) $(x + 1)^2 + (y + 5)^2 = 3$ 23) $x^2 + y^2 = 10$

25) $(x - 6)^2 + y^2 = 16$ 27) $x^2 + (y + 4)^2 = 8$

29) Group x- and y-terms separately.;
$(x^2 - 8x + 16) + (y^2 + 2y + 1) = -8 + 16 + 1$;
$(x - 4)^2 + (y + 1)^2 = 9$

31) $(x+1)^2 + (y+5)^2 = 9$; center: $(-1, -5)$; $r = 3$

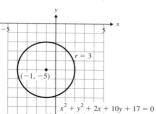

33) $(x+4)^2 + (y-1)^2 = 25$; center: $(-4, 1)$; $r = 5$

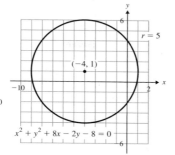

35) $(x-5)^2 + (y-7)^2 = 1$; center: $(5, 7)$; $r = 1$

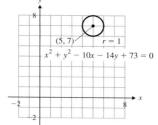

37) $x^2 + (y+3)^2 = 4$; center: $(0, -3)$; $r = 2$

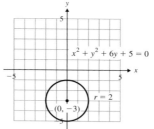

39) $(x-2)^2 + y^2 = 5$; center: $(2, 0)$; $r = \sqrt{5}$

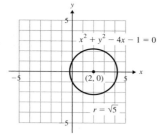

41) $(x-4)^2 + (y+4)^2 = 36$; center: $(4, -4)$; $r = 6$

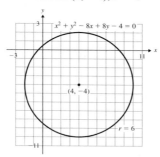

43) $\left(x - \frac{3}{2}\right)^2 + \left(y - \frac{1}{2}\right)^2 = 4$; center: $\left(\frac{3}{2}, \frac{1}{2}\right)$; $r = 2$

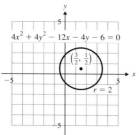

45) a) 128 m   b) 64 m   c) (0, 71)   d) $x^2 + (y - 71)^2 = 4096$

47) 11,127 mm²

49) $(x-5)^2 + (y-3)^2 = 1$

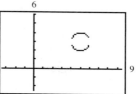

51) $(x+2)^2 + (y-4)^2 = 9$

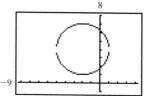

Section 12.3

1) false   3) true   5) true   7) false

9) center: $(-2, 1)$

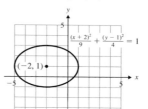

11) center: $(3, -2)$

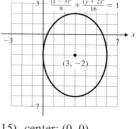

13) center: $(0, 0)$

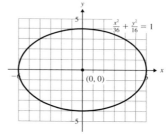

15) center: $(0, 0)$

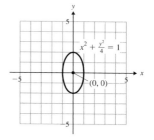

17) center: $(0, -4)$

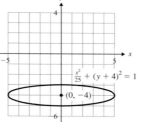

19) center: $(-1, -3)$

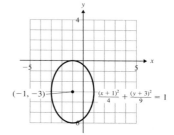

21) center: $(0, 0)$

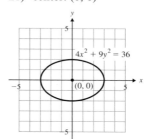

23) center: $(0, 0)$

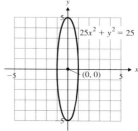

25) Group x- and y-terms separately.; $3(x^2 - 2x) + 2(y^2 + 2y) = 7$; Complete the square.; $3(x-1)^2 + 2(y+1)^2 = 12$; $\frac{(x-1)^2}{4} + \frac{(x+1)^2}{6} = 1$

27) $\dfrac{(x-1)^2}{16} + \dfrac{(y-3)^2}{4} = 1$ 29) $(x+4)^2 + \dfrac{(y+1)^2}{9} = 1$

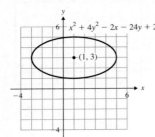
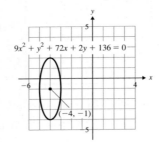

31) $\dfrac{(x-2)^2}{9} + \dfrac{(y-3)^2}{4} = 1$

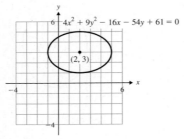

33) $\dfrac{(x+3)^2}{4} + \dfrac{y^2}{25} = 1$

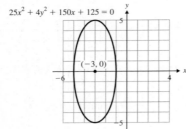

35) $\dfrac{x^2}{9} + \dfrac{y^2}{25} = 1$ 37) $\dfrac{x^2}{49} + y^2 = 1$

39) $\dfrac{(x-3)^2}{4} + \dfrac{(y-1)^2}{16} = 1$

41) Yes. If $a = b$ in the equation $\dfrac{(x-h)^2}{a^2} + \dfrac{(y-k)^2}{b^2} = 1$, then the ellipse is a circle.

43) $\dfrac{x^2}{9801} + \dfrac{y^2}{11{,}342.25} = 1$

45) $\dfrac{x^2}{36} + \dfrac{y^2}{16} = 1$

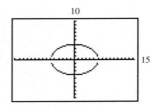

Section 12.4

1) false   3) true

5) center: (0, 0)   7) center: (0, 0)

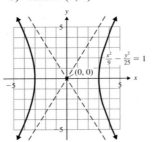
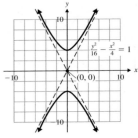

9) center: (2, −3)   11) center: (−4, −1)

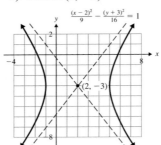
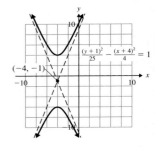

13) center: (1, 0)

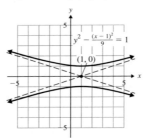

15) center: (1, 2)

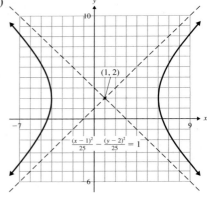

17) center: (0, 0)   19) center: (0, 0)

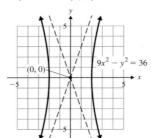
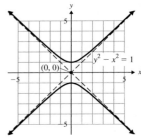

SA-54   Answers to Selected Exercises   www.mhhe.com/messersmith

21) $y = \dfrac{5}{3}x$ and $y = -\dfrac{5}{3}x$ 23) $y = 2x$ and $y = -2x$

25) $y = 3x$ and $y = -3x$ 27) $y = x$ and $y = -x$

29) 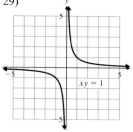   31)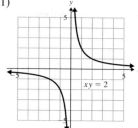

33) 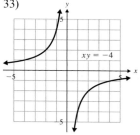   35)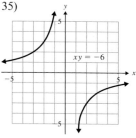

37) domain: $[-5, 5]$;
    range: $[0, 5]$
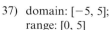

39) domain: $[-1, 1]$;
    range: $[-1, 0]$
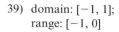

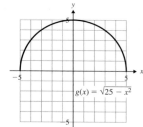   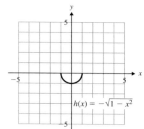

41) domain: $[-3, 3]$;
    range: $[-2, 0]$

43) domain:
    $(-\infty, -2] \cup [2, \infty)$;
    range: $(-\infty, 0]$

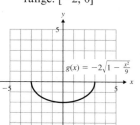   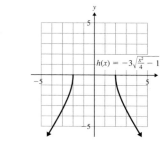

45)   47)

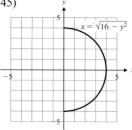   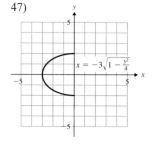

49) Group the x- and y-terms separately.;
    $4(x^2 - 2x) - 9(y^2 + 2y) = 41$;
    Complete the square.;
    $4(x-1)^2 - 9(y+1)^2 = 36$; $\dfrac{(x-1)^2}{9} - \dfrac{(y+1)^2}{4} = 1$

51) $\dfrac{(x-1)^2}{16} - \dfrac{(y+3)^2}{4} = 1$ 53) $\dfrac{(y-2)^2}{9} - \dfrac{(x-1)^2}{16} = 1$

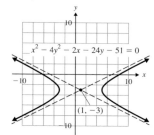   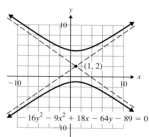

55) $y + 3 = \dfrac{4}{3}(x - 2)$ and $y + 3 = -\dfrac{4}{3}(x - 2)$

57) $y + 1 = \dfrac{5}{2}(x + 4)$ and $y + 1 = -\dfrac{5}{2}(x + 4)$

59) $y = \dfrac{1}{3}(x - 1)$ and $y = -\dfrac{1}{3}(x - 1)$

61) $y = x + 1$ and $y = -x + 3$ 63) $y^2 - \dfrac{x^2}{4} = 1$

65) $\dfrac{(x-2)^2}{9} - \dfrac{(y+3)^2}{16} = 1$

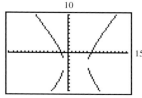

67) $\dfrac{(y+1)^2}{25} - \dfrac{(x+4)^2}{4} = 1$

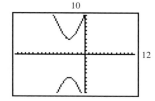

Chapter 12: Putting It All Together

1) parabola   2) circle

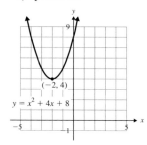   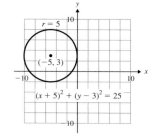

Answers to Selected Exercises **SA-55**

3) hyperbola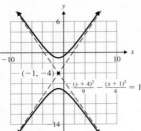

4) parabola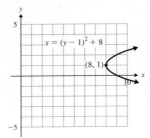

13) hyperbola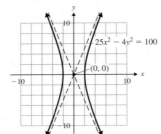

14) ellipse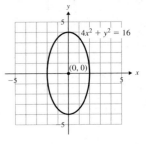

5) ellipse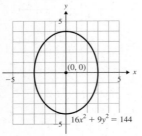

6) hyperbola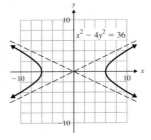

15) circle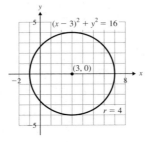

16) parabola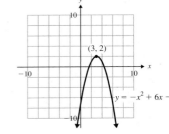

7) circle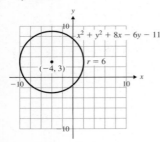

8) ellipse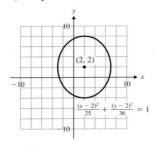

17) parabola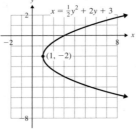

18) ellipse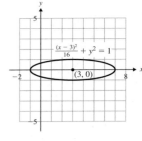

9) ellipse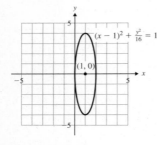

10) circle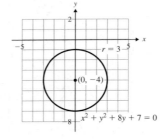

19) hyperbola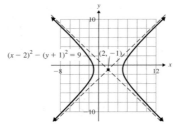

20) circle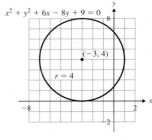

11) parabola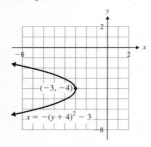

12) hyperbola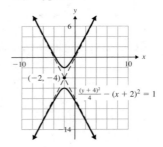

21)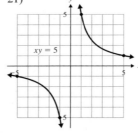

22) $\dfrac{(x+3)^2}{4} - \dfrac{y^2}{25} = 1$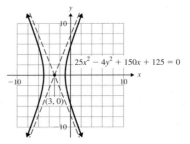

23) $(x - 3)^2 + \dfrac{(y + 2)^2}{9} = 1$

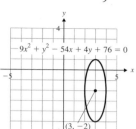

24)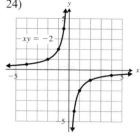

25) $\dfrac{(y - 1)^2}{4} - \dfrac{(x - 2)^2}{9} = 1$

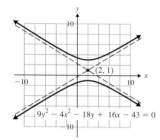

26) $\dfrac{(x - 1)^2}{9} + \dfrac{(y - 3)^2}{4} = 1$

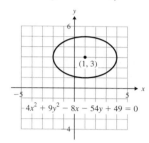

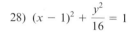

27) $x^2 + y^2 + 8x - 6y - 11 = 0$

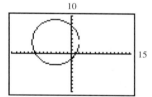

28) $(x - 1)^2 + \dfrac{y^2}{16} = 1$

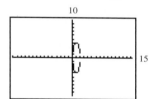

29) $x = -(y + 4)^2 - 3$

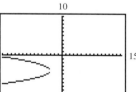

30) $25x^2 - 4y^2 = 100$

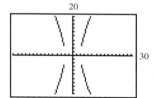

Section 12.5

1) c) 0, 1, or 2

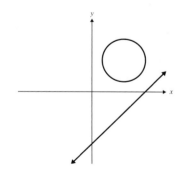

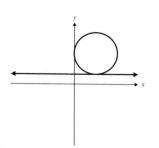
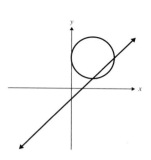

3) c) 0, 1, 2, 3, or 4

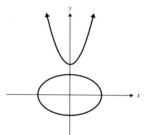

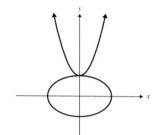
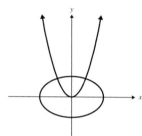

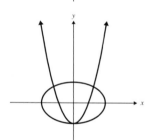
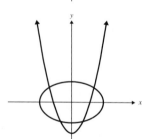

5) c) 0, 1, 2, 3, or 4

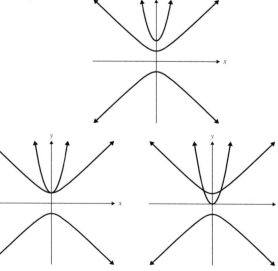

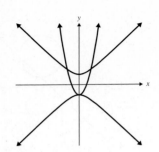 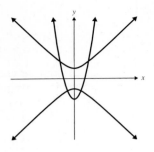

7) {(−4, −2), (6, −7)}   9) {(−1, 3), (3, 1)}

11) {(4, 2)}   13) {(3, 1), (3, −1), (−3, 1), (−3, −1)}

15) {(2, √2), (2, −√2), (−2, √2), (−2, −√2)}

17) {(3, −7), (−3, −7)}   19) {(0, −1)}

21) {(0, 2)}   23) ∅   25) ∅

27) $\left\{\left(\frac{\sqrt{2}}{2}, \frac{3\sqrt{2}}{2}\right), \left(\frac{\sqrt{2}}{2}, -\frac{3\sqrt{2}}{2}\right),\right.$
$\left.\left(-\frac{\sqrt{2}}{2}, \frac{3\sqrt{2}}{2}\right), \left(-\frac{\sqrt{2}}{2}, -\frac{3\sqrt{2}}{2}\right)\right\}$

29) {(0, 1)}   31) 8 and 5   33) 8 in. × 11 in.

35) 4000 basketballs; $240

Section 12.6

1) Three points that satisfy the inequality: (8, 0), (0, −8), and (6, 0); three that do not: (0, 0), (−3, −2), and (1, 1): answers may vary.

3) Three points that satisfy the inequality: (0, 0), (−4, −1), and (5, 2); three that do not: (0, 1), (0, −2), and (3, 3): answers may vary.

5) 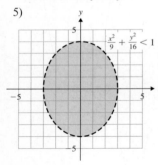   7)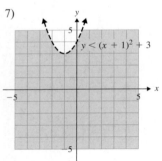

9) 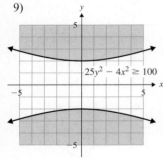   11)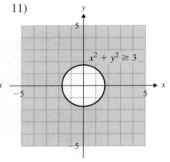

13) 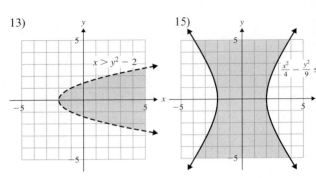 15)

17) 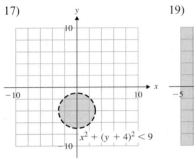   19)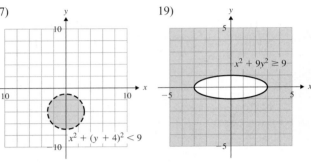

21) 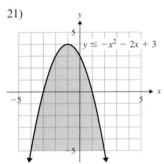   23)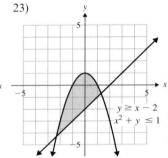

25)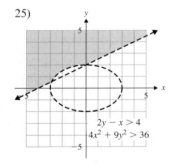

27)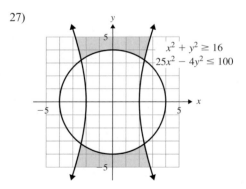

29)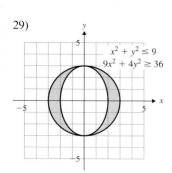

31)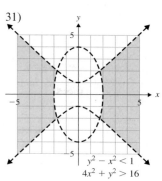

45) 47)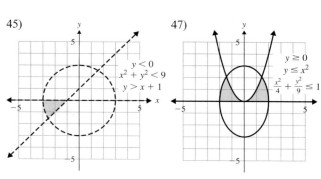

33)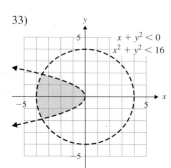

49)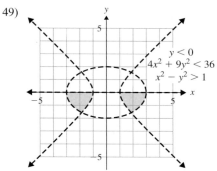

35)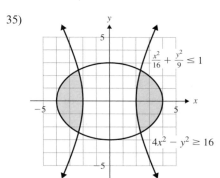

51)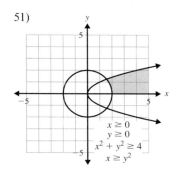

37)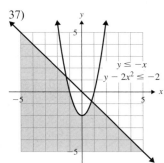

39)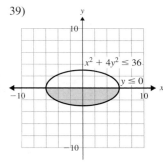

Chapter 12: Review Exercises

1) domain: $(-\infty, \infty)$; range: $[0, \infty)$

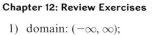

3) domain: $(-\infty, \infty)$; range: $[-4, \infty)$

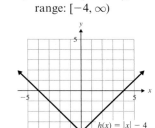

41)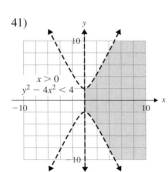

43)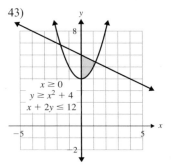

5) domain: $(-\infty, \infty)$; range: $(-\infty, 5]$

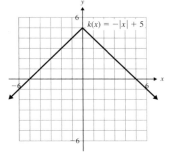

www.mhhe.com/messersmith

Answers to Selected Exercises SA-59

7)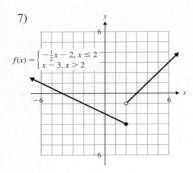

9) 7   11) −9   13) 0

15)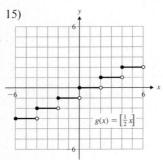

17) center: $(-3, 5)$; $r = 6$ 19) center: $(5, 2)$; $r = 4$

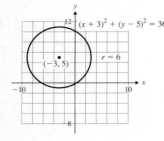   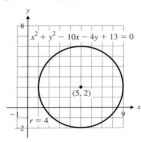

21) $(x - 3)^2 + y^2 = 16$

23) when $a = b$ in $\dfrac{(x - h)^2}{a^2} + \dfrac{(y - k)^2}{b^2} = 1$

25) center $(-3, 3)$ 27) center $(0, 0)$

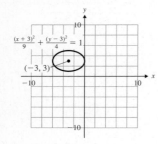   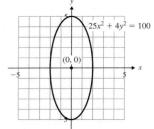

29) $\dfrac{(x - 2)^2}{4} + \dfrac{y^2}{25} = 1$

31) center $(0, 0)$ 33) center $(-1, -2)$

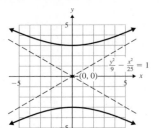   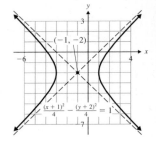

35)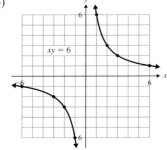

37) $(y + 3)^2 - \dfrac{(x - 1)^2}{16} = 1$

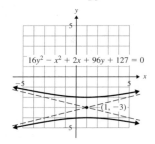

39) domain: $[-3, 3]$; range: $[0, 2]$ 41) ellipse

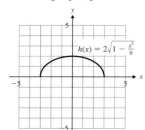   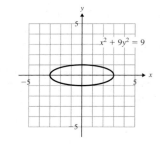

43) parabola   45) hyperbola

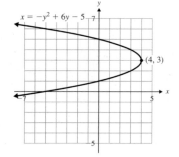   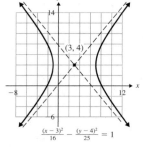

SA-60   Answers to Selected Exercises

47) circle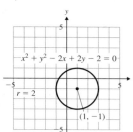

49) parabola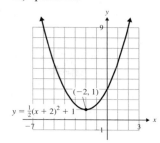

51) 0, 1, 2, 3, or 4   53) {(6, 7), (6, −7), (−6, 7), (−6, −7)}

55) {(1, 2), (−2, −1)}   57) ∅   59) 9 and 4

61)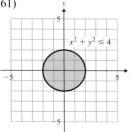

63)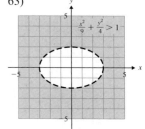

65)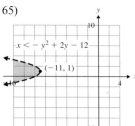

67)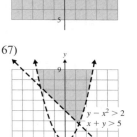

69)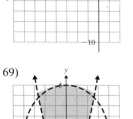

71)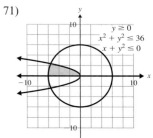

Chapter 12: Test

1)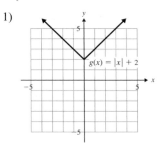

2)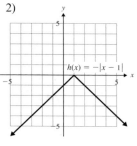

3)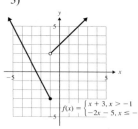

4) ellipse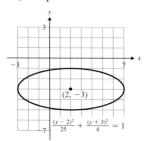

5) parabola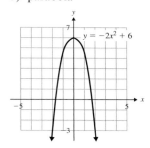

6) hyperbola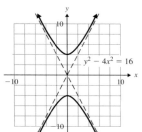

7) circle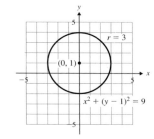

8) hyperbola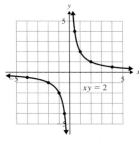

9) $(x + 1)^2 + (y - 3)^2 = 16$; center $(-1, 3)$; $r = 4$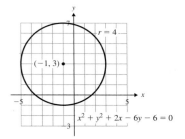

10) $(x - 5)^2 + (y - 2)^2 = 11$ 11) $\dfrac{x^2}{8836} + \dfrac{y^2}{6084} = 1$

12) domain: $[-5, 5]$; range: $[-5, 0]$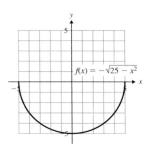

13) a)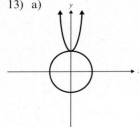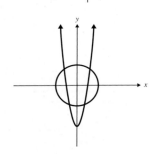

b) 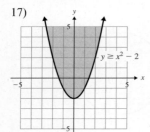   c) 0, 1, 2, 3, or 4

14) {(−1, 0), (7, −2)}

15) {(3, 1), (3, −1), (−3, 1), (−3, −1)}

16) 8 in. × 14 in.

17) 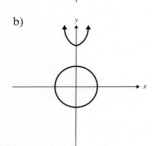   18)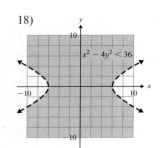

19) 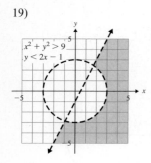   20)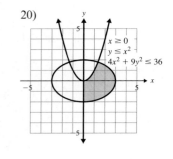

Chapter 12: Cumulative Review for Chapters 1–12

1) $-\dfrac{3}{4}$ 2) 12   3) $A = 15\text{ cm}^2$; $P = 18.5$ cm

4) $A = 128\text{ in}^2$; $P = 56$ in.   5) -1 6) 16

7) $\dfrac{1}{8a^{18}b^{15}}$ 8) $\{-40\}$ 9) $n = \dfrac{c-z}{a}$ 10) $[-4, \infty)$

11) 15, 17, 19   12) -1 13) 0

14)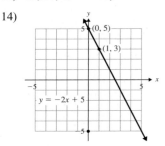

15) $y = -\dfrac{3}{4}x + 4$ 16) $\left(-1, \dfrac{3}{2}\right)$

17) 5 mL of 8%, 15 mL of 16%   18) $-3p^2 + 7p + 6$

19) $8w^3 + 30w^2 - 47w + 15$ 20) $x^2 + 4x + 9$

21) $2(3c-4)(c-1)$ 22) $(m-2)(m^2+2m+4)$

23) $\{-2, 6\}$ 24) $-\dfrac{5}{2(a+6)}$ 25) $6(t+3)$

26) $\left\{-6, -\dfrac{4}{3}\right\}$ 27) $(-\infty, -3) \cup \left(\dfrac{9}{5}, \infty\right)$

28) $5\sqrt{3}$ 29) $2\sqrt[3]{6}$ 30) $3ab^4\sqrt[3]{a^2b}$ 31) $\dfrac{1}{8}$

32) $3\sqrt{3}$ 33) $\dfrac{20 - 5\sqrt{3}}{13}$ 34) $\left\{\dfrac{1}{2} + 2i, \dfrac{1}{2} - 2i\right\}$

35) $\left\{-\dfrac{7}{2} + \dfrac{\sqrt{37}}{2}, -\dfrac{7}{2} - \dfrac{\sqrt{37}}{2}\right\}$ 36) $\left[-\dfrac{12}{5}, \dfrac{12}{5}\right]$

37) $\left(-\infty, -\dfrac{5}{2}\right) \cup (3, \infty)$

38) a) $\{0, 3, 4\}$ b) $\{-1, 0, 1, 2\}$ c) no

39)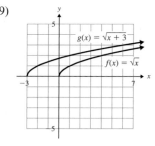

40) a) -3 b) $6x - 9$ c) 6   41) a) no   b) no

42) $f^{-1}(x) = 3x - 12$ 43) $\left\{-\dfrac{6}{13}\right\}$ 44) $\{5\}$ 45) 2

46)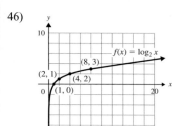

47) $\left\{\dfrac{\ln 8}{3}\right\}$; $\{0.6931\}$

48)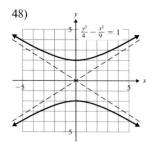

49)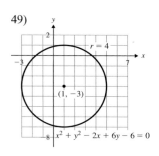

50) $(0, 3)$

Photo Credits

Page: 1: DreamPictures/VStock/Getty Images; **p. 2:** © Corbis RF; **p. 36:** © Manchan/Digital Vision/Getty RF; **p. 37:** © BananaStock/PictureQuest RF; **p. 53:** © Creatstock Photographic/Alamy RF; **p. 63(top):** © Vol. 25 PhotoDisc/Getty RF; **p. 63(bottom):** © BrandX/Punchstock RF; **p. 64:** © Corbis RF; **p. 73:** © Vol. 5 PhotoDisc/Getty RF; **p. 78:** © StockDisc/Getty RF; **p. 84:** © Comstock/Punchstock RF; **p. 85(left):** © Getty RF; **p. 85(right):** © Corbis RF; **p. 91:** © Adrian Sherratt/Alamy RF; **p. 94:** © Monty Rakusen/Getty RF; **p. 95:** © Insy Shah/Getty RF; **p. 106:** © Vol. 56 PhotoDisc/Getty RF; **p. 108:** © DynamicGraphics/Jupiter Images RF; **p. 125:** © The McGraw-Hill Companies, Inc./Andrew Resek, photographer; **p. 128:** © The McGraw-Hill Companies, Inc./Andrew Resek, photographer; **p. 132:** © Vol. 10 PhotoDisc/Getty RF; **p. 134:** © Digital Vision/Getty RF; **p. 135:** Stockbyte/Punchstock RF; **p. 136:** © Image Source/Getty RF; **p. 180:** © Stockbyte/Punchstock Images RF; **p. 197:** © BrandX 195/Getty RF; **p. 212 (left):** © The McGraw-Hill Companies, Inc./Gary He, photographer; **p. 212 (right):** © SS18 PhotoDisc/Getty RF; **p. 228:** © Rob Daly/Getty Images RF; **p. 229:** © Jose Luis Pelaez Inc./Getty RF; **p. 260:** © Bluestone Productions/Superstock RF; **p. 269(left):** © Royalty-Free/Corbis; **p. 269(right):** © BrandX/Punchstock RF; **p. 270:** © Corbis RF; **p. 271(top):** © Stockbyte/Punchstock RF; **p. 271(bottom):** © Corbis RF; **p. 287(left):** © RP020/Getty RF; **p. 287(top right):** © Ingram Publishing/Alamy RF; **p. 287(bottom right):** © BrandX Pictures/Punchstock RF; **p. 288:** © The McGraw-Hill Companies, Inc./Andrew Resek, photographer; **p. 290:** ©Jose Luis Pelaez Inc/Blend Images RF; **p. 291:** © DreamPictures/Getty RF; **p. 312:** © BananaStock/Punchstock RF; **p. 320:** © Vol. 21 PhotoDisc/Getty RF; **p. 349(left):** © Digital Vision RF; **p. 349(right):** © Stockbyte/PunchStock RF; **p. 353:** © Getty Images/OJO Images RF; **p. 354:** © franckreporter/Getty RF; **p. 409:** © SS14 PhotoDisc/Getty RF; **p. 417:** © BrandX/Punchstock RF; **p. 420:** © Blend Images/Hill Street Studios/Getty RF; **p. 421:** © Gary Conner/Photolibrary/Getty RF; **p. 479:** © Vol. 21 PhotoDisc/Getty RF; **p. 485(left):** © Corbis RF; **p. 485(right):** © Corbis RF; **p. 492:** © Comstock/JupiterImages RF; **p. 495(left):** © OS40 PhotoDisc/Getty RF; **p. 495(right):** © Creatas/PictureQuest RF; **p. 507:** © Klaus Tiedge/Blend Images RF; **p. 508:** © Dougal Waters/Digital Vision/Getty RF; **p. 524:** © The McGraw-Hill Companies, Inc., Mark Dierker, photographer; **p. 533:** © image100/Getty RF; **p. 586:** © Goodshoot/Alamy RF; **p. 607:** © Digital Vision/Getty RF; **p. 608:** © Image Source/Getty RF; **p. 644(left):** © Corbis/Superstock RF; **p. 644(right):** © Brand X Pictures/JupiterImages RF; **p. 649:** © Getty RF; **p. 678:** © Emma Lee/Life File/Getty RF; **p. 701:** © Antenna/Getty RF; **p. 702:** © Asia Images/Getty RF; **p. 729:** © Vol. 1 PhotoDisc/Getty RF; **p. 739:** © image100/PunchStock RF; **p. 742:** © Stockbyte/Getty RF; **p. 772:** © Creatas/PunchStock RF; **p. 784:** © Brand X Pictures/PunchStock RF; **p. 785:** © Vol. 10 PhotoDisc/Getty RF; **p. 788:** © SPL/Getty RF; **p. 789:** © AID/a.collection/Getty RF; **p. 797:** © OS31 PhotoDisc/Getty RF; **p. 809:** © Ingram Publishing/AGE Fotostock RF; **p. 810:** © Brand X Pictures/Punchstock; **p. 838:** © Comstock/PictureQuest.

Index

A

$a^{1/n}$, evaluating, 525–526
$a^{m/n}$, evaluating, 526–527
$a^{-m/n}$, evaluating, 527–528
Abscissa. *See* x-coordinate
Absolute value
   definition of, 9, 32
   finding, 9–10
Absolute value equations
   of form $|ax + b| = |cx + d|$, 120–121
   of form $|ax + b| = k$, 119–120
   understanding, 118, 131
Absolute value functions, graph of, 793–794, 846
Absolute value inequalities
   with < or ≤, 121–122, 131
   with > or ≥, 122–124, 132
   applied problems with, 125
   special cases of, 124–125
Addition
   applied problems with, 15
   of complex numbers, 590, 602
   in order of operations, 23–24
   of polynomial functions, 318–320, 347
   of polynomials, 315–317, 347
   of radical expressions, 554, 574, 600
      simplifying before, 555
   of rational expressions
      with common denominator, 441–443
      with denominator factors
         of $a - b$ and $b - a$, 446–447
      with different denominators, 443–445, 473, 498
   of real numbers, 13–14, 32
Addition property of equality
   definition of, 88
   equations solved with, 41–42
   linear inequalities in one variable solved with, 100–101
Additive inverse, definition of, 9, 32
Algebraic expressions
   definition of, 27, 33
   evaluating, 27–28
   functions evaluated for, 206
Angles, applied problems with
   complementary, 68–69
   measures of, 67–68
   supplementary, 68–69
Applied problems
   with absolute value inequalities, 125
   with addition, of real numbers, 15
   with angle measures, 67–68
   with area, 648–649
   with complementary angles, 68–69
   with cost, 262–263
   with equations with common logarithms, 755–756
   with exponential decay, 772–773
   with exponential functions, 726, 758–760, 770–772, 781
   with exponential growth, 772–773
   with geometry, 66–67, 89, 261–262
   with greatest integer function, 797
   with intersection of sets, 114
   with linear equations in one variable
      consecutive integers, 55–58
      distance, rate, and time, 81–83, 89–90
      general quantities, 53–54
      lengths, 54–55
      mixtures, 79–81, 89–90
      money, 76–79
      percent change, 58–59
      simple interest, 59–62
      steps for solving, 46–49, 88–89
   with linear equations in two variables, 147–149
   with linear functions, 207–208
   with linear inequalities in one variable, 105–106
   with logarithmic equations, 739–740
   with proportions, 478–480
   with quadratic equations
      consecutive integers, 404
      geometry, 403
      given an equation, 407–408
      maximum values, 669–672
      minimum values, 669–672
      Pythagorean theorem, 405–407, 416
      quadratic formula, 649–650, 694
      steps for solving, 403
   with quadratic formula, 629–630
   with rational equations
      distance, rate, and time, 480–482
      proportions, 478–480
      work, 482–484, 500–501
   with slope of a line, 160
   with subtraction, of real numbers, 15
   with supplementary angles, 68–69
   with systems of linear equations in three variables, 267–268, 285
   with systems of linear equations in two variables
      cost, 262–263
      distance, rate, and time, 265–266
      general, 260–261
      geometry, 261–262
      mixtures, 264–265
   with union of sets, 114
   with volume, 647–648
Approximately equal to (≈), 8
Area
   applied problems with, 648–649
   of rectangle, 66–67
   of trapezoid, 71–72
Associative properties, 28–29, 33
Augmented matrix, definition of, 274, 285
Axis of symmetry, of parabola, 654

B

Base, in exponential expressions, 294
Basic power rule for exponents, 295–297, 304–306, 346
Binomials
   definition of, 315
   factoring of
      difference of two squares, 380–383, 415
      sum and difference of two cubes, 383–385, 415
   multiplication of
      FOIL method for, 326–327, 347
      of form $(a + b)(a - b)$, 328–329, 347
      of other forms, 330–331
      squaring, 329–330, 347
   with radical expressions
      multiplication of, 556, 575
      squaring, 557–558
   substitution for, quadratic equations solved by, 641–642
Binomial square, 329–330, 347
Bounds
   of linear inequalities in three parts, 103
   of linear inequalities in two variables, 188
Brackets, in order of operations, 24

C

Cartesian coordinate system, review of, 140
Center
   of circle, 803
   of ellipse, 810
   of hyperbola, 818
Change-of-base formula, 760–761, 781
Circle
   as conic section, 802
   definition of, 803
   graph of
      in general form, 804–805, 828, 846
      in standard form, 803–804, 846
Coefficients, in algebraic expressions, 27
Combined variation, solving problems, 493–494, 501
Common logarithms
   equations with, solving, 755
      applied problems with, 755–756
   evaluating, 735–736, 780
      with calculator, 754–755
Commutative properties, 28–29, 33
Complementary angles, applied problems with, 68–69

Completing the square
    quadratic equations solved by, 619–621, 633–635, 692
    quadratic functions graphed by, 659
    rewriting $x = ay^2 + by + c$ by, 674–675
    for $x^2 + bx$, 616–618, 692
Complex fractions
    definition of, 450–451, 498
    simplifying
        with multiple terms in numerator and/or denominator, 452–456, 474, 499
        with one term in numerator and denominator, 451–452, 474, 499
Complex numbers
    addition of, 590, 602
    definition of, 588, 602
    division of, 592–593, 602
    multiplication of, 590–591
        by conjugate, 591–592, 602
    real numbers in, 588–589
    subtraction of, 590, 602
Composition functions
    definition of, 706, 778
    finding, 705–708
    using, 708
Compound inequality
    with *and,* 110–112, 131
    definition of, 103
    with *or,* 112–113, 131
    special, 113–114
    steps for solving, 111
    in three parts, 103–105
Compound interest, 758–759
Compound linear inequalities in two variables, graphing, 191–192, 220
Conic sections, definition of, 802
Conjugates of complex numbers, multiplication by, 591–592, 602
Consecutive integers, applied problems with
    linear equations in one variable, 55–58
    quadratic equations, 404
Consistent systems of linear equations in two variables, 234, 236, 282
Constant
    in algebraic expressions, 27
    in polynomials, 314
Constant of variation
    in direct variation, 488, 501
    in inverse variation, 491, 501
Continuous compounding, 759–760, 781
Cost, applied problems with, 262–263
Cross products, 465
Cube root, solving radical equations with, 583
Cube root functions
    domain of, 518–519, 598
    examples of, 517–518
    graph of, 519–520
Cubic equation, solving by factoring, 397–398

D

Daily to-do list, 95
Data, linear equations for modeling, 180–181
Decimals, linear equations in one variable with, 44–45
Decomposition of functions, 707–708
Degree of term, in polynomials, 314, 346

Denominator, of radical expressions, rationalizing
    definition of, 562, 601
    with one higher root, 565–567, 575–576
    with one square root, 562–565, 575
    with two terms, 567–569, 575–576, 601
Dependent systems of linear equations in two variables
    definition of, 236, 282, 283
    graphs of, 236
Dependent variables, 197
Diagonal, of matrix, 274
Difference of two squares, factoring, 380–383, 385, 415
Direct variation
    definition of, 488, 501
    solving problems with, 488–490
Discriminant, in determining quadratic equation solutions, 627–629, 693
Distance, rate, and time, applied problems with
    with linear equations in one variable, 81–83, 89–90
    rational equations, 480–482
    with systems of linear equations in two variables, 265–266
Distance formula, 615–616, 692
Distributive properties, 28–29, 34
Division
    complex fractions simplified by, 452–453, 455–456, 499
    of complex numbers, 592–593, 602
    of exponential expressions, with quotient rule for exponents, 300–301
    in order of operations, 23–24
    of polynomial functions, 341–342, 348
    of polynomials
        long division, 337–340
        by monomials, 336–337, 348
        by polynomials, 337–340, 348
    of radical expressions, with different indices, 550
    of rational expressions, 431–432, 473, 498
    of real numbers, 16–17, 32
    of square root expressions, with variables, 541–542
    of square roots, with negative numbers, 589–590
Division property of equality
    definition of, 88
    equations solved with, 41–42
Domain
    of cube root functions, 518–519, 598
    of rational functions, 425–427, 472, 497
    of relations
        definition of, 198, 220
        identifying, 198–201, 202–203, 221
    of square root functions, 517, 598

E

e, definition of, 723, 781
Element of matrix, definition of, 274
Elimination method
    nonlinear systems of equations solved by, 835–836
    systems of linear equations in two variables solved by, 241–243, 283

Ellipse
    as conic section, 802
    definition of, 810
    equation of
        with center at origin, 812
        in standard form, 810, 847
    graph of, 811–813, 830, 847
Empty set, 45
Equality, properties of, 88
    equations solved with, 41–42
    linear inequalities in one variable solved with, 100–103
Equation. *See specific types*
Equivalent rational expressions, 429–430
    with LCD as denominator, 439–441
Even integers, consecutive, 55–58
Exponent(s)
    in exponential expressions, 294
    negative, 297–300, 304–306, 346
    in order of operations, 23–24
    rational. *See* Rational exponents
    rules of
        basic power rule, 295–297, 304–306, 346
        combining, 304–306, 346, 528–529
        power rule for a product, 295–297, 304–306, 346
        power rule for a quotient, 295–297, 304–306, 346
        product rule, 294–295, 304–306, 346
        quotient rule, 300–301, 304–306, 346
        using, 20–21, 33
    variables as, 305–306
    zero as, 297–300, 304–306, 346
Exponential equations
    definition of, 724
    solving, 724–726, 766–767, 779, 782
    properties for, 765
Exponential expressions
    converting radical expressions to, 530
    definition of, 20, 294
    division of, with quotient rule for exponents, 300–301
    evaluating, 20–21
    multiplication of, with product rule for exponents, 294–295
    raising to a power, 295–297
    variable exponents in, 305–306
Exponential form
    conversion to logarithmic form, 732–733
    converted from logarithmic form, 732
Exponential functions
    applied problems with, 726, 758–760, 770–772, 781
    definition of, 720, 779
    graph of
        $f(x) = a^x$, 720–722
        $f(x) = a^{x+c}$, 722–723
        $f(x) = e^x$, 723–724

F

Factoring
    of binomials
        difference of two squares, 380–383, 385, 415
        sum and difference of two cubes, 383–385, 415

equations in quadratic form solved by, 638–640, 693
higher degree equations solved by, 397–398
of integers, 357
of polynomials
　definition of, 358, 414
　greatest common binomial factor, 360–361
　greatest common monomial factor, 359–360
　by grouping, 361–363, 414
　vs. multiplication, 358–359
　strategies for, 388–390
　by substitution, 374–375
　quadratic equations solved by, 394–397, 415–416, 633–635
of trinomials
　of form $ax^2 + bx + c$, 370–374, 414–415
　of form $x^2 + bx + c$, 367–369, 414
　by grouping, 370–371, 414–415
　perfect square trinomials, 378–380, 385, 415
　by trial and error, 371–374, 415
　with two variables, 369–370
Financial literacy, 608, 691
Focus (foci)
　of ellipse, 810
　of hyperbola, 818
FOIL method
　for multiplication of binomials, 326–327
　for multiplication of radical expressions, 557
Formulas
　change-of-base, 760–761, 781
　distance, 615–616, 692
　distance, rate, and time, 480
　midpoint, using, 146–147, 216
　perimeter of rectangle, 70–71
　quadratic. *See* Quadratic formula
　slope of a line, 158
　solving for specific variable, 70–72, 89, 645–646, 693
　vertex, 661–662
Fractions. *See also* Rational expressions
　complex. *See* Complex fractions
　equations with, quadratic equations from, 637–638
　linear equations in one variable with, solving, 44–45
　in linear inequalities, 104
　systems of linear equations in two variables with, solving by substitution, 240–241
Function(s), 702. *See also* Exponential functions; Linear function; Rational functions
　absolute value, graph of, 793–794, 846
　composition of
　　definition of, 706, 778
　　finding, 705–708
　　using, 708
　decomposition of, 707–708
　definition of, 198, 199, 220
　determining, 198–201
　　from given equation, 201–203
　domain of, finding, 201–203

evaluating
　for expressions, 206
　for variables, 206
function values for, finding, 204–206
greatest integer
　applied problems with, 797
　definition of, 795, 846
　graph of, 796–797
inverse of
　definition of, 711, 778
　finding, 711–713, 778
　graph of, 713–714
one-to-one
　definition of, 709
　determining, 709–711, 778
　horizontal line test for, 710–711, 778
piecewise, graph of, 794–795, 846
vertical line test for, 200–201, 220
Function notation, 203–204, 221

G

Gaussian elimination
　definition of, 274, 285
　systems of linear equations in three variables solved by, 275–278
　systems of linear equations in two variables solved by, 274–275, 285
Geometry, applied problems with, 66–67
　formulas, 89
　quadratic equations, 403
　systems of linear equations in two variables, 261–262
Graph
　of absolute value functions, 793–794, 846
　of circle
　　in general form, 804–805, 828, 846
　　in standard form, 803–804, 846
　of compound linear inequalities in two variables, 191–192, 220
　of cube root functions, 519–520
　of ellipse, 811–813, 830, 847
　of exponential functions
　　$f(x) = a^x$, 720–722
　　$f(x) = a^{x+c}$, 722–723
　　$f(x) = e^x$, 723–724
　of greatest integer function, 796–797
　of hyperbola
　　in nonstandard form, 821–822
　　in standard form, 818–821, 829, 847
　of inverse functions, 713–714
　of line
　　horizontal lines, 145–146, 216
　　with slope and point, 161–162, 216
　　with slope and y-intercept, 162–164, 217
　　summary of methods, 164
　　vertical lines, 145–146, 216
　of linear equations in two variables
　　by finding intercepts, 144–145, 215
　　horizontal lines, 145–146
　　by plotting points, 143–144, 215
　　vertical lines, 145–146
　of linear function, 207
　of linear inequalities in two variables, 188–191, 219
　of linear inequality in one variable, 98–99

of logarithmic function, 738–739
of natural logarithm functions, 757
of parabola
　finding vertex of, 675–676
　of form $x = a(y - k)^2 + h$, 672–674
　of form $x = ay^2 + by + c$, 674–675, 696, 829
of piecewise functions, 794–795, 846
of quadratic functions, 654–657
　of form $f(x) = a(x - h)^2 + k$, 657–658, 694
　of form $f(x) = ax^2 + bx + c$, 658–662, 695
　horizontal shift of, 655–656, 792–793
　reflection about the x-axis, 656–657, 792–793
　vertical shift of, 654–655, 792–793
quadratic inequality solved by, 681–682
of second-degree inequalities, 839–840, 848
of square root functions, 518–519, 598, 822–823
of systems of linear equations in two variables
　dependent, 236
　independent, 236
　solving with, 234–237, 282
of systems of nonlinear inequalities, 840–842
Graphing calculator
　absolute value equations on, 125–126
　circles on, 806
　complex numbers on, 594
　composition of functions on, 714–715
　cube root functions on, 521–522
　ellipses on, 814
　exponential equations on, 773–774
　hyperbola, 824
　linear equations in one variable on, 49
　linear inequalities in two variables on, 193–194
　logarithmic equations on, 773–774
　logarithmic functions on, 761
　nonlinear systems of equations on, 836–837
　parabolas on, 676
　perpendicular lines on, 181
　piecewise functions on, 798
　polynomial factoring on, 375
　quadratic equations on, 398–399, 662–663
　radical equations on, 583–584
　rational equations on, 468
　roots on, 520–521, 531
　scientific notation on, 310
　slope estimation on, 164–165
　square root functions on, 521–522
　systems of linear equations in three variables on, 278–279
　systems of linear equations in two variables on, 238
Greater than ($>$), 8, 122–124, 131
Greater than or equal to ($\geq$), 8, 122–124, 131
Greatest common factor (GCF)
　definition of, 357
　of monomials, finding, 357–358
　of polynomials, factoring out, 359

Greatest integer function
  applied problems with, 797
  definition of, 795, 846
  graph of, 796–797
Grouping, factoring by
  of polynomials, 361–363, 414
  of trinomials, 370–371, 414–415
Grouping symbols, in order of operations, 23–24

H

Higher roots
  of integers, simplifying, 546–547
  multiplication of, 545–546
  product rule for, 545–546, 600
  quotient rule for, 547–548, 600
Homework, 291, 345
Horizontal lines
  equations of
    graphing, 145–146, 216
    writing, 175
  slope of, 160–161, 216
Horizontal line test, 710–711, 778
Horizontal shift, of quadratic functions, 655–656, 792–793
Hyperbola
  as conic section, 802
  equation of
    with center at origin, 820, 847
    in standard form, 818, 847
  graph of
    in nonstandard form, 821–822, 847
    in standard form, 818–821, 829, 847
Hypotenuse, definition of, 405

I

Identity properties, 28–29, 34
Imaginary number
  definition of, 588
  simplifying powers of, 593, 602
Imaginary part, of complex numbers, 588, 602
Impedance, 596
Inconsistent systems of linear equations in two variables, definition of, 235, 282, 283
Independent systems of linear equations in two variables
  definition of, 234, 282
  graphs of, 236
Independent variables, 197
Index of radical, 513–515
Inequality symbols, comparing numbers with, 8
Integers
  consecutive, applied problems with, 55–58, 404
  definition of, 5
  graphing on number line, 5
  higher roots of, simplifying, 546–547
Intercepts, of linear equations in two variables, graphing with, 144–145, 215
Interest
  compound, applied problems with, 758–759
  compounded continuously, 759–760
  simple, applied problems with, 59–62

Internet, effective use of, 597
Intersection, of sets, 109–110, 131
  applied problems with, 114
Interval notation, 98
Inverse functions
  definition of, 711, 778
  finding, 711–713, 778
  graph of, 713–714
Inverse properties, 28–29, 34
Inverse variation
  definition of, 491, 501
  solving problems, 491–492
Irrational number
  definition of, 6, 32
  as square roots, 22, 513, 598

J

Joint variation
  definition of, 492, 501
  solving problems, 492–493

L

Least common denominator (LCD)
  complex fractions simplified by multiplication by, 454–456, 499
  for rational expressions, 437–439, 498
  rewriting, with LCD as denominator, 439–441
Legs of right triangle, 405
Length, applied problems with, 54–55
Less than (<), 8, 121–122, 132
Less than or equal to (≤), 8, 121–122, 132
Like radicals, definition of, 553
Like terms, combining, solving linear equations in one variable by, 43–44
Line(s)
  equation of
    horizontal lines, 175
    parallel lines, 178–179, 218
    perpendicular lines, 178–179, 218
    from slope and point, 172–173, 217
    from slope and y-intercept, 172, 217
    slope-intercept form for, 171
    standard form for, 171
    summary of methods, 175
    from two points, 174, 217–218
    vertical lines, 175
  graph of
    horizontal lines, 145–146, 216
    with slope and point, 161–162, 216
    with slope and y-intercept, 162–164, 217
    summary of methods, 164
    vertical lines, 145–146, 216
  parallel, slope of, 175–178, 218
  perpendicular, slope of, 175–178, 218
  slope of. See Slope of a line
Linear equations in one variable
  applied problems with
    consecutive integers, 55–58
    distance, rate, and time, 81–83, 89
    general quantities, 53–54
    lengths, 54–55
    mixtures, 79–81, 89
    money, 76–79
    percent change, 58–59
    simple interest, 59–62
    steps for solving, 46–49, 88–89

  with decimals, solving, 44–45
  definition of, 40–41
  with fractions, solving, 44–45
  with infinite solutions, solving, 45–46
  with no solution, solving, 45–46
  solving, 40–41, 88
    by combining like terms, 43–44
    with properties of equality, 41–42, 88
Linear equations in three variables, systems of. See Systems of linear equations in three variables
Linear equations in two variables. See also Line(s), equation of; Systems of linear equations in two variables
  applied problems with, 147–149
  graphing
    by finding intercepts, 144–145, 215
    of horizontal lines, 145–146
    by plotting points, 143–144, 215
    of vertical lines, 145–146
  modeling real-world data with, 180–181
  slope-intercept form for, 163
  solutions to, ordered pairs as, 141–142
  standard form for, 162, 215
Linear function
  applied problems with, 207–208
  definition of, 206, 221
  graph of, 207
Linear inequalities in one variable
  applied problems with, 105–106
  compound
    with *and*, 110–112, 131
    definition of, 103
    with *or*, 112–113, 131
    special, 113–114
    steps for solving, 111
    in three parts, 103–105
  definition of, 98
  fractions in, 104
  graphing, 98–99, 131
  solving, using properties of equality, 102–103
    addition property of equality, 100–101
    multiplication property of equality, 101–102
    subtraction property of equality, 100–101
Linear inequalities in two variables
  boundary lines of, 188
  compound, graph of, 191–192, 220
  definition of, 187, 219
  graphing, 188–191, 219
  solutions to, 187
Logarithm(s)
  change-of-base of, 760–761, 781
  common
    applied problems with, 755–756
    equations with, solving, 755
    evaluating, 735–736, 780
    evaluating with calculator, 754–755
  definition of, 731–732, 779
  evaluating, 734–735
  natural
    definition of, 756
    equations with, solving, 758
    evaluating, 756, 779

power rule for, 746–747
product rule for, 744–745
properties of, 736, 748, 780
    combining, 749–751
quotient rule for, 745–746
Logarithmic equations
  applied problems with, 739–740
  solving
    each term contains a
      logarithm, 768–769
    each term does not contain a
      logarithm, 769–770
    of form $\log_a b = c$, 733–734, 779–780
    of form $\log b = c$, 735–736
  properties for, 765, 767–770, 782
Logarithmic form
  conversion from exponential form, 732–733
  conversion to exponential form, 732
Logarithmic function
  characteristics of, 739, 780
  definition of, 737
  graph of, 738–739
  natural, 757
Long division, of polynomials, 337–340
Lower bounds, of linear inequalities in three parts, 103
Lowest terms, rational expressions in, 427–428, 497

M

Master calendar, 95
Math tests, taking, 229
Matrix
  augmented, 274, 285
  definition of, 274
  diagonal of, definition of, 274
  in row echelon form, 274, 285
  row operations on, 274, 285
Maximum values, of quadratic functions, 668–669, 695
  applied problems with, 669–672
Memory devices, 845
Memory improvement, 789, 845
Memory organization, 86–87
Midpoint formula, using, 146–147, 216
Minimum values, of quadratic functions, 668–669, 695
  applied problems with, 669–672
Mixtures, applied problems with, 79–81, 89–90, 264–265
Money, applied problems with, 76–79
Monomials
  definition of, 315
  division of polynomials by, 336–337, 348
  greatest common factor of, finding, 357–358
  multiplication of, by polynomials, 325, 347
  polynomials with radical expressions multiplied by, 556
Multiplication
  of binomials
    FOIL method for, 326–327, 347
    of form $(a + b)(a - b)$, 328–329, 347
    of other forms, 330–331
    with radical expressions, 556, 575
squaring, 329–330, 347
  of complex numbers, 590–591
    by conjugate, 591–592, 602
  of exponential expressions, with product rule for exponents, 294–295
  of monomials, by polynomials, 325
  of nth roots, 545–546
  in order of operations, 23–24
  of polynomial functions, 331–332, 348
  of polynomials
    vs. factoring, 358–359
    by monomials, 325, 347
    by multiple polynomials, 328
    by polynomials, 325–326, 347
  of radical expressions, 545–546, 601
    with different indices, 550
    of form $(a + b)(a - b)$, 558–559, 601
    using FOIL, 557
  of rational expressions, 430–431, 497
  of real numbers, 15–16, 32
  of square root expressions, with variables, 541–542
  of square roots, 534–535
    with negative numbers, 589–590
Multiplication property of equality
  definition of, 88
  equations solved with, 41–42
  linear inequality in one variable solved with, 101–102

N

Natural logarithm(s)
  equations with, solving, 758
  evaluating, 756, 781
Natural logarithm, definition of, 756
Natural logarithm functions, graph of, 757
Natural numbers, definition of, 5, 32
Negative exponents, 297–300, 304–306, 346
  rational expressions with, simplifying, 456–457
Negative nth roots, 513–515, 598
Negative numbers, square roots of, 588–589, 602
Negative reciprocals, 176
Negative square roots, finding, 512–513
Nonlinear systems of equations
  definition of, 832, 847
  solving
    by elimination method, 835–836
    by substitution, 832–834, 847–848
Nonlinear systems of inequalities
  graph of, 840–842
  solution set of, 840, 848
Not equal to ($\neq$), 8
Note-taking, 136, 214
nth roots
  finding, 513–515, 598
  multiplication of, 545–546
Null set, 45
Number(s)
  comparing, with inequality symbols, 8
  sets of, 5–7
Number line
  graphing integers on, 5
  linear inequality in one variable on, 98–99
Numerator, of radical expressions, rationalizing, 569

O

Odd integers, consecutive, 55–58
One-to-one functions
  definition of, 709
  determining, 709–711, 778
  horizontal line test for, 710–711, 778
Ordered pairs. *See also* Points
  definition of, 140
  plotting, 140–141
  as solution of linear equation in two variables, 141–142
  as solution of linear inequalities in two variables, 187
Ordered triples, 251, 284
Order of operations
  definition of, 23, 33
  using, 23–24
Ordinate. *See* y-coordinate
Origin, in Cartesian coordinate system, 140

P

Parabola
  as conic section, 802
  graph of
    finding vertex of, 675–676
    of form $x = a(y - k)^2 + h$, 672–674
    of form $x = ay^2 + by + c$, 674–675, 696, 829
  as graph of quadratic function, 654, 694
Parallel lines
  equations of, 178–179, 218
  slope of, 175–178
Parentheses, in order of operations, 23–24
Percent change, applied problems with, 58–59
Perfect square trinomials
  completing the square for, for $x^2 + bx$, 616–618
  definition of, 378, 415, 692
  examples of, 616
  factoring, 378–380, 385, 415
Perimeter, of rectangle, 70–71
Perpendicular lines
  equations of, 178–179, 218
  slope of, 175–178, 218
Piecewise functions, graph of, 794–795, 846
Plane, in Cartesian coordinate system, 140
Please Excuse My Dear Aunt Sally, 24, 845
Points. *See also* Ordered pairs
  equation of line from
    with slope, 172–173, 217
    with two points, 174, 217–218
  graphing linear equations in two variables from, 143–144
  graphing line from, with slope, 161–162, 216
  slope of a line from, 157–159
Point-slope formula, equation of a line from
  with slope and point, 172–173
  with two points, 174
Polynomial(s). *See also* Binomials; Monomials; Trinomials
  addition of, 315–317, 347
  constants in, 314
  definition of, 314
  degrees of, 314, 346

Polynomial(s)—(continued)
   degrees of terms in, 314, 346
   in descending powers, 314
   division of
      long division, 337–340
      by monomials, 336–337, 348
      by polynomials, 337–340, 348
   factoring of
      definition of, 358, 414
      greatest common binomial factor, 360–361
      greatest common monomial factor, 359–360
      by grouping, 361–363, 414
      vs. multiplication, 358–359
      strategies for, 388–390
      by substitution, 374–375
   greatest common factor of
      binomial, factoring out, 360–361
      monomial, factoring out, 359–360
      procedure for factoring out, 359
   identifying, 314–315
   in more than one variable, addition and subtraction of, 317
   multiplication of
      vs. factoring, 358–359
      by monomials, 325, 347
      by multiple polynomials, 328
      by polynomials, 325–326, 347
   subtraction of, 315–317, 347
   terms in, 314
Polynomial functions
   addition of, 318–320, 347
   definition of, 318, 347
   division of, 341–342, 348
   multiplication of, 331–332, 348
   subtraction of, 318–320, 347
Polynomial inequalities. See also Quadratic inequalities
   higher degree, solving, 685
P.O.W.E.R. Framework, 2
Power rule for logarithms, 746–747
Power rules for exponents, 295–297
   basic power rule, 295–297, 304–306
   power rule for a product, 295–297, 304–306, 346
   power rule for a quotient, 295–297, 304–306, 346
Powers of imaginary numbers, 602
   simplifying, 593
Prime factorization, definition of, 357
Prime trinomials, 368
Principal, in simple interest, 59
Principal nth roots, 513–515, 598
Principal square roots
   definition of, 22
   finding, 512–513
Product rule for exponents, 294–295, 304–306, 346
Product rule for higher roots, 545–546, 600
Product rule for logarithms, 744–745
Product rule for square roots, 534–535, 599
Progressive relaxation, 777
Properties of equality
   equations solved with, 41–42, 88
   linear inequalities in one variable solved with, 100–103

Proportion
   definition of, 465
   determining if true or false, 465
   solving, 465–466
Pythagorean theorem
   applied problems with, 405–407, 416
   definition of, 405, 416

Q

Quadrants, in Cartesian coordinate system, 140
Quadratic equations
   applied problems with
      consecutive integers, 404
      geometry, 403
      given an equation, 407–408
      Pythagorean theorem, 405–407, 416
      quadratic formula, 649–650, 694
      steps for solving, 403
   definition of, 392, 611
   vs. equations in quadratic form, 638
   from equations with fractions, solving, 637–638
   from equations with radicals, solving, 637–638
   solutions to, discriminant in determining, 627–629, 693
   solving
      choosing method for, 633–635
      by completing the square, 619–621, 692
      by factoring, 394–397, 415–416
      of form $ab = 0$, 393–394
      with quadratic formula, 626–627
      square root property and, 611–615
      by substitution for a binomial, 641–642
   standard form for, 393, 415
Quadratic form, equations in
   vs. quadratic equations, 638
   solving
      by factoring, 638–640, 693
      by substitution, 640–641
Quadratic formula
   applied problems with, 629–630
   definition of, 625, 693
   derivation of, 624–625
   quadratic equation solved by, 626–627, 633–635
Quadratic functions
   definition of, 653, 694
   graph of, 654–657, 681–682, 694
      of form $f(x) = a(x - h)^2 + k$, 657–658, 694
      of form $f(x) = ax^2 + bx + c$, 658–662, 695
      horizontal shift of, 655–656, 792–793
      reflection about the x-axis, 656–657, 792–793
      vertical shift of, 654–655, 792–793
   maximum values of, 668–669, 695
      applied problems with, 669–672
   minimum values of, 668–669, 695
      applied problems with, 669–672
Quadratic inequalities
   definition of, 681, 696

   solving
      by graphing, 681–682
      with special solutions, 684
      with test points, 682–684
Quotient rule for exponents, 300–301, 304–306, 346
Quotient rule for higher roots, 547–548, 600
Quotient rule for logarithms, 745–746
Quotient rule for square roots, 537–538, 599

R

Radical
   definition of, 22
   equations with, quadratic equations from, 637–638
   like, definition of, 553
   rational exponents and
      $a^{1/n}$, 525–526
      $a^{m/n}$, 526–527
      $a^{-m/n}$, 527–528
   simplifying, 546–547
Radical equations
   with cube roots, solving, 583
   solving
      with one square root, 578–580, 602
      with two square roots, 580–583
Radical expressions
   addition of, 554, 574, 600
      simplifying before, 555
   binomials with
      multiplication of, 556, 575
      squaring, 557–558, 601
   common factors in, dividing out, 570
   converting to exponential form, 530
   definition of, 516
   division of, with different indices, 550
   multiplication of, 545–546, 601
      with different indices, 550
      of form $(a + b)(a - b)$, 558–559, 601
      using FOIL, 557
   rationalizing denominator of
      definition of, 562, 601
      with one higher root, 565–567, 575–576
      with one square root, 562–565, 575
      with two terms, 567–569, 575–576, 601
   rationalizing numerator of, 569
   subtraction of, 554, 574, 600
      simplifying before, 555
   with variables
      division of, 541–542
      and even exponents, simplifying, 538–539
      multiplication of, 541–542
      and odd exponents, simplifying, 539–541
      simplifying, 548–550, 599, 600
Radical functions, definition of, 516
Radical sign, 22
Radicand, 22, 513
Radius, 803
Range, of relation
   definition of, 198, 220
   identifying, 198–201
Ratio, definition of, 465
Rational equations
   applied problems with
      distance, rate, and time, 480–482

proportions, 478–480
    work, 482–484, 500–501
proportions
    applied problems with, 478–480
    solving, 465–466
vs. rational expressions, 460–462, 475, 500
solving, 462–465, 500
    proportions, 465–466
    for specific variable, 466–467, 500
Rational exponents
    evaluating
        $a^{1/n}$, 525–526, 598
        $a^{m/n}$, 526–527, 574, 598
        $a^{-m/n}$, 526–527, 574, 599
    rules of exponents with, 528–529
Rational expressions. *See also* Complex fractions
    addition of
        with common denominator, 441–443
        with denominator factors
            of $a-b$ and $b-a$, 446–447
        with different denominators, 443–445, 473, 498
    definition of, 424, 497
    division of, 431–432, 473, 498
    equivalent, 429–430
        with LCD as denominator, 439–441
    fundamental property of, 427
    least common denominator for, 437–439, 498
    in lowest terms, 427–428, 497
    multiplication of, 430–431, 497
    with negative exponents, simplifying, 456–457
    vs. rational equations, 460–462, 475, 500
    rewriting with LCD as denominator, 439–441
    simplifying, 428–429, 497
        with negative exponents, 456–457
    subtraction of
        with common denominator, 441–443
        with denominator factors of $a-b$ and $b-a$, 446–447
        with different denominators, 443–445
Rational functions
    definition of, 425, 497
    domain of, 425–427, 472, 497
    undefined, 426
Rational inequalities, solving, 685–688, 696
Rational numbers, definition of, 6, 32, 424, 598
Real numbers
    addition of, 13–14, 32
    applied problems with, 15
    in complex numbers, 588–589
    definition of, 6, 32
    division of, 16–17, 32
    multiplication of, 15–16, 32
    properties of, 28–29, 33–34
    subtraction of, 14–15, 32
Real part, of complex numbers, 588, 602
Reciprocals, negative, 176
Rectangle
    area of, 66–67
    perimeter of, 70–71
Reflection, of quadratic functions, about *x*-axis, 656–657, 792–793
Relation. *See also* Function
    definition of, 197, 220

domain of
    definition of, 198, 220
    identifying, 198–201, 202–203, 221
range of
    definition of, 198, 220
    identifying, 198–201
    vertical line test for, 200–201, 220
Right triangle, definition of, 405
Rise, 157
Roots. *See also* Square root(s)
    $\sqrt[n]{a^n}$, evaluating, 515–516
    *n*th, finding, 513–515
    principal *n*th, 513–515
    principal square
        definition of, 22
        finding, 512–513
Row echelon form, 285
    matrix in, 274
Run, 157

S

Saving style, 691
Scientific notation
    converting from standard form, 308–309
    converting to standard form, 307–308, 346
    operations with, 309
    understanding, 306–307, 346
Second-degree inequalities
    definition of, 839, 848
    graph of, 839–840, 848
Set(s)
    intersection of, 109–110, 131
        applied problems with, 114
    union of, 109–110, 131
        applied problems with, 114
Set notation, 98
Signed numbers
    addition of, 13–14
    using, 8
Simple interest, applied problems with, 59–62
Simplification
    of complex fractions
        with multiple terms in numerator and/or denominator, 452–456, 474, 499
        with one term in numerator and denominator, 451–452, 474, 499
    of higher roots of integers, 546–547
    of powers of *i*, 593
    of radical expressions
        before addition and subtraction, 555
        with variables, 548–550, 599
    of rational expressions, 428–429, 497
        with negative exponents, 456–457
    of square root expressions
        with variables and even exponents, 538–539
        with variables and odd exponents, 539–541
    of square roots of whole number, 535–537
Slope-intercept form
    for equation of a line, 171, 217
    for linear equations in two variables, 163
Slope of a line
    applied problems with, 160
    definition of, 156, 216

equation of a line from
    with point, 172–173, 217
    with *y*-intercept, 172, 217
finding, given two points, 157–159
formula for, 158
graphing line from
    with one point, 161–162, 216
    with *y*-intercept, 162–164, 217
of horizontal lines, 160–161, 216
negative, 159
of parallel lines, 175–178, 218
of perpendicular lines, 175–178, 218
positive, 159
understanding, 156–157
of vertical lines, 160–161, 216
Square of a binomial, 329–330, 347
    with radical expressions, 557–558, 601
Square root(s)
    finding, 22–23, 33, 512–513
    irrational numbers as, 22, 513, 598
    multiplication of, 534–535
    negative, finding, 512–513
    of negative number, 588, 598, 602
    with negative numbers
        division of, 589–590
        multiplication of, 589–590
    principal
        definition of, 22
        finding, 512–513
    product rule for, 534–535, 599
    quotient rule for, 537–538, 599
    radical equation with one, solving, 578–580, 602
    radical equation with two, solving, 580–583
    of whole number, simplifying, 535–537
Square root expressions, with variables
    division of, 541–542
    and even exponents, simplifying, 538–539
    multiplication of, 541–542
    and odd exponents, simplifying, 539–541
Square root functions
    domain of, 517, 598
    examples of, 516
    graph of, 518–519, 598, 822–823
Square root property, 611–615, 633–635, 692
Square root symbol, 22
Standard form
    for equation of line, 171
    for linear equations in two variables, 162, 215
Stress, coping with, 702, 777
Study groups, 354, 413
Studying, 281
Study strategies
    financial literacy, 608
    homework, 291
    memory improvement, 789
    note-taking in class, 136
    P.O.W.E.R. Framework, 2
    reading textbooks, 37
    stress, coping with, 702
    study groups, 354
    test-taking, 229
    time management, 95
    working with technology, 508
    writing process, 421

Substitution method
  equations in quadratic form solved by, 640–641
  nonlinear systems of equations solved by, 832–834, 847–848
  polynomials factored by, 374–375
  systems of linear equations in two variables solved by, 238–241, 282
Subtraction
  applied problems with, 15
  of complex numbers, 590, 602
  in order of operations, 23–24
  of polynomial functions, 318–320, 347
  of polynomials, 315–317, 347
  of radical expressions, 554, 574, 600
    simplifying before, 555
  of rational expressions
    with common denominator, 441–443
    with denominator factors of $a - b$ and $b - a$, 446–447
    with different denominators, 443–445
  of real numbers, 14–15, 32
Subtraction property of equality
  definition of, 88
  equations solved with, 41–42
  linear inequalities in one variable solved with, 100–101
Sum and difference of two cubes, factoring, 383–385, 415
Supplementary angles, applied problems with, 68–69
Systems of linear equations in three variables
  applied problems with, 267–268, 285
  definition of, 251
  solutions of
    infinite solutions, 251–252
    no solution, 251–252
    one solution, 251–252
    ordered triples as, 251, 284
  solving, 252–254, 284
    by Gaussian elimination, 275–278
    with missing terms, 255–256
    special systems, 254–255
Systems of linear equations in two variables
  applied problems with
    cost, 262–263
    distance, rate, and time, 265–266
    general, 260–261
    geometry, 261–262
    mixtures, 264–265
  consistent, definition of, 234, 236, 282
  definition of, 233, 282
  dependent
    definition of, 236, 282, 283
    graphs of, 236
  with fractions, solving by substitution, 240–241
  inconsistent, definition of, 235, 282, 283
  independent
    definition of, 234, 282
    graphs of, 236
  solutions of
    definition of, 233, 282
    exactly one, 234, 282
    infinite solutions, 236, 282, 283

at least one, 234
no solution, 235, 282, 283
solving
  by elimination, 241–243, 283
  by Gaussian elimination, 274–275, 285
  by graphing, 234–237, 282
  special systems, 243–245
  by substitution, 238–241, 282
  without graphing, 237
Systems of nonlinear equations. See Nonlinear systems of equations
Systems of nonlinear inequalities
  graph of, 840–842
  solution set of, 840, 848

T

Technology, working with, 508
Terms
  in algebraic expressions, 27
  combining, solving linear equations in one variable by, 43–44
  in polynomials, 314
Test points
  linear inequalities graphed with, 188–189
  quadratic inequalities solved with, 682–684
Tests, taking, 229
Textbooks, reading, 37
Time log, 95
Time management, 95, 129–130
To-do list, daily, 95
Total impedance, 596
Trapezoid, area of, 71–72
Trial and error, factoring by, of trinomials, 371–374, 415
Trinomials
  definition of, 315
  factoring of
    of form $ax^2 + bx + c$, 370–374, 414–415
    of form $x^2 + bx + c$, 367–369, 414
    by grouping, 370–371, 414–415
    perfect square trinomials, 378–380, 385, 415
    by trial and error, 371–374, 415
    with two variables, 369–370
  perfect square trinomials
    completing the square for, 616–618
    definition of, 378, 415, 692
    examples of, 616
    factoring, 378–380, 385, 415
  prime, 368

U

Union, of sets, 109–110, 131
  applied problems with, 114
Unit circle, 804
Upper bounds, of linear inequalities in three parts, 103

V

Variables
  in algebraic expressions, 27
  dependent, 197

evaluating function for, 204–206
  as exponents, 305–306
  independent, 197
  solving formulas for, 70–72, 89, 645–646, 693
Variation
  combined, solving problems, 493–494, 501
  direct
    definition of, 488, 501
    solving problems with, 488–490
  inverse
    definition of, 491, 501
    solving problems, 491–492
  joint
    definition of, 492, 501
    solving problems, 492–493
Vertex, of parabola, 654
Vertex formula, quadratic functions graphed by, 661–662
Vertical lines
  equations of
    graphing, 145–146, 216
    writing, 175
  slope of, 160–161, 216
Vertical line test, 200–201, 220
Vertical shift, of quadratic functions, 654–655, 792–793
Volume, applied problems with, 647–648

W

Weekly timetable, 95
Whole numbers, definition of, 5, 32
Work, applied problems with, rational equations, 482–484, 500–501
Writing process, 421

X

x-axis
  in Cartesian coordinate system, 140
  reflection of quadratic functions about, 656–657, 792–793
x-coordinate, 140
x-intercept
  definition of, 144, 215
  of linear equations in two variables, graphing with, 144–145, 215

Y

y-axis, in Cartesian coordinate system, 140
y-coordinate, 140
y-intercept
  definition of, 144, 215
  equation of a line from, with slope, 172, 217
  graphing line from, with slope, 162–164, 217
  of linear equations in two variables, graphing with, 144–145, 215

Z

Zero, as exponent, 297–300, 304–306, 346
Zero product rule, 393, 415